CHAPTER 3 POLYNOMIAL AND RATIONAL FUNCTIONS

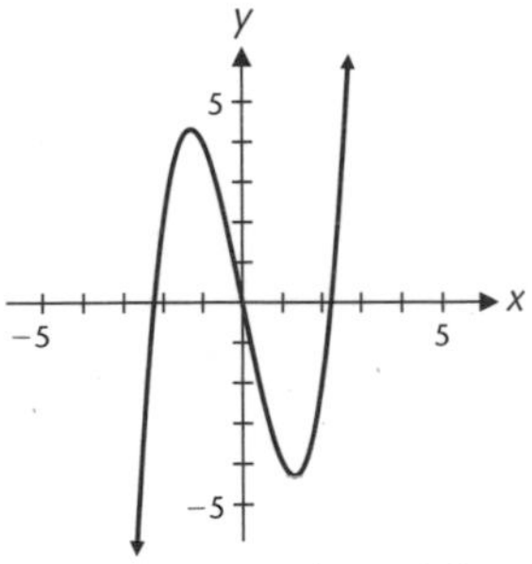

Third-degree polynomial
$g(x) = x^3 - 5x$

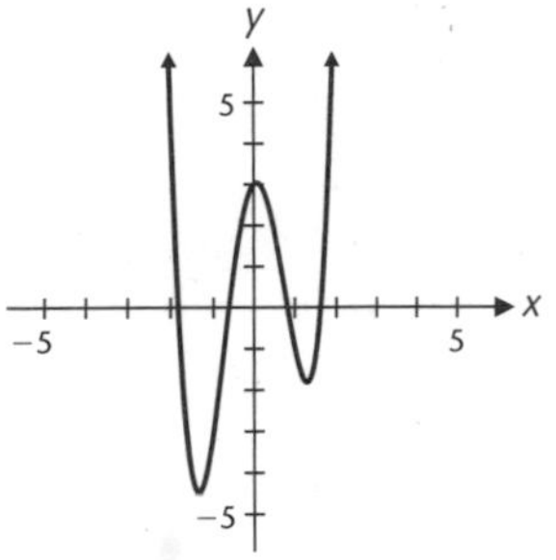

Fourth-degree polynomial
$G(x) = 2x^4 - 7x^2 + x + 3$

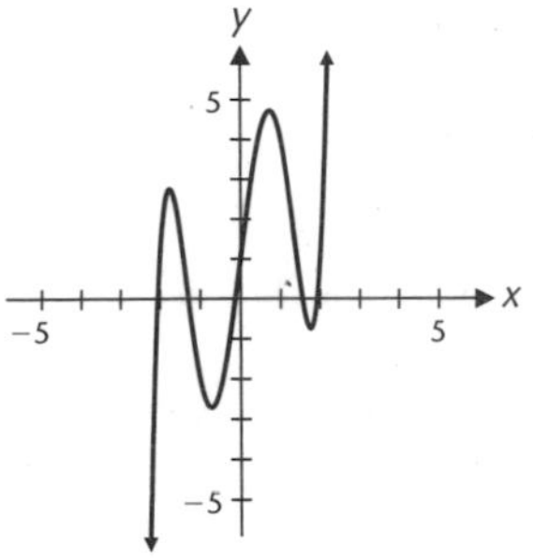

Fifth-degree polynomial
$h(x) = x^5 - 6x^3 + 8x + 1$

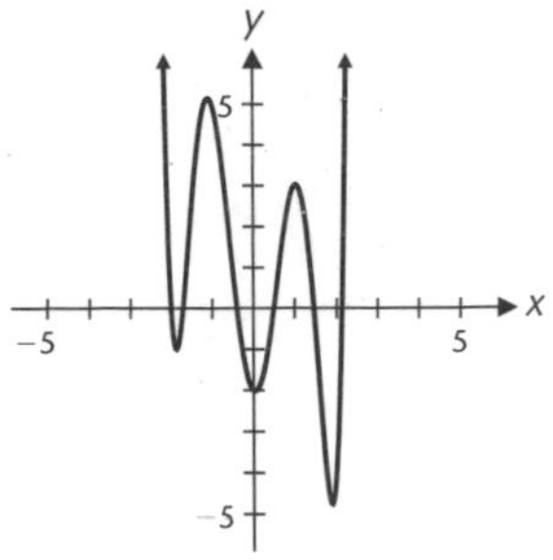

Sixth-degree polynomial
$H(x) = x^6 - 7x^4 + 12x^2 - x - 2$

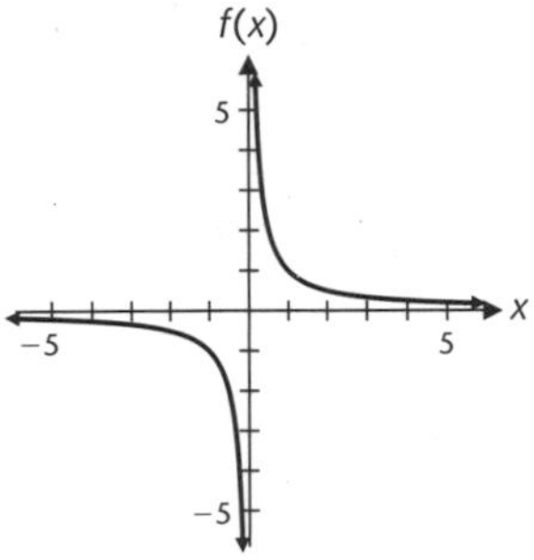

$f(x) = \dfrac{1}{x}$

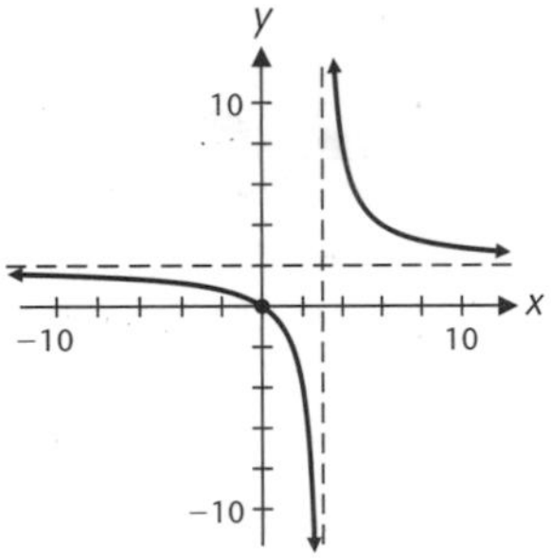

$f(x) = \dfrac{2x}{x - 3}$

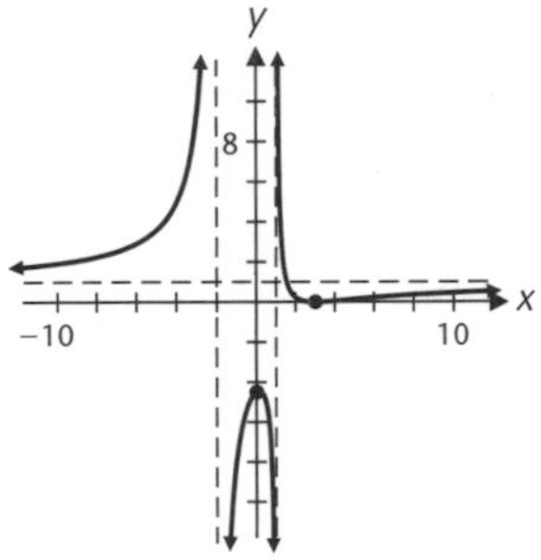

$f(x) = \dfrac{x^2 - 6x + 9}{x^2 + x - 2}$

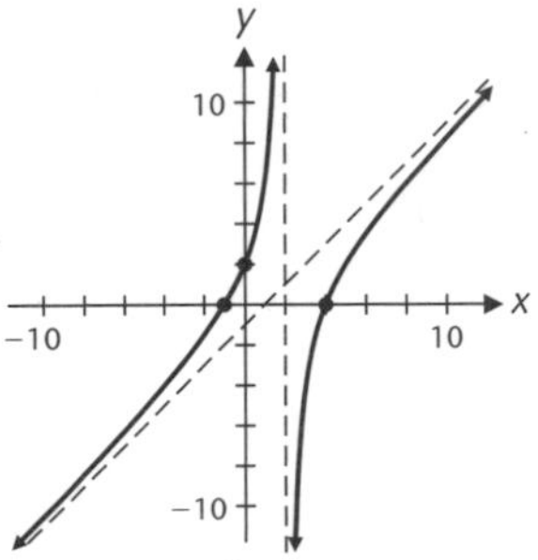

$f(x) = \dfrac{x^2 - 3x - 4}{x - 2}$

CHAPTER 4 EXPONENTIAL AND LOGARITHMIC FUNCTIONS

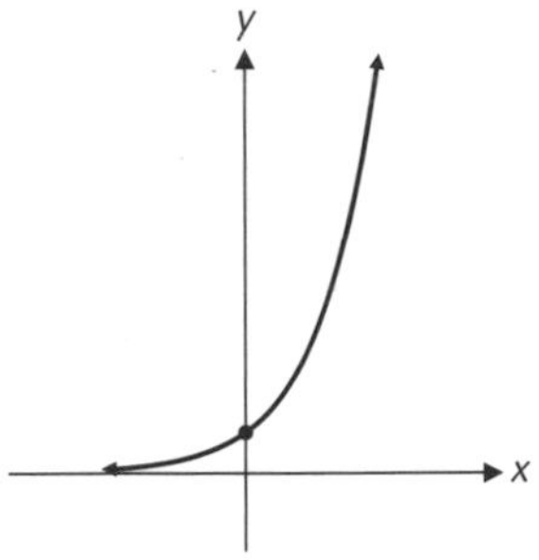

$y = b^x$
$b > 1$

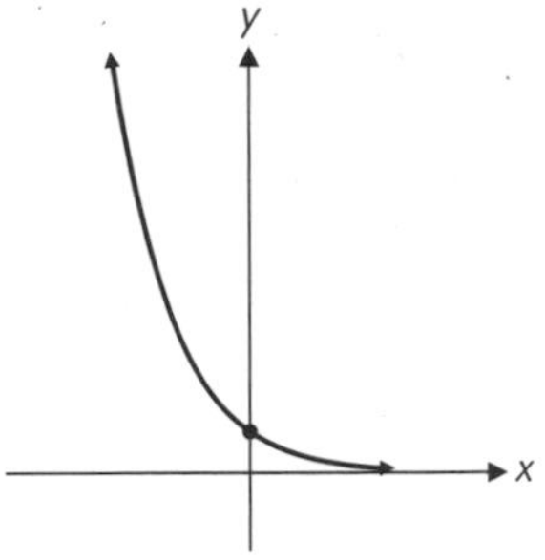

$y = b^x$
$0 < b < 1$

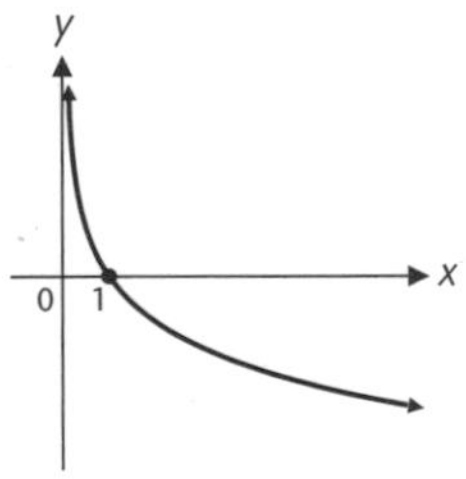

$y = \log_b x$
$0 < b < 1$

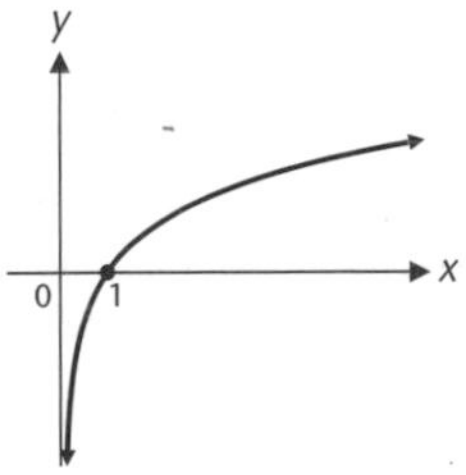

$y = \log_b x$
$b > 1$

COLLEGE ALGEBRA

A Graphing Approach

Barnett, Ziegler & Byleen's Precalculus Series

College Algebra, Sixth Edition
This book is the same as *College Algebra With Trigonometry* without the three chapters on trigonometry.
ISBN 0-07-006321-4

College Algebra With Trigonometry, Sixth Edition
This book is the same as *College Algebra* with three chapters of trigonometry added. Comparing *College Algebra With Trigonometry* with *Precalculus*, *College Algebra With Trigonometry* has more intermediate algebra review and starts trigonometry with angles and right triangles.
ISBN 0-07-006336-2

Precalculus: Functions and Graphs, Fourth Edition
This book differs from *College Algebra With Trigonometry* in that *Precalculus* starts at a higher level, placing intermediate algebra review in the appendix; and it starts trigonometry with the unit circle and circular functions.
ISBN 0-07-006341-9

College Algebra: A Graphing Approach
This book is the same as *Precalculus: A Graphing Approach* without the three chapters on trigonometry, but also includes coverage of probability. This text assumes the use of a graphing utility.
ISBN 0-07-005710-9

Precalculus: A Graphing Approach
This book is the same as *College Algebra: A Graphing Approach* with three additional chapters on trigonometry, but no probability coverage. This text assumes the use of a graphing utility.
ISBN 0-07-005717-6

COLLEGE ALGEBRA

A Graphing Approach

RAYMOND A. BARNETT
MERRITT COLLEGE

MICHAEL R. ZIEGLER
MARQUETTE UNIVERSITY

KARL E. BYLEEN
MARQUETTE UNIVERSITY

McGraw Hill

Boston Burr Ridge, IL Dubuque, IA Madison, WI New York San Francisco St. Louis
Bangkok Bogotá Caracas Lisbon London Madrid
Mexico City Milan New Delhi Seoul Singapore Sydney Taipei Toronto

McGraw-Hill Higher Education
A Division of The ***McGraw-Hill*** *Companies*

COLLEGE ALGEBRA: A GRAPHING APPROACH

This book is printed on acid-free paper.

1 2 3 4 5 6 7 8 9 0 VNH/VNH 9 0 9 8 7 6 5 4 3 2 1 0 9

ISBN 0-07-005710-9

Vice president and editorial director: *Kevin T. Kane*
Publisher: *JP Lenney*
Sponsoring editor: *Maggie Rogers*
Developmental editor: *Michelle Munn*
Marketing manager: *Mary K. Kittell*
Project manager: *Sheila M. Frank*
Senior production supervisor: *Sandra Hahn*
Design director: *Francis Owens*
Supplement coordinator: *Brenda A. Ernzen*
Compositor: *GTS Graphics, Inc.*
Typeface: *10.5/12 Times Roman*
Printer: *Von Hoffmann Press, Inc.*

Text/cover designer: *Andrew Ogus*
Cover photograph: © *PhotoDisc*
Chapter opener background photograph: © *Elizabeth Simpson/FPG International LLC*

Library of Congress Cataloging-in-Publication Data

Barnett, Raymond A.
College algebra : a graphing approach / Raymond A. Barnett, Michael R. Ziegler, Karl E. Byleen. — 1st ed.
p. cm.
ISBN 0-07-005710-9
1. Algebra. 2. Algebra—Graphic methods. I. Ziegler, Michael R. II. Byleen, Karl E. III. Title.

QA152.2 B3688 2000
512—dc21 99-049142
CIP

www.mhhe.com

About the Authors

Raymond A. Barnett, a native of and educated in California, received his B.A. in mathematical statistics from the University of California at Berkeley and his M.A. in mathematics from the University of Southern California. He has been a member of the Merritt College Mathematics Department and was chairman of the department for four years. Associated with four different publishers, Raymond Barnett has authored or co-authored eighteen textbooks in mathematics, most of which are still in use. In addition to international English editions, a number of the books have been translated into Spanish. Co-authors include Michael Ziegler, Marquette University; Thomas Kearns, Northern Kentucky University; Charles Burke, City College of San Francisco; John Fujii, Merritt College; and Karl Byleen, Marquette University.

Michael R. Ziegler received his B.S. from Shippensburg State College and his M.S. and Ph.D. from the University of Delaware. After completing postdoctoral work at the University of Kentucky, he was appointed to the faculty of Marquette University where he currently holds the rank of Professor in the Department of Mathematics, Statistics, and Computer Science. Dr. Ziegler has published over a dozen research articles in complex analysis and has co-authored over a dozen undergraduate mathematics textbooks with Raymond Barnett and Karl Byleen.

Karl E. Byleen received his B.S., M.A., and Ph.D. degrees in mathematics from the University of Nebraska. He is currently an Associate Professor in the Department of Mathematics, Statistics, and Computer Science of Marquette University. He has published a dozen research articles on the algebraic theory of semigroups, as well as co-authoring over a dozen undergraduate mathematics textbooks with Raymond Barnett and Michael Ziegler.

Preface

Objectives

Mathematical reform is the driving force behind the organization and development of this new college algebra text. The use of technology, primarily graphing utilities, is assumed throughout the text. The development of each topic proceeds from the concrete to the abstract and takes full advantage of technology, wherever appropriate.

The first major objective of this book is to encourage students to investigate mathematical ideas and processes graphically and numerically, as well as algebraically. Proceeding in this way, students gain a broader, deeper, and more useful understanding of a concept or process. Even though concept development and technology are emphasized, manipulative skills are not ignored, and plenty of opportunities to practice basic skills are present.

A brief look at the table of contents will reveal the importance of the function concept as a unifying theme. The second major objective of this book is the development of a library of elementary functions, including their important properties and uses. Having this library of elementary functions as a basic working tool in their mathematical tool boxes, students will be able to move into college algebra with greater confidence and understanding. In addition, a concise review of basic algebraic concepts is included in Appendix A for easy reference, or systematic review.

The third major objective of this book is to give the student substantial experience in solving and modeling real-world problems. Enough applications are included to convince even the most skeptical student that mathematics is really useful (see the Applications Index following the table of contents). Most of the applications are simplified versions of actual real-world problems taken from professional journals and professional books. No specialized experience is required to solve any of the applications.

Technology

The generic term *graphing utility* is used to refer to any of the various graphing calculators or computer software packages that might be available to a student using this book. We assume that each student has access to a graphing utility that can perform the following operations:

- Simultaneously display several graphs in a user-selected viewing window
- Explore graphs using zoom and trace
- Approximate roots
- Approximate intersection points
- Approximate maxima and minima
- Plot data sets and find associated regression equations
- Perform basic matrix operations, including row reduction and inversion

Most popular graphing calculators support all of these operations.

Features

CHAPTER OPENER:

Each chapter opens with a real-world application. Complete solutions to the chapter opener applications are located at the end of each chapter review, and require the use of graphical, numerical, and algebraic problem-solving techniques.

Polynomial and Rational Functions

3

Outline

3-1 Polynomial Functions

3-2 Finding Rational Zeros of Polynomials

3-3 Approximating Real Zeros of Polynomials

3-4 Rational Functions

Chapter 3 Group Activity: Interpolating Polynomials

Chapter 3 Review

Application

A manufacturer of camping supplies wants to produce an aluminum drinking cup in the shape of a right circular cylinder with an open top and a closed bottom. The cup will have a volume of 65 cubic inches. Find the dimensions of the cup that will use the minimum amount of aluminum and find the amount of aluminum used.

242 | 3 POLYNOMIAL AND RATIONAL FUNCTIONS

ANSWER TO APPLICATION

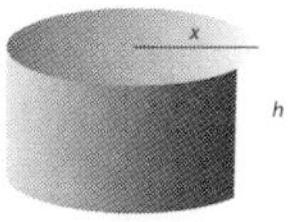

FIGURE 1

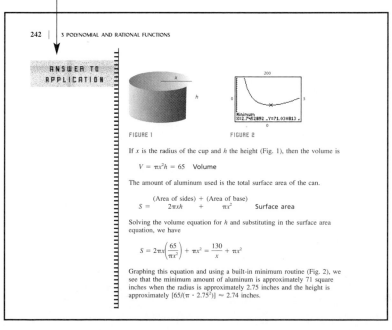

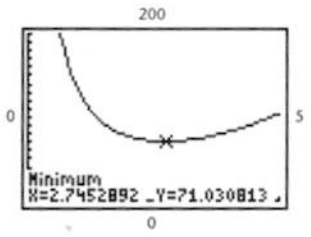

FIGURE 2

If x is the radius of the cup and h the height (Fig. 1), then the volume is

$$V = \pi x^2 h = 65 \quad \text{Volume}$$

The amount of aluminum used is the total surface area of the can.

$$S = \underset{2\pi x h}{(\text{Area of sides})} + \underset{\pi x^2}{(\text{Area of base})} \quad \text{Surface area}$$

Solving the volume equation for h and substituting in the surface area equation, we have

$$S = 2\pi x\left(\frac{65}{\pi x^2}\right) + \pi x^2 = \frac{130}{x} + \pi x^2$$

Graphing this equation and using a built-in minimum routine (Fig. 2), we see that the minimum amount of aluminum is approximately 71 square inches when the radius is approximately 2.75 inches and the height is approximately $[65/(\pi \cdot 2.75^2)] \approx 2.74$ inches.

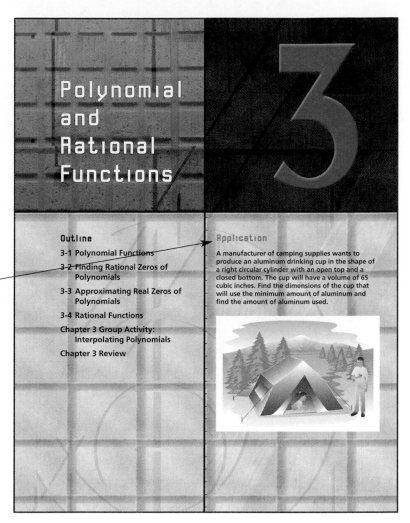

More formally, we define increasing, decreasing, and constant functions as follows:

DEFINITION 1

INCREASING, DECREASING, AND CONSTANT FUNCTIONS

Let I be an interval in the domain of a function f. Then,

1. **f is increasing** on I if $f(b) > f(a)$ whenever $b > a$ in I.
2. **f is decreasing** on I if $f(b) < f(a)$ whenever $b > a$ in I.
3. **f is constant** on I if $f(a) = f(b)$ for all a and b in I.

EXAMPLE 3 Describing a Graph

Use the terms increasing, decreasing, and constant to describe the graph of

$$f(x) = 60x - x^3 \quad \text{for} \quad -10 \le x \le 10$$

Solution We begin by graphing f in the standard viewing window [Fig. 10(a)]. Since this view of the graph is not very informative, we need to adjust the viewing window. The trace procedure can be used to determine appropriate values for Ymin and Ymax. (The trace procedure displays the coordinates of a point on the graph even if the point is not visible in the viewing window.) While tracing along the graph in Figure 10(a) from Xmin = −10 to Xmax = 10, notice that all the y values are between −400 and 400, inclusive. Using Ymin = −400 and Ymax = 400 produces the graph in Figure 10(b).

FIGURE 10

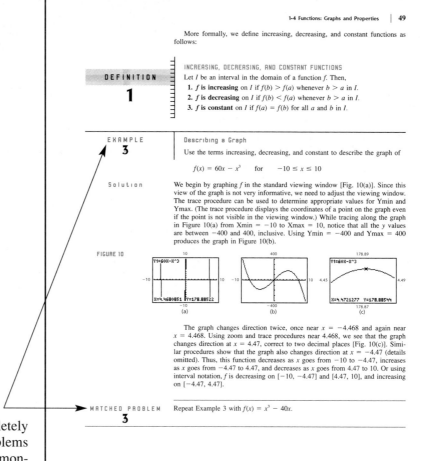

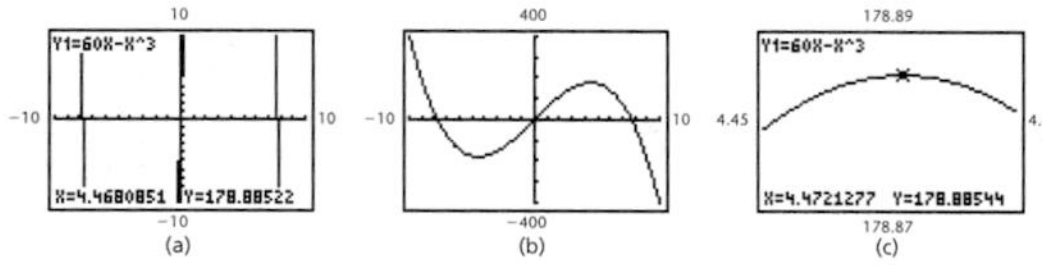

(a) (b) (c)

The graph changes direction twice, once near $x = -4.468$ and again near $x = 4.468$. Using zoom and trace procedures near 4.468, we see that the graph changes direction at $x = 4.47$, correct to two decimal places [Fig. 10(c)]. Similar procedures show that the graph also changes direction at $x = -4.47$ (details omitted). Thus, this function decreases as x goes from −10 to −4.47, increases as x goes from −4.47 to 4.47, and decreases as x goes from 4.47 to 10. Or using interval notation, f is decreasing on $[-10, -4.47]$ and $[4.47, 10]$, and increasing on $[-4.47, 4.47]$.

MATCHED PROBLEM 3 Repeat Example 3 with $f(x) = x^3 - 40x$.

EXAMPLES AND MATCHED PROBLEMS:

Integrated throughout the text, completely worked examples and practice problems are used to introduce concepts and demonstrate problem-solving techniques. Each example is followed by a similar Matched Problem for the student to work through while reading the material. Answers to the matched problems are located at the end of each section, for easy reference. This active involvement in the learning process helps students develop a more thorough understanding of algebraic concepts and processes.

EXPLORATION AND DISCUSSION:

Interspersed at appropriate places in every section, Explore/Discuss boxes encourage students to think critically about mathematics and to explore key concepts in more detail. Verbalization of mathematical concepts, results, and processes is encouraged in these Explore/Discuss boxes, as well as in some matched problems, and in particular problems in almost every exercise set. Explore/Discuss material can be used in class or as an out-of-class activity.

Tables 1 and 2 specify functions, since to each domain value there corresponds exactly one range value (for example, the cube of -2 is -8 and no other number). On the other hand, Table 3 does not specify a function, since to at least one domain value there corresponds more than one range value (for example, to the domain value 9 there corresponds -3 and 3, both square roots of 9).

Remark Some graphing utilities use the term *range* to refer to the window variables. In this book, *range* will always refer to the range of a function.

Explore/Discuss 1

Consider the set of students enrolled in a college and the set of faculty members of that college. Define a correspondence between the two sets by saying that a student corresponds to a faculty member if the student is currently enrolled in a course taught by the faculty member. Is this correspondence a function? Discuss.

Since a function is a rule that pairs each element in the domain with a corresponding element in the range, this correspondence can be illustrated by using ordered pairs of elements, where the first component represents a domain element

BALANCED EXERCISE SETS:

College Algebra: A Graphing Approach contains over 4,200 problems. Each Exercise Set is designed so that an average or below-average student will experience success and a very capable student will be challenged. Exercise Sets are found at the end of each section in the text, and are divided into A (routine, easy mechanics), B (more difficult mechanics), and C (difficult mechanics and some theory) levels of difficulty so that students at all levels can be challenged. Problem numbers that appear in blue indicate exercises that require the students to apply their reasoning and writing skills to the solution of the problem.

312 | 4 INVERSE FUNCTIONS: EXPONENTIAL AND LOGARITHMIC FUNCTIONS

EXERCISE 4-6

A

In Problems 1–8, evaluate to four decimal places.

1. log 82,734 **2.** log 843,250

3. log 0.001 439 **4.** log 0.035 604

5. ln 43.046 **6.** ln 2,843,100

7. ln 0.081 043 **8.** ln 0.000 032 4

In Problems 9–16, evaluate x to four significant digits, given:

9. $\log x = 5.3027$ **10.** $\log x = 1.9168$

11. $\log x = -3.1773$ **12.** $\log x = -2.0411$

13. $\ln x = 3.8655$ **14.** $\ln x = 5.0884$

15. $\ln x = -0.3916$ **16.** $\ln x = -4.1083$

B

In Problems 17–24, evaluate to three decimal places.

17. $n = \frac{\log 2}{\log 1.15}$ **18.** $n = \frac{\log 2}{\log 1.12}$

19. $n = \frac{\ln 3}{\ln 1.15}$ **20.** $n = \frac{\ln 4}{\ln 1.2}$

21. $x = \frac{\ln 0.5}{-0.21}$ **22.** $x = \frac{\ln 0.1}{-0.0025}$

23. $t = \frac{\ln 150}{\ln 3}$ **24.** $t = \frac{\log 200}{\log 2}$

In Problems 25–32, evaluate x to five significant digits.

25. $x = \log (5.3147 \times 10^{12})$

26. $x = \log (2.0991 \times 10^{17})$

27. $x = \ln (6.7917 \times 10^{-12})$

28. $x = \ln (4.0304 \times 10^{-8})$

29. $\log x = 32.068\ 523$ **30.** $\log x = -12.731\ 64$

31. $\ln x = -14.667\ 13$ **32.** $\ln x = 18.891\ 143$

In Problems 33–36, find f^{-1}. Check by graphing f, f^{-1}, and $y = x$ in the same viewing window on a graphing utility.

33. $f(x) = 2 \ln (x + 2)$ **34.** $f(x) = 2 \ln x + 2$

35. $f(x) = 4 \ln x - 3$ **36.** $f(x) = 4 \ln (x - 3)$

C

In Problems 37–40, find domain and range, x and y intercepts, and asymptotes. Round all approximate values to two decimal places.

37. $f(x) = -2 + \ln (1 + x^2)$ **38.** $f(x) = 2 - \ln (1 + |x|)$

39. $f(x) = 1 + \ln (1 - x^2)$ **40.** $f(x) = -1 + \ln (|1 - x^2|)$

41. Find the fallacy.

$1 < 3$

$\frac{1}{27} < \frac{3}{27}$ Dividc both sides by 27.

$\frac{1}{27} < \frac{1}{9}$

$(\frac{1}{3})^3 < (\frac{1}{3})^2$

$\log (\frac{1}{3})^3 < \log (\frac{1}{3})^2$

$3 \log \frac{1}{3} < 2 \log \frac{1}{3}$

$3 < 2$ Divide both sides by $\log \frac{1}{3}$.

42. Find the fallacy.

$3 > 2$

$3 \log \frac{1}{2} > 2 \log \frac{1}{2}$ Multiply both sides by $\log \frac{1}{2}$.

$\log (\frac{1}{2})^3 > \log (\frac{1}{2})^2$

$(\frac{1}{2})^3 > (\frac{1}{2})^2$

$\frac{1}{8} > \frac{1}{4}$

$1 > 2$ Multiply both sides by 8.

43. The function $f(x) = \log x$ increases extremely slowly as $x \to \infty$, but the composite function $g(x) = \log (\log x)$ increases still more slowly.
    (A) Illustrate this fact by computing the values of both functions for several large values of x.
    (B) Determine the domain and range of the function g.
    (C) Discuss the graphs of both functions.

44. The function $f(x) = \ln x$ increases extremely slowly as $x \to \infty$, but the composite function $g(x) = \ln (\ln x)$ increases still more slowly.
    (A) Illustrate this fact by computing the values of both functions for several large values of x.
    (B) Determine the domain and range of the function g.
    (C) Discuss the graphs of both functions.

In Problems 45–48, use a graphing utility to find the coordinates of all points of intersection to two decimal places.

45. $f(x) = \ln x, g(x) = 0.1x - 0.2$

46. $f(x) = \log x, g(x) = 4 - x^2$

APPLICATIONS:

One of the primary objectives of this book is to give the student substantial experience in modeling and solving real-world problems. Over 500 application exercises help convince even the most skeptical student that mathematics is relevant to everyday life. The most difficult application problems are marked with two stars (★★), the moderately difficult application problems with one star (★), and the easier application problems are not marked. An **Applications Index** is included following the table of contents to help locate particular applications.

78. Manufacturing. A box with a hinged lid is to be made out of a piece of cardboard that measures 20 by 40 inches. Six squares, x inches on a side, will be cut from each corner and the middle, and then the ends and sides will be folded up to form the box and its lid (see the figure). The volume of the box is given by

$$V(x) = 0.5x(40 - 3x)(20 - 2x) \qquad 0 \le x \le 10$$

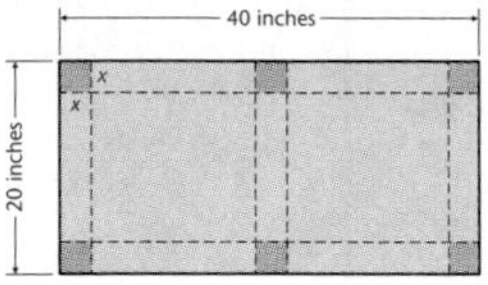

(A) Examine the values of the volume function $V(x)$ for $x = 0, 1, 2, \ldots, 10$ and estimate the size of the cut-out squares that will make the volume maximum. What is the estimated volume?

(B) Graph $y = V(x)$ and approximate to two decimal places the size of the cut-out squares that will make the volume maximum. Approximate the maximum volume to two decimal places also.

79. Construction. A freshwater pipe is to be run from a source on the edge of a lake to a small resort community on an island 8 miles offshore, as indicated in the figure. It costs \$10,000 per mile to lay the pipe on land and \$16,000 per mile to lay the pipe in the lake. The total cost $C(x)$ in thousands of dollars of laying the pipe is given by

$$C(x) = 10(20 - x) + 16\sqrt{x^2 + 64} \qquad 0 \le x \le 20$$

(A) Examine the values of the cost function $C(x)$ (rounded to the nearest thousand dollars) for $x = 0, 5, 10, 15,$ and 20 and estimate to the nearest mile the length of the land portion of the pipe that will make the production costs minimum. What is the estimated cost?

(B) Graph $y = C(x)$ and approximate to one decimal place the length of the land portion of the pipe that will make the production costs minimum. Approximate the minimum cost to the nearest thousand dollars also.

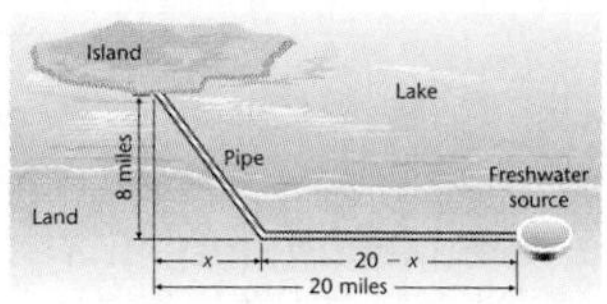

80. Transportation. The construction company laying the freshwater pipe in Problem 79 uses an amphibious vehicle to travel down the beach and then out to the island. The vehicle travels at 30 miles per hour on land and 7.5 miles per hour in water. The total time $T(x)$ in minutes for a trip from the freshwater source to the island is given by

$$T(x) = 2(20 - x) + 8\sqrt{x^2 + 64} \qquad 0 \le x \le 20$$

(A) Examine the values of the time function $T(x)$ (rounded to the nearest minute) for $x = 0, 5, 10, 15,$ and 20 and estimate the length of the land portion of the trip that will make the time minimum. What is the estimated time?

(B) Graph $y = T(x)$ and use zoom and trace procedures to approximate to one decimal place the length of the land portion of the trip that will make the time minimum. Approximate the minimum time to the nearest minute also.

81. Tire Mileage. An automobile tire manufacturer collected the data in the table relating tire pressure x, in pounds per square inch, and mileage in thousands of miles.

x	28	30	32	34	36
Mileage	45	52	55	51	47

A mathematical model for this data is given by

$$f(x) = -0.518x^2 + 33.3x - 481$$

(A) Complete the following table. Round values of $f(x)$ to one decimal place.

x	28	30	32	34	36
Mileage	45	52	55	51	47
$f(x)$					

(B) Sketch by hand the graph of f and the mileage data on the same axes.

(C) Use values of the modeling function rounded to two decimal places to estimate the mileage for a tire pressure of 31 lb/in.². For 35 lb/in.².

$$I = I_0 e^{-kd}$$

where I is the intensity d feet below the surface, I_0 is the intensity at the surface, and k is the coefficient of extinction. Measurements in the Sargasso Sea in the West Indies have indicated that half the surface light reaches a depth of 73.6 feet. Find k, and find the depth at which 1% of the surface light remains. Compute answers to three significant digits.

★ **95. Wildlife Management.** A lake formed by a newly constructed dam is stocked with 1,000 fish. Their population is expected to increase according to the logistic curve

$$N = \frac{30}{1 + 29e^{-1.35t}}$$

where N is the number of fish, in thousands, expected after t years. The lake will be open to fishing when the number of fish reaches 20,000. How many years, to the nearest year, will this take?

96. Medicare. The annual expenditures for Medicare (in billions of dollars) by the U.S. government for selected years since 1980 are shown in Table 1 (Bureau of the Census). Let x represent years since 1980.

TABLE 1 Medicare Expenditures

Year	Billion \$
1980	37
1985	72

(A) Find a logarithmic regression model of the form $y = a + b \ln x$ for this data. Estimate (to the nearest million) the total consumption in 1996 and in 2010.

(B) The actual consumption in 1996 was 1,583 million bushels. How does this compare with the estimated consumption in part A? What effect will this additional 1996 information have on the estimate for 2010? Explain.

CUMULATIVE REVIEW EXERCISE FOR CHAPTERS 3 AND 4

Work through all the problems in this cumulative review and check answers in the back of the book. Answers to all review problems are there, and following each answer is a number in italics indicating the section in which that type of problem is discussed. Where weaknesses show up, review appropriate sections in the text.

A

1. Let $P(x)$ be the polynomial whose graph is shown in the figure.

(A) Assuming that $P(x)$ has integer zeros and leading coefficient 1, find the lowest-degree equation that could produce this graph.

(B) Describe the left and right behavior of $P(x)$.

GROUP ACTIVITIES:

A Group Activity is located at the end of each chapter and involves many of the concepts discussed in that chapter. These activities strongly encourage the verbalization of mathematical concepts, results, and processes. All of these special activities are highlighted to emphasize their importance.

166 | 2 LINEAR AND QUADRATIC FUNCTIONS

Chapter 2 | Group Activity

Mathematical Modeling in Population Studies

In a study on population growth in California, Tulane University demographer Leon Bouvier recorded the past population totals for every 10 years starting at 1900. Then, using sophisticated demographic techniques, he made high, low, and medium projections to the year 2040. Table 1 shows actual populations up to 1990 and medium projections (to the nearest million) to 2040.

TABLE 1 California Population 1900–2040

Years after 1900	Date	Population (millions)	
0	1900	2	Actual
10	1910	3	
20	1920	4	
30	1930	5	
40	1940	5	
50	1950	10	
60	1960	15	
70	1970	20	
80	1980	23	
90	1990	30	
100	2000	35	Projected
110	2010	45	
120	2020	53	
130	2030	61	
140	2040	70	

1. Building a Mathematical Model.
   (A) Plot the first and last columns in Table 1 up to 1990 (actual populations). Would a linear or a quadratic function be the better model for this data? Why?
   (B) Use a graphing utility to compute a quadratic regression function to model the data you plotted in part A.
   (C) Graph this function and the data from part A for $0 \le x \le 150$. The results should look something like Figure 1.
2. Using the Mathematical Model for Projections.
   Use the quadratic regression model to answer the following questions.
   (A) Calculate projected populations for California at 10-year intervals, starting at 2000 and ending at 2040. Compare your projections with Professor Bouvier's projections, both numerically and graphically.
   (B) During what year would you project that the population will reach 40 million? 50 million?
   (C) For what years would you project the population to be between 34 million and 68 million, inclusive?

FIGURE 1

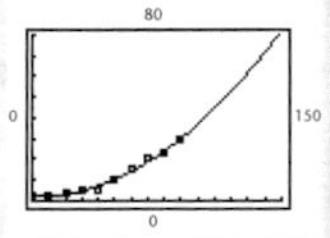

FOUNDATION FOR CALCULUS:

As many students will use this book to prepare for a calculus course, examples and exercises that are especially pertinent to calculus are marked with an icon C|∫.

102 | 2 LINEAR AND QUADRATIC FUNCTIONS

In Problems 29–32, write the slope–intercept form of the equation of the line with indicated slope and y intercept.

29. Slope = 1; y intercept = 0

30. Slope = -1; y intercept = 7

31. Slope = $-\frac{2}{3}$; y intercept = -4

32. Slope = $\frac{5}{3}$; y intercept = 6

B

In Problems 33–46, sketch a graph of the line that contains the indicated point(s) and/or has the indicated slope and/or has the indicated intercepts. Then write the equation of the line in the slope–intercept form $y = mx + b$ or in the form $x = c$ and check by graphing the equation on a graphing utility.

33. $(0, 4)$; $m = -3$ **34.** $(2, 0)$; $m = 2$

35. $(-5, 4)$; $m = -\frac{2}{5}$ **36.** $(-4, -2)$; $m = \frac{1}{2}$

37. $(1, 6)$; $(5, -2)$ **38.** $(-3, 4)$; $(6, 1)$

39. $(-4, 8)$; $(2, 0)$ **40.** $(2, -1)$; $(10, 5)$

41. $(-3, 4)$; $(5, 4)$ **42.** $(0, -2)$; $(4, -2)$

43. $(4, 6)$; $(4, -3)$ **44.** $(-3, 1)$; $(-3, -4)$

45. x intercept -4; y intercept 3

Problems 61–66 refer to the quadrilateral with vertices $A(0, 2)$, $B(4, -1)$, $C(1, -5)$, and $D(-3, -2)$.

61. Show that $AB \parallel DC$. **62.** Show that $DA \parallel CB$.

63. Show that $AB \perp BC$. **64.** Show that $AD \perp DC$.

65. Find an equation of the perpendicular bisector of AD. [*Hint:* First find the midpoint of AD. (See Problem 47 in Exercise 1-1.)]

66. Find an equation of the perpendicular bisector of AB.

C|∫ *Problems 67–72 are calculus-related. Recall that a line tangent to a circle at a point is perpendicular to the radius drawn to that point (see the figure). Find the equation of the line tangent to the circle at the indicated point. Write the final answer in the standard form $Ax + By = C$, $A \ge 0$. Graph the circle and the tangent line on the same coordinate system.*

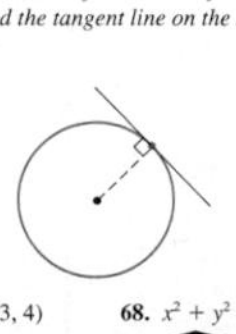

67. $x^2 + y^2 = 25$, $(3, 4)$ **68.** $x^2 + y^2 = 100$, $(-8, 6)$

GRAPHING CALCULATOR PROGRAMS (optional):

All of the graphing calculator programs found in this textbook are available for download at www.mhhe.com/barnett. The presence of an icon indicates the availability of downloadable TI 82, 83, 85, and 86 graphing calculator programs. Use of these programs is entirely optional.

TABLE 1 Synthetic Division on a Graphing Utility

Program SYNDIV

TI-82/TI-83	TI-85/TI-86	Output
Lbl A	Lbl A	{1, -6, 9, -3}→L1
Prompt R	Prompt R	{1 -6 9 -3}
2→I	2→I	prgmSYNDIV
dim(L1)→N	dimL L1→N	R=?0 {1 -6 9 -3}
{0}→L2	{0}→L2	R=?1 {1 -5 4 1}
N→dim(L2)	N→dimL L1	R=?2 {1 -4 1 -1}
L1(1)→L2(1)	L1(1)→L2(1)	R=?3 {1 -3 0 -3}
Lbl B	Lbl B	R=?4 {1 -2 1 1}
L2(I-1)*R+L1(I)→L2(I)	L2(I-1)*R+L1(I)→L2(I)	R=?
1+I→I	1+I→I	
If I≤N	If I≤N	
Goto B	Goto B	
Pause L2	Pause L2	
Goto A	Goto A	

Explore/Discuss 1

(A) If you have a TI-82, TI-83, TI-85, or TI-86 graphing calculator, enter the appropriate version of SYNDIV in your calculator exactly as shown in Table 1. If you have some other graphing utility that can store and execute programs, consult your manual and modify the statements in SYNDIV so that the program works on your graphing utility.

(B) Store the coefficients of $P(x) = x^3 - 6x^2 + 9x - 3$ in L1 (see the first line of output in Table 1) and execute the program. Type 0 at the "R=?" prompt and press ENTER to display the results of synthetic division with $r = 0$. To continue press ENTER again. Repeat this process for $r = 1, 2, 3$, and 4. Press QUIT at the "R=?" prompt to terminate the program. Compare the last number in each list with the values of $P(x)$ in Figure 3(a).

EXAMPLE 2 Bounding Real Zeros

Let $P(x) = x^4 - 2x^3 - 10x^2 + 40x - 90$. Find the smallest positive integer and the largest negative integer that, by Theorem 2, are upper and lower bounds, respectively, for the real zeros of $P(x)$. Also note the location of any zeros discovered in the process of searching for upper and lower bounds.

Solution We can use SYNDIV or hand calculations to perform synthetic division for $r = 1, 2, 3, \ldots$ until the quotient row turns nonnegative; then repeat this process

ond does no

EXAMPLE 1 Locating Real Zeros

Let $P(x) = x^3 - 6x^2 + 9x - 3$. Use a table to locate the zeros of $P(x)$ between successive integers and check graphically.

Solution We construct a table and look for sign changes [Fig. 3(a)]. The table in Figure 3(a) shows three sign changes. According to Theorem 1, $P(x)$ must have a real zero in each of the intervals (0, 1), (1, 2), and (3, 4). The graph in Figure 3(b) confirms this. Since $P(x)$ is a cubic polynomial, we have located all of its zeros.

FIGURE 3
$y_1 = P(x) = x^3 - 6x^2 + 9x - 3$.

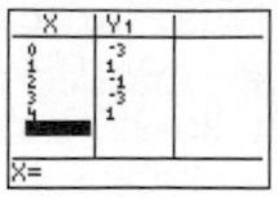
(a)

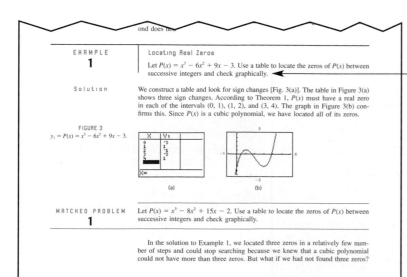

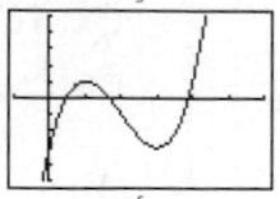

(b)

MATCHED PROBLEM 1 Let $P(x) = x^3 - 8x^2 + 15x - 2$. Use a table to locate the zeros of $P(x)$ between successive integers and check graphically.

In the solution to Example 1, we located three zeros in a relatively few number of steps and could stop searching because we knew that a cubic polynomial could not have more than three zeros. But what if we had not found three zeros?

TECHNOLOGY:

The generic term "graphing utility" is used to refer to any of the various graphing calculators or computer software packages that might be available to students using this book. The use of technology is integrated throughout the text for visualization, investigation, and verification.

A Comparison of Exponential Growth Phenomena

The equations and graphs given in Table 3 compare several widely used growth models. These are divided basically into two groups: unlimited growth and limited growth. Following each equation and graph is a short, incomplete list of areas in which the models are used. We have only touched on a subject that has been extensively developed and that you are likely to study in greater depth in the future.

TABLE 3 Exponential Growth and Decay

Description	Equation	Graph	Uses
Unlimited growth	$y = ce^{kt}$ $c, k > 0$	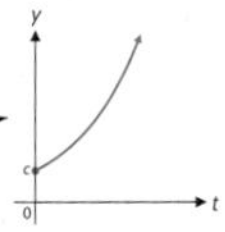	Short-term population growth (people, bacteria, etc.); growth of money at continuous compound interest
Exponential decay	$y = ce^{-kt}$ $c, k > 0$	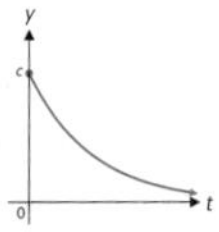	Radioactive decay; light absorption in water, glass, and the like; atmospheric pressure; electric circuits
Limited growth	$y = c(1 - e^{-kt})$ $c, k > 0$	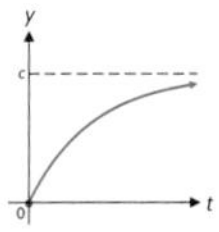	Learning skills; sales fads; company growth; electric circuits
Logistic growth	$y = \dfrac{M}{1 + ce^{-kt}}$ $c, k, M > 0$	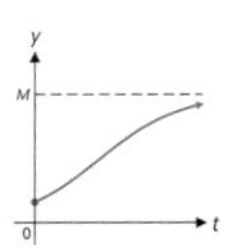	Long-term population growth; epidemics; sales of new products; company growth

GRAPHS AND ILLUSTRATIONS: All graphs in this text are computer generated to ensure mathematical accuracy. Graphing utility screens displayed in the text are actual output from a graphing calculator.

Explore/Discuss 1

An equation that is true for all permissible values of the variable is called an **identity.** An equation that is true for some values of the variable and false for others is called a **conditional equation.** Use algebraic and/or graphical techniques to solve each of the following and identify any identities.

(A) $2(x - 4) = 2x - 8$

(B) $2(x - 4) = 3x - 12$

(C) $2(x - 4) = 2x - 12$

EXAMPLE 2 Solving an Equation with a Variable in the Denominator

Solve algebraically and graphically: $\dfrac{7}{2x} - 3 = \dfrac{8}{3} - \dfrac{15}{x}$

Solution We begin with an algebraic solution. Note that 0 must be excluded from the permissible values of x because division by 0 is not permitted. To clear the fractions, we multiply both sides of the equation by $3(2x) = 6x$, the least common denominator (LCD) of all fractions in the equation. (For a discussion of LCDs and how to find them, see Section A-4.)

$$\frac{7}{2x} - 3 = \frac{8}{3} - \frac{15}{x} \qquad x \neq 0$$

$$6x\left(\frac{7}{2x} - 3\right) = 6x\left(\frac{8}{3} - \frac{15}{x}\right)$$

$$6x \cdot \frac{7}{2x} - 6x \cdot 3 = 6x \cdot \frac{8}{3} - 6x \cdot \frac{15}{x}$$

Multiply by $6x$, the LCD. This and the next step usually can be done mentally.

$$21 - 18x = 16x - 90$$

$$-34x = -111$$

The equation is now free of fractions.

$$x = \frac{111}{34}$$

The solution set is $\{\frac{111}{34}\}$.

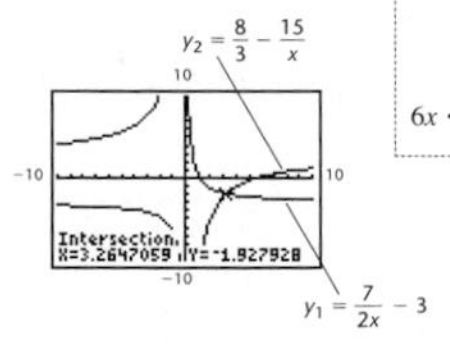

FIGURE 2 Graphical solution of $\frac{7}{2x} - 3 = \frac{8}{3} - \frac{15}{x}$.

Figure 2 shows the graphical solution. Note that $\frac{111}{34} \approx 3.2647059$, to seven decimal places.

STUDENT AIDS:

Annotation of examples and developments, in small colored type, is found throughout the text to help students through critical stages.

Think Boxes are dashed boxes used to enclose steps that are usually performed mentally.

136 | 2 LINEAR AND QUADRATIC FUNCTIONS

DEFINITION 3

EQUALITY AND BASIC OPERATIONS

1. **Equality:** $a + bi = c + di$ if and only if $a = c$ and $b = d$
2. **Addition:** $(a + bi) + (c + di) = (a + c) + (b + d)i$
3. **Multiplication:** $(a + bi)(c + di) = (ac - bd) + (ad + bc)i$

The basic properties of the real number system are discussed in Appendix A, Section A-1. Using the definition of addition and subtraction of complex numbers (Definition 3), it can be shown that the complex number system possesses the same properties. That is,

Screened Boxes are used to highlight important definitions, theorems, results, and step-by-step processes.

BASIC PROPERTIES OF THE COMPLEX NUMBER SYSTEM

1. Addition and multiplication of complex numbers are commutative and associative operations.
2. There is an additive identity and a multiplicative identity for complex numbers.
3. Every complex number has an additive inverse or negative.
4. Every nonzero complex number has a multiplicative inverse or reciprocal.
5. Multiplication distributes over addition.

As a consequence of these properties you will not have to memorize the definitions of addition and multiplication of complex numbers in Definition 3.

We can manipulate complex number symbols of the form $a + bi$ just like we manipulate real binomials of the form $a + bx$, as long as we remember that i is a special symbol for the imaginary unit, not for a real number.

The first arithmetic operation we consider is **addition.**

EXAMPLE 2 Addition of Complex Numbers

Carry out each operation and express the answer in standard form:

(A) $(2 - 3i) + (6 + 2i)$ (B) $(-5 + 4i) + (0 + 0i)$

Solutions

(A) We could apply the definition of addition directly, but it is easier to use complex number properties.

$$(2 - 3i) + (6 + 2i) = 2 - 3i + 6 + 2i \quad \text{Remove parentheses.}$$
$$= (2 + 6) + (-3 + 2)i \quad \text{Combine like terms.}$$
$$= 8 - i$$

Next we find the composition:

$$(f \circ g)(x) = f(g(x)) = f(\sqrt{3 - x})$$
$$= \sqrt{4 - (\sqrt{3 - x})^2}$$
$$= \sqrt{4 - (3 - x)} \qquad (\sqrt{t})^2 = t, t \geq 0$$
$$= \sqrt{1 + x}$$

Even though $\sqrt{1 + x}$ is defined for all $x \geq -1$, we must restrict the domain of $f \circ g$ to those values that also are in the domain of g. Thus,

Domain $f \circ g$: $x \geq -1$ and $x \leq 3$ or $[-1, 3]$

MATCHED PROBLEM 3 Find $(f \circ g)(x)$ and its domain for $f(x) = \sqrt{9 - x^2}$ and $g(x) = \sqrt{x - 1}$.

Caution Boxes appear throughout the text to indicate where student errors often occur.

CAUTION The domain of $f \circ g$ cannot always be determined simply by examining the final form of $(f \circ g)(x)$. Any numbers that are excluded from the domain of g must also be excluded from the domain of $f \circ g$.

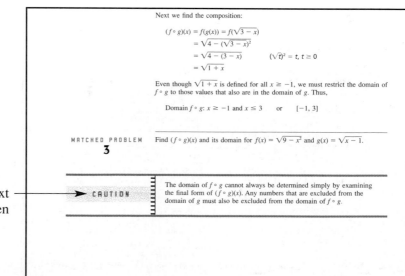

Boldface Type is used to introduce new terms and highlight important comments.

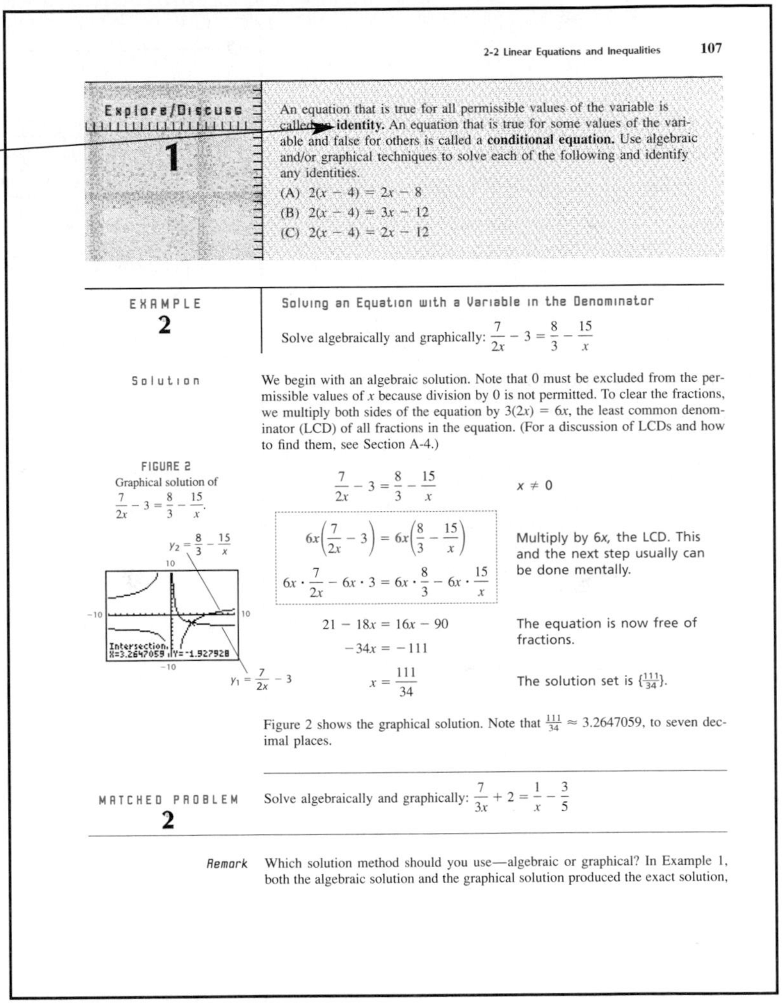

2-2 Linear Equations and Inequalities 107

Explore/Discuss 1

An equation that is true for all permissible values of the variable is called an **identity.** An equation that is true for some values of the variable and false for others is called a **conditional equation.** Use algebraic and/or graphical techniques to solve each of the following and identify any identities.

(A) $2(x - 4) = 2x - 8$

(B) $2(x - 4) = 3x - 12$

(C) $2(x - 4) = 2x - 12$

EXAMPLE 2 Solving an Equation with a Variable in the Denominator

Solve algebraically and graphically: $\frac{7}{2x} - 3 = \frac{8}{3} - \frac{15}{x}$

Solution We begin with an algebraic solution. Note that 0 must be excluded from the permissible values of x because division by 0 is not permitted. To clear the fractions, we multiply both sides of the equation by $3(2x) = 6x$, the least common denominator (LCD) of all fractions in the equation. (For a discussion of LCDs and how to find them, see Section A-4.)

FIGURE 2 Graphical solution of $\frac{7}{2x} - 3 = \frac{8}{3} - \frac{15}{x}$.

$$\frac{7}{2x} - 3 = \frac{8}{3} - \frac{15}{x} \qquad x \neq 0$$

$$6x\left(\frac{7}{2x} - 3\right) = 6x\left(\frac{8}{3} - \frac{15}{x}\right)$$

$$6x \cdot \frac{7}{2x} - 6x \cdot 3 = 6x \cdot \frac{8}{3} - 6x \cdot \frac{15}{x}$$

Multiply by 6x, the LCD. This and the next step usually can be done mentally.

$$21 - 18x = 16x - 90$$

$$-34x = -111$$

The equation is now free of fractions.

$$x = \frac{111}{34}$$

The solution set is $\{\frac{111}{34}\}$.

Figure 2 shows the graphical solution. Note that $\frac{111}{34} \approx 3.2647059$, to seven decimal places.

MATCHED PROBLEM 2 Solve algebraically and graphically: $\frac{7}{3x} + 2 = \frac{1}{x} - \frac{3}{5}$

Remark Which solution method should you use—algebraic or graphical? In Example 1, both the algebraic solution and the graphical solution produced the exact solution,

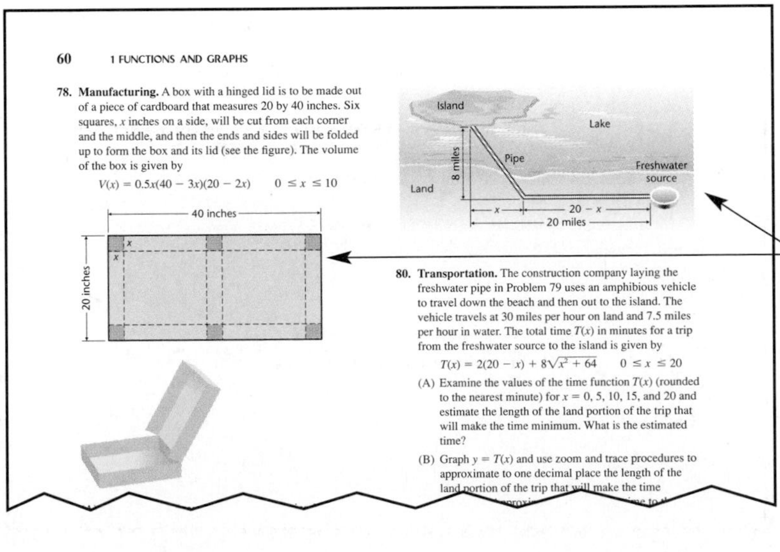

60 1 FUNCTIONS AND GRAPHS

78. Manufacturing. A box with a hinged lid is to be made out of a piece of cardboard that measures 20 by 40 inches. Six squares, x inches on a side, will be cut from each corner and the middle, and then the ends and sides will be folded up to form the box and its lid (see the figure). The volume of the box is given by

$$V(x) = 0.5x(40 - 3x)(20 - 2x) \qquad 0 \le x \le 10$$

80. Transportation. The construction company laying the freshwater pipe in Problem 79 uses an amphibious vehicle to travel down the beach and then out to the island. The vehicle travels at 30 miles per hour on land and 7.5 miles per hour in water. The total time $T(x)$ in minutes for a trip from the freshwater source to the island is given by

$$T(x) = 2(20 - x) + 8\sqrt{x^2 + 64} \qquad 0 \le x \le 20$$

(A) Examine the values of the time function $T(x)$ (rounded to the nearest minute) for $x = 0, 5, 10, 15$, and 20 and estimate the length of the land portion of the trip that will make the time minimum. What is the estimated time?

(B) Graph $y = T(x)$ and use zoom and trace procedures to approximate to one decimal place the length of the land portion of the trip that will make the time

Functional Use of Four Colors improves the clarity of many illustrations, graphs, and developments, and guides students through certain critical steps.

Chapter Review sections are provided at the end of each chapter and include a thorough review of all the important terms and symbols. This recap is followed by a comprehensive set of review exercises.

CHAPTER 2 | REVIEW

2-1 LINEAR FUNCTIONS

A function f is a **linear function** if $f(x) = mx + b$, $m \neq 0$, where m and b are real numbers. The **domain** is the set of all real numbers and the **range** is the set of all real numbers. If $m = 0$, then f is called a **constant function,** $f(x) = b$, which has the set of all real numbers as its *domain* and the constant b as its *range*. The **standard form** for the equation of a line is $Ax + By = C$, where A, B, and C are real constants, A and B not both 0. Every straight line in a Cartesian coordinate system is the graph of an equation of this type. The **slope** of the line through the points (x_1, y_1) and (x_2, y_2) is

$$m = \frac{y_2 - y_1}{x_2 - x_1} \qquad x_1 \neq x_2$$

The slope is not defined for a vertical line where $x_1 = x_2$. Two nonvertical lines with slopes m_1 and m_2 are **parallel** if and only if $m_1 = m_2$ and **perpendicular** if and only if $m_1 m_2 = -1$.

Equations of a Line

Standard Form	$Ax + By = C$	A and B not both 0
Slope–Intercept Form	$y = mx + b$	Slope: m; y intercept: b
Point–Slope Form	$y - y_1 = m(x - x_1)$	Slope: m; point: (x_1, y_1)
Horizontal Line	$y = b$	Slope: 0
Vertical Line	$x = a$	Slope: undefined

2-2 LINEAR EQUATIONS AND INEQUALITIES

The solution sets of linear equations can be found *algebraically* using the **properties of equality** to transform the given equation into an **equivalent equation** that has an obvious solution (see Section A-8) and approximated **graphically** using a graphing utility to find **intersection points.** An equation that is true for all permissible values of the variable is called an **identity.** An equation that is true for some values of the variable and false for others is called a **conditional equation.**

Linear inequalities in one variable are expressed using the inequality symbols $<$, $>$, $\leq$, and $\geq$. The **solution set of an inequality** is the set of all values of the variable that make the inequality a true statement. Each element of the solution set is called a **solution.** An equivalent inequality will result and the **sense will remain the same** if each side of the original inequality:

1. Has the same real number added to or subtracted from it.
2. Is multiplied or divided by the same positive number.

An equivalent inequality will result and the **sense will reverse** if each side of the original inequality:

3. Is multiplied or divided by the same negative number.

Note that multiplication by 0 and division by 0 are not permitted.

2-3 QUADRATIC FUNCTIONS

If a, b, and c are real numbers with $a \neq 0$, then the function $f(x) = ax^2 + bx + c$ is a **quadratic function** and its graph is a **parabola. Completing the square** of the quadratic expression $x^2 + bx$ produces a perfect square:

$$x^2 + bx + \left(\frac{b}{2}\right)^2 = \left(x + \frac{b}{2}\right)^2$$

Completing the square for $f(x) = ax^2 + bx + c$ produces the **standard form** $f(x) = a(x - h)^2 + k$ and the following properties:

1. The graph of f is a parabola:

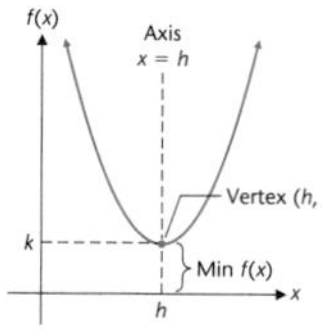

$a > 0$
Opens upward

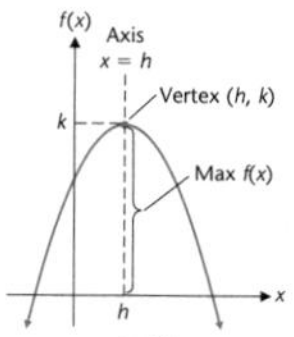

$a < 0$
Opens downward

2. Vertex: (h, k) (parabola increases on one side of the vertex and decreases on the other.)
3. Axis (of symmetry): $x = h$ (parallel to y axis)
4. $f(h) = k$ is the minimum if $a > 0$ and the maximum if $a < 0$.
5. Domain: All real numbers; range: $(-\infty, k]$ if $a < 0$ or $[k, \infty)$ if $a > 0$.
6. The graph of f is the graph of $g(x) = ax^2$ translated horizontally h units and vertically k units.

2-4 COMPLEX NUMBERS

A **complex number** in *standard form* is a number in the form $a + bi$, where a and b are real numbers and i is the **imaginary unit.** If $b \neq 0$, then $a + bi$ is also called an **imaginary number.** If $a = 0$ then $0 + bi = bi$ is also called a **pure imaginary number.** If $b = 0$, then $a + 0i = a$ is a **real number.** The complex

A Cumulative Review Exercise is provided after every second or third chapter, for additional reinforcement.

330 | 4 INVERSE FUNCTIONS: EXPONENTIAL AND LOGARITHMIC FUNCTIONS

$$I = I_0e^{-kd}$$

where I is the intensity d feet below the surface, I_0 is the intensity at the surface, and k is the coefficient of extinction. Measurements in the Sargasso Sea in the West Indies have indicated that half the surface light reaches a depth of 73.6 feet. Find k, and find the depth at which 1% of the surface light remains. Compute answers to three significant digits.

★ **95. Wildlife Management.** A lake formed by a newly constructed dam is stocked with 1,000 fish. Their population is expected to increase according to the logistic curve

$$N = \frac{30}{1 + 29e^{-1.35t}}$$

where N is the number of fish, in thousands, expected after t years. The lake will be open to fishing when the number of fish reaches 20,000. How many years, to the nearest year, will this take?

96. Medicare. The annual expenditures for Medicare (in billions of dollars) by the U.S. government for selected years since 1980 are shown in Table 1 (Bureau of the Census). Let x represent years since 1980.

TABLE 1 Medicare Expenditures

Year	Billion $
1980	37
1985	72
1990	111
1995	181

Source: U.S. Bureau of the Census.

(A) Find an exponential regression model of the form $y = ab^x$ for this data. Estimate (to the nearest billion) the total expenditures in 1996 and in 2010.

(B) When (to the nearest year) will the total expenditures reach 500 billion?

97. Agriculture. The total U.S. corn consumption (in millions of bushels) is shown in Table 2 for selected years since 1975. Let x represent years since 1900.

TABLE 2 Corn Consumption

Year	x	Total Consumption (million bushels)
1975	75	522
1980	80	659
1985	85	1,152
1990	90	1,373
1995	95	1,690

Source: U.S. Department of Agriculture.

(A) Find a logarithmic regression model of the form $y = a + b \ln x$ for this data. Estimate (to the nearest million) the total consumption in 1996 and in 2010.

(B) The actual consumption in 1996 was 1,583 million bushels. How does this compare with the estimated consumption in part A? What effect will this additional 1996 information have on the estimate for 2010? Explain.

CUMULATIVE REVIEW EXERCISE FOR CHAPTERS 3 AND 4

Work through all the problems in this cumulative review and check answers in the back of the book. Answers to all review problems are there, and following each answer is a number in italics indicating the section in which that type of problem is discussed. Where weaknesses show up, review appropriate sections in the text.

A

1. Let $P(x)$ be the polynomial whose graph is shown in the figure.

(A) Assuming that $P(x)$ has integer zeros and leading coefficient 1, find the lowest-degree equation that could produce this graph.

(B) Describe the left and right behavior of $P(x)$.

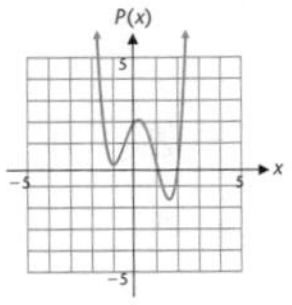

2. Match each equation with the graph of f, g, m, or n in the figure.

(A) $y = (\frac{3}{4})^x$ (B) $y = (\frac{4}{3})^x$

(C) $y = (\frac{3}{4})^x + (\frac{4}{3})^x$ (D) $y = (\frac{4}{3})^x - (\frac{3}{4})^x$

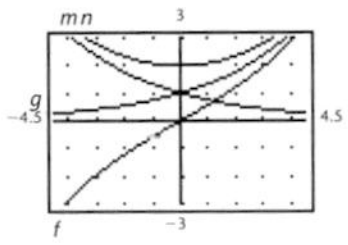

3. For $P(x) = 3x^3 + 5x^2 - 18x - 3$ and $D(x) = x + 3$, use synthetic division to divide $P(x)$ by $D(x)$, and write the answer in the form $P(x) = D(x)Q(x) + R$.

Supplements

A comprehensive set of ancillary materials for both the student and the instructor is available for use with this text.

Student's Solutions Manual: This supplement is available for sale to the student, and includes detailed solutions to all odd-numbered problems and most review exercises.

Instructor's Solutions Manual: This manual provides solutions to even-numbered problems and answers to all problems in the text.

Instructor's Resource Manual: This supplement provides transparency masters and sample tests for each chapter in the text.

Print and Computerized Testbanks: A Computerized Testbank is available that provides a variety of formats to enable the instructor to create tests using both algorithmically generated test questions and those from a static testbank. This testing system enables the instructor to choose questions either manually or randomly by section, question types, difficulty level, and other criteria. This testing software is available for PC and Macintosh computers. A softcover print version of the testbank includes the static questions found in the computerized version.

Barnett/Ziegler/Byleen Video Series: Course videotapes, created new for this series, provide students with additional reinforcement of the topics presented in the book. These videos are keyed specifically to the text and feature an effective combination of learning techniques, including personal instruction, state-of-the-art graphics, and real-world applications.

Interactive Diagrams CD-ROM: This software package is available for sale to the student. This CD contains 44 Interactive Diagrams that are designed for use with this textbook. Each Interactive Diagram (ID) is a separate Java Applet that contains an illustration that can be manipulated by the user for further conceptual understanding of the topic presented. For each section of the text where an ID has been created, an icon ID has been placed in the margin.

Course Solutions: Fully integrated multimedia, a full-scale Online Learning Center, and a Course Integration Guide are designed specifically to help you with your individual college algebra course needs. Assembled by an expert in your field of study, this printed manual fully integrates the numerous products available to accompany *College Algebra: A Graphing Approach.* The Course Integration Guide will also contain detailed solutions for the Group Activities found in the text, a description of each Interactive Diagram, and detailed solutions to the exploratory questions that accompany each Interactive Diagram.

In addition to the supplements listed above, a number of other technology and Web-based ancillaries are under development; they will support the ever-changing technology needs in college algebra and precalculus. For further information about these or any supplements, please contact your local McGraw-Hill sales representative, or visit our website at www.mhhe.com/barnett.

Accuracy

This book has been carefully checked by a number of mathematicians. If any errors are detected, the authors would be grateful if they were sent to: Karl E. Byleen, 9322 W. Garden Court, Hales Corners, WI 53130; or, by e-mail, to byleen@execpc.com.

Acknowledgments

In addition to the authors, many others are involved in the successful publication of this book. We wish to thank personally:

Precalculus: A Graphing Approach: Barbara Blythin, University of Nevada at Las Vegas; Jack Bookman, Duke University; Maria Brunett, Montgomery College; Bettyann Daley, University of Delaware; Patricia Dueck, Arizona State University; Joseph R. Ediger, Portland State University; Frances Gulick, University of Maryland–College Park Campus; Peg Hovde, Grossmont College; Giles W. Maloof, Boise State University; Carl Miller, San Diego City College; Michael Montano, Riverside Community College; Jim Paige, Wayne State College; Sandra Rucker, Clark Atlanta University; Jack C. Sharp, Floyd College.

College Algebra: A Graphing Approach: Jeffrey V. Berg, Arapahoe Community College; Paul Blankenship, Lexington Community College; Marc D. Campbell, Daytona Beach Community College; Janis M. Cimperman, St. Cloud State University; Dick J. Clark, Portland Community College; John Hamm, University of New Mexico; Jean M. Horn, Northern Virginia Community College–Woodbridge Campus; Adam Lewenberg, University of Akron; Lyn Miller, Western Kentucky University; Karen Mitchell, Rowan Cabarrus Community College; Robert Pearce, South Plains College; Mary Rice, University of Nevada at Las Vegas; Kathy V. Rodgers, University of Southern Indiana; JoAnn Royster, University of Central Arkansas; Bryan Stewart, Tarrant County Junior College; Tom Williams, Rowan Cabarrus Community College.

We also wish to thank:

Janis Cimperman, Margaret Donlan, Gitta Safier, Hossein Hamedani, and Caroline Woods for providing a careful and thorough check of all the mathematical calculations in this text, the Student's Solutions Manual, and the Instructor's Solutions Manual (a tedious but extremely important job).

Bettyann Daly, Joan Van Glabek, Norma James, Jare Confrey, Alan Maloney, Carolyn Meitler, Karen Davis, and Fred Safier, for developing the supplemental materials that are so important to the success of a text.

Jeanne Wallace for accurately and efficiently producing the Instructor's and Student's Solutions Manuals to supplement the text.

George and Brian Morris and their staff at Scientific Illustrators for their effective illustrations and accurate graphs.

All the people at McGraw-Hill who contributed their efforts to the production of this book, especially Sheila Frank, Nina Kreiden, J.P. Lenney, and Maggie Rogers.

Producing this new series with the help of all these extremely competent people has been a most satisfying experience.

R. A. Barnett
M. R. Ziegler
K. E. Byleen

Chapter Dependencies

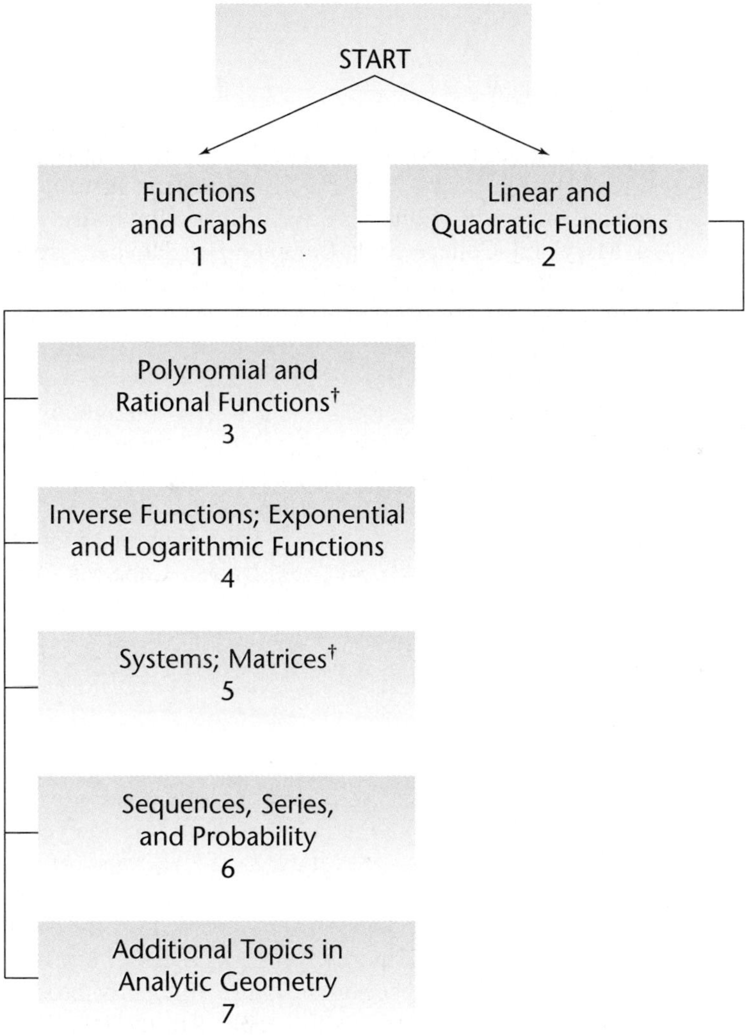
START
Functions and Graphs 1
Linear and Quadratic Functions 2
Polynomial and Rational Functions† 3
Inverse Functions; Exponential and Logarithmic Functions 4
Systems; Matrices† 5
Sequences, Series, and Probability 6
Additional Topics in Analytic Geometry 7

†Appendix B, Partial Fractions, can be covered in conjunction with this chapter.

Contents

Applications Index

MCGRAW-HILL IS PROUD TO OFFER AN EXCITING NEW SUITE OF MULTIMEDIA PRODUCTS AND SERVICES CALLED COURSE SOLUTIONS.

Designed specifically to help you with your individual course needs, **Course Solutions** will assist you in integrating your syllabus with our premier titles and state-of-the-art new media tools that support them.

AT THE HEART OF COURSE SOLUTIONS YOU'LL FIND:

- Fully integrated multimedia
- A full-scale Online Learning Center
- A Course Integration Guide

AS WELL AS THESE UNPARALLELED SERVICES:

- McGraw-Hill Learning Architecture
- McGraw-Hill Course Consultant Service
- McGraw-Hill Student Tutorial Service
- McGraw-Hill Instructor Syllabus Service
- PageOut Lite
- PageOut: The Course Web Site Development Center
- Other Delivery Options

COURSE SOLUTIONS truly has the solutions to your every teaching need. Read on to learn how we can specifically help you with your classroom challenges.

SPECIAL ATTENTION
to your specific needs.

These "perks" are all part of the extra service delivered through McGraw-Hill's **Course Solutions:**

MCGRAW-HILL LEARNING ARCHITECTURE

Each McGraw-Hill *Online Learning Center* is ready to be ported into our *McGraw-Hill Learning Architecture*—a full course management software system for Local Area Networks and Distance Learning Classes. Developed in conjunction with Top Class software, *McGraw-Hill Learning Architecture* is a powerful course management system available upon special request.

MCGRAW-HILL COURSE CONSULTANT SERVICE

In addition to the *Course Integration Guide*, instructors using **Course Solutions** textbooks can access a special curriculum-based *Course Consultant Service* within each *Online Learning Center*. A **McGraw-Hill Course Solutions Consultant** will personally help you—as a text adopter—integrate this text and media into your course to fit your specific needs. This content-based service is offered in addition to our usual software support services.

MCGRAW-HILL INSTRUCTOR SYLLABUS SERVICE

For *new* adopters of **Course Solutions** textbooks, McGraw-Hill will help correlate all text, supplement, and appropriate materials and services to your course syllabus. Simply call your McGraw-Hill sales representative for assistance.

PAGEOUT LITE

Free to **Course Solutions** textbook adopters, *PageOut Lite* is perfect for instructors who want to create their own Web site. In just a few minutes, even novices can turn their syllabus into a Web site using *PageOut Lite*.

PAGEOUT: THE COURSE WEB SITE DEVELOPMENT CENTER

For those who want the benefits of *PageOut Lite's* no-hassle approach to site development, but with even more features, we offer *PageOut: The Course Web Site Development Center.*

PageOut shares many of *PageOut Lite's* features, but also enables you to create links that will take your students to your original material, other Web site addresses, and to *McGraw-Hill Online Learning Center* content. This means you can assign *Online Learning Center* content within your syllabus-based Web site. *PageOut* also features a discussion board list where you and your students can exchange questions and post announcements, as well as an area for students to build personal Web pages.

OTHER DELIVERY OPTIONS

Online Learning Centers are also compatible with a number of full-service online course delivery systems or outside educational service providers. For a current list of compatible delivery systems, contact your McGraw-Hill sales representative.

And for your students . . .
MCGRAW-HILL STUDENT TUTORIAL SERVICE

Within each *Online Learning Center* resides a **FREE** *Student Tutorial Service*. This web-based "homework hotline" features guaranteed, 24-hour response time on weekdays. Students can also receive immediate tutorial assistance via an Internet whiteboard during regularly scheduled Net Tutor office hours.

www.mhhe.com/barnett

To the Student

The following suggestions are made to help you get the most out of this book and your efforts.

As you study the text we suggest a five-step approach. For each section,

1. Read the mathematical development.
2. Work through the illustrative examples.
3. Work the matched problem.
4. Review the main ideas in the section.
5. Work the assigned exercises at the end of the section.

Repeat the 1-2-3 cycle until the section is finished.

All of this should be done with graphing utility, paper, and pencil at hand. In fact, no mathematics text should be read without pencil and paper in hand; mathematics is not a spectator sport. Just as you cannot learn to swim by watching someone else swim, you cannot learn mathematics by simply reading worked examples—you must work problems, lots of them. In writing this text, we have assumed that you always have a graphing utility at hand. We have included many screen shots from a popular graphing utility, the Texas Instruments TI-83, but any standard graphing utility will do. As you read the text, you should reproduce each screen that is displayed in the text on your graphing utility. This will ensure that you are developing the graphing utility skills you will need to solve the matched problems and the exercises at the end of the section.

A graphing utility is a powerful tool for analyzing graphs. Like any good tool, it must be used properly. In particular, the display on most graphing utilities produces rough approximations to graphs. One of the skills you must develop as you study from this book is the ability to visualize the correct appearance of a graph, based on the rough graph produced by a graphing utility and your understanding of the mathematics under discussion. For example, consider the two figures shown in Figure 1.

FIGURE 1

$f(x) = \dfrac{x}{x^2 - 4}$.

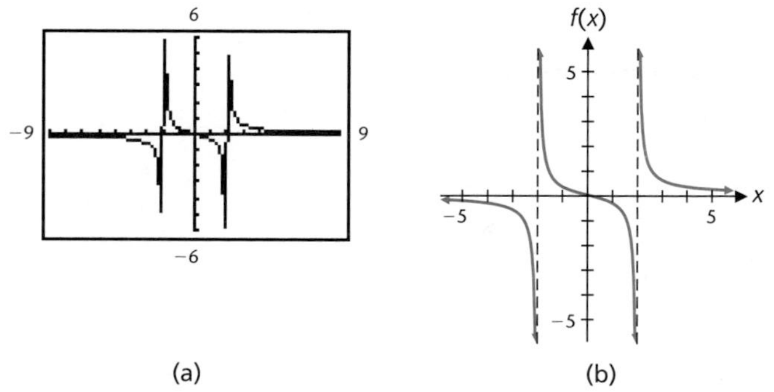

The graphing utility graph in Figure 1(a) seems to indicate that the graph is one continuous curve. However, the more accurate graph in Figure 1(b) clearly shows that the graph consists of three separate curves. When you study functions of this type in the text, you will learn that f is not defined at $x = -2$ and $x = 2$, but has vertical asymptotes at these points. Armed with this knowledge, you will be able to look at the graphing utility graph in Figure 1(a), visualize the more accurate graph in Figure 1(b), and sketch a good representation of the graph by hand.

If you have difficulty with the course, then, in addition to doing the regular assignments, spend more time on the examples and matched problems and work more A exercises, even if they are not assigned. If you find the course too easy, then work more C exercises and applied problems, even if they are not assigned. If you have difficulty with your graphing utility, consult our website (www.mhhe.com/barnett) or consult your calculator's user manual.

R. A. Barnett
M. R. Ziegler
K. E. Byleen

Functions and Graphs

1

Outline

Application

The data shown in the chart represent the yearly totals for personal income and savings in the United States. Use linear regression on a graphing utility to predict personal income when personal savings grow to $300 billion.

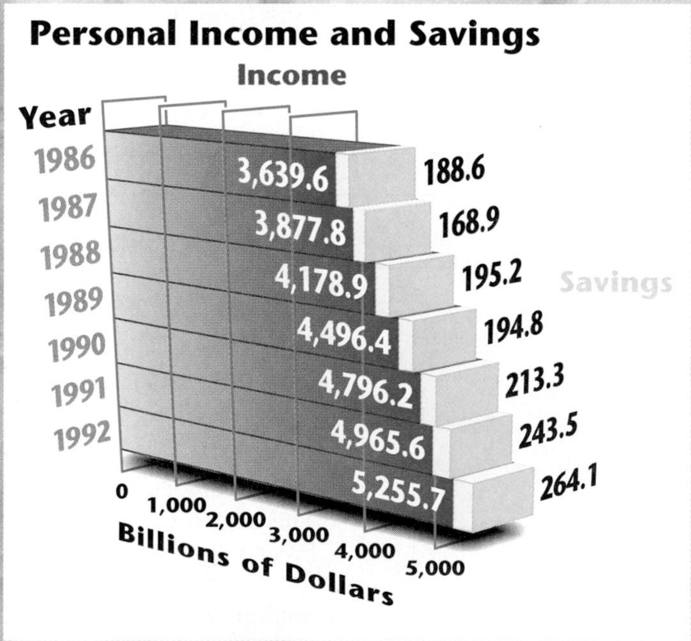

The function concept is one of the most important ideas in mathematics. The study of either the theory or the applications of mathematics beyond the most elementary level requires a firm understanding of functions and their graphs. The first two sections of this chapter are concerned with basic graphing techniques, including point-by-point plotting, graphing circles, and using an electronic graphing device such as a graphing calculator or a computer. In the remaining sections, we introduce the important concept of a function, discuss basic properties of functions and their graphs, and examine specific types of functions. Much of the remainder of this book is concerned with applying the ideas introduced in this chapter to a variety of different types of functions, as is evidenced by the chapter titles following this chapter (check the table of contents). Efforts made to understand and use the function concept correctly from the beginning will be rewarded many times in this course and in most future courses that involve mathematics.

Preparing for This Chapter

Before getting started on this chapter, review the following concepts:

Set Notation (Appendix A, Section A-1)

Polynomials (Appendix A, Sections A-2 and A-3)

Rational Expressions (Appendix A, Section A-4)

Square Root Radicals (Appendix A, Section A-7)

Interval Notation (Appendix A, Section A-8)

Pythagorean Theorem (Appendix D)

Section 1-1 Cartesian Coordinate System

- Cartesian Coordinate System
- Graphing: Point by Point
- Distance between Two Points
- Circles

Analytic geometry is concerned with the relationship between geometric forms, such as circles and lines, and algebraic forms, such as equations and inequalities. The key to this relationship is the Cartesian coordinate system, named after the French mathematician and philosopher René Descartes (1596–1650) who was the first to combine the study of algebra and geometry into a single discipline. In this section, we develop some of the basic tools used in analytic geometry and apply these tools to the graphing of equations and to the derivation of the equation of a circle.

Cartesian Coordinate System

FIGURE 1
Cartesian coordinate system.

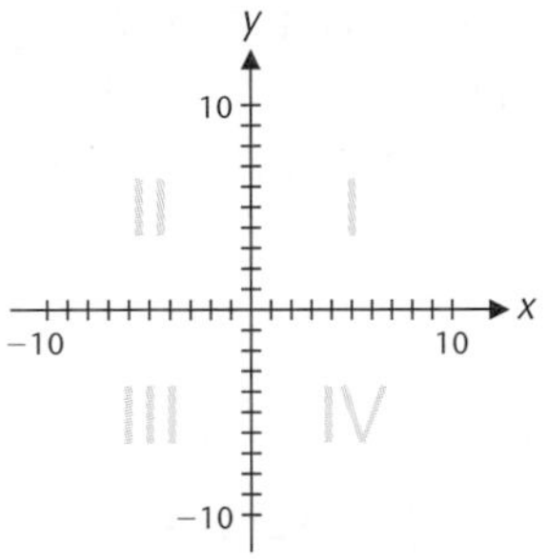

FIGURE 2
Coordinates in a plane.

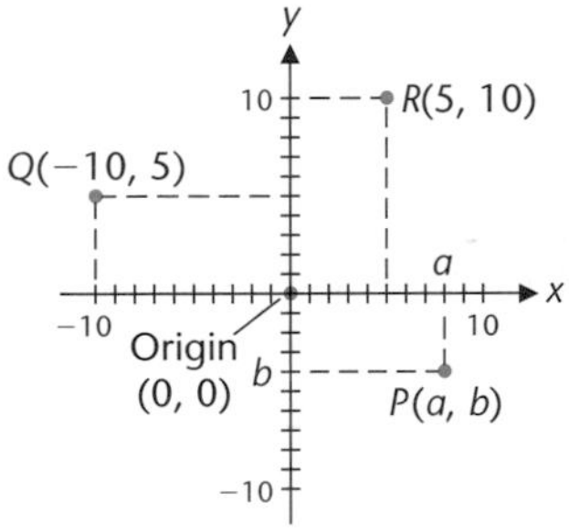

Just as a real number line establishes a one-to-one correspondence between the points on a line and the elements in the set of real numbers, we can form a **real plane** by establishing a one-to-one correspondence between the points in a plane and elements in the set of all ordered pairs of real numbers. This can be done by means of a Cartesian coordinate system.

Recall that to form a **Cartesian** or **rectangular coordinate system,** we select two real number lines, one horizontal and one vertical, and let them cross through their origins as indicated in Figure 1. Up and to the right are the usual choices for the positive directions. These two number lines are called the **horizontal axis** and the **vertical axis,** or together, the **coordinate axes.** The horizontal axis is usually referred to as the ***x* axis** and the vertical axis as the ***y* axis,** and each is labeled accordingly. Other labels may be used in certain situations. The coordinate axes divide the plane into four parts called **quadrants,** which are numbered counterclockwise from I to IV (see Fig. 1).

Now we want to assign *coordinates* to each point in the plane. Given an arbitrary point P in the plane, pass horizontal and vertical lines through the point (Fig. 2). The vertical line will intersect the horizontal axis at a point with coordinate a, and the horizontal line will intersect the vertical axis at a point with coordinate b. These two numbers written as the ordered pair (a, b) form the **coordinates** of the point P. The first coordinate a is called the **abscissa** of P; the second coordinate b is called the **ordinate** of P. The abscissa of Q in Figure 2 is -10, and the ordinate of Q is 5. The coordinates of a point can also be referenced in terms of the axis labels. The ***x* coordinate** of R in Figure 2 is 5, and the ***y* coordinate** of R is 10. The point with coordinates (0, 0) is called the **origin.**

The procedure we have just described assigns to each point P in the plane a unique pair of real numbers (a, b). Conversely, if we are given an ordered pair of real numbers (a, b), then, reversing this procedure, we can determine a unique point P in the plane. Thus:

There is a one-to-one correspondence between the points in a plane and the elements in the set of all ordered pairs of real numbers.

One set of ordered pairs

This is often referred to as the **fundamental theorem of analytic geometry.** Given any set of ordered pairs S, the **graph** of S is the set of points in the plane corresponding to the ordered pairs in S.

Graphing: Point by Point

The fundamental theorem of analytic geometry enables us to look at algebraic forms geometrically and to look at geometric forms algebraically. We begin by considering an algebraic form, an equation in two variables:

$$y = x^2 - 4 \tag{1}$$

A **solution** to equation (1) is an ordered pair of real numbers (a, b) such that

$$b = a^2 - 4$$

The **solution set** of equation (1) is the set of all these ordered pairs. More formally,

$$\text{Solution set of equation (1): } \{(x, y) \mid y = x^2 - 4\} \tag{2}$$

To find a solution for equation (1) we simply replace x with a number and calculate the value of y. For example, if $x = 2$, then $y = 2^2 - 4 = 0$, and the ordered pair (2, 0) is a solution. Similarly, if $x = -3$, then $y = (-3)^2 - 4 = 5$, and the ordered pair $(-3, 5)$ is also a solution of equation (1). In fact, any real number substituted for x in equation (1) will produce a solution to the equation. Thus, the solution set shown in (2) must have an infinite number of elements. We now use a rectangular coordinate system to provide a geometric representation of this set.

The **graph of an equation** is the graph of its solution set. To *sketch the graph of an equation,* we plot enough points from its solution set so that the total graph is apparent and then connect these points with a smooth curve, proceeding from left to right. This process is called **point-by-point plotting.**

EXAMPLE 1

Graphing an Equation Using Point-by-Point Plotting

Sketch a graph of $y = x^2 - 4$.

Solution

We make up a table of solutions—ordered pairs of real numbers that satisfy the given equation.

FIGURE 3

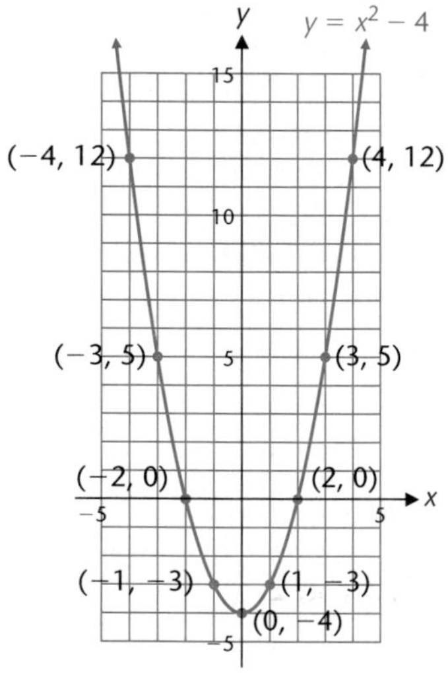

x	-4	-3	-2	-1	0	1	2	3	4
y	12	5	0	-3	-4	-3	0	5	12

After plotting these solutions, if there are any portions of the graph that are unclear, we plot additional points until the shape of the graph is apparent. Then we join all these plotted points with a smooth curve, as shown in Figure 3. Arrowheads are used to indicate that the graph continues beyond the portion shown here with no significant changes in shape.

The resulting figure is called a *parabola.* Notice that if we fold the paper along the y axis, the right side will match the left side. We say that the graph is *symmetric with respect to the y axis* and call the y axis the *axis of the parabola.* More will be said about parabolas later in the text.

MATCHED PROBLEM 1*

Sketch a graph of $y = 8 - x^2$ using point-by-point plotting.

EXAMPLE 2

Graphing an Equation Using Point-by-Point Plotting

Sketch a graph of $y = \sqrt{x}$.

Solution

Proceeding as before, we make up a table of solutions:

x	0	1	2	3	4	5	6	7	8	9
y	0	1	$\sqrt{2} \approx 1.4$	$\sqrt{3} \approx 1.7$	2	$\sqrt{5} \approx 2.2$	$\sqrt{6} \approx 2.4$	$\sqrt{7} \approx 2.6$	$\sqrt{8} \approx 2.8$	3

*Answers to matched problems in a given section are found near the end of the section, before the exercise set.

FIGURE 4

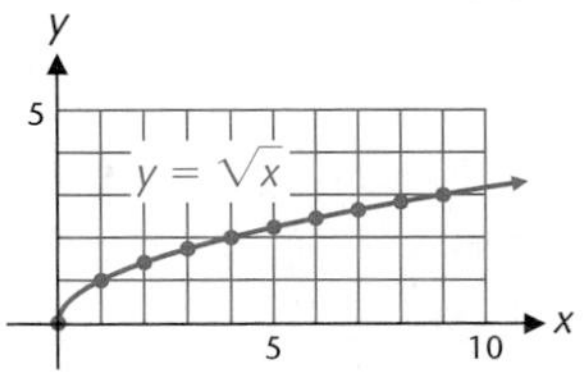

For graphing purposes, the irrational numbers in the table were evaluated on a calculator and rounded to one decimal place. Plotting these points and connecting them with a smooth curve produces the graph in Figure 4.

Notice that we did not include any negative values of x in the table. If x is a negative real number, then $\sqrt{x}$ is not a real number. Since the coordinates of a point in a rectangular coordinate system must be real numbers, *when graphing an equation, we consider only those values of the variables that produce real solutions to the equation.* We will have more to say about numbers of the form $\sqrt{x}$, where x is negative, later in this book.

MATCHED PROBLEM **2**

Sketch a graph of $y = 4 - \sqrt{x}$.

Explore/Discuss 1

To graph the equation $y = -x^3 + 2x$, we use point-by-point plotting to obtain the graph in Figure 5.

x	y
−1	−1
0	0
1	1

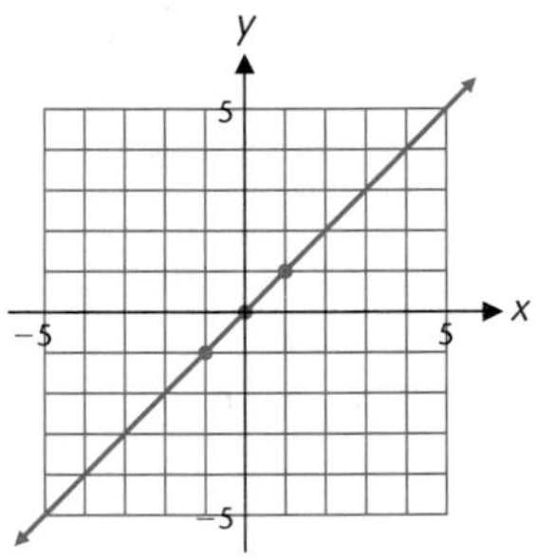

FIGURE 5

(A) Do you think this is the correct graph of the equation? If so, why? If not, why?

(B) Add points on the graph for $x = -2, -0.5, 0.5$, and 2.

(C) Now, what do you think the graph looks like? Sketch your version of the graph, adding more points as necessary.

(D) Write a short statement explaining any conclusions you might draw from parts (A), (B), and (C).

As Explore/Discuss 1 illustrates, sometimes it can be difficult to determine the apparent shape of a graph by simply plotting a few points. One of the major objectives of this book is to develop mathematical tools that will help us analyze graphs.

The use of graphs to illustrate relationships between quantities is commonplace. Estimating the coordinates of points on a graph provides specific examples of this relationship, even if no equation for the graph is available. The next example illustrates this process.

EXAMPLE **3**

Ozone Levels

The ozone level is measured in parts per billion (ppb). The ozone level during a 12-hour period in a suburb of Milwaukee, Wisconsin, on a particular summer day is given in Figure 6. Use this graph to estimate the following ozone levels to the nearest integer and times to the nearest quarter hour:

(A) The ozone level at 6 P.M.

(B) The highest ozone level and the time when it occurs.

(C) The time(s) when the ozone level is 90 ppb.

FIGURE 6

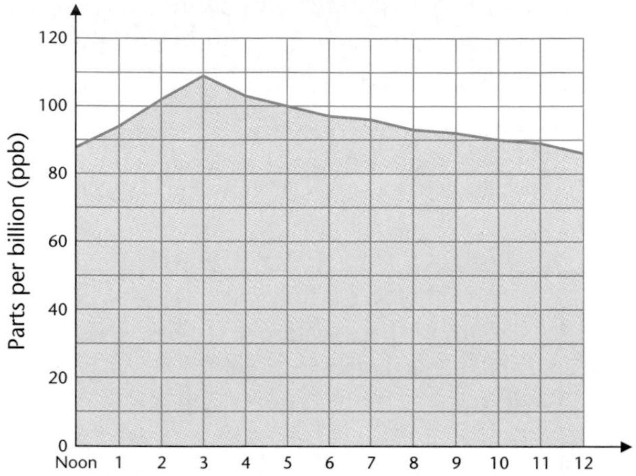

Source: Wisconsin Department of Natural Resources.

Solutions

(A) The ordinate of the point on the graph with abscissa 6 is approximately 97 ppb (see Fig. 7).

(B) The highest ozone level is approximately 109 ppb at 3 P.M.

(C) The ozone level is 90 ppb at about 12:30 P.M. and again at 10 P.M.

FIGURE 7

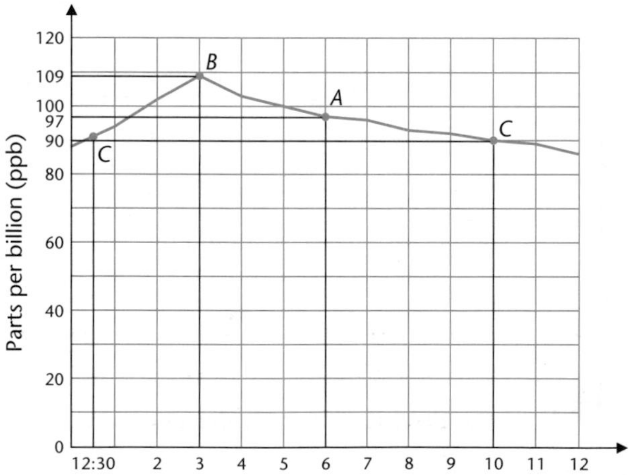

MATCHED PROBLEM 3

Use the graph in Figure 6 to estimate the following ozone level to the nearest integer and times to the nearest quarter hour.

(A) The ozone level at 7 P.M.

(B) The time(s) when the ozone level is 100 ppb.

Distance between Two Points

Analytic geometry is concerned with two basic problems:

1. Given an equation, find its graph.
2. Given a figure (line, circle, parabola, ellipse, etc.) in a coordinate system, find its equation.

So far we have concentrated on the first problem. We now introduce a basic tool that is used extensively in solving the second problem. This basic tool is the *distance-between-two-points formula,* which is easily derived using the Pythagorean theorem (see Appendix D). Let $P_1(x_1, y_1)$ and $P_2(x_2, y_2)$ be two points in a rectangular coordinate system and let $d(P_1, P_2)$ represent the distance between these two points. Then referring to Figure 8, we see that

$$[d(P_1, P_2)]^2 = |x_2 - x_1|^2 + |y_2 - y_1|^2$$
$$= (x_2 - x_1)^2 + (y_2 - y_1)^2 \quad \text{Since } |N|^2 = N^2.$$

FIGURE 8
Distance between two points.

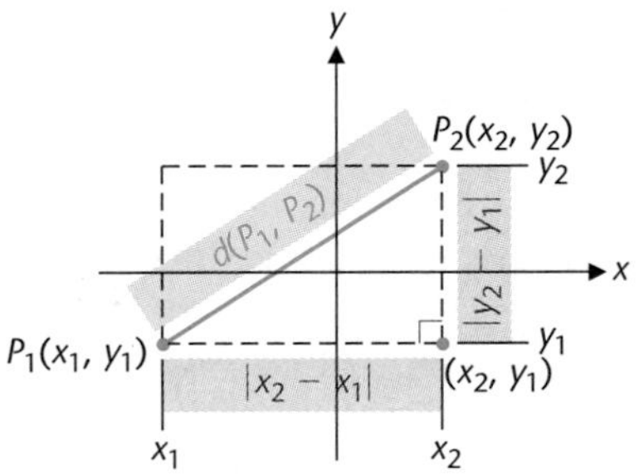

Thus:

THEOREM 1

DISTANCE BETWEEN $P_1(x_1, y_1)$ AND $P_2(x_2, y_2)$

$$d(P_1, P_2) = \sqrt{(x_2 - x_1)^2 + (y_2 - y_1)^2}$$

EXAMPLE 4

Using the Distance-between-Two-Points Formula

Find the distance between the points $(-3, 5)$ and $(-2, -8)$.*

Solution

It doesn't matter which point we designate as P_1 or P_2 because of the squaring in the formula. Let $(x_1, y_1) = (-3, 5)$ and $(x_2, y_2) = (-2, -8)$. Then,

$$d = \sqrt{[(-2) - (-3)]^2 + [(-8) - 5]^2}$$
$$= \sqrt{(-2 + 3)^2 + (-8 - 5)^2} = \sqrt{1^2 + (-13)^2} = \sqrt{1 + 169} = \sqrt{170}$$

*We often speak of the point (a, b) when we are referring to the point with coordinates (a, b). This shorthand, though not accurate, causes little trouble, and we will continue the practice.

Notice that if we choose $(x_1, y_1) = (-2, -8)$ and $(x_2, y_2) = (-3, 5)$, then

$$d = \sqrt{[(-3) - (-2)]^2 + [5 - (-8)^2]} = \sqrt{1 + 169} = \sqrt{170}$$

so it doesn't matter which point we designate as P_1 or P_2.

MATCHED PROBLEM 4

Find the distance between the points $(6, -3)$ and $(-7, -5)$.

Circles

The distance-between-two-points formula would still be helpful if its only use were to find actual distances between points, such as in Example 4. However, its more important use is in finding equations of figures in a rectangular coordinate system. We will use it to derive the standard equation of a circle. We start with a coordinate-free definition of a circle.

DEFINITION 1

CIRCLE

A **circle** is the set of all points in a plane equidistant from a fixed point. The fixed distance is called the **radius,** and the fixed point is called the **center.**

FIGURE 9
Circle.

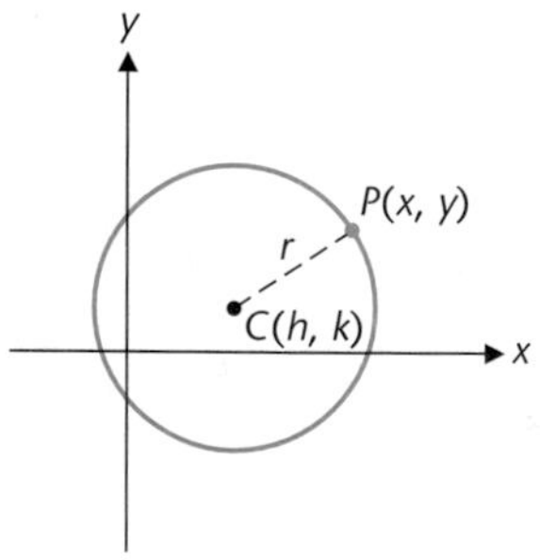

Let's find the equation of a circle with radius r ($r > 0$) and center C at (h, k) in a rectangular coordinate system (Fig. 9). The circle consists of all points $P(x, y)$ satisfying $d(P, C) = r$; that is, all points satisfying

$$\sqrt{(x - h)^2 + (y - k)^2} = r \quad r > 0$$

or, equivalently,

$$(x - h)^2 + (y - k)^2 = r^2 \quad r > 0$$

THEOREM 2

STANDARD EQUATION OF A CIRCLE

Circle with radius r and center at (h, k):

$$(x - h)^2 + (y - k)^2 = r^2 \quad r > 0$$

EXAMPLE 5

Equations and Graphs of Circles

Find the equation of a circle with radius 4 and center at:

(A) $(0, 0)$ (B) $(-3, 6)$

Graph each equation.

Solutions

(A) $(h, k) = (0, 0)$ and $r = 4$:

$$x^2 + y^2 = r^2$$
$$x^2 + y^2 = 4^2$$
$$x^2 + y^2 = 16$$

To graph the equation, locate the center at the origin and draw a circle of radius 4 (Fig. 10).

(B) $(h, k) = (-3, 6)$ and $r = 4$:

$$(x - h)^2 + (y - k)^2 = r^2$$
$$[x - (-3)]^2 + (y - 6)^2 = 4^2$$
$$(x + 3)^2 + (y - 6)^2 = 16$$

To graph the equation, locate the center $C(-3, 6)$ and draw a circle of radius 4 (Fig. 11).

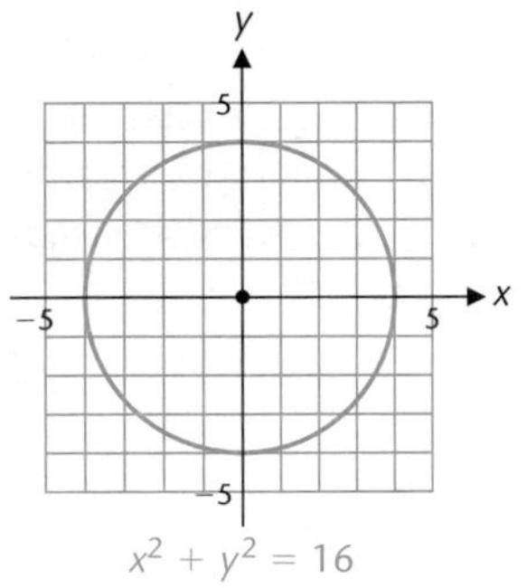

FIGURE 10

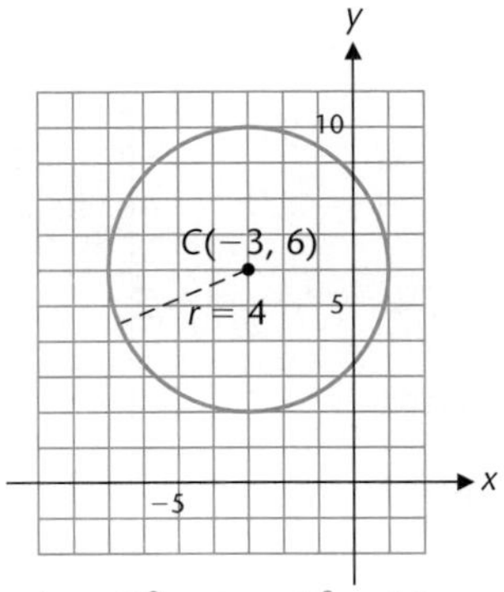

FIGURE 11

MATCHED PROBLEM 5

Find the equation of a circle with radius 3 and center at:

(A) (0, 0) (B) (3, −2)

Graph each equation.

Explore/Discuss 2

Each of the following statements is false. Indicate why, then modify each equation to make the statement true.

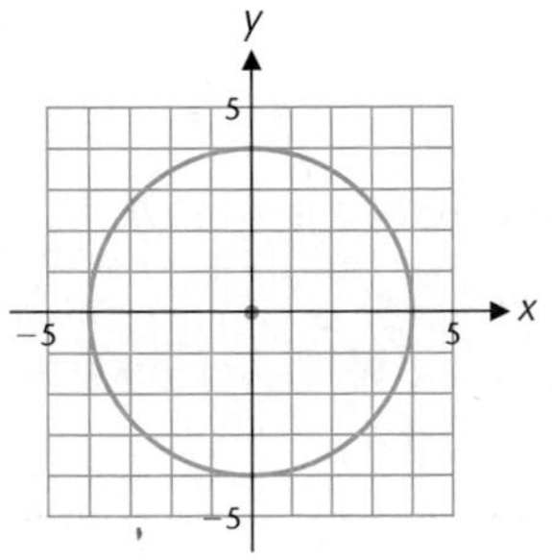

is the graph of $x^2 + y^2 = 4$.

Explore/Discuss 2 continued

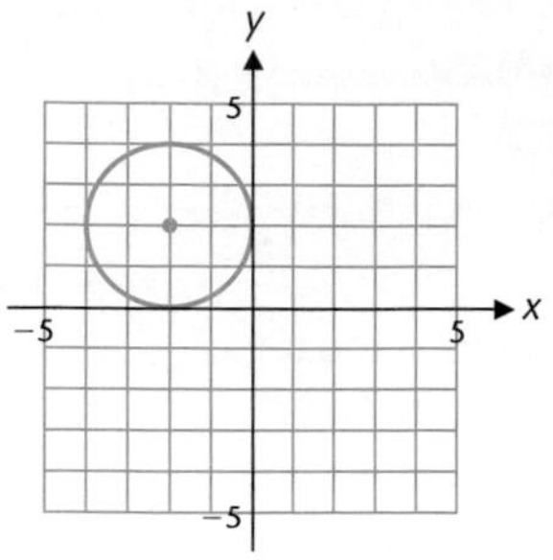

is the graph of $(x - 2)^2 + (y + 2)^2 = 4$.

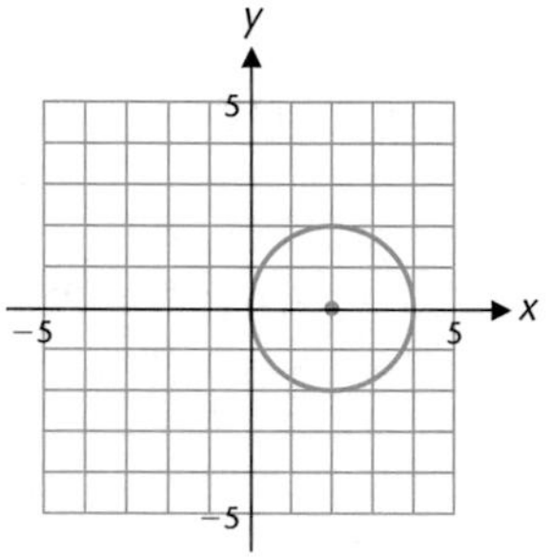

is the graph of $x^2 + (y - 2)^2 = 4$.

Answers to Matched Problems

1.

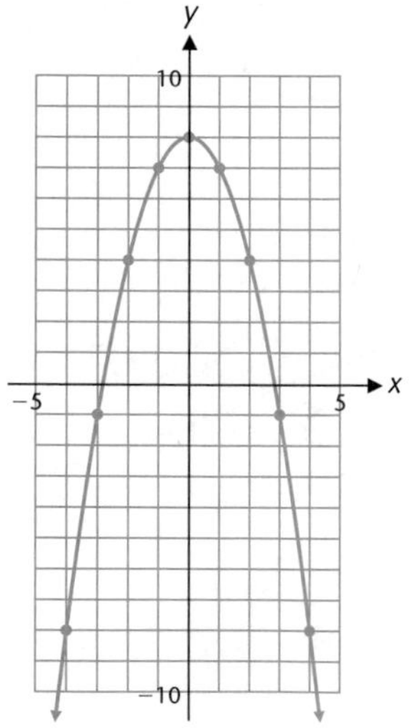

2.

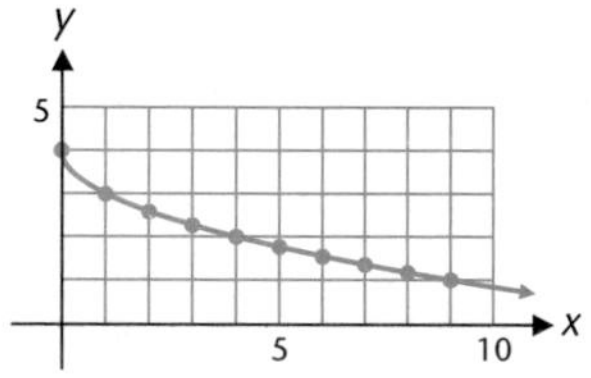

3. (A) 96 ppb (B) 1:45 P.M. and 5 P.M.

4. $d = \sqrt{173}$

5. (A) $x^2 + y^2 = 9$

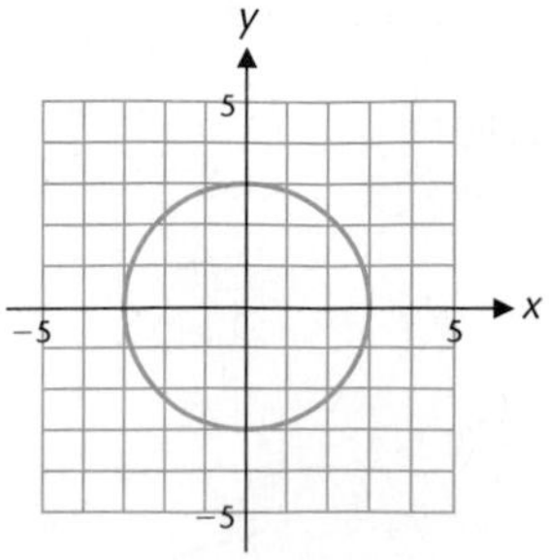

(B) $(x - 3)^2 + (y + 2)^2 = 9$

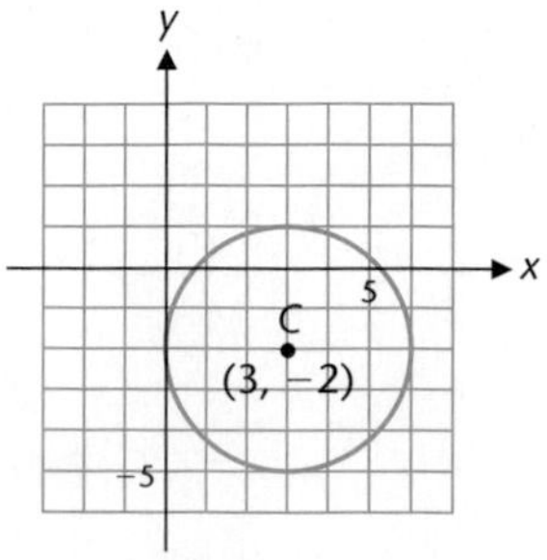

EXERCISE 1-1

A

In Problems 1–4, plot the given points in a rectangular coordinate system.

1. (5, 0), (3, −2), (−4, 2), (4, 4)

2. (0, 4), (−3, 2), (5, −1), (−2, −4)

3. (0, −2), (−1, −3), (4, −5), (−2, 1)

4. (−2, 0), (3, 2), (1, −4), (−3, 5)

In Problems 5–8, find the coordinates of points A, B, C, and D.

5.

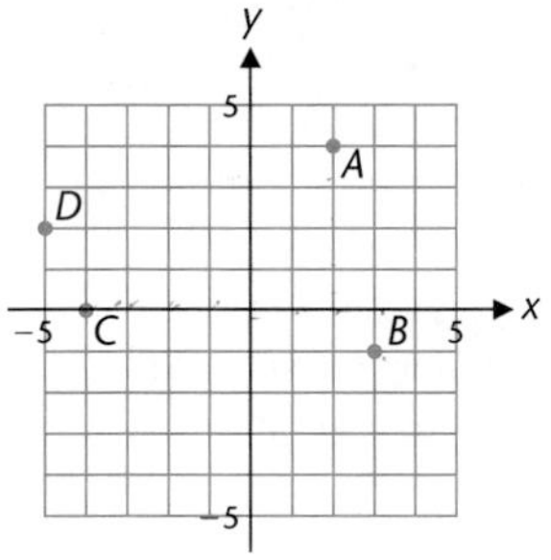

6.

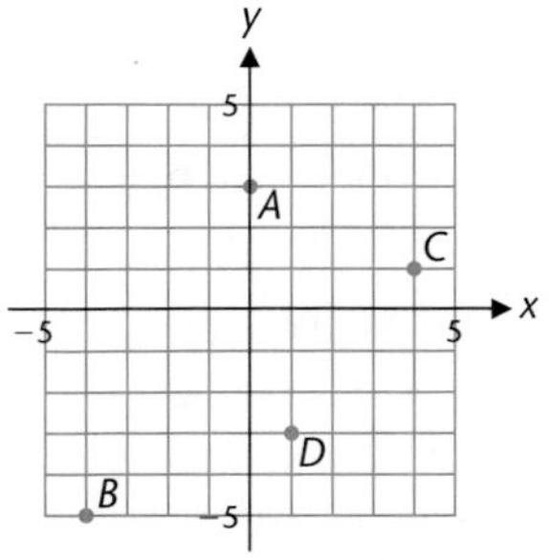

7.

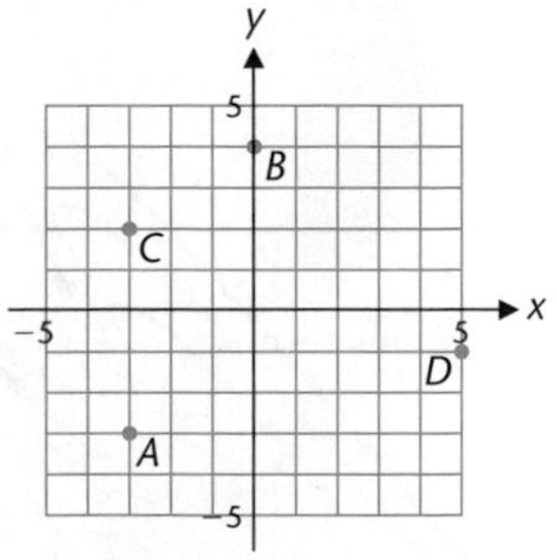

8.

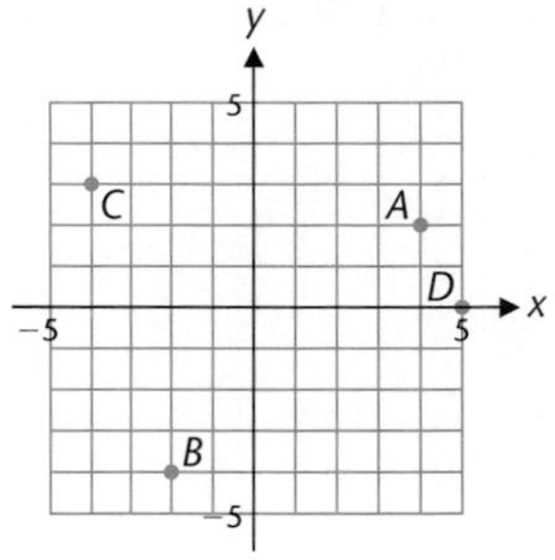

Find the distance between the indicated points in Problems 9–12. Leave the answer in radical form.

9. (−6, −4), (3, 4)

10. (−5, 4), (6, −1)

11. (6, 6), (4, −2)

12. (5, −3), (−1, 4)

In Problems 13–20, write the equation of a circle with the indicated center and radius.

13. $C(0, 0), r = 7$

14. $C(0, 0), r = 5$

15. $C(2, 3), r = 6$

16. $C(5, 6), r = 2$

17. $C(-4, 1), r = \sqrt{7}$

18. $C(-5, 6), r = \sqrt{11}$

19. $C(-3, -4), r = \sqrt{2}$

20. $C(4, -1), r = \sqrt{5}$

B

For each equation in Problems 21–26, make up a table of solutions using x = −3, −2, −1, 0, 1, 2, and 3. Plot these solutions and graph the equation.

21. $y = x + 1$

22. $y = 2 - x$

23. $y = x^2 - 5$

24. $y = 4 - x^2$

25. $y = 3 + x - 0.5x^2$

26. $y = 4 - x - 0.5x^2$

In Problems 27–30, use the graph to estimate to the nearest integer the missing coordinates of the indicated points. (Be sure you find all possible answers.)

27. (A) (8, ?) (B) (−5, ?) (C) (0, ?)
(D) (?, 6) (E) (?, −5) (F) (?, 0)

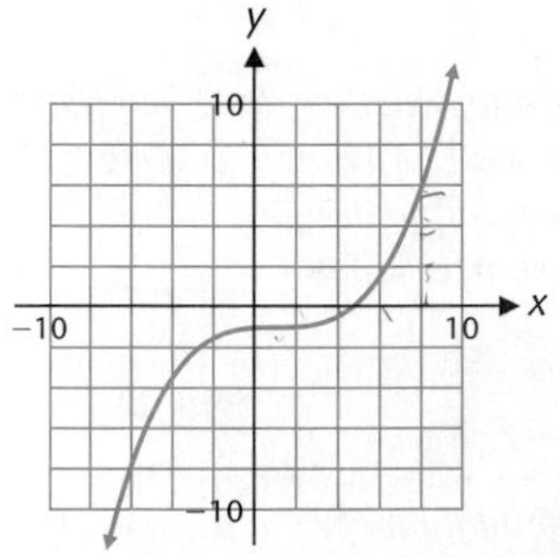

28. (A) (3, ?) (B) (−5, ?) (C) (0, ?)
(D) (?, 3) (E) (?, −4) (F) (?, 0)

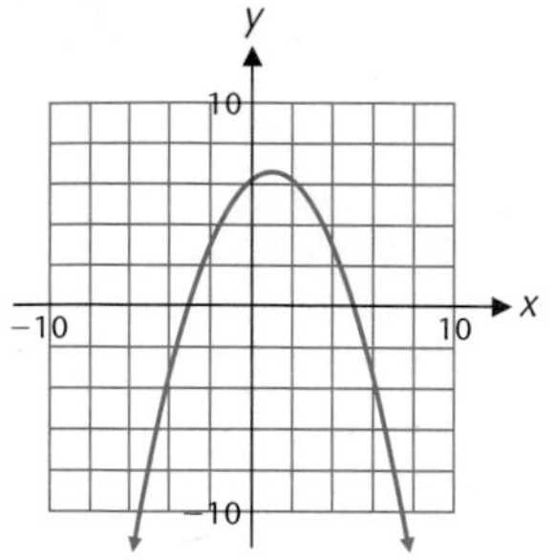

29. (A) (1, ?) (B) (−8, ?) (C) (0, ?)
(D) (?, −6) (E) (?, 4) (F) (?, 0)

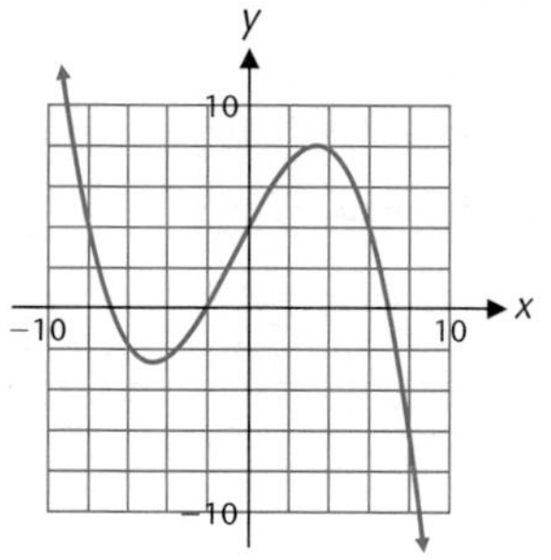

30. (A) (6, ?) (B) (−6, ?) (C) (0, ?)
(D) (?, −2) (E) (?, 1) (F) (?, 0)

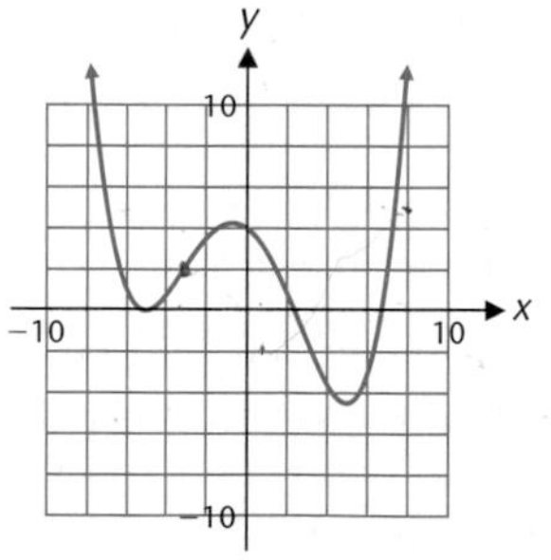

In Problems 31 and 32, determine whether the given points are vertices of a right triangle. (Recall, a triangle is a right triangle if and only if the square of the longest side is equal to the sum of the squares of the shorter sides.)

31. (−3, 2), (1, −2), (8, 5) **32.** (−4, −1), (0, 7), (6, −6)

Find the perimeter (to two decimal places) of the triangle with the vertices indicated in Problems 33 and 34.

33. (−3, 1), (1, −2), (4, 3) **34.** (−2, 4), (3, 1), (−3, −2)

In Problems 35–42, graph each equation using point-by-point plotting.

35. $y = x^{1/3}$ **36.** $y = x^{2/3}$

37. $y = x^3$ **38.** $y = x^4$

39. $y = \sqrt{x - 1}$ **40.** $y = \sqrt{5 - x}$

41. $y = \sqrt{1 + x^2}$ **42.** $y = x\sqrt{1 + x^2}$

43. (A) Graph the triangle with vertices $A(1, 1)$, $B(7, 2)$, and $C(4, 6)$.

(B) Now graph the triangle with vertices A′(1, −1), B′(7, −2), and C′(4, −6) in the same coordinate system.

(C) How are these two triangles related? How would you describe the effect of changing the sign of the y coordinate of all the points on a graph?

44. (A) Graph the triangle with vertices $A(1, 1)$, $B(7, 2)$, and $C(4, 6)$.

(B) Now graph the triangle with vertices $A'(-1, 1)$, $B'(-7, 2)$, and $C'(-4, 6)$ in the same coordinate system.

(C) How are these two triangles related? How would you describe the effect of changing the sign of the x coordinate of all the points on a graph?

45. (A) Graph the triangle with vertices $A(1, 1)$, $B(7, 2)$, and $C(4, 6)$.

(B) Now graph the triangle with vertices $A'(-1, -1)$, $B'(-7, -2)$, and $C'(-4, -6)$ in the same coordinate system.

(C) How are these two triangles related? How would you describe the effect of changing the signs of the x and y coordinates of all the points on a graph?

46. (A) Graph the triangle with vertices $A(1, 2)$, $B(1, 4)$, and $C(3, 4)$.

(B) Now graph $y = x$ and the triangle obtained by reversing the coordinates for each vertex of the original triangle: $A'(2, 1)$, $B'(4, 1)$, $C'(4, 3)$.

(C) How are these two triangles related? How would you describe the effect of reversing the coordinates of each point on a graph?

C

47. Use the distance-between-two-points formula to show that the point

$$\left(\frac{x_1 + x_2}{2}, \frac{y_1 + y_2}{2}\right)$$

is the **midpoint** of the line segment joining (x_1, y_1) and (x_2, y_2).

48. Use the midpoint formula from Problem 47 to find the midpoint of the line segment joining $(-3, 2)$ and $(5, -2)$.

Find the equation of a circle that has a diameter with the end points given in Problems 49 and 50. [Hint: See Problem 47.]

49. $(7, -3), (1, 7)$

50. $(-3, 2), (7, -4)$

51. Find the equation of a circle with center $(2, 2)$ whose graph passes through the point $(3, -5)$.

52. Find the equation of a circle with center $(-5, 4)$ whose graph passes through the point $(2, -3)$.

APPLICATIONS

53. **Price and Demand.** The quantity of a product that consumers are willing to buy during some period of time depends on its price. The price p and corresponding weekly demand q for a particular brand of diet soda in a city are shown in the figure. Use this graph to estimate the following demands to the nearest 100 cases.

(A) What is the demand when the price is \$6.00 per case?

(B) Does the demand increase or decrease if the price is increased to \$6.30 per case? By how much?

(C) Does the demand increase or decrease if the price is decreased to \$5.70? By how much?

(D) Write a brief description of the relationship between price and demand illustrated by this graph.

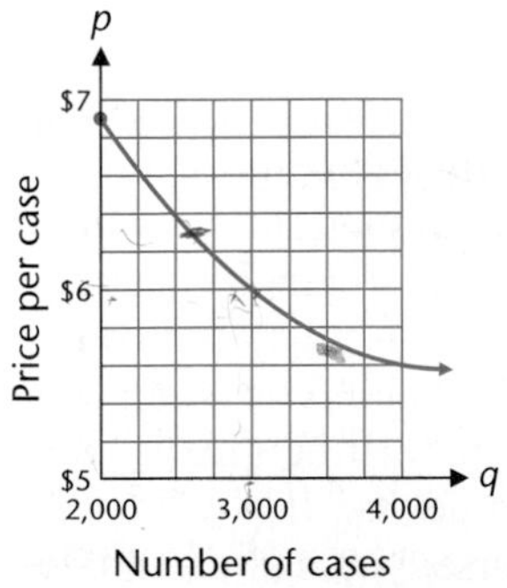

54. **Price and Supply.** The quantity of a product that suppliers are willing to sell during some period of time depends on its price. The price p and corresponding weekly supply q for a particular brand of diet soda in a city are shown in the figure. Use this graph to estimate the following supplies to the nearest 100 cases.

(A) What is the supply when the price is \$5.60 per case?

(B) Does the supply increase or decrease if the price is increased to \$5.80 per case? By how much?

(C) Does the supply increase or decrease if the price is decreased to \$5.40 per case? By how much?

(D) Write a brief description of the relationship between price and supply illustrated by this graph.

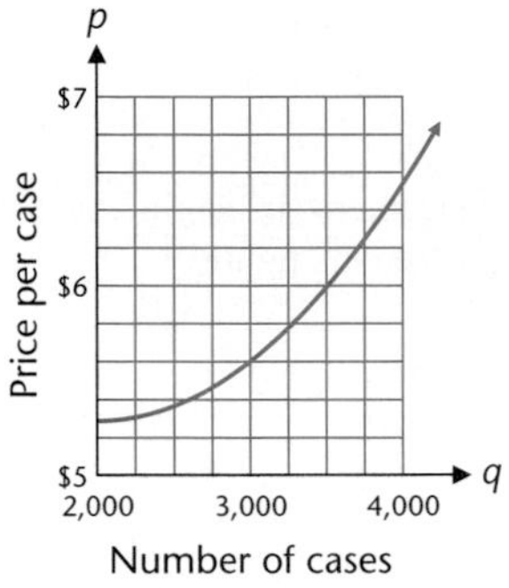

55. **Temperature.** The temperature (in degrees Fahrenheit) during a spring day in the Midwest is given in the figure. Use this graph to estimate the following temperatures to the nearest degree and times to the nearest hour.

(A) The temperature at 9:00 A.M.

(B) The highest temperature and the time when it occurs.

(C) The time(s) when the temperature is 49°.

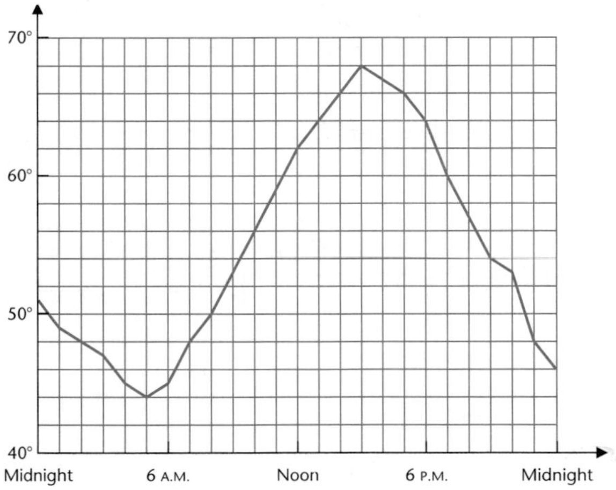

56. **Temperature.** Use the figure for Problem 55 to estimate the following temperatures to the nearest degree and times to the nearest half hour.

(A) The temperature at 7:00 P.M.

(B) The lowest temperature and the time when it occurs.

(C) The time(s) when the temperature is 52°.

57. After extensive surveys, the marketing research department of a producer of popular cassette tapes developed the demand equation

$$n = 10 - p \qquad 5 \le p \le 10$$

where n is the number of units (in thousands) retailers are willing to buy per day at \$$p$ per tape. The company's daily revenue R (in thousands of dollars) is given by

$$R = np = (10 - p)p \qquad 5 \le p \le 10$$

Graph the revenue equation for the indicated values of p.

58. Business. Repeat Problem 57 for the demand equation

$$n = 8 - p \qquad 4 \le p \le 8$$

59. Physics. The speed (in meters per second) of a ball swinging at the end of a pendulum is given by

$$v = 0.5\sqrt{2 - x}$$

where x is the vertical displacement (in centimeters) of the ball from its position at rest (see the figure).

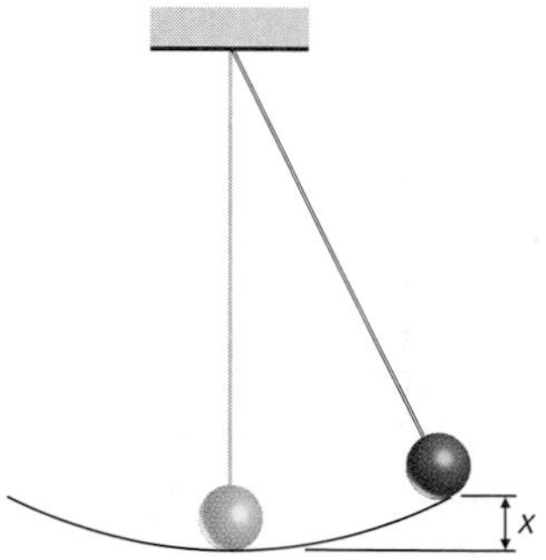

(A) Graph $v = 0.5\sqrt{2 - x}$ for $0 \le x \le 2$.

(B) Describe the relationship between this graph and the physical behavior of the ball as it swings back and forth.

60. Physics. The speed (in meters per second) of a ball oscillating at the end of a spring is given by

$$v = 4\sqrt{25 - x^2}$$

where x is the vertical displacement (in centimeters) of the ball from its position at rest (positive displacement measured downward—see the figure).

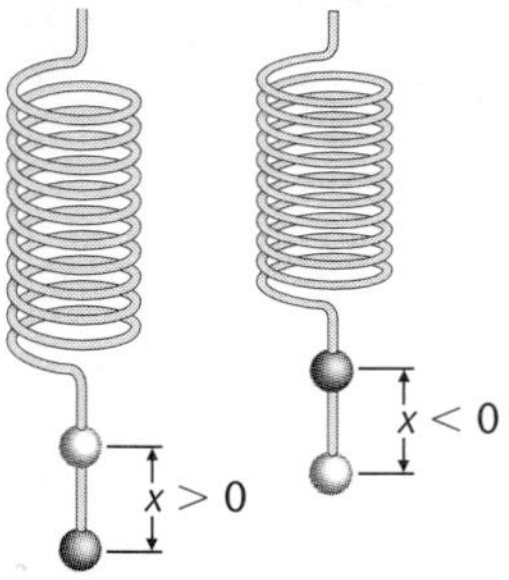

(A) Graph $v = 4\sqrt{25 - x^2}$ for $-5 \le x \le 5$.

(B) Describe the relationship between this graph and the physical behavior of the ball as it oscillates up and down.

Section 1-2 Using Graphing Utilities

- Graphing Utilities
- Screen Coordinates
- The Trace and Zoom Features

In the previous section, we sketched the graphs of equations by plotting points and then drawing by hand a smooth curve that passes through these points. Now we want to explore the use of electronic graphing devices to graph equations. The use of technology to aid in drawing and analyzing graphs is revolutionizing mathematics education and is the reason for this book. Your ability to interpret mathematical concepts and to discover patterns of behavior will be greatly increased as you become proficient with an electronic graphing device. If you have already used an electronic graphing device in a previous course, you can use this section to quickly review basic concepts.

Graphing Utilities

We now turn to the use of electronic graphing devices to graph equations. We will refer to any electronic device capable of displaying graphs as a **graphing utility.** The two most common graphing utilities are handheld graphing calculators and computers with appropriate software. You should have such a device as you proceed through this book.

We will discuss graphing utilities only in general terms. Refer to the manual or to the graphing utility supplement accompanying this text for specific details relative to your own graphing utility.

An image on the screen of a graphing utility is made up of darkened rectangles called **pixels** (see Fig. 1). The pixel rectangles are the same size, and don't change in size during any application. Graphing utilities use pixel-by-pixel plotting to produce graphs.

FIGURE 1
Pixel-by-pixel plotting on a graphing utility.

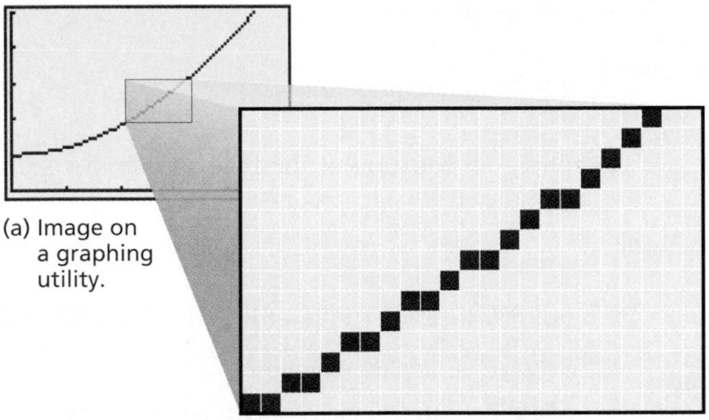

(a) Image on a graphing utility.

(b) Magnification to show pixels.

The accuracy of the graph depends on the resolution of the graphing utility. Most graphing utilities have screen resolutions of between 50 and 75 pixels per inch, which results in fairly rough but very useful graphs. Some computer systems can print very high quality graphs with resolutions over 1,000 pixels per inch.

Most graphing utility screens are rectangular in shape. The graphing screen on a graphing utility represents a portion of the plane in the rectangular coordinate system. But this representation is an approximation, since pixels are not really points, as is clearly shown in Figure 1. Points are geometric objects without dimensions, whereas a pixel has dimensions. The coordinates of a pixel are usually taken at the center of the pixel and represent all the infinitely many geometric points within the pixel. This will not cause much of a problem, as we will see.

The portion of a rectangular coordinate system displayed on the graphing screen is called a **viewing window** and is determined by assigning values to six **window variables:** the lower limit, upper limit, and scale for the x axis and the lower limit, upper limit, and scale for the y axis. Figure 2(a) illustrates the names and values of **standard window variables,** and Figure 2(b) shows the resulting **standard viewing window.**

FIGURE 2
A standard viewing window and its dimensions.

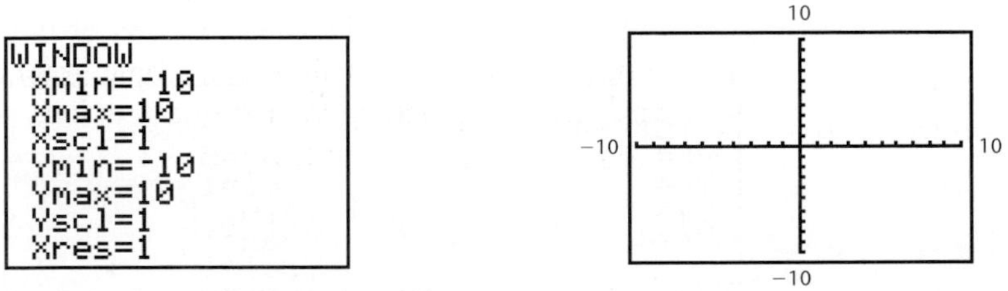

(a) Standard window variable values

(b) Standard viewing window

The names **Xmin, Xmax, Xscl, Ymin, Ymax,** and **Yscl** will be used for the six window variables. Xscl and Yscl determine the distance between tick marks on the x and y axes, respectively. Xres is a seventh window variable on some graphing utilities that controls the screen resolution; we will always leave this variable set to the default value 1. The window variables may be displayed slightly differently by your graphing utility. In this book, when a viewing window of a graphing utility is pictured in a figure, the values of Xmin, Xmax, Ymin, and Ymax are indicated by labels to make the graph easier to read [see Fig. 2(b)]. These labels are always centered on the sides of the viewing window, irrespective of the location of the axes.

We now turn to the use of a graphing utility to graph equations that can be written in the form

$$y = \text{(some expression in } x\text{)} \tag{1}$$

Graphing an equation of the type shown in equation (1) using a graphing utility is a simple three-step process:

GRAPHING EQUATIONS USING A GRAPHING UTILITY

Step 1. Enter the equation.

Step 2. Enter values for the window variables. (A rule of thumb for choosing Xscl and Yscl, unless there are reasons to the contrary, is to choose each about one-tenth the corresponding variable range.)

Step 3. Press the graph command.

The following example illustrates this procedure for graphing the equation $y = x^2 - 4$, the equation we graphed using point-by-point plotting in Example 1 of Section 1-1.

EXAMPLE 1

Graphing an Equation with a Graphing Utility

Use a graphing utility to graph $y = x^2 - 4$ for $-5 \le x \le 5$ and $-5 \le y \le 15$.

Solution

Enter the equation [Fig. 3(a)] and the values for the window variables [Fig. 3(b)] in the graphing utility; then press the graph command to obtain the graph in Figure 3(c). (The form of the screens in Fig. 3 may differ slightly, depending on the graphing utility used.)

FIGURE 3
Graphing is a three-step process.

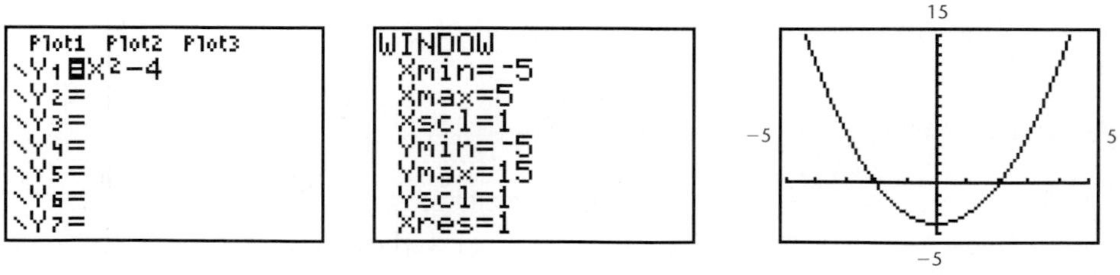

(a) Enter equation. (b) Enter window variables. (c) Press the graph command.

MATCHED PROBLEM 1

Use a graphing utility to graph $y = 8 - x^2$ for $-5 \le x \le 5$ and $-10 \le y \le 10$.

Remark For Example 1, we displayed a viewing window for each step in the graphing procedure. Generally, we will show only the final results, as illustrated in Figure 3(c).

The next example illustrates how a graphing utility can be used as an aid to sketching the graph of an equation by hand. The example illustrates the use of algebraic, numeric, and graphic approaches, which add considerably to the understanding of a problem.

EXAMPLE 2 Using a Graphing Utility as an Aid to Hand Graphing—Net Cash Flow

The net cash flow y in millions of dollars of a small high-tech company from 1991–1999 is given approximately by the following equation

$$y = 0.4x^3 - 2x + 1 \qquad -4 \le x \le 4 \tag{2}$$

where x represents the number of years before or after 1995, when the board of directors appointed a new CEO.

(A) Construct a table of values for equation (2) for each year from 1991 to 1999, inclusive. Compute y to one decimal place.

(B) Obtain a graph of equation (2) in the viewing window of your graphing utility. Plot the table values from part A by hand on graph paper, then join these points with a smooth curve using the graph in the viewing window as an aid.

Solutions

(A) After entering the given equation as y_1, we can find the value of y for a given value of x by storing the value of x in the variable X and simply typing y_1, as shown in Figure 4(a). To speed up this process, many graphing utilities can compute an entire table of values directly, as shown in Figure 4(b). We organize these results in Table 1.

FIGURE 4

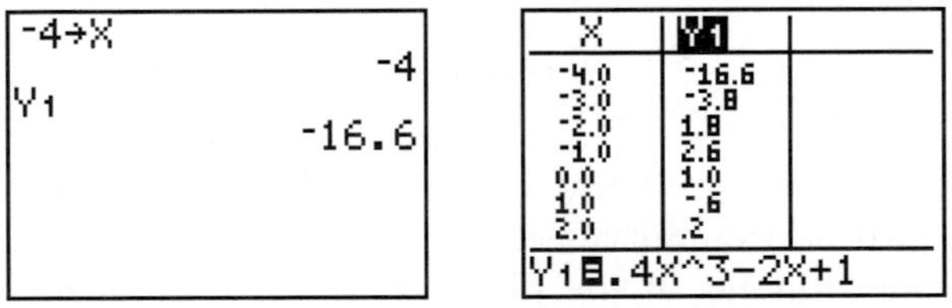

(a) (b)

TABLE 1 Net Cash Flow

Year	1991	1992	1993	1994	1995	1996	1997	1998	1999
x	−4	−3	−2	−1	0	1	2	3	4
y (million $)	−16.6	−3.8	1.8	2.6	1	−0.6	0.2	5.8	18.6

FIGURE 5

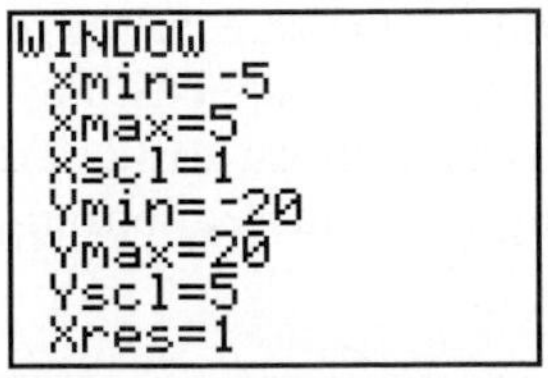

(B) To create a graph of equation (2) in the viewing window of a graphing utility, we select values for the viewing window variables that cover a little more than the values shown in Table 1, as shown in Figure 5. We add a grid to the viewing window to obtain the graphing utility graph shown in Figure 6(a). The corresponding hand sketch is shown in Figure 6(b).

FIGURE 6
Net cash flow.

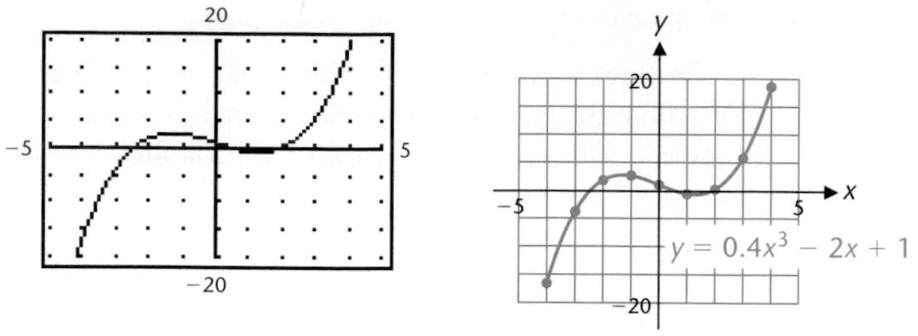

(a) Graphing utility graph (b) Hand sketch

Remark Table 1 gives us specific detail and the equation with its graph gives us an overview. Each viewpoint has its specific use.

MATCHED PROBLEM 2

Given the equation $y = 1 + 1.9x - 0.2x^3$, complete a table of values for the integers from -4 to 4, plot these points by hand, and then hand sketch the graph of the equation with the aid of a graphing utility.

Explore/Discuss 1

The choice of the viewing window has a pronounced effect on the shape of a graph. Graph $y = -x^3 + 2x$ in each of the following viewing windows:

(A) $-1 \le x \le 1,\ -1 \le y \le 1$

(B) $-10 \le x \le 10,\ -10 \le y \le 10$

(C) $-100 \le x \le 100,\ -100 \le y \le 100$

Which window gives the best view of the graph of this equation, and why?

Screen Coordinates

We now take a closer look at *screen coordinates* of pixels. Earlier we indicated that the coordinates of the center point of a pixel are usually used as the **screen coordinates of the pixel,** and these coordinates represent all points within the pixel. As expected, screen coordinates of pixels change as you change values of window variables.

To find screen coordinates of various pixels, move a *cursor* around the viewing window and observe the coordinates displayed on the screen. A **cursor** is a special symbol, such as a plus (+) or times (×) sign, that locates one pixel on the screen at a time. As the cursor is moved around the screen, it moves from pixel to pixel. To illustrate this, set the window variables in your graphing utility so that $-5 \le x \le 5$ and $-5 \le y \le 5$, and activate a grid for the screen. Move the cursor as close as you can to the point (2, 2) and observe what happens. Figure 7 shows the screen coordinates of the four pixels that are closest to (2, 2). The coordinates displayed on your screen may vary slightly from these, depending on the graphing utility used.

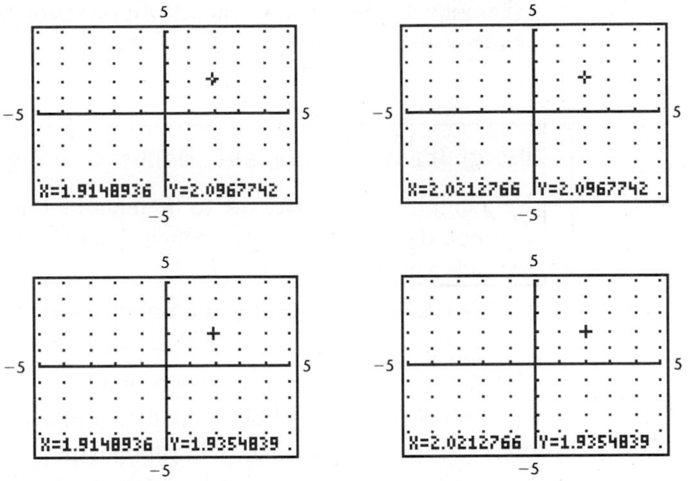

FIGURE 7
Screen coordinates of pixels near (2, 2).

Any of the four pixels in Figure 7 can be used to approximate the point (2, 2), but it is not possible to find a pixel in this viewing window whose screen coordinates are exactly (2, 2). It is instructive to repeat this exercise with different window variables, say, $-7 \le x \le 7$ and $-7 \le y \le 7$.

Remark We think it is important that actual output from existing graphing utilities be used in this book. You may not always be able to produce an exact replica of a text figure on your graphing utility, but the differences will be minor and should cause no difficulties.

The Trace and Zoom Features

Analyzing a graph of an equation frequently involves finding coordinates of points on the graph. Using the **trace feature** on a graphing utility is one way to accomplish this. The trace feature places a cursor directly on the graph and only permits movement left and right along the graph. The coordinates displayed during the tracing movement are coordinates of points that satisfy the equation. In most cases, these coordinates are not the same as the pixel screen coordinates displayed using the unrestricted cursor movement discussed earlier.

Explore/Discuss 2

Graph the equation $y = x$ in a standard viewing window.

(A) Without selecting the trace feature, move the cursor to a point on the screen that appears to lie on the graph of $y = x$ and is as close to (5, 5) as possible. Record these coordinates. Do these coordinates satisfy the equation $y = x$?

(B) Now select the trace feature and move the cursor along the graph of $y = x$ to a point that has the same x coordinate found in part A. Is the y coordinate of this point the same as you found in part A? Do the coordinates of the point using trace satisfy the equation $y = x$?

(C) Explain the difference in using trace along a curve and trying to use unrestricted movement of a cursor along a curve.

Example 3 illustrates a practical application of the trace feature in approximating coordinates of points on a graph.

EXAMPLE 3

Using Trace Techniques—Manufacturing

Rectangular open boxes are to be manufactured from 11- by 17-inch sheets of cardboard by cutting x- by x-inch squares out of the corners and folding up the sides, as shown in Figure 8.

(A) Write an equation for the volume y of the resulting box in terms of the size of the square x. Indicate appropriate restrictions on x.

(B) Graph the equation for appropriate values of x. Adjust the window variables for y to include the entire graph of interest.

(C) Use trace to estimate x to one decimal place for the smallest square that can be cut out to produce a box of volume 150 cubic inches.

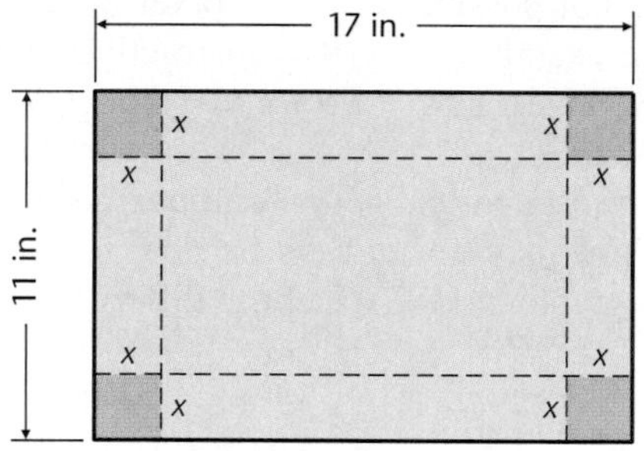

FIGURE 8
Template for boxes.

Solutions

(A) The box is shown in Figure 9 with the sides folded up and dimensions added. From this figure, we can write the equation of the volume in terms of x and establish the restrictions on x. No dimension can be negative; that is, $x \geq 0$, $11 - 2x \geq 0$, and $17 - 2x \geq 0$. These three inequalities imply that $0 \leq x \leq 5.5$. Thus, the volume of the box is given by

$$y = x(17 - 2x)(11 - 2x) \qquad 0 \leq x \leq 5.5 \tag{3}$$

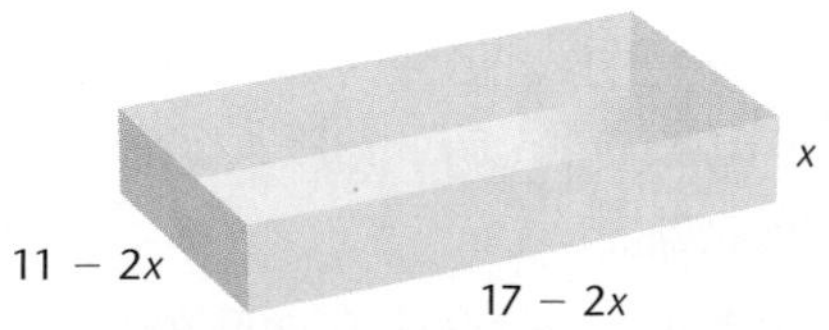

FIGURE 9
Actual box with dimensions added.

(B) Entering this equation in a graphing utility (it does not need to be multiplied out) and evaluating it for several integers between 0 and 5 [Fig. 10(a)], we see that a good choice for the window dimensions of y are $0 \leq y \leq 200$. This choice can easily be changed if too much space is above the graph or if part of the graph we are interested in is out of the viewing window. Figure 10(b) on the next page shows the graph of equation (3) in the viewing window selected.

FIGURE 10

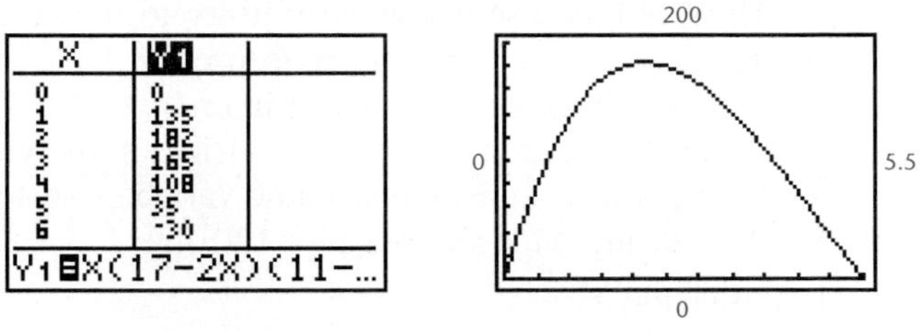

(a) (b)

(C) Select the trace command and move the cursor along the whole curve to observe the changes in the volume y as x moves from 0 to 5.5. Notice that y passes through 150 twice, once to the left side of the high point in the curve and once to the right. Thus, there are two solutions to the problem, but we are interested only in the smaller value of x. Tracing along the curve from left to right, starting at $x = 0$, we observe the two points with y coordinates closest to 150. These are shown in Figures 11(a) and 11(b). (Note that different graphing utilities may produce slightly different values for the two closest points.)

FIGURE 11

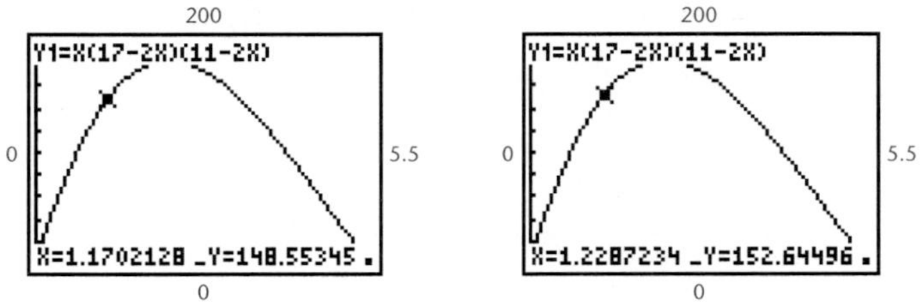

(a) Lower estimate (b) Upper estimate

If x_1 is the exact value of x that produces a volume of 150 cubic inches, then the point $(x_1, 150)$ must lie on the graph somewhere between the two points found in Figures 11(a) and 11(b). Thus, we have

$$1.1702128 < x_1 < 1.2287234$$

which implies that $x = 1.2$, correct to one decimal place. Evaluating y at $x = 1.2$ produces 150.672. The value of y is not exact because 1.2 is an approximation of the size of the square that will produce a volume of 150 cubic inches.

MATCHED PROBLEM 3

Use the methods outlined in Example 3 to approximate to one decimal place the size of the larger square that can be cut out to produce the volume 150 cubic inches.

The results and methods we used in Example 3 are very significant. If we had attempted to solve the problem algebraically, we would have been confronted with solving

$$150 = x(17 - 2x)(11 - 2x)$$

an equation that is very difficult to solve algebraically.

How can we use a graphing utility to obtain a more accurate approximation to the value of x_1? One way is to use the built-in **zoom** feature on most graphing utilities. In general, **zooming in** on a graph reduces the window variables and magnifies the portion of the graph visible in the viewing window [see Fig. 12(a)]. **Zooming out** enlarges the window variables so that more of the graph is visible in the viewing window [see Fig. 12(b)].

FIGURE 12
The zoom operation.

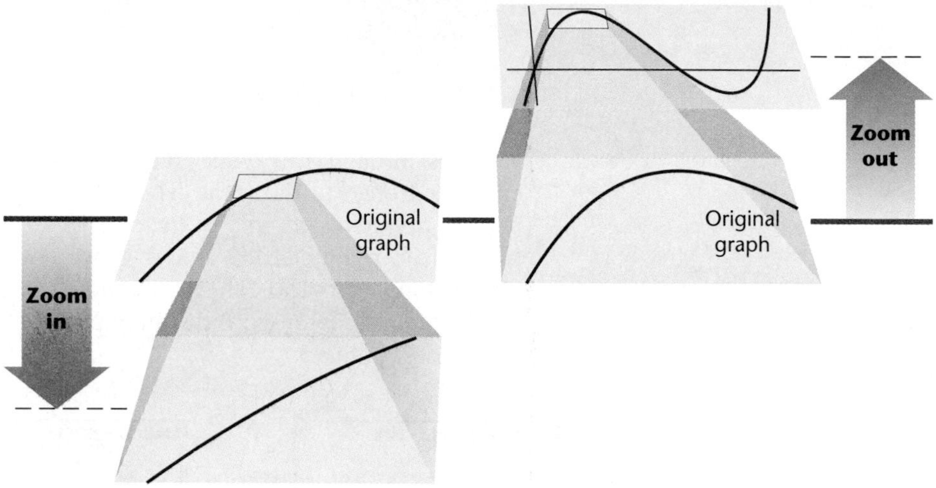

(a) Zooming in (b) Zooming out

Explore/Discuss 3

Figure 13 shows the zoom menu on a particular graphing utility. We want to explore the effects of some of these options on the graph of $y_1 = x$. Enter this equation in the equation editor and select ZStandard from the zoom menu. What are the window variables? In each of the following, position the cursor at the origin and select the indicated zoom option. Observe the changes in the window variables and examine the coordinates displayed by tracing along the curve.

(A) ZSquare (B) ZDecimal (C) ZInteger (D) ZoomFit

FIGURE 13
The zoom menu on a TI-83.

Most graphing utilities allow the user to choose the zoom factor that controls the amount of magnification or reduction produced by the zoom operation. The figures in the text were produced using a zoom factor of 5. Now we will use the zoom feature to improve the accuracy of the approximation we obtained in Example 3.

EXAMPLE 4 **Using Zoom Techniques—Manufacturing**

Use zoom techniques to find x_1 in Example 3 to two-decimal-place accuracy.

Solution Place the cursor near the point $(x_1, 150)$ [see Fig. 11(a)], select Zoom In from the zoom menu, and use trace to find the two points closest to $(x_1, 150)$ in the new graph [Figs. 14(b) and 14(c)]. Due to the decimal form of the window variables, in this example we choose to display the window variable screen, rather than placing these values on the graphs.

FIGURE 14

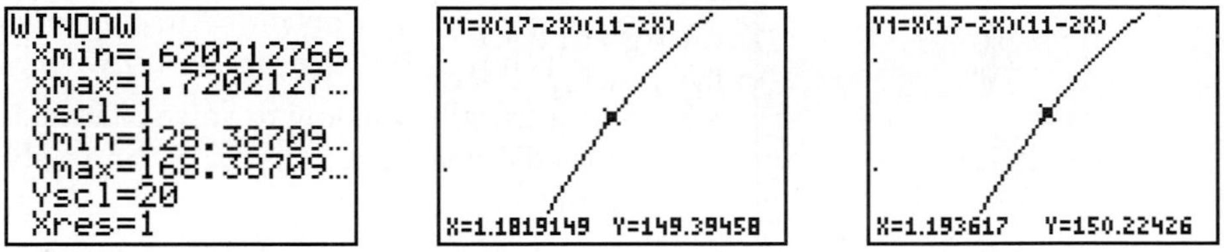

(a) New window variables (b) Lower estimate (c) Upper estimate

Since x_1 must be between the x coordinates of these points, we can conclude that

$$1.1819149 < x_1 < 1.193617$$

This is not quite enough to determine x_1 to two decimal places, so we zoom in on the graph in Figure 14(b) to produce Figure 15.

FIGURE 15

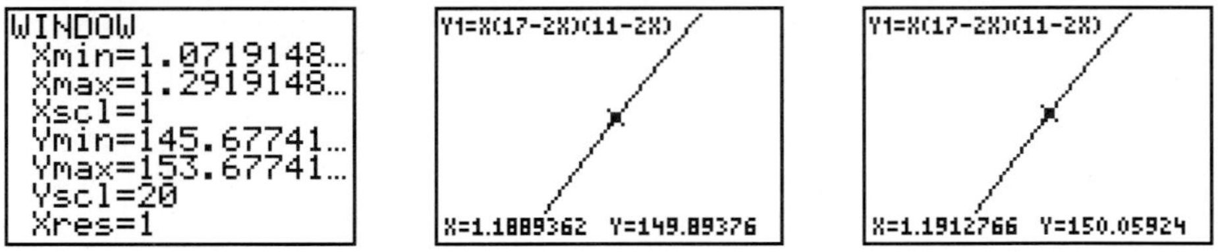

(a) New window variables (b) Lower estimate (c) Upper estimate

Now we can conclude that

$$1.1889362 < x_1 < 1.1912766$$

which implies that $x_1 = 1.19$, correct to two decimal places.

MATCHED PROBLEM 4

Proceeding as in Example 4, compute the value of x_1 for the larger square to two decimal places.

As Examples 3 and 4 illustrate, tracing along a graph usually produces coordinates with long decimal expansions. But in some windows, tracing produces coordinates with much simpler decimal expansions (see parts B and C in Explore/Discuss 3). Windows that display simple coordinates are sometimes called **friendly windows.** Friendly windows are formed using the relationship

$$\text{Xmax} = \text{Xmin} + m * h$$

where m is a number that depends on the particular graphing utility and h is selected by the user. Values of m for some popular Texas Instruments graphing calculators are given in Table 2. If you have some other graphing utility, you can determine m by counting the number of times you have to press the right arrow key to move the cursor from the left side of the screen to the right side.

TABLE 2

Calculator	m
TI-81	95
TI-82/83	94
TI-85/86	126
TI-89	158

Explore/Discuss 4

(A) Let Xmin = 0 and Xmax = $m * h$, where m is the machine dependent number described in the preceding paragraph. Describe the type of coordinates you obtain by tracing along the graph of $y = x$ if $h = 1$, if $h = 0.1$, and if $h = 0.25$.

(B) Use a friendly window to solve Matched Problem 4. How many times did you have to use zoom in to obtain two-decimal-place accuracy?

Answers to Matched Problems

1.

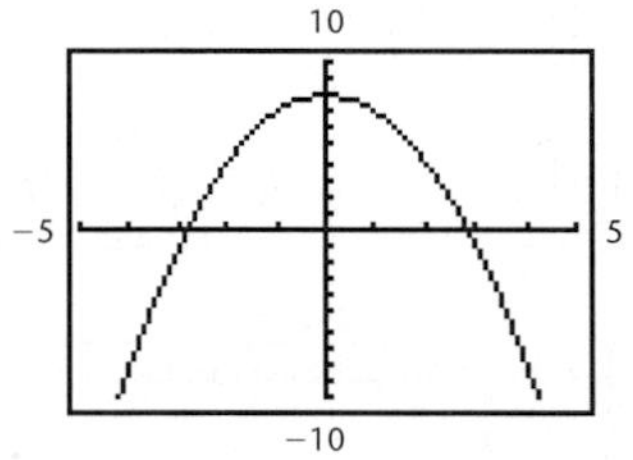

2.

x	−4	−3	−2	−1	0	1	2	3	4
y	6.2	0.7	−1.2	−0.7	1	2.7	3.2	1.3	−4.2

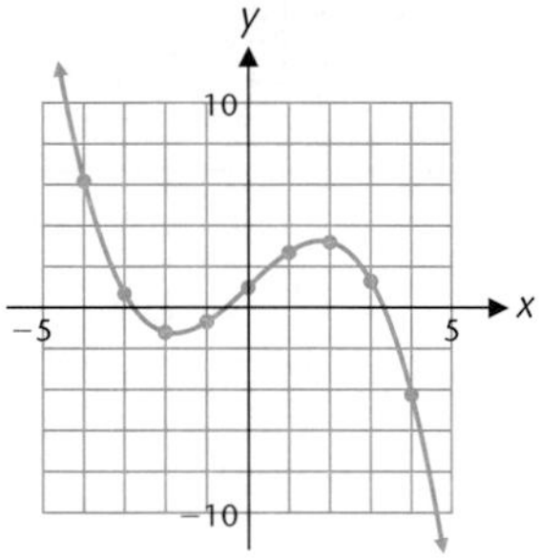

3. $x \approx 3.3$ in. 4. $x \approx 3.32$ in.

EXERCISE 1-2

A

In Problems 1–6, determine if the indicated point lies in the viewing window defined by

Xmin = −7, Xmax = 9, Ymin = −4, Ymax = 11

1. (0, 0) **2.** (0, 10) **3.** (10, 0)

4. (−3, −5) **5.** (−5, −3) **6.** (−8, 12)

7. Consider the points in the following table:

x	3	6	−7	−4	0
y	−9	−4	14	0	2

(A) Find the smallest rectangle in a Cartesian coordinate system that will contain all the points in the table. State your answers in terms of the window variables Xmin, Xmax, Ymin, and Ymax.

(B) Enter the window variables you determined in part A and display the corresponding viewing window. Can you use the cursor to display the coordinates of the points in the table on the graphing utility screen? Discuss the differences between the rectangle in the plane and the pixels displayed on the screen.

8. Repeat Problem 7 for the following table.

x	−4	0	−2	7	4
y	2	−4	0	−2	3

In Problems 9–14, graph each equation in a standard viewing window.

9. $y = -x$ **10.** $y = 0.5x$

11. $y = 9 - 0.4x^2$

12. $y = 0.3x^2 - 4$

13. $y = 2\sqrt{x + 5}$

14. $y = -2\sqrt{x + 5}$

B

For each equation in Problems 15–20, construct a table of values over the indicated interval, computing y values to the nearest tenth of a unit. Plot these points on graph paper, then with the aid of a graph on a graphing utility, complete the hand sketch of the graph.

15. $y = 4 + 4x - x^2$, $-2 \le x \le 6$ (Use even integers for the table.)

16. $y = 2x^2 + 12x + 5$, $-7 \le x \le 1$ (Use odd integers for the table.)

17. $y = 2\sqrt{2x + 10}$, $-5 \le x \le 5$ (Use odd integers for the table.)

18. $y = \sqrt{8 - 2x}$, $-4 \le x \le 4$ (Use even integers for the table.)

19. $y = 0.5x(4 - x)(x + 2)$, $-3 \le x \le 5$ (Use odd integers for the table.)

20. $y = 0.5x(x + 3.5)(2.8 - x)$, $-4 \le x \le 4$ (Use even integers for the table.)

In Problems 21–24, graph the equation in a standard viewing window. Use zoom and trace to approximate to two decimal places the x coordinates of the points in this window that are on the graph of the equation and have the indicated y coordinates.

21. $y = 4 - 3\sqrt[3]{x + 4}$
    (A) $(x, 8)$ (B) $(x, -1)$ (C) $(x, 0)$

22. $y = 3 + 4\sqrt[3]{x - 4}$
    (A) $(x, 8)$ (B) $(x, -6)$ (C) $(x, 0)$

23. $y = 3 + x + 0.1x^3$
    (A) $(x, 4)$ (B) $(x, -7)$ (C) $(x, 0)$

24. $y = 2 - 0.5x - 0.1x^3$
    (A) $(x, 7)$ (B) $(x, -5)$ (C) $(x, 0)$

The graph of each equation in Problems 25–28 is a parabola (see Example 1 in Section 1-1). Graph the equation in each of the viewing windows listed below and indicate which window gives the best view of the graph.

W_1: $-1 \le x \le 1, -1 \le y \le 1$

W_2: $-10 \le x \le 10, -10 \le y \le 10$

W_3: $-100 \le x \le 100, -100 \le y \le 100$

25. $y = x^2 - 2x$

26. $y = x^2 + 2x - 1$

27. $y = 0.8x - x^2$

28. $y = 20x - x^2$

C

In Problems 29–32, use trace and zoom to find all real solutions of each equation to two decimal places.

29. $2x^3 - 5x^2 + 2 = 0$

30. $3x^3 - 7x^2 + 3 = 0$

31. $0.01x^4 - 0.45x^2 + 2 = 0$

32. $-0.01x^4 + 0.56x^2 - 3 = 0$

33. The point $(\sqrt{2}, 2)$ is on the graph of $y = x^2$. Use trace and zoom to approximate $\sqrt{2}$ to four decimal places. Compare your result with the direct calculator evaluation of $\sqrt{2}$.

34. The point $(\sqrt[3]{4}, 4)$ is on the graph of $y = x^3$. Use trace and zoom to approximate $\sqrt[3]{4}$ to four decimal places. Compare your result with the direct calculator evaluation of $\sqrt[3]{4}$.

35. Find a set of window variables for your graphing utility containing the interval $0 \le x \le 90$ that will produce pixels with integer x coordinates.

36. Find a set of window variables for your graphing utility containing the interval $0 \le x \le 9$ that will produce pixels with one-decimal-place x coordinates.

37. In a few sentences, discuss the difference between the mathematical coordinates of a point and the screen coordinates of a pixel.

38. In a few sentences, discuss the difference between the coordinates displayed during unrestricted cursor movement and those displayed during the trace procedure.

APPLICATIONS

39. **Physics: Position of a Falling Object.** The Sears Tower in Chicago is 443 meters high, making it one of the world's tallest buildings. A construction worker drops a hammer from the top of the tower. After x seconds the hammer will be y meters above the ground, as given approximately by the equation

$$y = 443 - 4.877x^2$$

Graph this equation in the viewing window

$$0 \le x \le 10, -50 \le y \le 450$$

Use zoom and trace to approximate the time (to the nearest tenth of a second) it takes the hammer to fall to the ground.

40. **Physics: Position of a Moving Object.** A baseball fan throws a ball from a point in the stands 25 feet above the playing field (see the figure). After x seconds the ball will be y feet above the ground, as given approximately by the equation

$$y = 25 + 50x - 16x^2$$

Graph this equation in the viewing window

$$0 \le x \le 4, \ -10 \le y \le 70$$

Use zoom and trace to approximate the time (to the nearest tenth of a second) the ball is in the air.

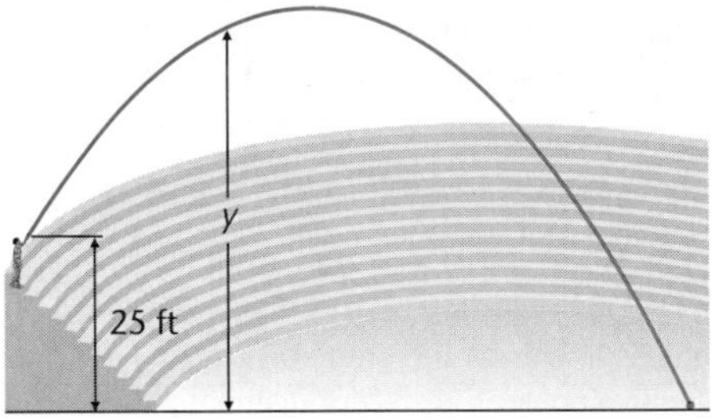

41. **Manufacturing.** A rectangular open-top box is to be constructed out of an 8.5-in. by 11-in. sheet of thin cardboard by cutting x-in. squares out of each corner and bending the sides up, as in Figures 8 and 9 in Example 3. What size squares to two decimal places should be cut out to produce a box with volume 55 cubic inches? Give the dimensions to two decimal places of all possible boxes with the given volume. Check your answers.

42. **Manufacturing.** A rectangular open-top box is to be constructed out of a 9-in. by 12-in. sheet of thin cardboard by cutting x-in. squares out of each corner and bending the sides up as shown in Figures 7 and 8 in Example 3. What size squares to two decimal places should be cut out to produce a box with volume 72 cubic inches? Give the dimensions to two decimal places of all possible boxes with the given volume. Check your answers.

43. **Manufacturing.** A box with a lid is to be cut out of a 12-in. by 24-in. sheet of thin cardboard by cutting out six x-in. squares and folding as indicated in the figure. What are the dimensions to two decimal places of all possible boxes that will have a volume of 100 cubic inches? Check your answers.

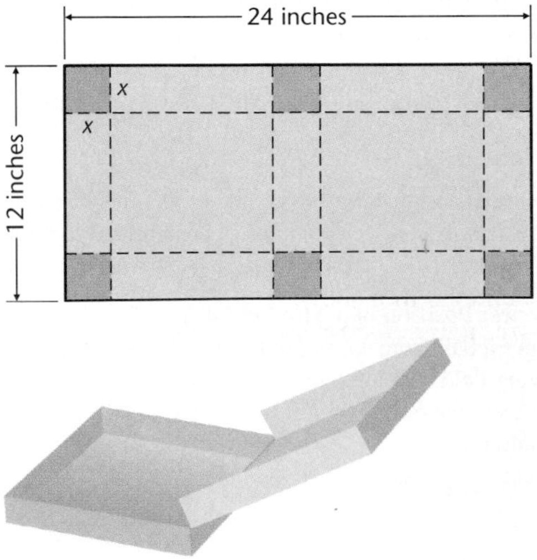

44. **Manufacturing.** A box with a lid is to be cut out of a 10-in. by 20-in. sheet of thin cardboard by cutting out six x-in. squares and folding as indicated in the figure. What are the dimensions to two decimal places of all possible boxes that will have a volume of 75 cubic inches? Check your answers. (Refer to the figure for Problem 43.)

45. **Price and Demand.** A nationwide office supply company sells high-grade paper for laser printers. The price per case y (in dollars) and the weekly demand x for this paper are related approximately by the equation

$$y = 100 - 0.6\sqrt{x} \qquad 5{,}000 \le x \le 20{,}000$$

(A) Use a graph of this equation along with zoom and trace to complete the following table. Approximate each value of x to the nearest hundred cases.

x			
y	20	25	30

(B) Does the demand increase or decrease if the price is increased from \$25 to \$30? By how much?

(C) Does the demand increase or decrease if the price is decreased from \$25 to \$20? By how much?

46. **Price and Demand.** Refer to the relationship between price and demand given in Problem 45.

(A) Use a graph of the equation along with zoom and trace to complete the following table. Approximate each value of x to the nearest hundred cases.

x			
y	35	40	45

(B) Does the demand increase or decrease if the price is increased from \$40 to \$45? By how much?

(C) Does the demand increase or decrease if the price is decreased from \$40 to \$35? By how much?

47. **Price and Revenue.** Refer to Problem 45. The revenue from the sale of x cases of paper at \$$y$ per case is given by the product $R = xy$.

(A) Use the results from Problem 45 to complete the following table of revenues.

y	20	25	30
R			

(B) Does the revenue increase or decrease if the price is increased from \$25 to \$30? By how much?

(C) Does the revenue increase or decrease if the price is decreased from \$25 to \$20? By how much?

(D) If the current price of paper is \$25 per case and the company wants to increase revenue, should they raise the price \$5, lower the price \$5, or leave the price unchanged?

48. Price and Revenue. Refer to Problem 46. The revenue from the sale of x cases of paper at $\$y$ per case is given by the product $R = xy$.

(A) Use the results from Problem 46 to complete the following table of revenues.

y	35	40	45
R			

(B) Does the revenue increase or decrease if the price is increased from \$40 to \$45? By how much?

(C) Does the revenue increase or decrease if the price is decreased from \$40 to \$35? By how much?

(D) If the current price of paper is \$40 per case and the company wants to increase revenue, should they raise the price \$5, lower the price \$5, or leave the price unchanged?

Section 1-3 | Functions

- Definition of a Function
- Functions Defined by Equations
- Function Notation
- Application
- A Brief History of the Function Concept

The idea of correspondence plays a central role in the formulation of the function concept. You have already had experiences with correspondences in everyday life. For example:

To each person there corresponds an age.

To each item in a store there corresponds a price.

To each automobile there corresponds a license number.

To each circle there corresponds an area.

To each number there corresponds its cube.

One of the most important aspects of any science (managerial, life, social, physical, computer, etc.) is the establishment of correspondences among various types of phenomena. Once a correspondence is known, predictions can be made. A chemist can use a gas law to predict the pressure of an enclosed gas, given its temperature. An engineer can use a formula to predict the deflections of a beam subject to different loads. A computer scientist can use formulas to compare the efficiency of algorithms for sorting data stored in a computer. An economist would like to be able to predict interest rates, given the rate of change of the money supply. And so on.

Definition of a Function

What do all the preceding examples have in common? Each describes the matching of elements from one set with elements in a second set. Consider Tables 1–3, which list values for the cube, square, and square root, respectively.

TABLE 1

Domain (number)	Range (cube)
−2	−8
−1	−1
0	0
1	1
2	8

TABLE 2

Domain (number)	Range (square)
−2	4
−1	1
0	0
1	1
2	4

TABLE 3

Domain (number)	Range (square root)
0	0
1	1, −1
4	2, −2
9	3, −3

Tables 1 and 2 specify functions, but Table 3 does not. Why not? The definition of the term *function* will explain.

DEFINITION 1

RULE FORM OF THE DEFINITION OF A FUNCTION

A **function** is a rule that produces a correspondence between two sets of elements such that to each element in the first set there corresponds *one and only one* element in the second set.

The first set is called the **domain** and the set of all corresponding elements in the second set is called the **range.**

Tables 1 and 2 specify functions, since to each domain value there corresponds exactly one range value (for example, the cube of -2 is -8 and no other number). On the other hand, Table 3 does not specify a function, since to at least one domain value there corresponds more than one range value (for example, to the domain value 9 there corresponds -3 and 3, both square roots of 9).

Remark Some graphing utilities use the term *range* to refer to the window variables. In this book, *range* will always refer to the range of a function.

Explore/Discuss 1

Consider the set of students enrolled in a college and the set of faculty members of that college. Define a correspondence between the two sets by saying that a student corresponds to a faculty member if the student is currently enrolled in a course taught by the faculty member. Is this correspondence a function? Discuss.

Since a function is a rule that pairs each element in the domain with a corresponding element in the range, this correspondence can be illustrated by using ordered pairs of elements, where the first component represents a domain element and the second component represents the corresponding range element. Thus, the functions defined in Tables 1 and 2 can be written as follows:

$$\text{Function 1} = \{(-2, -8), (-1, -1), (0, 0), (1, 1), (2, 8)\}$$

$$\text{Function 2} = \{(-2, 4), (-1, 1), (0, 0), (1, 1), (2, 4)\}$$

In both cases, notice that no two ordered pairs have the same first component and different second components. On the other hand, if we list the set A of ordered pairs determined by Table 3, we have

$$A = \{(0, 0), (1, 1), (1, -1), (4, 2), (4, -2), (9, 3), (9, -3)\}$$

In this case, there are ordered pairs with the same first component and different second components; for example, $(1, 1)$ and $(1, -1)$ both belong to the set A. Once again, we see that Table 3 does not define a function.

This suggests an alternative but equivalent way of defining functions that produces additional insight into this concept.

DEFINITION 2

SET FORM OF THE DEFINITION OF A FUNCTION

A **function** is a set of ordered pairs with the property that no two ordered pairs have the same first component and different second components.

The set of all first components in a function is called the **domain** of the function, and the set of all second components is called the **range.**

EXAMPLE 1

Functions Defined as Sets of Ordered Pairs

(A) The set $S = \{(1, 4), (2, 3), (3, 2), (4, 3), (5, 4)\}$ defines a function since no two ordered pairs have the same first component and different second components. The domain and range are

$$\text{Domain} = \{1, 2, 3, 4, 5\} \quad \text{Set of first components}$$

$$\text{Range} = \{2, 3, 4\} \quad \text{Set of second components}$$

(B) The set $T = \{(1, 4), (2, 3), (3, 2), (2, 4), (1, 5)\}$ does not define a function since there are ordered pairs with the same first component and different second components [for example, $(1, 4)$ and $(1, 5)$].

MATCHED PROBLEM 1

Determine whether each set defines a function. If it does, then state the domain and range.

(A) $S = \{(-2, 1), (-1, 2), (0, 0), (-1, 1), (-2, 2)\}$

(B) $T = \{(-2, 1), (-1, 2), (0, 0), (1, 2), (2, 1)\}$

Functions Defined by Equations

Both versions of the definition of a function are quite general, with no restrictions on the type of elements that make up the domain or range. In this text, unless otherwise indicated, **the domain and range of a function will be sets of real numbers.**

Defining a function by displaying the rule of correspondence in a table or listing all the ordered pairs in the function only works if the domain and range are relatively small finite sets. Functions with finite domains and ranges are used extensively in certain specialized areas, such as computer science, but most applications of functions involve infinite domains and ranges. If the domain and range of a function are infinite sets, then the rule of correspondence cannot be displayed in a table, and it is not possible to actually list all the ordered pairs belonging to the function. For most functions, we use an equation in two variables to specify both the rule of correspondence and the set of ordered pairs.

Consider the equation

$$y = x^2 + 2x \qquad x \text{ any real number} \tag{1}$$

This equation assigns to each domain value x exactly one range value y. For example,

$$\text{If } x = 4, \qquad \text{then} \qquad y = (4)^2 + 2(4) = 24$$

$$\text{If } x = -\tfrac{1}{3}, \qquad \text{then} \qquad y = (-\tfrac{1}{3})^2 + 2(-\tfrac{1}{3}) = -\tfrac{5}{9}$$

Thus, we can view equation (1) as a function with rule of correspondence

$$y = x^2 + 2x \qquad x^2 + 2x \text{ corresponds to } x$$

or, equivalently, as a function with set of ordered pairs

$$\{(x, y) \mid y = x^2 + 2x, x \text{ a real number}\}$$

The variable x is called an *independent variable,* indicating that values can be assigned "independently" to x from the domain. The variable y is called a *dependent variable,* indicating that the value of y "depends" on the value assigned to x and on the given equation. In general, any variable used as a placeholder for domain values is called an **independent variable;** any variable used as a placeholder for range values is called a **dependent variable.**

Which equations can be used to define functions?

FUNCTIONS DEFINED BY EQUATIONS

In an equation in two variables, if to each value of the independent variable there corresponds exactly one value of the dependent variable, then the equation defines a function.

If there is any value of the independent variable to which there corresponds more than one value of the dependent variable, then the equation does not define a function.

Notice that we have used the phrase "an equation defines a function" rather than "an equation is a function." This is a somewhat technical distinction, but it is employed consistently in mathematical literature and we will adhere to it in this text.

Explore/Discuss 2

(A) Graph $y = x^2 - 4$ for $-5 \le x \le 5$ and $-5 \le y \le 5$ and trace along this graph. Discuss the relationship between the coordinates displayed while tracing and the function defined by this equation.

(B) The graph of the equation $x^2 + y^2 = 16$ is a circle. Since most graphing utilities will accept only equations that have been solved for y, we must graph the equations $y_1 = \sqrt{16 - x^2}$ and $y_2 = -\sqrt{16 - x^2}$ to produce a graph of the circle. Graph these equations for $-5 \le x \le 5$ and $-5 \le y \le 5$. Then try different values for Xmin and Xmax until the graph looks more like a circle. Use the trace feature to find two points on this circle with the same x coordinate and different y coordinates.

(C) Is it possible to graph a single equation of the form $y =$ (expression in x) on your graphing utility and obtain a graph that is not the graph of a function? Explain your answer.

Remark If we want the graph of a circle to actually appear to be circular, we must choose window variables so that a unit length on the x axis is the same number of pixels as a unit length on the y axis. Such a window is often referred to as a **squared viewing window.** Most graphing utilities have an option under the zoom menu that will do this automatically.

Not all equations determine functions. One way to determine if an equation does determine a function is to examine its graph. The graphs of the equations

$$y = x^2 - 4 \qquad \text{and} \qquad x^2 + y^2 = 16$$

are shown in Figure 1.

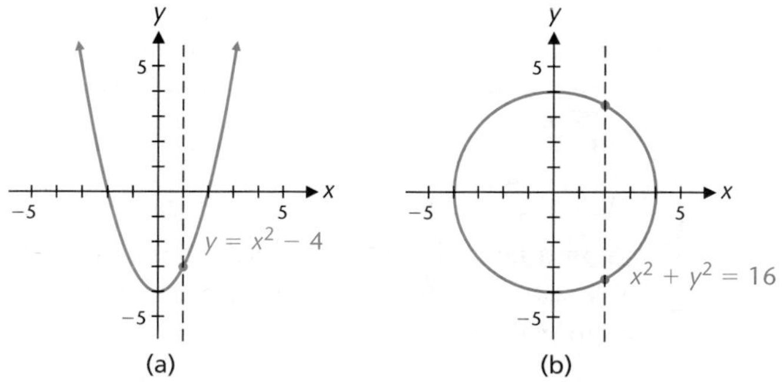

FIGURE 1
Graphs of equations and the vertical line test.

The graph in Figure 1(a) is a parabola and the graph in Figure 1(b) is a circle. Any vertical line intersects the parabola in exactly one point. This shows that each value of the independent variable x corresponds to exactly one value of the dependent variable y. Thus, the equation $y = x^2 - 4$ defines a function. On the other hand, there are vertical lines that intersect the circle in Figure 1(b) in two points. This indicates that there are values of the independent variable x that correspond to two different values of the dependent variable y. Consequently, the equation $x^2 + y^2 = 16$ does not define a function. These observations are generalized in Theorem 1.

THEOREM 1

VERTICAL LINE TEST FOR A FUNCTION

An equation defines a function if each vertical line in the rectangular coordinate system passes through at most one point on the graph of the equation.

If any vertical line passes through two or more points on the graph of an equation, then the equation does not define a function.

Since the expression $x^2 - 4$ represents a real number for all real values of x, the function defined by the equation $y = x^2 - 4$ is defined for all real numbers. Thus, its domain is the set of all real numbers, often denoted by the letter R or the interval* $(-\infty, \infty)$. On the other hand, the expression $\sqrt{16 - x^2}$ represents a real number only if $16 - x^2 \geq 0$. This inequality is equivalent to $x^2 \leq 16$ or $-4 \leq x \leq 4$. Thus, the domain of the function $y = \sqrt{16 - x^2}$ is $\{x \mid -4 \leq x \leq 4\}$ or $[-4, 4]$. Unless stated to the contrary, we will adhere to the following convention regarding domains and ranges for functions defined by equations.

AGREEMENT ON DOMAINS AND RANGES

If a function is defined by an equation and the domain is not indicated, then we assume that the domain is the set of all real number replacements of the independent variable that produce *real values* for the dependent variable. The range is the set of all values of the dependent variable corresponding to these domain values.

EXAMPLE 2 **Finding the Domain of a Function**

Find the domain of the function defined by the equation $y = 4 + \sqrt{x}$, assuming x is the independent variable.

Solution For $\sqrt{x}$ to be real, x must be greater than or equal to 0. Thus,

$$\text{Domain:} \quad \{x \mid x \geq 0\} \quad \text{or} \quad [0, \infty)$$

Note that in many cases we will dispense with set notation and simply write $x \geq 0$ instead of $\{x \mid x \geq 0\}$.

*See Appendix A, Section A-8, for a discussion of interval notation.

MATCHED PROBLEM 2

Find the domain of the function defined by the equation $y = 3 + \sqrt{-x}$, assuming x is the independent variable.

Function Notation

We will use letters to name functions and to provide a very important and convenient notation for defining functions. For example, if f is the name of the function defined by the equation $y = 2x + 1$, then instead of the more formal representations

$f: y = 2x + 1$ Rule of correspondence

or

$f: \{(x, y) \mid y = 2x + 1\}$ Set of ordered pairs

we simply write

$f(x) = 2x + 1$ Function notation

The symbol $f(x)$ is read "f of x," "f at x," or "the value of f at x" and represents the number in the range of the function f to which the domain value x is paired. Thus, $f(3)$ is the range value for the function f associated with the domain value 3. We find this range value by replacing x with 3 wherever x occurs in the function definition

$$f(x) = 2x + 1$$

and evaluating the right side,

$$\begin{aligned} f(3) &= 2 \cdot 3 + 1 \\ &= 6 + 1 \\ &= 7 \end{aligned}$$

The statement $f(3) = 7$ indicates in a concise way that the function f assigns the range value 7 to the domain value 3 or, equivalently, that the ordered pair (3, 7) belongs to f.

The symbol $f: x \to f(x)$, read "f maps x into $f(x)$," is also used to denote the relationship between the domain value x and the range value $f(x)$ (see Fig. 2). Whenever we write $y = f(x)$, we assume that x is an independent variable and that y and $f(x)$ both represent the dependent variable.

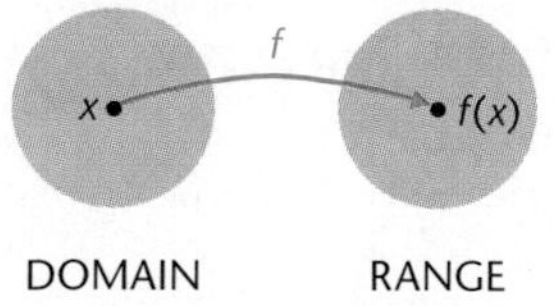

FIGURE 2
Function notation.

The function f "maps" the domain value x into the range value $f(x)$.

Letters other than f and x can be used to represent functions and independent variables. For example,

$$g(t) = t^2 - 3t + 7$$

defines g as a function of the independent variable t. To find $g(-2)$, we replace t by -2 wherever t occurs in

$$g(t) = t^2 - 3t + 7$$

and evaluate the right side:

$$\begin{aligned} g(-2) &= (-2)^2 - 3(-2) + 7 \\ &= 4 + 6 + 7 \\ &= 17 \end{aligned}$$

Thus, the function g assigns the range value 17 to the domain value -2; the ordered pair $(-2, 17)$ belongs to g.

It is important to understand and remember the definition of the symbol $f(x)$:

DEFINITION 3

THE SYMBOL $f(x)$

The symbol $f(x)$ represents the real number in the range of the function f corresponding to the domain value x. Symbolically, $f: x \to f(x)$. The ordered pair $(x, f(x))$ belongs to the function f. If x is a real number that is not in the domain of f, then f is **not defined** at x and $f(x)$ **does not exist.**

EXAMPLE 3

Evaluating Functions

For

$$f(x) = \frac{15}{x-3} \qquad g(x) = 16 + 3x - x^2 \qquad h(x) = \frac{6}{\sqrt{x} - 1}$$

find:

(A) $f(6)$ (B) $g(-7)$ (C) $h(-2)$ (D) $f(0) + g(4) - h(16)$

Solutions

(A) $f(6) \boxed{= \dfrac{15}{6-3}}^{*} = \dfrac{15}{3} = 5$

(B) $g(-7) \boxed{= 16 + 3(-7) - (-7)^2} = 16 - 21 - 49 = -54$

(C) $h(-2) = \dfrac{6}{\sqrt{-2} - 1}$

*Throughout the book, dashed boxes—called **think boxes**—are used to represent steps that are usually performed mentally.

But $\sqrt{-2}$ is not a real number. Since we have agreed to restrict the domain of a function to values of x that produce real values for the function, -2 is not in the domain of h and $h(-2)$ is not defined.

(D) $f(0) + g(4) - h(16)$

$$= \frac{15}{0 - 3} + [16 + 3(4) - 4^2] - \frac{6}{\sqrt{16} - 1}$$

$$= \frac{15}{-3} + 12 - \frac{6}{3}$$

$$= -5 + 12 - 2 = 5$$

MATCHED PROBLEM 3

Use the functions in Example 3 to find

(A) $f(-2)$ (B) $g(6)$ (C) $h(-8)$ (D) $\dfrac{f(8)}{h(9)}$

EXAMPLE 4

Finding Domains of Functions

Find the domains of functions f, g, and h:

$$f(x) = \frac{15}{x - 3} \qquad g(x) = 16 + 3x - x^2 \qquad h(x) = \frac{6}{\sqrt{x} - 1}$$

Solution

Domain of f

The fraction $15/(x - 3)$ represents a real number for all replacements of x by real numbers except $x = 3$, since division by 0 is not defined. Thus, $f(3)$ does not exist, and the domain of f is the set of all real numbers except 3. That is,

Domain of $f = \{x \mid x \neq 3)$ Set notation

$= (-\infty, 3) \cup (3, \infty)$ Interval notation

We often simplify this by writing

$$f(x) = \frac{15}{x - 3} \qquad x \neq 3$$

Domain of g

The domain is R, the set of all real numbers, since $16 + 3x - x^2$ represents a real number for all replacements of x by real numbers. To express this domain in interval notation, we write

Domain of $g = (-\infty, \infty)$

Domain of h

Since $\sqrt{x}$ is not a real number for negative real numbers x, x must be a nonnegative real number. But since $\sqrt{1} = 1$, evaluating $h(1)$ would result in division by 0. Thus, the domain of h is all nonnegative real numbers except 1. This can be written as

$$\text{Domain of } h = \{x \mid x \geq 0, x \neq 1\} \quad \text{Set notation}$$
$$= [0, 1) \cup (1, \infty) \quad \text{Interval notation}$$

or, more informally, as

$$h(x) = \frac{6}{\sqrt{x} - 1} \qquad x \geq 0, x \neq 1$$

MATCHED PROBLEM **4**

Find the domains of functions F, G, and H:

$$F(x) = x^2 + 5x - 2 \qquad G(x) = \frac{\sqrt{x}}{x - 3} \qquad H(x) = \frac{1}{x} + \frac{x}{x + 2}$$

Explore/Discuss 3

Let x and h be any real numbers.

(A) If $f(x) = 3x + 2$, which of the following is correct?
   (i) $f(x + h) = 3x + 2 + h$
   (ii) $f(x + h) = 3x + 3h + 2$
   (iii) $f(x + h) = 3x + 3h + 4$

(B) If $f(x) = x^2$, which of the following is correct?
   (i) $f(x + h) = x^2 + h^2$
   (ii) $f(x + h) = x^2 + h$
   (iii) $f(x + h) = x^2 + 2xh + h^2$

(C) If $f(x) = x^2 + 3x + 2$, write a verbal description of the operations that must be performed to evaluate $f(x + h)$.

In addition to evaluating functions at specific numbers, it is important to be able to evaluate functions at expressions that involve one or more variables. For example, the **difference quotient**

$$\frac{f(x + h) - f(x)}{h} \qquad x \text{ and } x + h \text{ in the domain of } f,\ h \neq 0$$

is studied extensively in a calculus course.

EXAMPLE **5** CL∫*

Evaluating and Simplifying a Difference Quotient

For $f(x) = x^2 + 4x + 5$, find and simplify:

(A) $f(x + h)$ (B) $\dfrac{f(x + h) - f(x)}{h}$, $h \neq 0$

*The symbol CL∫ denotes problems that are related to calculus.

Solutions (A) To find $f(x + h)$, we replace x with $x + h$ everywhere it appears in the equation that defines f and simplify:

$$\begin{aligned} f(x + h) &= (x + h)^2 + 4(x + h) + 5 \\ &= x^2 + 2xh + h^2 + 4x + 4h + 5 \end{aligned}$$

(B) Using the result of part (A), we get

$$\begin{aligned} \frac{f(x + h) - f(x)}{h} &= \frac{x^2 + 2xh + h^2 + 4x + 4h + 5 - (x^2 + 4x + 5)}{h} \\ &= \frac{x^2 + 2xh + h^2 + 4x + 4h + 5 - x^2 - 4x - 5}{h} \\ &= \frac{2xh + h^2 + 4h}{h} = \frac{h(2x + h + 4)}{h} = 2x + h + 4 \end{aligned}$$

MATCHED PROBLEM 5 Repeat Example 5 for $f(x) = x^2 + 3x + 7$.

CAUTION

1. If f is a function, then the symbol $f(x + h)$ represents the value of f at the number $x + h$ and must be evaluated by replacing the independent variable in the equation that defines f with the expression $x + h$, as we did in Example 5. Do not confuse this notation with the familiar algebraic notation for multiplication:

$f(x + h) \neq fx + fh$ — $f(x + h)$ is function notation.

$4(x + h) = 4x + 4h$ — $4(x + h)$ is algebraic multiplication notation.

2. There is another common incorrect interpretation of the symbol $f(x + h)$. If f is an arbitrary function, then

$$f(x + h) \neq f(x) + f(h)$$

It is possible to find some particular functions for which $f(x + h) = f(x) + f(h)$ is a true statement, but in general these two expressions are not equal.

Application

The next example explores the relationship between the *algebraic* definition of a function, the numeric values of the function, and a *graphic* representation of the function. The interplay between the algebraic, numeric, and graphic aspects of a function is one of the central themes of this book. In this example, we also see how a function can be used to describe data from the real world, a process that is generally referred to as *mathematical modeling*.

EXAMPLE 6

Consumer Debt

Revolving-credit debt (in billions of dollars) in the United States over a 20-year period is given in Table 4. A financial analyst used statistical techniques to produce a mathematical model for this data:

$$f(x) = 0.62x^2 - x + 5.1$$

where $x = 0$ corresponds to 1970.

(A) To compare the data in Table 4 and the values produced by the modeling function f, use a graphing utility to complete Table 5.

TABLE 4 Revolving-Credit Debt

Year	Total Debt (Billions)
1970	\$5.1
1975	\$15.0
1980	\$58.5
1985	\$128.9
1990	\$234.8

Source: Federal Reserve System.

TABLE 5

x	0	5	10	15	20
Debt	5.1	15.0	58.5	128.9	234.8
$f(x)$					

(B) Sketch by hand the graph of the modeling function f and the original data using the same axes.

(C) Use the modeling function f to estimate the debt to the nearest tenth of a billion in 1988 and in 1992.

Solutions

(A) As we mentioned earlier, most graphing utilities have a built-in routine for computing a table of values (Fig. 3). If yours does not, then simply evaluate the function at each value of x given in the table.

FIGURE 3

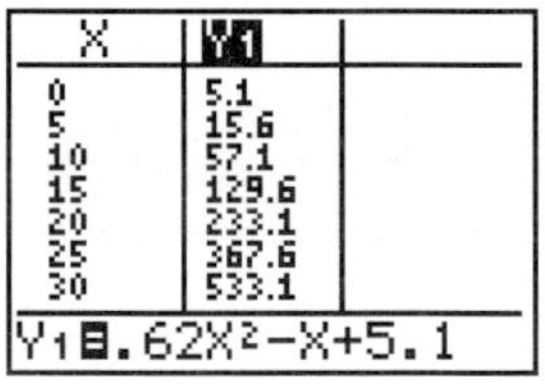

x	0	5	10	15	20
Debt	5.1	15.0	58.5	128.9	234.8
$f(x)$	5.1	15.6	57.1	129.6	233.1

(B) Figure 4 shows a sketch of the graph of $y = f(x)$ and the original data points in Table 4 with 0 corresponding to 1970.

FIGURE 4

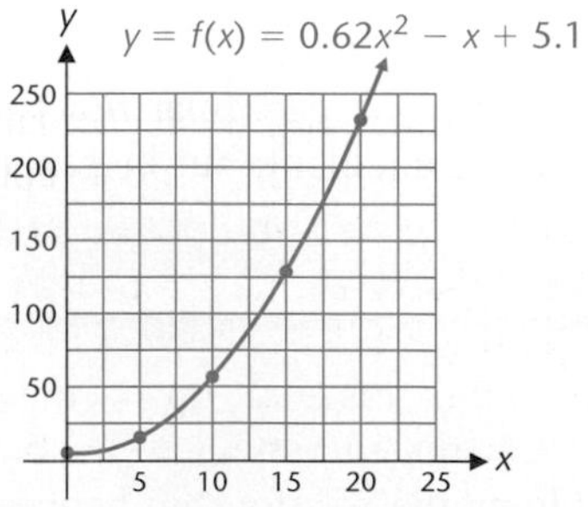

(C) Evaluate $f(x)$ at 18 and at 22:

$$f(18) = 188.0 \qquad f(22) = 283.2.$$

Thus, the revolving-credit debt should be \$188 billion in 1988 and \$283.2 billion in 1992.

MATCHED PROBLEM **6**

Credit union debt (in billions of dollars) in the United States is given in Table 6. Repeat Example 6 using this data and the modeling function.

$$y = f(x) = 0.5x^2 + 5.6x + 46.6$$

TABLE 6 Credit Union Debt

Year	Total Debt (Billions)
1970	\$48.7
1975	\$82.9
1980	\$147.0
1985	\$245.1
1990	\$347.1

Source: Federal Reserve System.

Remarks

1. Modeling functions like the function f in Example 6 provide reasonable and useful representations of the given data, but they do not always correctly predict future behavior. For example, the model in Example 6 indicated that the revolving-credit debt in 1992 should be about \$283.2 billion. But the actual debt for 1992 turned out to be \$267.9 billion, which differs from the predicted value by over \$18 billion. Proper use of mathematical models requires both an understanding of the techniques used to develop the model and frequent reevaluation, modification, and interpretation of the results produced by the model.

2. Later in this chapter we will discuss methods for finding a function f that models a given set of data. It turns out that this is easy to do with a graphing utility.

A Brief History of the Function Concept

The history of the use of functions in mathematics illustrates the tendency of mathematicians to extend and generalize a concept. The word "function" appears to have been first used by Leibniz in 1694 to stand for any quantity associated with a curve. By 1718, Johann Bernoulli considered a function any expression made up of constants and a variable. Later in the same century, Euler came to regard a function as any equation made up of constants and variables. Euler made extensive use of the extremely important notation $f(x)$, although its origin is generally attributed to Clairaut (1734).

The form of the definition of function that has been used until well into this century (many texts still contain this definition) was formulated by Dirichlet (1805–1859). He stated that, if two variables x and y are so related that for each value of x there corresponds exactly one value of y, then y is said to be a (single-valued) function of x. He called x, the variable to which values are assigned at will, the independent variable, and y, the variable whose values depend on the values assigned to x, the dependent variable. He called the values assumed by x the domain of the function, and the corresponding values assumed by y the range of the function.

Now, since set concepts permeate almost all mathematics, we have the more general definition of function presented in this section in terms of sets of ordered pairs of elements.

Answers to Matched Problems

1. (A) S does not define a function. (B) T defines a function with domain $\{-2, -1, 0, 1, 2\}$ and range $\{0, 1, 2\}$.

2. $x \leq 0$ **3.** (A) -3 (B) -2 (C) Does not exist (D) 1

4. Domain of F: all real numbers

Domain of G: $x \geq 0$, $x \neq 3$ or $[0, 3) \cup (3, \infty)$

Domain of H: all real numbers except 0 and -2 or $(-\infty, -2) \cup (-2, 0) \cup (0, 00)$

5. (A) $x^2 + 2xh + h^2 + 3x + 3h + 7$ (B) $2x + h + 3$

6. (A)

x	0	5	10	15	20
Debt	48.7	82.9	147	245.1	347.1
$f(x)$	46.6	87.1	152.6	243.1	358.6

(B)

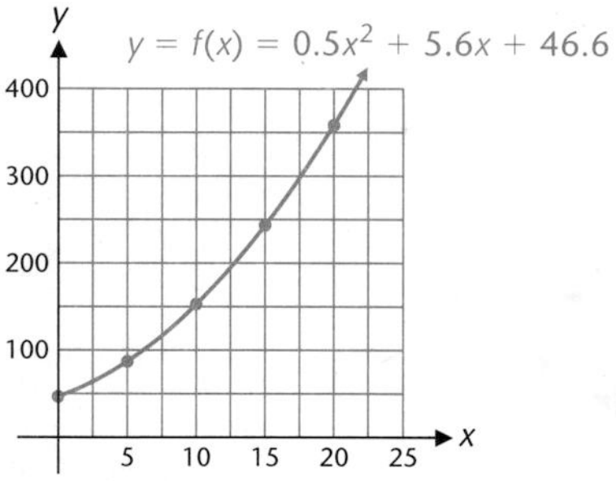

(C) \$309.4 billion; \$411.8 billion

EXERCISE 1-3

A

Indicate whether each table in Problems 1–6 defines a function.

1. **Domain** **Range**

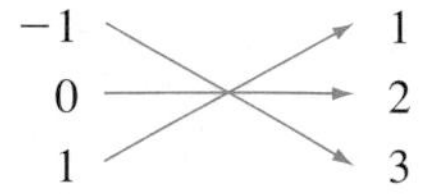

2. **Domain** **Range**

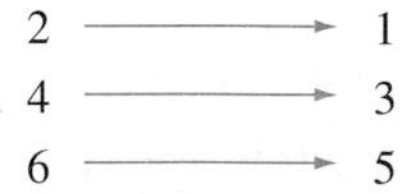

3. **Domain** **Range**

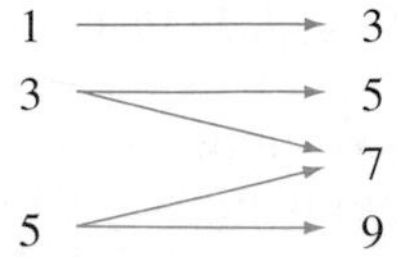

4. **Domain** **Range**

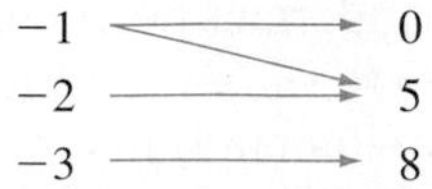

5. **Domain** **Range**

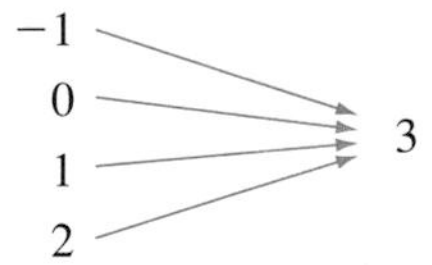

6. **Domain** **Range**

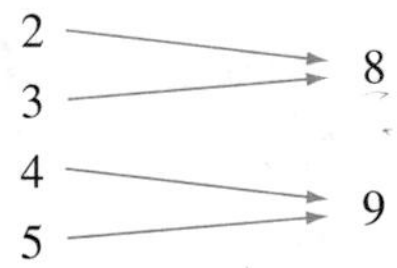

Indicate whether each set in Problems 7–12 defines a function. Find the domain and range of each function.

7. $\{(2, 4), (3, 6), (4, 8), (5, 10)\}$

8. $\{(-1, 4), (0, 3), (1, 2), (2, 1)\}$

9. $\{(10, -10), (5, -5), (0, 0), (5, 5)\ (10, 10)\}$

10. $\{(-10, 10), (-5, 5), (0, 0), (5, 5), (10, 10)\}$

11. $\{(0, 1), (1, 1), (2, 1), (3, 2), (4, 2), (5, 2)\}$

12. $\{(1, 1), (2, 1), (3, 1), (1, 2), (2, 2), (3, 2)\}$

Indicate whether each graph in Problems 13–18 is the graph of a function.

13.

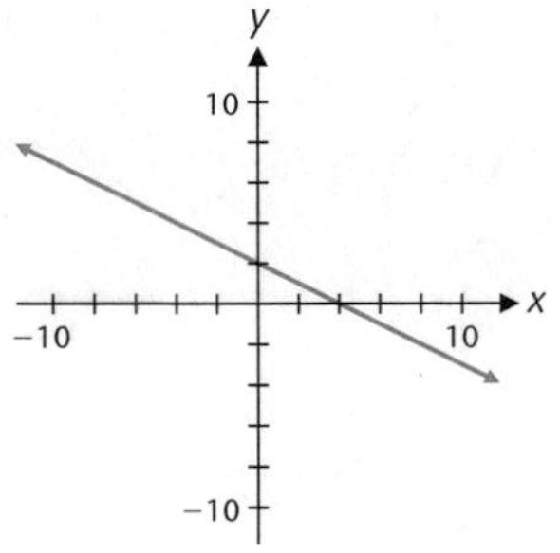

14.

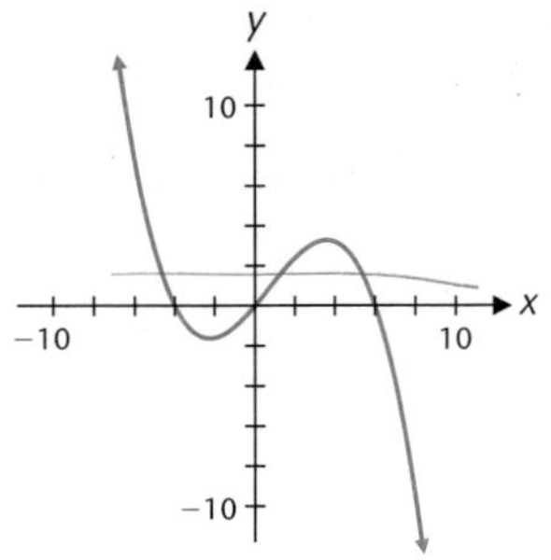

15.

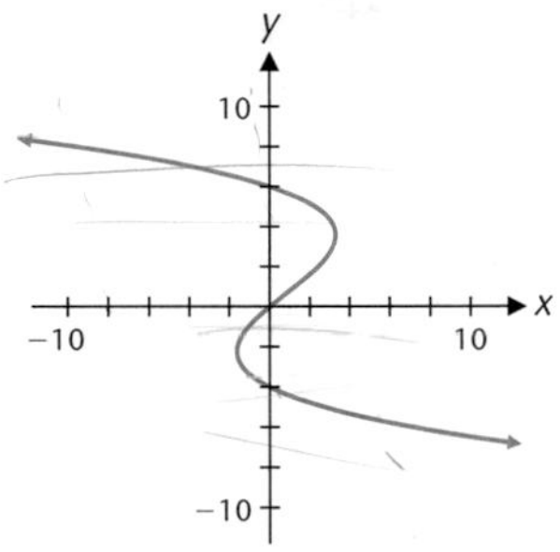

16.

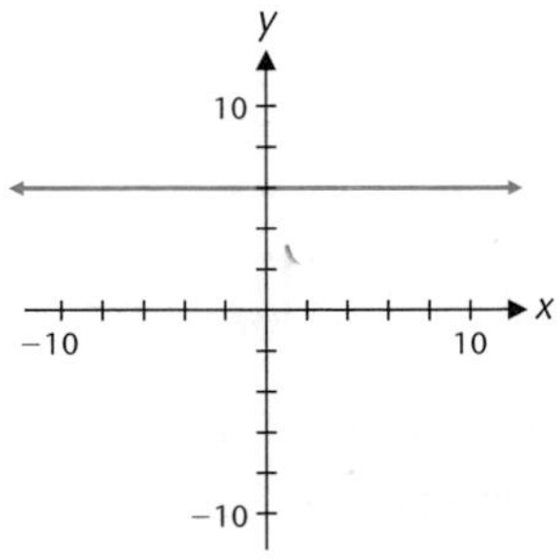

17.

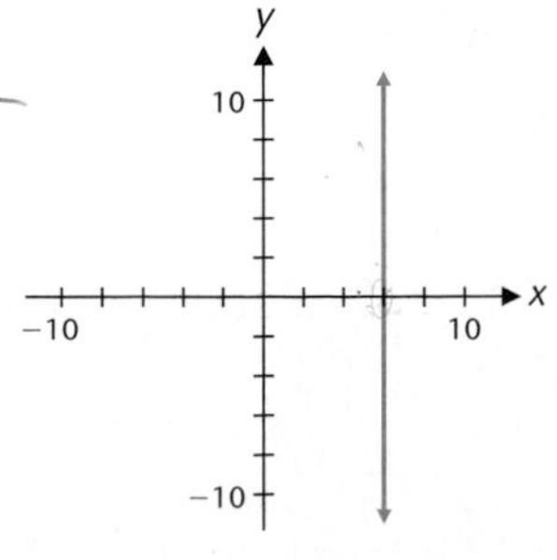

18.

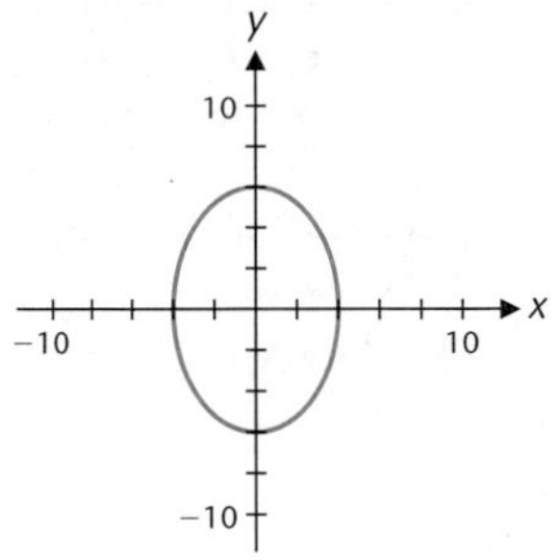

Problems 19–30 refer to the functions

$$f(x) = 3x - 5 \qquad g(t) = 4 - t$$

$$F(m) = 3m^2 + 2m - 4 \qquad G(u) = u - u^2$$

Evaluate as indicated.

19. $f(-1)$

20. $g(6)$

21. $G(-2)$

22. $F(-3)$

23. $F(-1) + f(3)$

24. $G(2) - g(-3)$

25. $2F(-2) - G(-1)$

26. $3G(-2) + 2F(-1)$

27. $\dfrac{f(0) \cdot g(-2)}{F(-3)}$

28. $\dfrac{g(4) \cdot f(2)}{G(1)}$

29. $\dfrac{f(4) - f(2)}{2}$

30. $\dfrac{g(5) - g(3)}{2}$

B

In Problems 31–42, find the domain of the indicated function. Express answers informally and formally using interval notation.

31. $f(x) = 4 - 9x + 3x^2$

32. $f(x) = 1 + 7x - 5x^2$

33. $f(x) = \dfrac{2}{4 - x}$

34. $f(x) = \dfrac{x}{x - 3}$

35. $f(x) = 2 - 3\sqrt{x}$

36. $f(x) = 4\sqrt{x} + 3$

37. $f(x) = 5 + \sqrt{-x}$

38. $f(x) = 2\sqrt{-x} - 1$

39. $f(x) = \dfrac{1}{x + 1} + \dfrac{1}{x - 1}$

40. $f(x) = \dfrac{x}{x - 2} - \dfrac{3}{x + 3}$

41. $f(x) = \dfrac{\sqrt{x}}{x - 5}$

42. $f(x) = \dfrac{\sqrt{-x}}{x + 4}$

In Problems 43–46, find a function f that makes all three equations true. [Hint: There may be more than one possible answer, but there is one obvious answer suggested by the pattern illustrated in the equations.]

43. $f(1) = 2(1) - 3$
$f(2) = 2(2) - 3$
$f(3) = 2(3) - 3$

44. $f(1) = 5(1)^2 - 6$
$f(2) = 5(2)^2 - 6$
$f(3) = 5(3)^2 - 6$

45. $f(1) = 4(1)^2 - 2(1) + 9$
$f(2) = 4(2)^2 - 2(2) + 9$
$f(3) = 4(3)^2 - 2(3) + 9$

46. $f(1) = -8 + 5(1) - 2(1)^2$
$f(2) = -8 + 5(2) - 2(2)^2$
$f(3) = -8 + 5(3) - 2(3)^2$

47. If $F(s) = 3s + 15$, find $\dfrac{F(2+h) - F(2)}{h}$.

48. If $K(r) = 7 - 4r$, find $\dfrac{K(1+h) - K(1)}{h}$.

49. If $g(x) = 2 - x^2$, find $\dfrac{g(3+h) - g(3)}{h}$.

50. If $P(m) = 2m^2 + 3$, find $\dfrac{P(2+h) - P(2)}{h}$.

51. If $L(w) = -2w^2 + 3w - 1$, find $\dfrac{L(-2+h) - L(-2)}{h}$.

52. If $D(p) = -3p^2 - 4p + 9$, find $\dfrac{D(-1+h) - D(-1)}{h}$.

The verbal statement "function f multiplies the square root of the domain element by 2 and then subtracts 5" and the algebraic statement $f(x) = 2\sqrt{x} - 5$ define the same function. In Problems 53–56, translate each verbal definition of the function into an algebraic definition.

53. Function g multiplies the domain element by 3 and then adds 1.

54. Function f multiplies the domain element by 7 and then adds the product of 5 and the cube of the domain element.

55. Function F divides the domain element by the sum of 8 and the square root of the domain element.

56. Function G takes the square root of the sum of 4 and the square of the domain element.

In Problems 57–60, translate each algebraic definition of the function into a verbal definition.

57. $f(x) = 2x - 3x^2$

58. $g(x) = 5x^3 - 8x$

59. $F(x) = \sqrt{x^4 + 9}$

60. $G(x) = \dfrac{x}{3x - 6}$

C

61. Find $f(x)$, given that $f(x + h) = 2(x + h)^2 - 4(x + h) + 6$

62. Find $g(x)$, given that $g(x + h) = 5 - 7(x + h)^2 + 8(x + h)$

63. Find $m(x)$, given that

$$m(x + h) = 4(x + h) - 3\sqrt{x + h} + 9$$

64. Find $s(x)$, given that

$$s(x + h) = 2\sqrt[3]{x + h} - 6(x + h) - 5$$

In Problems 65–72, find and simplify:

(A) $\dfrac{f(x+h) - f(x)}{h}$ (B) $\dfrac{f(x) - f(a)}{x - a}$

65. $f(x) = 3x - 4$

66. $f(x) = -2x + 5$

67. $f(x) = x^2 - 1$

68. $f(x) = x^2 + x - 1$

69. $f(x) = -3x^2 + 9x - 12$

70. $f(x) = -x^2 - 2x - 4$

71. $f(x) = x^3$

72. $f(x) = x^3 + x$

In Problems 73 and 74, $x = 1$ is not in the domain of the function f because the algebraic expression used to define f does not exist at $x = 1$. If you were to assign a numerical value to f at $x = 1$, what value would you choose? Support your choice with information obtained by exploring the graph of f near $x = 1$, by examining the numerical values of f near $x = 1$, and by algebraically simplifying the expression used to define f.

73. $f(x) = \dfrac{x^2 - 1}{x - 1}$

74. $f(x) = \dfrac{x^3 - 1}{x - 1}$

APPLICATIONS

75. Physics—Rate. The distance in feet that an object falls in a vacuum is given by $s(t) = 16t^2$, where t is time in seconds. Find

(A) $s(0)$, $s(1)$, $s(2)$, $s(3)$

(B) $\dfrac{s(2+h) - s(2)}{h}$

(C) What happens in part (B) when h tends to 0? Interpret physically.

76. Physics—Rate. An automobile starts from rest and travels along a straight and level road. The distance in feet traveled by the automobile is given by $s(t) = 10t^2$, where t is time in seconds. Find

(A) $s(8)$, $s(9)$, $s(10)$, $s(11)$

(B) $\dfrac{s(11+h) - s(11)}{h}$

(C) What happens in part B as h tends to 0? Interpret physically.

77. Boiling Point of Water. At sea level, water boils when it reaches a temperature of 212°F. At higher altitudes, the atmospheric pressure is lower and so is the temperature at which water boils. The boiling point $B(x)$ in degrees Fahrenheit at an altitude of x feet is given approximately by

$$B(x) = 212 - 0.0018x$$

(A) Complete the following table.

x	0	5,000	10,000	15,000	20,000	25,000	30,000
$B(x)$							

(B) Based on the information in the table, write a brief verbal description of the relationship between altitude and the boiling point of water.

78. Air Temperature. As dry air moves upward, it expands and cools. The air temperature $A(x)$ in degrees Celsius at an altitude of x kilometers is given approximately by

$$A(x) = 25 - 9x$$

(A) Complete the following table.

x	0	1	2	3	4	5
$A(x)$						

(B) Based on the information in the table, write a brief verbal description of the relationship between altitude and air temperature.

79. Car Rental. A car rental agency computes daily rental charges for compact cars with the function

$$D(x) = 20 + 0.25x$$

where $D(x)$ is the daily charge in dollars and x is the daily mileage. Translate this algebraic statement into a verbal statement that can be used to explain the daily charges to a customer.

80. Installation Charges. A telephone store computes charges for phone installation with the function

$$S(x) = 15 + 0.7x$$

where $S(x)$ is the installation charge in dollars and x is the time in minutes spent performing the installation. Translate this algebraic statement into a verbal statement that can be used to explain the installation charges to a customer.

Merck & Co., Inc. is the world's largest pharmaceutical company. Problems 81–84 refer to the data in Table 7 taken from the company's 1993 annual report.

TABLE 7 Selected Financial Data for Merck & Co., Inc. [$ in billions]

	1988	1989	1990	1991	1992
Sales	$5.9	$6.5	$7.7	$8.6	$9.7
R & D expenses	$0.66	$0.75	$0.85	$0.99	$1.1
Net income	$1.2	$1.5	$1.8	$2.1	$2.4

81. Sales Analysis. A mathematical model for Merck's sales is given by

$$S(t) = 5.74 + 0.97t$$

where $t = 0$ corresponds to 1988.

(A) Complete the following table. Round values of $S(t)$ to one decimal place.

t	0	1	2	3	4
Sales	5.9	6.5	7.7	8.6	9.7
$S(t)$					

(B) Sketch by hand the graph of S and the sales data on the same axes.

(C) Use the modeling function S to estimate the sales in 1993. In 2000.

(D) Write a brief verbal description of the company's sales from 1988 to 1992.

82. Income Analysis. A mathematical model for Merck's income is given by

$$I(t) = 1.2 + 0.3t$$

where $t = 0$ corresponds to 1988.

(A) Complete the following table. Round values of $I(t)$ to one decimal place.

t	0	1	2	3	4
Net income	1.2	1.5	1.8	2.1	2.4
$I(t)$					

(B) Sketch by hand the graph of I and the income data on the same axes.

(C) Use the modeling function I to estimate the income in 1993. In 2000.

(D) Write a brief verbal description of the company's income from 1988 to 1992.

83. Sales Analysis. A mathematical model for Merck's sales as a function of R & D (research & development) expenses is given by

$$S(r) = 0.2 + 8.6r$$

where r represents R & D expenditures.

(A) Complete the following table. Round values of $S(r)$ to one decimal place.

r (R & D)	0.66	0.75	0.85	0.99	1.1
Sales	5.9	6.5	7.7	8.6	9.7
$S(r)$					

(B) Sketch by hand the graph of S and the data on the same axes.

(C) Use the modeling function S to estimate the sales if the company spends $1.5 billion on research and development. $2 billion.

84. Income Analysis. A mathematical model for Merck's income as a function of R & D (research & development) expenses is given by

$I(r) = -0.5 + 2.7r$

where r represents R & D expenditures.

(A) Complete the following table. Round values of $I(r)$ to one decimal place.

r (R & D)	0.66	0.75	0.85	0.99	1.1
Net income	1.2	1.5	1.8	2.1	2.4
$I(r)$					

(B) Sketch by hand the graph of I and the data on the same axes.

(C) Use the modeling function I to estimate the income if the company spends \$1.5 billion on research and development. \$2 billion.

Section 1-4 | Functions: Graphs and Properties

- Basic Concepts
- Increasing and Decreasing Functions
- Local Maxima and Minima
- Piecewise-Defined Functions
- The Greatest Integer Function

One of the primary goals of this course is to provide you with a set of mathematical tools that can be used, in conjunction with a graphing utility, to analyze graphs that arise quite naturally in important applications. In this section, we discuss some basic concepts that are commonly used to describe graphs of functions.

Basic Concepts

Each function that has a real number domain and range has a graph—the graph of the ordered pairs of real numbers that constitute the function. When functions are graphed, domain values usually are associated with the horizontal axis and range values with the vertical axis. Thus, the **graph of a function f** is the same as the graph of the equation

$$y = f(x)$$

FIGURE 1
Graph of a function.

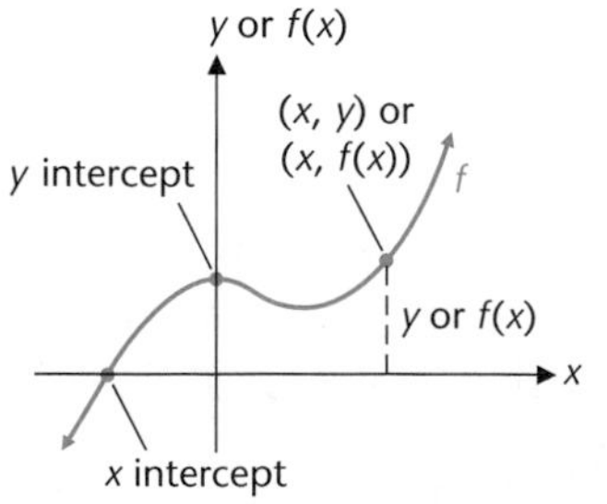

where x is the independent variable and the abscissa of a point on the graph of f. The variables y and $f(x)$ are dependent variables, and either is the ordinate of a point on the graph of f (see Fig. 1).

The abscissa of a point where the graph of a function intersects the x axis is called an **x intercept** or **zero** of the function. The x intercept is also a real solution or **root** of the equation $f(x) = 0$. The ordinate of a point where the graph of a function crosses the y axis is called the **y intercept** of the function. The y intercept is given by $f(0)$, provided 0 is in the domain of f. Note that a function can have more than one x intercept but can never have more than one y intercept—a consequence of the vertical line test discussed in the preceding section.

EXAMPLE 1

Finding x and y Intercepts

Find the x and y intercepts (correct to one decimal place) of $f(x) = x^3 + x - 3$.

Solution From the graph of f in Figure 2(a), we see that the y intercept is $f(0) = -3$ and that there is an x intercept between 1 and 2. Zooming in on the graph near $x = 1$ [Figs. 2(b) and 2(c)], we see that the x intercept (to one decimal place) is 1.2.

FIGURE 2

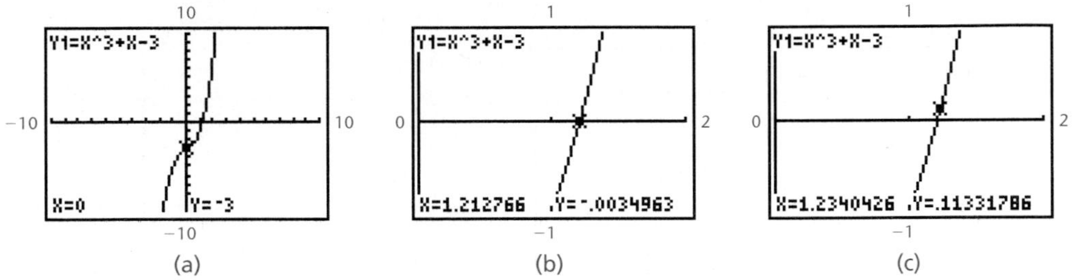

MATCHED PROBLEM 1 Find the x and y intercepts (correct to one decimal place) of $f(x) = x^3 + x + 5$.

Explore/Discuss 1

Suppose we want a more accurate approximation of the x intercept of the function $f(x) = x^3 + x - 3$ discussed in Example 1.

(A) Use zoom and trace to approximate the x intercept to five decimal places. How many times did you have to zoom in on the graph to obtain this accuracy?

(B) Consult the manual for your graphing utility to learn how to use its built-in routine for approximating x intercepts. **Root** and **zero** are two common names for this routine. Use your graphing utility to approximate the x intercept to five decimal places (Fig. 3).

FIGURE 3

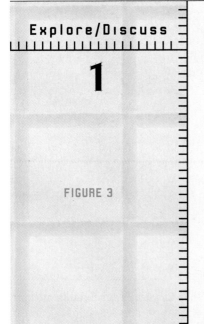

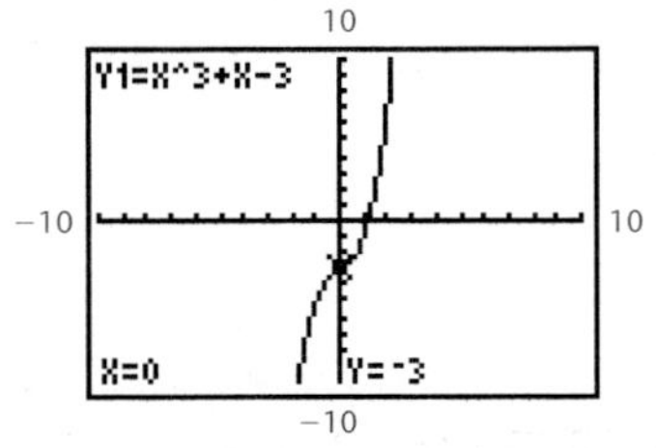

As you undoubtedly discovered in Explore/Discuss 1, using a built-in routine for approximating x intercepts is much faster than using zoom and trace. One of the great advantages of a graphing utility is the ability to approximate x intercepts or, equivalently, solutions to equations of the form $f(x) = 0$.

The domain of a function is the set of all the x coordinates of points on the graph of the function and the range is the set of all the y coordinates. It is instructive to view the domain and range as subsets of the coordinate axes as in Figure 4. Note the effective use of interval notation in describing the domain and range of the functions in this figure. In Figure 4(a) a solid dot is used to indicate that a point is on the graph of the function and in Figure 4(b) an open dot to indicate that a point is not on the graph of the function. An open or solid dot at the end of a graph indicates that the graph terminates there, while an arrowhead indicates that the graph continues beyond the portion shown with no significant changes in shape [see Fig. 4(b)].

FIGURE 4
Domain and range.

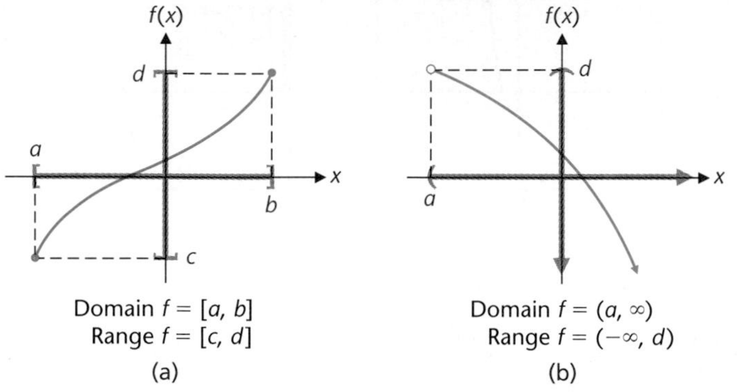

EXAMPLE
2

Finding the Domain and Range from a Graph

Find the domain and range of the function f whose graph is shown in Figure 5.

FIGURE 5

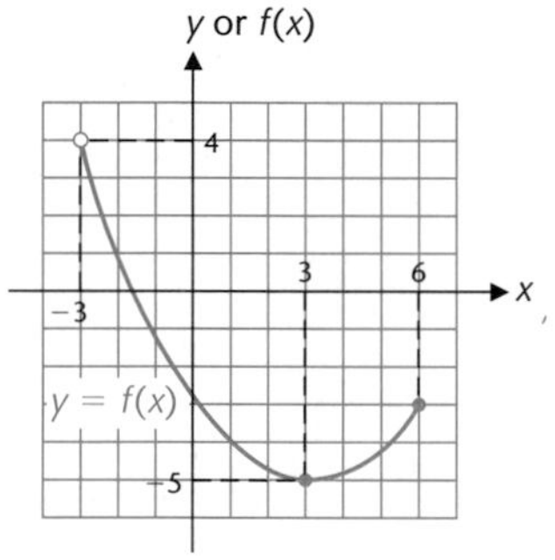

Solution

The dots at each end of the graph of f indicate that the graph terminates at these points. Thus, the x coordinates of the points on the graph are between -3 and 6. The open dot at $(-3, 4)$ indicates that -3 is not in the domain of f, while the closed dot at $(6, -3)$ indicates that 6 is in the domain of f. That is,

$$\text{Domain:} \quad -3 < x \leq 6 \quad \text{or} \quad (-3, 6]$$

The y coordinates are between -5 and 4 and, as before, the open dot at $(-3, 4)$ indicates that 4 is not in the range of f and the closed dot at $(3, -5)$ indicates that -5 is in the range of f. Thus,

$$\text{Range:} \quad -5 \leq y < 4 \quad \text{or} \quad [-5, 4)$$

MATCHED PROBLEM 2

Find the domain and range of the function f given by the graph in Figure 6.

FIGURE 6

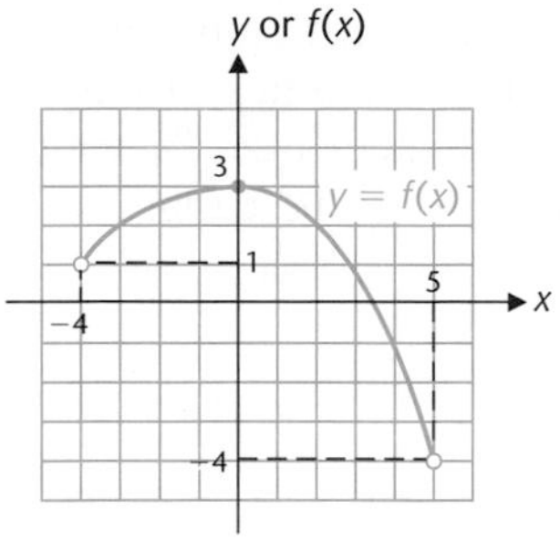

In Example 2, the domain and range for the function f were easily determined from the graph of f, since the entire graph of f was visible. More often we can only show part of a graph of a function defined by an equation, since the graph may continue to move indefinitely far away from the origin, and it may not be clear what happens to the graph as it moves further away. When a function is defined by an equation, such as

$$f(x) = 3\sqrt{x} \qquad \text{or} \qquad g(x) = 4\sqrt{x} - 0.4x \tag{1}$$

the domain is often easily determined from the equation, but the range is not. Since $\sqrt{x}$ represents a real number only for $x \geq 0$, functions f and g in equation (1) have the same domain, $[0, \infty)$. What about the range for f and for g? Looking at their graphs in standard viewing windows (Fig. 7), we might conclude that the range of each function is $[0, \infty)$. And we would be wrong!

FIGURE 7

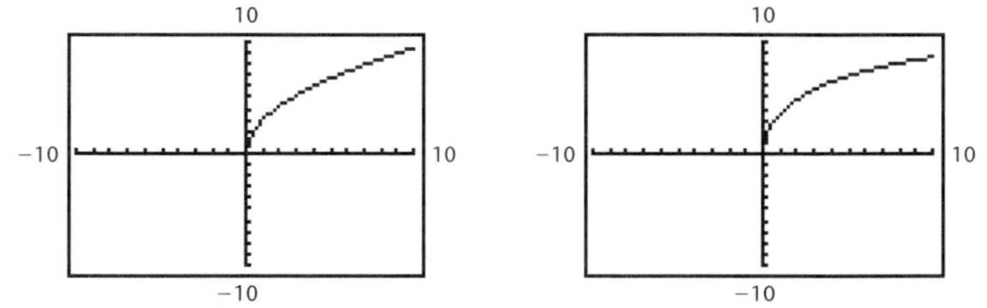

(a) $f(x) = 3\sqrt{x}$ (b) $g(x) = 4\sqrt{x} - 0.4x$

Using new window variables produces the graphs in Figure 8. The graph of f does appear to continue upward, but the graph of g changes direction and starts to go down. Using more advanced mathematics, it can be shown that $[0, \infty)$ is the correct range for f and the correct range for g is $(-\infty, 10]$.

FIGURE 8

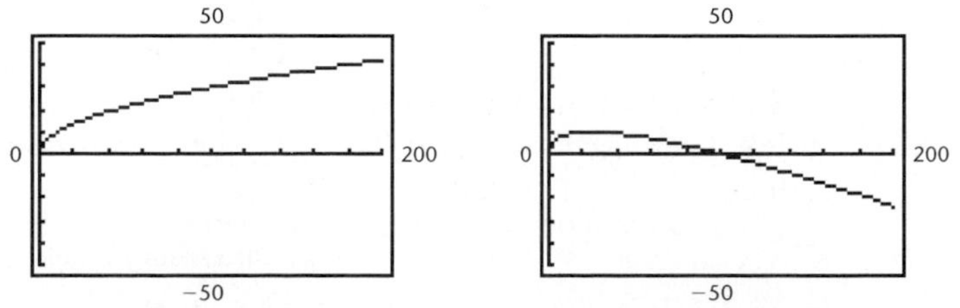

(a) $f(x) = 3\sqrt{x}$ (b) $g(x) = 4\sqrt{x} - 0.4x$

Before we continue, we must discuss another point about the graphs in Figure 7. Notice that the graphing utility we used to produce Figure 7 skipped over all the negative values of x because the functions were not defined as real numbers for $x < 0$. Most graphing utilities will disregard any values of x between Xmin and Xmax for which the function is not defined as a real number. But some do not. Instead, some graphing utilities display an error message if the interval [Xmin, Xmax] contains any values of x for which the function is not defined. If yours does this, you will have to be certain to exclude such x values from the viewing window.

Increasing and Decreasing Functions

Explore/Discuss 2

Graph each function in the standard viewing window, then write a verbal description of the behavior exhibited by the graph as x moves from left to right.

(A) $f(x) = 2 - x$ (B) $f(x) = x^3$

(C) $f(x) = 5$ (D) $f(x) = 9 - x^2$

We now take a look at *increasing* and *decreasing* properties of functions. Intuitively, a function is increasing over an interval I in its domain if its graph rises as the independent variable increases (moves from left to right) over I. A function is decreasing over I if its graph falls as the independent variable increases over I (Fig. 9).

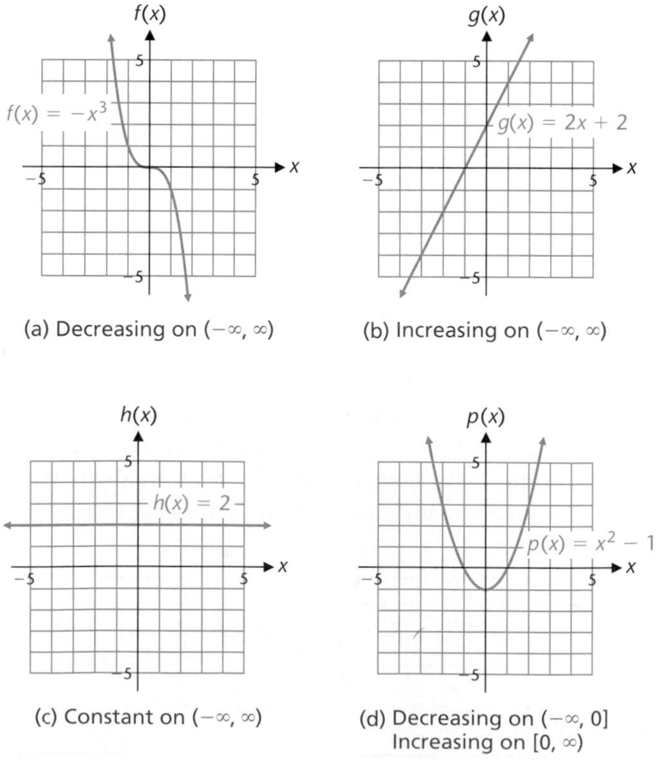

FIGURE 9
Increasing, decreasing, and constant functions.

(a) Decreasing on $(-\infty, \infty)$

(b) Increasing on $(-\infty, \infty)$

(c) Constant on $(-\infty, \infty)$

(d) Decreasing on $(-\infty, 0]$
Increasing on $[0, \infty)$

More formally, we define increasing, decreasing, and constant functions as follows:

DEFINITION 1

INCREASING, DECREASING, AND CONSTANT FUNCTIONS

Let I be an interval in the domain of a function f. Then,

1. **f is increasing** on I if $f(b) > f(a)$ whenever $b > a$ in I.
2. **f is decreasing** on I if $f(b) < f(a)$ whenever $b > a$ in I.
3. **f is constant** on I if $f(a) = f(b)$ for all a and b in I.

EXAMPLE 3

Describing a Graph

Use the terms increasing, decreasing, and constant to describe the graph of

$$f(x) = 60x - x^3 \qquad \text{for} \qquad -10 \le x \le 10$$

Solution

We begin by graphing f in the standard viewing window [Fig. 10(a)]. Since this view of the graph is not very informative, we need to adjust the viewing window. The trace procedure can be used to determine appropriate values for Ymin and Ymax. (The trace procedure displays the coordinates of a point on the graph even if the point is not visible in the viewing window.) While tracing along the graph in Figure 10(a) from Xmin $= -10$ to Xmax $= 10$, notice that all the y values are between -400 and 400, inclusive. Using Ymin $= -400$ and Ymax $= 400$ produces the graph in Figure 10(b).

FIGURE 10

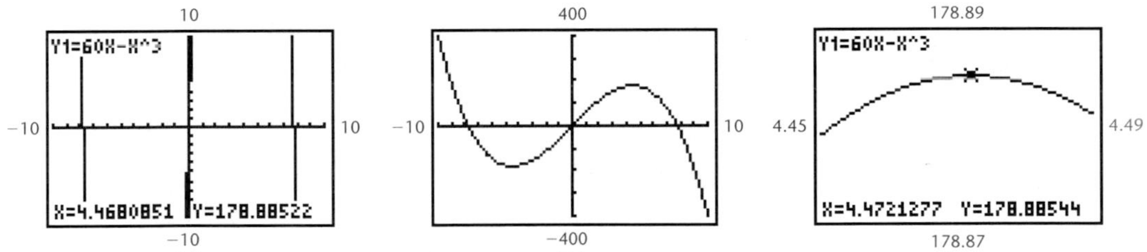

(a) (b) (c)

The graph changes direction twice, once near $x = -4.468$ and again near $x = 4.468$. Using zoom and trace procedures near 4.468, we see that the graph changes direction at $x = 4.47$, correct to two decimal places [Fig. 10(c)]. Similar procedures show that the graph also changes direction at $x = -4.47$ (details omitted). Thus, this function decreases as x goes from -10 to -4.47, increases as x goes from -4.47 to 4.47, and decreases as x goes from 4.47 to 10. Or using interval notation, f is decreasing on $[-10, -4.47]$ and $[4.47, 10]$, and increasing on $[-4.47, 4.47]$.

MATCHED PROBLEM 3

Repeat Example 3 with $f(x) = x^3 - 40x$.

In Example 3, we used zoom and trace procedures to approximate the points where the graph changes direction and we limited our discussion to the graphing interval [Xmin, Xmax]. Techniques for locating the exact points where graphs change direction and for analyzing increasing and decreasing behavior are discussed extensively in calculus.

Local Maxima and Minima

The points where the graph of a function changes direction play an essential role in the analysis of functions and graphs. These points are also critical to the solution of many applied problems involving the maximum or minimum values of a function. We define local maximum and minimum values of a function as follows:

LOCAL MAXIMA AND LOCAL MINIMA

The functional value $f(c)$ is called a **local maximum** if there is an interval (a, b) containing c such that

$$f(x) \le f(c) \text{ for all } x \text{ in } (a, b)$$

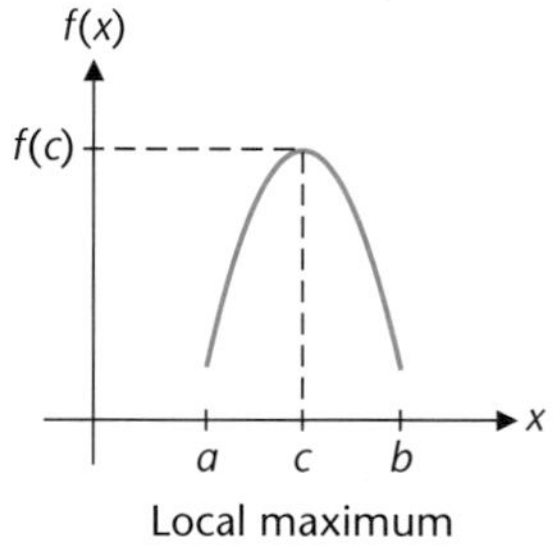

Local maximum

The functional value $f(c)$ is called a local minimum if there is an interval (a, b) containing c such that

$$f(x) \ge f(c) \text{ for all } x \text{ in } (a, b)$$

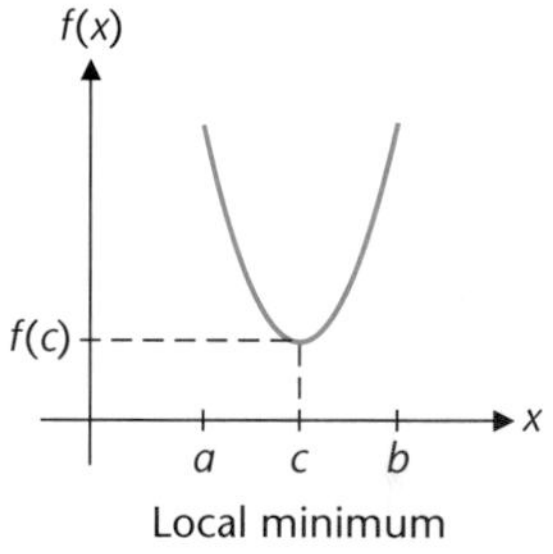

Local minimum

The functional value $f(c)$ is called a **local extremum** if it is either a local maximum or a local minimum.

Explore/Discuss 3

Most graphing utilities have built-in routines for approximating local maxima and minima that don't require you to zoom in on the graph (consult your manual for details). Graph $f(x) = 60x - x^3$ for $-10 \le x \le 10$, $-400 \le y \le 400$ and use a built-in routine to find the local maximum and local minimum [see Fig. 10(b)].

EXAMPLE 4 Maximizing Revenue

The revenue (in dollars) from the sale of x bicycle locks is given by

$$R(x) = 21x - 0.016x^2 \qquad 0 \le x \le 1{,}300$$

(A) Find the revenue for $x = 100, 200, \ldots, 1{,}300$. Use these values to estimate the number of locks that must be sold to maximize the revenue. What is the estimated maximum revenue?

(B) Graph $y = R(x)$ and find the exact number of locks that must be sold to maximize the revenue. What is the maximum revenue now?

Solutions

(A) Use a graphing utility to compute the required values of R (Fig. 11).

FIGURE 11

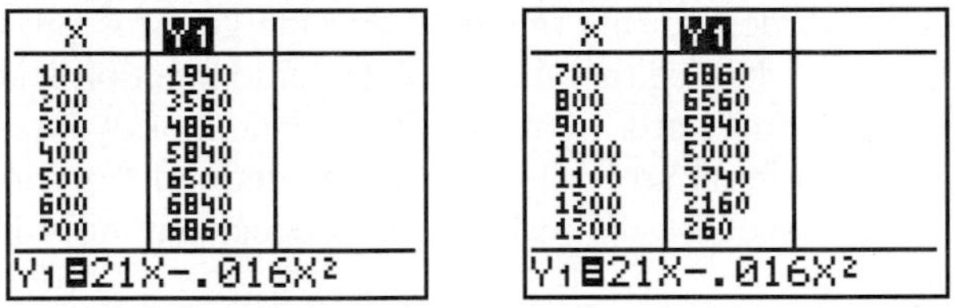

X	Y1
100	1940
200	3560
300	4860
400	5840
500	6500
600	6840
700	6860

Y1■21X−.016X²

X	Y1
700	6860
800	6560
900	5940
1000	5000
1100	3740
1200	2160
1300	260

Y1■21X−.016X²

Examining the values in Figure 11, we see that selling 700 locks will produce a revenue of \$6,860. This is the best estimate we can obtain from the values in these tables. It is likely that there exist values of x not in these tables that will produce larger values of $R(x)$.

(B) Using a built-in maximum routine (Fig. 12) and rounding to the nearest unit, we see that the maximum revenue of \$6,891 occurs when 656 locks are sold.

FIGURE 12

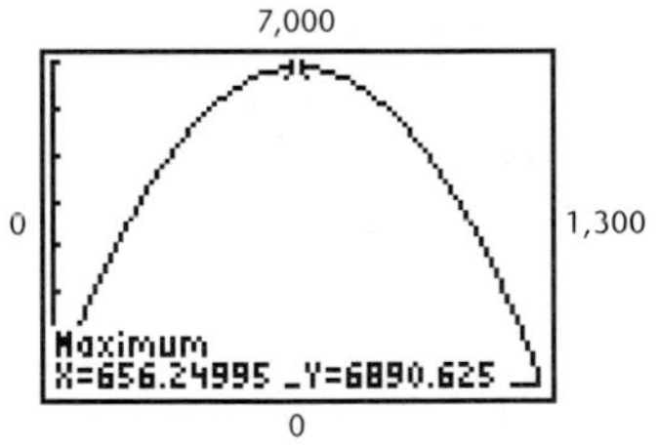

MATCHED PROBLEM 4

The profit (in dollars) from the sale of x bicycle locks is given by

$$P(x) = 17.5x - 0.016x^2 - 2{,}000 \qquad 0 \le x \le 1{,}300$$

(A) Find the profit for $x = 100, 200, \ldots, 1{,}300$. Use these values to estimate the number of locks that must be sold to maximize the profit. What is the estimated maximum profit?

(B) Graph $y = P(x)$ and find the number of locks that must be sold to maximize the profit. What is the maximum profit now?

Technically speaking, the revenue function R in Example 4 is defined only for integer values of x, $x = 0, 1, \ldots, 1{,}300$. However, for the purposes of mathematical analysis and as an aid in visualizing the behavior of the function R, we connect this discrete set of points with a continuous curve. There is nothing wrong with this, as long as you realize that it may not always be possible to realistically interpret values of R when x is not an integer. For example, the maximum value of R actually occurs at $x = 656.25$, but since it is not possible to manufacture one-fourth of a lock, we simply conclude that $x = 656$ is the number of locks that produces the maximum revenue.

FIGURE 13
Graph of $f(x) = |x| = \text{abs}(x)$.

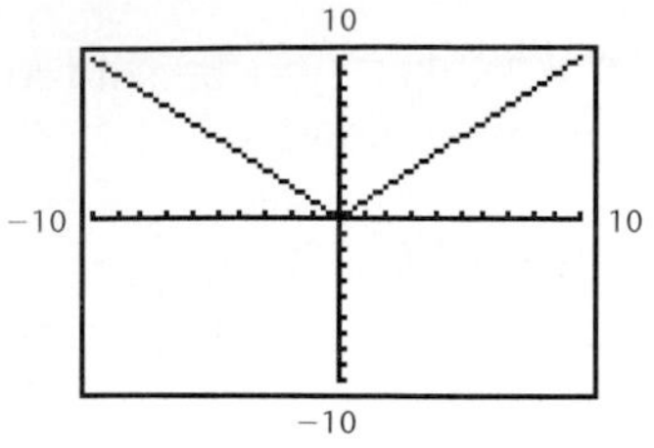

Piecewise-Defined Functions

The **absolute value function** is defined by

$$f(x) = |x| = \begin{cases} -x & \text{if } x < 0 \\ x & \text{if } x \geq 0 \end{cases}$$

The graph of $|x|$ is shown in Figure 13. Most graphing utilities use abs or ABS to denote this function, and the graph is produced directly using $y_1 = \text{abs}(x)$.

Notice that the absolute value function is defined by different formulas for different parts of its domain. Functions whose definitions involve more than one formula are called **piecewise-defined functions.** As the next example illustrates, piecewise-defined functions occur naturally in many applications.

EXAMPLE 5

Rental Charges

A car rental agency charges \$0.25 per mile if the mileage does not exceed 100. If the total mileage exceeds 100, the agency charges \$0.25 per mile for the first 100 miles and \$0.15 per mile for any additional mileage.

(A) If x represents the number of miles a rented vehicle is driven, express the mileage charge $C(x)$ as a function of x.

(B) Complete the following table.

x	0	50	100	150	200
$C(x)$					

(C) Sketch the graph of $y = C(x)$ by hand, using a graphing utility as an aid, and indicate the points in the table on the graph with solid dots.

Solutions

(A) If $0 \leq x \leq 100$, then

$$C(x) = 0.25x$$

If $x > 100$, then

$$\begin{aligned} C(x) &= \underbrace{0.25(100)}_{\text{Charge for the first 100 miles}} + \underbrace{0.15(x - 100)}_{\text{Charge for the additional mileage}} \\ &= 25 + 0.15x - 15 \\ &= 10 + 0.15x \end{aligned}$$

Thus, we see that C is a piecewise-defined function:

$$C(x) = \begin{cases} 0.25x & \text{if } 0 \leq x \leq 100 \\ 10 + 0.15x & \text{if } x > 100 \end{cases}$$

(B) Piecewise-defined functions are evaluated by first determining which rule applies and then using the appropriate rule to find the value of the function. To begin, we enter both rules in a graphing utility and use the table routine

FIGURE 14
$y_1 = 0.25x$, $y_2 = 10 + 0.15x$.

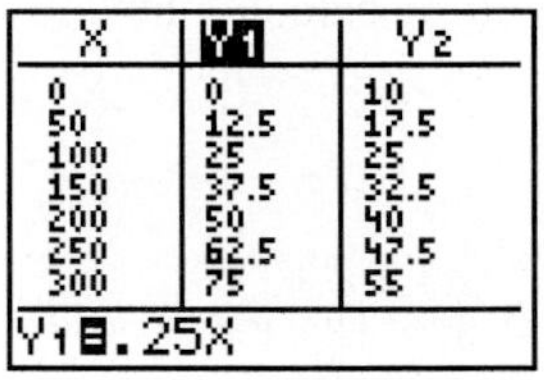

X	Y1	Y2
0	0	10
50	12.5	17.5
100	25	25
150	37.5	32.5
200	50	40
250	62.5	47.5
300	75	55

Y1=.25X

(Fig. 14). To complete the table, we use the values of $C(x)$ from the y_1 column if $0 \le x \le 100$, and from the y_2 column if $x > 100$.

x	0	50	100	150	200
$C(x)$	\$0	\$12.50	\$25	\$32.50	\$40

(C) Using a graph of both rules in the same viewing window as an aid (Fig. 15), we sketch the graph of $y = C(x)$ and add the points from the table to produce Figure 16.

FIGURE 15
$y_1 = 0.25x$, $y_2 = 10 + 0.15x$.

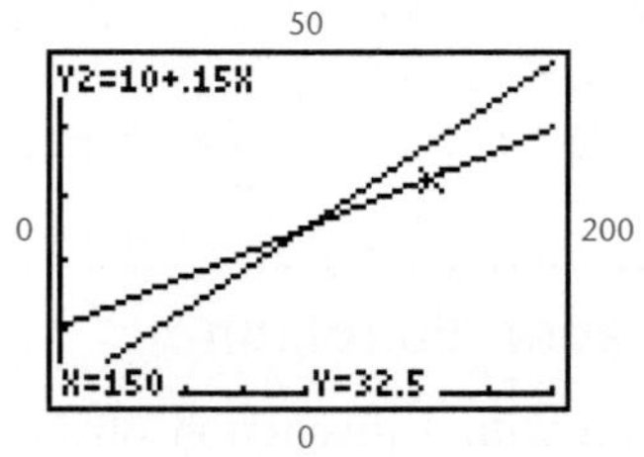

FIGURE 16
Hand sketch of the graph of $y = C(x)$.

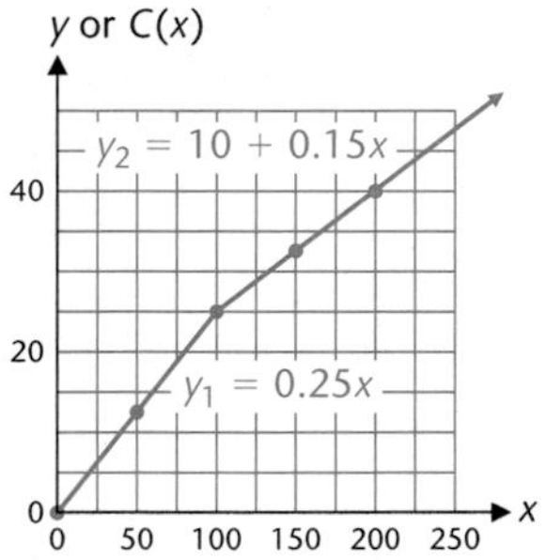

MATCHED PROBLEM 5

Another car rental agency charges \$0.30 per mile when the total mileage does not exceed 75, and \$0.30 per mile for the first 75 miles plus \$0.20 per mile for the additional mileage when the total mileage exceeds 75.

(A) If x represents the number of miles a rented vehicle is driven, express the mileage charge $C(x)$ as a function of x.

(B) Complete the following table.

x	0	50	75	100	150
$C(x)$					

(C) Sketch the graph of $y = C(x)$ by hand, using a graphing utility as an aid, and indicate the points in the table on the graph with solid dots.

Refer to Figures 14 and 16 in the solution to Example 5. Notice that the two formulas in the definition of C produce the same value at $x = 100$ and that the graph of C contains no breaks. Informally, a graph (or portion of a graph) is said to be **continuous** if it contains no breaks or gaps and can be drawn without lifting a pen from the paper. A graph is **discontinuous** at any points where there is a break or a gap. For example, the graph of the function in Figure 17 is discontinuous at $x = 1$. The entire graph cannot be drawn without lifting a pen from the paper. (A formal presentation of continuity can be found in calculus texts.)

FIGURE 17
Graph of

$$f(x) = \begin{cases} x^2 - 2 & \text{for } x \le 1 \\ x & \text{for } x > 1 \end{cases}$$

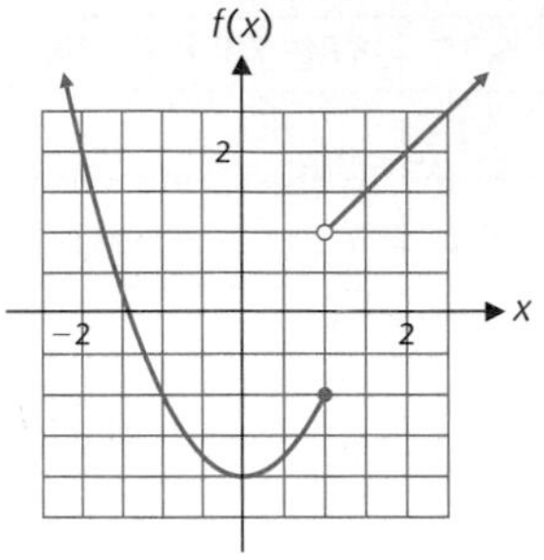

The Greatest Integer Function

We conclude this section with a discussion of an interesting and useful function called the *greatest integer function.*

The **greatest integer** of a real number x, denoted by $[\![x]\!]$, is the integer n such that $n \le x < n + 1$; that is, $[\![x]\!]$ is the largest integer less than or equal to x. For example,

$$[\![3.45]\!] = 3 \qquad [\![-2.13]\!] = -3 \quad \text{Not } -2$$
$$[\![7]\!] = 7 \qquad [\![-8]\!] = -8$$
$$[\![0]\!] = 0$$

The **greatest integer function** f is defined by the equation $f(x) = [\![x]\!]$. A piecewise definition of f for $-2 \le x < 3$ is shown below and a sketch of the graph of f for $-5 \le x \le 5$ is shown in Figure 18. Since the domain of f is all real numbers, the piecewise definition continues indefinitely in both directions, as does the stairstep pattern in the figure. Thus, the range of f is the set of all integers. The greatest integer function is an example of a more general class of functions called **step functions.**

FIGURE 18
Greatest integer function.

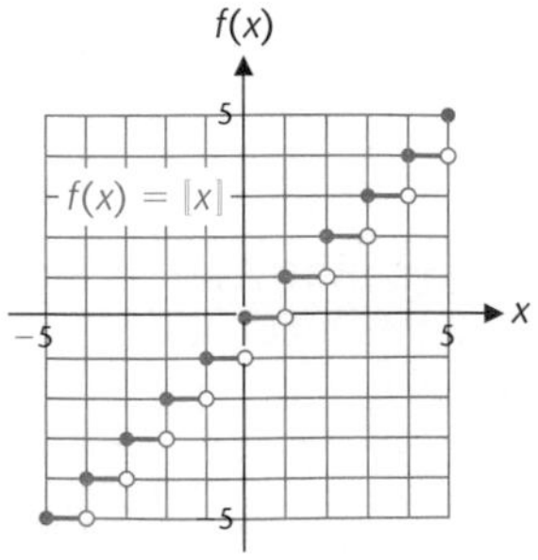

$$f(x) = [\![x]\!] = \begin{cases} \vdots & \\ -2 & \text{if } -2 \le x < -1 \\ -1 & \text{if } -1 \le x < 0 \\ 0 & \text{if } \quad 0 \le x < 1 \\ 1 & \text{if } \quad 1 \le x < 2 \\ 2 & \text{if } \quad 2 \le x < 3 \\ \vdots & \end{cases}$$

Notice in Figure 18 that at each integer value of x there is a break in the graph, and between integer values of x there is no break. Thus, the greatest integer function is discontinuous at each integer n and continuous on each interval of the form $[n, n + 1)$.

Most graphing utilities will graph the greatest integer function, usually denoted by int, but these graphs require careful interpretation. Comparing the sketch of $y = [\![x]\!]$ in Figure 18 with the graph of $y = \text{int}(x)$ in Figure 19(a), we see that the graphing utility has connected the end points of the horizontal line segments. This gives the appearance that the graph is continuous when it is not. To obtain a correct graph, consult the manual to determine how to change the graphing mode on your graphing utility from **connected mode** to **dot mode** [Fig. 19(b)].

FIGURE 19
Greatest integer function on a graphing utility.

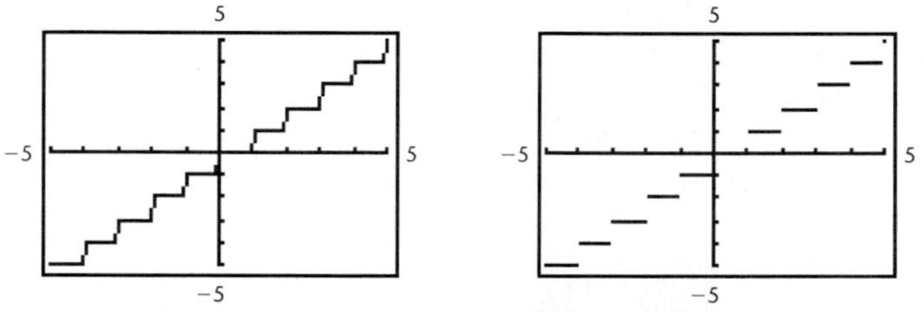

(a) Graph of $y = \text{int}(x)$ in the connected mode.

(b) Graph of $y = \text{int}(x)$ in the dot mode.

To avoid misleading graphs, use the dot mode on your graphing utility when graphing a function with discontinuities.

EXAMPLE 6

Computer Science

Let

$$f(x) = \frac{[\![10x + 0.5]\!]}{10}$$

Find

(A) $f(6)$ (B) $f(1.8)$ (C) $f(3.24)$ (D) $f(4.582)$ (E) $f(-2.68)$

What operation does this function perform?

Solutions

(A) $f(6) = \dfrac{[\![60.5]\!]}{10} = \dfrac{60}{10} = 6$

(B) $f(1.8) = \dfrac{[\![18.5]\!]}{10} = \dfrac{18}{10} = 1.8$

(C) $f(3.24) = \dfrac{[\![32.9]\!]}{10} = \dfrac{32}{10} = 3.2$

(D) $f(4.582) = \dfrac{[\![46.32]\!]}{10} = \dfrac{46}{10} = 4.6$

(E) $f(-2.68) = \dfrac{[\![-26.3]\!]}{10} = \dfrac{-27}{10} = -2.7$

x	$f[x]$
6	6
1.8	1.8
3.24	3.2
4.582	4.6
−2.68	−2.7

Comparing the values of x and $f(x)$ in the table, we conclude that this function rounds decimal fractions to the nearest tenth.

MATCHED PROBLEM 6

Let $f(x) = [\![x + 0.5]\!]$. Find

(A) $f(6)$ (B) $f(1.8)$ (C) $f(3.24)$ (D) $f(-4.3)$ (E) $f(-2.69)$

What operation does this function perform?

Answers to Matched Problems

1. x intercept: -1.5; y intercept: 5
2. Domain: $-4 < x < 5$ or $(-4, 5)$
   Range: $-4 < y \leq 3$ or $(-4, 3]$
3. f is increasing on $[-10, -3.65]$ and $[3.65, 10]$, and decreasing on $[-3.65, 3.65]$.
4. (A) The sale of 500 locks will produce an estimated maximum profit of \$2,750.

X	Y1	
100	-410	
200	860	
300	1810	
400	2440	
500	2750	
600	2740	
700	2410	

$Y_1 = 17.5X - .016X^2$...

X	Y1	
700	2410	
800	1760	
900	790	
1000	-500	
1100	-2110	
1200	-4040	
1300	-6290	

$Y_1 = 17.5X - .016X^2$...

   (B) The maximum profit of \$2,785 occurs when 547 locks are sold.

5. (A) $C(x) = \begin{cases} 0.3x & \text{if } 0 \leq x \leq 75 \\ 7.5 + 0.2x & \text{if } x > 75 \end{cases}$

   (B)

x	0	50	75	100	150
$C(x)$	\$0	\$15	\$22.50	\$27.50	\$37.50

   (C)

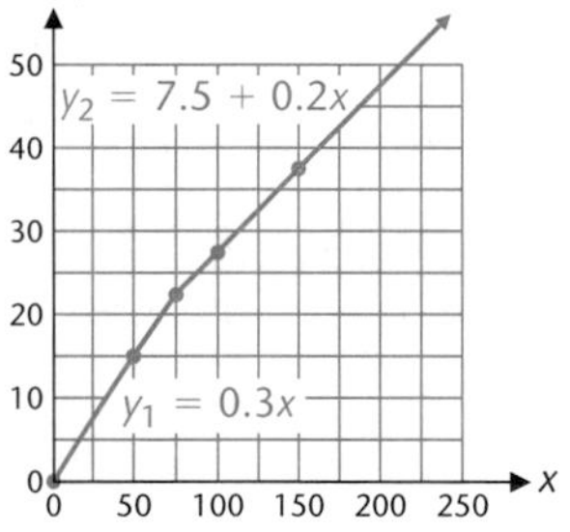

6. (A) 6 (B) 2 (C) 3 (D) -4 (E) -3; f rounds decimal fractions to the nearest integer.

EXERCISE 1-4

A

Problems 1–6 refer to functions f, g, h, k, p, and q given by the following graphs. (Assume the graphs continue as indicated beyond the parts shown.)

1. For the function f, find
   (A) Domain
   (B) Range
   (C) x intercepts
   (D) y intercept
   (E) Intervals over which f is increasing
   (F) Intervals over which f is decreasing
   (G) Intervals over which f is constant
   (H) Any points of discontinuity
2. Repeat Problem 1 for the function g.
3. Repeat Problem 1 for the function h.
4. Repeat Problem 1 for the function k.

5. Repeat Problem 1 for the function p.

6. Repeat Problem 1 for the function q.

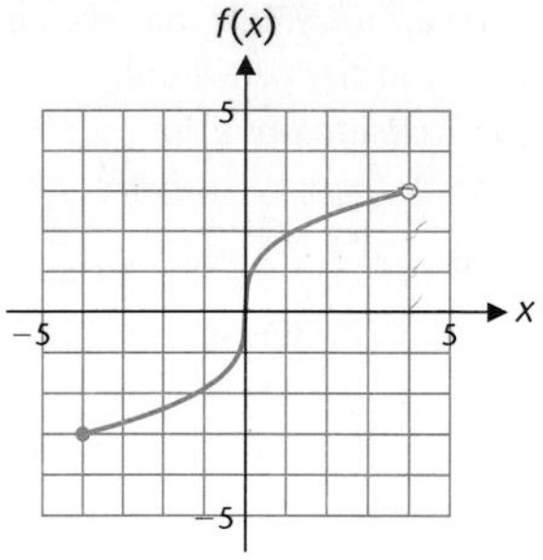

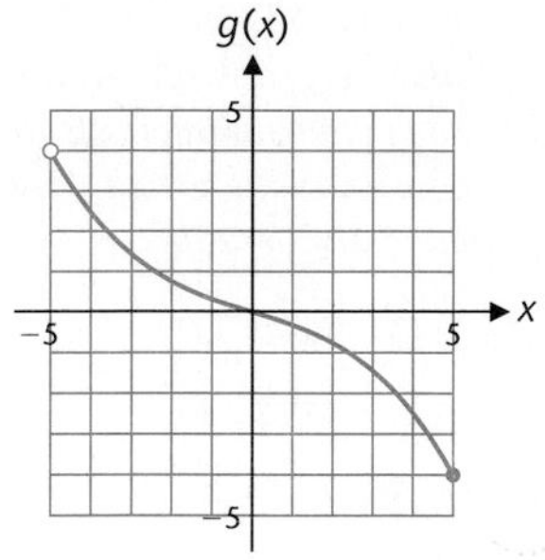

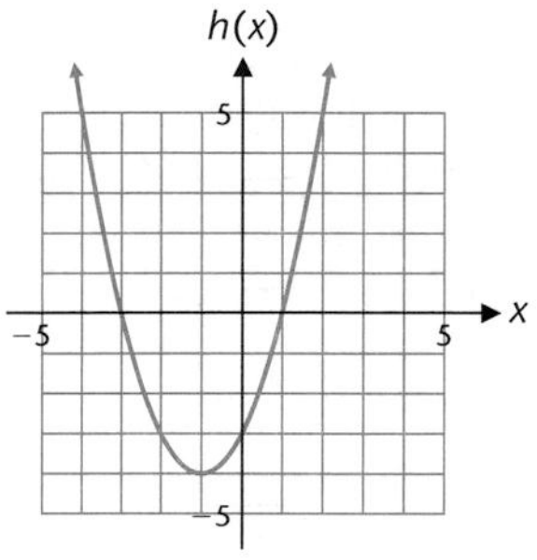

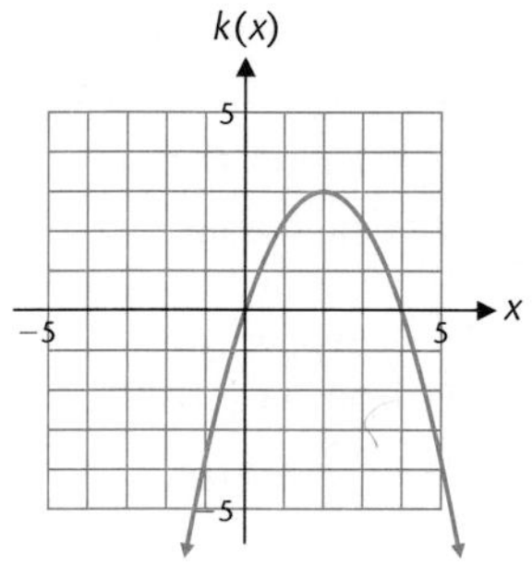

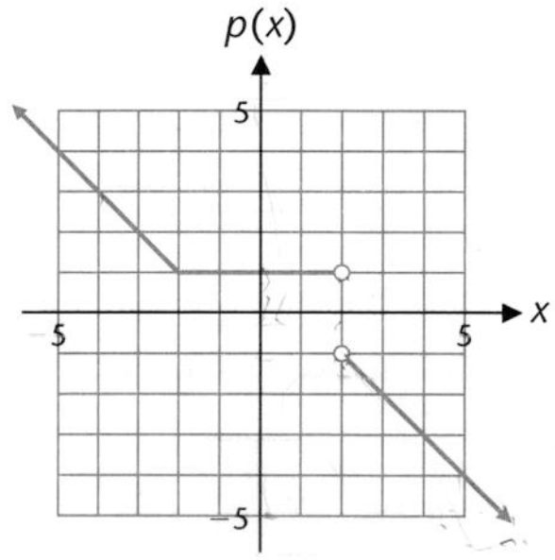

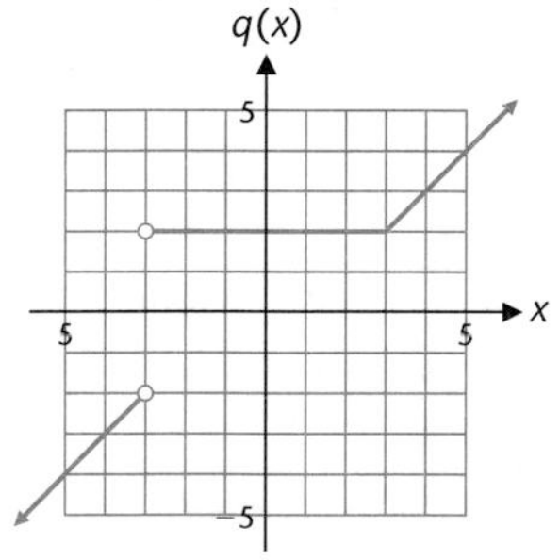

In Problems 7–16, examine the graph of the function on the interval [−10, 10] to determine the intervals over which the function is increasing, the intervals over which the function is decreasing, and the intervals over which the function is constant. Approximate the endpoints of the intervals to the nearest integer.

7. $f(x) = |x + 2| - 5$ **8.** $g(x) = 6 - |x - 3|$

9. $h(x) = x + |x + 3|$ **10.** $k(x) = |x - 2| - x$

11. $m(x) = |x - 3| - |x + 4|$

12. $n(x) = |x + 5| - |x - 2|$

13. $p(x) = |x - 1| + |x + 3|$

14. $q(x) = |x + 2| + |x - 4|$

15. $r(x) = |x + 4| - |x| + |x - 4|$

16. $s(x) = |x| - |x + 5| - |x - 3|$

Problems 17–22 describe the graph of a continuous function f over the interval [−5, 5]. Sketch by hand the graph of a function that is consistent with the given information.

17. The function f is increasing on $[-5, -2]$, constant on $[-2, 2]$, and decreasing on $[2, 5]$.

18. The function f is decreasing on $[-5, -2]$, constant on $[-2, 2]$, and increasing on $[2, 5]$.

19. The function f is decreasing on $[-5, -2]$, constant on $[-2, 2]$, and decreasing on $[2, 5]$.

20. The function f is increasing on $[-5, -2]$, constant on $[-2, 2]$, and increasing on $[2, 5]$.

21. The function f is decreasing on $[-5, -2]$, increasing on $[-2, 2]$, and decreasing on $[2, 5]$.

22. The function f is increasing on $[-5, -2]$, decreasing on $[-2, 2]$, and increasing on $[2, 5]$.

B

Each function f in Problems 23–28 has exactly one local extreme value f(c) and two x intercepts. Explore the graph of f with a graphing utility to locate the intercepts and to determine if the local extremum is a local maximum or a local minimum. Approximate c, f(c), and the x intercepts to one decimal place.

23. $f(x) = 0.7x^2 - 5x + 3$ **24.** $f(x) = 5 - 0.3x^2 - x$

25. $f(x) = 7\sqrt{x} - 2x$ **26.** $f(x) = 3x - 8\sqrt{x}$

27. $f(x) = x^2 - 4\sqrt[3]{x} - 4$ **28.** $f(x) = 2\sqrt[3]{x} - |x| + 2$

In Problems 29–36, sketch the graph by hand, using a graphing utility as an aid, and find the domain, range, and any points of discontinuity.

29. $f(x) = \begin{cases} x + 1 & \text{if } -1 \le x < 0 \\ -x + 1 & \text{if } 0 \le x \le 1 \end{cases}$

30. $f(x) = \begin{cases} x & \text{if } -2 \le x < 1 \\ -x + 2 & \text{if } 1 \le x \le 2 \end{cases}$

31. $f(x) = \begin{cases} -2 & \text{if } -3 \le x < -1 \\ 4 & \text{if } -1 < x \le 2 \end{cases}$

32. $f(x) = \begin{cases} 1 & \text{if } -2 \le x < 2 \\ -3 & \text{if } \ 2 < x \le 5 \end{cases}$

33. $f(x) = \begin{cases} x + 2 & \text{if } x < -1 \\ x - 2 & \text{if } x \ge -1 \end{cases}$

34. $f(x) = \begin{cases} -1 - x & \text{if } x \le 2 \\ 5 - x & \text{if } x > 2 \end{cases}$

35. $g(x) = \begin{cases} x^2 + 1 & \text{if } x < 0 \\ -x^2 - 1 & \text{if } x > 0 \end{cases}$

36. $h(x) = \begin{cases} -x^2 - 2 & \text{if } x < 0 \\ x^2 + 2 & \text{if } x > 0 \end{cases}$

In Problems 37–42, write a verbal description of the graph of the given function over the interval [−10, 10] using increasing and decreasing terminology, and indicating any local maximum and minimum values. Approximate to two decimal places the coordinates of any points used in your description.

37. $f(x) = x^2 + 4.3x - 32$

38. $g(x) = -x^2 + 6.9x + 25$

39. $h(x) = x^3 - x^2 - 74x + 60$

40. $k(x) = -x^3 + x^2 + 82x - 25$

41. $p(x) = |x^2 - x - 18|$

42. $q(x) = |x^2 - 2x - 30|$

Problems 43–48 describe the graph of a function f that is continuous on the interval [−5, 5], except as noted. Sketch by hand the graph of a function that is consistent with the given information.

43. The function f is increasing on $[-5, 0)$, discontinuous at $x = 0$, increasing on $(0, 5]$, $f(-2) = 0$, and $f(2) = 0$.

44. The function f is decreasing on $[-5, 0)$, discontinuous at $x = 0$, decreasing on $(0, 5]$, $f(-3) = 0$, and $f(3) = 0$.

45. The function f is discontinuous at $x = 0$, $f(-3) = -2$ is a local maximum, and $f(2) = 3$ is a local minimum.

46. The function f is discontinuous at $x = 0$, $f(-3) = 2$ is a local minimum, and $f(2) = -3$ is a local maximum.

47. The function f is discontinuous at $x = -2$ and $x = 2$, $f(-3) = -2$ and $f(3) = -2$ are local maxima, and $f(0) = 0$ is a local minimum.

48. The function f is discontinuous at $x = -2$ and $x = 2$, $f(-3) = 2$ and $f(3) = 2$ are local minima, and $f(0) = 0$ is a local maximum.

C

In Problems 49–54, graph $y = f(x)$ in a standard viewing window. Assuming that the graph continues as indicated beyond the part shown in this viewing window, find the domain, range, and any points of discontinuity. (Use the dot mode on your graphing utility.)

49. $f(x) = \dfrac{|5x - 10|}{x - 2}$

50. $f(x) = \dfrac{4x + 12}{|x + 3|}$

51. $f(x) = x + \dfrac{|4x - 4|}{x - 1}$

52. $f(x) = x + \dfrac{|2x + 2|}{x + 1}$

53. $f(x) = |x| - \dfrac{|9 - 3x|}{x - 3}$

54. $f(x) = |x| + \dfrac{|2x + 4|}{x + 2}$

In Problems 55–60, write a piecewise definition of f and sketch by hand the graph of f, using a graphing utility as an aid. Include sufficient intervals to clearly illustrate both the definition and the graph. Find the domain, range, and any points of discontinuity.

55. $f(x) = [\![x/2]\!]$

56. $f(x) = [\![x/3]\!]$

57. $f(x) = [\![3x]\!]$

58. $f(x) = [\![2x]\!]$

59. $f(x) = x - [\![x]\!]$

60. $f(x) = [\![x]\!] - x$

61. The function f is continuous and increasing on the interval $[1, 9]$ with $f(1) = -5$ and $f(9) = 4$.

(A) Sketch a graph of f that is consistent with the given information.

(B) How many times does your graph cross the x axis? Could the graph cross more times? Fewer times? Support your conclusions with additional sketches and/or verbal arguments.

62. Repeat Problem 61 if the function does not have to be continuous.

63. The function f is continuous on the interval $[-5, 5]$ with $f(-5) = -4$, $f(1) = 3$, and $f(5) = -2$.

(A) Sketch a graph of f that is consistent with the given information.

(B) How many times does your graph cross the x axis? Could the graph cross more times? Fewer times? Support your conclusions with additional sketches and/or verbal arguments.

64. Repeat Problem 63 if f is continuous on $[-8, 8]$ with $f(-8) = -6$, $f(-4) = 3$, $f(3) = -2$, and $f(8) = 5$.

65. The function f is continuous on $[0, 10]$, $f(5) = -5$ is a local minimum, and f has no other local extrema on this interval.

(A) Sketch a graph of f that is consistent with the given information.

(B) How many times does your graph cross the x axis? Could the graph cross more times? Fewer times? Support your conclusions with additional sketches and/or verbal arguments.

66. Repeat Problem 65 if $f(5) = 1$ and all other information is unchanged.

APPLICATIONS

67. Delivery Charges. A nationwide package delivery service charges $15 for overnight delivery of packages weighing

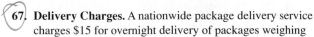

1 pound or less. Each additional pound (or fraction thereof) costs an additional \$3. Let $C(x)$ be the charge for overnight delivery of a package weighing x pounds.

(A) Write a piecewise definition of C for $0 < x \le 6$ and sketch the graph of C by hand.

(B) Can the function f defined by $f(x) = 15 + 3[\![x]\!]$ be used to compute the delivery charges for all x, $0 < x \le 6$? Justify your answer.

68. Telephone Charges. Calls to 900 numbers are charged to the caller. A 900 number hot line for tips and hints for video games charges \$4 for the first minute of the call and \$2 for each additional minute (or fraction thereof). Let $C(x)$ be the charge for a call lasting x minutes.

(A) Write a piecewise definition of C for $0 < x \le 6$ and sketch the graph of C by hand.

(B) Can the function f defined by $f(x) = 4 + 2[\![x]\!]$ be used to compute the charges for all x, $0 < x \le 6$? Justify your answer.

★ **69. Sales Commissions.** An appliance salesperson receives a base salary of \$200 a week and a commission of 4% on all sales over \$3,000 during the week. In addition, if the weekly sales are \$8,000 or more, the salesperson receives a \$100 bonus. If x represents weekly sales (in dollars), express the weekly earnings $E(x)$ as a function of x, and sketch its graph. Identify any points of discontinuity. Find $E(5{,}750)$ and $E(9{,}200)$.

★ **70. Service Charges.** On weekends and holidays, an emergency plumbing repair service charges \$2.00 per minute for the first 30 minutes of a service call and \$1.00 per minute for each additional minute. If x represents the duration of a service call in minutes, express the total service charge $S(x)$ as a function of x, and sketch its graph. Identify any points of discontinuity. Find $S(25)$ and $S(45)$.

71. Computer Science. Let $f(x) = 10[\![0.5 + x/10]\!]$. Evaluate f at 4, −4, 6, −6, 24, 25, 247, −243, −245, and −246. What operation does this function perform?

72. Computer Science. Let $f(x) = 100[\![0.5 + x/100]\!]$. Evaluate f at 40, −40, 60, −60, 740, 750, 7,551, −601, −649, and −651. What operation does this function perform?

★ **73. Computer Science.** Use the greatest integer function to define a function f that rounds real numbers to the nearest hundredth.

★ **74. Computer Science.** Use the greatest integer function to define a function f that rounds real numbers to the nearest thousandth.

75. Revenue. The revenue (in dollars) from the sale of x car seats for infants is given by

$$R(x) = 60x - 0.035x^2 \qquad 0 \le x \le 1{,}700$$

(A) Examine the values of the revenue function $R(x)$ for $x = 100, 200, \ldots, 1{,}700$ to estimate the number of car seats that must be sold to maximize the revenue. What is the estimated maximum revenue?

(B) Graph $y = R(x)$ and find the number of car seats that must be sold to maximize the revenue. What is the maximum revenue now?

76. Profit. The profit (in dollars) from the sale of x car seats for infants is given by

$$P(x) = 38x - 0.035x^2 - 4{,}000 \qquad 0 \le x \le 1{,}700$$

(A) Examine the values of the profit function $P(x)$ for $x = 100, 200, \ldots, 1{,}700$ to estimate the number of car seats that must be sold to maximize the profit. What is the estimated maximum profit?

(B) Graph $y = P(x)$ and find the number of car seats that must be sold to maximize the profit. What is the maximum profit now?

77. Manufacturing. A box is to be made out of a piece of cardboard that measures 18 by 24 inches. Squares, x inches on a side, will be cut from each corner and then the ends and sides will be folded up (see the figure). The volume of the box is given by

$$V(x) = x(24 - 2x)(18 - 2x) \qquad 0 \le x \le 9$$

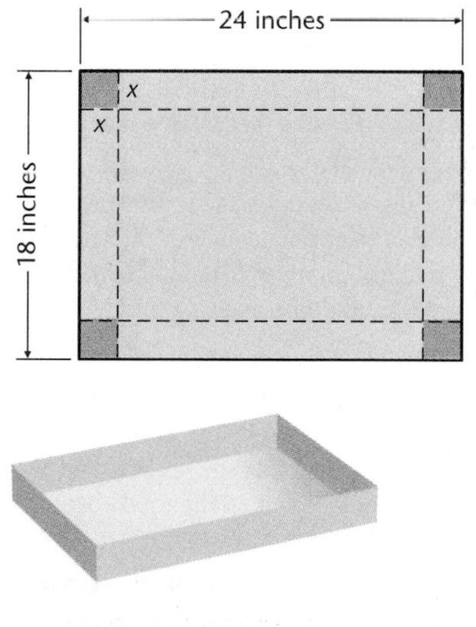

(A) Examine the values of the volume function $V(x)$ for $x = 0, 1, 2, \ldots, 9$ and estimate the size of the cut-out squares that will make the volume maximum. What is the estimated volume?

(B) Graph $y = V(x)$ and approximate to two decimal places the size of the cut-out squares that will make the volume maximum. Approximate the maximum volume to two decimal places also.

78. Manufacturing. A box with a hinged lid is to be made out of a piece of cardboard that measures 20 by 40 inches. Six squares, x inches on a side, will be cut from each corner and the middle, and then the ends and sides will be folded up to form the box and its lid (see the figure). The volume of the box is given by

$$V(x) = 0.5x(40 - 3x)(20 - 2x) \qquad 0 \le x \le 10$$

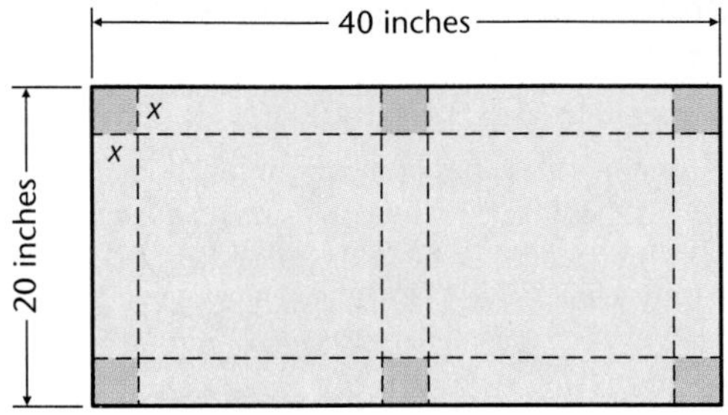

(A) Examine the values of the volume function $V(x)$ for $x = 0, 1, 2, \ldots, 10$ and estimate the size of the cut-out squares that will make the volume maximum. What is the estimated volume?

(B) Graph $y = V(x)$ and approximate to two decimal places the size of the cut-out squares that will make the volume maximum. Approximate the maximum volume to two decimal places also.

79. Construction. A freshwater pipe is to be run from a source on the edge of a lake to a small resort community on an island 8 miles offshore, as indicated in the figure. It costs \$10,000 per mile to lay the pipe on land and \$16,000 per mile to lay the pipe in the lake. The total cost $C(x)$ in thousands of dollars of laying the pipe is given by

$$C(x) = 10(20 - x) + 16\sqrt{x^2 + 64} \qquad 0 \le x \le 20$$

(A) Examine the values of the cost function $C(x)$ (rounded to the nearest thousand dollars) for $x = 0, 5, 10, 15,$ and 20 and estimate to the nearest mile the length of the land portion of the pipe that will make the production costs minimum. What is the estimated cost?

(B) Graph $y = C(x)$ and approximate to one decimal place the length of the land portion of the pipe that will make the production costs minimum. Approximate the minimum cost to the nearest thousand dollars also.

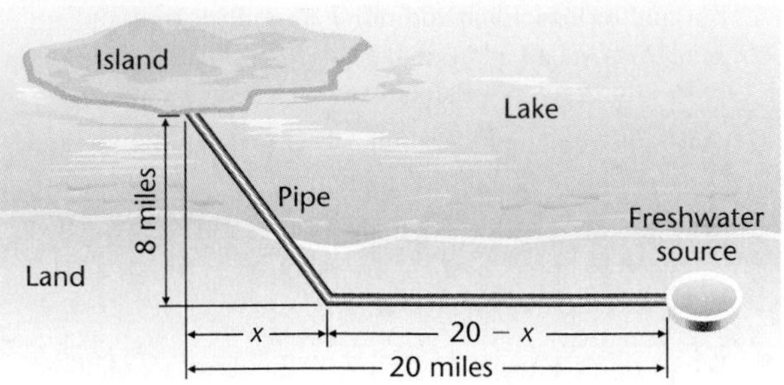

80. Transportation. The construction company laying the freshwater pipe in Problem 79 uses an amphibious vehicle to travel down the beach and then out to the island. The vehicle travels at 30 miles per hour on land and 7.5 miles per hour in water. The total time $T(x)$ in minutes for a trip from the freshwater source to the island is given by

$$T(x) = 2(20 - x) + 8\sqrt{x^2 + 64} \qquad 0 \le x \le 20$$

(A) Examine the values of the time function $T(x)$ (rounded to the nearest minute) for $x = 0, 5, 10, 15,$ and 20 and estimate the length of the land portion of the trip that will make the time minimum. What is the estimated time?

(B) Graph $y = T(x)$ and use zoom and trace procedures to approximate to one decimal place the length of the land portion of the trip that will make the time minimum. Approximate the minimum time to the nearest minute also.

81. Tire Mileage. An automobile tire manufacturer collected the data in the table relating tire pressure x, in pounds per square inch, and mileage in thousands of miles.

x	28	30	32	34	36
Mileage	45	52	55	51	47

A mathematical model for this data is given by

$$f(x) = -0.518x^2 + 33.3x - 481$$

(A) Complete the following table. Round values of $f(x)$ to one decimal place.

x	28	30	32	34	36
Mileage	45	52	55	51	47
$f(x)$					

(B) Sketch by hand the graph of f and the mileage data on the same axes.

(C) Use values of the modeling function rounded to two decimal places to estimate the mileage for a tire pressure of 31 lb/in.^{2}. For 35 lb/in.^{2}.

(D) Write a brief description of the relationship between tire pressure and mileage, using the terms increasing, decreasing, local maximum, and local minimum where appropriate.

82. Automobile Production. The table lists General Motor's total U.S. vehicle production in millions of units from 1989 to 1993.

Year	89	90	91	92	93
Production	4.7	4.1	3.5	3.7	5.0

A mathematical model for GM's production data is given by

$$f(x) = 0.33x^2 - 1.3x + 4.8$$

where $x = 0$ corresponds to 1989.

(A) Complete the following table. Round values of x to one decimal place.

x	0	1	2	3	4
Production	4.7	4.1	3.5	3.7	5.0
$f(x)$					

(B) Sketch by hand the graph of f and the production data on the same axes.

(C) Use values of the modeling function f rounded to two decimal places to estimate the production in 1994. In 1995.

(D) Write a brief verbal description of GM's production from 1989 to 1993, using increasing, decreasing, local maximum, and local minimum terminology where appropriate.

Section 1-5 Functions: Graphs and Transformations

- A Beginning Library of Elementary Functions
- Vertical and Horizontal Shifts
- Reflections, Expansions, and Contractions
- Even and Odd Functions

The functions

$$g(x) = x^2 + 2 \qquad h(x) = (x + 2)^2 \qquad k(x) = 2x^2$$

can be expressed in terms of the function $f(x) = x^2$ as follows:

$$g(x) = f(x) + 2 \qquad h(x) = f(x + 2) \qquad k(x) = 2f(x)$$

In this section we will see that the graphs of functions g, h, and k are closely related to the graph of function f. Insight gained by understanding these relationships will help us analyze and interpret the graphs of many different functions.

A Beginning Library of Elementary Functions

As we progress through this book, we will encounter a number of basic functions that we will want to add to our library of elementary functions. Figure 1 shows six basic functions that you will encounter frequently. You should know the definition, domain, and range of each of these functions, and be able to recognize their graphs. You should graph each basic function in Figure 1 on your graphing utility.

Most graphing utilities allow you to define a number of functions, usually denoted by $y_1, y_2, y_3, \ldots$ You can graph all of these functions simultaneously, or

FIGURE 1
Some basic functions and their graphs.
[*Note:* Letters used to designate these functions may vary from context to context; R is the set of all real numbers.]

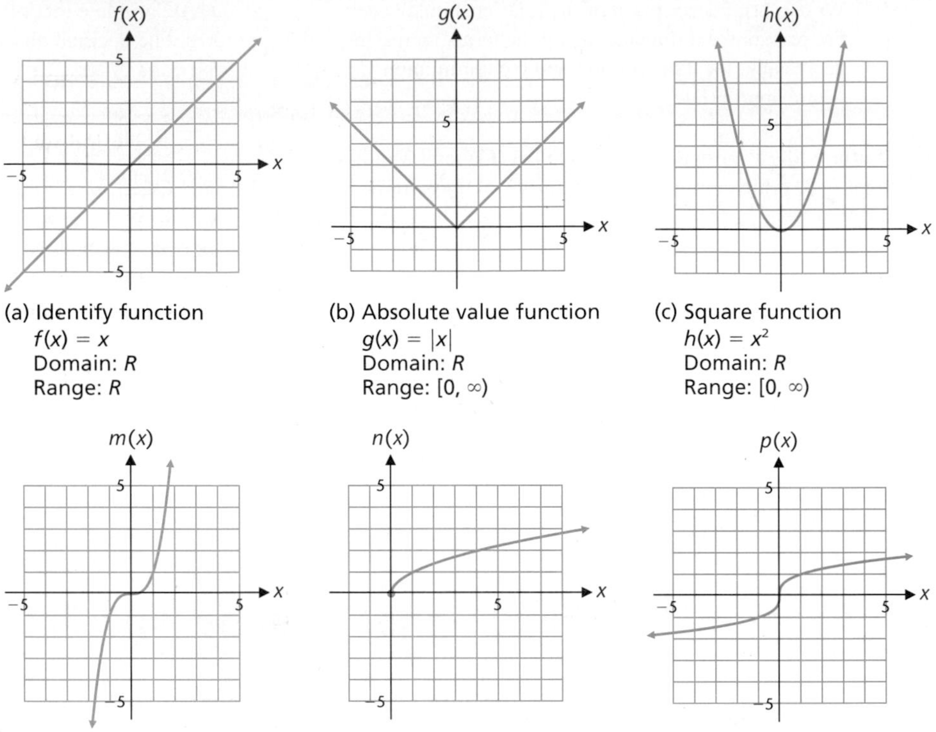

(a) Identify function
$f(x) = x$
Domain: R
Range: R

(b) Absolute value function
$g(x) = |x|$
Domain: R
Range: $[0, \infty)$

(c) Square function
$h(x) = x^2$
Domain: R
Range: $[0, \infty)$

(d) Cube function
$m(x) = x^3$
Domain: R
Range: R

(e) Square root function
$n(x) = \sqrt{x}$
Domain: $[0, \infty)$
Range: $[0, \infty)$

(f) Cube root function
$p(x) = \sqrt[3]{x}$
Domain: R
Range: R

you can select certain functions for graphing and suppress the graphs of the others. Consult your manual to determine how many functions can be stored in your graphing utility at one time and how to select particular functions for graphing. Many of our investigations in this section will involve graphing two or more functions at the same time.

Vertical and Horizontal Shifts

If a new function is formed by performing an operation on a given function, then the graph of the new function is called a **transformation** of the graph of the original function. For example, if we add a constant k to $f(x)$, then the graph of $y = f(x)$ is transformed into the graph of $y = f(x) + k$.

Explore/Discuss 1

Let $f(x) = |x|$.

(A) Graph $y = f(x) + k$ for $k = -2, 0$, and 1 simultaneously in the same viewing window. Describe the relationship between the graph of $y = f(x)$ and the graph of $y = f(x) + k$ for k any real number.

(B) Graph $y = f(x + h)$ for $h = -2, 0$, and 1 simultaneously in the same viewing window. Describe the relationship between the graph of $y = f(x)$ and the graph of $y = f(x + h)$ for h any real number.

EXAMPLE 1

Vertical and Horizontal Shifts

(A) How are the graphs of $y = x^2 + 2$ and $y = x^2 - 3$ related to the graph of $y = x^2$? Confirm your answer by graphing all three functions simultaneously in the same viewing window.

(B) How are the graphs of $y = (x + 2)^2$ and $y = (x - 3)^2$ related to the graph of $y = x^2$? Confirm your answer by graphing all three functions simultaneously in the same viewing window.

Solutions

(A) The graph of $y = x^2 + 2$ is the same as the graph of $y = x^2$ shifted upward 2 units, and the graph of $y = x^2 - 3$ is the same as the graph of $y = x^2$ shifted downward 3 units. Figure 2 confirms these conclusions. (It appears that the graph of $y = f(x) + k$ is the graph of $y = f(x)$ shifted up if k is positive and down if k is negative.)

FIGURE 2
Vertical shifts.

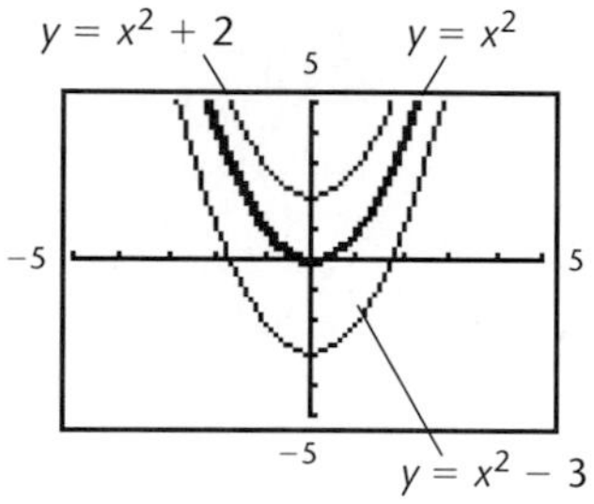

(B) The graph of $y = (x + 2)^2$ is the same as the graph of $y = x^2$ shifted to the left 2 units, and the graph of $y = (x - 3)^2$ is the same as the graph of $y = x^2$ shifted to the right 3 units. Figure 3 confirms these conclusions. [It appears that the graph of $y = f(x + h)$ is the graph of $y = f(x)$ shifted right if h is negative and left if h is positive—the opposite of what you might expect.]

FIGURE 3
Horizontal shifts.

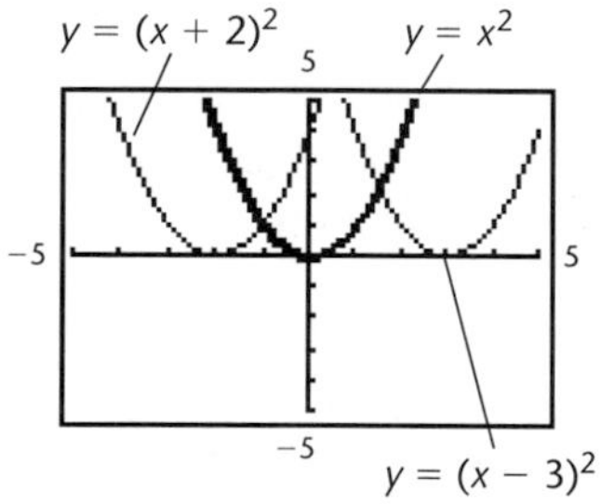

MATCHED PROBLEM 1

(A) How are the graphs of $y = \sqrt{x} + 3$ and $y = \sqrt{x} - 1$ related to the graph of $y = \sqrt{x}$? Confirm your answer by graphing all three functions simultaneously in the same viewing window.

(B) How are the graphs of $y = \sqrt{x + 3}$ and $y = \sqrt{x - 1}$ related to the graph of $y = \sqrt{x}$? Confirm your answer by graphing all three functions simultaneously in the same viewing window.

Comparing the graph of $y = f(x) + k$ with the graph of $y = f(x)$, we see that the graph of $y = f(x) + k$ can be obtained from the graph of $y = f(x)$ by **vertically translating** (shifting) the graph of the latter upward k units if k is positive and downward $|k|$ units if k is negative. Comparing the graph of $y = f(x + h)$ with the graph of $y = f(x)$, we see that the graph of $y = f(x + h)$ can be obtained from the graph of $y = f(x)$ by **horizontally translating** (shifting) the graph of the latter h units to the left if h is positive and $|h|$ units to the right if h is negative.

EXAMPLE 2

Vertical and Horizontal Translations (Shifts)

The graphs in Figure 4 are either horizontal or vertical shifts of the graph of $f(x) = |x|$. Write appropriate equations for functions G, H, M, and N in terms of f.

FIGURE 4
Vertical and horizontal shifts.

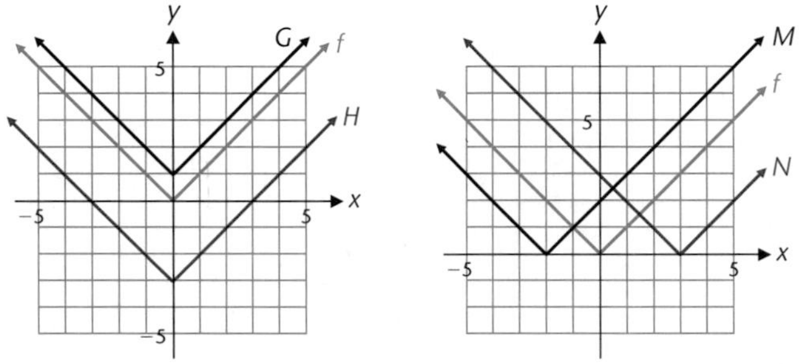

Solution

Functions H and G are vertical shifts given by

$$H(x) = |x| - 3 \qquad G(x) = |x| + 1$$

Functions M and N are horizontal shifts given by

$$M(x) = |x + 2| \qquad N(x) = |x - 3|$$

MATCHED PROBLEM 2

The graphs in Figure 5 are either horizontal or vertical shifts of the graph of $f(x) = x^3$. Write appropriate equations for functions H, G, M, and N in terms of f.

FIGURE 5
Vertical and horizontal shifts.

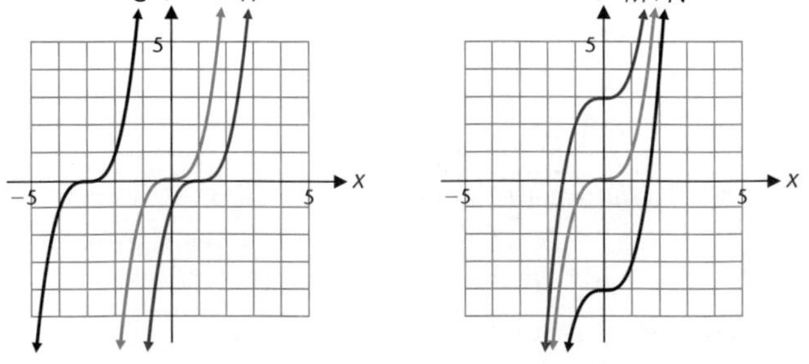

Reflections, Expansions, and Contractions

We now investigate how the graph of $y = Af(x)$ is related to the graph of $y = f(x)$ for different real numbers A.

Explore/Discuss 2

(A) Graph $y = A\sqrt{x}$ for $A = 1$, 2, and $\frac{1}{2}$ simultaneously in the same viewing window.

(B) Graph $y = A\sqrt{x}$ for $A = -1$, -2, and $-\frac{1}{2}$ simultaneously in the same viewing window.

(C) Describe the relationship between the graph of $h(x) = \sqrt{x}$ and the graph of $G(x) = A\sqrt{x}$ for A any nonzero real number.

Comparing the graph of $y = Af(x)$ with the graph of $y = f(x)$, we see that the graph of $y = Af(x)$ can be obtained from the graph of $y = f(x)$ by multiplying each ordinate value of the latter by A. The result is a **vertical expansion** of the graph of $y = f(x)$ if $A > 1$, a **vertical contraction** of the graph of $y = f(x)$ if $0 < A < 1$, and a **reflection in the x axis** if $A = -1$.

EXAMPLE 3 **Reflections, Expansions, and Contractions**

(A) How are the graphs of $y = 2\sqrt[3]{x}$ and $y = 0.5\sqrt[3]{x}$ related to the graph of $y = \sqrt[3]{x}$? Confirm your answer by graphing all three functions simultaneously in the same viewing window.

(B) How is the graph of $y = -2\sqrt[3]{x}$ related to the graph of $y = \sqrt[3]{x}$? Confirm your answer by graphing both functions simultaneously in the same viewing window.

Solutions

(A) The graph of $y = 2\sqrt[3]{x}$ is a vertical expansion of the graph of $y = \sqrt[3]{x}$ by a factor of 2, and the graph of $y = 0.5\sqrt[3]{x}$ is a vertical contraction of the graph of $y = \sqrt[3]{x}$ by a factor of 0.5. Figure 6 confirms this conclusion.

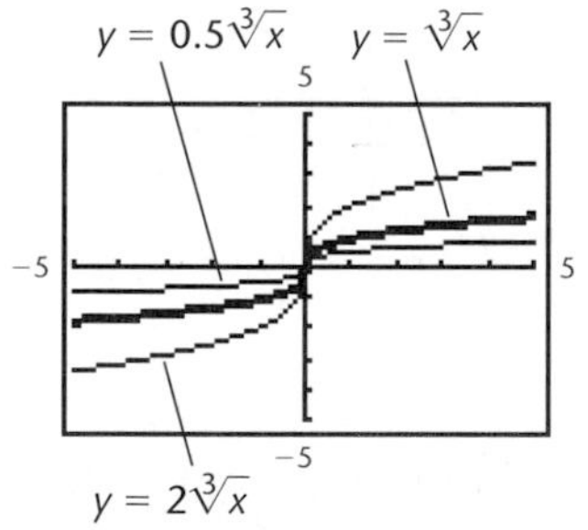

FIGURE 6 Vertical expansion and contraction.

(B) The graph of $y = -2\sqrt[3]{x}$ is a reflection in the x axis and a vertical expansion of the graph of $y = \sqrt[3]{x}$. Figure 7 confirms this conclusion.

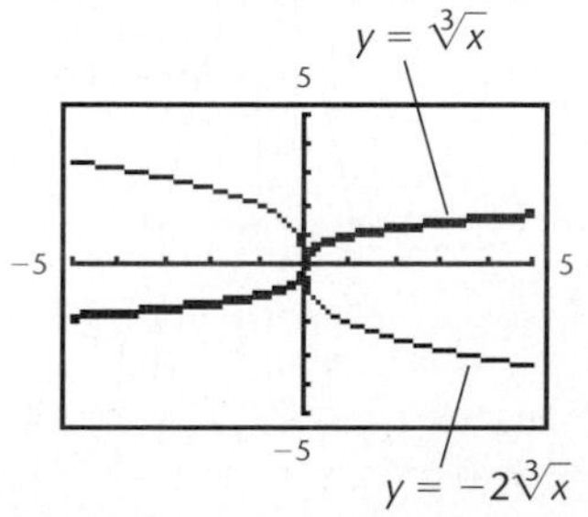

FIGURE 7 Reflection and vertical expansion.

MATCHED PROBLEM 3

(A) How are the graphs of $y = 2x$ and $y = 0.5x$ related to the graph of $y = x$? Confirm your answer by graphing all three functions simultaneously in the same viewing window.

(B) How is the graph of $y = -0.5x$ related to the graph of $y = x$? Confirm your answer by graphing both functions in the same viewing window.

The various transformations we have discussed are summarized in the following box for easy reference:

SUMMARY 1

GRAPH TRANSFORMATIONS apply to any Kind of a function

Vertical Translation [see Fig. 8(a)]:

$$y = f(x) + k \quad \begin{cases} k > 0 & \text{Shift graph of } y = f(x) \text{ up } k \text{ units.} \\ k < 0 & \text{Shift graph of } y = f(x) \text{ down } |k| \text{ units.} \end{cases}$$

Horizontal Translation [see Fig. 8(b)]:

$$y = f(x + h) \quad \begin{cases} h > 0 & \text{Shift graph of } y = f(x) \text{ left } h \text{ units.} \\ h < 0 & \text{Shift graph of } y = f(x) \text{ right } |h| \text{ units.} \end{cases}$$

Reflection [see Fig. 8(c)]:

$y = -f(x)$ Reflect the graph of $y = f(x)$ in the x axis.

Vertical Expansion and Contraction [see Fig. 8(d)]:

$$y = Af(x) \quad \begin{cases} A > 1 & \text{Vertically expand graph of } y = f(x) \text{ by multiplying each ordinate value by } A. \\ 0 < A < 1 & \text{Vertically contract graph of } y = f(x) \text{ by multiplying each ordinate value by } A. \end{cases}$$

FIGURE 8
Graph transformations.

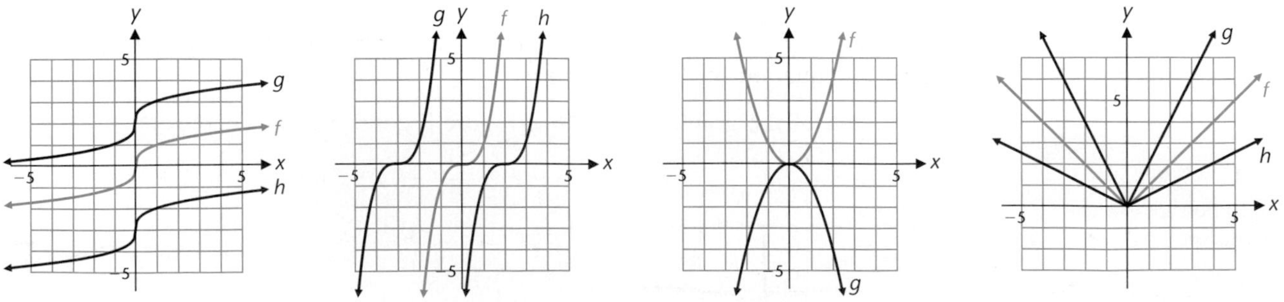

(a) Vertical Translation $g(x) = f(x) + 2$, $h(x) = f(x) - 3$

(b) Horizontal Translation $g(x) = f(x + 3)$, $h(x) = f(x - 2)$

(c) Reflection $g(x) = -f(x)$

(d) Expansion and Contraction $g(x) = 2f(x)$, $h(x) = 0.5f(x)$

Explore/Discuss 3

Use a graphing utility to explore the graph of $y = A(x + h)^2 + k$ for various values of the constants A, h, and k. Discuss how the graph of $y = A(x + h)^2 + k$ is related to the graph of $y = x^2$.

EXAMPLE 4

Combining Graph Transformations

The graph of $y = g(x)$ in Figure 9 is a transformation of the graph of $y = x^2$. Find an equation for the function g.

FIGURE 9

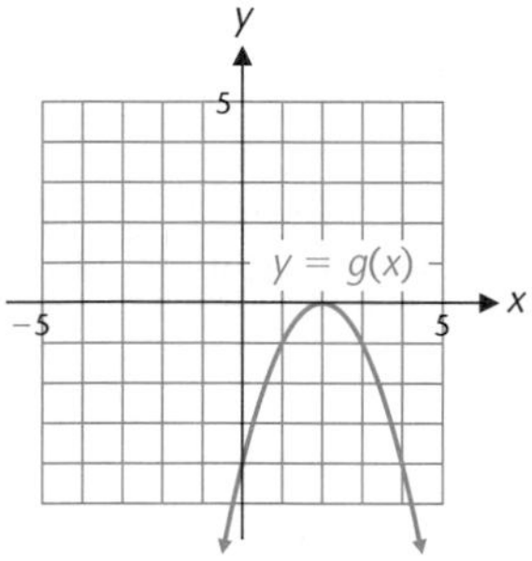

Solution

To transform the graph of $y = x^2$ [Fig. 10(a)] into the graph of $y = g(x)$, we first reflect the graph of $y = x^2$ in the x axis [Fig. 10(b)], then shift it to the right two units [Fig. 10(c)]. Thus, an equation for the function g is

$$g(x) = -(x - 2)^2$$

FIGURE 10

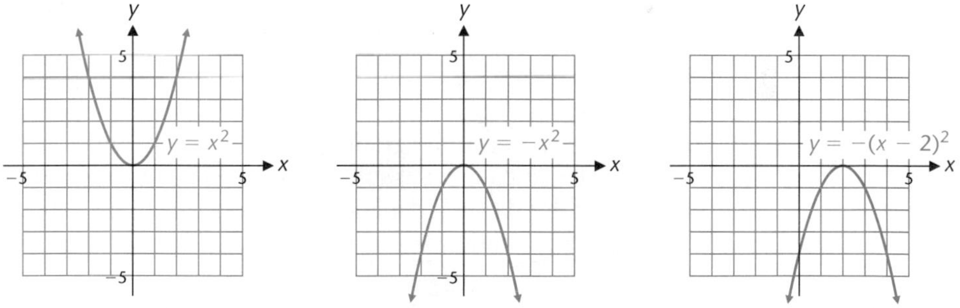

(a) $y = x^2$ (b) $y = -x^2$ (c) $y = -(x - 2)^2$

MATCHED PROBLEM 4

The graph of $y = h(x)$ in Figure 11 is a transformation of the graph of $y = x^3$. Find an equation for the function h.

FIGURE 11

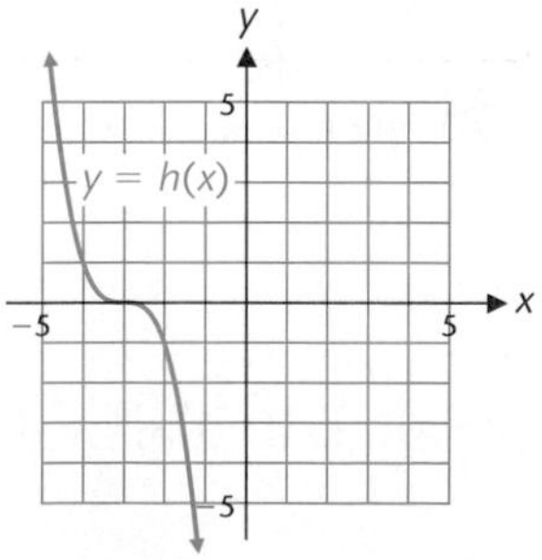

Even and Odd Functions

Certain transformations leave the graphs of some functions unchanged. For example, reflecting the graph of $y = x^2$ in the y axis does not change the graph. Functions with this property are called *even functions*. Similarly, reflecting the graph of $y = x^3$ in the x axis and then in the y axis does not change the graph. Functions with this property are called *odd functions*. More formally, we have the following definitions.

EVEN AND ODD FUNCTIONS

If $f(x) = f(-x)$ for all x in the domain of f, then f is an **even function.**
If $f(-x) = -f(x)$ for all x in the domain of f, then f is an **odd function.**

The graph of an even function is said to be **symmetric with respect to the *y* axis** and the graph of an odd function is said to be **symmetric with respect to the origin** (see Fig. 12).

FIGURE 12
Even and odd functions.

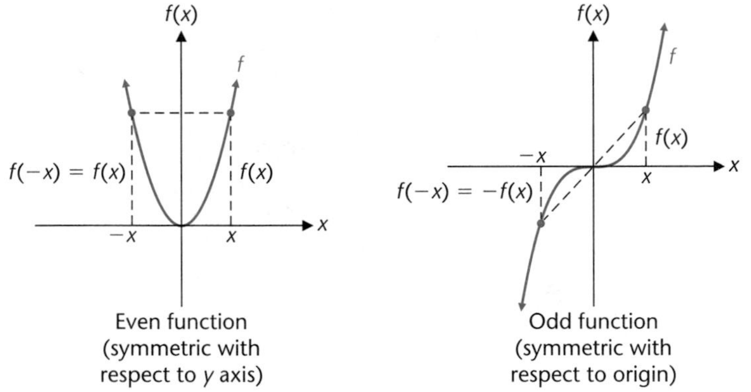

Refer to the graphs of the basic functions in Figure 1. These graphs show that the square and absolute value functions are even functions, and the identity, cube, and cube root functions are odd functions. Notice in Figure 1(e) that the square root function is not symmetric with respect to the y axis or the origin. Thus, the square root function is neither even nor odd.

In certain situations, it is useful to determine if a function is even or odd without graphing the function. The next example illustrates this process.

EXAMPLE 5

Testing for Even and Odd Functions

Without graphing, determine whether the functions f, g, and h are even, odd, or neither:

(A) $f(x) = x^4 + 1$ (B) $g(x) = x^3 + 1$ (C) $h(x) = x^5 + x$

Solutions

(A) $f(x) = x^4 + 1$

$$f(-x) = (-x)^4 + 1$$

$$= x^4 + 1 \qquad (-x)^4 = [(-1)x]^4 = (-1)^4x^4 = x^4 \text{ since } (-1)^4 = 1.$$

$$= f(x)$$

Therefore, f is even.

(B) $g(x) = x^3 + 1$

$$g(-x) = (-x)^3 + 1$$

$$= -x^3 + 1 \qquad (-x)^3 = (-1)^3x^3 = -x^3 \text{ since } (-1)^3 = -1.$$

$$-g(x) = -(x^3 + 1)$$

$$= -x^3 - 1$$

Since $g(-x) \neq g(x)$ and $g(-x) \neq -g(x)$, g is neither even nor odd.

(C) $h(x) = x^5 + x$

$$h(-x) = (-x)^5 + (-x) = -x^5 - x = -(x^5 + x) = -h(x)$$

Therefore, h is odd.

MATCHED PROBLEM 5

Without graphing, determine whether the functions F, G, and H are even, odd, or neither:

(A) $F(x) = x^3 - 2x$ odd   (B) $G(x) = x^2 + 1$ even   (C) $H(x) = 2x + 4$ niether

---

In the solution of Example 5, notice that we used the fact that

$$(-x)^n = \begin{cases} x^n & \text{if } n \text{ is an even integer} \\ -x^n & \text{if } n \text{ is an odd integer} \end{cases}$$

It is this property that motivates the use of the terms *even* and *odd* when describing symmetry properties of the graphs of functions. In addition to being an aid to graphing, certain problems and developments in calculus and more advanced mathematics are simplified if we recognize the presence of either an even or an odd function.

Answers to Matched Problems

1. (A) The graph of $y = \sqrt{x} + 3$ is the same as the graph of $y = \sqrt{x}$ shifted upward 3 units, and the graph of $y = \sqrt{x} - 1$ is the same as the graph of $y = \sqrt{x}$ shifted downward 1 unit. The figure confirms these conclusions.

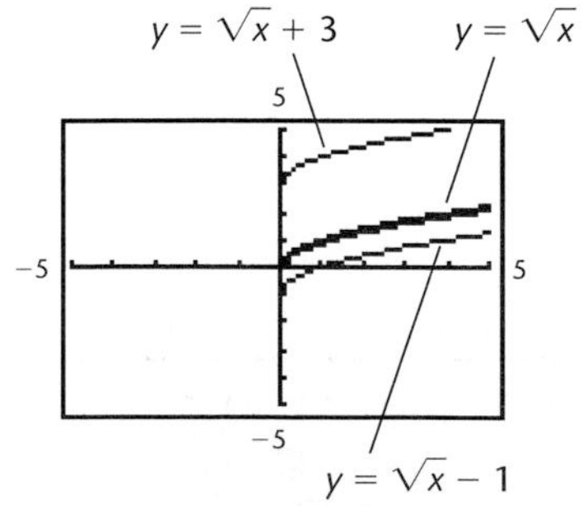

(B) The graph of $y = \sqrt{x + 3}$ is the same as the graph of $y = \sqrt{x}$ shifted to the left 3 units, and the graph of $y = \sqrt{x - 1}$ is the same as the graph of $y = \sqrt{x}$ shifted to the right 1 unit. The figure confirms these conclusions.

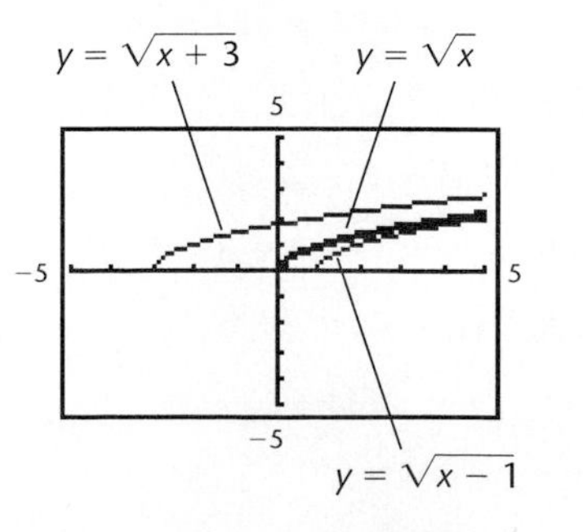

2. $G(x) = (x + 3)^3$, $H(x) = (x - 1)^3$, $M(x) = x^3 + 3$, $N(x) = x^3 - 4$

3. (A) The graph of $y = 2x$ is a vertical expansion of the graph of $y = x$, and the graph of $y = 0.5x$ is a vertical contraction of the graph of $y = x$. The figure confirms these conclusions.

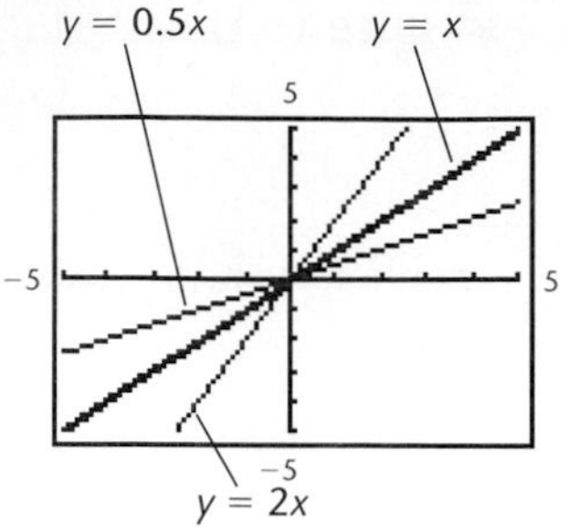

(B) The graph of $y = -0.5x$ is a vertical contraction and a reflection in the x axis of the graph of $y = x$. The figure confirms this conclusion.

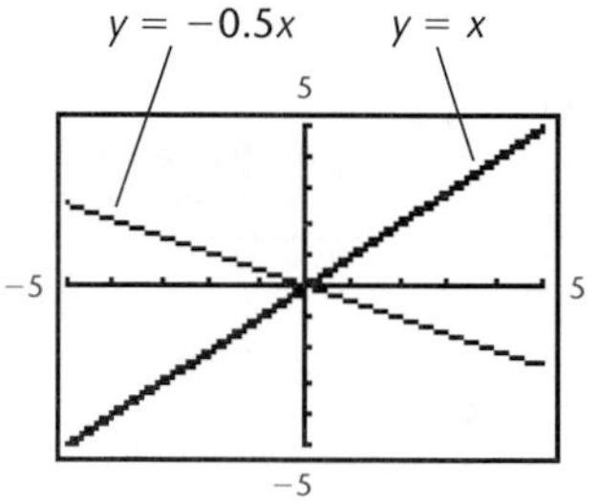

4. The graph of function h is a reflection in the x axis and a horizontal translation of 3 units to the left of the graph of $y = x^3$. An equation for h is $h(x) = -(x + 3)^3$.

5. (A) Odd (B) Even (C) Neither

EXERCISE 1-5

A

Without graphing, indicate whether each function in problems 1–10 is even, odd, or neither.

1. $g(x) = x^3 + x$
2. $f(x) = x^5 - x$ odd
3. $m(x) = x^4 + 3x^2$ even
4. $h(x) = x^4 - x^2$ even
5. $F(x) = x^5 + 1$
6. $f(x) = x^5 - 3$ niether
7. $G(x) = x^4 + 2$ even
8. $P(x) = x^4 - 4$ even
9. $q(x) = x^2 + x - 3$
10. $n(x) = 2x - 3$ niether

Problems 11–22 refer to the functions f and g given by the graphs below (the domain of each function is $[-2, 2]$). Use the graph of f or g, as required, to graph each given function.

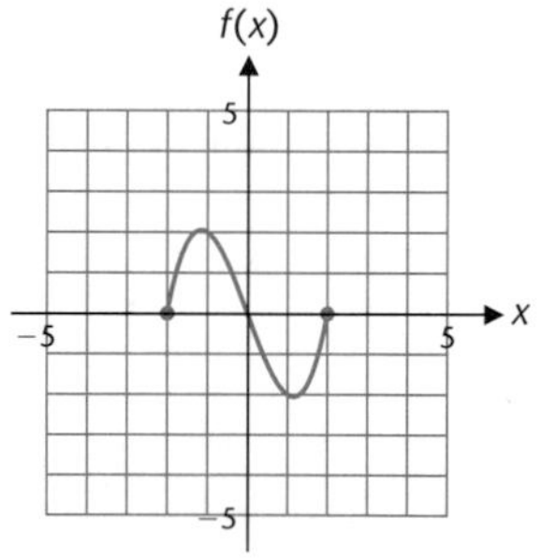

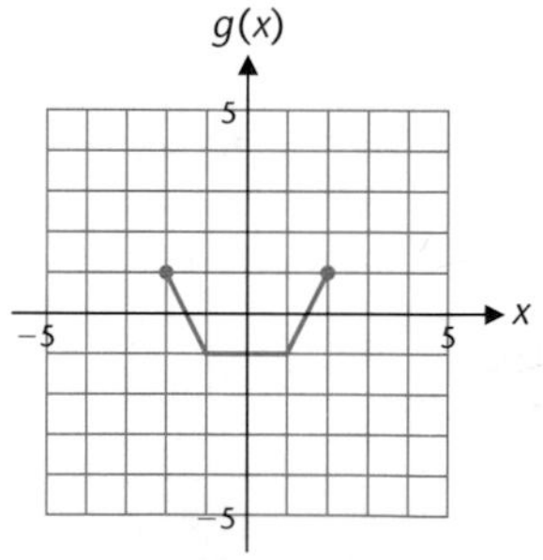

11. $f(x) + 2$
12. $g(x) - 1$
13. $g(x) + 2$
14. $f(x) - 1$
15. $f(x - 2)$
16. $g(x - 1)$
17. $g(x + 2)$
18. $f(x - 1)$
19. $-f(x)$
20. $-g(x)$ invert
21. $2g(x)$
22. $\frac{1}{2}f(x)$

B

Each graph in Problems 23–30 is the result of applying a transformation to the graph of one of the six basic functions in Figure 1. Identify the basic function, describe the transformation verbally, and find an equation for the given graph. Check by graphing the equation on a graphing utility.

23.

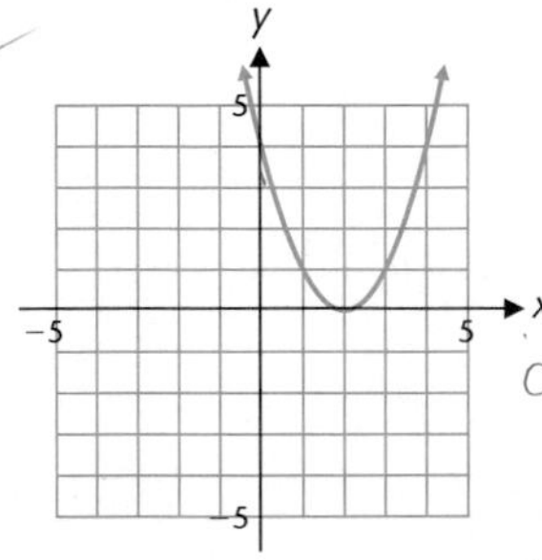

$g \to L$

$g(x+1) = |x+1|$

24.

25.

$m(x) - 2$

26.

$\sqrt[3]{x} + 3$

27.

28.

$p(x)$

$3 \cdot p(x)$

$3 \cdot \sqrt[3]{x}$

29.

30.

$-$ sign in front

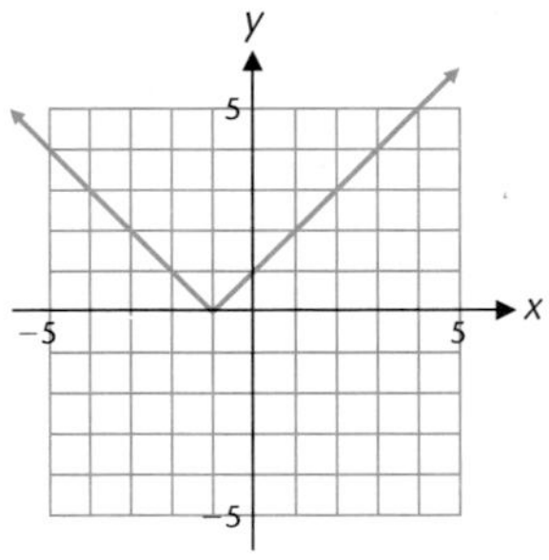

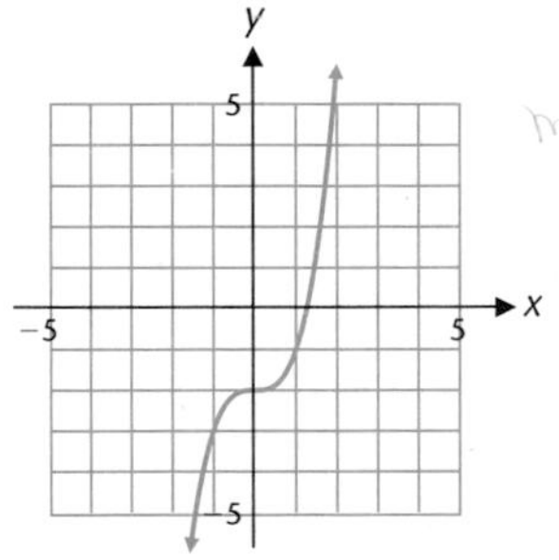

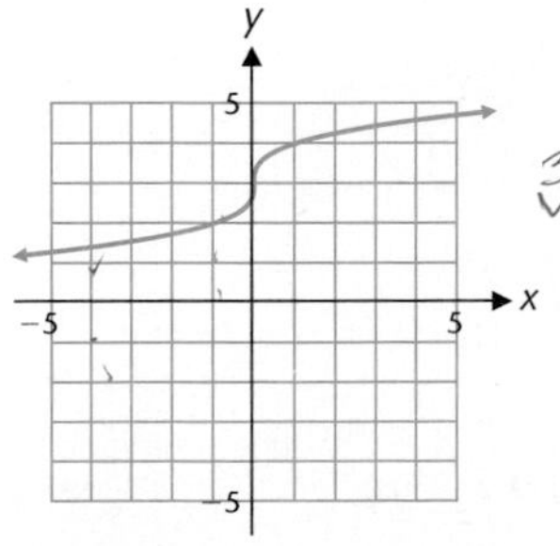

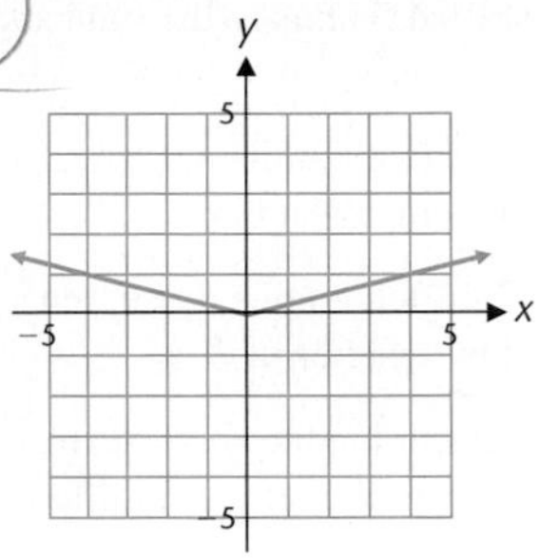

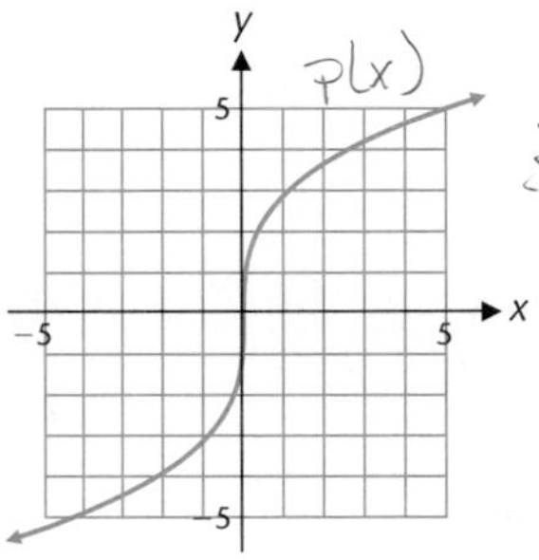

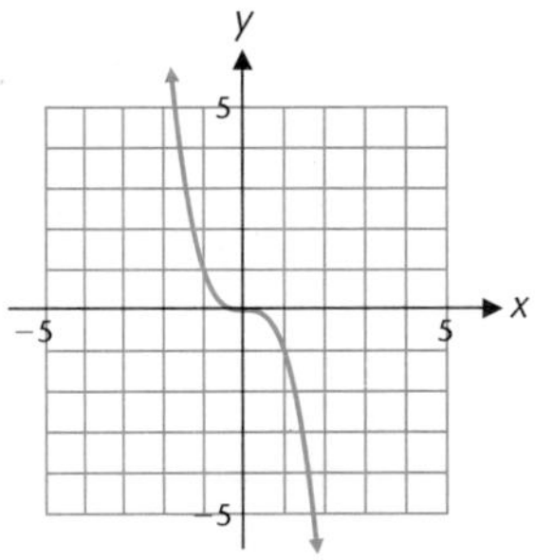

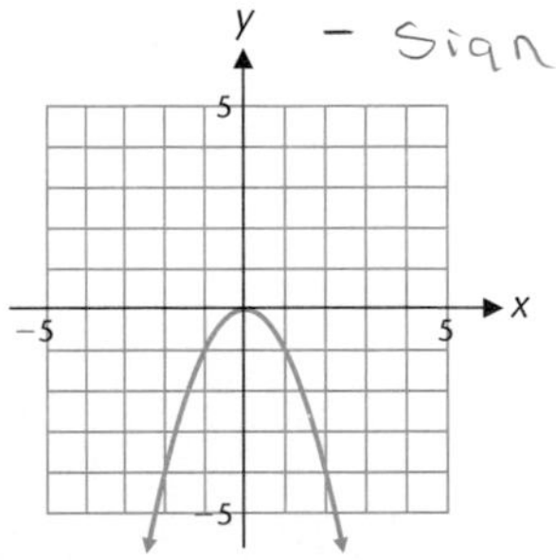

In Problems 31–38, the graph of the function g is formed by applying the indicated sequence of transformations to the given function f. Find an equation for the function g. Check your work by graphing f and g in a standard viewing window.

31. The graph of $f(x) = \sqrt[3]{x}$ is shifted 4 units to the left and 5 units down.

32. The graph of $f(x) = x^3$ is shifted 5 units to the right and 4 units up.

33. The graph of $f(x) = \sqrt{x}$ is shifted 6 units up, reflected in the x axis, and contracted by a factor of 0.5.

34. The graph of $f(x) = \sqrt{x}$ is shifted 2 units down, reflected in the x axis, and expanded by a factor of 4.

35. The graph of $f(x) = x^2$ is reflected in the x axis, expanded by a factor of 2, shifted 4 units to the left, and shifted 2 units down.

36. The graph of $f(x) = |x|$ is reflected in the x axis, contracted by a factor of 0.5, shifted 3 units to the right, and shifted 4 units up.

In Problems 37–44, indicate how the graph of each function is related to the graph of one of the six basic functions in Figure 1.

37. $f(x) = (x + 7)^2 + 9$ **38.** $g(x) = (x - 4)^2 - 6$ ®4↓6

39. $h(x) = -|x - 8|$ FLIP 9R8   **40.** $k(x) = -|x + 5|$ invert L5

41. $p(x) = 3 - \sqrt{x}$ **42.** $q(x) = -2 + \sqrt{x + 3}$

43. $r(x) = -4x^2$ **44.** $s(x) = -0.5|x|$

down 2 over L 3

Each graph in Problems 45–52 is the result of applying a sequence of transformations to the graph of one of the six basic functions in Figure 1. Find an equation for the given graph. Check by graphing the equation on a graphing utility.

45.
46.
4 R2
$g(x) = h(x-2) = |x + 2|$

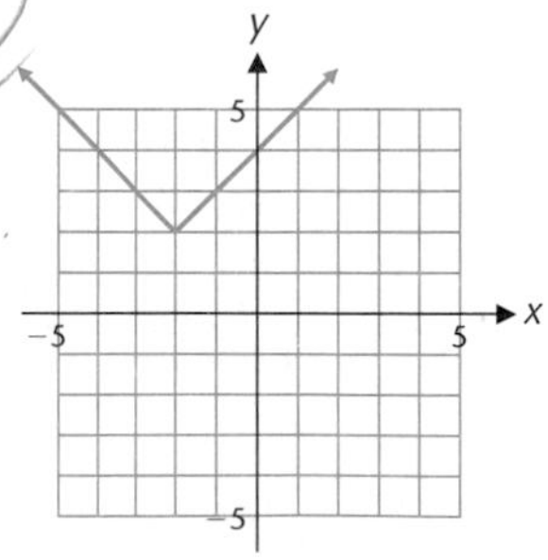

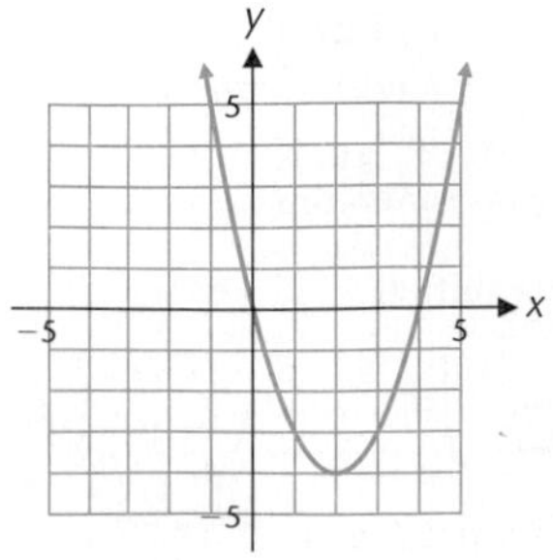

47.
48.
49.
50.
51.

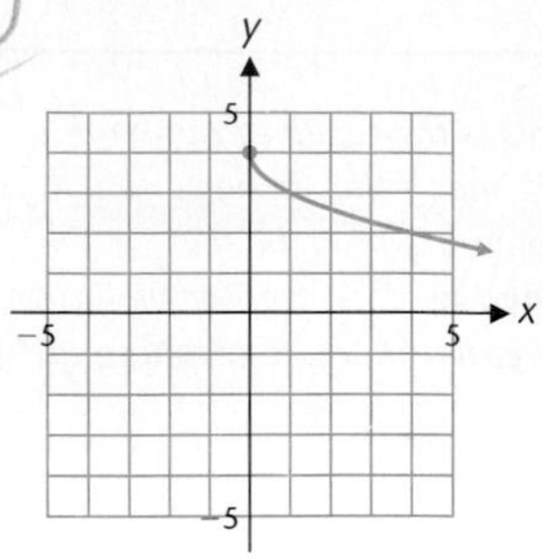

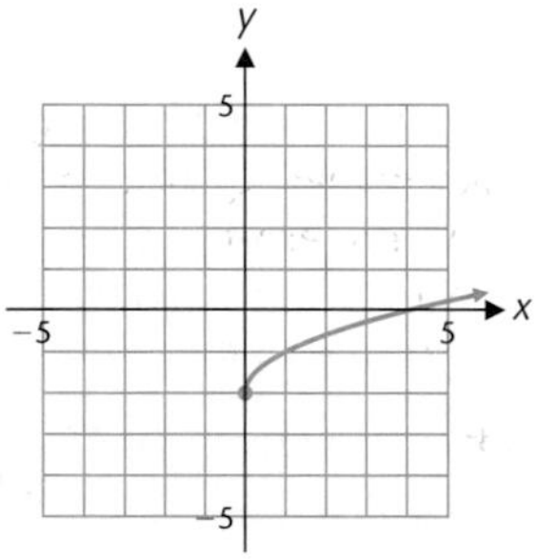

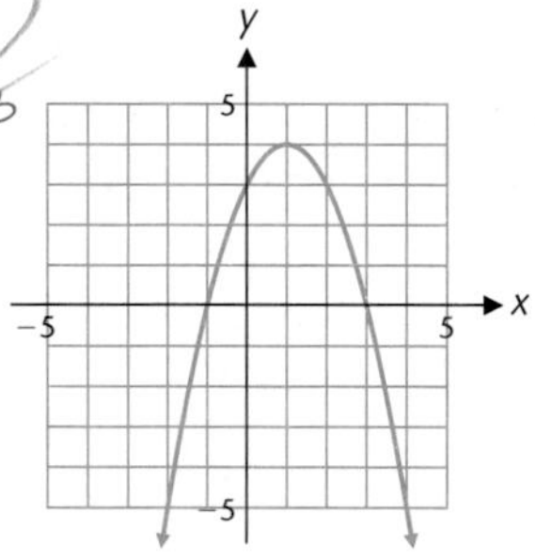

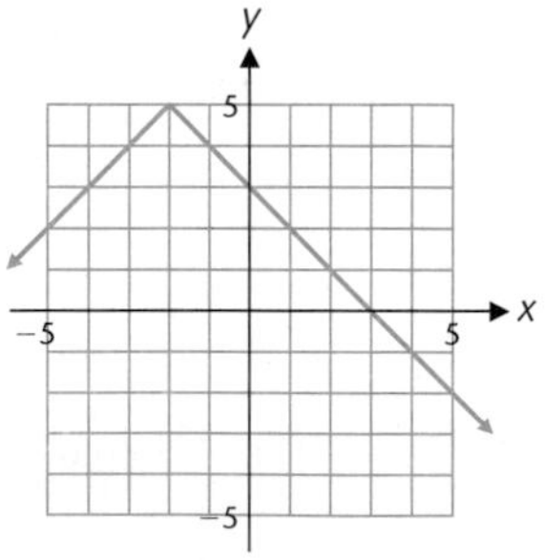

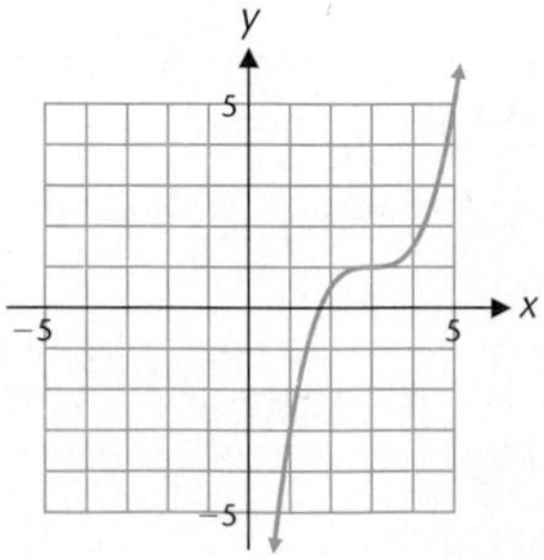

52.

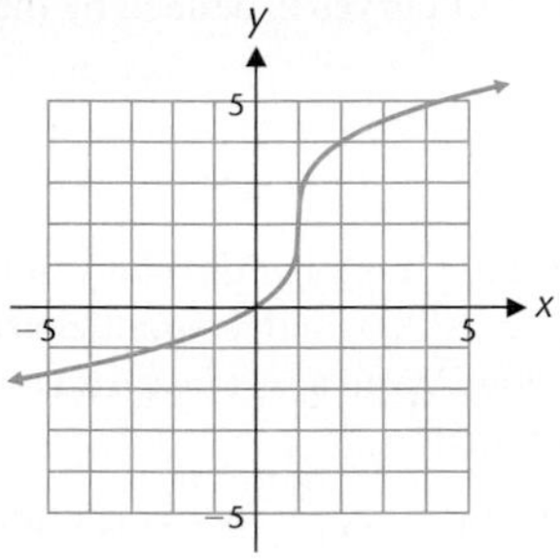

C

Changing the order in a sequence of transformations may change the final result. Investigate each pair of transformations in Problems 53–58 to determine if reversing their order can produce a different result. Support your conclusions with specific examples and/or mathematical arguments.

53. Vertical shift, horizontal shift

54. Vertical shift, reflection in y axis

55. Vertical shift, reflection in x axis

56. Vertical shift, expansion

57. Horizontal shift, reflection in x axis

58. Horizontal shift, contraction

Problems 59–62 refer to two functions f and g with domain $[-5, 5]$ and partial graphs as shown below.

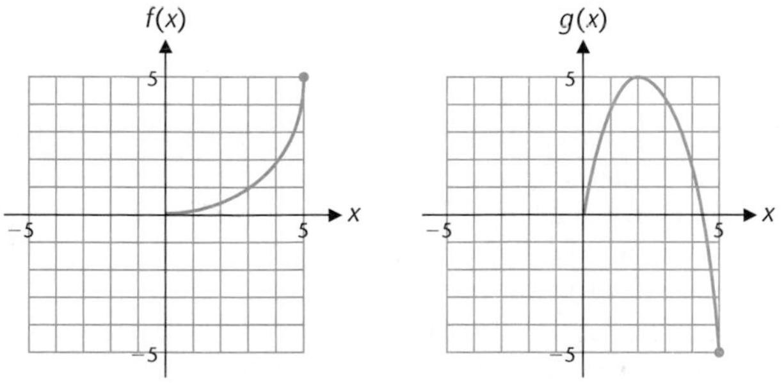

59. Complete the graph of f over the interval $[-5, 0]$, given that f is an even function.

60. Complete the graph of f over the interval $[-5, 0]$, given that f is an odd function.

61. Complete the graph of g over the interval $[-5, 0]$, given that g is an odd function.

62. Complete the graph of g over the interval $[-5, 0]$, given that g is an even function.

63. Let f be any function with the property that $-x$ is in the domain of f whenever x is in the domain of f, and let E and O be the functions defined by

$$E(x) = \tfrac{1}{2}[f(x) + f(-x)]$$

and

$$O(x) = \tfrac{1}{2}[f(x) - f(-x)]$$

(A) Show that E is always even.

(B) Show that O is always odd.

(C) Show that $f(x) = E(x) + O(x)$. What is your conclusion?

64. Let f be any function with the property that $-x$ is in the domain of f whenever x is in the domain of f, and let $g(x) = xf(x)$.

(A) If f is even, is g even, odd, or neither?

(B) If f is odd, is g even, odd, or neither?

In Problems 65–68, graph $f(x)$, $|f(x)|$, and $-|f(x)|$ in a standard viewing window. For purposes of comparison, it will be helpful to graph each function separately and make a hand sketch.

65. $f(x) = 0.2x^2 - 5$

66. $f(x) = 4 - 0.25x^2$

67. $f(x) = 4 - 0.1(x + 2)^3$

68. $f(x) = 0.25(x - 1)^3 - 1$

69. Describe the relationship between the graphs of $f(x)$ and $|f(x)|$ in Problems 65–68.

70. Describe the relationship between the graphs of $f(x)$ and $-|f(x)|$ in Problems 65–68.

APPLICATIONS

71. Production Costs. Total production costs for a product can be broken down into fixed costs, which do not depend on the number of units produced, and variable costs, which do depend on the number of units produced. Thus, the total cost of producing x units of the product can be expressed in the form

$$C(x) = K + f(x)$$

where K is a constant that represents the fixed costs and $f(x)$ is a function that represents the variable costs. Use the graph of the variable-cost function $f(x)$ shown in the figure to graph the total cost function if the fixed costs are \$30,000.

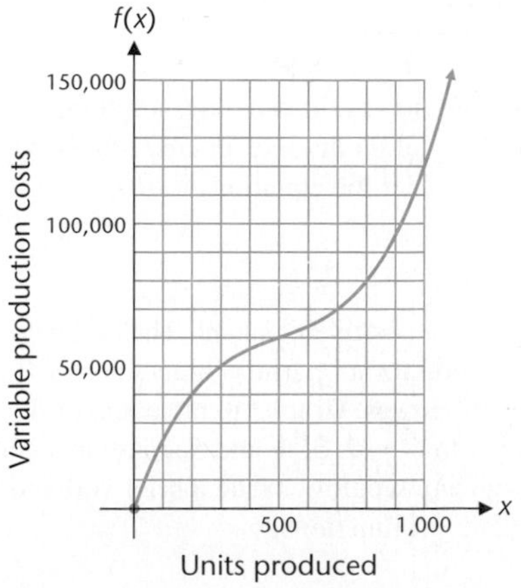

★ **72. Cost Function.** Refer to the variable-cost function $f(x)$ in Problem 71. Suppose construction of a new production facility results in a 25% decrease in the variable cost at all levels of output. If F is the new variable-cost function, use the graph of f to graph $y = F(x)$.

73. Timber Harvesting. To determine when a forest should be harvested, forest managers often use formulas to estimate the number of board feet a tree will produce. A board foot equals 1 square foot of wood, 1 inch thick. Suppose that the number of board feet y yielded by a tree can be estimated by

$$y = f(x) = C + 0.004(x - 10)^3$$

where x is the diameter of the tree in inches measured at a height of 4 feet above the ground and C is a constant that depends on the species being harvested. Graph $y = f(x)$ for $C = 10$, 15, and 20 simultaneously in the viewing window with Xmin = 10, Xmax = 25, Ymin = 10, and Ymax = 35. Write a brief verbal description of this collection of functions.

74. Safety Research. If a person driving a vehicle slams on the brakes and skids to a stop, the speed v in miles per hour at the time the brakes are applied is given approximately by

$$v = f(x) = C\sqrt{x}$$

where x is the length of the skid marks and C is a constant that depends on the road conditions and the weight of the vehicle. The table lists values of C for a midsize automobile and various road conditions. Graph $v = f(x)$ for the values of C in the table simultaneously in the viewing window with Xmin = 0, Xmax = 100, Ymin = 0, and Ymax = 60. Write a brief verbal description of this collection of functions.

Road Condition	C
Wet (concrete)	3.5
Wet (asphalt)	4
Dry (concrete)	5
Dry (asphalt)	5.5

75. Family of Curves. In calculus, solutions to certain types of problems often involve an unspecified constant. For example, consider the equation

$$y = \frac{1}{C}x^2 - C$$

where C is a positive constant. The collection of graphs of this equation for all permissible values of C is called a **family of curves.** Graph the members of this family corresponding to $C = 2$, 3, 4, and 5 simultaneously in a standard viewing window. Write a brief verbal description of this family of functions.

76. Family of Curves. A family of curves is defined by the equation

$$y = 2C - \frac{5}{C}x^2$$

where C is a positive constant. Graph the members of this family corresponding to $C = 1$, 2, 3, and 4 simultaneously in a standard viewing window. Write a brief verbal description of this family of functions.

77. Fluid Flow. A cubic tank is 4 feet on a side and is initially full of water. Water flows out an opening in the bottom of the tank at a rate proportional to the square root of the depth (see the figure). Using advanced concepts from mathematics and physics, it can be shown that the volume of the water in the tank t minutes after the water begins to flow is given by

$$V(t) = \frac{64}{C^2}(C - t)^2 \qquad 0 \le t \le C$$

where C is a constant that depends on the size of the opening. Sketch by hand the graphs of $y = V(t)$ for $C = 1$, 2, 4, and 8. Write a brief verbal description of this collection of functions.

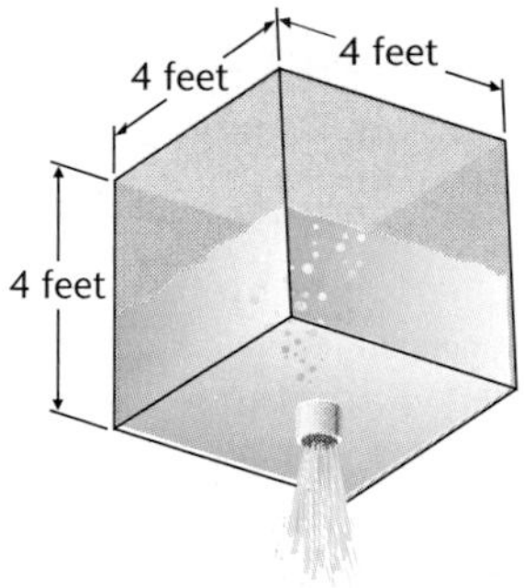

78. Evaporation. A water trough with triangular ends is 9 feet long, 4 feet wide, and 2 feet deep (see the figure). Initially, the trough is full of water, but due to evaporation, the volume of the water in the trough decreases at a rate proportional to the square root of the volume. Using advanced concepts from mathematics and physics, it can be shown that the volume after t hours is given by

$$V(t) = \frac{1}{C^2}(t + 6C)^2 \qquad 0 \le t \le 6|C|$$

where C is a constant. Sketch by hand the graphs of $y = V(t)$ for $C = -4$, -5, and -6. Write a brief verbal description of this collection of functions.

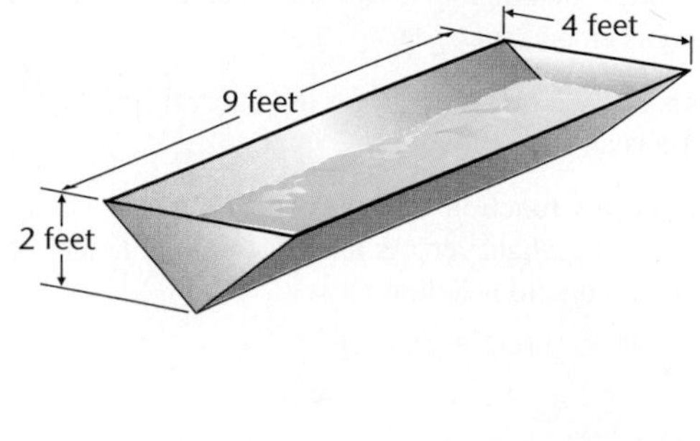

Chapter 1 Group Activity

Introduction to Regression Analysis

In real-world applications one often collects data in table form and then looks for a function to model the data. The very powerful mathematical tool *regression analysis* is often used for this purpose. A number of the modeling functions stated in examples and exercises in Chapter 1 were constructed using *regression techniques* and a graphing utility.

Regression analysis is the process of fitting a function to a set of data points. This process is also referred to as **curve fitting.** Table 1 lists the various types of regression equations that most graphing utilities are capable of producing. The name used to identify each regression equation is simply the name of the type of function that will be used to fit a curve to the data. As indicated in the table, each type of function will be discussed in subsequent chapters in the book.

TABLE 1 Regression Equations

Regression Process	Regression Equation	Location in Book
Linear regression	$y = ax + b$	Chapter 2
Quadratic regression	$y = ax^2 + bx + c$	Chapter 2
Cubic regression	$y = ax^3 + bx^2 + cx + d$	Chapter 3
Quartic regression	$y = ax^4 + bx^3 + cx^2 + dx + e$	Chapter 3
Logarithmic regression	$y = a + b \ln(x), x > 0$	Chapter 4
Exponential regression	$y = ab^x, y > 0$	Chapter 4
Power regression	$y = ax^b, x > 0$ and $y > 0$	Chapter 4
Sinusodial regression	$y = a \sin(bx + c) + d$	Chapter 5

We will not be interested in discussing the underlying mathematical methods used to construct a particular regression equation. Instead, we will concentrate on the mechanics of using a graphing utility to apply regression techniques to data sets.

In Problem 83, Exercise 1-3, we were given the data in Table 2 and the corresponding modeling function

$$S(r) = 0.2 + 8.6r \tag{1}$$

where r represents R & D expenditures and S represents sales (both in billions of dollars) for a large pharmaceutical company.

TABLE 2

r (R & D)	0.66	0.75	0.85	0.99	1.1
S (Sales)	5.9	6.5	7.7	8.6	9.7

CHAPTER 1 GROUP ACTIVITY continued

Now we want to see how the function f was constructed using linear regression on a graphing utility. On most graphing utilities, this process can be broken down into four steps:

Step 1. Enter the statistical data.

Step 2. Compute the desired regression equation.

Step 3. Transfer the regression equation to the equation editor.

Step 4. Graph the data set and the regression equation in the same viewing window.

The details for carrying out each step vary greatly from one graphing utility to another. Consult your manual. If you are using a Texas Instruments graphing calculator, you may also want to consult the graphing calculator supplement for this text (see Preface).

Figure 1 shows the results of applying this four-step process to the data in Table 2 using a linear regression equation.

FIGURE 1

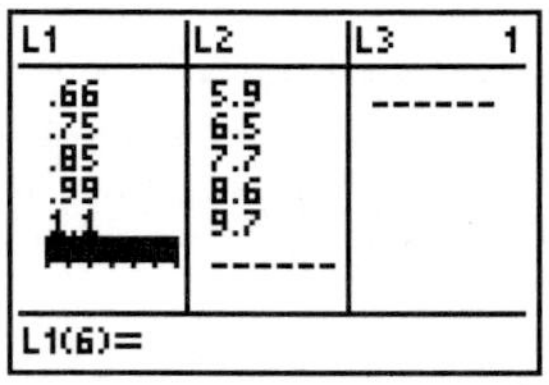

(a) Enter the data

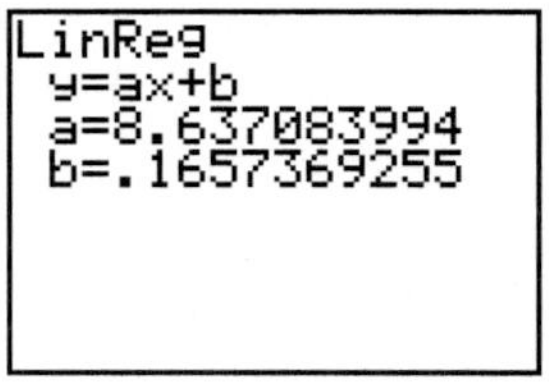

(b) Compute the regression equation.

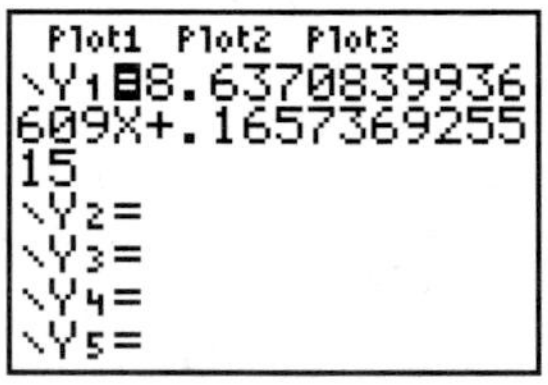

(c) Transfer the regression equation to the equation editor.

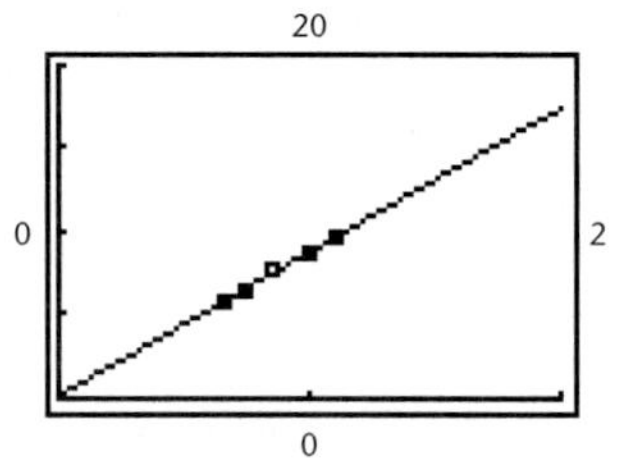

(d) Plot the data and graph the regression equation.

Notice how well the line in Figure 1(d) appears to fit this set of data. We will not discuss any mathematical techniques for determining how well a curve fits a given set of data, but will simply rely on an intuitive interpretation of graphs like Figure 1(d).

Most graphing utilities compute coefficients of regression equations to many more decimal places than we will need for our purposes. However, since it is a simple matter to transfer the regression equation directly to the equation editor, we might as well use all the available decimal places. [We did not enter the 15-digit numbers in Figure 1(c) by hand and neither should you!] Normally, any instructions to write a regression equation by hand will include rounding instructions. For example, if we round each coefficient in Figure 1(b) to one decimal place, we will obtain the modeling function S we used in Problem 83, Exercise 1-3 [see equation (1) on page 75].

(A) If you have not already done so, carry out steps 1–4 for the data in Table 2. Write a brief summary of the details for performing these steps on your graphing utility and keep it for future reference. Exercises in subsequent chapters will require you to compute regression equations.

CHAPTER 1 GROUP ACTIVITY continued

(B) Each of the following exercises from Chapter 1 contains a data set and an equation that models the data. In each case, use a graphing utility to compute a linear regression equation for the data. Plot the data and graph the equation in the same viewing window. Discuss the difference between the graphing utility's regression equation and the modeling function stated in the exercise.

1. Exercise 1-3, Problem 81

2. Exercise 1-3, Problem 82

3. Exercise 1-3, Problem 84

4. Exercise 1-4, Problem 81

5. Exercise 1-4, Problem 82

CHAPTER 1 | REVIEW

1-1 CARTESIAN COORDINATE SYSTEMS

A **real plane** is formed by establishing a one-to-one correspondence between the points in a plane and elements in the set of all ordered pairs of real numbers. A **Cartesian** or **rectangular coordinate system** is formed by the intersection of a horizontal real number line and a vertical real number line at their origins. These lines are called the **coordinate axes.** The **horizontal axis** is often referred to as the ***x* axis** and the **vertical axis** as the ***y* axis.** These axes divide the plane into four **quadrants.** Each point in the plane corresponds to its **coordinates**—an ordered pair (a, b) determined by passing horizontal and vertical lines through the point. The **abscissa** or ***x* coordinate** a is the coordinate of the intersection of the vertical line with the horizontal axis and the **ordinate** or ***y* coordinate** b is the coordinate of the intersection of the horizontal line with the vertical axis. The point $(0, 0)$ is called the **origin.** The **graph** of a set of ordered pairs is the set of corresponding points in the plane.

The **solution set** of an equation in two variables is the set of all ordered pairs of real numbers that make the equation a true statement. The **graph of an equation in two variables** is the graph of its solution set. Graphs are sketched by hand using **point-by-point plotting.** When graphing an equation, we only consider those values of the variables that produce real solutions to the equation. The **distance between the two points** $P_1(x_1, y_1)$ and $P_2(x_2, y_2)$ is

$$d(P_1, P_2) = \sqrt{(x_2 - x_1)^2 + (y_2 - y_1)^2}$$

The **standard equation for a circle** is:

$$(x - h)^2 + (y - k)^2 = r^2 \qquad \text{Radius: } r > 0, \quad \text{Center: } (h, k)$$

1-2 USING GRAPHING UTILITIES

A **graphing utility** is any electronic device capable of displaying the graph of an equation. The smallest darkened rectangular area that a graphing utility can display is called a **pixel.** The **window variables** for a **standard viewing window** are

$$\text{Xmin} = -10, \text{Xmax} = 10, \text{Xscl} = 1, \text{Ymin} = -10, \text{Ymax} = 10, \text{Yscl} = 1$$

Other viewing windows can be defined by assigning different values to these variables. Most graphing utilities will construct a table of ordered pairs that satisfy an equation. A **grid** can be added to a graph to aid in reading the graph. A **cursor** is used to locate a single pixel on the screen. The coordinates of the pixel at the cursor location, called **screen coordinates,** approximate the mathematical coordinates of all the points close to the pixel. The **trace** feature constrains cursor movement to the graph of an equation and displays coordinates of points that satisfy the equation. The **zoom** feature enlarges or reduces the viewing window. Zoom and trace can be used to approximate coordinates of points on a graph to any desired accuracy, up to the internal accuracy of the graphing utility.

1-3 FUNCTIONS

A **function** is a **rule** that produces a correspondence between two sets of elements such that to each element in the first set, there corresponds one and only one element in the second set. The first set is called the **domain** and the set of all corresponding elements in the second set is called the **range.** Equivalently, a *function* is a **set of ordered pairs** with the property that no two ordered pairs have the same first component and different second components. The *domain* is the set of all first components and the *range* is the set of all second components. An **equation** in two variables **defines a function** if to each value of the **independent variable,** the placeholder for domain values, there corresponds exactly one value of the **dependent variable,** the placeholder for range values. A **squared viewing window** is used on a graphing utility to improve the appearance of familiar graphs, such as circles. A **vertical line** will intersect the graph of a function in at most one point. Unless otherwise specified, the **domain of a function defined by an equation** is assumed to be the set of all real number replacements for the independent variable that produce real values for the dependent variable. The symbol $\boldsymbol{f(x)}$ represents the real number in the range of the function f corresponding to the domain value x. Equivalently, the ordered pair $(x, f(x))$ belongs to the function f.

1-4 FUNCTIONS: GRAPHS AND PROPERTIES

The **graph of a function** f is the graph of the equation $y = f(x)$. The abscissa of a point where the graph of a function intersects the x axis is called an **x intercept** or **zero** of the function. The x intercept is also a real solution or **root** of the equation $f(x) = 0$. The ordinate of a point where the graph of a function crosses the y axis is called the **y intercept** of the function. The y intercept is given by $f(0)$, provided 0 is in the domain of f. Most graphing utilities contain a built-in routine, usually called **root** or **zero,** for approximating x intercepts. A solid dot on a graph of a function indicates a point that belongs to the graph and an open dot indicates a point that does not belong to the graph. Dots are also used to indicate that a graph terminates at a point, and arrows are used to indicate that the graph continues with no significant changes.

Let I be an interval in the domain of a function f. Then,

1. **f is increasing** on I if $f(b) > f(a)$ whenever $b > a$ in I. (The graph of f rises as x moves from left to right over I.)
2. **f is decreasing** on I if $f(b) < f(a)$ whenever $b > a$ in I. (The graph of f falls as x moves from left to right over I.)
3. **f is constant** on I if $f(a) = f(b)$ for all a and b in I. (The graph of f is horizontal as x moves from left to right over I.)

The functional value $f(c)$ is called a **local maximum** if there is an interval (a, b) containing c such that $f(x) \le f(c)$ for all x in (a, b) and a local minimum if there is an interval (a, b) containing c such that $f(x) \ge f(c)$ for all x in (a, b). The functional value $f(c)$ is called a **local extremum** if it is either a local maximum or a local minimum. Most graphing utilities have built-in routines for approximating local maxima and minima.

A **piecewise-defined function** is a function whose definition involves more than one formula. The graph of a function is **continuous** if it has no holes or breaks and **discontinuous** at any point where it has a hole or break. Intuitively, the graph of a continuous function can be sketched without lifting a pen from the paper. The **greatest integer** of a real number x, denoted by $[\![x]\!]$, is the largest integer less than or equal to x; that is, $[\![x]\!] = n$, where n is an integer, $n \le x < n + 1$.

The **greatest integer function** f is defined by the equation $f(x) = [\![x]\!]$. Changing the mode on a graphing utility from **connected mode** to **dot mode** will make discontinuities on some graphs more apparent.

1-5 FUNCTIONS: GRAPHS AND TRANSFORMATIONS

The first six basic functions in a library of elementary functions are defined by $f(x) = x$ (identity function), $g(x) = |x|$ (absolute value function), $h(x) = x^2$ (square function), $m(x) = x^3$ (cube function), $n(x) = \sqrt{x}$ (square root function), and $p(x) = \sqrt[3]{x}$ (cube root function) (see Figure 1, Section 1-5). Performing an operation on a function produces a **transformation** of the graph of the function. The basic transformations are:

Vertical Translation:

$$y = f(x) + k \quad \begin{cases} k > 0 & \text{Shift graph of } y = f(x) \text{ up } k \text{ units} \\ k < 0 & \text{Shift graph of } y = f(x) \text{ down } |k| \text{ units} \end{cases}$$

Horizontal Translation:

$$y = f(x + h) \quad \begin{cases} h > 0 & \text{Shift graph of } y = f(x) \text{ left } h \text{ units} \\ h < 0 & \text{Shift graph of } y = f(x) \text{ right } |h| \text{ units} \end{cases}$$

Reflection:

$y = -f(x)$ Reflect the graph of $y = f(x)$ in the x axis

Vertical Expansion and Contraction:

$$y = Af(x) \quad \begin{cases} A > 1 & \text{Vertically expand graph of } y = f(x) \text{ by multiplying each ordinate value by } A \\ 0 < A < 1 & \text{Vertically contract graph of } y = f(x) \text{ by multiplying each ordinate value by } A \end{cases}$$

A function f is called an **even function** if $f(x) = f(-x)$ for all x in the domain of f and an **odd function** if $f(-x) = -f(x)$ for all x in the domain of f. The graph of an even function is said to be **symmetric with respect to the y axis** and the graph of an odd function is said to be **symmetric with respect to the origin.**

CHAPTER 1 | REVIEW EXERCISES

Work through all the problems in this review and check answers in the back of the book. Answers to most review problems are there, and following each answer is a number in italics indicating the section in which that type of problem is discussed. Where weaknesses show up, review appropriate sections in the text.

A

1. Plot the following points in a rectangular coordinate system: $(-4, 0)$, $(0, 3)$, $(-3, -4)$, $(4, -2)$.
2. Write the equation of a circle with radius $\sqrt{7}$ and center:
   (A) $(0, 0)$ (B) $(3, -2)$
3. Find the smallest viewing window that will contain all the points in the table. State your answer in terms of the window variables.

x	-3	5	-4	0	9
y	2	-6	7	-5	1

4. Indicate whether each set defines a function. Find the domain and range of each function.
(A) $\{(1, 1), (2, 4), (3, 9)\}$
(B) $\{(1, 1), (1, -1), (2, 2), (2, -2)\}$
(C) $\{(-2, 2), (-1, 2), (0, 2), (1, 2), (2, 2)\}$

5. Indicate whether each graph specifies a function.

(A)

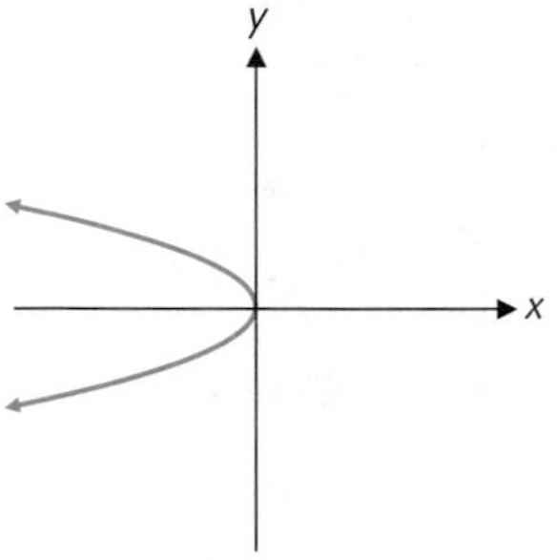

(B)

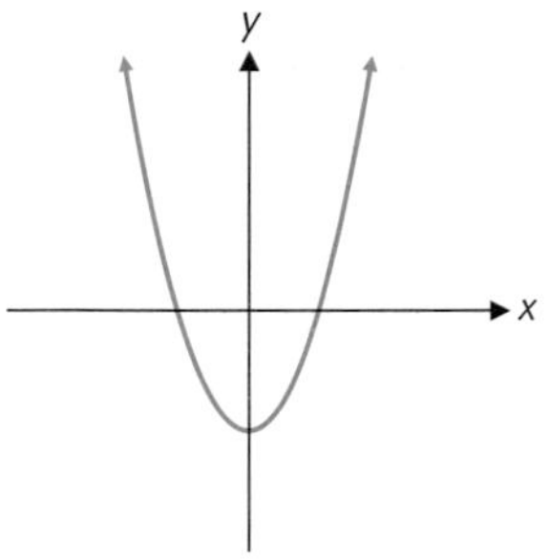

(C)

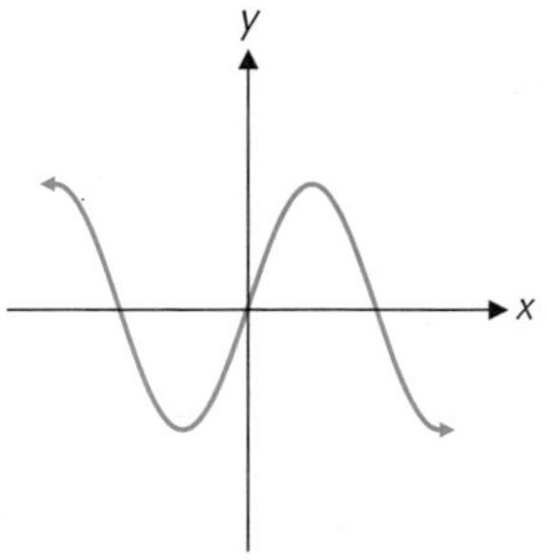

(D)

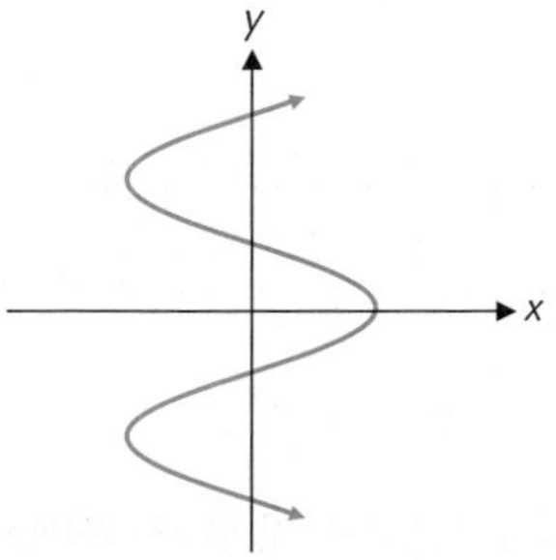

6. For $f(x) = x^2 - 2x$, find
(A) $f(1)$ (B) $f(-4)$
(C) $f(2) \cdot f(-1)$ (D) $\dfrac{f(0)}{f(3)}$

7. Without graphing, indicate whether each function is even, odd, or neither.
(A) $f(x) = x^5 + 6x$ (B) $g(t) = t^4 + 3t^2$
(C) $h(z) = z^5 + 4z^2$

Problems 8–16 refer to the function f given by the following graph.

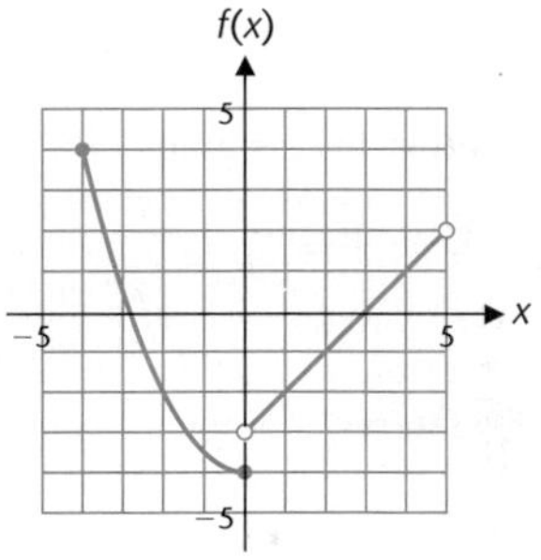

8. Find $f(-4)$, $f(0)$, $f(3)$, and $f(5)$.

9. Find all values of x for which $f(x) = -2$.

10. Find the domain and range of f.

11. Find the intervals over which f is increasing and decreasing.

12. Find any points of discontinuity.

Sketch the graph of each of the following:

13. $f(x) + 1$ **14.** $f(x + 1)$

15. $-f(x)$ **16.** $0.5f(x)$

17. Match each equation with a graph of one of the functions f, g, m, or n in the figure. Each graph is a graph of one of the equations and is assumed to continue without bound beyond the portion shown.
(A) $y = (x - 2)^2 - 4$ (B) $y = -(x + 2)^2 + 4$
(C) $y = -(x - 2)^2 + 4$ (D) $y = (x + 2)^2 - 4$

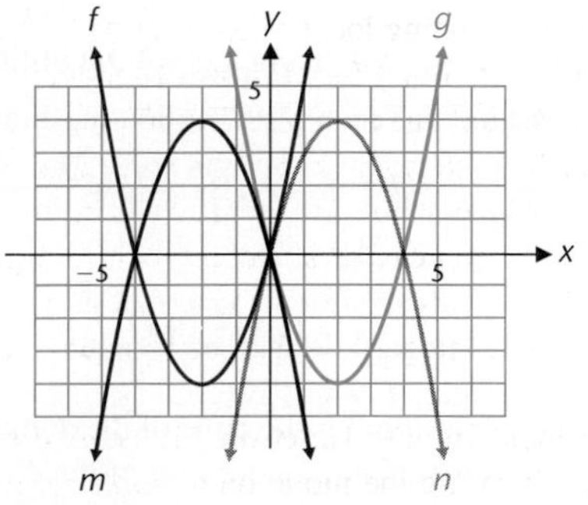

B

Problems 18–24 refer to the function q given by the following graph. (Assume the graph continues as indicated beyond the part shown.)

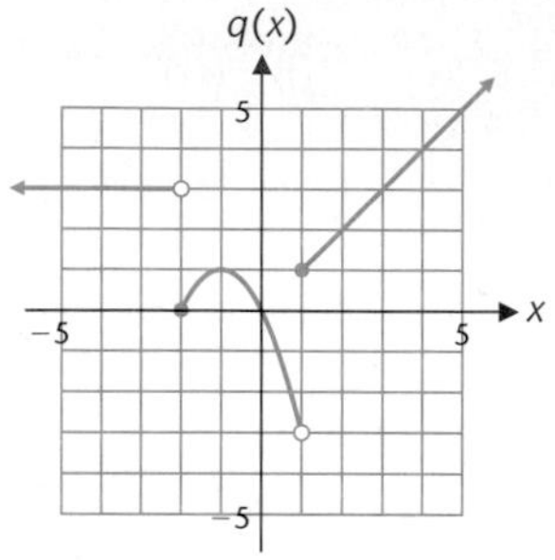

18. Find y to the nearest integer:
(A) $y = q(0)$ (B) $y = q(1)$
(C) $y = q(2)$ (D) $y = q(-2)$

19. Find x to the nearest integer:
(A) $q(x) = 0$ (B) $q(x) = 1$
(C) $q(x) = -3$ (D) $q(x) = 3$

20. Find the domain and range of q.

21. Find the intervals over which q is increasing.

22. Find the intervals over which q is decreasing.

23. Find the intervals over which q is constant.

24. Identify any points of discontinuity.

25. (A) Graph the triangle with vertices $(-1, -2)$, $(1, 4)$, and $(4, 3)$.
(B) Find the perimeter to two decimal places.
(C) Use the Pythagorean theorem to determine if the triangle is a right triangle.

26. Let $y = -x^2 + 6x - 4$.
(A) Complete the following table of solutions to this equation.

x	0	1	2	3	4	5	6
y							

(B) Plot the points in the table and sketch the graph of the equation by hand, using a graphing utility as an aid.

27. Find the domain of each of the following functions:
(A) $f(x) = x^2 - 4x + 5$ (B) $g(t) = \dfrac{t + 2}{t - 5}$
(C) $h(w) = 2 + 3\sqrt{w}$

28. If $g(t) = 2t^2 - 3t + 6$, find $\dfrac{g(2 + h) - g(2)}{h}$.

29. The function f multiplies the cube of the domain element by 4 and then subtracts the square root of the domain element. Write an algebraic definition of f.

30. Write a verbal description of the function $f(x) = 3x^2 + 4x - 6$.

31. The function $g(x) = 6\sqrt{x} - x^2$ has one local extreme value $g(c)$ and two x intercepts. Examine the graph of $y = g(x)$ to determine if this extremum is a local maximum or a local minimum. Approximate c, $g(c)$, and the x intercepts to one decimal place.

32. Let

$$f(x) = \begin{cases} -x - 5 & \text{for } -4 \le x < 0 \\ 0.2x^2 & \text{for } \quad 0 \le x \le 5 \end{cases}$$

(A) Sketch the graph of $y = f(x)$, using a graphing utility as an aid.
(B) Find the domain and range.
(C) Find any points of discontinuity.
(D) Find the intervals where f is increasing, those where f is decreasing, and those where f is constant.

33. Write a verbal description of the graph of $f(x) = 0.1x^3 - 6x + 5$ over the interval $[-10, 10]$ using increasing and decreasing terminology and indicating any local maximum and minimum values. Approximate to two decimal places the coordinates of any points used in your description.

34. How are the graphs of the following related to the graph of $y = x^2$?
(A) $y = -x^2$ (B) $y = x^2 - 3$ (C) $y = (x + 3)^2$

35. Each of the following graphs is the result of applying one or more transformations to the graph of one of the six basic functions in Figure 1, Section 1-5. Find an equation for the graph. Check by graphing the equation on a graphing utility.

(A)

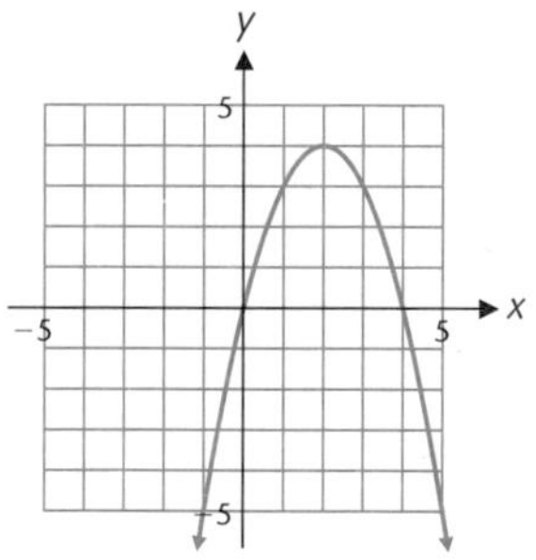

(B)

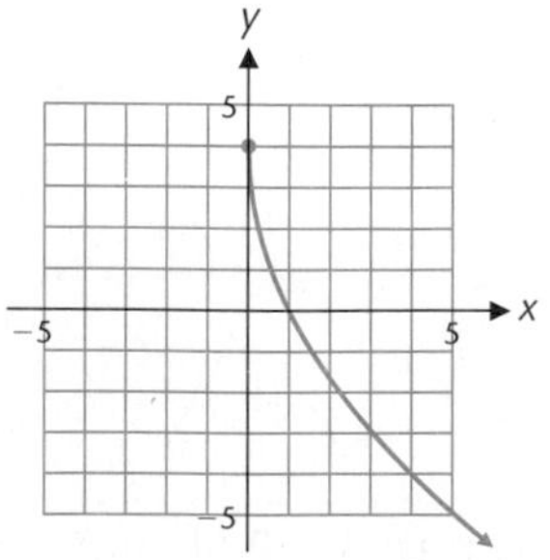

36. The graph of $f(x) = |x|$ is expanded by a factor of 3, reflected in the x axis, and shifted 4 units to the right and 8 units up to form the graph of the function g. Find an equation for the function g and graph g in a standard viewing window.

37. Find the center and radius, and sketch by hand the graph of the circle with equation

$$(x + 2)^2 + (y - 2)^2 = 4$$

38. Find the domain of $f(x) = \dfrac{x}{\sqrt{x} - 3}$.

39. Sketch by hand the graph of a function that is consistent with the given information.

(A) The function f is continuous on $[-5, 5]$, increasing on $[-5, -3]$, decreasing on $[-3, 1]$, constant on $[1, 3]$, and increasing on $[3, 5]$.

(B) The function f is continuous on $[-5, 1)$ and $[1, 5]$, $f(-2) = -1$ is a local maximum, and $f(3) = 2$ is a local minimum.

C

40. Find $g(t)$ given that

$$g(t + h) = 2(t + h)^2 - 4(t + h) + 5.$$

41. Graph in the standard viewing window:

$$f(x) = 0.1(x - 2)^2 + \frac{|3x - 6|}{x - 2}$$

Assuming the graph continues as indicated beyond the part shown in this viewing window, find the domain, range, and any points of discontinuity. [*Hint:* Use the dot mode on your graphing utility, if it has one.]

42. A partial graph of the function f is shown in the figure. Complete the graph of f over the interval $[0, 5]$ given that
(A) f is an even function. (B) f is an odd function.

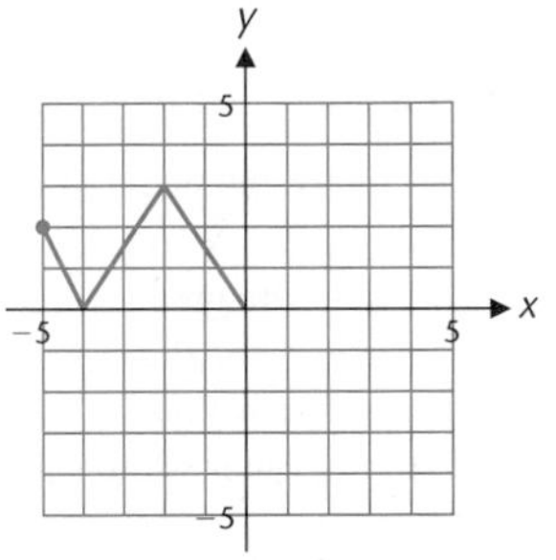

43. Find the equation of a circle with center $(4, -3)$ whose graph passes through the point $(1, 2)$.

44. For $f(x) = 3x^2 - 5x + 7$, find and simplify:

(A) $\dfrac{f(x + h) - f(x)}{h}$ (B) $\dfrac{f(x) - f(a)}{x - a}$

45. The function f is decreasing on $[-5, 5]$ with $f(-5) = 4$ and $f(5) = -3$.

(A) If f is continuous on $[-5, 5]$, how many times can the graph of f cross the x axis? Support your conclusion with examples and/or verbal arguments.

(B) Repeat part A if the function does not have to be continuous.

46. Let $f(x) = [\![|x|]\!]$.

(A) Write a piecewise definition of f. Include sufficient intervals to clearly illustrate the definition.

(B) Sketch by hand the graph of $y = f(x)$, using a graphing utility as an aid. Include sufficient intervals to clearly illustrate the graph.

(C) Find the range of f.

(D) Find any points of discontinuity.

(E) Indicate whether f is even, odd, or neither.

APPLICATIONS

47. Price and Demand. The price p and corresponding weekly demand q for a particular brand of liquid laundry detergent in a city are shown in the figure. Use this graph to estimate the following demands to the nearest 100 bottles.

(A) What is the demand when the price is \$5.00 per bottle?

(B) Does the demand increase or decrease if the price is increased to \$5.40 per bottle? By how much?

(C) Does the demand increase or decrease if the price is decreased to \$4.60 per bottle? By how much?

(D) Write a brief description of the relationship between price and demand illustrated by this graph.

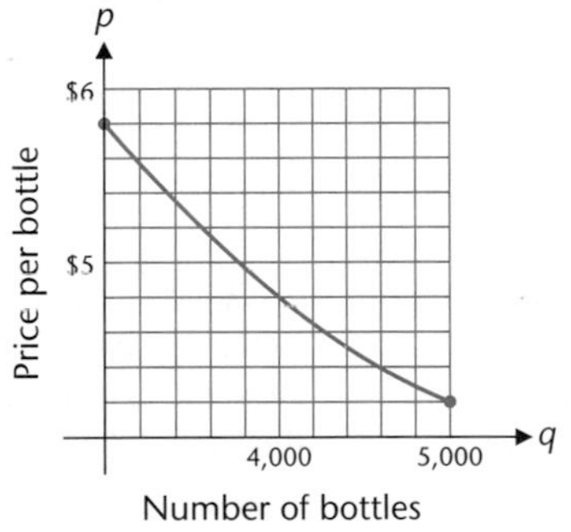

48. Price and Demand. Refer to Problem 47. An economic analyst determined that the price-demand equation for the liquid laundry detergent is given approximately by the equation

$$p = 11.3 - 0.1\sqrt{q} \qquad 3{,}000 \le q \le 5{,}000$$

Apply zoom and trace procedures to the graph of this equation to answer parts A, B, and C of Problem 47. Approximate each value of q to the nearest 10 bottles.

49. Physics: Position of a Moving Object. In flight shooting distance competitions, archers are capable of shooting arrows 600 meters or more. An archer standing on the ground shoots an arrow. After x seconds, the arrow is y meters above the ground as given approximately by

$$y = 55x - 4.88x^2$$

(A) Graph this equation in the viewing window $0 \le x \le 12$, $0 \le y \le 200$.

(B) Use a graphing utility to approximate the time (to the nearest tenth of a second) the arrow is airborne.
(C) Use a graphing utility to approximate the maximum altitude (to the nearest meter) the arrow reaches during its flight.

50. Manufacturing. A box with four flaps on each end is to be made out of a piece of cardboard that measures 48 by 72 inches. The width of each flap is x inches and the length of one pair of opposite flaps is $2x$ inches to ensure that the other pair of flaps will meet when folded over to close the box (see the figure). The volume of the box is given by

$$V(x) = 2x(48 - 2x)(36 - 2x) \qquad 0 \le x \le 9$$

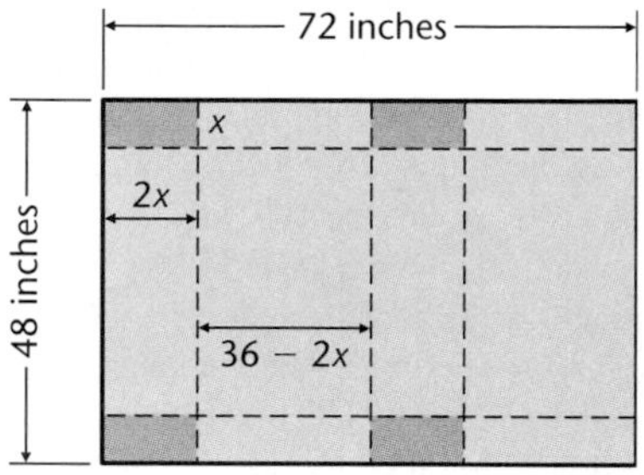

(A) Examine the values of the volume function $V(x)$ for $x = 0, 1, 2, \ldots, 9$ and estimate the width of the flap that will make the volume maximum. What is the estimated volume?
(B) Graph $y = V(x)$ and approximate to one decimal place the width of the flap that will make the volume maximum. Approximate the maximum volume to the nearest cubic inch also.

51. Demand. Egg consumption has been decreasing for some time, presumably because of increasing awareness of the high cholesterol in egg yolks. The table lists the annual per capita consumption of eggs in the United States.

1970	1975	1980	1985	1990
309	276	271	255	233

Source: U.S. Department of Agriculture

A mathematical model for this data is given by

$$f(x) = 303.4 - 3.46x$$

where $x = 0$ corresponds to 1970.
(A) Complete the following table. Round values of $f(x)$ to the nearest integer.

x	0	5	10	15	20
Consumption	309	276	271	255	233
$f(x)$					

(B) Sketch by hand the graph of $y = f(x)$ and the data in the table on the same set of axes.
(C) Use the modeling function f to estimate the per capita egg consumption in 1995 and in 2000.
(D) Use a graphing utility to find a linear regression equation for the data in the table and compare the results with the modeling function f (see Chapter 1 Group Activity).
(E) Based on the information in the table, write a brief verbal description of egg consumption from 1970 to 1990.

52. Medicine. Proscar is a drug produced by Merck & Co., Inc., to treat symptomatic benign prostate enlargement. One of the long-term effects of the drug is to increase urine flow rate. Results from a 3-year study show that

$$f(x) = 0.00005x^3 - 0.007x^2 + 0.255x$$

is a mathematical model for the average increase in urine flow rate in cubic centimeters per second where x is time in months.
(A) Graph this function for $0 \le x \le 36$ using a graphing utility.
(B) Write a brief verbal description of the graph using increasing, decreasing, local maximum, and local minimum as appropriate. Approximate to one decimal place the coordinates of any points used in your description.

53. Pricing. An office supply store sells ball point pens for \$0.49 each. For an order of three dozen or more pens, the price per pen for all pens ordered is reduced to \$0.44, and for an order of six dozen or more, the price per pen for all pens ordered is reduced to \$0.39.
(A) If $C(x)$ is the total cost in dollars for an order of x pens, write a piecewise definition for C.
(B) Sketch by hand the graph of $y = C(x)$ for $0 \le x \le 108$, using a graphing utility as an aid, and identify any points of discontinuity.

54. Computer Science. In computer programming, it is often necessary to check numbers for certain properties (even, odd, perfect square, etc.). The greatest integer function provides a convenient method for determining some of these properties. For example, the following function can be used to determine whether a number is the square of an integer:

$$f(x) = x - ([\![\sqrt{x}]\!])^2$$

(A) Find $f(1)$. (B) Find $f(2)$. (C) Find $f(3)$.
(D) Find $f(4)$. (E) Find $f(5)$.
(F) Find $f(n^2)$, where n is a positive integer.

ANSWER TO APPLICATION

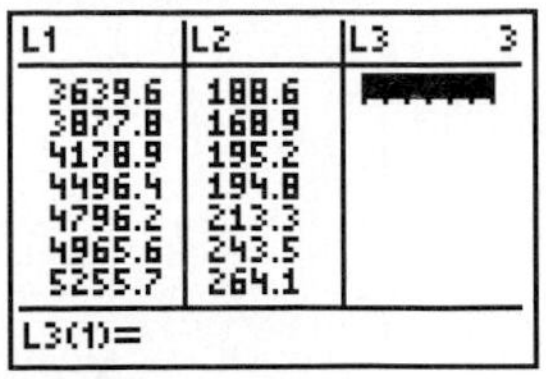

FIGURE 1

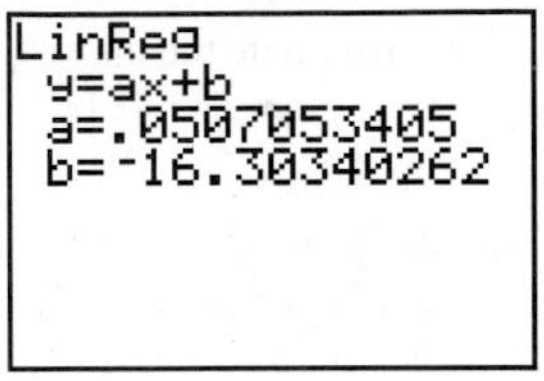

FIGURE 2

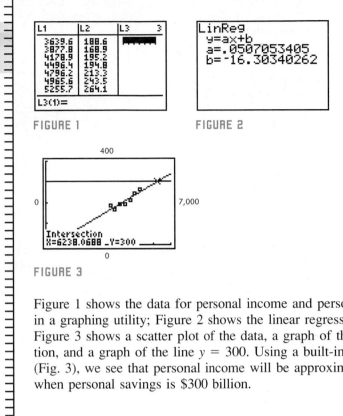

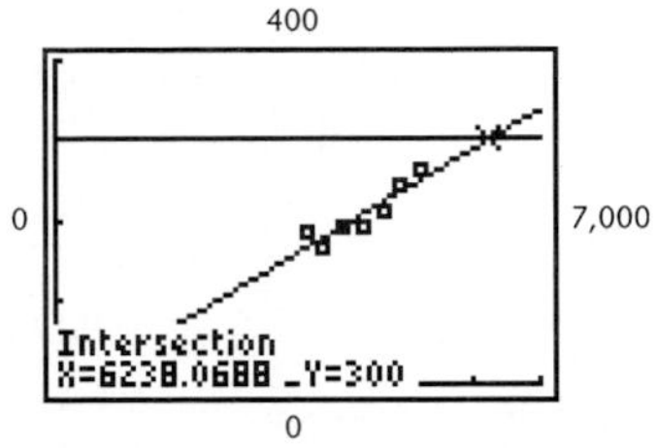

FIGURE 3

Figure 1 shows the data for personal income and personal savings entered in a graphing utility; Figure 2 shows the linear regression equation; and Figure 3 shows a scatter plot of the data, a graph of the regression equation, and a graph of the line $y = 300$. Using a built-in intersection routine (Fig. 3), we see that personal income will be approximately \$6,238 billion when personal savings is \$300 billion.

www.mhhe.com/barnett

Linear and Quadratic Functions

2

Outline

Application

Plans for a rectangular sunroom specify that the width of its interior is twice its depth and its walls are 7.5 inches thick. The home owners purchased 360 12-inch-square saltillo floor tiles while they were in Mexico. If the external area of the sunroom (including the walls) is 400 square feet, will there be enough tile to complete the floor of the interior?

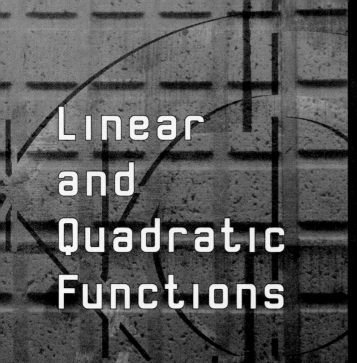

In the last chapter we investigated the general concept of function using graphs, tables, and algebraic equations. In this and subsequent chapters we investigate particular types of functions in more detail. By the time we finish, we will have a library of elementary functions that form a very important addition to our mathematical tool box. These elementary functions are used with great frequency in almost any place where mathematics is used: the physical, social, and life sciences; most technical fields; and most mathematical courses beyond this one. Take a few moments to look at the chapter titles in the table of contents and observe how the various types of elementary functions form the structure on which the course is organized.

In this chapter we investigate linear and quadratic functions. As you will see, many significant real-world problems require these functions in their representation and solution. In addition, to find all solutions to quadratic equations, we need to extend the real number system to include complex numbers.

Preparing for This Chapter

Before getting started on this chapter, review the following concepts:

Properties of Real Numbers (Appendix A, Section 1)

Polynomials (Appendix A, Sections 2 and 3)

Least Common Denominator (Appendix A, Section 4)

Rational Exponents (Appendix A, Section 6)

Square Root Radicals (Appendix A, Section 7)

Linear Equations and Inequalities (Appendix A, Section 8)

Set Operations (Appendix A, Section 8)

Cartesian Coordinate System (Chapter 1, Section 1)

Functions (Chapter 1, Section 3)

Graphs of Functions (Chapter 1, Section 4)

Elementary Functions (Chapter 1, Section 5)

Linear Regression (Chapter 1, Group Activity)

Section 2-1 Linear Functions

V

ID

- Constant and Linear Functions
- Graph of $Ax + By = C$
- Slope of a Line
- Equations of Lines—Special Forms
- Parallel and Perpendicular Lines
- Application: Linear Regression

The straight line is a fundamental geometric object and an important tool in mathematical modeling. In this section we will add linear functions to our library of elementary functions and explore the relationship between graphs of linear functions and straight lines. We will also determine how to find the equation of a line, given information about the line. And we will see how regression analysis can be used to find a line that is, in some sense, the "best" fit for a data set in an applied setting.

Constant and Linear Functions

One of the elementary functions introduced in Section 1-5 was the identity function $f(x) = x$ (Fig. 1).

FIGURE 1
Identity function: $f(x) = x$.

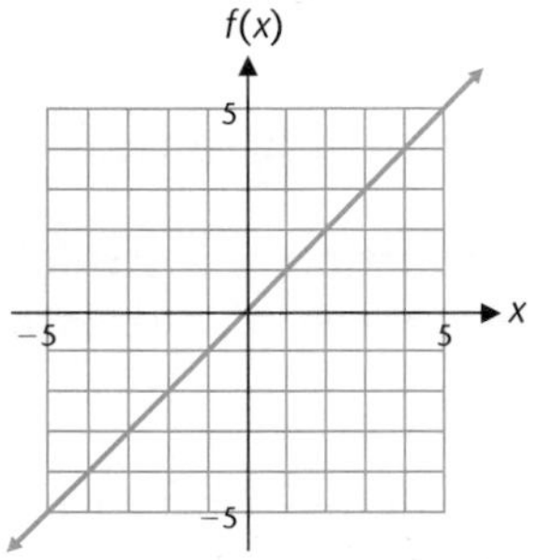

Explore/Discuss 1

Use the transformations discussed in Section 1-5 to describe verbally the relationship between the graph of $f(x) = x$ and each of the following functions. Graph each function.

(A) $g(x) = 3x + 1$ (B) $h(x) = 0.5x - 2$ (C) $k(x) = -x + 1$

If we apply a sequence of translations, reflections, expansions, and/or contractions to the identity function, the result is always a function whose graph is a straight line. Because of this, functions like g, h, and k in Explore/Discuss 1 are called *linear functions*. In general:

LINEAR AND CONSTANT FUNCTIONS

A function f is a **linear function** if

$$f(x) = mx + b \qquad m \neq 0$$

where m and b are real numbers. The **domain** is the set of all real numbers and the **range** is the set of all real numbers. If $m = 0$, then f is called a **constant function,**

$$f(x) = b$$

which has the set of all real numbers as its *domain* and the constant b as its *range*.

Figure 2 shows the graphs of two linear functions f and g, and a constant function h.

FIGURE 2
Two linear functions and a constant function.

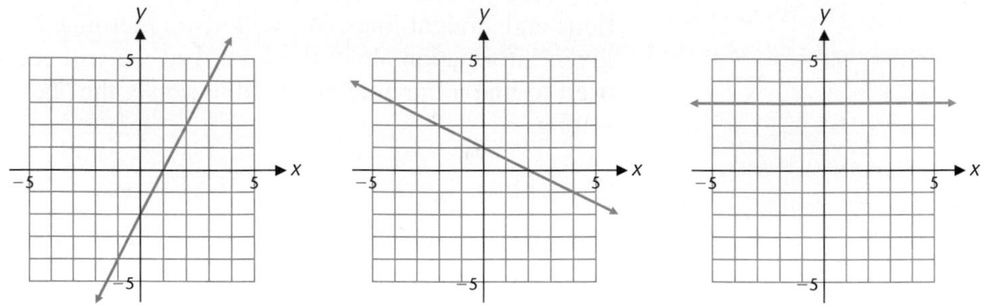

(a) $f(x) = 2x - 2$ (b) $g(x) = -0.5x + 1$ (c) $h(x) = 3$

It can be shown (see Problem 78 in Exercise 2-1 for a sketch of a proof) that

The graph of a linear function is a straight line that is neither horizontal nor vertical. The graph of a constant function is a horizontal straight line.

What about vertical lines? Recall from Chapter 1 that the graph of a function cannot contain two points with the same x coordinate and different y coordinates. Since *all* the points on a vertical line have the same x coordinate, the graph of a function can never be a vertical line. Later in this section we will discuss equations of vertical lines, but these equations never define functions.

Recall from Section 1-4 that the y intercept of a function f is $f(0)$, provided $f(0)$ exists, and the x intercepts are the solutions of the equation $f(x) = 0$.

Explore/Discuss 2

(A) Is it possible for a linear function to have two x intercepts? No x intercepts? If either of your answers is yes, give an example.

(B) Is it possible for a linear function to have two y intercepts? No y intercept? If either of your answers is yes, give an example.

(C) Discuss the possible number of x and y intercepts for a constant function.

EXAMPLE 1 Finding x and y Intercepts

Find the x and y intercepts for $f(x) = \frac{2}{3}x - 3$.

Solution The y intercept is $f(0) = -3$. The x intercept can be found algebraically using standard equation solving techniques, or graphically using a built-in routine on a graphing utility.

Algebraic Method

$$f(x) = 0$$

$$\frac{2}{3}x - 3 = 0$$

$$\frac{2}{3}x = 3$$

$$x = \frac{9}{2} = 4.5$$

Graphical Method

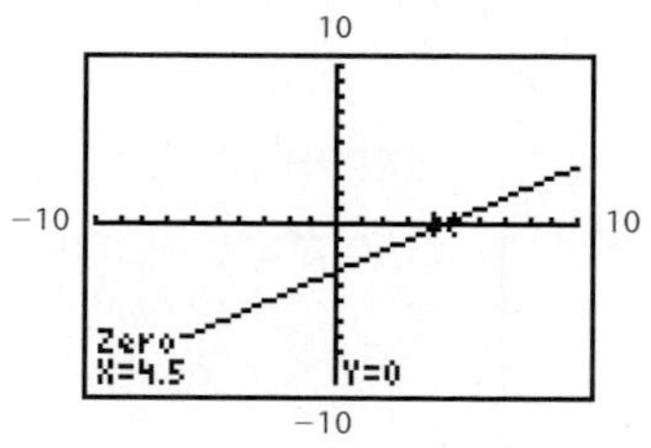

MATCHED PROBLEM 1

Find the x and y intercepts of $g(x) = -\frac{4}{3}x + 5$.

Graph of $Ax + By = C$

We now investigate graphs of linear equations in two variables:

$$Ax + By = C \tag{1}$$

where A and B are not both zero. Depending on the values of A and B, this equation defines a linear function, a constant function, or no function at all. If $A \neq 0$ and $B \neq 0$, then equation (1) can be written in the form

$$y = -\frac{A}{B}x + \frac{C}{B} \quad \text{Linear function (slanted line)} \tag{2}$$

which is in the form $f(x) = mx + b$, $m \neq 0$, hence is a linear function. If $A = 0$ and $B \neq 0$, then equation (1) can be written in the form

$$0x + By = C$$

$$y = \frac{C}{B} \quad \text{Constant function (horizontal line)} \tag{3}$$

which is in the form $g(x) = b$, hence is a constant function. If $A \neq 0$ and $B = 0$, then equation (1) can be written in the form

$$Ax + 0y = C$$

$$x = \frac{C}{A} \quad \text{Not a function (vertical line)} \tag{4}$$

We can see that the graph of equation (4) is a vertical line since the equation is satisfied for any value of y as long as x is the constant $\frac{C}{A}$. Hence this form does not define a function.

The following theorem is a generalization of the preceding discussion:

THEOREM 1

GRAPH OF A LINEAR EQUATION IN TWO VARIABLES

The graph of any equation of the form

$$Ax + By = C \quad \text{Standard form} \tag{5}$$

where A, B, and C are real constants (A and B not both 0) is a straight line. Every straight line in a Cartesian coordinate system is the graph of an equation of this type. Vertical and horizontal lines have particularly simple equations, which are special cases of equation (5):

Horizontal line with y intercept b: $y = b$

Vertical line with x intercept a: $x = a$

Explore/Discuss 3

Graph each of the following cases of $Ax + By = C$ in the same coordinate system:

1. $3x + 2y = 6$
2. $0x - 3y = 12$
3. $2x + 0y = 10$

Which cases define functions? Explain why or why not.

Graph each case using a graphing utility (check your manual on how to graph vertical lines).

Sketching the graph of an equation of the form

$$Ax + By = C \qquad \text{or} \qquad y = mx + b$$

is very easy, since the graph of each equation is a straight line. All that is necessary is to plot any two points from the solution set and use a straightedge to draw a line through these two points. The x and y intercepts are often the easiest to find.

EXAMPLE 2

Sketching Graphs of Lines

(A) Describe the graphs of $x = -2$ and $y = 3$. Graph simultaneously in the same rectangular coordinate system by hand and in the same viewing window on a graphing utility.

(B) Write the equations of the vertical and horizontal lines that pass through the point $(1, -4)$.

(C) Graph the equation $3x - 2y = 6$ by hand and on a graphing utility.

Solutions

(A) The graph of $x = -2$ is a vertical line with x intercept -2 and the graph of $y = 3$ is a horizontal line with y intercept 3 (Fig. 3).

FIGURE 3

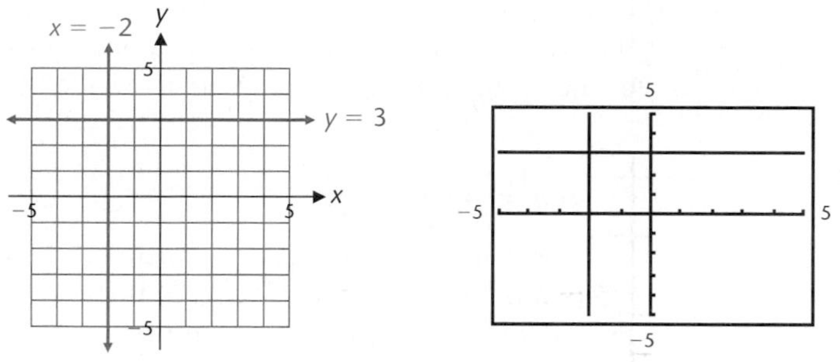

(a) By hand (b) On a graphing utility

(B) Horizontal line through $(1, -4)$: $y = -4$; vertical line through $(1, -4)$: $x = 1$

(C) To graph $3x - 2y = 6$ by hand, find the x intercept by substituting $y = 0$ and solving for x, and then find the y intercept by substituting $x = 0$ and solving for y. Then draw a line through the intercepts [Fig. 4(a)]. To use a graphing utility, solve $3x - 2y = 6$ for y, enter the result in the equation editor, and graph [Fig. 4(b)].

***x* intercept**	***y* intercept**	**Solve for *y***
$3x - 2(0) = 6$	$3(0) - 2y = 6$	$3x - 2y = 6$
$3x = 6$	$-2y = 6$	$-2y = -3x + 6$
$x = 2$	$y = -3$	$y = 1.5x - 3$

FIGURE 4

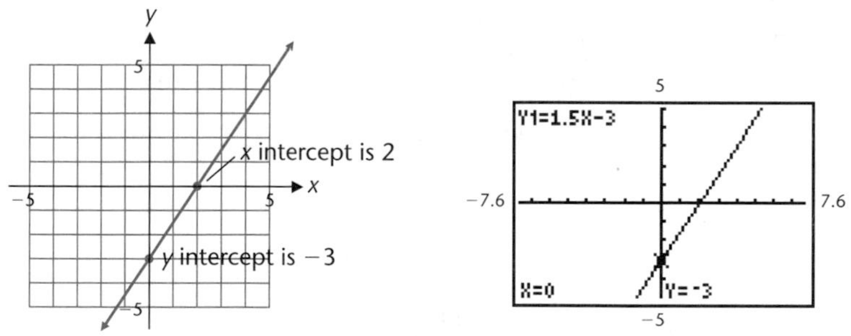

(a) By hand (b) On a graphing utility

Note that we used a *squared viewing window* in Figure 4(b) to produce units of the same length on both axes. This makes it easier to compare the hand sketch with the graphing utility graph.

MATCHED PROBLEM 2

(A) Describe the graphs of $x = 4$ and $y = -3$. Graph simultaneously in the same rectangular coordinate system by hand and in the same viewing window on a graphing utility.

(B) Write the equations of the vertical and horizontal lines that pass through the point $(-7, 5)$.

(C) Graph the equation $4x + 3y = 12$ by hand and on a graphing utility.

Slope of a Line

If we take two points $P_1(x_1, y_1)$ and $P_2(x_2, y_2)$ on a line, then the ratio of the change in y to the change in x as we move from point P_1 to point P_2 is called the **slope** of the line. Roughly speaking, slope is a measure of the "steepness" of a line. Sometimes the change in x is called the **run** and the change in y the **rise.**

SLOPE OF A LINE

If a line passes through two distinct points $P_1(x_1, y_1)$ and $P_2(x_2, y_2)$, then its slope m is given by the formula

$$m = \frac{y_2 - y_1}{x_2 - x_1} \qquad x_1 \neq x_2$$

$$= \frac{\text{Vertical change (rise)}}{\text{Horizontal change (run)}}$$

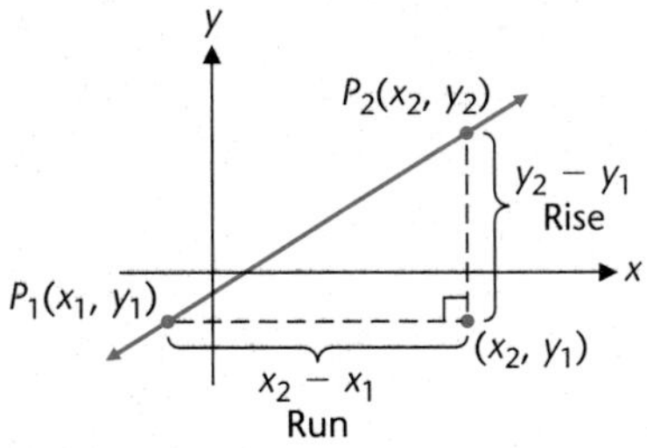

For a horizontal line, y doesn't change as x changes; hence, its slope is 0. For a vertical line, x doesn't change as y changes; hence, $x_1 = x_2$, the denominator in the slope formula is 0, and its slope is not defined. In general, the slope of a line may be positive, negative, zero, or not defined. Each case is illustrated geometrically as shown in Table 1.

TABLE 1 Geometric Interpretation of Slope

Line	Slope	Example
Rising as x moves from left to right	Positive	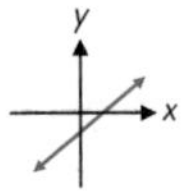
Falling as x moves from left to right	Negative	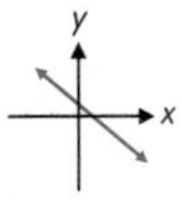
Horizontal	0	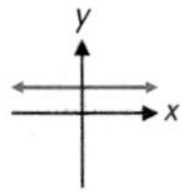
Vertical	Not defined	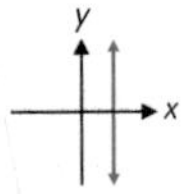

In using the formula to find the slope of the line through two points, it doesn't matter which point is labeled P_1 or P_2, since changing the labeling will change the sign in both the numerator and denominator of the slope formula:

$$\frac{y_2 - y_1}{x_2 - x_1} = \frac{y_1 - y_2}{x_1 - x_2}$$

FIGURE 5

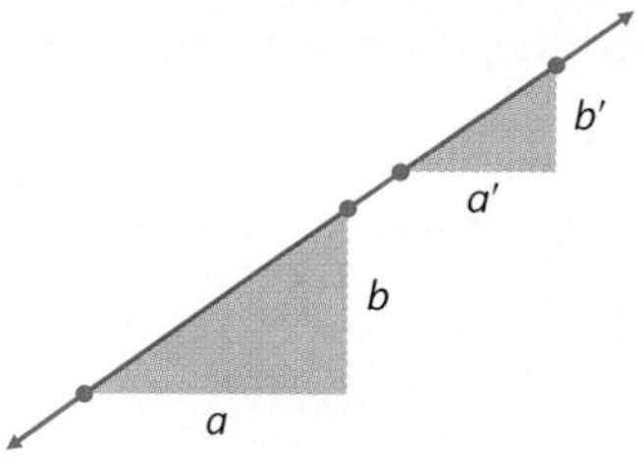

$$m = \frac{b}{a} = \frac{b'}{a'}$$

For example,

$$\frac{5-2}{7-3} = \frac{2-5}{3-7}$$

In addition, it is important to note that the definition of slope doesn't depend on the two points chosen on the line as long as they are distinct. This follows from the fact that the ratios of corresponding sides of similar triangles are equal (see Fig. 5).

EXAMPLE 3 Finding Slopes

Sketch a line through each pair of points and find the slope of each line.

(A) $(-3, -4), (3, 2)$ (B) $(-2, 3), (1, -3)$

(C) $(-4, 2), (3, 2)$ (D) $(2, 4), (2, -3)$

Solutions

(A)

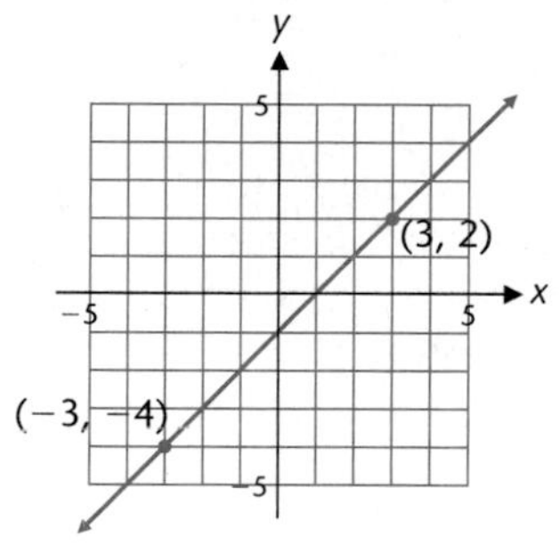

$$m = \frac{2-(-4)}{3-(-3)} = \frac{6}{6} = 1$$

(B)

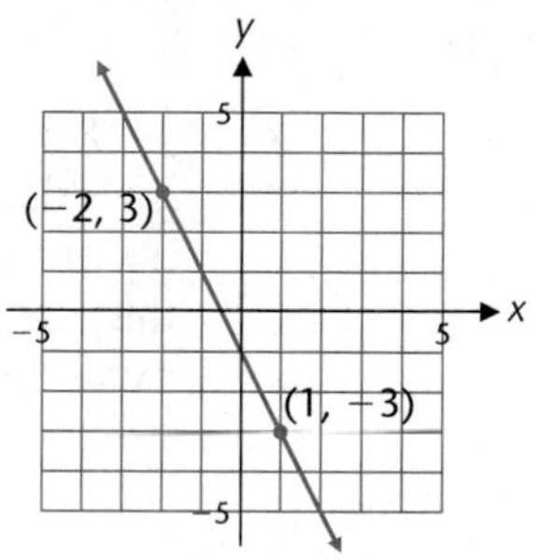

$$m = \frac{-3-3}{1-(-2)} = \frac{-6}{3} = -2$$

(C)

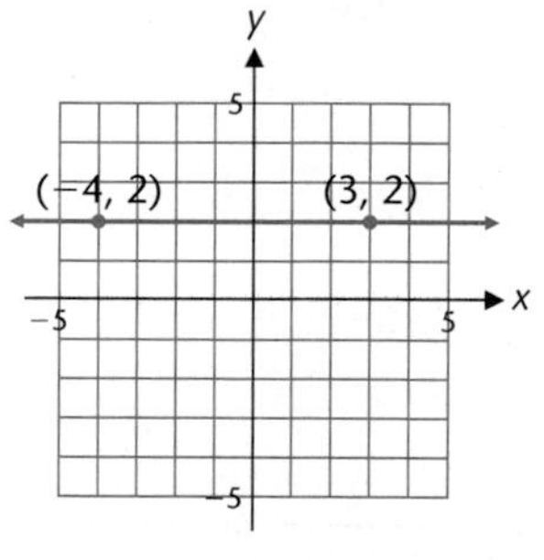

$$m = \frac{2-2}{3-(-4)} = \frac{0}{7} = 0$$

(D)

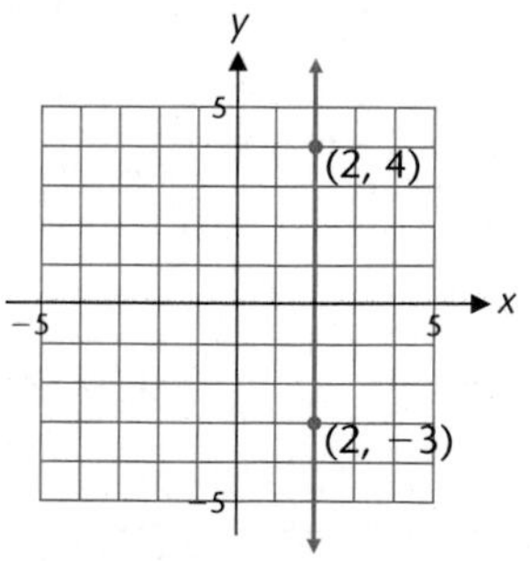

$$m = \frac{-3-4}{2-2} = \frac{-7}{0}$$

slope is not defined

MATCHED PROBLEM 3

Find the slope of the line through each pair of points. Do not graph.

(A) $(-3, -3), (2, -3)$ (B) $(-2, -1), (1, 2)$

(C) $(0, 4), (2, -4)$ (D) $(-3, 2), (-3, -1)$

Equations of Lines—Special Forms

Let us start by investigating why $y = mx + b$ is called the *slope–intercept form* for a line.

Explore/Discuss 4

(A) Using a graphing utility, graph $y = x + b$ for $b = -5, -3, 0, 3,$ and 5 simultaneously in a standard viewing window. Verbally describe the geometric significance of b.

(B) Using a graphing utility, graph $y = mx - 1$ for $m = -2, -1, 0, 1,$ and 2 simultaneously in a standard viewing window. Verbally describe the geometric significance of m.

As you see, constants m and b in $y = mx + b$ have special geometric significance, which we now explicitly state.

If we let $x = 0$, then $y = b$ and the graph of $y = mx + b$ crosses the y axis at $(0, b)$. Thus, the constant b is the y intercept. For example, the y intercept of the graph of $y = 2x - 7$ is -7.

We have already seen that the point $(0, b)$ is on the graph of $y = mx + b$. If we let $x = 1$, then it follows that the point $(1, m + b)$ is also on the graph. Since the graph of $y = mx + b$ is a line, we can use these two points to compute the slope:

$$\text{Slope} = \frac{y_2 - y_1}{x_2 - x_1} = \frac{(m + b) - b}{1 - 0} = m \qquad \begin{aligned}(x_1, y_1) &= (0, b)\\ (x_2, y_2) &= (1, m + b)\end{aligned}$$

Thus, m is the slope of the line with equation $y = mx + b$.

SLOPE–INTERCEPT FORM

The equation

$$y = mx + b \qquad m = \text{slope},\ b = y \text{ intercept}$$

is called the **slope–intercept form** of the equation of a line.

EXAMPLE 4 Using the Slope–Intercept Form

Sketch the graph of a line with y intercept -2 and slope $\frac{5}{4}$. Then write the equation of the line and graph on a graphing utility in a squared viewing window.

Solution If we start at the point $(0, -2)$ and move 4 units to the right (run), then the y coordinate of a point on the line must move up 5 units (rise) to the point $(4, 3)$. Drawing a line through these two points produces the graph shown in Figure 6(a). The equation of a line with y intercept -2 and slope $\frac{5}{4}$ is

$$y = \frac{5}{4}x - 2$$

Graphing this equation on a graphing utility produces the graph in Figure 6(b).

FIGURE 6

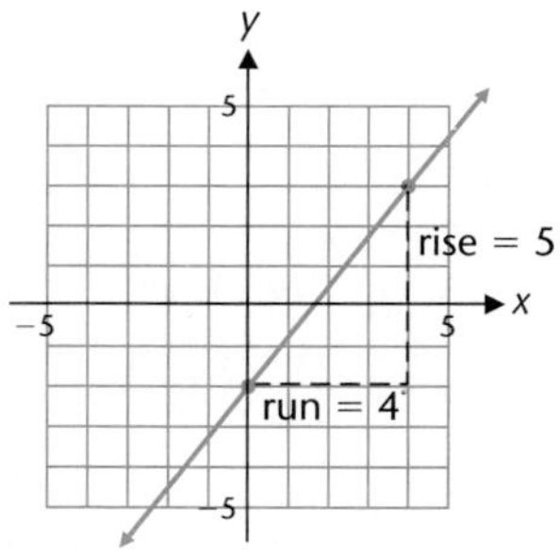

(a) By hand

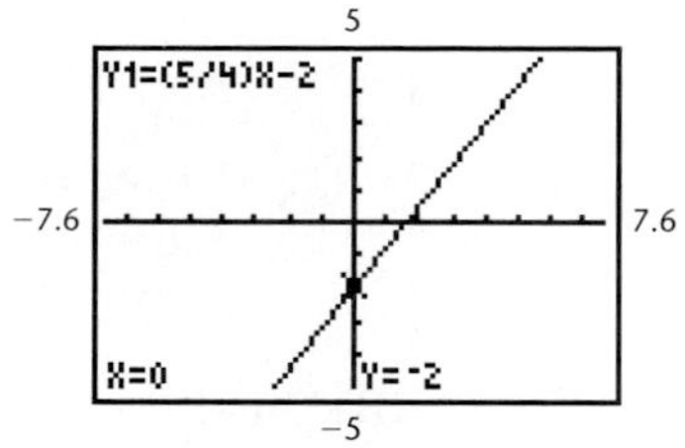

(b) On a graphing utility

MATCHED PROBLEM **4**

Sketch the graph of a line with y intercept 3 and slope $-\frac{3}{4}$. Then write the equation of the line and graph on a graphing utility in a squared viewing window.

FIGURE 7

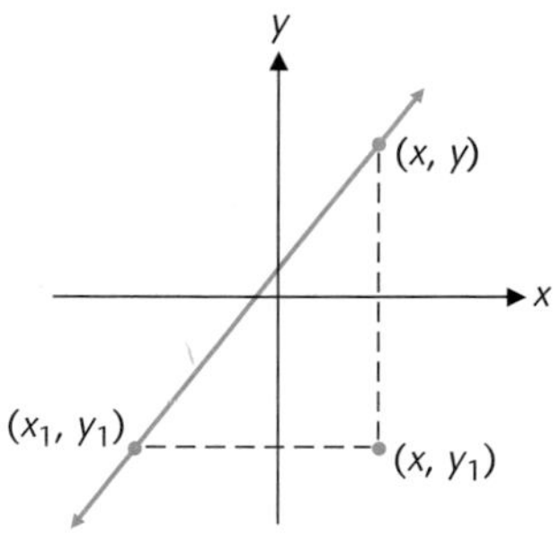

Suppose a line has slope m and passes through the point (x_1, y_1). If (x, y) is any other point on the line (Fig. 7), then

$$\frac{y - y_1}{x - x_1} = m$$

that is,

$$y - y_1 = m(x - x_1) \tag{6}$$

Since the point (x_1, y_1) also satisfies equation (6), we can conclude that equation (6) is an equation of a line with slope m that passes through (x_1, y_1).

POINT–SLOPE FORM

An equation of a line with slope m that passes through (x_1, y_1) is

$$y - y_1 = m(x - x_1)$$

which is called the **point–slope form** of an equation of a line.

If we are given the coordinates of two points on a line, we can use the given coordinates to find the slope and then use the point–slope form with either of the given points to find the equation of the line.

EXAMPLE **5**

Point–Slope Form

(A) Find an equation for the line that has slope $\frac{2}{3}$ and passes through the point $(-2, 1)$. Write the final answer in the form $Ax + By = C$.

(B) Find an equation for the line that passes through the two points $(4, -1)$ and $(-8, 5)$. Write the final answer in the form $y = mx + b$.

Solutions

(A) If $m = \frac{2}{3}$ and $(x_1, y_1) = (-2, 1)$, then

$$y - y_1 = m(x - x_1)$$

$$y - 1 = \frac{2}{3}[x - (-2)]$$

$$3y - 3 = 2x + 4$$

$$-2x + 3y = 7 \quad \text{or} \quad 2x - 3y = -7$$

(B) First use the slope formula to find the slope of the line:

$$m = \frac{y_2 - y_1}{x_2 - x_1} = \frac{5 - (-1)}{-8 - 4} = \frac{6}{-12} = -\frac{1}{2}$$

Now we choose $(x_1, y_1) = (4, -1)$ and proceed as in part A:

$$y - y_1 = m(x - x_1)$$

$$y - (-1) = -\frac{1}{2}(x - 4)$$

$$y + 1 = -\frac{1}{2}x + 2$$

$$y = -\frac{1}{2}x + 1$$

Verify that choosing $(x_1, y_1) = (-8, 5)$, the other given point, produces the same equation.

MATCHED PROBLEM 5

(A) Find an equation for the line that has slope $-\frac{2}{5}$ and passes through the point $(3, -2)$. Write the final answer in the form $Ax + By = C$.

(B) Find an equation for the line that passes through the two points $(-3, 1)$ and $(7, -3)$. Write the final answer in the form $y = mx + b$.

The various forms of the equation of a line that we have discussed are summarized in Table 2 for convenient reference.

TABLE 2

Equations of a Line

Standard Form	$Ax + By = C$	A and B not both 0
Slope–Intercept Form	$y = mx + b$	Slope: m; y intercept: b
Point–Slope Form	$y - y_1 = m(x - x_1)$	Slope: m; point: (x_1, y_1)
Horizontal Line	$y = b$	Slope: 0
Vertical Line	$x = a$	Slope: undefined

Parallel and Perpendicular Lines

Explore/Discuss 5

(A) Graph all of the following lines in the same viewing window. Discuss the relationship between these graphs and the slopes of the lines.

$$y = 2x - 5 \qquad y = 2x - 1 \qquad y = 2x + 3$$

(B) Graph each pair of lines in the same *squared* viewing window. Discuss the relationship between each pair of lines and their respective slopes.

$$y = 2x \quad \text{and} \quad y = -0.5x$$

$$y = -3x \quad \text{and} \quad y = \frac{1}{3}x$$

$$y = \frac{4}{5}x \quad \text{and} \quad y = -\frac{5}{4}x$$

From geometry, we know that two vertical lines are parallel and that a horizontal line and a vertical line are perpendicular to each other. How can we tell when two nonvertical lines are parallel or perpendicular to each other? Theorem 2, which we state without proof, provides a convenient test.

THEOREM 2

PARALLEL AND PERPENDICULAR LINES

Given two nonvertical lines L_1 and L_2, with slopes m_1 and m_2, respectively, then

$$L_1 \parallel L_2 \quad \text{if and only if } m_1 = m_2$$

$$L_1 \perp L_2 \quad \text{if and only if } m_1 m_2 = -1$$

The symbols $\parallel$ and $\perp$ mean, respectively, "is parallel to" and "is perpendicular to." In the case of perpendicularity, the condition $m_1 m_2 = -1$ can also be written as

$$m_2 = -\frac{1}{m_1} \quad \text{or} \quad m_1 = -\frac{1}{m_2}$$

Thus:

Two nonvertical lines are perpendicular if and only if their slopes are the negative reciprocals of each other.

EXAMPLE 6 **Parallel and Perpendicular Lines**

Given the line L with equation $3x - 2y = 5$ and the point P with coordinates $(-3, 5)$, find an equation of a line through P that is

(A) Parallel to L (B) Perpendicular to L

Solution First we write the equation for L in the slope–intercept form to find the slope of L:

$$3x - 2y = 5$$
$$-2y = -3x + 5$$
$$y = \tfrac{3}{2}x - \tfrac{5}{2}$$

Thus, the slope of L is $\frac{3}{2}$. The slope of a line parallel to L will also be $\frac{3}{2}$, and the slope of a line perpendicular to L will be $-\frac{2}{3}$. We now can find the equations of the two lines in parts A and B using the point–slope form.

(A) Parallel ($m = \frac{3}{2}$):

$$y - y_1 = m(x - x_1)$$
$$y - 5 = \tfrac{3}{2}(x + 3)$$
$$y - 5 = \tfrac{3}{2}x + \tfrac{9}{2}$$
$$y = \tfrac{3}{2}x + \tfrac{19}{2}$$

(B) Perpendicular ($m = -\frac{2}{3}$):

$$y - y_1 = m(x - x_1)$$
$$y - 5 = -\tfrac{2}{3}(x + 3)$$
$$y - 5 = -\tfrac{2}{3}x - 2$$
$$y = -\tfrac{2}{3}x + 3$$

MATCHED PROBLEM 6 Given the line L with equation $4x + 2y = 3$ and the point P with coordinates $(2, -3)$, find an equation of a line through P that is

(A) Parallel to L (B) Perpendicular to L

Application: Linear Regression

Regression analysis was introduced in the group activity for Chapter 1. The next example shows how the properties of linear functions developed in this section can be combined with linear regression techniques to analyze data in an applied setting.

EXAMPLE 7 **Cost Analysis**

A street corner vendor sells hot dogs from a pushcart. The vendor's daily costs (in dollars) at various sales levels are given in Table 3.

(A) Use regression analysis on a graphing utility to find a linear daily cost function $C(x) = ax + b$ that models the data in Table 3. (Round the constants a and b to two decimal places.)

(B) Discuss the interpretation of the constants a and b in terms of this data.

(C) Use the linear regression function (without rounding the constants a and b) to estimate the daily costs (to the nearest cent) on a day when 125 hot dogs are sold.

TABLE 3

Hot Dogs Sold	Daily Costs ($)
53	153.47
61	191.34
97	269.81
109	308.25
165	335.63

Solution (A) Figure 8(a) shows the data from Table 3 entered in a graphing utility and Figure 8(b) shows the resulting linear regression equation. Rounding the constants to two decimal places, we conclude that the linear regression equation for this data is

$$C(x) = 1.61x + 95.52 \tag{7}$$

Figure 8(c) shows the regression equation transferred to the equation editor and Figure 8(d) shows a graph of the data and the regression equation. It is clear from Figure 8(d) that there is no one line that will pass through all five data points, but that all five points are clustered around the graph of the regression equation. The linear regression equation is generally accepted as the "best" linear equation for modeling a given set of data.

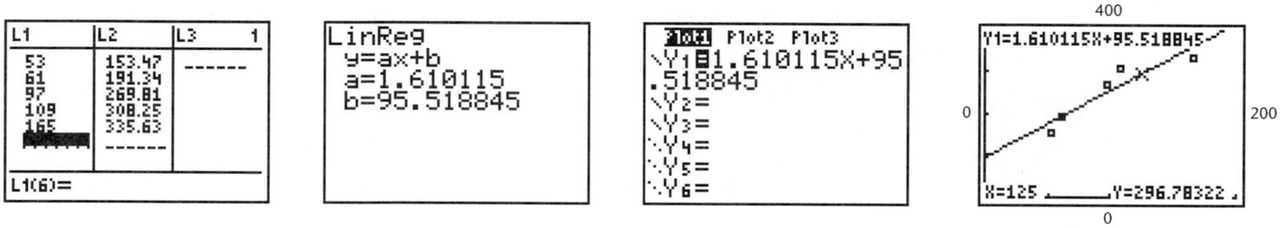

(a) Data

(b) Regression equation

(c) Regression equation transferred to equation editor

(d) Graph of data and regression equation

FIGURE 8

(B) The constant $b = 95.52$ is the y intercept. Thus, if no hot dogs are sold, the daily costs are $C(0) = \$95.52$. These are referred to as the vendor's **fixed costs,** that is, costs that do not depend on the number of hot dogs sold. Fixed costs can include permit fees, pushcart rental fees, insurance, and the like. The term $1.61x$ represents the **variable costs,** that is, costs that do depend on the number of hot dogs sold. Variable costs can include the cost of hot dogs, buns, condiments, and so on. The constant $a = 1.61$ is the slope. Each time an additional hot dog is sold, x increases by 1 (run) and daily costs increase by \$1.61 (rise). Thus, selling one more hot dog increases the daily cost by \$1.61. This quantity is often referred to as the **marginal cost.** (Marginal costs for nonlinear cost functions are more complicated to analyze and are usually discussed in more advanced courses.)

(C) The daily costs for selling 125 hot dogs are given by $C(125)$. If we evaluate this on the graphing utility, using the six-decimal place constants a and b stored in the calculator, we get $C(125) = \$296.78$ [see Figs. 8(c) and 8(d)]. On the other hand, if we use the rounded constants as shown in equation (7), we get $C(125) = \$296.77$. Although the difference is not very significant, it can cause confusion, especially when you compare your answer with one in the back of the book. Thus, whenever possible, we will use the unrounded constants, as stored in the graphing utility, to calculate quantities related to a regression equation and will only round the final answer.

MATCHED PROBLEM 7

Another street corner vendor sells tacos from a pushcart. The vendor's daily costs (in dollars) at various sales levels are given in Table 4.

(A) Use regression analysis on a graphing utility to find a linear daily cost function $C(x) = ax + b$ that models the data in Table 3. (Round the constants a and b to two decimal places.)

(B) Discuss the interpretation of the constants a and b in terms of this data.

(C) Use the linear regression function (without rounding the constants a and b) to estimate the daily costs (to the nearest cent) on a day when 90 tacos are sold.

TABLE 4

Tacos Sold	Daily Costs ($)
60	167.55
116	228.64
128	264.90
164	283.42
172	321.79

Answers to Matched Problems

1. x intercept: $\frac{15}{4} = 3.75$; y intercept: 5

2. (A) The graph of $x = 4$ is a vertical line with x intercept 4. The graph of $y = -3$ is a horizontal line with y intercept -3

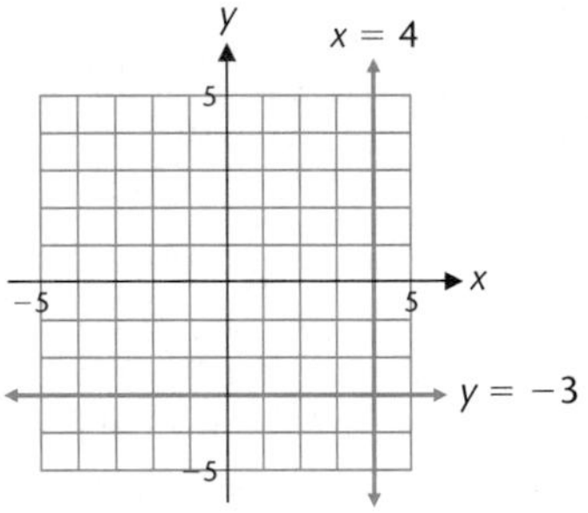

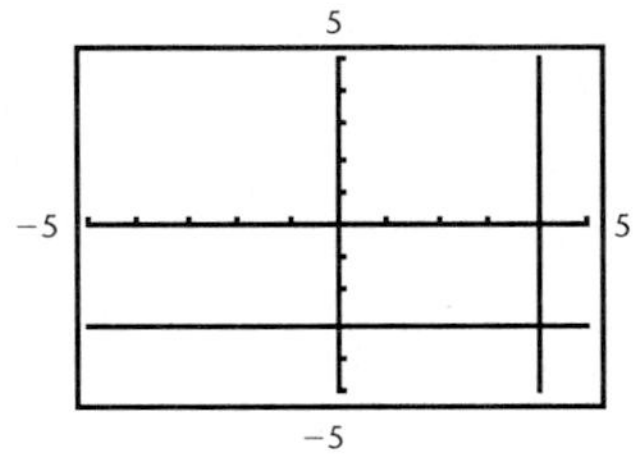

(B) Vertical: $x = -7$; horizontal: $y = 5$

(C)

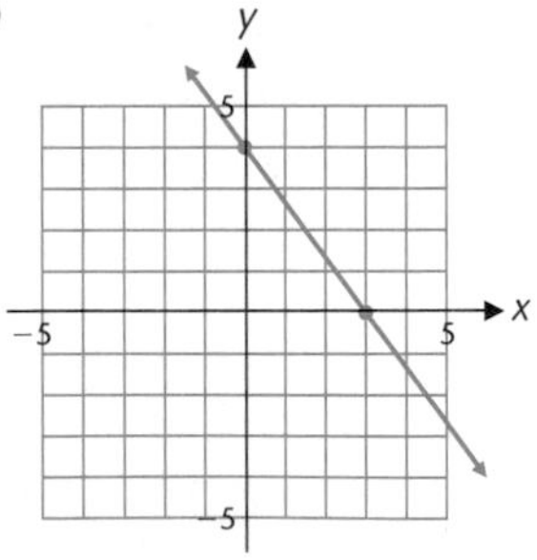

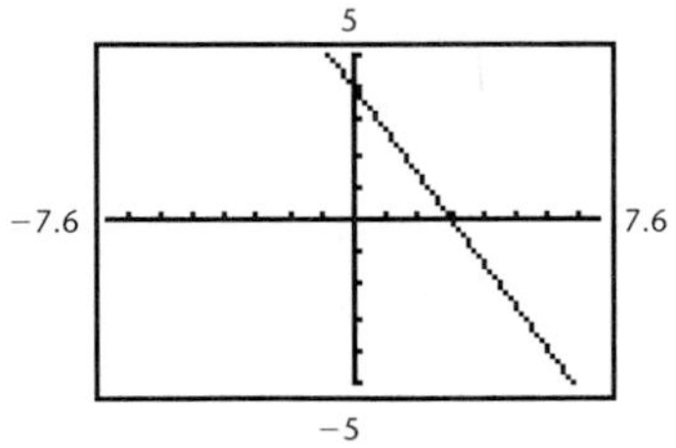

3. (A) $m = 0$ (B) $m = 1$ (C) $m = -4$ (D) m is not defined

4. $y = -\frac{3}{4}x + 3$

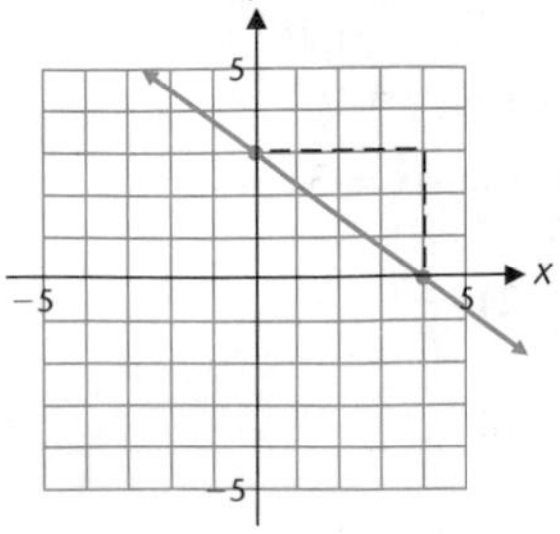

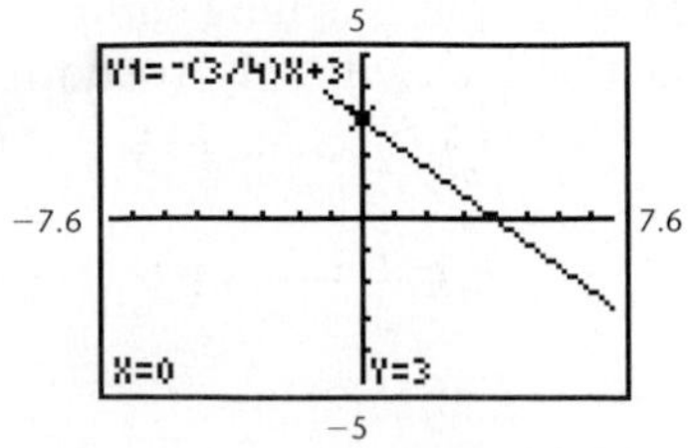

5. (A) $2x + 5y = -4$ (B) $y = -\frac{2}{5}x - \frac{1}{5}$ **6.** (A) $y = -2x + 1$ (B) $y = \frac{1}{2}x - 4$

7. (A) $C(x) = 1.28x + 89.66$

(B) The fixed costs are \$89.66, the variable costs are $1.28x$, and the cost of selling an additional taco is \$1.28.

(C) $C(90) = \$204.69$

EXERCISE 2-1

A

In Problems 1–6, use the graph of each linear function to find the x intercept, y intercept, and slope.

1.

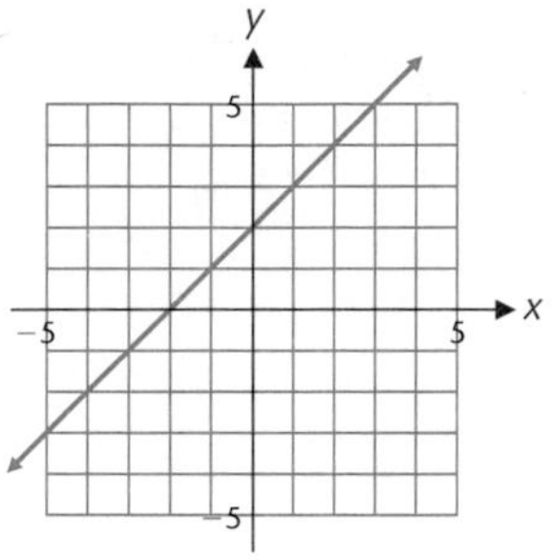

2.

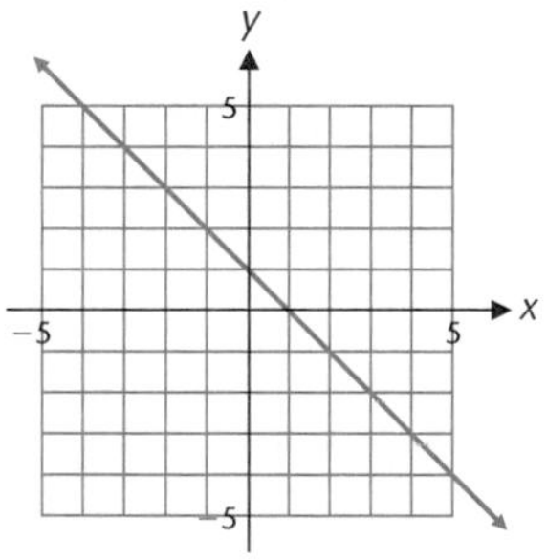

3.

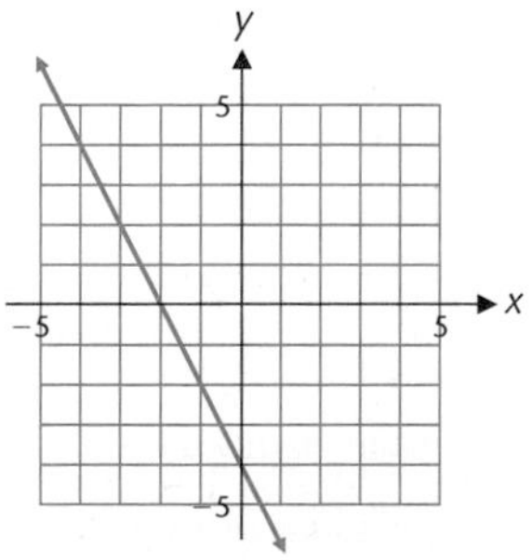

4.

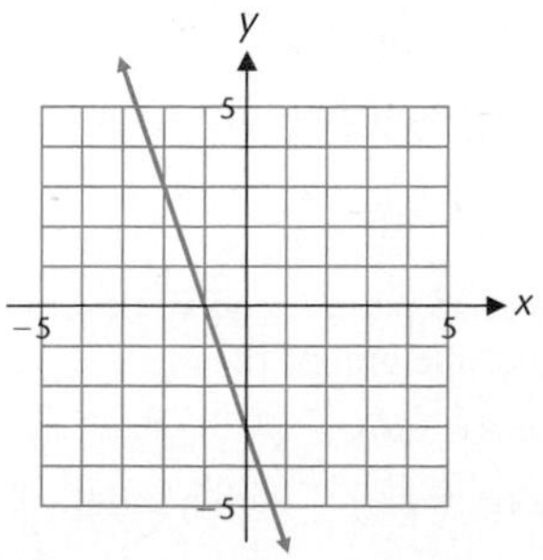

5.

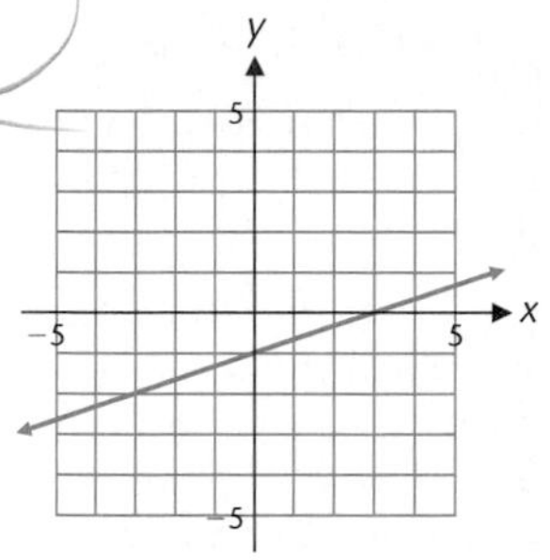

6.

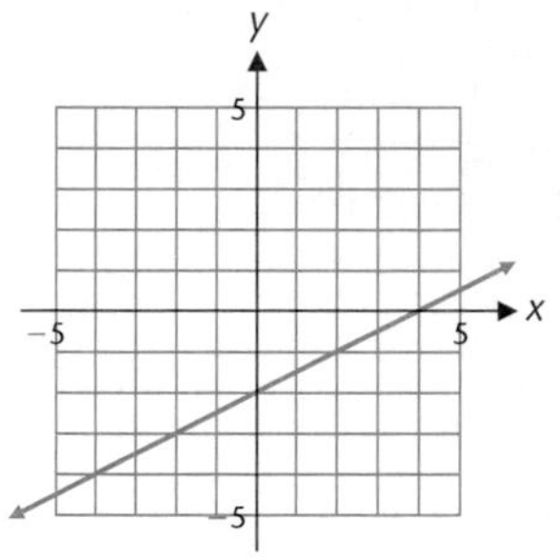

Which equations in Problems 7–16 define a linear function?

7. $y = 2x^2$

8. $y = 5 - 3x^3$

9. $y = 7 - \frac{1}{2}x$

10. $y = -2 + \sqrt{7}x$

11. $y = \dfrac{x - 5}{3}$

12. $y = \dfrac{3 - x}{2}$

13. $y = \frac{2}{3}(x - 7) - \frac{1}{2}(3 - x)$

14. $y = -\frac{1}{5}(2 - 3x) + \frac{2}{7}(x + 8)$

15. $y = \dfrac{3}{x - 5}$

16. $y = \dfrac{2}{3 - x}$

In Problems 17–28, find the x intercept, y intercept, and slope, if they exist, and graph each equation.

17. $y = -\frac{3}{5}x + 4$

18. $y = -\frac{3}{2}x + 6$

19. $y = -\frac{3}{4}x$

20. $y = \frac{2}{3}x - 3$

21. $2x - 3y = 15$

22. $4x + 3y = 24$

23. $\dfrac{y}{8} - \dfrac{x}{4} = 1$

24. $\dfrac{y}{6} - \dfrac{x}{5} = 1$

25. $x = -3$

26. $y = -2$

27. $y = 3.5$

28. $x = 2.5$

In Problems 29–32, write the slope–intercept form of the equation of the line with indicated slope and y intercept.

29. Slope $= 1$; y intercept $= 0$

30. Slope $= -1$; y intercept $= 7$

31. Slope $= -\frac{2}{3}$; y intercept $= -4$

32. Slope $= \frac{5}{3}$; y intercept $= 6$

B

In Problems 33–46, sketch a graph of the line that contains the indicated point(s) and/or has the indicated slope and/or has the indicated intercepts. Then write the equation of the line in the slope–intercept form $y = mx + b$ or in the form $x = c$ and check by graphing the equation on a graphing utility.

33. $(0, 4)$; $m = -3$

34. $(2, 0)$; $m = 2$

35. $(-5, 4)$; $m = -\frac{2}{5}$

36. $(-4, -2)$; $m = \frac{1}{2}$

37. $(1, 6)$; $(5, -2)$

38. $(-3, 4)$; $(6, 1)$

39. $(-4, 8)$; $(2, 0)$

40. $(2, -1)$; $(10, 5)$

41. $(-3, 4)$; $(5, 4)$

42. $(0, -2)$; $(4, -2)$

43. $(4, 6)$; $(4, -3)$

44. $(-3, 1)$; $(-3, -4)$

45. x intercept -4; y intercept 3

46. x intercept -4; y intercept -5

In Problems 47–58, write an equation of the line that contains the indicated point and meets the indicated condition(s). Write the final answer in standard form $Ax + By = C$, $A \geq 0$.

47. $(-3, 4)$; parallel to $y = 3x - 5$

48. $(-4, 0)$; parallel to $y = -2x + 1$

49. $(2, -3)$; perpendicular to $y = -\frac{1}{3}x$

50. $(-2, -4)$; perpendicular to $y = \frac{2}{3}x - 5$

51. $(2, 5)$; parallel to y axis

52. $(7, 3)$; parallel to x axis

53. $(3, -2)$; vertical

54. $(-2, -3)$; horizontal

55. $(5, 0)$; parallel to $3x - 2y = 4$

56. $(3, 5)$; parallel to $3x + 4y = 8$

57. $(0, -4)$; perpendicular to $x + 3y = 9$

58. $(-2, 4)$; perpendicular to $4x + 5y = 0$

59. Discuss the relationship between the graphs of the lines with equation $y = mx + 2$, where m is any real number.

60. Discuss the relationship between the graphs of the lines with equation $y = -0.5x + b$, where b is any real number.

Problems 61–66 refer to the quadrilateral with vertices $A(0, 2)$, $B(4, -1)$, $C(1, -5)$, and $D(-3, -2)$.

61. Show that $AB \parallel DC$.

62. Show that $DA \parallel CB$.

63. Show that $AB \perp BC$.

64. Show that $AD \perp DC$.

65. Find an equation of the perpendicular bisector of AD. [*Hint:* First find the midpoint of AD. (See Problem 47 in Exercise 1-1.)]

66. Find an equation of the perpendicular bisector of AB.

C∫ *Problems 67–72 are calculus-related. Recall that a line tangent to a circle at a point is perpendicular to the radius drawn to that point (see the figure). Find the equation of the line tangent to the circle at the indicated point. Write the final answer in the standard form $Ax + By = C$, $A \geq 0$. Graph the circle and the tangent line on the same coordinate system.*

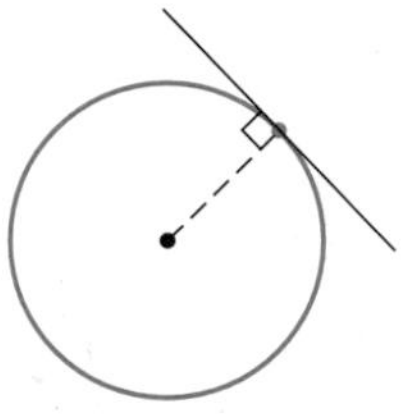

67. $x^2 + y^2 = 25$, $(3, 4)$

68. $x^2 + y^2 = 100$, $(-8, 6)$

69. $x^2 + y^2 = 50$, $(5, -5)$

70. $x^2 + y^2 = 80$, $(-4, -8)$

71. $(x - 3)^2 + (y + 4)^2 = 169$, $(8, -16)$

72. $(x + 5)^2 + (y - 9)^2 = 289$, $(-13, -6)$

C

73. (A) Graph the following equations in a squared viewing window:

$$3x + 2y = 6 \qquad 3x + 2y = 3$$
$$3x + 2y = -6 \qquad 3x + 2y = -3$$

(B) From your observations in part A, describe the family of lines obtained by varying C in $Ax + By = C$ while holding A and B fixed.

(C) Verify your conclusions in part B with a proof.

74. (A) Graph the following two equations in a squared viewing window:

$$3x + 4y = 12 \qquad 4x - 3y = 12$$

(B) Graph the following two equations in a squared viewing window:

$$2x + 3y = 12 \qquad 3x - 2y = 12$$

(C) From your observations in parts A and B, describe the apparent relationship of the graphs of

$$Ax + By = C \text{ and } Bx - Ay = C.$$

(D) Verify your conclusions in part C with a proof.

75. Describe the relationship between the graphs of $f(x) = mx + b$ and $g(x) = |mx + b|$, $m \neq 0$, and illustrate with examples. Is $g(x)$ always, sometimes, or never a linear function?

76. Describe the relationship between the graphs of $f(x) = mx + b$ and $g(x) = m|x| + b$, $m \neq 0$, and illustrate with examples. Is $g(x)$ always, sometimes, or never a linear function?

77. Prove that if a line L has x intercept $(a, 0)$ and y intercept $(0, b)$, then the equation of L can be written in the **intercept form**

$$\frac{x}{a} + \frac{y}{b} = 1 \qquad a, b \neq 0$$

78. Let

$$P_1(x_1, y_1) = P_1(x_1, mx_1 + b)$$
$$P_2(x_2, y_2) = P_2(x_2, mx_2 + b)$$
$$P_3(x_3, y_3) = P_3(x_3, mx_3 + b)$$

be three arbitrary points that satisfy $y = mx + b$ with $x_1 < x_2 < x_3$. Show that P_1, P_2, and P_3 are **collinear;** that is, they lie on the same line. [*Hint:* Use the distance formula and show that $d(P_1, P_2) + d(P_2, P_3) = d(P_1, P_3)$.] This proves that the graph of $y = mx + b$ is a straight line.

APPLICATIONS

79. **Physics.** The two temperature scales Fahrenheit (F) and Celsius (C) are linearly related. It is known that water freezes at 32°F or 0°C and boils at 212°F or 100°C.

(A) Find a linear equation that expresses F in terms of C.

(B) If a European family sets its house thermostat at 20°C, what is the setting in degrees Fahrenheit? If the outside temperature in Milwaukee is 86°F, what is the temperature in degrees Celsius?

(C) What is the slope of the graph of the linear equation found in part A? [The slope indicates the change in Fahrenheit degrees per unit change in Celsius degrees.]

80. **Physics.** Hooke's law states that the relationship between the stretch s of a spring and the weight w causing the stretch is linear (a principle on which all spring scales are constructed). For a particular spring, a 5-pound weight causes a stretch of 2 inches, while with no weight the stretch of the spring is 0.

(A) Find a linear equation that expresses s in terms of w.

(B) What weight will cause a stretch of 3.6 inches?

(C) What is the slope of the graph of the equation? [The slope indicates the amount of stretch per pound increase in weight.]

81. **Business–Depreciation.** A copy machine was purchased by a law firm for \$8,000 and is assumed to have a depreciated value of \$0 after 5 years. The firm takes straight-line depreciation over the 5-year period.

(A) Find a linear equation that expresses value V in dollars in terms of time t in years.

(B) What is the depreciated value after 3 years?

(C) What is the slope of the graph of the equation found in part A? Interpret verbally.

82. **Business–Markup Policy.** A clothing store sells a shirt costing \$20 for \$33 and a jacket costing \$60 for \$93.

(A) If the markup policy of the store for items costing over \$10 is assumed to be linear, write an equation that expresses retail price R in terms of cost C (wholesale price).

(B) What does a store pay for a suit that retails for \$240?

(C) What is the slope of the graph of the equation found in part A? Interpret verbally.

83. **Flight Conditions.** In stable air, the air temperature drops about 5°F for each 1,000-foot rise in altitude.

(A) If the temperature at sea level is 70°F and a commercial pilot reports a temperature of −20°F at 18,000 feet, write a linear equation that expresses temperature T in terms of altitude A (in thousands of feet).

(B) How high is the aircraft if the temperature is 0°F?

(C) What is the slope of the graph of the equation found in part A? Interpret verbally.

★ 84. **Flight Navigation.** An airspeed indicator on some aircraft is affected by the changes in atmospheric pressure at different altitudes. A pilot can estimate the true airspeed by observing the indicated airspeed and adding to it about 2% for every 1,000 feet of altitude.

(A) If a pilot maintains a constant reading of 200 miles per hour on the airspeed indicator as the aircraft climbs from sea level to an altitude of 10,000 feet, write a linear equation that expresses true airspeed T (miles per hour) in terms of altitude A (thousands of feet).

(B) What would be the true airspeed of the aircraft at 6,500 feet?

(C) What is the slope of the graph of the equation found in part A? Interpret verbally.

★ 85. **Oceanography.** After about 9 hours of a steady wind, the height of waves in the ocean is approximately linearly related to the duration of time the wind has been blowing. During a storm with 50-knot winds, the wave height after 9 hours was found to be 23 feet, and after 24 hours it was 40 feet.

(A) If t is time after the 50-knot wind started to blow and h is the wave height in feet, write a linear equation that expresses height h in terms of time t.

(B) How long will the wind have been blowing for the waves to be 50 feet high?

Express all calculated quantities to three significant digits.

86. Oceanography. As a diver descends into the ocean, pressure increases linearly with depth. The pressure is 15 pounds per square inch on the surface and 30 pounds per square inch 33 feet below the surface.

(A) If p is the pressure in pounds per square inch and d is the depth below the surface in feet, write an equation that expresses p in terms of d.

(B) How deep can a scuba diver go if the safe pressure for his equipment and experience is 40 pounds per square inch?

★ **87. Medicine.** Cardiovascular research has shown that above the 210 cholesterol level, each 1% increase in cholesterol level increases coronary risk 2%. For a particular age group, the coronary risk at a 210 cholesterol level is found to be 0.160 and at a level of 231 the risk is found to be 0.192.

(A) Find a linear equation that expresses risk R in terms of cholesterol level C.

(B) What is the risk for a cholesterol level of 260?

(C) What is the slope of the graph of the equation found in part A? Interpret verbally.

Express all calculated quantities to three significant digits.

★ **88. Demographics.** The average number of persons per household in the United States has been shrinking steadily for as long as statistics have been kept and is approximately linear with respect to time. In 1900, there were about 4.76 persons per household and in 1990, about 2.5.

(A) If N represents the average number of persons per household and t represents the number of years since 1900, write a linear equation that expresses N in terms of t.

(B) What is the predicted household size in the year 2000?

Express all calculated quantities to three significant digits.

89. Cost Analysis. The management of a company that manufactures surfboards produced the cost data in the table where C is the daily cost (in dollars) for a daily output of x surfboards.

Surfboards x	Daily Costs C ($)
7	940
10	1,780
12	2,280
19	2,660
28	3,930

(A) Use regression analysis to find a linear daily cost function $C(x) = ax + b$ that models the data in the table. (Round the constants a and b to the nearest unit.)

(B) Discuss the interpretation of the constants a and b in terms of this data.

(C) Use the linear regression function (without rounding the constants a and b) to estimate the daily costs (to the nearest dollar) on a day when 25 surfboards are produced.

90. Cost Analysis. The management of a company that manufactures roller skates produced the cost data in the table where C is the daily cost (in dollars) for a daily output of x pairs of skates.

Roller Skates x	Daily Costs C ($)
50	2,250
70	3,375
130	5,400
150	6,125
200	8,150

(A) Use regression analysis to find a linear daily cost function $C(x) = ax + b$ that models the data in the table. (Round the constants a and b to the nearest unit.)

(B) Discuss the interpretation of the constants a and b in terms of this data.

(C) Use the linear regression function (without rounding the constants a and b) to estimate the daily costs (to the nearest dollar) on a day when 175 pairs of skates are produced.

91. Energy Consumption. Analyzing data from the United States Energy Department for the period between 1920 and 1960 reveals that coal consumption as a percentage of all energy consumed (wood, coal, petroleum, natural gas, hydro, and nuclear) decreased almost linearly. Percentages for this period are given in the table.

Year	Consumption (%)
1920	72
1930	60
1940	50
1950	37
1960	22

(A) Use regression analysis to find a linear regression function $f(x)$ for this data, where x is the number of years since 1900.

(B) Use $f(x)$ to estimate (to the nearest 1%) the percentage of coal consumption in 1927. In 1953.

(C) If we assume that $f(x)$ continues to provide a good description of the percentage of coal consumption after 1960, when would $f(x)$ indicate that the percentage of coal consumption has reached 0? Did this really happen? (Consult some reference books if you are not certain.) If not, what are some reasons for the percentage of coal consumption to level off or even to increase at some point in time after 1960?

92. Petroleum Consumption. Analyzing data from the United States Energy Department for the period between 1920 and 1960 reveals that petroleum consumption as a percentage of all energy consumed (wood, coal, petroleum, natural gas, hydro, and nuclear) increased almost linearly. Percentages for this period are given in the table.

Year	Consumption (%)
1920	11
1930	22
1940	29
1950	37
1960	44

(A) Use regression analysis to find a linear regression function $f(x)$ for this data, where x is the number of years since 1900.

(B) Use $f(x)$ to estimate (to the nearest one percent) the percent of petroleum consumption in 1932. In 1956.

(C) If we assume that $f(x)$ continues to provide a good description of the percentage of petroleum consumption after 1960, when would this percentage reach 100%? Is this likely to happen? Explain.

Section 2-2 | Linear Equations and Inequalities

- Solving Linear Equations
- Solving Linear Inequalities
- Solving Equations and Inequalities Involving Absolute Value
- Application

In this section we discuss methods for solving equations and inequalities that involve linear functions. Some problems are best solved using algebraic techniques, while others benefit from a graphical approach. Since graphs often give additional insight into relationships, especially in applications, we will usually emphasize graphical techniques over algebraic methods. But you must be certain to master both. There are problems in this section that can only be solved algebraically. Later we will also encounter problems that can only be solved graphically.

Solving Linear Equations

In the preceding section we found the x intercept of a linear function $f(x) = mx + b$ by solving the equation $f(x) = 0$. Now we want to apply the same ideas to some more complicated equations.

EXAMPLE 1 **Solving an Equation Algebraically**

Solve $5x - 8 = 2x + 1$ and check.

Solution We use the familiar **properties of equality** to transform the given equation into an **equivalent equation** that has an obvious solution (see Section A-8).

$$5x - 8 = 2x + 1 \qquad \text{Original equation}$$

$$5x - 8 + 8 = 2x + 1 + 8 \qquad \text{Add 8 to both sides.}$$

$5x = 2x + 9$	Combine like terms.
$5x - 2x = 2x + 9 - 2x$	Subtract $2x$ from both sides.
$3x = 9$	Combine like terms.
$\frac{3x}{3} = \frac{9}{3}$	Divide both sides by 3.
$x = 3$	Simplify.

The solution set for this last equation is obvious:

Solution set: $\{3\}$

It follows from the properties of equality that $\{3\}$ is also the solution set of all the preceding equations in our solution. [*Note:* If an equation has only one element in its solution set, we generally use the last equation (in this case, $x = 3$) rather than set notation to represent the solution.]

Check

$5x - 8 = 2x + 1$	Original equation
$5(3) - 8 \stackrel{?}{=} 2(3) + 1$	Substitute $x = 3$.
$15 - 8 \stackrel{?}{=} 6 + 1$	Simplify each side.
$7 \stackrel{\checkmark}{=} 7$	A true statement

MATCHED PROBLEM 1

Solve and check: $2x + 1 = 4x + 5$

We can also use a graphing utility to solve equations of this type. From a graphical viewpoint, a solution to an equation of the form $f(x) = g(x)$ is an **intersection point** of the graphs of f and g. Figure 1 shows a graphical solution to Example 1 using a built-in intersection routine. Most graphing utilities have such a routine (consult your owner's manual or the graphing utility manual for this text; see the Preface). If yours does not, then use zoom and trace to approximate intersection points.

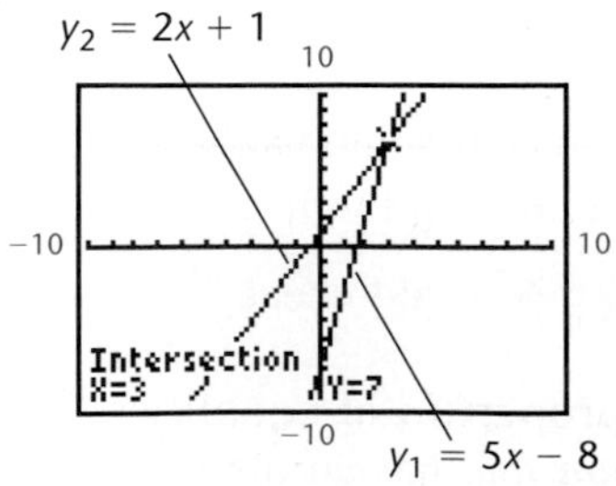

FIGURE 1
Graphical solution of $5x - 8 = 2x + 1$.

Explore/Discuss 1

An equation that is true for all permissible values of the variable is called an **identity.** An equation that is true for some values of the variable and false for others is called a **conditional equation.** Use algebraic and/or graphical techniques to solve each of the following and identify any identities.

(A) $2(x - 4) = 2x - 8$

(B) $2(x - 4) = 3x - 12$

(C) $2(x - 4) = 2x - 12$

EXAMPLE 2 Solving an Equation with a Variable in the Denominator

Solve algebraically and graphically: $\dfrac{7}{2x} - 3 = \dfrac{8}{3} - \dfrac{15}{x}$

Solution We begin with an algebraic solution. Note that 0 must be excluded from the permissible values of x because division by 0 is not permitted. To clear the fractions, we multiply both sides of the equation by $3(2x) = 6x$, the least common denominator (LCD) of all fractions in the equation. (For a discussion of LCDs and how to find them, see Section A-4.)

$$\frac{7}{2x} - 3 = \frac{8}{3} - \frac{15}{x} \qquad x \neq 0$$

$$6x\left(\frac{7}{2x} - 3\right) = 6x\left(\frac{8}{3} - \frac{15}{x}\right) \qquad \text{Multiply by } 6x\text{, the LCD. This and the next step usually can be done mentally.}$$

$$6x \cdot \frac{7}{2x} - 6x \cdot 3 = 6x \cdot \frac{8}{3} - 6x \cdot \frac{15}{x}$$

$$21 - 18x = 16x - 90 \qquad \text{The equation is now free of fractions.}$$

$$-34x = -111$$

$$x = \frac{111}{34} \qquad \text{The solution set is } \{\tfrac{111}{34}\}.$$

FIGURE 2
Graphical solution of $\dfrac{7}{2x} - 3 = \dfrac{8}{3} - \dfrac{15}{x}$.

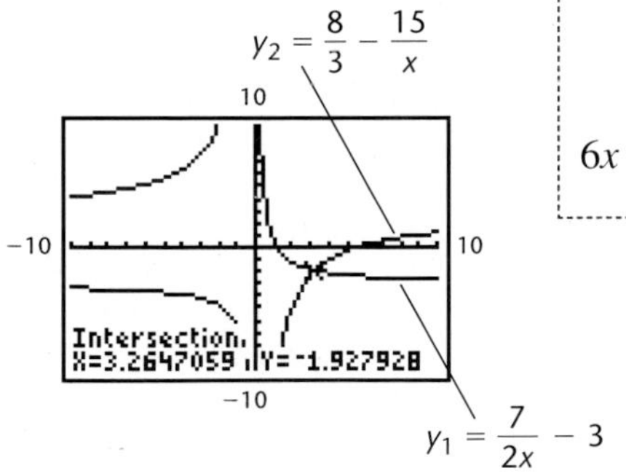

Figure 2 shows the graphical solution. Note that $\frac{111}{34} \approx 3.2647059$, to seven decimal places.

MATCHED PROBLEM 2 Solve algebraically and graphically: $\dfrac{7}{3x} + 2 = \dfrac{1}{x} - \dfrac{3}{5}$

Remark Which solution method should you use—algebraic or graphical? In Example 1, both the algebraic solution and the graphical solution produced the exact solution,

$x = 3$. In Example 2, the algebraic solution again produced the exact solution, $x = 111/34$, while the graphical solution produced $x = 3.2647059$, a seven-decimal-place approximation to the solution. Some like to argue that this makes the algebraic method superior to the graphical method. But exact solutions have little relevance to most applications of mathematics and decimal approximations are usually quite satisfactory.

We encourage you to choose the method that seems best to you, and when possible, use the other method to confirm your answer. In a simple problem, like Example 1, choose either method. In Example 2, we would recommend the algebraic method over the graphical method because of the complexity of the graphs. We have not yet studied graphs of functions involving fractions with x in the denominator. It was a fortunate accident that the intersection point was visible in a standard viewing window.

We frequently encounter equations involving more than one variable. For example, if L and W are the length and width of a rectangle, respectively, the area of the rectangle is given by (see Fig. 3):

FIGURE 3
Area of a rectangle.

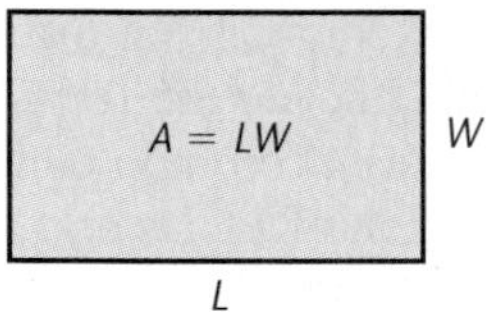

$$A = LW$$

Depending on the situation, we may want to solve this equation for L or W. To solve for W, we simply consider A and L to be constants and W to be a variable. Then the equation $A = LW$ becomes a linear equation in W, which can be solved easily by dividing both sides by L:

$$W = \frac{A}{L} \qquad L \neq 0$$

EXAMPLE 3

Solving an Equation with More than One Variable

Solve for P in terms of the other variables: $A = P + Prt$

Solution

$$A = P + Prt \qquad \text{Think of } A, r, \text{ and } t \text{ as constants.}$$

$$A = P(1 + rt) \qquad \text{Factor to isolate } P.$$

$$\frac{A}{1 + rt} = P \qquad \text{Divide both sides by } 1 + rt.$$

$$P = \frac{A}{1 + rt} \qquad \text{Restriction: } 1 + rt \neq 0$$

MATCHED PROBLEM 3

Solve for r in terms of the other variables: $A = P + Prt$

Solving Linear Inequalities

Now we want to turn our attention to inequalities. Any inequality that can be reduced to one of the four forms in (1) is called a **linear inequality in one variable.**

$$\left.\begin{array}{l} mx + b > 0 \\ mx + b \geq 0 \\ mx + b < 0 \\ mx + b \leq 0 \end{array}\right\} \quad \text{Linear inequalities} \qquad (1)$$

As was the case with equations, the **solution set of an inequality** is the set of all values of the variable that make the inequality a true statement. Each element of the solution set is called a **solution.** Two inequalities are said to be **equivalent** if they have the same solution set.

Explore/Discuss 2

Associated with the linear equation and inequalities

$$3x - 12 = 0 \qquad 3x - 12 < 0 \qquad 3x - 12 > 0$$

is the linear function

$$f(x) = 3x - 12$$

(A) Graph the function f.
(B) From the graph of f determine the values of x for which

$$f(x) = 0 \qquad f(x) < 0 \qquad f(x) > 0$$

(C) How are the answers to part B related to the solutions of

$$3x - 12 = 0 \qquad 3x - 12 < 0 \qquad 3x - 12 > 0$$

As you discovered in Explore/Discuss 2, solving inequalities graphically is both intuitive and efficient. On the other hand, algebraic methods can become quite complicated. So we will emphasize the graphical approach when solving inequalities.

EXAMPLE 4 **Solving a Linear Inequality**

Solve and graph on a number line: $0.5x + 1 \leq 0$

Solution The graph of $f(x) = 0.5x + 1$ is shown in Figure 4. It is clear from the graph that $f(x)$ is negative to the left of $x = -2$ and positive to the right. Thus, the solution set of the inequality

$$0.5x + 1 \leq 0$$

is

$$x \leq -2 \qquad \text{or} \qquad (-\infty, -2]$$

FIGURE 4 $f(x) = 0.5x + 1$.

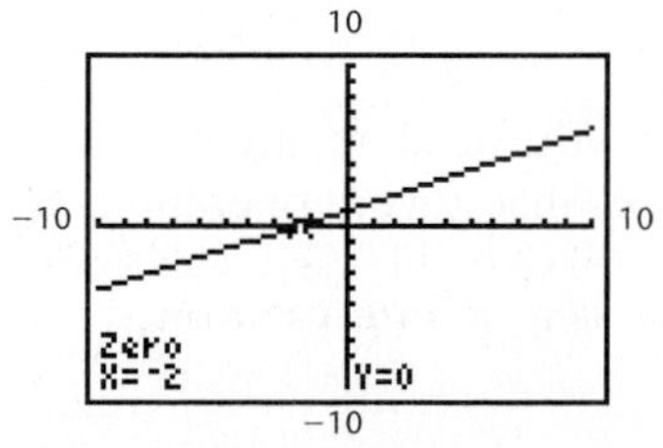

Figure 5 shows a graph of the solution set on a number line. A similar graph can be produced on most graphing utilities by entering $y_1 = 0.5x + 1 \leq 0$ (Fig. 6). The expression $0.5x + 1 \leq 0$ is assigned the value 1 for those values of x that make it a true statement and the value 0 for those values of x that make it a false statement.

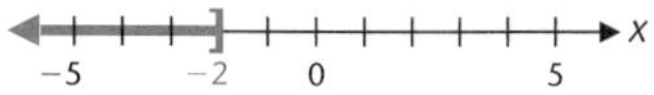

FIGURE 5

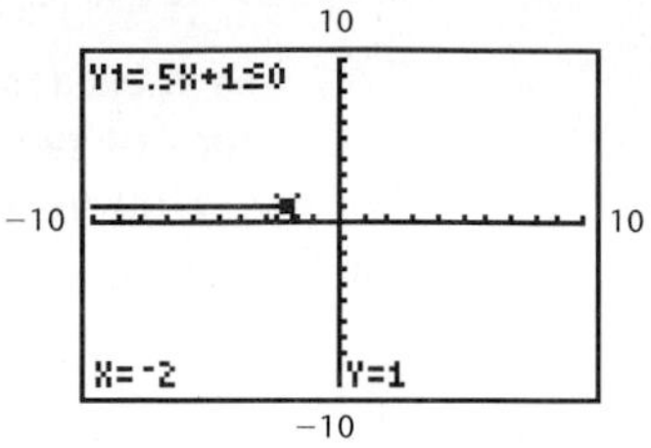

FIGURE 6

MATCHED PROBLEM 4

Solve and graph on a number line: $2x - 6 \geq 0$

Explore/Discuss 3

Associated with the following equations and inequalities

$$-\tfrac{1}{2}x + 4 = x - 2$$
$$-\tfrac{1}{2}x + 4 < x - 2$$
$$-\tfrac{1}{2}x + 4 > x - 2$$

are the two linear functions

$$f(x) = -\tfrac{1}{2}x + 4 \text{ and } g(x) = x - 2$$

(A) Graph both f and g in the same viewing window.

(B) From the graph in part A determine the value(s) of x for which

$$f(x) = g(x) \qquad f(x) < g(x) \qquad f(x) > g(x)$$

(C) How are the answers to part B related to the solutions of

$$-\tfrac{1}{2}x + 4 = x - 2$$
$$-\tfrac{1}{2}x + 4 < x - 2$$
$$-\tfrac{1}{2}x + 4 > x - 2$$

Most inequalities can be solved graphically. If you need to algebraically manipulate an inequality, Theorem 1 lists the properties that govern operations on inequalities.

THEOREM 1

INEQUALITY PROPERTIES

An equivalent inequality will result and the **sense will remain the same,** if each side of the original inequality

1. Has the same real number added to or subtracted from it.
2. Is multiplied or divided by the same positive number.

An equivalent inequality will result and the **sense will reverse,** if each side of the original inequality

3. Is multiplied or divided by the same negative number.

Note: Multiplication by 0 and division by 0 are not permitted.

To gain some experience with these properties, we will solve the next example two ways, algebraically and graphically.

EXAMPLE 5

Solving a Double Inequality

Solve and graph on a number line: $-3 \le 4 - 7x < 18$

Solution

To solve algebraically, we perform operations on the double inequality until we have isolated x in the middle with a coefficient of 1.

$$-3 \le 4 - 7x < 18$$

$$-3 - 4 \le 4 - 7x - 4 < 18 - 4 \qquad \text{Subtract 4 from each member.}$$

$$-7 \le -7x < 14$$

$$\frac{-7}{-7} \ge \frac{-7x}{-7} > \frac{14}{-7} \qquad \text{Divide each member by } -7 \text{ and reverse each inequality.}$$

$$1 \ge x > -2 \quad \text{or} \quad -2 < x \le 1 \quad \text{or} \quad (-2, 1] \tag{2}$$

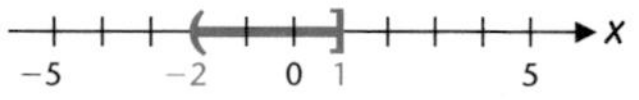

To solve graphically, enter $y_1 = -3$, $y_2 = 4 - 7x$, $y_3 = 18$, graph [Fig. 7(a)], and find the intersection points [Figs. 7(b) and 7(c)]. It is clear from the graph that y_2 is between y_1 and y_3 for x between -2 and 1. Since $y_2 = -3$ at $x = 1$, we include 1 in the solution set, obtaining the same solution, as shown in equation (2).

FIGURE 7

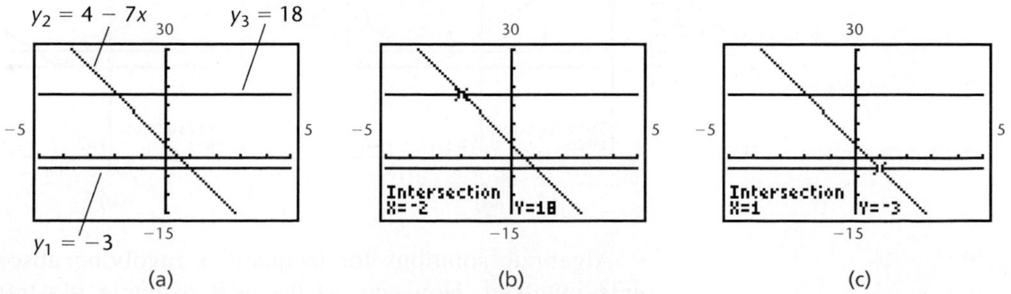

MATCHED PROBLEM 5

Solve algebraically and graphically and graph on a number line: $-3 < 7 - 2x \le 7$

Solving Equations and Inequalities Involving Absolute Value

Explore/Discuss 4

Recall the definition of the absolute value function (see Section 1-4):

$$f(x) = |x| = \begin{cases} -x & \text{if } x < 0 \\ x & \text{if } x \ge 0 \end{cases}$$

(A) Graph the absolute value function $f(x) = |x|$ and the constant function $g(x) = 3$ in the same viewing window.

(B) From the graph in part A, determine the values of x for which

$$|x| < 3 \qquad |x| = 3 \qquad |x| > 3$$

(C) Discuss methods for using the definition of $|x|$ and algebraic techniques to solve part B.

The algebraic solution of an equation or inequality involving the absolute value function usually must be broken down into two or more cases. For example, to solve the equation

$$|x - 4| = 2 \tag{3}$$

we consider two cases:

$$x - 4 = 2 \quad \text{or} \quad x - 4 = -2$$
$$x = 6 \qquad\qquad x = 2$$

We can also solve equation (3) graphically by graphing $y_1 = |x - 4|$ and $y_2 = 2$, and finding their intersection points, as shown in Figure 8.

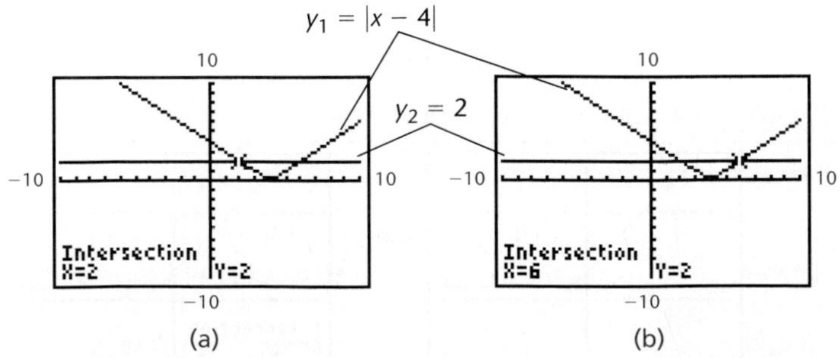

FIGURE 8 Graphical solution of $|x - 4| = 2$.

Algebraic solutions for inequalities involving absolute values can become quite involved. However, as the next example illustrates, even problems that appear to be complicated are easily solved with a graphing utility.

EXAMPLE 6

Solving Absolute Value Problems Graphically

Solve graphically. Write solutions in both inequality and interval notation and graph on a number line.

(A) $|2x - 5| > 4$ (B) $|0.5x + 2| \geq 3x - 5$

Solutions

(A) Graph $y_1 = |2x - 5|$ and $y_2 = 4$ in the same viewing window and find the intersection points (Fig. 9). Examining the graphs in Figure 9, we see if $x < 0.5$ or $x > 4.5$, then the graph of y_1 is above the graph of y_2. Thus, the solution is

$x < 0.5$ or $x > 4.5$ Inequality notation

$(-\infty, 0.5) \cup (4.5, \infty)$* Interval notation

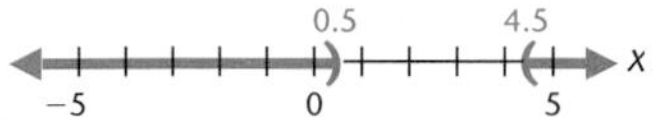

FIGURE 9

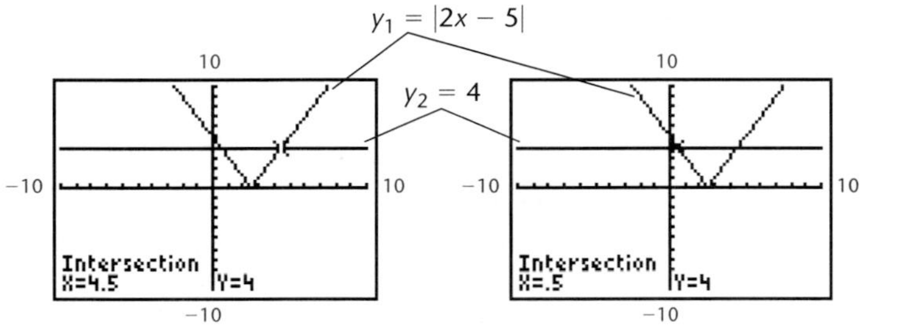

(a) (b)

FIGURE 10

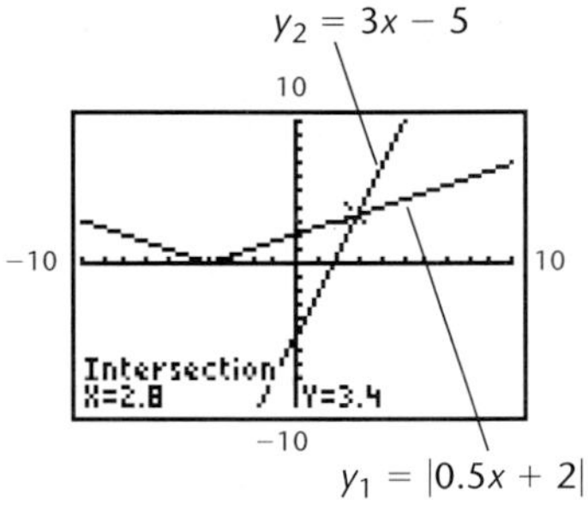

(B) Figure 10 shows the appropriate graphs for the inequality $|0.5x + 2| \geq 3x - 5$. The graph in Figure 10 shows that $y_1 > y_2$ for $x < 2.8$. Since $y_1 = y_2$ for $x = 2.8$, we must include this value of x in the solution set:

$x \leq 2.8$ or $(-\infty, 2.8]$

MATCHED PROBLEM 6

Solve graphically and write solutions in both inequality and interval notation.

(A) $|\frac{2}{3}x + 1| \geq 2$ (B) $|2x - 5| < 0.4x + 2$

Application

EXAMPLE 7

Break-Even, Profit, and Loss

A recording company produces compact discs (CDs). One-time fixed costs for a particular CD are $24,000, which include costs such as recording, album

*The symbol $\cup$ denotes the union operation for sets. See Section A-8 for a discussion of interval notation and set operations.

design, and promotion. Variable costs amount to \$5.50 per CD and include the manufacturing, distribution, and royalty costs for each disc actually manufactured and sold to a retailer. The CD is sold to retail outlets for \$8.00 each.

(A) Find the level of sales for which the company will break even. Describe verbally and graphically the sales levels that result in a profit and those that result in a loss.

(B) Find the sales level that will produce a profit of \$20,000.

Solutions (A) Let

x = Number of CDs sold

C = Total cost for producing x CDs

R = Revenue (return) on sales of x CDs

Now form the cost and revenue functions.

$$C(x) = \text{Fixed costs} + \text{Variable costs}$$

$$= 24{,}000 + 5.5x \quad \text{Cost function}$$

$$R(x) = 8x \quad \text{Revenue function}$$

The company will break even when revenue = cost; that is, when $R(x) = C(x)$. The solution to this equation is often referred to as the **break-even point.** Graphs of both functions and their intersection point are shown in Figure 11. Examining this graph, we see that the company will break even if they sell 9,600 CDs. If they sell more than 9,600 CDs, then revenue is greater than cost, and the company will make a profit. If they sell fewer than 9,600 CDs, then cost is greater than revenue and the company will lose money. These sales levels are illustrated in Figure 12.

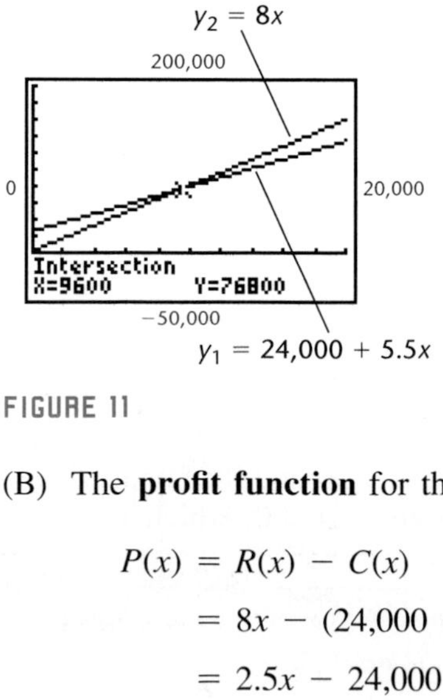

FIGURE 11

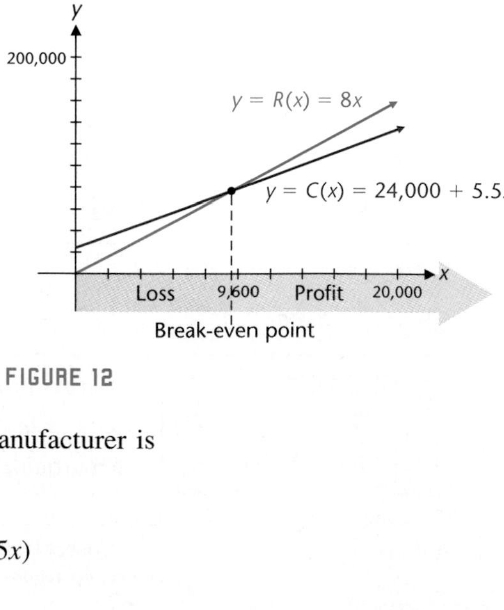

FIGURE 12

(B) The **profit function** for this manufacturer is

$$P(x) = R(x) - C(x)$$

$$= 8x - (24{,}000 + 5.5x)$$

$$= 2.5x - 24{,}000$$

The sales level x that will produce a profit of \$20,000 is the solution of the equation $P(x) = 20{,}000$. Figure 13 shows a graphical solution of this linear equation. Thus, we see that the company will make a profit of \$20,000 when they sell 17,600 CDs.

FIGURE 13

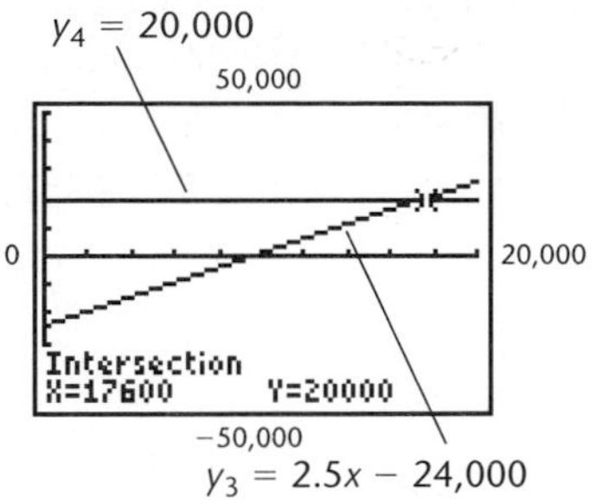

MATCHED PROBLEM 7

Repeat Example 7 if fixed costs are \$28,000, variable costs are \$6.60 per CD, and the CDs are sold for \$9.80 each.

Explore/Discuss 5

(A) Find the x intercept of the profit function in Example 7 (see Fig. 13).

(B) Discuss the relationship between the x intercept of the profit function and the sales levels for which the company incurs a loss, breaks even, or makes a profit.

(C) In general, compare the graphical solutions of the inequalities

$$f(x) > g(x) \quad \text{and} \quad f(x) - g(x) > 0$$

Answers to Matched Problems

1. -2 **2.** $-\frac{20}{39} \approx -0.5128205$ **3.** $r = \frac{A - P}{Pt}$

4. $x \geq 3$ or $[3, \infty)$

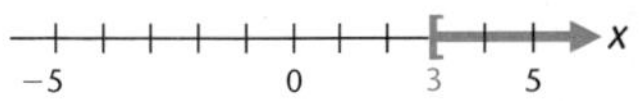

5. $0 \leq x < 5$ or $[0, 5)$

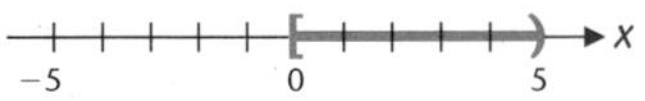

6. (A) $x \leq -4.5$ or $x \geq 1.5$; $(-\infty, -4.5] \cup [1.5, \infty)$

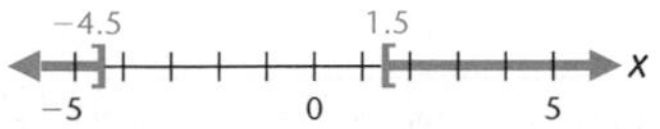

(B) $1.25 < x < 4.375$; $(1.25, 4.375)$

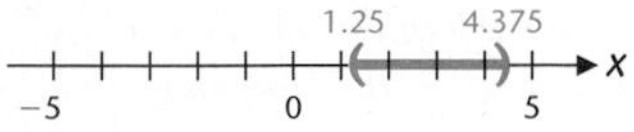

7. (A) The company breaks even if they sell 8,750 CDs, makes a profit if they sell more than 8,750 CDs, and loses money if they sell less than 8,750 CDs.

(B) The company must sell 15,000 CDs to make a profit of \$20,000.

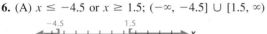

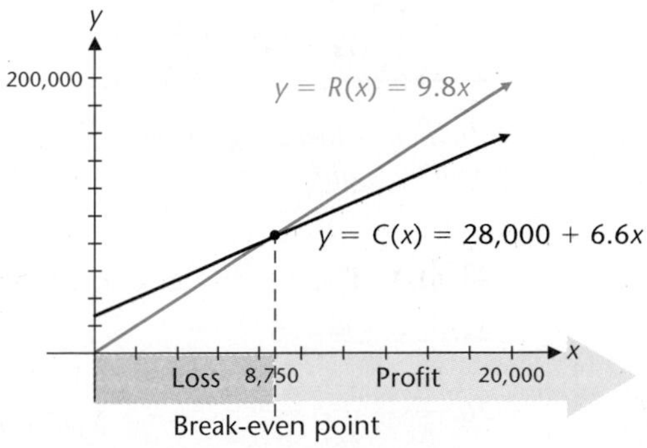

EXERCISE 2-2

A

Use the graphs of functions u and v in the figure to solve the equations and inequalities in Problems 1–8. (Assume the graphs continue as indicated beyond the portions shown here.) Express solutions to inequalities in interval notation.

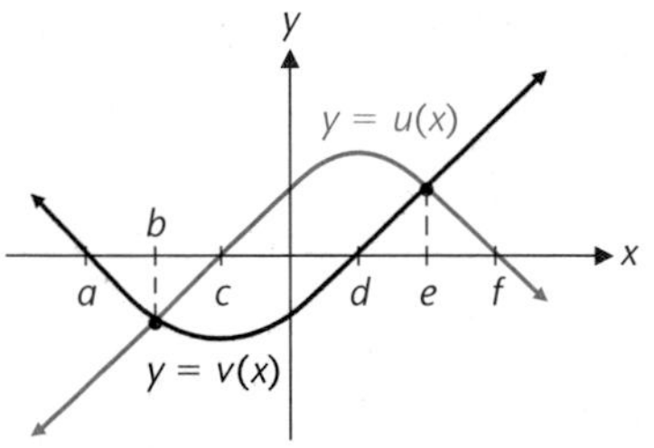

1. $u(x) = 0$

2. $v(x) = 0$

3. $u(x) = v(x)$

4. $u(x) - v(x) = 0$

5. $u(x) > 0$

6. $v(x) \geq 0$

7. $v(x) \geq u(x)$

8. $v(x) < 0$

Solve Problems 9–14 algebraically and check graphically.

9. $3(x + 2) = 5(x - 6)$

10. $5x + 10(x - 2) = 40$

11. $5 + 4(t - 2) = 2(t + 7) + 1$

12. $5w - (7w - 4) - 2 = 5 - (3w + 2)$

13. $5 - \dfrac{2x - 1}{4} = \dfrac{x + 2}{3}$

14. $\dfrac{x + 3}{4} - \dfrac{x - 4}{2} = \dfrac{3}{8}$

Solve Problems 15–20 algebraically and check graphically. Represent each solution using inequality notation, interval notation, and a graph on a real number line.

15. $7x - 8 < 4x + 7$

16. $4x + 8 \geq x - 1$

17. $-5t < -10$

18. $-7n \geq 21$

19. $-4 < 5t + 6 \leq 21$

20. $2 \leq 3m - 7 < 14$

B

In Problems 21–36, solve each equation or inequality. When applicable, write answers using both inequality notation and interval notation.

21. $|y - 5| = 3$

22. $|x + 1| = 5$

23. $|5t + 3| \leq 7$

24. $|2w - 9| < 6$

25. $\dfrac{1}{m} - \dfrac{1}{9} = \dfrac{4}{9} - \dfrac{2}{3m}$

26. $\dfrac{2}{3x} + \dfrac{1}{2} = \dfrac{4}{x} + \dfrac{4}{3}$

27. $-12 < \dfrac{3}{4}(2 - x) \leq 24$

28. $24 \leq \dfrac{2}{3}(x - 5) < 36$

29. $|3t - 7| = \dfrac{4}{3}t + \dfrac{1}{2}$

30. $|2s + 3| = 6 - 0.5s$

31. $|1.5x + 6| > 0.3x + 7.5$

32. $|7 - 2x| \geq x - 0.8$

33. $\dfrac{2x}{x - 3} = 7 + \dfrac{4}{x - 3}$

34. $\dfrac{2x}{x + 4} = 7 - \dfrac{6}{x + 4}$

35. $6 < |x - 2| + |x + 1| < 12$

36. $|x + 1| - |x - 2| < 0.4x$

In Problems 37–44, solve for the indicated variable in terms of the other variables.

37. $a_n = a_1 + (n - 1)d$ for d (arithmetic progressions)

38. $F = \frac{9}{5}C + 32$ for C (temperature scale)

39. $\dfrac{1}{f} = \dfrac{1}{d_1} + \dfrac{1}{d_2}$ for f (simple lens formula)

40. $\dfrac{1}{R} = \dfrac{1}{R_1} + \dfrac{1}{R_2}$ for R_1 (electric circuit)

41. $A = 2ab + 2ac$ for a (surface area of a rectangular solid)

42. $A = 2ab + 2ac + 2bc$ for c

43. $y = \dfrac{2x - 3}{3x + 5}$ for x

44. $x = \dfrac{3y + 2}{y - 3}$ for y

45. Discuss the relationship between the graphs of $y_1 = x$ and $y_2 = \sqrt{x^2}$.

46. Discuss the relationship between the graphs of $y_1 = |x|$ and $y_2 = \sqrt{x^2}$.

47. Discuss the possible signs of the numbers a and b given that

(A) $ab > 0$

(B) $ab < 0$

(C) $\dfrac{a}{b} > 0$

(D) $\dfrac{a}{b} < 0$

48. Discuss the possible signs of the numbers a, b, and c given that

(A) $abc > 0$

(B) $\dfrac{ab}{c} < 0$

(C) $\dfrac{a}{bc} > 0$

(D) $\dfrac{a^2}{bc} < 0$

In Problems 49–52, replace each question mark with < or > and explain why your choice makes the statement true.

49. If $a - b = 1$, then a ? b.

50. If $u - v = -2$, then u ? v.

51. If $a < 0$, $b < 0$, and $\dfrac{b}{a} > 1$, then a ? b.

52. If $a > 0$, $b > 0$, and $\dfrac{b}{a} > 1$, then a ? b.

C

C *Problems 53–56 are calculus-related. Solve and graph. Write each solution using interval notation.*

53. $0 < |x - 3| < 0.1$ **54.** $0 < |x - 5| < 0.01$

55. $0 < |x - c| < d$ **56.** $0 < |x - 4| < d$

C **57.** What are the possible values of $\dfrac{x}{|x|}$?

C **58.** What are the possible values of $\dfrac{|x - 1|}{x - 1}$?

APPLICATIONS

59. Sales Commissions. One employee of a computer store is paid a base salary of \$2,150 a month plus an 8% commission on all sales over \$7,000 during the month. How much must the employee sell in 1 month to earn a total of \$3,170 for the month?

60. Sales Commissions. A second employee of the computer store in Problem 59 is paid a base salary of \$1,175 a month plus a 5% commission on all sales during the month.

(A) How much must this employee sell in 1 month to earn a total of \$3,170 for the month?

(B) Determine the sales level where both employees receive the same monthly income. If employees can select either of these payment methods, how would you advise an employee to make this selection?

C **61. Approximation.** The area A of a region is approximately equal to 12.436. The error in this approximation is less than 0.001. Describe the possible values of this area both with an absolute value inequality and with interval notation.

C **62. Approximation.** The volume V of a solid is approximately equal to 6.94. The error in this approximation is less than 0.02. Describe the possible values of this volume both with an absolute value inequality and with interval notation.

63. Break-Even Analysis. An electronics firm is planning to market a new graphing calculator. The fixed costs are \$650,000 and the variable costs are \$47 per calculator. The wholesale price of the calculator will be \$63. For the company to make a profit, it is clear that revenues must be greater than costs.

(A) How many calculators must be sold for the company to make a profit?

(B) How many calculators must be sold for the company to break even?

(C) Discuss the relationship between the results in parts A and B.

64. Break-Even Analysis. A video game manufacturer is planning to market a 64-bit version of its game machine. The fixed costs are \$550,000 and the variable costs are \$120 per machine. The wholesale price of the machine will be \$140.

(A) How many game machines must be sold for the company to make a profit?

(B) How many game machines must be sold for the company to break even?

(C) Discuss the relationship between the results in parts A and B.

65. Break-Even Analysis. The electronics firm in Problem 63 finds that rising prices for parts increases the variable costs to \$50.5 per calculator.

(A) Discuss possible strategies the company might use to deal with this increase in costs.

(B) If the company continues to sell the calculators for \$63, how many must they sell now to make a profit?

(C) If the company wants to start making a profit at the same production level as before the cost increase, how much should they increase the wholesale price?

66. Break-Even Analysis. The video game manufacturer in Problem 64 finds that unexpected programming problems increases the fixed costs to \$660,000.

(A) Discuss possible strategies the company might use to deal with this increase in costs.

(B) If the company continues to sell the game machines for \$140, how many must they sell now to make a profit?

(C) If the company wants to start making a profit at the same production level as before the cost increase, how much should they increase the wholesale price?

★ **67. Significant Digits.** If $N = 2.37$ represents a measurement, then we assume an accuracy of 2.37 ± 0.005. Express the accuracy assumption using an absolute value inequality.

★ **68. Significant Digits.** If $N = 3.65 \times 10^{-3}$ is a number from a measurement, then we assume an accuracy of $3.65 \times 10^{-3} \pm 5 \times 10^{-6}$. Express the accuracy assumption using an absolute value inequality.

★ **69. Finance.** If an individual aged 65–69 continues to work after Social Security benefits start, benefits will be reduced when earnings exceed an earnings limitation. In 1989, benefits were reduced by \$1 for every \$2 earned in excess of \$8,880. Find the range of benefit reductions for individuals earning between \$13,000 and \$16,000.

★ **70. Finance.** Refer to Problem 69. In 1990 the law was changed so that benefits were reduced by \$1 for every \$3 earned in excess of \$8,880. Find the range of benefit reductions for individuals earning between \$13,000 and \$16,000.

71. Celsius/Fahrenheit. A formula for converting Celsius degrees to Fahrenheit degrees is given by the linear function

$$F = \frac{9}{5}C + 32$$

Determine to the nearest degree the Celsius range in temperature that corresponds to the Fahrenheit range of 60°F to 80°F.

72. Celsius/Fahrenheit. A formula for converting Fahrenheit degrees to Celsius degrees is given by the linear function

$$C = \frac{5}{9}(F - 32)$$

Determine to the nearest degree the Fahrenheit range in temperature that corresponds to a Celsius range of 20°C to 30°C.

★ **73. Earth Science.** In 1984, the Soviets led the world in drilling the deepest hole in the Earth's crust—more than 12 kilometers deep. They found that below 3 kilometers the temperature T increased 2.5°C for each additional 100 meters of depth.

(A) If the temperature at 3 kilometers is 30°C and x is the depth of the hole in kilometers, write an equation using x that will give the temperature T in the hole at any depth beyond 3 kilometers.

(B) What would the temperature be at 15 kilometers? [The temperature limit for their drilling equipment was about 300°C.]

(C) At what interval of depths will the temperature be between 200°C and 300°C, inclusive?

★ **74. Aeronautics.** Because air is not as dense at high altitudes, planes require a higher ground speed to become airborne. A rule of thumb is 3% more ground speed per 1,000 feet of elevation, assuming no wind and no change in air temperature. (Compute numerical answers to 3 significant digits.)

(A) Let

V_s = Takeoff ground speed at sea level for a particular plane (in miles per hour)

A = Altitude above sea level (in thousands of feet)

V = Takeoff ground speed at altitude A for the same plane (in miles per hour)

Write a formula relating these three quantities.

(B) What takeoff ground speed would be required at Lake Tahoe airport (6,400 feet), if takeoff ground speed at San Francisco airport (sea level) is 120 miles per hour?

(C) If a landing strip at a Colorado Rockies hunting lodge (8,500 feet) requires a takeoff ground speed of 125 miles per hour, what would be the takeoff ground speed in Los Angeles (sea level)?

(D) If the takeoff ground speed at sea level is 135 miles per hour and the takeoff ground speed at a mountain resort is 155 miles per hour, what is the altitude of the mountain resort in thousands of feet?

Section 2-3 | Quadratic Functions

- Quadratic Functions
- Completing the Square
- Properties of Quadratic Functions and Their Graphs
- Applications

Quadratic Functions

FIGURE 1
Square function $h(x) = x^2$.

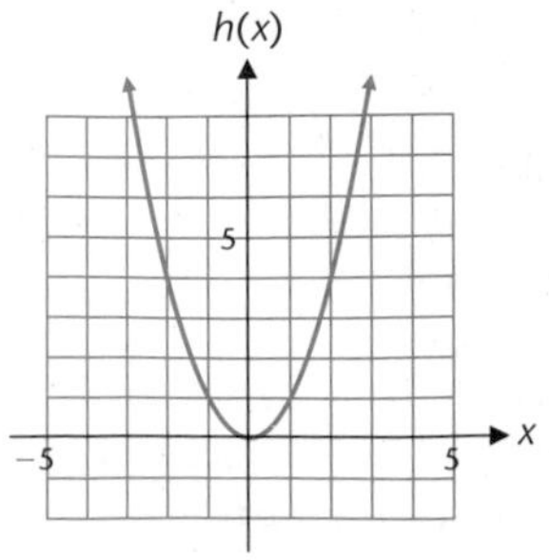

The graph of the square function, $h(x) = x^2$, is shown in Figure 1. Notice that the graph is symmetric with respect to the y axis and that (0, 0) is the lowest point on the graph. Let's explore the effect of applying a sequence of basic transformations to the graph of h. (A brief review of Section 1-5 might prove helpful at this point.)

Explore/Discuss 1

Indicate how the graph of each function is related to the graph of $h(x) = x^2$. Discuss the symmetry of the graphs and find the highest or lowest point, whichever exists, on each graph.

(A) $f(x) = (x - 3)^2 - 7 = x^2 - 6x + 2$

(B) $g(x) = 0.5(x + 2)^2 + 3 = 0.5x^2 + 2x + 5$

(C) $m(x) = -(x - 4)^2 + 8 = -x^2 + 8x - 8$

(D) $n(x) = -3(x + 1)^2 - 1 = -3x^2 - 6x - 4$

Graphing the functions in Explore/Discuss 1 produces figures similar in shape to the graph of the square function in Figure 1. These figures are called *parabolas*. The functions that produced these parabolas are examples of the important class of *quadratic functions*, which we now define.

QUADRATIC FUNCTIONS

If a, b, and c are real numbers with $a \neq 0$, then the function

$$f(x) = ax^2 + bx + c$$

is a **quadratic function** and its graph is a **parabola.***

Since the expression $ax^2 + bx + c$ represents a real number for all real number replacements of x,

the domain of a quadratic function is the set of all real numbers.

We will discuss methods for determining the range of a quadratic function later in this section. Typical graphs of quadratic functions are illustrated in Figure 2.

FIGURE 2 Graphs of quadratic functions.

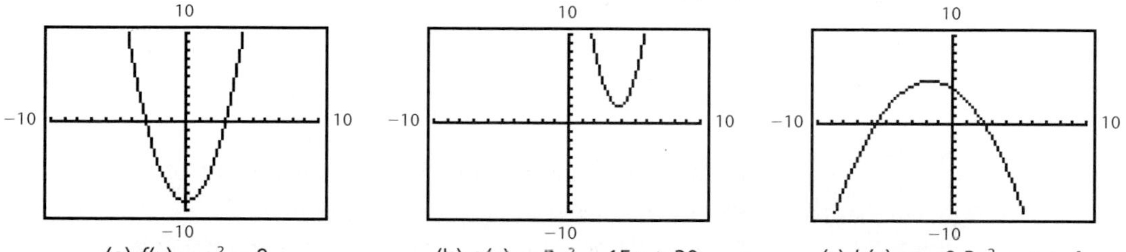

(a) $f(x) = x^2 - 9$ (b) $g(x) = 2x^2 - 15x + 30$ (c) $h(x) = -0.3x^2 - x + 4$

Completing the Square

In Explore/Discuss 1 we wrote each function as two different, but equivalent, expressions. For example,

$$f(x) = (x - 3)^2 - 7 = x^2 - 6x + 2$$

*A more general definition of a parabola that is independent of any coordinate system is given in Section 7-1.

It is easy to verify that these two expressions are equivalent by expanding the first expression. The first expression is more useful than the second for analyzing the graph of f. If we are given only the second expression, how can we determine the first? It turns out that this is a routine process, called *completing the square*, that is another useful tool to be added to our mathematical toolbox.

Explore/Discuss 2

Replace ? in each of the following with a number that makes the equation valid.

(A) $(x + 1)^2 = x^2 + 2x + ?$ (B) $(x + 2)^2 = x^2 + 4x + ?$

(C) $(x + 3)^2 = x^2 + 6x + ?$ (D) $(x + 4)^2 = x^2 + 8x + ?$

Replace ? in each of the following with a number that makes the expression a perfect square of the form $(x + h)^2$.

(E) $x^2 + 10x + ?$ (F) $x^2 + 12x + ?$ (G) $x^2 + bx + ?$

Given the quadratic expression

$$x^2 + bx$$

what must be added to this expression to make it a perfect square? To find out, consider the square of the following expression:

$$(x + m)^2 = x^2 + \underbrace{2mx} + \underbrace{m^2}$$

m^2 is the square of one-half the coefficient of x.

We see that the third term on the right side of the equation is the square of one-half the coefficient of x in the second term on the right; that is, m^2 is the square of $\frac{1}{2}(2m)$. This observation leads to the following rule:

COMPLETING THE SQUARE

To **complete the square** of the quadratic expression

$$x^2 + bx$$

add the square of one-half the coefficient of x; that is, add

$$\left(\frac{b}{2}\right)^2 \quad \text{or} \quad \frac{b^2}{4}$$

The resulting expression can be factored as a perfect square:

$$x^2 + bx + \left(\frac{b}{2}\right)^2 = \left(x + \frac{b}{2}\right)^2$$

EXAMPLE 1

Completing the Square

Complete the square for each of the following:

(A) $x^2 - 3x$ (B) $x^2 - 6bx$

Solutions

(A) $x^2 - 3x$

$$x^2 - 3x + \frac{9}{4} = \left(x - \frac{3}{2}\right)^2 \qquad \text{Add } \left(\frac{-3}{2}\right)^2\text{; that is, } \frac{9}{4}.$$

(B) $x^2 - 6bx$

$$x^2 - 6bx + 9b^2 = (x - 3b)^2 \qquad \text{Add } \left(\frac{-6b}{2}\right)^2\text{; that is, } 9b^2.$$

MATCHED PROBLEM 1

Complete the square for each of the following:

(A) $x^2 - 5x$ (B) $x^2 + 4mx$

It is important to note that the rule for completing the square applies to only quadratic expressions in which the coefficient of x^2 is 1. This causes little trouble, however, as you will see.

Properties of Quadratic Functions and Their Graphs

We now use the process of completing the square to transform the quadratic function

$$f(x) = ax^2 + bx + c$$

into the **standard form**

$$f(x) = a(x - h)^2 + k$$

Many important features of the graph of a quadratic function can be determined by examining the standard form. We begin with a specific example and then generalize the results.

Consider the quadratic function given by

$$f(x) = 2x^2 - 8x + 4 \qquad (1)$$

We use completing the square to transform this function into standard form:

$$f(x) = 2x^2 - 8x + 4$$

$$= 2(x^2 - 4x) + 4 \qquad \text{Factor the coefficient of } x^2 \text{ out of the first two terms.}$$

$$= 2(x^2 - 4x + ?) + 4$$

$$= 2(x^2 - 4x + 4) + 4 - 8$$

We add 4 to complete the square inside the parentheses. But because of the 2 outside the parentheses, we have actually added 8, so we must subtract 8.

$$= 2(x - 2)^2 - 4$$

The transformation is complete and can be checked by expanding.

Thus, the standard form is

$$f(x) = 2(x - 2)^2 - 4 \tag{2}$$

If $x = 2$, then $2(x - 2)^2 = 0$ and $f(2) = -4$. For any other value of x, the positive number $2(x - 2)^2$ is added to -4, making $f(x)$ larger. Therefore,

$$f(2) = -4$$

is the *minimum value* of $f(x)$ for all x—a very important result! Furthermore, if we choose any two values of x that are equidistant from $x = 2$, we will obtain the same value for the function. For example, $x = 1$ and $x = 3$ are each one unit from $x = 2$ and their functional values are

$$f(1) = 2(-1)^2 - 4 = -2$$

$$f(3) = 2(1)^2 - 4 = -2$$

Thus, the vertical line $x = 2$ is a line of symmetry—if the graph of equation (1) is drawn on a piece of paper and the paper folded along the line $x = 2$, then the two sides of the parabola will match exactly.

The above results are illustrated by graphing equation (1) or (2) and the line $x = 2$ in a suitable viewing window (Fig. 3).

FIGURE 3

Graph of a quadratic function.

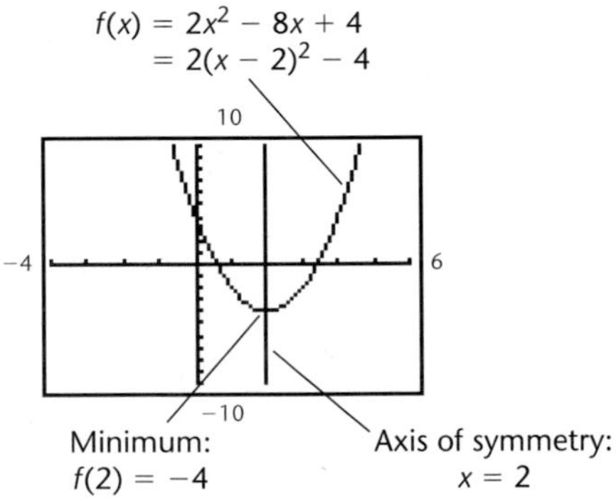

From the analysis of equation (2), illustrated by the graph in Figure 3, we conclude that $f(x)$ is decreasing on $(-\infty, 2]$ and increasing on $[2, \infty)$. Furthermore, $f(x)$ can assume any value greater than or equal to -4, but no values less than -4. Thus,

Range of f: $y \geq -4$ or $[-4, \infty)$

In general, the graph of a quadratic function is a parabola with line of symmetry parallel to the vertical axis. The lowest or highest point on the parabola, whichever exists, is called the **vertex.** The maximum or minimum value of a quadratic function always occurs at the vertex of the graph. The vertical line of symmetry through the vertex is called the **axis** of the parabola. Thus, for $f(x) = 2x^2 - 8x + 4$, the vertical line $x = 2$ is the axis of the parabola and $(2, -4)$ is its vertex.

From equation (2), we can see that the graph of f is simply the graph of $g(x) = 2x^2$ translated to the right 2 units and down 4 units, as shown in Figure 4.

FIGURE 4
Graph of f is the graph of g translated.

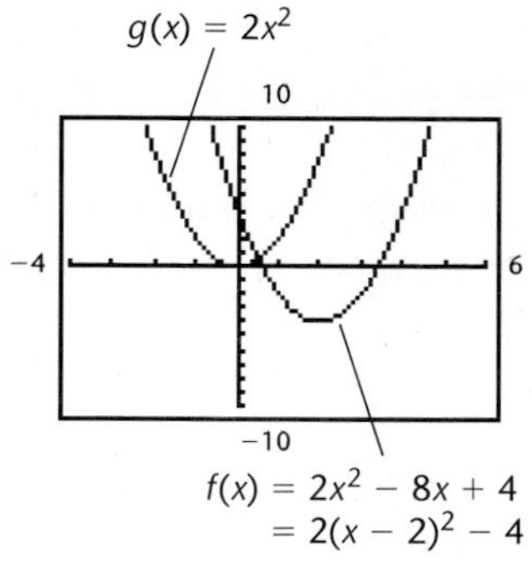

Notice the important results we have obtained from the standard form of the quadratic function f:

→ The vertex of the parabola

→ The axis of the parabola

→ The minimum value of $f(x)$

→ The range of f

→ A relationship between the graph of f and the graph of g

Explore/Discuss 3

Explore the effect of changing the constants a, h, and k on the graph of $f(x) = a(x - h)^2 + k$.

(A) Let $a = 1$ and $h = 5$. Graph function f for $k = -4$, 0, and 3 simultaneously in the same viewing window. Explain the effect of changing k on the graph of f.

(B) Let $a = 1$ and $k = 2$. Graph function f for $h = -4$, 0, and 5 simultaneously in the same viewing window. Explain the effect of changing h on the graph of f.

(C) Let $h = 5$ and $k = -2$. Graph function f for $a = 0.25$, 1, and 3 simultaneously in the same viewing window. Graph function f for $a = 1$, -1, and -0.25 simultaneously in the same viewing window. Explain the effect of changing a on the graph of f.

(D) Can all quadratic functions of the form $y = ax^2 + bx + c$ be rewritten as $a(x + h)^2 + k$?

We generalize the above discussion in the following box:

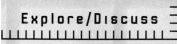

PROPERTIES OF A QUADRATIC FUNCTION AND ITS GRAPH

Given a quadratic function and the standard form obtained by completing the square

$$f(x) = ax^2 + bx + c = a(x - h)^2 + k \qquad a \neq 0$$

we summarize general properties as follows:

1. The graph of f is a parabola:

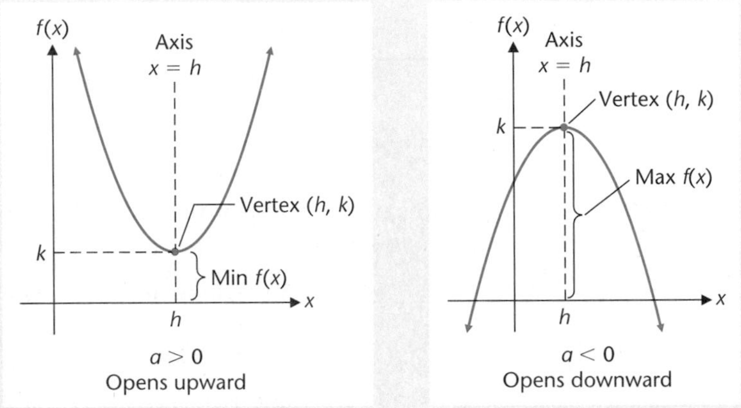

2. Vertex: (h, k) (parabola increases on one side of the vertex and decreases on the other).
3. Axis (of symmetry): $x = h$ (parallel to y axis).
4. $f(h) = k$ is the minimum if $a > 0$ and the maximum if $a < 0$.
5. Domain: all real numbers; range: $(-\infty, k]$ if $a < 0$ or $[k, \infty)$ if $a > 0$.
6. The graph of f is the graph of $g(x) = ax^2$ translated horizontally h units and vertically k units.

EXAMPLE 2

Analyzing a Quadratic Function

Find the standard form for the following quadratic function, analyze the graph, and check your results with a graphing utility:

$$f(x) = -0.5x^2 - x + 5$$

Solution

We complete the square to find the standard form:

$$\begin{aligned} f(x) &= -0.5x^2 - x + 5 \\ &= -0.5(x^2 + 2x + ?) + 5 \\ &= -0.5(x^2 + 2x + 1) + 5 + 0.5 \\ &= -0.5(x + 1)^2 + 5.5 \end{aligned}$$

FIGURE 5

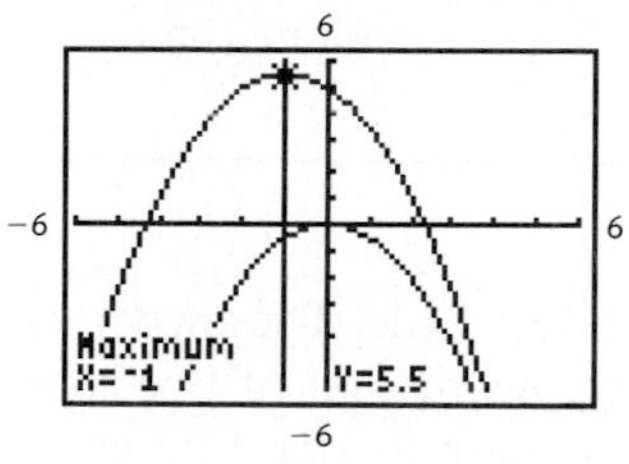

From the standard form we see that $h = -1$ and $k = 5.5$. Thus, the vertex is $(-1, 5.5)$, the axis of symmetry is $x = -1$, the maximum value is $f(-1) = 5.5$, and the range is $(-\infty, 5.5]$. The function f is increasing on $(-\infty, -1]$ and decreasing on $[-1, \infty)$. The graph of f is the graph of $g(x) = -0.5x^2$ shifted to the left 1 unit and upward 5.5 units. To check these results, we graph f and g simultaneously in the same viewing window, use the built-in maximum routine to locate the vertex, and add the graph of the axis of symmetry (Fig. 5).

MATCHED PROBLEM **2**

Find the standard form for the following quadratic function, analyze the graph, and check your results with a graphing utility:

$$f(x) = -x^2 + 3x - 1$$

EXAMPLE **3**

Finding the Equation of a Parabola

Find an equation for the parabola whose graph is shown in Figure 6.

FIGURE 6

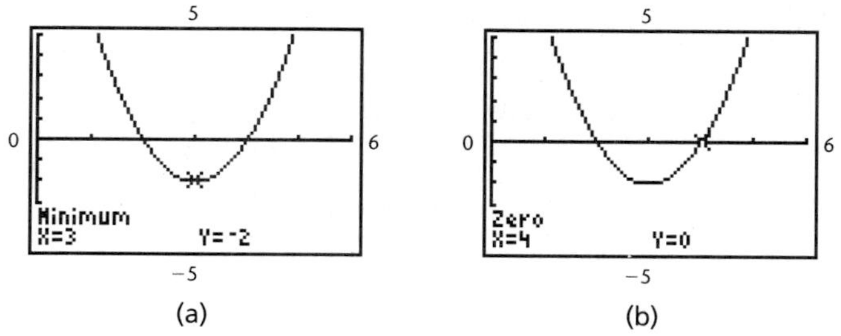

Solution

Figure 6(a) shows that the vertex of the parabola is $(h, k) = (3, -2)$. Thus, the standard equation must have the form

$$f(x) = a(x - 3)^2 - 2 \tag{3}$$

Figure 6(b) shows that $f(4) = 0$. Substituting in equation (3) and solving for a, we have

$$f(4) = a(4 - 3)^2 - 2 = 0$$
$$a = 2$$

Thus, the equation for the parabola is

$$f(x) = 2(x - 3)^2 - 2 = 2x^2 - 12x + 16$$

MATCHED PROBLEM **3**

Find the equation of the parabola with vertex (2, 4) and y intercept (0, 2).

Applications

We now look at several applications that can be modeled using quadratic functions.

EXAMPLE 4

Maximum Area

A dairy farm has a barn that is 150 feet long and 75 feet wide. The owner has 240 ft of fencing and wishes to use all of it in the construction of two identical adjacent outdoor pens with the long side of the barn as one side of the pens and a common fence between the two (Fig. 7). The owner wants the pens to be as large as possible.

FIGURE 7

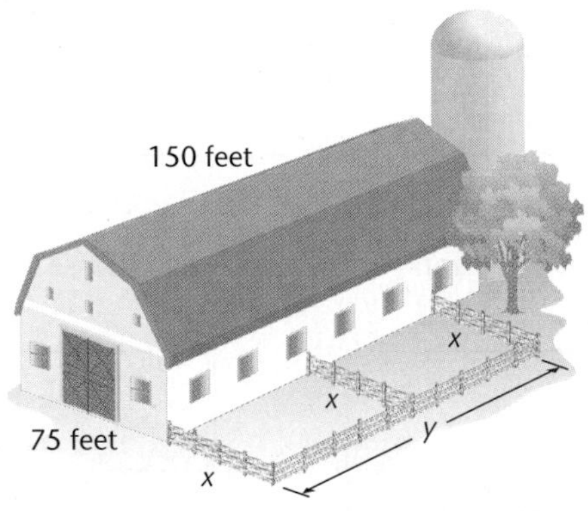

(A) Construct a mathematical model for the combined area of both pens in the form of a function $A(x)$ (see Fig. 7) and state the domain of A.

(B) Find the value of x that produces the maximum combined area.

(C) Find the dimensions and the area of each pen.

Solutions

(A) Since $y = 240 - 3x$,

$$A(x) = (240 - 3x)x = 240x - 3x^2$$

The distances x and y must be nonnegative. Since $y = 240 - 3x$, it follows that x cannot exceed 80. Thus, a model for this problem is

$$A(x) = 240x - 3x^2, \; 0 \le x \le 80$$

FIGURE 8
$A(x) = 240x - 3x^2$.

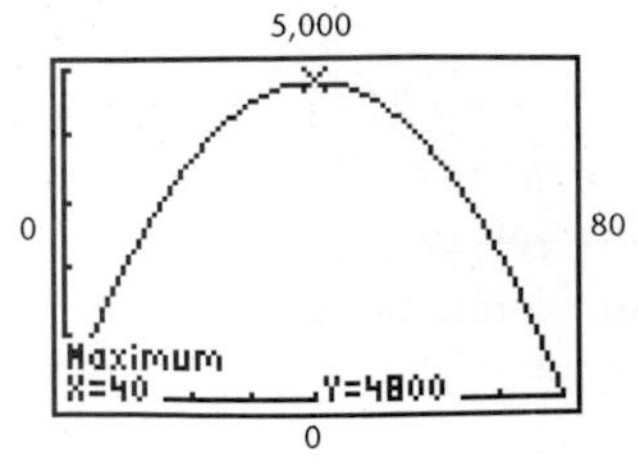

(B) Omitting the details, the standard form for A is

$$A(x) = -3(x - 40)^2 + 4{,}800$$

Thus, the maximum combined area of 4,800 ft^2 occurs at $x = 40$. This result is confirmed in Figure 8.

(C) Each pen is x by $y/2$ or 40 ft by 60 ft. The area of each pen is 40 ft $\times$ 60 ft $=$ 2,400 ft^2.

MATCHED PROBLEM 4

Repeat Example 4 with the owner constructing three identical adjacent pens instead of two.

Now that we have added quadratic functions to our mathematical toolbox, we can use this new tool in conjunction with another tool discussed previously—regression analysis. In the next example, we use both of these tools to investigate the effect of recycling efforts on solid waste disposal.

EXAMPLE 5

Solid Waste Disposal

Franklin Associates Ltd. of Prairie Village, Kansas, reported the data in Table 1 to the U.S. Environmental Protection Agency.

TABLE 1 Municipal Solid Waste Disposal

Year	Annual Landfill Disposal (millions of tons)	Per Person Per Day (pounds)
1970	88.2	2.37
1980	123.3	2.97
1985	136.4	3.13
1987	140.0	3.15
1990	131.6	2.90
1993	127.6	2.70
1995	118.4	2.50

(A) Let x represent time in years with $x = 0$ corresponding to 1960, and let y represent the corresponding annual landfill disposal. Use regression analysis on a graphing utility to find a quadratic function of the form $y = ax^2 + bx + c$ that models this data. (Round the constants a, b, and c to three significant digits* when reporting your results.)

(B) If landfill disposal continues to follow the trend exhibited in Table 1, when (to the nearest year) would the annual landfill disposal return to the 1970 level?

(C) Is it reasonable to expect the annual landfill disposal to follow this trend indefinitely? Explain.

Solutions

(A) Since the values of y increase from 1970 to 1987 and then begin to decrease, a quadratic model seems a better choice than a linear one. Figure 9 shows the details of constructing the model on a graphing utility.

*For those not familiar with the meaning of *significant digits*, see Appendix C for a brief discussion of this concept.

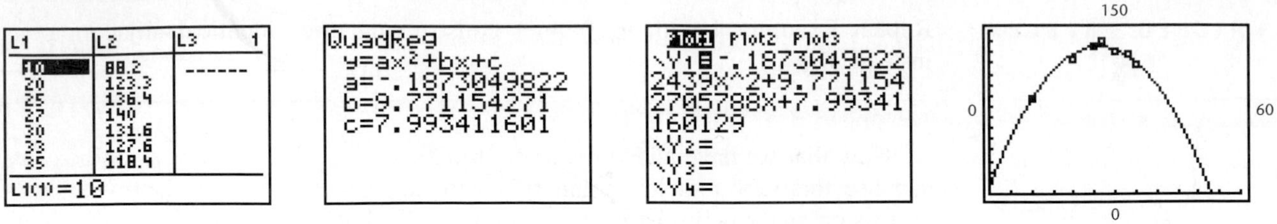

(a) Data

(b) Regression equation

(c) Regression equation transferred to equation editor

(d) Graph of data and regression equation

FIGURE 9

Rounding the constants to three significant digits, a quadratic regression equation for this data is

$$y_1 = -0.187x^2 + 9.77x + 7.99$$

The graph in Figure 9(d) indicates that this is a reasonable model for this data. It is, in fact, the "best" quadratic equation for this data.

(B) To determine when the annual landfill disposal returns to the 1970 level, we add the graph of $y_2 = 88.2$ to the graph [Fig. 10(a)]. The graphs of y_1 and y_2 intersect twice, once at $x = 10$ (1970), and again at a later date. Using a built-in intersection routine [Fig. 10(b)] shows that the x coordinate of the second intersection point (to the nearest integer) is 42. Thus, the annual landfill disposal returns to the 1970 level of 88.2 million tons in 2002. [*Note:* You will obtain slightly different results if you round the constants a, b, and c before finding the intersection point. As we stated before, we will always use the unrounded constants in calculations and only round the final answer.]

FIGURE 10

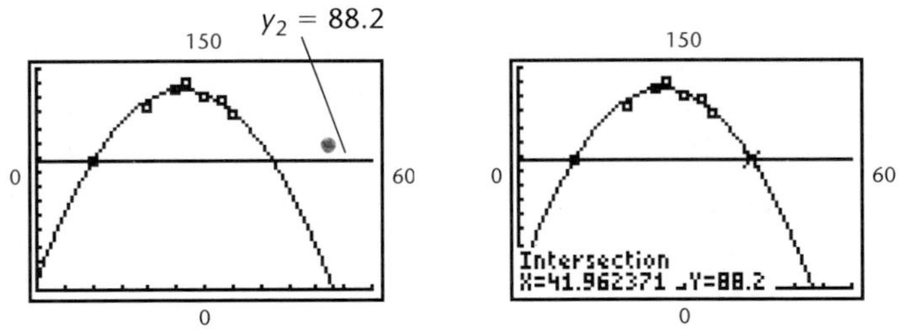

(a)

(b)

(C) The graph of y_1 continues to decrease and reaches 0 somewhere between 2110 and 2115. It is highly unlikely that the annual landfill disposal will ever reach 0. As time goes by and more data becomes available, new models will have to be constructed to better predict future trends.

MATCHED PROBLEM **5**

Refer to Table 1.

(A) Let x represent time in years with $x = 0$ corresponding to 1960, and let y represent the corresponding landfill disposal per person per day. Use regression analysis on a graphing utility to find a quadratic function of the form $y = ax^2 + bx + c$ that models this data. (Round the constants a, b, and c to three significant digits when reporting your results.)

(B) If landfill disposal per person per day continues to follow the trend exhibited in Table 1, when (to the nearest year) would it fall below 1.5 pounds per person per day?

(C) Is it reasonable to expect the landfill disposal per person per day to follow this trend indefinitely? Explain.

Answers to Matched Problems

1. (A) $x^2 - 5x + \frac{25}{4} = \left(x - \frac{5}{2}\right)^2$ (B) $x^2 + 4mx + 4m^2 = (x + 2m)^2$

2. Standard form: $f(x) = -(x - 1.5)^2 + 1.25$. The vertex is $(1.5, 1.25)$, the axis of symmetry is $x = 1.5$, the maximum value of $f(x)$ is 1.25, and the range of f is $(-\infty, 1.25]$. The function f is increasing on $(-\infty, 1.5]$ and decreasing on $[1.5, \infty)$. The graph of f is the graph of $g(x) = -x^2$ shifted 1.5 units to the right and 1.25 units upward.

3. $f(x) = -0.5(x - 2)^2 + 4 = -0.5x^2 + 2x + 2$

4. (A) $A(x) = (240 - 4x)x$, $0 \le x \le 60$ (B) The maximum combined area of 3,600 ft^2 occurs at $x = 30$ ft.
(C) Each pen is 30 ft by 40 ft with area 1,200 ft^2.

5. (A) $y = -0.00434x^2 + 0.202x + 0.759$ (B) 2003

EXERCISE 2-3

A

In Problems 1–6, complete the square and find the standard form of each quadratic function.

1. $f(x) = x^2 - 4x + 5$ **2.** $g(x) = -x^2 - 2x - 3$

3. $h(x) = -x^2 - 2x - 1$ **4.** $k(x) = x^2 - 4x + 4$

5. $m(x) = x^2 - 4x + 1$ **6.** $n(x) = -x^2 - 2x + 3$

In Problems 7–12, write a brief verbal description of the relationship between the graph of the indicated function (from Problems 1–6) and the graph of $y = x^2$.

7. $f(x) = x^2 - 4x + 5$ **8.** $g(x) = -x^2 - 2x - 3$

9. $h(x) = -x^2 - 2x - 1$ **10.** $k(x) = x^2 - 4x + 4$

11. $m(x) = x^2 - 4x + 1$ **12.** $n(x) = -x^2 - 2x + 3$

In Problems 13–18, match each graph with one of the functions in Problems 1–6.

13.

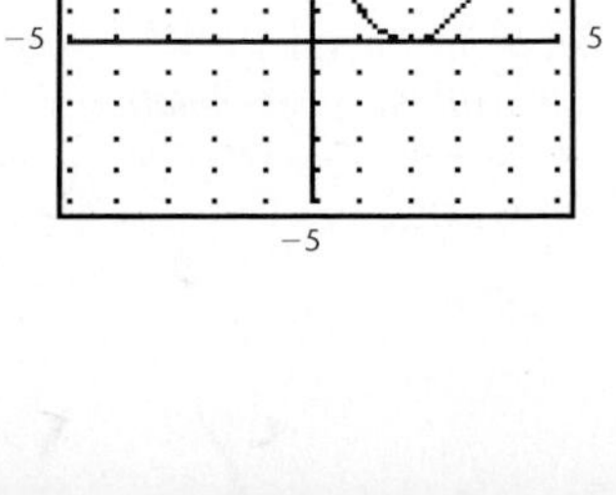

14.

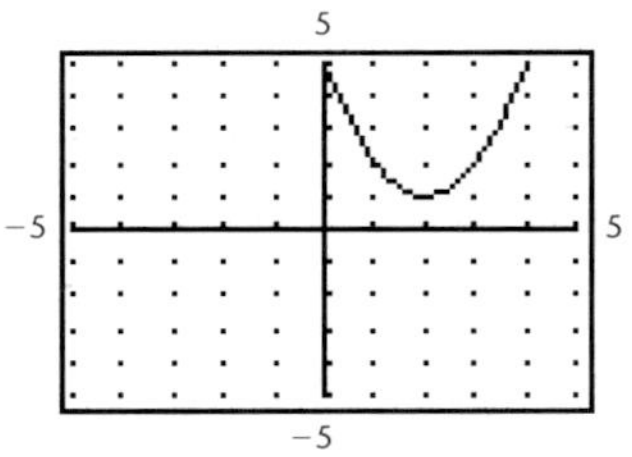

15.

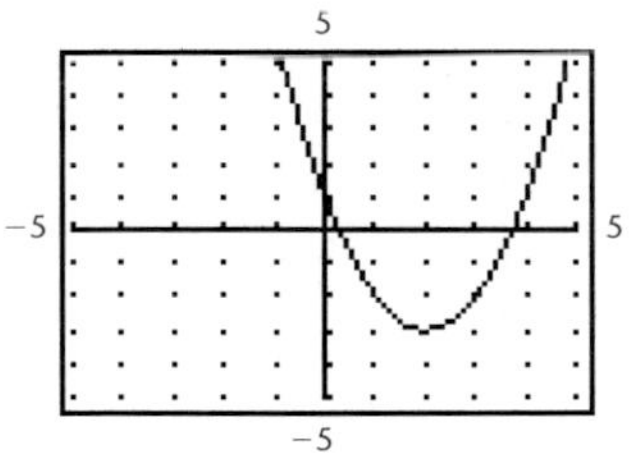

16.

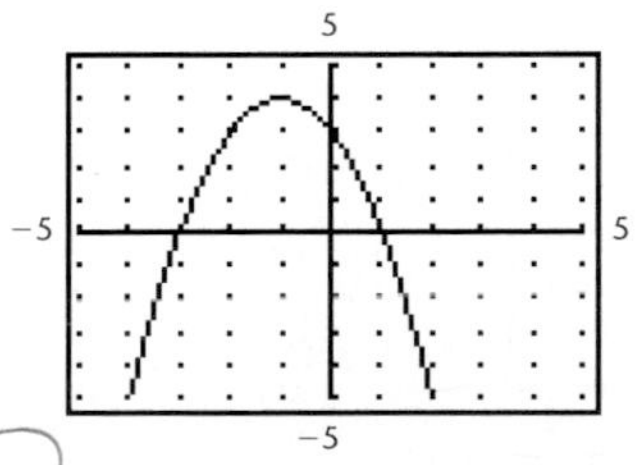

17.

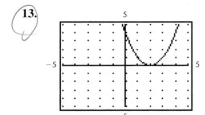

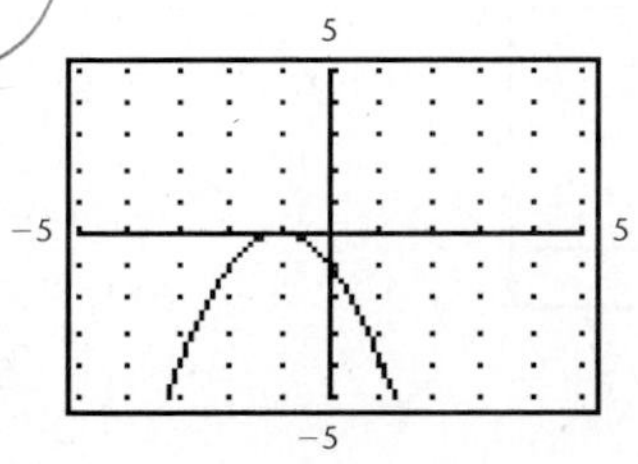

18.

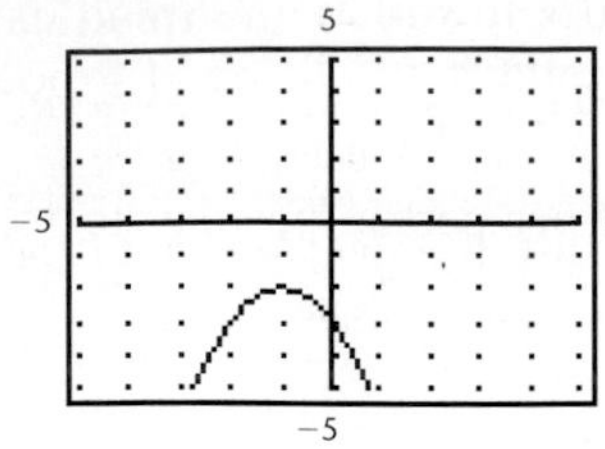

B

For each quadratic function in Problems 19–24, sketch a graph of the function and label the axis and the vertex.

19. $f(x) = 2x^2 - 24x + 90$

20. $f(x) = 3x^2 + 24x + 30$

21. $f(x) = -x^2 - 6x - 4$

22. $f(x) = -x^2 + 10x - 30$

23. $f(x) = 0.5x^2 - 2x - 7$

24. $f(x) = 0.4x^2 + 4x + 4$

In Problems 25–28, find the intervals where f is increasing, the intervals where f is decreasing, and the range. Express answers in interval notation.

25. $f(x) = 4x^2 - 18x + 25$

26. $f(x) = 5x^2 + 29x - 17$

27. $f(x) = -10x^2 + 44x + 12$

28. $f(x) = -8x^2 - 20x + 16$

In Problems 29–32, use the graph of the parabola to find the equation of the corresponding quadratic function.

29.

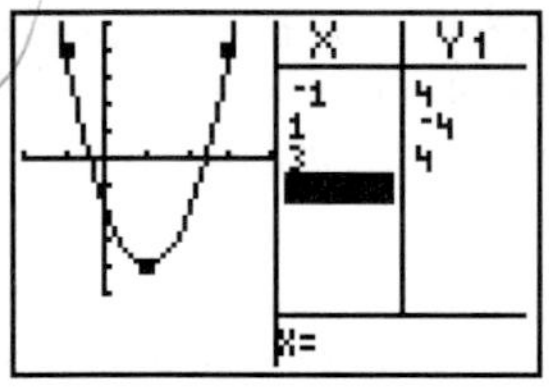

30.

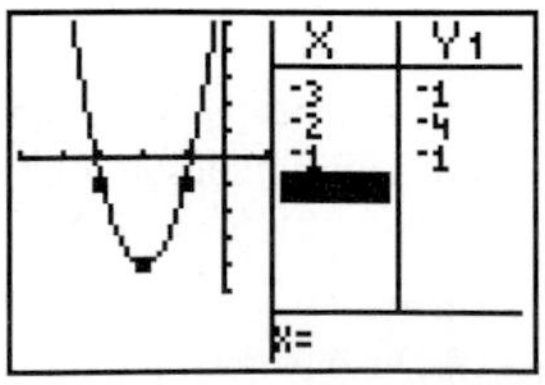

31.

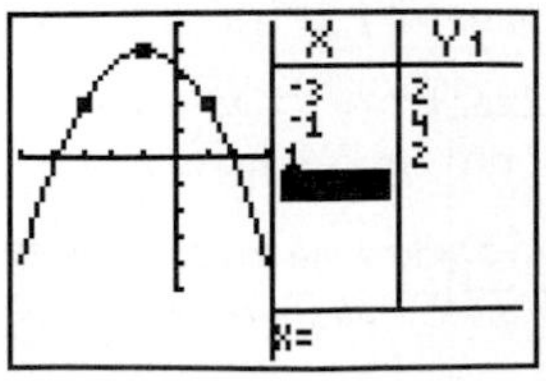

32.

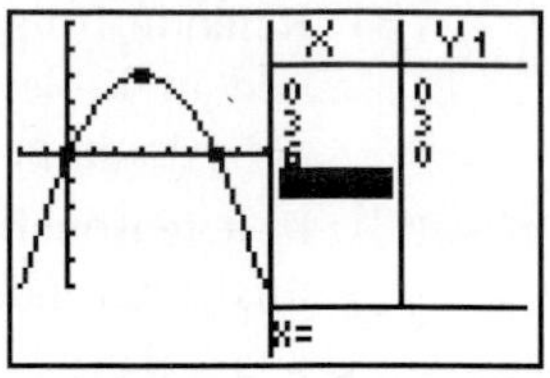

In Problems 33–38, find the equation of a quadratic function whose graph satisfies the given conditions.

33. Vertex: (4, 8); x intercept: 6

34. Vertex: $(-2, -12)$; x intercept: -4

35. Vertex: $(-4, 12)$; y intercept: 4

36. Vertex: (5, 8); y intercept: -2

37. Vertex: $(-5, -25)$; additional point on graph: $(-2, 20)$

38. Vertex: $(6, -40)$; additional point on graph: (3, 50)

39. Graph the line $y = 0.5x + 3$. Choose any two distinct points on this line and find the linear regression model for the data set consisting of the two points you chose. Experiment with other lines of your choosing. Discuss the relationship between a linear regression model for two points and the line that goes through the two points.

40. Graph the parabola $y = x^2 - 5x$. Choose any three distinct points on this parabola and find the quadratic regression model for the data set consisting of the three points you chose. Experiment with other parabolas of your choice. Discuss the relationship between a quadratic regression model for three noncollinear points and the parabola that goes through the three points.

41. Let $f(x) = (x - 1)^2 + k$. Discuss the relationship between the values of k and the number of x intercepts for the graph of f. Generalize your comments to any function of the form

$$f(x) = a(x - h)^2 + k,\ a > 0$$

42. Let $f(x) = -(x - 2)^2 + k$. Discuss the relationship between the values of k and the number of x intercepts for the graph of f. Generalize your comments to any function of the form

$$f(x) = a(x - h)^2 + k,\ a < 0$$

C

Recall that the standard equation of a circle with radius r and center (h, k) is

$$(x - h)^2 + (y - k)^2 = r^2$$

In Problems 43–46, use completing the square twice to find the center and radius of the circle with the given equation.

43. $x^2 + y^2 - 6x - 4y = 36$

44. $x^2 + y^2 - 2x - 10y = 55$

45. $x^2 + y^2 + 8x - 2y = 8$

46. $x^2 + y^2 - 4x + 12y = 24$

47. Let $f(x) = a(x - h)^2 + k$. Compare the values of $f(h + r)$ and $f(h - r)$ for any real number r. Interpret the results in terms of the graph of f.

48. Let $f(x) = ax^2 + bx + c$, $a \neq 0$. Express each of the following in terms of a, b, and c:

(A) The axis of symmetry

(B) The vertex

(C) The maximum or minimum value of f, whichever exists.

Problems 49–52 are calculus-related. In geometry, a line that intersects a circle in two distinct points is called a secant line, as shown in figure (a). In calculus, the line through the points $(x_1, f(x_1))$ and $(x_2, f(x_2))$ is called a ***secant line*** *for the graph of the function f, as shown in figure (b).*

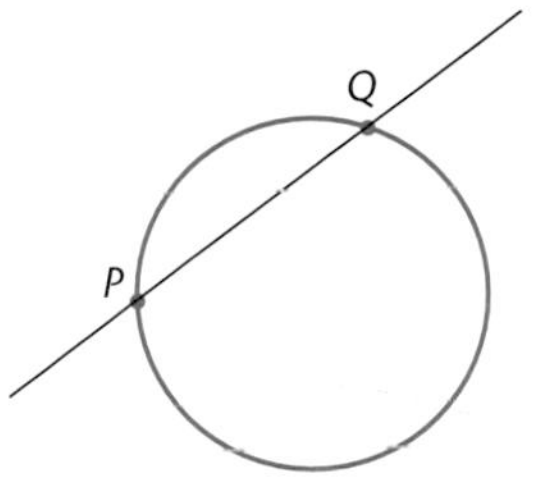

Secant line for a circle
(a)

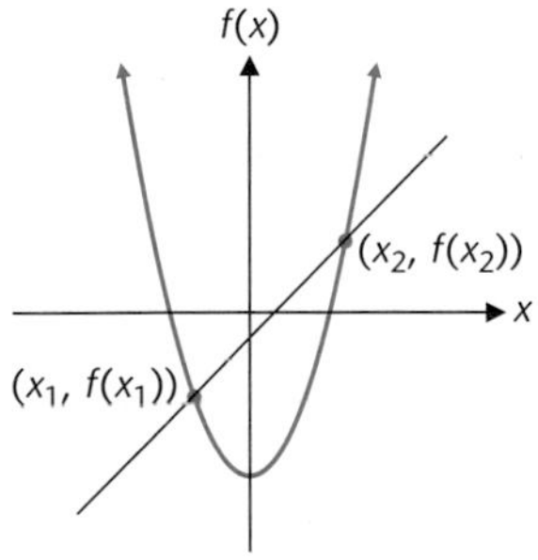

Secant line for the graph of a function
(b)

In Problems 49 and 50, find the equation of the secant line through the indicated points on the graph of f. Graph f and the secant line on the same coordinate system.

49. $f(x) = x^2 - 4$; $(-1, -3)$, $(3, 5)$

50. $f(x) = 9 - x^2$; $(-2, 5)$, $(4, -7)$

51. Let $f(x) = x^2 - 3x + 5$. If h is a nonzero real number, then $(2, f(2))$ and $(2 + h, f(2 + h))$ are two distinct points on the graph of f.

(A) Find the slope of the secant line through these two points.

(B) Evaluate the slope of the secant line for $h = 1$, $h = 0.1$, $h = 0.01$, and $h = 0.001$. What value does the slope seem to be approaching?

52. Repeat Problem 51 for $f(x) = x^2 + 2x - 6$.

53. Find the minimum product of two numbers whose difference is 30. Is there a maximum product? Explain.

54. Find the maximum product of two numbers whose sum is 60. Is there a minimum product? Explain.

APPLICATIONS

55. Construction. A horse breeder wants to construct a corral next to a horse barn 50 feet long, using all of the barn as one side of the corral (see the figure). He has 250 feet of fencing available and wants to use all of it.

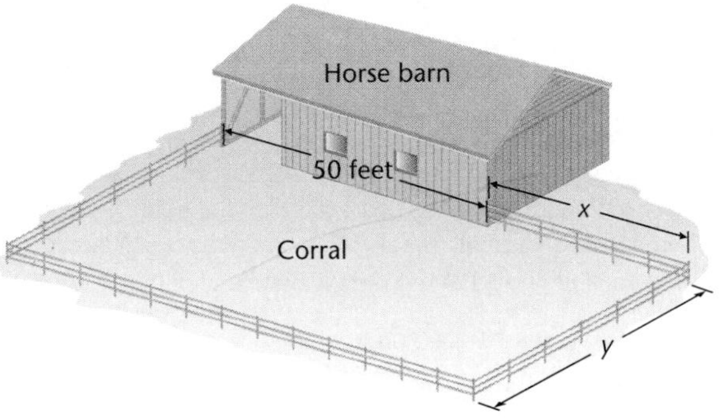

(A) Express the area $A(x)$ of the corral as a function of x and indicate its domain.

(B) Find the value of x that produces the maximum area.

(C) What are the dimensions of the corral with the maximum area?

56. Construction. Repeat Problem 55 if the horse breeder has only 140 feet of fencing available for the corral. Does the maximum value of the area function still occur at the vertex? Explain.

57. Projectile Flight. An arrow shot vertically into the air from a cross bow reaches a maximum height of 484 feet after 5.5 seconds of flight. Let the quadratic function $d(t)$ represent the distance above ground (in feet) t seconds after the arrow is released. (If air resistance is neglected, a quadratic model provides a good approximation for the flight of a projectile.)

(A) Find $d(t)$ and state its domain.

(B) At what times (to two decimal places) will the arrow be 250 feet above the ground?

58. Projectile Flight. Repeat Problem 57 if the arrow reaches a maximum height of 324 feet after 4.5 seconds of flight.

59. Engineering. The arch of a bridge is in the shape of a parabola 14 feet high at the center and 20 feet wide at the base (see the figure).

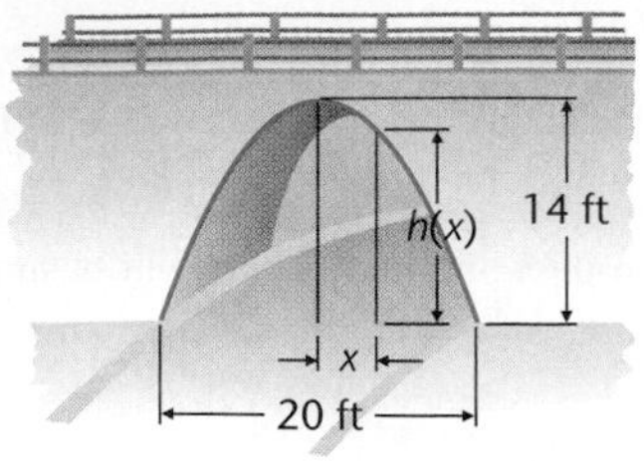

(A) Express the height of the arch $h(x)$ in terms of x and state its domain.

(B) Can a truck that is 8 feet wide and 12 feet high pass through the arch?

(C) What is the tallest 8-foot-wide truck that can pass through the arch?

(D) What (to two decimal places) is the widest 12-foot-high truck that can pass through the arch?

60. Engineering. The roadbed of one section of a suspension bridge is hanging from a large cable suspended between two towers that are 200 feet apart (see the figure). The cable forms a parabola that is 60 feet above the roadbed at the towers and 10 feet above the roadbed at the lowest point.

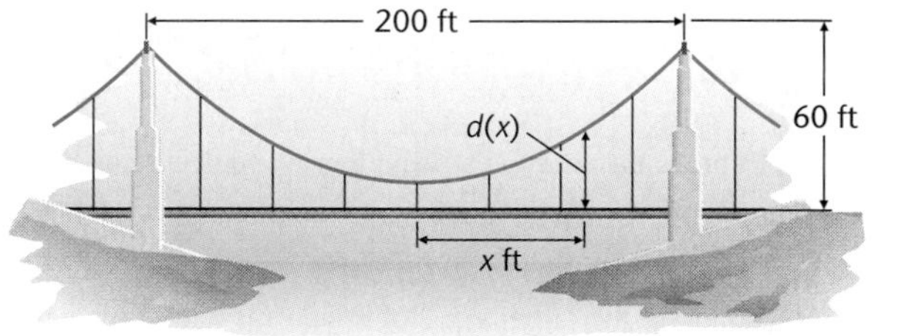

(A) Express the vertical distance $d(x)$ (in feet) from the roadbed to the suspension cable in terms of x and state the domain of d.

(B) The roadbed is supported by seven equally spaced vertical cables (see the figure). Find the combined total length of these supporting cables.

61. Break-Even Analysis. Table 1 contains revenue and cost data for the production of lawn mowers where R is the total revenue (in dollars) from the sale of x lawn mowers and C is the total cost (in dollars) of producing x lawn mowers.

TABLE 1

x	R ($)	C ($)
200	95,000	145,000
650	275,000	160,000
1,000	290,000	210,000
1,350	260,000	230,000
1,700	140,000	270,000

(A) Find a quadratic regression model for the revenue data using x as the independent variable.

(B) Find a linear regression model for the cost data using x as the independent variable.

(C) Use the regression models from parts A and B to estimate the x coordinates (to the nearest integer) of the break-even points.

62. Profit Analysis. Use the regression models computed in Problem 61 to estimate the indicated quantities.

(A) How many lawn mowers (to the nearest integer) must be produced and sold to realize a profit of $50,000?

(B) How many lawn mowers (to the nearest integer) must be produced and sold to realize the maximum profit? What is the maximum profit (to the nearest dollar)?

63. Water Consumption. Table 2 contains data related to the water consumption in the United States for selected years from 1960 to 1990. This data is based on U.S. Geological Survey, *Estimated Use of Water in the United States in 1990,* circular 1081, and previous quinquennial issues.

TABLE 2 Daily Water Consumption

Year	Total (billion gallons)	Irrigation (billion gallons)
1960	61	52
1965	77	66
1970	87	73
1975	96	80
1980	100	83
1985	92	74
1990	94	76

(A) Let the independent variable x represent years since 1960. Find a quadratic regression model for the total daily water consumption.

(B) If daily water consumption continues to follow the trend exhibited in Table 2, when (to the nearest year) would the total consumption return to the 1960 level?

64. Water Consumption. Refer to Problem 63.

(A) Let the independent variable x represent years since 1960. Find a quadratic regression model for the daily water consumption for irrigation.

(B) If daily water consumption continues to follow the trend exhibited in Table 2, when (to the nearest year) would the consumption for irrigation return to the 1960 level?

Section 2-4 | Complex Numbers

V
- Introductory Remarks
- The Complex Number System
- Complex Numbers and Radicals

Introductory Remarks

The Pythagoreans (c. 500 B.C.) found that the simple quadratic equation

$$x^2 - 2 = 0 \tag{1}$$

had no rational number solution—it is not possible to find the ratio of two integers whose square is 2. If equation (1) were to have a solution, then a new kind of number had to be invented—an irrational number. The irrational numbers $\sqrt{2}$ and $-\sqrt{2}$ are both solutions to equation (1). Irrational numbers were not put on a firm mathematical foundation until the nineteenth century. The rational and irrational numbers together constitute the real number system. Is there any reason to invent any other kinds of numbers?

Explore/Discuss 1

Graph $g(x) = x^2 - 1$ in a standard viewing window and discuss the relationship between the real zeros of the function and the x intercepts of its graph. Do the same for $f(x) = x^2 + 1$.

Does the simple quadratic equation

$$x^2 + 1 = 0 \tag{2}$$

have a solution? If equation (2) is to have a solution, x^2 must be negative. But the square of a real number is never negative. Thus, equation (2) cannot have any real number solutions. Once again, a new type of number must be invented—a number whose square can be negative. These new numbers are among the numbers called *complex numbers*. The complex numbers evolved over a long period of time, but, like the real numbers, it was not until the nineteenth century that they were given a firm mathematical foundation. Table 1 gives a brief history of the evolution of complex numbers.

TABLE 1 Brief History of Complex Numbers

Approximate Date	Person	Event
50	Heron of Alexandria	First recorded encounter of a square root of a negative number.
850	Mahavira of India	Said that a negative has no square root, since it is not a square.
1545	Cardano of Italy	Solutions to cubic equations involved square roots of negative numbers.
1637	Descartes of France	Introduced the terms *real* and *imaginary*.
1748	Euler of Switzerland	Used i for $\sqrt{-1}$.
1832	Gauss of Germany	Introduced the term *complex number*.

The Complex Number System

We start the development of the complex number system by defining a complex number and several special types of complex numbers. We then define equality, addition, and multiplication in this system, and from these definitions the important special properties and operational rules for addition, subtraction, multiplication, and division will follow.

DEFINITION 1

COMPLEX NUMBER

A **complex number** is a number of the form

$$a + bi \qquad \textbf{Standard Form}$$

where a and b are real numbers and i is called the **imaginary unit.**

The imaginary unit i introduced in Definition 1 is not a real number. It is a special symbol used in the representation of the elements in this new complex number system.

Some examples of complex numbers are

$$3 - 2i \qquad \tfrac{1}{2} + 5i \qquad 2 - \tfrac{1}{3}i$$
$$0 + 3i \qquad 5 + 0i \qquad 0 + 0i$$

Particular kinds of complex numbers are given special names as follows:

DEFINITION 2

NAMES FOR PARTICULAR KINDS OF COMPLEX NUMBERS

Imaginary Unit:	i	
Complex Number:	$a + bi$	a and b real numbers
Imaginary Number:	$a + bi$	$b \neq 0$
Pure Imaginary Number:	$0 + bi = bi$	$b \neq 0$
Real Number:	$a + 0i = a$	
Zero:	$0 + 0i = 0$	
Conjugate of $a + bi$:	$a - bi$	

EXAMPLE 1

Special Types of Complex Numbers

Given the list of complex numbers:

$$3 - 2i \qquad \tfrac{1}{2} + 5i \qquad 2 - \tfrac{1}{3}i$$
$$0 + 3i = 3i \qquad 5 + 0i = 5 \qquad 0 + 0i = 0$$

(A) List all the imaginary numbers, pure imaginary numbers, real numbers, and zero.

(B) Write the conjugate of each.

Solutions

(A) Imaginary numbers: $3 - 2i, \frac{1}{2} + 5i, 2 - \frac{1}{3}i, 3i$

Pure imaginary numbers: $0 + 3i = 3i$

Real numbers: $5 + 0i = 5, 0 - 0i = 0$

Zero: $0 + 0i = 0$

(B) $3 + 2i$ $\quad \frac{1}{2} - 5i$ $\quad 2 + \frac{1}{3}i$

$0 - 3i = -3i$ $\quad 5 - 0i = 5$ $\quad 0 - 0i = 0$

MATCHED PROBLEM 1

Given the list of complex numbers:

$6 + 7i$ $\quad \sqrt{2} - \frac{1}{3}i$ $\quad 0 - i = -i$

$0 + \frac{2}{3}i = \frac{2}{3}i$ $\quad -\sqrt{3} + 0i = -\sqrt{3}$ $\quad 0 - 0i = 0$

(A) List all the imaginary numbers, pure imaginary numbers, real numbers, and zero.

(B) Write the conjugate of each.

In Definition 2, notice that we identify a complex number of the form $a + 0i$ with the real number a, a complex number of the form $0 + bi$, $b \neq 0$, with the **pure imaginary number** bi, and the complex number $0 + 0i$ with the real number 0. Thus, a real number is also a complex number, just as a rational number is also a real number. Any complex number that is not a real number is called an **imaginary number.** If we combine the set of all real numbers with the set of all imaginary numbers, we obtain ***C*, the set of complex numbers.** The relationship of the complex number system to the other number systems we have studied is shown in Figure 1.

FIGURE 1

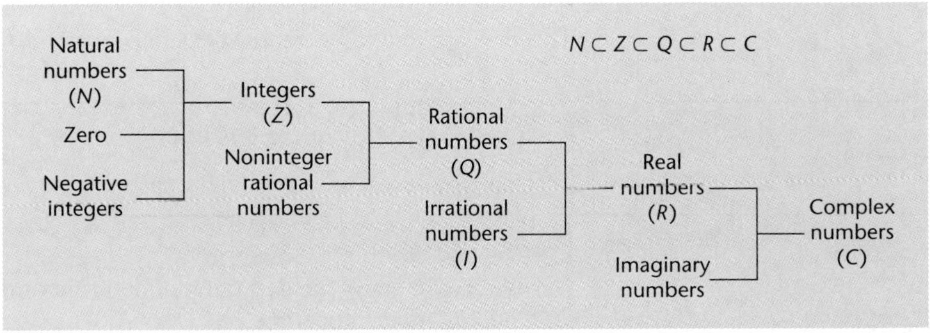

To use complex numbers, we must know how to add, subtract, multiply, and divide them. We start by defining equality, addition, and multiplication.

DEFINITION 3

EQUALITY AND BASIC OPERATIONS

1. **Equality:** $a + bi = c + di$ if and only if $a = c$ and $b = d$
2. **Addition:** $(a + bi) + (c + di) = (a + c) + (b + d)i$
3. **Multiplication:** $(a + bi)(c + di) = (ac - bd) + (ad + bc)i$

The basic properties of the real number system are discussed in Appendix A, Section A-1. Using the definition of addition and subtraction of complex numbers (Definition 3), it can be shown that the complex number system possesses the same properties. That is,

BASIC PROPERTIES OF THE COMPLEX NUMBER SYSTEM

1. Addition and multiplication of complex numbers are commutative and associative operations.
2. There is an additive identity and a multiplicative identity for complex numbers.
3. Every complex number has an additive inverse or negative.
4. Every nonzero complex number has a multiplicative inverse or reciprocal.
5. Multiplication distributes over addition.

As a consequence of these properties you will not have to memorize the definitions of addition and multiplication of complex numbers in Definition 3.

We can manipulate complex number symbols of the form $a + bi$ just like we manipulate real binomials of the form $a + bx$, as long as we remember that i is a special symbol for the imaginary unit, not for a real number.

The first arithmetic operation we consider is **addition.**

EXAMPLE 2

Addition of Complex Numbers

Carry out each operation and express the answer in standard form:

(A) $(2 - 3i) + (6 + 2i)$ (B) $(-5 + 4i) + (0 + 0i)$

Solutions

(A) We could apply the definition of addition directly, but it is easier to use complex number properties.

$$(2 - 3i) + (6 + 2i) = 2 - 3i + 6 + 2i \quad \text{Remove parentheses.}$$
$$= (2 + 6) + (-3 + 2)i \quad \text{Combine like terms.}$$
$$= 8 - i$$

(B) $(-5 + 4i) + (0 + 0i) = -5 + 4i + 0 + 0i$

$= -5 + 4i$

MATCHED PROBLEM 2

Carry out each operation and express the answer in standard form:
(A) $(3 + 2i) + (6 - 4i)$ (B) $(0 + 0i) + (7 - 5i)$

Example 2, part B, and Matched Problem 2, part B, illustrate the following general result: For any complex number $a + bi$,

$$(a + bi) + (0 + 0i) = (0 + 0i) + (a + bi) = a + bi$$

Thus, $0 + 0i$ is the **additive identity** or **zero** for the complex numbers. We anticipated this result in Definition 1 when we identified the complex number $0 + 0i$ with the real number 0.

Your graphing utility may be able to perform complex number arithmetic (check your manual). Figure 2(a) shows the solution of Example 2 on a TI-83 using the standard $a + bi$ notation. Figure 2(b) shows the same operations on a TI-86.

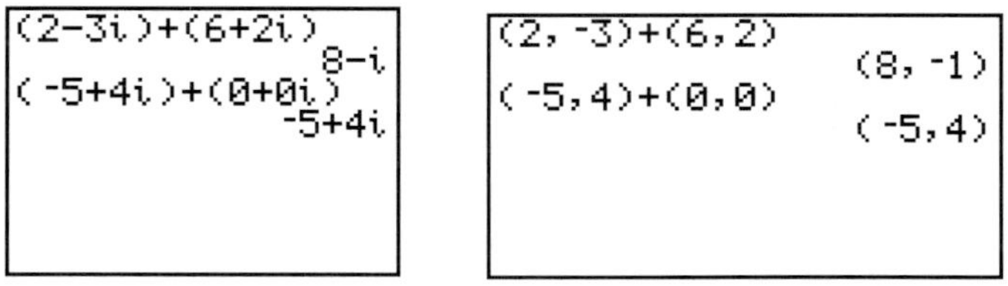

(a) TI-83 (b) TI-86

FIGURE 2
Complex number arithmetic on some typical graphing utilities.

Note that the TI-86 does not support the $a + bi$ notation for complex numbers. Instead, the complex number $a + bi$ is entered as the ordered pair (a, b). For now, if your graphing utility uses ordered pair notation for complex numbers, just translate each $a + bi$ form into an ordered pair, perform the operation, and translate back to $a + bi$ form.

We now turn to **negatives** and **subtraction,** which can be defined in terms of the additive inverse of a complex number. However, because of the already stated properties of complex numbers, we can manipulate $a + bi$ in the same way we manipulate the real binomial form $a + bx$.

EXAMPLE 3

Negation and Subtraction

Carry out each operation and express the answer in standard form:
(A) $-(4 - 5i)$ (B) $(7 - 3i) - (6 + 2i)$ (C) $(-2 + 7i) + (2 - 7i)$

Solutions

(A) $-(4 - 5i) = (-1)(4 - 5i) = -4 + 5i$

(B) $(7 - 3i) - (6 + 2i) = 7 - 3i - 6 - 2i$

$= 1 - 5i$

(C) $(-2 + 7i) + (2 - 7i) = -2 + 7i + 2 - 7i = 0$

MATCHED PROBLEM 3

Carry out each operation and express the answer in standard form:

(A) $-(-3 + 2i)$ (B) $(3 - 5i) - (1 - 3i)$ (C) $(-4 + 9i) + (4 - 9i)$

Figure 3 shows a check of the calculations in Example 3 on two different graphing utilities.

FIGURE 3

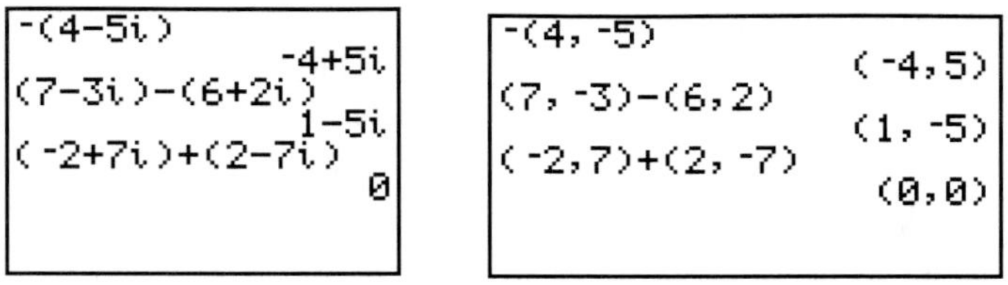

(a) TI-83 (b) TI-86

In general, the **additive inverse** or **negative** of $a + bi$ is $-a - bi$ since

$$(a + bi) + (-a - bi) = (-a - bi) + (a + bi) = 0$$

(see Example 3, part C, and Matched Problem 3, part C).

Now we turn our attention to multiplication. First, we use the definition of multiplication to see what happens to the complex unit i when it is squared:

$$\begin{aligned} & \qquad (a + bi)(c + di) \\ i^2 &= (0 + 1i)(0 + 1i) \\ & \qquad (ac - bd) + (ad + bc)i \\ &= (0 \cdot 0 - 1 \cdot 1) + (0 \cdot 1 + 1 \cdot 0)i \\ &= -1 + 0i \\ &= -1 \end{aligned}$$

Thus, we have proved that

$$i^2 = -1$$

That is, the square of i is a negative real number and i is a solution to $x^2 + 1 = 0$. Since $i^2 = -1$, we define $\sqrt{-1}$ to be the imaginary unit i. Thus,

$$i = \sqrt{-1} \quad \text{and} \quad -i = -\sqrt{-1}$$

Just as in the case of addition and subtraction, **multiplication of complex numbers** can be carried out using the properties of complex numbers stated above. That is, we can manipulate $a + bi$ in the same way we manipulate the real binomial form $a + bx$. The key difference is that we replace i^2 with -1 each time it occurs.

EXAMPLE 4

Multiplying Complex Numbers

Carry out each operation and express the answer in standard form:

(A) $(2 - 3i)(6 + 2i)$ (B) $1(3 - 5i)$

(C) $i(1 + i)$ (D) $(3 + 4i)(3 - 4i)$

Solutions

(A) $(2 - 3i)(6 + 2i) = 2(6 + 2i) - 3i(6 + 2i)$

$= 12 + 4i - 18i - 6i^2$

$= 12 - 14i - 6(-1)$ Replace i^2 with -1.

$= 18 - 14i$

(B) $1(3 - 5i) = 1 \cdot 3 - 1 \cdot 5i = 3 - 5i$

(C) $i(1 + i) = i + i^2 = i - 1 = -1 + i$

(D) $(3 + 4i)(3 - 4i) = 9 - 12i + 12i - 16i^2$

$= 9 + 16 = 25$

MATCHED PROBLEM 4

Carry out each operation and express the answer in standard form:

(A) $(5 + 2i)(4 - 3i)$ (B) $3(-2 + 6i)$ (C) $i(2 - 3i)$ (D) $(2 + 3i)(2 - 3i)$

A graphing utility check of Example 4 is shown in Figure 4.

FIGURE 4

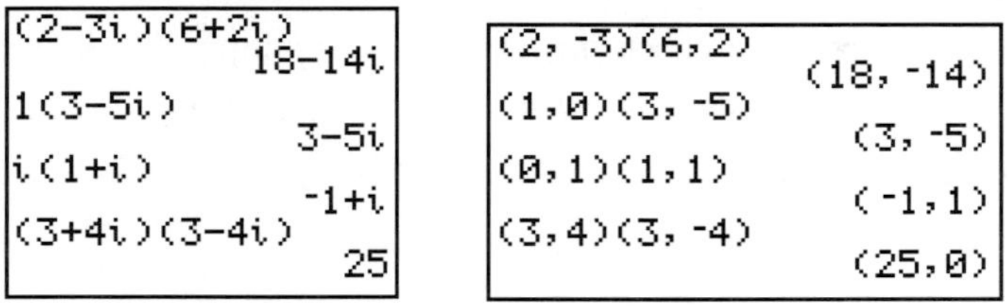

(a) TI-83 (b) TI-86

For any complex number $a + bi$,

$$1(a + bi) = (a + bi)1 = a + bi$$

(see Example 4, part B). Thus, 1 is the **multiplicative identity** for complex numbers, just as it is for real numbers.

Earlier we stated that every nonzero complex number has a multiplicative inverse or reciprocal. We will denote this as a fraction, just as we do with real numbers. Thus,

$$\frac{1}{a + bi} \quad \text{is the \textbf{reciprocal} of} \quad a + bi \qquad a + bi \neq 0$$

The following important property of the conjugate of a complex number is used to express reciprocals and quotients in standard form.

THEOREM 1

PRODUCT OF A COMPLEX NUMBER AND ITS CONJUGATE

$$(a + bi)(a - bi) = a^2 + b^2 \quad \text{A real number}$$

We now turn to the fourth arithmetic operation, **division of complex numbers.** Division can be performed by making direct use of Theorem 1. As before, we can manipulate $a + bi$ in the same way we manipulate the real binomial form $a + bx$, except we replace i^2 with -1 each time it occurs. Example 5 should make the process clear.

EXAMPLE 5

Reciprocals and Quotients

Carry out each operation and express the answer in standard form:

(A) $\dfrac{1}{2 + 3i}$ (B) $\dfrac{7 - 3i}{1 + i}$

Solutions

(A) Multiply numerator and denominator by the conjugate of the denominator:

$$\frac{1}{2 + 3i} = \frac{1}{2 + 3i} \cdot \frac{2 - 3i}{2 - 3i} \boxed{= \frac{2 - 3i}{4 - 9i^2} = \frac{2 - 3i}{4 + 9}}$$

$$= \frac{2 - 3i}{13} = \frac{2}{13} - \frac{3}{13}i$$

To check this answer, we multiply the divisor by the quotient:

Check

$$(2 + 3i)\left(\frac{2}{13} - \frac{3}{13}i\right) = \frac{4}{13} - \frac{6}{13}i + \frac{6}{13}i - \frac{9}{13}i^2$$

$$= \frac{4}{13} + \frac{9}{13} = 1$$

$$\text{(B)} \quad \frac{7 - 3i}{1 + i} = \frac{7 - 3i}{1 + i} \cdot \frac{1 - i}{1 - i} \boxed{= \frac{7 - 7i - 3i + 3i^2}{1 - i^2}}$$

$$= \frac{4 - 10i}{2} = 2 - 5i$$

Check

$$(1 + i)(2 - 5i) = 2 - 5i + 2i - 5i^2 = 7 - 3i$$

MATCHED PROBLEM 5

Carry out each operation and express the answer in standard form:

(A) $\dfrac{1}{4 + 2i}$ (B) $\dfrac{6 + 7i}{2 - i}$

Figure 5 shows the result of carrying out the operations in Example 5 on two graphing utilities. In the answer to Example 5, part A, note that we used a built-in routine to convert the decimal forms to fraction forms.

FIGURE 5

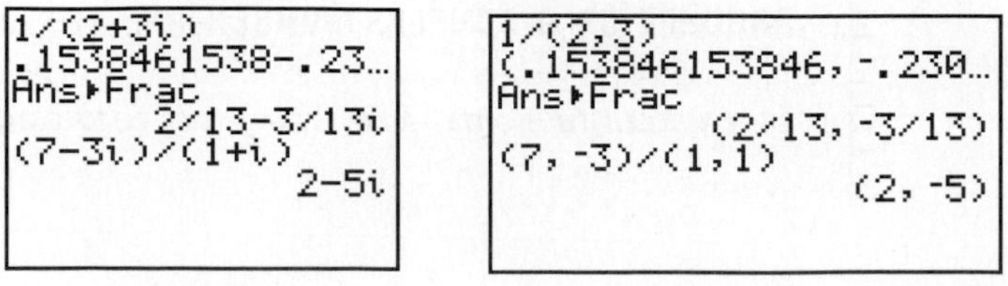

(a) TI-83 (b) TI-86

EXAMPLE 6 **Combined Operations**

Carry out the indicated operations and write each answer in standard form:

(A) $(3 - 2i)^2 - 6(3 - 2i) + 13$ (B) $\dfrac{2 - 3i}{2i}$

Solutions

(A) $$\begin{aligned}(3 - 2i)^2 - 6(3 - 2i) + 13 &= 9 - 12i + 4i^2 - 18 + 12i + 13\\ &= 9 - 12i - 4 - 18 + 12i + 13\\ &= 0\end{aligned}$$

(B) If a complex number is divided by a pure imaginary number, we can make the denominator real by multiplying numerator and denominator by i.

$$\frac{2 - 3i}{2i} \cdot \frac{i}{i} = \frac{2i - 3i^2}{2i^2} = \frac{2i + 3}{-2} = -\frac{3}{2} - i$$

MATCHED PROBLEM 6

Carry out the indicated operations and write each answer in standard form:

(A) $(3 + 2i)^2 - 6(3 + 2i) + 13$ (B) $\dfrac{4 - i}{3i}$

Explore/Discuss 2

Natural number powers of i take on particularly simple forms:

i	$i^5 = i^4 \cdot i = (1)i = i$
$i^2 = -1$	$i^6 = i^4 \cdot i^2 = 1(-1) = -1$
$i^3 = i^2 \cdot i = (-1)i = -i$	$i^7 = i^4 \cdot i^3 = 1(-i) = -i$
$i^4 = i^2 \cdot i^2 = (-1)(-1) = 1$	$i^8 = i^4 \cdot i^4 = 1 \cdot 1 = 1$

In general, what are the possible values for i^n, n a natural number? Explain how you could easily evaluate i^n for any natural number n. Then evaluate each of the following:

(A) i^{17} (B) i^{24} (C) i^{38} (D) i^{47}

If your graphing utility can perform complex arithmetic, use it to check your calculations in parts A–D.

Complex Numbers and Radicals

Recall that we say that a is a square root of b if $a^2 = b$. If x is a positive real number, then x has two square roots, the principal square root, denoted by $\sqrt{x}$, and its negative, $-\sqrt{x}$ (Section A-7). If x is a negative real number, then x still has two square roots, but now these square roots are imaginary numbers.

DEFINITION 4

PRINCIPAL SQUARE ROOT OF A NEGATIVE REAL NUMBER

The **principal square root of a negative real number,** denoted by $\sqrt{-a}$, where a is positive, is defined by

$$\sqrt{-a} = i\sqrt{a} \qquad \sqrt{-3} = i\sqrt{3} \qquad \sqrt{-9} = i\sqrt{9} = 3i$$

The other square root of $-a$, $a > 0$, is $-\sqrt{-a} = -i\sqrt{a}$.

Note in Definition 4 that we wrote $i\sqrt{a}$ and $i\sqrt{3}$ in place of the standard forms $\sqrt{a}i$ and $\sqrt{3}i$. We follow this convention whenever it appears that i might accidentally slip under a radical sign ($\sqrt{a}i \neq \sqrt{ai}$, but $\sqrt{a}i = i\sqrt{a}$). Definition 4 is motivated by the fact that

$$(i\sqrt{a})^2 = i^2a = -a$$

EXAMPLE 7

Complex Numbers and Radicals

Write in standard form:

(A) $\sqrt{-4}$ (B) $4 + \sqrt{-5}$ (C) $\dfrac{-3 - \sqrt{-5}}{2}$ (D) $\dfrac{1}{1 - \sqrt{-9}}$

Solutions

(A) $\sqrt{-4} = i\sqrt{4} = 2i$ (B) $4 + \sqrt{-5} = 4 + i\sqrt{5}$

(C) $\dfrac{-3 - \sqrt{-5}}{2} = \dfrac{-3 - i\sqrt{5}}{2} = -\dfrac{3}{2} - \dfrac{\sqrt{5}}{2}i$

(D) $\dfrac{1}{1 - \sqrt{-9}} = \dfrac{1}{1 - 3i} = \dfrac{1 \cdot (1 + 3i)}{(1 - 3i) \cdot (1 + 3i)}$

$$= \frac{1 + 3i}{1 - 9i^2} = \frac{1 + 3i}{10} = \frac{1}{10} + \frac{3}{10}i$$

MATCHED PROBLEM 7

Write in standard form:

(A) $\sqrt{-16}$ (B) $5 + \sqrt{-7}$ (C) $\dfrac{-5 - \sqrt{-2}}{2}$ (D) $\dfrac{1}{3 - \sqrt{-4}}$

Explore/Discuss 3

From Theorem 1 in Section A-7, we know that if a and b are positive real numbers, then

$$\sqrt{a}\sqrt{b} = \sqrt{ab} \tag{3}$$

Thus, we can evaluate expressions like $\sqrt{9}\sqrt{4}$ two ways:

$$\sqrt{9}\sqrt{4} = \sqrt{(9)(4)} = \sqrt{36} = 6 \quad \text{and} \quad \sqrt{9}\sqrt{4} = (3)(2) = 6$$

Explore/Discuss 3 continued

Evaluate each of the following two ways. Is equation (3) a valid property to use in all cases?

(A) $\sqrt{9}\sqrt{-4}$ (B) $\sqrt{-9}\sqrt{4}$ (C) $\sqrt{-9}\sqrt{-4}$

If your graphing utility can perform complex arithmetic, use it to check your calculations in parts A–C. If entering $\sqrt{-4}$ produces an error message, try entering $\sqrt{-4 + 0i}$.

CAUTION

Note that in Example 7, part D, we wrote $1 - \sqrt{-9} = 1 - 3i$ before proceeding with the simplification. This is a necessary step because some of the properties of radicals that are true for real numbers turn out not to be true for complex numbers. In particular, for positive real numbers a and b,

$$\sqrt{a}\sqrt{b} = \sqrt{ab} \quad \text{but} \quad \sqrt{-a}\sqrt{-b} \neq \sqrt{(-a)(-b)}$$

(See Explore/Discuss 3.)

Early resistance to these new numbers is suggested by the words used to name them: *complex* and *imaginary.* In spite of this early resistance, complex numbers have come into widespread use in both pure and applied mathematics. They are used extensively, for example, in electrical engineering, physics, chemistry, statistics, and aeronautical engineering. Our first use of them will be in connection with solutions of second-degree equations in the next section.

Answers to Matched Problems

1. (A) Imaginary numbers: $6 + 7i$, $\sqrt{2} - \frac{1}{3}i$, $0 - i = -i$, $0 + \frac{2}{3}i = \frac{2}{3}i$
Pure imaginary numbers: $0 - i = -i$, $0 + \frac{2}{3}i = \frac{2}{3}i$
Real numbers: $-\sqrt{3} + 0i = -\sqrt{3}$, $0 - 0i = 0$
Zero: $0 - 0i = 0$
(B) $6 - 7i$, $\sqrt{2} + \frac{1}{3}i$, $0 + i = i$, $0 - \frac{2}{3}i = -\frac{2}{3}i$, $-\sqrt{3} - 0i = -\sqrt{3}$, $0 + 0i = 0$

2. (A) $9 - 2i$ (B) $7 - 5i$

3. (A) $3 - 2i$ (B) $2 - 2i$ (C) 0

4. (A) $26 - 7i$ (B) $-6 + 18i$ (C) $3 + 2i$ (D) 13

5. (A) $\frac{1}{5} - \frac{1}{10}i$ (B) $1 + 4i$

6. (A) 0 (B) $-\frac{1}{3} - \frac{4}{3}i$

7. (A) $4i$ (B) $5 + i\sqrt{7}$ (C) $-\frac{5}{2} - (\sqrt{2}/2)i$ (D) $\frac{3}{13} + \frac{2}{13}i$

EXERCISE 2-4

A

In Problems 1–26, perform the indicated operations and write each answer in standard form.

1. $(2 + 4i) + (5 + i)$

2. $(3 + i) + (4 + 2i)$

3. $(-2 + 6i) + (7 - 3i)$

4. $(6 - 2i) + (8 - 3i)$

5. $(6 + 7i) - (4 + 3i)$

6. $(9 + 8i) - (5 + 6i)$

7. $(3 + 5i) - (-2 - 4i)$

8. $(8 - 4i) - (11 - 2i)$

9. $(4 - 5i) + 2i$

10. $6 + (3 - 4i)$

11. $(4i)(6i)$

12. $(3i)(8i)$

13. $-3i(2 - 4i)$

14. $-2i(5 - 3i)$

15. $(3 + 3i)(2 - 3i)$

16. $(-2 - 3i)(3 - 5i)$

17. $(2 - 3i)(7 - 6i)$

18. $(3 + 2i)(2 - i)$

19. $(7 + 4i)(7 - 4i)$

20. $(5 + 3i)(5 - 3i)$

21. $\dfrac{1}{2 + i}$

22. $\dfrac{1}{3 - i}$

23. $\dfrac{3 + i}{2 - 3i}$

24. $\dfrac{2 - i}{3 + 2i}$

25. $\dfrac{13 + i}{2 - i}$

26. $\dfrac{15 - 3i}{2 - 3i}$

B

In Problems 27–36, convert imaginary numbers to standard form, perform the indicated operations, and express answers in standard form.

27. $(2 - \sqrt{-4}) + (5 - \sqrt{-9})$

28. $(3 - \sqrt{-4}) + (-8 + \sqrt{-25})$

29. $(9 - \sqrt{-9}) - (12 - \sqrt{-25})$

30. $(-2 - \sqrt{-36}) - (4 + \sqrt{-49})$

31. $(3 - \sqrt{-4})(-2 + \sqrt{-49})$

32. $(2 - \sqrt{-1})(5 + \sqrt{-9})$

33. $\dfrac{5 - \sqrt{-4}}{7}$

34. $\dfrac{6 - \sqrt{-64}}{2}$

35. $\dfrac{1}{2 - \sqrt{-9}}$

36. $\dfrac{1}{3 - \sqrt{-16}}$

Write Problems 37–42 in standard form.

37. $\dfrac{2}{5i}$

38. $\dfrac{1}{3i}$

39. $\dfrac{1 + 3i}{2i}$

40. $\dfrac{2 - i}{3i}$

41. $(2 - 3i)^2 - 2(2 - 3i) + 9$

42. $(2 - i)^2 + 3(2 - i) - 5$

43. Let $f(x) = x^2 - 2x + 2$.

(A) Show that the conjugate complex numbers $1 + i$ and $1 - i$ are both zeros of f.

(B) Does f have any real zeros? Any x intercepts? Explain.

44. Let $g(x) = -x^2 + 4x - 5$.

(A) Show that the conjugate complex numbers $2 + i$ and $2 - i$ are both zeros of g.

(B) Does g have any real zeros? Any x intercepts? Explain.

45. Simplify: i^{18}, i^{32}, and i^{67}.

46. Simplify: i^{21}, i^{43}, and i^{52}.

In Problems 47–50, solve for x and y.

47. $(2x - 1) + (3y + 2)i = 5 - 4i$

48. $3x + (y - 2)i = (5 - 2x) + (3y - 8)i$

49. $\dfrac{(1 + x) + (y - 2)i}{1 + i} = 2 - i$

50. $\dfrac{(2 + x) + (y + 3)i}{1 - i} = -3 + i$

In Problems 51–54, solve for z. Express answers in standard form.

51. $(2 + i)z + i = 4i$

52. $(3 - i)z + 2 = i$

53. $3iz + (2 - 4i) = (1 + 2i)z - 3i$

54. $(2 - i)z + (1 - 4i) = (-1 + 3i)z + (4 + 2i)$

55. Explain what is wrong with the following "proof" that $-1 = 1$:

$$-1 = i^2 = \sqrt{-1}\sqrt{-1} = \sqrt{(-1)(-1)} = \sqrt{1} = 1$$

56. Explain what is wrong with the following "proof" that $1/i = i$. What is the correct value of $1/i$?

$$\frac{1}{i} = \frac{1}{\sqrt{-1}} = \frac{\sqrt{1}}{\sqrt{-1}} = \sqrt{\frac{1}{-1}} = \sqrt{-1} = i$$

C

In Problems 57–62, perform the indicated operations, and write each answer in standard form.

57. $(a + bi) + (c + di)$

58. $(a + bi) - (c + di)$

59. $(a + bi)(a - bi)$

60. $(u - vi)(u + vi)$

61. $(a + bi)(c + di)$

62. $\dfrac{a + bi}{c + di}$

63. Show that $i^{4k} = 1$, k a natural number.

64. Show that $i^{4k+1} = i$, k a natural number.

65. Show that $2 - i$ and $-2 + i$ are square roots of $3 - 4i$.

66. Show that $-3 + 2i$ and $3 - 2i$ are square roots of $5 - 12i$.

67. Describe how you could find the square roots of $8 - 6i$ without using a graphing utility. What are the square roots of $8 - 6i$?

68. Describe how you could find the square roots of $2i$ without using a graphing utility. What are the square roots of $2i$?

69. Let $S_n = i + i^2 + i^3 + \cdots + i^n$, $n \geq 1$. Describe the possible values of S_n.

70. Let $T_n = i^2 + i^4 + i^6 + \cdots + i^{2n}$, $n \geq 1$. Describe the possible values of T_n.

Supply the reasons in the proofs for the theorems stated in Problems 71 and 72.

71. *Theorem:* The complex numbers are commutative under addition.

Proof: Let $a + bi$ and $c + di$ be two arbitrary complex numbers; then,

Statement

1. $(a + bi) + (c + di) = (a + c) + (b + d)i$

2. $= (c + a) + (d + b)i$

3. $= (c + di) + (a + bi)$

Reason

1.

2.

3.

72. *Theorem:* The complex numbers are commutative under multiplication.

Proof: Let $a + bi$ and $c + di$ be two arbitrary complex numbers; then,

Statement

1. $(a + bi) \cdot (c + di) = (ac - bd) + (ad + bc)i$

2. $= (ca - db) + (da + cb)i$

3. $= (c + di)(a + bi)$

Reason

1.

2.

3.

Letters z and w are often used as complex variables, where $z = x + yi$, $w = u + vi$, and x, y, u, and v are real numbers. The conjugates of z and w, denoted by $\bar{z}$ and $\bar{w}$, respectively, are given by $\bar{z} = x - yi$ and $\bar{w} = u - vi$. In Problems 73–80, express each property of conjugates verbally and then prove the property.

73. $z\bar{z}$ is a real number.

74. $z + \bar{z}$ is a real number.

75. $\bar{z} = z$ if and only if z is real.

76. $\bar{\bar{z}} = z$

77. $\overline{z + w} = \bar{z} + \bar{w}$

78. $\overline{z - w} = \bar{z} - \bar{w}$

79. $\overline{zw} = \bar{z} \cdot \bar{w}$

80. $\overline{z/w} = \bar{z}/\bar{w}$

Section 2-5 | Quadratic Equations and Inequalities

- Introduction
- Solution by Factoring
- Solution by Completing the Square
- Solution by Quadratic Formula
- Solving Quadratic Inequalities

Introduction

In this book we are primarily interested in functions with real number domains and ranges. However, if we want to fully understand the nature of the zeros of a function or the roots of an equation, it is necessary to extend some of the definitions in Section 1-4 to include complex numbers. A complex number r is a **zero** of the function $f(x)$ and a **root** of the equation $f(x) = 0$ if $f(r) = 0$. As before, if r is a real number, then r is also an x intercept of the graph of f. An imaginary zero can never be an x intercept.

If a, b, and c are real numbers, $a \neq 0$, then associated with the quadratic function

$$f(x) = ax^2 + bx + c$$

is the **quadratic equation**

$$ax^2 + bx + c = 0$$

Explore/Discuss 1

Match the zeros of each function with one of the sets A, B, or C:

Function	Zeros
$f(x) = x^2 - 1$	$A = \{1\}$
$g(x) = x^2 + 1$	$B = \{-1, 1\}$
$h(x) = (x - 1)^2$	$C = \{-i, i\}$

Which of these sets of zeros can be found using graphical approximation techniques? Which cannot?

A graphing utility can be used to approximate the real roots of an equation, but not the imaginary roots. In this section we will develop algebraic techniques for finding the exact value of the roots of a quadratic equation, real or imaginary. In the process, we will derive the well-known *quadratic formula,* another important tool for our mathematical toolbox.

Solution by Factoring

If $ax^2 + bx + c$ can be written as the product of two first-degree factors, then the quadratic equation can be quickly and easily solved. The method of solution by factoring rests on the zero property of complex numbers, which is a generalization of the zero property of real numbers introduced in Section A-1.

ZERO PROPERTY

If m and n are complex numbers, then

$$m \cdot n = 0 \quad \text{if and only if} \quad m = 0 \text{ or } n = 0 \text{ (or both)}$$

EXAMPLE 1 **Solving Quadratic Equations by Factoring**

Solve by factoring:

(A) $6x^2 - 19x - 7 = 0$ (B) $x^2 - 6x + 5 = -4$ (C) $2x^2 = 3x$

Solutions

(A) $6x^2 - 19x - 7 = 0$

$(2x - 7)(3x + 1) = 0$ Factor left side.

$2x - 7 = 0$ or $3x + 1 = 0$

$x = \frac{7}{2}$ $\qquad x = -\frac{1}{3}$

The solution set is $\{-\frac{1}{3}, \frac{7}{2}\}$.

(B) $x^2 - 6x + 5 = -4$

$x^2 - 6x + 9 = 0$ Write in standard form.

$(x - 3)^2 = 0$ Factor left side.

$x = 3$

The solution set is $\{3\}$. The equation has one root, 3. But since it came from two factors, we call 3 a **double root** or a **root of multiplicity 2.**

(C) $2x^2 = 3x$

$2x^2 - 3x = 0$

$x(2x - 3) = 0$

$x = 0$ or $2x - 3 = 0$

$x = \frac{3}{2}$

Solution set: $\{0, \frac{3}{2}\}$

MATCHED PROBLEM 1

Solve by factoring:

(A) $3x^2 + 7x - 20 = 0$ (B) $4x^2 + 12x + 9 = 0$ (C) $4x^2 = 5x$

CAUTION

1. One side of an equation must be 0 before the zero property can be applied. Thus

$$x^2 - 6x + 5 = -4$$

$$(x - 1)(x - 5) = -4$$

does not imply that $x - 1 = -4$ or $x - 5 = -4$. See Example 1, part B, for the correct solution of this equation.

2. The equations

$$2x^2 = 3x \qquad \text{and} \qquad 2x = 3$$

are not equivalent. The first has solution set $\{0, \frac{3}{2}\}$, while the second has solution set $\{\frac{3}{2}\}$. The root $x = 0$ is lost when each member of the first equation is divided by the variable x. See Example 1, part C, for the correct solution of this equation.

Do not divide both members of an equation by an expression containing the variable for which you are solving. You may be dividing by 0.

Remark It is common practice to represent solutions of quadratic equations informally by the last equation rather than by writing a solution set using set notation. From now on, we will follow this practice unless a particular emphasis is desired.

Solution by Completing the Square

Factoring is a specialized method that is very efficient if the factors can be quickly identified. However, not all quadratic equations are easy to factor. We now turn to a more general process that is guaranteed to work in all cases. This process is based on completing the square, discussed in Section 2-3, and the following square root property:

SQUARE ROOT PROPERTY

For any complex numbers r and s, if $r^2 = s$, then $r = \pm\sqrt{s}$.

EXAMPLE 2 Solution by Completing the Square

Use completing the square and the square root property to solve each of the following:

(A) $(x + \frac{1}{2})^2 - \frac{5}{4} = 0$ (B) $x^2 + 6x - 2 = 0$ (C) $2x^2 - 4x + 3 = 0$

Solutions

(A) This quadratic expression is already written in standard form. We solve for the squared term and then use the square root property:

$$(x + \tfrac{1}{2})^2 - \tfrac{5}{4} = 0$$

$$(x + \tfrac{1}{2})^2 = \tfrac{5}{4} \qquad \text{Apply the square root property.}$$

$$x + \tfrac{1}{2} = \pm\sqrt{\tfrac{5}{4}} \qquad \text{Solve for } x.$$

$$x = -\frac{1}{2} \pm \frac{\sqrt{5}}{2}$$

$$= \frac{-1 \pm \sqrt{5}}{2}$$

(B) We can speed up the process of completing the square by taking advantage of the fact that we are working with a quadratic equation, not a quadratic expression.

$$x^2 + 6x - 2 = 0$$

$$x^2 + 6x = 2$$

$$x^2 + 6x + 9 = 2 + 9 \qquad \text{Complete the square on the left side, and add the same number to the right side.}$$

$$(x + 3)^2 = 11$$

$$x + 3 = \pm\sqrt{11}$$

$$x = -3 \pm \sqrt{11}$$

(C) $2x^2 - 4x + 3 = 0$

$x^2 - 2x + \frac{3}{2} = 0$ Make the leading coefficient 1 by dividing by 2.

$x^2 - 2x = -\frac{3}{2}$

$x^2 - 2x + 1 = -\frac{3}{2} + 1$ Complete the square on the left side and add the same number to the right side.

$(x - 1)^2 = -\frac{1}{2}$ Factor the left side.

$x - 1 = \pm\sqrt{-\frac{1}{2}}$

$x = 1 \pm i\sqrt{\frac{1}{2}}$

$= 1 \pm \frac{\sqrt{2}}{2}i$ Answer in $a + bi$ form.

MATCHED PROBLEM 2

Solve by completing the square:

(A) $(x + \frac{1}{3})^2 - \frac{2}{9} = 0$ (B) $x^2 + 8x - 3 = 0$ (C) $3x^2 - 12x + 13 = 0$

Explore/Discuss 2

Graph the quadratic functions associated with the three quadratic equations in Example 2. Approximate the x intercepts of each function and compare with the roots found in Example 2. Which of these equations has roots that cannot be approximated graphically?

Solution by Quadratic Formula

Now consider the general quadratic equation with unspecified coefficients:

$$ax^2 + bx + c = 0 \qquad a \neq 0$$

We can solve it by completing the square exactly as we did in Example 2, part C. To make the leading coefficient 1, we must multiply both sides of the equation by $1/a$. Thus,

$$x^2 + \frac{b}{a}x + \frac{c}{a} = 0$$

Adding $-c/a$ to both sides of the equation and then completing the square of the left side, we have

$$x^2 + \frac{b}{a}x + \frac{b^2}{4a^2} = \frac{b^2}{4a^2} - \frac{c}{a}$$

We now factor the left side and solve using the square root property:

$$\left(x + \frac{b}{2a}\right)^2 = \frac{b^2 - 4ac}{4a^2}$$

$$x + \frac{b}{2a} = \pm\sqrt{\frac{b^2 - 4ac}{4a^2}}$$

$$x = -\frac{b}{2a} \pm \frac{\sqrt{b^2 - 4ac}}{2a} \qquad \text{See Problem 77.}$$

$$= \frac{-b \pm \sqrt{b^2 - 4ac}}{2a}$$

We have thus derived the well-known and widely used **quadratic formula:**

THEOREM 1

QUADRATIC FORMULA

If $ax^2 + bx + c = 0$, $a \neq 0$, then

$$x = \frac{-b \pm \sqrt{b^2 - 4ac}}{2a}$$

The quadratic formula and completing the square are equivalent methods. Either can be used to find the exact value of the roots of any quadratic equation.

EXAMPLE 3

Using the Quadratic Formula

Solve $2x + \frac{3}{2} = x^2$ by use of the quadratic formula. Leave the answer in simplest radical form.

Solution

$$2x + \tfrac{3}{2} = x^2$$

$$4x + 3 = 2x^2 \qquad \text{Multiply both sides by 2.}$$

$$2x^2 - 4x - 3 = 0 \qquad \text{Write in standard form.}$$

$$x = \frac{-b \pm \sqrt{b^2 - 4ac}}{2a} \qquad a = 2,\ b = -4,\ c = -3$$

$$= \frac{-(-4) \pm \sqrt{(-4)^2 - 4(2)(-3)}}{2(2)}$$

$$= \frac{4 \pm \sqrt{40}}{4} = \frac{4 \pm 2\sqrt{10}}{4} = \frac{2 \pm \sqrt{10}}{2}$$

CAUTION

1. $-4^2 \neq (-4)^2 \qquad -4^2 = -16$ and $(-4)^2 = 16$
2. $2 + \frac{\sqrt{10}}{2} \neq \frac{2 + \sqrt{10}}{2} \qquad 2 + \frac{\sqrt{10}}{2} = \frac{4 + \sqrt{10}}{2}$
3. $\frac{\not{4} \pm 2\sqrt{10}}{\not{4}} \neq \pm 2\sqrt{10} \qquad \frac{4 \pm 2\sqrt{10}}{4} = \frac{2(2 \pm \sqrt{10})}{4} = \frac{2 \pm \sqrt{10}}{2}$

MATCHED PROBLEM 3

Solve $x^2 - \frac{5}{2} = -3x$ using the quadratic formula. Leave the answer in simplest radical form.

Explore/Discuss 3

Given the quadratic function $f(x) = ax^2 + bx + c$, let $D = b^2 - 4ac$. How many real zeros does f have if

(A) $D > 0$ (B) $D = 0$ (C) $D < 0$

In each of these three cases, what type of roots does the quadratic equation $f(x) = 0$ have?

The quantity $\boldsymbol{b^2 - 4ac}$ in the quadratic formula is called the **discriminant** and gives us information about the roots of the corresponding equation and the zeros of the associated quadratic function. This information is summarized in Table 1.

TABLE 1 Discriminants, Roots, and Zeros

Discriminant $b^2 - 4ac$	Roots of* $ax^2 + bx + c = 0$	Number of Real Zeros of* $f(x) = ax^2 + bx + c$
Positive	Two distinct real roots	2
0	One real root (a double root)	1
Negative	Two imaginary roots, one the conjugate of the other	0

*a, b, and c are real numbers with $a \neq 0$.

EXAMPLE 4

Design

A picture frame of uniform width has outer dimensions of 12 inches by 18 inches. How wide (to the nearest tenth of an inch) must the frame be to display an area of 140 square inches?

Solution

We begin by drawing and labeling a figure:

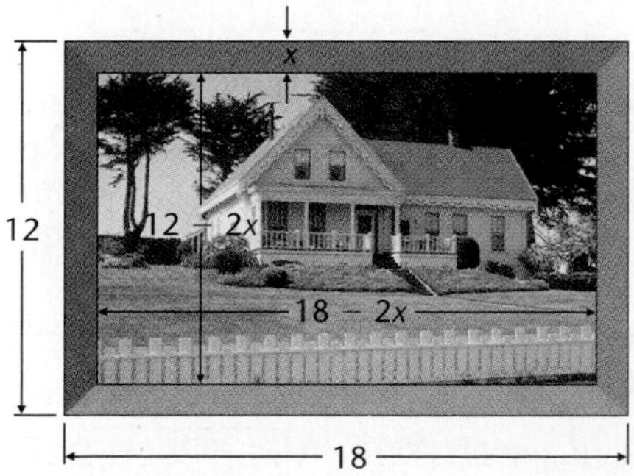

If x is the width of the frame, then x must satisfy the equation

$$(18 - 2x)(12 - 2x) = 140 \qquad (1)$$

Note that x must satisfy $0 \leq x \leq 6$ to insure that both $12 - 2x$ and $18 - 2x$ are nonnegative. The roots of this quadratic equation can be found algebraically or approximated graphically. Using both methods will confirm that we have the correct answer. We begin with the algebraic solution:

$$\begin{aligned}(18 - 2x)(12 - 2x) &= 140 \\ 216 - 36x - 24x + 4x^2 &= 140 \\ 4x^2 - 60x + 76 &= 0 \\ x^2 - 15x + 19 &= 0 \\ x &= \frac{15 \pm \sqrt{149}}{2}\end{aligned}$$

FIGURE 1

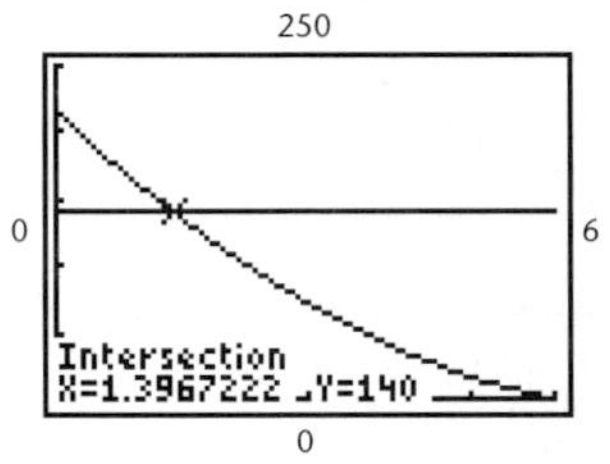

Thus, the quadratic equation has two solutions (rounded to one decimal place):

$$x = \frac{15 + \sqrt{149}}{2} = 13.6 \quad \text{and} \quad x = \frac{15 - \sqrt{149}}{2} = 1.4$$

The first must be discarded as being much too large. So the width of the frame is 1.4 inches.

Graphing both sides of equation (1) for $0 \leq x \leq 6$ and using an intersection routine confirms that this answer is correct (Fig. 1).

MATCHED PROBLEM 4

A 1,200 square foot garden is enclosed with 150 feet of fencing. Find the dimensions of the garden to the nearest tenth of a foot.

Solving Quadratic Inequalities

Explore/Discuss 4

Graph $f(x) = (x + 2)(x - 3)$ and examine the graph to determine the solutions of the following inequalities:

(A) $f(x) > 0$ (B) $f(x) < 0$ (C) $f(x) \geq 0$ (D) $f(x) \leq 0$

The simplest method for solving inequalities involving a function is to find the zeros of the function and then examine the graph to determine where the function is positive and where it is negative. Inequalities involving quadratic functions are handled routinely by this method, as the following examples illustrate.

CIL **EXAMPLE 5**

Finding the Domain of a Function

Find the domain of $f(x) = \sqrt{x^2 - 4x - 9}$. Express answer in interval notation using exact values.

Solution The domain of this function is the set of all real numbers x that produce real values for $f(x)$ (Section 1-3).

This is precisely the solution set of the quadratic inequality

$$x^2 - 4x - 9 \geq 0 \tag{2}$$

FIGURE 2
$y = x^2 - 4x - 9.$

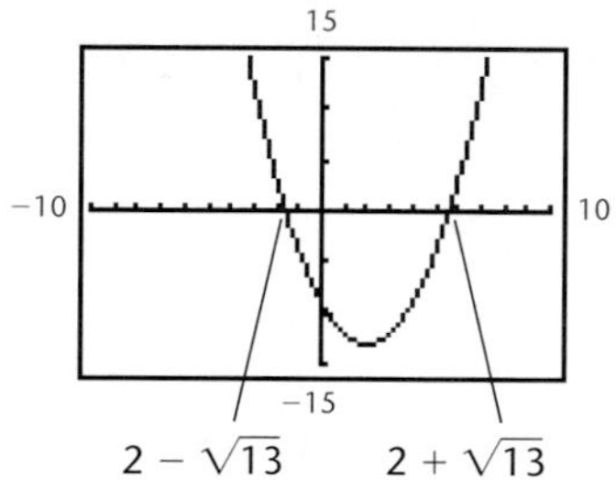

The solution of this inequality consists of all values of x for which the graph of $y = x^2 - 4x - 9$ is on or above the x axis. Using either completing the square or the quadratic formula, we find that the x intercepts are

$$x = \frac{4 \pm \sqrt{52}}{2} = 2 \pm \sqrt{13}$$

Examining the graph in Figure 2, we see that the solution of inequality (2) and, hence, the domain of f, is

$$(-\infty, 2 - \sqrt{13}] \cup [2 + \sqrt{13}, \infty)$$

MATCHED PROBLEM 5

Find the domain of $g(x) = \sqrt{2 + 2x - x^2}$. Express answer in interval notation using exact values.

EXAMPLE 6

Projectile Motion

If a projectile is shot straight upward from the ground with an initial velocity of 160 feet per second, its distance d (in feet) above the ground at the end of t seconds (neglecting air resistance) is given approximately by

$$d(t) = 160t - 16t^2$$

(A) What is the domain of d?

(B) At what times (to two decimal places) will the projectile be more than 200 feet above the ground?

Express answers in inequality notation.

Solutions

(A) Factoring $d(t)$, we have

$$d(t) = 160t - 16t^2 = 16t(10 - t)$$

Thus, $d(0) = 0$ and $d(10) = 0$. The projectile is released at $t = 0$ seconds and returns to the ground at $t = 10$ seconds, so the domain of d is $0 \leq t \leq 10$.

(B) Since we are asked for two-decimal-place accuracy, we can solve this problem graphically. Graph d and the horizontal line $y = 200$ and find the intersection points (Fig. 3).

FIGURE 3

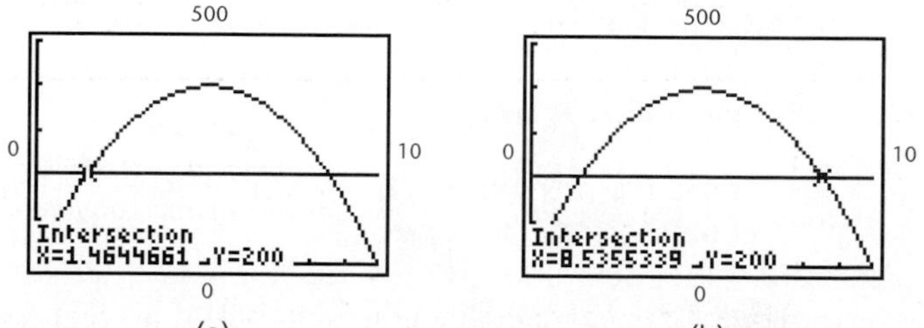

(a) (b)

From Figure 3 we see that the projectile will be above 200 feet for $1.46 < t < 8.54$.

MATCHED PROBLEM 6

Refer to the projectile equation in Example 6. At what times (to two decimal places) during its flight will the projectile be less than 250 feet above the ground? Express answer in inequality notation.

Answers to Matched Problems

1. (A) $\{-4, \frac{5}{3}\}$ (B) $\{-\frac{3}{2}\}$ (a double root) (C) $\{0, \frac{5}{4}\}$
2. (A) $x = (-1 \pm \sqrt{2})/3$ (B) $x = -4 \pm \sqrt{19}$ (C) $x = (6 \pm i\sqrt{3})/3$ or $2 \pm (\sqrt{3}/3)i$
3. $x = (-3 \pm \sqrt{19})/2$ **4.** 23.1 ft by 51.9 ft **5.** $[1 - \sqrt{3}, 1 + \sqrt{3}]$ **6.** $0 \le t < 1.94$ or $8.06 < t \le 10$

EXERCISE 2-5

A

In Problems 1–6, solve by factoring.

1. $4u^2 = 8u$
2. $3A^2 = -12A$
3. $9y^2 = 12y - 4$
4. $16x^2 + 8x = -1$
5. $11x = 2x^2 + 12$
6. $8 - 10x = 3x^2$

In Problems 7–18, solve by using the square root property.

7. $m^2 - 12 = 0$
8. $y^2 - 45 = 0$
9. $x^2 + 25 = 0$
10. $x^2 + 16 = 0$
11. $9y^2 - 16 = 0$
12. $4x^2 - 9 = 0$
13. $4x^2 + 25 = 0$
14. $16a^2 + 9 = 0$
15. $(n + 5)^2 = 9$
16. $(m - 3)^2 = 25$
17. $(d - 3)^2 = -4$
18. $(t + 1)^2 = -9$

In Problems 19–26, solve using the quadratic formula.

19. $x^2 - 10x - 3 = 0$
20. $x^2 - 6x - 3 = 0$
21. $x^2 + 8 = 4x$
22. $y^2 + 3 = 2y$
23. $2x^2 + 1 = 4x$
24. $2m^2 + 3 = 6m$
25. $5x^2 + 2 = 2x$
26. $7x^2 + 6x + 4 = 0$

In Problems 27–34, solve and graph. Express answers in both inequality and interval notation.

27. $x^2 < 10 - 3x$
28. $x^2 + x < 12$
29. $x^2 + 21 > 10x$
30. $x^2 + 7x + 10 > 0$
31. $x^2 \le 8x$
32. $x^2 + 6x \ge 0$
33. $x^2 + 5x \le 0$
34. $x^2 \le 4x$

B

In Problems 35–44, find exact answers and check with a graphing utility, if possible.

35. $x^2 - 6x - 3 = 0$
36. $y^2 - 10y - 3 = 0$
37. $2y^2 - 6y + 3 = 0$
38. $2d^2 - 4d + 1 = 0$
39. $3x^2 - 2x - 2 = 0$
40. $3x^2 + 5x - 4 = 0$
41. $12x^2 + 7x = 10$
42. $9x^2 + 9x = 4$
43. $x^2 = 3x + 1$
44. $x^2 + 2x = 2$

In Problems 45–48, solve for the indicated variable in terms of the other variables. Use positive square roots only.

45. $s = \frac{1}{2}gt^2$ for t
46. $a^2 + b^2 = c^2$ for a
47. $P = EI - RI^2$ for I
48. $A = P(1 + r)^2$ for r

In Problems 49–52, solve to two decimal places. Express answers in inequality notation.

49. $2.07x^2 - 3.79x + 1.34 > 0$
50. $0.61x^2 - 4.28x + 2.93 < 0$
51. $4.83x^2 + 2.04x - 3.18 \le 0$
52. $5.13x^2 + 7.27x - 4.32 \ge 0$

C

In Problems 53–60, find the domain of each function. Express answers in interval notation using exact values.

53. $f(x) = \sqrt{x^2 - 9}$
54. $g(x) = \sqrt{4 - x^2}$

55. $h(x) = \sqrt[4]{2x^2 + x - 6}$

56. $k(x) = \sqrt[6]{3x^2 - 7x - 6}$

57. $F(x) = \dfrac{1}{\sqrt{6x - x^2 - 4}}$

58. $G(x) = \dfrac{1}{\sqrt{8x - x^2 - 14}}$

59. Consider the quadratic equation

$$x^2 + 4x + c = 0$$

where c is a real number. Discuss the relationship between the values of c and the three types of roots listed in Table 1.

60. Consider the quadratic equation

$$x^2 - 2x + c = 0$$

where c is a real number. Discuss the relationship between the values of c and the three types of roots listed in Table 1.

In Problems 61–64, use the given information concerning the roots of the quadratic equation $ax^2 + bx + c = 0$, $a \neq 0$, to describe the possible solution sets for the indicated inequality. Illustrate your conclusions with specific examples.

61. $ax^2 + bx + c > 0$, given distinct real roots r_1 and r_2 with $r_1 < r_2$.

62. $ax^2 + bx + c \leq 0$, given distinct real roots r_1 and r_2 with $r_1 < r_2$.

63. $ax^2 + bx + c \geq 0$, given one (double) real root r.

64. $ax^2 + bx + c < 0$, given one (double) real root r.

65. Give an example of a quadratic inequality whose solution set is the entire real line.

66. Give an example of a quadratic inequality whose solution set is the empty set.

C

Solve Problems 67–70 and express answer in $a + bi$ form.

67. $x^2 + 3ix - 2 = 0$

68. $x^2 - 7ix - 10 = 0$

69. $x^2 + 2ix = 3$

70. $x^2 = 2ix - 3$

In Problems 71 and 72, find all solutions.

71. $x^3 - 1 = 0$

72. $x^4 - 1 = 0$

73. Can a quadratic equation with rational coefficients have one rational root and one irrational root? Explain.

74. Can a quadratic equation with real coefficients have one real root and one imaginary root? Explain.

75. Show that if r_1 and r_2 are the two roots of $ax^2 + bx + c = 0$, then $r_1r_2 = c/a$.

76. For r_1 and r_2 in Problem 75, show that $r_1 + r_2 = -b/a$.

77. In one stage of the derivation of the quadratic formula, we replaced the expression

$$\pm\sqrt{(b^2 - 4ac)/4a^2}$$

with

$$\pm\sqrt{b^2 - 4ac}/2a$$

What justifies using $2a$ in place of $|2a|$?

78. Find the error in the following "proof" that two arbitrary numbers are equal to each other: Let a and b be arbitrary numbers such that $a \neq b$. Then

$$(a - b)^2 = a^2 - 2ab + b^2 = b^2 - 2ab + a^2$$
$$(a - b)^2 = (b - a)^2$$
$$a - b = b - a$$
$$2a = 2b$$
$$a = b$$

APPLICATIONS

79. Numbers. Find two numbers such that their sum is 21 and their product is 104.

80. Numbers. Find all numbers with the property that when the number is added to itself the sum is the same as when the number is multiplied by itself.

81. Numbers. Find two consecutive positive even integers whose product is 168.

82. Numbers. Find two consecutive positive integers whose product is 600.

83. Profit Analysis. A screen printer produces custom silk-screen apparel. The cost $C(x)$ of printing x custom T-shirts and the revenue $R(x)$ from the sale of x T-shirts (both in dollars) are given by

$$C(x) = 200 + 2.25x$$
$$R(x) = 10x - 0.05x^2$$

Determine the production levels x (to the nearest integer) that will result in the printer showing a profit.

84. Profit Analysis. Refer to Problem 83. Determine the production levels x (to the nearest integer) that will result in the printer showing a profit of at least \$60.

85. Air Search. A search plane takes off from an airport at 6:00 A.M. and travels due north at 200 miles per hour. A second plane takes off at 6:30 A.M. and travels due east at 170 miles per hour. The planes carry radios with a maximum range of 500 miles. When (to the nearest minute) will these planes no longer be able to communicate with each other?

86. Projectile Flight. If a projectile is shot straight upward from the ground with an initial velocity of 176 feet per second, its distance d (in feet) above the ground at the end of t seconds (neglecting air resistance) is given approximately by

$$d(t) = 176t - 16t^2$$

(A) What is the domain of d?

(B) At what times (to two decimal places) will the projectile be more than 200 feet above the ground?

Express answers in inequality notation.

87. Construction. A gardener has a 30 foot by 20 foot rectangular plot of ground. She wants to build a brick walkway of uniform width on the border of the plot (see the figure). If the gardener wants to have 400 square feet of ground left for planting, how wide (to two decimal places) should she build the walkway?

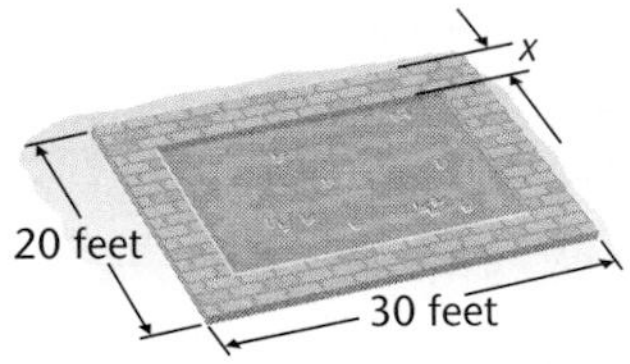

88. Construction. Refer to Problem 87. The gardener buys enough brick to build 160 square feet of walkway. Is this sufficient to build the walkway determined in Problem 87? If not, how wide (to two decimal places) can she build the walkway with these bricks?

★ **89. Architecture.** A developer wants to erect a rectangular building on a triangular-shaped piece of property that is 200 feet wide and 400 feet long (see the figure).

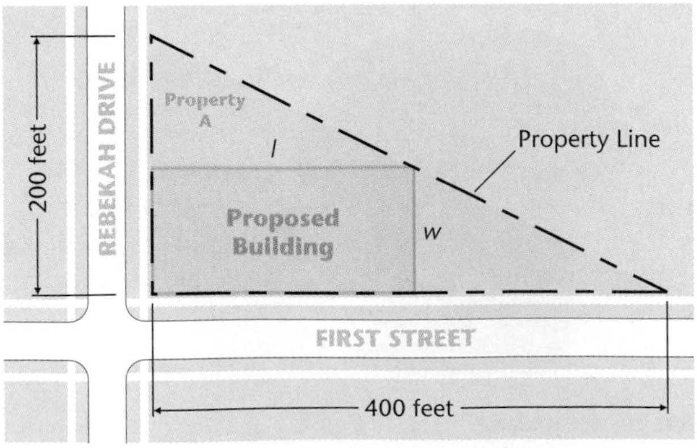

(A) Express the cross-sectional area $A(w)$ of the building as a function of the width w and state the domain of this function. [*Hint:* Use Euclid's theorem* to find a relationship between the length l and width w.]

(B) Building codes require that this building have a cross-sectional area of at least 15,000 square feet. What are the widths of the buildings that will satisfy the building codes?

(C) Can the developer construct a building with a cross-sectional area of 25,000 square feet? What is the maximum cross-sectional area of a building constructed in this manner?

★ **90. Architecture.** An architect is designing a small A-frame cottage for a resort area. A cross-section of the cottage is an isosceles triangle with a base of 5 meters and an altitude of 4 meters. The front wall of the cottage must accommodate a sliding door positioned as shown in the figure.

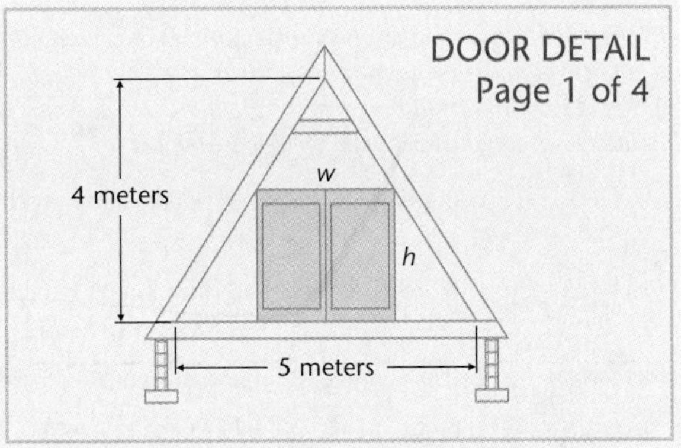

(A) Express the area $A(w)$ of the door as a function of the width w and state the domain of this function. [See the hint for Problem 89.]

(B) A provision of the building code requires that doorways must have an area of at least 4.2 square meters. Find the width of the doorways that satisfy this provision.

(C) A second provision of the building code requires all doorways to be at least 2 meters high. Discuss the effect of this requirement on the answer to part B.

91. Transportation. A delivery truck leaves a warehouse and travels north to factory A. From factory A the truck travels east to factory B and then returns directly to the warehouse (see the figure). The driver recorded the truck's odometer reading at the warehouse at both the beginning and the end of the trip and also at factory B, but forgot to record it at factory A (see the table). The driver does recall that it was further from the warehouse to factory A than it was from factory A to factory B. Since delivery charges are based on distance from the warehouse, the driver needs to know how far factory A is from the warehouse. Find this distance.

**Euclid's theorem:* If two triangles are similar, their corresponding sides are proportional:

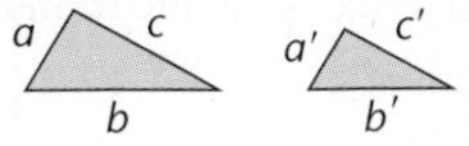

$$\frac{a}{a'} = \frac{b}{b'} = \frac{c}{c'}$$

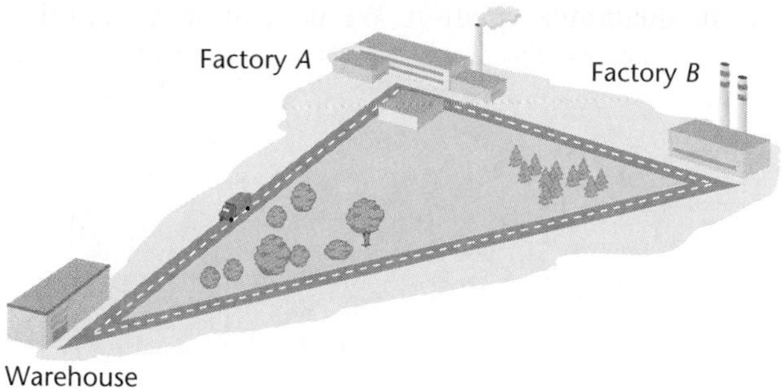

	Odometer Readings
Warehouse	52846
Factory A	52???
Factory B	52937
Warehouse	53002

★★ **92. Construction.** A $\frac{1}{4}$-mile track for racing stock cars consists of two semicircles connected by parallel straight-aways (see the figure). To provide sufficient room for pit crews, emergency vehicles, and spectator parking, the track must enclose an area of 100,000 square feet. Find the length of the straightaways and the diameter of the semicircles to the nearest foot. [*Recall:* The area A and circumference C of a circle of diameter d are given by $A = \pi d^2/4$ and $C = \pi d$.]

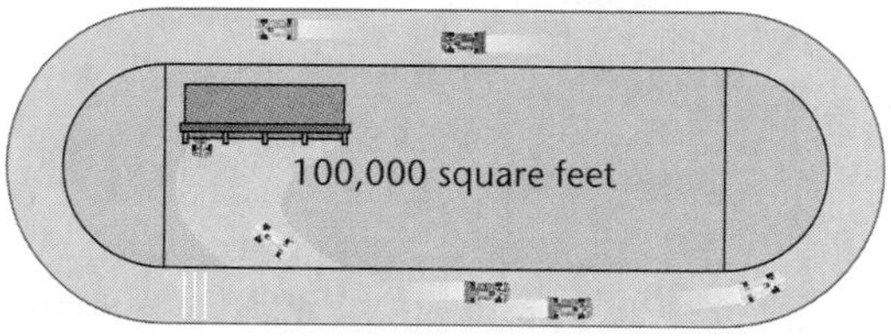

Section 2-6 | Additional Equation Solving Techniques

- Equations Involving Radicals
- Equations of the Form $ax^{2p} + bx^p + c = 0$

In this section we examine equations that can be transformed into quadratic equations by various algebraic manipulations. With proper interpretation, the solutions of the resulting quadratic equations will lead to the solutions of the original equations.

Equations Involving Radicals

Consider the equation

$$x = \sqrt{x + 2} \tag{1}$$

Graphing both sides of the equation and using an intersection routine shows that $x = 2$ is a solution to the equation (Fig. 1). Is it the only solution?

FIGURE 1 $y_1 = x$, $y_2 = \sqrt{x + 2}$.

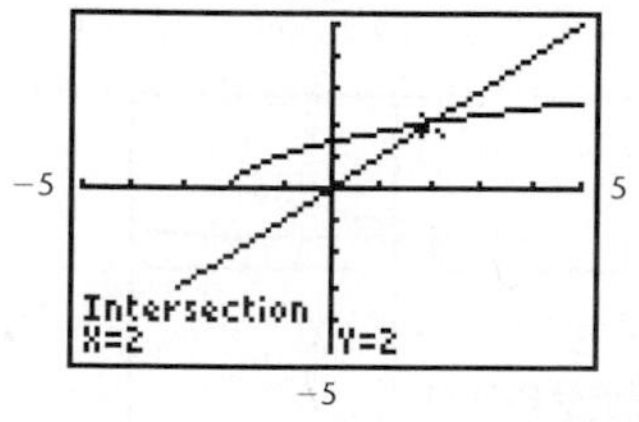

There may be other solutions not visible in this viewing window. Or there may be imaginary solutions (remember, graphical approximation applies only to real solutions). To solve this equation algebraically, we square each side of equation (1) and then proceed to solve the resulting quadratic equation. Thus,

$$\begin{aligned} x^2 &= (\sqrt{x + 2})^2 \qquad (2) \\ x^2 &= x + 2 \\ x^2 - x - 2 &= 0 \\ (x - 2)(x + 1) &= 0 \\ x &= 2, -1 \end{aligned}$$

These are the only solutions to the quadratic equation. We have already seen that $x = 2$ is a solution to the original equation. To check if -1 is a solution, we substitute in equation (1):

$$x = \sqrt{x + 2}$$
$$-1 \stackrel{?}{=} \sqrt{-1 + 2}$$
$$-1 \stackrel{?}{=} \sqrt{1}$$
$$-1 \neq 1$$

Thus, -1 is not a solution to equation (1). What have we gained by performing these algebraic manipulations? If we can be certain that all solutions of equation (1) must be among the solutions of equation (2), then we can rule out the possibility of any additional solutions to equation (1). Theorem 1 provides the necessary tool to do this.

THEOREM 1

POWER OPERATION ON EQUATIONS

If both sides of an equation are raised to the same natural number power, then the solution set of the original equation is a subset of the solution set of the new equation.

Equation	Solution Set
$x = 3$	$\{3\}$
$x^2 = 9$	$\{-3, 3\}$

Referring to equations (1) and (2) on page 157, we know that 2 and -1 are the only solutions to the quadratic equation (2). And we checked that -1 is not a solution to equation (1). Theorem 1 now implies that 2 must be the *only* solution to equation (1). We call -1 an *extraneous solution*. In general, an **extraneous solution** is a solution introduced during the solution process that does not satisfy the original equation.

Every solution of the new equation must be checked in the original equation to eliminate extraneous solutions.

Explore/Discuss 1

Figure 2 shows that $x = -1$ is a solution of the equation

$$\sqrt{x + 2} = 0.01x + 1.01$$

Are there any other solutions? Find any additional solutions both algebraically and graphically. What are some advantages and disadvantages of each of these solution methods?

FIGURE 2
$y_1 = \sqrt{x + 2}$, $y_2 = 0.01x + 1.01$.

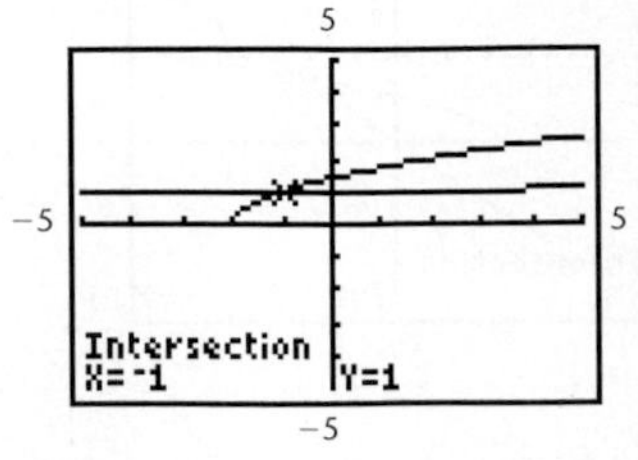

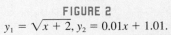

EXAMPLE 1 **Solving Equations Involving Radicals**

Solve algebraically:

(A) $\sqrt{4x^2 + 8x + 7} - x = 1$ (B) $\sqrt{2x + 3} - \sqrt{x - 2} = 2$

Solutions

(A)
$$\sqrt{4x^2 + 8x + 7} - x = 1$$
$$\sqrt{4x^2 + 8x + 7} = x + 1 \qquad \text{Isolate radical on one side.}$$
$$4x^2 + 8x + 7 = x^2 + 2x + 1 \qquad \text{Square both sides.}$$
$$3x^2 + 6x + 6 = 0$$
$$x^2 + 2x + 2 = 0 \qquad \text{Use the quadratic formula.}$$
$$x = \frac{-2 \pm \sqrt{-4}}{2}$$
$$= -1 + i, -1 - i$$

Check

$$x = -1 + i$$
$$\sqrt{4x^2 + 8x + 7} - x = 1$$
$$\sqrt{4(-1 + i)^2 + 8(-1 + i) + 7} - (-1 + i) \stackrel{?}{=} 1$$
$$\sqrt{4 - 8i - 4 - 8 + 8i + 7} + 1 - i \stackrel{?}{=} 1$$
$$1 \stackrel{\checkmark}{=} 1$$

$$x = -1 - i$$
$$\sqrt{4x^2 + 8x + 7} - x = 1$$
$$\sqrt{4(-1 - i)^2 + 8(-1 - i) + 7} - (-1 - i) \stackrel{?}{=} 1$$
$$\sqrt{4 + 8i - 4 - 8 - 8i + 7} + 1 + i \stackrel{?}{=} 1$$
$$1 + 2i \neq 1$$

FIGURE 3

$y_1 = 1$,
$y_2 = \sqrt{4x^2 + 8x + 7} - x$.

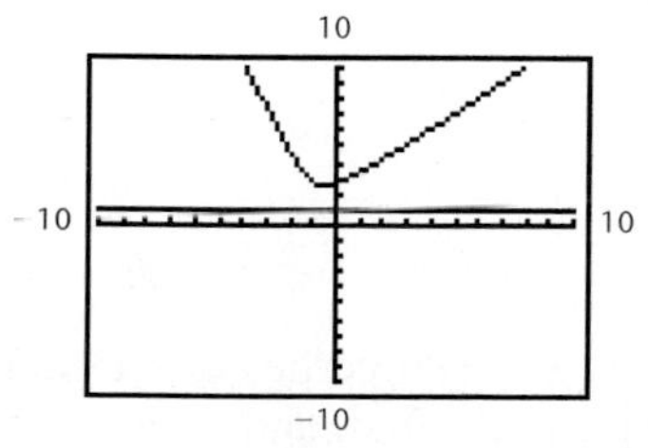

The check shows that $-1 + i$ is a solution to the original equation and $-1 - i$ is extraneous. Thus, the only solution is the imaginary number

$$x = -1 + i$$

Graphing both sides of the equation illustrates that there are no intersection points in a standard viewing window (Fig. 3). The algebraic solution shows that the equation has no real solutions, hence there cannot be any intersection points anywhere in the plane.

(B) To solve an equation that contains more than one radical, isolate one radical at a time and square both sides to eliminate the isolated radical. Repeat this process until all the radicals are eliminated.

$$\sqrt{2x+3} - \sqrt{x-2} = 2$$

$$\sqrt{2x+3} = \sqrt{x-2} + 2 \qquad \text{Isolate one of the radicals.}$$

$$2x+3 = x - 2 + 4\sqrt{x-2} + 4 \qquad \text{Square both sides.}$$

$$x + 1 = 4\sqrt{x-2} \qquad \text{Isolate the remaining radical.}$$

$$x^2 + 2x + 1 = 16(x-2) \qquad \text{Square both sides.}$$

$$x^2 - 14x + 33 = 0$$

$$(x-3)(x-11) = 0$$

$$x = 3, 11$$

Check

$x = 3$	$x = 11$
$\sqrt{2x+3} - \sqrt{x-2} = 2$	$\sqrt{2x+3} - \sqrt{x-2} = 2$
$\sqrt{2(3)+3} - \sqrt{3-2} \stackrel{?}{=} 2$	$\sqrt{2(11)+3} - \sqrt{11-2} \stackrel{?}{=} 2$
$2 \stackrel{\checkmark}{=} 2$	$2 \stackrel{\checkmark}{=} 2$

FIGURE 4
$y_1 = 2,$
$y_2 = \sqrt{2x+3} - \sqrt{x-2}.$

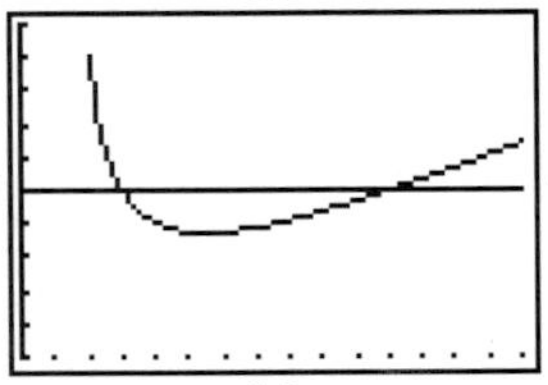

Both solutions check. Thus,

$x = 3, 11$ Two solutions

These results are illustrated in Figure 4, which shows the two real solutions.

MATCHED PROBLEM 1

Solve algebraically

(A) $\sqrt{x^2 - 2x - 2} + 2x = 2$ (B) $\sqrt{2x+5} + \sqrt{x+2} = 5$

Equations Involving Rational Exponents

To solve the equation

$$x^{2/3} - x^{1/3} - 6 = 0$$

write it in the form

$$(x^{1/3})^2 - x^{1/3} - 6 = 0$$

You can now recognize that the equation is quadratic in $x^{1/3}$. (Properties of rational exponents are reviewed in Section A-6.) So, we solve for $x^{1/3}$ first, and then solve for x. We can solve the equation directly or make the substitution $u = x^{1/3}$, solve for u, and then solve for x. Both methods of solution are shown on the next page.

Method I. Direct solution:

$$(x^{1/3})^2 - x^{1/3} - 6 = 0$$

$$(x^{1/3} - 3)(x^{1/3} + 2) = 0 \qquad \text{Factor left side.}$$

$$x^{1/3} = 3 \qquad \text{or} \qquad x^{1/3} = -2$$

$$(x^{1/3})^3 = 3^3 \qquad (x^{1/3})^3 = (-2)^3 \qquad \text{Cube both sides.}$$

$$x = 27 \qquad x = -8$$

Solution set: $\{-8, 27\}$

Method II. Using substitution:
Let $u = x^{1/3}$, solve for u, and then solve for x.

$$u^2 - u - 6 = 0$$

$$(u - 3)(u + 2) = 0$$

$$u = 3, -2$$

Replacing u with $x^{1/3}$, we obtain

$$x^{1/3} = 3 \qquad \text{or} \qquad x^{1/3} = -2$$

$$x = 27 \qquad x = -8$$

Solution set: $\{-8, 27\}$

FIGURE 5
$y_1 = x^{2/3} - x^{1/3} - 6.$

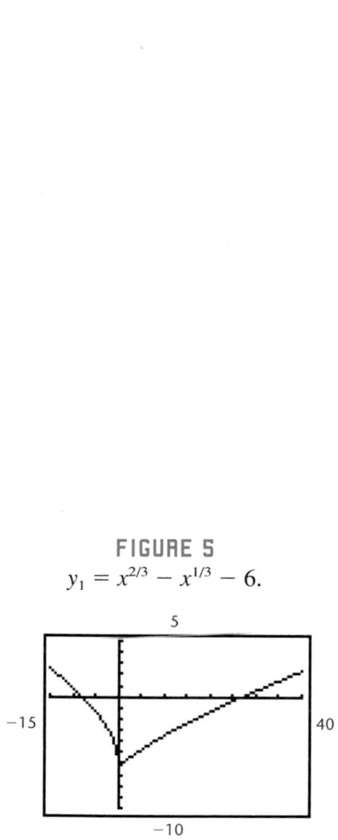

The graph in Figure 5 confirms these results. [*Note:* In some graphing utilities you may have to enter the left side of the equation in the form $y_1 = (x^2)^{1/3} - x^{1/3} - 6$ rather than $y_1 = x^{2/3} - x^{1/3} - 6$. Try both forms to see what happens.]

In general, if an equation that is not quadratic can be transformed to the form

$$au^2 + bu + c = 0$$

where u is an expression in some other variable, then the equation is called an **equation of quadratic type.** Equations of quadratic type often can be solved using quadratic methods.

Explore/Discuss 2

Which of the following can be transformed into an equation of quadratic type by making a substitution of the form $u = x^n$?

(A) $3x^{-4} + 2x^{-2} + 7$ (B) $7x^5 - 3x^2 + 3$

(C) $2x^5 + 4x^2\sqrt{x} - 6$ (D) $8x^{-2}\sqrt{x} - 5x^{-1}\sqrt{x} - 2$

In general, if a, b, c, m, and n are nonzero real numbers, when can an expression of the form $ax^m + bx^n + c$ be transformed into an equation of quadratic type?

EXAMPLE 2 Solving Equations of Quadratic Type

Solve algebraically:

(A) $x^4 - 3x^2 - 4 = 0$ (B) $3x^{-2/5} - 6x^{-1/5} + 2 = 0$

Solutions

(A) The equation is quadratic in x^2. We solve for x^2 and then for x:

$$(x^2)^2 - 3x^2 - 4 = 0$$
$$(x^2 - 4)(x^2 + 1) = 0$$
$$x^2 = 4 \quad \text{or} \quad x^2 = -1$$
$$x = \pm 2 \quad \text{or} \quad x = \pm i$$

Solution set: $\{-2, 2, -i, i\}$

Since we did not raise each side of the equation to a natural number power, we do not have to check for extraneous solutions. (You should still check the accuracy of the solutions.) Figure 6 shows the two real solutions. The imaginary solutions cannot be illustrated graphically.

FIGURE 6
$y_1 = x^4 - 3x^2 - 4$.

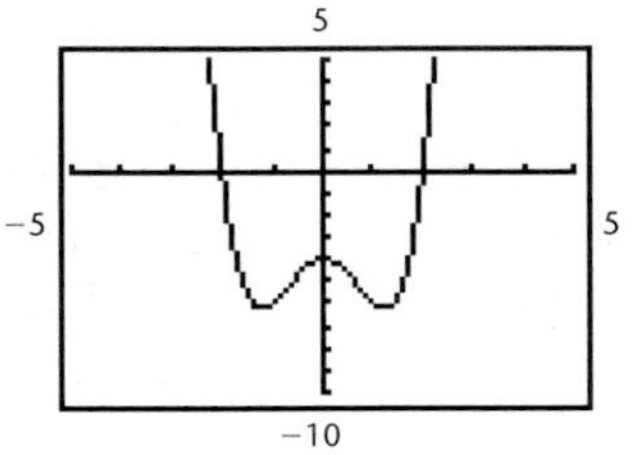

(B) The equation $3x^{-2/5} - 6x^{-1/5} + 2 = 0$ is quadratic in $x^{-1/5}$. We substitute $u = x^{-1/5}$ and solve for u:

$$3u^2 - 6u + 2 = 0$$
$$u = \frac{6 \pm \sqrt{12}}{6} = \frac{3 \pm \sqrt{3}}{3} \qquad \text{Use the quadratic formula.}$$
$$x = u^{-5}$$
$$= \left(\frac{3 \pm \sqrt{3}}{3}\right)^{-5}$$

Thus, the two solutions are

$$x = \left(\frac{3}{3 \pm \sqrt{3}}\right)^5 \qquad \text{Use a calculator.}$$
$$\approx 0.102\,414,\ 74.147\,586$$

The thick curve in Figure 7(a) is the graph of $y = 3x^{-2/5} - 6x^{-1/5} + 2$. The graph crosses the x axis near $x = 0$ and again near $x = 75$. The solution near $x = 75$ is easily approximated in this viewing window. The solution near the origin can be approximated in the same viewing window, but more insight is gained by zooming in on the graph [Fig. 7(b)].

FIGURE 7
$y = 3x^{-2/5} - 6x^{-1/5} + 2$.

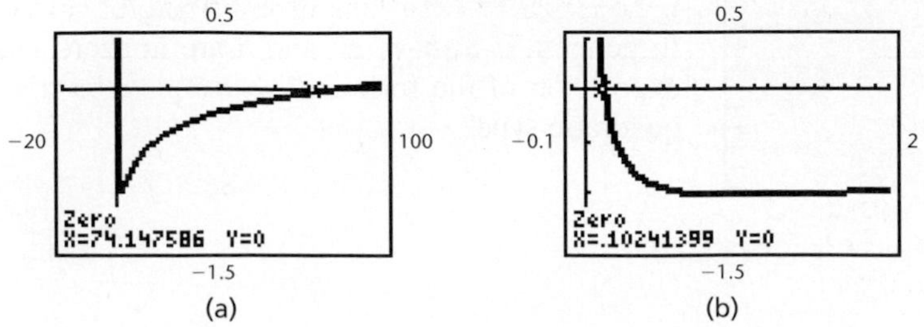

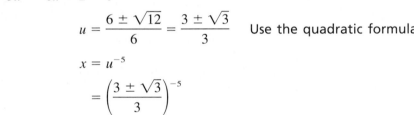

MATCHED PROBLEM 2

Solve algebraically:

(A) $x^4 + 3x^2 - 4 = 0$ (B) $3x^{-2/5} - x^{-1/5} - 2 = 0$

EXAMPLE 3

Solving an Equation Graphically

The diagonal of a rectangle is 10 inches and the area is 45 square inches. Find the dimensions of the rectangle, correct to one decimal place.

Solution

Draw a rectangle and label the dimensions as shown in Figure 8. From the Pythagorean theorem,

FIGURE 8

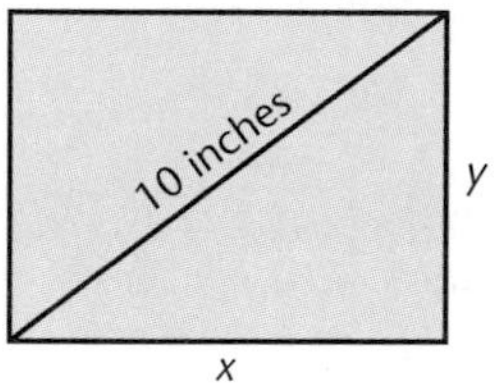

$$x^2 + y^2 = 10^2$$

Thus,

$$y = \sqrt{100 - x^2} \qquad 0 \le x \le 10$$

Since the area of the rectangle is given by xy, we have

$$x\sqrt{100 - x^2} = 45 \qquad \text{Area of the rectangle}$$

Squaring both sides of this equation and combining like terms will produce a quadratic equation in x^2, which can be solved using the techniques we have discussed. However, since we only need to find approximate real solutions, a graphical solution will suffice. Figure 9 shows that the equation $x\sqrt{100 - x^2} = 45$ has two solutions, $x = 5.3$ and $x = 8.5$ (rounded to one decimal place). The corresponding values of y are

$$y = \sqrt{100 - 5.3^2} = 8.5 \qquad \text{when } x = 5.3$$
$$y = \sqrt{100 - 8.5^2} = 5.3 \qquad \text{when } x = 8.5$$

In either case, the dimensions of the rectangle are 5.3 inches by 8.5 inches.

FIGURE 9
$y_1 = x\sqrt{100 - x^2}$, $y_2 = 45$.

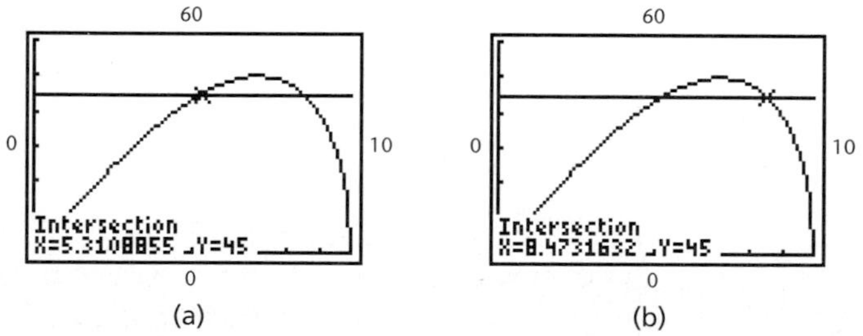

MATCHED PROBLEM 3

If the area of a right triangle is 24 square inches and the hypotenuse is 12 inches, find the lengths of the legs of the triangle, correct to one decimal place.

Answers to Matched Problems

1. (A) $x = 1 - i$ (B) $x = 2$
2. (A) $x = \pm 1, \pm 2i$ (B) $x = 1, -\frac{243}{32}$
3. 4.3 inches and 11.2 inches

EXERCISE 2-6

A

In Problems 1–6, determine the validity of each statement. If a statement is false, explain why.

1. If $x^2 = 5$, then $x = \pm\sqrt{5}$ **2.** $\sqrt{25} = \pm 5$

3. $(\sqrt{x-1} + 1)^2 = x$ **4.** $(\sqrt{x-1})^2 + 1 = x$

5. If $x^3 = 2$, then $x = 8$ **6.** If $x^{1/3} = 2$, then $x = 8$

Solve Problems 7–22 algebraically. Illustrate all real solutions on a graphing utility.

7. $\sqrt[3]{x+5} = 3$ **8.** $\sqrt[4]{x-3} = 2$

9. $\sqrt{5n+9} = n - 1$ **10.** $m - 13 = \sqrt{m+7}$

11. $\sqrt{x+5} + 7 = 0$ **12.** $3 + \sqrt{2x-1} = 0$

13. $\sqrt{3x+4} = 2 + \sqrt{x}$ **14.** $\sqrt{3w-2} - \sqrt{w} = 2$

15. $y^4 - 2y^2 - 8 = 0$ **16.** $x^4 - 7x^2 - 18 = 0$

17. $3x = \sqrt{x^2 - 2}$ **18.** $x = \sqrt{5x^2 + 9}$

19. $2x^{2/3} + 3x^{1/3} - 2 = 0$ **20.** $x^{2/3} - 3x^{1/3} - 10 = 0$

21. $(m^2 - m)^2 - 4(m^2 - m) = 12$

22. $(x^2 + 2x)^2 - (x^2 + 2x) = 6$

B

Solve Problems 23–38 algebraically. Illustrate all real solutions on a graphing utility.

23. $\sqrt{u-2} = 2 + \sqrt{2u+3}$

24. $\sqrt{3t+4} + \sqrt{t} = -3$

25. $\sqrt{3y-2} = 3 - \sqrt{3y+1}$

26. $\sqrt{2x-1} - \sqrt{x-4} = 2$

27. $\sqrt{7x-2} - \sqrt{x+1} = \sqrt{3}$

28. $\sqrt{3x+6} - \sqrt{x+4} = \sqrt{2}$

29. $\sqrt{4x^2 + 12x + 1} - 6x = 9$

30. $6x - \sqrt{4x^2 - 20x + 17} = 15$

31. $3n^{-2} - 11n^{-1} - 20 = 0$ **32.** $6x^{-2} - 5x^{-1} - 6 = 0$

33. $9y^{-4} - 10y^{-2} + 1 = 0$ **34.** $4x^{-4} - 17x^{-2} + 4 = 0$

35. $y^{1/2} - 3y^{1/4} + 2 = 0$ **36.** $4x^{-1} - 9x^{-1/2} + 2 = 0$

37. $(m-5)^4 + 36 = 13(m-5)^2$

38. $(x-3)^4 + 3(x-3)^2 = 4$

C

Solve Problems 39–42 algebraically. Illustrate all real solutions on a graphing utility.

39. $\sqrt{5-2x} - \sqrt{x+6} = \sqrt{x+3}$

40. $\sqrt{2x+3} - \sqrt{x-2} = \sqrt{x+1}$

41. $2 + 3y^{-4} = 6y^{-2}$

42. $4m^{-2} = 2 + m^{-4}$

Solve Problems 43–46 two ways: by squaring and by substitution.

43. $m - 7\sqrt{m} + 12 = 0$ **44.** $y - 6 + \sqrt{y} = 0$

45. $t - 11\sqrt{t} + 18 = 0$ **46.** $x = 15 - 2\sqrt{x}$

In Problems 47–50, solve algebraically and graphically. Discuss the advantages and disadvantages of each method.

47. $2\sqrt{x+5} = 0.01x + 2.04$

48. $3\sqrt{x-1} = 0.05x + 2.9$

49. $2x^{-2/5} - 5x^{-1/5} + 1 = 0$

50. $x^{-2/5} - 3x^{-1/5} + 1 = 0$

APPLICATIONS

51. Manufacturing. A lumber mill cuts rectangular beams from circular logs (see the figure). If the diameter of the log is 16 inches and the cross-sectional area of the beam is 120 square inches, find the dimensions of the cross section of the beam correct to one decimal place.

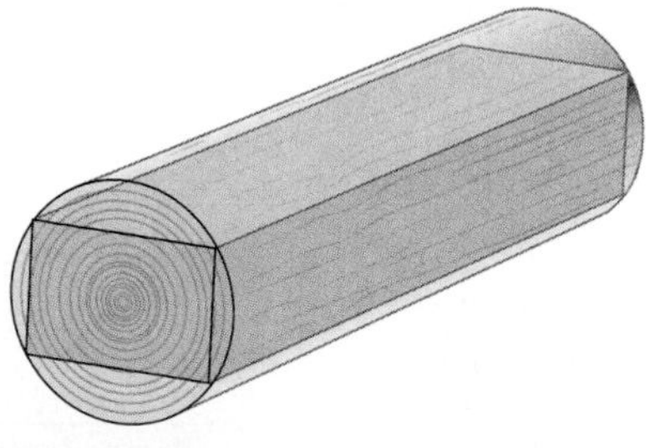

52. Design. A food-processing company packages an assortment of their products in circular metal tins 12 inches in diameter. Four identically sized rectangular boxes are used to divide the tin into eight compartments (see the figure on the next page). If the cross-sectional area of each box is 15 square inches, find the dimensions of the boxes correct to one decimal place.

★ **53. Construction.** A water trough is constructed by bending a 4- by 6-foot rectangular sheet of metal down the middle and attaching triangular ends (see the figure). If the volume of the trough is 9 cubic feet, find the width correct to two decimal places.

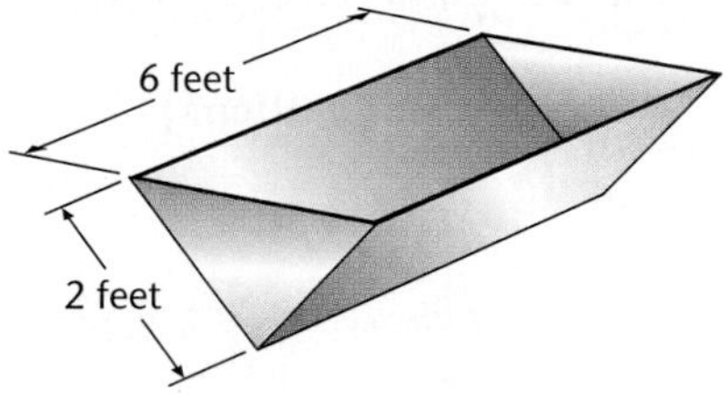

★ **54. Design.** A paper drinking cup in the shape of a right circular cone is constructed from 125 square centimeters of paper (see the figure). If the height of the cone is 10 centimeters, find the radius correct to two decimal places.

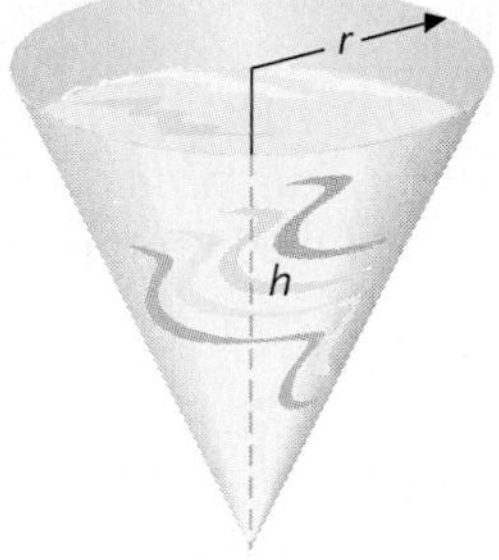

Lateral surface area:

$$S = \pi r\sqrt{r^2 + h^2}$$

Chapter 2 | Group Activity

Mathematical Modeling in Population Studies

In a study on population growth in California, Tulane University demographer Leon Bouvier recorded the past population totals for every 10 years starting at 1900. Then, using sophisticated demographic techniques, he made high, low, and medium projections to the year 2040. Table 1 shows actual populations up to 1990 and medium projections (to the nearest million) to 2040.

TABLE 1 California Population 1900–2040

Years after 1900	Date	Population (millions)	
0	1900	2	Actual
10	1910	3	
20	1920	4	
30	1930	5	
40	1940	5	
50	1950	10	
60	1960	15	
70	1970	20	
80	1980	23	
90	1990	30	
100	2000	35	Projected
110	2010	45	
120	2020	53	
130	2030	61	
140	2040	70	

1. Building a Mathematical Model.
   (A) Plot the first and last columns in Table 1 up to 1990 (actual populations). Would a linear or a quadratic function be the better model for this data? Why?
   (B) Use a graphing utility to compute a quadratic regression function to model the data you plotted in part A.
   (C) Graph this function and the data from part A for $0 \le x \le 150$. The results should look something like Figure 1.

FIGURE 1

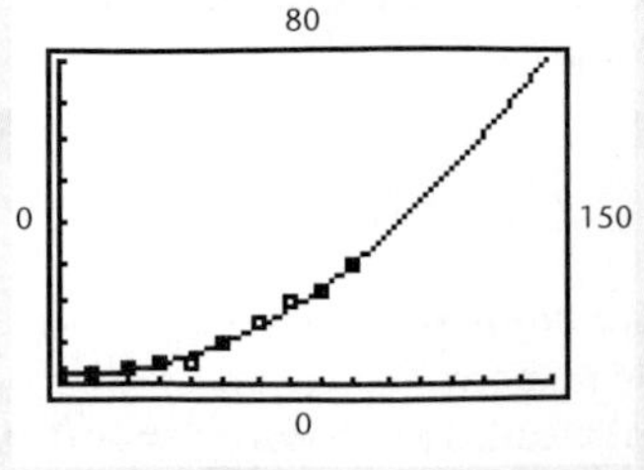

2. Using the Mathematical Model for Projections.
   Use the quadratic regression model to answer the following questions.
   (A) Calculate projected populations for California at 10-year intervals, starting at 2000 and ending at 2040. Compare your projections with Professor Bouvier's projections, both numerically and graphically.
   (B) During what year would you project that the population will reach 40 million? 50 million?
   (C) For what years would you project the population to be between 34 million and 68 million, inclusive?

CHAPTER 2 | REVIEW

2-1 LINEAR FUNCTIONS

A function f is a **linear function** if $f(x) = mx + b$, $m \neq 0$, where m and b are real numbers. The **domain** is the set of all real numbers and the **range** is the set of all real numbers. If $m = 0$, then f is called a **constant function,** $f(x) = b$, which has the set of all real numbers as its *domain* and the constant b as its *range*. The **standard form** for the equation of a line is $Ax + By = C$, where A, B, and C are real constants, A and B not both 0. Every straight line in a Cartesian coordinate system is the graph of an equation of this type. The **slope** of the line through the points (x_1, y_1) and (x_2, y_2) is

$$m = \frac{y_2 - y_1}{x_2 - x_1} \qquad x_1 \neq x_2$$

The slope is not defined for a vertical line where $x_1 = x_2$. Two nonvertical lines with slopes m_1 and m_2 are **parallel** if and only if $m_1 = m_2$ and **perpendicular** if and only if $m_1 m_2 = -1$.

Equations of a Line

Standard Form	$Ax + By = C$	A and B not both 0
Slope–Intercept Form	$y = mx + b$	Slope: m; y intercept: b
Point–Slope Form	$y - y_1 = m(x - x_1)$	Slope: m; point: (x_1, y_1)
Horizontal Line	$y = b$	Slope: 0
Vertical Line	$x = a$	Slope: undefined

2-2 LINEAR EQUATIONS AND INEQUALITIES

The solution sets of linear equations can be found *algebraically* using the **properties of equality** to transform the given equation into an **equivalent equation** that has an obvious solution (see Section A-8) and approximated **graphically** using a graphing utility to find **intersection points.** An equation that is true for all permissible values of the variable is called an **identity.** An equation that is true for some values of the variable and false for others is called a **conditional equation.**

Linear inequalities in one variable are expressed using the inequality symbols $<$, $>$, $\leq$, and $\geq$. The **solution set of an inequality** is the set of all values of the variable that make the inequality a true statement. Each element of the solution set is called a **solution.** An equivalent inequality will result and the **sense will remain the same** if each side of the original inequality:

1. Has the same real number added to or subtracted from it.
2. Is multiplied or divided by the same positive number.

An equivalent inequality will result and the **sense will reverse** if each side of the original inequality:

3. Is multiplied or divided by the same negative number.

Note that multiplication by 0 and division by 0 are not permitted.

2-3 QUADRATIC FUNCTIONS

If a, b, and c are real numbers with $a \neq 0$, then the function $f(x) = ax^2 + bx + c$ is a **quadratic function** and its graph is a **parabola. Completing the square** of the quadratic expression $x^2 + bx$ produces a perfect square:

$$x^2 + bx + \left(\frac{b}{2}\right)^2 = \left(x + \frac{b}{2}\right)^2$$

Completing the square for $f(x) = ax^2 + bx + c$ produces the **standard form** $f(x) = a(x - h)^2 + k$ and the following properties:

1. The graph of f is a parabola:

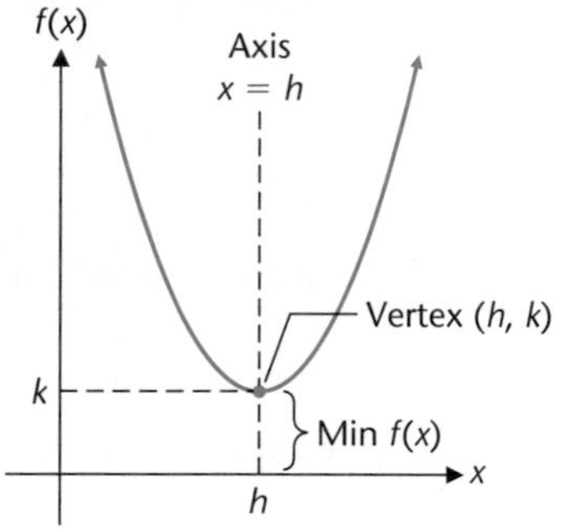

$a > 0$
Opens upward

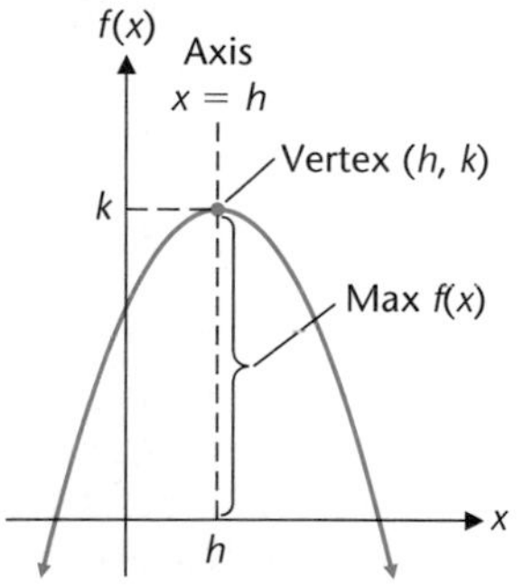

$a < 0$
Opens downward

2. Vertex: (h, k) (parabola increases on one side of the vertex and decreases on the other.)
3. Axis (of symmetry): $x = h$ (parallel to y axis)
4. $f(h) = k$ is the minimum if $a > 0$ and the maximum if $a < 0$.
5. Domain: All real numbers; range: $(-\infty, k]$ if $a < 0$ or $[k, \infty)$ if $a > 0$.
6. The graph of f is the graph of $g(x) = ax^2$ translated horizontally h units and vertically k units.

2-4 COMPLEX NUMBERS

A **complex number** in *standard form* is a number in the form $a + bi$, where a and b are real numbers and i is the **imaginary unit.** If $b \neq 0$, then $a + bi$ is also called an **imaginary number.** If $a = 0$ then $0 + bi = bi$ is also called a **pure imaginary number.** If $b = 0$, then $a + 0i = a$ is a **real number.** The complex

zero is $0 + 0i = 0$. The **conjugate** of $a + bi$ is $a - bi$. **Equality, addition,** and **multiplication** are defined as follows:

1. $a + bi = c + di$ if and only if $a = c$ and $b = d$
2. $(a + bi) + (c + di) = (a + c) + (b + d)i$
3. $(a + bi)(c + di) = (ac - bd) + (ad + bc)i$

Since complex numbers obey the same commutative, associative, and distributive properties as real numbers, most operations with complex numbers are performed by using these properties and the fact that $i^2 = -1$. The **property of conjugates,**

$$(a + bi)(a - bi) = a^2 + b^2$$

can be used to find **reciprocals** and **quotients.** If $a > 0$, then the **principal square root of the negative real number** $-a$ is $\sqrt{-a} = i\sqrt{a}$.

2-5 QUADRATIC EQUATIONS AND INEQUALITIES

A **quadratic equation** is an equation that can be written in the form

$$ax^2 + bx + c = 0 \qquad a \neq 0$$

where x is a variable and a, b, and c are constants. Methods of algebraic solution include:

1. **Factoring** and using the **zero property:** $m \cdot n = 0$ if and only if $m = 0$ or $n = 0$ (or both)
2. **Completing the square** and using the **square root property:**

   If $r^2 = s$, then $r = \pm\sqrt{s}$
3. Using the **quadratic formula:**

$$x = \frac{-b \pm \sqrt{b^2 - 4ac}}{2a}$$

If the **discriminant $b^2 - 4ac$** is positive, the equation has two distinct **real roots;** if the discriminant is 0, the equation has one real **double root;** and if the discriminant is negative, the equation has two **imaginary roots,** each the conjugate of the other.

2-6 ADDITIONAL EQUATION SOLVING TECHNIQUES

A **radical** can be eliminated from an equation by isolating the radical on one side of the equation and raising both sides of the equation to the same power producing a new equation whose solution set always contains the solution set of the original equation. The new equation may have **extraneous solutions** that are not solutions of the original equation. Consequently, **every solution of the new equation must be checked in the original equation to eliminate extraneous solutions.** If an equation contains more than one radical, then the process of isolating a radical and squaring both sides can be repeated until all radicals are eliminated. If a substitution transforms an equation into the form $au^2 + bu + c = 0$, where u is an expression in some other variable, then the equation is an **equation of quadratic type,** which can be solved by quadratic methods.

CHAPTER 2 | REVIEW EXERCISES

Work through all the problems in this chapter review and check answers in the back of the book. Answers to most review problems are there, and following each answer is a number in italics indicating the section in which that type of problem is discussed. Where weaknesses show up, review appropriate sections in the text.

A

1. Graph $3x + 2y = 9$ and indicate its slope.
2. Write an equation of a line with x intercept 6 and y intercept 4. Write the final answer in the form $Ax + By = C$, where A, B, and C are integers with $A > 0$.
3. Write the slope intercept form of the equation of the line with slope $-\frac{2}{3}$ and y intercept 2.
4. Write the equations of the vertical and horizontal lines passing through the point $(-3, 4)$. What is the slope of each?
5. Solve algebraically and check graphically:
   (A) $0.05x + 0.25(30 - x) = 3.3$
   (B) $\frac{5x}{3} - \frac{4 + x}{2} = \frac{x - 2}{4} + 1$

In Problems 6 and 7,

(A) Complete the square and find the standard form of the function.

(B) Write a brief verbal description of the relationship between the graph of the function and the graph of $y = x^2$.

(C) Find the x intercepts algebraically and check graphically.

6. $f(x) = -x^2 - 2x + 3$ 7. $f(x) = x^2 - 3x - 2$

8. Perform the indicated operations and write the answers in standard form:
   (A) $(-3 + 2i) + (6 - 8i)$ (B) $(3 - 3i)(2 + 3i)$
   (C) $\frac{13 - i}{5 - 3i}$

Solve Problems 9–14 by any algebraic method and check graphically, if possible.

9. $x^2 - 2x + 3 = x^2 + 3x - 7$
10. $2x^2 - 7 = 0$
11. $2x^2 = 4x$
12. $2x^2 = 7x - 3$
13. $m^2 + m + 1 = 0$
14. $y^2 = \frac{3}{2}(y + 1)$
15. $\sqrt{5x - 6} - x = 0$

In Problems 16–18, solve and graph. Express answers in inequality and interval notation using exact values.

16. $3(2 - x) - 2 \le 2x - 1$
17. $x^2 + x < 20$ 18. $x^2 > 4x + 12$

19. Discuss the use of the terms rising, falling, increasing, and decreasing as they apply to the descriptions of the following:
(A) A line with positive slope
(B) A line with negative slope
(C) A parabola that opens upward
(D) A parabola that opens downward

B

20. Find an equation of the line through the points $(-4, 3)$ and $(0, -3)$. Write the final answer in the form $Ax + By = C$, where A, B, and C are integers with $A > 0$.

21. Write the slope–intercept form of the equation of the line that passes through the point $(-2, 1)$ and is
(A) parallel to the line $6x + 3y = 5$
(B) perpendicular to the line $6x + 3y = 5$

In Problems 22–24, solve each inequality. Write answers in inequality notation.

22. $|y + 9| < 5$

23. $|2x - 8| \geq 3$

24. $\sqrt{(1 - 2m)^2} \leq 3$

25. Find the domain of each of the following functions. Express answers in interval notation using exact values.
(A) $f(x) = \sqrt{2 - x}$
(B) $g(x) = \sqrt{4 - 2x - x^2}$
(C) $h(x) = \dfrac{1}{\sqrt{\frac{1}{2}x^2 - 3x + 3}}$

26. Let $f(x) = 0.5x^2 - 4x + 5$.
(A) Sketch the graph of f and label the axis and the vertex.
(B) Where is f increasing? Decreasing? What is the range? (Express answers in interval notation.)

27. Find the equations of the linear function g and the quadratic function f whose graphs are shown in the figure. This line is called the tangent line to the graph of f at the point $(-1, 0)$.

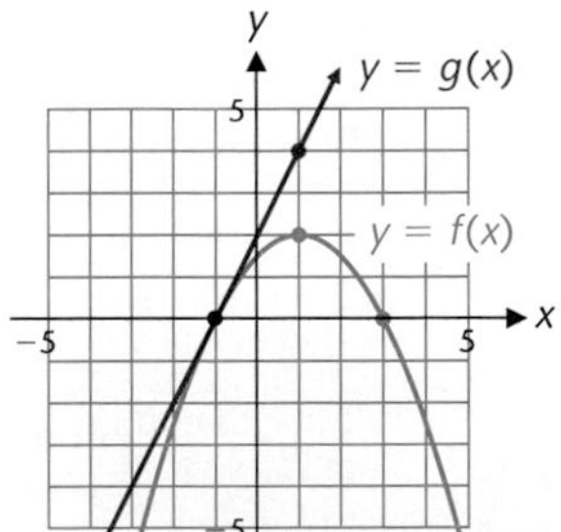

28. Perform the indicated operations and write the final answers in standard form:
(A) $(3 + i)^2 - 2(3 + i) + 3$
(B) i^{27}

29. Convert to $a + bi$ forms, perform the indicated operations, and write the final answers in standard form:
(A) $(2 - \sqrt{-4}) - (3 - \sqrt{-9})$
(B) $\dfrac{2 - \sqrt{-1}}{3 + \sqrt{-4}}$ (C) $\dfrac{4 + \sqrt{-25}}{\sqrt{-4}}$

Solve Problems 30–34 algebraically. Illustrate all real solutions on a graphing utility.

30. $(x + \frac{5}{2})^2 = \frac{5}{4}$ **31.** $1 + \dfrac{3}{u^2} = \dfrac{2}{u}$

32. $2x + 3\sqrt{4x^2 - 4x + 9} = 1$

33. $2x^{2/3} - 5x^{1/3} - 12 = 0$

34. $m^4 + 5m^2 - 36 = 0$

35. $\sqrt{y - 2} - \sqrt{5y + 1} = -3$

36. Can a quadratic function have only imaginary zeros? If not, explain why. If so, give an example and discuss any special relationship between the zeros.

37. If a quadratic function has only imaginary zeros, can the function be graphed? If not, explain why. If so, what is the graph's relationship to the x axis?

Solve Problems 38 and 39 to two decimal places. Express answers in inequality notation.

38. $3.57x - 4.23 < 6.47 - 1.83x$

39. $1.44x^2 - 3.45x - 2.98 \geq 0$

40. Consider the quadratic equation

$$x^2 - 6x + c = 0$$

where c is a real number. Discuss the relationship between the values of c and the three types of roots listed in Table 1 in Section 2-5.

Solve Problems 41 and 42 for the indicated variable in terms of the other variables.

41. $P = M - Mdt$ for M (mathematics of finance)

42. $P = EI - RI^2$ for I (electrical engineering)

C

43. For what values of a and b is the following inequality true?

$$a + b < b - a$$

44. If a and b are negative numbers and $a > b$, then is a/b greater than 1 or less than 1?

45. Solve and graph. Write answer using interval notation: $0 < |x - 6| < d$

46. Evaluate: $(a + bi)\left(\dfrac{a}{a^2 + b^2} - \dfrac{b}{a^2 + b^2}i\right)$, $a, b, \neq 0$

47. Are the graphs of $mx - y = b$ and $x + my = b$ parallel, perpendicular, or neither? Justify your answer.

48. Use completing the square to find the center and radius of the circle with equation:

$$x^2 - 4x + y^2 - 2y - 3 = 0$$

49. Refer to Problem 48. Find the equation of the line tangent to the circle at the point $(4, 3)$. Graph the circle and the line on the same coordinate system.

50. Solve $3x^{-2/5} - 4x^{-1/5} + 1 = 0$ algebraically and graphically.

51. Find all solutions of $x^3 + 1 = 0$.

APPLICATIONS

52. Cost Analysis. Cost equations for manufacturing companies are often quadratic in nature—costs are high at low and high production levels. The weekly cost $C(x)$ (in dollars) for manufacturing x inexpensive calculators is

$$C(x) = 0.001x^2 - 9.5x + 30{,}000$$

Find the production level(s) (to the nearest integer) that

(A) Produces the minimum weekly cost. What is the minimum weekly cost (to the nearest cent)?

(B) Produces a weekly cost of \$12,000.

(C) Produces a weekly cost of \$6,000.

53. Break-Even Analysis. The manufacturing company in Problem 52 sells its calculators to wholesalers for \$3 each. How many calculators (to the nearest integer) must the company sell to break even?

54. Profit Analysis. Refer to Problems 52 and 53. Find the production levels that produce a profit. A loss. (Express answers in inequality notation.)

55. Linear Depreciation. A computer system was purchased by a small company for \$12,000 and is assumed to have a depreciated value of \$2,000 after 8 years. If the value is depreciated linearly from \$12,000 to \$2,000:

(A) Find the linear equation that relates value V (in dollars) to time t (in years).

(B) What would be the depreciated value of the system after 5 years?

56. Business–Pricing. A sporting goods store sells tennis shorts that cost \$30 for \$48 and sunglasses that cost \$20 for \$32.

(A) If the markup policy of the store for items that cost over \$10 is assumed to be linear and is reflected in the pricing of these two items, write an equation that expresses retail price R as a function of cost C.

(B) What should be the retail price of a pair of skis that cost \$105?

★ **57. Income.** A salesperson receives a base salary of \$200 per week and a commission of 10% on all sales over \$3,000 during the week. If x represents the salesperson's weekly sales, express the total weekly earnings $E(x)$ as a function of x. Find $E(2{,}000)$ and $E(5{,}000)$.

58. Construction. A farmer has 120 feet of fencing to be used in the construction of two identical rectangular pens sharing a common side (see the figure).

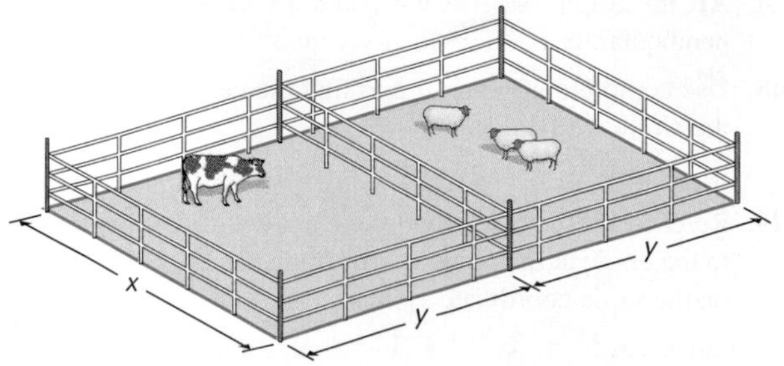

(A) Express the total area $A(x)$ enclosed by both pens as a function of the width x.

(B) From physical considerations, what is the domain of the function A?

(C) Find the dimensions of the pens that will make the total enclosed area maximum.

59. Sports Medicine. The following quotation was found in a sports medicine handout: "The idea is to raise and sustain your heart rate to 70% of its maximum safe rate for your age. One way to determine this is to subtract your age from 220 and multiply by 0.7."

(A) If H is the maximum safe sustained heart rate (in beats per minute) for a person of age A (in years), write a formula relating H and A.

(B) What is the maximum safe sustained heart rate for a 20-year-old?

(C) If the maximum safe sustained heart rate for a person is 126 beats per minute, how old is the person?

★ **60. Design.** The pages of a textbook have uniform margins of 2 centimeters on all four sides (see the figure). If the area of the entire page is 480 square centimeters and the area of the printed portion is 320 square centimeters, find the dimensions of the page.

★ **61. Design.** A landscape designer uses 8-foot timbers to form a pattern of isosceles triangles along the wall of a building (see the figure). If the area of each triangle is 24 square feet, find the base correct to two decimal places.

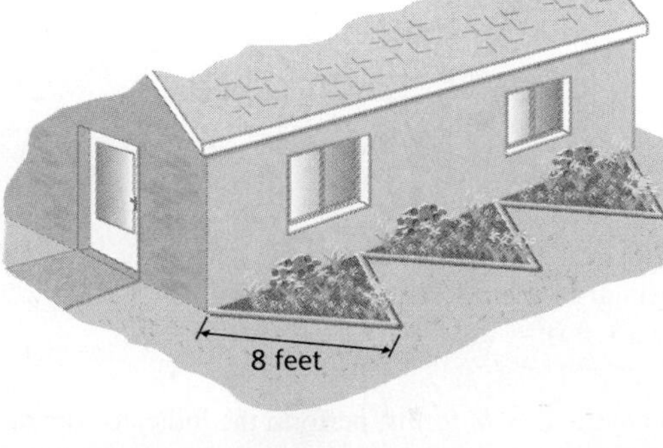

★ **62. Architecture.** An entrance way in the shape of a parabola 12 feet wide and 12 feet high must enclose a rectangular door that is 8.4 feet high. What is the widest doorway (to the nearest tenth of a foot) that can be installed in the entrance way?

63. Drug Use. The use of marijuana by teenagers declined throughout the 80s, but began to increase during the 90s. Table 1 gives the percentage of 12- to 17-year-olds who have ever used marijuana for selected years from 1979 to 1995. (Source: National Household Survey on Drug Abuse.)

TABLE 1 Marijuana Use: 12 to 17 Years Old

Year	Ever Used (%)
1979	26.7
1985	20.1
1990	12.7
1994	13.6
1995	16.2

(A) Find a quadratic regression model for the percentage of 12- to 17-year-olds who have ever used marijuana, using years since 1970 for the independent variable.

(B) Use your model to predict the year during which the percentage of marijuana users will return to the 1979 level.

64. Political Science. Association of economic class and party affiliation did not start with Roosevelt's New Deal; it goes back to the time of Andrew Jackson (1767–1845). Paul Lazarsfeld of Columbia University published an article in the November 1950 issue of *Scientific American* in which he discusses statistical investigations of the relationships between economic class and party affiliation. The data in the table are taken from this article.

Political Affiliations in 1836

Ward	Average Assessed Value per Person (in \$100)	Democratic Votes (%)
12	1.7	51
3	2.1	49
1	2.3	53
5	2.4	36
2	3.6	65
11	3.7	35
10	4.7	29
4	6.2	40
6	7.1	34
9	7.4	29
8	8.7	20
7	11.9	23

(A) Find a linear regression model for the data in the second and third columns of the table, using the average assessed value as the independent variable.

(B) Use the linear regression model to predict (to two decimal places) the percentage of votes for democrats in a ward with an average assessed value of \$300.

CUMULATIVE REVIEW EXERCISES FOR CHAPTERS 1 AND 2

Work through all the problems in this chapter review and check answers in the back of the book. Answers to most review problems are there, and following each answer is a number in italics indicating the section in which that type of problem is discussed. Where weaknesses show up, review appropriate sections in the text.

A

1. (A) Plot the points in the table in a rectangular coordinate system.

(B) Find the smallest viewing window that will contain all of these points. State your answer in terms of the window variables.

(C) Does this set of points define a function? Explain.

x	-3	1	-2	1	3
y	4	-2	4	-4	4

2. Given points $A(3, 2)$ and $B(5, 6)$, find

(A) Distance between A and B.

(B) Slope–intercept form of the equation of the line through A and B.

(C) Slope–intercept form of the equation of the line through B and perpendicular to the line through A and B.

(D) Standard form of the equation of the circle with center at A and passing through B.

(E) Graph the lines from parts B and C and the circle from part D on the same coordinate system.

3. Graph $2x - 3y = 6$ and indicate its slope and intercepts.

4. For $f(x) = x^2 - 2x + 5$ and $g(x) = 3x - 2$, find

(A) $f(-2) + g(-3)$

(B) $f(1) \cdot g(1)$

(C) $\dfrac{g(0)}{f(0)}$

5. How are the graphs of the following related to the graph of $y = |x|$?

(A) $y = 2|x|$

(B) $y = |x - 2|$

(C) $y = |x| - 2$

Problems 6–8 refer to the function f given by the following graph:

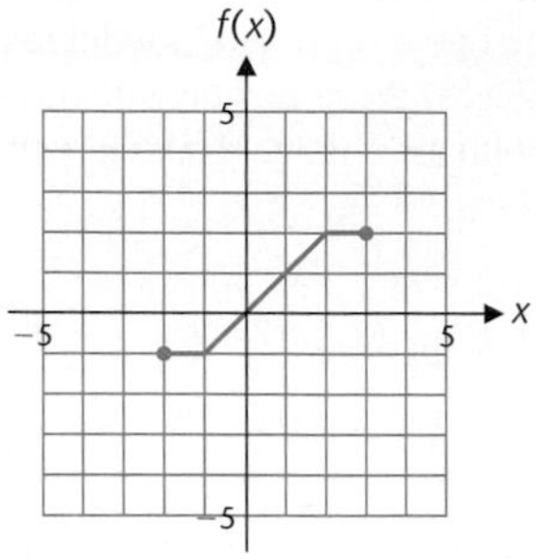

6. Find the domain and range of f. Express answers in interval notation.

7. Is f an even function, an odd function, or neither? Explain.

8. Use the graph of f to sketch a graph of the following:

(A) $y = -f(x + 1)$ (B) $y = 2f(x) - 2$

Solve Problems 9–13 algebraically and check graphically.

9. $\frac{7x}{5} - \frac{3 + 2x}{2} = \frac{x - 10}{3} + 2$

10. $3x^2 = -12x$ **11.** $4x^2 - 20 = 0$

12. $x^2 - 6x + 2 = 0$ **13.** $x - \sqrt{12 - x} = 0$

In Problems 14–16, solve and graph. Express answers in inequality and interval notation.

14. $2(3 - y) + 4 \le 5 - y$ **15.** $|x - 2| < 7$

16. $x^2 + 3x \ge 10$

17. Let $f(x) = x^2 - 4x - 1$.

(A) Find the standard form of f.

(B) How is the graph of f related to the graph of $y = x^2$?

(C) Find the x intercepts algebraically and check graphically.

18. Perform the indicated operations and write the answer in standard form:

(A) $(2 - 3i) - (-5 + 7i)$ (B) $(1 + 4i)(3 - 5i)$

(C) $\frac{5 + i}{2 + 3i}$

B

19. Find each of the following for the function f given by the graph shown below.

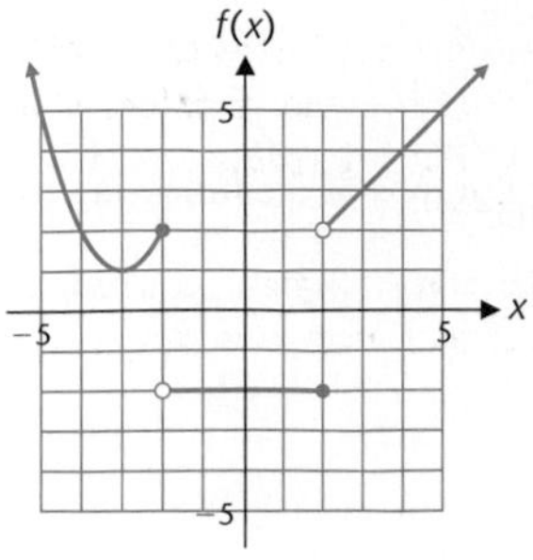

(A) The domain of f

(B) The range of f

(C) $f(-3) + f(-2) + f(2)$

(D) The intervals over which f is increasing.

(E) The x coordinates of any points of discontinuity.

20. Write the slope–intercept form of the equation of the line passing through the point $(-6, 1)$ that is

(A) parallel to the line $3x + 2y = 12$.

(B) perpendicular to the line $3x + 2y = 12$.

21. Graph $f(x) = x^2 - 2x - 8$. Label the axis of symmetry and the coordinates of the vertex, and find the range, intercepts, and maximum or minimum value of $f(x)$.

In Problems 22 and 23, solve and graph. Express answers in inequality and interval notation using exact values.

22. $|4x - 9| > 3$

23. $\sqrt{(3m - 4)^2} \le 2$

In Problems 24 and 25, find the domain of each function. Express answers in interval notation using exact values.

24. $f(x) = \frac{1}{\sqrt{9 - 7x}}$ **25.** $g(x) = \sqrt{7x - x^2 - 11}$

26. Perform the indicated operations and write the final answers in standard form.

(A) $(2 - 3i)^2 - (4 - 5i)(2 - 3i) - (2 + 10i)$

(B) $\frac{3}{5} + \frac{4}{5}i + \frac{1}{\frac{3}{5} + \frac{4}{5}i}$ (C) i^{35}

27. Convert to $a + bi$ forms, perform the indicated operations, and write the final answers in standard form.

(A) $(5 + 2\sqrt{-9}) - (2 - 3\sqrt{-16})$

(B) $\frac{2 + 7\sqrt{-25}}{3 - \sqrt{-1}}$ (C) $\frac{12 - \sqrt{-64}}{\sqrt{-4}}$

28. Graph, finding the domain, range, and any points of discontinuity.

$$f(x) = \begin{cases} x - 1 & \text{if } x < 0 \\ x^2 + 1 & \text{if } x \ge 0 \end{cases}$$

29. Find the center and radius of the circle given by $x^2 - 6x + y^2 + 2y = 0$. Graph the circle and show the center and the radius.

30. The graph in the figure is the result of applying a sequence of transformations to the graph of $y = |x|$. Describe the transformations verbally and write an equation for the graph in the figure.

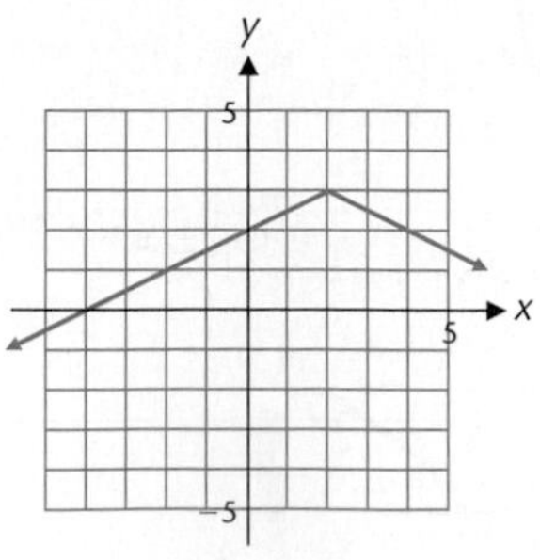

31. Find the standard form of the quadratic function whose graph is shown in the figure.

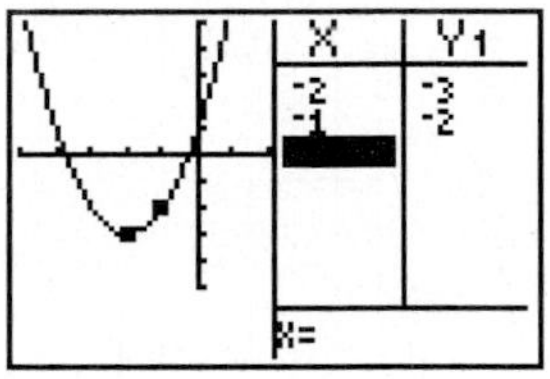

Solve Problems 32–36 algebraically. Check graphically, if possible.

32. $1 + \frac{14}{y^2} = \frac{6}{y}$

33. $4x^{2/3} - 4x^{1/3} - 3 = 0$

34. $u^4 + u^2 - 12 = 0$

35. $\sqrt{8t - 2} - 2\sqrt{t} = 1$

36. $6x = \sqrt{9x^2 - 48}$

Solve Problems 37 and 38 to two decimal places. Express answers in inequality and interval notation, when appropriate.

37. $-3.45 < 1.86 - 0.33x \leq 7.92$

38. $2.35x^2 + 10.44x - 16.47 = 0$

39. Consider the quadratic equation

$$x^2 + bx + 1 = 0$$

where b is a real number. Discuss the relationship between the values of b and the three types of roots listed in Table 1 in Section 2-5.

40. Give a sample example of an odd function. Of an even function. Can a function be both even and odd? Explain.

41. Can a quadratic equation with real coefficients have one imaginary root and one real root? One double imaginary root? Explain.

42. If $g(x) = -2x^2 + 3x - 1$, find $\frac{g(2 + h) - g(2)}{h}$.

43. The graph below is the result of applying one or more transformations to the graph of one of the six basic functions in Figure 1, Section 1-5. Find an equation for the graph.

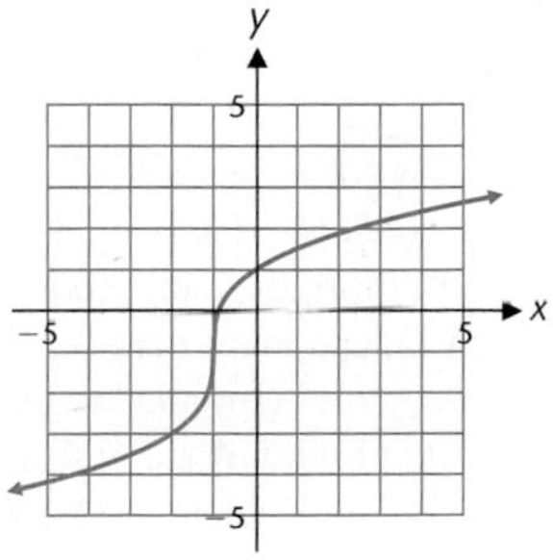

44. The total surface area of a right circular cylinder with radius r and height h is given by

$$A = 2\pi r(r + h) \qquad r > 0, h > 0$$

(A) Solve for h in terms of the other variables.

(B) Solve for r in terms of the other variables. Why is there only one solution?

C

45. Evaluate $x^2 - x + 2$ for $x = \frac{1}{2} - \frac{i}{2}\sqrt{7}$

46. For what values of a and b is the inequality $a - b < b - a$ true?

47. Write in standard form: $\frac{a + bi}{a - bi}$, $a, b \neq 0$

Solve Problems 48–51 algebraically and check graphically, if possible.

48. $3x^2 = 2\sqrt{2}x - 1$

49. $1 + 13x^{-2} + 36x^{-4} = 0$

50. $\sqrt{16x^2 + 48x + 39} - 2x = 3$

51. $3x^{-2/5} - x^{-1/5} - 1 = 0$

52. Show that $5 + i$ and $-5 - i$ are the square roots of $24 + 10i$. Describe how you could find these square roots algebraically.

53. For $f(x) = 0.5^2 - 3x - 7$, find

(A) $\frac{f(x + h) - f(x)}{h}$ (B) $\frac{f(x) - f(a)}{x - a}$

54. The function f is continuous for all real numbers and its graph passes through the points $(0, 4)$, $(5, -3)$, $(10, 2)$. Discuss the minimum and maximum number of x intercepts for f.

55. Let $f(x) = |x + 2| + |x - 2|$. Find a piecewise definition of f that does not involve the absolute value function. Graph f and find the domain and range.

56. Let $f(x) = 2x - [\![2x]\!]$. Write a piecewise definition for f and sketch the graph of f. Include sufficient intervals to clearly illustrate both the definition and the graph. Find the domain, range, and any point of discontinuity.

57. Find all solutions of $x^3 - 8 = 0$.

APPLICATIONS

58. Break-Even Analysis. The publisher's fixed costs for the production of a new cookbook are $41,800. Variable costs are $4.90 per book. If the book is sold to bookstores for $9.65, how many must be sold for the publisher to break even?

59. Finance. An investor instructs a broker to purchase a certain stock whenever the price per share p of the stock is within $10 of $200. Express this instruction as an absolute value inequality.

60. Profit and Loss Analysis. At a price of $\$p$ per unit, the marketing department in a company estimates that the weekly cost C and the weekly revenue R, in thousands of dollars, will be given by the equations

$C = 88 - 12p$ Cost equation

$R = 15p - 2p^2$ Revenue equation

Find the prices for which the company has

(A) A profit (B) A loss

★ **61. Shipping.** A ship leaves Port A, sails east to Port B, and then north to Port C, a total distance of 115 miles. The next day the ship sails directly form Port C back to Port A, a distance of 85 miles. Find the distance between Ports A and B and between Ports B and C.

62. Price and Demand. The weekly demand for mouthwash in a chain of drug stores is 1,160 bottles at a price of \$3.79 each. If the price is lowered to \$3.59, the weekly demand increases to 1,340 bottles. Assuming the relationship between the weekly demand x and the price per bottle p is linear, express x as a function of p. How many bottles would the store sell each week if the price were lowered to \$3.29?

63. Business–Pricing. A telephone company begins a new pricing plan that charges customers for local calls as follows: The first 60 calls each month are 6 cents each, the next 90 are 5 cents each, the next 150 are 4 cents each, and any additional calls are 3 cents each. If C is the cost, in dollars, of placing x calls per month, write a piecewise definition of C as a function of x and graph.

64. Construction. A home owner has 80 feet of chain-link fencing to be used to construct a dog pen adjacent to a house (see the figure).

(A) Express the area $A(x)$ enclosed by the pen as a function of the width x.

(B) From physical considerations, what is the domain of the function A?

(C) Graph A and determine the dimensions of the pen that will make the area maximum.

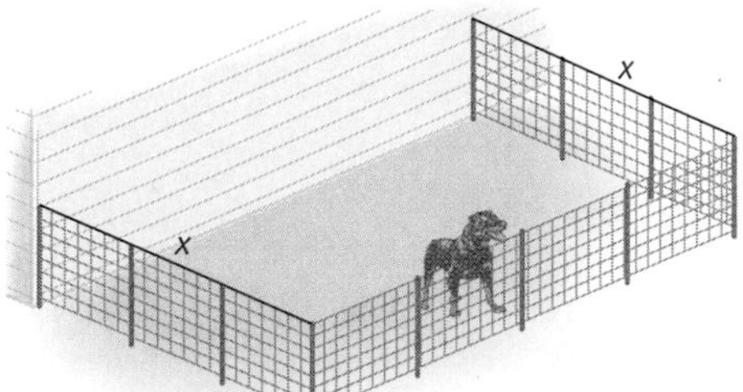

65. Computer Science. Let $f(x) = x - 2[\![x/2]\!]$. This function can be used to determine if an integer is odd or even.

(A) Find $f(1), f(2), f(3), f(4)$.

(B) Find $f(n)$ for any integer n. [*Hint:* Consider two cases, $n = 2k$ and $n = 2k + 1$, k an integer.]

66. Price and Demand. The demand for barley q (in thousands of bushels) and the corresponding price p (in cents) at a midwestern grain exchange are shown in the figure.

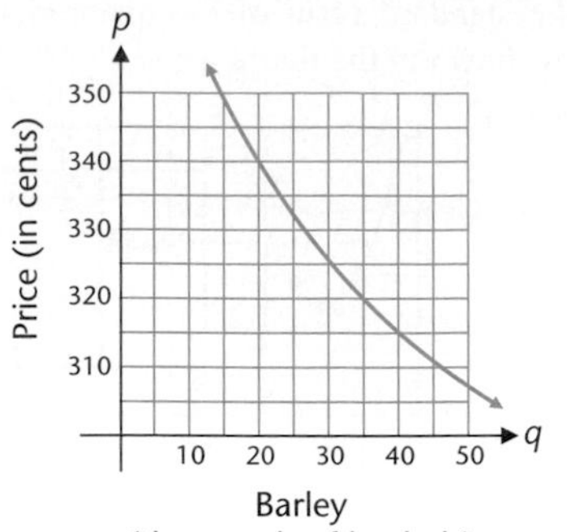

(A) What is the demand (to the nearest thousand bushels) when the price is 325 cents per bushel?

(B) Does the demand increase or decrease if the price is increased to 340 cents per bushel? By how much?

(C) Does the demand increase or decrease if the price is decreased to 315 cents per bushel? By how much?

(D) Write a brief description of the relationship between price and demand illustrated by this graph.

(E) Use the graph to estimate the price (to the nearest cent) when the demand is 20, 25, 30, 35, and 40 thousand bushels. Use this data to find a quadratic regression model for the price of barley using the demand as the independent variable.

67. Engineering. The July 1994 issue of *SEA* magazine contained a review of a 33-foot twin-drive power cruiser. Included in the review was the following information on fuel consumption at various speeds, where MPH represents miles per hour and GPH represents gallons per hour.

Fuel Consumption Versus Speed

MPH	GPH
15	8.3
20	11.8
25	15.7
30	20.3

(A) Find a linear regression model for this data using MPH as the independent variable.

(B) Find a quadratic regression model for this data using MPH as the independent variable.

(C) Compare the values (rounded to one decimal place) of the models you found in A and B with the values in the table. Which model seems to best fit the data?

(D) The fuel tank of the cruiser holds 186 gallons. Use the model you chose in C to determine the cruising range without refueling at 17 MPH. At 28 MPH. Give answers in hours (to the nearest hour) and in miles (to the nearest 10 miles).

ANSWER TO APPLICATION

The interior and exterior dimensions of the sunroom are shown in Figure 1. (Note that 7.5 inches equals 0.625 feet.) Since the external area is 400 square feet, we have the following equation:

$$(x + 0.625)(2x + 1.25) = 400$$

FIGURE 1

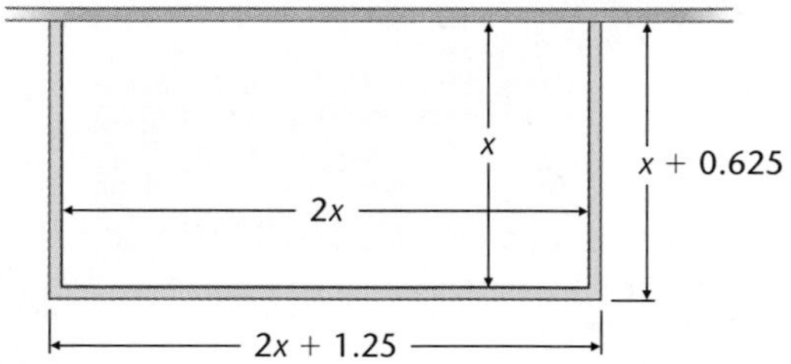

Graphing both sides of this equation and using a built-in intersection routine (Fig. 2), we see that $x \approx 13.5$. The interior area of the sunroom is

$$(2x)x \approx (27)(13.5) = 364.5$$

FIGURE 2

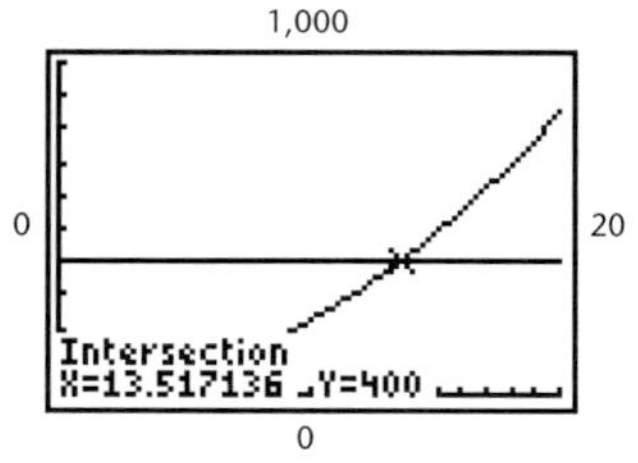

Thus, 360 square feet of tile will not be sufficient to cover the floor of the interior.

In Section 3-3 we discuss methods for locating the real zeros of a polynomial with real coefficients. Once located, the real zeros are easily approximated with a graphing utility. Section 3-4 deals with rational functions and their graphs.

Preparing for This Chapter

Before getting started on this chapter, review the following concepts:
Polynomials (Appendix A, Sections 2 and 3)
Rational Expressions (Appendix A, Section 5)
Graphs of Functions (Chapter 1, Section 4)
Linear Regression (Chapter 1, Group Activity)
Linear Functions (Chapter 2, Section 1)
Quadratic Functions (Chapter 2, Section 3)
Complex Numbers (Chapter 2, Section 4)
Quadratic Formula (Chapter 2, Section 5)

Section 3-1 Polynomial Functions

- Polynomial Functions
- Polynomial Division
- Division Algorithm
- Remainder Theorem
- Graphing Polynomial Functions
- Polynomial Regression

Polynomial Functions

In Chapters 1 and 2 you were introduced to the basic functions

$f(x) = b$ Constant function

$f(x) = ax + b \quad a \neq 0$ Linear function

$f(x) = ax^2 + bx + c \quad a \neq 0$ Quadratic function

as well as some special cases of more complicated functions such as

$f(x) = ax^3 + bx^2 + cx + d \quad a \neq 0$ Cubic function

Notice the evolving pattern going from the constant function to the cubic function—the terms in each equation are of the form ax^n, where n is a nonnegative integer and a is a real number. All these functions are special cases of the general class of functions called *polynomial functions*. The function

$$P(x) = a_nx^n + a_{n-1}x^{n-1} + \cdots + a_1x + a_0 \qquad a_n \neq 0$$

is called an **nth-degree polynomial function.** We will also refer to $P(x)$ as a **polynomial of degree n** or, more simply, as a **polynomial.** The numbers $a_n, a_{n-1}, \ldots, a_1, a_0$ are called the **coefficients of the function.** A nonzero constant function is a zero-degree polynomial, a linear function is a first-degree polynomial, and a quadratic function is a second-degree polynomial. The zero function $Q(x) = 0$ is also considered to be a polynomial but is not assigned a degree. The coefficients of a polynomial function may be complex numbers, or may be restricted to real numbers, rational numbers, or integers, depending on our interests. The domain of a polynomial function can be the set of complex numbers, the set of real numbers, or appropriate subsets of either, depending on our interests. In general, the context will dictate the choice of coefficients and domain.

The number r is said to be a **zero of the function P,** or a **zero of the polynomial $P(x)$,** or a **solution or root of the equation $P(x) = 0$,** if

$$P(r) = 0$$

A zero of a polynomial may or may not be the number 0. A zero of a polynomial is any number that makes the value of the polynomial 0. If the coefficients of a polynomial $P(x)$ are real numbers, then a real zero is simply an x intercept for the graph of $y = P(x)$. Consider the polynomial

$$P(x) = x^2 - 4x + 3$$

The graph of P is shown in Figure 1.

FIGURE 1
Zeros, roots, and x intercepts.

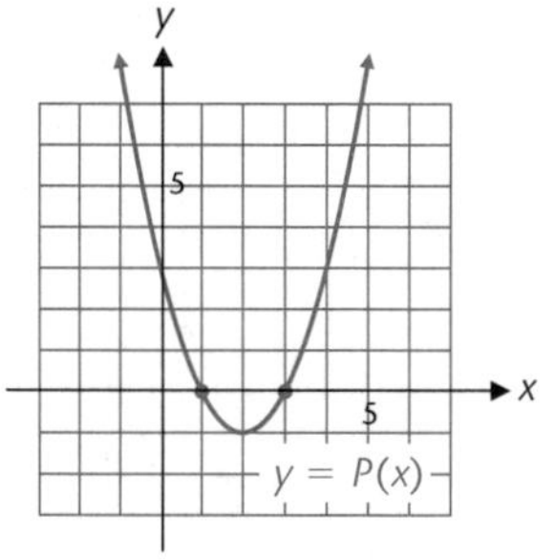

The x intercepts 1 and 3 are zeros of $P(x) = x^2 - 4x + 3$, since $P(1) = 0$ and $P(3) = 0$. The x intercepts 1 and 3 are also solutions or roots for the equation $x^2 - 4x + 3 = 0$.

In general:

ZEROS AND ROOTS

If the coefficients of a polynomial $P(x)$ are real, then the x intercepts of the graph of $y = P(x)$ are real **zeros** of P and $P(x)$ and real **solutions,** or **roots,** for the equation $P(x) = 0$.

Polynomial Division

We can find quotients of polynomials by a long-division process similar to that used in arithmetic. An example will illustrate the process.

EXAMPLE 1 **Algebraic Long Division**

Divide $5 + 4x^3 - 3x$ by $2x - 3$.

Solution

$$\begin{array}{r} 2x^2 + 3x + 3 \\ 2x - 3\overline{)4x^3 + 0x^2 - 3x + 5} \\ \underline{4x^3 - 6x^2} \qquad\qquad \\ 6x^2 - 3x \qquad \\ \underline{6x^2 - 9x} \qquad \\ 6x + 5 \\ \underline{6x - 9} \\ 14 = R \\ \text{Remainder} \end{array}$$

Arrange the dividend and the divisor in descending powers of the variable. Insert, with 0 coefficients, any missing terms of degree less than 3. Divide the first term of the divisor into the first term of the dividend. Multiply the divisor by $2x^2$, line up like terms, subtract (change the sign and add) as in arithmetic, and bring down $-3x$. Repeat the process until the degree of the remainder is less than that of the divisor.

Thus,

$$\frac{4x^3 - 3x + 5}{2x - 3} = 2x^2 + 3x + 3 + \frac{14}{2x - 3}$$

Check

$$(2x - 3)\left[(2x^2 + 3x + 3) + \frac{14}{2x - 3}\right] = (2x - 3)(2x^2 + 3x + 3) + 14$$
$$= 4x^3 - 3x + 5$$

MATCHED PROBLEM 1

Divide $6x^2 - 30 + 9x^3$ by $3x - 4$.

Being able to divide a polynomial $P(x)$ by a linear polynomial of the form $x - r$ quickly and accurately will be of great help in the search for zeros of higher-degree polynomial functions. This kind of division can be carried out more efficiently by a method called **synthetic division.** The method is most easily understood through an example. Let's start by dividing $P(x) = 2x^4 + 3x^3 - x - 5$ by $x + 2$, using ordinary long division. The critical parts of the process are indicated in color.

$$\begin{array}{lr l} & 2x^3 - 1x^2 + 2x - 5 & \text{Quotient} \\ \text{Divisor} & x + 2\overline{)2x^4 + 3x^3 + 0x^2 - 1x - 5} & \text{Dividend} \\ & \underline{2x^4 + 4x^3} \qquad\qquad\qquad\qquad & \\ & -1x^3 + 0x^2 \qquad\qquad\quad & \\ & \underline{-1x^3 - 2x^2} \qquad\qquad\quad & \\ & 2x^2 - 1x \qquad\quad & \\ & \underline{2x^2 + 4x} \qquad\quad & \\ & -5x - 5 & \\ & \underline{-5x - 10} & \\ & 5 & \text{Remainder} \end{array}$$

The numerals printed in color, which represent the essential part of the division process, are arranged more conveniently as follows:

Dividend coefficients

$$\begin{array}{r|rrrrr} & 2 & 3 & 0 & -1 & -5 \\ & & 4 & -2 & 4 & -10 \\ \hline 2 & 2 & -1 & 2 & -5 & 5 \end{array}$$

Quotient coefficients   Remainder

Mechanically, we see that the second and third rows of numerals are generated as follows. The first coefficient, 2, of the dividend is brought down and multiplied by 2 from the divisor; and the product, 4, is placed under the second dividend coefficient, 3, and subtracted. The difference, -1, is again multiplied by the 2 from the divisor; and the product is placed under the third coefficient from the dividend and subtracted. This process is repeated until the remainder is reached. The process can be made a little faster, and less prone to sign errors, by changing $+2$ from the divisor to -2 and adding instead of subtracting. Thus

Dividend coefficients

$$\begin{array}{r|rrrrr} & 2 & 3 & 0 & -1 & -5 \\ & & -4 & 2 & -4 & 10 \\ \hline -2 & 2 & -1 & 2 & -5 & 5 \end{array}$$

Quotient coefficients   Remainder

KEY STEPS IN THE SYNTHETIC DIVISION PROCESS

To divide the polynomial $P(x)$ by $x - r$:

Step 1. Arrange the coefficients of $P(x)$ in order of descending powers of x. Write 0 as the coefficient for each missing power.

Step 2. After writing the divisor in the form $x - r$, use r to generate the second and third rows of numbers as follows. Bring down the first coefficient of the dividend and multiply it by r; then add the product to the second coefficient of the dividend. Multiply this sum by r, and add the product to the third coefficient of the dividend. Repeat the process until a product is added to the constant term of $P(x)$.

Step 3. The last number to the right in the third row of numbers is the remainder. The other numbers in the third row are the coefficients of the quotient, which is of degree 1 less than $P(x)$.

EXAMPLE **2**

Synthetic Division

Use synthetic division to find the quotient and remainder resulting from dividing $P(x) = 4x^5 - 30x^3 - 50x - 2$ by $x + 3$. Write the answer in the form $Q(x) + R/(x - r)$, where R is a constant.

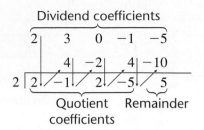

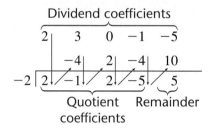

Solution Since $x + 3 = x - (-3)$, we have $r = -3$, and

$$\begin{array}{r|rrrrrr} & 4 & 0 & -30 & 0 & -50 & -2 \\ & & -12 & 36 & -18 & 54 & -12 \\ \hline -3 & 4 & -12 & 6 & -18 & 4 & -14 \end{array}$$

The quotient is $4x^4 - 12x^3 + 6x^2 - 18x + 4$ with a remainder of -14. Thus,

$$\frac{P(x)}{x+3} = 4x^4 - 12x^3 + 6x^2 - 18x + 4 + \frac{-14}{x+3}$$

MATCHED PROBLEM 2 Repeat Example 2 with $P(x) = 3x^4 - 11x^3 - 18x + 8$ and divisor $x - 4$.

A calculator is a convenient tool for performing synthetic division. Any type of calculator can be used, although one with a memory will save some keystrokes. The flowchart in Figure 2 shows the repetitive steps in the synthetic division process, and Figure 3 illustrates the results of applying this process to Example 2 on a graphing calculator.

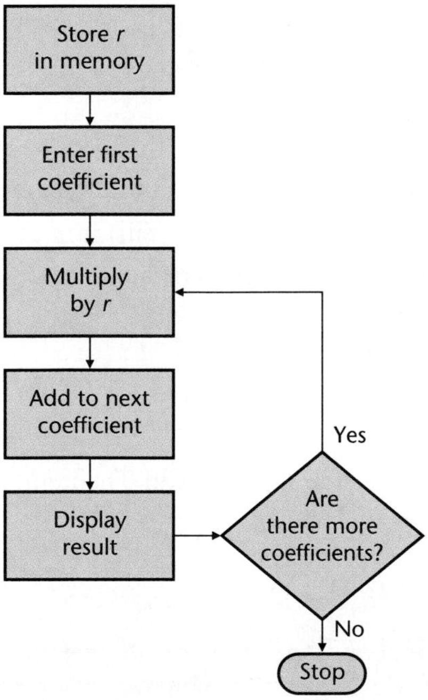

FIGURE 2 Synthetic division.

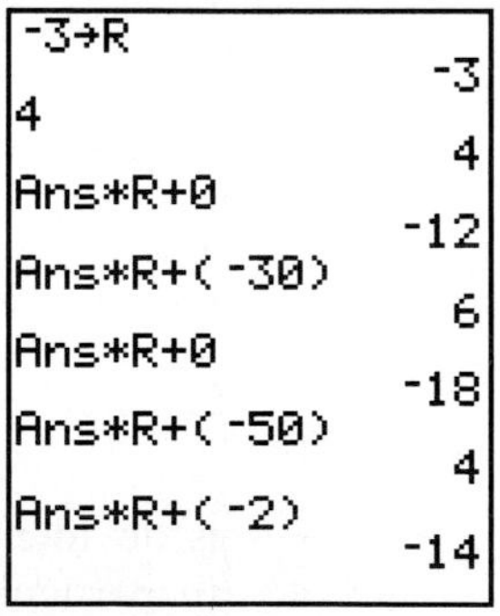

FIGURE 3

Division Algorithm

If we divide $P(x) = 2x^4 - 5x^3 - 4x^2 + 13$ by $x - 3$, we

$$\frac{2x^4 - 5x^3 - 4x^2 + 13}{x - 3} = 2x^3 + x^2 - x - 3 + \frac{4}{x}$$

If we multiply both sides of this equation by $x - 3$, then we get

$$2x^4 - 5x^3 - 4x^2 + 13 = (x - 3)(2x^3 + x^2 - x - 3) + 4$$

This last equation is an identity in that the left side is equal to the right side for *all* replacements of x by real or imaginary numbers, including $x = 3$. This example suggests the important **division algorithm,** which we state as Theorem 1 without proof.

THEOREM 1

DIVISION ALGORITHM

For each polynomial $P(x)$ of degree greater than 0 and each number r, there exists a unique polynomial $Q(x)$ of degree 1 less than $P(x)$ and a unique number R such that

$$P(x) = (x - r)Q(x) + R$$

The polynomial $Q(x)$ is called the **quotient,** $x - r$ is the **divisor,** and R is the **remainder.** Note that R may be 0.

Explore/Discuss 1

Let $P(x) = x^3 - 3x^2 - 2x + 8$.

(A) Evaluate $P(x)$ for

(i) $x = -2$ (ii) $x = 1$ (iii) $x = 3$

(B) Use synthetic division to find the remainder when $P(x)$ is divided by

(i) $x + 2$ (ii) $x - 1$ (iii) $x - 3$

What conclusion does a comparison of the results in parts A and B suggest?

Remainder Theorem

We now use the division algorithm in Theorem 1 to prove the *remainder theorem.* The equation in Theorem 1,

$$P(x) = (x - r)Q(x) + R$$

is an identity; that is, it is true for all real or imaginary replacements for x. In particular, if we let $x = r$, then we observe a very interesting and useful relationship:

$$\begin{aligned} P(r) &= (r - r)Q(r) + R \\ &= 0 \cdot Q(r) + R \\ &= 0 + R \\ &= R \end{aligned}$$

In words, the value of a polynomial $P(x)$ at $x = r$ is the same as the remainder R obtained when we divide $P(x)$ by $x - r$. We have proved the well-known remainder theorem:

THEOREM 2

REMAINDER THEOREM

If R is the remainder after dividing the polynomial $P(x)$ by $x - r$, then

$$P(r) = R$$

EXAMPLE 3

Two Methods for Evaluating Polynomials

If $P(x) = 4x^4 + 10x^3 + 19x + 5$, find $P(-3)$ by

(A) Using the remainder theorem and synthetic division

(B) Evaluating $P(-3)$ directly

Solutions

(A) Use synthetic division to divide $P(x)$ by $x - (-3)$.

$$\begin{array}{r|rrrrr} & 4 & 10 & 0 & 19 & 5 \\ & & -12 & 6 & -18 & -3 \\ \hline -3 & 4 & -2 & 6 & 1 & 2 = R = P(-3) \end{array}$$

(B) $P(-3) = 4(-3)^4 + 10(-3)^3 + 19(-3) + 5$

$= 2$

MATCHED PROBLEM 3

Repeat Example 3 for $P(x) = 3x^4 - 16x^2 - 3x + 7$ and $x = -2$.

You might think the remainder theorem is not a very effective tool for evaluating polynomials. But let's consider the number of operations performed in parts A and B of Example 3. Synthetic division requires only 4 multiplications and 4 additions to find $P(-3)$, while the direct evaluation requires 10 multiplications and 4 additions. [Note that evaluating $4(-3)^4$ actually requires 5 multiplications.] The difference becomes even larger as the degree of the polynomial increases. Computer programs that involve numerous polynomial evaluations often use synthetic division because of its efficiency. We will find synthetic division and the remainder theorem to be useful tools later in this chapter.

Graphing Polynomial Functions

The shape of the graph of a polynomial function is connected to the degree of the polynomial. The shapes of odd-degree polynomial functions have something in common, and the shapes of even-degree polynomial functions have something in common. Figure 4 shows graphs of representative polynomial functions from degrees 1 to 6 and suggests some general properties of graphs of polynomial functions.

FIGURE 4
Graphs of polynomial functions.

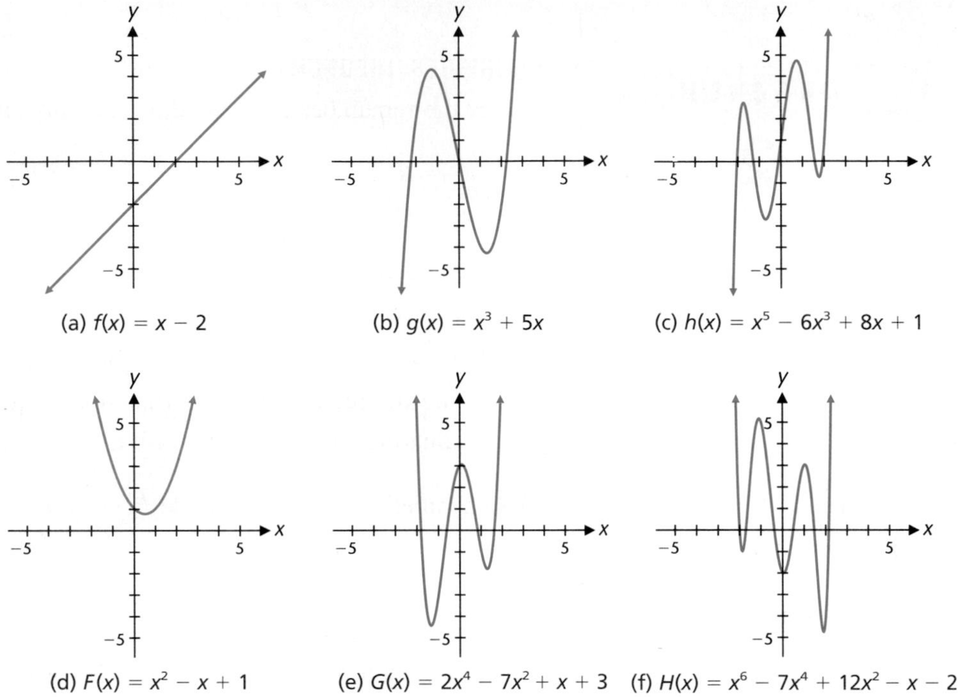

Notice that the odd-degree polynomial graphs start negative, end positive, and cross the x axis at least once. The even-degree polynomial graphs start positive, end positive, and may not cross the x axis at all. In all cases in Figure 4, the coefficient of the highest-degree term was chosen positive. If any leading coefficient had been chosen negative, then we would have a similar graph but reflected in the x axis.

Explore/Discuss 2

Using a graphing utility, discuss the shape of the graph of each of the following functions.

(A) $y_1 = x^2 + x$

(B) $y_2 = x^3 + x^2$

(C) $y_3 = x^4 + x^3$

(D) $y_4 = x^5 + x^4$

In each case, which term seems to most influence the shape of the graph?

The shape of the graph of a polynomial is also related to the shape of the graph of the term with highest degree or **leading term** of the polynomial. Figure 5 compares the graph of one of the polynomials from Figure 4 with the graph of its leading term. Although quite dissimilar for points close to the origin, as we "zoom out" to points distant from the origin, the graphs become quite similar. The leading term in the polynomial dominates all other terms combined.

FIGURE 5

$p(x) = x^5$,
$h(x) = x^5 - 6x^3 + 8x + 1$.

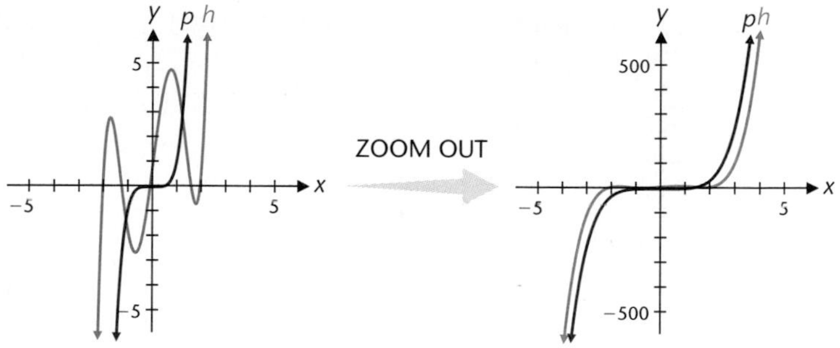

In general, the behavior of the graph of a polynomial function as x decreases without bound to the left or as x increases without bound to the right is determined by its leading term. We often use the symbols $-\infty$ and ∞ to help describe this left and right behavior.* The various possibilities are summarized in Theorem 3.

THEOREM 3

LEFT AND RIGHT BEHAVIOR OF A POLYNOMIAL

$P(x) = a_n x^n + a_{n-1}x^{n-1} + \cdots + a_1 x + a_0 \qquad a_n \neq 0$

1. $a_n > 0$ and n even
Graph of $P(x)$ increases without bound as x decreases to the left and as x increases to the right.

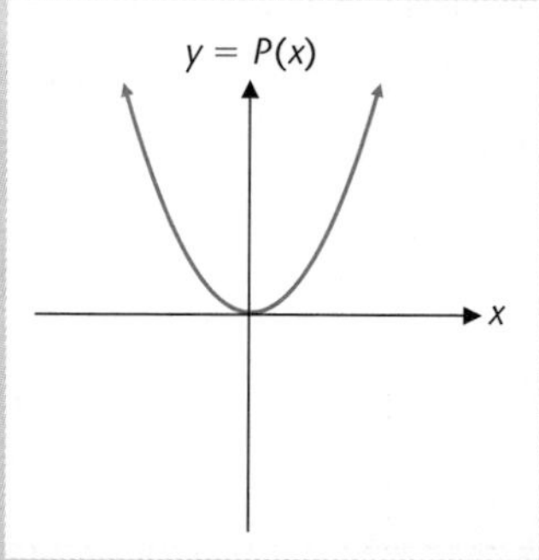

$$P(x) \to \begin{cases} \infty & \text{as } x \to -\infty \\ \infty & \text{as } x \to \infty \end{cases}$$

2. $a_n > 0$ and n odd
Graph of $P(x)$ decreases without bound as x decreases to the left and increases without bound as x increases to the right.

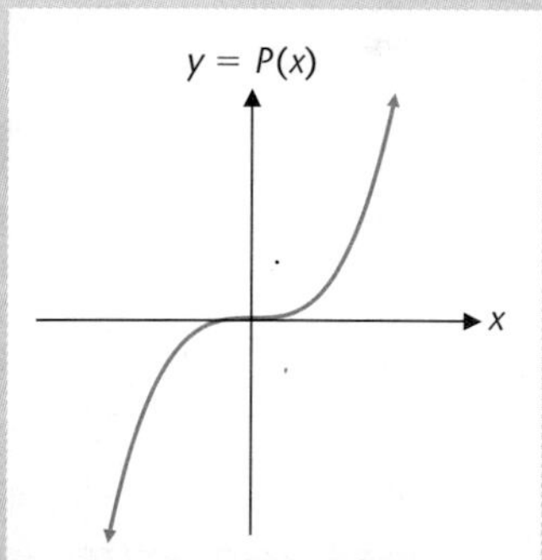

$$P(x) \to \begin{cases} -\infty & \text{as } x \to -\infty \\ \infty & \text{as } x \to \infty \end{cases}$$

*Remember, the symbol ∞ does not represent a real number. Earlier, we used ∞ to denote unbounded intervals, such as $[0, \infty)$. Now we are using it to describe quantities that are growing with no upper limit on their size.

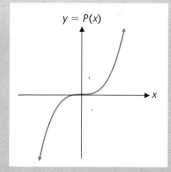

THEOREM 3 continued

3. $a_n < 0$ and n even
Graph of $P(x)$ decreases without bound as x decreases to the left and as x increases to the right.

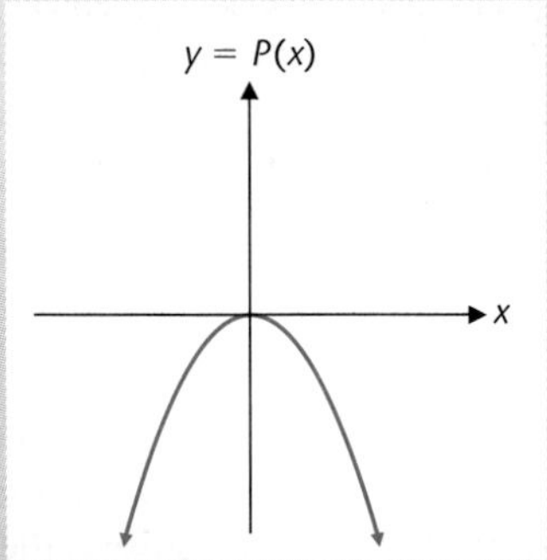

$$P(x) \to \begin{cases} -\infty & \text{as } x \to -\infty \\ -\infty & \text{as } x \to \infty \end{cases}$$

4. $a_n < 0$ and n odd
Graph of $P(x)$ increases without bound as x decreases to the left and decreases without bound as x increases to the right.

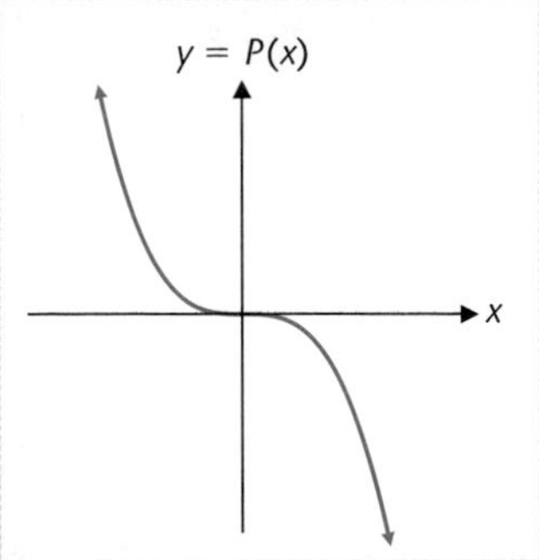

$$P(x) \to \begin{cases} \infty & \text{as } x \to -\infty \\ -\infty & \text{as } x \to \infty \end{cases}$$

Figure 4 shows examples of polynomial functions with graphs containing the maximum number of local extrema. The points on a continuous graph where the local extrema occur are sometimes referred to as **turning points.** Listed in Theorem 4 are useful properties of polynomial functions we accept without proof. Property 3 is discussed in detail later in this chapter. The other properties are established in calculus.

THEOREM 4

GRAPH PROPERTIES OF POLYNOMIAL FUNCTIONS

Let P be an nth-degree polynomial function with real coefficients.

1. P is continuous for all real numbers.
2. The graph of P is a smooth curve.
3. The graph of P has at most n x intercepts.
4. P has at most $n - 1$ local extrema or turning points.

Explore/Discuss 3

(A) What is the least number of local extrema an odd-degree polynomial function can have? An even-degree polynomial function?

(B) What is the maximum number of x intercepts the graph of a polynomial function of degree n can have?

(C) What is the maximum number of real solutions an nth-degree polynomial equation can have?

(D) What is the least number of x intercepts the graph of a polynomial function of odd degree can have? Of even degree?

(E) What is the least number of real zeros a polynomial function of odd degree can have? Of even degree?

EXAMPLE 4

Analyzing the Graph of a Polynomial

Approximate to two decimal places the zeros and local extrema for

$$P(x) = x^3 - 14x^2 + 27x - 12$$

Solution

Examining the graph of P in a standard viewing window [Fig. 6(a)], we see two zeros and a local maximum near $x = 1$. Zooming in shows these points more clearly [Fig. 6(b)]. Using built-in routines (details omitted), we find that $P(x) = 0$ for $x \approx 0.66$ and $x \approx 1.54$, and that $P(1.09) \approx 2.09$ is a local maximum value.

FIGURE 6
$P(x) = x^3 - 14x^2 + 27x - 12.$

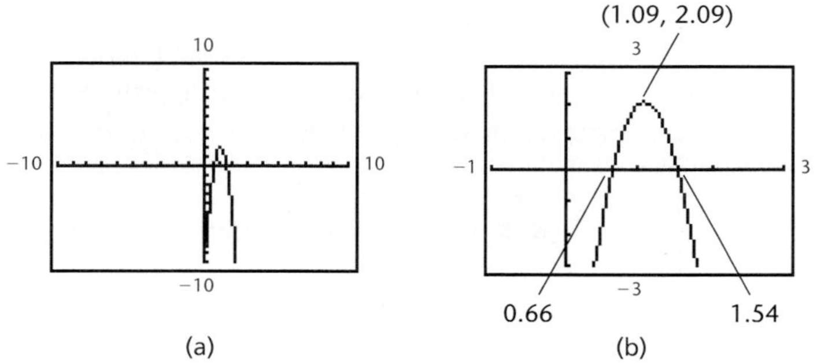

Have we found all the zeros and local extrema? The graph in Figure 6(a) seems to indicate that $P(x)$ is decreasing as x decreases to the left and as x increases to the right. However, the leading term for $P(x)$ is x^3. Since x^3 increases without bound as x increases to the right without bound, $P(x)$ must change direction at some point and become increasing. Thus, there must exist a local minimum and another zero that are not visible in this viewing window. Tracing along the graph in Figure 6(a) to the right and observing the coordinates on the screen we discover a local minimum near $x = 8$ and a zero near $x = 12$. Adjusting the window variables produces the graph in Figure 7. Using built-in routines (details omitted) we find that $P(x) = 0$ for $x \approx 11.80$ and that $P(8.24) \approx -180.61$ is a local minimum. Since a third-degree polynomial can have at most three zeros and two local extrema, we have found all the zeros and local extrema for this polynomial.

FIGURE 7
$P(x) = x^3 - 14x^2 + 27x - 12.$

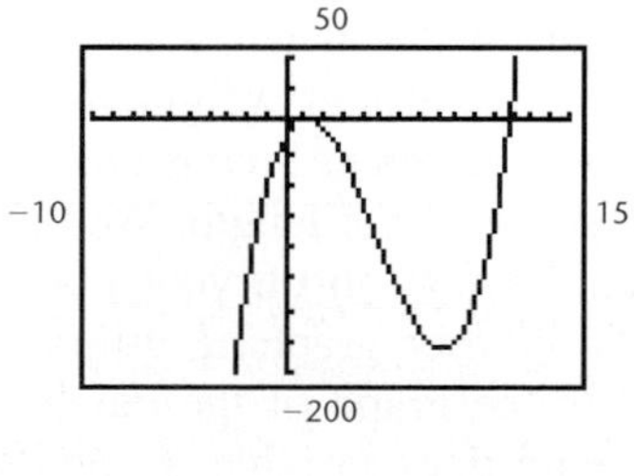

MATCHED PROBLEM 4

Approximate to two decimal places the zeros and the coordinates of the local extrema for

$$P(x) = -x^3 - 14x^2 - 15x + 5$$

Polynomial Regression

In the first two chapters, we saw that regression techniques can be used to construct a linear or quadratic model for a set of data. Most graphing utilities have the ability to use a variety of functions for modeling data. We discuss polynomial regression models in this section and other types of regression models in later sections.

EXAMPLE 5

Estimating the Weight of Fish

Using the length of a fish to estimate its weight is of interest to both scientists and sport anglers. The data in Table 1 give the average weight of North American sturgeon for certain lengths. Use these data and regression techniques to find a cubic polynomial model that can be used to estimate the weight of a sturgeon for any length. Estimate (to the nearest ounce) the weights of sturgeon of lengths 45, 46, 47, 48, 49, and 50 inches, respectively.

TABLE 1 Sturgeon

Length (in.) x	Weight (oz.) y
18	13
22	26
26	46
30	75
34	115
38	166
44	282
52	492
60	796

Source: www.thefishernet.com

Solution

The graph of the data in Table 1 [Fig. 8(a)] indicates that a linear regression model would not be appropriate for this data. And, in fact, we would not expect a linear relationship between length and weight. Instead, since weight is associated with volume, which involves three dimensions, it is more likely that the weight would be related to the cube of the length. We use a cubic regression polynomial to model this data [Fig. 8(b)]. (Consult your manual for the details of calculating regression polynomials on your graphing utility.) Figure 8(c) adds the graph of the polynomial model to the graph of the data. The graph in Figure 8(c) shows that this cubic polynomial does provide a good fit for the data. (We will have more to say about the choice of functions and the accuracy of the fit provided by regression analysis later in the text.) Figure 8(d) shows the estimated weights for the requested lengths.

FIGURE 8

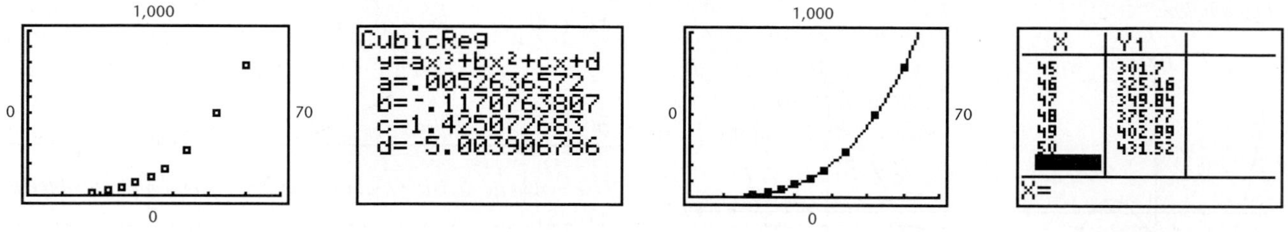

1,000

0   70

0

(a)

CubicReg
y=ax³+bx²+cx+d
a=.0052636572
b=-.1170763807
c=1.425072683
d=-5.003906786

(b)

1,000

0   70

0

(c)

X	Y1
45	301.7
46	325.16
47	349.84
48	375.77
49	402.99
50	431.52

X=

(d)

MATCHED PROBLEM
5

Find a quadratic regression model for the data in Table 1 and compare it with the cubic regression model found in Example 5. Which model appears to provide a better fit for this data? Use numerical and/or graphical comparisons to support your choice.

Answers to Matched Problems

1. $3x^2 + 6x + 8 + \dfrac{2}{3x - 4}$ 2. $\dfrac{P(x)}{x - 4} = 3x^3 + x^2 + 4x - 2 + \dfrac{0}{x - 4} = 3x^3 + x^2 + 4x - 2$

3. $P(-2) = -3$ for both parts, as it should

4. Zeros: -12.80, -1.47, 0.27; local maximum: $P(-0.57) \approx 9.19$; local minimum: $P(-8.76) \approx -265.71$

5. The cubic regression model provides a better model for this data, especially for $18 \le x \le 26$.

QuadReg
y=ax²+bx+c
a=.4941296639
b=-20.47687445
c=234.5707386

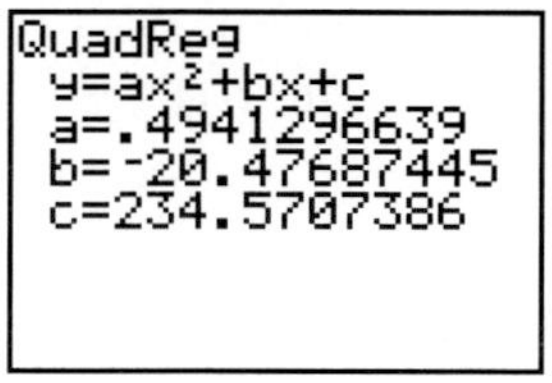

EXERCISE 3-1

A

In Problems 1–4, a is a positive real number. Match each function with one of graphs (a)–(d).

1. $f(x) = -ax^3$

2. $g(x) = -ax^4$

3. $h(x) = ax^6$

4. $k(x) = -ax^5$

(c)   (d)

(a)

(b)

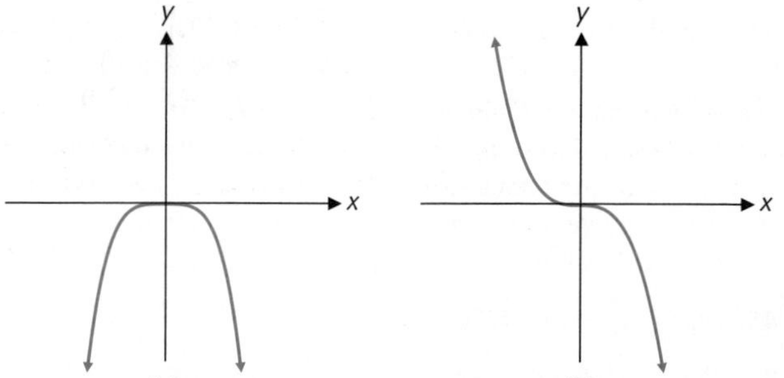

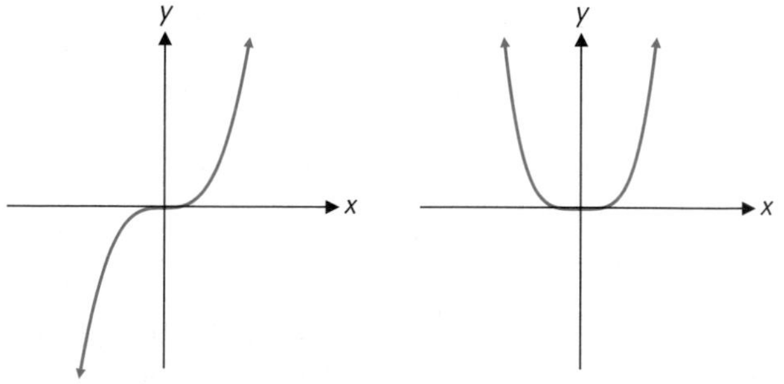

Problems 5–8 refer to the graphs of functions f, g, h, and k shown below.

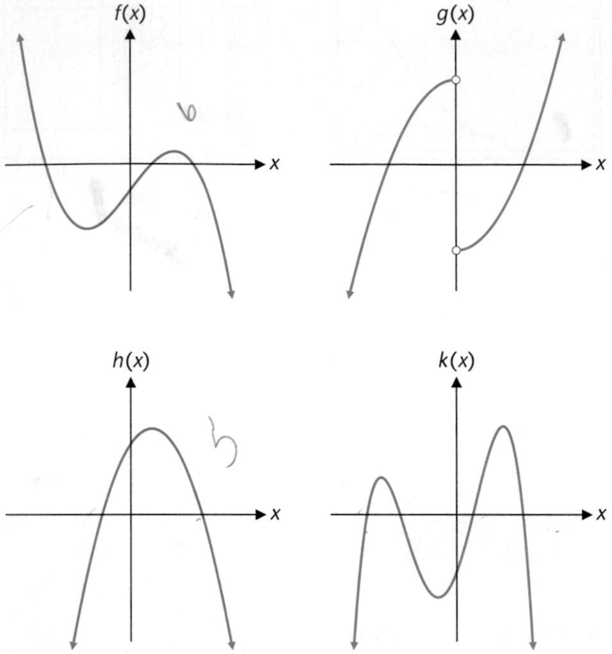

5. Which of these functions could be a second-degree polynomial?
6. Which of these functions could be a third-degree polynomial?
7. Which of these functions could be a fourth-degree polynomial?
8. Which of these functions is not a polynomial?

In Problems 9–16, divide, using algebraic long division. Write the quotient, and indicate the remainder.

9. $(4m^2 - 1) \div (2m - 1)$
10. $(y^2 - 9) \div (y + 3)$
11. $(6 - 6x + 8x^2) \div (2x + 1)$
12. $(11x - 2 + 12x^2) \div (3x + 2)$
13. $(x^3 - 1) \div (x - 1)$
14. $(a^3 + 27) \div (a + 3)$
15. $(3y - y^2 + 2y^3 - 1) \div (y + 2)$
16. $(3 + x^3 - x) \div (x - 3)$

In Problems 17–22, use synthetic division to write the quotient $P(x) \div (x - r)$ in the form $P(x)/(x - r) = Q(x) + R/(x - r)$, where R is a constant.

17. $(x^2 + 3x - 7) \div (x - 2)$
18. $(x^2 + 3x - 3) \div (x - 3)$
19. $(4x^2 + 10x - 9) \div (x + 3)$
20. $(2x^2 + 7x - 5) \div (x + 4)$
21. $(2x^3 - 3x + 1) \div (x - 2)$
22. $(x^3 + 2x^2 - 3x - 4) \div (x + 2)$

B

Use synthetic division and the remainder theorem in Problems 23–28.

23. Find $P(-2)$, given $P(x) = 3x^2 - x - 10$.
24. Find $P(-3)$, given $P(x) = 4x^2 + 10x - 8$.
25. Find $P(2)$, given $P(x) = 2x^3 - 5x^2 + 7x - 7$.
26. Find $P(5)$, given $P(x) = 2x^3 - 12x^2 - x + 30$.
27. Find $P(-4)$, given $P(x) = x^4 - 10x^2 + 25x - 2$.
28. Find $P(-7)$, given $P(x) = x^4 + 5x^3 - 13x^2 - 30$.

In Problems 29–44, divide, using synthetic division. Write the quotient, and indicate the remainder. As coefficients get more involved, a calculator should prove helpful. Do not round off—all quantities are exact.

29. $(3x^4 - x - 4) \div (x + 1)$
30. $(5x^4 - 2x^2 - 3) \div (x - 1)$
31. $(x^5 + 1) \div (x + 1)$
32. $(x^4 - 16) \div (x - 2)$
33. $(3x^4 + 2x^3 - 4x - 1) \div (x + 3)$
34. $(x^4 - 3x^3 - 5x^2 + 6x - 3) \div (x - 4)$
35. $(2x^6 - 13x^5 + 75x^3 + 2x^2 - 50) \div (x - 5)$
36. $(4x^6 + 20x^5 - 24x^4 - 3x^2 - 13x + 30) \div (x + 6)$
37. $(4x^4 + 2x^3 - 6x^2 - 5x + 1) \div (x + \frac{1}{2})$
38. $(2x^3 - 5x^2 + 6x + 3) \div (x - \frac{1}{2})$
39. $(4x^3 + 4x^2 - 7x - 6) \div (x + \frac{3}{2})$
40. $(3x^3 - x^2 + x + 2) \div (x + \frac{2}{3})$
41. $(3x^4 - 2x^3 + 2x^2 - 3x + 1) \div (x - 0.4)$
42. $(4x^4 - 3x^3 + 5x^2 + 7x - 6) \div (x - 0.7)$
43. $(3x^5 + 2x^4 + 5x^3 - 7x - 3) \div (x + 0.8)$
44. $(7x^5 - x^4 + 3x^3 - 2x^2 - 5) \div (x + 0.9)$

For each polynomial function in Problems 45–50:
(A) State the left and right behavior, the maximum number of x intercepts, and the maximum number of local extrema.
(B) Approximate (to two decimal places) the x intercepts and the local extrema.

45. $P(x) = x^3 - 5x^2 + 2x + 6$
46. $P(x) = x^3 + 2x^2 - 5x - 3$

47. $P(x) = -x^3 + 4x^2 + x + 5$

48. $P(x) = -x^3 - 3x^2 + 4x - 4$

49. $P(x) = x^4 + x^3 - 5x^2 - 3x + 12$

50. $P(x) = -x^4 + 6x^2 - 3x - 16$

In Problems 51–54, either give an example of a polynomial with real coefficients that satisfies the given conditions or explain why such a polynomial cannot exist.

51. $P(x)$ is a third-degree polynomial with one x intercept.

52. $P(x)$ is a fourth-degree polynomial with no x intercepts.

53. $P(x)$ is a third-degree polynomial with no x intercepts.

54. $P(x)$ is a fourth-degree polynomial with no turning points.

C

In Problems 55 and 56, divide, using algebraic long division. Write the quotient, and indicate the remainder.

55. $(16x - 5x^3 - 8 + 6x^4 - 8x^2) \div (2x - 4 + 3x^2)$

56. $(8x^2 - 7 - 13x + 24x^4) \div (3x + 5 + 6x^2)$

In Problems 57 and 58, divide, using synthetic division. Do not use a calculator.

57. $(x^3 - 3x^2 + x - 3) \div (x - i)$

58. $(x^3 - 2x^2 + x - 2) \div (x + i)$

59. Let $P(x) = x^2 + 2ix - 10$. Find

(A) $P(2 - i)$ (B) $P(5 - 5i)$

(C) $P(3 - i)$ (D) $P(-3 - i)$

60. Let $P(x) = x^2 - 4ix - 13$. Find

(A) $P(5 + 6i)$ (B) $P(1 + 2i)$

(C) $P(3 + 2i)$ (D) $P(-3 + 2i)$

In Problems 61–68, approximate (to two decimal places) the x intercepts and the local extrema.

61. $P(x) = 40 + 50x - 9x^2 - x^3$

62. $P(x) = 40 + 70x + 18x^2 + x^3$

63. $P(x) = 0.04x^3 - 10x + 5$

64. $P(x) = -0.01x^3 + 2.8x - 3$

65. $P(x) = 0.1x^4 + 0.3x^3 - 23x^2 - 23x + 90$

66. $P(x) = 0.1x^4 + 0.2x^3 - 19x^2 + 17x + 100$

67. $P(x) = x^4 - 24x^3 + 167x^2 - 275x + 131$

68. $P(x) = x^4 + 20x^3 + 118x^2 + 178x + 79$

69. (A) Divide $P(x) = a_2x^2 + a_1x + a_0$ by $x - r$, using both synthetic division and the long-division process, and compare the coefficients of the quotient and the remainder produced by each method.

(B) Expand the expression representing the remainder. What do you observe?

70. Repeat Problem 69 for

$$P(x) = a_3x^3 + a_2x^2 + a_1x + a_0$$

71. Polynomials also can be evaluated conveniently using a "nested factoring" scheme. For example, the polynomial $P(x) = 2x^4 - 3x^3 + 2x^2 - 5x + 7$ can be written in a nested factored form as follows:

$$\begin{aligned} P(x) &= 2x^4 - 3x^3 + 2x^2 - 5x + 7 \\ &= (2x - 3)x^3 + 2x^2 - 5x + 7 \\ &= [(2x - 3)x + 2]x^2 - 5x + 7 \\ &= \{[(2x - 3)x + 2]x - 5\}x + 7 \end{aligned}$$

Use the nested factored form to find $P(-2)$ and $P(1.7)$. [*Hint:* To evaluate $P(-2)$, store -2 in your calculator's memory and proceed from left to right recalling -2 as needed.]

72. Find $P(-2)$ and $P(1.3)$ for $P(x) = 3x^4 + x^3 - 10x^2 + 5x - 2$ using the nested factoring scheme presented in Problem 71.

APPLICATIONS

73. Revenue. The price-demand equation for 8,000 BTU window air conditioners is given by

$$p = 0.0004x^2 - x + 569 \qquad 0 \le x \le 800$$

where x is the number of air conditioners that can be sold at a price of p dollars each.

(A) Find the revenue function.

(B) Find the number of air conditioners that must be sold to maximize the revenue, the corresponding price to the nearest dollar, and the maximum revenue to the nearest dollar.

74. Profit. Refer to Problem 73. The cost of manufacturing 8,000 BTU window air conditioners is given by

$$C(x) = 10{,}000 + 90x$$

where $C(x)$ is the total cost in dollars of producing x air conditioners.

(A) Find the profit function.

(B) Find the number of air conditioners that must be sold to maximize the profit, the corresponding price to the nearest dollar, and the maximum profit to the nearest dollar.

75. Construction. A rectangular container measuring 1 foot by 2 feet by 4 feet is covered with a layer of lead shielding of uniform thickness (see the figure).

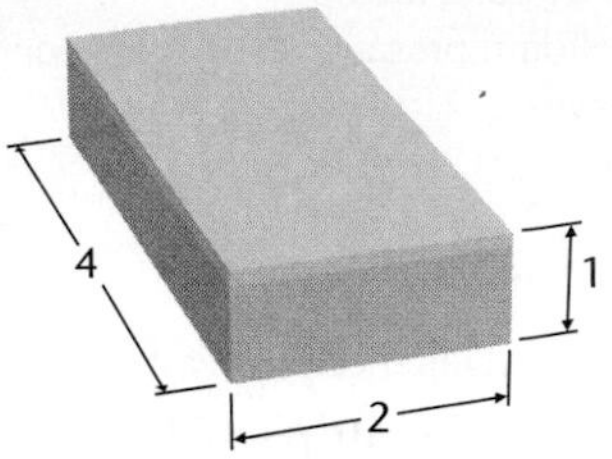

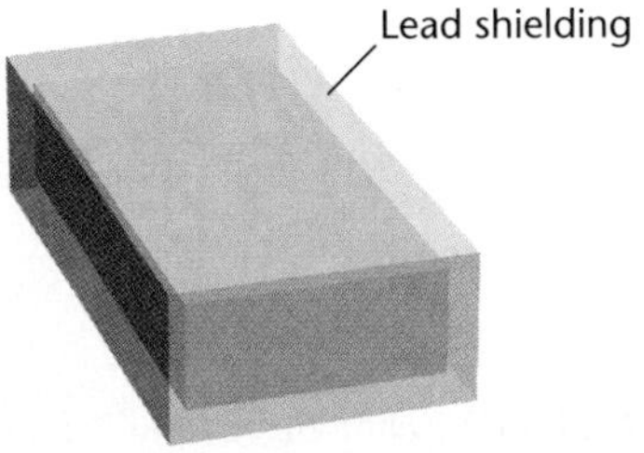

(A) Find the volume of lead shielding V as a function of the thickness x (in feet) of the shielding.

(B) Find the thickness of the lead shielding to three decimal places if the volume of the shielding is 3 cubic feet.

76. Manufacturing. A rectangular storage container measuring 2 feet by 2 feet by 3 feet is coated with a protective coating of plastic of uniform thickness.

(A) Find the volume of plastic V as a function of the thickness x (in feet) of the coating.

(B) Find the thickness of the plastic coating to four decimal places if the volume of the shielding is 0.1 cubic feet.

77. Health Care. Table 2 shows the total national expenditures (in billion dollars) and the per capita expenditures (in dollars) for selected years since 1960.

TABLE 2 National Health Expenditures

Year	Total Expenditures (billion $)	Per Capita Expenditures ($)
1960	27.1	143
1965	41.6	204
1970	74.4	346
1975	132.9	592
1980	247.2	1,002
1985	428.2	1,666
1990	697.5	2,588

Source: U.S. Census Bureau.

(A) Let x represent the number of years since 1960 and find a cubic regression polynomial for the total national expenditures.

(B) Use the polynomial model from part A to estimate the total national expenditures (to the nearest tenth of a billion) for 1995.

78. Health Care. Refer to Table 2.

(A) Let x represent the number of years since 1960 and find a cubic regression polynomial for the per capita expenditures.

(B) Use the polynomial model from part A to estimate the per capita expenditures (to the nearest dollar) for 1995.

79. Marriage. Table 3 shows the marriage and divorce rates per 1,000 population for selected years since 1950.

TABLE 3 Marriages and Divorces (per 1,000 population)

Year	Marriages	Divorces
1950	11.1	2.6
1955	9.3	2.3
1960	8.5	2.2
1965	9.3	2.5
1970	10.6	3.5
1975	10.0	4.8
1980	10.6	5.2
1985	10.1	5.0
1990	9.8	4.7

Source: U.S. Census Bureau.

(A) Let x represent the number of years since 1950 and find a cubic regression polynomial for the marriage rate.

(B) Use the polynomial model from part A to estimate the marriage rate (to one decimal place) for 1995.

80. Divorce. Refer to Table 3.

(A) Let x represent the number of years since 1950 and find a cubic regression polynomial for the divorce rate.

(B) Use the polynomial model from part A to estimate the divorce rate (to one decimal place) for 1995.

Section 3-2 Finding Rational Zeros of Polynomials

- Factor Theorem
- Fundamental Theorem of Algebra
- Imaginary Zeros
- Rational Zeros

In this section we develop some important properties of polynomials with arbitrary coefficients. Then we consider the problem of finding all the rational zeros of a polynomial with rational coefficients. In some cases, this process will also enable us to find irrational or imaginary zeros.

Explore/Discuss 1

Let $P(x) = x^3 - 5x^2 + 2x + 8$

(A) (i) Find $P(2)$.

(ii) What is the remainder if $P(x)$ is divided by $x - 2$?

(iii) Is $x - 2$ a factor of $P(x)$?

(B) Repeat part A for $x = -1$, $x = 1$, $x = 3$, and $x = 4$.

What relationship between factors and zeros is suggested by these results?

Factor Theorem

The division algorithm (Theorem 1 in Section 3-1)

$$P(x) = (x - r)Q(x) + R$$

may, because of the remainder theorem (Theorem 2 in Section 3-1), be written in a form where R is replaced by $P(r)$:

$$P(x) = (x - r)Q(x) + P(r)$$

It is easy to see that $x - r$ is a factor of $P(x)$ if and only if $P(r) = 0$; that is, if and only if r is a zero of the polynomial $P(x)$. This result is known as the **factor theorem:**

THEOREM 1

FACTOR THEOREM

If r is a zero of the polynomial $P(x)$, then $x - r$ is a factor of $P(x)$. Conversely, if $x - r$ is a factor of $P(x)$, then r is a zero of $P(x)$.

The relationship between zeros, roots, factors, and x intercepts is fundamental to the study of polynomials. With the addition of the factor theorem, we now know that the following statements are equivalent for any polynomial $P(x)$:

1. r is a root of the equation $P(x) = 0$.

2. r is a zero of $P(x)$.

3. $x - r$ is a factor of $P(x)$.

If, in addition, the coefficients of $P(x)$ are real numbers and r is a real number, then we can add a fourth statement to this list:

4. r is an x intercept of the graph of $P(x)$.

EXAMPLE 1

Factors, Zeros, Roots, and Intercepts

(A) Use the factor theorem to show that $x + 1$ is a factor of $P(x) = x^{25} + 1$.

(B) What are the zeros of $P(x) = 3(x - 5)(x + 2)(x - 3)$?

(C) What are the roots of $x^4 - 1 = 0$?

(D) What are the x intercepts of the graph of $P(x) = x^4 - 1$?

Solutions

(A) Since $x + 1 = x - (-1)$, we have $r = -1$ and

$$P(r) = P(-1) = (-1)^{25} + 1 = -1 + 1 = 0$$

Hence, -1 is a zero of $P(x) = x^{25} + 1$. By the factor theorem, $x - (-1) = x + 1$ is a factor of $x^{25} + 1$.

(B) Since $(x - 5)$, $(x + 2)$, and $(x - 3)$ are all factors of $P(x)$, 5, -2, and 3 are zeros of $P(x)$.

FIGURE 1
x intercepts of $P(x) = x^4 - 1$.

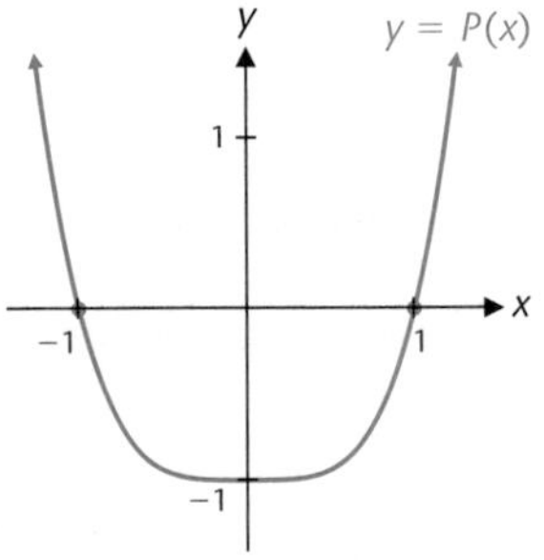

(C) Factoring the left side, we have

$$x^4 - 1 = 0$$

$$(x^2 - 1)(x^2 + 1) = 0$$

$$(x - 1)(x + 1)(x - i)(x + i) = 0$$

Thus, the roots of $x^4 - 1 = 0$ are 1, -1, i, and $-i$.

(D) From part C, the zeros of $P(x)$ are 1, -1, i, and $-i$. However, x intercepts must be real numbers. Thus the x intercepts of the graph of $P(x) = x^4 - 1$ are 1 and -1 (see Fig. 1).

MATCHED PROBLEM 1

(A) Use the factor theorem to show that $x - 1$ is a factor of $P(x) = x^{54} - 1$.

(B) What are the zeros of

$$P(x) = 2(x + 3)(x + 7)(x - 8)(x + 1)?$$

(C) What are the roots of $x^2 + 4 = 0$?

(D) What are the x intercepts of $P(x) = x^2 + 4$?

Fundamental Theorem of Algebra

Theorem 2, often referred to as the **fundamental theorem of algebra,** requires verification that is beyond the scope of this book, so we state it without proof.

THEOREM 2

FUNDAMENTAL THEOREM OF ALGEBRA

Every polynomial $P(x)$ of degree $n > 0$ has at least one zero.

If $P(x) = a_nx^n + a_{n-1}x^{n-1} + \cdots + a_1x + a_0$ is a polynomial of degree $n > 0$ with complex coefficients, then, according to Theorem 2, it has at least one zero, say r_1. According to the factor theorem, $x - r_1$ is a factor of $P(x)$. Thus,

$$P(x) = (x - r_1)Q_1(x)$$

where $Q_1(x)$ is a polynomial of degree $n - 1$. If $n - 1 = 0$, then $Q_1(x) = a_n$. If $n - 1 > 0$, then, by Theorem 2, $Q_1(x)$ has at least one zero, say r_2. And

$$Q_1(x) = (x - r_2)Q_2(x)$$

where $Q_2(x)$ is a polynomial of degree $n - 2$. Thus,

$$P(x) = (x - r_1)(x - r_2)Q_2(x)$$

If $n - 2 = 0$, then $Q_2(x) = a_n$. If $n - 2 > 0$, then $Q_2(x)$ has at least one zero, say r_3. And

$$Q_2(x) = (x - r_3)Q_3(x)$$

where $Q_3(x)$ is a polynomial of degree $n - 3$.

We continue in this way until $Q_k(x)$ is of degree zero—that is, until $k = n$. At this point, $Q_n(x) = a_n$, and we have

$$P(x) = (x - r_1)(x - r_2) \cdot \cdots \cdot (x - r_n)a_n$$

Thus, $r_1, r_2, \ldots, r_n$ are n zeros, not necessarily distinct, of $P(x)$. Is it possible for $P(x)$ to have more than these n zeros? Let's assume that r is a number different from the zeros above. Then

$$P(r) = a_n(r - r_1)(r - r_2) \cdot \cdots \cdot (r - r_n) \neq 0$$

since r is not equal to any of the zeros. Hence, r is not a zero, and we conclude that $r_1, r_2, \ldots, r_n$ are the only zeros of $P(x)$. We have just sketched a proof of Theorem 3.

THEOREM 3

n ZEROS THEOREM

Every polynomial $P(x)$ of degree $n > 0$ can be expressed as the product of n linear factors. Hence, $P(x)$ has exactly n zeros—not necessarily distinct.

Theorems 2 and 3 were first proved in 1797 by Carl Friedrich Gauss, one of the greatest mathematicians of all time, at the age of 20.

If $P(x)$ is represented as the product of linear factors and $x - r$ occurs m times, then r is called a **zero of multiplicity *m*.** For example, if

$$P(x) = 4(x - 5)^3(x + 1)^2(x - i)(x + i)$$

then this seventh-degree polynomial has seven zeros, not all distinct. That is, 5 is a zero of multiplicity 3, or a triple zero; -1 is a zero of multiplicity 2, or a double zero; and i and $-i$ are zeros of multiplicity 1, or simple zeros. Thus, this seventh-degree polynomial has exactly seven zeros if we count 5 and -1 with their respective multiplicities.

Explore/Discuss 2

Graph each of the following polynomials one at a time for $-10 \le x \le 10$ and an appropriate interval of y values:

$$y_1 = (x + 2)^2(3 - x)$$
$$y_2 = (x + 2)(3 - x)^2$$
$$y_3 = (x + 2)^2(3 - x)^2$$
$$y_4 = (x + 2)^2(3 - x)^3$$
$$y_5 = (x + 2)^3(3 - x)^2$$
$$y_6 = (x + 2)^3(3 - x)^3$$

Discuss the various possibilities for the behavior of the graph of a polynomial at a zero of odd multiplicity and at a zero of even multiplicity.

EXAMPLE 2 **Factoring a Polynomial**

If -2 is a double zero of $P(x) = x^4 - 7x^2 + 4x + 20$, write $P(x)$ as a product of first-degree factors.

Solution Since -2 is a double zero of $P(x)$, we can write

$$P(x) = (x + 2)^2Q(x)$$
$$= (x^2 + 4x + 4)Q(x)$$

and find $Q(x)$ by dividing $P(x)$ by $x^2 + 4x + 4$. Carrying out the algebraic long division, we obtain

$$Q(x) = x^2 - 4x + 5$$

The zeros of $Q(x)$ are found, using the quadratic formula, to be $2 - i$ and $2 + i$. Thus, $P(x)$ written as a product of linear factors is

$$P(x) = (x + 2)^2[x - (2 - i)][x - (2 + i)]$$

[*Note:* Any time $Q(x)$ is a quadratic polynomial, its zeros can be found using the quadratic formula.]

MATCHED PROBLEM 2

If 3 is a double zero of $P(x) = x^4 - 12x^3 + 55x^2 - 114x + 90$, write $P(x)$ as a product of first-degree factors.

FIGURE 2

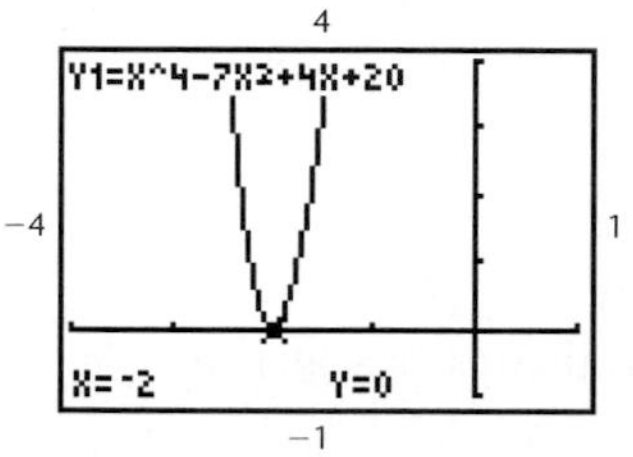

The graph of $y = P(x)$ from Example 2 is shown in Figure 2. Notice that $P(x)$ has four zeros, taking multiplicity into account, but only one x intercept—the double zero at $x = -2$. Since an nth degree polynomial has exactly n zeros (Theorem 3), it can have a maximum of n x intercepts, but may have fewer, due to imaginary zeros and multiple real zeros. Also note that the two imaginary zeros of $P(x)$ are not visible on the graph in Figure 2. The real zeros of a polynomial with real coefficients can be approximated graphically on a graphing utility, but the imaginary zeros cannot.

CAUTION

Approximating real zeros of even multiplicity with a built-in root approximation routine can result in an error message. The maximum and minimum routines on a graphing utility can be used to approximate real zeros of even multiplicity if the root approximation routine fails.

Imaginary Zeros

Something interesting happens if we restrict the coefficients of a polynomial to real numbers. Let's use the quadratic formula to find the zeros of the polynomial

$$P(x) = x^2 - 6x + 13$$

To find the zeros of $P(x)$, we solve $P(x) = 0$:

$$x^2 - 6x + 13 = 0$$

$$x = \frac{6 \pm \sqrt{36 - 52}}{2}$$

$$= \frac{6 \pm \sqrt{-16}}{2} = \frac{6 \pm 4i}{2} = 3 \pm 2i$$

The zeros of $P(x)$ are $3 - 2i$ and $3 + 2i$, conjugate imaginary numbers (see Section 2-4). Also observe that the imaginary zeros in Example 2 are the conjugate imaginary numbers $2 - i$ and $2 + i$.

This is generalized in the following theorem:

THEOREM 4

IMAGINARY ZEROS THEOREM

Imaginary zeros of polynomials with real coefficients, if they exist, occur in conjugate pairs.

As a consequence of Theorems 3 and 4, we also know (think this through) the following:

THEOREM 5

REAL ZEROS AND ODD-DEGREE POLYNOMIALS

A polynomial of odd degree with real coefficients always has at least one real zero.

Explore/Discuss 3

(A) Let $P(x)$ be a third-degree polynomial with real coefficients. Indicate which of the following statements are true and which are false. Justify your conclusions.
  (i) $P(x)$ has at least one real zero.
  (ii) $P(x)$ has three zeros.
  (iii) $P(x)$ can have two real zeros and one imaginary zero.

(B) Let $P(x)$ be a fourth-degree polynomial with real coefficients. Indicate which of the following statements are true and which are false. Justify your conclusions.
  (i) $P(x)$ has four zeros.
  (ii) $P(x)$ has at least two real zeros.
  (iii) If we know $P(x)$ has three real zeros, then the fourth zero must be real.

Rational Zeros

First note that a polynomial with rational coefficients can always be written as a constant times a polynomial with integer coefficients. For example,

$$\begin{aligned} P(x) &= \tfrac{1}{2}x^3 - \tfrac{2}{3}x^2 + \tfrac{7}{4}x + 5 \\ &= \tfrac{1}{12}(6x^3 - 8x^2 + 21x + 60) \end{aligned}$$

Thus, it is sufficient to confine our attention to polynomials with integer coefficients.

We introduce the rational zero theorem by examining the following quadratic polynomial whose zeros can be found easily by factoring:

$$P(x) = 6x^2 - 13x - 5 = (2x - 5)(3x + 1)$$

Zeros of $P(x)$: $\dfrac{5}{2}$ and $-\dfrac{1}{3} = \dfrac{-1}{3}$

Notice that the numerators, 5 and -1, of the zeros are both integer factors of -5, the constant term in $P(x)$. The denominators 2 and 3 of the zeros are both integer factors of 6, the coefficient of the highest-degree term in $P(x)$. These observations are generalized in Theorem 6.

THEOREM

6

RATIONAL ZERO THEOREM

If the rational number b/c, in lowest terms, is a zero of the polynomial

$$P(x) = a_n x^n + a_{n-1} x^{n-1} + \cdots + a_1 x + a_0 \qquad a_n \neq 0$$

with integer coefficients, then b must be an integer factor of a_0 and c must be an integer factor of a_n.

$$P(x) = a_n x^n + a_{n-1} x^{n-1} + \cdots + a_1 x + a_0$$

$\dfrac{b}{c}$

c must be a factor of a_n

b must be a factor of a_0

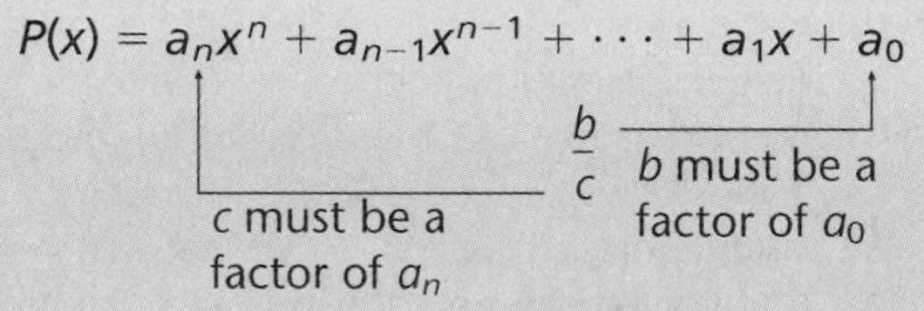

The proof of Theorem 6 is not difficult and is instructive, so we sketch it here.

Proof

Since b/c is a zero of $P(x)$,

$$a_n\left(\frac{b}{c}\right)^n + a_{n-1}\left(\frac{b}{c}\right)^{n-1} + \cdots + a_1\left(\frac{b}{c}\right) + a_0 = 0 \qquad (1)$$

If we multiply both sides of equation (1) by c^n, we obtain

$$a_n b^n + a_{n-1} b^{n-1} c + \cdots + a_1 b c^{n-1} + a_0 c^n = 0 \qquad (2)$$

which can be written in the form

$$a_n b^n = c(-a_{n-1} b^{n-1} - \cdots - a_0 c^{n-1}) \qquad (3)$$

Since both sides of equation (3) are integers, c must be a factor of $a_n b^n$. And since the rational number b/c is given to be in lowest terms, b and c can have no common factors other than ± 1. That is, b and c are **relatively prime.** This implies that b^n and c also are relatively prime. Hence, c must be a factor of a_n.

Now, if we solve equation (2) for $a_0 c^n$ and factor b out of the right side, we have

$$a_0 c^n = b(-a_n b^{n-1} - \cdots - a_1 c^{n-1})$$

We see that b is a factor of a_0c^n and, hence, a factor of a_0, since b and c are relatively prime.

Explore/Discuss 4

Let $P(x) = a_3x^3 + a_2x^2 + a_1x + a_0$, where a_3, a_2, a_1, and a_0 are integers.

1. If $P(2) = 0$, there is one coefficient that must be an even integer. Identify this coefficient and explain why it must be even.
2. If $P(\frac{1}{2}) = 0$, there is one coefficient that must be an even integer. Identify this coefficient and explain why it must be even.
3. If $a_3 = a_0 = 1$, $P(-1) \neq 0$, and $P(1) \neq 0$, does $P(x)$ have any rational zeros? Support your conclusion with verbal arguments and/or examples.

It is important to understand that Theorem 6 does not say that a polynomial *P(x)* with integer coefficients must have rational zeros.

It simply states that if $P(x)$ does have a rational zero, then the numerator of the zero must be an integer factor of a_0 and the denominator of the zero must be an integer factor of a_n. Since every integer has a finite number of integer factors, Theorem 6 enables us to construct a finite list of possible rational zeros. Finding any rational zeros then becomes a routine, although sometimes tedious, process of elimination.

As the next example illustrates, a graphing utility can greatly reduce the effort required to locate rational zeros.

EXAMPLE 3 Finding Rational Zeros

Find all the rational zeros for $P(x) = 2x^3 + 9x^2 + 7x - 6$.

Solution If b/c in lowest terms is a rational zero of $P(x)$, then b must be a factor of -6 and c must be a factor of 2.

$$\text{Possible values of } b \text{ are the integer factors of } -6\text{: } \pm1, \pm2, \pm3, \pm6 \qquad (4)$$

$$\text{Possible values of } c \text{ are the integer factors of } 2\text{: } \pm1, \pm2 \qquad (5)$$

Writing all possible fractions b/c where b is from (4) and c is from (5), we have

$$\text{Possible rational zeros for } P(x)\text{: } \pm1, \pm2, \pm3, \pm6, \pm\tfrac{1}{2}, \pm\tfrac{3}{2} \qquad (6)$$

[Note that all fractions are in lowest terms and duplicates like $\pm6/\pm2 = \pm3$ are not repeated.] If $P(x)$ has any rational zeros, they must be in list (6). We can test each number r in this list simply by evaluating $P(r)$. However, exploring the graph of $y = P(x)$ first will usually indicate which numbers in the list are the most likely candidates for zeros. Examining a graph of $P(x)$, we see that there are zeros near -3, near -2, and between 1 and 2, so we begin by evaluating $P(x)$ at -3, -2, and $\frac{1}{2}$ (Fig. 3).

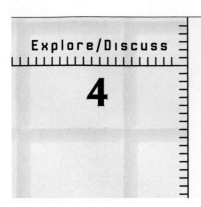

FIGURE 3

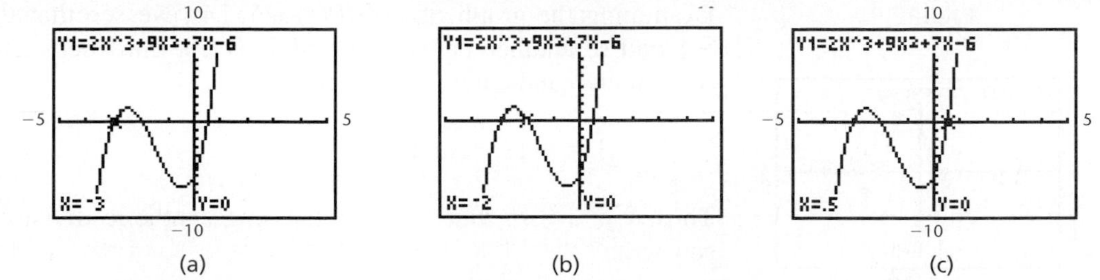

Thus -3, -2, and $\frac{1}{2}$ are rational zeros of $P(x)$. Since a third-degree polynomial can have at most three zeros, we have found all the rational zeros. There is no need to test the remaining candidates in list (6).

MATCHED PROBLEM 3

Find all rational zeros for $P(x) = 2x^3 + x^2 - 11x - 10$.

As we saw in the solution of Example 3, rational zeros can be located by simply evaluating the polynomial. However, if we want to find multiple zeros, imaginary zeros, or exact values of irrational zeros, we need to consider *reduced polynomials*. If r is a zero of a polynomial $P(x)$, then we can write

$$P(x) = (x - r)Q(x)$$

where $Q(x)$ is a polynomial of degree one less than the degree of $P(x)$. The quotient polynomial $Q(x)$ is called the **reduced polynomial** for $P(x)$. In Example 3, after determining that -3 is a zero of $P(x)$, we can write

$$\begin{array}{r|rrrr} & 2 & 9 & 7 & -6 \\ & & -6 & -9 & 6 \\ \hline -3 & 2 & 3 & -2 & 0 \end{array}$$

$$\begin{aligned} P(x) &= 2x^3 + 9x^2 + 7x - 6 \\ &= (x + 3)(2x^2 + 3x - 2) \\ &= (x + 3)Q(x) \end{aligned}$$

Since the reduced polynomial $Q(x) = 2x^2 + 3x - 2$ is a quadratic, we can find its zeros by factoring or the quadratic formula. Thus,

$$P(x) = (x + 3)(2x^2 + 3x - 2) = (x + 3)(x + 2)(2x - 1)$$

and we see that the zeros of $P(x)$ are -3, -2, and $\frac{1}{2}$, as before.

EXAMPLE 4 **Finding Rational and Irrational Zeros**

Find all zeros exactly for $P(x) = 2x^3 - 7x^2 + 4x + 3$.

Solution First, list the possible rational zeros:

$$\pm 1, \pm 3, \pm \tfrac{1}{2}, \pm \tfrac{3}{2}$$

FIGURE 4

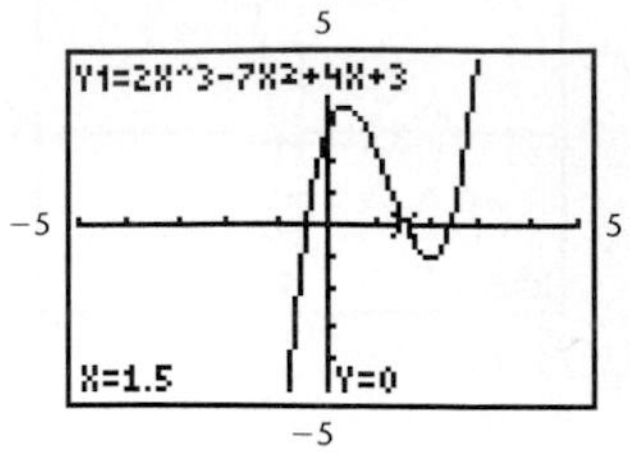

Examining the graph of $y = P(x)$ (Fig. 4), we see that there is a zero between -1 and 0, another between 1 and 2, and a third between 2 and 3. We test the only likely candidates, $-\frac{1}{2}$ and $\frac{3}{2}$:

$$P(-\tfrac{1}{2}) = -1 \qquad \text{and} \qquad P(\tfrac{3}{2}) = 0$$

Thus, $\frac{3}{2}$ is a zero, but $-\frac{1}{2}$ is not. Using synthetic division (details omitted), we can write

$$P(x) = (x - \tfrac{3}{2})(2x^2 - 4x - 2)$$

Since the reduced polynomial is quadratic, we can use the quadratic formula to find the exact values of the remaining zeros:

$$2x^2 - 4x - 2 = 0$$

$$x^2 - 2x - 1 = 0$$

$$x = \frac{2 \pm \sqrt{4 - 4(1)(-1)}}{2}$$

$$= \frac{2 \pm 2\sqrt{2}}{2} = 1 \pm \sqrt{2}$$

Thus, the exact zeros of $P(x)$ are $\frac{3}{2}$ and $1 \pm \sqrt{2}$.

MATCHED PROBLEM 4

Find all zeros exactly for $P(x) = 3x^3 - 10x^2 + 5x + 4$.

EXAMPLE 5

Finding Rational and Imaginary Zeros

Find all zeros exactly for $P(x) = x^4 - 6x^3 + 14x^2 - 14x + 5$.

Solution

FIGURE 5

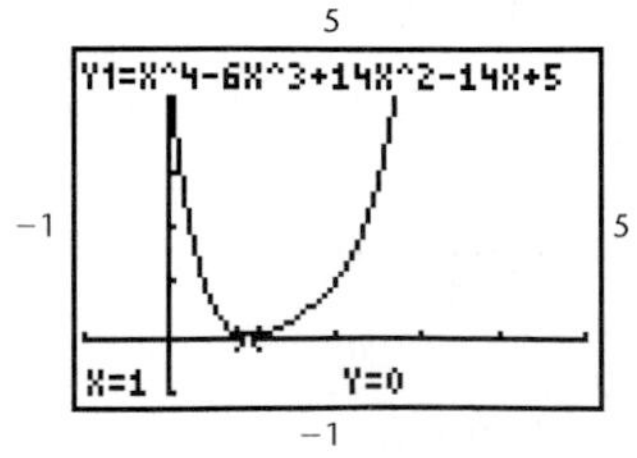

The possible rational zeros are ± 1 and ± 5. Examining the graph of $P(x)$ (Fig. 5), we see that 1 is a zero. Since the graph of $P(x)$ does not appear to change sign at 1, this may be a multiple root (see Explore/Discuss 2). Using synthetic division (details omitted), we find that

$$P(x) = (x - 1)(x^3 - 5x^2 + 9x - 5)$$

The possible rational zeros of the reduced polynomial

$$Q(x) = x^3 - 5x^2 + 9x - 5$$

are ± 1 and ± 5. Examining the graph of $Q(x)$ (Fig. 6), we see that 1 is a rational zero. After a division, we have a quadratic reduced polynomial:

$$Q(x) = (x - 1)Q_1(x) = (x - 1)(x^2 - 4x + 5)$$

FIGURE 6

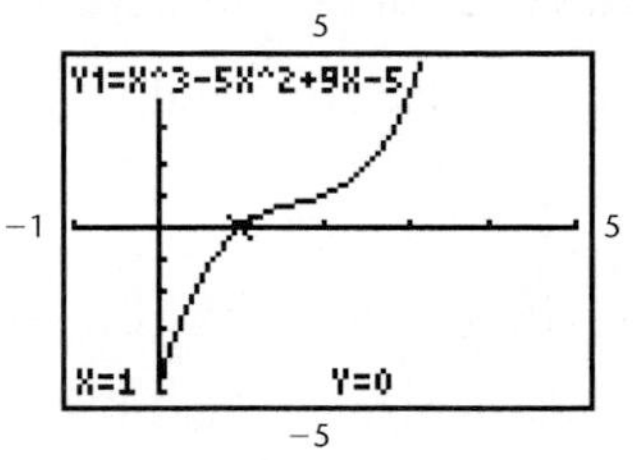

We use the quadratic formula to find the zeros of $Q_1(x)$:

$$x^2 - 4x + 5 = 0$$

$$x = \frac{4 \pm \sqrt{16 - 4(1)(5)}}{2}$$

$$= \frac{4 \pm \sqrt{-4}}{2} = 2 \pm i$$

Thus, the exact zeros of $P(x)$ are 1 (multiplicity 2), $2 - i$, and $2 + i$.

MATCHED PROBLEM 5

Find all zeros exactly for $P(x) = x^4 + 4x^3 + 10x^2 + 12x + 5$.

Remark We were successful in finding all the zeros of the polynomials in Examples 4 and 5 because we could find sufficient rational zeros to reduce the original polynomial to a quadratic. This is not always possible. For example, the polynomial

$$P(x) = x^3 + 6x - 2$$

has no rational zeros, but does have an irrational zero at $x \approx 0.32748$ (Fig. 7). The other two zeros are imaginary. The techniques we have developed will not find the exact value of these roots.

FIGURE 7
$P(x) = x^3 + 6x - 2.$

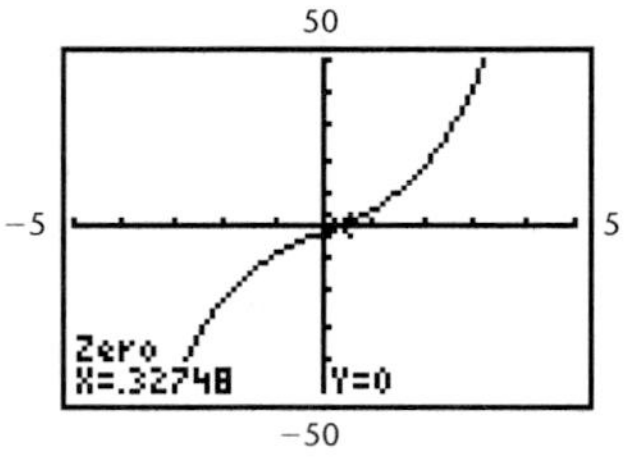

Explore/Discuss 5

There is a technique for finding the exact zeros of cubic polynomials, usually referred to as Cardano's formula.* This formula shows that the exact value of the irrational zero of $P(x) = x^3 + 6x - 2$ (see Fig. 7) is

$$x = \sqrt[3]{4} - \sqrt[3]{2}$$

(A) Verify that this is correct by expanding and simplifying

$$P(\sqrt[3]{4} - \sqrt[3]{2}).$$

(B) Cardano's formula also shows that the two imaginary zeros are

$$-\tfrac{1}{2}(\sqrt[3]{4} - \sqrt[3]{2}) \pm \tfrac{1}{2}i\sqrt{3}(\sqrt[3]{4} - \sqrt[3]{2})$$

If you like algebraic manipulation, you can also verify that these are correct.

---

*Girolamo Cardano (1501–1576), an Italian mathematician and physician, was the first to publish a formula for the solution to cubic equations of the form $x^3 + ax + b = 0$ and the first to realize that this technique could be used to solve other cubic equations. Having predicted that he would live to the age of 75, Cardano committed suicide in 1576.

Explore/Discuss 5 continued

(C) Find a reference for Cardano's formula in a library or on the Internet. Use this formula to find the exact value of the irrational zero of

$$P(x) = x^3 + 9x - 6$$

Check your answer by comparing it with the approximate value obtained on a graphing utility.

Answers to Matched Problems

1. (A) $P(1) = 1^{54} - 1 = 0$ implies that $x = 1$ is a factor of $P(x)$.
   (B) $-3, -7, 8, -1$ (C) $-2i, 2i$ (D) No x intercepts
2. $P(x) = (x - 3)^2[x - (3 - i)][x - (3 + i)]$
3. $-2, -1, \frac{5}{2}$
4. $\frac{4}{3}, 1 - \sqrt{2}, 1 + \sqrt{2}$
5. -1 (multiplicity 2), $-1 - 2i$, $-1 + 2i$

EXERCISE 3-2

A

Write the zeros of each polynomial in Problems 1–4, and indicate the multiplicity of each if over 1. What is the degree of each polynomial?

1. $P(x) = (x + 8)^3(x - 6)^2$
2. $P(x) = (x - 5)(x + 7)^2$
3. $P(x) = 3(x + 4)^3(x - 3)^2(x + 1)$
4. $P(x) = 5(x - 2)^3(x + 3)^2(x - 1)$

In Problems 5–10, find a polynomial P(x) of lowest degree, with leading coefficient 1, that has the indicated set of zeros. Leave the answer in a factored form. Indicate the degree of the polynomial.

5. 3 (multiplicity 2) and -4
6. -2 (multiplicity 3) and 1 (multiplicity 2)
7. -7 (multiplicity 3), $-3 + \sqrt{2}$, $-3 - \sqrt{2}$
8. $\frac{1}{3}$ (multiplicity 2), $5 + \sqrt{7}$, $5 - \sqrt{7}$
9. $(2 - 3i)$, $(2 + 3i)$, -4 (multiplicity 2)
10. $i\sqrt{3}$ (multiplicity 2), $-i\sqrt{3}$ (multiplicity 2), and 4 (multiplicity 3)

In Problems 11–16, find a polynomial of lowest degree, with leading coefficient 1, that has the indicated graph. Assume all zeros are integers. Leave the answer in a factored form. Indicate the degree of each polynomial.

11.

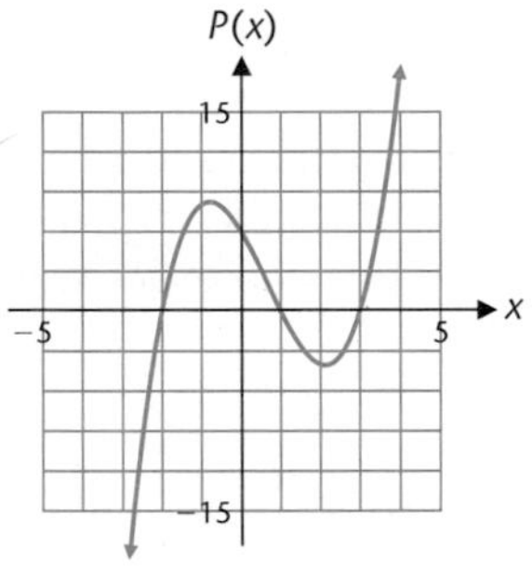

12.

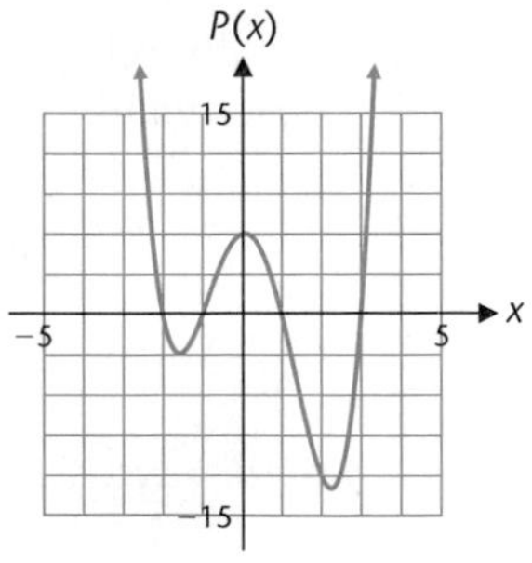

13.

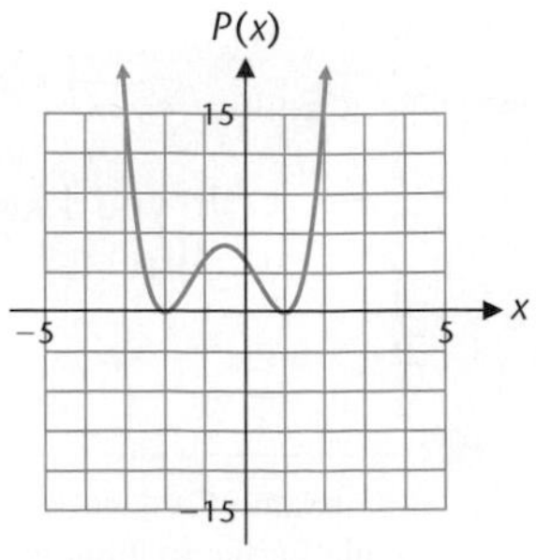

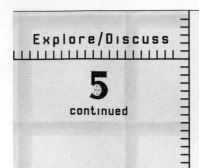

14.

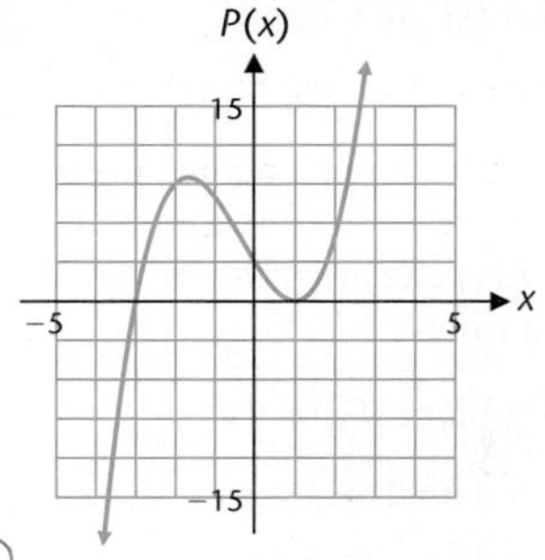

15.

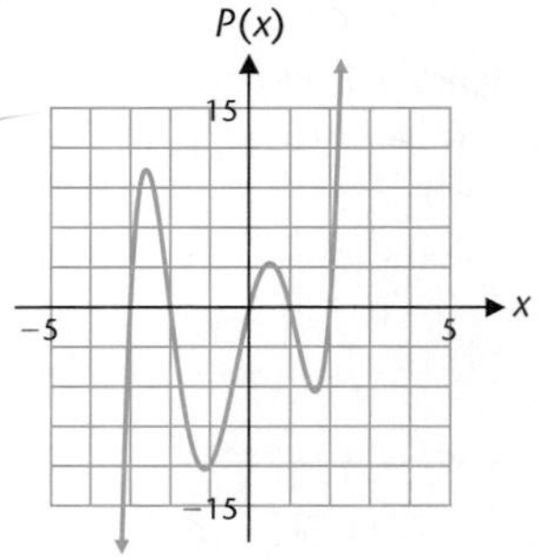

16.

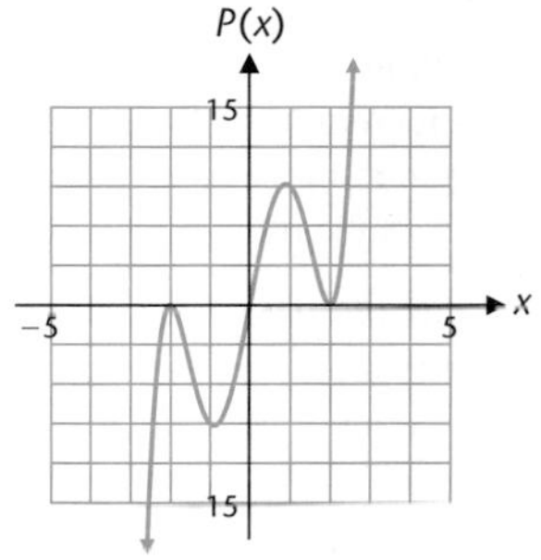

In Problems 17–20, determine whether the second polynomial is a factor of the first polynomial without dividing or using synthetic division. [Hint: Evaluate directly and use the factor theorem.]

17. $x^{18} - 1$; $x - 1$

18. $x^{18} - 1$; $x + 1$

19. $3x^3 - 7x^2 - 8x + 2$; $x + 1$

20. $3x^4 - 2x^3 + 5x - 6$; $x - 1$

B

For each polynomial in problems 21–26, list all possible rational zeros (Theorem 6).

21. $P(x) = x^3 - 2x^2 - 5x + 6$

22. $P(x) = x^3 + 3x^2 - 6x - 8$

23. $P(x) = 3x^3 - 11x^2 + 8x + 4$

24. $P(x) = 2x^3 + x^2 - 4x - 3$

25. $P(x) = 12x^3 - 16x^2 - 5x + 3$

26. $P(x) = 2x^3 - 9x^2 + 14x - 5$

In Problems 27–32, write P(x) as a product of linear terms.

27. $P(x) = x^3 + 9x^2 + 24x + 16$; -1 is a zero

28. $P(x) = x^3 - 4x^2 - 3x + 18$; 3 is a double zero

29. $P(x) = x^4 - 1$; 1 and -1 are zeros

30. $P(x) = x^4 + 2x^2 + 1$; i is a double zero

31. $P(x) = 2x^3 - 17x^2 + 90x - 41$; $\frac{1}{2}$ is a zero

32. $P(x) = 3x^3 - 10x^2 + 31x + 26$; $-\frac{2}{3}$ is a zero

In Problems 33–40, find all roots exactly (rational, irrational, and imaginary) for each polynomial equation.

33. $2x^3 - 5x^2 + 1 = 0$

34. $2x^3 - 10x^2 + 12x - 4 = 0$

35. $x^4 + 4x^3 - x^2 - 20x - 20 = 0$

36. $x^4 - 4x^2 - 4x - 1 = 0$

37. $x^4 - 2x^3 - 5x^2 + 8x + 4 = 0$

38. $x^4 - 2x^2 - 16x - 15 = 0$

39. $2x^5 - 3x^4 - 2x + 3 = 0$

40. $2x^5 + x^4 - 6x^3 - 3x^2 - 8x - 4 = 0$

In Problems 41–48, find all zeros exactly (rational, irrational, and imaginary) for each polynomial.

41. $P(x) = x^3 - 19x + 30$

42. $P(x) = x^3 - 7x^2 + 36$

43. $P(x) = x^4 - \frac{21}{10}x^3 + \frac{2}{5}x$

44. $P(x) = x^4 + \frac{7}{6}x^3 - \frac{7}{3}x^2 - \frac{5}{2}x$

45. $P(x) = x^4 - 5x^3 + \frac{15}{2}x^2 - 2x - 2$

46. $P(x) = x^4 - \frac{13}{4}x^2 - \frac{5}{2}x - \frac{1}{4}$

47. $P(x) = 3x^5 - 5x^4 - 8x^3 + 16x^2 + 21x + 5$

48. $P(x) = 2x^5 - 3x^4 - 6x^3 + 23x^2 - 26x + 10$

In Problems 49–54, write each polynomial as a product of linear factors.

49. $P(x) = 6x^3 + 13x^2 - 4$

50. $P(x) = 6x^3 - 17x^2 - 4x + 3$

51. $P(x) = x^3 + 2x^2 - 9x - 4$

52. $P(x) = x^3 - 8x^2 + 17x - 4$

53. $P(x) = 4x^4 - 4x^3 - 9x^2 + x + 2$

54. $P(x) = 2x^4 + 3x^3 - 4x^2 - 3x + 2$

In Problems 55–60, multiply.

55. $[x - (4 - 5i)][x - (4 + 5i)]$

56. $[x - (2 - 3i)][x - (2 + 3i)]$

57. $[x - (3 + 4i)][x - (3 - 4i)]$

58. $[x - (5 + 2i)][x - (5 - 2i)]$

59. $[x - (a + bi)][x - (a - bi)]$

60. $(x - bi)(x + bi)$

C

In Problems 61–66, find all other zeros of P(x), given the indicated zero.

61. $P(x) = x^3 - 5x^2 + 4x + 10$; $3 - i$ is one zero

62. $P(x) = x^3 + x^2 - 4x + 6$; $1 + i$ is one zero

63. $P(x) = x^3 - 3x^2 + 25x - 75$; $-5i$ is one zero

64. $P(x) = x^3 + 2x^2 + 16x + 32$; $4i$ is one zero

65. $P(x) = x^4 - 4x^3 + 3x^2 + 8x - 10$; $2 + i$ is one zero

66. $P(x) = x^4 - 2x^3 + 7x^2 - 18x - 18$; $-3i$ is one zero

Prove that each of the real numbers in Problems 67–70 is not rational by writing an appropriate polynomial and making use of Theorem 6.

67. $\sqrt{6}$

68. $\sqrt{12}$

69. $\sqrt[3]{5}$

70. $\sqrt[5]{8}$

In Problems 71–76, find all zeros (rational, irrational, and imaginary) exactly.

71. $P(x) = 3x^3 - 37x^2 + 84x - 24$

72. $P(x) = 2x^3 - 9x^2 - 2x + 30$

73. $P(x) = 4x^4 + 4x^3 + 49x^2 + 64x - 240$

74. $P(x) = 6x^4 + 35x^3 + 2x^2 - 233x - 360$

75. $P(x) = 4x^4 - 44x^3 + 145x^2 - 192x + 90$

76. $P(x) = x^5 - 6x^4 + 6x^3 + 28x^2 - 72x + 48$

77. The solutions to the equation $x^3 - 1 = 0$ are all the cube roots of 1.

(A) How many cube roots of 1 are there?

(B) 1 is obviously a cube root of 1; find all others.

78. The solutions to the equation $x^3 - 8 = 0$ are all the cube roots of 8.

(A) How many cube roots of 8 are there?

(B) 2 is obviously a cube root of 8; find all others.

79. If P is a polynomial function with real coefficients of degree n, with n odd, then what is the maximum number of times the graph of $y = P(x)$ can cross the x axis? What is the minimum number of times?

80. Answer the questions in Problem 79 for n even.

81. Given $P(x) = x^2 + 2ix - 5$ with $2 - i$ a zero, show that $2 + i$ is not a zero of $P(x)$. Does this contradict Theorem 4? Explain.

82. If $P(x)$ and $Q(x)$ are two polynomials of degree n, and if $P(x) = Q(x)$ for more than n values of x, then how are $P(x)$ and $Q(x)$ related?

APPLICATIONS

Find all rational solutions exactly, and find irrational solutions to two decimal places.

83. Storage. A rectangular storage unit has dimensions 1 by 2 by 3 feet. If each dimension is increased by the same amount, how much should this amount be to create a new storage unit with volume 10 times the old?

84. Construction. A rectangular box has dimensions 1 by 1 by 2 feet. If each dimension is increased by the same amount, how much should this amount be to create a new box with volume six times the old?

★ **85. Packaging.** An open box is to be made from a rectangular piece of cardboard that measures 8 by 5 inches, by cutting out squares of the same size from each corner and bending up the sides (see the figure). If the volume of the box is to be 14 cubic inches, how large a square should be cut from each corner? [*Hint:* Determine the domain of x from physical considerations before starting.]

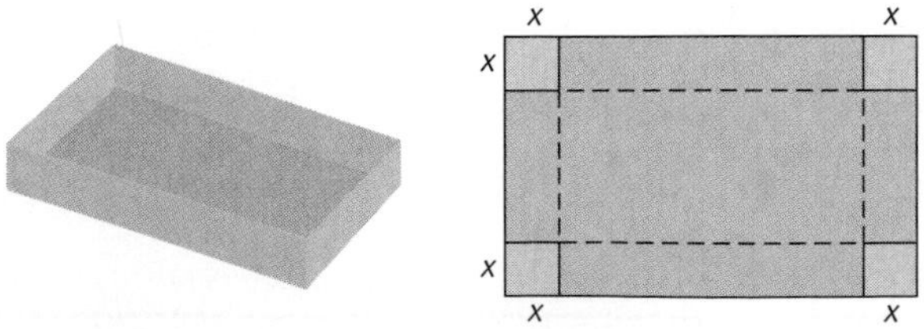

★ **86. Fabrication.** An open metal chemical tank is to be made from a rectangular piece of stainless steel that measures 10 by 8 feet, by cutting out squares of the same size from each corner and bending up the sides (see the figure). If the volume of the tank is to be 48 cubic feet, how large a square should be cut from each corner?

Section 3-3 Approximating Real Zeros of Polynomials

- Locating Real Zeros
- The Bisection Method
- Approximating Multiple Zeros
- Application

The methods for finding zeros discussed in the preceding section are designed to find as many exact real and imaginary zeros as possible. But there are zeros that cannot be found by using these methods. For example, the polynomial

$$P(x) = x^5 + x - 1$$

must have at least one real zero (Theorem 5 in Section 3-2). Since the only possible rational zeros are ± 1, and neither of these turns out to be a zero, $P(x)$ must have at least one irrational zero. We cannot find the exact value of this zero, but it can be approximated using various well-known methods.

In this section we develop two important tools for locating real zeros, *the location theorem* and *the upper and lower bound theorem*. We will see that the upper and lower bound theorem is a useful tool for approximating real zeros on a graphing utility. Next we discuss how the location theorem leads to the *bisection method*, a simple approximation technique that forms the basis for the zero approximation routines on most graphing utilities. Finally, we investigate some of the problems that can be encountered when a polynomial has multiple zeros.

In this section, we restrict our attention to the real zeros of polynomials with real coefficients.

Locating Real Zeros

FIGURE 1
$P(x) = x^5 + x - 1.$

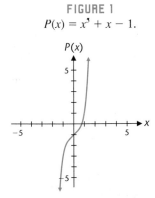

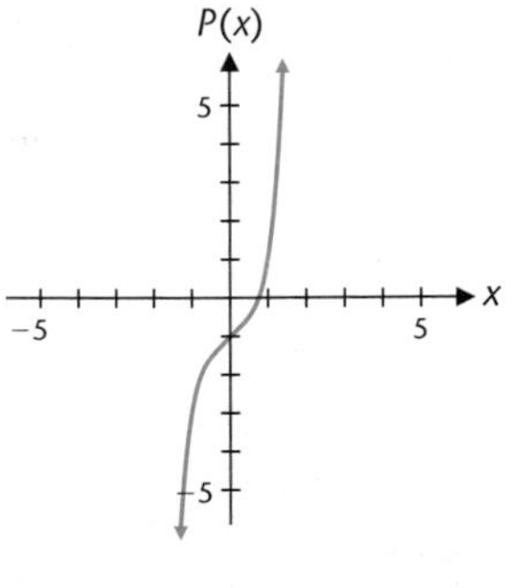

Let us return to the polynomial function

$$P(x) = x^5 + x - 1$$

As we have found, $P(x)$ has no rational zeros and at least one irrational zero. The graph of $P(x)$ is shown in Figure 1.

Note that $P(0) = -1$ and $P(1) = 1$. Since the graph of a polynomial function is continuous, the graph of $P(x)$ must cross the x axis at least once between $x = 0$ and $x = 1$. This observation is the basis for Theorem 1 and leads to a simple method for approximating zeros.

THEOREM 1

LOCATION THEOREM

If f is continuous on an interval I, a and b are two numbers in I, and $f(a)$ and $f(b)$ are of opposite sign, then there is at least one x intercept between a and b.

We will find Theorem 1 very useful when we are searching for real zeros, hence the name *location theorem.* It is important to remember that "at least" in Theorem 1 means "one or more." Notice in Figure 2(a) that $f(-3) = -15 < 0$, $f(3) = 15 > 0$, and f has one zero between -3 and 3. In Figure 2(b), $f(-3) = -15$ and $f(3) = 15$, but this time there are three zeros between -3 and 3.

FIGURE 2
The location theorem.

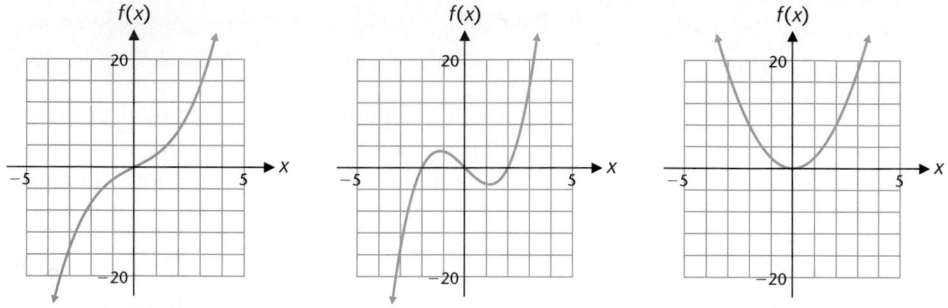

(a) $f(x) = \frac{1}{3}x^3 + 2x$ (b) $f(x) = x^3 - 4x$ (c) $f(x) = 2x^2$

The converse to the location theorem (Theorem 1) is false; that is, if c is a zero of f, then f may or may not change sign at c. Compare Figures 2(a) and 2(c). Both functions have a zero at $x = 0$, but the first changes sign at 0 and the second does not.

EXAMPLE 1

Locating Real Zeros

Let $P(x) = x^3 - 6x^2 + 9x - 3$. Use a table to locate the zeros of $P(x)$ between successive integers and check graphically.

Solution

We construct a table and look for sign changes [Fig. 3(a)]. The table in Figure 3(a) shows three sign changes. According to Theorem 1, $P(x)$ must have a real zero in each of the intervals (0, 1), (1, 2), and (3, 4). The graph in Figure 3(b) confirms this. Since $P(x)$ is a cubic polynomial, we have located all of its zeros.

FIGURE 3
$y_1 = P(x) = x^3 - 6x^2 + 9x - 3.$

X	Y1	
0	-3	
1	1	
2	-1	
3	-3	
4	1	

X=

(a)

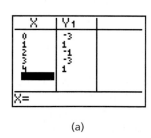

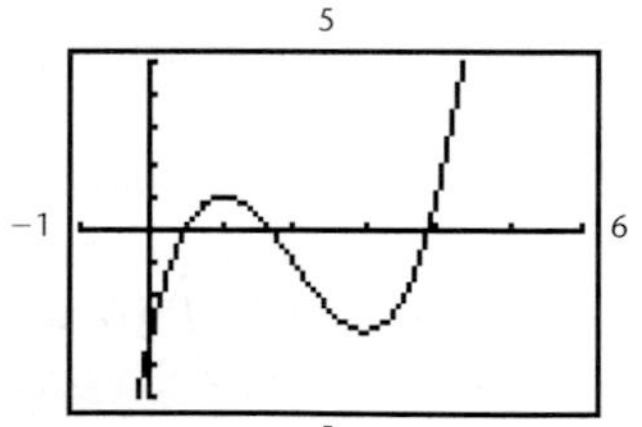

(b)

MATCHED PROBLEM 1

Let $P(x) = x^3 - 8x^2 + 15x - 2$. Use a table to locate the zeros of $P(x)$ between successive integers and check graphically.

In the solution to Example 1, we located three zeros in a relatively few number of steps and could stop searching because we knew that a cubic polynomial could not have more than three zeros. But what if we had not found three zeros?

Some cubic polynomials have only one real zero. How can we tell when we have searched far enough? The next theorem tells us how to find *upper* and *lower bounds* for the real zeros of a polynomial. Any number that is greater than or equal to the largest zero of a polynomial is called an **upper bound** of the zeros of the polynomial. Similarly, any number that is less than or equal to the smallest zero of the polynomial is called a **lower bound** of the zeros of the polynomial. Theorem 2, based on the synthetic division process, enables us to determine upper and lower bounds of the real zeros of a polynomial with real coefficients.

THEOREM 2

UPPER AND LOWER BOUNDS OF REAL ZEROS

Given an nth-degree polynomial $P(x)$ with real coefficients, $n > 0$, $a_n > 0$, and $P(x)$ divided by $x - r$ using synthetic division:

1. **Upper Bound.** If $r > 0$ and all numbers in the quotient row of the synthetic division, including the remainder, are nonnegative, then r is an upper bound of the real zeros of $P(x)$.
2. **Lower Bound.** If $r < 0$ and all numbers in the quotient row of the synthetic division, including the remainder, alternate in sign, then r is a lower bound of the real zeros of $P(x)$.

[*Note:* In the lower-bound test, if 0 appears in one or more places in the quotient row, including the remainder, the sign in front of it can be considered either positive or negative, but not both. For example, the numbers 1, 0, 1 can be considered to alternate in sign, while 1, 0, −1 cannot.]

We sketch a proof of part 1 of Theorem 2. The proof of part 2 is similar, only a little more difficult.

Proof If all the numbers in the quotient row of the synthetic division are nonnegative after dividing $P(x)$ by $x - r$, then

$$P(x) = (x - r)Q(x) + R$$

where the coefficients of $Q(x)$ are nonnegative and R is nonnegative. If $x > r > 0$, then $x - r > 0$ and $Q(x) > 0$; hence,

$$P(x) = (x - r)Q(x) + R > 0$$

Thus, $P(x)$ cannot be 0 for any x greater than r, and r is an upper bound for the real zeros of $P(x)$.

Theorem 2 requires performing synthetic division repeatedly until the desired patterns occur in the quotient row. This is a simple, but tedious, operation to carry out by hand. Table 1 shows a simple program for a graphing calculator that will perform synthetic division.

TABLE 1 Synthetic Division on a Graphing Utility

Program SYNDIV

TI-82/TI-83	TI-85/TI-86
Lbl A	Lbl A
Prompt R	Prompt R
2→I	2→I
dim(L1)→N	dimL L1→N
{0}→L2	{0}→L2
N→dim(L2)	N→dimL L1
L1(1)→L2(1)	L1(1)→L2(1)
Lbl B	Lbl B
L2(I−1)*R+L1(I)→L2(I)	L2(I−1)*R+L1(I)→L2(I)
1+I→I	1+I→I
If I≤N	If I≤N
Goto B	Goto B
Pause L2	Pause L2
Goto A	Goto A

Output

```
{1,-6,9,-3}→L1
        {1 -6 9 -3}
prgmSYNDIV
R=?0
        {1 -6 9 -3}
R=?1
         {1 -5 4 1}
R=?2
        {1 -4 1 -1}
R=?3
        {1 -3 0 -3}
R=?4
         {1 -2 1 1}
R=?
```

Explore/Discuss 1

(A) If you have a TI-82, TI-83, TI-85, or TI-86 graphing calculator, enter the appropriate version of SYNDIV in your calculator exactly as shown in Table 1. If you have some other graphing utility that can store and execute programs, consult your manual and modify the statements in SYNDIV so that the program works on your graphing utility.

(B) Store the coefficients of $P(x) = x^3 - 6x^2 + 9x - 3$ in L1 (see the first line of output in Table 1) and execute the program. Type 0 at the "R=?" prompt and press ENTER to display the results of synthetic division with $r = 0$. To continue press ENTER again. Repeat this process for $r = 1, 2, 3$, and 4. Press QUIT at the "R=?" prompt to terminate the program. Compare the last number in each list with the values of $P(x)$ in Figure 3(a).

EXAMPLE 2 Bounding Real Zeros

Let $P(x) = x^4 - 2x^3 - 10x^2 + 40x - 90$. Find the smallest positive integer and the largest negative integer that, by Theorem 2, are upper and lower bounds, respectively, for the real zeros of $P(x)$. Also note the location of any zeros discovered in the process of searching for upper and lower bounds.

Solution We can use SYNDIV or hand calculations to perform synthetic division for $r = 1, 2, 3, \ldots$ until the quotient row turns nonnegative; then repeat this process for $r = -1, -2, -3, \ldots$ until the quotient row alternates in sign. We organize these results in the *synthetic division table* shown below. In a **synthetic division table** we dispense with writing the product of r with each coefficient in the quotient and simply list the results in the table. It is also useful to include $r = 0$ in the table to detect any sign changes between $r = 0$ and $r = \pm 1$.

		1	−2	−10	40	−90	
	0	1	−2	−10	40	−90	
	1	1	−1	−11	29	−61	
	2	1	0	−10	20	−50	
	3	1	1	−7	19	−33	
	4	1	2	−2	32	38	
UB	5	1	3	5	65	235	← This quotient row is nonnegative; hence, 5 is an upper bound (UB).
	−1	1	−3	−7	47	−137	
	−2	1	−4	−2	44	−178	
	−3	1	−5	5	25	−165	
	−4	1	−6	14	−16	−26	
LB	−5	1	−7	25	−85	335	← This quotient row alternates in sign; hence, −5 is a lower bound (LB).

FIGURE 4
$P(x) = x^4 - 2x^3 - 10x^2 + 40x - 90$.

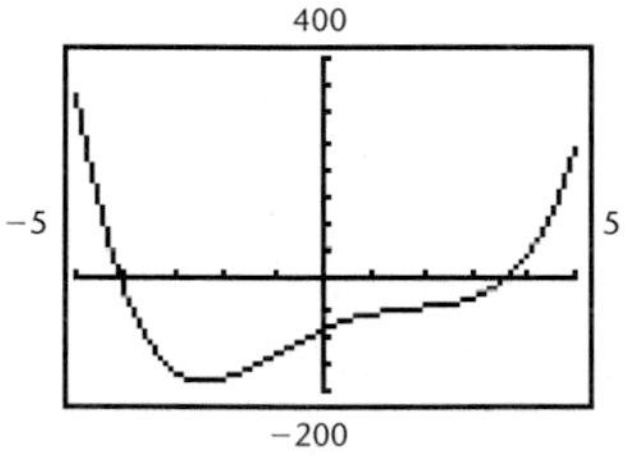

The graph of $P(x) = x^4 - 2x^3 - 10x^2 + 40x - 90$ for $-5 \le x \le 5$ is shown in Figure 4. Theorem 2 implies that all the real zeros of $P(x)$ are between -5 and 5. We can be certain that the graph does not change direction and cross the x axis somewhere outside the viewing window in Figure 4. We also note that there is at least one zero in (3, 4) and at least one in $(-5, -4)$, as indicated by the sign changes in the values of $P(x)$ shown in the last column of the synthetic division table.

MATCHED PROBLEM 2

Let $P(x) = x^4 - 5x^3 - x^2 + 40x - 70$. Find the smallest positive integer and the largest negative integer that, by Theorem 2, are upper and lower bounds, respectively, for the real zeros of $P(x)$. Also note the location of any zeros discovered in the process of searching for upper and lower bounds.

EXAMPLE 3

Approximating Real Zeros with a Graphing Utility

Let $P(x) = x^3 - 30x^2 + 275x - 720$.

(A) Find the smallest positive integer multiple of 10 and the largest negative integer multiple of 10 that, by Theorem 2, are upper and lower bounds, respectively, for the real zeros of $P(x)$.

(B) Approximate the real zeros of $P(x)$ to two decimal places.

Solutions

(A) We construct a synthetic division table to search for bounds for the zeros of $P(x)$. The size of the coefficients in $P(x)$ indicates that we can speed up this search by choosing larger increments between test values.

		1	−30	275	−720
	10	1	−20	75	30
	20	1	−10	75	780
UB	30	1	0	275	7,530
LB	−10	1	−40	675	−7,470

Thus, all real zeros of $P(x) = x^3 - 30x^2 + 275x - 720$ must lie between -10 and 30.

(B) Graphing $P(x)$ for $-10 \le x \le 30$ (Fig. 5) shows that $P(x)$ has three zeros. The approximate values of these zeros (details omitted) are 4.48, 11.28, and 14.23.

FIGURE 5
$P(x) = x^3 - 30x^2 + 275x - 720.$

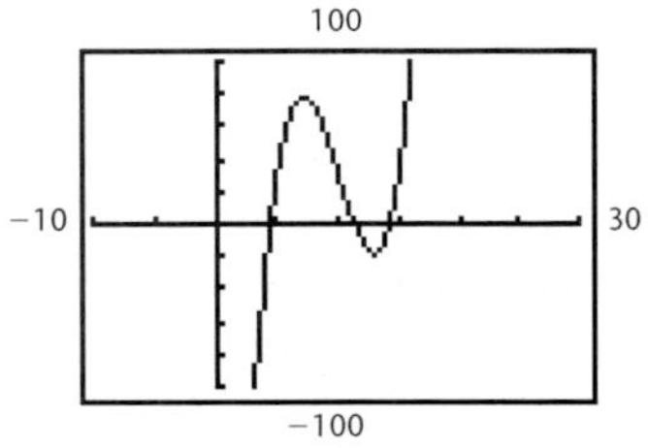

MATCHED PROBLEM 3

Let $P(x) = x^3 - 25x^2 + 170x - 170$.

(A) Find the smallest positive integer multiple of 10 and the largest negative integer multiple of 10 that, by Theorem 2, are upper and lower bounds, respectively, for the real zeros of $P(x)$.

(B) Approximate the real zeros of $P(x)$ to two decimal places.

Remark One of the most frequently asked questions concerning graphing utilities is how to determine the correct viewing window. The upper and lower bound theorem provides an answer to this question for polynomial functions. As Example 3 illustrates, the upper and lower bound theorem and the zero approximation routine on a graphing utility are two important mathematical tools that work very well together.

The Bisection Method

Now that we know more about locating real zeros of a polynomial, we want to discuss a technique that can be used to approximate real zeros. Explore/Discuss 2 provides an introduction to the repeated systematic application of the location theorem (Theorem 1) called the *bisection method*. This method forms the basis for the zero approximation routines in many graphing utilities.

Explore/Discuss 2

Let $P(x) = x^3 + x - 1$. Since $P(0) = -1$ and $P(1) = 1$, the location theorem implies that $P(x)$ must have at least one zero in (0, 1).

(A) Is $P(0.5)$ positive or negative? Is there a zero in (0, 0.5) or in (0.5, 1)?

Explore/Discuss 2 continued

(B) Let m be the midpoint of the interval from part A that contains a zero. Is $P(m)$ positive or negative? What does this tell you about the location of the zero?

(C) Explain how this process could be used repeatedly to approximate a zero to any desired accuracy.

The **bisection method** used to approximate real zeros is straightforward: Let $P(x)$ be a polynomial with real coefficients. If $P(x)$ has opposite signs at the endpoints of the interval (a, b), then a real zero r lies in this interval. We bisect this interval [find the midpoint $m = (a + b)/2$], check the sign of $P(m)$, and choose the interval (a, m) or (m, b) on which $P(x)$ has opposite signs at the endpoints. We repeat this bisecting process (producing a set of "nested" intervals, each half the size of the preceding one and each containing the real zero r) until we get the desired decimal accuracy for the zero approximation. At any point in the process if $P(m) = 0$, we stop, since m is a real zero. An example will help clarify the process.

EXAMPLE 4

Approximating Real Zeros by Bisection

For the polynomial $P(x) = x^4 - 2x^3 - 10x^2 + 40x - 90$ in Example 2, we found that all the real zeros lie between -5 and 5 and that each of the intervals $(-5, -4)$ and $(3, 4)$ contained at least one zero. Use bisection to approximate a real zero on the interval $(3, 4)$ to one-decimal-place accuracy.

Solution

We organize the results of our calculations in a table. Since the sign of $P(x)$ changes at the endpoints of the interval $(3.5625, 3.625)$, we conclude that a real zero lies in this interval and is given by $r = 3.6$ to one-decimal-place accuracy (each endpoint rounds to 3.6).

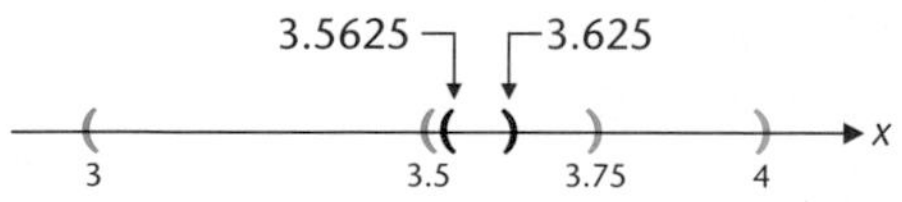

FIGURE 6
Nested intervals produced by the bisection method in Table 2.

Figure 6 illustrates the nested intervals produced by the bisection method in Table 2. Match each step in Table 2 with an interval in Figure 6. Note how each interval that contains a zero gets smaller and smaller and is contained in the preceding interval that contained the zero.

TABLE 2 Bisection Approximation

Sign Change Interval (a, b)	Midpoint m	Sign of P: $P(a)$	$P(m)$	$P(b)$
(3, 4)	3.5	−	−	+
(3.5, 4)	3.75	−	+	+
(3.5, 3.75)	3.625	−	+	+
(3.5, 3.625)	3.5625	−	−	+
(3.5625, 3.625)	We stop here	−		+

If we had wanted two-decimal-place accuracy, we would have continued the process in Table 2 until the endpoints of a sign change interval rounded to the same two-decimal-place number.

MATCHED PROBLEM
4

Use the bisection method to approximate to one-decimal-place accuracy a zero on the interval $(-5, -4)$ for the polynomial in Example 4.

The bisection method is easy to implement on a graphing utility. Table 3 shows a program that computes the sequence of nested intervals shown in Table 2.

T A B L E   3   Bisection on a Graphing Utility

Program BISECT

TI 82/83	TI 85/86	Output
{0}→L₁	{0}→L1	Plot1 Plot2 Plot3
Input " LEFT BOUND: ",L	Input " LEFT BOUND: ",L	\Y₁■X^4-2X^3-10X
Input "RIGHT BOUND: ",R	Input "RIGHT BOUND: ",R	²+40X-90
If L≥R	If L≥R	\Y₂=
Then	Then	\Y₃=
Disp "BOUND ERROR"	Disp "BOUND ERROR"	\Y₄=
Stop	Stop	\Y₅=
End	End	\Y₆=
If Y₁(L)Y₁(R)≥0	If evalF(y1,x,L)evalF(y1,x,R)≥0	(a)
Then	Then	
Disp "NO SIGN CHANGE"	Disp "NO SIGN CHANGE"	LEFT BOUND: 3
Stop	Stop	RIGHT BOUND: 4
Else	Else	(3 4)
Lbl A	Lbl A	(3.5 4)
L→L₁(1):R→L₁(2)	L→L1(1):R→L1(2)	(3.5 3.75)
Pause L₁	Pause L1	(3.5 3.625)
(L+R)/2→M	(L+R)/2→M	(3.5625 3.625)
If Y₁(M)=0	If evalF(y1,x,M)==0	(b)
Then	Then	
Disp "Y = 0 AT X =",M	Disp "Y = 0 AT X =",M	ERR:BREAK
Stop	Stop	1■Quit
End	End	2:Goto
If Y₁(L)Y₁(M)<0	If evalF(y1,x,L)evalF(y1,x,M)<0	
Then	Then	(c)
M→R	M→R	
Else	Else	FIGURE 7
M→L	M→L	
End	End	
Goto A	Goto A	

Explore/Discuss
3

(A) If you have a TI-82, TI-83, TI-85, or TI-86 graphing calculator, enter the appropriate version of BISECT in your calculator exactly as shown in Table 3. If you have some other graphing utility that can store and execute programs, consult your manual and modify the statements in BISECT so that the program works on your graphing utility.

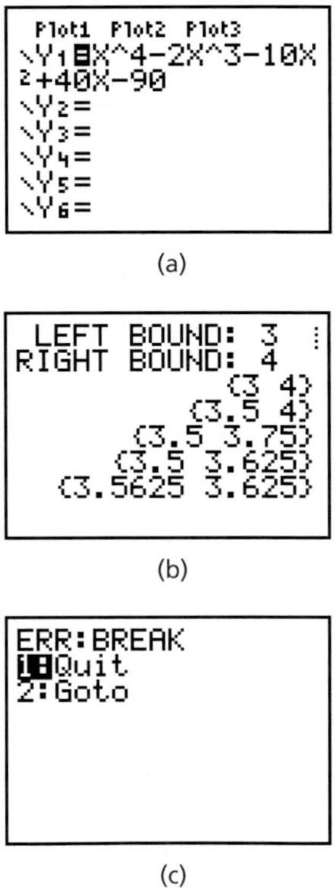

Explore/Discuss 3 continued

(B) To use the program, first enter $P(x) = x^4 - 2x^3 - 10x^2 + 40x - 90$ as Y1 in the equation editor of the graphing utility [Fig. 7(a)]. Then enter 3 and 4 at the prompts and press ENTER repeatedly to generate the sequence of nested intervals [Fig. 7(b)]. Press ON and select Quit to terminate execution [Fig. 7(c)].

Approximating Multiple Zeros

Consider the polynomial

FIGURE 8
$P(x) = x(x - 1)^2(x + 1)^4$.

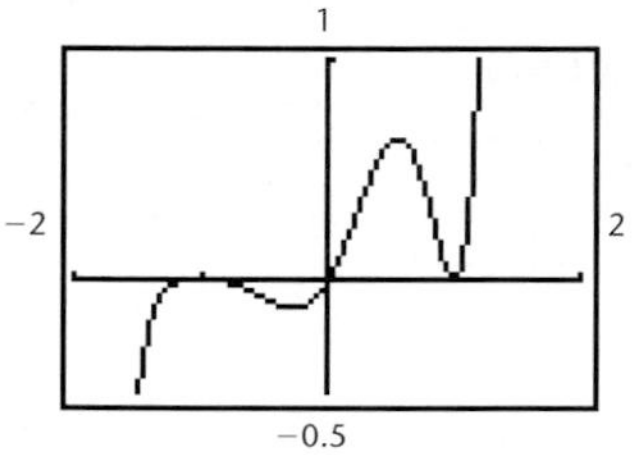

$$P(x) = x(x - 1)^2(x + 1)^4$$

which has a simple zero at 0, a zero of multiplicity 2 at 1 and a zero of multiplicity 4 at -1 (see Fig. 8). Notice that $P(x)$ has a local maximum at $x = -1$ and does not change sign at $x = -1$. Also, x has a local minimum at $x = 1$ and does not change sign there either. Both of these are zeros of even multiplicity. On the other hand, 0 is a zero of odd multiplicity, $P(x)$ does change sign at $x = 0$, and does not have a local extremum at $x = 0$. Theorem 3, which we state without proof, generalizes these observations.

THEOREM 3

ZEROS OF EVEN AND ODD MULTIPLICITY

If $P(x)$ is a polynomial with real coefficients, then

1. If r is a zero of odd multiplicity, then $P(x)$ changes sign at r and does not have a local extremum at $x = r$.
2. If r is a zero of even multiplicity, then $P(x)$ does not change sign at r and has a local extremum at $x = r$.

Explore/Discuss 4

Let $P(x) = x(x - 1)^2(x + 1)^4$.

(A) Use BISECT to approximate the zeros of $P(x)$. Where does it fail to work? Why?

(B) Use the zero approximation routine on your graphing utility to approximate the zeros of $P(x)$. Does this routine ever fail to work?

The bisection method shown in Table 3 requires that the function change sign at a zero to approximate that zero. Thus, this method will always fail at a zero of even multiplicity. Most graphing utilities use a more complicated routine that may or may not work at zeros of even multiplicity. For example, the TI-83 graphing calculator was able to approximate the zero of $P(x) = x(x - 1)^2(x + 1)^4$ at 1, but not the zero at -1. What are we to do if the zero routine fails? The ideas introduced in Theorem 3 provide the answer:

Zeros of even multiplicity can be approximated by using a maximum or minimum approximation routine, whichever applies.

The next example illustrates this approach.

EXAMPLE 5

Approximating Multiple Zeros

Let $P(x) = x^5 + 6x^4 + 4x^3 - 24x^2 - 16x + 32$. Approximate the zeros of $P(x)$ to two decimal places. Use maximum and minimum routines to approximate any zeros of even multiplicity. Determine the multiplicity of each zero.

Solution

Examining the graph of $P(x)$, we see that there is a zero of odd multiplicity at $x = -2$ [Fig. 9(a)]. It appears that there may be zeros of even multiplicity near $x = -3$ and $x = 1$. Using a maximum routine near $x = -3$ [Fig. 9(b)] and a minimum routine near $x = 1$ [Fig. 9(c)], we find that -3.24 and 1.24 are zeros of even multiplicity. Since $P(x)$ is a fifth-degree polynomial, these zeros must be double zeros and -2 must be a simple zero.

FIGURE 9
Zeros of $P(x) = x^5 + 6x^4 + 4x^3 - 24x^2 - 16x + 32$.

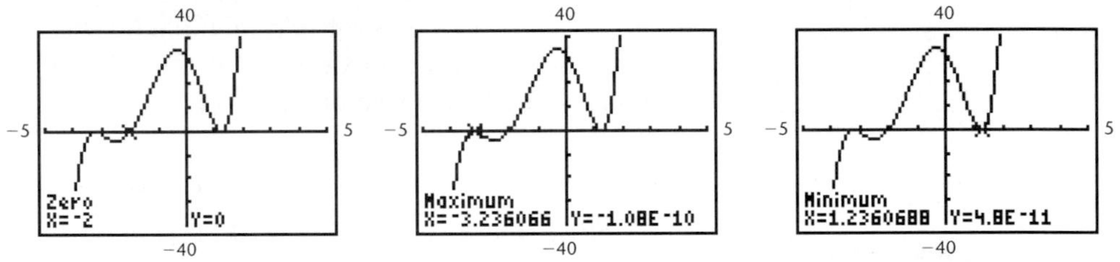

(a) Simple zero (b) Double zero (c) Double zero

MATCHED PROBLEM 5

Let $P(x) = x^5 - 6x^4 + 40x^2 - 12x - 72$. Approximate the zeros of $P(x)$ to two decimal places. Use maximum and minimum routines to approximate any zeros of even multiplicity. Determine the multiplicity of each zero.

Application

EXAMPLE 6

Construction

An oil tank is in the shape of a right circular cylinder with a hemisphere at each end (see Fig. 10). The cylinder is 55 inches long, and the volume of the tank is $11{,}000\pi$ cubic inches (approximately 20 cubic feet). Let x denote the common radius of the hemispheres and the cylinder.

FIGURE 10

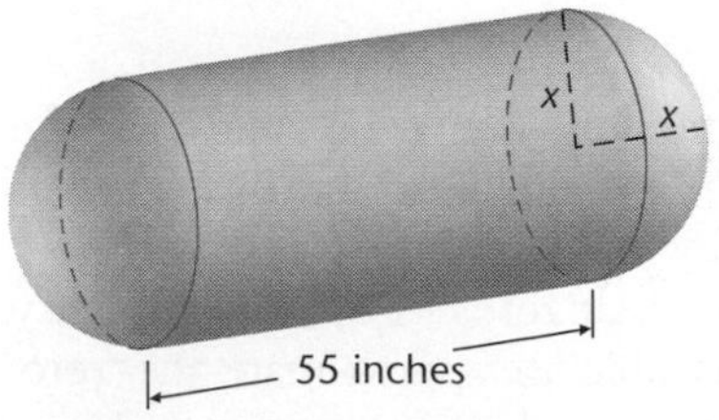

(A) Find a polynomial equation that x must satisfy.

(B) Approximate x to one decimal place.

Solutions (A) If x is the common radius of the hemispheres and the cylinder in inches, then

$$\begin{pmatrix}\text{Volume}\\\text{of}\\\text{tank}\end{pmatrix} = \begin{pmatrix}\text{Volume}\\\text{of two}\\\text{hemispheres}\end{pmatrix} + \begin{pmatrix}\text{Volume}\\\text{of}\\\text{cylinder}\end{pmatrix}$$

$$11{,}000\pi = \tfrac{4}{3}\pi x^3 + 55\pi x^2 \qquad \text{Multiply by } 3/\pi.$$

$$33{,}000 = 4x^3 + 165x^2$$

$$0 = 4x^3 + 165x^2 - 33{,}000$$

Thus, x must be a positive zero of

$$P(x) = 4x^3 + 165x^2 - 33{,}000$$

(B) Since the coefficients of $P(x)$ are large, we use larger increments in the synthetic division table:

		4	165	0	−33,000
	10	4	205	2,050	−12,500
UB	20	4	245	4,900	65,000

Graphing $y = P(x)$ for $0 \le x \le 20$ (Fig. 11), we see that $x = 12.4$ inches (to one decimal place).

FIGURE 11
$P(x) = 4x^3 + 165x^2 - 33{,}000.$

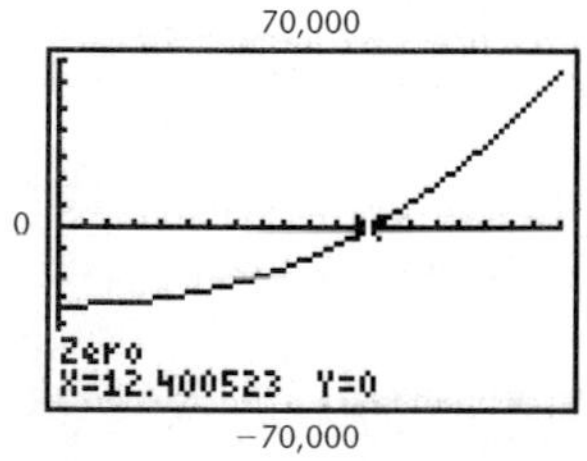

MATCHED PROBLEM 6 Repeat Example 6 if the volume of the tank is $44{,}000\pi$ cubic inches.

Answers to Matched Problems

1. Intervals containing zeros: (0, 1), (2, 3), (5, 6)
2. Lower bound: −3; upper bound: 6; intervals containing zeros: (−3, −2), (3, 4)
3. (A) Lower bound: −10; upper bound: 30 (B) Real zeros: 1.20, 11.46, 12.34
4. $x = -4.1$ **5.** −1.65 (double zero), 2 (simple zero), 3.65 (double zero)
6. (A) $P(x) = 4x^3 + 165x^2 - 132{,}000 = 0$ (B) 22.7 inches

EXERCISE 3-3

A

In Problems 1–4, use the table of values for the polynomial function P to discuss the possible locations of the x intercepts of the graph of $y = P(x)$.

1.

x	−7	−5	−1	3	5	8
$P(x)$	9	4	−3	6	4	−2

2.

x	−8	−2	0	2	4	9
$P(x)$	−3	4	5	2	−5	6

3.

x	−6	−4	0	2	4	7
$P(x)$	−5	3	−4	−6	3	−5

4.

x	−5	−3	−1	0	2	5
$P(x)$	7	4	2	−1	3	−6

In Problems 5–8, use a synthetic division table and Theorem 1 to locate each real zero between successive integers.

5. $P(x) = x^3 - 9x^2 + 23x - 14$

6. $P(x) = x^3 - 12x^2 + 44x - 49$

7. $P(x) = x^3 + 3x^2 - x - 5$

8. $P(x) = x^3 + x^2 - 4x - 3$

Find the smallest positive integer and largest negative integer that, by Theorem 2, are upper and lower bounds, respectively, for the real zeros of each of the polynomials given in Problems 9–14.

9. $P(x) = x^3 - 3x + 1$

10. $P(x) = x^3 - 4x^2 + 4$

11. $P(x) = x^4 - 3x^3 + 4x^2 + 2x - 9$

12. $P(x) = x^4 - 4x^3 + 6x^2 - 4x - 7$

13. $P(x) = x^5 - 3x^3 + 3x^2 + 2x - 2$

14. $P(x) = x^5 - 3x^4 + 3x^2 + 2x - 1$

B

In Problems 15–22,

(A) Find the smallest positive integer and largest negative integer that, by Theorem 2, are upper and lower bounds, respectively, for the real zeros of P(x). Also note the location of any zeros between successive integers.

(B) If (k, k + 1) is the interval determined in part A that contains the largest real zero of P(x), determine the number of additional intervals required by the bisection method to obtain a one-decimal-place approximation to this zero and state the approximate value of the zero.

15. $P(x) = x^3 - 2x^2 - 5x + 4$

16. $P(x) = x^3 + x^2 - 4x - 1$

17. $P(x) = x^3 - 2x^2 - x + 5$

18. $P(x) = x^3 - 3x^2 - x - 2$

19. $P(x) = x^4 - 2x^3 - 7x^2 + 9x + 7$

20. $P(x) = x^4 - x^3 - 9x^2 + 9x + 4$

21. $P(x) = x^4 - x^3 - 4x^2 + 4x + 3$

22. $P(x) = x^4 - 3x^3 - x^2 + 3x + 3$

In Problems 23–30,

(A) Find the smallest positive integer and largest negative integer that, by Theorem 2, are upper and lower bounds, respectively, for the real zeros of P(x).

(B) Approximate the real zeros of each polynomial to two decimal places.

23. $P(x) = x^3 - 2x^2 + 3x - 8$

24. $P(x) = x^3 + 3x^2 + 4x + 5$

25. $P(x) = x^4 + x^3 - 5x^2 + 7x - 22$

26. $P(x) = x^4 - x^3 - 8x^2 - 12x - 25$

27. $P(x) = x^5 - 3x^3 - 4x + 4$

28. $P(x) = x^5 - x^4 - 2x^2 - 4x - 5$

29. $P(x) = x^5 + x^4 + 3x^3 + x^2 + 2x - 5$

30. $P(x) = x^5 - 2x^4 - 6x^2 - 9x + 10$

Problems 31–34 refer to the polynomial

$$P(x) = (x - 1)^2(x - 2)(x - 3)^4$$

31. Can the zero at $x = 1$ be approximated by the bisection method? Explain.

32. Can the zero at $x = 2$ be approximated by the bisection method? Explain.

33. Can the zero at $x = 3$ be approximated by the bisection method? Explain.

34. Which of the zeros can be approximated by a maximum approximation routine? By a minimum approximation routine? By the zero approximation routine on your graphing utility?

For each polynomial P(x) in Problems 35–40, use a maximum or minimum routine to approximate zeros of even multiplicity and a zero approximation routine to approximate zeros of odd multiplicity, all to two decimal places. State the multiplicity of each zero.

35. $P(x) = x^4 - 4x^3 - 10x^2 + 28x + 49$

36. $P(x) = x^4 + 4x^3 - 4x^2 - 16x + 16$

37. $P(x) = x^5 - 6x^4 + 4x^3 + 24x^2 - 16x - 32$

38. $P(x) = x^5 - 6x^4 + 2x^3 + 28x^2 - 15x + 2$

39. $P(x) = x^5 - 6x^4 + 11x^3 - 4x^2 - 3.75x - 0.5$

40. $P(x) = x^5 + 12x^4 + 47x^3 + 56x^2 - 15.75x + 1$

C

In Problems 41–50,

(A) Find the smallest positive integer multiple of 10 and largest negative integer multiple of 10 that, by Theorem 2, are upper and lower bounds, respectively, for the real zeros of each polynomial.

(B) Approximate the real zeros of each polynomial to two decimal places.

41. $P(x) = x^3 - 24x^2 - 25x + 10$

42. $P(x) = x^3 - 37x^2 + 70x - 20$

43. $P(x) = x^4 + 12x^3 - 900x^2 + 5{,}000$

44. $P(x) = x^4 - 12x^3 - 425x^2 + 7{,}000$

45. $P(x) = x^4 - 100x^2 - 1{,}000x - 5{,}000$

46. $P(x) = x^4 - 5x^3 - 50x^2 - 500x + 7{,}000$

47. $P(x) = 4x^4 - 40x^3 - 1{,}475x^2 + 7{,}875x - 10{,}000$

48. $P(x) = 9x^4 + 120x^3 - 3{,}083x^2 - 25{,}674x - 48{,}400$

49. $P(x) = 0.01x^5 - 0.1x^4 - 12x^3 + 9{,}000$

50. $P(x) = 0.1x^5 + 0.7x^4 - 18.775x^3 - 340x^2 - 1{,}645x - 2{,}450$

APPLICATIONS

Express the solutions to Problems 51–56 as the roots of a polynomial equation of the form P(x) = 0 and approximate these solutions to three decimal places. Use a graphing utility, if available; otherwise, use the bisection method.

★ **51. Geometry.** Find all points on the graph of $y = x^2$ that are 1 unit away from the point (1, 2). [*Hint:* Use the distance-between-two-points formula from Section 1-1.]

★ **52. Geometry.** Find all points on the graph of $y = x^2$ that are 1 unit away from the point (2, 1).

★ **53. Manufacturing.** A box is to be made out of a piece of cardboard that measures 18 by 24 inches. Squares, x inches on a side, will be cut from each corner, and then the ends and sides will be folded up (see the figure). Find the value of x that would result in a box with a volume of 600 cubic inches.

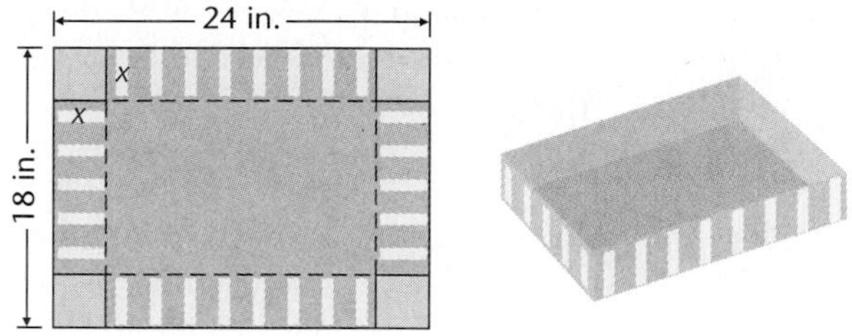

★ **54. Manufacturing.** A box with a hinged lid is to be made out of a piece of cardboard that measures 20 by 40 inches. Six squares, x inches on a side, will be cut from each corner and the middle, and then the ends and sides will be folded up to form the box and its lid (see the figure). Find the value of x that would result in a box with a volume of 500 cubic inches.

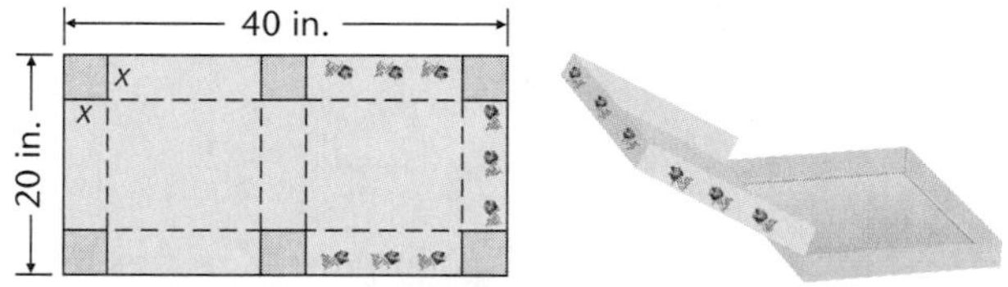

★ **55. Construction.** A propane gas tank is in the shape of a right circular cylinder with a hemisphere at each end (see the figure). If the overall length of the tank is 10 feet and the volume is 20π cubic feet, find the common radius of the hemispheres and the cylinder.

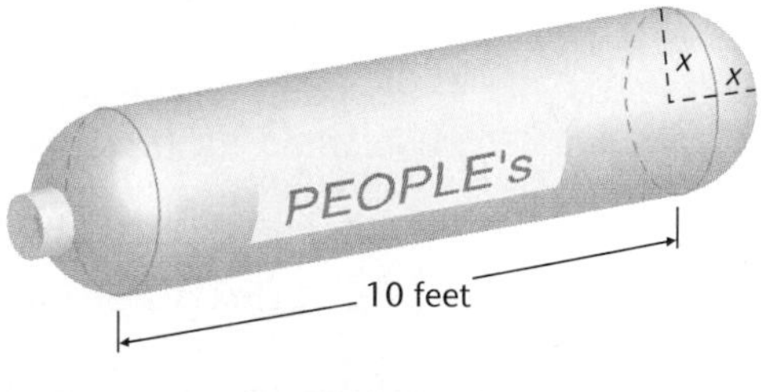

★ **56. Shipping.** A shipping box is reinforced with steel bands in all three directions (see the figure). A total of 20.5 feet of steel tape is to be used, with 6 inches of waste because of a 2-inch overlap in each direction. If the box has a square base and a volume of 2 cubic feet, find its dimensions.

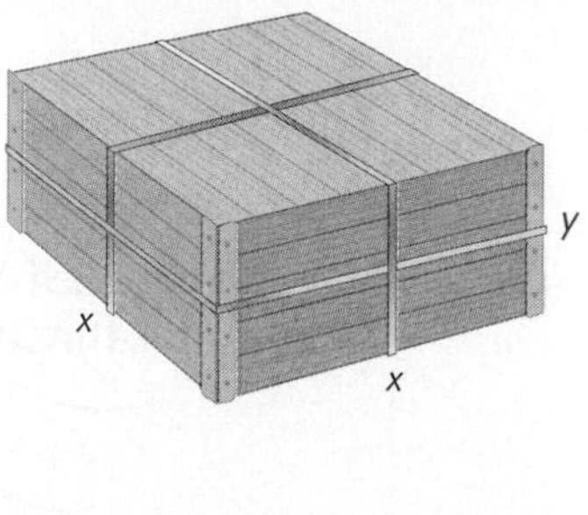

Section 3-4 Rational Functions

- Rational Functions
- Vertical and Horizontal Asymptotes
- Graphing Rational Functions

In this section we develop the tools necessary to analyze the graph of a rational function. A graphing utility will be an important aid in this analysis, but a firm understanding of the properties of rational functions is necessary to correctly interpret results obtained from a graphing utility. The final goal is to produce a hand sketch showing all the important features of the graph.

Rational Functions

Just as rational numbers are defined in terms of quotients of integers, rational functions are defined in terms of quotients of polynomials. The following equations define rational functions:

$$f(x) = \frac{x-1}{x^2-x-6} \qquad g(x) = \frac{1}{x} \qquad h(x) = \frac{x^3-1}{x}$$

$$p(x) = 2x^2 - 3 \qquad q(x) = 3 \qquad r(x) = 0$$

In general, a function f is a **rational function** if

$$f(x) = \frac{n(x)}{d(x)} \qquad d(x) \neq 0$$

where $n(x)$ and $d(x)$ are polynomials. The **domain of f** is the set of all real numbers x such that $d(x) \neq 0$.

If $x = a$ and $d(a) = 0$, then f is not defined at $x = a$ and there can be no point on the graph of f with abscissa $x = a$. Remember, division by 0 is never allowed. It can be shown that

If $f(x) = n(x)/d(x)$ and $d(a) = 0$, then f is discontinuous at $x = a$ and the graph of f has a hole or break at $x = a$.

If $x = a$ is in the domain of $f(x)$ and $n(a) = 0$, then the graph of f crosses the x axis at $x = a$. Thus,

If $f(x) = n(x)/d(x)$, $n(a) = 0$, and $d(a) \neq 0$, then $x = a$ is an x intercept for the graph of f.

What happens if both $n(a) = 0$ and $d(a) = 0$? In this case, we know that $x - a$ is a factor of both $n(x)$ and $d(x)$, and thus, $f(x)$ is not in lowest terms (see Section A-4).

Unless specifically stated to the contrary, we assume that all the rational functions we consider are reduced to lowest terms.

EXAMPLE 1

Finding the Domain and x Intercepts for a Rational Function

Find the domain and x intercepts for $f(x) = \dfrac{2x^2 - 2x - 4}{x^2 - 9}$

Solution

$$f(x) = \frac{n(x)}{d(x)} = \frac{2x^2 - 2x - 4}{x^2 - 9} = \frac{2(x - 2)(x + 1)}{(x - 3)(x + 3)}$$

Since $d(3) = 0$ and $d(-3) = 0$, the domain of f is

$$x \neq \pm 3 \qquad \text{or} \qquad (-\infty, -3) \cup (-3, 3) \cup (3, \infty)$$

Since $n(2) = 0$ and $n(-1) = 0$, the graph of f crosses the x axis at $x = 2$ and $x = -1$.

MATCHED PROBLEM 1

Find the domain and x intercepts for: $f(x) = \dfrac{3x^2 - 12}{x^2 + 2x - 3}$

Vertical and Horizontal Asymptotes

Even though a rational function f may be discontinuous at $x = a$ (no graph for $x = a$), it is still useful to know what happens to the graph of f when x is close to a. For example, consider the very simple rational function f defined by

$$f(x) = \frac{1}{x}$$

It is clear that the function f is discontinuous at $x = 0$. But what happens to $f(x)$ when x approaches 0 from either side of 0? A numerical approach will give us an idea of what happens to $f(x)$ when x gets close to 0. From Table 1, we see that as x approaches 0 from the right, $1/x$ gets larger and larger—that is, $1/x$ increases without bound. We write this symbolically* as

$$\frac{1}{x} \to \infty \qquad \text{as} \qquad x \to 0^+$$

TABLE 1 Behavior of $1/x$ as $x \to 0^+$

x	1	0.1	0.01	0.001	0.0001	0.000 01	0.000 001	. . . x approaches 0 from the right ($x \to 0^+$)
$1/x$	1	10	100	1,000	10,000	100,000	1,000,000	. . . $1/x$ increases without bound ($1/x \to \infty$)

*Remember, the symbol ∞ does not represent a real number. In this context, ∞ is used to indicate that the values of $1/x$ increase without bound. That is, $1/x$ exceeds any given number N no matter how large N is chosen.

If x approaches 0 from the left, then x and $1/x$ are both negative and the values of $1/x$ decrease without bound (see Table 2). This is denoted as

$$\frac{1}{x} \to -\infty \quad \text{as} \quad x \to 0^-$$

TABLE 2 Behavior of $1/x$ as $x \to 0^-$

x	−1	−0.1	−0.01	−0.001	−0.0001	−0.000 01	−0.000 001	. . .	x approaches 0 from the left ($x \to 0^-$)
$1/x$	−1	−10	−100	−1,000	−10,000	−100,000	−1,000,000	. . .	$1/x$ decreases without bound ($1/x \to -\infty$)

The graph of $f(x) = 1/x$ for $-1 \le x \le 1$, $x \ne 0$, is shown in Figure 1. The behavior of f as x approaches 0 from the right is illustrated on the graph by drawing a curve that becomes almost vertical and placing an arrow on the curve to indicate that the values of $1/x$ continue to increase without bound as x approaches 0 from the right. The behavior as x approaches 0 from the left is illustrated in a similar manner.

FIGURE 1

$f(x) = \frac{1}{x}$ near $x = 0$.

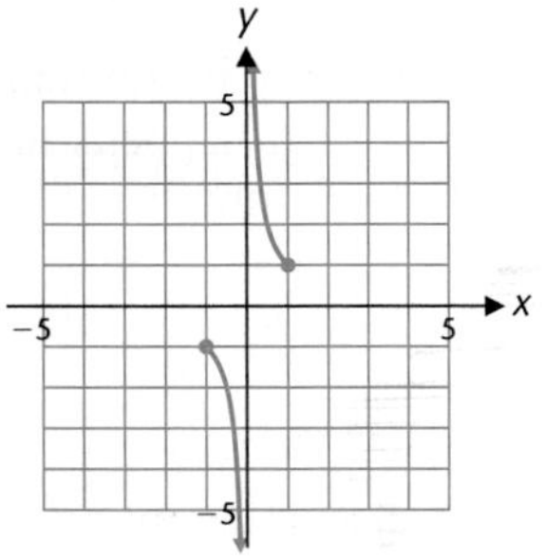

Explore/Discuss 1

Construct tables similar to Tables 1 and 2 for $g(x) = 1/x^2$, and discuss the behavior of the graph of $g(x)$ near $x = 0$.

The preceding discussion suggests that vertical asymptotes are associated with the zeros of the denominator of a rational function. Using the same kind of reasoning, we state the following general method of locating vertical asymptotes for rational functions.

THEOREM 1

VERTICAL ASYMPTOTES AND RATIONAL FUNCTIONS

Let f be a rational function defined by

$$f(x) = \frac{n(x)}{d(x)}$$

THEOREM 1 continued

where $n(x)$ and $d(x)$ are polynomials. If a is a real number such that $d(a) = 0$ and $n(a) \neq 0$, then the line $x = a$ is a vertical asymptote of the graph of $y = f(x)$.

Now we look at the behavior of $f(x) = 1/x$ as $|x|$ gets very large—that is, as $x \to \infty$ and as $x \to -\infty$. Consider Tables 3 and 4. As x increases without bound, $1/x$ is positive and approaches 0 from above. As x decreases without bound, $1/x$ is negative and approaches 0 from below. For our purposes, it is not necessary to distinguish between $1/x$ approaching 0 from above and from below. Thus, we will describe this behavior by writing

$$\frac{1}{x} \to 0 \qquad \text{as} \qquad x \to \infty \text{ and as } x \to -\infty$$

TABLE 3 Behavior of $1/x$ as $x \to \infty$

x	1	10	100	1,000	10,000	100,000	1,000,000	. . .	x increases without bound ($x \to \infty$)
$1/x$	1	0.1	0.01	0.001	0.0001	0.000 01	0.000 001	. . .	$1/x$ approaches 0 ($1/x \to 0$)

TABLE 4 Behavior of $1/x$ as $x \to -\infty$

x	−1	−10	−100	−1,000	−10,000	−100,000	−1,000,000	. . .	x decreases without bound ($x \to -\infty$)
$1/x$	−1	−0.1	−0.01	−0.001	−0.0001	−0.000 01	−0.000 001	. . .	$1/x$ approaches 0 ($1/x \to 0$)

The graph of $f(x) = 1/x$ on a graphing utility is shown in Figure 2(a). Note that this graph confirms the behavior at the vertical and horizontal asymptotes indicated in Tables 1–4. A hand sketch of the graph is shown in Figure 2(b). The behavior as $x \to -\infty$ and as $x \to \infty$ is illustrated by drawing a curve that is almost horizontal and adding arrows at the end.

FIGURE 2
$f(x) = \frac{1}{x}, x \neq 0.$

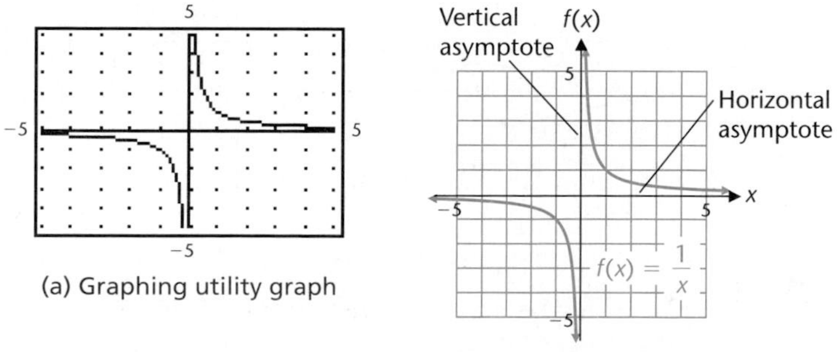

(a) Graphing utility graph

(b) Hand sketch

DEFINITION 1

HORIZONTAL AND VERTICAL ASYMPTOTES

The line $x = a$ is a **vertical asymptote** for the graph of $y = f(x)$ if $f(x)$ either increases or decreases without bound as x approaches a from the right or from the left. Symbolically,

$$f(x) \to \infty \quad \text{or} \quad f(x) \to -\infty \quad \text{as} \quad x \to a^{+} \quad \text{or} \quad x \to a^{-}$$

The line $y = b$ is a **horizontal asymptote** for the graph of $y = f(x)$ if $f(x)$ approaches b as x increases without bound or as x decreases without bound. Symbolically,

$$f(x) \to b \quad \text{as} \quad x \to \infty \quad \text{or} \quad x \to -\infty$$

Explore/Discuss 2

Construct tables similar to Tables 3 and 4 for each of the following functions and discuss the behavior of each as $x \to \infty$ and as $x \to -\infty$:

(A) $f(x) = \dfrac{3x}{x^2 + 1}$ (B) $g(x) = \dfrac{3x^2}{x^2 + 1}$ (C) $h(x) = \dfrac{3x^3}{x^2 + 1}$

In Section 3-1, we saw that the behavior of a polynomial

$$P(x) = a_n x^n + \cdots + a_1 x + a_0$$

as $x \to \pm\infty$ is determined by its leading term, $a_n x^n$. In a similar manner, the behavior of a rational function is determined by the ratio of the leading terms of its numerator and denominator; that is, the graphs of

$$f(x) = \frac{a_m x^m + \cdots + a_1 x + a_0}{b_n x^n + \cdots + b_1 x + b_0} \quad \text{and} \quad g(x) = \frac{a_m x^m}{b_n x^n}$$

exhibit the same behavior as $x \to \infty$ and as $x \to -\infty$ (see Fig. 3).

FIGURE 3
Graphs of rational functions as $x \to \pm\infty$.

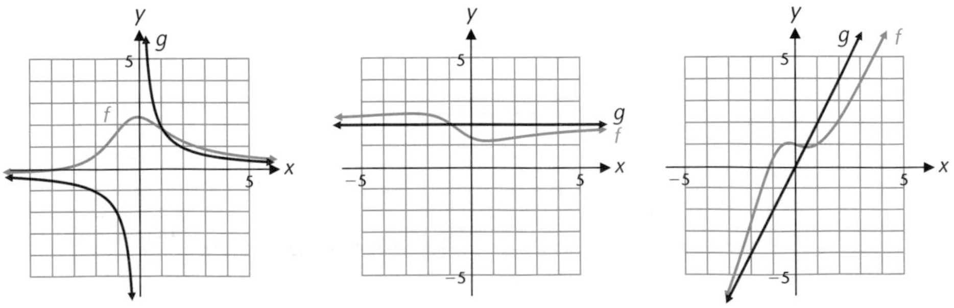

(a) $f(x) = \dfrac{2x + 7}{x^2 + x + 3}$, $g(x) = \dfrac{2x}{x^2} = \dfrac{2}{x}$

(b) $f(x) = \dfrac{2x^2 + 4}{x^2 + x + 3}$, $g(x) = \dfrac{2x^2}{x^2} = 2$

(c) $f(x) = \dfrac{2x^3 + 3}{x^2 + x + 3}$, $g(x) = \dfrac{2x^3}{x^2} = 2x$

In Figure 3(a), the degree of the numerator is less than the degree of the denominator and the x axis is a horizontal asymptote. In Figure 3(b), the degree of the numerator equals the degree of the denominator and the line $y = 2$ is a horizontal asymptote. In Figure 3(c), the degree of the numerator is greater than the degree of the denominator and there are no horizontal asymptotes. These ideas are generalized in Theorem 2 to provide a simple way to locate the horizontal asymptotes of any rational function.

THEOREM 2

HORIZONTAL ASYMPTOTES AND RATIONAL FUNCTIONS

Let f be a rational function defined by the quotient of two polynomials as follows:

$$f(x) = \frac{a_m x^m + \cdots + a_1 x + a_0}{b_n x^n + \cdots + b_1 x + b_0}$$

1. For $m < n$, the line $y = 0$ (the x axis) is a horizontal asymptote.
2. For $m = n$, the line $y = a_m/b_n$ is a horizontal asymptote.
3. For $m > n$, the graph will increase or decrease without bound as $x \to -\infty$ and $x \to \infty$, depending on m, n, a_m, and b_n, and there are no horizontal asymptotes.

EXAMPLE 2

Finding Vertical and Horizontal Asymptotes for a Rational Function

Find all vertical and horizontal asymptotes for

$$f(x) = \frac{n(x)}{d(x)} = \frac{2x^2 - 2x - 4}{x^2 - 9}$$

Solution

Since $d(x) = x^2 - 9 = (x - 3)(x + 3)$, the graph of $f(x)$ has vertical asymptotes at $x = 3$ and $x = -3$ (Theorem 1). Since $n(x)$ and $d(x)$ have the same degree, the line

$$y = \frac{a_2}{b_2} = \frac{2}{1} = 2 \qquad a_2 = 2,\ b_2 = 1$$

is a horizontal asymptote (Theorem 2, part 2).

MATCHED PROBLEM 2

Find all vertical and horizontal asymptotes for

$$f(x) = \frac{3x^2 - 12}{x^2 + 2x - 3}$$

Graphing Rational Functions

We now use the techniques for locating asymptotes, along with other graphing aids discussed in the text, to graph several rational functions. First, we outline a systematic approach to the problem of graphing rational functions.

ANALYZING AND SKETCHING THE GRAPH OF A RATIONAL FUNCTION: $f(x) = n(x)/d(x)$

Step 1. *Intercepts.* Find the real solutions of the equation $n(x) = 0$ and use these solutions to plot any x intercepts of the graph of f. Evaluate $f(0)$, if it exists, and plot the y intercept.

Step 2. *Vertical Asymptotes.* Find the real solutions of the equation $d(x) = 0$ and use these solutions to determine the domain of f, the points of discontinuity, and the vertical asymptotes. Sketch any vertical asymptotes as dashed lines.

Step 3. *Horizontal Asymptotes.* Determine whether there is a horizontal asymptote and if so, sketch it as a dashed line.

Step 4. *Complete the Sketch.* Using a graphing utility graph as an aid and the information determined in steps 1–3, sketch the graph.

EXAMPLE 3 Graphing a Rational Function

Graph: $y = f(x) = \dfrac{2x}{x - 3}$

Solution

$$f(x) = \frac{2x}{x-3} = \frac{n(x)}{d(x)}$$

Step 1. *Intercepts.* Find real zeros of $n(x) = 2x$ and find $f(0)$:

$$2x = 0$$

$$x = 0 \quad x \text{ intercept}$$

$$f(0) = 0 \quad y \text{ intercept}$$

The graph crosses the coordinate axes only at the origin. Plot this intercept, as shown in Figure 4.

FIGURE 4

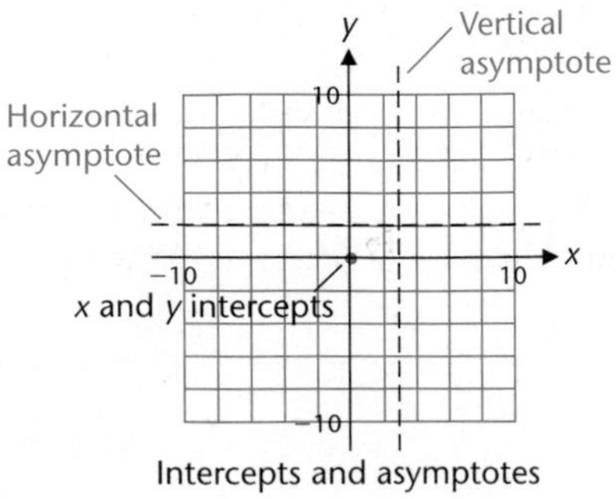

Intercepts and asymptotes

Step 2. *Vertical Asymptotes.* Find real zeros of $d(x) = x - 3$:

$$x - 3 = 0$$
$$x = 3$$

The domain of f is $(-\infty, 3) \cup (3, \infty)$, f is discontinuous at $x = 3$, and the graph has a vertical asymptote at $x = 3$. Sketch this asymptote, as shown in Figure 4.

Step 3. *Horizontal Asymptote.* Since $n(x)$ and $d(x)$ have the same degree, the line $y = 2$ is a horizontal asymptote, as shown in Figure 4.

Step 4. *Complete the Sketch.* Using the graphing utility graph in Figure 5(a), we obtain the graph in Figure 5(b). Notice that the graph is a smooth continuous curve over the interval $(-\infty, 3)$ and over the interval $(3, \infty)$. As expected, there is a break in the graph at $x = 3$.

FIGURE 5

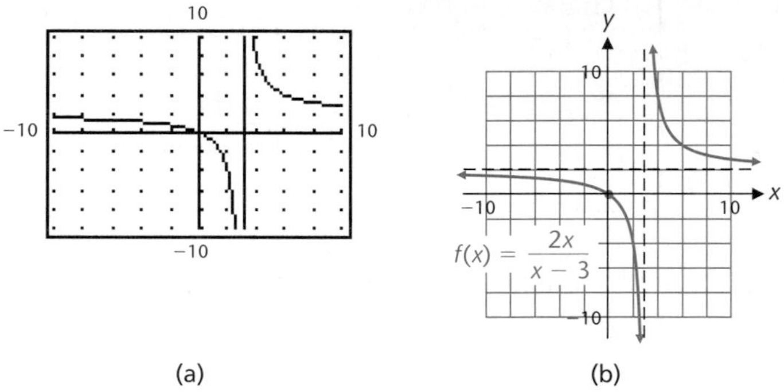

MATCHED PROBLEM 3

Proceed as in Example 3 and sketch the graph of $y = f(x) = \dfrac{3x}{x + 2}$

Remark Refer to Example 3. When $f(x) = 2x/(x - 3)$ is graphed on a graphing utility [Fig. 5(a)], it appears that the graphing utility has also drawn the vertical asymptote at $x = 3$, but this is not the case. Most graphing utilities, when set in *connected mode*, calculate points on a graph and connect these points with line segments. The last point plotted to the left of the asymptote and the first plotted to the right of the asymptote will usually have very large y coordinates. If these y coordinates have opposite signs, then the graphing utility may connect the two points with a nearly vertical line segment, which gives the appearance of an asymptote. If you wish, you can set the calculator in *dot mode* to plot the points without the connecting line segments [Fig. 6(a)].

Depending on the scale, a graph may even appear to be continuous at a vertical asymptote [Fig. 6(b)]. It is important to always locate the vertical asymptotes as we did in step 2 before turning to the graphing utility graph to complete the sketch.

FIGURE 6
Graphing utility graphs of $f(x) = \frac{2x}{x - 3}$.

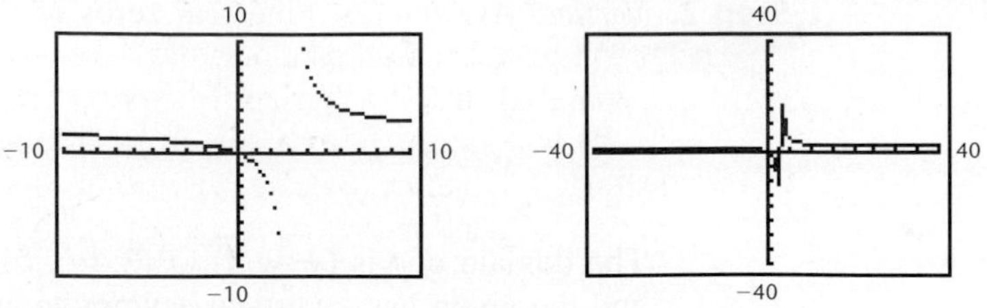

(a) Dot mode (b) Connected mode

In the remaining examples we will just list the results of each step in the graphing strategy and omit the computational details.

EXAMPLE 4

Graphing a Rational Function

Graph: $y = f(x) = \dfrac{x^2 - 6x + 9}{x^2 + x - 2}$

Solution

$$f(x) = \frac{x^2 - 6x + 9}{x^2 + x - 2} = \frac{(x - 3)^2}{(x + 2)(x - 1)}$$

x intercept: $x = 3$

y intercept: $y = f(0) = -\frac{9}{2} = -4.5$

Domain: $(-\infty, -2) \cup (-2, 1) \cup (1, \infty)$

Points of discontinuity: $x = -2$ and $x = 1$

Vertical asymptotes: $x = -2$ and $x = 1$

Horizontal asymptote: $y = 1$

Using the graphing utility graph in Figure 7 as an aid, sketch in the intercepts and asymptotes, then sketch the graph of f (Fig. 8).

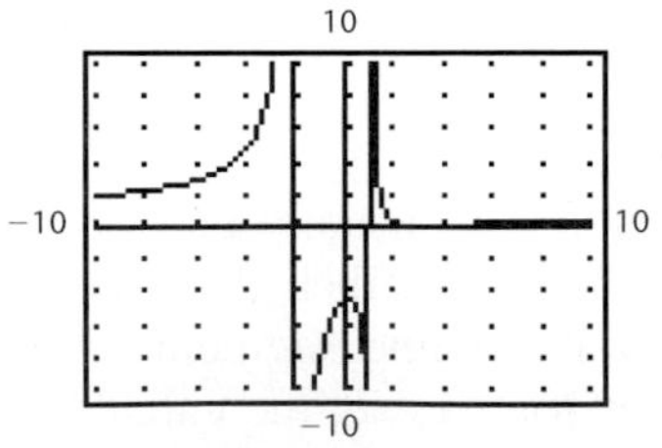

FIGURE 7

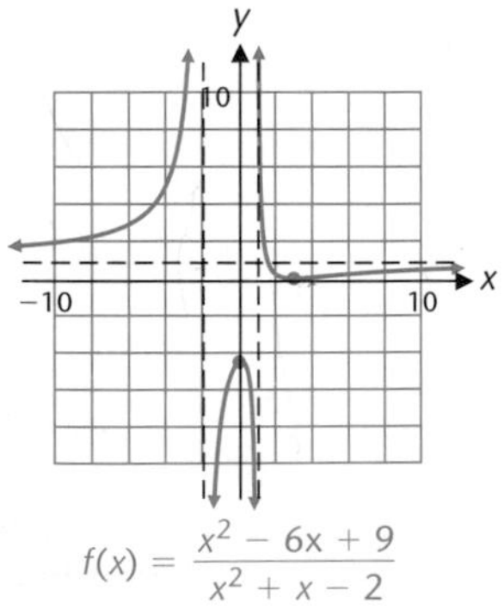

FIGURE 8

MATCHED PROBLEM 4

Graph: $y = f(x) = \dfrac{x^2}{x^2 - 7x + 10}$

CAUTION

The graph of a function cannot cross a vertical asymptote, but the same statement is not true for horizontal asymptotes. The graph in Example 4 clearly shows that **the graph of a function can cross a horizontal asymptote.** The definition of a horizontal asymptote requires $f(x)$ to approach b as x increases or decreases without bound, but it does not preclude the possibility that $f(x) = b$ for one or more values of x. In fact, using the cosine function from trigonometry, it is possible to construct a function whose graph crosses a horizontal asymptote an infinite number of times (see Fig. 9).

FIGURE 9
Multiple intersections of a graph and a horizontal asymptote.

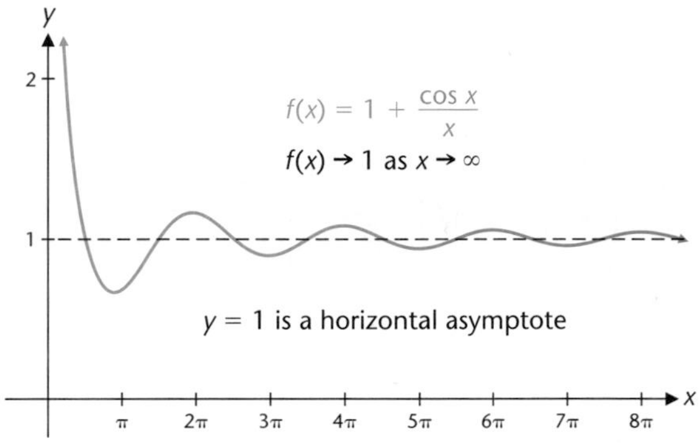

EXAMPLE 5

Graphing a Rational Function

Graph: $y = f(x) = \dfrac{x^2 - 3x - 4}{x - 2}$

Solution

$$f(x) = \frac{x^2 - 3x - 4}{x - 2} = \frac{(x + 1)(x - 4)}{x - 2}$$

x intercepts: $x = -1$ and $x = 4$

y intercept: $y = f(0) = 2$

Domain: $(-\infty, 2) \cup (2, \infty)$

Points of discontinuity: $x = 2$

Vertical asymptote: $x = 2$

No horizontal asymptote

Even though the graph of f does not have a horizontal asymptote, we can still gain some useful information about the behavior of the graph as $x \to -\infty$ and as $x \to \infty$ if we first perform a long division:

$$\begin{array}{r l} x - 1 & \quad \text{Quotient} \\ x - 2\overline{)x^2 - 3x - 4} & \\ \underline{x^2 - 2x} \phantom{{}-4} & \\ -x - 4 & \\ \underline{-x + 2} & \\ -6 & \quad \text{Remainder} \end{array}$$

Thus,

$$f(x) = \frac{x^2 - 3x - 4}{x - 2} = x - 1 - \frac{6}{x - 2}$$

As $x \to -\infty$ or $x \to \infty$, $6/(x - 2) \to 0$ and the graph of f approaches the line $y = x - 1$. This line is called an **oblique asymptote** for the graph of f. A graphing utility graph, including the oblique asymptote, is shown in Figure 10, and the graph of f is sketched in Figure 11.

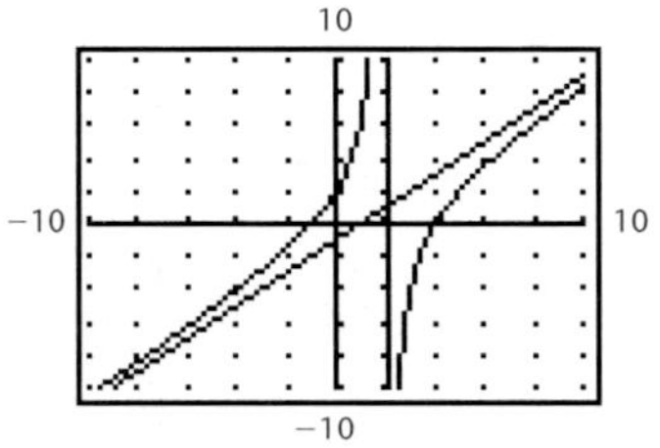

FIGURE 10

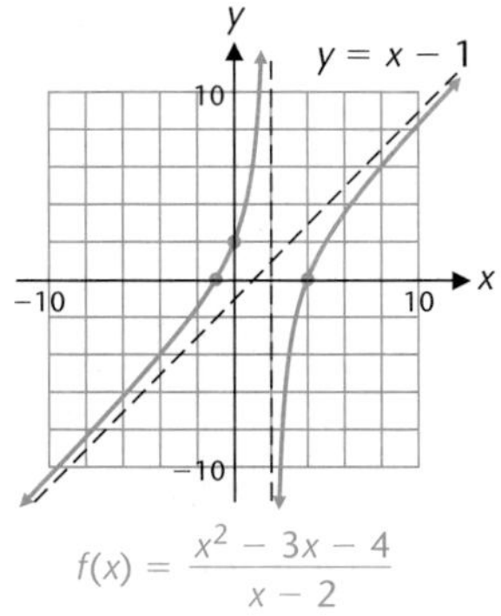

FIGURE 11

Generalizing the results of Example 5, we have Theorem 3.

THEOREM 3

OBLIQUE ASYMPTOTES AND RATIONAL FUNCTIONS

If $f(x) = n(x)/d(x)$, where $n(x)$ and $d(x)$ are polynomials and the degree of $n(x)$ is 1 more than the degree of $d(x)$, then $f(x)$ can be expressed in the form

$$f(x) = mx + b + \frac{r(x)}{d(x)}$$

where the degree of $r(x)$ is less than the degree of $d(x)$. The line

$$y = mx + b$$

is an oblique asymptote for the graph of f. That is,

$$[f(x) - (mx + b)] \to 0 \quad \text{as} \quad x \to -\infty \quad \text{or} \quad x \to \infty$$

MATCHED PROBLEM 5

Graph, including any oblique asymptotes: $y = f(x) = \dfrac{x^2 + 5}{x + 1}$

Answers to Matched Problems

1. Domain: $(-\infty, -3) \cup (-3, 1) \cup (1, \infty)$; x intercepts: $x = -2$, $x = 2$
2. Vertical asymptotes: $x = -3$, $x = 1$; horizontal asymptote: $y = 3$

3.

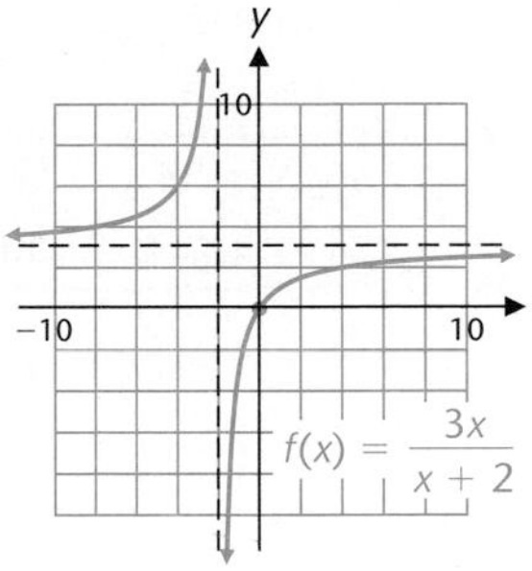

4.

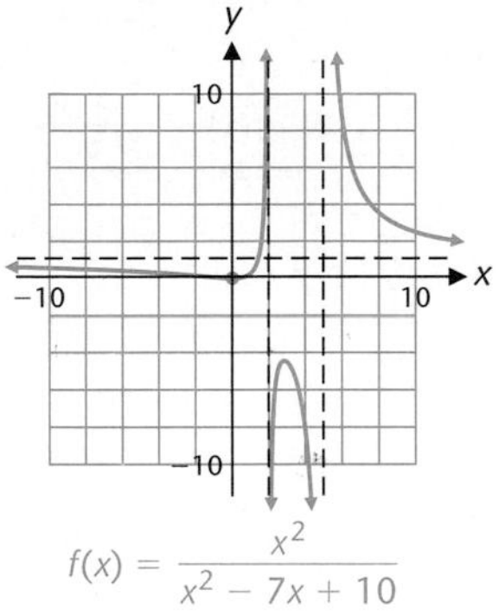

5.

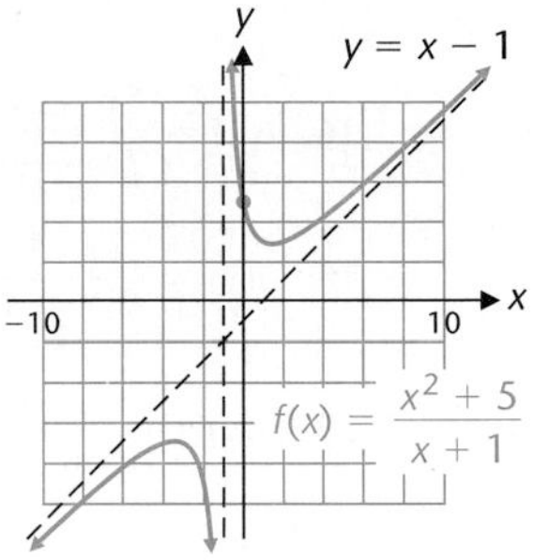

EXERCISE 3-4

A

In Problems 1–4, match each graph with one of the following functions:

$$f(x) = \frac{2x - 4}{x + 2} \qquad g(x) = \frac{2x + 4}{2 - x}$$

$$h(x) = \frac{2x + 4}{x - 2} \qquad k(x) = \frac{4 - 2x}{x + 2}$$

1.

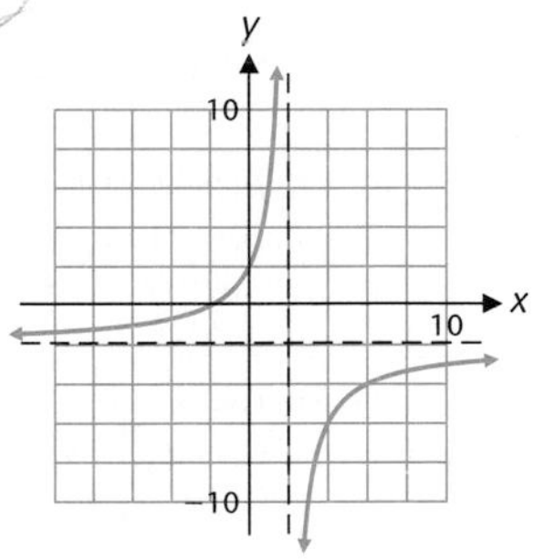

2.

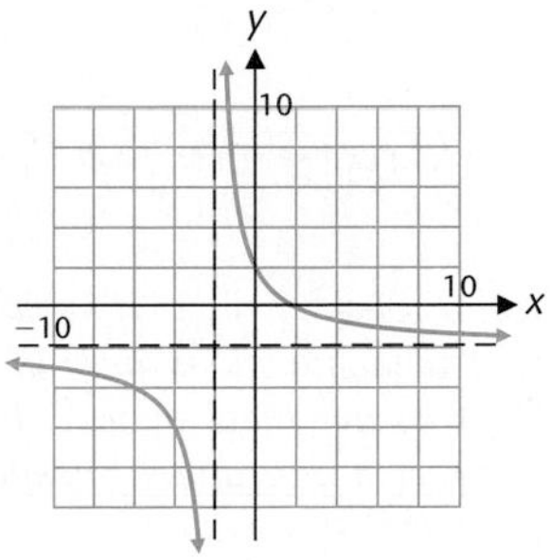

3.

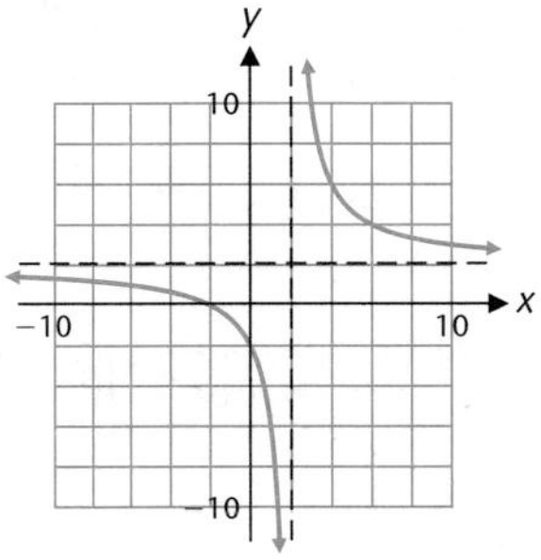

4.

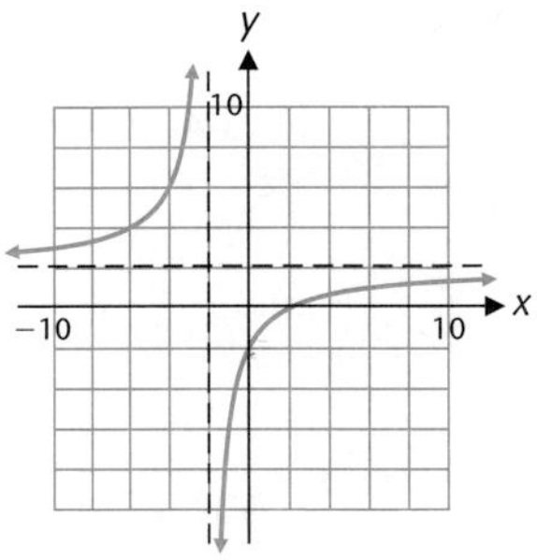

In Problems 5–12, find the domain and x intercepts.

5. $f(x) = \dfrac{2x - 4}{x + 1}$

6. $g(x) = \dfrac{3x + 6}{x - 1}$

7. $h(x) = \dfrac{x^2 - 1}{x^2 - 16}$

8. $k(x) = \dfrac{x^2 - 36}{x^2 - 25}$

9. $r(x) = \dfrac{x^2 - x - 6}{x^2 - x - 12}$

10. $s(x) = \dfrac{x^2 + x - 12}{x^2 + x - 6}$

11. $F(x) = \dfrac{x}{x^2 + 4}$

12. $G(x) = \dfrac{x^2}{x^2 + 16}$

In Problems 13–20, find all vertical and horizontal asymptotes.

13. $f(x) = \dfrac{2x}{x - 4}$ **14.** $h(x) = \dfrac{3x}{x + 5}$

15. $s(x) = \dfrac{2x^2 + 3x}{3x^2 - 48}$ **16.** $r(x) = \dfrac{5x^2 - 7x}{2x^2 - 50}$

17. $p(x) = \dfrac{2x}{x^4 + 1}$ **18.** $q(x) = \dfrac{5x^4}{2x^2 + 3x - 2}$

19. $t(x) = \dfrac{6x^4}{3x^2 - 2x - 5}$ **20.** $g(x) = \dfrac{3x}{x^4 + 2x^2 + 1}$

B

In Problems 21–40, use the graphing strategy outlined in the text to sketch the graph of each function.

21. $f(x) = \dfrac{1}{x - 4}$ **22.** $g(x) = \dfrac{1}{x + 3}$

23. $f(x) = \dfrac{x}{x + 1}$ **24.** $f(x) = \dfrac{3x}{x - 3}$

25. $h(x) = \dfrac{x}{2x - 2}$ **26.** $p(x) = \dfrac{3x}{4x + 4}$

27. $f(x) = \dfrac{2x - 4}{x + 3}$ **28.** $f(x) = \dfrac{3x + 3}{2 - x}$

29. $g(x) = \dfrac{1 - x^2}{x^2}$ **30.** $f(x) = \dfrac{x^2 + 1}{x^2}$

31. $f(x) = \dfrac{9}{x^2 - 9}$ **32.** $g(x) = \dfrac{6}{x^2 - x - 6}$

33. $f(x) = \dfrac{x}{x^2 - 1}$ **34.** $p(x) = \dfrac{x}{1 - x^2}$

35. $g(x) = \dfrac{2}{x^2 + 1}$ **36.** $f(x) = \dfrac{x}{x^2 + 1}$

37. $f(x) = \dfrac{12x^2}{(3x + 5)^2}$ **38.** $f(x) = \dfrac{7x^2}{(2x - 3)^2}$

39. $f(x) = \dfrac{x^2 - 1}{x^2 + 7x + 10}$ **40.** $f(x) = \dfrac{x^2 + 6x + 8}{x^2 - x - 2}$

41. If $f(x) = n(x)/d(x)$, where $n(x)$ and $d(x)$ are quadratic functions, what is the maximum number of x intercepts $f(x)$ can have? What is the minimum number? Illustrate both cases with examples.

42. If $f(x) = n(x)/d(x)$, where $n(x)$ and $d(x)$ are quadratic functions, what is the maximum number of vertical asymptotes $f(x)$ can have? What is the minimum number? Illustrate both cases with examples.

In Problems 43–48, find all vertical, horizontal, and oblique asymptotes.

43. $f(x) = \dfrac{2x^2}{x - 1}$ **44.** $g(x) = \dfrac{3x^2}{x + 2}$

45. $p(x) = \dfrac{x^3}{x^2 + 1}$ **46.** $q(x) = \dfrac{x^5}{x^3 - 8}$

47. $r(x) = \dfrac{2x^2 - 3x + 5}{x}$ **48.** $s(x) = \dfrac{-3x^2 + 5x + 9}{x}$

In Problems 49–52, investigate the behavior of each function as $x \to \infty$ and as $x \to -\infty$, and find any horizontal asymptotes.

49. $f(x) = \dfrac{5x}{\sqrt{x^2 + 1}}$ **50.** $f(x) = \dfrac{2x}{\sqrt{x^2 - 1}}$

51. $f(x) = \dfrac{4\sqrt{x^2 - 4}}{x}$ **52.** $f(x) = \dfrac{3\sqrt{x^2 + 1}}{x - 1}$

C

In Problems 53–58, use the graphing strategy outlined in the text to sketch the graph of each function. Include any oblique asymptotes.

53. $f(x) = \dfrac{x^2 + 1}{x}$ **54.** $g(x) = \dfrac{x^2 - 1}{x}$

55. $k(x) = \dfrac{x^2 - 4x + 3}{2x - 4}$ **56.** $h(x) = \dfrac{x^2 + x - 2}{2x - 4}$

57. $F(x) = \dfrac{8 - x^3}{4x^2}$ **58.** $G(x) = \dfrac{x^4 + 1}{x^3}$

If $f(x) = n(x)/d(x)$, where the degree of $n(x)$ is greater than the degree of $d(x)$, then long division can be used to write $f(x) = p(x) + q(x)/d(x)$, where $p(x)$ and $q(x)$ are polynomials with the degree of $q(x)$ less than the degree of $d(x)$. In Problems 59–62, perform the long division and discuss the relationship between the graphs of $f(x)$ and $p(x)$ as $x \to \infty$ and as $x \to -\infty$.

59. $f(x) = \dfrac{x^4}{x^2 + 1}$ **60.** $f(x) = \dfrac{x^5}{x^2 + 1}$

61. $f(x) = \dfrac{x^5}{x^2 - 1}$ **62.** $f(x) = \dfrac{x^5}{x^3 - 1}$

CU *In calculus, it is often necessary to consider rational functions that are not in lowest terms, such as the functions given in Problems 63–66. For each function, state the domain, reduce the function to lowest terms, and sketch its graph. Remember to exclude from the graph any points with x values that are not in the domain.*

63. $f(x) = \dfrac{x^2 - 4}{x - 2}$ **64.** $g(x) = \dfrac{x^2 - 1}{x + 1}$

65. $r(x) = \dfrac{x + 2}{x^2 - 4}$

66. $s(x) = \dfrac{x - 1}{x^2 - 1}$

APPLICATIONS

67. Employee Training. A company producing electronic components used in television sets has established that on the average, a new employee can assemble $N(t)$ components per day after t days of on-the-job training, as given by

$$N(t) = \frac{50t}{t + 4} \qquad t \geq 0$$

Sketch the graph of N, including any vertical or horizontal asymptotes. What does N approach as $t \to \infty$?

68. Physiology. In a study on the speed of muscle contraction in frogs under various loads, researchers W. O. Fems and J. Marsh found that the speed of contraction decreases with increasing loads. More precisely, they found that the relationship between speed of contraction S (in centimeters per second) and load w (in grams) is given approximately by

$$S(w) = \frac{26 + 0.06w}{w} \qquad w \geq 5$$

Sketch the graph of S, including any vertical or horizontal asymptotes. What does S approach as $w \to \infty$?

69. Retention. An experiment on retention is conducted in a psychology class. Each student in the class is given 1 day to memorize the same list of 40 special characters. The lists are turned in at the end of the day, and for each succeeding day for 20 days each student is asked to turn in a list of as many of the symbols as can be recalled. Averages are taken, and it is found that a good approximation of the average number of symbols, $N(t)$, retained after t days is given by

$$N(t) = \frac{5t + 30}{t} \qquad t \geq 1$$

Sketch the graph of N, including any vertical or horizontal asymptotes. What does N approach as $t \to \infty$?

70. Learning Theory. In 1917, L. L. Thurstone, a pioneer in quantitative learning theory, proposed the function

$$f(x) = \frac{a(x + c)}{(x + c) + b}$$

to describe the number of successful acts per unit time that a person could accomplish after x practice sessions. Suppose that for a particular person enrolling in a typing class,

$$f(x) = \frac{50(x + 1)}{x + 5} \qquad x \geq 0$$

where $f(x)$ is the number of words per minute the person is able to type after x weeks of lessons. Sketch the graph of f, including any vertical or horizontal asymptotes. What does f approach as $x \to \infty$?

CIC *Problems 71–74 are calculus related.*

★ **71. Replacement Time.** A desktop office copier has an initial price of \$2,500. A maintenance/service contract costs \$200 for the first year and increases \$50 per year thereafter. It can be shown that the total cost of the copier after n years is given by

$$C(n) = 2{,}500 + 175n + 25n^2$$

The average cost per year for n years is $\overline{C}(n) = C(n)/n$.

(A) Find the rational function $\overline{C}$.

(B) When is the average cost per year minimum? (This is frequently referred to as the *replacement time* for this piece of equipment.)

(C) Sketch the graph of $\overline{C}$, including any asymptotes.

★ **72. Average Cost.** The total cost of producing x units of a certain product is given by

$$C(x) = \tfrac{1}{5}x^2 + 2x + 2{,}000$$

The average cost per unit for producing x units is $\overline{C}(x) = C(x)/x$.

(A) Find the rational function $\overline{C}$.

(B) At what production level will the average cost per unit be minimal?

(C) Sketch the graph of $\overline{C}$, including any asymptotes.

★ **73. Construction.** A rectangular dog pen is to be made to enclose an area of 225 square feet.

(A) If x represents the width of the pen, express the total length $L(x)$ of the fencing material required for the pen in terms of x.

(B) Considering the physical limitations, what is the domain of the function L?

(C) Find the dimensions of the pen that will require the least amount of fencing material.

(D) Graph the function L, including any asymptotes.

★ **74. Construction.** Rework Problem 73 with the added assumption that the pen is to be divided into two sections, as shown in the figure. (Approximate dimensions to three decimal places.)

Chapter 3 Group Activity

Interpolating Polynomials

Given two points in the plane, we can use the point–slope form of the equation of a line to find a polynomial whose graph passes through these two points. How can we proceed if we are given more than two points? For example, how can we find the equation of a polynomial $P(x)$ whose graph passes through the points listed in Table 1 and graphed in Figure 1?

TABLE 1

x	1	2	3	4
$P(x)$	1	3	−3	1

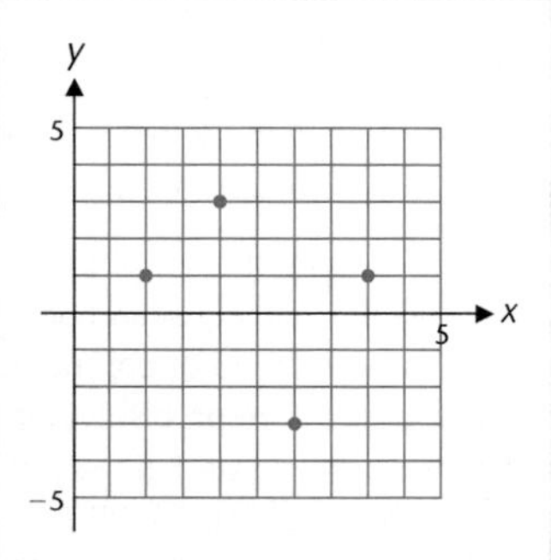

FIGURE 1

The key to solving this problem is to write the unknown polynomial $P(x)$ in the following special form:

$$P(x) = a_0 + a_1(x - 1) + a_2(x - 1)(x - 2) + a_3(x - 1)(x - 2)(x - 3) \quad (1)$$

Since the graph of $P(x)$ is to pass through each point in Table 1, we can substitute each value of x in equation (1) to determine the coefficients a_0, a_1, a_2, and a_3. First we evaluate equation (1) at $x = 1$ to determine a_0:

$$1 = P(1)$$
$$= a_0 \qquad \text{All other terms in equation (1) are 0 when } x = 1.$$

Using this value for a_0 in equation (1) and evaluating at $x = 2$, we have

$$3 = P(2) = 1 + a_1(1) \qquad \text{All other terms are 0.}$$
$$2 = a_1$$

Continuing in this manner, we have

$$-3 = P(3) = 1 + 2(2) + a_2(2)(1)$$
$$-8 = 2a_2$$
$$-4 = a_2$$
$$1 = P(4) = 1 + 2(3) - 4(3)(2) + a_3(3)(2)(1)$$
$$18 = 6a_3$$
$$3 = a_3$$

We have now evaluated all the coefficients in equation (1) and can write

$$P(x) = 1 + 2(x - 1) - 4(x - 1)(x - 2) + 3(x - 1)(x - 2)(x - 3) \quad (2)$$

If we expand the products in equation (2) and collect like terms, we can express $P(x)$ in the more conventional form (verify this):

$$P(x) = 3x^3 - 22x^2 + 47x - 27$$

(A) To check these calculations, evaluate $P(x)$ at $x = 1, 2, 3$, and 4 and compare the results with Table 1. Then add the graph of $P(x)$ to Figure 1.

(B) Write a verbal description of the special form of $P(x)$ in equation (1).

In general, given a set of $n + 1$ points:

x	x_0	x_1	$\cdots$	x_n
y	y_0	y_1	$\cdots$	y_n

the **interpolating polynomial** for these points is the polynomial $P(x)$ of degree less than or equal to n that satisfies $P(x_k) = y_k$ for $k = 0, 1, \ldots, n$. The **general form** of the interpolating polynomial is

$$P(x) = a_0 + a_1(x - x_0) + a_2(x - x_0)(x - x_1) + \cdots + a_n(x - x_0)(x - x_1) \cdot \cdots \cdot (x - x_{n-1})$$

(C) Summarize the procedure for using the points in the table to find the coefficients in the general form.

(D) Give an example to show that the interpolating polynomial can have degree strictly less than n.

(E) Could there be two different polynomials of degree less than or equal to n whose graph passes through the given $n + 1$ points? Justify your answer.

(F) Find the interpolating polynomial for each of Tables 2 and 3. Check your answers by evaluating the polynomial, and illustrate by graphing the points in the table and the polynomial in the same viewing window.

TABLE 2

x	-1	0	1	2
y	5	3	3	11

TABLE 3

x	-2	-1	0	1	2
y	-3	0	5	0	-3

A surprisingly short program on a graphing utility can be used to calculate the coefficients in the general form of an interpolating polynomial. Table 4 shows such a program for a Texas Instruments graphing calculator and the output generated when we use the program to find the coefficients of the interpolating polynomial for Table 1.

TABLE 4 Interpolating Polynomial Coefficients on a Graphing Utility

Program INTERP

```
L2→L3
dimL L3→M
For(I,2,M,1)
For(J,M,I,-1)
(L3(J)-L3(J-1))/(L1(J)-L1(J-I+1))→L3(J)
End
End
Disp L3
```

Output

```
{1,2,3,4}→L1
                {1 2 3 4}
{1,3,-3,1}→L2
                {1 3 -3 1}
INTERP
                {1 2 -4 3}
                      Done
```

(G) If you have a TI-85 or TI-86 graphing calculator, enter INTERP in your calculator exactly as shown in Table 4. To use the program, enter the x values in L1 and the corresponding y values in L2 (see the output in Table 4) and then execute the program. If you have some other graphing utility that can store and execute programs, consult your manual and modify the statements in INTERP so that the program works on your graphing utility. Use INTERP to check your answers to part F.

CHAPTER 3 | REVIEW

3-1 POLYNOMIAL FUNCTIONS AND GRAPHS

In this chapter, unless indicated otherwise, the coefficients of the ***n*th-degree polynomial function** $P(x) = a_nx^n + a_{n-1}x^{n-1} + \cdots + a_1x + a_0$ are complex numbers and the domain is the set of complex numbers. The number r is said to be a **zero of the function *P*,** or a **zero of the polynomial *P*(*x*),** or a **solution or root of the equation *P*(*x*) = 0,** if $P(r) = 0$. If the coefficients of $P(x)$ are real numbers, then the x intercepts of the graph of $y = P(x)$ are real **zeros** of P and $P(x)$ and real **solutions** or **roots** for the equation $P(x) = 0$.

Synthetic division is an efficient method for dividing polynomials by linear terms of the form $x - r$ that is well-suited to calculator use.

Let $P(x)$ be a polynomial of degree greater than 0 and let r be a real number. Then we have the following important theorems:

Division Algorithm. $P(x) = (x - r)Q(x) + R$, where $x - r$ is the **divisor;** $Q(x)$, a unique polynomial of degree 1 less than $P(x)$, is the **quotient;** and R, a unique real number, is the **remainder.**

Remainder Theorem. $P(r) = R$.

The left and right behavior of an nth-degree polynomial $P(x)$ with real coefficients is determined by its highest degree or **leading term.** As $x \to \pm\infty$, a_nx^n and $P(x)$ both approach $\pm\infty$, depending on n and the sign of a_n. The points on a continuous graph where local extrema occur are called **turning points.** Important graph properties are

1. P is continuous for all real numbers.
2. The graph of P is a smooth curve.
3. The graph of P has at most n x intercepts.
4. P has at most $n - 1$ turning points.

3-2 FINDING RATIONAL ZEROS OF POLYNOMIALS

If $P(x)$ is a polynomial of degree $n > 0$, then we have the following important theorems:

Factor Theorem. The number r is a zero of $P(x)$ if and only if $(x - r)$ is a factor of $P(x)$.

Fundamental Theorem of Algebra. $P(x)$ has at least one zero.

***n* Zeros Theorem.** $P(x)$ can be expressed as a product of n linear factors and has n zeros, not necessarily distinct.

If $P(x)$ is represented as the product of linear factors and $x - r$ occurs m times, then r is called a **zero of multiplicity *m*.**

Imaginary Zeros Theorem. If $P(x)$ has real coefficients, then imaginary zeros of $P(x)$, if they exist, must occur in conjugate pairs.

Real Zeros and Odd-Degree Polynomials. If $P(x)$ has real coefficients and is of odd degree, then $P(x)$ always has at least one real zero.

Rational Zero Theorem. If the rational number b/c, in lowest terms, is a zero of the polynomial

$$P(x) = a_nx^n + a_{n-1}x^{n-1} + \cdots + a_1x + a_0 \qquad a_n \neq 0$$

with integer coefficients, then b must be an integer factor of a_0 and c must be an integer factor of a_n.

If $P(x) = (x - r)Q(x)$, then $Q(x)$ is called a **reduced polynomial** for $P(x)$.

3-3 APPROXIMATING REAL ZEROS OF POLYNOMIALS

The following theorems are important tools for locating the real zeros of a polynomial with real coefficients. Once located, a graphing utility can be used to approximate the zeros.

Location Theorem. If f is continuous on an interval I, a and b are two numbers in I, and $f(a)$ and $f(b)$ are of opposite sign, then there is at least one x intercept between a and b.

Upper and Lower Bounds of Real Zeros. If $n > 0$, $a_n > 0$, and $P(x)$ is divided by $x - r$ using synthetic division:

1. If $r > 0$ and all numbers in the quotient row of the synthetic division, including the remainder, are nonnegative, then r is greater than or equal to the largest zero of $P(x)$ and is called an **upper bound** of the real zeros of $P(x)$.
2. If $r < 0$ and all numbers in the quotient row of the synthetic division, including the remainder, alternate in sign, then r is less than or equal to the smallest zero of $P(x)$ and is called a **lower bound** of the real zeros of $P(x)$.

Zeros of Even and Odd Multiplicity. If $P(x)$ is a polynomial with real coefficients, then

1. If r is a zero of odd multiplicity, then $P(x)$ changes sign at r and does not have a local extremum at $x = r$.
2. If r is a zero of even multiplicity, then $P(x)$ does not change sign at r and has a local extremum at $x = r$.

3-4 RATIONAL FUNCTIONS

A function of the form $f(x) = n(x)/d(x)$, where $n(x)$ and $d(x)$ are polynomials is a **rational function.** The line $x = a$ is a **vertical asymptote** for the graph of $y = f(x)$ if $f(x) \to \infty$ or $f(x) \to -\infty$ as $x \to a^+$ or $x \to a^-$. If $d(a) = 0$ and $n(a) \neq 0$, then the line $x = a$ is a vertical asymptote. The line $y = b$ is a **horizontal asymptote** for the graph of $y = f(x)$ if $f(x) \to b$ as $x \to \infty$ or $x \to -\infty$. The line $y = mx + b$ is an **oblique asymptote** if $[f(x) - (mx + b)] \to 0$ as $x \to \infty$ or $x \to -\infty$.

$$\text{Let } f(x) = \frac{a_mx^m + \cdots + a_1x + a_0}{b_nx^n + \cdots + b_1x + b_0}, \quad a_m, b_n \neq 0.$$

The behavior of the graph of f as $x \to \infty$ or $x \to -\infty$ is determined by the ratio of the leading terms of the numerator and denominator, a_mx^m/b_nx^n.

1. If $m < n$, then the x axis is a horizontal asymptote.
2. If $m = n$, then the line $y = a_m/b_n$ is a horizontal asymptote.
3. If $m > n$, then there are no horizontal asymptotes.

Analyzing and Sketching the Graph of a Rational Function: $f(x) = n(x)/d(x)$

Step 1. *Intercepts.* Find the real solutions of the equation $n(x) = 0$ and use these solutions to plot any x intercepts of the graph of f. Evaluate $f(0)$, if it exists, and plot the y intercept.

Step 2. *Vertical Asymptotes.* Find the real solutions of the equation $d(x) = 0$ and use these solutions to determine the domain of f, the points of discontinuity, and the vertical asymptotes. Sketch any vertical asymptotes as dashed lines.

Step 3. *Horizontal Asymptotes.* Determine whether there is a horizontal asymptote and if so, sketch it as a dashed line.

Step 4. *Complete the Sketch.* Using a graphing utility graph as an aid and the information determined in steps 1–3, sketch the graph.

CHAPTER 3 | REVIEW EXERCISES

Work through all the problems in this chapter review, and check answers in the back of the book. Answers to all review problems are there, and following each answer is a number in italics indicating the section in which that type of problem is discussed. Where weaknesses show up, review appropriate sections in the text.

A

1. Use synthetic division to divide $P(x) = 2x^3 + 3x^2 - 1$ by $D(x) = x + 2$, and write the answer in the form $P(x) = D(x)Q(x) + R$.
2. If $P(x) = x^5 - 4x^4 + 9x^2 - 8$, find $P(3)$ using the remainder theorem and synthetic division.
3. What are the zeros of $P(x) = 3(x - 2)(x + 4)(x + 1)$?
4. If $P(x) = x^2 - 2x + 2$ and $P(1 + i) = 0$, find another zero of $P(x)$.
5. Let $P(x)$ be the polynomial whose graph is shown in the figure.
   (A) Assuming that $P(x)$ has integer zeros and leading coefficient 1, find the lowest-degree equation that could produce this graph.
   (B) Describe the left and right behavior of $P(x)$.

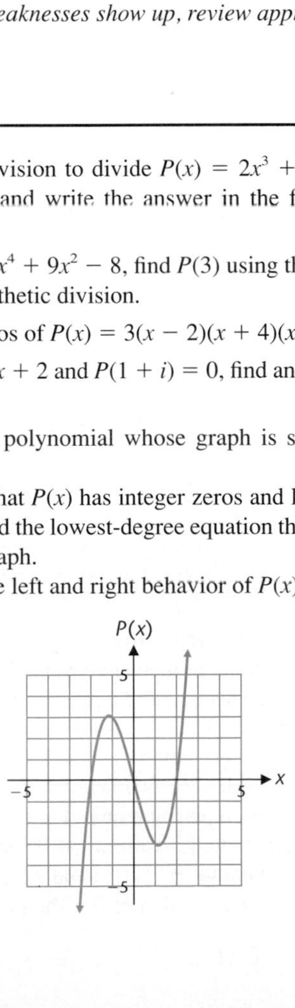

6. According to the upper and lower bound theorem, which of the following are upper or lower bounds of the zeros of $P(x) = x^3 - 4x^2 + 2$?

$-2, -1, 3, 4$

7. How do you know that $P(x) = 2x^3 - 3x^2 + x - 5$ has at least one real zero between 1 and 2?

8. Write the possible rational zeros for $P(x) = x^3 - 4x^2 + x + 6$.

9. Find all rational zeros for $P(x) = x^3 - 4x^2 + x + 6$.

10. Find the domain and x intercept(s) for:

(A) $f(x) = \dfrac{2x - 3}{x + 4}$ (B) $g(x) = \dfrac{3x}{x^2 - x - 6}$

11. Find the horizontal and vertical asymptotes for the functions in Problem 10.

B

12. Let $P(x) = x^3 - 3x^2 - 3x + 4$.
(A) Graph $P(x)$ and describe the graph verbally, including the number of x intercepts, the number of turning points, and the left and right behavior.
(B) Approximate the largest x intercept to two decimal places.

13. If $P(x) = 8x^4 - 14x^3 - 13x^2 - 4x + 7$, find $Q(x)$ and R such that $P(x) = (x - \frac{1}{4})Q(x) + R$. What is $P(\frac{1}{4})$?

14. If $P(x) = 4x^3 - 8x^2 - 3x - 3$, find $P(-\frac{1}{2})$ using the remainder theorem and synthetic division.

15. Use the quadratic formula and the factor theorem to factor $P(x) = x^2 - 2x - 1$.

16. Is $x + 1$ a factor of $P(x) = 9x^{26} - 11x^{17} + 8x^{11} - 5x^4 - 7$? Explain, without dividing or using synthetic division.

17. Determine all rational zeros of $P(x) = 2x^3 - 3x^2 - 18x - 8$.

18. Factor the polynomial in Problem 17 into linear factors.

19. Find all rational zeros of $P(x) = x^3 - 3x^2 + 5$.

20. Find all zeros (rational, irrational, and imaginary) exactly for $P(x) = 2x^4 - x^3 + 2x - 1$.

21. Factor the polynomial in Problem 20 into linear factors.

22. Let $P(x) = x^5 - 10x^4 + 30x^3 - 20x^2 - 15x - 2$.
(A) Approximate the zeros of $P(x)$ to two decimal places and state the multiplicity of each zero.
(B) Can any of these zeros be approximated with the bisection method? A maximum routine? A minimum routine? Explain.

23. Let $P(x) = x^4 - 2x^3 - 30x^2 - 25$.
(A) Find the smallest positive and largest negative integers that, by Theorem 2 in Section 3-3, are upper and lower bounds, respectively, for the real zeros of $P(x)$.
(B) If $(k, k + 1)$, k an integer, is the interval containing the largest real zero of $P(x)$, determine how many additional intervals are required in the bisection method to approximate this zero to one decimal place.
(C) Approximate the real zeros of $P(x)$ to two decimal places.

24. Let $f(x) = \dfrac{x - 1}{2x + 2}$
(A) Find the domain and the intercepts for f.
(B) Find the vertical and horizontal asymptotes for f.
(C) Sketch a graph of f. Draw vertical and horizontal asymptotes with dashed lines.

C

25. Use synthetic division to divide $P(x) = x^3 + 3x + 2$ by $[x - (1 + i)]$, and write the answer in the form $P(x) = D(x)Q(x) + R$.

26. Find a polynomial of lowest degree with leading coefficient 1 that has zeros $-\frac{1}{2}$ (multiplicity 2), -3, and 1 (multiplicity 3). (Leave the answer in factored form.) What is the degree of the polynomial?

27. Repeat Problem 26 for a polynomial $P(x)$ with zeros -5, $2 - 3i$, and $2 + 3i$.

28. Find all zeros (rational, irrational, and imaginary) exactly for $P(x) = 2x^5 - 5x^4 - 8x^3 + 21x^2 - 4$.

29. Factor the polynomial in Problem 28 into linear factors.

30. Let $P(x) = x^4 + 16x^3 + 47x^2 - 137x + 73$. Approximate (to two decimal places) the x intercepts and the local extrema.

31. What is the minimal degree of a polynomial $P(x)$, given that $P(-1) = -4$, $P(0) = 2$, $P(1) = -5$, and $P(2) = 3$? Justify your conclusion.

32. If $P(x)$ is a cubic polynomial with integer coefficients and if $1 + 2i$ is a zero of $P(x)$, can $P(x)$ have an irrational zero? Explain.

33. The solutions to the equation $x^3 - 27 = 0$ are the cube roots of 27.
(A) How many cube roots of 27 are there?
(B) 3 is obviously a cube root of 27; find all others.

34. Let $P(x) = x^4 + 2x^3 - 500x^2 - 4{,}000$.
(A) Find the smallest positive integer multiple of 10 and the largest negative integer multiple of 10 that, by Theorem 2 in Section 3-3, are upper and lower bounds, respectively, for the real zeros of $P(x)$.
(B) Approximate the real zeros of $P(x)$ to two decimal places.

35. Graph

$$f(x) = \frac{x^2 + 2x + 3}{x + 1}$$

Indicate any vertical, horizontal, or oblique asymptotes with dashed lines.

36. Use a graphing utility to find any horizontal asymptotes for

$$f(x) = \frac{2x}{\sqrt{x^2 + 3x + 4}}$$

APPLICATIONS

In Problems 37–40, express the solutions as the roots of a polynomial equation of the form $P(x) = 0$. Find rational solutions exactly and irrational solutions to three decimal places.

37. Architecture. An entryway is formed by placing a rectangular door inside an arch in the shape of the parabola with graph $y = 16 - x^2$, x and y in feet (see the figure). If the area of the door is 48 square feet, find the dimensions of the door.

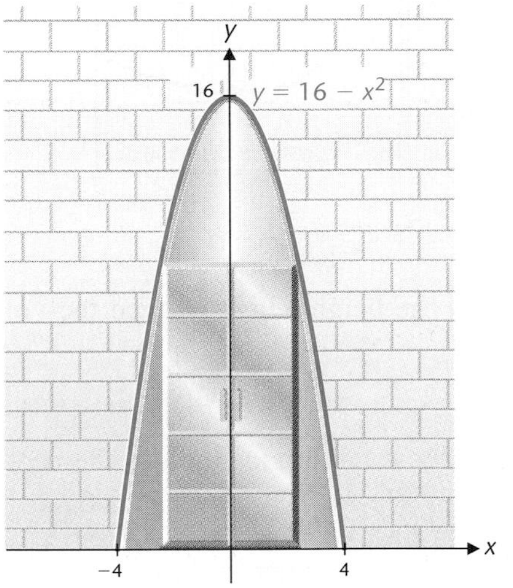

38. Construction. A grain silo is formed by attaching a hemisphere to the top of a right circular cylinder (see the figure). If the cylinder is 18 feet high and the volume of the silo is 486π cubic feet, find the common radius of the cylinder and the hemisphere.

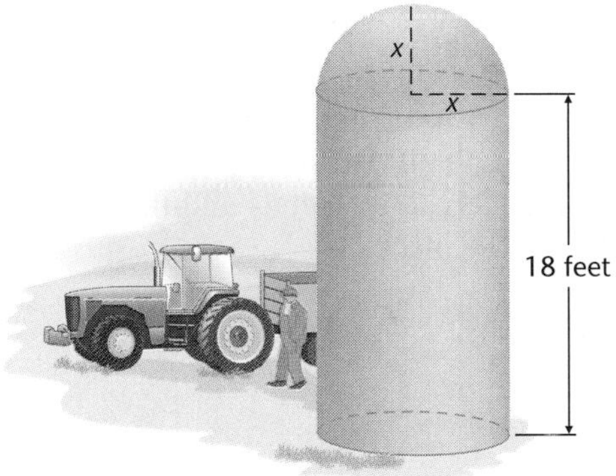

★ **39. Manufacturing.** A box is to be made out of a piece of cardboard that measures 15 by 20 inches. Squares, x inches on a side, will be cut from each corner, and then the ends and sides will be folded up (see the figure). Find the value of x that would result in a box with a volume of 300 cubic inches.

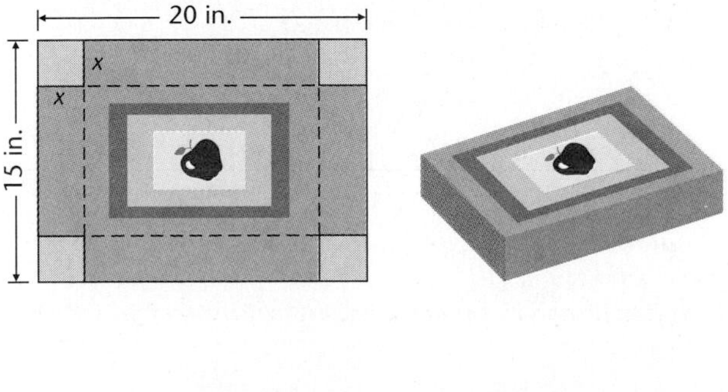

★ **40. Geometry.** Find all points on the graph of $y = x^2$ that are 3 units from the point (1, 4).

41. Advertising. A chain of appliance stores uses television ads to promote the sale of refrigerators. Analyzing past records produced the data in the table, where x is the number of ads placed monthly and y is the number of refrigerators sold that month.

(A) Find a cubic regression equation for this data using the number of ads as the independent variable.

(B) Estimate (to the nearest integer) the number of refrigerators that would be sold if 15 ads are placed monthly.

(C) Estimate (to the nearest integer) the number of ads that should be placed to sell 750 refrigerators monthly.

Number of Ads x	Number of Refrigerators y
10	270
20	430
25	525
30	630
45	890
48	915

42. Women in the Workforce. It is reasonable to conjecture from the data given in the table that many Japanese women tend to leave the work force to marry and have children, but then reenter the workforce when the children are grown.

(A) Explain why you might expect cubic regression to provide a better fit to the data than linear or quadratic regression.

(B) Find a cubic regression for this data using age as the independent variable.

(C) Use the regression equation to estimate (to the nearest year) the ages at which 65% of the women are in the workforce.

Women in the Workforce in Japan (1993)

Age	Percentage of Women Employed
22	75
27	64
32	52
37	61
42	70
47	72
52	66
57	57
62	40

ANSWER TO APPLICATION

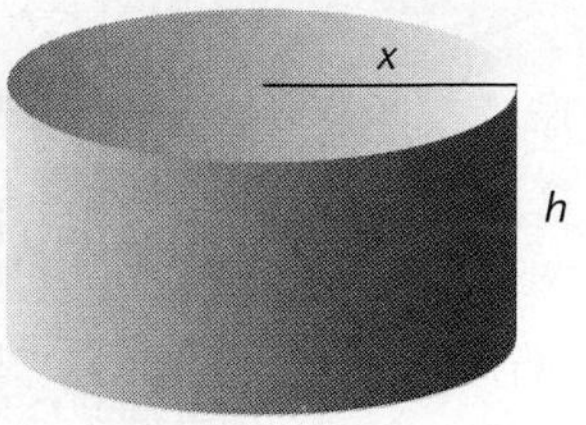

FIGURE 1

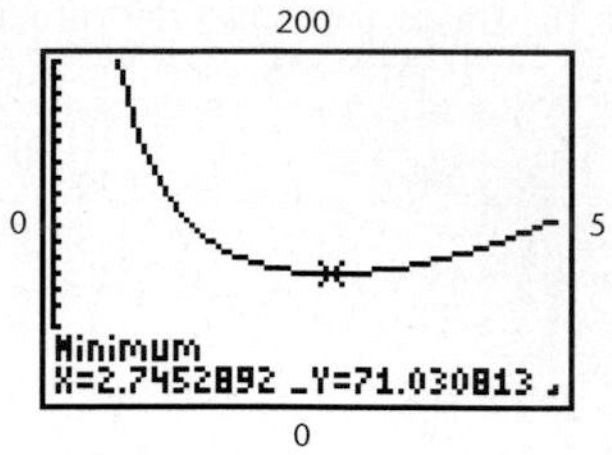

FIGURE 2

If x is the radius of the cup and h the height (Fig. 1), then the volume is

$$V = \pi x^2 h = 65 \quad \text{Volume}$$

The amount of aluminum used is the total surface area of the can.

$$\begin{array}{lccc} & \text{(Area of sides)} & + & \text{(Area of base)} \\ S = & 2\pi x h & + & \pi x^2 \end{array} \quad \text{Surface area}$$

Solving the volume equation for h and substituting in the surface area equation, we have

$$S = 2\pi x\left(\frac{65}{\pi x^2}\right) + \pi x^2 = \frac{130}{x} + \pi x^2$$

Graphing this equation and using a built-in minimum routine (Fig. 2), we see that the minimum amount of aluminum is approximately 71 square inches when the radius is approximately 2.75 inches and the height is approximately $[65/(\pi \cdot 2.75^2)] \approx 2.74$ inches.

Inverse Functions; Exponential and Logarithmic Functions

4

Outline

Application

You have just inherited a trust fund of $18,000 that you cannot access for 8 more years. You are hoping that this investment will double by then, as you have plans to purchase a sports car for $35,000. The *Rule of 72* can be used to quickly determine how long it will take an investment to double at a given interest rate. If r is the annual rate, then the Rule of 72 states that an investment at this rate will double in approximately $72/(100r)$ years. If your trust fund is invested at 10.4% compounded annually, how accurate is the Rule of 72 approximation?

Most of the functions we have considered so far have been polynomial and rational functions, with a few others involving roots or powers of polynomial or rational functions. The general class of functions defined by means of the algebraic operations of addition, subtraction, multiplication, division, and the taking of powers and roots on variables and constants are called *algebraic functions.*

In the first two sections of this chapter, we discuss some operations that can be performed on functions to produce new functions, including the very important concept of an *inverse function.* Next, we define and investigate the properties of two new and important types of functions called *exponential functions* and *logarithmic functions.* These functions are not algebraic, but are members of another class of functions called *transcendental functions.* The exponential functions and logarithmic functions are used in describing and solving a wide variety of real-world problems, including growth of populations of people, animals, and bacteria; radioactive decay; growth of money at compound interest; absorption of light as it passes through air, water, or glass; and magnitudes of sounds and earthquakes. We consider applications in these areas plus many more in the sections that follow.

Preparing for This Chapter

Before getting started on this chapter, review the following concepts:
Exponents (Appendix A, Sections 5 and 6)
Functions (Chapter 1, Section 3)
Graphs of Functions (Chapter 1, Section 4)
Quadratic Equations (Chapter 2, Section 5)
Equation Solving Techniques (Chapter 2, Section 6)

Section 4-1 | Operations on Functions; Composition

V

- Operations on Functions
- Composition
- Applications

If two functions f and g are both defined at a real number x, and if $f(x)$ and $g(x)$ are both real numbers, then it is possible to perform real number operations such as addition, subtraction, multiplication, or division with $f(x)$ and $g(x)$. Furthermore, if $g(x)$ is a number in the domain of f, then it is also possible to evaluate f at $g(x)$. In this section we see how operations on the values of functions can be used to define operations on the functions themselves.

Operations on Functions

The functions f and g given by

$$f(x) = 2x + 3 \qquad \text{and} \qquad g(x) = x^2 - 4$$

are defined for all real numbers. Thus, for any real x we can perform the following operations:

$$f(x) + g(x) = 2x + 3 + x^2 - 4 = x^2 + 2x - 1$$
$$f(x) - g(x) = 2x + 3 - (x^2 - 4) = -x^2 + 2x + 7$$
$$f(x)g(x) = (2x + 3)(x^2 - 4) = 2x^3 + 3x^2 - 8x - 12$$

For $x \neq \pm 2$ we can also form the quotient

$$\frac{f(x)}{g(x)} = \frac{2x + 3}{x^2 - 4} \qquad x \neq \pm 2$$

Notice that the result of each operation is a new function. Thus, we have

$$(f + g)(x) = f(x) + g(x) = x^2 + 2x - 1 \qquad \text{Sum}$$
$$(f - g)(x) = f(x) - g(x) = -x^2 + 2x + 7 \qquad \text{Difference}$$
$$(fg)(x) = f(x)g(x) = 2x^3 + 3x^2 - 8x - 12 \qquad \text{Product}$$
$$\left(\frac{f}{g}\right)(x) = \frac{f(x)}{g(x)} = \frac{2x + 3}{x^2 - 4} \qquad x \neq \pm 2 \qquad \text{Quotient}$$

Notice that the sum, difference, and product functions are defined for all values of x, as were f and g, but the domain of the quotient function must be restricted to exclude those values where $g(x) = 0$.

DEFINITION 1

OPERATIONS ON FUNCTIONS

The **sum, difference, product,** and **quotient** of the functions f and g are the functions defined by

$$(f + g)(x) = f(x) + g(x) \qquad \text{Sum function}$$
$$(f - g)(x) = f(x) - g(x) \qquad \text{Difference function}$$
$$(fg)(x) = f(x)g(x) \qquad \text{Product function}$$
$$\left(\frac{f}{g}\right)(x) = \frac{f(x)}{g(x)} \qquad g(x) \neq 0 \qquad \text{Quotient function}$$

Each function is defined on the intersection of the domains of f and g, with the exception that the values of x where $g(x) = 0$ must be excluded from the domain of the quotient function.

Explore/Discuss 1

Enter $y_1 = \sqrt{4 - x}$ and $y_2 = \sqrt{3 + x}$ in the equation editor of a graphing utility [Fig. 1(a)], graph y_1 and y_2 in the same viewing window* [Fig. 1(b)], and use Trace to determine the values of x for which each function is defined. Use Trace in Figures 1(c) through 1(f) to determine the domains of the corresponding functions.

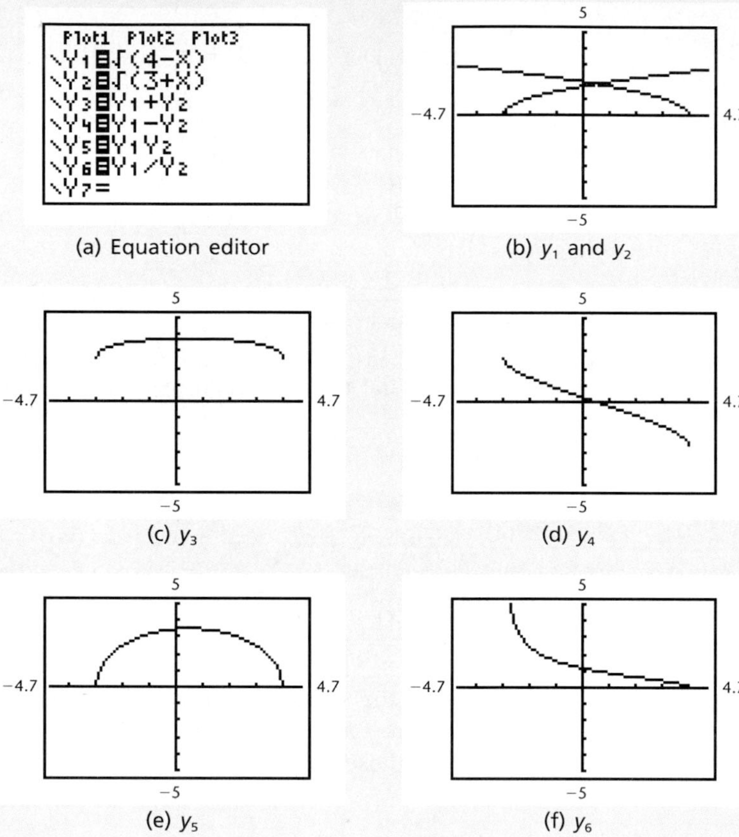

FIGURE 1
Graphing sum, difference, product, and quotient functions.

EXAMPLE 1 Finding the Sum, Difference, Product, and Quotient Functions

Let $f(x) = \sqrt{4 - x}$ and $g(x) = \sqrt{3 + x}$. Find the functions $f + g, f - g, fg,$ and f/g, and find their domains.

Solution

$$(f + g)(x) = f(x) + g(x) = \sqrt{4 - x} + \sqrt{3 + x}$$
$$(f - g)(x) = f(x) - g(x) = \sqrt{4 - x} - \sqrt{3 + x}$$
$$(fg)(x) = f(x)g(x) = \sqrt{4 - x}\sqrt{3 + x}$$
$$= \sqrt{(4 - x)(3 + x)}$$
$$= \sqrt{12 + x - x^2}$$
$$\left(\frac{f}{g}\right)(x) = \frac{f(x)}{g(x)} = \frac{\sqrt{4 - x}}{\sqrt{3 + x}} = \sqrt{\frac{4 - x}{3 + x}}$$

*It is convenient to choose Xmin and Xmax so that the pixels have one-decimal-place screen coordinates. See Problems 35 and 36 in Exercise 1-2 or consult your manual.

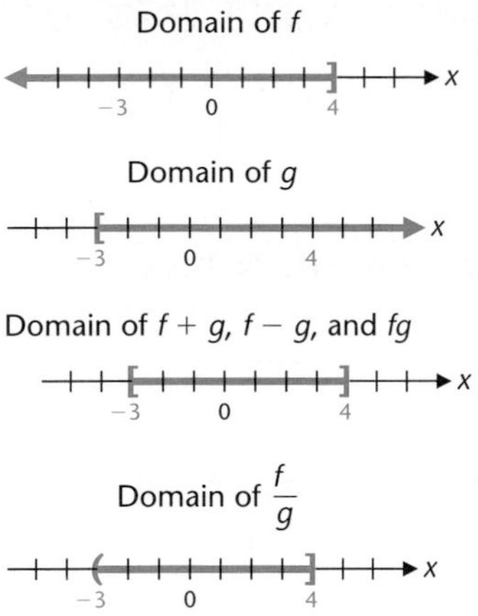

The domains of f and g are

$$\text{Domain of } f\text{: } x \leq 4 \text{ or } (-\infty, 4]$$

$$\text{Domain of } g\text{: } x \geq -3 \text{ or } [-3, \infty)$$

The intersection of these domains is

$$(-\infty, 4] \cap [-3, \infty) = [-3, 4]$$

This is the domain of the functions $f + g$, $f - g$, and fg. Since $g(-3) = 0$, $x = -3$ must be excluded from the domain of the quotient function. Thus,

$$\text{Domain of } \frac{f}{g}\text{: } (-3, 4]$$

MATCHED PROBLEM 1

Let $f(x) = \sqrt{x}$ and $g(x) = \sqrt{10 - x}$. Find the functions $f + g$, $f - g$, fg, and f/g, and find their domains.

Composition

Consider the function h given by the equation

$$h(x) = \sqrt{2x + 1}$$

Inside the radical is a first-degree polynomial that defines a linear function. So the function h is really a combination of a square root function and a linear function. We can see this more clearly as follows. Let

$$u = 2x + 1 = g(x)$$

$$y = \sqrt{u} = f(u)$$

Then

$$h(x) = f(g(x))$$

The function h is said to be the *composite* of the two functions f and g. (Loosely speaking, we can think of h as a function of a function.) What can we say about the domain of h given the domains of f and g? In forming the composite $h(x) = f(g(x))$: **x must be restricted so that x is in the domain of g and $g(x)$ is in the domain of f.** Since the domain of f, where $f(u) = \sqrt{u}$, is the set of nonnegative real numbers, we see that $g(x)$ must be nonnegative; that is,

$$g(x) \geq 0$$

$$2x + 1 \geq 0$$

$$x \geq -\tfrac{1}{2}$$

Thus, the domain of h is this restricted domain of g.

A special function symbol is often used to represent the *composite of two functions*, which we define in general terms below.

DEFINITION 2

COMPOSITE FUNCTIONS

Given functions f and g, then $f \circ g$ is called their **composite** and is defined by the equation

$$(f \circ g)(x) = f(g(x))$$

The domain of $f \circ g$ is the set of all real numbers x in the domain of g where $g(x)$ is in the domain of f.

As an immediate consequence of Definition 2, we have (see Fig. 2):

The domain of $f \circ g$ is always a subset of the domain of g, and the range of $f \circ g$ is always a subset of the range of f.

FIGURE 2
Composite functions.

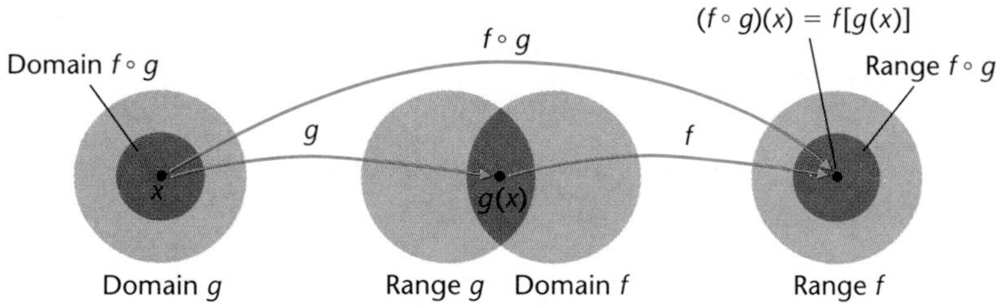

EXAMPLE 2

Finding the Composition of Two Functions

Find $(f \circ g)(x)$ and $(g \circ f)(x)$ and their domains for $f(x) = x^{10}$ and $g(x) = 3x^4 - 1$.

Solution

$$(f \circ g)(x) = f(g(x)) = f(3x^4 - 1) = (3x^4 - 1)^{10}$$
$$(g \circ f)(x) = g(f(x)) = g(x^{10}) = 3(x^{10})^4 - 1 = 3x^{40} - 1$$

The functions f and g are both defined for all real numbers. If x is any real number, then x is in the domain of g, $g(x)$ is in the domain of f, and, consequently, x is in the domain of $f \circ g$. Thus, the domain of $f \circ g$ is the set of all real numbers. Using similar reasoning, the domain of $g \circ f$ also is the set of all real numbers.

MATCHED PROBLEM 2

Find $(f \circ g)(x)$ and $(g \circ f)(x)$ and their domains for $f(x) = 2x + 1$ and $g(x) = (x - 1)/2$.

If two functions are both defined for all real numbers, then so is their composition.

Explore/Discuss 2

Verify that if $f(x) = 1/(1 - 2x)$ and $g(x) = 1/x$, then $(f \circ g)(x) = x/(x - 2)$. Clearly, $f \circ g$ is not defined at $x = 2$. Are there any other values of x where $f \circ g$ is not defined? Explain.

If either function in a composition is not defined for some real numbers, then, as Example 3 illustrates, the domain of the composition may not be what you first think it should be.

EXAMPLE 3 **Finding the Composition of Two Functions**

Find $(f \circ g)(x)$ and its domain for $f(x) = \sqrt{4 - x^2}$ and $g(x) = \sqrt{3 - x}$.

Solution We begin by stating the domains of f and g, a good practice in any composition problem:

$$\text{Domain } f\text{: } -2 \le x \le 2 \qquad \text{or} \qquad [-2, 2]$$

$$\text{Domain } g\text{: } x \le 3 \qquad \text{or} \qquad (-\infty, 3]$$

Next we find the composition:

$$\begin{aligned}(f \circ g)(x) &= f(g(x)) = f(\sqrt{3 - x})\\ &= \sqrt{4 - (\sqrt{3 - x})^2}\\ &= \sqrt{4 - (3 - x)} \qquad (\sqrt{t})^2 = t,\ t \ge 0\\ &= \sqrt{1 + x}\end{aligned}$$

Even though $\sqrt{1 + x}$ is defined for all $x \ge -1$, we must restrict the domain of $f \circ g$ to those values that also are in the domain of g. Thus,

$$\text{Domain } f \circ g\text{: } x \ge -1 \text{ and } x \le 3 \qquad \text{or} \qquad [-1, 3]$$

MATCHED PROBLEM 3

Find $(f \circ g)(x)$ and its domain for $f(x) = \sqrt{9 - x^2}$ and $g(x) = \sqrt{x - 1}$.

CAUTION

The domain of $f \circ g$ cannot always be determined simply by examining the final form of $(f \circ g)(x)$. Any numbers that are excluded from the domain of g must also be excluded from the domain of $f \circ g$.

Explore/Discuss 3

Refer to Example 3. Enter $y_1 = \sqrt{4 - x^2}$, $y_2 = \sqrt{3 - x}$, and $y_3 = y_1(y_2(x))$ in the equation editor of your graphing utility and graph y_3. Does this graph agree with the answer we found in Example 3? Does your graphing utility seem to handle composition correctly? (Not all do!)

Applications

In calculus, it is not only important to be able to find the composition of two functions, but also to recognize when a given function is the composition of two simpler functions.

CI EXAMPLE 4 **Recognizing Composition Forms**

Express h as a composition of two simpler functions for

$$h(x) = (3x + 5)^5$$

Solution If we let $f(x) = x^5$ and $g(x) = 3x + 5$, then

$$h(x) = (3x + 5)^5 = f(3x + 5) = f(g(x)) = (f \circ g)(x)$$

and we have expressed h as the composition of f and g.

MATCHED PROBLEM 4 Express h as a composition of the square root function and a linear function for $h(x) = \sqrt{4x - 7}$.

You will encounter the operations discussed in this section in many different situations. The next example shows how these operations are used in economic analysis.

EXAMPLE 5 **Market Research**

The research department for an electronics firm estimates that the weekly demand for a certain brand of audiocassette players is given by

$$x = f(p) = 20{,}000 - 1{,}000p \qquad \text{Demand function}$$

where x is the number of cassette players retailers are likely to buy per week at $\$p$ per player. The research department also has determined that the total cost (in dollars) of producing x cassette players per week is given by

$$C(x) = 75{,}000 + 4x \qquad \text{Cost function}$$

and the total weekly revenue (in dollars) obtained from the sale of these cassette players is given by

$$R(x) = 20x - \frac{1}{1{,}000}x^2 \quad \text{Revenue function}$$

Express the firm's weekly profit as a function of the price p.

Solution

Since profit is revenue minus cost, the profit function is the difference of the revenue and cost functions, $P = R - C$. Since R and C are given as functions of x, we first express P as a function of x:

$$\begin{aligned} P(x) &= (R - C)(x) \\ &= R(x) - C(x) \\ &= 20x - \frac{1}{1{,}000}x^2 - (75{,}000 + 4x) \\ &= 16x - \frac{1}{1{,}000}x^2 - 75{,}000 \end{aligned}$$

Next, we use composition to express P as a function of the price p:

$$\begin{aligned} (P \circ f)(p) &= P(f(p)) = P(20{,}000 - 1{,}000p) \\ &= 16(20{,}000 - 1{,}000p) - \frac{1}{1{,}000}(20{,}000 - 1{,}000p)^2 - 75{,}000 \\ &= 320{,}000 - 16{,}000p - 400{,}000 + 40{,}000p - 1{,}000p^2 - 75{,}000 \\ &= -155{,}000 + 24{,}000p - 1{,}000p^2 \end{aligned}$$

Technically, $P \circ f$ and P are different functions since the first has independent variable p and the second has independent variable x. However, since both functions represent the same quantity, it is customary to use the same symbol to name each function. Thus,

$$P(p) = -155{,}000 + 24{,}000p - 1{,}000p^2$$

expresses the weekly profit P as a function of price p.

MATCHED PROBLEM 5

Repeat Example 5 for the functions

$$x = f(p) = 10{,}000 - 1{,}000p$$

$$C(x) = 90{,}000 + 5x \qquad R(x) = 10x - \frac{1}{1{,}000}x^2$$

Answers to Matched Problems

1. $(f+g)(x) = \sqrt{x} + \sqrt{10-x}$; $(f-g)(x) = \sqrt{x} - \sqrt{10-x}$; $(fg)(x) = \sqrt{10x - x^2}$; $(f/g)(x) = \sqrt{x/(10-x)}$; the functions $f+g$, $f-g$, and fg have domain $[0, 10]$, the domain of f/g is $[0, 10)$
2. $(f \circ g)(x) = x$, domain $= (-\infty, \infty)$; $(g \circ f)(x) = x$, domain $= (-\infty, \infty)$
3. $(f \circ g)(x) = \sqrt{10-x}$; domain: $x \geq 1$ and $x \leq 10$ or $[1, 10]$
4. $h(x) = (f \circ g)(x)$, where $f(x) = \sqrt{x}$ and $g(x) = 4x - 7$
5. $P(p) = -140{,}000 + 15{,}000p - 1{,}000p^2$

EXERCISE 4-1

A

Problems 1–10 refer to the graphs of f and g shown below.

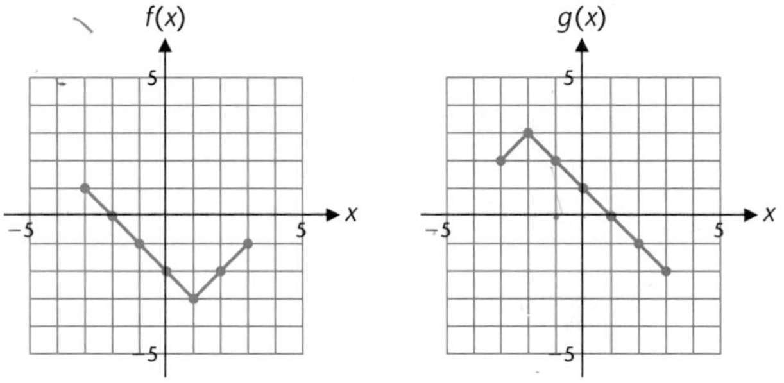

1. Construct a table of values of $(f+g)(x)$ for $x = -3, -2, -1, 0, 1, 2$, and 3, and sketch the graph of $f+g$.

2. Construct a table of values of $(g-f)(x)$ for $x = -3, -2, -1, 0, 1, 2$, and 3, and sketch the graph of $g-f$.

Use the graphs of f and g to find each of the following:

3. $(f \circ g)(-1)$
4. $(f \circ g)(2)$
5. $(g \circ f)(-2)$
6. $(g \circ f)(3)$
7. $f(g(1))$
8. $f(g(0))$
9. $g(f(2))$
10. $g(f(-3))$

In Problems 11–16, for the indicated functions f and g, find the functions $f+g$, $f-g$, fg, and f/g, and find their domains.

11. $f(x) = 4x$; $g(x) = x + 1$
12. $f(x) = 3x$; $g(x) = x - 2$
13. $f(x) = 2x^2$; $g(x) = x^2 + 1$
14. $f(x) = 3x$; $g(x) = x^2 + 4$
15. $f(x) = 3x + 5$; $g(x) = x^2 - 1$
16. $f(x) = 2x - 7$; $g(x) = 9 - x^2$

In Problems 17–22, for the indicated functions f and g, find the functions $f \circ g$ and $g \circ f$, and find their domains.

17. $f(x) = x^3$; $g(x) = x^2 - x + 1$
18. $f(x) = x^2$; $g(x) = x^3 + 2x + 4$
19. $f(x) = |x + 1|$; $g(x) = 2x + 3$
20. $f(x) = |x - 4|$; $g(x) = 3x + 2$
21. $f(x) = x^{1/3}$; $g(x) = 2x^3 + 4$
22. $f(x) = x^{2/3}$; $g(x) = 8 - x^3$

B

In Problems 23–26, find $f \circ g$ and $g \circ f$. Graph f, g, $f \circ g$, and $g \circ f$ in a squared viewing window and describe any apparent symmetry between these graphs.

23. $f(x) = \frac{1}{2}x + 1$; $g(x) = 2x - 2$
24. $f(x) = 3x + 2$; $g(x) = \frac{1}{3}x - \frac{2}{3}$
25. $f(x) = -\frac{2}{3}x - \frac{5}{3}$; $g(x) = -\frac{3}{2}x - \frac{5}{2}$
26. $f(x) = -2x + 3$; $g(x) = -\frac{1}{2}x + \frac{3}{2}$

In Problems 27–32, for the indicated functions f and g, find the functions $f+g$, $f-g$, fg, and f/g, and find their domains.

27. $f(x) = \sqrt{2 - x}$; $g(x) = \sqrt{x + 3}$
28. $f(x) = \sqrt{x + 4}$; $g(x) = \sqrt{3 - x}$
29. $f(x) = \sqrt{x} + 2$; $g(x) = \sqrt{x} - 4$
30. $f(x) = 1 - \sqrt{x}$; $g(x) = 2 - \sqrt{x}$
31. $f(x) = \sqrt{x^2 + x - 6}$; $g(x) = \sqrt{7 + 6x - x^2}$
32. $f(x) = \sqrt{8 + 2x - x^2}$; $g(x) = \sqrt{x^2 - 7x + 10}$

In Problems 33–38, for the indicated functions f and g, find the functions $f \circ g$ and $g \circ f$, and find their domains.

33. $f(x) = \sqrt{x}$; $g(x) = x - 4$
34. $f(x) = \sqrt{x}$; $g(x) = 2x + 5$
35. $f(x) = x + 2$; $g(x) = \dfrac{1}{x}$
36. $f(x) = x - 3$; $g(x) = \dfrac{1}{x^2}$
37. $f(x) = |x|$; $g(x) = \dfrac{1}{x - 1}$
38. $f(x) = |x - 1|$; $g(x) = \dfrac{1}{x}$

Use the graphs of functions f and g shown below to match each function in Problems 39–44 with one of graphs (a)–(f).

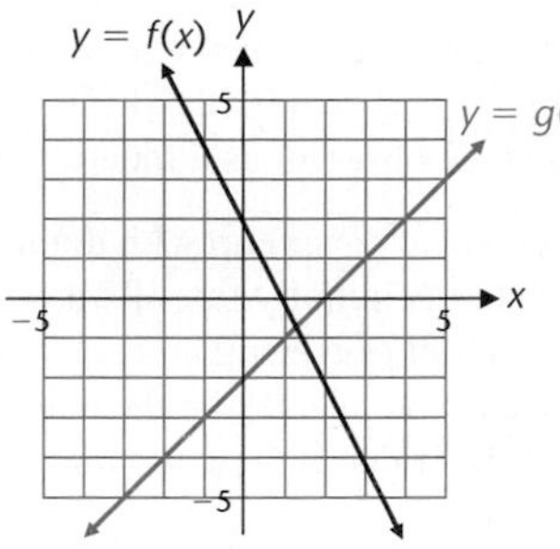

39. $(f + g)(x)$ **40.** $(f - g)(x)$ **41.** $(g - f)(x)$

42. $(fg)(x)$ **43.** $\left(\frac{f}{g}\right)(x)$ **44.** $\left(\frac{g}{f}\right)(x)$

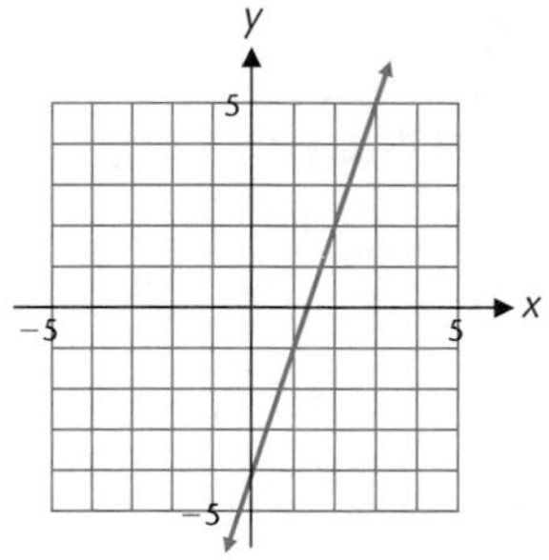

(a)

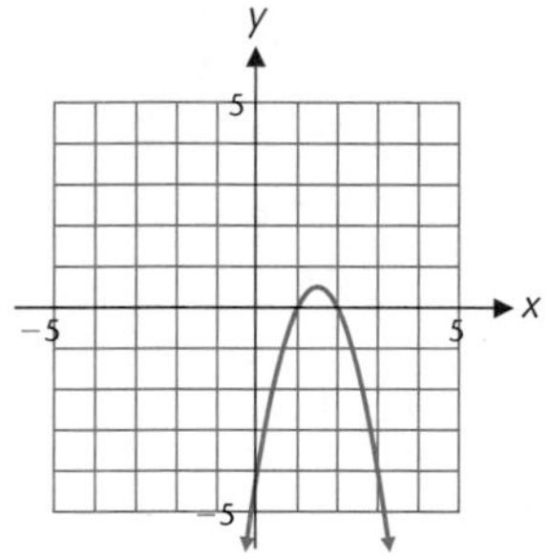

(b)

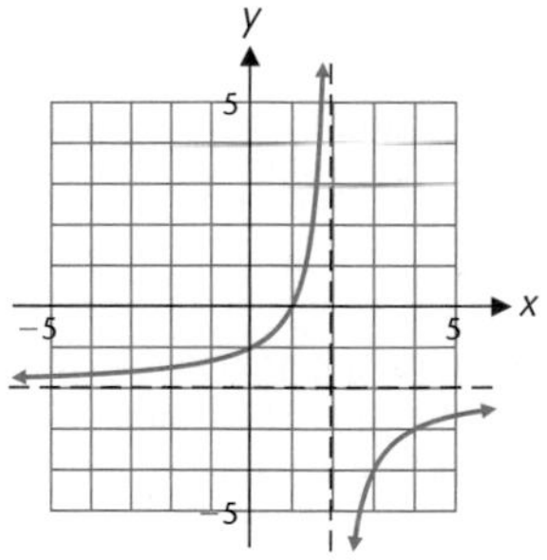

(c)

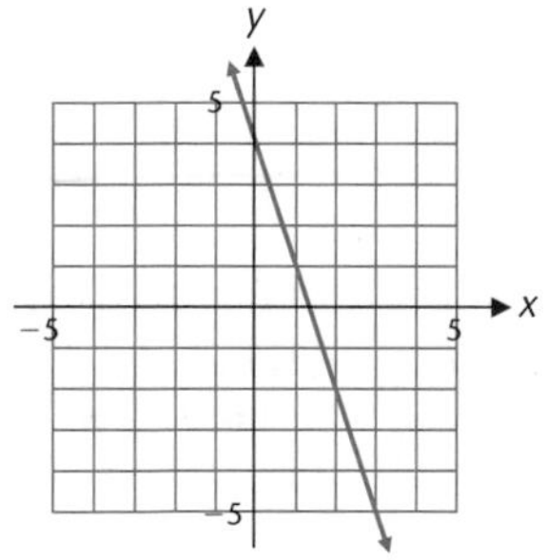

(d)

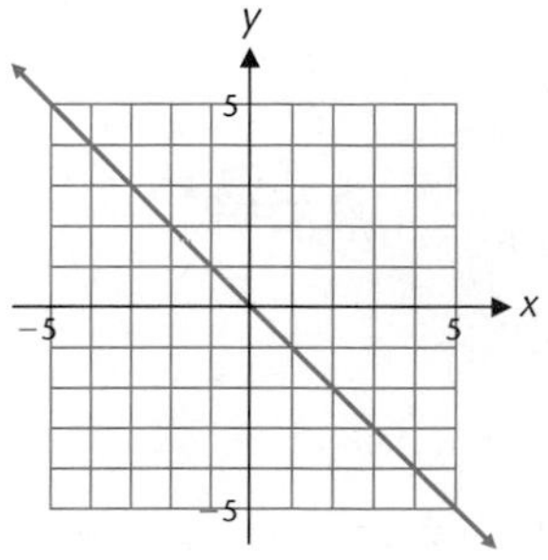

(e)

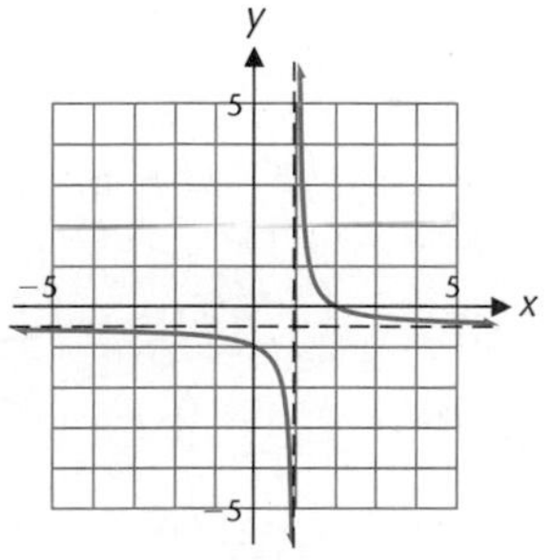

(f)

C|[*In Problems 45–52, express h as a composition of two simpler functions f and g of the form $f(x) = x^n$ and $g(x) = ax + b$, where n is a rational number and a and b are integers.*

45. $h(x) = (2x - 7)^4$ **46.** $h(x) = (3 - 5x)^7$

47. $h(x) = \sqrt{4 + 2x}$ **48.** $h(x) = \sqrt{3x - 11}$

49. $h(x) = 3x^7 - 5$ **50.** $h(x) = 5x^6 + 3$

51. $h(x) = \frac{4}{\sqrt{x}} + 3$ **52.** $h(x) = -\frac{2}{\sqrt{x}} + 1$

53. Are the functions fg and gf identical? Justify your answer.

54. Are the functions $f \circ g$ and $g \circ f$ identical? Justify your answer.

55. Is there a function g that satisfies $f \circ g = g \circ f = f$ for all functions f? If so, what is it?

56. Is there a function g that satisfies $fg = gf = f$ for all functions f? If so, what is it?

In Problems 57–60, for the indicated functions f and g, find the functions $f + g$, $f - g$, fg, and f/g, and find their domains.

57. $f(x) = x + \frac{1}{x}$; $g(x) = x - \frac{1}{x}$

58. $f(x) = x - 1$; $g(x) = x - \frac{6}{x - 1}$

59. $f(x) = 1 - \frac{x}{|x|}$; $g(x) = 1 + \frac{x}{|x|}$

60. $f(x) = x + |x|$; $g(x) = x - |x|$

In Problems 61–66, for the indicated functions f and g, find the functions $f \circ g$ and $g \circ f$, and find their domains.

61. $f(x) = \sqrt{4 - x}$; $g(x) = x^2$

62. $f(x) = \sqrt{x - 1}$; $g(x) = x^2$

63. $f(x) = \frac{x + 5}{x}$; $g(x) = \frac{x}{x - 2}$

64. $f(x) = \frac{x}{x - 1}$; $g(x) = \frac{2x - 4}{x}$

65. $f(x) = \sqrt{25 - x^2}$; $g(x) = \sqrt{9 + x^2}$

66. $f(x) = \sqrt{x^2 - 9}$; $g(x) = \sqrt{x^2 + 25}$

In Problems 67–72, enter the given expression for $(f \circ g)(x)$ exactly as it is written and graph on a graphing utility for $-10 \le x \le 10$. Then simplify the expression, enter the result, and graph in a new viewing window, again for $-10 \le x \le 10$. Find the domain of $f \circ g$. Which is the correct graph of $f \circ g$?

67. $f(x) = \sqrt{5 - x^2}$; $g(x) = \sqrt{3 - x}$;
$(f \circ g)(x) = \sqrt{5 - (\sqrt{3 - x})^2}$

68. $f(x) = \sqrt{6 - x^2}$; $g(x) = \sqrt{x - 1}$;
$(f \circ g)(x) = \sqrt{6 - (\sqrt{x - 1})^2}$

69. $f(x) = \sqrt{x^2 + 5};\quad g(x) = \sqrt{x^2 - 4};$

$(f \circ g)(x) = \sqrt{(\sqrt{x^2 - 4})^2 + 5}$

70. $f(x) = \sqrt{x^2 + 5};\quad g(x) = \sqrt{4 - x^2};$

$(f \circ g) = \sqrt{(\sqrt{4 - x^2})^2 + 5}$

71. $f(x) = \sqrt{x^2 + 7};\quad g(x) = \sqrt{9 - x^2};$

$(f \circ g)(x) = \sqrt{(\sqrt{9 - x^2})^2 + 7}$

72. $f(x) = \sqrt{x^2 + 7};\quad g(x) = \sqrt{x^2 - 9};$

$(f \circ g)(x) = \sqrt{(\sqrt{x^2 - 9})^2 + 7}$

APPLICATIONS

73. Market Research. The demand x and the price p (in dollars) for a certain product are related by

$$x = f(p) = 4{,}000 - 200p$$

The revenue (in dollars) from the sale of x units is given by

$$R(x) = 20x - \frac{1}{200}x^2$$

and the cost (in dollars) of producing x units is given by

$$C(x) = 10x + 30{,}000$$

Express the profit as a function of the price p.

74. Market Research. The demand x and the price p (in dollars) for a certain product are related by

$$x = f(p) = 5{,}000 - 100p$$

The revenue (in dollars) from the sale of x units and the cost (in dollars) of producing x units are given, respectively, by

$$R(x) = 50x - \frac{1}{100}x^2 \quad \text{and} \quad C(x) = 20x + 40{,}000$$

Express the profit as a function of the price p.

75. Pollution. An oil tanker aground on a reef is leaking oil that forms a circular oil slick about 0.1 foot thick (see the figure). The radius of the slick (in feet) t minutes after the leak first occurred is given by

$$r(t) = 0.4t^{1/3}$$

Express the volume of the oil slick as a function of t.

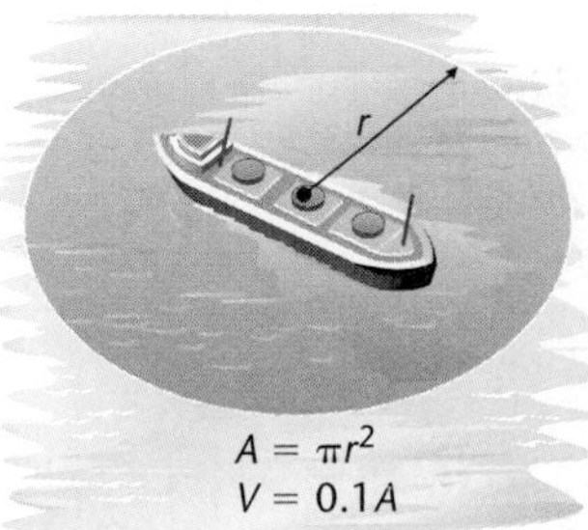

76. Weather Balloon. A weather balloon is rising vertically. An observer is standing on the ground 100 meters from the point where the weather balloon was released.

(A) Express the distance d between the balloon and the observer as a function of the balloon's distance h above the ground.

(B) If the balloon's distance above ground after t seconds is given by $h = 5t$, express the distance d between the balloon and the observer as a function of t.

★ **77. Fluid Flow.** A conical paper cup with diameter 4 inches and height 4 inches is initially full of water. A small hole is made in the bottom of the cup and the water begins to flow out of the cup. Let h and r be the height and radius, respectively, of the water in the cup t minutes after the water begins to flow.

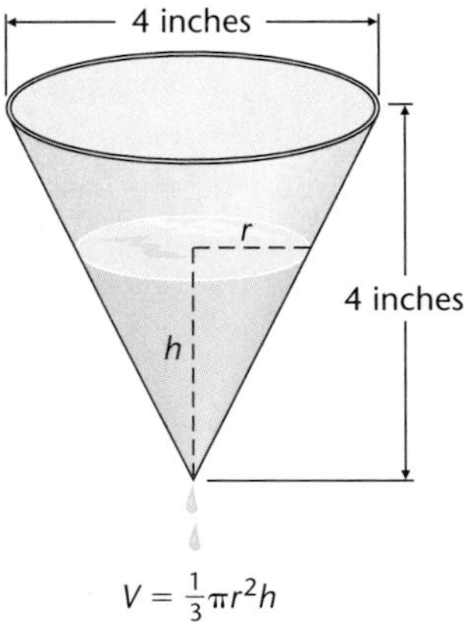

(A) Express r as a function of h.

(B) Express the volume V as a function of h.

(C) If the height of the water after t minutes is given by

$$h(t) = 0.5\sqrt{t}$$

express V as a function of t.

★ **78. Evaporation.** A water trough with triangular ends is 6 feet long, 4 feet wide, and 2 feet deep. Initially, the trough is full of water, but due to evaporation, the volume of the water is decreasing. Let h and w be the height and width, respectively, of the water in the tank t hours after it began to evaporate.

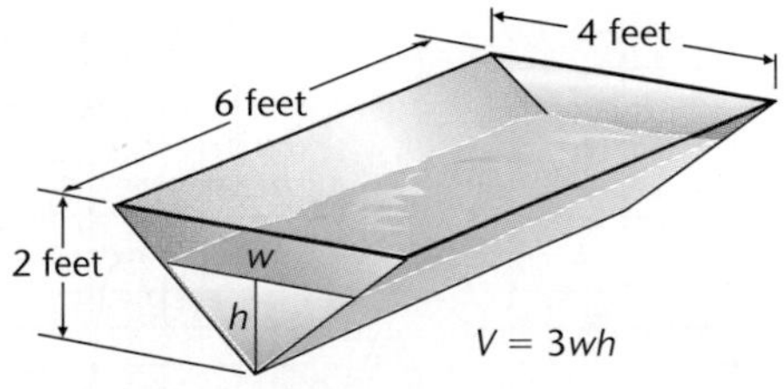

(A) Express w as a function of h.

(B) Express V as a function of h.

(C) If the height of the water after t hours is given by

$$h(t) = 2 - 0.2\sqrt{t}$$

express V as a function of t.

Section 4-2 Inverse Functions

- One-to-One Functions
- Inverse Functions

Many important mathematical relationships can be expressed in terms of functions. For example,

$C = \pi d = f(d)$	The circumference of a circle is a function of the diameter *d*.
$V = s^3 = g(s)$	The volume of a cube is a function of the edge *s*.
$d = 1{,}000 - 100p = h(p)$	The demand for a product is a function of the price *p*.
$F = \frac{9}{5}C + 32$	Temperature measured in °F is a function of temperature in °C.

In many cases, we are interested in *reversing* the correspondence determined by a function. Thus,

$d = \frac{C}{\pi} = m(C)$	The diameter of a circle is a function of the circumference *C*.
$s = \sqrt[3]{V} = n(V)$	The edge of a cube is a function of the volume *V*.
$p = 10 - \frac{1}{100}d = r(d)$	The price of a product is a function of the demand *d*.
$C = \frac{5}{9}(F - 32)$	Temperature measured in °C is a function of temperature in °F.

As these examples illustrate, reversing the relationship between two quantities often produces a new function. This new function is called the *inverse* of the original function. Later in this text we will see that many important functions (for example, logarithmic functions) are actually defined as the inverses of other functions.

In this section, we develop techniques for determining whether the inverse function exists, some general properties of inverse functions, and methods for finding the rule of correspondence that defines the inverse function. A review of Section 1-3 will prove very helpful at this point.

One-to-One Functions

Recall the set form of the definition of a function:

A function is a set of ordered pairs with the property that no two ordered pairs have the same first component and different second components.

However, it is possible that two ordered pairs in a function could have the same second component and different first components. If this does not happen, then we call the function a *one-to-one function*. It turns out that one-to-one functions are the only functions that have inverse functions.

DEFINITION 1

ONE-TO-ONE FUNCTION

A function is one-to-one if no two ordered pairs in the function have the same second component and different first components.

To illustrate this concept, consider the following three sets of ordered pairs:

$$f = \{(0, 3), (0, 5)\ (4, 7)\}$$

$$g = \{(0, 3), (2, 3), (4, 7)\}$$

$$h = \{(0, 3), (2, 5), (4, 7)\}$$

Set f is not a function because the ordered pairs (0, 3) and (0, 5) have the same first component and different second components. Set g is a function, but it is not a one-to-one function because the ordered pairs (0, 3) and (2, 3) have the same second component and different first components. But set h is a function, and it is one-to-one. Representing these three sets of ordered pairs as rules of correspondence provides some additional insight into this concept.

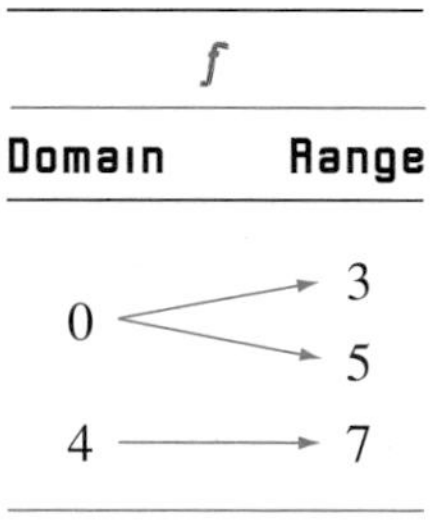

f is not a function.

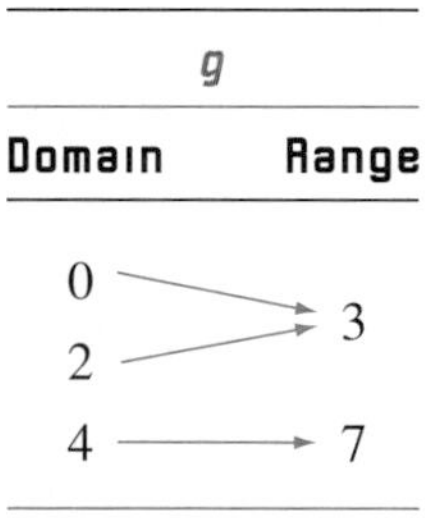

g is a function but is not one-to-one.

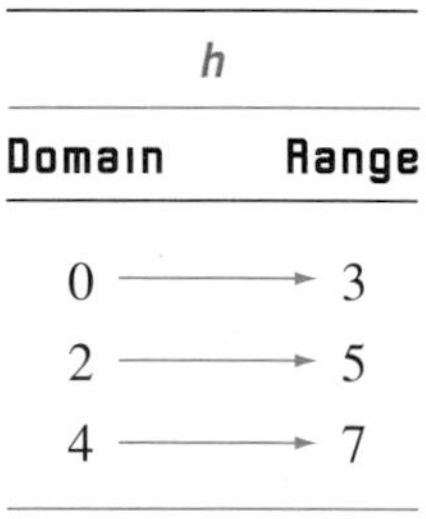

h is a one-to-one function.

EXAMPLE 1

Determining Whether a Function Is One-to-One

Determine whether f is a one-to-one function for

(A) $f(x) = x^2$ (B) $f(x) = 2x - 1$

Solutions

(A) To show that a function is not one-to-one, all we have to do is find two different ordered pairs in the function with the same second component and different first components. Since

$$f(2) = 2^2 = 4 \quad \text{and} \quad f(-2) = (-2)^2 = 4$$

the ordered pairs (2, 4) and (−2, 4) both belong to f and f is not one-to-one.

(B) To show that a function is one-to-one, we have to show that no two ordered pairs have the same second component and different first components. To do this, we assume there are two ordered pairs $(a, f(a))$ and $(b, f(b))$ in f with the same second components and then show that the first components must also be the same. That is, we show that $f(a) = f(b)$ implies $a = b$. We proceed as follows:

$f(a) = f(b)$	Assume second components are equal.
$2a - 1 = 2b - 1$	Evaluate $f(a)$ and $f(b)$.
$2a = 2b$	Simplify.
$a = b$	Conclusion: f is one-to-one.

Thus, by Definition 1, f is a one-to-one function.

MATCHED PROBLEM 1

Determine whether f is a one-to-one function for

(A) $f(x) = 4 - x^2$ (B) $f(x) = 4 - 2x$

The methods used in the solution of Example 1 can be stated as a theorem.

THEOREM 1

ONE-TO-ONE FUNCTIONS

1. If $f(a) = f(b)$ for at least one pair of domain values a and b, $a \neq b$, then f is not one-to-one.
2. If the assumption $f(a) = f(b)$ always implies that the domain values a and b are equal, then f is one-to-one.

Applying Theorem 1 is not always easy—try testing $f(x) = x^3 + 2x + 3$, for example. However, if we are given the graph of a function, then there is a simple graphic procedure for determining if the function is one-to-one. If a horizontal line intersects the graph of a function in more than one point, then the function is not one-to-one, as shown in Figure 1(a). However, if each horizontal line intersects the graph in one point, or not at all, then the function is one-to-one, as shown in Figure 1(b). These observations form the basis for the *horizontal line test.*

FIGURE 1 Intersections of graphs and horizontal lines.

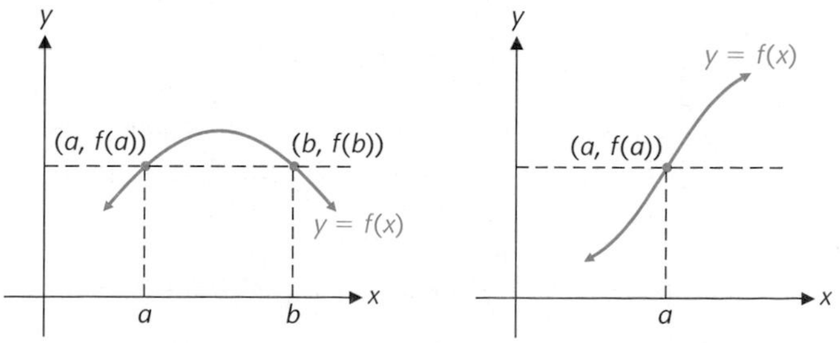

$f(a) = f(b)$ for $a \neq b$
f is not one-to-one
(a)

Only one point has ordinate $f(a)$; f is one-to-one
(b)

A is true B is true an also true reversed

THEOREM 2

HORIZONTAL LINE TEST

A function is one-to-one if and only if each horizontal line intersects the graph of the function in at most one point.

The graphs of the function considered in Example 1 are shown in Figure 2. Applying the horizontal line test to each graph confirms the results we obtained in Example 1.

FIGURE 2
Applying the horizontal line test.

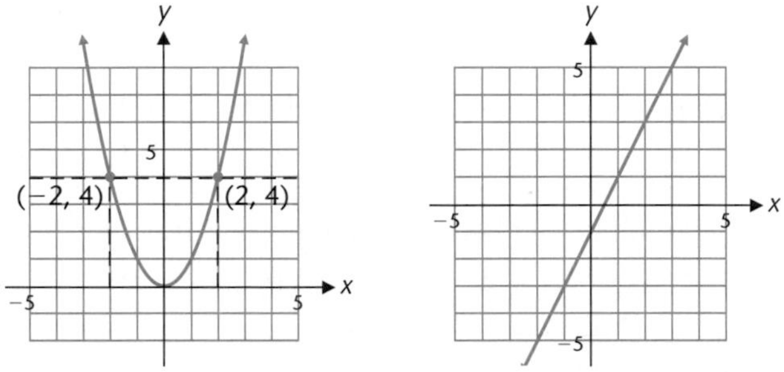

$f(x) = x^2$ does not pass the horizontal line test; f is not one-to-one
(a)

$f(x) = 2x - 1$ passes the horizontal line test; f is one-to-one
(b)

A function that is increasing throughout its domain or decreasing throughout its domain will always pass the horizontal line test [see Figs. 3(a) and 3(b)]. Thus, we have the following theorem.

THEOREM 3

INCREASING AND DECREASING FUNCTIONS

If a function f is increasing throughout its domain or decreasing throughout its domain, then f is a one-to-one function.

FIGURE 3
Increasing, decreasing, and one-to-one functions.

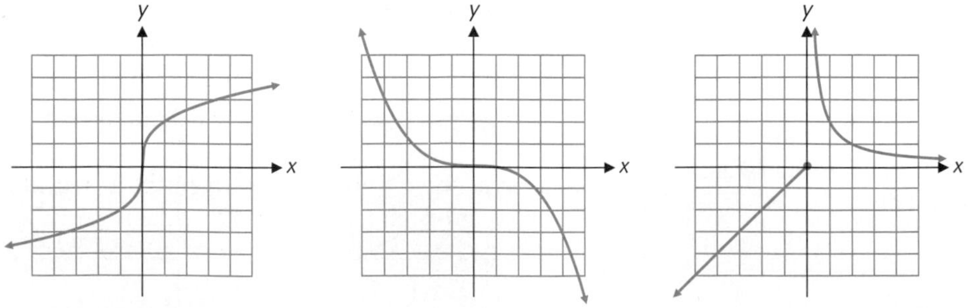

An increasing function is always one-to-one
(a)

A decreasing function is always one-to-one
(b)

A one-to-one function is not always increasing or decreasing
(c)

The converse of Theorem 3 is false. To see this, consider the function graphed in Figure 3(c). This function is increasing on $(-\infty, 0]$ and decreasing on $(0, \infty)$, yet the graph passes the horizontal line test. Thus, this is a one-to-one function that is neither an increasing function nor a decreasing function.

Inverse Functions

Now we want to see how we can form a new function by reversing the correspondence determined by a given function. Let g be the function defined as follows:

$$g = \{(-3, 9), (0, 0), (3, 9)\} \qquad g \text{ is not one-to-one.}$$

Notice that g is not one-to-one because the domain elements -3 and 3 both correspond to the range element 9. We can reverse the correspondence determined by function g simply by reversing the components in each ordered pair in g, producing the following set:

$$G = \{(9, -3), (0, 0), (9, 3)\} \qquad G \text{ is not a function.}$$

But the result is not a function because the domain element 9 corresponds to two different range elements, -3 and 3. On the other hand, if we reverse the ordered pairs in the function

$$f = \{(1, 2), (2, 4), (3, 9)\} \qquad f \text{ is one-to-one.}$$

we obtain

$$F = \{(2, 1), (4, 2), (9, 3)\} \qquad F \text{ is a function.}$$

This time f is a one-to-one function, and the set F turns out to be a function also. This new function F, formed by reversing all the ordered pairs in f, is called the *inverse* of f and is usually denoted* by f^{-1}. Thus,

$$f^{-1} = \{(2, 1) (4, 2), (9, 3)\} \qquad \text{The inverse of } f$$

Notice that f^{-1} is also a one-to-one function and that the following relationships hold:

$$\text{Domain of } f^{-1} = \{2, 4, 9\} = \text{Range of } f$$
$$\text{Range of } f^{-1} = \{1, 2, 3\} = \text{Domain of } f$$

Thus, reversing all the ordered pairs in a one-to-one function forms a new one-to-one function and reverses the domain and range in the process. We are now ready to present a formal definition of the inverse of a function.

---

*f^{-1}, read "f inverse," is a special symbol used here to represent the inverse of the function f. It does *not* mean $1/f$.

DEFINITION 2

INVERSE OF A FUNCTION

If f is a one-to-one function, then the **inverse** of f, denoted f^{-1}, is the function formed by reversing all the ordered pairs in f. Thus,

$$f^{-1} = \{(y, x) \mid (x, y) \text{ is in } f\}$$

If f is not one-to-one, then f **does not have an inverse** and f^{-1} **does not exist.**

The following properties of inverse functions follow directly from the definition.

THEOREM 4

PROPERTIES OF INVERSE FUNCTIONS

If f^{-1} exists, then

1. f^{-1} is a one-to-one function.
2. Domain of f^{-1} = Range of f
3. Range of f^{-1} = Domain of f

Explore/Discuss 1

Most graphing utilities have a routine, usually denoted by Draw Inverse (or an abbreviation of this phrase—consult your manual), that will draw the graph formed by reversing the ordered pairs of all the points on the graph of a function. For example, Figure 4(a) shows the graph of $f(x) = 2x - 1$ along with the graph obtained by using the Draw Inverse routine. Figure 4(b) does the same for $f(x) = x^2$.

(A) Is the graph produced by Draw Inverse in Figure 4(a) the graph of a function? Does f^{-1} exist? Explain.

(B) Is the graph produced by Draw Inverse in Figure 4(b) the graph of a function? Does f^{-1} exist? Explain.

(C) If your graphing utility has a Draw Inverse routine, apply it to the graphs of $y = \sqrt{x - 1}$ and $y = 4x - x^2$ to determine if the result is the graph of a function and if the inverse of the original function exists.

FIGURE 4

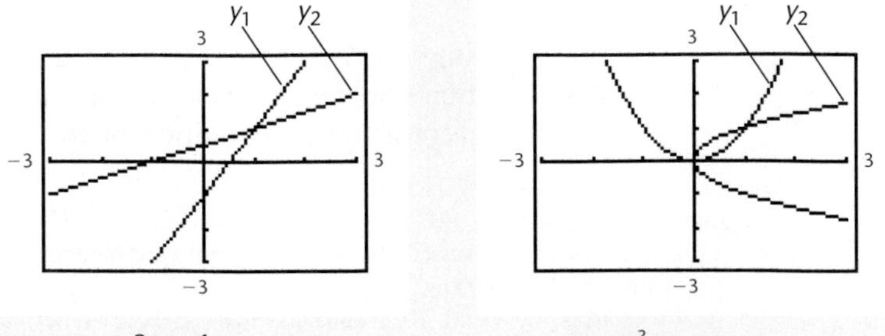

$y_1 = 2x - 1$
y_2 = Draw Inverse y_1
(a)

$y_1 = x^2$
y_2 = Draw Inverse y_1
(b)

Finding the inverse of a function defined by a finite set of ordered pairs is easy; just reverse each ordered pair. But how do we find the inverse of a function defined by an equation? Consider the one-to-one function f defined by

$$f(x) = 2x - 1$$

To find f^{-1}, we let $y = f(x)$ and solve for x:

$$y = 2x - 1$$
$$y + 1 = 2x$$
$$\tfrac{1}{2}y + \tfrac{1}{2} = x$$

Since the ordered pair (x, y) is in f if and only if the reversed ordered pair (y, x) is in f^{-1}, this last equation defines f^{-1}:

$$x = f^{-1}(y) = \tfrac{1}{2}y + \tfrac{1}{2} \qquad (1)$$

Something interesting happens if we form the composition* of f and f^{-1} in either of the two possible orders.

$$f^{-1}(f(x)) = f^{-1}(2x - 1) = \tfrac{1}{2}(2x - 1) + \tfrac{1}{2} = x - \tfrac{1}{2} + \tfrac{1}{2} = x$$

and

$$f(f^{-1}(y)) = f(\tfrac{1}{2}y + \tfrac{1}{2}) = 2(\tfrac{1}{2}y + \tfrac{1}{2}) - 1 = y + 1 - 1 = y$$

These compositions indicate that if f maps x into y, then f^{-1} maps y back into x and if f^{-1} maps y into x, then f maps x back into y. This is interpreted schematically in Figure 5.

FIGURE 5
Composition of f and f^{-1}.

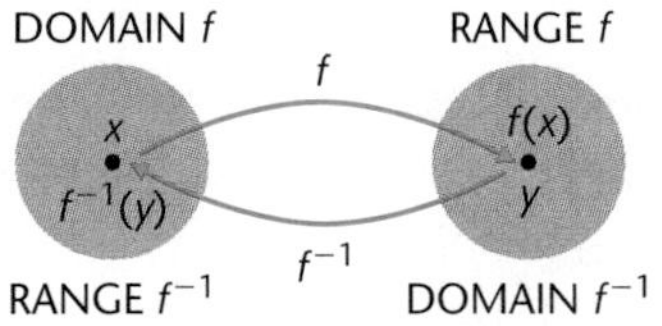

Finally, we note that we usually use x to represent the independent variable and y the dependent variable in an equation that defines a function. It is customary to do this for inverse functions also. Thus, interchanging the variables x and y in equation (1), we can state that the inverse of

$$y = f(x) = 2x - 1$$

is

$$y = f^{-1}(x) = \tfrac{1}{2}x + \tfrac{1}{2}$$

*When working with inverse functions, it is customary to write compositions as $f(g(x))$ rather than as $(f \circ g)(x)$.

In general, we have the following result:

THEOREM 5

RELATIONSHIP BETWEEN f AND f^{-1}

If f^{-1} exists, then

1. $x = f^{-1}(y)$ if and only if $y = f(x)$.
2. $f^{-1}(f(x)) = x$ for all x in the domain of f.
3. $f(f^{-1}(y)) = y$ for all y in the domain of f^{-1} or, if x and y have been interchanged, $f(f^{-1}(x)) = x$ for all x in the domain of f^{-1}.

If f and g are one-to-one functions satisfying

$$f(g(x)) = x \qquad \text{for all } x \text{ in the domain of } g$$

$$g(f(x)) = x \qquad \text{for all } x \text{ in the domain of } f$$

then it can be shown that $g = f^{-1}$ and $f = g^{-1}$. Thus, the inverse function is the only function that satisfies both these compositions. We can use this fact to check that we have found the inverse correctly.

Explore/Discuss 2

Find $f(g(x))$ and $g(f(x))$ for

$$f(x) = (x - 1)^3 + 2 \qquad \text{and} \qquad g(x) = (x - 2)^{1/3} + 1$$

How are f and g related?

The procedure for finding the inverse of a function defined by an equation is given in the next box. This procedure can be applied whenever it is possible to solve $y = f(x)$ for x in terms of y.

FINDING THE INVERSE OF A FUNCTION f

Step 1. Find the domain of f and verify that f is one-to-one. If f is not one-to-one, then stop, since f^{-1} does not exist.

Step 2. Solve the equation $y = f(x)$ for x. The result is an equation of the form $x = f^{-1}(y)$.

Step 3. Interchange x and y in the equation found in step 2. This expresses f^{-1} as a function of x.

Step 4. Find the domain of f^{-1}. Remember, the domain of f^{-1} must be the same as the range of f.

Check your work by verifying that

$$f^{-1}(f(x)) = x \qquad \text{for all } x \text{ in the domain of } f$$

and

$$f(f^{-1}(x)) = x \qquad \text{for all } x \text{ in the domain of } f^{-1}$$

EXAMPLE 2

Finding the Inverse of a Function

Find f^{-1} for $f(x) = \sqrt{x - 1}$

Solution

Step 1. Find the domain of f and verify that f is one-to-one. The domain of f is $[1, \infty)$. The graph of f in Figure 6 shows that f is one-to-one, hence f^{-1} exists.

FIGURE 6

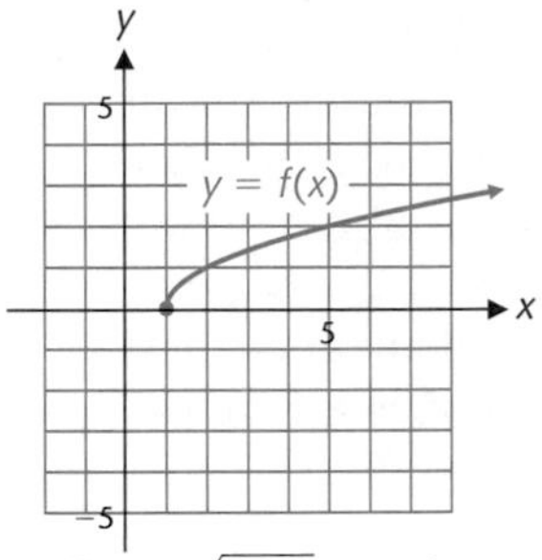

$f(x) = \sqrt{x - 1}$, $x \geq 1$

Step 2. Solve the equation $y = f(x)$ for x.

$$y = \sqrt{x - 1}$$
$$y^2 = x - 1$$
$$x = y^2 + 1$$

Thus,

$$x = f^{-1}(y) = y^2 + 1$$

Step 3. Interchange x and y.

$$y = f^{-1}(x) = x^2 + 1$$

Step 4. Find the domain of f^{-1}. The equation $f^{-1}(x) = x^2 + 1$ is defined for all values of x, but this does not tell us what the domain of f^{-1} is. Remember, the domain of f^{-1} must equal the range of f. From the graph of f, we see that the range of f is $[0, \infty)$. Thus, the domain of f^{-1} is also $[0, \infty)$. That is,

$$f^{-1}(x) = x^2 + 1 \qquad x \geq 0$$

Check

For x in $[1, \infty)$, the domain of f, we have

$$\begin{aligned} f^{-1}(f(x)) &= f^{-1}(\sqrt{x - 1}) \\ &= (\sqrt{x - 1})^2 + 1 \\ &= x - 1 + 1 \\ &\stackrel{\checkmark}{=} x \end{aligned}$$

For x in $[0, \infty)$, the domain of f^{-1}, we have

$$\begin{aligned} f(f^{-1}(x)) &= f(x^2 + 1) \\ &= \sqrt{(x^2 + 1) - 1} \\ &= \sqrt{x^2} \\ &= |x| && \sqrt{x^2} = |x| \text{ for any real number } x. \\ &\stackrel{\checkmark}{=} x && |x| = x \text{ for } x \geq 0. \end{aligned}$$

MATCHED PROBLEM 2

Find f^{-1} for $f(x) = \sqrt{x + 2}$.

Explore/Discuss 3

Most basic arithmetic operations can be reversed by performing a second operation: subtraction reverses addition, division reverses multiplication, squaring reverses taking the square root, and so on. Viewing a function as a sequence of reversible operations gives additional insight into the inverse function concept. For example, the function $f(x) = 2x - 1$ can be described verbally as a function that multiplies each domain element by 2 and then subtracts 1. Reversing this sequence describes a function g that adds 1 to each domain element and then divides by 2, or $g(x) = (x + 1)/2$, which is the inverse of the function f. For each of the following functions, write a verbal description of the function, reverse your description, and write the resulting algebraic equation. Verify that the result is the inverse of the original function.

(A) $f(x) = 3x + 5$ (B) $f(x) = \sqrt{x - 1}$ (C) $f(x) = \dfrac{1}{x + 1}$

There is an important relationship between the graph of any function and its inverse that is based on the following observation: In a rectangular coordinate system, the points (a, b) and (b, a) are symmetric with respect to the line $y = x$ [see Fig. 7(a)]. Theorem 6 is an immediate consequence of this observation.

FIGURE 7 Symmetry with respect to the line $y = x$.

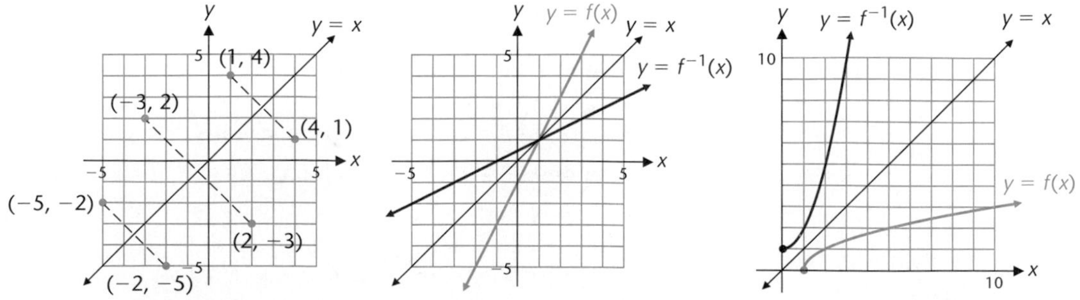

(a, b) and (b, a) are symmetric with respect to the line $y = x$
(a)

$f(x) = 2x - 1$
$f^{-1}(x) = \frac{1}{2}x + \frac{1}{2}$
(b)

$f(x) = \sqrt{x - 1}$
$f^{-1}(x) = x^2 + 1, x \geq 0$
(c)

THEOREM 6

SYMMETRY PROPERTY FOR THE GRAPHS OF f AND f^{-1}

The graphs of $y = f(x)$ and $y = f^{-1}(x)$ are symmetric with respect to the line $y = x$.

Knowledge of this symmetry property makes it easy to graph f^{-1} if the graph of f is known, and vice versa. Figures 7(b) and 7(c) illustrate this property for the two inverse functions we found earlier in this section.

If a function is not one-to-one, we usually can restrict the domain of the function to produce a new function that is one-to-one. Then we can find an inverse for the restricted function. Suppose we start with $f(x) = x^2 - 4$. Since f is not one-to-one, f^{-1} does not exist [Fig. 8(a)]. But there are many ways the domain of f can be restricted to obtain a one-to-one function. Figures 8(b) and 8(c) illustrate two such restrictions.

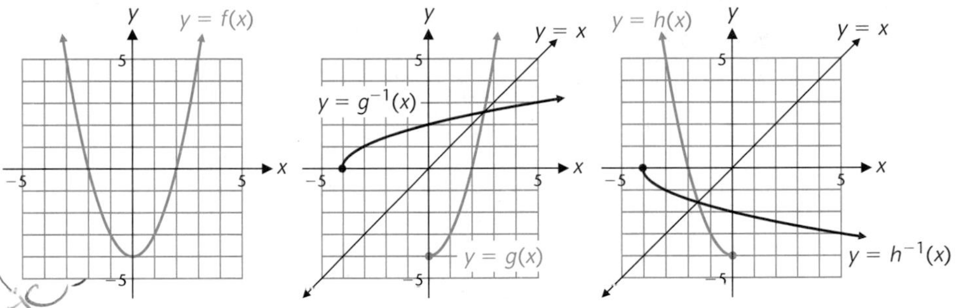

$f(x) = x^2 - 4$
f^{-1} does not exist
(a)

$g(x) = x^2 - 4,\ x \geq 0$
$g^{-1}(x) = \sqrt{x + 4},\ x \geq -4$
(b)

$h(x) = x^2 - 4,\ x \leq 0$
$h^{-1}(x) = -\sqrt{x + 4},\ x \geq -4$
(c)

FIGURE 8 Restricting the domain of a function.

Explore/Discuss 4

To graph the function

$$g(x) = 4x - x^2, \qquad x \geq 0$$

on a graphing utility, enter

$$y_1 = (4x - x^2)/(x \geq 0)$$

(A) The Boolean expression $(x \geq 0)$ is assigned the value 1 if the inequality is true and 0 if it is false. How does this result in restricting the graph of $4x - x^2$ to just those values of x satisfying $x \geq 0$?

(B) Use this concept to reproduce Figures 8(b) and 8(c) on a graphing utility.

(C) Do your graphs appear to be symmetric with respect to the line $y = x$? What happens if you use a squared window for your graph?

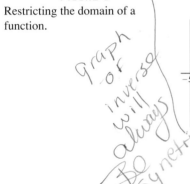

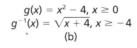

Recall from Theorem 3 that increasing and decreasing functions are always one-to-one. This provides the basis for a convenient and popular method of restricting the domain of a function:

If the domain of a function *f* is restricted to an interval on the *x* axis over which *f* is increasing (or decreasing), then the new function determined by this restriction is one-to-one and has an inverse.

We used this method to form the functions g and h in Figure 8.

EXAMPLE **3**

Finding the Inverse of a Function

Find the inverse of $f(x) = 4x - x^2$, $x \le 2$. Graph f, f^{-1}, and $y = x$ in a squared viewing window on a graphing utility and then sketch the graph by hand, adding appropriate labels.

Solution

Step 1. Find the domain of f and verify that f is one-to-one. We are given that the domain of f is $(-\infty, 2]$. Using the Boolean expression $x \le 2$ [Fig. 9(a)] to restrict the graph to this domain produces the graph of f in Figure 9(b). This graph shows that f is one-to-one.

FIGURE 9

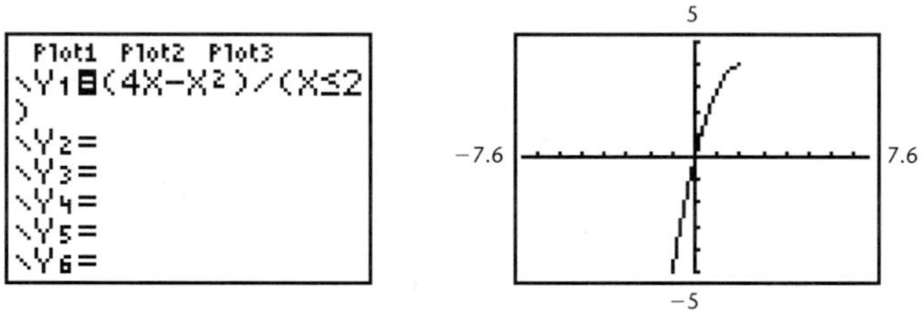

(a) (b)

Step 2. Solve the equation $y = f(x)$ for x.

$$y = 4x - x^2$$

$$x^2 - 4x = -y \qquad \text{Rearrange terms.}$$

$$x^2 - 4x + 4 = -y + 4 \qquad \text{Add 4 to complete the square on the left side.}$$

$$(x - 2)^2 = 4 - y$$

Taking the square root of both sides of this last equation, we obtain two possible solutions:

$$x - 2 = \pm\sqrt{4 - y}$$

The restricted domain of f tells us which solution to use. Since $x \le 2$ implies $x - 2 \le 0$, we must choose the negative square root. Thus,

$$x - 2 = -\sqrt{4 - y}$$

$$x = 2 - \sqrt{4 - y}$$

and we have found

$$x = f^{-1}(y) = 2 - \sqrt{4 - y}$$

Step 3. Interchange x and y.

$$y = f^{-1}(x) = 2 - \sqrt{4 - x}$$

Step 4. Find the domain of f^{-1}. The equation $f^{-1}(x) = 2 - \sqrt{4 - x}$ is defined for $x \leq 4$. From the graph in Figure 9(b), the range of f also is $(-\infty, 4]$. Thus,

$$f^{-1}(x) = 2 - \sqrt{4 - x} \qquad x \leq 4$$

The check is left for the reader.

The graphs of f, f^{-1}, and $y = x$ on a graphing utility are shown in Figure 10 and a hand sketch is shown in Figure 11. Note that we plotted several points on the graph of f and their reflections on the graph of f^{-1} to aid in preparing the hand sketch.

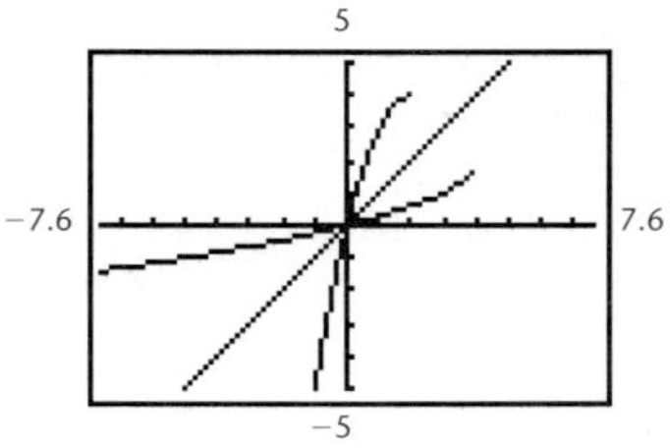

FIGURE 10

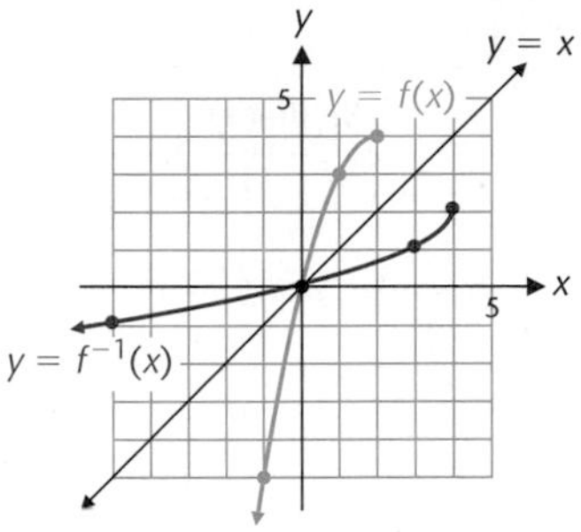

FIGURE 11

MATCHED PROBLEM 3

Find the inverse of $f(x) = 4x - x^2$, $x \geq 2$. Graph f, f^{-1}, and $y = x$ in the same coordinate system.

Answers to Matched Problems

1. (A) Not one-to-one (B) One-to-one **2.** $f^{-1}(x) = x^2 - 2$, $x \geq 0$

3. $f^{-1}(x) = 2 + \sqrt{4 - x}$, $x \leq 4$

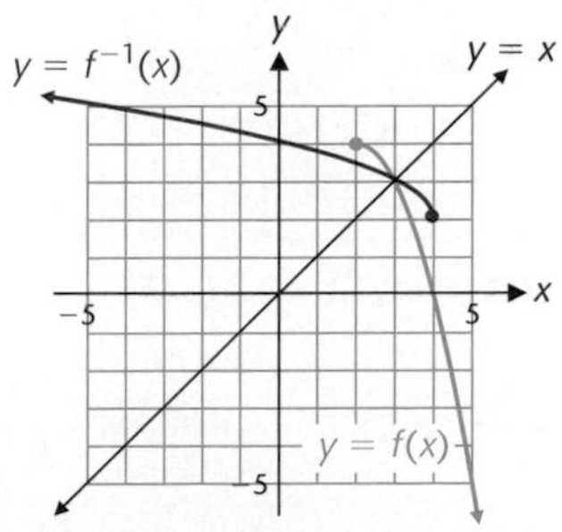

EXERCISE 4-2

A

Which of the functions in Problems 1–16 are one-to-one?

1. {(1, 2), (2, 1), (3, 4), (4, 3)}

2. {(−1, 0), (0, 1), (1, −1), (2, 1)}

3. {(5, 4), (4, 3), (3, 3), (2, 4)}

4. {(5, 4), (4, 3), (3, 2), (2, 1)}

5.

Domain	Range
−2 →	−4
−1 →	−2
0 →	0
1 →	1
2 →	5

6.

Domain	Range
−2 →	−3
−1 →	−3
0 →	7
1 →	9
2 →	9

7.

Domain	Range
1 →	7
2 →	7
3 →	7
4 →	7
5 →	7

8.

Domain	Range
1 →	5
2 →	3
3 →	1
4 →	2
5 →	4

9.

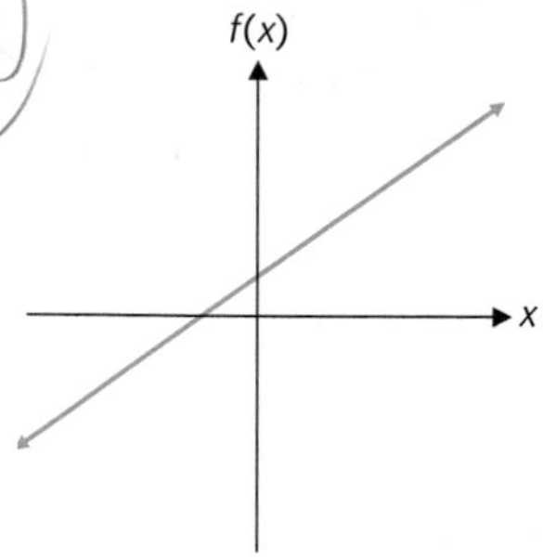

10.

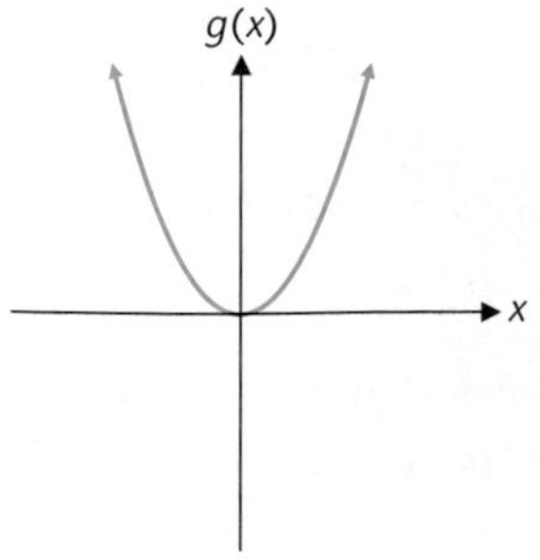

11.

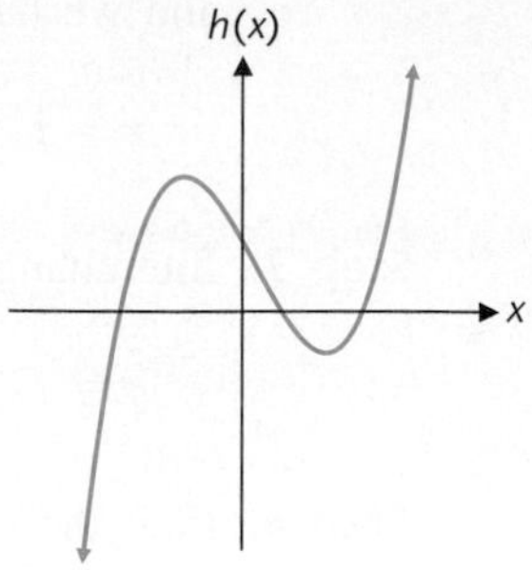

12.

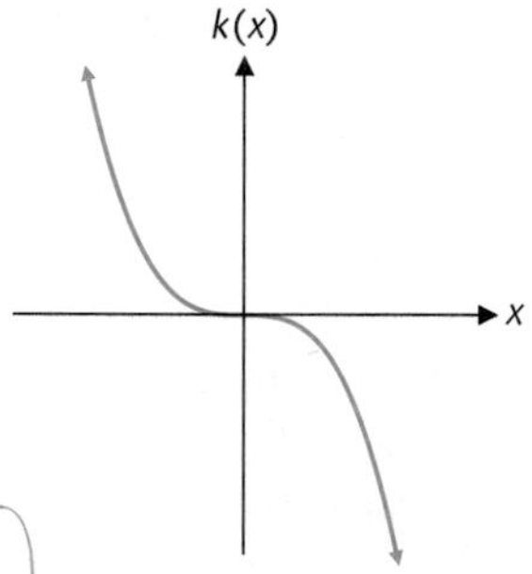

13.

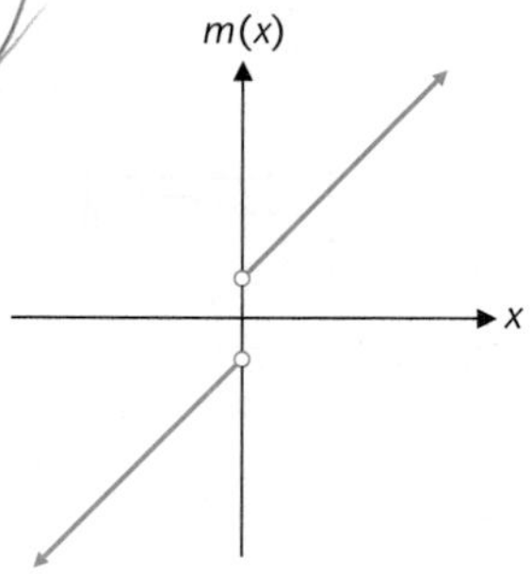

14.

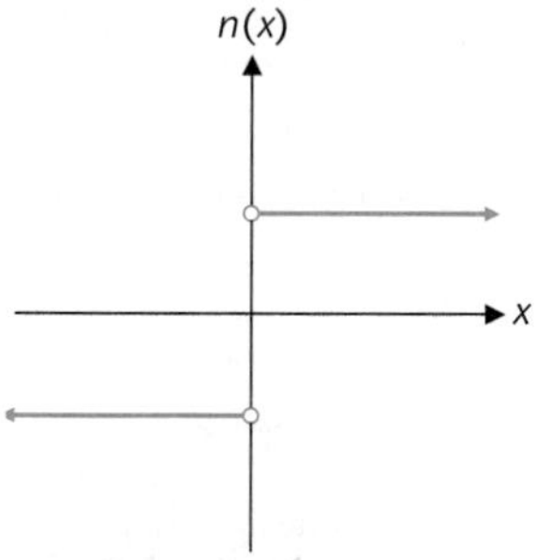

15.

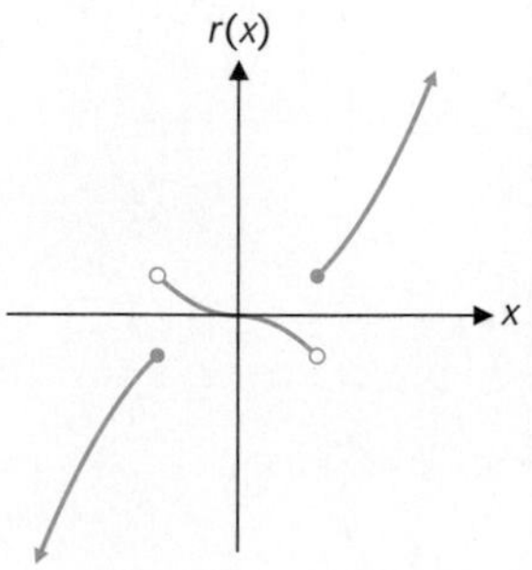

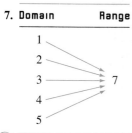

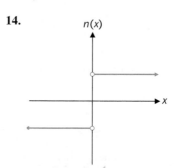

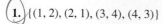

16.

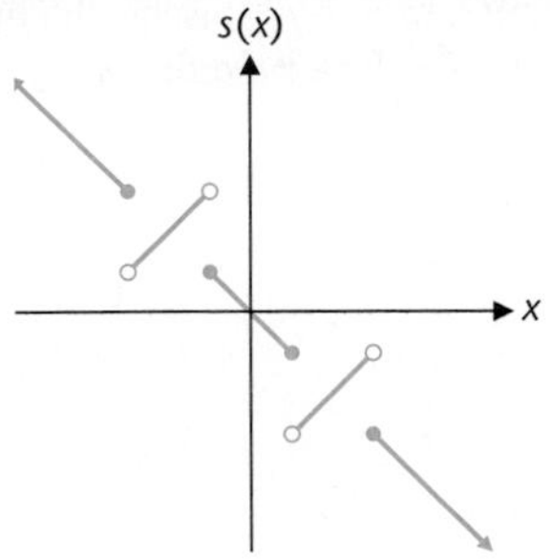

B

In Problems 17–22, use Theorem 1 to determine which functions are one-to-one.

17. $F(x) = \frac{1}{2}x + 2$

18. $G(x) = -\frac{1}{3}x + 1$

19. $H(x) = 4 - x^2$

20. $K(x) = \sqrt{4 - x}$

21. $M(x) = \sqrt{x + 1}$

22. $N(x) = x^2 - 1$

In Problems 23–30, use the horizontal line test (Theorem 2) to determine which functions are one-to-one.

23. $f(x) = \dfrac{x^2 + |x|}{x}$

24. $f(x) = \dfrac{x^2 - |x|}{x}$

25. $f(x) = \dfrac{x^3 + |x|}{x}$

26. $f(x) = \dfrac{|x|^3 + |x|}{x}$

27. $f(x) = \dfrac{x^2 - 4}{|x - 2|}$

28. $f(x) = \dfrac{1 - x^2}{|x + 1|}$

29. $f(x) = \dfrac{x^3 - 9x}{|x^2 - 9|}$

30. $f(x) = \dfrac{4x - x^3}{|x^2 - 4|}$

In Problems 31–34, use the graph of the one-to-one function f to sketch the graph of f^{-1}. State the domain and range of f^{-1}.

31.

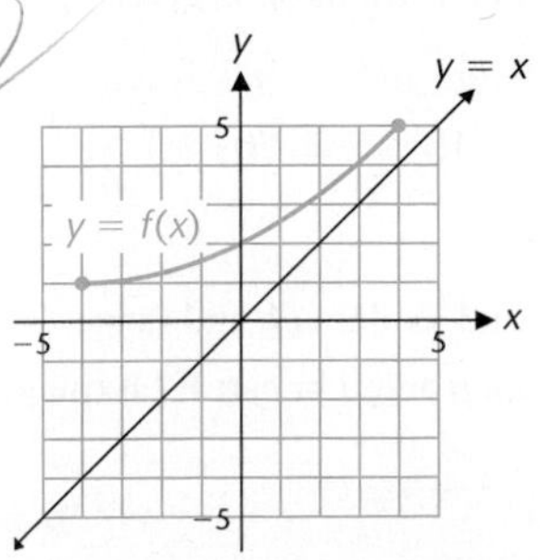

32.

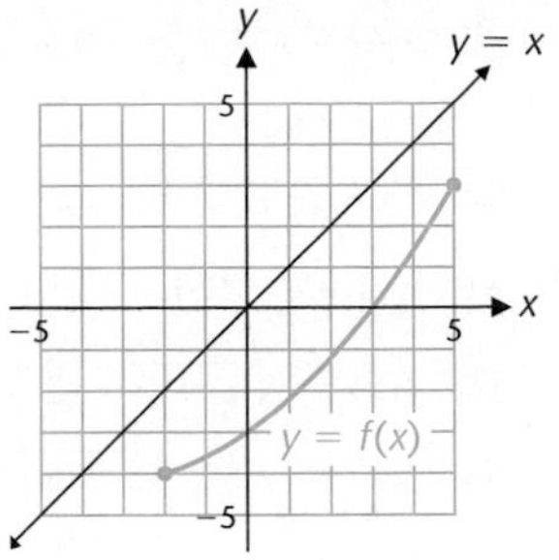

33.

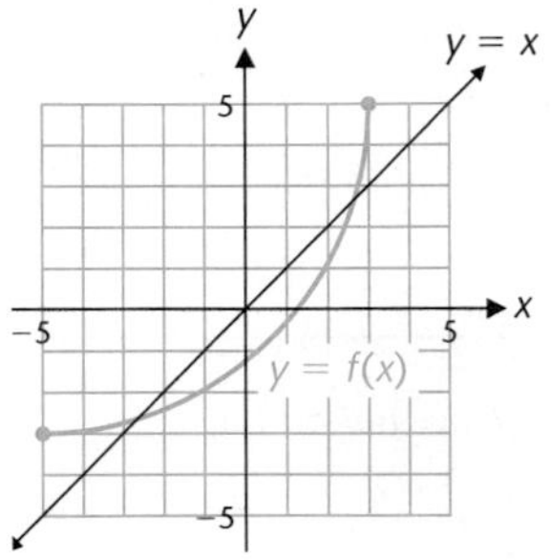

34.

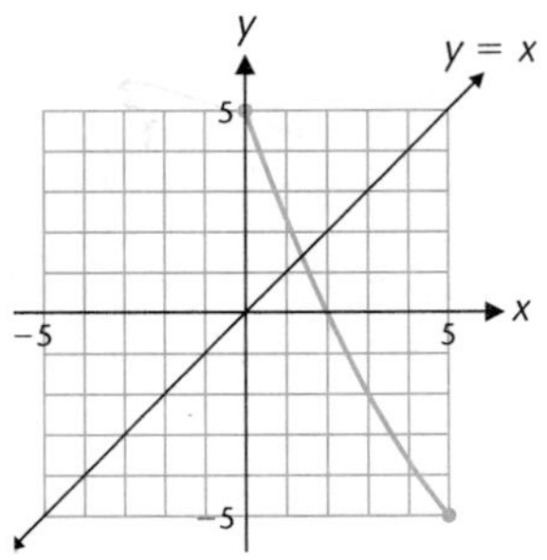

In Problems 35–40, verify that g is the inverse of the one-to-one function f by showing that $g(f(x)) = x$ and $f(g(x)) = x$. Sketch the graphs of f, g, and $y = x$ in the same coordinate system and identify each graph.

35. $f(x) = 3x + 6$; $g(x) = \frac{1}{3}x - 2$

36. $f(x) = -\frac{1}{2}x + 2$; $g(x) = -2x + 4$

37. $f(x) = 4 + x^2, x \geq 0$; $g(x) = \sqrt{x - 4}$

38. $f(x) = \sqrt{x + 2}$; $g(x) = x^2 - 2, x \geq 0$

39. $f(x) = -\sqrt{x - 2}$; $g(x) = x^2 + 2, x \leq 0$

40. $f(x) = 6 - x^2, x \leq 0$; $g(x) = -\sqrt{6 - x}$

The functions in Problems 41–60 are one-to-one. Find f^{-1}.

41. $f(x) = 3x$

42. $f(x) = \frac{1}{2}x$

43. $f(x) = 4x - 3$

44. $f(x) = -\frac{1}{3}x + \frac{5}{3}$

45. $f(x) = \frac{1}{10}x + \frac{3}{5}$

46. $f(x) = -2x - 7$

47. $f(x) = \dfrac{2}{x - 1}$

48. $f(x) = \dfrac{3}{x + 4}$

49. $f(x) = \dfrac{x}{x + 2}$

50. $f(x) = \dfrac{x - 3}{x}$

51. $f(x) = \dfrac{2x + 5}{3x - 4}$

52. $f(x) = \dfrac{5 - 3x}{7 - 4x}$

53. $f(x) = x^3 + 1$

54. $f(x) = x^5 - 2$

55. $f(x) = 4 - \sqrt[5]{x + 2}$

56. $f(x) = \sqrt[3]{x + 3} - 2$

57. $f(x) = \frac{1}{2}\sqrt{16 - x}$

58. $f(x) = -\frac{1}{3}\sqrt{36 - x}$

59. $f(x) = 3 - \sqrt{x - 2}$

60. $f(x) = 4 + \sqrt{5 - x}$

61. How are the x and y intercepts of a function and its inverse related?

62. Does a constant function have an inverse? Explain.

C

The functions in Problems 63–66 are one-to-one. Find f^{-1}.

63. $f(x) = (x - 1)^2 + 2, x \geq 1$

64. $f(x) = 3 - (x - 5)^2, x \leq 5$

65. $f(x) = x^2 + 2x - 2, x \leq -1$

66. $f(x) = x^2 + 8x + 7, x \geq -4$

In Problems 67–74, find f^{-1}, find the domain and range of f^{-1}, sketch the graphs of f, f^{-1}, and $y = x$ in the same coordinate system, and identify each graph.

67. $f(x) = -\sqrt{9 - x^2}, 0 \leq x \leq 3$

68. $f(x) = \sqrt{9 - x^2}, 0 \leq x \leq 3$

69. $f(x) = \sqrt{9 - x^2}, -3 \leq x \leq 0$

70. $f(x) = -\sqrt{9 - x^2}, -3 \leq x \leq 0$

71. $f(x) = 1 + \sqrt{1 - x^2}, 0 \leq x \leq 1$

72. $f(x) = 1 - \sqrt{1 - x^2}, 0 \leq x \leq 1$

73. $f(x) = 1 - \sqrt{1 - x^2}, -1 \leq x \leq 0$

74. $f(x) = 1 + \sqrt{1 - x^2}, -1 \leq x \leq 0$

75. Find $f^{-1}(x)$ for $f(x) = ax + b, a \neq 0$.

76. Find $f^{-1}(x)$ for $f(x) = \sqrt{a^2 - x^2}, a > 0, 0 \leq x \leq a$.

77. Refer to Problem 75. For which a and b is f its own inverse?

78. How could you recognize the graph of a function that is its own inverse?

79. Show that the line through the points (a, b) and (b, a), $a \neq b$, is perpendicular to the line $y = x$ (see the figure at the top of the next column).

80. Show that the point $((a + b)/2, (a + b)/2)$ bisects the line segment from (a, b) to (b, a), $a \neq b$ (see the figure).

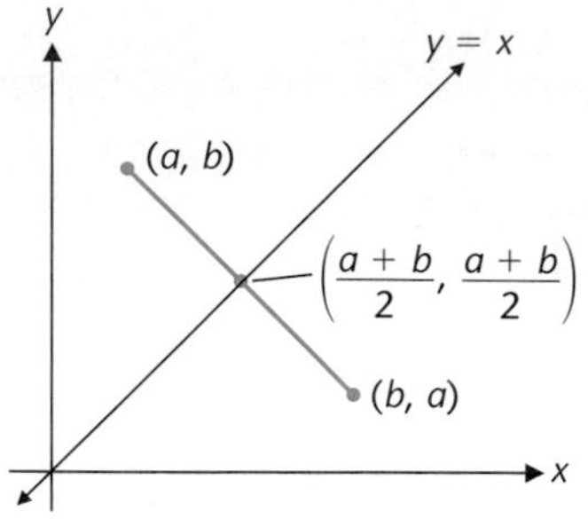

In Problems 81–84, the function f is not one-to-one. Find the inverses of the functions formed by restricting the domain of f as indicated. Check by graphing f, f^{-1}, and the line $y = x$ in a squared viewing window on a graphing utility. [Hint: To restrict the graph of $y = f(x)$ to an interval of the form $a \leq x \leq b$, enter $y = f(x)/((a \leq x)*(x \leq b))$.]

81. $f(x) = (2 - x)^2$:
(A) $x \leq 2$ (B) $x \geq 2$

82. $f(x) = (1 + x)^2$:
(A) $x \leq -1$ (B) $x \geq -1$

83. $f(x) = \sqrt{4x - x^2}$:
(A) $0 \leq x \leq 2$ (B) $2 \leq x \leq 4$

84. $f(x) = \sqrt{6x - x^2}$:
(A) $0 \leq x \leq 3$ (B) $3 \leq x \leq 6$

APPLICATIONS

85. Price and Demand. The number q of CD players consumers are willing to buy per week from a retail chain at a price of \$$p$ is given approximately by

$$q = d(p) = \frac{3{,}000}{0.2p + 1} \qquad 10 \leq p \leq 70$$

(A) Find the range of d.
(B) Find $p = d^{-1}(q)$, and find its domain and range.
(C) Should you interchange p and q in part B? Explain.

86. Price and Supply. The number q of CD players a retail chain is willing to supply at a price of \$$p$ is given approximately by

$$q = s(p) = \frac{900p}{p + 20} \qquad 10 \leq p \leq 70$$

(A) Find the range of s.
(B) Find $p = s^{-1}(q)$, and find its domain and range.
(C) Should you interchange p and q in part B? Explain.

87. Business—Markup Policy. A bookstore sells a book with a wholesale price of \$6 for \$9.95 and one with a wholesale price of \$10 for \$14.95.

(A) If the markup policy for the store is assumed to be linear, find a function $r = m(w)$ that expresses the retail price r as a function of the wholesale price w.

(B) Describe the function m verbally.

(C) Find $w = m^{-1}(r)$.

(D) Describe the function m^{-1} verbally.

88. Flight Conditions. In stable air, the air temperature drops about 5°F for each 1,000-foot rise in altitude.

(A) If the temperature at sea level is 63°F and a pilot reports a temperature of -12°F at 15,000 feet, find a linear function $t = d(a)$ that expresses the temperature t in terms of the altitude a (in thousands of feet).

(B) Find $a = d^{-1}(t)$.

Section 4-3 | Exponential Functions

- Exponential Functions
- Basic Exponential Graphs
- Additional Exponential Properties
- Applications

In this section we define exponential functions, look at some of their important properties—including graphs—and consider several significant applications.

Exponential Functions

Let's start by noting that the functions f and g given by

$$f(x) = 2^x \qquad \text{and} \qquad g(x) = x^2$$

are not the same function. Whether a variable appears as an exponent with a constant base or as a base with a constant exponent makes a big difference. The function g is a quadratic function, which we have already discussed. The function f is a new type of function called an *exponential function.*

The values of the exponential function $f(x) = 2^x$ for x an integer are easy to compute [Fig. 1(a)]. If $x = m/n$ is a rational number, then $f(m/n) = \sqrt[n]{2^m}$, which can be evaluated on almost any calculator [Fig. 1(b)]. Finally, a graphing utility can graph the function $f(x) = 2^x$ [Fig. 1(c)] for any given interval of x values.

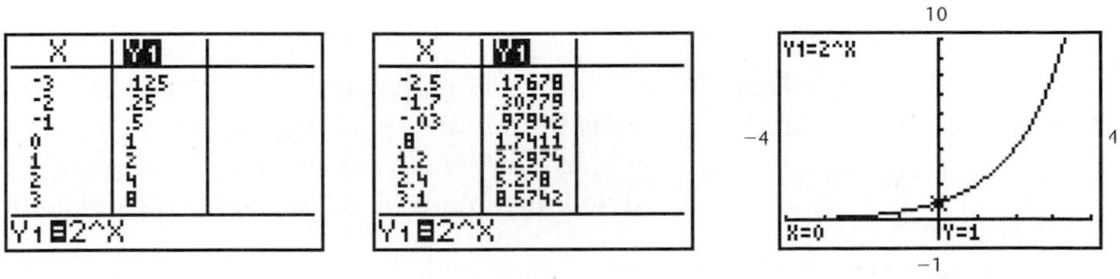

(a) (b) (c)

FIGURE 1 $f(x) = 2^x$.

The only catch is that we have not yet defined 2^x for *all* real numbers. For example, what does

$$2^{\sqrt{2}}$$

mean? The question is not easy to answer at this time. In fact, a precise definition of $2^{\sqrt{2}}$ must wait for more advanced courses, where we can show that, if b is a positive real number and x is any real number, then

$$b^x$$

names a real number, and the graph of $f(x) = 2^x$ is as indicated in Figure 1. We also can show that for x irrational, b^x can be approximated as closely as we like by using rational number approximations for x. Since $\sqrt{2} = 1.414213\ldots$, for example, the sequence

$$2^{1.4}, 2^{1.41}, 2^{1.414}, \ldots$$

approximates $2^{\sqrt{2}}$, and as we use more decimal places, the approximation improves.

DEFINITION 1

EXPONENTIAL FUNCTION

The equation

$$f(x) = b^x \qquad b > 0, b \neq 1$$

defines an **exponential function** for each different constant b, called the **base.** The independent variable x may assume any real value.

Thus, the **domain of f** is the set of all real numbers, and it can be shown that the **range of f** is the set of all positive real numbers. We require the base b to be positive to avoid imaginary numbers such as $(-2)^{1/2}$.

Basic Exponential Graphs

Explore/Discuss 1

Compare the graphs of $f(x) = 3^x$ and $g(x) = 2^x$ by graphing both functions in the same viewing window. Find all points of intersection of the graphs. For which values of x is the graph of f above the graph of g? Below the graph of g? Are the graphs of f and g close together as $x \to \infty$? As $x \to -\infty$? Discuss.

It is useful to compare the graphs of $y = 2^x$ and $y = (\frac{1}{2})^x = 2^{-x}$ by plotting both on the same coordinate system, as shown in Figure 2(a). The graph of

$$f(x) = b^x \qquad b > 1 \qquad \text{Fig. 2(b)}$$

looks very much like the graph of the particular case $y = 2^x$, and the graph of

$$f(x) = b^x \qquad 0 < b < 1 \qquad \text{Fig. 2(b)}$$

looks very much like the graph of $y = (\frac{1}{2})^x$. Note in both cases that the x axis is a *horizontal asymptote* for the graph.

FIGURE 2
Basic exponential graphs.

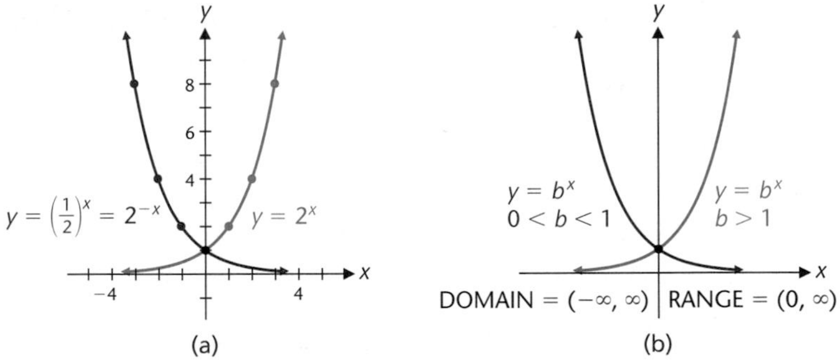

The graphs in Figure 2 suggest the following important general properties of exponential functions, which we state without proof:

BASIC PROPERTIES OF THE GRAPH OF $f(x) = b^x$, $b > 0$, $b \neq 1$

1. All graphs pass through the point (0, 1). $b^0 = 1$ for any permissible base b.
2. All graphs are continuous, with no holes or jumps.
3. The x axis is a horizontal asymptote.
4. If $b > 1$, then b^x increases as x increases.
5. If $0 < b < 1$, then b^x decreases as x increases.
6. The function f is one-to-one.

Property 6 implies that an exponential function has an inverse, called a *logarithmic function,* which we discuss in Section 4-5.

Graphing exponential functions on a graphing utility is routine, but interpreting the results requires an understanding of the preceding properties.

EXAMPLE 1

Graphing Exponential Functions

Let $f(x) = \frac{1}{2}(4^x)$. Construct a table of values (rounded to two decimal places) for $f(x)$ using integer values from -3 to 3. Graph f on a graphing utility and then sketch a graph by hand.

Solution

Set the graphing utility in two-decimal-place mode, construct the table [Fig. 3(a)], and graph the function [Fig. 3(b)]. The points on the graph of $f(x)$ for $x < 0$ are indistinguishable from the x axis in Figure 3(b). However, from the properties of an exponential function, we know that $f(x) > 0$ for all real numbers x and that

$f(x) \to 0$ as $x \to -\infty$. The hand sketch in Figure 3(c) illustrates the behavior for $x < 0$ more clearly. Of course, zooming in on the graphing utility will also illustrate this behavior.

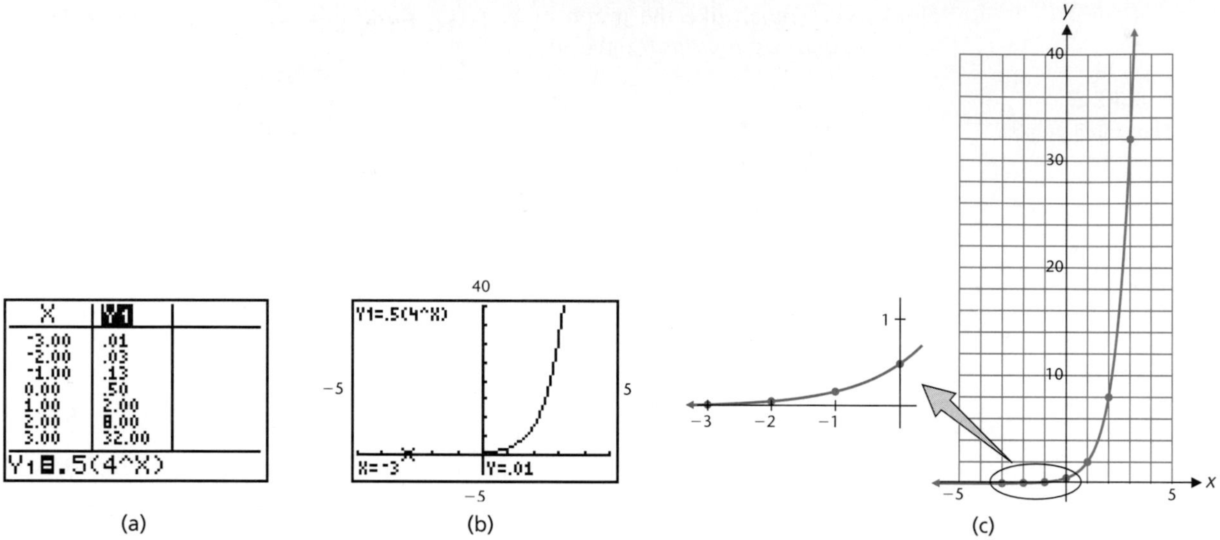

FIGURE 3

MATCHED PROBLEM 1

Repeat Example 1 for $y = \frac{1}{2}(\frac{1}{4})^x = \frac{1}{2}(4^{-x})$.

Additional Exponential Properties

Exponential functions whose domains include irrational numbers obey the familiar laws of exponents for rational exponents (see Appendix A). We summarize these exponent laws here and add two other important and useful properties.

EXPONENTIAL FUNCTION PROPERTIES

For a and b positive, $a \neq 1$, $b \neq 1$, and x and y real:

1. Exponent laws:

$$a^x a^y = a^{x+y} \qquad (a^x)^y = a^{xy} \qquad (ab)^x = a^x b^x$$

$$\left(\frac{a}{b}\right)^x = \frac{a^x}{b^x} \qquad \frac{a^x}{a^y} = a^{x-y} \qquad \frac{2^{5x}}{2^{7x}} = 2^{5x-7x} = 2^{-2x}$$

2. $a^x = a^y$ if and only if $x = y$. If $6^{4x} = 6^{2x+4}$, then $4x = 2x + 4$, and $x = 2$.

3. For $x \neq 0$, $a^x = b^x$ if and only if $a = b$. If $a^4 = 3^4$, then $a = 3$.

EXAMPLE 2 **Using Exponential Function Properties**

Solve $4^{x-3} = 8$ for x.

Solution Express both sides in terms of the same base, and use property 2 to equate exponents.

$$4^{x-3} = 8$$

$$(2^2)^{x-3} = 2^3 \qquad \text{Express 4 and 8 as powers of 2.}$$

$$2^{2x-6} = 2^3 \qquad (a^x)^y = a^{xy}$$

$$2x - 6 = 3 \qquad \text{Property 2}$$

$$2x = 9$$

$$x = \tfrac{9}{2}$$

Check $$4^{(9/2)-3} = 4^{3/2} = (\sqrt{4})^3 = 2^3 \stackrel{\checkmark}{=} 8$$

MATCHED PROBLEM 2 Solve $27^{x+1} = 9$ for x.

Applications

We now consider three applications that utilize exponential functions in their analysis: population growth, radioactive decay, and compound interest. Population growth and compound interest are examples of exponential growth, while radioactive decay is an example of negative exponential growth.

Our first example involves the growth of populations, such as people, animals, insects, and bacteria. Populations tend to grow exponentially and at different rates. A convenient and easily understood measure of growth rate is the **doubling time**—that is, the time it takes for a population to double. Over short periods of time the **doubling time growth model** is often used to model population growth:

$$P = P_0 2^{t/d}$$

where P = Population at time t

P_0 = Population at time $t = 0$

d = Doubling time

Note that when $t = d$,

$$P = P_0 2^{d/d} = P_0 2$$

and the population is double the original, as it should be. We use this model to solve a population growth problem in Example 3.

EXAMPLE 3 **Population Growth**

Mexico has a population of around 100 million people, and it is estimated that the population will double in 21 years. If population growth continues at the same rate, what will be the population:

(A) 15 years from now? (B) 30 years from now?

Calculate answers to 3 significant digits.

Solutions We use the doubling time growth model:

$$P = P_0 2^{t/d}$$

Substituting $P_0 = 100$ and $d = 21$, we obtain

$$P = 100(2^{t/21}) \quad \text{See Figure 4.}$$

FIGURE 4 $P = 100(2^{t/21})$.

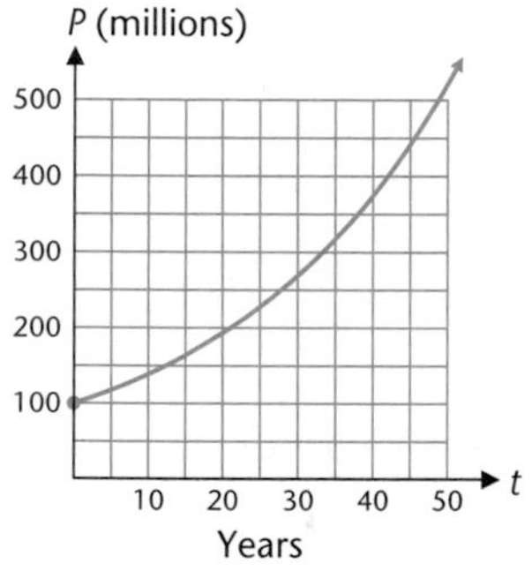

(A) Find P when $t = 15$ years:

$$P = 100(2^{15/21})$$
$$\approx 164 \text{ million people}$$

(B) Find P when $t = 30$ years:

$$P = 100(2^{30/21})$$
$$\approx 269 \text{ million people}$$

MATCHED PROBLEM 3

The bacterium *Escherichia coli* (*E. coli*) is found naturally in the intestines of many mammals. In a particular laboratory experiment, the doubling time for *E. coli* is found to be 25 minutes. If the experiment starts with a population of 1,000 *E. coli* and there is no change in the doubling time, how many bacteria will be present:

(A) In 10 minutes? (B) In 5 hours?

Write answers to three significant digits.

Explore/Discuss 2

The doubling time growth model would *not* be expected to give accurate results over long periods of time. According to the doubling time growth model of Example 3, what was the population of Mexico 500 years ago at the height of Aztec civilization? What will the population of Mexico be 200 years from now? Explain why these results are unrealistic. Discuss factors that affect human populations which are not taken into account by the doubling time growth model.

Our second application involves radioactive decay, which is often referred to as negative growth. Radioactive materials are used extensively in medical diagnosis and therapy, as power sources in satellites, and as power sources in many countries. If we start with an amount A_0 of a particular radioactive isotope, the amount declines exponentially in time. The rate of decay varies from isotope to isotope. A convenient and easily understood measure of the rate of decay is the **half-life** of the isotope—that is, the time it takes for half of a particular material to decay. In this section we use the following **half-life decay model:**

$$A = A_0(\tfrac{1}{2})^{t/h} = A_0 2^{-t/h}$$

where A = Amount at time t

A_0 = Amount at time $t = 0$

h = Half-life

Note that when $t = h$,

$$A = A_0 2^{-h/h} = A_0 2^{-1} = \frac{A_0}{2}$$

and the amount of isotope is half the original amount, as it should be.

EXAMPLE 4

Radioactive Decay

The radioactive isotope gallium 67 (^{67}Ga), used in the diagnosis of malignant tumors, has a biological half-life of 46.5 hours. If we start with 100 milligrams of the isotope, how many milligrams will be left after

(A) 24 hours? (B) 1 week?

Compute answers to three significant digits.

Solutions

We use the half-life decay model:

$$A = A_0(\tfrac{1}{2})^{t/h} = A_0 2^{-t/h}$$

Using $A_0 = 100$ and $h = 46.5$, we obtain

$$A = 100(2^{-t/46.5}) \quad \text{See Figure 5.}$$

FIGURE 5 $A = 100(2^{-t/46.5})$.

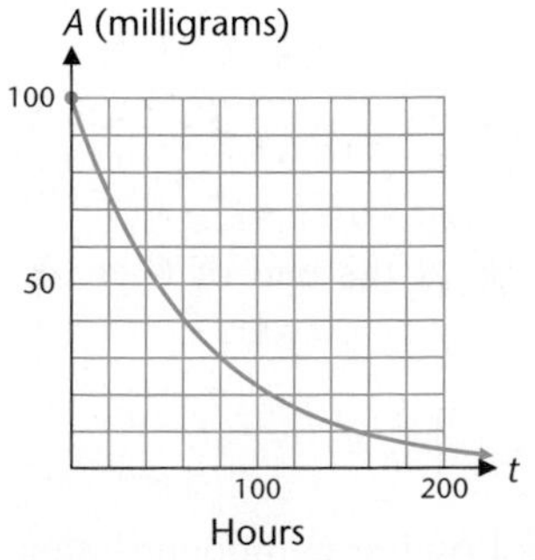

(A) Find A when $t = 24$ hours:

$$A = 100(2^{-24/46.5}) = 69.9 \text{ milligrams}$$

(B) Find A when $t = 168$ hours (1 week = 168 hours):

$$A = 100(2^{-168/46.5}) = 8.17 \text{ milligrams}$$

MATCHED PROBLEM 4

Radioactive gold 198 (^{198}Au), used in imaging the structure of the liver, has a half-life of 2.67 days. If we start with 50 milligrams of the isotope, how many milligrams will be left after:

(A) $\frac{1}{2}$ day? (B) 1 week?

Compute answers to three significant digits.

Our third application deals with the growth of money at compound interest. This topic is important to most people and is fundamental to many topics in the mathematics of finance.

The fee paid to use another's money is called **interest.** It is usually computed as a percentage, called the **interest rate,** of the principal over a given period of time. If, at the end of a payment period, the interest due is reinvested at the same rate, then the interest earned as well as the principal will earn interest during the next payment period. Interest paid on interest reinvested is called **compound interest.**

Suppose you deposit \$1,000 in a savings and loan that pays 8% compounded semiannually. How much will the savings and loan owe you at the end of 2 years? Compounded semiannually means that interest is paid to your account at the end of each 6-month period, and the interest will in turn earn interest. The **interest rate per period** is the annual rate, 8% = 0.08, divided by the number of compounding periods per year, 2. If we let A_1, A_2, A_3, and A_4 represent the new amounts due at the end of the first, second, third, and fourth periods, respectively, then

$$A_1 = \$1{,}000 + \$1{,}000\left(\frac{0.08}{2}\right)$$

$$= \$1{,}000(1 + 0.04) \qquad P\left(1 + \frac{r}{n}\right)$$

$$A_2 = A_1(1 + 0.04)$$

$$= [\$1{,}000(1 + 0.04)](1 + 0.04)$$

$$= \$1{,}000(1 + 0.04)^2 \qquad P\left(1 + \frac{r}{n}\right)^2$$

$$A_3 = A_2(1 + 0.04)$$

$$= [\$1{,}000(1 + 0.04)^2](1 + 0.04)$$

$$= \$1{,}000(1 + 0.04)^3 \qquad P\left(1 + \frac{r}{n}\right)^3$$

$$A_4 = A_3(1 + 0.04)$$

$$= [\$1{,}000(1 + 0.04)^3](1 + 0.04)$$

$$= \$1{,}000(1 + 0.04)^4 \qquad P\left(1 + \frac{r}{n}\right)^4$$

What do you think the savings and loan will owe you at the end of 6 years? If you guessed

$$A = \$1{,}000(1 + 0.04)^{12}$$

you have observed a pattern that is generalized in the following compound interest formula:

COMPOUND INTEREST

If a **principal** P is invested at an annual **rate** r compounded m times a year, then the **amount** A in the account at the end of n compounding periods is given by

$$A = P\left(1 + \frac{r}{m}\right)^n$$

The annual rate r is expressed in decimal form.

EXAMPLE 5 **Compound Interest**

If you deposit \$5,000 in an account paying 9% compounded daily, how much will you have in the account in 5 years? Compute the answer to the nearest cent.

Solution We use the compound interest formula with $P = 5{,}000$, $r = 0.09$, $m = 365$, and $n = 5(365) = 1{,}825$:

$$A = P\left(1 + \frac{r}{m}\right)^n$$

$$= 5{,}000\left(1 + \frac{0.09}{365}\right)^{1825}$$

$$= \$7{,}841.13$$

FIGURE 6

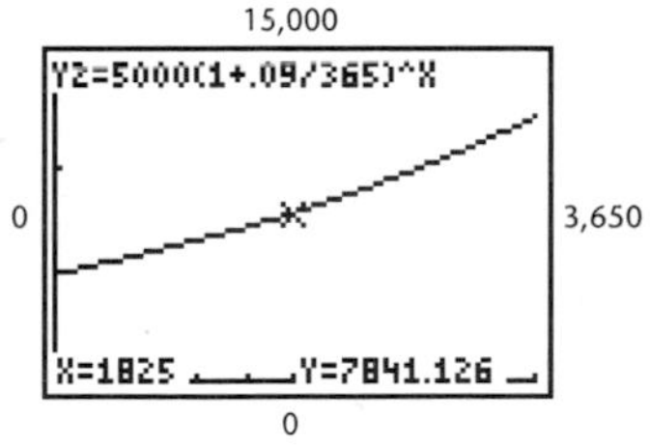

The graph of

$$A = 5{,}000\left(1 + \frac{0.09}{365}\right)^x$$

is shown in Figure 6.

MATCHED PROBLEM 5

If \$1,000 is invested in an account paying 10% compounded monthly, how much will be in the account at the end of 10 years? Compute the answer to the nearest cent.

EXAMPLE 6 **Comparing Investments**

If \$1,000 is deposited into an account earning 10% compounded monthly and, at the same time, \$2,000 is deposited into an account earning 4% compounded monthly, will the first account ever be worth more than the second? If so, when?

Solution Let y_1 and y_2 represent the amounts in the first and second accounts, respectively, then

$$y_1 = 1{,}000(1 + 0.10/12)^x$$

$$y_2 = 2{,}000(1 + 0.04/12)^x$$

where x is the number of compounding periods (months). Examining the graphs of y_1 and y_2 [Fig. 7(a)], we see that the graphs intersect at $x \approx 139.438$ months. Since compound interest is paid at the end of each compounding period, we compare the amount in the accounts after 139 months and after 140 months [Fig. 7(b)]. Thus, the first account is worth more than the second for $x \geq 140$ months or 11 years and 8 months.

FIGURE 7

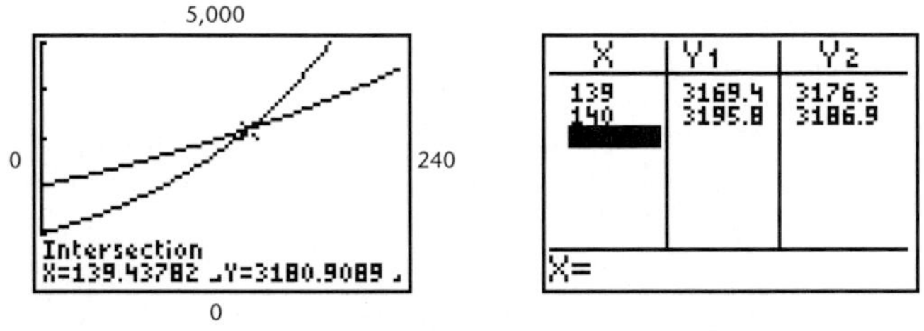

(a) (b)

MATCHED PROBLEM 6 If \$4,000 is deposited into an account earning 10% compounded quarterly and, at the same time, \$5,000 is deposited into an account earning 6% compounded quarterly, when will the first account be worth more than the second?

Answers to Matched Problems

1. $y = \frac{1}{2}(4^{-x})$

x	y
−3	32.00
−2	8.00
−1	2.00
0	0.50
1	0.13
2	0.03
3	0.01

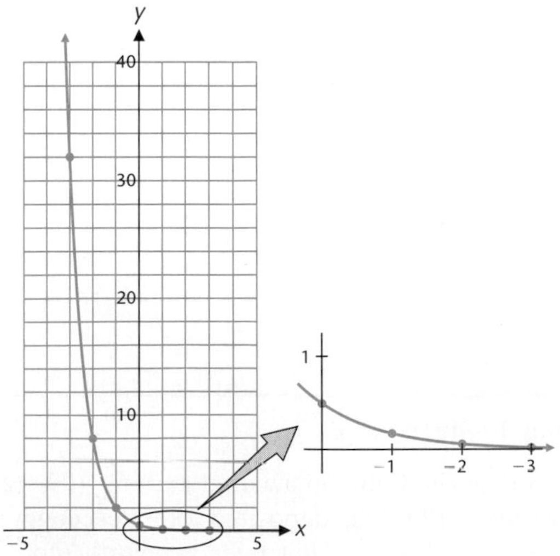

2. $x = -\frac{1}{3}$ **3.** (A) 1,320 (B) $4{,}100{,}000 = 4.10 \times 10^6$ **4.** (A) 43.9 mg (B) 8.12 mg
5. \$2,707.04 **6.** After 23 quarters

EXERCISE 4-3

A

1. Match each equation with the graph of *f*, *g*, *m*, or *n* in the figure.

(A) $y = (0.2)^x$ (B) $y = 2^x$

(C) $y = (\frac{1}{3})^x$ (D) $y = 4^x$

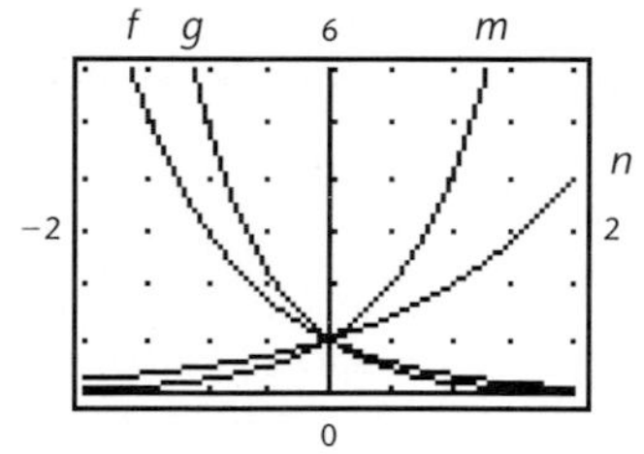

2. Match each equation with the graph of *f*, *g*, *m*, or *n* in the figure.

(A) $y = 5^x$ (B) $y = (0.5)^x$

(C) $y = 3^x$ (D) $y = (\frac{1}{4})^x$

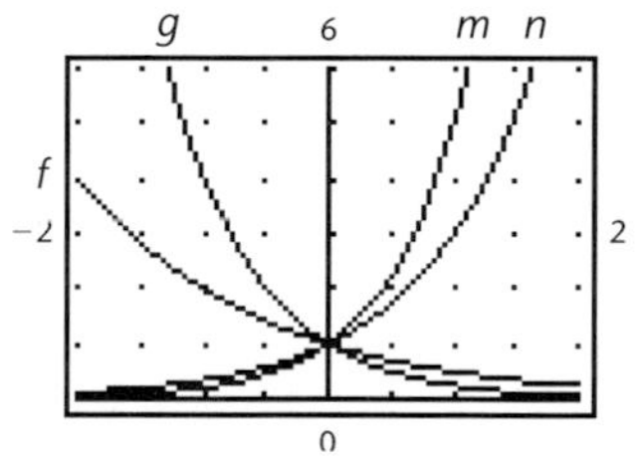

In Problems 3–8, compute answers to four significant digits.

3. $5^{\sqrt{3}}$ **4.** $3^{-\sqrt{2}}$

5. $\pi^{\sqrt{2}}$ **6.** $\pi^{-\sqrt{3}}$

7. $\dfrac{2^{\pi} + 2^{-\pi}}{2}$ **8.** $\dfrac{3^{\pi} - 3^{-\pi}}{2}$

Before graphing the functions in Problems 9–18, classify each function as increasing or decreasing, find the y intercept, and identify any asymptotes. Then examine the graph to check your answers.

9. $y = 3^x$ **10.** $y = 5^x$

11. $y = (\frac{1}{3})^x = 3^{-x}$ **12.** $y = (\frac{1}{5})^x = 5^{-x}$

13. $g(x) = -3^{-x}$ **14.** $f(x) = -5^x$

15. $h(x) = 5(3^x)$ **16.** $f(x) = 4(5^x)$

17. $y = 3^{x+3} - 5$ **18.** $y = 5^{x+2} + 4$

In Problems 19–24, simplify.

19. $10^{3x-1}10^{4-x}$ **20.** $(4^{3x})^{2y}$ **21.** $\dfrac{3^x}{3^{1-x}}$

22. $\dfrac{5^{x-3}}{5^{x-4}}$ **23.** $\left(\dfrac{4^x}{5^y}\right)^{3z}$ **24.** $(2^x3^y)^z$

B

In Problems 25–36, solve for x.

25. $5^{3x} = 5^{4x-2}$ **26.** $10^{2-3x} = 10^{5x-6}$

27. $7^{x^2} = 7^{2x+3}$ **28.** $4^{5x-x^2} = 4^{-6}$

29. $(1 - x)^5 = (2x - 1)^5$ **30.** $5^3 = (x + 2)^3$

31. $2^x = 4^{x+1}$ **32.** $9^{x-1} = 3^x$

33. $25^{x+1} = 125^{2x}$ **34.** $100^{x-1} = 1{,}000^{2x}$

35. $9^{x^2} = 3^{3x-1}$ **36.** $4^{x^2} = 2^{x+3}$

37. Find all real numbers a such that $a^2 = a^{-2}$. Explain why this does not violate the second exponential function property in the box on page 274.

38. Find real numbers a and b such that $a \neq b$ but $a^4 = b^4$. Explain why this does not violate the third exponential function property in the box on page 274.

39. Examine the graph of $y = 1^x$ on a graphing utility and explain why 1 cannot be the base for an exponential function.

40. Examine the graph of $y = 0^x$ on a graphing utility and explain why 0 cannot be the base for an exponential function. [*Hint:* Turn the axes off before graphing.]

Graph each function in Problems 41–48 using the graph of f shown in the figure.

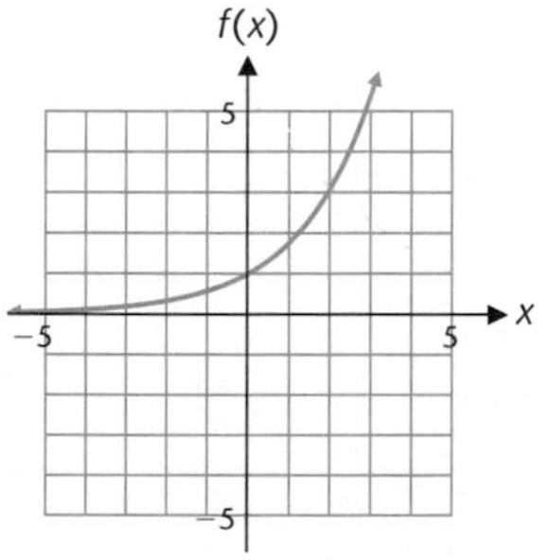

41. $y = f(x) - 2$ **42.** $y = f(x) + 1$

43. $y = f(x - 2)$ **44.** $y = f(x + 1)$

45. $y = 2f(x) - 4$ **46.** $y = 3 - 5f(x)$

47. $y = 2 - 3f(x - 4)$ **48.** $y = 2f(x + 1) - 1$

In Problems 49–56, sketch the graph of a function of the form $f(x) = a(b^x) + c$ that satisfies the given conditions.

49. f is an increasing function, asymptotic to the x axis, and satisfying $f(-5) = 0.25, f(0) = 1$, and $f(10) = 16$.

50. f is an increasing function, asymptotic to the x axis, and satisfying $f(-1) = 0.5, f(0) = 1$, and $f(3) = 8$.

51. f is a decreasing function, asymptotic to the x axis, and satisfying $f(-1) = 24, f(0) = 6$, and $f(2) = 0.375$.

52. f is a decreasing function, asymptotic to the x axis, and satisfying $f(-1) = 32, f(0) = 4$, and $f(1) = 0.5$.

53. f is a decreasing function, asymptotic to the line $y = 3$, and satisfying $f(-20) = 2.75, f(0) = 2$, and $f(20) = -1$.

54. f is an increasing function, asymptotic to the line $y = 4$, and satisfying $f(-10) = 0, f(0) = 3$, and $f(10) = 3.75$.

55. f is an increasing function, asymptotic to the line $y = -2$, and satisfying $f(-2) = -1.75, f(0) = -1$, and $f(2) = 2$.

56. f is a decreasing function, asymptotic to the line $y = -5$, and satisfying $f(-4) = 11, f(0) = -4$, and $f(2) = -4.75$.

C

In Problems 57–60, simplify.

57. $(6^x + 6^{-x})(6^x - 6^{-x})$

58. $(3^x - 3^{-x})(3^x + 3^{-x})$

59. $(6^x + 6^{-x})^2 - (6^x - 6^{-x})^2$

60. $(3^x - 3^{-x})^2 + (3^x + 3^{-x})^2$

In Problems 61–64, use a graphing utility to approximate local extrema and x intercepts to two decimal places. Investigate the behavior as $x \to \infty$ and as $x \to -\infty$ and identify any horizontal asymptotes.

61. $m(x) = 2x(3^{-x}) + 2$

62. $h(x) = 3x(2^{-x}) - 1$

63. $f(x) = \dfrac{2^x + 2^{-x}}{2}$

64. $g(x) = \dfrac{3^x + 3^{-x}}{2}$

APPLICATIONS

65. Gaming. A person bets on red and black on a roulette wheel using a *Martingale strategy*. That is, a \$2 bet is placed on red, and the bet is doubled each time until a win occurs. The process is then repeated. If black occurs n times in a row, then $L = 2^n$ dollars is lost on the nth bet. Graph this function for $1 \le n \le 10$. Even though the function is defined only for positive integers, points on this type of graph are usually joined with a smooth curve as a visual aid.

66. Bacterial Growth. If bacteria in a certain culture double every $\frac{1}{2}$ hour, write an equation that gives the number of bacteria N in the culture after t hours, assuming the culture has 100 bacteria at the start. Graph the equation for $0 \le t \le 5$.

67. Population Growth. Because of its short life span and frequent breeding, the fruit fly *Drosophila* is used in some genetic studies. Raymond Pearl of Johns Hopkins University, for example, studied 300 successive generations of descendants of a single pair of *Drosophila* flies. In a laboratory situation with ample food supply and space, the doubling time for a particular population is 2.4 days. If we start with 5 male and 5 female flies, how many flies should we expect to have in

(A) 1 week? (B) 2 weeks?

68. Population Growth. If Kenya has a population of about 30,000,000 people and a doubling time of 19 years and if the growth continues at the same rate, find the population in

(A) 10 years (B) 30 years

Compute answers to two significant digits.

69. Insecticides. The use of the insecticide DDT is no longer allowed in many countries because of its long-term adverse effects. If a farmer uses 25 pounds of active DDT, assuming its half-life is 12 years, how much will still be active after

(A) 5 years? (B) 20 years?

Compute answers to two significant digits.

70. Radioactive Tracers. The radioactive isotope technetium 99m (^{99m}Tc) is used in imaging the brain. The isotope has a half-life of 6 hours. If 12 milligrams are used, how much will be present after

(A) 3 hours? (B) 24 hours?

Compute answers to three significant digits.

71. Finance. Suppose \$4,000 is invested at 11% compounded weekly. How much money will be in the account in

(A) $\frac{1}{2}$ year? (B) 10 years?

Compute answers to the nearest cent.

72. Finance. Suppose \$2,500 is invested at 7% compounded quarterly. How much money will be in the account in

(A) $\frac{3}{4}$ year? (B) 15 years?

Compute answers to the nearest cent.

★ **73. Finance.** A couple just had a new child. How much should they invest now at 8.25% compounded daily in order to have $40,000 for the child's education 17 years from now? Compute the answer to the nearest dollar.

★ **74. Finance.** A person wishes to have $15,000 cash for a new car 5 years from now. How much should be placed in an account now if the account pays 9.75% compounded weekly? Compute the answer to the nearest dollar.

★ **75. Finance.** If $3,000 is deposited into an account earning 8% compounded daily and, at the same time, $5,000 is deposited into an account earning 5% compounded daily, will the first account be worth more than the second? If so, when?

★ **76. Finance.** If $4,000 is deposited into an account earning 9% compounded weekly and, at the same time, $6,000 is deposited into an account earning 7% compounded weekly, will the first account be worth more than the second? If so, when?

★ **77. Finance.** Will an investment of $10,000 at 8.9% compounded daily ever be worth more at the end of a quarter than an investment of $10,000 at 9% compounded quarterly? Explain.

★ **78. Finance.** A sum of $5,000 is invested at 13% compounded semiannually. Suppose that a second investment of $5,000 is made at interest rate r compounded daily. For which values of r, to the nearest tenth of a percent, is the second investment better than the first? Discuss.

Problems 79 and 80 require a graphing utility that can compute exponential regression equations of the form $y = ab^x$ (consult your manual).

79. Depreciation. Table 1 gives the market value of a minivan (in dollars) x years after its purchase. Find an exponential regression model of the form $y = ab^x$ for this data set. Estimate the purchase price of the van. Estimate the value of the van 10 years after its purchase. Round answers to the nearest dollar.

TABLE 1

x	Value ($)
1	12,575
2	9,455
3	8,115
4	6,845
5	5,225
6	4,485

80. Depreciation. Table 2 gives the market value of a luxury sedan (in dollars) x years after its purchase. Find an exponential regression model of the form $y = ab^x$ for this data set. Estimate the purchase price of the sedan. Estimate the value of the sedan 10 years after its purchase. Round answers to the nearest dollar.

TABLE 2

x	Value ($)
1	23,125
2	19,050
3	15,625
4	11,875
5	9,450
6	7,125

Section 4-4 | The Exponential Function with Base e

V ID

- Base e Exponential Function
- Growth and Decay Applications Revisited
- Continuous Compound Interest
- A Comparison of Exponential Growth Phenomena

Until now the number π has probably been the most important irrational number you have encountered. In this section we will introduce another irrational number, e, that is just as important in mathematics and its applications.

Base *e* Exponential Function

The following expression is important to the study of calculus and, as we will see later in this section, also is closely related to the compound interest formula discussed in the preceding section:

$$\left(1 + \frac{1}{x}\right)^x \tag{1}$$

Explore/Discuss 1

(A) Calculate the values of $[1 + (1/x)]^x$ for $x = 1, 2, 3, 4$, and 5. Are the values increasing or decreasing as x gets larger?

(B) Graph $y = [1 + (1/x)]^x$ and discuss the behavior of the graph as x increases without bound.

TABLE 1

x	$\left(1 + \frac{1}{x}\right)^x$
1	2
10	2.593 74 . . .
100	2.704 81 . . .
1,000	2.716 92 . . .
10,000	2.718 14 . . .
100,000	2.718 27 . . .
1,000,000	2.718 28 . . .

Interestingly, by calculating the value of expression (1) for larger and larger values of x (see Table 1), it appears that $[1 + (1/x)]^x$ approaches a number close to 2.7183. In a calculus course we can show that as x increases without bound, the value of $[1 + (1/x)]^x$ approaches an irrational number that we call e. Just as irrational numbers such as π and $\sqrt{2}$ have unending, nonrepeating decimal representations (see Section A-1), e also has an unending, nonrepeating decimal representation. To 12 decimal places,

$$e = 2.718\ 281\ 828\ 459$$

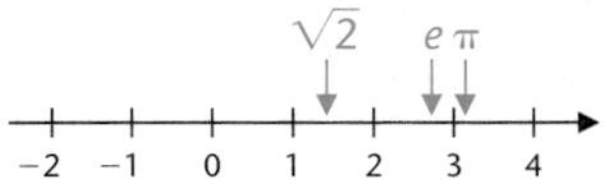

Exactly who discovered e is still being debated. It is named after the great Swiss mathematician Leonhard Euler (1707–1783), who computed e to 23 decimal places using $[1 + (1/x)]^x$.

The constant e turns out to be an ideal base for an exponential function because in calculus and higher mathematics many operations take on their simplest form using this base. This is why you will see e used extensively in expressions and formulas that model real-world phenomena.

DEFINITION 1

EXPONENTIAL FUNCTION WITH BASE *e*

For x a real number, the equation

$$f(x) = e^x$$

defines the **exponential function with base *e*.**

The exponential function with base e is used so frequently that it is often referred to as *the* exponential function. The graphs of $y = e^x$ and $y = e^{-x}$ are shown in Figure 1.

FIGURE 1
Exponential functions with base e.

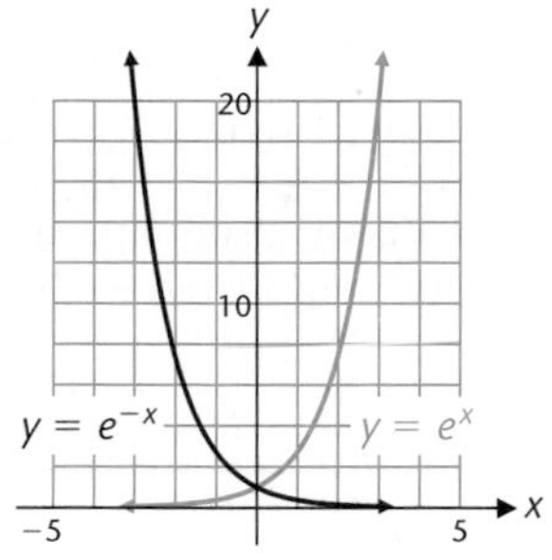

Explore/Discuss 2

(A) Graph $y_1 = e^x$, $y_2 = e^{0.5x}$, and $y_3 = e^{2x}$ in the same viewing window. How do these graphs compare with the graph of $y = b^x$ for $b > 1$?

(B) Graph $y_1 = e^{-x}$, $y_2 = e^{-0.5x}$, and $y_3 = e^{-2x}$ in the same viewing window. How do these graphs compare with the graph of $y = b^x$ for $0 < b < 1$?

(C) Use the properties of exponential functions to show that all of these functions are exponential functions.

EXAMPLE 1 Analyzing a Graph

Describe the graph of $f(x) = 4 - e^{x/2}$, including x and y intercepts, increasing and decreasing properties, and horizontal asymptotes. Round any approximate values to two decimal places.

Solution The graph of f is shown in Figure 2(a). The y intercept is $f(0) = 4 - 1 = 3$ and the x intercept is 2.77 (to two decimal places). The graph shows that f is decreasing for all x. Since the exponential function $e^{x/2} \to 0$ as $x \to -\infty$, it follows that $f(x) = 4 - e^{x/2} \to 4$ as $x \to -\infty$. The table in Figure 2(b) confirms this. Thus, the line $y = 4$ is a horizontal asymptote for the graph.

FIGURE 2
$f(x) = 4 - e^{x/2}$.

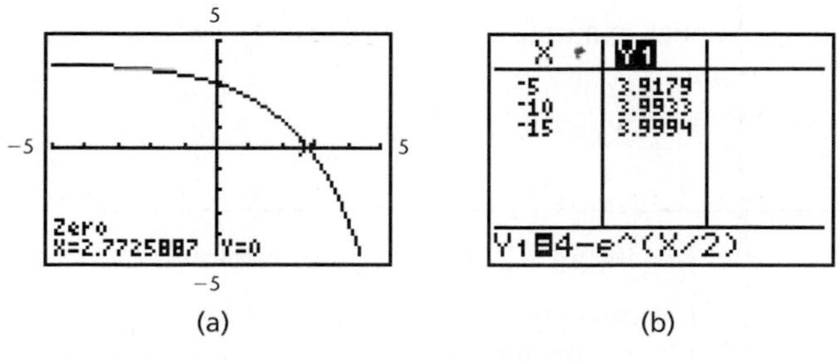

MATCHED PROBLEM 1

Describe the graph of $f(x) = 2e^{x/2} - 5$, including x and y intercepts, increasing and decreasing properties, and horizontal asymptotes. Round any approximate values to two decimal places.

Growth and Decay Applications Revisited

Most exponential growth and decay problems are modeled using base e exponential functions. We present two applications here and many more in Exercise 4-4.

EXAMPLE 2

Medicine—Bacteria Growth

Cholera, an intestinal disease, is caused by a cholera bacterium that multiplies exponentially by cell division as modeled by

$$N = N_0e^{1.386t}$$

where N is the number of bacteria present after t hours and N_0 is the number of bacteria present at $t = 0$. If we start with 1 bacterium, how many bacteria will be present in

(A) 5 hours? (B) 12 hours?

Compute the answers to three significant digits.

Solutions

(A) Use $N_0 = 1$ and $t = 5$:

$$\begin{aligned} N &= N_0e^{1.386t} \\ &= e^{1.386(5)} \\ &= 1{,}020 \end{aligned}$$

(B) Use $N_0 = 1$ and $t = 12$:

$$\begin{aligned} N &= N_0e^{1.386t} \\ &= e^{1.386(12)} \\ &= 16{,}700{,}000 \end{aligned}$$

MATCHED PROBLEM 2

Repeat example 2 if $N = N_0e^{0.783t}$ and all other information remains the same.

EXAMPLE 3

Carbon 14 Dating

Cosmic-ray bombardment of the atmosphere produces neutrons, which in turn react with nitrogen to produce radioactive carbon 14. Radioactive carbon 14 enters all living tissues through carbon dioxide, which is first absorbed by plants. As long as a plant or animal is alive, carbon 14 is maintained in the living organism at a constant level. Once the organism dies, however, carbon 14 decays according to the equation

$$A = A_0 e^{-0.000124t}$$

where A is the amount of carbon 14 present after t years and A_0 is the amount present at time $t = 0$. If 1,000 milligrams of carbon 14 are present at the start, how many milligrams will be present in

(A) 10,000 years? (B) 50,000 years?

Compute answers to three significant digits.

Solutions

Substituting $A_0 = 1{,}000$ in the decay equation, we have

$$A = 1{,}000e^{-0.000124t} \quad \text{See Figure 3.}$$

FIGURE 3

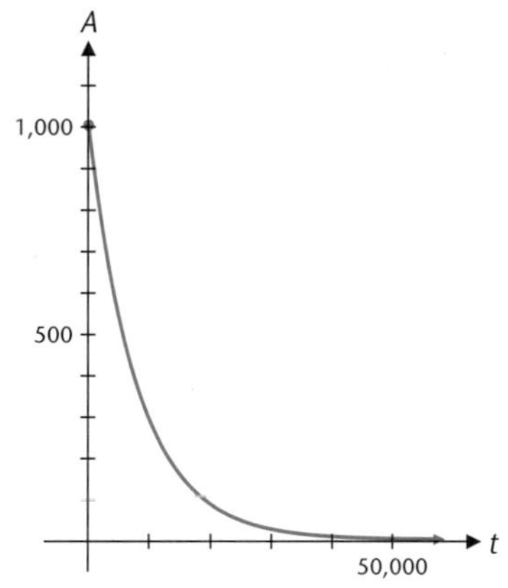

(A) Solve for A when $t = 10{,}000$:

$$A = 1{,}000e^{-0.000124(10{,}000)}$$
$$= 289 \text{ milligrams}$$

(B) Solve for A when $t = 50{,}000$:

$$A = 1{,}000e^{-0.000124(50{,}000)}$$
$$= 2.03 \text{ milligrams}$$

More will be said about carbon 14 dating in Exercise 4-4, where we will be interested in solving for t after being given information about A and A_0.

MATCHED PROBLEM 3

Referring to Example 3, how many milligrams of carbon 14 would have to be present at the beginning in order to have 10 milligrams present after 20,000 years? Compute the answer to four significant digits.

EXAMPLE 4

Limited Growth in an Epidemic

A community of 1,000 individuals is assumed to be homogeneously mixed. One individual who has just returned from another community has influenza. Assume the community has not had influenza shots and all are susceptible. The spread of the disease in the community is predicted to be given by the logistic curve

$$N(t) = \frac{1{,}000}{1 + 999e^{-0.3t}}$$

where N is the number of people who have contracted influenza after t days.

(A) How many people have contracted influenza after 10 days? After 20 days? Round answers to the nearest integer.

(B) How many days will it take until half the community has contracted influenza? Round answer to the nearest integer.

(C) Does N approach a limiting value as t increases without bound? Explain.

Solutions

(A) The table in Figure 4(a) shows that $N(10) \approx 20$ individuals and $N(20) \approx 288$ individuals.

FIGURE 4
Logistic growth.

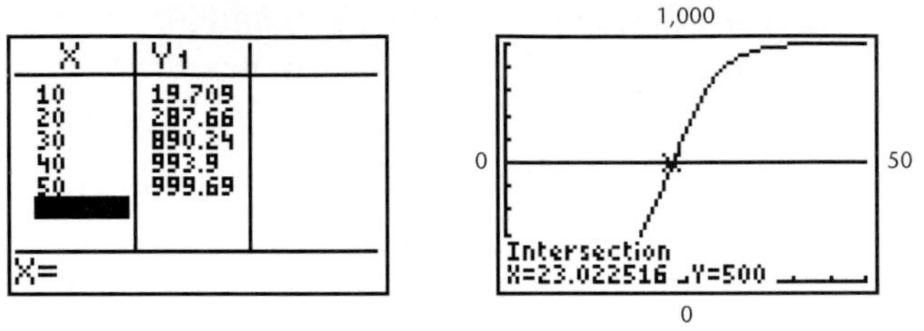

(a) (b)

(B) Figure 4(b) shows that the graph of $N(t)$ intersects the line $y = 500$ after approximately 23 days.

(C) The values in Figure 4(a) and the graph in Figure 4(b) both indicate that N approaches 1,000 as t increases without bound. We can confirm this algebraically by noting that since $999e^{-0.3t} \to 0$ as t increases without bound,

$$N(t) = \frac{1{,}000}{1 + 999e^{-0.3t}} \to \frac{1{,}000}{1 + 0} = 1{,}000$$

Thus, the upper limit on the growth of N is 1,000, the total number of people in the community.

MATCHED PROBLEM 4

A group of 400 parents, relatives, and friends are waiting anxiously at Kennedy Airport for a charter flight returning students after a year in Europe. It is stormy and the plane is late. A particular parent thought he had heard that the plane's radio had gone out and related this news to some friends, who in turn passed it on to others. The propagation of this rumor is predicted to be given by

$$N(t) = \frac{400}{1 + 399e^{-0.4t}}$$

where N is the number of people who have heard the rumor after t minutes.

(A) How many people have heard the rumor after 10 minutes? After 20 minutes? Round answers to the nearest integer.

(B) How many minutes will it take until half the group has heard the rumor? Round answer to the nearest integer.

(C) Does N approach a limiting value as t increases without bound? Explain.

Continuous Compound Interest

The constant e occurs naturally in the study of compound interest. Returning to the compound interest formula discussed in Section 4-3,

$$A = P\left(1 + \frac{r}{m}\right)^n \quad \text{Compound interest}$$

recall that P is the principal invested at an annual rate compounded m times a year and A is the amount in the account after n compounding periods. For the purposes of the discussion here, it is convenient to let $n = mt$, where t is time in years, so that

$$A = P\left(1 + \frac{r}{m}\right)^{mt}$$

is the amount in the account after t years. Suppose P, r, and t are held fixed, and m, the number of compounding periods in 1 year, is increased without bound. Will the amount A increase without bound or will it tend to some limiting value? Let's examine a specific case numerically before we attack the general problem. If $P = \$100$, $r = 0.08$, and $t = 2$ years, then

$$A = 100\left(1 + \frac{0.08}{m}\right)^{2m}$$

The amount A is computed for several values of m in Table 2. Notice that the largest gain appears in going from annually to semiannually. Then, the gains slow down as m increases. In fact, it appears that A might be tending to something close to \$117.35 as m gets larger and larger.

TABLE 2 Effect of Compounding Frequency

Compounding Frequency	n	$A = 100\left(1 + \frac{0.08}{m}\right)^{2m}$
Annually	1	\$116.6400
Semiannually	2	116.9859
Quarterly	4	117.1659
Weekly	52	117.3367
Daily	365	117.3490
Hourly	8,760	117.3501

We now return to the general problem to see if we can determine what happens to $A = P[1 + (r/m)]^{mt}$ as m increases without bound. A little algebraic manipulation of the compound interest formula will lead to an answer and a significant result in the mathematics of finance:

$$A = P\left(1 + \frac{r}{m}\right)^{mt}$$

$$= P\left(1 + \frac{1}{m/r}\right)^{(m/r)rt} \quad \text{Change algebraically.}$$

$$= \left[\left(1 + \frac{1}{x}\right)^{x}\right]^{rt} \quad \text{Let } x = m/r.$$

The expression within the square brackets should look familiar. Recall from the first part of this section that

$$\left(1 + \frac{1}{x}\right)^x \to e \qquad \text{as} \qquad x \to \infty$$

Since r is fixed, $x = m/r \to \infty$ as $m \to \infty$. Thus,

$$P\left(1 + \frac{r}{m}\right)^{mt} \to Pe^{rt} \qquad \text{as} \qquad m \to \infty$$

and we have arrived at the **continuous compound interest formula,** a very important and widely used formula in business, banking, and economics.

CONTINUOUS COMPOUND INTEREST FORMULA

If a principal P is invested at an annual rate r compounded continuously, then the amount A in the account at the end of t years is given by

$$A = Pe^{rt}$$

The annual rate r is expressed as a decimal.

EXAMPLE 5

Continuous Compound Interest

If \$100 is invested at an annual rate of 8% compounded continuously, what amount, to the nearest cent, will be in the account after 2 years?

Solution

Use the continuous compound interest formula to find A when $P = \$100$, $r = 0.08$, and $t = 2$:

$$\begin{aligned} A &= Pe^{rt} \\ &= \$100e^{(0.08)(2)} \qquad \text{8\% is equivalent to } r = 0.08. \\ &= \$117.35 \end{aligned}$$

Compare this result with the values calculated in Table 2.

MATCHED PROBLEM 5

What amount will an account have after 5 years if \$100 is invested at an annual rate of 12% compounded annually? Quarterly? Continuously? Compute answers to the nearest cent.

The continuous compound interest formula may also be used to model short-term population growth. If a population P is assumed to grow continuously at an annual rate r, then the population A at the end of t years is given by $A = Pe^{rt}$.

A Comparison of Exponential Growth Phenomena

The equations and graphs given in Table 3 compare several widely used growth models. These are divided basically into two groups: unlimited growth and limited growth. Following each equation and graph is a short, incomplete list of areas in which the models are used. We have only touched on a subject that has been extensively developed and that you are likely to study in greater depth in the future.

TABLE 3 Exponential Growth and Decay

Description	Equation	Graph	Uses
Unlimited growth	$y = ce^{kt}$ $c, k > 0$	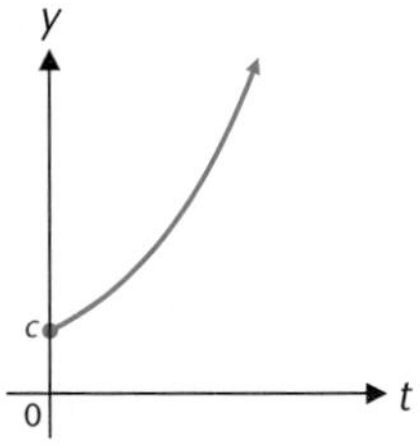	Short-term population growth (people, bacteria, etc.); growth of money at continuous compound interest
Exponential decay	$y = ce^{-kt}$ $c, k > 0$	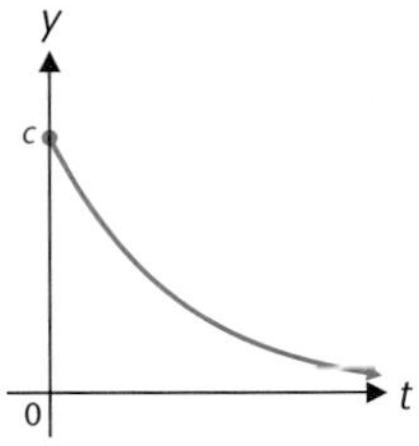	Radioactive decay; light absorption in water, glass, and the like; atmospheric pressure; electric circuits
Limited growth	$y = c(1 - e^{-kt})$ $c, k > 0$	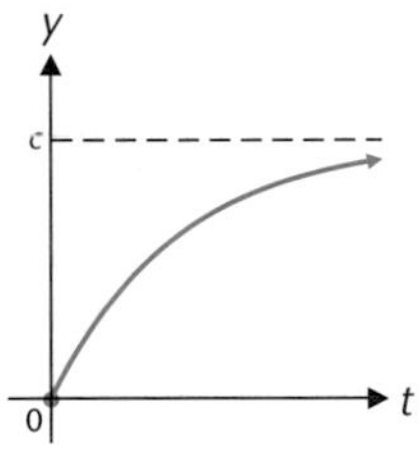	Learning skills; sales fads; company growth; electric circuits
Logistic growth	$y = \dfrac{M}{1 + ce^{-kt}}$ $c, k, M > 0$	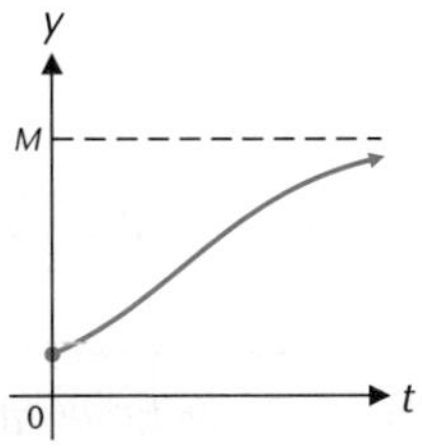	Long-term population growth; epidemics; sales of new products; company growth

Answers to Matched Problems

1. y intercept: -3; x intercept: 1.83; increasing for all x; horizontal asymptote: $y = -5$
2. (A) 50 bacteria (B) 12,000 bacteria **3.** 119.4 mg
4. (A) 48 individuals; 353 individuals (B) 15 minutes
(C) N approaches an upper limit of 400, the number of people in the entire group.
5. Annually: \$176.23; quarterly: \$180.61; continuously: \$182.21

EXERCISE 4-4

A

1. Match each equation with the graph of f, g, m, or n in the figure.

(A) $y = e^{1.2x}$ (B) $y = e^{-0.6x}$

(C) $y = e^{0.4x}$ (D) $y = e^{-1.5x}$

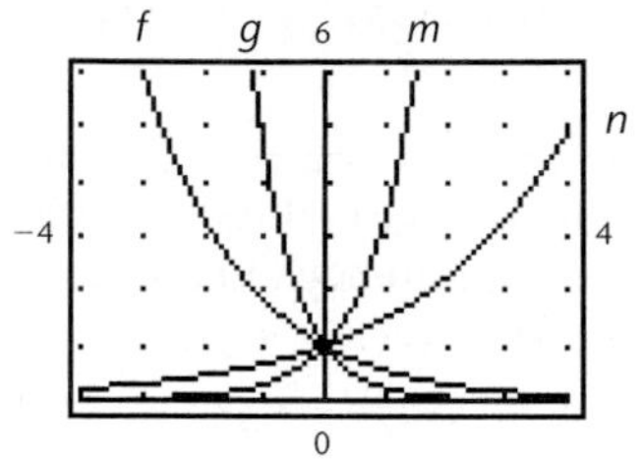

2. Match each equation with the graph of f, g, m, or n in the figure.

(A) $y = e^{-1.2x}$ (B) $y = e^{0.7x}$

(C) $y = e^{-0.4x}$ (D) $y = e^{1.3x}$

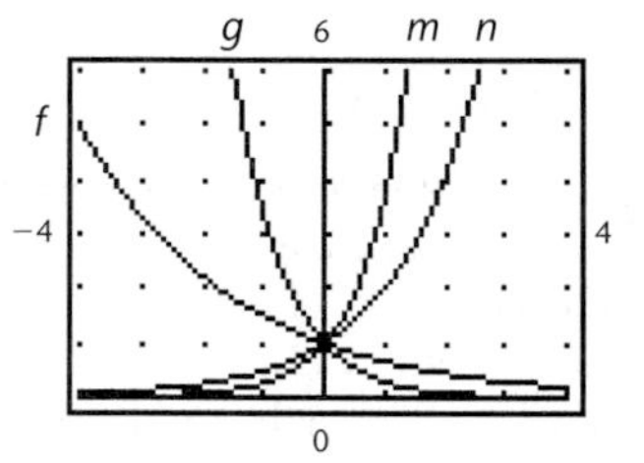

In Problems 3–8, compute answers to four significant digits.

3. $e^2 + e^{-2}$ **4.** $e - e^{-1}$

5. $\sqrt{e}$ **6.** $e^{\sqrt{2}}$

7. e^e **8.** e^{-e}

In Problems 9–14, simplify.

9. $e^{2x}e^{-3x}$ **10.** $(e^{-x})^4$ **11.** $(e^x)^3$

12. $e^{-4x}e^{6x}$ **13.** $\dfrac{e^{5x}}{e^{2x+1}}$ **14.** $\dfrac{e^{4-3x}}{e^{2-5x}}$

15. (A) Explain what is wrong with the following reasoning about the expression $[1 + (1/x)]^x$: As x gets large, $1 + (1/x)$ approaches 1 because $1/x$ approaches 0, and 1 raised to any power is 1, so $[1 + 1/x]^x$ approaches 1.

(B) Which number does $[1 + (1/x)]^x$ approach as x approaches ∞?

16. (A) Explain what is wrong with the following reasoning about the expression $[1 + (1/x)]^x$: If $b > 1$, then the exponential function b^x approaches ∞ as x approaches ∞, and $1 + (1/x)$ is greater than 1, so $[1 + (1/x)]^x$ approaches infinity as $x \to \infty$.

(B) Which number does $[1 + (1/x)]^x$ approach as x approaches ∞?

B

Before graphing the functions in Problems 17–26, classify each function as increasing or decreasing, find the x and y intercepts, and identify any asymptotes. Round any approximate values to two decimal places. Examine the graph to check your answers.

17. $y = -e^x$ **18.** $y = -e^{-x}$

19. $y = 10e^{0.2x}$ **20.** $y = 100e^{0.1x}$

21. $f(t) = 100e^{-0.1t}$ **22.** $g(t) = 10e^{-0.2t}$

23. $F(x) = 2 - e^{-x}$ **24.** $G(x) = e^{2x} - 3$

25. $m(t) = e^{3t} - 2$ **26.** $n(t) = 3 + e^{-2t}$

In Problems 27–32, describe the transformations that can be used to obtain the graph of g from the graph of $f(x) = e^x$ (see Section 1-5). Check your answers by graphing f and g in the same viewing window.

27. $g(x) = e^{x-2}$ **28.** $g(x) = e^{x+3}$

29. $g(x) = e^x + 2$ **30.** $g(x) = e^x - 1$

31. $g(x) = 2e^{-(x+2)}$ **32.** $g(x) = 0.5e^{-(x-1)}$

C

In Problems 33–38, simplify.

33. $\dfrac{-2x^3e^{-2x} - 3x^2e^{-2x}}{x^6}$ **34.** $\dfrac{5x^4e^{5x} - 4x^3e^{5x}}{x^8}$

35. $(e^x + e^{-x})^2 + (e^x - e^{-x})^2$

36. $e^x(e^{-x} + 1) - e^{-x}(e^x + 1)$

37. $\dfrac{e^{-x}(e^x - e^{-x}) + e^{-x}(e^x + e^{-x})}{e^{-2x}}$

38. $\dfrac{e^x(e^x + e^{-x}) - (e^x - e^{-x})e^x}{e^{2x}}$

In Problems 39–42, solve each equation. [Remember: $e^{-x} \neq 0$ for any real number x.]

39. $2xe^{-x} = 0$ **40.** $(x - 3)e^x = 0$

41. $x^2e^x - 5xe^x = 0$ **42.** $3xe^{-x} + x^2e^{-x} = 0$

In Problems 43–50, use a graphing utility to find local extrema, y intercepts, and x intercepts. Investigate the behavior as $x \to \infty$ and as $x \to -\infty$ and identify any horizontal *asymptotes. Round any approximate values to two decimal places.*

43. $f(x) = 2 + e^{x-2}$

44. $g(x) = -3 + e^{1+x}$

45. $m(x) = e^{|x|}$

46. $n(x) = e^{-|x|}$

47. $s(x) = e^{-x^2}$

48. $r(x) = e^{x^2}$

49. $F(x) = \dfrac{200}{1 + 3e^{-x}}$

50. $G(x) = \dfrac{100}{1 + e^{-x}}$

C

51. Use a graphing utility to investigate the behavior of $f(x) = (1 + x)^{1/x}$ as x approaches 0.

52. Use a graphing utility to investigate the behavior of $f(x) = (1 + x)^{1/x}$ as x approaches ∞.

It is common practice in many applications of mathematics to approximate nonpolynomial functions with appropriately selected polynomials. For example, the polynomials in Problems 53–56, called ***Taylor polynomials,*** *can be used to approximate the exponential function $f(x) = e^x$. To illustrate this approximation graphically, in each problem graph $f(x) = e^x$ and the indicated polynomial in the same viewing window, $-4 \le x \le 4$ and $-5 \le y \le 50$.*

53. $P_1(x) = 1 + x + \frac{1}{2}x^2$

54. $P_2(x) = 1 + x + \frac{1}{2}x^2 + \frac{1}{6}x^3$

55. $P_3(x) = 1 + x + \frac{1}{2}x^2 + \frac{1}{6}x^3 + \frac{1}{24}x^4$

56. $P_4(x) = 1 + x + \frac{1}{2}x^2 + \frac{1}{6}x^3 + \frac{1}{24}x^4 + \frac{1}{120}x^5$

57. Investigate the behavior of the functions $f_1(x) = x/e^x$, $f_2(x) = x^2/e^x$, and $f_3(x) = x^3/e^x$ as $x \to \infty$ and as $x \to -\infty$, and find any horizontal asymptotes. Generalize to functions of the form $f_n(x) = x^n/e^x$, where n is any positive integer.

58. Investigate the behavior of the functions $g_1(x) = xe^x$, $g_2(x) = x^2e^x$, and $g_3(x) = x^3e^x$ as $x \to \infty$ and as $x \to -\infty$, and find any horizontal asymptotes. Generalize to functions of the form $g_n(x) = x^ne^x$, where n is any positive integer.

APPLICATIONS

59. Population Growth. If the world population is about 6 billion people now and if the population grows continuously at an annual rate of 1.7%, what will the population be in 10 years? Compute the answer to two significant digits.

60. Population Growth. If the population in Mexico is around 100 million people now and if the population grows continuously at an annual rate of 2.3%, what will the population be in 8 years? Compute the answer to two significant digits.

61. Population Growth. In 1996 the population of Russia was 148 million and the population of Nigeria was 104 million. If the populations of Russia and Nigeria grow continuously at annual rates of -0.62% and 3.0%, respectively, when will Nigeria have a greater population than Russia?

62. Population Growth. In 1996 the population of Germany was 84 million and the population of Egypt was 64 million. If the populations of Germany and Egypt grow continuously at annual rates of -0.15% and 1.9%, respectively, when will Egypt have a greater population than Germany?

63. Space Science. Radioactive isotopes, as well as solar cells, are used to supply power to space vehicles. The isotopes gradually lose power because of radioactive decay. On a particular space vehicle the nuclear energy source has a power output of P watts after t days of use as given by

$$P = 75e^{-0.0035t}$$

Graph this function for $0 \le t \le 100$.

64. Earth Science. The atmospheric pressure P, in pounds per square inch, decreases exponentially with altitude h, in miles above sea level, as given by

$$P = 14.7e^{-0.21h}$$

Graph this function for $0 \le h \le 10$.

65. Marine Biology. Marine life is dependent upon the microscopic plant life that exists in the *photic zone,* a zone that goes to a depth where about 1% of the surface light still remains. Light intensity I relative to depth d, in feet, for one of the clearest bodies of water in the world, the Sargasso Sea in the West Indies, can be approximated by

$$I = I_0e^{-0.00942d}$$

where I_0 is the intensity of light at the surface. What percentage of the surface light will reach a depth of

(A) 50 feet? (B) 100 feet?

66. Marine Biology. Refer to Problem 65. In some waters with a great deal of sediment, the photic zone may go down only 15 to 20 feet. In some murky harbors, the intensity of light d feet below the surface is given approximately by

$$I = I_0e^{-0.23d}$$

What percentage of the surface light will reach a depth of

(A) 10 feet? (B) 20 feet?

67. Money Growth. If you invest \$5,250 in an account paying 11.38% compounded continuously, how much money will be in the account at the end of

(A) 6.25 years? (B) 17 years?

68. Money Growth. If you invest \$7,500 in an account paying 8.35% compounded continuously, how much money will be in the account at the end of

(A) 5.5 years? (B) 12 years?

69. **Money Growth.** *Barron's*, a national business and financial weekly, published the following "Top Savings Deposit Yields" for $2\frac{1}{2}$-year certificate of deposit accounts:

Gill Savings	8.30% (CC)
Richardson Savings and Loan	8.40% (CQ)
USA Savings	8.25% (CD)

where CC represents compounded continuously, CQ compounded quarterly, and CD compounded daily. Compute the value of \$1,000 invested in each account at the end of $2\frac{1}{2}$ years.

70. **Money Growth.** Refer to Problem 69. In another issue of *Barron's*, 1-year certificate of deposit accounts included:

Alamo Savings	8.25% (CQ)
Lamar Savings	8.05% (CC)

Compute the value of \$10,000 invested in each account at the end of 1 year.

★ 71. **Present Value.** A promissory note will pay \$30,000 at maturity 10 years from now. How much should you be willing to pay for the note now if the note gains value at a rate of 9% compounded continuously?

★ 72. **Present Value.** A promissory note will pay \$50,000 at maturity $5\frac{1}{2}$ years from now. How much should you be willing to pay for the note now if the note gains value at a rate of 10% compounded continuously?

73. **AIDS Epidemic.** In June 1996 the World Health Organization estimated that 7.7 million cases of AIDS (acquired immunodeficiency syndrome) had occurred worldwide since the beginning of the epidemic. Assuming that the disease spreads continuously at an annual rate of 17%, estimate the total number of AIDS cases that have occurred by June of the year

(A) 2000 (B) 2004

74. **AIDS Epidemic.** In June 1996 the World Health Organization estimated that 28 million people worldwide had been infected with HIV (human immunodeficiency virus) since the beginning of the AIDS epidemic. Assuming that HIV infection spreads continuously at an annual rate of 19%, estimate the total number of people who have been infected with HIV by June of the year

(A) 2000 (B) 2004

75. **Learning Curve.** People assigned to assemble circuit boards for a computer manufacturing company undergo on-the-job training. From past experience, it was found that the learning curve for the average employee is given by

$$N = 40(1 - e^{-0.12t})$$

where N is the number of boards assembled per day after t days of training.

(A) How many boards can an average employee produce after 3 days of training? After 5 days of training? Round answers to the nearest integer.

(B) How many days of training will it take until an average employee can assemble 25 boards a day? Round answer to the nearest integer.

(C) Does N approach a limiting value as t increases without bound? Explain.

76. **Advertising.** A company is trying to expose a new product to as many people as possible through television advertising in a large metropolitan area with 2 million potential viewers. A model for the number of people N, in millions, who are aware of the product after t days of advertising was found to be

$$N = 2(1 - e^{-0.037t})$$

(A) How many viewers are aware of the product after 2 days? After 10 days? Express answers as integers, rounded to three significant digits.

(B) How many days will it take until half of the potential viewers will become aware of the product? Round answer to the nearest integer.

(C) Does N approach a limiting value as t increases without bound? Explain.

77. **Newton's Law of Cooling.** This law states that the rate at which an object cools is proportional to the difference in temperature between the object and its surrounding medium. The temperature T of the object t hours later is given by

$$T = T_m + (T_0 - T_m)e^{-kt}$$

where T_m is the temperature of the surrounding medium and T_0 is the temperature of the object at $t = 0$. Suppose a bottle of wine at a room temperature of 72°F is placed in the refrigerator to cool before a dinner party. If the temperature in the refrigerator is kept at 40°F and $k = 0.4$, find the temperature of the wine, to the nearest degree, after 3 hours. (In Exercise 4-7 we will find out how to determine k.)

78. **Newton's Law of Cooling.** Refer to Problem 77. What is the temperature, to the nearest degree, of the wine after 5 hours in the refrigerator?

79. **Photography.** An electronic flash unit for a camera is activated when a capacitor is discharged through a filament of wire. After the flash is triggered, and the capacitor is discharged, the circuit (see the figure) is connected and the battery pack generates a current to recharge the capacitor. The time it takes for the capacitor to recharge is called the *recycle time*. For a particular flash unit using a 12-volt battery pack, the charge q, in coulombs, on the capacitor t seconds after recharging has started is given by

$$q = 0.0009(1 - e^{-0.2t})$$

Find the value that q approaches as t increases without bound and interpret.

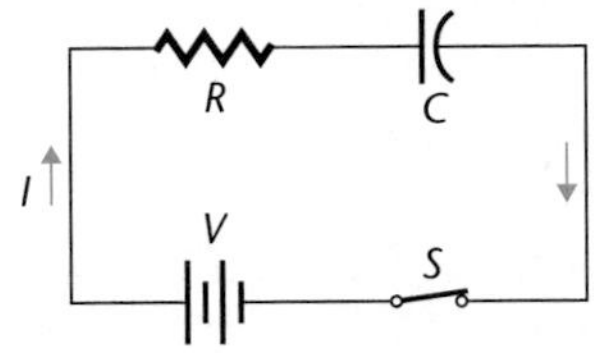

80. **Medicine.** An electronic heart pacemaker utilizes the same type of circuit as the flash unit in Problem 79, but it is designed so that the capacitor discharges 72 times a minute. For a particular pacemaker, the charge on the capacitor t seconds after it starts recharging is given by

$$q = 0.000\,008(1 - e^{-2t})$$

Find the value that q approaches as t increases without bound and interpret.

81. **Wildlife Management.** A herd of 20 white-tailed deer is introduced to a coastal island where there had been no deer before. Their population is predicted to increase according to the logistic curve

$$N = \frac{100}{1 + 4e^{-0.14t}}$$

where N is the number of deer expected in the herd after t years.

(A) How many deer will be present after 2 years? After 6 years? Round answers to the nearest integer.

(B) How many years will it take for the herd to grow to 50 deer? Round answer to the nearest integer.

(C) Does N approach a limiting value as t increases without bound? Explain.

82. **Training.** A trainee is hired by a computer manufacturing company to learn to test a particular model of a personal computer after it comes off the assembly line. The learning curve for an average trainee is given by

$$N = \frac{200}{4 + 21e^{-0.1t}}$$

(A) How many computers can an average trainee be expected to test after 3 days of training? After 6 days? Round answers to the nearest integer.

(B) How many days will it take until an average trainee can test 30 computers per day? Round answer to the nearest integer.

(C) Does N approach a limiting value as t increases without bound? Explain.

Section 4-5 | Logarithmic Functions

- Definition of Logarithmic Function
- From Logarithmic Form to Exponential Form, and Vice Versa
- Properties of Logarithmic Functions

Definition of Logarithmic Function

We now define a new class of functions, called **logarithmic functions,** as inverses of exponential functions. Since exponential functions are one-to-one, their inverses exist. Here you will see why we placed special emphasis on the general concept of inverse functions in Section 4-2. If you know quite a bit about a function, then, based on a knowledge of inverses in general, you will automatically know quite a bit about its inverse. For example, the graph of f^{-1} is the graph of f reflected across the line $y = x$, and the domain and range of f^{-1} are, respectively, the range and domain of f.

If we start with the exponential function

$$f\colon y = 2^x$$

and interchange the variables x and y, we obtain the inverse of f:

$$f^{-1}\colon x = 2^y$$

The graphs of f, f^{-1}, and the line $y = x$ are shown in Figure 1. This new function is given the name **logarithmic function with base 2.** Since we cannot solve the equation $x = 2^y$ for y using the algebra properties discussed so far, we introduce a new symbol to represent this inverse function:

$$y = \log_2 x \quad \text{Read "log to the base 2 of } x\text{."}$$

Thus,

$$y = \log_2 x \qquad \text{is equivalent to} \qquad x = 2^y$$

that is, $\log_2 x$ is the exponent to which 2 must be raised to obtain x. Symbolically, $x = 2^y = 2^{\log_2 x}$.

FIGURE 1
Logarithmic function with base 2.

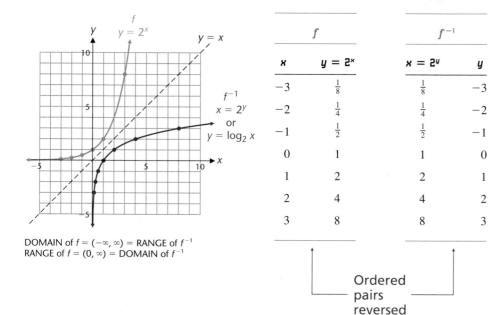

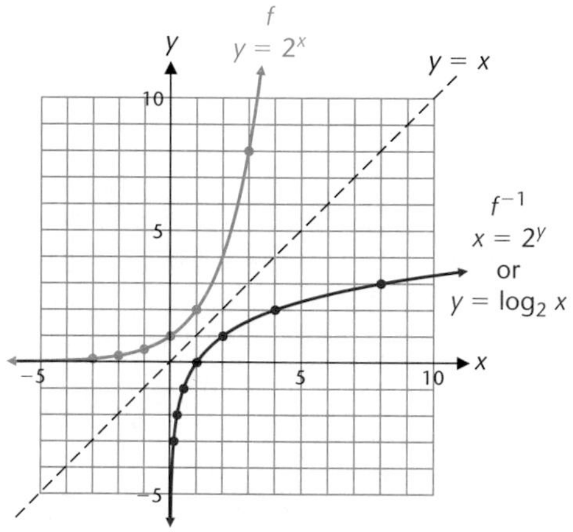

DOMAIN of $f = (-\infty, \infty)$ = RANGE of f^{-1}
RANGE of $f = (0, \infty)$ = DOMAIN of f^{-1}

f		f^{-1}	
x	$y = 2^x$	$x = 2^y$	y
-3	$\frac{1}{8}$	$\frac{1}{8}$	-3
-2	$\frac{1}{4}$	$\frac{1}{4}$	-2
-1	$\frac{1}{2}$	$\frac{1}{2}$	-1
0	1	1	0
1	2	2	1
2	4	4	2
3	8	8	3

Ordered pairs reversed

In general, we define the **logarithmic function with base *b*** to be the inverse of the exponential function with base b ($b > 0$, $b \neq 1$).

DEFINITION 1

DEFINITION OF LOGARITHMIC FUNCTION

For $b > 0$ and $b \neq 1$,

Logarithmic form		Exponential form
$y = \log_b x$	is equivalent to	$x = b^y$

The log to the base b of x is the exponent to which b must be raised to obtain x.

$$y = \log_{10} x \qquad \text{is equivalent to} \qquad x = 10^y$$

$$y = \log_e x \qquad \text{is equivalent to} \qquad x = e^y$$

Remember: A logarithm is an exponent.

It is very important to remember that $y = \log_b x$ and $x = b^y$ define the same function, and as such can be used interchangeably.

Since the domain of an exponential function includes all real numbers and its range is the set of positive real numbers, the **domain** of a logarithmic function is the set of all positive real numbers and its **range** is the set of all real numbers. Thus, $\log_{10} 3$ is defined, but $\log_{10} 0$ and $\log_{10} (-5)$ are not defined. That is, 3 is a logarithmic domain value, but 0 and -5 are not. Typical logarithmic curves are shown in Figure 2.

FIGURE 2
Typical logarithmic graphs.

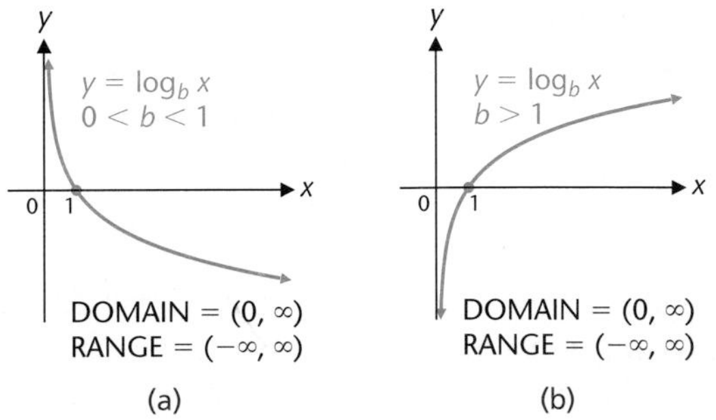

Explore/Discuss 1

For the exponential function $f = \{(x, y) \mid y = (\frac{2}{3})^x\}$, graph f and $y = x$ on the same coordinate system. Then sketch the graph of f^{-1}. Use the Draw Inverse routine on a graphing utility to check your work. Discuss the domains and ranges of f and its inverse. By what other name is f^{-1} known?

From Logarithmic Form to Exponential Form, and Vice Versa

We now look into the matter of converting logarithmic forms to equivalent exponential forms, and vice versa.

EXAMPLE 1 **Logarithmic–Exponential Conversions**

Change each logarithmic form to an equivalent exponential form.

(A) $\log_2 8 = 3$ (B) $\log_{25} 5 = \frac{1}{2}$ (C) $\log_2 (\frac{1}{4}) = -2$

Solutions

(A) $\log_2 8 = 3$ is equivalent to $8 = 2^3$.
(B) $\log_{25} 5 = \frac{1}{2}$ is equivalent to $5 = 25^{1/2}$.
(C) $\log_2 (\frac{1}{4}) = -2$ is equivalent to $\frac{1}{4} = 2^{-2}$.

MATCHED PROBLEM 1

Change each logarithmic form to an equivalent exponential form.

(A) $\log_3 27 = 3$ (B) $\log_{36} 6 = \frac{1}{2}$ (C) $\log_3 (\frac{1}{9}) = -2$

EXAMPLE 2

Logarithmic–Exponential Conversions

Change each exponential form to an equivalent logarithmic form.

(A) $49 = 7^2$ (B) $3 = \sqrt{9}$ (C) $\frac{1}{5} = 5^{-1}$

Solutions

(A) $49 = 7^2$ is equivalent to $\log_7 49 = 2$.

(B) $3 = \sqrt{9}$ is equivalent to $\log_9 3 = \frac{1}{2}$.

(C) $\frac{1}{5} = 5^{-1}$ is equivalent to $\log_5 (\frac{1}{5}) = -1$.

MATCHED PROBLEM 2

Change each exponential form to an equivalent logarithmic form.

(A) $64 = 4^3$ (B) $2 = \sqrt[3]{8}$ (C) $\frac{1}{16} = 4^{-2}$

To gain a little deeper understanding of logarithmic functions and their relationship to the exponential functions, we consider a few problems where we want to find x, b, or y in $y = \log_b x$, given the other two values. All values were chosen so that the problems can be solved without tables or a calculator.

EXAMPLE 3

Solutions of the Equation $y = \log_b x$

Find x, b, or y as indicated.

(A) Find y: $y = \log_4 8$. (B) Find x: $\log_3 x = -2$.

(C) Find b: $\log_b 1{,}000 = 3$.

Solutions

(A) Write $y = \log_4 8$ in equivalent exponential form.

$$8 = 4^y$$

$$2^3 = 2^{2y} \quad \text{Write each number to the same base 2.}$$

$$2y = 3 \quad \text{Recall that } b^m = b^n \text{ if and only if } m = n.$$

$$y = \tfrac{3}{2}$$

Thus, $\frac{3}{2} = \log_4 8$.

(B) Write $\log_3 x = -2$ in equivalent exponential form.

$$x = 3^{-2}$$

$$= \frac{1}{3^2} = \frac{1}{9}$$

Thus, $\log_3 (\frac{1}{9}) = -2$.

(C) Write $\log_b 1{,}000 = 3$ in equivalent exponential form:

$$1{,}000 = b^3$$

$$10^3 = b^3 \quad \text{Write 1,000 as a third power.}$$

$$b = 10$$

Thus, $\log_{10} 1{,}000 = 3$.

MATCHED PROBLEM 3

Find x, b, or y as indicated.

(A) Find y: $y = \log_9 27$. (B) Find x: $\log_2 x = -3$.

(C) Find b: $\log_b 100 = 2$.

Properties of Logarithmic Functions

The familiar properties of exponential functions imply corresponding properties of logarithmic functions.

Explore/Discuss 2

Discuss the connection between the exponential equation and the logarithmic equation, and explain why each equation is valid.

(A) $2^4 2^7 = 2^{11}$; $\log_2 2^4 + \log_2 2^7 = \log_2 2^{11}$

(B) $2^{13}/2^5 = 2^8$; $\log_2 2^{13} - \log_2 2^5 = \log_2 2^8$

(C) $(2^6)^9 = 2^{54}$; $9 \log_2 2^6 = \log_2 2^{54}$

Several of the powerful and useful properties of logarithmic functions are listed in Theorem 1.

THEOREM 1

PROPERTIES OF LOGARITHMIC FUNCTIONS

If b, M, and N are positive real numbers, $b \neq 1$, and p and x are real numbers, then

1. $\log_b 1 = 0$
2. $\log_b b = 1$
3. $\log_b b^x = x$
4. $b^{\log_b x} = x, x > 0$
5. $\log_b MN = \log_b M + \log_b N$
6. $\log_b \dfrac{M}{N} = \log_b M - \log_b N$
7. $\log_b M^p = p \log_b M$
8. $\log_b M = \log_b N$ if and only if $M = N$

The first two properties in Theorem 1 follow directly from the definition of a logarithmic function:

$$\log_b 1 = 0 \quad \text{since} \quad b^0 = 1$$

$$\log_b b = 1 \quad \text{since} \quad b^1 = b$$

The third and fourth properties look more complicated than they are. They follow directly from the fact that exponential and logarithmic functions are inverses of each other. Recall from Section 4-2 that if f is one-to-one, then f^{-1} is a one-to-one function satisfying

$$f^{-1}(f(x)) = x \quad \text{and} \quad f(f^{-1}(x)) = x$$

Applying these general properties to $f(x) = b^x$ and $f^{-1}(x) = \log_b x$, we see that

$$f^{-1}(f(x)) = x \qquad f(f^{-1}(x)) = x$$
$$\log_b (f(x)) = x \qquad b^{f^{-1}(x)} = x$$
$$\log_b b^x = x \qquad b^{\log_b x} = x$$

Properties 5 to 7 enable us to convert multiplication into addition, division into subtraction, and power and root problems into multiplication. The proofs of these properties are based on properties of exponents. A sketch of a proof of the fifth property follows: To bring exponents into the proof, we let

$$u = \log_b M \qquad \text{and} \qquad v = \log_b N$$

and convert these to the equivalent exponential forms

$$M = b^u \qquad \text{and} \qquad N = b^v$$

Now, see if you can provide the reasons for each of the following steps:

$$\log_b MN = \log_b b^u b^v = \log_b b^{u+v} = u + v = \log_b M + \log_b N$$

The other properties are established in a similar manner (see Problems 111 and 112 in Exercise 4-5.)

Finally, the eighth property follows from the fact that logarithmic functions are one-to-one.

We now illustrate the use of these properties in several examples.

EXAMPLE 4 **Using Logarithmic Properties**

Simplify using the properties in Theorem 1.

(A) $\log_e 1$ (B) $\log_{10} 10$ (C) $\log_e e^{2x+1}$

(D) $\log_{10} 0.01$ (E) $10^{\log_{10} 7}$ (F) $e^{\log_e x^2}$

Solutions

(A) $\log_e 1 = 0$ (B) $\log_{10} 10 = 1$

(C) $\log_e e^{2x+1} = 2x + 1$ (D) $\log_{10} 0.01 = \log_{10} 10^{-2} = -2$

(E) $10^{\log_{10} 7} = 7$ (F) $e^{\log_e x^2} = x^2$

MATCHED PROBLEM 4

Simplify using the properties in Theorem 1.

(A) $\log_{10} 10^{-5}$ (B) $\log_5 25$ (C) $\log_{10} 1$

(D) $\log_e e^{m+n}$ (E) $10^{\log_{10} 4}$ (F) $e^{\log_e (x^4+1)}$

EXAMPLE 5 **Using Logarithmic Properties**

Write in terms of simpler logarithmic forms.

(A) $\log_b 3x$ (B) $\log_b \dfrac{x}{5}$ (C) $\log_b x^7$

(D) $\log_b \dfrac{mn}{pq}$ (E) $\log_b (mn)^{2/3}$ (F) $\log_b \dfrac{x^8}{y^{1/5}}$

Solutions

(A) $\log_b 3x = \log_b 3 + \log_b x$ — $\log_b MN = \log_b M + \log_b N$

(B) $\log_b \frac{x}{5} = \log_b x - \log_b 5$ — $\log_b \frac{M}{N} = \log_b M - \log_b N$

(C) $\log_b x^7 = 7 \log_b x$ — $\log_b M^p = p \log_b M$

(D) $\log_b \frac{mn}{pq} = \log_b mn - \log_b pq$ — $\log_b \frac{M}{N} = \log_b M - \log_b N$

$= \log_b m + \log_b n - (\log_b p + \log_b q)$ — $\log_b MN = \log_b M + \log_b N$

$= \log_b m + \log_b n - \log_b p - \log_b q$

(E) $\log_b (mn)^{2/3} = \frac{2}{3} \log_b mn$ — $\log_b M^p = p \log_b M$

$= \frac{2}{3}(\log_b m + \log_b n)$ — $\log_b MN = \log_b M + \log_b N$

(F) $\log_b \frac{x^8}{y^{1/5}} = \log_b x^8 - \log_b y^{1/5}$ — $\log_b \frac{M}{N} = \log_b M - \log_b N$

$= 8 \log_b x - \frac{1}{5} \log_b y$ — $\log_b M^p = p \log_b M$

MATCHED PROBLEM 5

Write in terms of simpler logarithmic forms, as in Example 5.

(A) $\log_b \frac{r}{uv}$ (B) $\log_b \left(\frac{m}{n}\right)^{3/5}$ (C) $\log_b \frac{u^{1/3}}{v^5}$

EXAMPLE 6

Using Logarithmic Properties

If $\log_e 3 = 1.10$ and $\log_e 7 = 1.95$, find

(A) $\log_e (\frac{7}{3})$ (B) $\log_e \sqrt[3]{21}$

Solutions

(A) $\log_e (\frac{7}{3}) = \log_e 7 - \log_e 3 = 1.95 - 1.10 = 0.85$

(B) $\log_e \sqrt[3]{21} = \log_e (21)^{1/3} = \frac{1}{3} \log_e (3 \cdot 7) = \frac{1}{3}(\log_e 3 + \log_e 7)$

$= \frac{1}{3}(1.10 + 1.95) = 1.02$

MATCHED PROBLEM 6

If $\log_e 5 = 1.609$ and $\log_e 8 = 2.079$, find

(A) $\log_e \frac{5^{10}}{8}$ (B) $\log_e \sqrt[4]{\frac{8}{5}}$

The following example and problem, though somewhat artificial, will give you additional practice in using the properties in Theorem 1.

EXAMPLE 7

Using Logarithmic Properties

Find x so that $\log_b x = \frac{2}{3} \log_b 27 + 2 \log_b 2 - \log_b 3$ without using a calculator or table.

Solution First we use properties from Theorem 1 to express the right side as the logarithm of a single number.

$$\begin{aligned}\log_b x &= \tfrac{2}{3}\log_b 27 + 2\log_b 2 - \log_b 3\\ &= \log_b 27^{2/3} + \log_b 2^2 - \log_b 3\\ &= \log_b 9 + \log_b 4 - \log_b 3 && 27^{2/3} = 9;\ 2^2 = 4\\ &= \log_b \frac{9\cdot 4}{3} = \log_b 12 && \text{Properties 5 and 6 of Theorem 1}\end{aligned}$$

Thus,

$$\log_b x = \log_b 12$$

Now we use property 8 of Theorem 1 to find x:

$$x = 12$$

MATCHED PROBLEM 7 Find x so that $\log_b x = \frac{2}{3}\log_b 8 + \frac{1}{2}\log_b 9 - \log_b 6$ without using a calculator or table.

CAUTION

We conclude this section by noting two common errors:

1. $\dfrac{\log_b M}{\log_b N} \neq \log_b M - \log_b N$ — $\log_b M - \log_b N = \log_b \dfrac{M}{N}$; $\dfrac{\log_b M}{\log_b N}$ cannot be simplified.
2. $\log_b (M + N) \neq \log_b M + \log_b N$ — $\log_b M + \log_b N = \log_b MN$; $\log_b (M + N)$ cannot be simplified.

Answers to Matched Problems

1. (A) $27 = 3^3$ (B) $6 = 36^{1/2}$ (C) $\frac{1}{9} = 3^{-2}$ **2.** (A) $\log_4 64 = 3$ (B) $\log_8 2 = \frac{1}{3}$ (C) $\log_4 (\frac{1}{16}) = -2$
3. (A) $y = \frac{3}{2}$ (B) $x = \frac{1}{8}$ (C) $b = 10$ **4.** (A) -5 (B) 2 (C) 0 (D) $m + n$ (E) 4 (F) $x^4 + 1$
5. (A) $\log_b r - \log_b u - \log_b v$ (B) $\frac{3}{5}(\log_b m - \log_b n)$ (C) $\frac{1}{3}\log_b u - 5\log_b v$
6. (A) 14.01 (to four significant digits) (B) 0.1175 (to four significant digits) **7.** $x = 2$

EXERCISE 4-5

A

Rewrite Problems 1–8 in equivalent exponential form.

1. $\log_3 81 = 4$

2. $\log_5 125 = 3$

3. $\log_{10} 0.001 = -3$

4. $\log_{10} 1{,}000 = 3$

5. $\log_{81} 3 = \frac{1}{4}$

6. $\log_4 2 = \frac{1}{2}$

7. $\log_{1/2} 16 = -4$

8. $\log_{1/3} 27 = -3$

Rewrite Problems 9–16 in equivalent logarithmic form.

9. $0.0001 = 10^{-4}$

10. $10{,}000 = 10^4$

11. $8 = 4^{3/2}$

12. $9 = 27^{2/3}$

13. $\frac{1}{2} = 32^{-1/5}$

14. $\frac{1}{8} = 2^{-3}$

15. $7 = \sqrt{49}$

16. $4 = \sqrt[3]{64}$

In Problems 17–30, simplify each expression using Theorem 1.

17. $\log_{16} 1$

18. $\log_{25} 1$

19. $\log_{0.5} 0.5$

20. $\log_7 7$

21. $\log_e e^4$

22. $\log_{10} 10^5$

23. $\log_{10} 0.01$

24. $\log_{10} 100$

25. $\log_5 \sqrt[3]{5}$

26. $\log_2 \sqrt{8}$

27. $e^{\log_e \sqrt{x}}$

28. $e^{\log_e (x-1)}$

29. $e^{2 \log_e x}$

30. $10^{-3 \log_{10} u}$

B

Find x, y, or b, as indicated in Problems 31–44.

31. $\log_2 x = 2$

32. $\log_3 x = 3$

33. $\log_4 16 = y$

34. $\log_8 64 = y$

35. $\log_b 16 = 2$

36. $\log_b 10^{-3} = -3$

37. $\log_b 1 = 0$

38. $\log_b b = 1$

39. $\log_4 x = \frac{1}{2}$

40. $\log_8 x = \frac{1}{3}$

41. $\log_{1/3} 9 = y$

42. $\log_{49} (\frac{1}{7}) = y$

43. $\log_b 1{,}000 = \frac{3}{2}$

44. $\log_b 4 = \frac{2}{3}$

Write Problems 45–58 in terms of simpler logarithmic forms (see Example 5).

45. $\log_b u^2 v^7$

46. $\log_b u^{1/2} v^{1/3}$

47. $\log_b \dfrac{m^{2/3}}{n^{1/2}}$

48. $\log_b \dfrac{u^3}{v^5}$

49. $\log_b \dfrac{u}{vw}$

50. $\log_b \dfrac{uv}{w}$

51. $\log_b \dfrac{1}{a^2}$

52. $\log_b \dfrac{1}{M^5}$

53. $\log_b \sqrt[3]{x^2 - y^2}$

54. $\log_b \sqrt{u^2 + 1}$

55. $\log_b \dfrac{\sqrt[3]{N}}{p^2 q^3}$

56. $\log_b \dfrac{m^5 n^3}{\sqrt{p}}$

57. $\log_b \sqrt[4]{\dfrac{x^2 y^3}{\sqrt{z}}}$

58. $\log_b \sqrt[5]{\left(\dfrac{x}{y^4 z^9}\right)^3}$

In Problems 59–68, write each expression in terms of a single logarithm with a coefficient of 1. Example: $\log_b u^2 - \log_b v = \log_b (u^2/v)$.

59. $2 \log_b x - \log_b y$

60. $\log_b m - \frac{1}{2} \log_b n$

61. $\log_b w - \log_b x - \log_b y$

62. $\log_b w + \log_b x - \log_b y$

63. $3 \log_b x + 2 \log_b y - \frac{1}{4} \log_b z$

64. $\frac{1}{3} \log_b w - 3 \log_b x - 5 \log_b y$

65. $5(\frac{1}{2} \log_b u - 2 \log_b v)$

66. $7(4 \log_b m + \frac{1}{3} \log_b n)$

67. $\frac{1}{5} (2 \log_b x + 3 \log_b y)$

68. $\frac{1}{3}(4 \log_b x - 2 \log_b y)$

C

In Problems 69–76, write each expression in terms of logarithms of first-degree polynomials. Example:

$$\log_b \frac{(2x + 1)^3}{(3x - 5)^4} = 3 \log_b (2x + 1) - 4 \log_b (3x - 5)$$

69. $\log_b [(x + 3)^5 (2x - 7)^2]$

70. $\log_b [(5x - 4)^3 (3x + 2)^4]$

71. $\log_b \dfrac{(x + 10)^7}{(1 + 10x)^2}$

72. $\log_b \dfrac{(x - 3)^5}{(5 + x)^3}$

73. $\log_b \dfrac{x^2}{\sqrt{x + 1}}$

74. $\log_b \dfrac{\sqrt{x - 1}}{x^3}$

75. $\log_b (x^4 + x^3 - 20x^2)$

76. $\log_b (x^5 + 5x^4 - 14x^3)$

In Problems 77–86, solve for x without using a calculator or table.

77. $\log_2 (x + 5) = 2 \log_2 3$

78. $\log_{10} (5 - x) = 3 \log_{10} 2$

79. $2 \log_5 x = \log_5 (x^2 - 6x + 2)$

80. $\log_{10} (x^2 - 2x - 2) = 2 \log_{10} (x - 2)$

81. $\log_e (x + 8) - \log_e x = 3 \log_e 2$

82. $\log_7 4x - \log_7 (x + 1) = \frac{1}{2} \log_7 4$

83. $2 \log_3 x = \log_3 2 + \log_3 (4 - x)$

84. $\log_4 x + \log_4 (x + 2) = \frac{1}{2} \log_4 9$

85. $3 \log_b 2 + \frac{1}{2} \log_b 25 - \log_b 20 = \log_b x$

86. $\frac{3}{2} \log_b 4 - \frac{2}{3} \log_b 8 + 2 \log_b 2 = \log_b x$

If $\log_b 2 = 0.69$, $\log_b 3 = 1.10$, and $\log_b 5 = 1.61$, find the value of each expression in Problems 87–96.

87. $\log_b 30$ **88.** $\log_b 12$ **89.** $\log_b \frac{2}{5}$

90. $\log_b \frac{5}{3}$ **91.** $\log_b 27$ **92.** $\log_b 16$

93. $\log_b \sqrt[3]{2}$ **94.** $\log_b \sqrt{3}$ **95.** $\log_b \sqrt{0.9}$

96. $\log_b \sqrt[3]{1.5}$

C

In Problems 97–100,

(A) Use the graph of $y = \log_2 x$ (Fig. 1) and graph transformations to sketch the graph of f.

(B) Find f^{-1} and use the Draw Inverse routine on a graphing utility to check the graph in part A.

97. $f(x) = \log_2 (x - 2)$

98. $f(x) = \log_2 (x + 3)$

99. $f(x) = \log_2 x - 2$

100. $f(x) = \log_2 x + 3$

101. (A) For $f = \{(x, y) \mid y = (\frac{1}{2})^x = 2^{-x}\}$, graph f, f^{-1}, and $y = x$ on the same coordinate system.

(B) Indicate the domain and range of f and f^{-1}.

(C) What other name can you use for the inverse of f?

102. (A) For $f = \{(x, y) \mid y = (\frac{1}{3})^x = 3^{-x}\}$, graph f, f^{-1}, and $y = x$ on the same coordinate system.

(B) Indicate the domain and range of f and f^{-1}.

(C) What other name can you use for the inverse of f?

Find the inverse of each function in Problems 103–106.

103. $f(x) = 5^{3x-1} + 4$

104. $g(x) = 3^{2x-3} - 2$

105. $g(x) = 3 \log_e (5x - 2)$

106. $f(x) = 2 + \log_e (5x - 3)$

107. Explain why the graph of the reflection of the function $y = 3^{x^2}$ in the line $y = x$ is not the graph of a function.

108. Explain why the graph of the reflection of the function $y = 2^{|x|}$ in the line $y = x$ is not the graph of a function.

109. Write $\log_e x - \log_e 100 = -0.08t$ in an exponential form that is free of logarithms.

110. Write $\log_e x - \log_e C + kt = 0$ in an exponential form that is free of logarithms.

111. Prove that $\log_b (M/N) = \log_b M - \log_b N$ under the hypotheses of Theorem 1.

112. Prove that $\log_b M^p = p \log_b M$ under the hypotheses of Theorem 1.

Section 4-6 | Common and Natural Logarithms

- Common and Natural Logarithmic Functions
- Applications

John Napier (1550–1617) is credited with the invention of logarithms, which evolved out of an interest in reducing the computational strain in research in astronomy. This new computational tool was immediately accepted by the scientific world. Now, with the availability of inexpensive calculators, logarithms have lost most of their importance as a computational device. However, the logarithmic concept has been greatly generalized since its conception, and logarithmic functions are used widely in both theoretical and applied sciences.

Of all possible logarithmic bases, the base e and the base 10 are used almost exclusively. Before we can use logarithms in certain practical problems, we need to be able to approximate the logarithm of any positive number to either base 10 or base e. And conversely, if we are given the logarithm of a number to base 10 or base e, we need to be able to approximate the number. Historically, tables were used for this purpose, but now calculators are used since they are faster and can find far more values than any table can possibly include.

Common and Natural Logarithmic Functions

Common logarithms, also called **Briggsian logarithms,** are logarithms with base 10. **Natural logarithms,** also called **Naperian logarithms,** are logarithms with base *e*. Most calculators have a function key labeled "log" and a function key labeled "ln." The former represents the common logarithmic function and the latter the natural logarithmic function. In fact, "log" and "ln" are both used extensively in mathematical literature, and whenever you see either used in this book without a base indicated, they should be interpreted as in the following box.

LOGARITHMIC FUNCTIONS

$y = \log x = \log_{10} x$ Common logarithmic function

$y = \ln x = \log_e x$ Natural logarithmic function

Explore/Discuss 1

(A) Sketch the graph of $y = 10^x$, $y = \log x$, and $y = x$ in the same coordinate system and state the domain and range of the common logarithmic function.

(B) Sketch the graph of $y = e^x$, $y = \ln x$, and $y = x$ in the same coordinate system and state the domain and range of the natural logarithmic function.

EXAMPLE 1 Calculator Evaluation of Logarithms

Use a calculator to evaluate each to six decimal places.

(A) log 3,184 (B) ln 0.000 349 (C) log (−3.24)

Solutions

(A) $\log 3{,}184 = 3.502\ 973$

(B) $\ln 0.000\ 349 = -7.960\ 439$

(C) $\log(-3.24) =$ Error

Why is an error indicated in part C? Because −3.24 is not in the domain of the log function. [*Note:* Calculators display error messages in various ways. Some calculators use a more advanced definition of logarithmic functions that involves complex numbers. They will display an ordered pair, representing a complex number, as the value of log (−3.24), rather than an error message. You should interpret such a display as indicating that the number entered is not in the domain of the logarithmic function as we have defined it.]

MATCHED PROBLEM 1

Use a calculator to evaluate each to six decimal places.

(A) log 0.013 529 (B) ln 28.693 28 (C) ln (−0.438)

When working with common and natural logarithms, we follow the common practice of using the equal sign "=" where it might be more appropriate to use the approximately equal sign "≈." No harm is done as long as we keep in mind that in a statement such as $\log 3.184 = 0.503$, the number on the right is only assumed accurate to three decimal places and is not exact.

Explore/Discuss 2

Graphs of the functions $f(x) = \log x$ and $g(x) = \ln x$ are shown in the graphing utility display of Figure 1. Which graph belongs to which function? It appears from the display that one of the functions may be a constant multiple of the other. Is that true? Find and discuss the evidence for your answer.

FIGURE 1

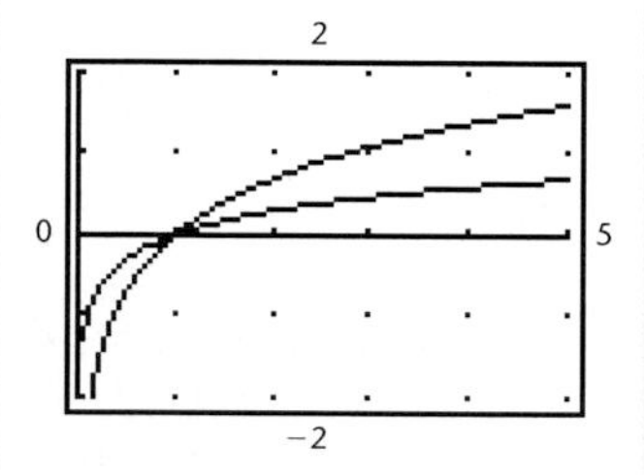

EXAMPLE 2

Calculator Evaluation of Logarithms

Use a calculator to evaluate each expression to three decimal places.

(A) $\dfrac{\log 2}{\log 1.1}$ (B) $\log \dfrac{2}{1.1}$ (C) $\log 2 - \log 1.1$

Solutions

(A) $\dfrac{\log 2}{\log 1.1} = 7.273$

(B) $\log \dfrac{2}{1.1} = 0.260$

(C) $\log 2 - \log 1.1 = 0.260$. Note that $\dfrac{\log 2}{\log 1.1} \neq \log 2 - \log 1.1$, but $\log \dfrac{2}{1.1} = \log 2 - \log 1.1$ (see Theorem 1, Section 4-5).

MATCHED PROBLEM 2

Use a calculator to evaluate each to three decimal places.

(A) $\dfrac{\ln 3}{\ln 1.08}$ (B) $\ln \dfrac{3}{1.08}$ (C) $\ln 3 - \ln 1.08$

We now turn to the second problem: Given the logarithm of a number, find the number. To solve this problem, we make direct use of the logarithmic–exponential relationships discussed in Section 4-5.

LOGARITHMIC–EXPONENTIAL RELATIONSHIPS

$\log x = y$ is equivalent to $x = 10^y$

$\ln x = y$ is equivalent to $x = e^y$

EXAMPLE 3

Solving $\log_b x = y$ for x

Find x to three significant digits, given the indicated logarithms.

(A) $\log x = -9.315$ (B) $\ln x = 2.386$

Solutions

(A) $\log x = -9.315$

$x = 10^{-9.315}$ Change to equivalent exponential form.

$= 4.84 \times 10^{-10}$

Notice that the answer is displayed in scientific notation in the calculator.

(B) $\ln x = 2.386$

$x = e^{2.386}$ Change to equivalent exponential form.

$= 10.9$

MATCHED PROBLEM 3

Find x to four significant digits, given the indicated logarithms.

(A) $\ln x = -5.062$ (B) $\log x = 12.0821$

Explore/Discuss 3

Example 3 was solved algebraically using the logarithmic–exponential relationships. Use the intersection routine on a graphing utility to solve this problem graphically. Discuss the relative merits of the two approaches.

Applications

We now consider three applications that are solved using common and natural logarithms. The first application concerns sound intensity; the second, earthquake intensity; and the third, rocket flight theory.

Sound Intensity The human ear is able to hear sound over an incredible range of intensities. The loudest sound a healthy person can hear without damage to the eardrum has an intensity 1 trillion (1,000,000,000,000) times that of the softest sound a person can hear. Working directly with numbers over such a wide range is very cumbersome. Since the logarithm, with base greater than 1, of a number increases much more slowly than the number itself, logarithms are often used to create more convenient compressed scales. The decibel scale for sound intensity is an example of such a scale. The **decibel,** named after the inventor of the telephone, Alexander Graham Bell (1847–1922), is defined as follows:

$$D = 10 \log \frac{I}{I_0} \qquad \text{Decibel scale} \tag{1}$$

where D is the **decibel level** of the sound, I is the **intensity** of the sound measured in watts per square meter (W/m^2), and I_0 is the intensity of the least audible sound that an average healthy young person can hear. The latter is standardized to be $I_0 = 10^{-12}$ watt per square meter. Table 1 lists some typical sound intensities from familiar sources.

TABLE 1 Typical Sound Intensities

Sound Intensity (W/m^2)	Sound
1.0×10^{-12}	Threshold of hearing
5.2×10^{-10}	Whisper
3.2×10^{-6}	Normal conversation
8.5×10^{-4}	Heavy traffic
3.2×10^{-3}	Jackhammer
1.0×10^{0}	Threshold of pain
8.3×10^{2}	Jet plane with afterburner

EXAMPLE 4

Sound Intensity

Find the number of decibels from a whisper with sound intensity 5.20×10^{-10} watt per square meter. Compute the answer to two decimal places.

Solution

We use the decibel formula (1):

$$\begin{aligned} D &= 10 \log \frac{I}{I_0} \\ &= 10 \log \frac{5.2 \times 10^{-10}}{10^{-12}} \\ &= 10 \log 520 \\ &= 27.16 \text{ decibels} \end{aligned}$$

MATCHED PROBLEM 4

Find the number of decibels from a jackhammer with sound intensity 3.2×10^{-3} watt per square meter. Compute the answer to two decimal places.

Explore/Discuss 4

Imagine using a large sheet of graph paper, ruled with horizontal and vertical lines $\frac{1}{8}$ inch apart, to plot the sound intensities of Table 1 on the x axis and the corresponding decibel levels on the y axis. Suppose that each $\frac{1}{8}$-inch unit on the x axis represents the intensity of the least audible sound (10^{-12} W/m^2), and each $\frac{1}{8}$-inch unit on the y axis represents 1 decibel. If the point corresponding to a jet plane with afterburner is plotted on the graph paper, how far is it from the x axis? From the y axis? (Give the first answer in inches and the second in miles!) Discuss.

Earthquake Intensity The energy released by the largest earthquake recorded, measured in joules, is about 100 billion (100,000,000,000) times the energy released by a small earthquake that is barely felt. Over the past 150 years several people from various countries have devised different types of measures of earthquake magnitudes so that their severity could be easily compared. In 1935 the California seismologist Charles Richter devised a logarithmic scale that bears his name and is still widely used in the United States. The **magnitude** M on the **Richter scale*** is given as follows:

$$M = \frac{2}{3}\log\frac{E}{E_0} \qquad \text{Richter scale} \qquad (2)$$

where E is the energy released by the earthquake, measured in joules, and E_0 is the energy released by a very small reference earthquake which has been standardized to be

$$E_0 = 10^{4.40} \text{ joules}$$

The destructive power of earthquakes relative to magnitudes on the Richter scale is indicated in Table 2.

TABLE 2 The Richter Scale

Magnitude on Richter Scale	Destructive Power
$M < 4.5$	Small
$4.5 < M < 5.5$	Moderate
$5.5 < M < 6.5$	Large
$6.5 < M < 7.5$	Major
$7.5 < M$	Greatest

*Originally, Richter defined the magnitude of an earthquake in terms of logarithms of the maximum seismic wave amplitude, in thousandths of a millimeter, measured on a standard seismograph. Formula (2) gives essentially the same magnitude that Richter obtained for a given earthquake but in terms of logarithms of the energy released by the earthquake.

EXAMPLE 5 **Earthquake Intensity**

The 1906 San Francisco earthquake released approximately 5.96×10^{16} joules of energy. What was its magnitude on the Richter scale? Compute the answer to two decimal places.

Solution We use the magnitude formula (2):

$$\begin{aligned} M &= \frac{2}{3} \log \frac{E}{E_0} \\ &= \frac{2}{3} \log \frac{5.96 \times 10^{16}}{10^{4.40}} \\ &= 8.25 \end{aligned}$$

MATCHED PROBLEM 5 The 1985 earthquake in central Chile released approximately 1.26×10^{16} joules of energy. What was its magnitude on the Richter scale? Compute the answer to two decimal places.

EXAMPLE 6 **Earthquake Intensity**

If the energy release of one earthquake is 1,000 times that of another, how much larger is the Richter scale reading of the larger than the smaller?

Solution Let

$$M_1 = \frac{2}{3} \log \frac{E_1}{E_0} \qquad \text{and} \qquad M_2 = \frac{2}{3} \log \frac{E_2}{E_0}$$

be the Richter equations for the smaller and larger earthquakes, respectively. Substituting $E_2 = 1{,}000E_1$ into the second equation, we obtain

$$\begin{aligned} M_2 &= \frac{2}{3} \log \frac{1{,}000E_1}{E_0} \\ &= \frac{2}{3} \left(\log 10^3 + \log \frac{E_1}{E_0} \right) \\ &= \frac{2}{3}(3) + \frac{2}{3} \log \frac{E_1}{E_0} \\ &= 2 + M_1 \end{aligned}$$

Thus, an earthquake with 1,000 times the energy of another has a Richter scale reading of 2 more than the other.

MATCHED PROBLEM 6 If the energy release of one earthquake is 10,000 times that of another, how much larger is the Richter scale reading of the larger than the smaller?

Rocket Flight Theory The theory of rocket flight uses advanced mathematics and physics to show that the velocity v of a rocket at burnout (depletion of fuel supply) is given by

$$v = c \ln \frac{W_t}{W_b} \qquad \text{Rocket equation} \qquad (3)$$

where c is the exhaust velocity of the rocket engine, W_t is the takeoff weight (fuel, structure, and payload), and W_b is the burnout weight (structure and payload).

Because of the Earth's atmospheric resistance, a launch vehicle velocity of at least 9.0 kilometers per second is required to achieve the minimum altitude needed for a stable orbit. It is clear that to increase velocity v, either the weight ratio W_t/W_b must be increased or the exhaust velocity c must be increased. The weight ratio can be increased by the use of solid fuels, and the exhaust velocity can be increased by improving the fuels, solid or liquid.

EXAMPLE 7 **Rocket Flight Theory**

A typical single-stage, solid-fuel rocket may have a weight ratio $W_t/W_b = 18.7$ and an exhaust velocity $c = 2.38$ kilometers per second. Would this rocket reach a launch velocity of 9.0 kilometers per second?

Solution We use the rocket equation (3):

$$\begin{aligned} v &= c \ln \frac{W_t}{W_b} \\ &= 2.38 \ln 18.7 \\ &= 6.97 \text{ kilometers per second} \end{aligned}$$

The velocity of the launch vehicle is far short of the 9.0 kilometers per second required to achieve orbit. This is why multiple-stage launchers are used—the deadweight from a preceding stage can be jettisoned into the ocean when the next stage takes over.

MATCHED PROBLEM 7 A launch vehicle using liquid fuel, such as a mixture of liquid hydrogen and liquid oxygen, can produce an exhaust velocity of $c = 4.7$ kilometers per second. However, the weight ratio W_t/W_b must be low—around 5.5 for some vehicles—because of the increased structural weight to accommodate the liquid fuel. How much more or less than the 9.0 kilometers per second required to reach orbit will be achieved by this vehicle?

Answers to Matched Problems

1. (A) $-1.868\ 734$ (B) $3.356\ 663$ (C) Not possible **2.** (A) 14.275 (B) 1.022 (C) 1.022
3. (A) $x = 0.006\ 333$ (B) $x = 1.21 \times 10^{12}$ **4.** 95.05 decibels **5.** 7.80 **6.** 2.67 **7.** 1 km/s less

EXERCISE 4-6

A

In Problems 1–8, evaluate to four decimal places.

1. $\log 82{,}734$
2. $\log 843{,}250$
3. $\log 0.001\,439$
4. $\log 0.035\,604$
5. $\ln 43.046$
6. $\ln 2{,}843{,}100$
7. $\ln 0.081\,043$
8. $\ln 0.000\,032\,4$

In Problems 9–16, evaluate x to four significant digits, given:

9. $\log x = 5.3027$
10. $\log x = 1.9168$
11. $\log x = -3.1773$
12. $\log x = -2.0411$
13. $\ln x = 3.8655$
14. $\ln x = 5.0884$
15. $\ln x = -0.3916$
16. $\ln x = -4.1083$

B

In Problems 17–24, evaluate to three decimal places.

17. $n = \dfrac{\log 2}{\log 1.15}$
18. $n = \dfrac{\log 2}{\log 1.12}$
19. $n = \dfrac{\ln 3}{\ln 1.15}$
20. $n = \dfrac{\ln 4}{\ln 1.2}$
21. $x = \dfrac{\ln 0.5}{-0.21}$
22. $x = \dfrac{\ln 0.1}{-0.0025}$
23. $t = \dfrac{\ln 150}{\ln 3}$
24. $t = \dfrac{\log 200}{\log 2}$

In Problems 25–32, evaluate x to five significant digits.

25. $x = \log (5.3147 \times 10^{12})$
26. $x = \log (2.0991 \times 10^{17})$
27. $x = \ln (6.7917 \times 10^{-12})$
28. $x = \ln (4.0304 \times 10^{-8})$
29. $\log x = 32.068\,523$
30. $\log x = -12.731\,64$
31. $\ln x = -14.667\,13$
32. $\ln x = 18.891\,143$

In Problems 33–36, find f^{-1}. Check by graphing f, f^{-1}, and $y = x$ in the same viewing window on a graphing utility.

33. $f(x) = 2 \ln (x + 2)$
34. $f(x) = 2 \ln x + 2$
35. $f(x) = 4 \ln x - 3$
36. $f(x) = 4 \ln (x - 3)$

C

In Problems 37–40, find domain and range, x and y intercepts, and asymptotes. Round all approximate values to two decimal places.

37. $f(x) = -2 + \ln (1 + x^2)$
38. $f(x) = 2 - \ln (1 + |x|)$
39. $f(x) = 1 + \ln (1 - x^2)$
40. $f(x) = -1 + \ln (|1 - x^2|)$

41. Find the fallacy.

$$1 < 3$$
$$\tfrac{1}{27} < \tfrac{3}{27} \qquad \text{Divide both sides by 27.}$$
$$\tfrac{1}{27} < \tfrac{1}{9}$$
$$(\tfrac{1}{3})^3 < (\tfrac{1}{3})^2$$
$$\log (\tfrac{1}{3})^3 < \log (\tfrac{1}{3})^2$$
$$3 \log \tfrac{1}{3} < 2 \log \tfrac{1}{3}$$
$$3 < 2 \qquad \text{Divide both sides by } \log \tfrac{1}{3}.$$

42. Find the fallacy.

$$3 > 2$$
$$3 \log \tfrac{1}{2} > 2 \log \tfrac{1}{2} \qquad \text{Multiply both sides by } \log \tfrac{1}{2}.$$
$$\log (\tfrac{1}{2})^3 > \log (\tfrac{1}{2})^2$$
$$(\tfrac{1}{2})^3 > (\tfrac{1}{2})^2$$
$$\tfrac{1}{8} > \tfrac{1}{4}$$
$$1 > 2 \qquad \text{Multiply both sides by 8.}$$

43. The function $f(x) = \log x$ increases extremely slowly as $x \to \infty$, but the composite function $g(x) = \log (\log x)$ increases still more slowly.
   (A) Illustrate this fact by computing the values of both functions for several large values of x.
   (B) Determine the domain and range of the function g.
   (C) Discuss the graphs of both functions.

44. The function $f(x) = \ln x$ increases extremely slowly as $x \to \infty$, but the composite function $g(x) = \ln (\ln x)$ increases still more slowly.
   (A) Illustrate this fact by computing the values of both functions for several large values of x.
   (B) Determine the domain and range of the function g.
   (C) Discuss the graphs of both functions.

In Problems 45–48, use a graphing utility to find the coordinates of all points of intersection to two decimal places.

45. $f(x) = \ln x,\ g(x) = 0.1x - 0.2$
46. $f(x) = \log x,\ g(x) = 4 - x^2$

47. $f(x) = \ln x,\ g(x) = x^{1/3}$

48. $f(x) = 3 \ln (x - 2),\ g(x) = 4e^{-x}$

The polynomials in Problems 49–52, called ***Taylor polynomials,*** *can be used to approximate the function $g(x) = \ln (1 + x)$. To illustrate this approximation graphically, in each problem, graph $g(x) = \ln (1 + x)$ and the indicated polynomial in the same viewing window, $-1 \le x \le 3$ and $-2 \le y \le 2$.*

49. $P_1(x) = x - \frac{1}{2}x^2$

50. $P_2(x) = x - \frac{1}{2}x^2 + \frac{1}{3}x^3$

51. $P_3(x) = x - \frac{1}{2}x^2 + \frac{1}{3}x^3 - \frac{1}{4}x^4$

52. $P_4(x) = x - \frac{1}{2}x^2 + \frac{1}{3}x^3 - \frac{1}{4}x^4 + \frac{1}{5}x^5$

APPLICATIONS

53. Sound. What is the decibel level of

(A) The threshold of hearing, 1.0×10^{-12} watt per square meter?

(B) The threshold of pain, 1.0 watt per square meter?

Compute answers to two significant digits.

54. Sound. What is the decibel level of

(A) A normal conversation, 3.2×10^{-6} watt per square meter?

(B) A jet plane with an afterburner, 8.3×10^{2} watts per square meter?

Compute answers to two significant digits.

55. Sound. If the intensity of a sound from one source is 1,000 times that of another, how much more is the decibel level of the louder sound than the quieter one?

56. Sound. If the intensity of a sound from one source is 10,000 times that of another, how much more is the decibel level of the louder sound than the quieter one?

57. Earthquakes. The largest recorded earthquake to date was in Colombia in 1906, with an energy release of 1.99×10^{17} joules. What was its magnitude on the Richter scale? Compute the answer to one decimal place.

58. Earthquakes. Anchorage, Alaska, had a major earthquake in 1964 that released 7.08×10^{16} joules of energy. What was its magnitude on the Richter scale? Compute the answer to one decimal place.

★★ **59. Earthquakes.** The 1933 Long Beach, California, earthquake had a Richter scale reading of 6.3, and the 1964 Anchorage, Alaska, earthquake had a Richter scale reading of 8.3. How many times more powerful was the Anchorage earthquake than the Long Beach earthquake?

★★ **60. Earthquakes.** Generally, an earthquake requires a magnitude of over 5.6 on the Richter scale to inflict serious damage. How many times more powerful than this was the great 1906 Colombia earthquake, which registered a magnitude of 8.6 on the Richter scale?

61. Space Vehicles. A new solid-fuel rocket has a weight ratio $W_t/W_b = 19.8$ and an exhaust velocity $c = 2.57$ kilometers per second. What is its velocity at burnout? Compute the answer to two decimal places.

62. Space Vehicles. A liquid-fuel rocket has a weight ratio $W_t/W_b = 6.2$ and an exhaust velocity $c = 5.2$ kilometers per second. What is its velocity at burnout? Compute the answer to two decimal places.

63. Chemistry. The hydrogen ion concentration of a substance is related to its acidity and basicity. Because hydrogen ion concentrations vary over a very wide range, logarithms are used to create a compressed **pH scale,** which is defined as follows:

$$\text{pH} = -\log [\text{H}^+]$$

where $[\text{H}^+]$ is the hydrogen ion concentration, in moles per liter. Pure water has a pH of 7, which means it is neutral. Substances with a pH less than 7 are acidic, and those with a pH greater than 7 are basic. Compute the pH of each substance listed, given the indicated hydrogen ion concentration.

(A) Seawater, 4.63×10^{-9}

(B) Vinegar, 9.32×10^{-4}

Also, indicate whether it is acidic or basic. Compute answers to one decimal place.

64. Chemistry. Refer to Problem 63. Compute the pH of each substance below, given the indicated hydrogen ion concentration. Also, indicate whether it is acidic or basic. Compute answers to one decimal place.

(A) Milk, 2.83×10^{-7}

(B) Garden mulch, 3.78×10^{-6}

★ **65. Ecology.** Refer to Problem 63. Many lakes in Canada and the United States will no longer sustain some forms of wildlife because of the increase in acidity of the water from acid rain and snow caused by sulfur dioxide emissions from industry. If the pH of a sample of rainwater is 5.2, what is its hydrogen ion concentration in moles per liter? Compute the answer to two significant digits.

★ **66. Ecology.** Refer to Problem 63. If normal rainwater has a pH of 5.7, what is its hydrogen ion concentration in moles per liter? Compute the answer to two significant digits.

Section 4-7 Exponential and Logarithmic Equations

- Exponential Equations
- Logarithmic Equations
- Change of Base

Equations involving exponential and logarithmic functions, such as

$$2^{3x-2} = 5 \qquad \text{and} \qquad \log(x + 3) + \log x = 1$$

are called **exponential** and **logarithmic equations,** respectively. Logarithmic properties play a central role in their solution. Of course, a graphing utility can be used to find approximate solutions for many exponential and logarithmic equations. However, there are situations where the algebraic solution is necessary. In this section, we emphasize algebraic solutions and use a graphing utility as a check, when appropriate.

Exponential Equations

The following examples illustrate the use of logarithmic properties in solving exponential equations.

EXAMPLE 1 **Solving an Exponential Equation**

Solve $2^{3x-2} = 5$ for x to four decimal places.

Solution How can we get x out of the exponent? Use logs! Since the logarithm function is one-to-one, if two positive quantities are equal, their logs are equal. See Theorem 1 in Section 4-5.

$$2^{3x-2} = 5$$

$$\log 2^{3x-2} = \log 5 \qquad \text{Take the common or natural log of both sides.}$$

$$(3x - 2) \log 2 = \log 5 \qquad \text{Use } \log_b N^p = p \log_b N \text{ to get } 3x - 2 \text{ out of the exponent position.}$$

$$3x - 2 = \frac{\log 5}{\log 2}$$

$$x = \frac{1}{3}\left(2 + \frac{\log 5}{\log 2}\right) \qquad \textit{Remember: } \frac{\log 5}{\log 2} \neq \log 5 - \log 2.$$

$$= 1.4406 \qquad \text{To four decimal places.}$$

FIGURE 1
$y_1 = 2^{3x-2}$, $y_2 = 5$.

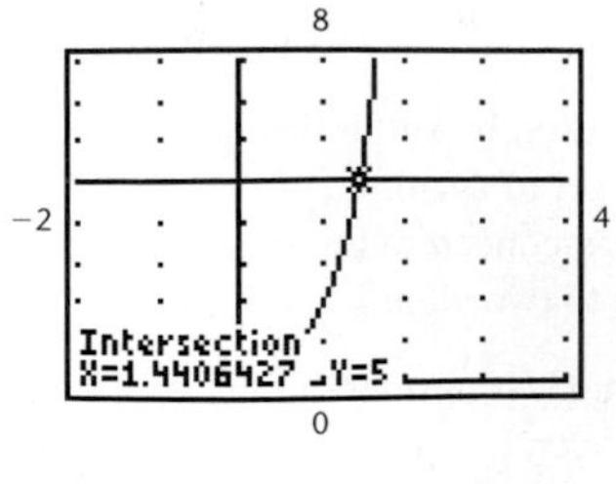

Figure 1 shows a graphical solution that confirms this result.

MATCHED PROBLEM 1

Solve $35^{1-2x} = 7$ for x to four decimal places.

EXAMPLE 2

Compound Interest

A certain amount of money P (principal) is invested at an annual rate r compounded annually. The amount of money A in the account after t years, assuming no withdrawals, is given by

$$A = P\left(1 + \frac{r}{m}\right)^n = P(1 + r)^n \qquad m = 1 \text{ for annual compounding.}$$

How many years to the nearest year will it take the money to double if it is invested at 6% compounded annually?

Solution

To find the doubling time, we replace A in $A = P(1.06)^n$ with $2P$ and solve for n.

$$2P = P(1.06)^n$$

$$2 = 1.06^n \qquad \text{Divide both sides by } P.$$

$$\log 2 = \log 1.06^n \qquad \text{Take the common or natural log of both sides.}$$

$$= n \log 1.06 \qquad \text{Note how log properties are used to get } n \text{ out of the exponent position.}$$

$$n = \frac{\log 2}{\log 1.06}$$

$$= 12 \text{ years} \qquad \text{To the nearest year.}$$

Figure 2 confirms this result.

FIGURE 2
$y_1 = 1.06^x$, $y_2 = 2$.

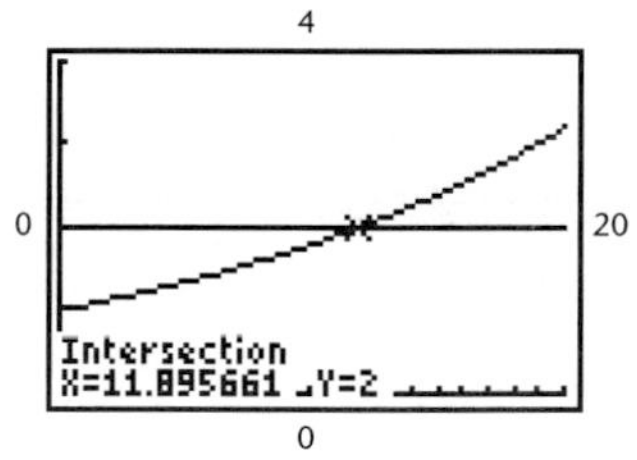

MATCHED PROBLEM 2

Repeat Example 2, changing the interest rate to 9% compounded annually.

EXAMPLE 3

Atmospheric Pressure

The atmospheric pressure P, in pounds per square inch, at x miles above sea level is given approximately by

$$P = 14.7e^{-0.21x}$$

At what height will the atmospheric pressure be half the sea-level pressure? Compute the answer to two significant digits.

Solution

Sea-level pressure is the pressure at $x = 0$. Thus,

$$P = 14.7e^0 = 14.7$$

One-half of sea-level pressure is $14.7/2 = 7.35$. Now our problem is to find x so that $P = 7.35$; that is, we solve $7.35 = 14.7e^{-0.21x}$ for x:

$$7.35 = 14.7e^{-0.21x}$$

$$0.5 = e^{-0.21x} \qquad \text{Divide both sides by 14.7 to simplify.}$$

$$\ln 0.5 = \ln e^{-0.21x} \qquad \text{Since the base is } e\text{, take the natural log of both sides.}$$

$$= -0.21x \qquad \ln e = 1$$

$$x = \frac{\ln 0.5}{-0.21}$$

$$= 3.3 \text{ miles} \qquad \text{To two significant digits.}$$

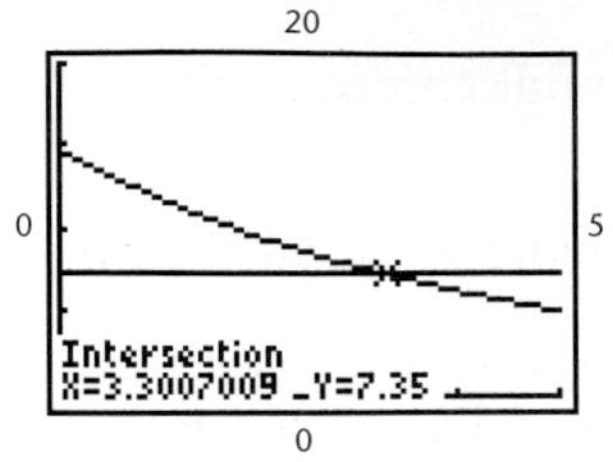

FIGURE 3
$y_1 = 14.7e^{-0.21x}$, $y_2 = 7.35$.

Figure 3 shows that this answer is correct.

MATCHED PROBLEM 3

Using the formula in Example 3, find the altitude in miles so that the atmospheric pressure will be one-eighth that at sea level. Compute the answer to two significant digits.

The graph of

$$y = \frac{e^x + e^{-x}}{2} \tag{1}$$

is a curve called a **catenary** (Fig. 4). A uniform cable suspended between two fixed points is a physical example of such a curve.

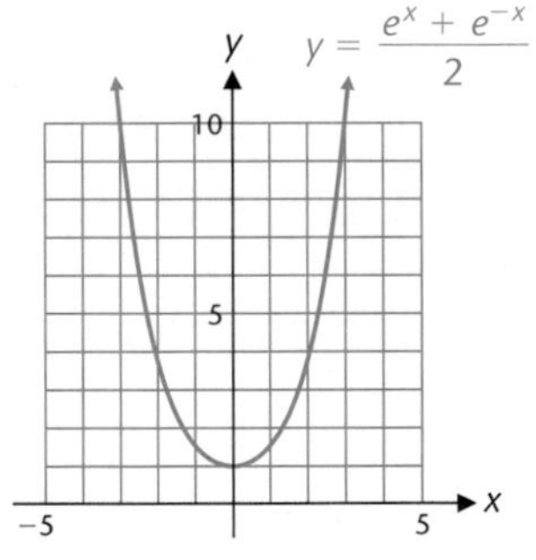

FIGURE 4
Catenary.

EXAMPLE 4 Solving an Exponential Equation

Given equation (1), find x for $y = 2.5$. Compute the answer to four decimal places.

Solution

$$y = \frac{e^x + e^{-x}}{2}$$

$$2.5 = \frac{e^x + e^{-x}}{2}$$

$$5 = e^x + e^{-x}$$

$$5e^x = e^{2x} + 1 \qquad \text{Multiply both sides by } e^x.$$

$$e^{2x} - 5e^x + 1 = 0 \qquad \text{This is a quadratic in } e^x.$$

Let $u = e^x$, then

$$u^2 - 5u + 1 = 0$$

$$u = \frac{5 \pm \sqrt{25 - 4(1)(1)}}{2}$$

$$= \frac{5 \pm \sqrt{21}}{2}$$

$$e^x = \frac{5 \pm \sqrt{21}}{2}$$ Replace u with e^x and solve for x.

$$\ln e^x = \ln \frac{5 \pm \sqrt{21}}{2}$$ Take the natural log of both sides (both values on the right are positive).

$$x = \ln \frac{5 \pm \sqrt{21}}{2}$$ $\log_b b^x = x$.

$$= -1.5668, 1.5668$$

FIGURE 5

$y_1 = \dfrac{e^x + e^{-x}}{2}$, $y_2 = 2.5$.

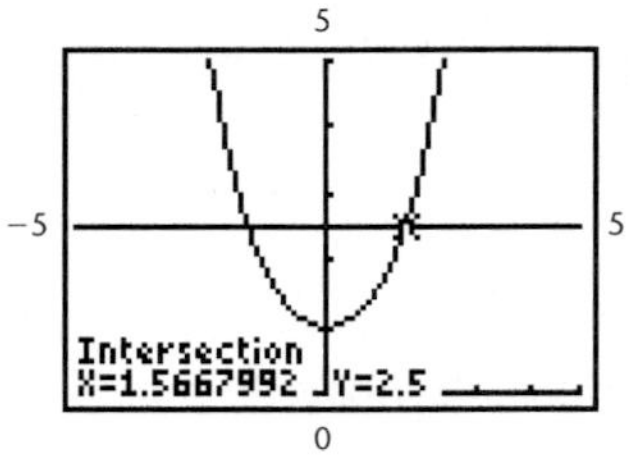

Figure 5 confirms the positive solution. Note that the algebraic method also produced exact solutions, an important consideration in certain calculus applications (see Problems 57–60 in Exercise 4-7).

MATCHED PROBLEM
4

Given $y = (e^x - e^{-x})/2$, find x for $y = 1.5$. Compute the answer to three decimal places.

Explore/Discuss 1

Let $y = e^{2x} + 3e^x + e^{-x}$

(A) Try to find x when $y = 7$ using the method of Example 4. Explain the difficulty that arises.

(B) Use a graphing utility to find x when $y = 7$.

Logarithmic Equations

We now illustrate the solution of several types of logarithmic equations.

EXAMPLE
5

Solving a Logarithmic Equation

Solve $\log (x + 3) + \log x = 1$, and check.

Solution

First use properties of logarithms to express the left side as a single logarithm, then convert to exponential form and solve for x.

$$\log(x+3) + \log x = 1$$

$\log[x(x+3)] = 1$ — Combine left side using $\log M + \log N = \log MN$.

$x(x+3) = 10^1$ — Change to equivalent exponential form.

$x^2 + 3x - 10 = 0$ — Write in $ax^2 + bx + c = 0$ form and solve.

$$(x+5)(x-2) = 0$$

$$x = -5, 2$$

Check $x = -5$: $\log(-5+3) + \log(-5)$ is not defined because the domain of the log function is $(0, \infty)$.

$x = 2$: $\log(2+3) + \log 2 = \log 5 + \log 2$

$$= \log(5 \cdot 2) = \log 10 \overset{\checkmark}{=} 1$$

FIGURE 6
$y_1 = \log(x+3) + \log x$, $y_2 = 1$.

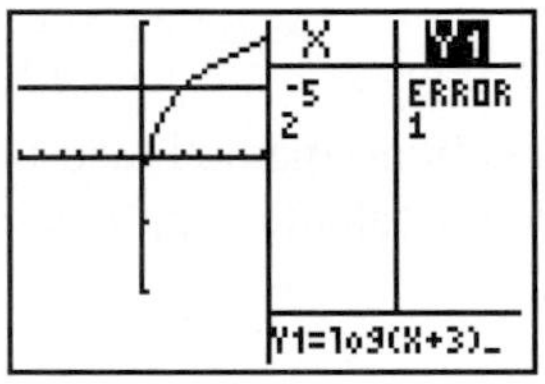

Thus, the only solution to the original equation is $x = 2$. Remember, answers should be checked in the original equation to see whether any should be discarded.

Figure 6 shows the solution at $x = 2$ and also shows that the left side of the equation is not defined at $x = -5$, the extraneous solution produced by the algebraic method.

MATCHED PROBLEM 5

Solve $\log(x - 15) = 2 - \log x$, and check.

EXAMPLE 6

Solving a Logarithmic Equation

Solve $(\ln x)^2 = \ln x^2$.

Solution There are no logarithmic properties for simplifying $(\ln x)^2$. However, we can simplify $\ln x^2$, obtaining an equation involving $\ln x$ and $(\ln x)^2$.

$$(\ln x)^2 = \ln x^2$$

$= 2 \ln x$ — This is a quadratic equation in $\ln x$. Move all nonzero terms to the left and factor.

$$(\ln x)^2 - 2\ln x = 0$$

$$(\ln x)(\ln x - 2) = 0$$

$\ln x = 0$ or $\ln x - 2 = 0$

$x = e^0$ $\ln x = 2$

$= 1$ $x = e^2$

FIGURE 7
$y_1 = (\ln x)^2$, $y_2 = \ln x^2$.

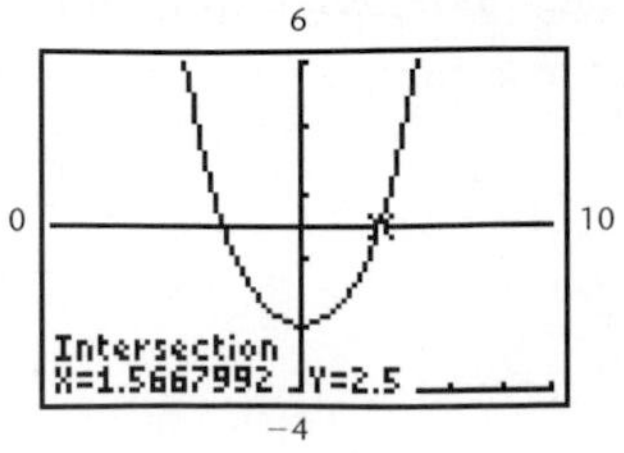

Checking that both $x = 1$ and $x = e^2$ are solutions to the original equation is left to you.

Figure 7 confirms the solution at $e^2 \approx 7.3890561$.

MATCHED PROBLEM 6

Solve $\log x^2 = (\log x)^2$.

CAUTION

Note that

$$(\log_b x)^2 \neq \log_b x^2 \qquad (\log_b x)^2 = (\log_b x)(\log_b x)$$
$$\log_b x^2 = 2 \log_b x$$

EXAMPLE 7

Earthquake Intensity

Recall from Section 4-6 that the magnitude of an earthquake on the Richter scale is given by

$$M = \frac{2}{3} \log \frac{E}{E_0}$$

Solve for E in terms of the other symbols.

Solution

$$M = \frac{2}{3} \log \frac{E}{E_0}$$

$$\log \frac{E}{E_0} = \frac{3M}{2} \qquad \text{Multiply both sides by } \tfrac{3}{2}.$$

$$\frac{E}{E_0} = 10^{3M/2} \qquad \text{Change to exponential form.}$$

$$E = E_0 10^{3M/2}$$

MATCHED PROBLEM 7

Solve the rocket equation from Section 4-6 for W_b in terms of the other symbols:

$$v = c \ln \frac{W_t}{W_b}$$

Change of Base

How would you find the logarithm of a positive number to a base other than 10 or e? For example, how would you find $\log_3 5.2$? In Example 8 we evaluate this logarithm using a direct process. Then we develop a change-of-base formula to find such logarithms in general. You may find it easier to remember the process than the formula.

EXAMPLE 8 **Evaluating a Base 3 Logarithm**

Evaluate $\log_3 5.2$ to four decimal places.

Solution Let $y = \log_3 5.2$ and proceed as follows:

$$\log_3 5.2 = y$$

$$5.2 = 3^y \qquad \text{Change to exponential form.}$$

$$\ln 5.2 = \ln 3^y \qquad \text{Take the natural log (or common log) of each side.}$$

$$= y \ln 3 \qquad \log_b M^p = p \log_b M$$

$$y = \frac{\ln 5.2}{\ln 3} \qquad \text{Solve for } y.$$

Replace y with $\log_3 5.2$ from the first step, and use a calculator to evaluate the right side:

$$\log_3 5.2 = \frac{\ln 5.2}{\ln 3} = 1.5007$$

MATCHED PROBLEM 8 Evaluate $\log_{0.5} 0.0372$ to four decimal places.

To develop a change-of-base formula for arbitrary positive bases, with neither base equal to 1, we proceed as above. Let $y = \log_b N$, where N and b are positive and $b \neq 1$. Then

$$\log_b N = y$$

$$N = b^y \qquad \text{Write in exponential form.}$$

$$\log_a N = \log_a b^y \qquad \text{Take the log of each side to another positive base } a,\ a \neq 1.$$

$$= y \log_a b \qquad \log_b M^p = p \log_b M$$

$$y = \frac{\log_a N}{\log_a b} \qquad \text{Solve for } y.$$

Replacing y with $\log_b N$ from the first step, we obtain the **chain-of-base formula:**

$$\log_b N = \frac{\log_a N}{\log_a b}$$

In words, this formula states that the logarithm of a number to a given base is the logarithm of that number to a new base divided by the logarithm of the old base to the new base. In practice, we usually choose either e or 10 for the new base so that a calculator can be used to evaluate the necessary logarithms (see Example 8).

Explore/Discuss 2

If b is any positive real number different from 1, the change-of-base formula implies that the function $y = \log_b x$ is a constant multiple of the natural logarithmic function; that is, $\log_b x = k \ln x$ for some k.

(A) Graph the functions $y = \ln x$, $y = 2 \ln x$, $y = 0.5 \ln x$, and $y = -3 \ln x$.

(B) Write each function of part A in the form $y = \log_b x$ by finding the base b to two decimal places.

(C) Is every exponential function $y = b^x$ a constant multiple of $y = e^x$? Explain.

Answers to Matched Problems

1. $x = 0.2263$ **2.** More than double in 9 years, but not quite double in 8 years **3.** 9.9 miles **4.** $x = 1.195$
5. $x = 20$ **6.** $x = 1{,}100$ **7.** $W_b = W_t e^{-v/c}$ **8.** 4.7486

EXERCISE 4-7

A

Solve Problems 1–12 algebraically and check graphically. Round answers to three significant digits.

1. $10^{-x} = 0.0347$ **2.** $10^x = 14.3$ **3.** $10^{3x+1} = 92$

4. $10^{5x-2} = 348$ **5.** $e^x = 3.65$ **6.** $e^{-x} - 0.0142$

7. $e^{2x-1} = 405$ **8.** $e^{3x+5} = 23.8$ **9.** $5^x = 18$

10. $3^x = 4$ **11.** $2^{-x} = 0.238$ **12.** $3^{-x} = 0.074$

Solve Problems 13–18 exactly.

13. $\log 5 + \log x = 2$ **14.** $\log x - \log 8 = 1$

15. $\log x + \log (x - 3) = 1$

16. $\log (x - 9) + \log 100x = 3$

17. $\log (x + 1) - \log (x - 1) = 1$

18. $\log (2x + 1) = 1 + \log (x - 2)$

B

Solve Problems 19–26 algebraically and check graphically. Round answers to three significant digits.

19. $2 = 1.05^x$ **20.** $3 = 1.06^x$

21. $e^{-1.4x} = 13$ **22.** $e^{0.32x} = 632$

23. $123 = 500e^{-0.12x}$ **24.** $438 = 200e^{0.25x}$

25. $e^{-x^2} = 0.23$ **26.** $e^{x^2} = 125$

Solve Problems 27–38 exactly.

27. $\log x - \log 5 = \log 2 - \log (x - 3)$

28. $\log (6x + 5) - \log 3 = \log 2 - \log x$

29. $\ln x = \ln (2x - 1) - \ln (x - 2)$

30. $\ln (x + 1) = \ln (3x + 1) - \ln x$

31. $\log (2x + 1) = 1 - \log (x - 1)$

32. $1 - \log (x - 2) = \log (3x + 1)$

33. $(\ln x)^3 = \ln x^4$ **34.** $(\log x)^3 = \log x^4$

35. $\ln (\ln x) = 1$ **36.** $\log (\log x) = 1$

37. $x^{\log x} = 100x$ **38.** $3^{\log x} = 3x$

In Problems 39–40,

(A) Explain the difficulty in solving the equation exactly.

(B) Determine the number of solutions by graphing the functions on each side of the equation.

39. $e^{x/2} = 5 \ln x$ **40.** $\ln (\ln x) + \ln x = 2$

In Problems 41–42,

(A) Explain the difficulty in solving the equation exactly.

(B) Use a graphing utility to find all solutions to three decimal places.

41. $3^x + 2 = 7 + x - e^{-x}$ **42.** $e^{x/4} = 5 \log x + 4 \ln x$

Evaluate Problems 43–48 to four decimal places.

43. $\log_5 372$ **44.** $\log_4 23$ **45.** $\log_8 0.0352$

46. $\log_2 0.005\,439$ **47.** $\log_3 0.1483$ **48.** $\log_{12} 435.62$

C

Solve Problems 49–56 for the indicated variable in terms of the remaining symbols. Use the natural log for solving exponential equations.

49. $A = Pe^{rt}$ for r (finance)

50. $A = P\left(1 + \dfrac{r}{n}\right)^{nt}$ for t (finance)

51. $D = 10 \log \dfrac{I}{I_0}$ for I (sound)

52. $t = \dfrac{-1}{k}(\ln A - \ln A_0)$ for A (decay)

53. $M = 6 - 2.5 \log \dfrac{I}{I_0}$ for I (astronomy)

54. $L = 8.8 + 5.1 \log D$ for D (astronomy)

55. $I = \dfrac{E}{R}(1 - e^{-Rt/L})$ for t (circuitry)

56. $S = R\dfrac{(1 + i)^n - 1}{i}$ for n (annuity)

The following combinations of exponential functions define four of six ***hyperbolic functions,*** *an important class of functions in calculus and higher mathematics. Solve Problems 57–60 for x in terms of y. The results are used to define* ***inverse hyperbolic functions,*** *another important class of functions in calculus and higher mathematics.*

57. $y = \dfrac{e^x + e^{-x}}{2}$ **58.** $y = \dfrac{e^x - e^{-x}}{2}$

59. $y = \dfrac{e^x - e^{-x}}{e^x + e^{-x}}$ **60.** $y = \dfrac{e^x + e^{-x}}{e^x - e^{-x}}$

In Problems 61–64, use a graphing utility to graph each function. [Hint: Use the change-of-base formula first.]

61. $y = 3 + \log_2 (2 - x)$ **62.** $y = \log_3 (4 + x) - 5$

63. $y = \log_3 x - \log_2 x$ **64.** $y = \log_3 x - \log_2 x$

In Problems 65–76, use a graphing utility to approximate to two decimal places any solutions of the equation in the interval $0 \le x \le 1$. None of these equations can be solved exactly using any step-by-step algebraic process.

65. $2^{-x} - 2x = 0$ **66.** $3^{-x} - 3x = 0$

67. $x3^x - 1 = 0$ **68.** $x2^x - 1 = 0$

69. $e^{-x} - x = 0$ **70.** $xe^{2x} - 1 = 0$

71. $xe^x - 2 = 0$ **72.** $e^{-x} - 2x = 0$

73. $\ln x + 2x = 0$ **74.** $\ln x + x^2 = 0$

75. $\ln x + e^x = 0$ **76.** $\ln x + x = 0$

APPLICATIONS

Solve Problems 77–90 algebraically or graphically, whichever seems more appropriate.

77. Compound Interest. How many years, to the nearest year, will it take a sum of money to double if it is invested at 15% compounded annually?

78. Compound Interest. How many years, to the nearest year, will it take money to quadruple if it is invested at 20% compounded annually?

79. Compound Interest. At what annual rate compounded continuously will \$1,000 have to be invested to amount to \$2,500 in 10 years? Compute the answer to three significant digits.

80. Compound Interest. How many years will it take \$5,000 to amount to \$8,000 if it is invested at an annual rate of 9% compounded continuously? Compute the answer to three significant digits.

★★ **81. Astronomy.** The brightness of stars is expressed in terms of magnitudes on a numerical scale that increases as the brightness decreases. The magnitude m is given by the formula

$$m = 6 - 2.5 \log \frac{L}{L_0}$$

where L is the light flux of the star and L_0 is the light flux of the dimmest stars visible to the naked eye.

(A) What is the magnitude of the dimmest stars visible to the naked eye?

(B) How many times brighter is a star of magnitude 1 than a star of magnitude 6?

82. Astronomy. An optical instrument is required to observe stars beyond the sixth magnitude, the limit of ordinary vision. However, even optical instruments have their limitations. The limiting magnitude L of any optical telescope with lens diameter D, in inches, is given by

$$L = 8.8 + 5.1 \log D$$

(A) Find the limiting magnitude for a homemade 6-inch reflecting telescope.

(B) Find the diameter of a lens that would have a limiting magnitude of 20.6.

Compute answers to three significant digits.

83. World Population. A mathematical model for world population growth over short periods of time is given by

$$P = P_0e^{rt}$$

where P is the population after t years, P_0 is the population at $t = 0$, and the population is assumed to grow continuously at the annual rate r. How many years, to the nearest year, will it take the world population to double if it grows continuously at an annual rate of 2%?

★ **84. World Population.** Refer to Problem 83. Starting with a world population of 4 billion people and assuming that the population grows continuously at an annual rate of 2%, how many years, to the nearest year, will it be before there is only 1 square yard of land per person? Earth contains approximately 1.7×10^{14} square yards of land.

★ **85. Archaeology—Carbon 14 Dating.** As long as a plant or animal is alive, carbon 14 is maintained in a constant amount in its tissues. Once dead, however, the plant or animal ceases taking in carbon, and carbon 14 diminishes by radioactive decay according to the equation

$$A = A_0e^{-0.000124t}$$

where A is the amount after t years and A_0 is the amount when $t = 0$. Estimate the age of a skull uncovered in an archaeological site if 10% of the original amount of carbon 14 is still present. Compute the answer to three significant digits.

★ **86. Archaeology—Carbon 14 Dating.** Refer to Problem 85. What is the half-life of carbon 14? That is, how long will it take for half of a sample of carbon 14 to decay? Compute the answer to three significant digits.

★ **87. Photography.** An electronic flash unit for a camera is activated when a capacitor is discharged through a filament of wire. After the flash is triggered and the capacitor is discharged, the circuit (see the figure) is connected and the battery pack generates a current to recharge the capacitor. The time it takes for the capacitor to recharge is called the *recycle time*. For a particular flash unit using a 12-volt battery pack, the charge q, in coulombs, on the capacitor t seconds after recharging has started is given by

$$q = 0.0009(1 - e^{-0.2t})$$

How many seconds will it take the capacitor to reach a charge of 0.0007 coulomb? Compute the answer to three significant digits.

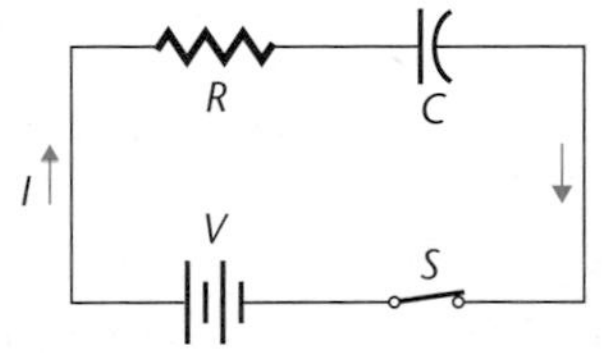

★ **88. Advertising.** A company is trying to expose a new product to as many people as possible through television advertising in a large metropolitan area with 2 million possible viewers. A model for the number of people N, in millions, who are aware of the product after t days of advertising was found to be

$$N = 2(1 - e^{-0.037t})$$

How many days, to the nearest day, will the advertising campaign have to last so that 80% of the possible viewers will be aware of the product?

★★ **89. Newton's Law of Cooling.** This law states that the rate at which an object cools is proportional to the difference in temperature between the object and its surrounding medium. The temperature T of the object t hours later is given by

$$T = T_m + (T_0 - T_m)e^{-kt}$$

where T_m is the temperature of the surrounding medium and T_0 is the temperature of the object at $t = 0$. Suppose a bottle of wine at a room temperature of 72°F is placed in a refrigerator at 40°F to cool before a dinner party. After an hour the temperature of the wine is found to be 61.5°F. Find the constant k, to two decimal places, and the time, to one decimal place, it will take the wine to cool from 72 to 50°F.

★ **90. Marine Biology.** Marine life is dependent upon the microscopic plant life that exists in the *photic zone*, a zone that goes to a depth where about 1% of the surface light still remains. Light intensity is reduced according to the exponential function

$$I = I_0e^{-kd}$$

where I is the intensity d feet below the surface and I_0 is the intensity at the surface. The constant k is called the *coefficient of extinction*. At Crystal Lake in Wisconsin it was found that half the surface light remained at a depth of 14.3 feet. Find k, and find the depth of the photic zone. Compute answers to three significant digits.

91. Agriculture. Table 1 shows the yield (bushels per acre) and the total production (millions of bushels) for corn in the United States for selected years since 1950. Let x represent years since 1900.

TABLE 1 United States Corn Production

Year	x	Yield (bushels per acre)	Total Production (million bushels)
1950	50	37.6	2,782
1960	60	55.6	3,479
1970	70	81.4	4,802
1980	80	97.7	6,867
1990	90	115.6	7,802

Source: U.S. Department of Agriculture.

(A) Find a logarithmic regression model ($y = a + b \ln x$) for the yield. Estimate (to one decimal place) the yield in 1996 and in 2010.

(B) The actual yield in 1996 was 127.1 bushels per acre. How does this compare with the estimated yield in part A? What effect will this additional 1996 information have on the estimate for 2010? Explain.

92. Agriculture. Refer to Table 1.

(A) Find a logarithmic regression model ($y = a + b \ln x$) for the total production. Estimate (to the nearest million) the production in 1996 and in 2010.

(B) The actual production in 1996 was 7,949 billion bushels. How does this compare with the estimated production in part A? What effect will this 1996 production information have on the estimate for 2010? Explain.

Chapter 4 | Group Activity

Comparing Regression Models

We have used polynomial, exponential, and logarithmic regression models to fit curves to data sets. And there are other equations that can be used for curve fitting (the TI-83 graphing calculator has 12 different equations on its STAT-CALC menu). How can we determine which equation provides the best fit for a given set of data? There are two principal ways to select models. The first is to use information about the type of data to help make a choice. For example, we expect the weight of a fish to be related to the cube of its length. And we expect most populations to grow exponentially, at least over the short term. The second method for choosing among equations involves developing a measure of how closely an equation fits a given data set. This is best introduced through an example. Consider the data set in Figure 1, where L1 represents the x coordinates and L2 represents the y coordinates. The graph of this data set is shown in Figure 2. Suppose we arbitrarily choose the equation $y_1 = 0.6x + 1$ to model this data (Fig. 3).

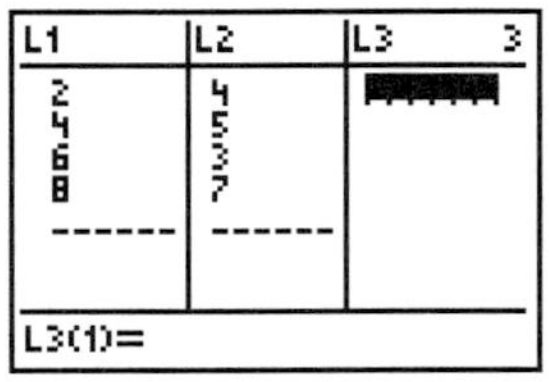

FIGURE 1

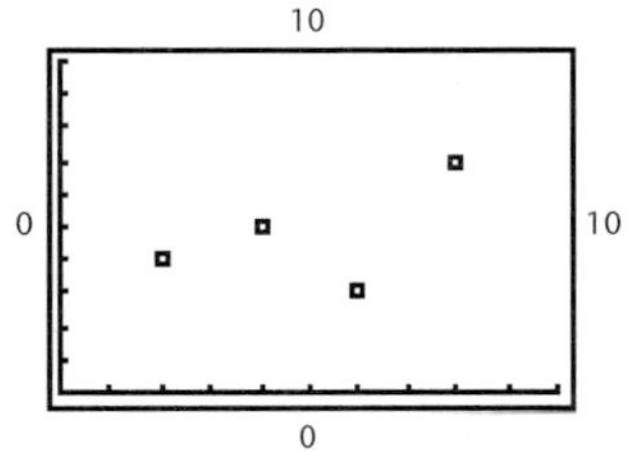

FIGURE 2

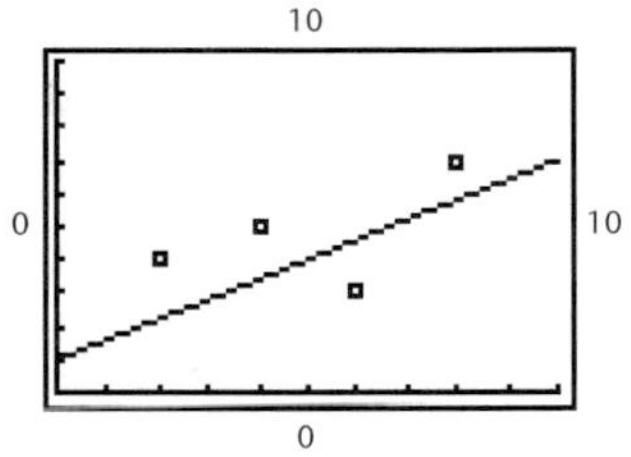

FIGURE 3
$y_1 = 0.6x + 1$.

To measure how well the graph of y_1 fits this data, we examine the difference between the y coordinates in the data set and the corresponding y coordinates on the graph of y_1 (L3 in Figs. 4 and 5). Each of these differences is called a **residual.** The most commonly accepted measure of the fit provided by a given model is the **sum of the squares of the residuals (SSR).** Computing this quantity is a simple matter on a graphing utility (Fig. 6).

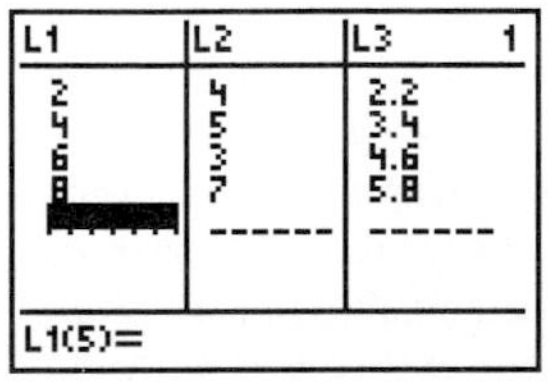

FIGURE 4

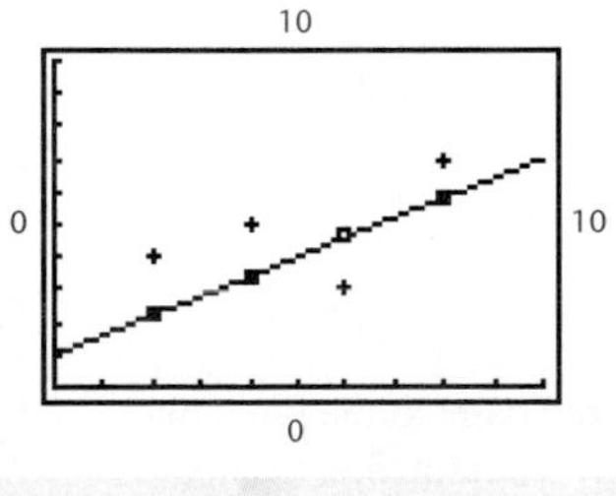

FIGURE 5
Here + is L2 and □ is L3.

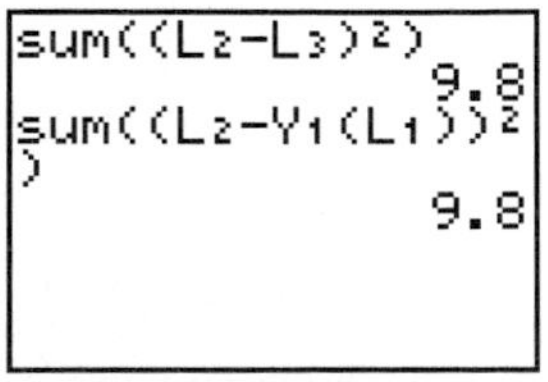

FIGURE 6
Two ways to calculate SSR.

(A) Find the linear regression model for the data in Figure 1, compute the SSR for this equation and compare it with the one we computed for y_1.

It turns out that among all possible linear polynomials, **the linear regression model minimizes the sum of the squares of the residuals.** For this reason, the linear regression model is often called the **least-squares line.** A similar statement

CHAPTER 4 GROUP ACTIVITY continued

can be made for polynomials of any fixed degree. That is, the quadratic regression model minimizes the SSR over all quadratic polynomials, the cubic regression model minimizes the SSR over all cubic polynomials, and so on. The same statement cannot be made for exponential or logarithmic regression models. Nevertheless, the SSR can still be used to compare exponential, logarithmic, and polynomial models.

(B) Find the exponential and logarithmic regression models for the data in Figure 1, compute their SSRs, and compare with the linear model.

(C) National annual advertising expenditures for selected years since 1950 are shown in Table 1 where x is years since 1950 and y is total expenditures in billions of dollars. Which regression model would fit this data best: a quadratic model, a cubic model, or an exponential model? Use the SSRs to support your choice.

TABLE 1 Annual Advertising Expenditures, 1950–1995

x (years)	0	5	10	15	20	25	30	35	40	45
y (billion \$)	5.7	9.2	12.0	15.3	19.6	27.9	53.6	94.8	128.6	160.9

Source: U.S. Bureau of the Census.

CHAPTER 4 | REVIEW

4-1 OPERATIONS ON FUNCTIONS; COMPOSITION

The **sum, difference, product,** and **quotient** of the functions f and g are defined by

$$(f + g)(x) = f(x) + g(x) \qquad (f - g)(x) = f(x) - g(x)$$

$$(fg)(x) = f(x)g(x) \qquad \left(\frac{f}{g}\right)(x) = \frac{f(x)}{g(x)} \qquad g(x) \neq 0$$

The **domain** of each function is the intersection of the domains of f and g, with the exception that values of x where $g(x) = 0$ must be excluded from the domain of f/g.

The **composition** of functions f and g is defined by $(f \circ g)(x) = f(g(x))$. The *domain* of $f \circ g$ is the set of all real numbers x in the domain of g where $g(x)$ is in the domain of f. The domain of $f \circ g$ is always a subset of the domain of g.

4-2 INVERSE FUNCTIONS

A function is **one-to-one** if no two ordered pairs in the function have the same second component and different first components. A **horizontal line** will intersect the graph of a one-to-one function in at most one point. A function that is increasing (decreasing) throughout its domain is one-to-one. The **inverse** of the one-to-one function f is the function f^{-1} formed by reversing all the ordered pairs in f. If f is not one-to-one, then f^{-1} **does not exist.**

Assuming that f^{-1} exists, then:

1. f^{-1} is one-to-one.
2. Domain of f^{-1} = Range of f.
3. Range of f^{-1} = Domain of f.
4. $x = f^{-1}(y)$ if and only if $y = f(x)$.
5. $f^{-1}(f(x)) = x$ for all x in the domain of f.
6. $f(f^{-1}(x)) = x$ for all x in the domain of f^{-1}.
7. To find f^{-1}, solve the equation $y = f(x)$ for x and then interchange x and y.
8. The graphs of $y = f(x)$ and $y = f^{-1}(x)$ are symmetric with respect to the line $y = x$.

4-3 EXPONENTIAL FUNCTIONS

The equation $f(x) = b^x$, $b > 0$, $b \neq 1$, defines an **exponential function** with **base** b. The domain of f is $(-\infty, \infty)$ and the range is $(0, \infty)$. The **graph** of an exponential function is a continuous curve that always passes through the point $(0, 1)$ and has the x axis as a horizontal asymptote. If $b > 1$, then b^x increases as x increases, and if $0 < b < 1$, then b^x decreases as x increases. The function f is one-to-one and has an inverse. We have the following **exponential function properties:**

1. $a^x a^y = a^{x+y} \qquad (a^x)^y = a^{xy} \qquad (ab)^x = a^x b^x$

   $\left(\frac{a}{b}\right)^x = \frac{a^x}{b^x} \qquad \frac{a^x}{a^y} = a^{x-y}$
2. $a^x = a^y$ if and only if $x = y$.
3. For $x \neq 0$, then $a^x = b^x$ if and only if $a = b$.

Exponential functions are used to describe various types of **growth.**

1. **Population growth** can be modeled by using the **doubling time growth model** $P = P_0 2^{t/d}$, where P is population at time t, P_0 is the population at time $t = 0$, and d is the **doubling time**—the time it takes for the population to double.
2. **Radioactive decay** can be modeled by using the **half-life decay model** $A = A_0(\frac{1}{2})^{t/h} = A_0 2^{-t/h}$, where A is the amount at time t, A_0 is the amount at time $t = 0$, and h is the **half-life**—the time it takes for half the material to decay.
3. The growth of money in an account paying **compound interest** is described by $A = P(1 + r/m)^n$, where P is the **principal,** r is the annual **rate,** m is the number of compounding periods in 1 year, and A is the **amount** in the account after n compounding periods.

4-4 THE EXPONENTIAL FUNCTION WITH BASE *e*

As x approaches ∞, the expression $[1 + (1/x)]^x$ approaches the irrational number $e \approx 2.718\ 281\ 828\ 459$. The function $f(x) = e^x$ is called the **exponential function with base *e*.** Exponential functions with base e are used to model a variety of different types of exponential growth and decay, including growth of money in accounts that pay **continuous compound interest.** If a principal P is invested at an annual rate r compounded continuously, then the amount A in the account after t years is given by $A = Pe^{rt}$.

4-5 LOGARITHMIC FUNCTIONS

The **logarithmic function with base *b*** is defined to be the inverse of the exponential function with base b and is denoted by $y = \log_b x$. Thus, $y = \log_b x$ if and only if $x = b^y$, $b > 0$, $b \neq 1$. The domain of a logarithmic function is $(0, \infty)$ and the range is $(-\infty, \infty)$. The graph of a logarithmic function is a continuous curve that always passes through the point (1, 0) and has the y axis as a vertical asymptote. We have the following **properties of logarithmic functions:**

1. $\log_b 1 = 0$
2. $\log_b b = 1$
3. $\log_b b^x = x$
4. $b^{\log_b x} = x, x > 0$
5. $\log_b MN = \log_b M + \log_b N$
6. $\log_b \frac{M}{N} = \log_b M - \log_b N$
7. $\log_b M^p = p \log_b M$
8. $\log_b M = \log_b N$ if and only if $M = N$

4-6 COMMON AND NATURAL LOGARITHMS

Logarithms to the base 10 are called **common logarithms** and are denoted by $\log x$. Logarithms to the base e are called **natural logarithms** and are denoted by $\ln x$. Thus, $\log x = y$ is equivalent to $x = 10^y$, and $\ln x = y$ is equivalent to $x = e^y$.

The following applications involve logarithms:

1. The **decibel** is defined by $D = 10 \log (I/I_0)$, where D is the **decibel level** of the sound, I is the **intensity** of the sound, and $I_0 = 10^{-12}$ watt per square meter is a standardized sound level.
2. The **magnitude** M of an earthquake on the **Richter scale** is given by $M = \frac{2}{3} \log (E/E_0)$, where E is the energy released by the earthquake and $E_0 = 10^{4.40}$ joules is a standardized energy level.
3. The **velocity** v of a rocket at burnout is given by the **rocket equation** $v = c \ln (W_t/W_b)$, where c is the exhaust velocity, W_t is the takeoff weight, and W_b is the burnout weight.

4-7 EXPONENTIAL AND LOGARITHMIC EQUATIONS

Various techniques for solving **exponential equations,** such as $2^{3x-2} = 5$, and **logarithmic equations,** such as $\log (x + 3) + \log x = 1$, are illustrated by examples. The **change-of-base formula,** $\log_b N = (\log_a N)/(\log_a b)$, relates logarithms to two different bases and can be used, along with a calculator, to evaluate logarithms to bases other than e or 10.

CHAPTER 4 | REVIEW EXERCISES

Work through all the problems in this chapter review and check answers in the back of the book. Answers to all review problems are there, and following each answer is a number in italics indicating the section in which that type of problem is discussed. Where weaknesses show up, review appropriate sections in the text.

A

Problems 1–7 refer to the graphs of f and g shown in the figures.

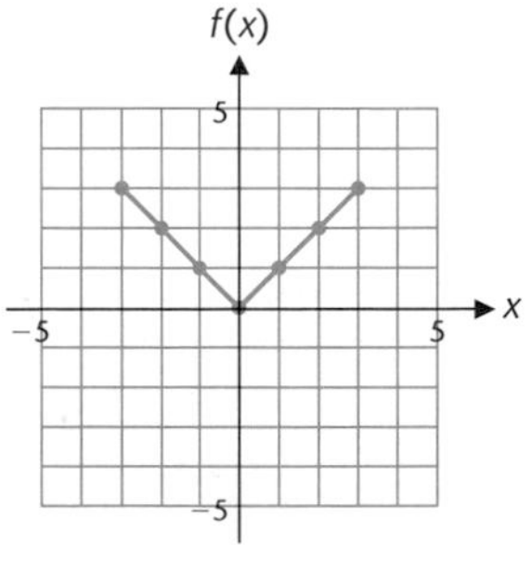

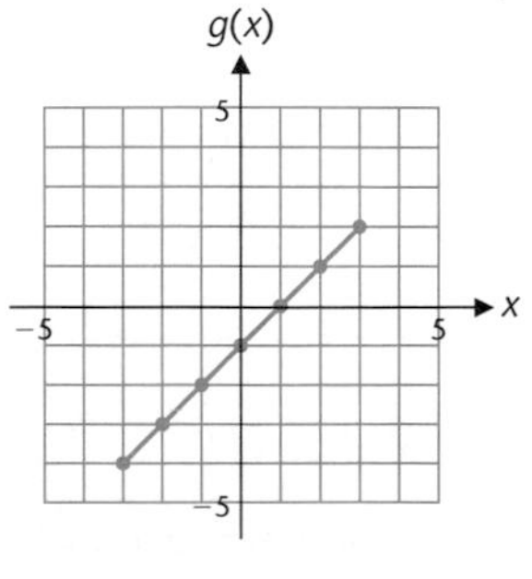

1. Construct a table of values of $(f - g)(x)$ for $x = -3, -2, -1, 0, 1, 2$, and 3, and sketch the graph of $f - g$.

Use the graphs of f and g to find each of the following:

2. $(f \circ g)(-1)$
3. $(g \circ f)(-2)$
4. $f(g(1))$
5. $g(f(-3))$
6. Is f a one-to-one function?

7. Is g a one-to-one function?

8. Match each equation with the graph of f, g, m, or n in the figure.
(A) $y = \log_2 x$ (B) $y = 0.5^x$
(C) $y = \log_{0.5} x$ (D) $y = 2^x$

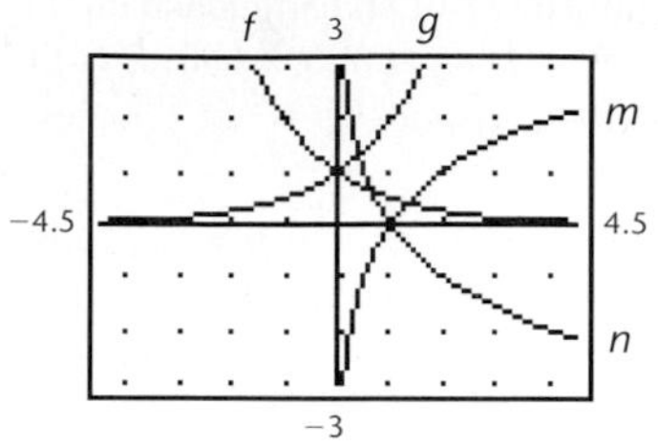

9. Write in logarithmic form using base 10: $m = 10^n$.

10. Write in logarithmic form using base e: $x = e^y$.

Write Problems 11 and 12 in exponential form.

11. $\log x = y$ **12.** $\ln y = x$

In Problems 13 and 14, simplify.

13. $\dfrac{7^{x+2}}{7^{2-x}}$ **14.** $\left(\dfrac{e^x}{e^{-x}}\right)^x$

Solve Problems 15–17 for x exactly. Do not use a calculator or table.

15. $\log_2 x = 3$ **16.** $\log_x 25 = 2$ **17.** $\log_3 27 = x$

Solve Problems 18–21 for x to three significant digits.

18. $10^x = 17.5$ **19.** $e^x = 143{,}000$

20. $\ln x = -0.015\,73$ **21.** $\log x = 2.013$

Evaluate Problems 22–25 to four significant digits using a calculator.

22. $\ln \pi$ **23.** $\log(-e)$

24. $\pi^{\ln 2}$ **25.** $\dfrac{e^{\pi} + e^{-\pi}}{2}$

26. Let $f(x) = x^2 - 4$ and $g(x) = x + 3$. Find each of the following functions and find their domains.
(A) f/g (B) g/f (C) $f \circ g$ (D) $g \circ f$

B

Solve Problems 27–37 for x exactly. Do not use a calculator or table.

27. $\ln(2x - 1) = \ln(x + 3)$

28. $\log(x^2 - 3) = 2\log(x - 1)$

29. $e^{x^2-3} = e^{2x}$ **30.** $4^{x-1} = 2^{1-x}$

31. $2x^2e^{-x} = 18e^{-x}$ **32.** $\log_{1/4} 16 = x$

33. $\log_x 9 = -2$ **34.** $\log_{16} x = \frac{3}{2}$

35. $\log_x e^5 = 5$ **36.** $10^{\log_{10} x} = 33$

37. $\ln x = 0$

Solve Problems 38–47 for x to three significant digits.

38. $x = 2(10^{1.32})$ **39.** $x = \log_5 23$

40. $\ln x = -3.218$ **41.** $x = \log(2.156 \times 10^{-7})$

42. $x = \dfrac{\ln 4}{\ln 2.31}$ **43.** $25 = 5(2^x)$

44. $4{,}000 = 2{,}500(e^{0.12x})$ **45.** $0.01 = e^{-0.05x}$

46. $5^{2x-3} = 7.08$ **47.** $\dfrac{e^x - e^{-x}}{2} = 1$

48. Given $f(x) = \sqrt{x} - 8$ and $g(x) = |x|$,
(A) Find $f \circ g$ and $g \circ f$.
(B) Find the domains of $f \circ g$ and $g \circ f$.

49. Given $f(x) = e^{x-1}$ and $g(x) = \ln(x + 1)$,
(A) Find $f \circ g$ and $g \circ f$.
(B) Find the domains of $f \circ g$ and $g \circ f$.

50. Which of the following functions are one-to-one?
(A) $f(x) = x^3$
(B) $g(x) = (x - 2)^2$
(C) $h(x) = 2x - 3$
(D) $F(x) = (x + 3)^2, x \geq -3$
(E) $G(x) = e^{1-x^2}$
(F) $H(x) = \ln(4 - x)$

In Problems 51–55, find f^{-1}, find the domain and range of f^{-1}, sketch the graphs of f, f^{-1}, and $y = x$ in the same coordinate system, and identify each graph.

51. $f(x) = 3x - 7$ **52.** $f(x) = \sqrt{x - 1}$

53. $f(x) = x^2 - 1, x \geq 0$ **54.** $f(x) = e^x + 1$

55. $f(x) = 2\ln(x - 1)$

Solve Problems 56–61 for x exactly. Do not use a calculator or table.

56. $\log 3x^2 - \log 9x = 2$

57. $\log x - \log 3 = \log 4 - \log(x + 4)$

58. $\ln(x + 3) - \ln x = 2\ln 2$

59. $\ln(2x + 1) - \ln(x - 1) = \ln x$

60. $(\log x)^3 = \log x^9$ **61.** $\ln(\log x) = 1$

In Problems 62 and 63, simplify.

62. $(e^x + 1)(e^{-x} - 1) - e^x(e^{-x} - 1)$

63. $(e^x + e^{-x})(e^x - e^{-x}) - (e^x - e^{-x})^2$

In Problems 64–67, find domain and range, intercepts, and asymptotes. Round all approximate values to two decimal places.

64. $y = 2^{x-1}$ **65.** $f(t) = 10e^{-0.08t}$

66. $y = \ln(x - 1)$ **67.** $N = \dfrac{100}{1 + 3e^{-t}}$

68. If the graph of $y = e^x$ is reflected in the line $y = x$, the graph of the function $y = \ln x$ is obtained. Discuss the functions that are obtained by reflecting the graph of $y = e^x$ in the x axis and the y axis.

69. (A) Explain why the equation $e^{-x/3} = 4 \ln (x + 1)$ has exactly one solution.
(B) Find the solution of the equation to three decimal places.

70. Approximate all real zeros of $f(x) = 4 - x^2 + \ln x$ to three decimal places.

71. Find the coordinates of the points of intersection of $f(x) = 10^{x-3}$ and $g(x) = 8 \log x$ to three decimal places.

C

72. Given $f(x) = x^2$ and $g(x) = \sqrt{1 - x}$, find each function and its domain.
(A) fg (B) f/g (C) $f \circ g$ (D) $g \circ f$

Solve Problems 73–76 for the indicated variable in terms of the remaining symbols.

73. $D = 10 \log \dfrac{I}{I_0}$ for I (sound intensity)

74. $y = \dfrac{1}{\sqrt{2\pi}} e^{-x^2/2}$ for x (probability)

75. $x = -\dfrac{1}{k} \ln \dfrac{I}{I_0}$ for I (X-ray intensity)

76. $r = P \dfrac{i}{1 - (1 + i)^{-n}}$ for n (finance)

Find the inverse of each function in Problems 77 and 78.

77. $f(x) = \dfrac{x + 2}{x - 3}$

78. $f(x) = \dfrac{e^x \quad e^{-x}}{2}$

79. Write $\ln y = -5t + \ln c$ in an exponential form free of logarithms; then solve for y in terms of the remaining symbols.

80. For $f = \{(x, y) \mid y = \log_2 x\}$, graph f and f^{-1} on the same coordinate system. What are the domains and ranges for f and f^{-1}?

81. Explain why 1 cannot be used as a logarithmic base.

82. Prove that $\log_b (M/N) = \log_b M - \log_b N$.

APPLICATIONS

Solve these application problems algebraically or graphically, whichever seems more appropriate.

83. **Price and Demand.** The number q of hot dogs that can be sold during a baseball game at a price of \$$p$ is given approximately by

$$p = d(q) = \frac{9}{1 + 0.002q} \qquad 1{,}000 \le q \le 4{,}000$$

(A) Find the range of d.
(B) Find $q = d^{-1}(p)$ and find its domain and range.

★ 84. **Market Research.** If x units of a product are produced each week and sold for a price of \$$p$ per unit, then the weekly demand, revenue, and cost equations are, respectively,

$$x = 500 - 10p$$
$$R(x) = 50x - \tfrac{1}{10}x^2$$
$$C(x) = 20x + 4{,}000$$

Express the weekly profit as a function of the price p.

85. **Population Growth.** Many countries have a population growth rate of 3% (or more) per year. At this rate, how many years will it take a population to double? Use the annual compounding growth model $P = P_0(1 + r)^t$. Compute the answer to three significant digits.

86. **Population Growth.** Repeat Problem 85 using the continuous compounding growth model $P = P_0e^{rt}$.

87. **Carbon 14 Dating.** How many years will it take for carbon 14 to diminish to 1% of the original amount after the death of a plant or animal? Use the formula $A = A_0e^{-0.000124t}$. Compute the answer to three significant digits.

★ 88. **Medicine.** One leukemic cell injected into a healthy mouse will divide into two cells in about $\frac{1}{2}$ day. At the end of the day these two cells will divide into four. This doubling continues until 1 billion cells are formed; then the animal dies with leukemic cells in every part of the body.
(A) Write an equation that will give the number N of leukemic cells at the end of t days.
(B) When, to the nearest day, will the mouse die?

89. **Money Growth.** Assume \$1 had been invested at an annual rate of 3% compounded continuously at the birth of Christ. What would be the value of the account in the year 2000? Compute the answer to two significant digits.

90. **Present Value.** Solving $A = Pe^{rt}$ for P, we obtain $P = Ae^{-rt}$, which is the **present value** of the amount A due in t years if money is invested at a rate r compounded continuously.
(A) Graph $P = 1{,}000(e^{-0.08t})$, $0 \le t \le 30$.
(B) What does it appear that P tends to as t tends to infinity? [*Conclusion:* The longer the time until the amount A is due, the smaller its present value, as we would expect.]

91. **Earthquakes.** The 1971 San Fernando, California, earthquake released 1.99×10^{14} joules of energy. Compute its magnitude on the Richter scale using the formula $M = \frac{2}{3} \log (E/E_0)$, where $E_0 = 10^{4.40}$ joules. Compute the answer to one decimal place.

92. **Earthquakes.** Refer to Problem 91. If the 1906 San Francisco earthquake had a magnitude of 8.3 on the Richter scale, how much energy was released? Compute the answer to three significant digits.

★ 93. **Sound.** If the intensity of a sound from one source is 100,000 times that of another, how much more is the decibel level of the louder sound than the softer one? Use the formula $D = 10 \log (I/I_0)$.

★★ 94. **Marine Biology.** The intensity of light entering water is reduced according to the exponential function

$$I = I_0 e^{-kd}$$

where I is the intensity d feet below the surface, I_0 is the intensity at the surface, and k is the coefficient of extinction. Measurements in the Sargasso Sea in the West Indies have indicated that half the surface light reaches a depth of 73.6 feet. Find k, and find the depth at which 1% of the surface light remains. Compute answers to three significant digits.

★ **95. Wildlife Management.** A lake formed by a newly constructed dam is stocked with 1,000 fish. Their population is expected to increase according to the logistic curve

$$N = \frac{30}{1 + 29e^{-1.35t}}$$

where N is the number of fish, in thousands, expected after t years. The lake will be open to fishing when the number of fish reaches 20,000. How many years, to the nearest year, will this take?

96. Medicare. The annual expenditures for Medicare (in billions of dollars) by the U.S. government for selected years since 1980 are shown in Table 1 (Bureau of the Census). Let x represent years since 1980.

TABLE 1 Medicare Expenditures

Year	Billion $
1980	37
1985	72
1990	111
1995	181

Source: U.S. Bureau of the Census.

(A) Find an exponential regression model of the form $y = ab^x$ for this data. Estimate (to the nearest billion) the total expenditures in 1996 and in 2010.

(B) When (to the nearest year) will the total expenditures reach 500 billion?

97. Agriculture. The total U.S. corn consumption (in millions of bushels) is shown in Table 2 for selected years since 1975. Let x represent years since 1900.

TABLE 2 Corn Consumption

Year	x	Total Consumption (million bushels)
1975	75	522
1980	80	659
1985	85	1,152
1990	90	1,373
1995	95	1,690

Source: U.S. Department of Agriculture.

(A) Find a logarithmic regression model of the form $y = a + b \ln x$ for this data. Estimate (to the nearest million) the total consumption in 1996 and in 2010.

(B) The actual consumption in 1996 was 1,583 million bushels. How does this compare with the estimated consumption in part A? What effect will this additional 1996 information have on the estimate for 2010? Explain.

CUMULATIVE REVIEW EXERCISE FOR CHAPTERS 3 AND 4

Work through all the problems in this cumulative review and check answers in the back of the book. Answers to all review problems are there, and following each answer is a number in italics indicating the section in which that type of problem is discussed. Where weaknesses show up, review appropriate sections in the text.

A

1. Let $P(x)$ be the polynomial whose graph is shown in the figure.

(A) Assuming that $P(x)$ has integer zeros and leading coefficient 1, find the lowest-degree equation that could produce this graph.

(B) Describe the left and right behavior of $P(x)$.

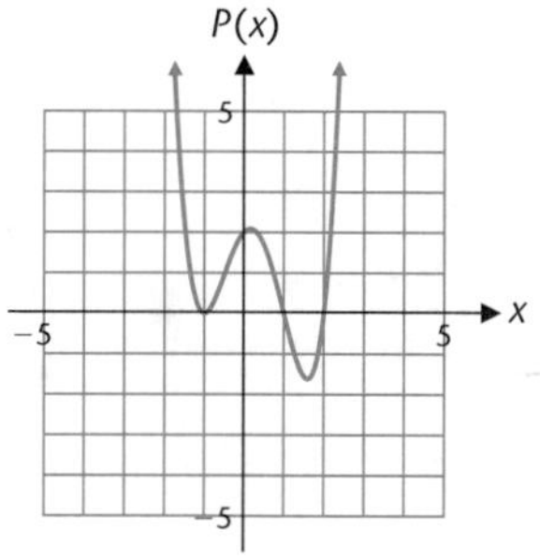

2. Match each equation with the graph of f, g, m, or n in the figure.

(A) $y = (\frac{3}{4})^x$ (B) $y = (\frac{4}{3})^x$

(C) $y = (\frac{3}{4})^x + (\frac{4}{3})^x$ (D) $y = (\frac{4}{3})^x - (\frac{3}{4})^x$

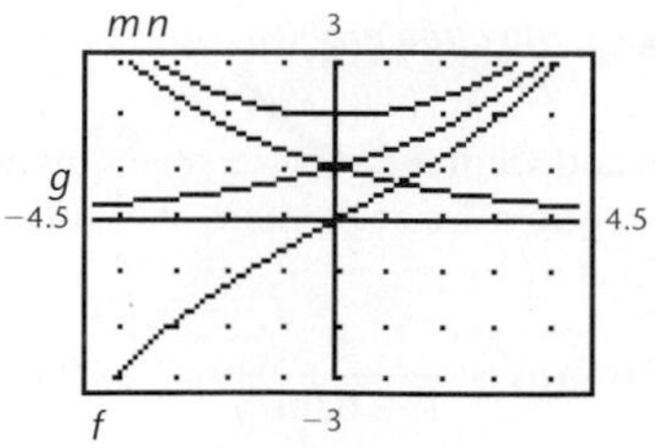

3. For $P(x) = 3x^3 + 5x^2 - 18x - 3$ and $D(x) = x + 3$, use synthetic division to divide $P(x)$ by $D(x)$, and write the answer in the form $P(x) = D(x)Q(x) + R$.

4. Let $P(x) = 2(x + 2)(x - 3)(x - 5)$. What are the zeros of $P(x)$?

5. Let $P(x) = 4x^3 - 5x^2 - 3x - 1$. How do you know that $P(x)$ has at least one real zero between 1 and 2?

6. Let $P(x) = x^3 + x^2 - 10x + 8$. Find all rational zeros for $P(x)$.

7. Solve for x.
(A) $y = 10^x$ (B) $y = \ln x$

8. Simplify.
(A) $(2e^x)^3$ (B) $\dfrac{e^{3x}}{e^{-2x}}$

9. Solve for x exactly. Do not use a calculator or a table.
(A) $\log_3 x = 2$
(B) $\log_3 81 = x$
(C) $\log_x 4 = -2$

10. Solve for x to three significant digits.
(A) $10^x = 2.35$ (B) $e^x = 87{,}500$
(C) $\log x = -1.25$ (D) $\ln x = 2.75$

B

11. Given $f(x) = 1/(x - 2)$ and $g(x) = (x + 3)/x$, find $f \circ g$. What is the domain of $f \circ g$?

12. Find $f^{-1}(x)$ for $f(x) = 2x + 5$.

13. The function f subtracts the square root of the domain element from three times the natural log of the domain element. Write an algebraic definition of f.

14. Write a verbal description of the function $f(x) = 100e^{0.5x} - 50$.

15. Let $f(x) = \dfrac{2x + 8}{x + 2}$.
(A) Find the domain and the intercepts for f.
(B) Find the vertical and horizontal asymptotes for f.
(C) Sketch the graph of f. Draw vertical and horizontal asymptotes with dashed lines.

16. Let $f(x) = \sqrt{x + 4}$
(A) Find $f^{-1}(x)$.
(B) Find the domain and range of f and f^{-1}.
(C) Graph f, f^{-1}, and $y = x$ on the same coordinate system and identify each graph.

17. Which of the following functions is one-to-one?
(A) $f(x) = x^3 + x$ (B) $g(x) = x^3 + x^2$
(C) $h(x) = e^x + \ln x$ (D) $k(x) = e^x - \ln x$

18. If $P(x) = 2x^3 - 5x^2 + 3x + 2$, find $P(\frac{1}{2})$ using the remainder theorem and synthetic division.

19. Which of the following is a factor of

$$P(x) = x^{25} - x^{20} + x^{15} + x^{10} - x^5 + 1$$

(A) $x - 1$ (B) $x + 1$

20. Let $P(x) = x^4 - 8x^2 + 3$.
(A) Graph $P(x)$ and describe the graph verbally, including the number of x intercepts, the number of turning points, and the left and right behavior.
(B) Approximate the largest x intercept to two decimal places.

21. Let $P(x) = x^5 - 8x^4 + 17x^3 + 2x^2 - 20x - 8$.
(A) Approximate the zeros of $P(x)$ to two decimal places and state the multiplicity of each zero.
(B) Can any of these zeros be approximated with the bisection method? A maximum routine? A minimum routine? Explain.

22. Let $P(x) = x^4 + 2x^3 - 20x^2 - 30$.
(A) Find the smallest positive and largest negative integers that, by Theorem 2 in Section 3-3, are upper and lower bounds, respectively, for the real zeros of $P(x)$.
(B) If $(k, k + 1)$, k an integer, is the interval containing the largest real zero of $P(x)$, determine how many additional intervals are required in the bisection method to approximate this zero to one decimal place.
(C) Approximate the real zeros of $P(x)$ to two decimal places.

23. Find all zeros (rational, irrational, and imaginary) exactly for $P(x) = 4x^3 - 20x^2 + 29x - 15$.

24. Find all zeros (rational, irrational, and imaginary) exactly for $P(x) = x^4 + 5x^3 + x^2 - 15x - 12$, and factor $P(x)$ into linear factors.

Solve Problems 25–34 for x exactly. Do not use a calculator or a table.

25. $2^{x^2} = 4^{x+4}$

26. $2x^2e^{-x} + xe^{-x} = e^{-x}$

27. $e^{\ln x} = 2.5$

28. $\log_x 10^4 = 4$

29. $\log_9 x = -\frac{3}{2}$

30. $\ln (x + 4) - \ln (x - 4) = 2 \ln 3$

31. $\ln (2x^2 + 2) = 2 \ln (2x - 4)$

32. $\log x + \log (x + 15) = 2$

33. $\log (\ln x) = -1$

34. $4 (\ln x)^2 = \ln x^2$

Solve Problems 35–39 for x to three significant digits.

35. $x = \log_3 41$

36. $\ln x = 1.45$

37. $4(2^x) = 20$

38. $10e^{-0.5x} = 1.6$

39. $\dfrac{e^x - e^{-x}}{e^x + e^{-x}} = \dfrac{1}{2}$

In Problems 40–44, find domain, range, intercepts, and asymptotes. Round all approximate values to two decimal places.

40. $f(x) = 3^{1-x}$

41. $g(x) = \ln (2 - x)$

42. $A(t) = 100e^{-0.3t}$

43. $h(x) = -2e^{-x} + 3$

44. $N(t) = \dfrac{6}{2 + e^{-0.1t}}$

45. If the graph of $y = \ln x$ is reflected in the line $y = x$, the graph of the function $y = e^x$ is obtained. Discuss the functions that are obtained by reflecting the graph of $y = \ln x$ in the x axis and in the y axis.

46. (A) Explain why the equation $e^{-x} = \ln x$ has exactly one solution.
    (B) Approximate the solution of the equation to two decimal places.

C

47. Given $f(x) = x^2$ and $g(x) = \sqrt{4 - x^2}$, find
    (A) Domain of g
    (B) f/g and its domain
    (C) $f \circ g$ and its domain

48. Let $f(x) = x^2 - 2x - 3$, $x \geq 1$.
    (A) Find $f^{-1}(x)$.
    (B) Find the domain and range of f^{-1}.
    (C) Graph, f, f^{-1}, and $y = x$ on the same coordinate system.

49. Graph f and indicate any horizontal, vertical, or oblique asymptotes with dashed lines:

$$f(x) = \frac{x^2 + 4x + 8}{x + 2}$$

50. Let $P(x) = x^4 - 28x^3 + 262x^2 - 922x + 1{,}083$. Approximate (to two decimal places) the x intercepts and the local extrema.

51. Find a polynomial of lowest degree with leading coefficient 1 that has zeros -1 (multiplicity 2), 0 (multiplicity 3), $3 + 5i$, and $3 - 5i$. Leave the answer in factored form. What is the degree of the polynomial?

52. If $P(x)$ is a fourth-degree polynomial with integer coefficients and if i is a zero of $P(x)$, can $P(x)$ have any irrational zeros? Explain.

53. Let $P(x) = x^4 + 9x^3 - 500x^2 + 20{,}000$.
    (A) Find the smallest positive integer multiple of 10 and the largest negative integer multiple of 10 that, by Theorem 2 in Section 3-3, are upper and lower bounds, respectively, for the real zeros of $P(x)$.
    (B) Approximate the real zeros of $P(x)$ to two decimal places.

54. Find all zeros (rational, irrational, and imaginary) exactly for

$$P(x) = x^5 - 4x^4 + 3x^3 + 10x^2 - 10x - 12$$

and factor $P(x)$ into linear factors.

55. Find rational roots exactly and irrational roots to two decimal places for

$$P(x) = x^5 + 4x^4 + x^3 - 11x^2 - 8x + 4$$

56. Let $f(x) = 3 \ln (x - 2)$.
    (A) Find $f^{-1}(x)$.
    (B) Find the domain and range of f and f^{-1}.
    (C) Graph f, f^{-1}, and $y = x$ on the same coordinate system and identify each graph.

57. Use natural logarithms to solve for n.

$$A = P\frac{(1 + i)^n - 1}{i}$$

58. Solve $\ln y = 5x + \ln A$ for y. Express the answer in a form that is free of logarithms.

59. Solve for x.

$$y = \frac{e^x - 2e^{-x}}{2}$$

APPLICATIONS

60. **Weather Balloon.** A spherical weather balloon is being inflated. The radius of the balloon t seconds after inflation begins is given by $r = 0.3t^{1/4}$ meters. Express the volume of the balloon as a function of t. [Volume of a sphere: $V = \frac{4}{3}\pi r^3$.]

61. **Depreciation.** Office equipment was purchased for \$20,000 and is assumed to depreciate linearly to a scrap value of \$4,000 after 8 years.
    (A) Find a linear function $v = d(t)$ that relates value v in dollars to time t in years.
    (B) Find $t = d^{-1}(v)$.

62. **Shipping.** A mailing service provides customers with rectangular shipping containers. The length plus the girth of one of these containers is 10 feet (see the figure). If the end of the container is square and the volume is 8 cubic feet, find the dimensions. Find rational solutions exactly and irrational solutions to two decimal places.

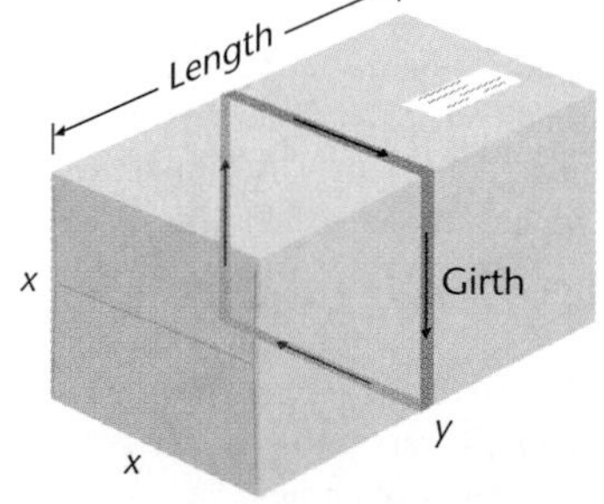

63. **Geometry.** The diagonal of a rectangle is 2 feet longer than one of the sides, and the area of the rectangle is 6 square feet. Find the dimensions of the rectangle. Find rational solutions exactly and irrational solutions to two decimal places.

64. **Population Growth.** If the Republic of the Congo has a population of about 40 million people and a doubling time of 22 years, find the population in
    (A) 5 years (B) 30 years
    Compute answers to three significant digits.

65. Compound Interest. How long will it take money invested in an account earning 7% compounded annually to double? Use the annual compounding growth model $P = P_0(1 + r)^t$, and compute the answer to three significant digits.

66. Compound Interest. Repeat Problem 66 using the continuous compound interest model $P = P_0e^{rt}$.

67. Earthquakes. If the 1906 and 1989 San Francisco earthquakes registered 8.3 and 7.1, respectively, on the Richter scale, how many times more powerful was the 1906 earthquake than the 1989 earthquake? Use the formula $M = \frac{2}{3}\log(E/E_0)$, where $E_0 = 10^{4.40}$ joules, and compute the answer to one decimal place.

68. Sound. If the decibel level at a rock concert is 88, find the intensity of the sound at the concert. Use the formula $D = 10\log(I/I_0)$, where $I_0 = 10^{-12}$ watt per square meter, and compute the answer to two significant digits.

69. Table 1 shows the life expectancy (in years) at birth for residents of the United States from 1970 to 1995. Let x represent years since 1970. Use the indicated regression model to estimate the life expectancy (to the nearest tenth of a year) for a U.S. resident born in 2010.

(A) Linear regression
(B) Quadratic regression
(C) Cubic regression
(D) Exponential regression

TABLE 1

Year	Life Expectancy
1970	70.8
1975	72.6
1980	73.7
1985	74.7
1990	75.4
1995	75.9

Source: U.S. Census Bureau.

70. Refer to Problem 69. The Census Bureau projected the life expectancy for a U.S. resident born in 2010 to be 77.6 years. Which of the models in Problem 69 is closest to the Census Bureau projection?

ANSWER TO APPLICATION

According to the Rule of 72, your investment will double in

$$\frac{72}{100(0.104)} \approx 6.9 \text{ years}$$

To compute the exact doubling time, we use the compound interest formula $A = P(1 + r/m)^n$ with $A = 36{,}000$, $P = 18{,}000$, $r = 0.104$, $m = 1$, and solve for n:

$$36{,}000 = 18{,}000(1 + 0.104)^n$$

Graphing both sides and finding the point of intersection (Fig. 1), we see that the exact doubling time is 7.0 years. Thus, for this interest rate, the Rule of 72 provides a reasonable approximation.

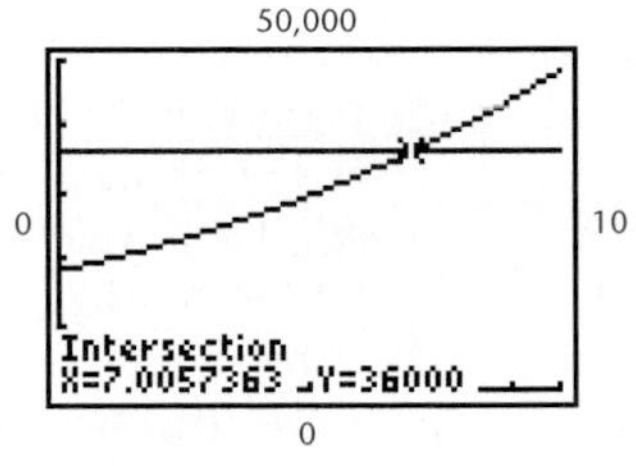

FIGURE 1

Systems; Matrices

5

Outline

5-1 Systems of Linear Equations in Two Variables

5-2 Systems of Linear Equations and Augmented Matrices

5-3 Gauss–Jordan Elimination

5-4 Matrix Operations

5-5 Inverse of a Square Matrix

5-6 Matrix Equations and Systems of Linear Equations

5-7 Systems of Linear Inequalities

5-8 Linear Programming

Chapter 5 Group Activity: Modeling with Systems of Linear Equations

Chapter 5 Review

Application

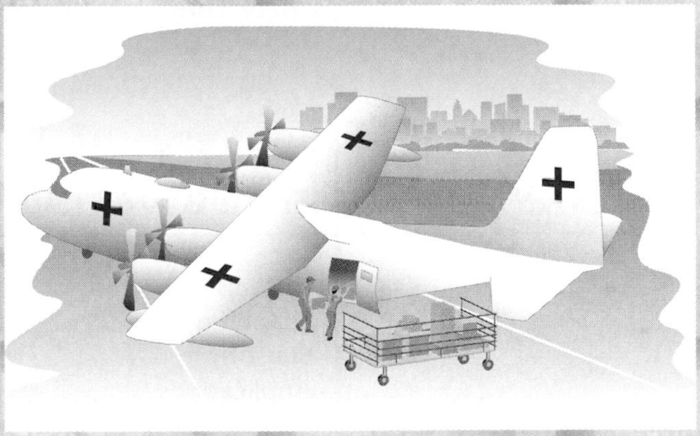

A Red Cross plane is being loaded with bottled water and dehydrated food for transport to an earthquake region. Each bottle of water weighs 18 pounds, occupies 1 cubic foot of space, and will supply 15 people. Each package of dehydrated food weighs 9 pounds, occupies 0.75 cubic feet of space, and will supply 11 people. The plane has space for 4,500 cubic feet of cargo weighing at most 64,800 pounds. How many bottles of water and packages of dehydrated food should be loaded into the plane to maximize the total number of people supplied with either food or water by this shipment?

In this chapter we first discuss how systems of linear equations involving two variables are solved graphically and algebraically. Because these techniques are not suitable for linear systems involving larger numbers of equations and variables, we then turn to a different method of solution involving the concept of an *augmented matrix,* which arises quite naturally when dealing with larger linear systems. We then study *matrices* and *matrix operations* in their own right as a new mathematical form. With these new operations added to our mathematical toolbox, we return to systems of equations from a fresh point of view. Finally, we discuss systems of linear inequalities and linear programming. Throughout the chapter we use these new mathematical tools to solve a variety of interesting and important applied problems.

Preparing for This Chapter

Before getting started on this chapter, review the following concepts:

Properties of Real Numbers (Appendix A, Section 1)

Linear Equations and Inequalities (Appendix A, Section 8, and Chapter 2, Section 2)

Linear Functions (Chapter 2, Section 1)

Section 5-1 | Systems of Linear Equations in Two Variables

- Systems of Equations
- Graphing
- Substitution
- Applications

In this section we discuss both graphical and algebraic methods for solving systems of linear equations in two variables. Then we use systems of this type to construct and solve mathematical models for several applications.

Systems of Equations

To establish basic concepts, consider the following example. At a computer fair, student tickets cost \$2 and general admission tickets cost \$3. If a total of 7 tickets are purchased for a total cost of \$18, how many of each type were purchased?

Let

x = Number of student tickets

y = Number of general admission tickets

Then

$$x + y = 7 \quad \text{Total number of tickets purchased}$$

$$2x + 3y = 18 \quad \text{Total purchase cost}$$

We now have a system of two linear equations in two variables. Thus, we can solve this problem by finding all pairs of numbers x and y that satisfy both equations.

In general, we are interested in solving linear systems of the type

$$ax + by = h \quad \text{System of two linear equations in two variables}$$
$$cx + dy = k$$

where x and y are variables, a, b, c, and d are real numbers called the **coefficients** of x and y, and h and k are real numbers called the **constant terms** in the equations. A pair of numbers $x = x_0$ and $y = y_0$ is a **solution** of this system if each equation is satisfied by the pair. The set of all such pairs of numbers is called the **solution set** for the system. To **solve** a system is to find its solution set.

Graphing

Recall that the graph of a linear equation is the line consisting of all ordered pairs that satisfy the equation. To solve the ticket problem by graphing, we graph both equations in the same coordinate system. The coordinates of any points that the lines have in common must be solutions to the system, since they must satisfy both equations.

EXAMPLE 1

Solving a System by Graphing

Solve the ticket problem by graphing:
$$x + y = 7$$
$$2x + 3y = 18$$

Solution

From Figure 1 we see that

FIGURE 1

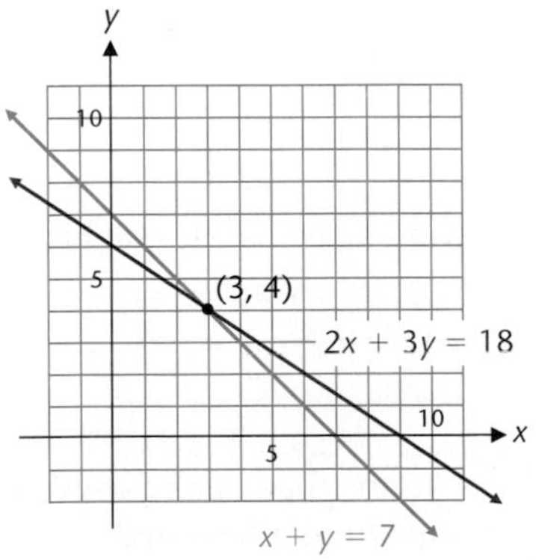

$$x = 3 \quad \text{Student tickets}$$
$$y = 4 \quad \text{General admission tickets}$$

Check

$x + y = 7$	$2x + 3y = 18$
$3 + 4 \overset{?}{=} 7$	$2(3) + 3(4) \overset{?}{=} 18$
$7 \overset{\checkmark}{=} 7$	$18 \overset{\checkmark}{=} 18$

MATCHED PROBLEM 1

Solve by graphing and check:
$$x - y = 3$$
$$x + 2y = -3$$

It is clear that the preceding example has exactly one solution, since the lines have exactly one point of intersection. In general, lines in a rectangular coordinate system are related to each other in one of three ways, as illustrated in the next example.

EXAMPLE 2

Solving Three Important Types of Systems by Graphing

Solve each of the following systems by graphing:

(A) $2x - 3y = 2$, $x + 2y = 8$ (B) $4x + 6y = 12$, $2x + 3y = -6$ (C) $2x - 3y = -6$, $-x + \frac{3}{2}y = 3$

Solutions

(A)

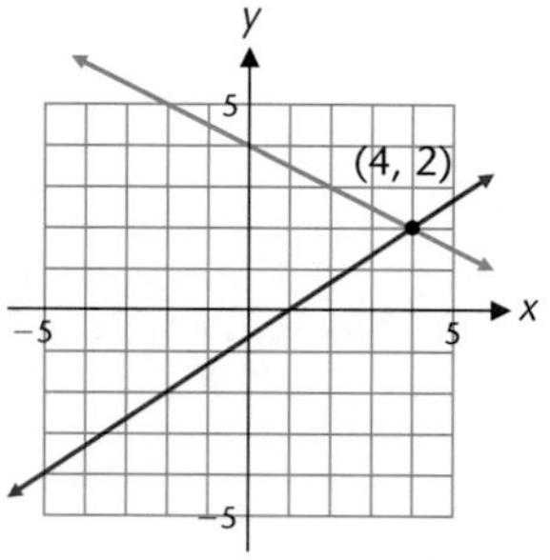

Lines intersect at one point only.
Exactly one solution: $x = 4$, $y = 2$

(B)

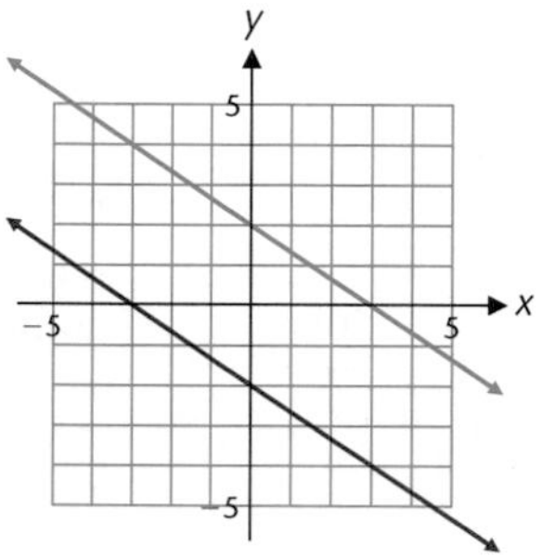

Lines are parallel (each has slope $-\frac{2}{3}$). No solution.

(C)

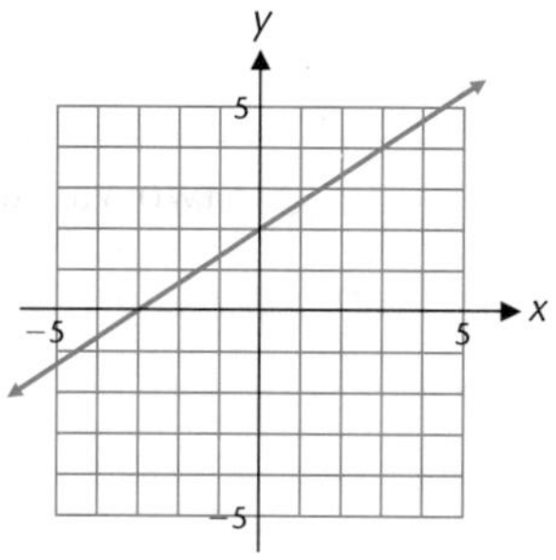

Lines coincide.
Infinitely many solutions.

MATCHED PROBLEM 2

Solve each of the following systems by graphing:

(A) $2x + 3y = 12$, $x - 3y = -3$ (B) $x - 3y = -3$, $-2x + 6y = 12$ (C) $2x - 3y = 12$, $-x + \frac{3}{2}y = -6$

We now define some terms that can be used to describe the different types of solutions to systems of equations illustrated in Example 2.

SYSTEMS OF LINEAR EQUATIONS: BASIC TERMS

A system of linear equations is **consistent** if it has one or more solutions and **inconsistent** if no solutions exist. Furthermore, a consistent system is said to be **independent** if it has exactly one solution (often referred to as the **unique solution**) and **dependent** if it has more than one solution.

Referring to the three systems in Example 2, the system in part A is consistent and independent, with the unique solution $x = 4$ and $y = 2$. The system in part B is inconsistent, with no solution. And the system in part C is consistent and dependent, with an infinite number of solutions: all the points on the two coinciding lines.

Explore/Discuss 1

Can a consistent and dependent system have exactly two solutions? Exactly three solutions? Explain.

By geometrically interpreting a system of two linear equations in two variables, we gain useful information about what to expect in the way of solutions to the system. In general, any two lines in a rectangular coordinate plane must intersect in exactly one point, be parallel, or coincide (have identical graphs). Thus, the systems in Example 2 illustrate the only three possible types of solutions for systems of two linear equations in two variables. These ideas are summarized in Theorem 1.

THEOREM 1

POSSIBLE SOLUTIONS TO A LINEAR SYSTEM

The linear system

$$ax + by = h$$
$$cx + dy = k$$

must have

1. Exactly one solution — Consistent and independent

or

2. No solution — Inconsistent

or

3. Infinitely many solutions — Consistent and dependent

There are no other possibilities.

One drawback of finding a solution by graphing is the inaccuracy of hand-drawn graphs. Graphic solutions performed on a graphing utility, however, provide both a useful geometric interpretation and an accurate approximation of the solution to a system of linear equations in two variables.

EXAMPLE 3 **Solving a System Using a Graphing Utility**

Solve to two decimal places using a graphing utility:

$$\begin{aligned} 5x - 3y &= 13 \\ 2x + 4y &= 15 \end{aligned}$$

Solution First solve each equation for y:

$$\begin{aligned} 5x - 3y &= 13 \\ -3y &= -5x + 13 \\ y &= \tfrac{5}{3}x - \tfrac{13}{3} \end{aligned} \qquad \begin{aligned} 2x + 4y &= 15 \\ 4y &= -2x + 15 \\ y &= -0.5x + 3.75 \end{aligned}$$

Next, enter each equation in a graphing utility [Fig. 2(a)], graph in an appropriate viewing window, and approximate the intersection point [Fig. 2(b)].

FIGURE 2

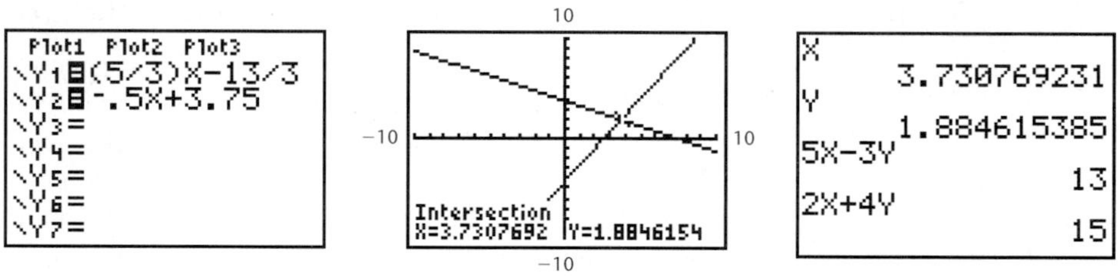

(a) Equation definitions (b) Intersection point (c) Check

Rounding the values in Figure 2(b) to two decimal places, we see that the solution is

$$x = 3.73 \text{ and } y = 1.88 \qquad \text{or} \qquad (3.73, 1.88)$$

Figure 2(c) shows a check of this solution.

MATCHED PROBLEM 3 Solve to two decimal places using a graphing utility:

$$\begin{aligned} 2x - 5y &= -25 \\ 4x + 3y &= 5 \end{aligned}$$

Remark In the solution to Example 3, you might wonder why we checked a solution produced by a graphing utility. After all, we don't expect a graphing utility to make an error. But the equations in the original system and the equations entered in Figure 2(a) are not identical. We might have made an error when solving the original equations for y. The check in Figure 2(c) eliminates this possibility.

Graphic methods help us visualize a system and its solutions, frequently reveal relationships that might otherwise be hidden, and, with the assistance of a graphing utility, provide very accurate approximations to solutions.

Substitution

There are a number of different algebraic techniques that can also be used to solve systems of linear equations in two variables. One of the simplest is the *substitution method.* To solve a system by **substitution,** we first choose one of the two equations in a system and solve for one variable in terms of the other. (We make a choice that avoids fractions, if possible.) Then we substitute the result in the other equation and solve the resulting linear equation in one variable. Finally, we substitute this result back into the expression obtained in the first step to find the second variable. We return to the ticket problem stated at the beginning of the section to illustrate this process.

EXAMPLE **4**

Solving a System by Substitution

Use substitution to solve the ticket problem:

$$\begin{aligned} x + y &= 7 \\ 2x + 3y &= 18 \end{aligned}$$

Solution

Solve either equation for one variable and substitute into the remaining equation. We choose to solve the first equation for y in terms of x:

$x + y = 7$ Solve the first equation for y in terms of x.

$y = 7 - x$ Substitute into the second equation.

$$\begin{aligned} 2x + 3y &= 18 \\ 2x + 3(7 - x) &= 18 \\ 2x + 21 - 3x &= 18 \\ -x &= -3 \\ x &= 3 \end{aligned}$$

Now, replace x with 3 in $y = 7 - x$:

$$\begin{aligned} y &= 7 - x \\ y &= 7 - 3 \\ y &= 4 \end{aligned}$$

Thus the solution is 3 student tickets and 4 general admission tickets.

Check

$$\begin{aligned} x + y &= 7 \\ 3 + 4 &\overset{?}{=} 7 \\ 7 &\overset{\checkmark}{=} 7 \end{aligned} \qquad \begin{aligned} 2x + 3y &= 18 \\ 2(3) + 3(4) &\overset{?}{=} 18 \\ 18 &\overset{\checkmark}{=} 18 \end{aligned}$$

MATCHED PROBLEM 4

Solve by substitution and check:
$$\begin{aligned} x - y &= 3 \\ x + 2y &= -3 \end{aligned}$$

EXAMPLE 5

Solving a System by Substitution

Solve by substitution and check:
$$\begin{aligned} 2x - 3y &= 7 \\ 3x - y &= 7 \end{aligned}$$

Solution

To avoid fractions, we choose to solve the second equation for y:

$$\begin{aligned}
3x - y &= 7 && \text{Solve for } y \text{ in terms of } x. \\
-y &= -3x + 7 \\
y &= 3x - 7 && \text{Substitute into first equation.} \\
2x - 3y &= 7 && \text{First equation} \\
2x - 3(3x - 7) &= 7 && \text{Solve for } x. \\
2x - 9x + 21 &= 7 \\
-7x &= -14 \\
x &= 2 && \text{Substitute } x = 2 \text{ in } y = 3x - 7. \\
y &= 3x - 7 \\
y &= 3(2) - 7 \\
y &= -1
\end{aligned}$$

Thus, the solution is $x = 2$ and $y = -1$.

Check

$$\begin{aligned} 2x - 3y &= 7 \\ 2(2) - 3(-1) &\stackrel{?}{=} 7 \\ 7 &\stackrel{\checkmark}{=} 7 \end{aligned} \qquad \begin{aligned} 3x - y &= 7 \\ 3(2) - (-1) &\stackrel{?}{=} 7 \\ 7 &\stackrel{\checkmark}{=} 7 \end{aligned}$$

MATCHED PROBLEM 5

Solve by substitution and check:
$$\begin{aligned} 3x - 4y &= 18 \\ 2x + y &= 1 \end{aligned}$$

Explore/Discuss 2

Use substitution to solve each of the following systems. Discuss the nature of the solution sets you obtain.

$$\begin{aligned} x + 3y &= 4 \\ 2x + 6y &= 7 \end{aligned} \qquad \begin{aligned} x + 3y &= 4 \\ 2x + 6y &= 8 \end{aligned}$$

Applications

The following examples illustrate the use of systems of linear equations to construct models for applied problems. Each model can be solved by either graphing or substitution—the choice is really a matter of personal preference.

EXAMPLE 6

Food Processing

A food manufacturer produces regular and lite smoked sausages. A regular sausage is 72% pork and 28% turkey and a lite sausage is 22% pork and 78% turkey. The company has just received a shipment of 2,000 pounds of pork and 2,000 pounds of turkey. How many pounds of each type of sausage should be produced to use all the meat in this shipment?

Solution

First we define the relevant variables:

x = Pounds of regular sausage

y = Pounds of lite sausage

Next we summarize the given information in Table 1. It is convenient to organize the table so that the quantities represented by variables correspond to columns in the table (rather than to rows), as shown.

TABLE 1

	Regular Sausage	Lite Sausage	Total
Pork	72%	22%	2,000
Turkey	28%	78%	2,000

Now we use the information in the table to form equations involving x and y:

$$\begin{pmatrix}\text{Pork in } x \text{ pounds}\\ \text{of regular sausage}\end{pmatrix} + \begin{pmatrix}\text{Pork in } y \text{ pounds}\\ \text{of lite sausage}\end{pmatrix} = \begin{pmatrix}\text{Total}\\ \text{pork}\end{pmatrix}$$

$$0.72x + 0.22y = 2{,}000$$

$$\begin{pmatrix}\text{Turkey in } x \text{ pounds}\\ \text{of regular sausage}\end{pmatrix} + \begin{pmatrix}\text{Turkey in } y \text{ pounds}\\ \text{of lite sausage}\end{pmatrix} = \begin{pmatrix}\text{Total}\\ \text{turkey}\end{pmatrix}$$

$$0.28x + 0.78y = 2{,}000$$

We will solve this system graphically. Figure 3(a) shows the equations after they have been solved for y and entered in the equation editor of a graphing utility. From Figure 3(b), we conclude that producing 2,240 pounds of regular sausage and 1,760 pounds of lite sausage will use all the available pork and turkey.

FIGURE 3

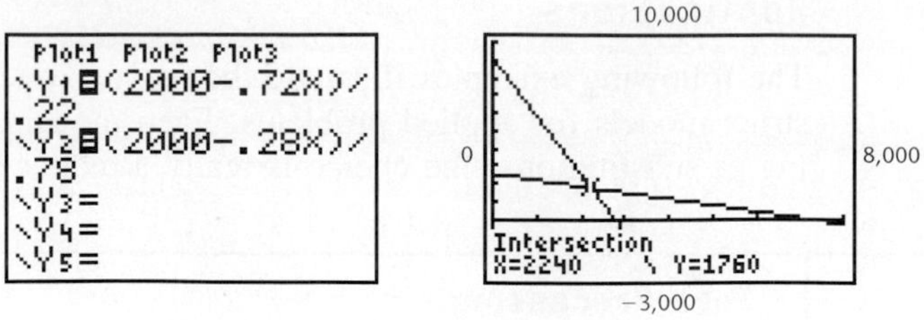

(a) (b)

MATCHED PROBLEM 6

A food manufacturer produces regular and deluxe rice mixtures by mixing wild rice with long-grain rice. The regular rice mixture is 5% wild rice and 95% long-grain rice and the deluxe rice mixture is 10% wild rice and 90% long-grain rice. The company has just received a shipment of 120 pounds of wild rice and 1,500 pounds of long-grain rice. How many pounds of each type of rice mixture should be produced to use all the rice in this shipment?

EXAMPLE 7

Airspeed

An airplane makes the 2,400-mile trip from Washington, D.C., to San Francisco in 7.5 hours and makes the return trip in 6 hours. Assuming that the plane travels at a constant airspeed and that the wind blows at a constant rate from west to east, find the plane's airspeed and the wind rate.

Solution

Let x represent the airspeed of the plane and let y represent the rate at which the wind is blowing (both in miles per hour). The ground speed of the plane is determined by combining these two rates; that is,

$x - y$ = Ground speed flying east to west (headwind)

$x + y$ = Ground speed flying west to east (tailwind)

Applying the familiar formula $D = RT$ to each leg of the trip leads to the following system of equations:

$2{,}400 = 7.5(x - y)$ From Washington to San Francisco

$2{,}400 = 6(x + y)$ From San Francisco to Washington

After simplification, we have

$$x - y = 320$$
$$x + y = 400$$

Solve using substitution:

$x = y + 320$ Solve first equation for x.

$y + 320 + y = 400$ Substitute in second equation.

$2y = 80$

$y = 40$ **mph** Wind rate

$$x = 40 + 320$$

$$x = \mathbf{360\ mph} \qquad \text{Airspeed}$$

Check

$$2{,}400 = 7.5(x - y) \qquad 2{,}400 = 6(x + y)$$

$$2{,}400 \stackrel{?}{=} 7.5(360 - 40) \qquad 2{,}400 \stackrel{?}{=} 6(360 + 40)$$

$$2{,}400 \stackrel{\checkmark}{=} 2{,}400 \qquad 2{,}400 \stackrel{\checkmark}{=} 2{,}400$$

MATCHED PROBLEM 7

A boat takes 8 hours to travel 80 miles upstream and 5 hours to return to its starting point. Find the speed of the boat in still water and the speed of the current.

EXAMPLE 8

Supply and Demand

The quantity of a product that people are willing to buy during some period of time depends on its price. Generally, the higher the price, the less the demand; the lower the price, the greater the demand. Similarly, the quantity of a product that a supplier is willing to sell during some period of time also depends on the price. Generally, a supplier will be willing to supply more of a product at higher prices and less of a product at lower prices. This example uses linear models to analyze the relationship between supply and demand.

Solution

Suppose we are interested in analyzing the sale of cherries each day in a particular city. An analyst arrives at the following price–demand and price–supply equations:

$$p = -0.3q + 5 \qquad \text{Demand equation (consumer)}$$

$$p = 0.06q + 0.68 \qquad \text{Supply equation (supplier)}$$

where q represents the quantity in thousands of pounds and p represents the price in dollars. The graphs of these equations are shown in Figure 4(a), where we have substituted x for q.

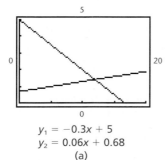

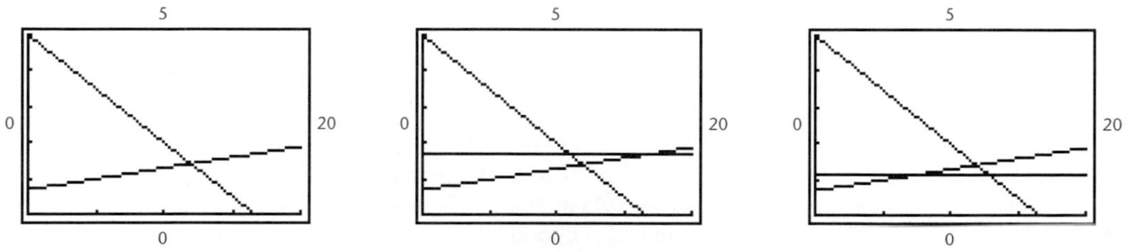

$y_1 = -0.3x + 5$
$y_2 = 0.06x + 0.68$
(a)

$y_3 = 1.70$
Supply exceeds demand
(b)

$y_4 = 1.10$
Demand exceeds supply
(c)

FIGURE 4

Suppose that cherries are selling for \$1.70 per pound. Using a built-in intersection routine (details omitted), we find that the horizontal line $p = 1.70$ intersects the demand equation at $q = 11$ and the supply equation at $q = 17$ [see

FIGURE 5

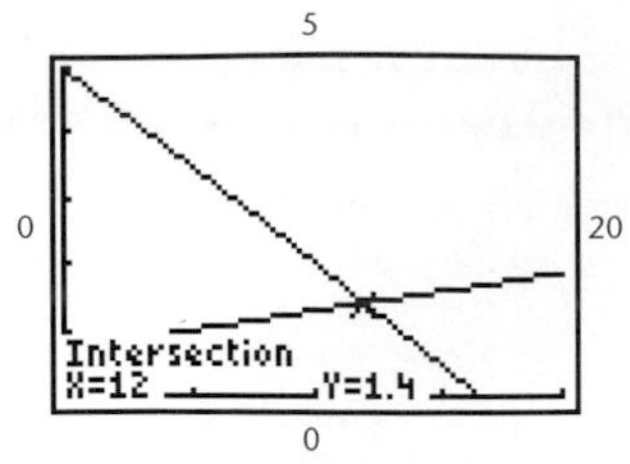

Equilibrium point

Fig. 4(b)]. Thus, at a price of \$1.70 per pound, consumers will purchase 11,000 pounds of cherries and suppliers are willing to supply 17,000 pounds. The supply exceeds the demand at this price, and the price will come down. Now suppose that the price drops to \$1.10 per pound [Fig. 4(c)]. Proceeding as before (details omitted), we find that at this price consumers will purchase 13,000 pounds of cherries, but suppliers will supply only 7,000 pounds. Thus, at \$1.10 per pound the demand exceeds the supply and the price will go up. At what price will cherries stabilize for the day? That is, at what price will supply equal demand? This price, if it exists, is called the **equilibrium price,** the quantity sold at that price is called the **equilibrium quantity,** and the point of intersection of the supply and demand equations is called the **equilibrium point.** Using a built-in intersection routine (Fig. 5), we see that the equilibrium quantity is 12,000 pounds and the equilibrium price is \$1.40.

MATCHED PROBLEM **8**

The price–demand and price–supply equations for strawberries in a certain city are

$$p = -0.2q + 4 \qquad \text{Demand equation}$$

$$p = 0.04q + 1.84 \qquad \text{Supply equation}$$

where q represents the quantity in thousands of pounds and p represents the price in dollars.

Find the equilibrium quantity and the equilibrium price.

Answers to Matched Problems

1.

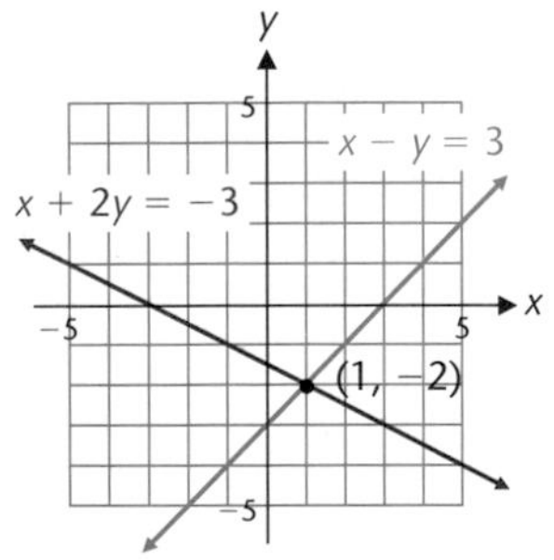

$x = 1$, $y = -2$

Check:

$$x - y = 3$$
$$1 - (-2) \stackrel{?}{=} 3$$
$$3 \stackrel{\checkmark}{=} 3$$
$$x + 2y = -3$$
$$1 + 2(-2) \stackrel{?}{=} -3$$
$$-3 \stackrel{\checkmark}{=} -3$$

2. (A) (3, 2) or $x = 3$ and $y = 2$ (B) No solutions (C) Infinite number of solutions
3. $(-1.92, 4.23)$ or $x = -1.92$ and $y = 4.23$ **4.** $x = 1$, $y = -2$ **5.** $x = 2$, $y = -3$
6. 840 pounds of regular mix, 780 pounds of deluxe mix **7.** Boat: 13 mph, current: 3 mph
8. Equilibrium quantity = 9 thousand pounds; Equilibrium price = \$2.20 per pound

EXERCISE 5-1

A

Match each system in Problems 1–4 with one of the following graphs, and use the graph to solve the system.

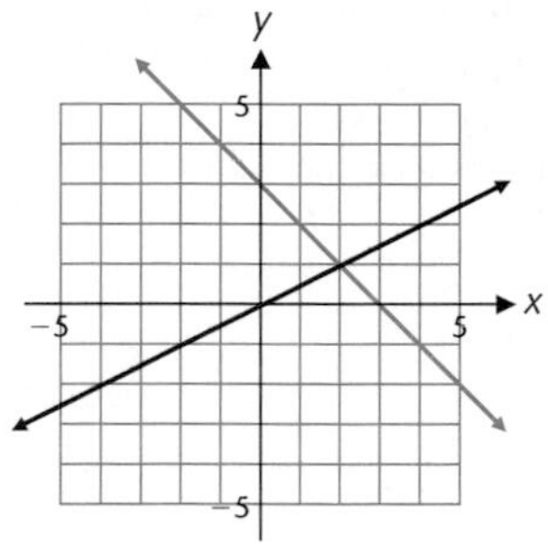

(a)

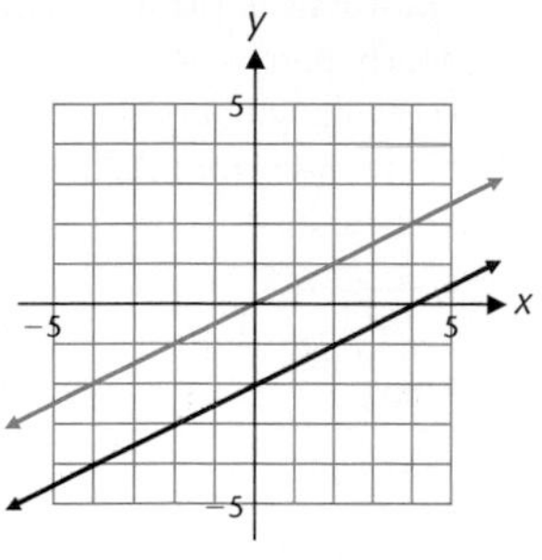

(b)

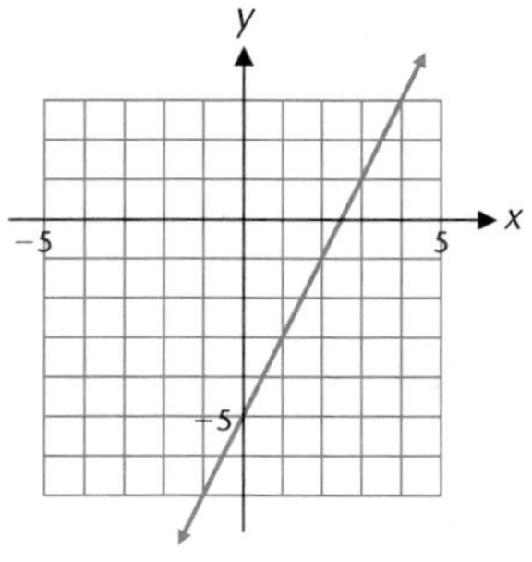

(c)

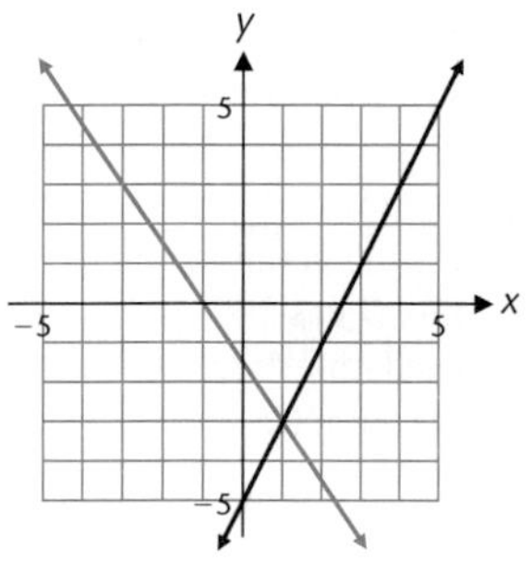

(d)

1. $2x - 4y = 8$
$x - 2y = 0$

2. $x + y = 3$
$x - 2y = 0$

3. $2x - y = 5$
$3x + 2y = -3$

4. $4x - 2y = 10$
$2x - y = 5$

Solve Problems 5–10 by graphing.

5. $x + y = 7$
$x - y = 3$

6. $x - y = 2$
$x + y = 4$

7. $3x - 2y = 12$
$7x + 2y = 8$

8. $3x - y = 2$
$x + 2y = 10$

9. $3u + 5v = 15$
$6u + 10v = -30$

10. $m + 2n = 4$
$2m + 4n = -8$

Solve Problems 11–16 by substitution.

11. $y = 2x + 3$
$y = 3x - 5$

12. $y = x + 4$
$y = 5x - 8$

13. $x - y = 4$
$x + 3y = 12$

14. $2x - y = 3$
$x + 2y = 14$

15. $3x - y = 7$
$2x + 3y = 1$

16. $2x + y = 6$
$x - y = -3$

B

Solve Problems 17–30 by either method. Round any approximate values to two decimal places.

17. $4x + 3y = 26$
$3x - 11y = -7$

18. $9x - 3y = 24$
$11x + 2y = 1$

19. $7m + 12n = -1$
$5m - 3n = 7$

20. $3p + 8q = 4$
$15p + 10q = -10$

21. $y = 0.08x$
$y = 100 + 0.04x$

22. $y = 0.07x$
$y = 80 + 0.05x$

23. $0.2u - 0.5v = 0.07$
$0.8u - 0.3v = 0.79$

24. $0.3s - 0.6t = 0.18$
$0.5s - 0.2t = 0.54$

25. $\frac{2}{5}x + \frac{3}{2}y = 2$
$\frac{7}{3}x - \frac{5}{4}y = -5$

26. $\frac{7}{2}x - \frac{5}{6}y = 10$
$\frac{2}{5}x + \frac{4}{3}y = 6$

27. $2x - 3y = -5$
$3x + 4y = 13$

28. $7x - 3y = 20$
$5x + 2y = 8$

29. $3.5x - 2.4y = 0.1$
$2.6x - 1.7y = -0.2$

30. $5.4x + 4.2y = -12.9$
$3.7x + 6.4y = -4.5$

31. In the process of solving a system by substitution, suppose you encounter a contradiction, such as $0 = 1$. How would you describe the solutions to such a system? Illustrate your ideas with the system

$$x - 2y = -3$$
$$-2x + 4y = 7$$

32. In the process of solving a system by substitution, suppose you encounter an identity, such as $0 = 0$. How would you describe the solutions to such a system? Illustrate your ideas with the system

$$x - 2y = -3$$
$$-2x + 4y = 6$$

C

In Problems 33 and 34, solve each system for p and q in terms of x and y. Explain how you could check your solution and then perform the check.

33. $x = 2 + p - 2q$
$y = 3 - p + 3q$

34. $x = -1 + 2p - q$
$y = 4 - p + q$

Problems 35 and 36 refer to the system

$$ax + by = h$$
$$cx + dy = k$$

where x and y are variables and a, b, c, d, h, and k are real constants.

35. Solve the system for x and y in terms of the constants a, b, c, d, h, and k. Clearly state any assumptions you must make about the constants during the solution process.

36. Discuss the nature of solutions to systems that do not satisfy the assumptions you made in Problem 35.

APPLICATIONS

37. **Airspeed.** It takes a private airplane 8.75 hours to make the 2,100-mile flight from Atlanta to Los Angeles and 5 hours to make the return trip. Assuming that the wind blows at a constant rate from Los Angeles to Atlanta, find the airspeed of the plane and the wind rate.

38. **Airspeed.** A plane carries enough fuel for 20 hours of flight at an airspeed of 150 miles per hour. How far can it fly into a 30 mph headwind and still have enough fuel to return to its starting point? (This distance is called the *point of no return.*)

39. **Rate–Time.** A crew of eight can row 20 kilometers per hour in still water. The crew rows upstream and then returns to its starting point in 15 minutes. If the river is flowing at 2 kilometers per hour, how far upstream did the crew row?

40. **Rate–Time.** It takes a boat 2 hours to travel 20 miles down a river and 3 hours to return upstream to its starting point. What is the rate of the current in the river?

41. **Chemistry.** A chemist has two solutions of hydrochloric acid in stock: a 50% solution and an 80% solution. How much of each should be used to obtain 100 milliliters of a 68% solution?

42. **Business.** A jeweler has two bars of gold alloy in stock, one of 12 carats and the other of 18 carats (24-carat gold is pure gold, 12-carat is $\frac{12}{24}$ pure, 18-carat gold is $\frac{18}{24}$ pure, and so on). How many grams of each alloy must be mixed to obtain 10 grams of 14-carat gold?

43. **Finance.** Suppose you have \$12,000 to invest. If part is invested at 10% and the rest at 15%, how much should be invested at each rate to yield 12% on the total amount invested?

44. **Finance.** An investor has \$20,000 to invest. If part is invested at 8% and the rest at 12%, how much should be invested at each rate to yield 11% on the total amount invested?

45. **Production.** A supplier for the electronics industry manufactures keyboards and screens for graphing calculators at plants in Mexico and Taiwan. The hourly production rates at each plant are given in the table. How many hours should each plant be operated to exactly fill an order for 4,000 keyboards and screens?

Plant	Keyboards	Screens
Mexico	40	32
Taiwan	20	32

46. **Production.** A company produces Italian sausages and bratwursts at plants in Green Bay and Sheboygan. The hourly production rates at each plant are given in the table. How many hours should each plant be operated to exactly fill an order for 62,250 Italian sausages and 76,500 bratwursts?

Plant	Italian Sausage	Bratwurst
Green Bay	800	800
Sheboygan	500	1,000

47. **Nutrition.** Animals in an experiment are to be kept on a strict diet. Each animal is to receive, among other things, 20 grams of protein and 6 grams of fat. The laboratory technician is able to purchase two food mixes of the following compositions: Mix A has 10% protein and 6% fat; mix B has 20% protein and 2% fat. How many grams of each mix should be used to obtain the right diet for a single animal?

48. **Nutrition.** A fruit grower can use two types of fertilizer in an orange grove, brand A and brand B. Each bag of brand A contains 8 pounds of nitrogen and 4 pounds of phosphoric acid. Each bag of brand B contains 7 pounds of nitrogen and 7 pounds of phosphoric acid. Tests indicate that the grove needs 720 pounds of nitrogen and 500 pounds of phosphoric acid. How many bags of each brand should be used to provide the required amounts of nitrogen and phosphoric acid?

49. **Supply and Demand.** Suppose the supply and demand equations for printed T-shirts in a resort town for a particular week are

$$p = 0.007q + 3 \quad \text{Supply equation}$$
$$p = -0.018q + 15 \quad \text{Demand equation}$$

where p is the price in dollars and q is the quantity.

(A) Find the supply and the demand (to the nearest unit) if T-shirts are priced at \$4 each. Discuss the stability of the T-shirt market at this price level.

(B) Find the supply and the demand (to the nearest unit) if T-shirts are priced at \$8 each. Discuss the stability of the T-shirt market at this price level.

(C) Find the equilibrium price and quantity.

(D) Graph the two equations in the same coordinate system and identify the equilibrium point, supply curve, and demand curve.

50. Supply and Demand. Suppose the supply and demand for printed baseball caps in a resort town for a particular week are

$$p = 0.006q + 2 \qquad \text{Supply equation}$$
$$p = -0.014q + 13 \qquad \text{Demand equation}$$

where p is the price in dollars and q is the quantity in hundreds.

(A) Find the supply and the demand (to the nearest unit) if baseball caps are priced at \$4 each. Discuss the stability of the baseball cap market at this price level.

(B) Find the supply and the demand (to the nearest unit) if baseball caps are priced at \$8 each. Discuss the stability of the baseball cap market at this price level.

(C) Find the equilibrium price and quantity.

(D) Graph the two equations in the same coordinate system and identify the equilibrium point, supply curve, and demand curve.

★ **51. Supply and Demand.** At \$0.60 per bushel, the daily supply for wheat is 450 bushels and the daily demand is 645 bushels. When the price is raised to \$0.90 per bushel, the daily supply increases to 750 bushels and the daily demand decreases to 495 bushels. Assume that the supply and demand equations are linear.

(A) Find the supply equation. [*Hint:* Write the supply equation in the form $p = aq + b$ and solve for a and b.]

(B) Find the demand equation.

(C) Find the equilibrium price and quantity.

★ **52. Supply and Demand.** At \$1.40 per bushel, the daily supply for soybeans is 1,075 bushels and the daily demand is 580 bushels. When the price falls to \$1.20 per bushel, the daily supply decreases to 575 bushels and the daily demand increases to 980 bushels. Assume that the supply and demand equations are linear.

(A) Find the supply equation. [See the hint in Problem 51.]

(B) Find the demand equation.

(C) Find the equilibrium price and quantity.

★ **53. Physics.** An object dropped off the top of a tall building falls vertically with constant acceleration. If s is the distance of the object above the ground (in feet) t seconds after its release, then s and t are related by an equation of the form

$$s = a + bt^2$$

where a and b are constants. Suppose the object is 180 feet above the ground 1 second after its release and 132 feet above the ground 2 seconds after its release.

(A) Find the constants a and b.

(B) How high is the building?

(C) How long does the object fall?

★ **54. Physics.** Repeat Problem 53 if the object is 240 feet above the ground after 1 second and 192 feet above the ground after 2 seconds.

★ **55. Earth Science.** An earthquake emits a primary wave and a secondary wave. Near the surface of the Earth the primary wave travels at about 5 miles per second and the secondary wave at about 3 miles per second. From the time lag between the two waves arriving at a given receiving station, it is possible to estimate the distance to the quake. (The *epicenter* can be located by obtaining distance bearings at three or more stations.) Suppose a station measured a time difference of 16 seconds between the arrival of the two waves. How long did each wave travel, and how far was the earthquake from the station?

★ **56. Earth Science.** A ship using sound-sensing devices above and below water recorded a surface explosion 6 seconds sooner by its underwater device than its above-water device. Sound travels in air at about 1,100 feet per second and in seawater at about 5,000 feet per second.

(A) How long did it take each sound wave to reach the ship?

(B) How far was the explosion from the ship?

Section 5-2 Systems of Linear Equations and Augmented Matrices

- Elimination by Addition
- Matrices
- Solving Linear Systems Using Augmented Matrices

Most real-world applications of linear systems involve a large number of variables and equations. Computers are usually used to solve these larger systems. Although very effective for systems involving two variables, the graphing and substitution methods discussed in the preceding section are not well-suited for computer use in the solution of larger systems. In this section, we begin the development of a method that will work for systems of any size and that lends itself to computer implementation.

Elimination by Addition

We begin with an algebraic solution method called **elimination by addition.** As we will see, this method is readily generalized to larger systems. The method involves the replacement of systems of equations with simpler *equivalent systems*, by performing appropriate operations, until we obtain a system with an obvious solution. **Equivalent systems** of equations are systems that have the same solution set. Theorem 1 lists operations that produce equivalent systems.

THEOREM 1

ELEMENTARY EQUATION OPERATIONS PRODUCING EQUIVALENT SYSTEMS

A system of linear equations is transformed into an equivalent system if

1. Two equations are interchanged.
2. An equation is multiplied by a nonzero constant.
3. A constant multiple of another equation is added to a given equation.

Any one of the three operations in Theorem 1 can be used to produce an equivalent system, but operations 2 and 3 will be of most use to us now. Operation 1 becomes more important later in the section. The use of Theorem 1 is best illustrated by examples.

EXAMPLE 1 **Solving a System Using Elimination by Addition**

Solve using elimination by addition:

$$\begin{aligned} 3x - 2y &= 8 \\ 2x + 5y &= -1 \end{aligned}$$

Solution We use Theorem 1 to eliminate one of the variables and thus obtain a system with an obvious solution.

$$\begin{aligned} 3x - 2y &= 8 \\ 2x + 5y &= -1 \end{aligned}$$

If we multiply the top equation by 5, the bottom by 2, and then add, we can eliminate y.

$$\begin{aligned} 15x - 10y &= 40 \\ 4x + 10y &= -2 \\ \hline 19x \qquad &= 38 \\ x &= 2 \end{aligned}$$

The equation $x = 2$ paired with either of the two original equations produces an equivalent system. Thus, we can substitute $x = 2$ back into either of the two original equations to solve for y. We choose the second equation.

$$\begin{aligned} 2(2) + 5y &= -1 \\ 5y &= -5 \\ y &= -1 \end{aligned}$$

Solution: $x = 2$, $y = -1$ or $(2, -1)$.

Check

$$\begin{aligned} 3x - 2y &= 8 \\ 3(2) - 2(-1) &\stackrel{?}{=} 8 \\ 8 &\stackrel{\checkmark}{=} 8 \end{aligned} \qquad \begin{aligned} 2x + 5y &= -1 \\ 2(2) + 5(-1) &\stackrel{?}{=} -1 \\ -1 &\stackrel{\checkmark}{=} -1 \end{aligned}$$

MATCHED PROBLEM 1

Solve using elimination by addition: $6x + 3y = 3$
$5x + 4y = 7$

Explore/Discuss 1

In each of the following systems, compare the results of applying elimination by addition with the graphical solution and discuss the nature of the solution sets.

(A) $\begin{aligned} 3x + 2y &= 6 \\ 6x + 4y &= 12 \end{aligned}$ (B) $\begin{aligned} 3x + 2y &= 6 \\ 6x + 4y &= 13 \end{aligned}$

Let's see what happens in the elimination process when a system either has no solution or has infinitely many solutions. Consider the following system:

$$\begin{aligned} 2x + 6y &= -3 \\ x + 3y &= 2 \end{aligned}$$

Multiplying the second equation by -2 and adding, we obtain

$$\begin{aligned} 2x + 6y &= -3 \\ -2x - 6y &= -4 \\ \hline 0 &= -7 \end{aligned}$$

We have obtained a contradiction. An assumption that thc original system has solutions must be false, otherwise, we have proved that $0 = -7$! Thus, the system has no solution. The graphs of the equations are parallel and the system is inconsistent.

Now consider the system

$$\begin{aligned} x - \tfrac{1}{2}y &= 4 \\ -2x + y &= -8 \end{aligned}$$

If we multiply the top equation by 2 and add the result to the bottom equation, we get

$$\begin{aligned} 2x - y &= 8 \\ -2x + y &= -8 \\ \hline 0 &= 0 \end{aligned}$$

Obtaining $0 = 0$ by addition implies that the two original equations are equivalent. That is, their graphs coincide and the system is dependent. If we let $x = t$, where t is any real number, and solve either equation for y, we obtain $y = 2t - 8$. Thus,

$$(t, 2t - 8) \qquad t \text{ a real number}$$

describes the solution set for the system. The variable t is called a **parameter,** and replacing t with a real number produces a **particular solution** to the system. For example, some particular solutions to this system are

$t = -1$	$t = 2$	$t = 5$	$t = 9.4$
$(-1, -10)$	$(2, -4)$	$(5, 2)$	$(9.4, 10.8)$

The next example illustrates that elimination by addition provides an efficient method for solving applied problems.

EXAMPLE 2

Diet

An individual wants to use milk and orange juice to increase the amount of calcium and vitamin A in her daily diet. An ounce of milk contains 41 milligrams of calcium and 59 micrograms* of vitamin A. An ounce of orange juice contains 5 milligrams of calcium and 75 micrograms of vitamin A. How many ounces of milk and orange juice should she drink each day to provide exactly 550 milligrams of calcium and 1,300 micrograms of vitamin A?

Solution

First we define the relevant variables:

x = Number of ounces of milk

y = Number of ounces of orange juice

Next we summarize the given information in Table 1.

TABLE 1

	Milk	Orange Juice	Total Needed
Calcium (mg)	41	5	550
Vitamin A (μg)	59	75	1,300

*A microgram (μg) is one-millionth (10^{-6}) of a gram.

Now we use the information in the table to form equations involving x and y:

$$\begin{pmatrix}\text{Calcium in } x \text{ oz}\\ \text{of milk}\end{pmatrix} + \begin{pmatrix}\text{Calcium in } y \text{ oz}\\ \text{of orange juice}\end{pmatrix} = \begin{pmatrix}\text{Total calcium}\\ \text{needed (mg)}\end{pmatrix}$$

$$41x \quad + \quad 5y \quad = \quad 550$$

$$\begin{pmatrix}\text{Vitamin A in } x \text{ oz}\\ \text{of milk}\end{pmatrix} + \begin{pmatrix}\text{Vitamin A in } y \text{ oz}\\ \text{of orange juice}\end{pmatrix} = \begin{pmatrix}\text{Total vitamin A}\\ \text{needed } (\mu\text{g})\end{pmatrix}$$

$$59x \quad + \quad 75y \quad = \quad 1{,}300$$

Solve using elimination by addition:

$$\begin{aligned} -615x - 75y &= -8{,}250 \\ 59x + 75y &= 1{,}300 \\ \hline -556x \qquad &= -6{,}950 \\ x &= \mathbf{12.5} \end{aligned} \qquad \begin{aligned} 41(\mathbf{12.5}) + 5y &= 550 \\ 5y &= 37.5 \\ y &= \mathbf{7.5} \end{aligned}$$

Drinking 12.5 ounces of milk and 7.5 ounces of orange juice each day will provide the required amounts of calcium and vitamin A.

Check

$$\begin{aligned} 41x + 5y &= 550 \\ 41(12.5) + 5(7.5) &\stackrel{?}{=} 550 \\ 550 &\stackrel{\checkmark}{=} 550 \end{aligned} \qquad \begin{aligned} 59x + 75y &= 1{,}300 \\ 59(12.5) + 75(7.5) &\stackrel{?}{=} 1{,}300 \\ 1{,}300 &\stackrel{\checkmark}{=} 1{,}300 \end{aligned}$$

MATCHED PROBLEM 2

An individual wants to use cottage cheese and yogurt to increase the amount of protein and calcium in his daily diet. An ounce of cottage cheese contains 3 grams of protein and 15 milligrams of calcium. An ounce of yogurt contains 1 gram of protein and 41 milligrams of calcium. How many ounces of cottage cheese and yogurt should he eat each day to provide exactly 62 grams of protein and 760 milligrams of calcium?

Matrices

In solving systems of equations using elimination by addition, the coefficients of the variables and the constant terms played a central role. The process can be made more efficient for generalization and computer work by the introduction of a mathematical form called a *matrix*. A **matrix** is a rectangular array of numbers written within brackets. Two examples are

$$A = \begin{bmatrix} 1 & -3 & 7 \\ 5 & 0 & -4 \end{bmatrix} \qquad B = \begin{bmatrix} -5 & 4 & 11 \\ 0 & 1 & 6 \\ -2 & 12 & 8 \\ -3 & 0 & -1 \end{bmatrix}$$

Each number in a matrix is called an **element** of the matrix. Matrix A has six elements arranged in two rows and three columns. Matrix B has 12 elements arranged in four rows and three columns. If a matrix has m rows and n columns, it is called an $\boldsymbol{m \times n}$ **matrix** (read "m by n matrix"). The expression $m \times n$ is called the **size** of the matrix, and the numbers m and n are called the **dimensions** of the matrix. It is important to note that the number of rows is always given first. Referring to matrices A and B on page 353, A is a 2×3 matrix and B is a 4×3 matrix. A matrix with n rows and n columns is called a **square matrix of order $\boldsymbol{n}$.** A matrix with only one column is called a **column matrix,** and a matrix with only one row is called a **row matrix.** These definitions are illustrated by the following:

3×3	4×1	1×4
$\begin{bmatrix} 0.5 & 0.2 & 1.0 \\ 0.0 & 0.3 & 0.5 \\ 0.7 & 0.0 & 0.2 \end{bmatrix}$	$\begin{bmatrix} 3 \\ -2 \\ 1 \\ 0 \end{bmatrix}$	$\begin{bmatrix} 2 & \frac{1}{2} & 0 & -\frac{2}{3} \end{bmatrix}$
Square matrix of order 3	Column matrix	Row matrix

The **position** of an element in a matrix is the row and column containing the element. This is usually denoted using **double subscript notation** a_{ij}, where i is the row and j is the column containing the element a_{ij}, as illustrated below:

$$A = \begin{bmatrix} 1 & 5 & -3 \\ 6 & 0 & -4 \end{bmatrix} \qquad \begin{aligned} a_{11} &= 1, a_{12} = 5, a_{13} = -3 \\ a_{21} &= 6, a_{22} = 0, a_{23} = -4 \end{aligned}$$

Note that a_{12} is read "a one two," not "a twelve." The elements $a_{11} = 1$ and $a_{22} = 0$ make up the *principal diagonal* of A. In general, the **principal diagonal** of a matrix A consists of the elements $a_{11}, a_{22}, a_{33}, \ldots$.

Remark Most graphing utilities are capable of storing and manipulating matrices. Figure 1 shows matrix A displayed in the editing screen of a particular graphing calculator. The size of the matrix is given at the top of the screen. The position and the value of the currently selected element is given at the bottom. Notice that a comma is used in the notation of the position. This is common practice on graphing utilities but not in mathematical literature.

FIGURE 1
Matrix notation on a graphing utility.

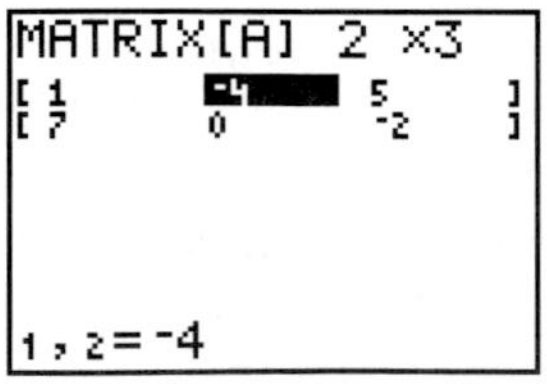

The coefficients and constant terms in a system of linear equations can be used to form several matrices of interest to our work. Related to the system

$$\begin{aligned} 2x - 3y &= 5 \\ x + 2y &= -3 \end{aligned} \tag{1}$$

are the following matrices:

Coefficient matrix	Constant matrix	Augmented coefficient matrix
$\begin{bmatrix} 2 & -3 \\ 1 & 2 \end{bmatrix}$	$\begin{bmatrix} 5 \\ -3 \end{bmatrix}$	$\left[\begin{array}{cc\|c} 2 & -3 & 5 \\ 1 & 2 & -3 \end{array}\right]$

The augmented coefficient matrix will be used in this section. The other matrices will be used in later sections. The augmented coefficient matrix contains the essential parts of the system—both the coefficients and the constants. The vertical bar is included only as a visual aid to help us separate the coefficients from the constant terms. (Matrices entered and displayed on a graphing utility will not display this line.)

For ease of generalization to the larger systems in the following sections, we are now going to change the notation for the variables in system (1) to a subscript form (we would soon run out of letters, but we will not run out of subscripts). That is, in place of x and y, we will use x_1 and x_2, respectively, and (1) will be written as

$$\begin{aligned} 2x_1 - 3x_2 &= 5 \\ x_1 + 2x_2 &= -3 \end{aligned}$$

In general, associated with each linear system of the form

$$\begin{aligned} a_{11}x_1 + a_{12}x_2 &= k_1 \\ a_{21}x_1 + a_{22}x_2 &= k_2 \end{aligned} \qquad (2)$$

where x_1 and x_2 are variables, is the **augmented matrix** of the system:

Column 1 (C_1)
Column 2 (C_2)
Column 3 (C_3)

$$\left[\begin{array}{cc|c} a_{11} & a_{12} & k_1 \\ a_{21} & a_{22} & k_2 \end{array}\right] \begin{array}{l} \leftarrow \text{Row 1 } (R_1) \\ \leftarrow \text{Row 2 } (R_2) \end{array}$$

This matrix contains the essential parts of system (2). Our objective is to learn how to manipulate augmented matrices in such a way that a solution to system (2) will result, if a solution exists.

In our earlier discussion of using elimination by addition, we said that two systems were equivalent if they had the same solution. And we used the operations in Theorem 1 to transform a system into an equivalent system. Paralleling this approach, we now say that two augmented matrices are **row-equivalent,** denoted by the symbol $\sim$ between the two matrices, if they are augmented matrices of equivalent systems of equations. And we use the operations listed in Theorem 2 below to transform augmented matrices into row-equivalent matrices. Note that Theorem 2 is a direct consequence of Theorem 1.

THEOREM 2

ELEMENTARY ROW OPERATIONS PRODUCING ROW-EQUIVALENT MATRICES

An augmented matrix is transformed into a row-equivalent matrix if any of the following **row operations** is performed:

1. Two rows are interchanged ($R_i \leftrightarrow R_j$).

2. A row is multiplied by a nonzero constant ($kR_i \rightarrow R_i$).

3. A constant multiple of one row is added to another row ($kR_j + R_i \rightarrow R_i$).

[*Note:* The arrow means "replaces."]

Solving Linear Systems Using Augmented Matrices

The use of Theorem 2 in solving systems in the form of (2) is best illustrated by examples.

EXAMPLE 3

Solving a System Using Augmented Matrix Methods

Solve, using augmented matrix methods:

$$\begin{aligned} 3x_1 + 4x_2 &= 1 \\ x_1 - 2x_2 &= 7 \end{aligned} \qquad (3)$$

Solution We start by writing the augmented matrix corresponding to system (3):

$$\left[\begin{array}{rr|r} 3 & 4 & 1 \\ 1 & -2 & 7 \end{array}\right] \qquad (4)$$

Our objective is to use row operations from Theorem 2 to try to transform matrix (4) into the form

$$\left[\begin{array}{rr|r} 1 & 0 & m \\ 0 & 1 & n \end{array}\right] \qquad (5)$$

where m and n are real numbers. The solution to system (3) will then be obvious, since matrix (5) will be the augmented matrix of the following system:

$$\begin{aligned} x_1 &= m \qquad & x_1 + 0x_2 &= m \\ x_2 &= n \qquad & 0x_1 + x_2 &= n \end{aligned}$$

We now proceed to use row operations to transform (4) into form (5).

Step 1. To get a 1 in the upper left corner, we interchange rows 1 and 2—Theorem 2, part 1:

$$\left[\begin{array}{rr|r} 3 & 4 & 1 \\ 1 & -2 & 7 \end{array}\right] \quad R_1 \leftrightarrow R_2 \quad \left[\begin{array}{rr|r} 1 & -2 & 7 \\ 3 & 4 & 1 \end{array}\right]$$

Now you see why we wanted Theorem 2, part 1.

Step 2. To get a 0 in the lower left corner, we multiply R_1 by -3 and add to R_2—Theorem 2, part 3. This changes R_2 but not R_1. Some people find it useful to write $(-3)R_1$ outside the matrix to help reduce errors in arithmetic:

$$\left[\begin{array}{rr|r} 1 & -2 & 7 \\ 3 & 4 & 1 \end{array}\right] \quad \underbrace{(-3)R_1}_{} + R_2 \to R_2 \quad \left[\begin{array}{rr|r} 1 & -2 & 7 \\ 0 & 10 & -20 \end{array}\right]$$

$$-3 \quad 6 \quad -21$$

Step 3. To get a 1 in the second row, second column, we multiply R_2 by $\frac{1}{10}$—Theorem 2, part 2:

$$\left[\begin{array}{rr|r} 1 & -2 & 7 \\ 0 & 10 & -20 \end{array}\right] \quad \tfrac{1}{10}R_2 \to R_2 \quad \left[\begin{array}{rr|r} 1 & -2 & 7 \\ 0 & 1 & -2 \end{array}\right]$$

Step 4. To get a 0 in the first row, second column, we multiply R_2 by 2 and add the result to R_1—Theorem 2, part 3. This changes R_1 but not R_2.

$$\begin{array}{ccc} 0 & 2 & -4 \end{array}$$

$$\left[\begin{array}{cc|c} 1 & -2 & 7 \\ 0 & 1 & -2 \end{array}\right] \underset{\sim}{2R_2 + R_1 \to R_1} \left[\begin{array}{cc|c} 1 & 0 & 3 \\ 0 & 1 & -2 \end{array}\right]$$

We have accomplished our objective! The last matrix is the augmented matrix for the system

$$\begin{aligned} x_1 &= 3 \\ x_2 &= -2 \end{aligned} \qquad (6)$$

Since system (6) is equivalent to the original system (3), we have solved system (3). That is, $x_1 = 3$ and $x_2 = -2$.

Check

$$\begin{aligned} 3x_1 + 4x_2 &= 1 \\ 3(3) + 4(-2) &\stackrel{?}{=} 1 \\ 1 &\stackrel{\checkmark}{=} 1 \end{aligned} \qquad \begin{aligned} x_1 - 2x_2 &= 7 \\ 3 - 2(-2) &\stackrel{?}{=} 7 \\ 7 &\stackrel{\checkmark}{=} 7 \end{aligned}$$

The above process is written more compactly as follows:

Step 1. Need a 1 here

$$\left[\begin{array}{cc|c} 3 & 4 & 1 \\ 1 & -2 & 7 \end{array}\right] \quad R_1 \leftrightarrow R_2$$

Step 2. Need a 0 here

$$\sim \left[\begin{array}{cc|c} 1 & -2 & 7 \\ 3 & 4 & 1 \end{array}\right] \quad (-3)R_1 + R_2 \to R_2$$

$$\begin{array}{ccc} -3 & 6 & -21 \end{array}$$

Step 3. Need a 1 here

$$\sim \left[\begin{array}{cc|c} 1 & -2 & 7 \\ 0 & 10 & -20 \end{array}\right] \quad \tfrac{1}{10}R_2 \to R_2$$

$$\begin{array}{ccc} 0 & 2 & -4 \end{array}$$

Step 4. Need a 0 here

$$\sim \left[\begin{array}{cc|c} 1 & -2 & 7 \\ 0 & 1 & -2 \end{array}\right] \quad 2R_2 + R_1 \to R_1$$

$$\sim \left[\begin{array}{cc|c} 1 & 0 & 3 \\ 0 & 1 & -2 \end{array}\right]$$

Therefore, $x_1 = 3$ and $x_2 = -2$.

MATCHED PROBLEM 3

Solve, using augmented matrix methods:

$$\begin{aligned} 2x_1 - x_2 &= -7 \\ x_1 + 2x_2 &= 4 \end{aligned}$$

Explore/Discuss 2

The summary at the end of Example 3 shows five augmented coefficient matrices. Write the linear system that each matrix represents, solve each system graphically, and discuss the relationship between these solutions.

EXAMPLE 4 Solving a System Using Augmented Matrix Methods

Solve, using augmented matrix methods:
$$\begin{aligned} 2x_1 - 3x_2 &= 7 \\ 3x_1 + 4x_2 &= 2 \end{aligned}$$

Solution

Step 1. Need a 1 here

$$\left[\begin{array}{cc|c} 2 & -3 & 7 \\ 3 & 4 & 2 \end{array}\right] \quad \tfrac{1}{2}R_1 \to R_1$$

Step 2. Need a 0 here

$$\sim \left[\begin{array}{cc|c} 1 & -\frac{3}{2} & \frac{7}{2} \\ 3 & 4 & 2 \end{array}\right] \quad (-3)R_1 + R_2 \to R_2$$

$$\begin{array}{ccc} -3 & \frac{9}{2} & -\frac{21}{2} \end{array}$$

Step 3. Need a 1 here

$$\sim \left[\begin{array}{cc|c} 1 & -\frac{3}{2} & \frac{7}{2} \\ 0 & \frac{17}{2} & -\frac{17}{2} \end{array}\right] \quad \tfrac{2}{17}R_2 \to R_2$$

$$\begin{array}{ccc} 0 & \frac{3}{2} & -\frac{3}{2} \end{array}$$

Step 4. Need a 0 here

$$\sim \left[\begin{array}{cc|c} 1 & -\frac{3}{2} & \frac{7}{2} \\ 0 & 1 & -1 \end{array}\right] \quad \tfrac{3}{2}R_2 + R_1 \to R_1$$

$$\sim \left[\begin{array}{cc|c} 1 & 0 & 2 \\ 0 & 1 & -1 \end{array}\right]$$

Thus, $x_1 = 2$ and $x_2 = -1$. You should check this solution in the original system.

MATCHED PROBLEM 4

Solve, using augmented matrix methods:
$$\begin{aligned} 5x_1 - 2x_2 &= 12 \\ 2x_1 + 3x_2 &= 1 \end{aligned}$$

EXAMPLE 5 Solving a System Using Augmented Matrix Methods

Solve, using augmented matrix methods:
$$\begin{aligned} 2x_1 - x_2 &= 4 \\ -6x_1 + 3x_2 &= -12 \end{aligned} \tag{7}$$

Solution

$$\left[\begin{array}{cc|c} 2 & -1 & 4 \\ -6 & 3 & -12 \end{array}\right]$$

$\tfrac{1}{2}R_1 \to R_1$ (This produces a 1 in the upper left corner.)
$\tfrac{1}{3}R_2 \to R_2$ (This simplifies R_2.)

$$\sim \left[\begin{array}{cc|c} 1 & -\frac{1}{2} & 2 \\ -2 & 1 & -4 \end{array}\right]$$

$$\begin{array}{ccc} 2 & -1 & 4 \end{array}$$

$2R_1 + R_2 \to R_2$ (This produces a 0 in the lower left corner.)

$$\sim \left[\begin{array}{cc|c} 1 & -\frac{1}{2} & 2 \\ 0 & 0 & 0 \end{array}\right]$$

The last matrix corresponds to the system

$$x_1 - \tfrac{1}{2}x_2 = 2 \qquad x_1 - \tfrac{1}{2}x_2 = 2$$
$$0 = 0 \qquad 0x_1 + 0x_2 = 0$$

Thus, $x_1 = \frac{1}{2}x_2 + 2$. Hence, for any real number t, if $x_2 = t$, then $x_1 = \frac{1}{2}t + 2$. That is, the solution set is described by

$$(\tfrac{1}{2}t + 2, t) \qquad t \text{ a real number} \tag{8}$$

For example, if $t = 6$, then $(5, 6)$ is a particular solution; if $t = -2$, then $(1, -2)$ is another particular solution; and so on. Geometrically, the graphs of the two original equations coincide and there are infinitely many solutions.

In general, if we end up with a row of 0s in an augmented matrix for a two-equation–two-variable system, the system is dependent and there are infinitely many solutions.

Check

The following is a check that (8) provides a solution for system (7) for any real number t:

$$2x_1 - x_2 = 4 \qquad -6x_1 + 3x_2 = -12$$
$$2(\tfrac{1}{2}t + 2) - t \stackrel{?}{=} 4 \qquad -6(\tfrac{1}{2}t + 2) + 3t \stackrel{?}{=} -12$$
$$t + 4 - t \stackrel{?}{=} 4 \qquad -3t - 12 + 3t \stackrel{?}{=} -12$$
$$4 \stackrel{\checkmark}{=} 4 \qquad -12 \stackrel{\checkmark}{=} -12$$

MATCHED PROBLEM 5

Solve, using augmented matrix methods:

$$-2x_1 + 6x_2 = 6$$
$$3x_1 - 9x_2 = -9$$

Explore/Discuss 3

Most graphing utilities can perform row operations. Figure 2 shows the solution to Example 5 on a particular graphing calculator. Consult your manual to see how to perform row operations, and solve Matched Problem 5 on your graphing utility.

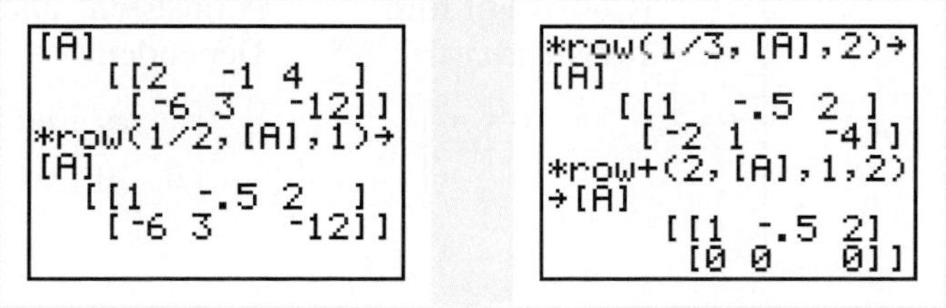

FIGURE 2
Performing row operations on a graphing utility.

EXAMPLE 6 **Solving a System Using Augmented Matrix Methods**

Solve, using augmented matrix methods:

$$\begin{aligned} 2x_1 + 6x_2 &= -3 \\ x_1 + 3x_2 &= 2 \end{aligned}$$

Solution

$$\left[\begin{array}{cc|c} 2 & 6 & -3 \\ 1 & 3 & 2 \end{array}\right] \quad R_1 \leftrightarrow R_2$$

$$\sim \left[\begin{array}{cc|c} 1 & 3 & 2 \\ 2 & 6 & -3 \end{array}\right] \quad (-2)R_1 + R_2 \to R_2$$
$$\begin{array}{ccc} -2 & -6 & -4 \end{array}$$

$$\sim \left[\begin{array}{cc|c} 1 & 3 & 2 \\ 0 & 0 & -7 \end{array}\right] \quad R_2 \text{ implies the contradiction: } 0 = -7$$

This is the augmented matrix of the system

$$\begin{aligned} x_1 + 3x_2 &= 2 \qquad & x_1 + 3x_2 &= 2 \\ 0 &= -7 \qquad & 0x_1 + 0x_2 &= -7 \end{aligned}$$

The second equation is not satisfied by any ordered pair of real numbers. Hence, the original system is inconsistent and has no solution. Otherwise, we have proved that $0 = -7$!

Thus, if we obtain all 0s to the left of the vertical bar and a nonzero number to the right of the bar in a row of an augmented matrix, then the system is inconsistent and there are no solutions.

MATCHED PROBLEM 6

Solve, using augmented matrix methods:

$$\begin{aligned} 2x_1 - x_2 &= 3 \\ 4x_1 - 2x_2 &= -1 \end{aligned}$$

SUMMARY

For m, n, p real numbers, $p \neq 0$:

Form 1: A Unique Solution (Consistent and Independent)	**Form 2: Infinitely Many Solutions (Consistent and Dependent)**	**Form 3: No Solution (Inconsistent)**
$\left[\begin{array}{cc\|c} 1 & 0 & m \\ 0 & 1 & n \end{array}\right]$	$\left[\begin{array}{cc\|c} 1 & m & n \\ 0 & 0 & 0 \end{array}\right]$	$\left[\begin{array}{cc\|c} 1 & m & n \\ 0 & 0 & p \end{array}\right]$

The process of solving systems of equations described in this section is referred to as *Gauss–Jordan elimination.* We will use this method to solve larger-scale systems in the next section, including systems where the number of equations and the number of variables are not the same.

Answers to Matched Problems

1. $(-1, 3)$, or $x = -1$ and $y = 3$ **2.** 16.5 oz of cottage cheese, 12.5 oz of yogurt
3. $x_1 = -2, x_2 = 3$ **4.** $x_1 = 2, x_2 = -1$
5. The system is dependent. For t any real number, $x_2 = t, x_1 = 3t - 3$ is a solution.
6. Inconsistent—no solution

EXERCISE 5-2

A

Solve Problems 1–4 using elimination by addition.

1. $\begin{aligned} 2x + 3y &= 1 \\ 3x - y &= 7 \end{aligned}$

2. $\begin{aligned} 2m - n &= 10 \\ m - 2n &= -4 \end{aligned}$

3. $\begin{aligned} 4x - 3y &= 15 \\ 3x + 4y &= 5 \end{aligned}$

4. $\begin{aligned} 5x + 2y &= 1 \\ 2x - 3y &= -11 \end{aligned}$

Problems 5–14 refer to the following matrices:

$$A = \begin{bmatrix} 3 & -2 & 0 \\ 4 & 1 & -6 \end{bmatrix} \qquad B = \begin{bmatrix} -2 & 8 & 0 \\ -3 & 6 & 9 \\ 4 & 2 & 0 \end{bmatrix}$$

$$C = [3 \quad -2 \quad 0] \qquad D = \begin{bmatrix} -4 \\ 7 \end{bmatrix}$$

5. What is the size of A? Of C?

6. What is the size of B? Of D?

7. Identify all row matrices.

8. Identify all column matrices.

9. Identify all square matrices.

10. How many additional rows would matrix A need to be a square matrix?

11. For matrix A, find a_{12} and a_{23}.

12. For matrix A, find a_{21} and a_{13}.

13. Find the elements on the principal diagonal of matrix B.

14. Find the elements on the principal diagonal of matrix A.

Perform each of the row operations indicated in Problems 15–26 on the following matrix:

$$\left[\begin{array}{rr|r} 1 & -3 & 2 \\ 4 & -6 & -8 \end{array}\right]$$

15. $R_1 \leftrightarrow R_2$

16. $\frac{1}{2}R_2 \rightarrow R_2$

17. $-4R_1 \rightarrow R_1$

18. $-2R_1 \rightarrow R_1$

19. $2R_2 \rightarrow R_2$

20. $-1R_2 \rightarrow R_2$

21. $(-4)R_1 + R_2 \rightarrow R_2$

22. $(-\frac{1}{2})R_2 + R_1 \rightarrow R_1$

23. $(-2)R_1 + R_2 \rightarrow R_2$

24. $(-3)R_1 + R_2 \rightarrow R_2$

25. $(-1)R_1 + R_2 \rightarrow R_2$

26. $1R_1 + R_2 \rightarrow R_2$

B

Each of the matrices in Problems 27–32 is the result of performing a single row operation on the matrix A shown below. Identify the row operation. Check your work by performing the row operation you identified on a graphing utility.

$$A = \left[\begin{array}{rr|r} -1 & 2 & -3 \\ 6 & -3 & 12 \end{array}\right]$$

27. $\left[\begin{array}{rr|r} -1 & 2 & -3 \\ 2 & -1 & 4 \end{array}\right]$

28. $\left[\begin{array}{rr|r} -2 & 4 & -6 \\ 6 & -3 & 12 \end{array}\right]$

29. $\left[\begin{array}{rr|r} -1 & 2 & -3 \\ 0 & 9 & -6 \end{array}\right]$

30. $\left[\begin{array}{rr|r} 3 & 0 & 5 \\ 6 & -3 & 12 \end{array}\right]$

31. $\left[\begin{array}{rr|r} 1 & 1 & 1 \\ 6 & -3 & 12 \end{array}\right]$

32. $\left[\begin{array}{rr|r} -1 & 2 & -3 \\ 2 & 5 & 0 \end{array}\right]$

Solve Problems 33 and 34 using augmented matrix methods. Write the linear system represented by each augmented matrix in your solution, and solve each of these systems graphically. Discuss the relationship between the solutions of these systems.

33. $\begin{aligned} x_1 + x_2 &= 7 \\ x_1 - x_2 &= 1 \end{aligned}$

34. $\begin{aligned} x_1 + x_2 &= 5 \\ x_1 - x_2 &= -3 \end{aligned}$

Solve Problems 35–46 using augmented matrix methods:

35. $\begin{aligned} x_1 - 4x_2 &= -2 \\ -2x_1 + x_2 &= -3 \end{aligned}$

36. $\begin{aligned} x_1 - 3x_2 &= -5 \\ -3x_1 - x_2 &= 5 \end{aligned}$

37. $\begin{aligned} 3x_1 - x_2 &= 2 \\ x_1 + 2x_2 &= 10 \end{aligned}$

38. $\begin{aligned} 2x_1 + x_2 &= 0 \\ x_1 - 2x_2 &= -5 \end{aligned}$

39. $\begin{aligned} x_1 + 2x_2 &= 4 \\ 2x_1 + 4x_2 &= -8 \end{aligned}$

40. $\begin{aligned} 2x_1 - 3x_2 &= -2 \\ -4x_1 + 6x_2 &= 7 \end{aligned}$

41. $\begin{aligned} 2x_1 + x_2 &= 6 \\ x_1 - x_2 &= -3 \end{aligned}$

42. $\begin{aligned} 3x_1 - x_2 &= -5 \\ x_1 + 3x_2 &= 5 \end{aligned}$

43. $\begin{aligned} 3x_1 - 6x_2 &= -9 \\ -2x_1 + 4x_2 &= 6 \end{aligned}$

44. $\begin{aligned} 2x_1 - 4x_2 &= -2 \\ -3x_1 + 6x_2 &= 3 \end{aligned}$

45. $\begin{aligned} 4x_1 - 2x_2 &= 2 \\ -6x_1 + 3x_2 &= -3 \end{aligned}$

46. $\begin{aligned} -6x_1 + 2x_2 &= 4 \\ 3x_1 - x_2 &= -2 \end{aligned}$

C

47. The coefficients of the three systems below are very similar. One might guess that the solution sets to the three systems would also be nearly identical. Develop evidence for or against this guess by considering graphs of the systems and solutions obtained using elimination by addition.

(A) $4x + 5y = 4$
$9x + 11y = 4$

(B) $4x + 5y = 4$
$8x + 11y = 4$

(C) $4x + 5y = 4$
$8x + 10y = 4$

48. Repeat Problem 47 for the following systems.

(A) $5x - 6y = -10$
$11x - 13y = -20$

(B) $5x - 6y = -10$
$10x - 13y = -20$

(C) $5x - 6y = -10$
$10x - 12y = -20$

Solve Problems 49–52 using augmented matrix methods. Use a graphing utility to perform the row operations.

49. $0.8x_1 + 2.88x_2 = 4$
$1.25x_1 + 4.34x_2 = 5$

50. $2.7x_1 - 15.12x_2 = 27$
$3.25x_1 - 18.52x_2 = 33$

51. $4.8x_1 - 40.32x_2 = 295.2$
$-3.75x_1 + 28.7x_2 = -211.2$

52. $5.7x_1 - 8.55x_2 = -35.91$
$4.5x_1 + 5.73x_2 = 76.17$

APPLICATIONS

53. Puzzle. A friend of yours came out of the post office having spent \$19.50 on 32¢ and 23¢ stamps. If she bought 75 stamps in all, how many of each type did she buy?

54. Puzzle. A parking meter contains only nickels and dimes worth \$6.05. If there are 89 coins in all, how many of each type are there?

55. Investments. Bond *A* pays 6% compounded annually and bond *B* pays 9% compounded annually. If a \$200,000 investment in a combination of the two bonds returns \$14,775 annually, how much is invested in each bond?

56. Investments. Past history indicates that mutual fund *A* will earn 14.6% annually and mutual fund *B* will earn 9.8% annually. How should an investment be divided between the two funds to produce an expected return of 11%?

57. Chemistry. A chemist has two solutions of sulfuric acid: a 20% solution and an 80% solution. How much of each should be used to obtain 100 liters of a 62% solution?

58. Chemistry. A chemist has two solutions: one containing 40% alcohol and another containing 70% alcohol. How much of each should be used to obtain 80 liters of a 49% solution?

59. Nutrition. Animals in an experiment are to be kept on a strict diet. Each animal is to receive, among other things, 54 grams of protein and 24 grams of fat. The laboratory technician is able to purchase two food mixes of the following compositions: Mix *A* has 15% protein and 10% fat; mix *B* has 30% protein and 5% fat. How many grams of each mix should be used to obtain the right diet for a single animal?

60. Nutrition—Plants. A fruit grower can use two types of fertilizer in his orange grove, brand *A* and brand *B*. Each bag of brand *A* contains 9 pounds of nitrogen and 5 pounds of phosphoric acid. Each bag of brand *B* contains 8 pounds of nitrogen and 6 pounds of phosphoric acid. Tests indicate that the grove needs 770 pounds of nitrogen and 490 pounds of phosphoric acid. How many bags of each brand should be used to provide the required amounts of nitrogen and phosphoric acid?

61. Delivery Charges. United Express, a nationwide package delivery service, charges a base price for overnight delivery of packages weighing 1 pound or less and a surcharge for each additional pound (or fraction thereof). A customer is billed \$27.75 for shipping a 5-pound package and \$64.50 for shipping a 20-pound package. Find the base price and the surcharge for each additional pound.

62. Delivery Charges. Refer to Problem 61. Federated Shipping, a competing overnight delivery service, informs the customer in Problem 61 that it would ship the 5-pound package for \$29.95 and the 20-pound package for \$59.20.

(A) If Federated Shipping computes its cost in the same manner as United Express, find the base price and the surcharge for Federated Shipping.

(B) Devise a simple rule that the customer can use to choose the cheaper of the two services for each package shipped. Justify your answer.

63. Resource Allocation. A coffee manufacturer uses Colombian and Brazilian coffee beans to produce two blends, robust and mild. A pound of the robust blend requires 12 ounces of Colombian beans and 4 ounces of Brazilian beans. A pound of the mild blend requires 6 ounces of Colombian beans and 10 ounces of Brazilian beans. Coffee is shipped in 132-pound burlap bags. The company has 50 bags of Colombian beans and 40 bags of Brazilian beans on hand. How many pounds of each blend should it produce in order to use all the available beans?

64. **Resource Allocation.** Refer to Problem 63.

(A) If the company decides to discontinue production of the robust blend and only produce the mild blend, how many pounds of the mild blend can it produce and how many beans of each type will it use? Are there any beans that are not used?

(B) Repeat part A if the company decides to discontinue production of the mild blend and only produce the robust blend.

Section 5-3 Gauss–Jordan Elimination

- Reduced Matrices
- Solving Systems by Gauss–Jordan Elimination
- Application

Now that you have had some experience with row operations on simple augmented matrices, we will consider systems involving more than two variables. In addition, we will not require that a system have the same number of equations as variables. It turns out that the results for two-variable–two-equation linear systems, stated in Theorem 1 in Section 5-1, actually hold for linear systems of any size.

POSSIBLE SOLUTIONS TO A LINEAR SYSTEM

It can be shown that any linear system must have exactly one solution, no solution, or an infinite number of solutions, regardless of the number of equations or the number of variables in the system. The terms *unique*, *consistent*, *inconsistent*, *dependent*, and *independent* are used to describe these solutions, just as they are for systems with two variables.

Reduced Matrices

In the last section we used row operations to transform the augmented coefficient matrix for a system of two equations in two variables

$$\left[\begin{array}{cc|c} a_{11} & a_{12} & k_1 \\ a_{21} & a_{22} & k_2 \end{array}\right] \qquad \begin{aligned} a_{11}x_1 + a_{12}x_2 &= k_1 \\ a_{21}x_1 + a_{22}x_2 &= k_2 \end{aligned}$$

into one of the following simplified forms:

$$\begin{array}{ccc} \text{Form 1} & \text{Form 2} & \text{Form 3} \\ \left[\begin{array}{cc|c} 1 & 0 & m \\ 0 & 1 & n \end{array}\right] & \left[\begin{array}{cc|c} 1 & m & n \\ 0 & 0 & 0 \end{array}\right] & \left[\begin{array}{cc|c} 1 & m & n \\ 0 & 0 & p \end{array}\right] \end{array} \qquad (1)$$

where m, n, and p are real numbers, $p \neq 0$. Each of these reduced forms represents a system that has a different type of solution set, and no two of these forms are row-equivalent. Thus, we consider each of these to be a different simplified form. Now we want to consider larger systems with more variables and more equations.

Explore/Discuss 1

Forms 1, 2, and 3 above represent systems that have, respectively, a unique solution, an infinite number of solutions, and no solution. Discuss the number of solutions for the systems of three equations in three variables represented by the following augmented coefficient matrices.

$$\text{(A)} \left[\begin{array}{ccc|c} 1 & 1 & 1 & 2 \\ 0 & 0 & 0 & 3 \\ 0 & 0 & 0 & 0 \end{array}\right] \quad \text{(B)} \left[\begin{array}{ccc|c} 1 & 1 & 1 & 2 \\ 0 & 0 & 0 & 0 \\ 0 & 0 & 0 & 0 \end{array}\right] \quad \text{(C)} \left[\begin{array}{ccc|c} 1 & 0 & 0 & 2 \\ 0 & 1 & 0 & 3 \\ 0 & 0 & 1 & 4 \end{array}\right]$$

Since there is no upper limit on the number of variables or the number of equations in a linear system, it is not feasible to explicitly list all possible "simplified forms" for larger systems, as we did for systems of two equations in two variables. Instead, we state a general definition of a simplified form called a *reduced matrix* that can be applied to all matrices and systems, regardless of size.

DEFINITION 1

REDUCED MATRIX

A matrix is in **reduced form*** if

1. Each row consisting entirely of 0s is below any row having at least one nonzero element.
2. The leftmost nonzero element in each row is 1.
3. The column containing the leftmost 1 of a given row has 0s above and below the 1.
4. The leftmost 1 in any row is to the right of the leftmost 1 in the preceding row.

EXAMPLE 1

Reduced Forms

The matrices below are not in reduced form. Indicate which condition in the definition is violated for each matrix. State the row operation(s) required to transform the matrix to reduced form, and find the reduced form.

(A) $\left[\begin{array}{cc|c} 0 & 1 & -2 \\ 1 & 0 & 3 \end{array}\right]$ (B) $\left[\begin{array}{ccc|c} 1 & 2 & -2 & 3 \\ 0 & 0 & 1 & -1 \end{array}\right]$

(C) $\left[\begin{array}{cc|c} 1 & 0 & -3 \\ 0 & 0 & 0 \\ 0 & 1 & -2 \end{array}\right]$ (D) $\left[\begin{array}{ccc|c} 1 & 0 & 0 & -1 \\ 0 & 2 & 0 & 3 \\ 0 & 0 & 1 & -5 \end{array}\right]$

Solutions

(A) Condition 4 is violated: The leftmost 1 in row 2 is not to the right of the leftmost 1 in row 1. Perform the row operation $R_1 \leftrightarrow R_2$ to obtain the reduced form:

$$\left[\begin{array}{cc|c} 1 & 0 & 3 \\ 0 & 1 & -2 \end{array}\right]$$

(B) Condition 3 is violated: The column containing the leftmost 1 in row 2 does not have a 0 above the 1. Perform the row operation $2R_2 + R_1 \rightarrow R_1$ to obtain the reduced form:

$$\left[\begin{array}{ccc|c} 1 & 2 & 0 & 1 \\ 0 & 0 & 1 & -1 \end{array}\right]$$

*The reduced form we have defined here is often referred to as the **reduced row-echelon form** to distinguish it from other reduced forms (see Problems 57–62 in Exercise 5-3). Most graphing utilities use the abbreviation **rref** to refer to this reduced form.

(C) Condition 1 is violated: The second row contains all 0s, and it is above a row having at least one nonzero element. Perform the row operation $R_2 \leftrightarrow R_3$ to obtain the reduced form:

$$\left[\begin{array}{cc|c} 1 & 0 & -3 \\ 0 & 1 & -2 \\ 0 & 0 & 0 \end{array}\right]$$

(D) Condition 2 is violated: The leftmost nonzero element in row 2 is not a 1. Perform the row operation $\frac{1}{2}R_2 \to R_2$ to obtain the reduced form:

$$\left[\begin{array}{ccc|c} 1 & 0 & 0 & -1 \\ 0 & 1 & 0 & \frac{3}{2} \\ 0 & 0 & 1 & -5 \end{array}\right]$$

MATCHED PROBLEM 1

The matrices below are not in reduced form. Indicate which condition in the definition is violated for each matrix. State the row operation(s) required to transform the matrix to reduced form and find the reduced form.

(A) $\left[\begin{array}{cc|c} 1 & 0 & 2 \\ 0 & 3 & -6 \end{array}\right]$ (B) $\left[\begin{array}{ccc|c} 1 & 5 & 4 & 3 \\ 0 & 1 & 2 & -1 \\ 0 & 0 & 0 & 0 \end{array}\right]$

(C) $\left[\begin{array}{ccc|c} 0 & 1 & 0 & -3 \\ 1 & 0 & 0 & 0 \\ 0 & 0 & 1 & 2 \end{array}\right]$ (D) $\left[\begin{array}{ccc|c} 1 & 2 & 0 & 3 \\ 0 & 0 & 0 & 0 \\ 0 & 0 & 1 & 4 \end{array}\right]$

Solving Systems by Gauss–Jordan Elimination

We are now ready to outline the Gauss–Jordan elimination method for solving systems of linear equations. The method systematically transforms an augmented matrix into a reduced form. The system corresponding to a reduced augmented coefficient matrix is called a **reduced system.** As we will see, reduced systems are easy to solve.

The Gauss–Jordan elimination method is named after the German mathematician Carl Friedrich Gauss (1777–1885) and the German geodesist Wilhelm Jordan (1842–1899). Gauss, one of the greatest mathematicians of all time, used a method of solving systems of equations that was later generalized by Jordan to solve problems in large-scale surveying.

EXAMPLE 2

Solving a System Using Gauss–Jordan Elimination

Solve by Gauss–Jordan elimination:

$$\begin{aligned} 2x_1 - 2x_2 + x_3 &= 3 \\ 3x_1 + x_2 - x_3 &= 7 \\ x_1 - 3x_2 + 2x_3 &= 0 \end{aligned}$$

Solution Write the augmented matrix and follow the steps indicated at the right to produce a reduced form.

Need a 1 here.

$$\left[\begin{array}{rrr|r} 2 & -2 & 1 & 3 \\ 3 & 1 & -1 & 7 \\ 1 & -3 & 2 & 0 \end{array}\right] \quad R_1 \leftrightarrow R_3$$

Step 1. Choose the leftmost nonzero column and get a 1 at the top.

Need 0s here.

$$\sim \left[\begin{array}{rrr|r} 1 & -3 & 2 & 0 \\ 3 & 1 & -1 & 7 \\ 2 & -2 & 1 & 3 \end{array}\right] \quad \begin{array}{l} (-3)R_1 + R_2 \rightarrow R_2 \\ (-2)R_1 + R_3 \rightarrow R_3 \end{array}$$

Step 2. Use multiples of the row containing the 1 from step 1 to get zeros in all remaining places in the column containing this 1.

Need a 1 here.

$$\sim \left[\begin{array}{rrr|r} 1 & -3 & 2 & 0 \\ 0 & 10 & -7 & 7 \\ 0 & 4 & -3 & 3 \end{array}\right] \quad 0.1R_2 \rightarrow R_2$$

Step 3. Repeat step 1 with the *submatrix* formed by (mentally) deleting the top row.

Need 0s here.

$$\sim \left[\begin{array}{rrr|r} 1 & -3 & 2 & 0 \\ 0 & 1 & -0.7 & 0.7 \\ 0 & 4 & -3 & 3 \end{array}\right] \quad \begin{array}{l} 3R_2 + R_1 \rightarrow R_1 \\ (-4)R_2 + R_3 \rightarrow R_3 \end{array}$$

Step 4. Repeat step 2 with the *entire matrix.*

Need a 1 here.

$$\sim \left[\begin{array}{rrr|r} 1 & 0 & -0.1 & 2.1 \\ 0 & 1 & -0.7 & 0.7 \\ 0 & 0 & -0.2 & 0.2 \end{array}\right] \quad (-5)R_3 \rightarrow R_3$$

Step 5. Repeat step 1 with the *submatrix* formed by (mentally) deleting the top two rows.

Need 0s here.

$$\sim \left[\begin{array}{rrr|r} 1 & 0 & -0.1 & 2.1 \\ 0 & 1 & -0.7 & 0.7 \\ 0 & 0 & 1 & -1 \end{array}\right] \quad \begin{array}{l} 0.1R_3 + R_1 \rightarrow R_1 \\ 0.7R_3 + R_2 \rightarrow R_2 \end{array}$$

Step 6. Repeat step 2 with the *entire matrix.*

$$\sim \left[\begin{array}{rrr|r} 1 & 0 & 0 & 2 \\ 0 & 1 & 0 & 0 \\ 0 & 0 & 1 & -1 \end{array}\right]$$

The matrix is now in reduced form, and we can proceed to solve the corresponding reduced system.

$$\begin{aligned} x_1 \quad\quad &= 2 \\ x_2 \quad &= 0 \\ x_3 &= -1 \end{aligned}$$

The solution to this system is $x_1 = 2$, $x_2 = 0$, $x_3 = -1$. You should check this solution in the original system.

GAUSS–JORDAN ELIMINATION

Step 1. Choose the leftmost nonzero column and use appropriate row operations to get a 1 at the top.

Step 2. Use multiples of the row containing the 1 from step 1 to get zeros in all remaining places in the column containing this 1.

Step 3. Repeat step 1 with the **submatrix** formed by (mentally) deleting the row used in step 2 and all rows above this row.

Step 4. Repeat step 2 with the **entire matrix,** including the mentally deleted rows. Continue this process until it is impossible to go further.

[*Note:* If at any point in this process we obtain a row with all zeros to the left of the vertical line and a nonzero number to the right, we can stop, since we will have a contradiction: $0 = n$, $n \neq 0$. We can then conclude that the system has no solution.]

Remarks

1. Even though each matrix has a unique reduced form, the sequence of steps (algorithm) presented here for transforming a matrix into a reduced form is not unique. That is, other sequences of steps (using row operations) can produce a reduced matrix. (For example, it is possible to use row operations in such a way that computations involving fractions are minimized.) But we emphasize again that we are not interested in the most efficient hand methods for transforming small matrices into reduced forms. Our main interest is in giving you a little experience with a method that is suitable for solving large-scale systems on a computer or graphing utility.

2. Most graphing utilities have the ability to find reduced forms, either directly or with some programming. Figure 1 illustrates the solution of Example 2 on a graphing utility that has a built-in routine for finding reduced forms. Notice that in row 2 and column 4 of the reduced form the graphing utility has displayed the very small number $-3.5E\text{-}13$ instead of the exact value 0. This is a common occurrence on a graphing utility and causes no problems. Just replace any very small numbers displayed in scientific notation with 0.

FIGURE 1
Gauss–Jordan elimination on a graphing calculator.

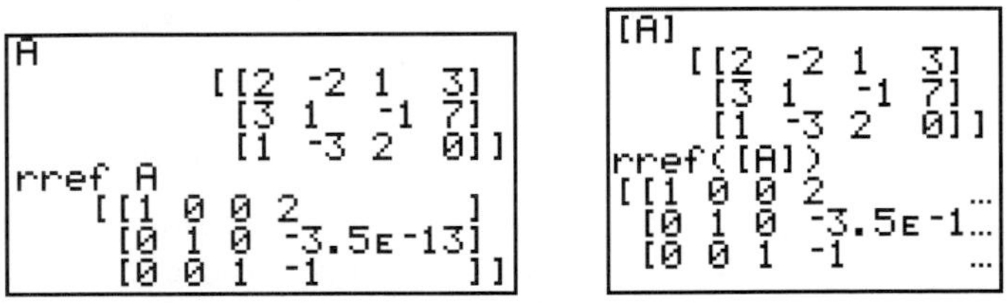

(a) TI-85/86 (b) TI-83

Remark Most of the graphing calculator screens displayed in the text were produced on a Texas Instruments TI-83. However, in this chapter there are a few screens that were produced on a TI-85/86 to display more of the matrix (see Fig. 1).

MATCHED PROBLEM **2**

Solve by Gauss–Jordan elimination:

$$\begin{aligned} 3x_1 + x_2 - 2x_3 &= 2 \\ x_1 - 2x_2 + x_3 &= 3 \\ 2x_1 - x_2 - 3x_3 &= 3 \end{aligned}$$

EXAMPLE 3

Solving a System Using Gauss–Jordan Elimination

Solve by Gauss–Jordan elimination:

$$\begin{aligned} 2x_1 - 4x_2 + x_3 &= -4 \\ 4x_1 - 8x_2 + 7x_3 &= 2 \\ -2x_1 + 4x_2 - 3x_3 &= 5 \end{aligned}$$

Solution

$$\left[\begin{array}{rrr|r} 2 & -4 & 1 & -4 \\ 4 & -8 & 7 & 2 \\ -2 & 4 & -3 & 5 \end{array}\right] \quad 0.5R_1 \to R_1$$

$$\sim \left[\begin{array}{rrr|r} 1 & -2 & 0.5 & -2 \\ 4 & -8 & 7 & 2 \\ -2 & 4 & -3 & 5 \end{array}\right] \quad \begin{array}{l} (-4)R_1 + R_2 \to R_2 \\ 2R_1 + R_3 \to R_3 \end{array}$$

$$\sim \left[\begin{array}{rrr|r} 1 & -2 & 0.5 & -2 \\ 0 & 0 & 5 & 10 \\ 0 & 0 & -2 & 1 \end{array}\right] \quad 0.2R_2 \to R_2$$

Note that column 3 is the leftmost nonzero column in this submatrix.

$$\sim \left[\begin{array}{rrr|r} 1 & -2 & 0.5 & -2 \\ 0 & 0 & 1 & 2 \\ 0 & 0 & -2 & 1 \end{array}\right] \quad \begin{array}{l} (-0.5)R_2 + R_1 \to R_1 \\ \\ 2R_2 + R_3 \to R_3 \end{array}$$

$$\sim \left[\begin{array}{rrr|r} 1 & -2 & 0 & -3 \\ 0 & 0 & 1 & 2 \\ 0 & 0 & 0 & 5 \end{array}\right]$$

We stop the Gauss–Jordan elimination, even though the matrix is not in reduced form, since the last row produces a contradiction.

The system is inconsistent and has no solution.

MATCHED PROBLEM 3

Solve by Gauss–Jordan elimination:

$$\begin{aligned} 2x_1 - 4x_2 - x_3 &= -8 \\ 4x_1 - 8x_2 + 3x_3 &= 4 \\ -2x_1 + 4x_2 + x_3 &= 11 \end{aligned}$$

CAUTION

Figure 2 shows the solution to Example 3 on a graphing utility with a built-in reduced form routine. Notice that the graphing utility does not stop when a contradiction first occurs, as we did in the solution to Example 3, but continues on to find the reduced form. Nevertheless, the last row in the reduced form still produces a contradiction, indicating that the system has no solution.

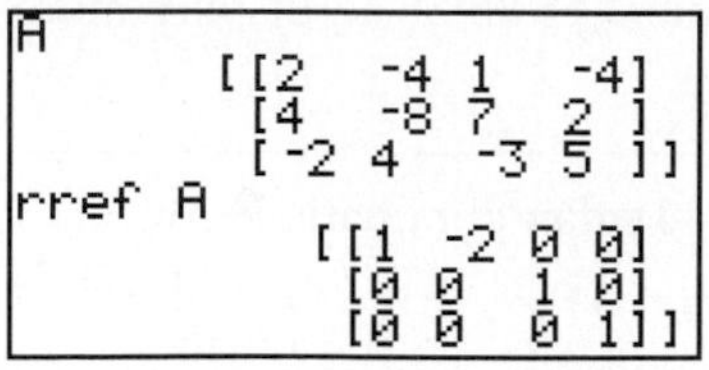

FIGURE 2
Recognizing contradictions on a graphing utility.

EXAMPLE 4 **Solving a System Using Gauss–Jordan Elimination**

Solve by Gauss–Jordan elimination:

$$\begin{aligned} 3x_1 + 6x_2 - 9x_3 &= 15 \\ 2x_1 + 4x_2 - 6x_3 &= 10 \\ -2x_1 - 3x_2 + 4x_3 &= -6 \end{aligned}$$

Solution

$$\left[\begin{array}{rrr|r} 3 & 6 & -9 & 15 \\ 2 & 4 & -6 & 10 \\ -2 & -3 & 4 & -6 \end{array}\right] \quad \tfrac{1}{3}R_1 \to R_1$$

$$\sim \left[\begin{array}{rrr|r} 1 & 2 & -3 & 5 \\ 2 & 4 & -6 & 10 \\ -2 & -3 & 4 & -6 \end{array}\right] \quad \begin{array}{l} (-2)R_1 + R_2 \to R_2 \\ 2R_1 + R_3 \to R_3 \end{array}$$

$$\sim \left[\begin{array}{rrr|r} 1 & 2 & -3 & 5 \\ 0 & 0 & 0 & 0 \\ 0 & 1 & -2 & 4 \end{array}\right] \quad R_2 \leftrightarrow R_3$$

Note that we must interchange rows 2 and 3 to obtain a nonzero entry at the top of the second column of this submatrix.

$$\sim \left[\begin{array}{rrr|r} 1 & 2 & -3 & 5 \\ 0 & 1 & -2 & 4 \\ 0 & 0 & 0 & 0 \end{array}\right] \quad (-2)R_2 + R_1 \to R_1$$

FIGURE 3

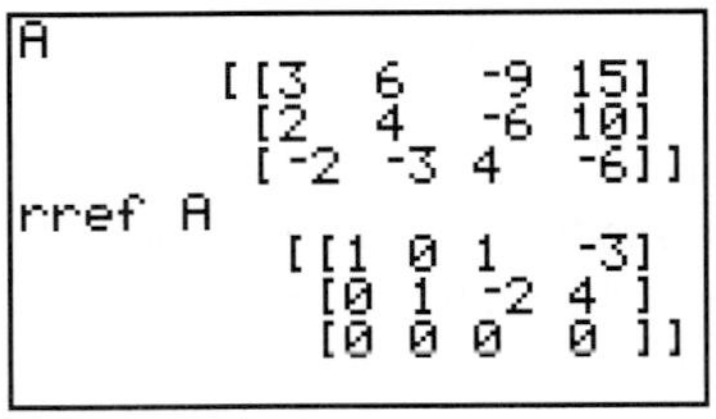

$$\sim \left[\begin{array}{rrr|r} 1 & 0 & 1 & -3 \\ 0 & 1 & -2 & 4 \\ 0 & 0 & 0 & 0 \end{array}\right]$$

This matrix is now in reduced form. (See Fig. 3 for a graphing utility solution.) Write the corresponding reduced system and solve.

$$\begin{aligned} x_1 \quad + x_3 &= -3 \\ x_2 - 2x_3 &= 4 \end{aligned}$$

We discard the equation corresponding to the third (all 0) row in the reduced form, since it is satisfied by all values of x_1, x_2, and x_3.

Note that the leftmost variable in each equation appears in one and only one equation. We solve for the leftmost variables x_1 and x_2 in terms of the remaining variable x_3:

$$\begin{aligned} x_1 &= -x_3 - 3 \\ x_2 &= 2x_3 + 4 \end{aligned}$$

This dependent system has an infinite number of solutions. We will use a parameter to represent all the solutions. If we let $x_3 = t$, then for any real number t,

$$\begin{aligned} x_1 &= -t - 3 \\ x_2 &= 2t + 4 \\ x_3 &= t \end{aligned}$$

You should check that $(-t - 3, 2t + 4, t)$ is a solution of the original system for any real number t. Some particular solutions are

$t = 0$	$t = -2$	$t = 3.5$
$(-3, 4, 0)$	$(-1, 0, -2)$	$(-6.5, 11, 3.5)$

MATCHED PROBLEM 4

Solve by Gauss–Jordan elimination:

$$\begin{aligned} 2x_1 - 2x_2 - 4x_3 &= -2 \\ 3x_1 - 3x_2 - 6x_3 &= -3 \\ -2x_1 + 3x_2 + x_3 &= 7 \end{aligned}$$

In general,

> **If the number of leftmost 1s in a reduced augmented coefficient matrix is less than the number of variables in the system and there are no contradictions, then the system is dependent and has infinitely many solutions.**

There are many different ways to use the reduced augmented coefficient matrix to describe the infinite number of solutions of a dependent system. We will always proceed as follows: Solve each equation in a reduced system for its leftmost variable and then introduce a different parameter for each remaining variable. As the solution to Example 4 illustrates, this method produces a concise and useful representation of the solutions to a dependent system. Example 5 illustrates a dependent system where two parameters are required to describe the solution.

Explore/Discuss 2

Explain why the definition of reduced form ensures that each leftmost variable in a reduced system appears in one and only one equation and no equation contains more than one leftmost variable. Discuss methods for determining if a consistent system is independent or dependent by examining the reduced form.

EXAMPLE 5

Solving a System Using Gauss–Jordan Elimination

Solve by Gauss–Jordan elimination:

$$\begin{aligned} x_1 + 2x_2 + 4x_3 + x_4 - x_5 &= 1 \\ 2x_1 + 4x_2 + 8x_3 + 3x_4 - 4x_5 &= 2 \\ x_1 + 3x_2 + 7x_3 + 3x_5 &= -2 \end{aligned}$$

Solution The augmented coefficient matrix and its reduced form are shown in Figure 4. Write the corresponding reduced system and solve.

FIGURE 4

```
A
     [[1 2 4 1 -1 1 ]
      [2 4 8 3 -4 2 ]
      [1 3 7 0 3  -2]]
rref A
    [[1 0 -2 0 -3 7 ]
     [0 1 3  0 2  -3]
     [0 0 0  1 -2 0 ]]
```

$$\begin{aligned} x_1 - 2x_3 - 3x_5 &= 7 \\ x_2 + 3x_3 + 2x_5 &= -3 \\ x_4 - 2x_5 &= 0 \end{aligned}$$

Solve for the leftmost variables x_1, x_2, and x_4 in terms of the remaining variables x_3 and x_5.

$$\begin{aligned} x_1 &= 2x_3 + 3x_5 + 7 \\ x_2 &= -3x_3 - 2x_5 - 3 \\ x_4 &= 2x_5 \end{aligned}$$

If we let $x_3 = s$ and $x_5 = t$, then for any real numbers s and t,

$$\begin{aligned} x_1 &= 2s + 3t + 7 \\ x_2 &= -3s - 2t - 3 \\ x_3 &= s \\ x_4 &= 2t \\ x_5 &= t \end{aligned}$$

is a solution. The check is left for you to perform.

MATCHED PROBLEM 5

Solve by Gauss–Jordan elimination:
$$\begin{aligned} x_1 - x_2 + 2x_3 - 2x_5 &= 3 \\ -2x_1 + 2x_2 - 4x_3 - x_4 + x_5 &= -5 \\ 3x_1 - 3x_2 + 7x_3 + x_4 - 4x_5 &= 6 \end{aligned}$$

Application

Dependent systems of linear equations provide an excellent opportunity to discuss mathematical modeling in a little more detail. The process of using mathematics to solve real-world problems can be broken down into three steps (Fig. 5):

Step 1. *Construct* a mathematical model whose solution will provide information about the real-world problem.

Step 2. *Solve* the mathematical model.

Step 3. *Interpret* the solution to the mathematical model in terms of the original real-world problem.

FIGURE 5

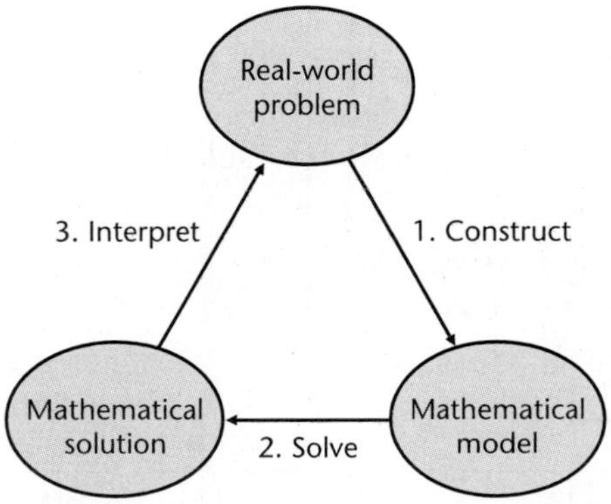

In more complex problems, this cycle may have to be repeated several times to obtain the required information about the real-world problem.

EXAMPLE 6 **Purchasing**

A chemical manufacturer wants to purchase a fleet of 24 railroad tank cars with a combined carrying capacity of 250,000 gallons. Tank cars with three different carrying capacities are available: 6,000 gallons, 8,000 gallons, and 18,000 gallons. How many of each type of tank car should be purchased?

Solution Let

x_1 = Number of 6,000-gallon tank cars

x_2 = Number of 8,000-gallon tank cars

x_3 = Number of 18,000-gallon tank cars

Then

$$x_1 + x_2 + x_3 = 24 \qquad \text{Total number of tank cars}$$

$$6{,}000x_1 + 8{,}000x_2 + 18{,}000x_3 = 250{,}000 \qquad \text{Total carrying capacity}$$

FIGURE 6

```
A
[[1    1    1     24...
 [6000 8000 18000 25...
rref A
    [[1 0 -5 -29]
     [0 1 6   53 ]]
```

The augmented coefficient matrix and its reduced form are shown in Figure 6. Write the corresponding reduced system and solve.

$$x_1 - 5x_3 = -29 \quad \text{or} \quad x_1 = 5x_3 - 29$$

$$x_2 + 6x_3 = 53 \quad \text{or} \quad x_2 = -6x_3 + 53$$

Let $x_3 = t$. Then for t any real number,

$$x_1 = 5t - 29$$

$$x_2 = -6t + 53$$

$$x_3 = t$$

is a solution—or is it? Since the variables in this system represent the number of tank cars purchased, the values of x_1, x_2, and x_3 must be nonnegative integers. Thus, the third equation requires that t must be a nonnegative integer. The first equation requires that $5t - 29 \geq 0$, so t must be at least 6. The middle equation requires that $-6t + 53 \geq 0$, so t can be no larger than 8. Thus, 6, 7, and 8 are the only possible values for t. There are only three possible combinations that meet the company's specifications of 24 tank cars with a total carrying capacity of 250,000 gallons, as shown in Table 1.

TABLE 1

t	6,000-Gallon Tank Cars x_1	8,000-Gallon Tank Cars x_2	18,000-Gallon Tank Cars x_3
6	1	17	6
7	6	11	7
8	11	5	8

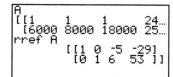

The final choice would probably be influenced by other factors. For example, the company might want to minimize the cost of the 24 tank cars.

MATCHED PROBLEM 6

A commuter airline wants to purchase a fleet of 30 airplanes with a combined carrying capacity of 960 passengers. The three available types of planes carry 18, 24, and 42 passengers, respectively. How many of each type of plane should be purchased?

Answers to Matched Problems

1. (A) Condition 2 is violated: the 3 in row 2 and column 2 should be a 1. Perform the operation $\frac{1}{3}R_2 \to R_2$ to obtain

$$\left[\begin{array}{cc|c} 1 & 0 & 2 \\ 0 & 1 & -2 \end{array}\right]$$

(B) Condition 3 is violated: the 5 in row 1 and column 2 should be a 0. Perform the operation $(-5)R_2 + R_1 \to R_1$ to obtain

$$\left[\begin{array}{ccc|c} 1 & 0 & -6 & 8 \\ 0 & 1 & 2 & -1 \\ 0 & 0 & 0 & 0 \end{array}\right]$$

(C) Condition 4 is violated: the leftmost 1 in the second row is not to the right of the leftmost 1 in the first row. Perform the operation $R_1 \leftrightarrow R_2$ to obtain

$$\left[\begin{array}{ccc|c} 1 & 0 & 0 & 0 \\ 0 & 1 & 0 & -3 \\ 0 & 0 & 1 & 2 \end{array}\right]$$

(D) Condition 1 is violated: the all-zero second row should be at the bottom. Perform the operation $R_2 \leftrightarrow R_3$ to obtain

$$\left[\begin{array}{ccc|c} 1 & 2 & 0 & 3 \\ 0 & 0 & 1 & 4 \\ 0 & 0 & 0 & 0 \end{array}\right]$$

2. $x_1 = 1, x_2 = -1, x_3 = 0$ **3.** Inconsistent; no solution **4.** $x_1 = 5t + 4, x_2 = 3t + 5, x_3 = t$, t any real number
5. $x_1 = s + 7, x_2 = s, x_3 = t - 2, x_4 = -3t - 1, x_5 = t$, s and t any real numbers
6.

t	18-Passenger Planes x_1	24-Passenger Planes x_2	42-Passenger Planes x_3
14	2	14	14
15	5	10	15
16	8	6	16
17	11	2	17

EXERCISE 5-3

A

In Problems 1–10, if a matrix is in reduced form, say so. If not, explain why and indicate the row operation(s) necessary to transform the matrix into reduced form.

1. $\left[\begin{array}{cc|c} 1 & 0 & 3 \\ 0 & 1 & -2 \end{array}\right]$

2. $\left[\begin{array}{cc|c} 0 & 1 & 3 \\ 1 & 0 & -2 \end{array}\right]$

3. $\left[\begin{array}{ccc|c} 1 & 0 & 3 & 2 \\ 0 & 0 & 0 & 0 \\ 0 & 1 & -1 & 5 \end{array}\right]$

4. $\left[\begin{array}{ccc|c} 1 & 0 & 0 & -4 \\ 0 & 1 & 0 & 0 \\ 0 & 0 & 1 & 1 \end{array}\right]$

5. $\left[\begin{array}{ccc|c} 0 & 1 & 0 & 4 \\ 0 & 0 & 3 & -2 \\ 0 & 0 & 0 & 0 \end{array}\right]$

6. $\left[\begin{array}{ccc|c} 1 & 1 & -3 & 2 \\ 0 & 0 & 1 & 5 \\ 0 & 0 & 0 & 0 \end{array}\right]$

7. $\left[\begin{array}{ccc|c} 1 & 1 & 0 & -1 \\ 0 & 0 & 1 & 1 \\ 0 & 0 & 0 & 0 \end{array}\right]$

8. $\left[\begin{array}{ccc|c} 1 & 0 & -1 & 4 \\ 0 & 2 & 1 & 3 \\ 0 & 0 & 0 & 0 \end{array}\right]$

9. $\left[\begin{array}{cccc|c} 1 & 0 & -3 & 0 & 1 \\ 0 & 0 & 1 & 1 & 0 \end{array}\right]$ **10.** $\left[\begin{array}{cccc|c} 1 & -3 & 0 & 0 & 1 \\ 0 & 0 & 1 & 1 & 0 \end{array}\right]$

In Problems 11–18, write the linear system corresponding to each reduced augmented matrix and solve.

11. $\left[\begin{array}{ccc|c} 1 & 0 & 0 & -2 \\ 0 & 1 & 0 & 3 \\ 0 & 0 & 1 & 0 \end{array}\right]$ **12.** $\left[\begin{array}{cccc|c} 1 & 0 & 0 & 0 & -2 \\ 0 & 1 & 0 & 0 & 0 \\ 0 & 0 & 1 & 0 & 1 \\ 0 & 0 & 0 & 1 & 3 \end{array}\right]$

13. $\left[\begin{array}{ccc|c} 1 & 0 & -2 & 3 \\ 0 & 1 & 1 & -5 \\ 0 & 0 & 0 & 0 \end{array}\right]$ **14.** $\left[\begin{array}{ccc|c} 1 & -2 & 0 & -3 \\ 0 & 0 & 1 & 5 \\ 0 & 0 & 0 & 0 \end{array}\right]$

15. $\left[\begin{array}{cc|c} 1 & 0 & 0 \\ 0 & 1 & 0 \\ 0 & 0 & 1 \end{array}\right]$ **16.** $\left[\begin{array}{cc|c} 1 & 0 & 5 \\ 0 & 1 & -3 \\ 0 & 0 & 1 \end{array}\right]$

17. $\left[\begin{array}{cccc|c} 1 & -2 & 0 & -3 & -5 \\ 0 & 0 & 1 & 3 & 2 \end{array}\right]$ **18.** $\left[\begin{array}{cccc|c} 1 & 0 & -2 & 3 & 4 \\ 0 & 1 & -1 & 2 & -1 \end{array}\right]$

B

Use row operations to change each matrix in Problems 19–24 to reduced form.

19. $\left[\begin{array}{cc|c} 1 & 2 & -1 \\ 0 & 1 & 3 \end{array}\right]$ **20.** $\left[\begin{array}{cc|c} 1 & 3 & 1 \\ 0 & 2 & -4 \end{array}\right]$

21. $\left[\begin{array}{ccc|c} 1 & 0 & -3 & 1 \\ 0 & 1 & 2 & 0 \\ 0 & 0 & 3 & -6 \end{array}\right]$ **22.** $\left[\begin{array}{ccc|c} 1 & 0 & 4 & 0 \\ 0 & 1 & -3 & -1 \\ 0 & 0 & -2 & 2 \end{array}\right]$

23. $\left[\begin{array}{ccc|c} 1 & 2 & -2 & -1 \\ 0 & 3 & -6 & 1 \\ 0 & -1 & 2 & -\frac{1}{3} \end{array}\right]$ **24.** $\left[\begin{array}{ccc|c} 0 & -2 & 8 & 1 \\ 2 & -2 & 6 & -4 \\ 0 & -1 & 4 & \frac{1}{2} \end{array}\right]$

Solve Problems 25–44 using Gauss–Jordan elimination.

25. $\begin{aligned} 2x_1 + 4x_2 - 10x_3 &= -2 \\ 3x_1 + 9x_2 - 21x_3 &= 0 \\ x_1 + 5x_2 - 12x_3 &= 1 \end{aligned}$ **26.** $\begin{aligned} 3x_1 + 5x_2 - x_3 &= -7 \\ x_1 + x_2 + x_3 &= -1 \\ 2x_1 \qquad + 11x_3 &= 7 \end{aligned}$

27. $\begin{aligned} 3x_1 + 8x_2 - x_3 &= -18 \\ 2x_1 + x_2 + 5x_3 &= 8 \\ 2x_1 + 4x_2 + 2x_3 &= -4 \end{aligned}$ **28.** $\begin{aligned} 2x_1 + 7x_2 + 15x_3 &= -12 \\ 4x_1 + 7x_2 + 13x_3 &= -10 \\ 3x_1 + 6x_2 + 12x_3 &= -9 \end{aligned}$

29. $\begin{aligned} 2x_1 - x_2 - 3x_3 &= 8 \\ x_1 - 2x_2 \qquad &= 7 \end{aligned}$ **30.** $\begin{aligned} 2x_1 + 4x_2 - 6x_3 &= 10 \\ 3x_1 + 3x_2 - 3x_3 &= 6 \end{aligned}$

31. $\begin{aligned} 2x_1 - x_2 &= 0 \\ 3x_1 + 2x_2 &= 7 \\ x_1 - x_2 &= -1 \end{aligned}$ **32.** $\begin{aligned} 2x_1 - x_2 &= 0 \\ 3x_1 + 2x_2 &= 7 \\ x_1 - x_2 &= -2 \end{aligned}$

33. $\begin{aligned} 3x_1 - 4x_2 - x_3 &= 1 \\ 2x_1 - 3x_2 + x_3 &= 1 \\ x_1 - 2x_2 + 3x_3 &= 2 \end{aligned}$ **34.** $\begin{aligned} 3x_1 + 7x_2 - x_3 &= 11 \\ x_1 + 2x_2 - x_3 &= 3 \\ 2x_1 + 4x_2 - 2x_3 &= 10 \end{aligned}$

35. $\begin{aligned} -2x_1 + x_2 + 3x_3 &= -7 \\ x_1 - 4x_2 + 2x_3 &= 0 \\ x_1 - 3x_2 + x_3 &= 1 \end{aligned}$ **36.** $\begin{aligned} 2x_1 + 5x_2 + 4x_3 &= -7 \\ -4x_1 - 5x_2 + 2x_3 &= 9 \\ -2x_1 - x_2 + 4x_3 &= 3 \end{aligned}$

37. $\begin{aligned} 2x_1 - 2x_2 - 4x_3 &= -2 \\ -3x_1 + 3x_2 + 6x_3 &= 3 \end{aligned}$ **38.** $\begin{aligned} 2x_1 + 8x_2 - 6x_3 &= 4 \\ -3x_1 - 12x_2 + 9x_3 &= -6 \end{aligned}$

39. $\begin{aligned} 4x_1 - x_2 + 2x_3 &= 3 \\ -4x_1 + x_2 - 3x_3 &= -10 \\ 8x_1 - 2x_2 + 9x_3 &= -1 \end{aligned}$

40. $\begin{aligned} 4x_1 - 2x_2 + 2x_3 &= 5 \\ -6x_1 + 3x_2 - 3x_3 &= -2 \\ 10x_1 - 5x_2 + 9x_3 &= 4 \end{aligned}$

41. $\begin{aligned} 2x_1 - 5x_2 - 3x_3 &= 7 \\ -4x_1 + 10x_2 + 2x_3 &= 6 \\ 6x_1 - 15x_2 - x_3 &= -19 \end{aligned}$

42. $\begin{aligned} -4x_1 + 8x_2 + 10x_3 &= -6 \\ 6x_1 - 12x_2 - 15x_3 &= 9 \\ -8x_1 + 14x_2 + 19x_3 &= -8 \end{aligned}$

43. $\begin{aligned} 5x_1 - 3x_2 + 2x_3 &= 13 \\ 2x_1 - x_2 - 3x_3 &= 1 \\ 4x_1 - 2x_2 + 4x_3 &= 12 \end{aligned}$ **44.** $\begin{aligned} 4x_1 - 2x_2 + 3x_3 &= 3 \\ 3x_1 - x_2 - 2x_3 &= -10 \\ 2x_1 + 4x_2 - x_3 &= -1 \end{aligned}$

45. Consider a consistent system of three linear equations in three variables. Discuss the nature of the solution set for the system if the reduced form of the augmented coefficient matrix has

(A) One leftmost 1 (B) Two leftmost 1s

(C) Three leftmost 1s (D) Four leftmost 1s

46. Consider a system of three linear equations in three variables. Give examples of two reduced forms that are not row equivalent if the system is

(A) Consistent and dependent

(B) Inconsistent

In Problems 47–50, discuss the relationship between the number of solutions of the system and the constant k.

47. $\begin{aligned} x_1 - x_2 &= 4 \\ 3x_1 + kx_2 &= 7 \end{aligned}$ **48.** $\begin{aligned} x_1 + 2x_2 &= 4 \\ -2x_1 + kx_2 &= -8 \end{aligned}$

49. $\begin{aligned} x_1 + kx_2 &= 3 \\ 2x_1 + 6x_2 &= 6 \end{aligned}$ **50.** $\begin{aligned} x_1 + kx_2 &= 3 \\ 2x_1 + 4x_2 &= 8 \end{aligned}$

C

In Problems 51–56, solve using Gauss–Jordan elimination. Use a row reduction routine on a graphing utility to find the reduced forms.

51. $\begin{aligned} x_1 + 2x_2 - 4x_3 - x_4 &= 7 \\ 2x_1 + 5x_2 - 9x_3 - 4x_4 &= 16 \\ x_1 + 5x_2 - 7x_3 - 7x_4 &= 13 \end{aligned}$

52. $$\begin{aligned} 2x_1 + 4x_2 + 5x_3 + 4x_4 &= 8 \\ x_1 + 2x_2 + 2x_3 + x_4 &= 3 \end{aligned}$$

53. $$\begin{aligned} x_1 - x_2 + 3x_3 - 2x_4 &= 1 \\ -2x_1 + 4x_2 - 3x_3 + x_4 &= 0.5 \\ 3x_1 - x_2 + 10x_3 - 4x_4 &= 2.9 \\ 4x_1 - 3x_2 + 8x_3 - 2x_4 &= 0.6 \end{aligned}$$

54. $$\begin{aligned} x_1 + x_2 + 4x_3 + x_4 &= 1.3 \\ -x_1 + x_2 - x_3 &= 1.1 \\ 2x_1 + x_3 + 3x_4 &= -4.4 \\ 2x_1 + 5x_2 + 11x_3 + 3x_4 &= 5.6 \end{aligned}$$

55. $$\begin{aligned} x_1 - 2x_2 + x_3 + x_4 + 2x_5 &= 2 \\ -2x_1 + 4x_2 + 2x_3 + 2x_4 - 2x_5 &= 0 \\ 3x_1 - 6x_2 + x_3 + x_4 + 5x_5 &= 4 \\ -x_1 + 2x_2 + 3x_3 + x_4 + x_5 &= 3 \end{aligned}$$

56. $$\begin{aligned} x_1 - 3x_2 + x_3 + x_4 + 2x_5 &= 2 \\ -x_1 + 5x_2 + 2x_3 + 2x_4 - 2x_5 &= 0 \\ 2x_1 - 6x_2 + 2x_3 + 2x_4 + 4x_5 &= 4 \\ -x_1 + 3x_2 - x_3 - x_5 &= -3 \end{aligned}$$

Most graphing utilities also have a routine that produces the row-echelon form (ref) of a matrix. Problems 57–62 require a graphing utility with this routine. In Problems 57–60 use the row-echelon form to solve the indicated system.

57. $$\begin{aligned} x_1 + 2x_2 &= 6 \\ 2x_1 - 5x_2 &= 3 \end{aligned}$$

58. $$\begin{aligned} 3x_1 - 4x_2 &= 10 \\ 4x_1 + 2x_2 &= 6 \end{aligned}$$

59. $$\begin{aligned} x_1 - 2x_2 + x_3 &= -7 \\ 2x_1 + x_2 - 2x_3 &= -6 \\ -2x_1 - x_2 + 4x_3 &= 14 \end{aligned}$$

60. $$\begin{aligned} 2x_1 - x_2 + 3x_3 &= 17 \\ -x_1 + 2x_2 - x_3 &= -12 \\ 4x_1 + x_2 - 5x_3 &= 3 \end{aligned}$$

61. Based on the results in Problems 57–60, discuss the differences between the row-echelon form and the reduced row-echelon form of a matrix.

62. Describe a general procedure for using the row-echelon form to find the solution of a linear system.

APPLICATIONS

Solve Problems 63–78 using Gauss–Jordan elimination.

★ **63. Puzzle.** A friend of yours came out of the post office after spending \$14.00 on 15¢, 20¢, and 35¢ stamps. If she bought 45 stamps in all, how many of each type did she buy?

★ **64. Puzzle.** A parking meter accepts only nickels, dimes, and quarters. If the meter contains 32 coins with a total value of \$6.80, how many of each type are there?

★★ **65. Chemistry.** A chemist can purchase a 10% saline solution in 500 cubic centimeter containers, a 20% saline solution in 500 cubic centimeter containers, and a 50% saline solution in 1,000 cubic centimeter containers. He needs 12,000 cubic centimeters of 30% saline solution. How many containers of each type of solution should he purchase to form this solution?

★★ **66. Chemistry.** Repeat Problem 65 if the 50% saline solution is available only in 1,500 cubic centimeter containers.

67. Geometry. Find a, b, and c so that the graph of the parabola with equation $y = a + bx + cx^2$ passes through the points $(-2, 3)$, $(-1, 2)$, and $(1, 6)$.

68. Geometry. Find a, b, and c so that the graph of the parabola with equation $y = a + bx + cx^2$ passes through the points $(1, 3)$, $(2, 2)$, and $(3, 5)$.

69. Geometry. Find a, b, and c so that the graph of the circle with equation $x^2 + y^2 + ax + by + c = 0$ passes through the points $(6, 2)$, $(4, 6)$, and $(-3, -1)$.

70. Geometry. Find a, b, and c so that the graph of the circle with equation $x^2 + y^2 + ax + by + c = 0$ passes through the points $(-4, 1)$, $(-1, 2)$, and $(3, -6)$.

71. Production Scheduling. A small manufacturing plant makes three types of inflatable boats: one-person, two-person, and four-person models. Each boat requires the services of three departments, as listed in the table. The cutting, assembly, and packaging departments have available a maximum of 380, 330, and 120 labor-hours per week, respectively.

	One-Person Boat	Two-Person Boat	Four-Person Boat
Cutting department	0.5 h	1.0 h	1.5 h
Assembly department	0.6 h	0.9 h	1.2 h
Packaging department	0.2 h	0.3 h	0.5 h

(A) How many boats of each type must be produced each week for the plant to operate at full capacity?

(B) How is the production schedule in part A affected if the packaging department is no longer used?

(C) How is the production schedule in part A affected if the four-person boat is no longer produced?

72. Production Scheduling. Repeat Problem 71 assuming the cutting, assembly, and packaging departments have available a maximum of 350, 330, and 115 labor-hours per week, respectively.

73. Nutrition. A dietitian in a hospital is to arrange a special diet using three basic foods. The diet is to include exactly 340 units of calcium, 180 units of iron, and 220 units of vitamin A. The number of units per ounce of each special ingredient for each of the foods is indicated in the table.

	Units Per Ounce		
	Food A	Food B	Food C
Calcium	30	10	20
Iron	10	10	20
Vitamin A	10	30	20

(A) How many ounces of each food must be used to meet the diet requirements?

(B) How is the diet in part A affected if food C is not used?

(C) How is the diet in part A affected if the vitamin A requirement is dropped?

74. Nutrition. Repeat Problem 73 if the diet is to include exactly 400 units of calcium, 160 units of iron, and 240 units of vitamin A.

75. Agriculture. A farmer can buy four types of plant food. Each barrel of mix A contains 30 pounds of phosphoric acid, 50 pounds of nitrogen, and 30 pounds of potash; each barrel of mix B contains 30 pounds of phosphoric acid, 75 pounds of nitrogen, and 20 pounds of potash; each barrel of mix C contains 30 pounds of phosphoric acid, 25 pounds of nitrogen, and 20 pounds of potash; and each barrel of mix D contains 60 pounds of phosphoric acid, 25 pounds of nitrogen, and 50 pounds of potash. Soil tests indicate that a particular field needs 900 pounds of phosphoric acid, 750 pounds of nitrogen, and 700 pounds of potash. How many barrels of each type of food should the farmer mix together to supply the necessary nutrients for the field?

76. Animal Nutrition. In a laboratory experiment, rats are to be fed five packets of food containing a total of 80 units of vitamin E. There are four different brands of food packets that can be used. A packet of brand A contains 5 units of vitamin E, a packet of brand B contains 10 units of vitamin E, a packet of brand C contains 15 units of vitamin E, and a packet of brand D contains 20 units of vitamin E. How many packets of each brand should be mixed and fed to the rats?

77. Sociology. Two sociologists have grant money to study school busing in a particular city. They wish to conduct an opinion survey using 600 telephone contacts and 400 house contacts. Survey company A has personnel to do 30 telephone and 10 house contacts per hour; survey company B can handle 20 telephone and 20 house contacts per hour. How many hours should be scheduled for each firm to produce exactly the number of contacts needed?

78. Sociology. Repeat Problem 77 if 650 telephone contacts and 350 house contacts are needed.

Section 5-4 Matrix Operations

- Addition and Subtraction
- Multiplication of a Matrix by a Number
- Matrix Product

Matrices are both a very ancient and a very current mathematical concept. References to matrices and systems of equations can be found in Chinese manuscripts dating back to around 200 B.C. Over the years, mathematicians and scientists have found many applications of matrices. More recently, the advent of personal and large-scale computers has increased the use of matrices in a wide variety of applications. In 1979 Dan Bricklin and Robert Frankston introduced VisiCalc, the first electronic spreadsheet program for personal computers. Simply put, a *spreadsheet* is a computer program that allows the user to enter and manipulate numbers, often using matrix notation and operations. Spreadsheets were initially used by businesses in areas such as budgeting, sales projections, and cost estimation. However, many other applications have begun to appear. For example, a scientist can use a spreadsheet to analyze the results of an experiment, or a teacher can use one to record and average grades. There are even spreadsheets that can be used to help compute an individual's income tax.

In Section 5-2 we introduced basic matrix terminology and solved systems of equations by performing row operations on augmented coefficient matrices. Matrices have many other useful applications and possess an interesting mathematical structure in their own right. As we will see, matrix addition and multiplication are similar to real number addition and multiplication in many respects, but there are some important differences. To help you understand the similarities and the differences, you should review the basic properties of real number operations discussed in Section A-1.

Addition and Subtraction

Before we can discuss arithmetic operations for matrices, we have to define equality for matrices. Two matrices are **equal** if they have the same size and their corresponding elements are equal. For example,

$$\underset{2 \times 3}{\begin{bmatrix} a & b & c \\ d & e & f \end{bmatrix}} = \underset{2 \times 3}{\begin{bmatrix} u & v & w \\ x & y & z \end{bmatrix}} \quad \text{if and only if} \quad \begin{matrix} a = u & b = v & c = w \\ d = x & e = y & f = z \end{matrix}$$

The **sum of two matrices** of the same size is a matrix with elements that are the sums of the corresponding elements of the two given matrices.

Addition is not defined for matrices of different sizes.

EXAMPLE 1

Matrix Addition

(A) $\begin{bmatrix} a & b \\ c & d \end{bmatrix} + \begin{bmatrix} w & x \\ y & z \end{bmatrix} = \begin{bmatrix} (a + w) & (b + x) \\ (c + y) & (d + z) \end{bmatrix}$

(B) $\begin{bmatrix} 2 & -3 & 0 \\ 1 & 2 & -5 \end{bmatrix} + \begin{bmatrix} 3 & 1 & 2 \\ -3 & 2 & 5 \end{bmatrix} = \begin{bmatrix} 5 & -2 & 2 \\ -2 & 4 & 0 \end{bmatrix}$

MATCHED PROBLEM 1

Add: $\begin{bmatrix} 3 & 2 \\ -1 & -1 \\ 0 & 3 \end{bmatrix} + \begin{bmatrix} -2 & 3 \\ 1 & -1 \\ 2 & -2 \end{bmatrix}$

FIGURE 1

Addition on a graphing utility.

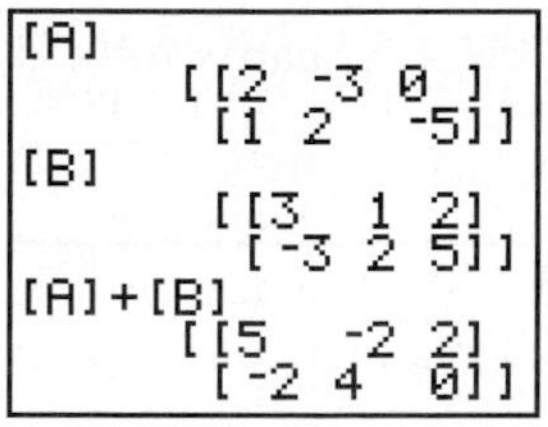

Graphing utilities can also be used to solve problems involving matrix operations. Figure 1 illustrates the solution to Example 1, Part B, on a graphing utility.

Because we add two matrices by adding their corresponding elements, it follows from the properties of real numbers that matrices of the same size are commutative and associative relative to addition. That is, if A, B, and C are matrices of the same size, then

$$A + B = B + A \qquad \text{Commutative}$$

$$(A + B) + C = A + (B + C) \qquad \text{Associative}$$

A matrix with elements that are all 0s is called a **zero matrix.** For example, the following are zero matrices of different sizes:

$$\begin{bmatrix} 0 & 0 & 0 \end{bmatrix} \quad \begin{bmatrix} 0 & 0 \\ 0 & 0 \end{bmatrix} \quad \begin{bmatrix} 0 \\ 0 \\ 0 \\ 0 \end{bmatrix} \quad \begin{bmatrix} 0 & 0 & 0 & 0 \\ 0 & 0 & 0 & 0 \\ 0 & 0 & 0 & 0 \end{bmatrix}$$

[*Note:* "0" can be used to denote the zero matrix of any size.]

The **negative of a matrix *M*,** denoted by $-M$, is a matrix with elements that are the negatives of the elements in M. Thus, if

$$M = \begin{bmatrix} a & b \\ c & d \end{bmatrix}$$

then

$$-M = \begin{bmatrix} -a & -b \\ -c & -d \end{bmatrix}$$

Note that $M + (-M) = 0$ (a zero matrix).

If A and B are matrices of the same size, then we define **subtraction** as follows:

$$A - B = A + (-B)$$

Thus, to subtract matrix B from matrix A, we simply subtract corresponding elements.

EXAMPLE 2

Matrix Subtraction

$$\begin{bmatrix} 3 & -2 \\ 5 & 0 \end{bmatrix} - \begin{bmatrix} -2 & 2 \\ 3 & 4 \end{bmatrix} = \begin{bmatrix} 3 & -2 \\ 5 & 0 \end{bmatrix} + \begin{bmatrix} 2 & -2 \\ -3 & -4 \end{bmatrix} = \begin{bmatrix} 5 & -4 \\ 2 & -4 \end{bmatrix}$$

MATCHED PROBLEM 2

Subtract: $\begin{bmatrix} 2 & -3 & 5 \end{bmatrix} - \begin{bmatrix} 3 & -2 & 1 \end{bmatrix}$

Multiplication of a Matrix by a Number

The **product of a number *k* and a matrix *M*,** denoted by ***kM*,** is a matrix formed by multiplying each element of M by k.

EXAMPLE 3

Multiplication of a Matrix by a Number

$$-2\begin{bmatrix} 3 & -1 & 0 \\ -2 & 1 & 3 \\ 0 & -1 & -2 \end{bmatrix} = \begin{bmatrix} -6 & 2 & 0 \\ 4 & -2 & -6 \\ 0 & 2 & 4 \end{bmatrix}$$

MATCHED PROBLEM 3

Find: $10\begin{bmatrix} 1.3 \\ 0.2 \\ 3.5 \end{bmatrix}$

Explore/Discuss 1

Multiplication of two numbers can be interpreted as repeated addition if one of the numbers is a positive integer. That is,

$$2a = a + a \qquad 3a = a + a + a \qquad 4a = a + a + a + a$$

and so on. Discuss this interpretation for the product of an integer k and a matrix M. Use specific examples to illustrate your remarks.

We now consider an application that uses various matrix operations.

EXAMPLE 4

Sales and Commissions

Ms. Fong and Mr. Petris are salespeople for a new car agency that sells only two models. August was the last month for this year's models, and next year's models were introduced in September. Gross dollar sales for each month are given in the following matrices:

$$\begin{array}{c} \\ \text{Fong} \\ \text{Petris} \end{array} \begin{array}{c} \text{AUGUST SALES} \\ \begin{array}{cc} \text{Compact} & \text{Luxury} \end{array} \\ \begin{bmatrix} \$36{,}000 & \$72{,}000 \\ \$72{,}000 & \$0 \end{bmatrix} \end{array} = A \qquad \begin{array}{c} \text{SEPTEMBER SALES} \\ \begin{array}{cc} \text{Compact} & \text{Luxury} \end{array} \\ \begin{bmatrix} \$144{,}000 & \$288{,}000 \\ \$180{,}000 & \$216{,}000 \end{bmatrix} \end{array} = B$$

For example, Ms. Fong had \$36,000 in compact sales in August and Mr. Petris had \$216,000 in luxury car sales in September.

(A) What are the combined dollar sales in August and September for each salesperson and each model?

(B) What was the increase in dollar sales from August to September?

(C) If both salespeople receive a 3% commission on gross dollar sales, compute the commission for each salesperson for each model sold in September.

Solutions

We use matrix addition for part A, matrix subtraction for part B, and multiplication of a matrix by a number for part C.

$$\text{(A)}\quad A + B = \begin{array}{c} \begin{array}{cc} \text{Compact} & \text{Luxury} \end{array} \\ \begin{bmatrix} \$180{,}000 & \$360{,}000 \\ \$252{,}000 & \$216{,}000 \end{bmatrix} \end{array} \begin{array}{l} \\ \text{Fong} \\ \text{Petris} \end{array}$$

$$\text{(B)}\quad B - A = \begin{bmatrix} \$108{,}000 & \$216{,}000 \\ \$108{,}000 & \$216{,}000 \end{bmatrix} \begin{array}{l} \text{Fong} \\ \text{Petris} \end{array}$$

$$\begin{array}{ccc} & \text{Compact} & \text{Luxury} \end{array}$$

$$\text{(C)}\quad 0.03B = \begin{bmatrix} (0.03)(\$144{,}000) & (0.03)(\$288{,}000) \\ (0.03)(\$180{,}000) & (0.03)(\$216{,}000) \end{bmatrix}$$

$$= \begin{bmatrix} \$4{,}320 & \$8{,}640 \\ \$5{,}400 & \$6{,}480 \end{bmatrix} \begin{matrix} \text{Fong} \\ \text{Petris} \end{matrix}$$

MATCHED PROBLEM 4

Repeat Example 4 with

$$A = \begin{bmatrix} \$72{,}000 & \$72{,}000 \\ \$36{,}000 & \$72{,}000 \end{bmatrix} \quad \text{and} \quad B = \begin{bmatrix} \$180{,}000 & \$216{,}000 \\ \$144{,}000 & \$216{,}000 \end{bmatrix}$$

Example 4 involved an agency with only two salespeople and two models. A more realistic problem might involve 20 salespeople and 15 models. Problems of this size are often solved with the aid of a spreadsheet on a personal computer. Figure 2 illustrates a computer spreadsheet solution for Example 4.

	A	B	C	D	E	F	G
1		Compact	Luxury	Compact	Luxury	Compact	Luxury
2		August Sales		September Sales		September Commissions	
3	Fong	$36,000	$72,000	$144,000	$288,000	$4,320	$8,640
4	Petris	$72,000	$0	$180,000	$216,000	$5,400	$6,480
5		Combined Sales		Sales Increases			
6	Fong	$180,000	$360,000	$108,000	$216,000		
7	Petris	$252,000	$216,000	$108,000	$216,000		

FIGURE 2

Matrix Product

Now we are going to introduce a matrix multiplication that may at first seem rather strange. In spite of its apparent strangeness, this operation is well-founded in the general theory of matrices and, as we will see, is extremely useful in many practical problems.

Historically, matrix multiplication was introduced by the English mathematician Arthur Cayley (1821–1895) in studies of linear equations and linear transformations. In Section 5-6, you will see how matrix multiplication is central to the process of expressing systems of equations as matrix equations and to the process of solving matrix equations. Matrix equations and their solutions provide us with an alternate method of solving linear systems with the same number of variables as equations.

We start by defining the product of two special matrices, a row matrix and a column matrix.

DEFINITION 1

PRODUCT OF A ROW MATRIX AND A COLUMN MATRIX

The **product** of a $1 \times n$ row matrix and an $n \times 1$ column matrix is a 1×1 matrix given by

$$\underset{1 \times n}{[a_1 \; a_2 \; \cdots \; a_n]} \underset{n \times 1}{\begin{bmatrix} b_1 \\ b_2 \\ \vdots \\ b_n \end{bmatrix}} = [a_1b_1 + a_2b_2 + \cdots + a_nb_n]$$

Note that the number of elements in the row matrix and in the column matrix must be the same for the product to be defined.

EXAMPLE 5 Product of a Row Matrix and a Column Matrix

$$[2 \quad -3 \quad 0]\begin{bmatrix} -5 \\ 2 \\ -2 \end{bmatrix} = [(2)(-5) + (-3)(2) + (0)(-2)]$$

$$= [-10 - 6 + 0] = [-16]$$

MATCHED PROBLEM 5

$$[-1 \quad 0 \quad 3 \quad 2]\begin{bmatrix} 2 \\ 3 \\ 4 \\ -1 \end{bmatrix} = ?$$

Refer to Example 5. The distinction between the real number -16 and the 1×1 matrix $[-16]$ is a technical one, and it is common to see 1×1 matrices written as real numbers without brackets. In the work that follows, we will frequently refer to 1×1 matrices as real numbers and omit the brackets whenever it is convenient to do so.

EXAMPLE 6 Production Scheduling

A factory produces a slalom water ski that requires 4 labor-hours in the fabricating department and 1 labor-hour in the finishing department. Fabricating personnel receive \$10 per hour, and finishing personnel receive \$8 per hour. Total labor cost per ski is given by the product

$$[4 \quad 1]\begin{bmatrix} 10 \\ 8 \end{bmatrix} = [(4)(10) + (1)(8)] = [40 + 8] = [48] \text{ or \$48 per ski}$$

MATCHED PROBLEM 6

If the factory in Example 6 also produces a trick water ski that requires 6 labor-hours in the fabricating department and 1.5 labor-hours in the finishing department, write a product between appropriate row and column matrices that gives the total labor cost for this ski. Compute the cost.

We now use the product of a $1 \times n$ row matrix and an $n \times 1$ column matrix to extend the definition of matrix product to more general matrices.

DEFINITION 2

MATRIX PRODUCT

If A is an $m \times p$ matrix and B is a $p \times n$ matrix, then the **matrix product** of A and B, denoted AB, is an $m \times n$ matrix whose element in the ith row and jth column is the real number obtained from the product of the ith row of A and the jth column of B. If the number of columns in A does not equal the number of rows in B, then the matrix product AB is **not defined.**

It is important to check sizes before starting the multiplication process. If A is an $a \times b$ matrix and B is a $c \times d$ matrix, then if $b = c$, the product AB will exist and will be an $a \times d$ matrix (see Fig. 3). If $b \neq c$, then the product AB does not exist.

FIGURE 3

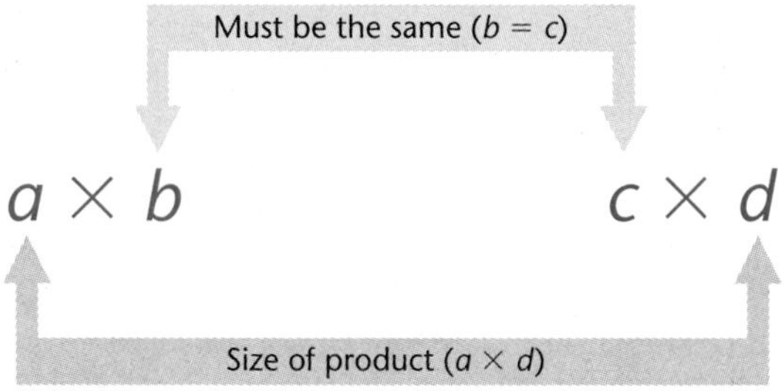

The definition is not as complicated as it might first seem. An example should help clarify the process. For

$$A = \begin{bmatrix} 2 & 3 & -1 \\ -2 & 1 & 2 \end{bmatrix} \quad \text{and} \quad B = \begin{bmatrix} 1 & 3 \\ 2 & 0 \\ -1 & 2 \end{bmatrix}$$

A is 2×3, B is 3×2, and so AB is 2×2. To find the first row of AB, we take the product of the first row of A with every column of B and write each result as a real number, not a 1×1 matrix. The second row of AB is computed in the same manner. The four products of row and column matrices used to produce the four elements in AB are shown in the dashed box at the top of the next page. These products are usually calculated mentally, or with the aid of a calculator, and need not be written out. The shaded portions highlight the steps involved in computing the element in the first row and second column of AB.

$$\overset{2\times 3}{\begin{bmatrix} 2 & 3 & -1 \\ -2 & 1 & 2 \end{bmatrix}} \overset{3\times 2}{\begin{bmatrix} 1 & 3 \\ 2 & 0 \\ -1 & 2 \end{bmatrix}} = \begin{bmatrix} [2 \quad 3 \quad -1]\begin{bmatrix} 1 \\ 2 \\ -1 \end{bmatrix} & [2 \quad 3 \quad -1]\begin{bmatrix} 3 \\ 0 \\ 2 \end{bmatrix} \\ [-2 \quad 1 \quad 2]\begin{bmatrix} 1 \\ 2 \\ -1 \end{bmatrix} & [-2 \quad 1 \quad 2]\begin{bmatrix} 3 \\ 0 \\ 2 \end{bmatrix} \end{bmatrix}$$

$$= \overset{2\times 2}{\begin{bmatrix} 9 & 4 \\ -2 & -2 \end{bmatrix}}$$

EXAMPLE 7

Matrix Product

(A) $$\overset{3\times 2}{\begin{bmatrix} 2 & 1 \\ 1 & 0 \\ -1 & 2 \end{bmatrix}} \overset{2\times 4}{\begin{bmatrix} 1 & -1 & 0 & 1 \\ 2 & 1 & 2 & 0 \end{bmatrix}} = \overset{3\times 4}{\begin{bmatrix} 4 & -1 & 2 & 2 \\ 1 & -1 & 0 & 1 \\ 3 & 3 & 4 & -1 \end{bmatrix}}$$

(B) $$\overset{2\times 4}{\begin{bmatrix} 1 & -1 & 0 & 1 \\ 2 & 1 & 2 & 0 \end{bmatrix}} \overset{3\times 2}{\begin{bmatrix} 2 & 1 \\ 1 & 0 \\ -1 & 2 \end{bmatrix}}$$

Product is not defined

(C) $$\begin{bmatrix} 2 & 6 \\ -1 & -3 \end{bmatrix}\begin{bmatrix} 1 & 2 \\ 3 & 6 \end{bmatrix} = \begin{bmatrix} 20 & 40 \\ -10 & -20 \end{bmatrix}$$

(D) $$\begin{bmatrix} 1 & 2 \\ 3 & 6 \end{bmatrix}\begin{bmatrix} 2 & 6 \\ -1 & -3 \end{bmatrix} = \begin{bmatrix} 0 & 0 \\ 0 & 0 \end{bmatrix}$$

(E) $$[2 \quad -3 \quad 0]\begin{bmatrix} -5 \\ 2 \\ -2 \end{bmatrix} = [-16]$$

(F) $$\begin{bmatrix} -5 \\ 2 \\ -2 \end{bmatrix}[2 \quad -3 \quad 0] = \begin{bmatrix} -10 & 15 & 0 \\ 4 & -6 & 0 \\ -4 & 6 & 0 \end{bmatrix}$$

MATCHED PROBLEM 7

Find each product, if it is defined:

(A) $$\begin{bmatrix} -1 & 0 & 3 & -2 \\ 1 & 2 & 2 & 0 \end{bmatrix}\begin{bmatrix} -1 & 1 \\ 2 & 3 \\ 1 & 0 \end{bmatrix}$$

(B) $$\begin{bmatrix} -1 & 1 \\ 2 & 3 \\ 1 & 0 \end{bmatrix}\begin{bmatrix} -1 & 0 & 3 & -2 \\ 1 & 2 & 2 & 0 \end{bmatrix}$$

(C) $\begin{bmatrix} 1 & 2 \\ -1 & -2 \end{bmatrix}\begin{bmatrix} -2 & 4 \\ 1 & -2 \end{bmatrix}$ (D) $\begin{bmatrix} -2 & 4 \\ 1 & -2 \end{bmatrix}\begin{bmatrix} 1 & 2 \\ -1 & -2 \end{bmatrix}$

(E) $\begin{bmatrix} 3 & -2 & 1 \end{bmatrix}\begin{bmatrix} 4 \\ 2 \\ 3 \end{bmatrix}$ (F) $\begin{bmatrix} 4 \\ 2 \\ 3 \end{bmatrix}\begin{bmatrix} 3 & -2 & 1 \end{bmatrix}$

FIGURE 4
Multiplication on a graphing utility.

```
[A]
        [[2  1]
         [1  0]
         [-1 2]]
[B]
     [[1 -1 0 1]
      [2 1  2 0]]
[A]*[B]
    [[4 -1 2 2 ]
     [1 -1 0 1 ]
     [3 3  4 -1]]
```

Figure 4 illustrates a graphing utility solution to Example 7, part A. What would you expect to happen if you tried to solve Example 7, part B, on a graphing utility?

In the arithmetic of real numbers it does not matter in which order we multiply; for example, $5 \times 7 = 7 \times 5$. In matrix multiplication, however, it does make a difference. That is, AB does not always equal BA, even if both multiplications are defined and both products are the same size (see Example 7, parts C and D). Thus,

Matrix multiplication is not commutative.

Also, AB may be zero with neither A nor B equal to zero (see Example 7, part D). Thus,

The zero property does not hold for matrix multiplication.

(See Section A-1 for a discussion of the zero property for real numbers.)

Just as we used the familiar algebraic notation AB to represent the product of matrices A and B, we use the notation A^2 for AA, the product of A with itself, A^3 for AAA, and so on.

Explore/Discuss 2

In addition to the commutative and zero properties, there are other significant differences between real number multiplication and matrix multiplication.

(A) In real number multiplication, the only real number whose square is 0 is the real number 0 ($0^2 = 0$). Find at least one 2×2 matrix A with all elements nonzero such that $A^2 = 0$, where 0 is the 2×2 zero matrix.

(B) In real number multiplication, the only nonzero real number that is equal to its square is the real number 1 ($1^2 = 1$). Find at least one 2×2 matrix A with all elements nonzero such that $A^2 = A$.

We will continue our discussion of properties of matrix multiplication later in this chapter. Now we consider an application of matrix multiplication.

EXAMPLE 8

Labor Costs

Let us combine the time requirements for slalom and trick water skis discussed in Example 6 and Matched Problem 6 into one matrix:

$$\begin{array}{l} \\ \\ \\ \text{Trick ski} \\ \text{Slalom ski} \end{array} \begin{array}{c} \text{Labor-hours per ski} \\ \begin{array}{cc} \text{Assembly} & \text{Finishing} \\ \text{department} & \text{department} \end{array} \\ \begin{bmatrix} 6\text{ h} & 1.5\text{ h} \\ 4\text{ h} & 1\text{ h} \end{bmatrix} \end{array} = L$$

Now suppose that the company has two manufacturing plants, X and Y, in different parts of the country and that the hourly wages for each department are given in the following matrix:

$$\begin{array}{l} \\ \\ \\ \text{Assembly department} \\ \text{Finishing department} \end{array} \begin{array}{c} \text{Hourly wages} \\ \begin{array}{cc} \text{Plant} & \text{Plant} \\ X & Y \end{array} \\ \begin{bmatrix} \$10 & \$12 \\ \$\ 8 & \$10 \end{bmatrix} \end{array} = H$$

Since H and L are both 2×2 matrices, we can take the product of H and L in either order and the result will be a 2×2 matrix:

$$HL = \begin{bmatrix} 10 & 12 \\ 8 & 10 \end{bmatrix} \begin{bmatrix} 6 & 1.5 \\ 4 & 1 \end{bmatrix} = \begin{bmatrix} 108 & 27 \\ 88 & 22 \end{bmatrix}$$

$$LH = \begin{bmatrix} 6 & 1.5 \\ 4 & 1 \end{bmatrix} \begin{bmatrix} 10 & 12 \\ 8 & 10 \end{bmatrix} = \begin{bmatrix} 72 & 87 \\ 48 & 58 \end{bmatrix}$$

How can we interpret the elements in these products? Let's begin with the product HL. The element 108 in the first row and first column of HL is the product of the first row matrix of H and the first column matrix of L:

$$\begin{array}{cc} \text{Plant} & \text{Plant} \\ X & Y \\ [10 & 12] \end{array} \begin{bmatrix} 6 \\ 4 \end{bmatrix} \begin{array}{l} \text{Trick} \\ \text{Slalom} \end{array} = 10(6) + 12(4) = 60 + 48 = 108$$

Notice that \$60 is the labor cost for assembling a trick ski at plant X and \$48 is the labor cost for assembling a slalom ski at plant Y. Although both numbers represent labor costs, it makes no sense to add them together. They do not pertain to the same type of ski or to the same plant. Thus, even though the product HL happens to be defined mathematically, it has no useful interpretation in this problem.

Now let's consider the product LH. The element 72 in the first row and first column of LH is given by the following product:

$$\begin{array}{cc} \text{Assembly} & \text{Finishing} \\ [6 & 1.5] \end{array} \begin{bmatrix} 10 \\ 8 \end{bmatrix} \begin{array}{l} \text{Assembly} \\ \text{Finishing} \end{array} = 6(10) + 1.5(8)$$

$$= 60 + 12 = 72$$

where \$60 is the labor cost for assembling a trick ski at plant X and \$12 is the labor cost for finishing a trick ski at plant X. Thus, the sum is the total labor cost for producing a trick ski at plant X. The other elements in LH also represent total labor costs, as indicated by the row and column labels shown below:

$$\begin{array}{c} \text{Labor costs per ski} \\ \begin{array}{cc} \text{Plant} & \text{Plant} \\ X & Y \end{array} \end{array}$$

$$LH = \begin{bmatrix} \$72 & \$87 \\ \$48 & \$58 \end{bmatrix} \begin{array}{l} \text{Trick ski} \\ \text{Slalom ski} \end{array}$$

MATCHED PROBLEM 8

Refer to Example 8. The company wants to know how many hours to schedule in each department to produce 1,000 trick skis and 2,000 slalom skis. These production requirements can be represented by either of the following matrices:

$$\begin{array}{cc} \text{Trick} & \text{Slalom} \\ \text{skis} & \text{skis} \end{array}$$

$$P = [1{,}000 \quad 2{,}000] \qquad Q = \begin{bmatrix} 1{,}000 \\ 2{,}000 \end{bmatrix} \begin{array}{l} \text{Trick skis} \\ \text{Slalom skis} \end{array}$$

Using the labor-hour matrix L from Example 8, find PL or LQ, whichever has a meaningful interpretation for this problem, and label the rows and columns accordingly.

Figure 5 shows a solution to Example 8 on a spreadsheet.

	A	B	C	D	E	F
1		Labor-hours per ski			Hourly wages	
2		Fabricating	Finishing		Plant X	Plant Y
3	Trick ski	6	1.5	Fabricating	\$10	\$12
4	Slalom ski	4	1	Finishing	\$8	\$10
5		Labor costs per ski				
6		Plant X	Plant Y			
7	Trick ski	\$72	\$87			
8	Slalom ski	\$48	\$58			

FIGURE 5 Matrix multiplication in a spreadsheet.

CAUTION

Example 8 and Problem 8 illustrate an important point about matrix multiplication. Even if you are using a graphing utility to perform the calculations in a matrix product, it is still necessary for you to know the definition of matrix multiplication so that you can interpret the results correctly.

Answers to Matched Problems

1. $\begin{bmatrix} 1 & 5 \\ 0 & -2 \\ 2 & 1 \end{bmatrix}$ **2.** $[-1 \quad -1 \quad 4]$ **3.** $\begin{bmatrix} 13 \\ 2 \\ 35 \end{bmatrix}$

4. (A) $\begin{bmatrix} \$252{,}000 & \$288{,}000 \\ \$180{,}000 & \$288{,}000 \end{bmatrix}$ (B) $\begin{bmatrix} \$108{,}000 & \$144{,}000 \\ \$108{,}000 & \$144{,}000 \end{bmatrix}$ (C) $\begin{bmatrix} \$5{,}400 & \$6{,}480 \\ \$4{,}320 & \$6{,}480 \end{bmatrix}$

5. $[8]$ **6.** $[6 \quad 1.5]\begin{bmatrix} 10 \\ 8 \end{bmatrix} = [72]$ or \$72

7. (A) Not defined (B) $\begin{bmatrix} 2 & 2 & -1 & 2 \\ 1 & 6 & 12 & -4 \\ -1 & 0 & 3 & -2 \end{bmatrix}$ (C) $\begin{bmatrix} 0 & 0 \\ 0 & 0 \end{bmatrix}$ (D) $\begin{bmatrix} -6 & -12 \\ 3 & 6 \end{bmatrix}$ (E) $[11]$ (F) $\begin{bmatrix} 12 & -8 & 4 \\ 6 & -4 & 2 \\ 9 & -6 & 3 \end{bmatrix}$

8. $PL = \overset{\text{Assembly}}{[14{,}000} \quad \overset{\text{Finishing}}{3{,}500]}$ Labor hours

EXERCISE 5-4

A

Perform the indicated operations in Problems 1–18, if possible.

1. $\begin{bmatrix} -1 & 4 \\ 2 & -6 \end{bmatrix} + \begin{bmatrix} 1 & -2 \\ 0 & 5 \end{bmatrix}$

2. $\begin{bmatrix} 2 & -1 \\ 3 & 0 \end{bmatrix} + \begin{bmatrix} -3 & 1 \\ 2 & -3 \end{bmatrix}$

3. $\begin{bmatrix} -3 & 5 \\ 2 & 0 \\ 1 & 4 \end{bmatrix} + \begin{bmatrix} 2 & 1 \\ -6 & 3 \\ 0 & -5 \end{bmatrix}$

4. $\begin{bmatrix} 4 & -1 & 0 \\ 2 & 1 & 3 \end{bmatrix} + \begin{bmatrix} -2 & 1 & 3 \\ 5 & 6 & -8 \end{bmatrix}$

5. $\begin{bmatrix} -3 & 5 \\ 2 & 0 \\ 1 & 4 \end{bmatrix} + \begin{bmatrix} -2 & 1 & 3 \\ 5 & 6 & -8 \end{bmatrix}$

6. $\begin{bmatrix} 4 & -1 & 0 \\ 2 & 1 & 3 \end{bmatrix} + \begin{bmatrix} 2 & 1 \\ -6 & 3 \\ 0 & -5 \end{bmatrix}$

7. $\begin{bmatrix} 6 & 2 & -3 \\ 0 & -4 & 5 \end{bmatrix} - \begin{bmatrix} 4 & -1 & 2 \\ -5 & 1 & -2 \end{bmatrix}$

8. $\begin{bmatrix} 4 & -5 \\ 1 & 0 \\ 1 & -3 \end{bmatrix} - \begin{bmatrix} -1 & 2 \\ 6 & -2 \\ 1 & -7 \end{bmatrix}$

9. $10\begin{bmatrix} 2 & -1 & 3 \\ 0 & -4 & 5 \end{bmatrix}$

10. $5\begin{bmatrix} 1 & -2 & 0 & 4 \\ -3 & 2 & -1 & 6 \end{bmatrix}$

11. $[2 \quad 4]\begin{bmatrix} 3 \\ 1 \end{bmatrix}$

12. $[1 \quad 5]\begin{bmatrix} 6 \\ 2 \end{bmatrix}$

13. $\begin{bmatrix} 3 & 4 \\ -1 & -2 \end{bmatrix}\begin{bmatrix} -1 \\ 2 \end{bmatrix}$

14. $\begin{bmatrix} -1 & 1 \\ 2 & -3 \end{bmatrix}\begin{bmatrix} 4 \\ -2 \end{bmatrix}$

15. $\begin{bmatrix} 2 & -3 \\ 1 & 2 \end{bmatrix}\begin{bmatrix} 1 & -1 \\ 0 & -2 \end{bmatrix}$

16. $\begin{bmatrix} -3 & 2 \\ 4 & -1 \end{bmatrix}\begin{bmatrix} -2 & 5 \\ -1 & 3 \end{bmatrix}$

17. $\begin{bmatrix} 1 & -1 \\ 0 & -2 \end{bmatrix}\begin{bmatrix} 2 & -3 \\ 1 & 2 \end{bmatrix}$

18. $\begin{bmatrix} -2 & 5 \\ -1 & 3 \end{bmatrix}\begin{bmatrix} -3 & 2 \\ 4 & -1 \end{bmatrix}$

B

Find the products in Problems 19–26.

19. $[4 \quad -2]\begin{bmatrix} -5 \\ -3 \end{bmatrix}$

20. $[2 \quad -1]\begin{bmatrix} 3 \\ -4 \end{bmatrix}$

21. $\begin{bmatrix} -5 \\ -3 \end{bmatrix}[4 \quad -2]$

22. $\begin{bmatrix} 3 \\ -4 \end{bmatrix}[2 \quad -1]$

23. $[3 \quad -2 \quad -4]\begin{bmatrix} 1 \\ 2 \\ -3 \end{bmatrix}$

24. $[1 \quad -2 \quad 2]\begin{bmatrix} 2 \\ -1 \\ 1 \end{bmatrix}$

25. $\begin{bmatrix} 1 \\ 2 \\ -3 \end{bmatrix}[3 \quad -2 \quad -4]$

26. $\begin{bmatrix} 2 \\ -1 \\ 1 \end{bmatrix}[1 \quad -2 \quad 2]$

Problems 27–44 refer to the following matrices.

$$A = \begin{bmatrix} 2 & -1 & 3 \\ 0 & 4 & -2 \end{bmatrix} \qquad B = \begin{bmatrix} -3 & 1 \\ 2 & 5 \end{bmatrix}$$

$$C = \begin{bmatrix} -1 & 0 & 2 \\ 4 & -3 & 1 \\ -2 & 3 & 5 \end{bmatrix} \qquad D = \begin{bmatrix} 3 & -2 \\ 0 & -1 \\ 1 & 2 \end{bmatrix}$$

Perform the indicated operations, if possible.

27. CA **28.** AC **29.** BA

30. AB **31.** C^2 **32.** B^2

33. $C + DA$ **34.** $B + AD$ **35.** $0.2CD$

36. $0.1DB$ **37.** $2DB + 5CD$ **38.** $3BA + 4AC$

39. $(-1)AC + 3DB$ **40.** $(-2)BA + 6CD$

41. CDA **42.** ACD

43. DBA **44.** BAD

In Problems 45 and 46, calculate $B, B^2, B^3, \ldots,$ and $AB, AB^2, AB^3, \ldots$. Describe any patterns you observe in each sequence of matrices.

45. $A = [0.3 \quad 0.7]$ and $B = \begin{bmatrix} 0.4 & 0.6 \\ 0.2 & 0.8 \end{bmatrix}$

46. $A = [0.4 \quad 0.6]$ and $B = \begin{bmatrix} 0.9 & 0.1 \\ 0.3 & 0.7 \end{bmatrix}$

47. Find a, b, c, and d so that

$$\begin{bmatrix} a & b \\ c & d \end{bmatrix} + \begin{bmatrix} 2 & -3 \\ 0 & 1 \end{bmatrix} = \begin{bmatrix} 1 & -2 \\ 3 & -4 \end{bmatrix}$$

48. Find w, x, y, and z so that

$$\begin{bmatrix} 4 & -2 \\ -3 & 0 \end{bmatrix} + \begin{bmatrix} w & x \\ y & z \end{bmatrix} = \begin{bmatrix} 2 & -3 \\ 0 & 5 \end{bmatrix}$$

49. Find x and y so that

$$\begin{bmatrix} 3x & 5 \\ -1 & 4x \end{bmatrix} + \begin{bmatrix} 2y & -3 \\ -6 & -y \end{bmatrix} = \begin{bmatrix} 7 & 2 \\ -7 & 2 \end{bmatrix}$$

50. Find x and y so that

$$\begin{bmatrix} 4 & 2x \\ -4x & -3 \end{bmatrix} + \begin{bmatrix} -5 & -3y \\ 5y & 3 \end{bmatrix} = \begin{bmatrix} -1 & 1 \\ 1 & 0 \end{bmatrix}$$

C

51. Find x and y so that

$$\begin{bmatrix} 1 & 3 \\ -2 & -2 \end{bmatrix} \begin{bmatrix} x & 1 \\ 3 & 2 \end{bmatrix} = \begin{bmatrix} y & 7 \\ y & -6 \end{bmatrix}$$

52. Find x and y so that

$$\begin{bmatrix} x & -1 \\ 1 & 0 \end{bmatrix} \begin{bmatrix} 2 & 1 \\ 4 & 1 \end{bmatrix} = \begin{bmatrix} y & y \\ 2 & 1 \end{bmatrix}$$

53. Find a, b, c, and d so that

$$\begin{bmatrix} 1 & 3 \\ 1 & 4 \end{bmatrix} \begin{bmatrix} a & b \\ c & d \end{bmatrix} = \begin{bmatrix} 6 & -5 \\ 7 & -7 \end{bmatrix}$$

54. Find a, b, c, and d so that

$$\begin{bmatrix} 1 & -2 \\ 2 & -3 \end{bmatrix} \begin{bmatrix} a & b \\ c & d \end{bmatrix} = \begin{bmatrix} 1 & 0 \\ 3 & 2 \end{bmatrix}$$

55. A square matrix is a **diagonal matrix** if all elements not on the principal diagonal are zero. Thus, a 2×2 diagonal matrix has the form

$$A = \begin{bmatrix} a & 0 \\ 0 & d \end{bmatrix}$$

where a and d are any real numbers. Discuss the validity of each of the following statements. If the statement is always true, explain why. If not, give examples.

(A) If A and B are 2×2 diagonal matrices, then $A + B$ is a 2×2 diagonal matrix.

(B) If A and B are 2×2 diagonal matrices, then $A + B = B + A$.

(C) If A and B are 2×2 diagonal matrices, then AB is a 2×2 diagonal matrix.

(D) If A and B are 2×2 diagonal matrices, then $AB = BA$.

56. A square matrix is an **upper triangular matrix** if all elements below the principal diagonal are zero. Thus, a 2×2 upper triangular matrix has the form

$$A = \begin{bmatrix} a & b \\ 0 & d \end{bmatrix}$$

where a, b, and d are any real numbers. Discuss the validity of each of the following statements. If the statement is always true, explain why. If not, give examples.

(A) If A and B are 2×2 upper triangular matrices, then $A + B$ is a 2×2 upper triangular matrix.

(B) If A and B are 2×2 upper triangular matrices, then $A + B = B + A$.

(C) If A and B are 2×2 upper triangular matrices, then AB is a 2×2 upper triangular matrix.

(D) If A and B are 2×2 upper triangular matrices, then $AB = BA$.

APPLICATIONS

57. **Cost Analysis.** A company with two different plants manufactures guitars and banjos. Its production costs for each instrument are given in the following matrices:

	Plant X Guitar	Plant X Banjo		Plant Y Guitar	Plant Y Banjo	
Materials	\$30	\$25	$= A$	\$36	\$27	$= B$
Labor	\$60	\$80		\$54	\$74	

Find $\frac{1}{2}(A + B)$, the average cost of production for the two plants.

58. **Cost Analysis.** If both labor and materials at plant X in Problem 57 are increased 20%, find $\frac{1}{2}(1.2A + B)$, the new average cost of production for the two plants.

59. **Markup.** An import car dealer sells three models of a car. Current dealer invoice price (cost) and the retail price for the basic models and the indicated options are given in the following two matrices (where "Air" means air conditioning):

Dealer invoice price

	Basic car	Air	AM/FM radio	Cruise control	
Model A	\$10,400	\$682	\$215	\$182	
Model B	\$12,500	\$721	\$295	\$182	$= M$
Model C	\$16,400	\$827	\$443	\$192	

Retail price

	Basic car	Air	AM/FM radio	Cruise control	
Model A	\$13,900	\$783	\$263	\$215	
Model B	\$15,000	\$838	\$395	\$236	$= N$
Model C	\$18,300	\$967	\$573	\$248	

We define the markup matrix to be $N - M$ (**markup** is the difference between the retail price and the dealer invoice price). Suppose the value of the dollar has had a sharp decline and the dealer invoice price is to have an across-the-board 15% increase next year. To stay competitive with domestic cars, the dealer increases the retail prices only 10%. Calculate a markup matrix for next year's models and the indicated options. (Compute results to the nearest dollar.)

60. Markup. Referring to Problem 59, what is the markup matrix resulting from a 20% increase in dealer invoice prices and an increase in retail prices of 15%? (Compute results to the nearest dollar.)

61. Labor Costs. A company with manufacturing plants located in different parts of the country has labor-hour and wage requirements for the manufacturing of three types of inflatable boats as given in the following two matrices:

Labor-hours per boat

$$M = \begin{array}{c} \text{Cutting department} \quad \text{Assembly department} \quad \text{Packaging department} \\ \begin{bmatrix} 0.6\text{ h} & 0.6\text{ h} & 0.2\text{ h} \\ 1.0\text{ h} & 0.9\text{ h} & 0.3\text{ h} \\ 1.5\text{ h} & 1.2\text{ h} & 0.4\text{ h} \end{bmatrix} \end{array} \begin{array}{l} \\ \text{One-person boat} \\ \text{Two-person boat} \\ \text{Four-person boat} \end{array}$$

Hourly wages

$$N = \begin{array}{c} \text{Plant I} \quad \text{Plant II} \\ \begin{bmatrix} \$8 & \$9 \\ \$10 & \$12 \\ \$5 & \$6 \end{bmatrix} \end{array} \begin{array}{l} \\ \text{Cutting department} \\ \text{Assembly department} \\ \text{Packaging department} \end{array}$$

(A) Find the labor costs for a one-person boat manufactured at plant I.

(B) Find the labor costs for a four-person boat manufactured at plant II.

(C) Discuss possible interpretations of the elements in the matrix products MN and NM.

(D) If either of the products MN or NM has a meaningful interpretation, find the product and label its rows and columns.

62. Inventory Value. A personal computer retail company sells five different computer models through three stores located in a large metropolitan area. The inventory of each model on hand in each store is summarized in matrix M. Wholesale (W) and retail (R) values of each model computer are summarized in matrix N.

Model

$$M = \begin{array}{c} A \quad B \quad C \quad D \quad E \\ \begin{bmatrix} 4 & 2 & 3 & 7 & 1 \\ 2 & 3 & 5 & 0 & 6 \\ 10 & 4 & 3 & 4 & 3 \end{bmatrix} \end{array} \begin{array}{l} \\ \text{Store 1} \\ \text{Store 2} \\ \text{Store 3} \end{array}$$

$$N = \begin{array}{c} W \qquad R \\ \begin{bmatrix} \$700 & \$840 \\ \$1{,}400 & \$1{,}800 \\ \$1{,}800 & \$2{,}400 \\ \$2{,}700 & \$3{,}300 \\ \$3{,}500 & \$4{,}900 \end{bmatrix} \end{array} \begin{array}{l} \\ A \\ B \\ C \\ D \\ E \end{array}$$

(A) What is the retail value of the inventory at store 2?

(B) What is the wholesale value of the inventory at store 3?

(C) Discuss possible interpretations of the elements in the matrix products MN and NM.

(D) If either of the products MN or NM has a meaningful interpretation, find the product and label its rows and columns.

(E) Discuss methods of matrix multiplication that can be used to find the total inventory of each model on hand at all three stores. State the matrices that can be used, and perform the necessary operations.

(F) Discuss methods of matrix multiplication that can be used to find the total inventory of all five models at each store. State the matrices that can be used, and perform the necessary operations.

63. Airfreight. A nationwide airfreight service has connecting flights between five cities, as illustrated in the figure. To represent this schedule in matrix form, we construct a 5×5 **incidence matrix** A, where the rows represent the origins of each flight and the columns represent the destinations. We place a 1 in the ith row and jth column of this matrix if there is a connecting flight from the ith city to the jth city and a 0 otherwise. We also place 0s on the principal diagonal, because a connecting flight with the same origin and destination does not make sense.

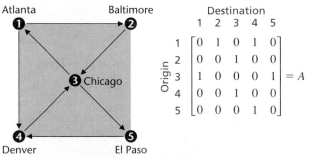

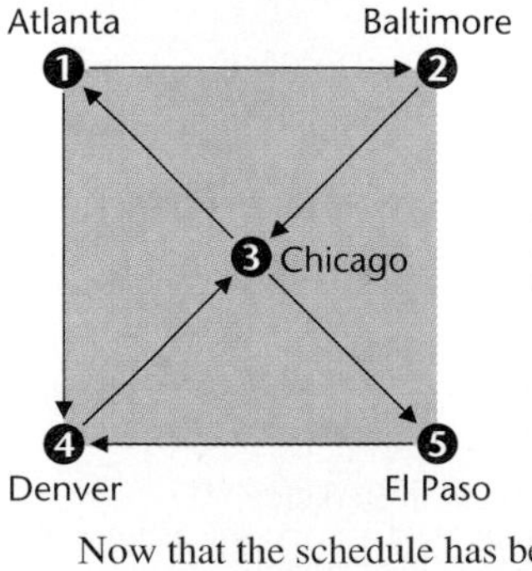

Destination

$$\text{Origin} \begin{array}{c} \\ 1 \\ 2 \\ 3 \\ 4 \\ 5 \end{array} \begin{array}{c} 1 \quad 2 \quad 3 \quad 4 \quad 5 \\ \begin{bmatrix} 0 & 1 & 0 & 1 & 0 \\ 0 & 0 & 1 & 0 & 0 \\ 1 & 0 & 0 & 0 & 1 \\ 0 & 0 & 1 & 0 & 0 \\ 0 & 0 & 0 & 1 & 0 \end{bmatrix} \end{array} = A$$

Now that the schedule has been represented in the mathematical form of a matrix, we can perform operations on this matrix to obtain information about the schedule.

(A) Find A^2. What does the 1 in row 2 and column 1 of A^2 indicate about the schedule? What does the 2 in row 1 and column 3 indicate about the schedule? In general, how would you interpret each element off the principal diagonal of A^2? [*Hint:* Examine the diagram for possible connections between the ith city and the jth city.]

(B) Find A^3. What does the 1 in row 4 and column 2 of A^3 indicate about the schedule? What does the 2 in row 1 and column 5 indicate about the schedule? In general, how would you interpret each element off the principal diagonal of A^3?

(C) Compute $A, A + A^2, A + A^2 + A^3, \ldots,$ until you obtain a matrix with no zero elements (except possibly on the principal diagonal), and interpret.

64. Airfreight. Find the incidence matrix A for the flight schedule illustrated in the figure. Compute $A, A + A^2,$ $A + A^2 + A^3, \ldots,$ until you obtain a matrix with no zero elements (except possibly on the principal diagonal), and interpret.

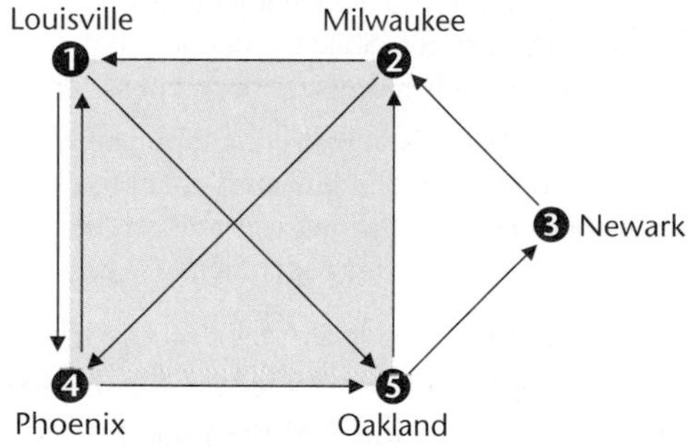

65. Politics. In a local election, a group hired a public relations firm to promote its candidate in three ways: telephone, house calls, and letters. The cost per contact is given in matrix M:

$$M = \begin{bmatrix} \$0.80 \\ \$1.50 \\ \$0.40 \end{bmatrix} \begin{matrix} \text{Telephone} \\ \text{House call} \\ \text{Letter} \end{matrix} \quad (\text{Cost per contact})$$

The number of contacts of each type made in two adjacent cities is given in matrix N:

$$\begin{matrix} & \text{Telephone} & \text{House call} & \text{Letter} \end{matrix}$$
$$N = \begin{bmatrix} 1{,}000 & 500 & 5{,}000 \\ 2{,}000 & 800 & 8{,}000 \end{bmatrix} \begin{matrix} \text{Berkeley} \\ \text{Oakland} \end{matrix}$$

(A) Find the total amount spent in Berkeley.

(B) Find the total amount spent in Oakland.

(C) Discuss possible interpretations of the elements in the matrix products MN and NM.

(D) If either of the products MN or NM has a meaningful interpretation, find the product and label its rows and columns.

(E) Discuss methods of matrix multiplication that can be used to find the total number of telephone calls, house calls, and letters. State the matrices that can be used, and perform the necessary operations.

(F) Discuss methods of matrix multiplication that can be used to find the total number of contacts in Berkeley and in Oakland. State the matrices that can be used, and perform the necessary operations.

66. Nutrition. A nutritionist for a cereal company blends two cereals in different mixes. The amounts of protein, carbohydrate, and fat (in grams per ounce) in each cereal are given by matrix M. The amounts of each cereal used in the three mixes are given by matrix N.

$$\begin{matrix} & \text{Cereal } A & \text{Cereal } B \end{matrix}$$
$$M = \begin{bmatrix} 4 \text{ g/oz} & 2 \text{ g/oz} \\ 20 \text{ g/oz} & 16 \text{ g/oz} \\ 3 \text{ g/oz} & 1 \text{ g/oz} \end{bmatrix} \begin{matrix} \text{Protein} \\ \text{Carbohydrate} \\ \text{Fat} \end{matrix}$$

$$\begin{matrix} & \text{Mix } X & \text{Mix } Y & \text{Mix } Z \end{matrix}$$
$$N = \begin{bmatrix} 15 \text{ oz} & 10 \text{ oz} & 5 \text{ oz} \\ 5 \text{ oz} & 10 \text{ oz} & 15 \text{ oz} \end{bmatrix} \begin{matrix} \text{Cereal } A \\ \text{Cereal } B \end{matrix}$$

(A) Find the amount of protein in mix X.

(B) Find the amount of fat in mix Z.

(C) Discuss possible interpretations of the elements in the matrix products MN and NM.

(D) If either of the products MN or NM has a meaningful interpretation, find the product and label its rows and columns.

67. Dominance Relation. To rank players for an upcoming tennis tournament, a club decides to have each player play one set with every other player. The results are given in the table.

Player	Defeated
1. Aaron	Charles, Dan, Elvis
2. Bart	Aaron, Dan, Elvis
3. Charles	Bart, Dan
4. Dan	Frank
5. Elvis	Charles, Dan, Frank
6. Frank	Aaron, Bart, Charles

(A) Express the outcomes as an incidence matrix A by placing a 1 in the ith row and jth column of A if player i defeated player j and a 0 otherwise (see Problem 63).

(B) Compute the matrix $B = A + A^2$.

(C) Discuss matrix multiplication methods that can be used to find the sum of the rows in B. State the matrices that can be used and perform the necessary operations.

(D) Rank the players from strongest to weakest. Explain the reasoning behind your ranking.

68. Dominance Relation. Each member of a chess team plays one match with every other player. The results are given in the table.

Player	Defeated
1. Anne	Diane
2. Bridget	Anne, Carol, Diane
3. Carol	Anne
4. Diane	Carol, Erlene
5. Erlene	Anne, Bridget, Carol

(A) Express the outcomes as an incidence matrix A by placing a 1 in the ith row and jth column of A if player i defeated player j and a 0 otherwise (see Problem 63).

(B) Compute the matrix $B = A + A^2$.

(C) Discuss matrix multiplication methods that can be used to find the sum of the rows in B. State the matrices that can be used and perform the necessary operations.

(D) Rank the players from strongest to weakest. Explain the reasoning behind your ranking.

Section 5-5 | Inverse of a Square Matrix

- Identity Matrix for Multiplication
- Inverse of a Square Matrix
- Application: Cryptography

In this section we introduce the identity matrix and the inverse of a square matrix. These matrix forms, along with matrix multiplication, are then used to solve some systems of equations written in matrix form in Section 5-6.

Identity Matrix for Multiplication

We know that for any real number a

$$(1)a = a(1) = a$$

The number 1 is called the *identity* for real number multiplication. Does the set of all matrices of a given dimension have an identity element for multiplication? That is, if M is an arbitrary $m \times n$ matrix, does M have an identity element I such that $IM = MI = M$? The answer in general is no. However, the set of all **square matrices of order n** (matrices with n rows and n columns) does have an identity.

DEFINITION 1

IDENTITY MATRIX

The **identity matrix for multiplication** for the set of all square matrices of order n is the square matrix of order n, denoted by I, with 1s along the principal diagonal (from upper left corner to lower right corner) and 0s elsewhere.

For example,

$$\begin{bmatrix} 1 & 0 \\ 0 & 1 \end{bmatrix} \quad \text{and} \quad \begin{bmatrix} 1 & 0 & 0 \\ 0 & 1 & 0 \\ 0 & 0 & 1 \end{bmatrix}$$

are the identity matrices for all square matrices of order 2 and 3, respectively.

Most graphing utilities have a built-in command for generating the identity matrix of a given order (see Fig. 1).

FIGURE 1
Identity matrices.

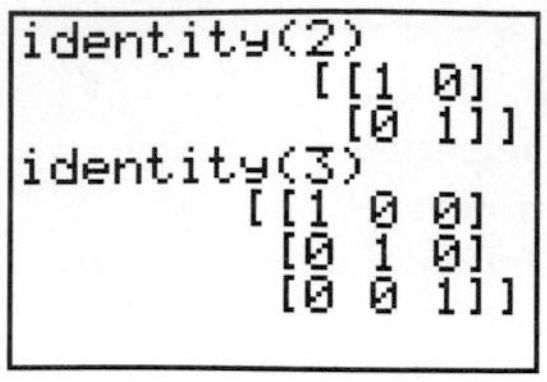

EXAMPLE 1

Identity Matrix Multiplication

(A) $\begin{bmatrix} 1 & 0 & 0 \\ 0 & 1 & 0 \\ 0 & 0 & 1 \end{bmatrix}\begin{bmatrix} a & b & c \\ d & e & f \\ g & h & i \end{bmatrix} = \begin{bmatrix} a & b & c \\ d & e & f \\ g & h & i \end{bmatrix}$

(B) $\begin{bmatrix} a & b & c \\ d & e & f \\ g & h & i \end{bmatrix}\begin{bmatrix} 1 & 0 & 0 \\ 0 & 1 & 0 \\ 0 & 0 & 1 \end{bmatrix} = \begin{bmatrix} a & b & c \\ d & e & f \\ g & h & i \end{bmatrix}$

(C) $\begin{bmatrix} 1 & 0 \\ 0 & 1 \end{bmatrix}\begin{bmatrix} a & b & c \\ d & e & f \end{bmatrix} = \begin{bmatrix} a & b & c \\ d & e & f \end{bmatrix}$

(D) $\begin{bmatrix} a & b & c \\ d & e & f \end{bmatrix}\begin{bmatrix} 1 & 0 & 0 \\ 0 & 1 & 0 \\ 0 & 0 & 1 \end{bmatrix} = \begin{bmatrix} a & b & c \\ d & e & f \end{bmatrix}$

MATCHED PROBLEM 1

Multiply:

(A) $\begin{bmatrix} 1 & 0 \\ 0 & 1 \end{bmatrix}\begin{bmatrix} 3 & -5 \\ 4 & 6 \end{bmatrix}$ and $\begin{bmatrix} 3 & -5 \\ 4 & 6 \end{bmatrix}\begin{bmatrix} 1 & 0 \\ 0 & 1 \end{bmatrix}$

(B) $\begin{bmatrix} 1 & 0 & 0 \\ 0 & 1 & 0 \\ 0 & 0 & 1 \end{bmatrix}\begin{bmatrix} 5 & -7 \\ 2 & 4 \\ 6 & -8 \end{bmatrix}$ and $\begin{bmatrix} 5 & -7 \\ 2 & 4 \\ 6 & -8 \end{bmatrix}\begin{bmatrix} 1 & 0 \\ 0 & 1 \end{bmatrix}$

In general, we can show that if M is a square matrix of order n and I is the identity matrix of order n, then

$$IM = MI = M$$

If M is an $m \times n$ matrix that is not square ($m \neq n$), then it is still possible to multiply M on the left and on the right by an identity matrix, but not with the same-size identity matrix (see Example 1, parts C and D). To avoid the complications involved with associating two different identity matrices with each non-square matrix, we restrict our attention in this section to square matrices.

Explore/Discuss 1

The only real number solutions to the equation $x^2 = 1$ are $x = 1$ and $x = -1$.

(A) Show that $A = \begin{bmatrix} 0 & 1 \\ 1 & 0 \end{bmatrix}$ satisfies $A^2 = I$, where I is the 2×2 identity.

(B) Show that $B = \begin{bmatrix} 0 & -1 \\ -1 & 0 \end{bmatrix}$ satisfies $B^2 = I$.

(C) Find a 2×2 matrix with all elements nonzero whose square is the 2×2 identity matrix.

Inverse of a Square Matrix

In the set of real numbers, we know that for each real number a, except 0, there exists a real number a^{-1} such that

$$a^{-1}a = 1$$

The number a^{-1} is called the *inverse* of the number a relative to multiplication, or the *multiplicative inverse* of a. For example, 2^{-1} is the multiplicative inverse of 2, since $2^{-1}(2) = 1$. We use this idea to define the *inverse of a square matrix.*

DEFINITION 2

INVERSE OF A SQUARE MATRIX

If M is a square matrix of order n and if there exists a matrix M^{-1} (read "M inverse") such that

$$M^{-1}M = MM^{-1} = I$$

then M^{-1} is called the **multiplicative inverse of *M*** or, more simply, the **inverse of *M*.**

The multiplicative inverse of a nonzero real number a also can be written as $1/a$. This notation is not used for matrix inverses.

Let's use Definition 2 to find M^{-1}, if it exists, for

$$M = \begin{bmatrix} 2 & 3 \\ 1 & 2 \end{bmatrix}$$

We are looking for

$$M^{-1} = \begin{bmatrix} a & c \\ b & d \end{bmatrix}$$

such that

$$MM^{-1} = M^{-1}M = I$$

Thus, we write

$$\overset{M}{\begin{bmatrix} 2 & 3 \\ 1 & 2 \end{bmatrix}} \overset{M^{-1}}{\begin{bmatrix} a & c \\ b & d \end{bmatrix}} = \overset{I}{\begin{bmatrix} 1 & 0 \\ 0 & 1 \end{bmatrix}}$$

and try to find a, b, c, and d so that the product of M and M^{-1} is the identity matrix I. Multiplying M and M^{-1} on the left side, we obtain

$$\begin{bmatrix} (2a + 3b) & (2c + 3d) \\ (a + 2b) & (c + 2d) \end{bmatrix} = \begin{bmatrix} 1 & 0 \\ 0 & 1 \end{bmatrix}$$

which is true only if

$$\begin{aligned} 2a + 3b &= 1 \qquad & 2c + 3d &= 0 \\ a + 2b &= 0 \qquad & c + 2d &= 1 \end{aligned}$$

Solving these two systems, we find that $a = 2$, $b = -1$, $c = -3$, and $d = 2$. Thus,

$$M^{-1} = \begin{bmatrix} 2 & -3 \\ -1 & 2 \end{bmatrix}$$

as is easily checked:

$$\overset{M}{\begin{bmatrix} 2 & 3 \\ 1 & 2 \end{bmatrix}} \overset{M^{-1}}{\begin{bmatrix} 2 & -3 \\ -1 & 2 \end{bmatrix}} = \overset{I}{\begin{bmatrix} 1 & 0 \\ 0 & 1 \end{bmatrix}} = \overset{M^{-1}}{\begin{bmatrix} 2 & -3 \\ -1 & 2 \end{bmatrix}} \overset{M}{\begin{bmatrix} 2 & 3 \\ 1 & 2 \end{bmatrix}}$$

Unlike nonzero real numbers, inverses do not always exist for nonzero square matrices. For example, if

$$N = \begin{bmatrix} 2 & 1 \\ 4 & 2 \end{bmatrix}$$

then, proceeding as before, we are led to the systems

$$\begin{aligned} 2a + b &= 1 \qquad & 2c + d &= 0 \\ 4a + 2b &= 0 \qquad & 4c + 2d &= 1 \end{aligned}$$

These systems are both inconsistent and have no solution. Hence, N^{-1} does not exist.

Being able to find inverses, when they exist, leads to direct and simple solutions to many practical problems. In the next section, for example, we will show how inverses can be used to solve systems of linear equations.

The method outlined above for finding the inverse, if it exists, gets very involved for matrices of order larger than 2. Now that we know what we are looking for, we can use augmented matrices, as in Section 5-3, to make the process more efficient. Details are illustrated in Example 2.

EXAMPLE
2

Finding an Inverse

Find the inverse, if it exists, of

$$M = \begin{bmatrix} 1 & -1 & 1 \\ 0 & 2 & -1 \\ 2 & 3 & 0 \end{bmatrix}$$

Solution

We start as before and write

$$\overset{M}{\begin{bmatrix} 1 & -1 & 1 \\ 0 & 2 & -1 \\ 2 & 3 & 0 \end{bmatrix}} \overset{M^{-1}}{\begin{bmatrix} a & d & g \\ b & e & h \\ c & f & i \end{bmatrix}} = \overset{I}{\begin{bmatrix} 1 & 0 & 0 \\ 0 & 1 & 0 \\ 0 & 0 & 1 \end{bmatrix}}$$

This is true only if

$$\begin{array}{lll} a - b + c = 1 & d - e + f = 0 & g - h + i = 0 \\ 2b - c = 0 & 2e - f = 1 & 2h - i = 0 \\ 2a + 3b = 0 & 2d + 3e = 0 & 2g + 3h = 1 \end{array}$$

Now we write augmented matrices for each of the three systems:

First

$$\begin{bmatrix} 1 & -1 & 1 & | & 1 \\ 0 & 2 & -1 & | & 0 \\ 2 & 3 & 0 & | & 0 \end{bmatrix}$$

Second

$$\begin{bmatrix} 1 & -1 & 1 & | & 0 \\ 0 & 2 & -1 & | & 1 \\ 2 & 3 & 0 & | & 0 \end{bmatrix}$$

Third

$$\begin{bmatrix} 1 & -1 & 1 & | & 0 \\ 0 & 2 & -1 & | & 0 \\ 2 & 3 & 0 & | & 1 \end{bmatrix}$$

Since each matrix to the left of the vertical bar is the same, exactly the same row operations can be used on each augmented matrix to transform it into a reduced form. We can speed up the process substantially by combining all three augmented matrices into the single augmented matrix form

$$\begin{bmatrix} 1 & -1 & 1 & | & 1 & 0 & 0 \\ 0 & 2 & -1 & | & 0 & 1 & 0 \\ 2 & 3 & 0 & | & 0 & 0 & 1 \end{bmatrix} = [M \mid I] \qquad (1)$$

We now try to perform row operations on matrix (1) until we obtain a row-equivalent matrix that looks like matrix (2):

$$\begin{bmatrix} \overset{I}{1} & 0 & 0 & | & \overset{B}{a} & d & g \\ 0 & 1 & 0 & | & b & e & h \\ 0 & 0 & 1 & | & c & f & i \end{bmatrix} = [I \mid B] \qquad (2)$$

If this can be done, then the new matrix to the right of the vertical bar is M^{-1}! Now let's try to transform matrix (1) into a form like that of matrix (2). We follow the same sequence of steps as in the solution of linear systems by Gauss–Jordan elimination (see Section 5-3):

$$\begin{array}{c} \quad M \qquad\qquad I \\ \left[\begin{array}{ccc|ccc} 1 & -1 & 1 & 1 & 0 & 0 \\ 0 & 2 & -1 & 0 & 1 & 0 \\ 2 & 3 & 0 & 0 & 0 & 1 \end{array}\right] \end{array} \quad (-2)R_1 + R_3 \to R_3$$

$$\sim \left[\begin{array}{ccc|ccc} 1 & -1 & 1 & 1 & 0 & 0 \\ 0 & 2 & -1 & 0 & 1 & 0 \\ 0 & 5 & -2 & -2 & 0 & 1 \end{array}\right] \quad \tfrac{1}{2}R_2 \to R_2$$

$$\sim \left[\begin{array}{ccc|ccc} 1 & -1 & 1 & 1 & 0 & 0 \\ 0 & 1 & -\frac{1}{2} & 0 & \frac{1}{2} & 0 \\ 0 & 5 & -2 & -2 & 0 & 1 \end{array}\right] \quad \begin{array}{l} R_2 + R_1 \to R_1 \\ \\ (-5)R_2 + R_3 \to R_3 \end{array}$$

$$\sim \left[\begin{array}{ccc|ccc} 1 & 0 & \frac{1}{2} & 1 & \frac{1}{2} & 0 \\ 0 & 1 & -\frac{1}{2} & 0 & \frac{1}{2} & 0 \\ 0 & 0 & \frac{1}{2} & -2 & -\frac{5}{2} & 1 \end{array}\right] \quad 2R_3 \to R_3$$

$$\sim \left[\begin{array}{ccc|ccc} 1 & 0 & \frac{1}{2} & 1 & \frac{1}{2} & 0 \\ 0 & 1 & -\frac{1}{2} & 0 & \frac{1}{2} & 0 \\ 0 & 0 & 1 & -4 & -5 & 2 \end{array}\right] \quad \begin{array}{l} (-\frac{1}{2})R_3 + R_1 \to R_1 \\ \frac{1}{2}R_3 + R_2 \to R_2 \end{array}$$

$$\sim \left[\begin{array}{ccc|ccc} 1 & 0 & 0 & 3 & 3 & -1 \\ 0 & 1 & 0 & -2 & -2 & 1 \\ 0 & 0 & 1 & -4 & -5 & 2 \end{array}\right] = [I \mid B]$$

Converting back to systems of equations equivalent to our three original systems (we won't have to do this step in practice), we have

$$\begin{array}{lll} a = 3 & d = 3 & g = -1 \\ b = -2 & e = -2 & h = 1 \\ c = -4 & f = -5 & i = 2 \end{array}$$

And these are just the elements of M^{-1} that we are looking for! Hence,

$$M^{-1} = \begin{bmatrix} 3 & 3 & -1 \\ -2 & -2 & 1 \\ -4 & -5 & 2 \end{bmatrix}$$

Note that this is the matrix to the right of the vertical line in the last augmented matrix.

Check Since the definition of matrix inverse requires that

$$M^{-1}M = I \quad \text{and} \quad MM^{-1} = I \tag{3}$$

it appears that we must compute both $M^{-1}M$ and MM^{-1} to check our work. However, it can be shown that if one of the equations in (3) is satisfied, then the other is also satisfied. Thus, for checking purposes it is sufficient to compute either $M^{-1}M$ or MM^{-1}—we don't need to do both.

$$M^{-1}M = \begin{bmatrix} 3 & 3 & -1 \\ -2 & -2 & 1 \\ -4 & -5 & 2 \end{bmatrix} \begin{bmatrix} 1 & -1 & 1 \\ 0 & 2 & -1 \\ 2 & 3 & 0 \end{bmatrix} = \begin{bmatrix} 1 & 0 & 0 \\ 0 & 1 & 0 \\ 0 & 0 & 1 \end{bmatrix} = I$$

MATCHED PROBLEM 2

Let $M = \begin{bmatrix} 3 & -1 & 1 \\ -1 & 1 & 0 \\ 1 & 0 & 1 \end{bmatrix}$

(A) Form the augmented matrix $[M \mid I]$.

(B) Use row operations to transform $[M \mid I]$ into $[I \mid B]$.

(C) Verify by multiplication that $B = M^{-1}$.

The procedure used in Example 2 can be used to find the inverse of any square matrix, if the inverse exists, and will also indicate when the inverse does not exist. These ideas are summarized in Theorem 1.

THEOREM 1

INVERSE OF A SQUARE MATRIX M

If $[M \mid I]$ is transformed by row operations into $[I \mid B]$, then the resulting matrix B is M^{-1}. If, however, we obtain all 0s in one or more rows to the left of the vertical line, then M^{-1} does not exist.

Explore/Discuss 2

(A) Suppose that the square matrix M has a row of all zeros. Explain why M has no inverse.

(B) Suppose that the square matrix M has a column of all zeros. Explain why M has no inverse.

EXAMPLE 3

Finding a Matrix Inverse

Find M^{-1}, given $M = \begin{bmatrix} 4 & -1 \\ -6 & 2 \end{bmatrix}$

Solution

$$\left[\begin{array}{cc|cc} 4 & -1 & 1 & 0 \\ -6 & 2 & 0 & 1 \end{array}\right] \quad \tfrac{1}{4}R_1 \to R_1$$

$$\sim \left[\begin{array}{cc|cc} 1 & -\frac{1}{4} & \frac{1}{4} & 0 \\ -6 & 2 & 0 & 1 \end{array}\right] \quad 6R_1 + R_2 \to R_2$$

$$\sim \left[\begin{array}{cc|cc} 1 & -\frac{1}{4} & \frac{1}{4} & 0 \\ 0 & \frac{1}{2} & \frac{3}{2} & 1 \end{array}\right] \quad 2R_2 \to R_2$$

$$\sim \left[\begin{array}{cc|cc} 1 & -\frac{1}{4} & \frac{1}{4} & 0 \\ 0 & 1 & 3 & 2 \end{array}\right] \quad \tfrac{1}{4}R_2 + R_1 \to R_1$$

$$\sim \left[\begin{array}{cc|cc} 1 & 0 & 1 & \frac{1}{2} \\ 0 & 1 & 3 & 2 \end{array}\right]$$

Thus,

$$M^{-1} = \begin{bmatrix} 1 & \frac{1}{2} \\ 3 & 2 \end{bmatrix} \qquad \text{Check by showing } M^{-1}M = I.$$

MATCHED PROBLEM 3

Find M^{-1}, given $M = \begin{bmatrix} 2 & -6 \\ 1 & -2 \end{bmatrix}$

EXAMPLE 4

Finding an Inverse

Find M^{-1}, if it exists, given $M = \begin{bmatrix} 10 & -2 \\ -5 & 1 \end{bmatrix}$

Solution

$$\left[\begin{array}{cc|cc} 10 & -2 & 1 & 0 \\ -5 & 1 & 0 & 1 \end{array}\right] \sim \left[\begin{array}{cc|cc} 1 & -\frac{1}{5} & \frac{1}{10} & 0 \\ -5 & 1 & 0 & 1 \end{array}\right]$$

$$\sim \left[\begin{array}{cc|cc} 1 & -\frac{1}{5} & \frac{1}{10} & 0 \\ 0 & 0 & \frac{1}{2} & 1 \end{array}\right]$$

We have all 0s in the second row to the left of the vertical line. Therefore, M^{-1} does not exist.

MATCHED PROBLEM 4

Find M^{-1}, if it exists, given $M = \begin{bmatrix} 6 & -3 \\ -2 & 1 \end{bmatrix}$

Most graphing utilities can compute matrix inverses and can identify those matrices that do not have inverses. A matrix that does not have an inverse is often referred to as a **singular matrix.** Figure 2 illustrates the procedure on a graphing utility. Note that the inverse operation is performed by pressing the x^{-1} key. Entering $[A]$^(-1) results in an error message.

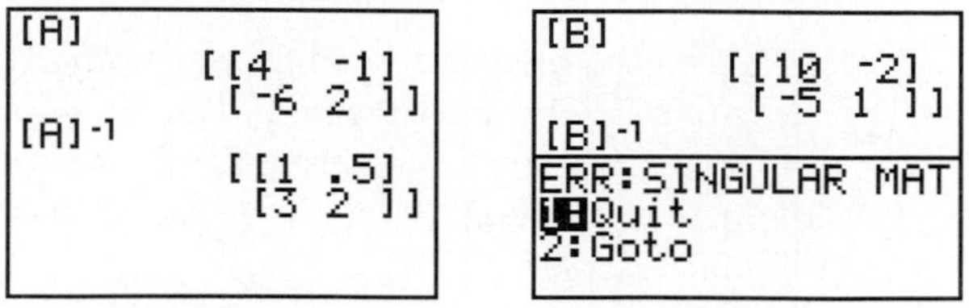

(a) Example 3 (b) Example 4

FIGURE 2
Finding matrix inverses on a graphing utility.

Application: Cryptography

Matrix inverses can be used to provide a simple and effective procedure for encoding and decoding messages. To begin, we assign the numbers 1 to 26 to the letters in the alphabet, as shown below. We also assign the number 27 to a blank to provide for space between words. (A more sophisticated code could include both uppercase and lowercase letters and punctuation symbols.)

A	B	C	D	E	F	G	H	I	J	K	L	M	N
1	2	3	4	5	6	7	8	9	10	11	12	13	14
O	P	Q	R	S	T	U	V	W	X	Y	Z	Blank	
15	16	17	18	19	20	21	22	23	24	25	26	27	

Thus, the message I LOVE MATH corresponds to the sequence

9 27 12 15 22 5 27 13 1 20 8

Any matrix whose elements are positive integers and whose inverse exists can be used as an **encoding matrix.** For example, to use the 2×2 matrix

$$A = \begin{bmatrix} 4 & 3 \\ 5 & 4 \end{bmatrix}$$

to encode the above message, first we divide the numbers in the sequence into groups of 2 and use these groups as the columns of a matrix B with two rows:

$$B = \begin{bmatrix} 9 & 12 & 22 & 27 & 1 & 8 \\ 27 & 15 & 5 & 13 & 20 & 27 \end{bmatrix}$$

Proceed down the columns, not across the rows.

(Notice that we added an extra blank at the end of the message to make the columns come out even.) Then we multiply this matrix on the left by A:

$$AB = \begin{bmatrix} 4 & 3 \\ 5 & 4 \end{bmatrix} \begin{bmatrix} 9 & 12 & 22 & 27 & 1 & 8 \\ 27 & 15 & 5 & 13 & 20 & 27 \end{bmatrix}$$

$$= \begin{bmatrix} 117 & 93 & 103 & 147 & 64 & 113 \\ 153 & 120 & 130 & 187 & 85 & 148 \end{bmatrix}$$

The coded message is

117 153 93 120 103 130 147 187 64 85 113 148

This message can be decoded simply by putting it back into matrix form and multiplying on the left by the **decoding matrix** A^{-1}. Since A^{-1} is easily determined if A is known, the encoding matrix A is the only key needed to decode messages encoded in this manner. Although simple in concept, codes of this type can be very difficult to crack.

EXAMPLE 5 **Cryptography**

The message

31 54 69 37 64 82 7 34 58 51 69 75 23 30 36 65 84 84

was encoded with the matrix A shown below. Decode this message.

$$A = \begin{bmatrix} 0 & 2 & 1 \\ 1 & 2 & 1 \\ 2 & 1 & 1 \end{bmatrix}$$

Solution We begin by entering the 3×3 encoding matrix A (Fig. 3). Then we enter the coded message in the columns of a matrix C with three rows (Fig. 3). If B is the matrix containing the uncoded message, then B and C are related by $C = AB$. To find B, we multiply both sides of the equation $C = AB$ by A^{-1} (Fig. 4).

```
A
               [[0 2 1]
                [1 2 1]
                [2 1 1]]
C
[[31 37 7  51 23 65]
 [54 64 34 69 30 84]
 [69 82 58 75 36 84]]
```

FIGURE 3

```
A⁻¹C→B
[[23 27 27 18 7  19]
 [8  9  3  12 1  19]
 [15 19 1  27 21 27]]
```

FIGURE 4

Writing the numbers in the columns of this matrix in sequence and using the correspondence between numbers and letters noted earlier produces the decoded message:

23	8	15	27	9	19	27	3	1	18	12	27	7	1	21	19	19	27
W	H	O		I	S		C	A	R	L		G	A	U	S	S	

The answer to this question can be found earlier in this chapter.

MATCHED PROBLEM 5

The message

46 84 85 55 101 100 59 95 132 25 42 53 52 91 90 43 71 83 19 37 25

was encoded with the matrix A shown below. Decode this message.

$$A = \begin{bmatrix} 1 & 1 & 1 \\ 2 & 1 & 2 \\ 2 & 3 & 1 \end{bmatrix}$$

Answers to Matched Problems

1. (A) $\begin{bmatrix} 3 & -5 \\ 4 & 6 \end{bmatrix}$ (B) $\begin{bmatrix} 5 & -7 \\ 2 & 4 \\ 6 & -8 \end{bmatrix}$

2. (A) $\left[\begin{array}{rrr|rrr} 3 & -1 & 1 & 1 & 0 & 0 \\ -1 & 1 & 0 & 0 & 1 & 0 \\ 1 & 0 & 1 & 0 & 0 & 1 \end{array}\right]$ (B) $\left[\begin{array}{rrr|rrr} 1 & 0 & 0 & 1 & 1 & -1 \\ 0 & 1 & 0 & 1 & 2 & -1 \\ 0 & 0 & 1 & -1 & -1 & 2 \end{array}\right]$ (C) $\begin{bmatrix} 1 & 1 & -1 \\ 1 & 2 & -1 \\ -1 & -1 & 2 \end{bmatrix}\begin{bmatrix} 3 & -1 & 1 \\ -1 & 1 & 0 \\ 1 & 0 & 1 \end{bmatrix} = \begin{bmatrix} 1 & 0 & 0 \\ 0 & 1 & 0 \\ 0 & 0 & 1 \end{bmatrix}$

3. $\begin{bmatrix} -1 & 3 \\ -\frac{1}{2} & 1 \end{bmatrix}$ **4.** Does not exist **5.** WHO IS WILHELM JORDAN

EXERCISE 5-5

A

Perform the indicated operations in Problems 1–8.

1. $\begin{bmatrix} 1 & 0 \\ 0 & 1 \end{bmatrix}\begin{bmatrix} 2 & -3 \\ 4 & 5 \end{bmatrix}$

2. $\begin{bmatrix} 1 & 0 \\ 0 & 1 \end{bmatrix}\begin{bmatrix} 4 & -3 \\ 0 & 2 \end{bmatrix}$

3. $\begin{bmatrix} 2 & -3 \\ 4 & 5 \end{bmatrix}\begin{bmatrix} 1 & 0 \\ 0 & 1 \end{bmatrix}$

4. $\begin{bmatrix} 4 & -3 \\ 0 & 2 \end{bmatrix}\begin{bmatrix} 1 & 0 \\ 0 & 1 \end{bmatrix}$

5. $\begin{bmatrix} 1 & 0 & 0 \\ 0 & 1 & 0 \\ 0 & 0 & 1 \end{bmatrix}\begin{bmatrix} -2 & 1 & 3 \\ 2 & 4 & -2 \\ 5 & 1 & 0 \end{bmatrix}$

6. $\begin{bmatrix} 1 & 0 & 0 \\ 0 & 1 & 0 \\ 0 & 0 & 1 \end{bmatrix}\begin{bmatrix} -3 & 0 & 2 \\ 1 & 1 & 5 \\ 2 & -1 & 7 \end{bmatrix}$

7. $\begin{bmatrix} -2 & 1 & 3 \\ 2 & 4 & -2 \\ 5 & 1 & 0 \end{bmatrix}\begin{bmatrix} 1 & 0 & 0 \\ 0 & 1 & 0 \\ 0 & 0 & 1 \end{bmatrix}$

8. $\begin{bmatrix} -3 & 0 & 2 \\ 1 & 1 & 5 \\ 2 & -1 & 7 \end{bmatrix}\begin{bmatrix} 1 & 0 & 0 \\ 0 & 1 & 0 \\ 0 & 0 & 1 \end{bmatrix}$

In Problems 9–18, examine the product of the two matrices to determine if each is the inverse of the other.

9. $\begin{bmatrix} 3 & -4 \\ -2 & 3 \end{bmatrix}; \begin{bmatrix} 3 & 4 \\ 2 & 3 \end{bmatrix}$

10. $\begin{bmatrix} -2 & -1 \\ -4 & 2 \end{bmatrix}; \begin{bmatrix} 1 & -1 \\ 2 & -2 \end{bmatrix}$

11. $\begin{bmatrix} 2 & 2 \\ -1 & -1 \end{bmatrix}; \begin{bmatrix} 1 & 1 \\ -1 & -1 \end{bmatrix}$

12. $\begin{bmatrix} 5 & -7 \\ -2 & 3 \end{bmatrix}; \begin{bmatrix} 3 & 7 \\ 2 & 5 \end{bmatrix}$

13. $\begin{bmatrix} -5 & 2 \\ -8 & 3 \end{bmatrix}; \begin{bmatrix} 3 & -2 \\ 8 & -5 \end{bmatrix}$

14. $\begin{bmatrix} 7 & 4 \\ -5 & -3 \end{bmatrix}; \begin{bmatrix} 3 & 4 \\ -5 & -7 \end{bmatrix}$

15. $\begin{bmatrix} 1 & 2 & 0 \\ 0 & 1 & 0 \\ -1 & -1 & 1 \end{bmatrix}; \begin{bmatrix} 1 & -2 & 0 \\ 0 & 1 & 0 \\ 1 & -1 & 0 \end{bmatrix}$

16. $\begin{bmatrix} 1 & 0 & 1 \\ -3 & 1 & -2 \\ 0 & 0 & 1 \end{bmatrix}; \begin{bmatrix} 1 & 0 & -1 \\ 3 & 1 & -1 \\ 0 & 0 & 1 \end{bmatrix}$

17. $\begin{bmatrix} 1 & -1 & 1 \\ 0 & 2 & -1 \\ 2 & 3 & 0 \end{bmatrix}; \begin{bmatrix} 3 & 3 & -1 \\ -2 & -2 & 1 \\ -4 & -5 & 2 \end{bmatrix}$

18. $\begin{bmatrix} 1 & 0 & -1 \\ 3 & 1 & -1 \\ 0 & 0 & 0 \end{bmatrix}; \begin{bmatrix} 1 & 0 & -1 \\ -3 & 1 & -2 \\ 0 & 0 & 1 \end{bmatrix}$

B

Given M in Problems 19–28, find M^{-1}, and show that $M^{-1}M = I$.

19. $\begin{bmatrix} 0 & -1 \\ 1 & 4 \end{bmatrix}$ **20.** $\begin{bmatrix} -1 & 5 \\ 0 & -1 \end{bmatrix}$ **21.** $\begin{bmatrix} 1 & 2 \\ 1 & 3 \end{bmatrix}$

22. $\begin{bmatrix} 2 & 1 \\ 5 & 3 \end{bmatrix}$ **23.** $\begin{bmatrix} 1 & 3 \\ 2 & 7 \end{bmatrix}$ **24.** $\begin{bmatrix} 2 & 1 \\ 1 & 1 \end{bmatrix}$

25. $\begin{bmatrix} 1 & -2 & 0 \\ 0 & 1 & 1 \\ 2 & -1 & 2 \end{bmatrix}$ **26.** $\begin{bmatrix} 1 & 3 & 0 \\ 1 & 2 & 3 \\ 0 & -1 & 2 \end{bmatrix}$

27. $\begin{bmatrix} 1 & 1 & 0 \\ 0 & 2 & -1 \\ 1 & 0 & 1 \end{bmatrix}$ **28.** $\begin{bmatrix} 1 & 0 & -1 \\ 2 & -1 & 0 \\ 1 & 1 & -4 \end{bmatrix}$

Find the inverse of each matrix in Problems 29–32, if it exists.

29. $\begin{bmatrix} 3 & 9 \\ 2 & 6 \end{bmatrix}$ **30.** $\begin{bmatrix} 2 & -4 \\ -3 & 6 \end{bmatrix}$

31. $\begin{bmatrix} 2 & 3 \\ 3 & 5 \end{bmatrix}$

32. $\begin{bmatrix} -5 & 4 \\ 4 & -3 \end{bmatrix}$

C

Find the inverse of each matrix in Problems 33–38, if it exists.

33. $\begin{bmatrix} 2 & 2 & -1 \\ 0 & 4 & -1 \\ -1 & -2 & 1 \end{bmatrix}$

34. $\begin{bmatrix} 4 & 2 & -1 \\ 1 & 1 & -1 \\ -3 & -1 & 1 \end{bmatrix}$

35. $\begin{bmatrix} 2 & 1 & 1 \\ 1 & 1 & 0 \\ -1 & -1 & 0 \end{bmatrix}$

36. $\begin{bmatrix} 1 & -1 & 0 \\ 2 & -1 & 1 \\ 0 & 1 & 1 \end{bmatrix}$

37. $\begin{bmatrix} 1 & 5 & 10 \\ 0 & 1 & 4 \\ 1 & 6 & 15 \end{bmatrix}$

38. $\begin{bmatrix} 1 & -5 & -10 \\ 0 & 1 & 6 \\ 1 & -4 & -3 \end{bmatrix}$

39. Discuss the existence of M^{-1} for 2×2 diagonal matrices of the form

$$M = \begin{bmatrix} a & 0 \\ 0 & d \end{bmatrix}$$

40. Discuss the existence of M^{-1} for 2×2 upper triangular matrices of the form

$$M = \begin{bmatrix} a & b \\ 0 & d \end{bmatrix}$$

41. Find A^{-1} and A^2 for each of the following matrices.

(A) $A = \begin{bmatrix} 3 & 2 \\ -4 & -3 \end{bmatrix}$ (B) $A = \begin{bmatrix} -2 & -1 \\ 3 & 2 \end{bmatrix}$

42. Based on your observations in Problem 41, if $A = A^{-1}$ for a square matrix A, what is A^2? Give a mathematical argument to support your conclusion.

43. Find $(A^{-1})^{-1}$ for each of the following matrices.

(A) $A = \begin{bmatrix} 4 & 2 \\ 1 & 3 \end{bmatrix}$ (B) $A = \begin{bmatrix} 5 & 5 \\ -1 & 3 \end{bmatrix}$

44. Based on your observations in Problem 43, if A^{-1} exists for a square matrix A, what is $(A^{-1})^{-1}$? Give a mathematical argument to support your conclusion.

45. Find $(AB)^{-1}$, $A^{-1}B^{-1}$, and $B^{-1}A^{-1}$ for each of the following pairs of matrices.

(A) $A = \begin{bmatrix} 3 & 4 \\ 2 & 3 \end{bmatrix}$ and $B = \begin{bmatrix} 3 & 7 \\ 2 & 5 \end{bmatrix}$

(B) $A = \begin{bmatrix} 1 & -1 \\ 2 & 3 \end{bmatrix}$ and $B = \begin{bmatrix} 6 & 2 \\ 2 & 1 \end{bmatrix}$

46. Based on your observations in Problem 45, which of the following is a true statement? Give a mathematical argument to support your conclusion.

(A) $(AB)^{-1} = A^{-1}B^{-1}$

(B) $(AB)^{-1} = B^{-1}A^{-1}$

APPLICATIONS

Problems 47–50 refer to the encoding matrix $A = \begin{bmatrix} 3 & 5 \\ 1 & 2 \end{bmatrix}$

47. **Cryptography.** Encode the message CAT IN THE HAT with the matrix A given above.

48. **Cryptography.** Encode the message FOX IN SOCKS with the matrix A given above.

49. **Cryptography.** The following message was encoded with the matrix A given above. Decode this message.

111 43 40 15 177 68 50 19 116 45 86
29 62 22 121 43 68 27

50. **Cryptography.** The following message was encoded with the matrix A given above. Decode this message.

99 38 154 58 115 43 121 43 20 7 149
56 86 29 196 73 99 38

Problems 51–54 refer to the encoding matrix

$$B = \begin{bmatrix} 1 & 0 & 1 & 0 & 1 \\ 0 & 1 & 1 & 0 & 3 \\ 2 & 1 & 1 & 1 & 1 \\ 0 & 0 & 1 & 0 & 2 \\ 1 & 1 & 1 & 2 & 1 \end{bmatrix}$$

51. **Cryptography.** Encode the message DWIGHT DAVID EISENHOWER with the matrix B given above.

52. **Cryptography.** Encode the message JOHN FITZGERALD KENNEDY with the matrix B given above.

53. **Cryptography.** The following message was encoded with the matrix B given above. Decode this message.

41 84 82 44 74 25 56 67 20 54 43
54 89 39 102 44 67 86 44 90 68 135
136 81 149

54. **Cryptography.** The following message was encoded with the matrix B given above. Decode this message.

22 15 57 5 47 54 58 89 45 84 46
80 87 53 96 51 68 116 39 113 68 135
136 81 149

Section 5-6 Matrix Equations and Systems of Linear Equations

- Matrix Equations
- Matrix Equations and Systems of Linear Equations
- Application

The identity matrix and inverse matrix discussed in the last section can be put to immediate use in the solving of certain simple matrix equations. Being able to solve a matrix equation gives us another important method of solving a system of equations having the same number of variables as equations. If the system either has fewer variables than equations or more variables than equations, then we must return to the Gauss–Jordan method of elimination.

Matrix Equations

Before we discuss the solution of matrix equations, you will probably find it helpful to briefly review the basic properties of real numbers and linear equations discussed in Sections A-1 and A-8.

Explore/Discuss 1

Let a, b, and c be real numbers, with $a \neq 0$. Solve each equation for x.

(A) $ax = b$ (B) $ax + b = c$

Solving simple matrix equations follows very much the same procedures used in solving real number equations. We have, however, less freedom with matrix equations, because matrix multiplication is not commutative. In solving matrix equations, we will be guided by the properties of matrices summarized in Theorem 1.

THEOREM 1

BASIC PROPERTIES OF MATRICES

Assuming all products and sums are defined for the indicated matrices A, B, C, I, and 0, then

Addition Properties

Associative:	$(A + B) + C = A + (B + C)$
Commutative:	$A + B = B + A$
Additive Identity:	$A + 0 = 0 + A = A$
Additive Inverse:	$A + (-A) = (-A) + A = 0$

Multiplication Properties

Associative Property:	$A(BC) = (AB)C$
Multiplicative Identity:	$AI = IA = A$
Multiplicative Inverse:	If A is a square matrix and A^{-1} exists, then $AA^{-1} = A^{-1}A = I$.

THEOREM 1 continued

Combined Properties	
Left Distributive:	$A(B + C) = AB + AC$
Right Distributive:	$(B + C)A = BA + CA$
Equality	
Addition:	If $A = B$, then $A + C = B + C$.
Left Multiplication:	If $A = B$, then $CA = CB$.
Right Multiplication:	If $A = B$, then $AC = BC$.

The process of solving certain types of simple matrix equations is best illustrated by an example.

EXAMPLE 1 Solving a Matrix Equation

Given an $n \times n$ matrix A and $n \times 1$ column matrices B and X, solve $AX = B$ for X. Assume all necessary inverses exist.

Solution We are interested in finding a column matrix X that satisfies the matrix equation $AX = B$. To solve this equation, we multiply both sides, on the left, by A^{-1}, assuming it exists, to isolate X on the left side.

$$AX = B$$

$$A^{-1}(AX) = A^{-1}B \quad \text{Use the left multiplication property.}$$

$$(A^{-1}A)X = A^{-1}B \quad \text{Associative property}$$

$$IX = A^{-1}B \quad A^{-1}A = I$$

$$X = A^{-1}B \quad IX = X$$

CAUTION

Do not mix the left multiplication property and the right multiplication property. If $AX = B$, then

$$A^{-1}(AX) \neq BA^{-1}$$

MATCHED PROBLEM 1 Given an $n \times n$ matrix A and $n \times 1$ column matrices B, C, and X, solve $AX + C = B$ for X. Assume all necessary inverses exist.

Matrix Equations and Systems of Linear Equations

We now show how independent systems of linear equations with the same number of variables as equations can be solved by first converting the system into a matrix equation of the form $AX = B$ and using $X = A^{-1}B$ as obtained in Example 1.

EXAMPLE 2

Using Inverses to Solve Systems of Equations

Use matrix inverse methods to solve the system

$$\begin{aligned} x_1 - x_2 + x_3 &= 1 \\ 2x_2 - x_3 &= 1 \\ 2x_1 + 3x_2 &= 1 \end{aligned} \qquad (1)$$

Solution

The inverse of the coefficient matrix

$$A = \begin{bmatrix} 1 & -1 & 1 \\ 0 & 2 & -1 \\ 2 & 3 & 0 \end{bmatrix}$$

provides an efficient method for solving this system. To see how, we convert system (1) into a matrix equation:

$$\overset{A}{\begin{bmatrix} 1 & -1 & 1 \\ 0 & 2 & -1 \\ 2 & 3 & 0 \end{bmatrix}} \overset{X}{\begin{bmatrix} x_1 \\ x_2 \\ x_3 \end{bmatrix}} = \overset{B}{\begin{bmatrix} 1 \\ 1 \\ 1 \end{bmatrix}} \qquad (2)$$

Check that matrix equation (2) is equivalent to system (1) by finding the product of the left side and then equating corresponding elements on the left with those on the right. Now you see another important reason for defining matrix multiplication as we did.

We are interested in finding a column matrix X that satisfies the matrix equation $AX = B$. In Example 1 we found that if $AX = B$ and if A^{-1} exists, then

$$X = A^{-1}B$$

The inverse of A was found in Example 2 in Section 5-5 to be

$$A^{-1} = \begin{bmatrix} 3 & 3 & -1 \\ -2 & -2 & 1 \\ -4 & -5 & 2 \end{bmatrix}$$

Thus,

$$\overset{X}{\begin{bmatrix} x_1 \\ x_2 \\ x_3 \end{bmatrix}} = \overset{A^{-1}}{\begin{bmatrix} 3 & 3 & -1 \\ -2 & -2 & 1 \\ -4 & -5 & 2 \end{bmatrix}} \overset{B}{\begin{bmatrix} 1 \\ 1 \\ 1 \end{bmatrix}} = \begin{bmatrix} 5 \\ -3 \\ -7 \end{bmatrix}$$

and we can conclude that $x_1 = 5$, $x_2 = -3$, and $x_3 = -7$. Check this result in system (1).

To solve this problem on a graphing utility, enter A and B (Fig. 1) and simply type $A^{-1}B$ (Fig. 2).

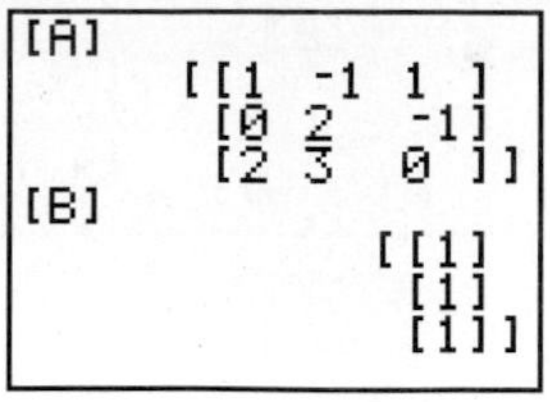

FIGURE 1

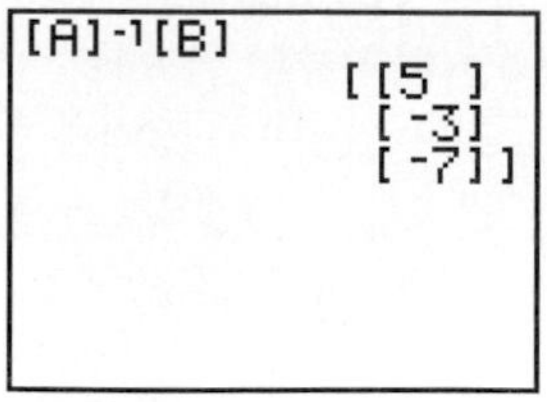

FIGURE 2

MATCHED PROBLEM 2

Use matrix inverse methods to solve the system:

$$\begin{aligned} 3x_1 - x_2 + x_3 &= 1 \\ -x_1 + x_2 \qquad &= 3 \\ x_1 \qquad + x_3 &= 2 \end{aligned}$$

At first glance, using matrix inverse methods seems to require the same amount of effort as using Gauss–Jordan elimination. In either case, row operations must be applied to an augmented matrix involving the coefficients of the system. The advantage of the inverse matrix method becomes readily apparent when solving a number of systems with a common coefficient matrix and different constant terms.

EXAMPLE 3

Using Inverses to Solve Systems of Equations

Use matrix inverse methods to solve each of the following systems:

(A) $\begin{aligned} x_1 - x_2 + x_3 &= 3 \\ 2x_2 - x_3 &= 1 \\ 2x_1 + 3x_2 \qquad &= 4 \end{aligned}$ (B) $\begin{aligned} x_1 - x_2 + x_3 &= -5 \\ 2x_2 - x_3 &= 2 \\ 2x_1 + 3x_2 \qquad &= -3 \end{aligned}$

Solutions

Notice that both systems have the same coefficient matrix A as system (1) in Example 2. Only the constant terms have been changed. Thus, we can use A^{-1} to solve these systems just as we did in Example 2.

(A) X, A^{-1}, B

$$\begin{bmatrix} x_1 \\ x_2 \\ x_3 \end{bmatrix} = \begin{bmatrix} 3 & 3 & -1 \\ -2 & -2 & 1 \\ -4 & -5 & 2 \end{bmatrix} \begin{bmatrix} 3 \\ 1 \\ 4 \end{bmatrix} = \begin{bmatrix} 8 \\ -4 \\ -9 \end{bmatrix}$$ See Figure 3.

Thus, $x_1 = 8$, $x_2 = -4$, and $x_3 = -9$

(B) X, A^{-1}, B

$$\begin{bmatrix} x_1 \\ x_2 \\ x_3 \end{bmatrix} = \begin{bmatrix} 3 & 3 & -1 \\ -2 & -2 & 1 \\ -4 & -5 & 2 \end{bmatrix} \begin{bmatrix} -5 \\ 2 \\ -3 \end{bmatrix} = \begin{bmatrix} -6 \\ 3 \\ 4 \end{bmatrix}$$ See Figure 4.

Thus, $x_1 = -6$, $x_2 = 3$, and $x_3 = 4$

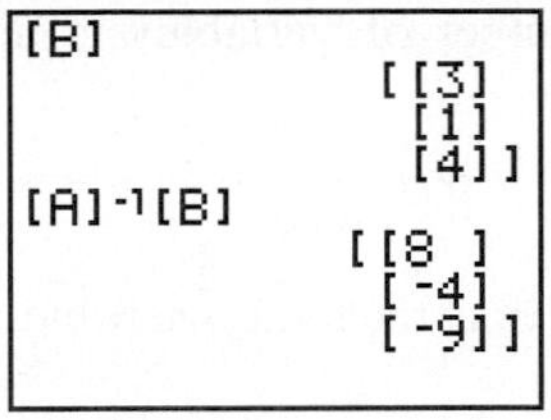

FIGURE 3

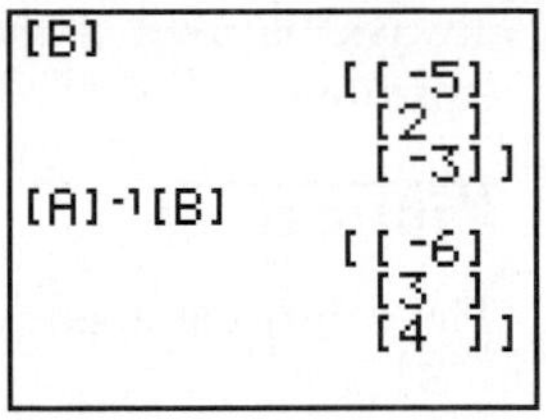

FIGURE 4

MATCHED PROBLEM 3

Use matrix inverse methods to solve each of the following systems (see Matched Problem 2):

(A)
$$\begin{aligned} 3x_1 - x_2 + x_3 &= 3 \\ -x_1 + x_2 &= -3 \\ x_1 + x_3 &= 2 \end{aligned}$$

(B)
$$\begin{aligned} 3x_1 - x_2 + x_3 &= -5 \\ -x_1 + x_2 &= 1 \\ x_1 + x_3 &= -4 \end{aligned}$$

Explore/Discuss 2

Use matrix inverse methods to solve each of the following systems, if possible, otherwise use Gauss–Jordan elimination. Describe the types of systems that can be solved by inverse methods and those that cannot. Are there any systems that cannot be solved by Gauss–Jordan elimination?

(A)
$$\begin{aligned} x_1 - x_2 &= 1 \\ x_1 + x_2 &= 7 \\ 3x_1 - x_2 &= 9 \end{aligned}$$

(B)
$$\begin{aligned} x_1 - x_2 + x_3 &= 1 \\ x_1 + x_2 - x_3 &= 7 \\ 3x_1 - x_2 + x_3 &= 9 \end{aligned}$$

(C)
$$\begin{aligned} x_1 - x_2 + x_3 &= 1 \\ x_1 + x_2 - x_3 &= 7 \\ 3x_1 - x_2 + x_3 &= 8 \end{aligned}$$

(D)
$$\begin{aligned} x_1 - x_2 + x_3 &= 1 \\ x_1 + x_2 - x_3 &= 7 \\ 3x_1 - x_2 + 2x_3 &= 8 \end{aligned}$$

USING INVERSE METHODS TO SOLVE SYSTEMS OF EQUATIONS

If the number of equations in a system equals the number of variables and the coefficient matrix has an inverse, then the system will always have a unique solution that can be found by using the inverse of the coefficient matrix to solve the corresponding matrix equation.

Matrix equation	Solution
$AX = B$	$X = A^{-1}B$

Remark What happens if the coefficient matrix does not have an inverse? In this case, it can be shown that the system does not have a unique solution and is either dependent or inconsistent. Gauss–Jordan elimination must be used to determine which is the case. Also, as we mentioned earlier, Gauss–Jordan elimination must always be used if the number of variables is not the same as the number of equations.

Application

The following application illustrates the usefulness of the inverse method.

EXAMPLE 4

Investment Allocation

An investment adviser currently has two types of investments available for clients: an investment *M* that pays 10% per year and an investment *N* of higher risk that pays 20% per year. Clients may divide their investments between the two to achieve any total return desired between 10 and 20%. However, the higher the desired return, the higher the risk. How should each client listed in the table invest to achieve the indicated return?

	Client			
	1	2	3	*k*
Total investment	\$20,000	\$50,000	\$10,000	k_1
Annual return desired	\$2,400	\$7,500	\$1,300	k_2
	(12%)	(15%)	(13%)	

Solution

We first solve the problem for an arbitrary client *k* using inverses, and then apply the result to the three specific clients.

Let

$$x_1 = \text{Amount invested in } M$$
$$x_2 = \text{Amount invested in } N$$

Then

$$\begin{aligned} x_1 + \quad x_2 &= k_1 \quad \text{Total invested} \\ 0.1x_1 + 0.2x_2 &= k_2 \quad \text{Total annual return} \end{aligned}$$

Write as a matrix equation:

$$\overset{A}{\begin{bmatrix} 1 & 1 \\ 0.1 & 0.2 \end{bmatrix}} \overset{X}{\begin{bmatrix} x_1 \\ x_2 \end{bmatrix}} = \overset{B}{\begin{bmatrix} k_1 \\ k_2 \end{bmatrix}}$$

FIGURE 5

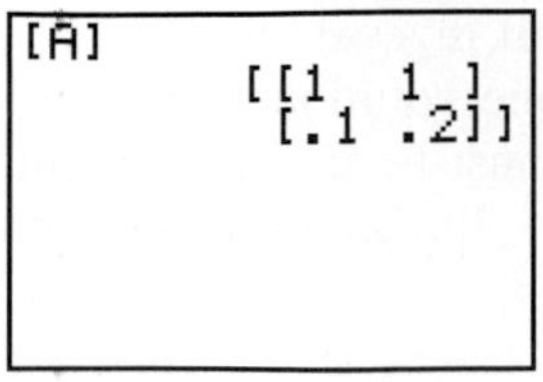

If A^{-1} exists, then

$$X = A^{-1}B$$

To solve each client's investment problem on a graphing utility, first we enter *A* (Fig. 5), then we enter the appropriate values for *B* and compute $A^{-1}B$ (Fig. 6).

FIGURE 6

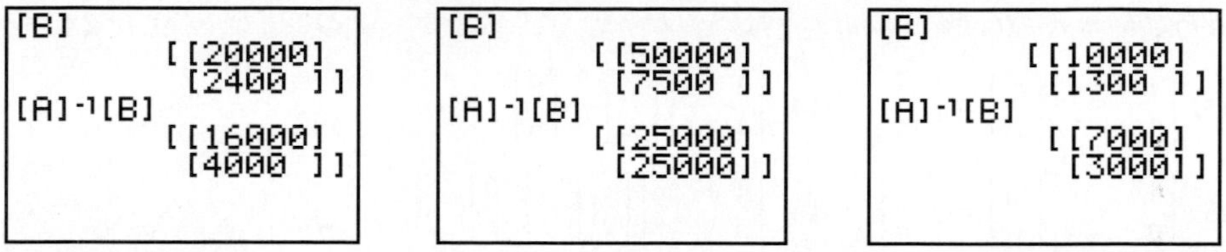

(a) Client 1 (b) Client 2 (c) Client 3

From Figure 6, we see that Client 1 should invest \$16,000 in investment M and \$4,000 in investment N, Client 2 should invest \$25,000 in investment M and \$25,000 in investment N, and Client 3 should invest \$7,000 in investment M and \$3,000 in investment N.

MATCHED PROBLEM 4

Repeat Example 4 with investment M paying 8% and investment N paying 24%.

Answers to Matched Problems

1.
$$AX + C = B$$
$$(AX + C) - C = B - C$$
$$AX + (C - C) = B - C$$
$$AX + 0 = B - C$$
$$AX = B - C$$
$$A^{-1}(AX) = A^{-1}(B - C)$$
$$(A^{-1}A)X = A^{-1}(B - C)$$
$$IX = A^{-1}(B - C)$$
$$X = A^{-1}(B - C)$$

2. $x_1 = 2, x_2 = 5, x_3 = 0$ 3. (A) $x_1 = -2, x_2 = -5, x_3 = 4$ (B) $x_1 = 0, x_2 = 1, x_3 = -4$

4. $A^{-1} = \begin{bmatrix} 1.5 & -6.25 \\ -0.5 & 6.25 \end{bmatrix}$; Client 1: \$15,000 in M and \$5,000 in N; Client 2: \$28,125 in M and \$21,875 in N; Client 3: \$6,875 in M and \$3,125 in N

EXERCISE 5-6

A

Write Problems 1–4 as systems of linear equations without matrices.

1. $\begin{bmatrix} 2 & -1 \\ 1 & 3 \end{bmatrix}\begin{bmatrix} x_1 \\ x_2 \end{bmatrix} = \begin{bmatrix} 3 \\ -2 \end{bmatrix}$

2. $\begin{bmatrix} -3 & 1 \\ -1 & 2 \end{bmatrix}\begin{bmatrix} x_1 \\ x_2 \end{bmatrix} = \begin{bmatrix} -2 \\ 5 \end{bmatrix}$

3. $\begin{bmatrix} -2 & 0 & 1 \\ 1 & 2 & 1 \\ 0 & 1 & -1 \end{bmatrix}\begin{bmatrix} x_1 \\ x_2 \\ x_3 \end{bmatrix} = \begin{bmatrix} 3 \\ -4 \\ 2 \end{bmatrix}$

4. $\begin{bmatrix} 1 & -2 & 0 \\ -3 & 1 & -1 \\ 2 & 0 & 4 \end{bmatrix}\begin{bmatrix} x_1 \\ x_2 \\ x_3 \end{bmatrix} = \begin{bmatrix} 3 \\ -2 \\ 5 \end{bmatrix}$

Write each system in Problems 5–8 as a matrix equation of the form $AX = B$.

5. $\begin{aligned} 4x_1 - 3x_2 &= 2 \\ x_1 + 2x_2 &= 1 \end{aligned}$

6. $\begin{aligned} x_1 - 2x_2 &= 7 \\ -3x_1 + x_2 &= -3 \end{aligned}$

7. $\begin{aligned} x_1 - 2x_2 + x_3 &= -1 \\ -x_1 + x_2 &= 2 \\ 2x_1 + 3x_2 + x_3 &= -3 \end{aligned}$

8. $\begin{aligned} 2x_1 + 3x_3 &= 5 \\ x_1 - 2x_2 + x_3 &= -4 \\ -x_1 + 3x_2 &= 2 \end{aligned}$

In Problems 9–12, find x_1 and x_2.

9. $\begin{bmatrix} x_1 \\ x_2 \end{bmatrix} = \begin{bmatrix} 3 & -2 \\ 1 & 4 \end{bmatrix}\begin{bmatrix} -2 \\ 1 \end{bmatrix}$

10. $\begin{bmatrix} x_1 \\ x_2 \end{bmatrix} = \begin{bmatrix} -2 & 1 \\ -1 & 2 \end{bmatrix}\begin{bmatrix} 3 \\ -2 \end{bmatrix}$

11. $\begin{bmatrix} x_1 \\ x_2 \end{bmatrix} = \begin{bmatrix} -2 & 3 \\ 2 & -1 \end{bmatrix}\begin{bmatrix} 3 \\ 2 \end{bmatrix}$

12. $\begin{bmatrix} x_1 \\ x_2 \end{bmatrix} = \begin{bmatrix} 3 & -1 \\ 0 & 2 \end{bmatrix}\begin{bmatrix} -2 \\ 1 \end{bmatrix}$

In Problems 13–16, find x_1 and x_2.

13. $\begin{bmatrix} 1 & -1 \\ 1 & -2 \end{bmatrix}\begin{bmatrix} x_1 \\ x_2 \end{bmatrix} = \begin{bmatrix} 5 \\ 7 \end{bmatrix}$

14. $\begin{bmatrix} 1 & 3 \\ 1 & 4 \end{bmatrix}\begin{bmatrix} x_1 \\ x_2 \end{bmatrix} = \begin{bmatrix} 9 \\ 6 \end{bmatrix}$

15. $\begin{bmatrix} 1 & 1 \\ 2 & -3 \end{bmatrix}\begin{bmatrix} x_1 \\ x_2 \end{bmatrix} = \begin{bmatrix} 15 \\ 10 \end{bmatrix}$

16. $\begin{bmatrix} 1 & 1 \\ 3 & -2 \end{bmatrix}\begin{bmatrix} x_1 \\ x_2 \end{bmatrix} = \begin{bmatrix} 10 \\ 20 \end{bmatrix}$

B

Write each system in Problems 17–24 as a matrix equation and solve using inverses.

17. $x_1 + 2x_2 = k_1$
$x_1 + 3x_2 = k_2$
(A) $k_1 = 1, k_2 = 3$
(B) $k_1 = 3, k_2 = 5$
(C) $k_1 = -2, k_2 = 1$

18. $2x_1 + x_2 = k_1$
$5x_1 + 3x_2 = k_2$
(A) $k_1 = 2, k_2 = 13$
(B) $k_1 = -2, k_2 = 4$
(C) $k_1 = 1, k_2 = -3$

19. $x_1 + 3x_2 = k_1$
$2x_1 + 7x_2 = k_2$
(A) $k_1 = 2, k_2 = -1$
(B) $k_1 = 1, k_2 = 0$
(C) $k_1 = 3, k_2 = -1$

20. $2x_1 + x_2 = k_1$
$x_1 + x_2 = k_2$
(A) $k_1 = -1, k_2 = -2$
(B) $k_1 = 2, k_2 = 3$
(C) $k_1 = 2, k_2 = 0$

21. $x_1 - 2x_2 = k_1$
$x_2 + x_3 = k_2$
$2x_1 - x_2 + 2x_3 = k_3$
(A) $k_1 = 1, k_2 = 0, k_3 = 2$
(B) $k_1 = -1, k_2 = 1, k_3 = 0$
(C) $k_1 = 2, k_2 = -2, k_3 = 1$

22. $x_1 + 3x_2 = k_1$
$x_1 + 2x_2 + 3x_3 = k_2$
$-x_2 + 2x_3 = k_3$
(A) $k_1 = 0, k_2 = 2, k_3 = 1$
(B) $k_1 = -2, k_2 = 0, k_3 = 1$
(C) $k_1 = 3, k_2 = 1, k_3 = 0$

23. $x_1 + x_2 = k_1$
$2x_2 - x_3 = k_2$
$x_1 + x_3 = k_3$
(A) $k_1 = 2, k_2 = 0, k_3 = 4$
(B) $k_1 = 0, k_2 = 4, k_3 = -2$
(C) $k_1 = 4, k_2 = 2, k_3 = 0$

24. $x_1 - x_3 = k_1$
$2x_1 - x_2 = k_2$
$x_1 + x_2 - 4x_3 = k_3$
(A) $k_1 = 4, k_2 = 8, k_3 = 0$
(B) $k_1 = 4, k_2 = 0, k_3 = -4$
(C) $k_1 = 0, k_2 = 8, k_3 = -8$

In Problems 25–30, explain why the system cannot be solved by matrix inverse methods. Discuss methods that could be used and then solve the system.

25. $-2x_1 + 4x_2 = -5$
$6x_1 - 12x_2 = 15$

26. $-2x_1 + 4x_2 = 5$
$6x_1 - 12x_2 = 15$

27. $x_1 - 3x_2 - 2x_3 = -1$
$-2x_1 + 6x_2 + 4x_3 = 3$

28. $x_1 - 3x_2 - 2x_3 = -1$
$-2x_1 + 7x_2 + 3x_3 = 3$

29. $x_1 - 2x_2 + 3x_3 = 1$
$2x_1 - 3x_2 - 2x_3 = 3$
$x_1 - x_2 - 5x_3 = 2$

30. $x_1 - 2x_2 + 3x_3 = 1$
$2x_1 - 3x_2 - 2x_3 = 3$
$x_1 - x_2 - 5x_3 = 4$

C

For $n \times n$ matrices A and B and $n \times 1$ matrices C, D, and X, solve each matrix equation in Problems 31–36 for X. Assume all necessary inverses exist.

31. $AX - BX = C$

32. $AX + BX = C$

33. $AX + X = C$

34. $AX - X = C$

35. $AX - C = D - BX$

36. $AX + C = BX + D$

37. Use matrix inverse methods to solve the following system for the indicated values of k_1 and k_2.

$x_1 + 2.001x_2 = k_1$
$x_1 + 2x_2 = k_2$

(A) $k_1 = 1, k_2 = 1$
(B) $k_1 = 1, k_2 = 0$
(C) $k_1 = 0, k_2 = 1$

Discuss the effect of small changes in the constant terms on the solution set of this system.

38. Repeat Problem 37 for the following system:

$x_1 - 3.001x_2 = k_1$
$x_1 - 3x_2 = k_2$

APPLICATIONS

Solve using systems of equations and matrix inverse methods.

39. Resource Allocation. A concert hall has 10,000 seats and two categories of ticket prices, $4 and $8. Assume all seats in each category can be sold.

	Concert		
	1	2	3
Tickets sold	10,000	10,000	10,000
Return required	$56,000	$60,000	$68,000

(A) How many tickets of each category should be sold to bring in each of the returns indicated in the table?

(B) Is it possible to bring in a return of $9,000? Of $3,000? Explain.

(C) Describe all the possible returns.

40. Production Scheduling. Labor and material costs for manufacturing two guitar models are given in the following table:

Guitar Model	Labor Cost	Material Cost
A	\$30	\$20
B	\$40	\$30

(A) If a total of \$3,000 a week is allowed for labor and material, how many of each model should be produced each week to exactly use each of the allocations of the \$3,000 indicated in the following table?

	Weekly Allocation		
	1	2	3
Labor	\$1,800	\$1,750	\$1,720
Material	\$1,200	\$1,250	\$1,280

(B) Is it possible to use an allocation of \$1,600 for labor and \$1,400 for material? Of \$2,000 for labor and \$1,000 for material? Explain.

★ **41. Circuit Analysis.** A direct current electric circuit consisting of conductors (wires), resistors, and batteries is diagrammed in the figure.

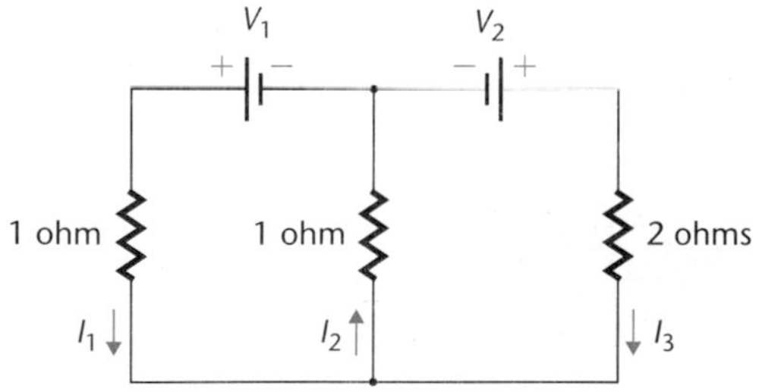

If I_1, I_2, and I_3 are the currents (in amperes) in the three branches of the circuit and V_1 and V_2 are the voltages (in volts) of the two batteries, then *Kirchhoff's* laws* can be used to show that the currents satisfy the following system of equations:

$$I_1 - I_2 + I_3 = 0$$
$$I_1 + I_2 = V_1$$
$$I_2 + 2I_3 = V_2$$

Solve this system for

(A) $V_1 = 10$ volts, $V_2 = 10$ volts

(B) $V_1 = 10$ volts, $V_2 = 15$ volts

(C) $V_1 = 15$ volts, $V_2 = 10$ volts

*Gustav Kirchhoff (1824–1887), a German physicist, was among the first to apply theoretical mathematics to physics. He is best-known for his development of certain properties of electric circuits, which are now known as **Kirchhoff's laws.**

★ **42. Circuit Analysis.** Repeat Problem 41 for the electric circuit shown in the figure.

$$I_1 - I_2 + I_3 = 0$$
$$I_1 + 2I_2 = V_1$$
$$2I_2 + 2I_3 = V_2$$

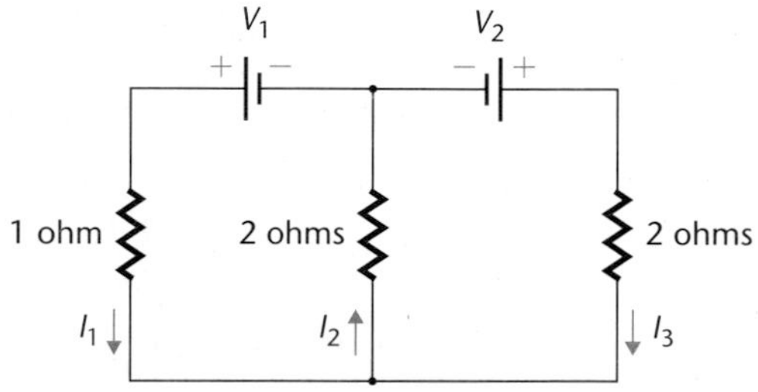

★★ **43. Geometry.** The graph of $f(x) = ax^2 + bx + c$ passes through the points $(1, k_1)$, $(2, k_2)$, and $(3, k_3)$. Determine a, b, and c for

(A) $k_1 = -2, k_2 = 1, k_3 = 6$

(B) $k_1 = 4, k_2 = 3, k_3 = -2$

(C) $k_1 = 8, k_2 = -5, k_3 = 4$

★★ **44. Geometry.** Repeat Problem 43 if the graph passes through the points $(-1, k_1)$, $(0, k_2)$, and $(1, k_3)$.

Check your answers in Problems 43 and 44 by graphing $y = f(x)$ on a graphing utility and verifying that the graph passes through the indicated points.

45. Diets. A biologist has available two commercial food mixes with the following percentages of protein and fat:

Mix	Protein (%)	Fat (%)
A	20	2
B	10	6

(A) How many ounces of each mix should be used to prepare each of the diets listed in the following table?

	Diet		
	1	2	3
Protein	20 oz	10 oz	10 oz
Fat	6 oz	4 oz	6 oz

(B) Is it possible to prepare a diet consisting of 20 ounces of protein and 14 ounces of fat? Of 20 ounces of protein and 1 ounce of fat? Explain.

Section 5-7 Systems of Linear Inequalities

- Graphing Linear Inequalities in Two Variables
- Solving Systems of Linear Inequalities Graphically
- Application

Many applications of mathematics involve systems of inequalities rather than systems of equations. A graph is often the most convenient way to represent the solutions of a system of inequalities in two variables. In this section, we discuss techniques for graphing both a single linear inequality in two variables and a system of linear inequalities in two variables.

Graphing Linear Inequalities in Two Variables

We know how to graph first-degree equations such as

$$y = 2x - 3 \qquad \text{and} \qquad 2x - 3y = 5$$

but how do we graph first-degree inequalities such as

$$y \le 2x - 3 \qquad \text{and} \qquad 2x - 3y > 5$$

Actually, graphing these inequalities is almost as easy as graphing the equations. But before we begin, we must discuss some important subsets of a plane in a rectangular coordinate system.

A line divides a plane into two halves called **half-planes.** A vertical line divides a plane into **left** and **right half-planes** [Fig. 1(a)]; a nonvertical line divides a plane into **upper** and **lower half-planes** [Fig. 1(b)].

FIGURE 1
Half-planes.

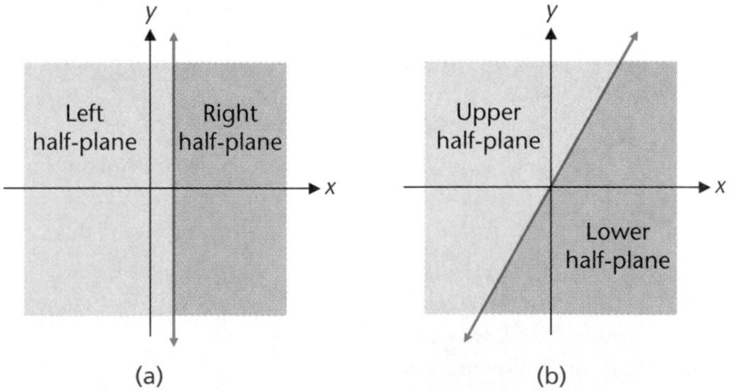

Explore/Discuss 1

Consider the following linear equation and related linear inequalities:

(1) $2x - 3y = 12$ (2) $2x - 3y < 12$ (3) $2x - 3y > 12$

(A) Graph the line with equation (1).

(B) Find the point on this line with x coordinate 3 and draw a vertical line through this point. Discuss the relationship between the y coordinates of the points on this line and statements (1), (2), and (3).

Explore/Discuss 1 continued

(C) Repeat part B for $x = -3$. For $x = 9$.

(D) Based on your observations in parts B and C, write a verbal description of all the points in the plane that satisfy equation (1), those that satisfy inequality (2), and those that satisfy inequality (3).

Now let's investigate the half-planes determined by the linear equation $y = 2x - 3$. We start by graphing $y = 2x - 3$ (Fig. 2). For any given value of x, there is exactly one value for y such that (x, y) lies on the line. For the same x, if the point (x, y) is below the line, then $y < 2x - 3$. Thus, the lower half-plane corresponds to the solution of the inequality $y < 2x - 3$. Similarly, the upper half-plane corresponds to the solution of the inequality $y > 2x - 3$, as shown in Figure 2.

FIGURE 2

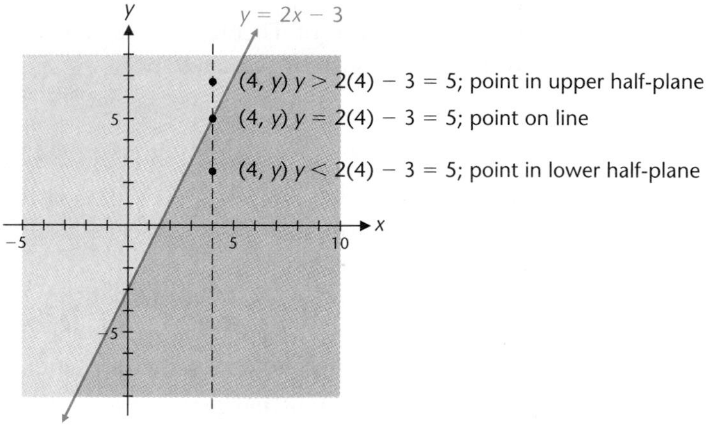

The four inequalities formed from $y = 2x - 3$ by replacing the = sign by $\geq$, $>$, $\leq$, and $<$, respectively, are

$$y \geq 2x - 3 \qquad y > 2x - 3 \qquad y \leq 2x - 3 \qquad y < 2x - 3$$

The graph of each is a half-plane. The line $y = 2x - 3$, called the **boundary line** for the half-plane, is included for $\geq$ and $\leq$ and excluded for $>$ and $<$. In Figure 3, the half-planes are indicated with small arrows on the graph of $y = 2x - 3$ and then graphed as shaded regions. Included boundary lines are shown as solid lines, and excluded boundary lines are shown as dashed lines.

FIGURE 3

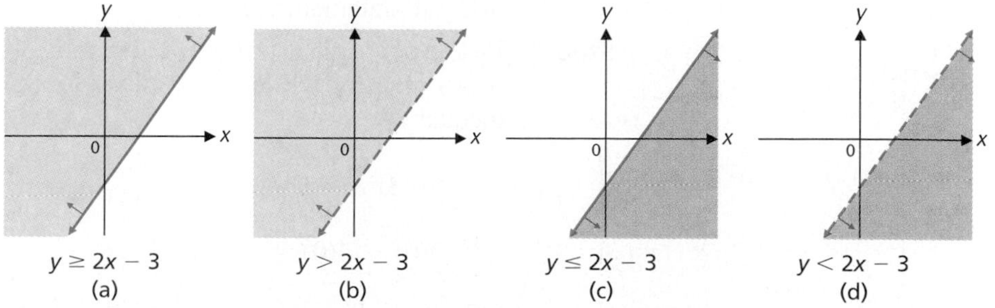

THEOREM 1

GRAPHS OF LINEAR INEQUALITIES IN TWO VARIABLES

The graph of a linear inequality

$$Ax + By < C \qquad \text{or} \qquad Ax + By > C$$

with $B \neq 0$, is either the upper half-plane or the lower half-plane (but not both) determined by the line $Ax + By = C$.

If $B = 0$, then the graph of

$$Ax < C \qquad \text{or} \qquad Ax > C$$

is either the left half-plane or the right half-plane (but not both) determined by the line $Ax = C$.

As a consequence of Theorem 1, we state a simple and fast mechanical procedure for graphing linear inequalities.

PROCEDURE FOR GRAPHING LINEAR INEQUALITIES IN TWO VARIABLES

Step 1. Graph $Ax + By = C$ as a dashed line if equality is not included in the original statement or as a solid line if equality is included.

Step 2. Choose a test point anywhere in the plane not on the line and substitute the coordinates into the inequality. The origin (0, 0) often requires the least computation.

Step 3. The graph of the original inequality includes the half-plane containing the test point if the inequality is satisfied by that point, or the half-plane not containing that point if the inequality is not satisfied by that point.

EXAMPLE 1

Graphing a Linear Inequality

Graph: $3x - 4y \leq 12$
Check on a graphing utility.

Solution

FIGURE 4

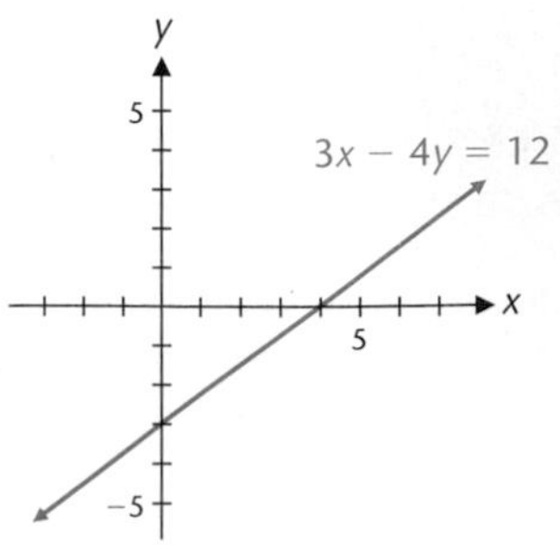

Step 1. Graph $3x - 4y = 12$ as a solid line, since equality is included in the original statement (Fig. 4).

Step 2. Pick a convenient test point above or below the line. The origin (0, 0) requires the least computation. Substituting (0, 0) into the inequality

$$3x - 4y \leq 12$$

$$3(0) - 4(0) = 0 \leq 12$$

produces a true statement; therefore, (0, 0) is in the solution set.

Step 3. The line $3x - 4y = 12$ and the half-plane containing the origin form the graph of $3x - 4y \le 12$ (Fig. 5).

FIGURE 5

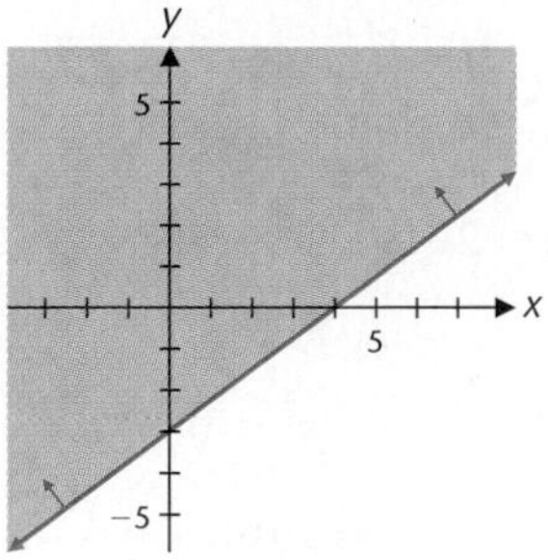

Check

Figure 6 shows a check of this solution on a graphing utility. In Figure 6(a), the small triangle to the left of y_1 indicates that the option to shade above the graph was selected. Consult the manual to see how to shade graphs on your graphing utility.

FIGURE 6

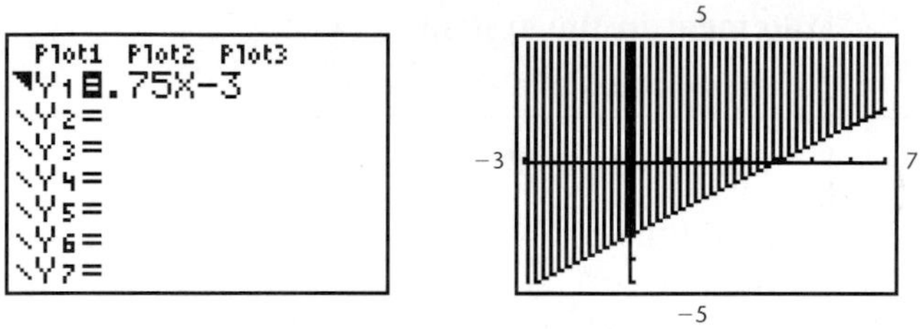

(a) (b)

MATCHED PROBLEM 1

Graph: $2x + 3y < 6$
Check on a graphing utility.

EXAMPLE 2

Graphing a Linear Inequality

Graph: (A) $y > -3$ (B) $2x \le 5$

Solutions

(A) The graph of $y > -3$ is shown in Figure 7.

(B) The graph of $2x \le 5$ is shown in Figure 8.

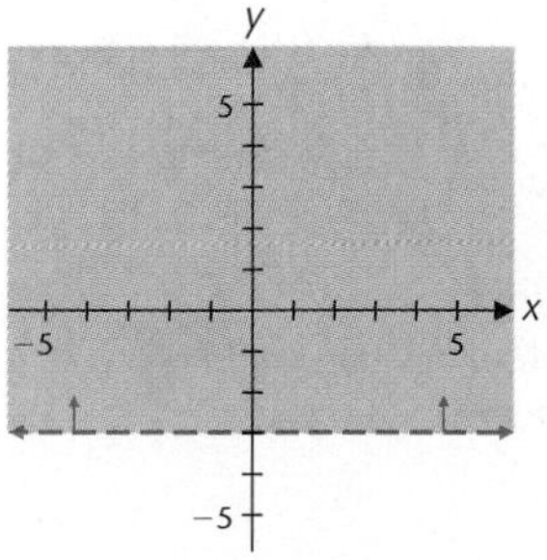

FIGURE 7

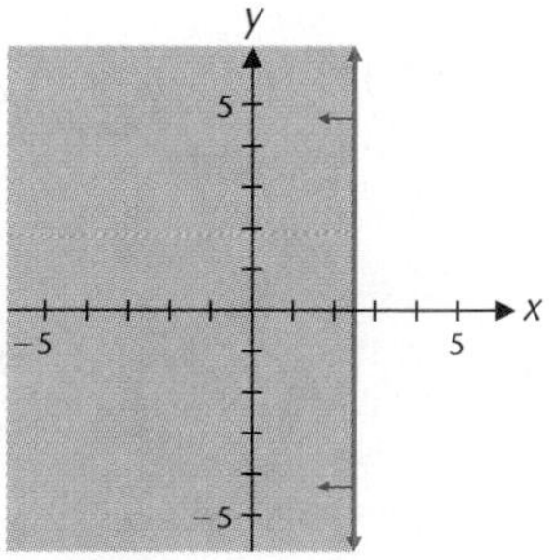

FIGURE 8

MATCHED PROBLEM 2

Graph: (A) $y \le 2$ (B) $3x > -8$

Solving Systems of Linear Inequalities Graphically

We now consider systems of linear inequalities such as

$$\begin{aligned} x + y &\geq 6 \\ 2x - y &\geq 0 \end{aligned} \qquad \text{and} \qquad \begin{aligned} 2x + y &\leq 22 \\ x + y &\leq 13 \\ 2x + 5y &\leq 50 \\ x &\geq 0 \\ y &\geq 0 \end{aligned}$$

We wish to **solve** such systems **graphically**—that is, to find the graph of all ordered pairs of real numbers (x, y) that simultaneously satisfy all the inequalities in the system. The graph is called the **solution region** for the system. To find the solution region, we graph each inequality in the system and then take the intersection of all the graphs. To simplify the discussion that follows, **we will consider only systems of linear inequalities where equality is included in each statement in the system.**

EXAMPLE 3

Solving a System of Linear Inequalities Graphically

Solve the following system of linear inequalities graphically:

$$\begin{aligned} x + y &\geq 6 \\ 2x - y &\geq 0 \end{aligned}$$

Solution

First, graph the line $x + y = 6$ and shade the region that satisfies the inequality $x + y \geq 6$. This region is shaded in blue in Figure 9(a). Next, graph the line $2x - y = 0$ and shade the region that satisfies the inequality $2x - y \geq 0$. This region is shaded in red in Figure 9(a). The solution region for the system of inequalities is the intersection of these two regions. This is the region shaded in both red and blue in Figure 9(a), which is redrawn in Figure 9(b) with only the solution region shaded for clarity. The coordinates of any point in the shaded region of Figure 9(b) specify a solution to the system. For example, the points (2, 4), (6, 3), and (7.43, 8.56) are three of infinitely many solutions, as can be easily checked. The intersection point (2, 4) can be obtained by solving the equations $x + y = 6$ and $2x - y = 0$ simultaneously.

FIGURE 9

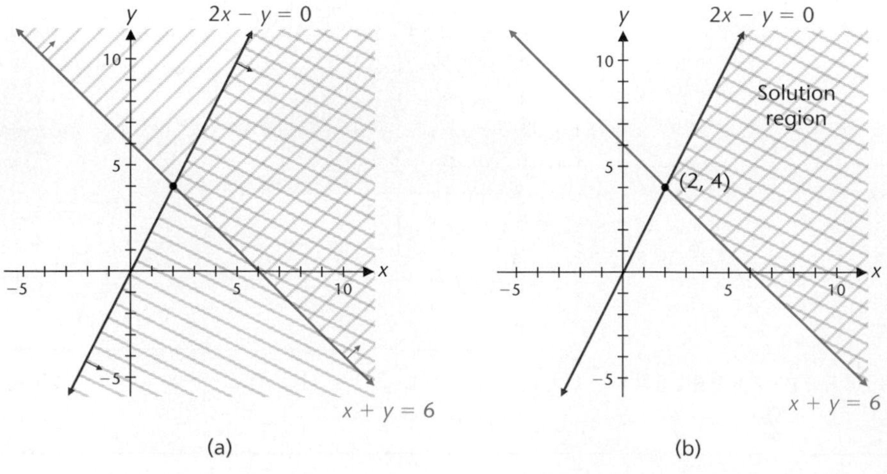

MATCHED PROBLEM 3

Solve the following system of linear inequalities graphically:

$$\begin{aligned} 3x + y &\le 21 \\ x - 2y &\le 0 \end{aligned}$$

Explore/Discuss 2

Refer to Example 3. Graph each boundary line and shade the regions obtained by reversing each inequality. That is, shade the region of the plane that corresponds to the inequality $x + y < 6$ and then shade the region that corresponds to the inequality $2x - y < 0$. What portion of the plane is left unshaded? Compare this method with the one used in the solution to Example 3.

The method of solving inequalities investigated in Explore/Discuss 2 works very well on a graphing utility that allows the user to shade above and below a graph. Referring to Example 3, the unshaded region in Figure 10(b) corresponds to the solution region in Figure 9(b).

FIGURE 10

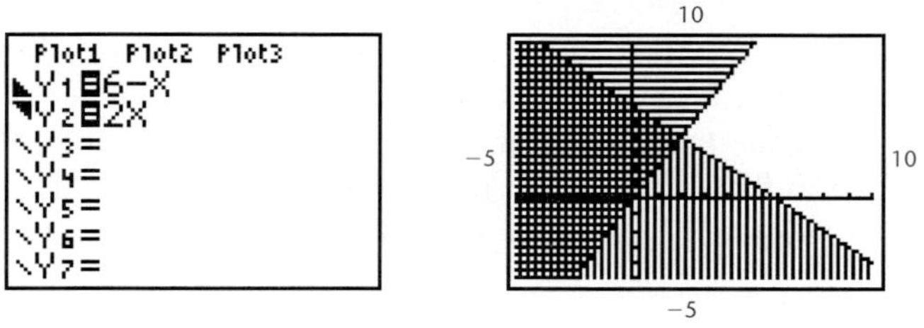

(a) (b)

The points of intersection of the lines that form the boundary of a solution region play a fundamental role in the solution of linear programming problems, which are discussed in the next section.

DEFINITION 1

CORNER POINT

A **corner point** of a solution region is a point in the solution region that is the intersection of two boundary lines.

The point (2, 4) is the only corner point of the solution region in Example 3; see Figure 9(b).

EXAMPLE 4

Solving a System of Linear Inequalities Graphically

Solve the following system of linear inequalities graphically, and find the corner points.

$$\begin{aligned} 2x + y &\leq 22 \\ x + y &\leq 13 \\ 2x + 5y &\leq 50 \\ x &\geq 0 \\ y &\geq 0 \end{aligned}$$

Solution The inequalities $x \geq 0$ and $y \geq 0$, called **nonnegative restrictions,** occur frequently in applications involving systems of inequalities since x and y often represent quantities that can't be negative—number of units produced, number of hours worked, and the like. The solution region lies in the first quadrant, and we can restrict our attention to that portion of the plane. First we graph the line associated with each inequality on a graphing utility (Fig. 11).

FIGURE 11

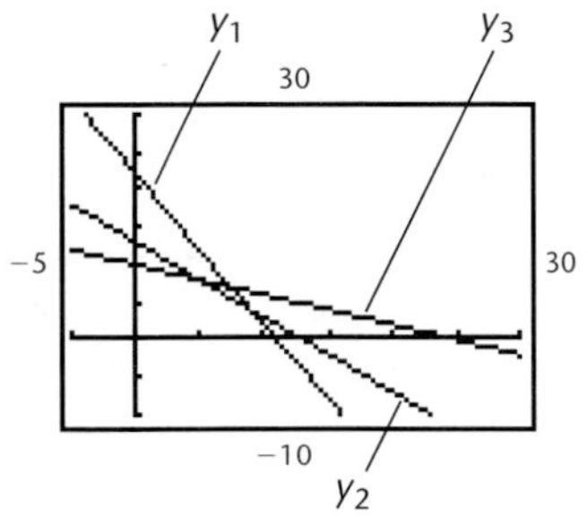

$2x + y = 22$ Enter $y_1 = 22 - 2x$.

$x + y = 13$ Enter $y_2 = 13 - x$.

$2x + 5y = 50$ Enter $y_3 = 10 - 0.4x$.

Next, choosing (0, 0) as a test point, we see that the graph of each of the first three inequalities in the system consists of its corresponding line and the half-plane lying below it. Thus, the solution region of the system consists of the points in the first quadrant that simultaneously lie below all three of these lines (Fig. 12). Figure 13 provides a check on a graphing utility.

FIGURE 12

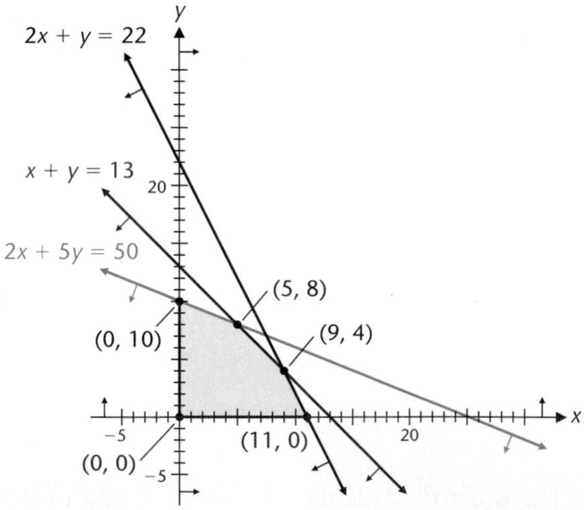

FIGURE 13

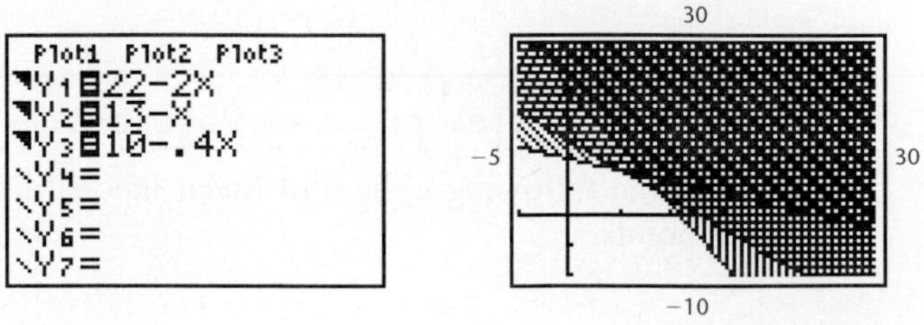

(a) (b)

The corner points (0, 0), (0, 10), and (11, 0) are easily read from the graph. Using an intersection routine (details omitted), the other two corner points are

(9, 4) Intersection of y_1 and y_2

and

(5, 8) Intersection of y_2 and y_3

Note that lines y_1 and y_3 also intersect, but the intersection point is not part of the solution region, and hence, is not a corner point. This can be checked by finding this intersection point and testing it in the system of inequalities.

MATCHED PROBLEM 4

Solve the following system of linear inequalities graphically, and find the corner points:

$$\begin{aligned} 5x + y &\geq 20 \\ x + y &\geq 12 \\ x + 3y &\geq 18 \\ x &\geq 0 \\ y &\geq 0 \end{aligned}$$

If we compare the solution regions of Examples 3 and 4, we see that there is a fundamental difference between these two regions. We can draw a circle around the solution region in Example 4. However, it is impossible to include all the points in the solution region in Example 3 in any circle, no matter how large we draw it. This leads to the following definition.

DEFINITION 2

BOUNDED AND UNBOUNDED SOLUTION REGIONS

A solution region of a system of linear inequalities is **bounded** if it can be enclosed within a circle. If it cannot be enclosed within a circle, then it is **unbounded.**

Thus, the solution region for Example 4 is bounded and the solution region for Example 3 is unbounded. This definition will be important in the next section.

Application

EXAMPLE 5

Production Scheduling

A manufacturer of surfboards makes a standard model and a competition model. Each standard board requires 6 labor-hours for fabricating and 1 labor-hour for finishing. Each competition board requires 8 labor-hours for fabricating and 3 labor-hours for finishing. The maximum labor-hours available per week in the fabricating and finishing departments are 120 and 30, respectively. What combinations of boards can be produced each week so as not to exceed the number of labor-hours available in each department per week?

Solution To clarify relationships, we summarize the information in the following table:

	Standard Model (labor-hours per board)	Competition Model (labor-hours per board)	Maximum Labor-Hours Available per Week
Fabricating	6	8	120
Finishing	1	3	30

Let

x = Number of standard boards produced per week

y = Number of competition boards produced per week

These variables are restricted as follows:

Fabricating department restriction:

$$\begin{pmatrix}\text{Weekly fabricating}\\ \text{time for } x\\ \text{standard boards}\end{pmatrix} + \begin{pmatrix}\text{Weekly fabricating}\\ \text{time for } y\\ \text{competition boards}\end{pmatrix} \le \begin{pmatrix}\text{Maximum labor-hours}\\ \text{available per week}\end{pmatrix}$$

$$6x \quad + \quad 8y \quad \le \quad 120$$

Finishing department restriction:

$$\begin{pmatrix}\text{Weekly finishing}\\ \text{time for } x\\ \text{standard boards}\end{pmatrix} + \begin{pmatrix}\text{Weekly finishing}\\ \text{time for } y\\ \text{competition boards}\end{pmatrix} \le \begin{pmatrix}\text{Maximum labor-hours}\\ \text{available per week}\end{pmatrix}$$

$$1x \quad + \quad 3y \quad \le \quad 30$$

Since it is not possible to manufacture a negative number of boards, x and y also must satisfy the nonnegative restrictions

$$x \ge 0$$
$$y \ge 0$$

Thus, x and y must satisfy the following system of linear inequalities:

$$\begin{aligned} 6x + 8y &\le 120 && \text{Fabricating department restriction}\\ x + 3y &\le 30 && \text{Finishing department restriction}\\ x &\ge 0 && \text{Nonnegative restriction}\\ y &\ge 0 && \text{Nonnegative restriction} \end{aligned}$$

Graphing this system of linear inequalities, we obtain the set of **feasible solutions,** or the **feasible region,** as shown in Figure 14. For problems of this type and for the linear programming problems we consider in the next section, solution regions are often referred to as feasible regions. Any point within the shaded area, including the boundary lines, represents a possible production schedule. Any point outside the shaded area represents an impossible schedule. For example, it would be possible to produce 12 standard boards and 5 competition boards per week, but it would not be possible to produce 12 standard boards and 7 competition boards per week (see the figure).

FIGURE 14

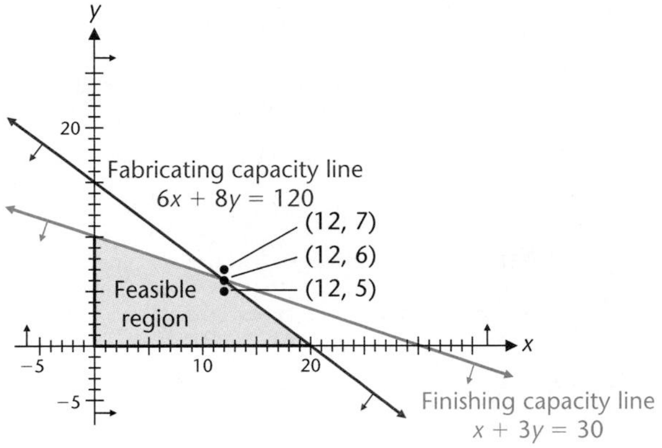

MATCHED PROBLEM 5

Repeat Example 5 using 5 hours for fabricating a standard board and a maximum of 27 labor-hours for the finishing department.

Remark Refer to Example 5. How do we interpret a production schedule of 10.5 standard boards and 4.3 competition boards? It is not possible to manufacture a fraction of a board. But it is possible to *average* 10.5 standard and 4.3 competition boards per week. In general, we will assume that all points in the feasible region represent acceptable solutions, even though noninteger solutions might require special interpretation.

Answers to Matched Problems

1.

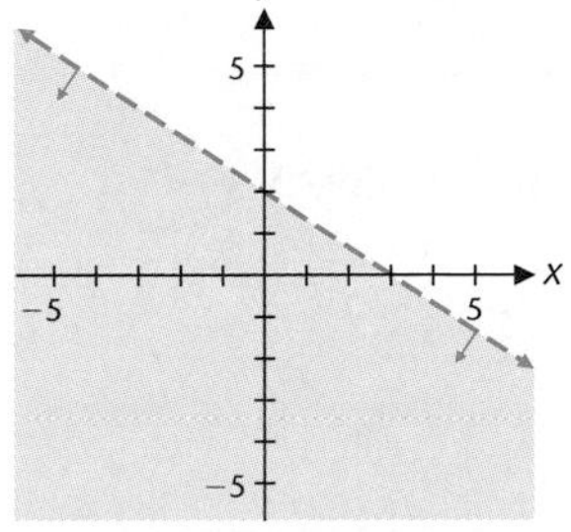

2. (A)

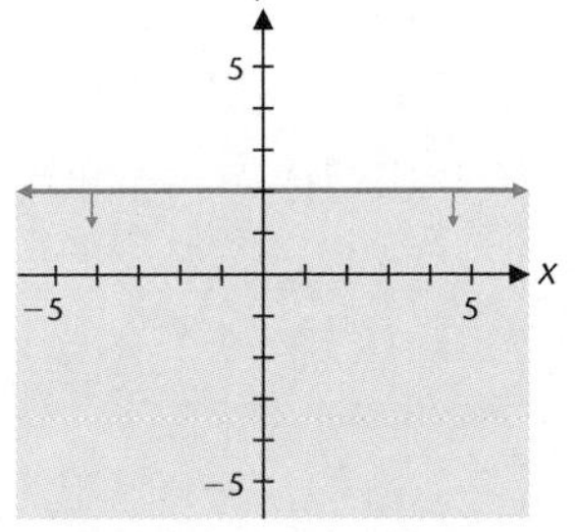

(B)

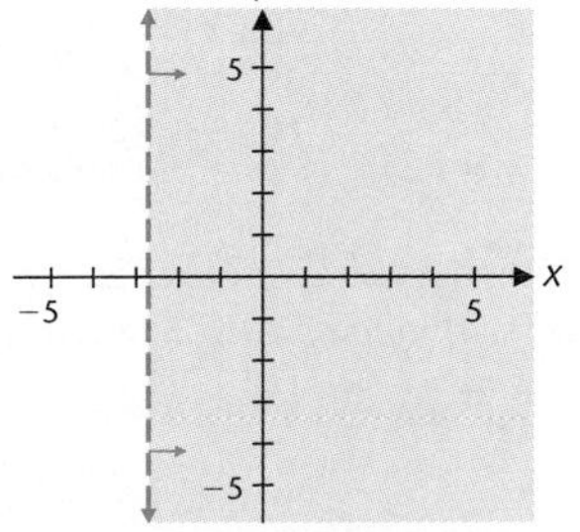

3.

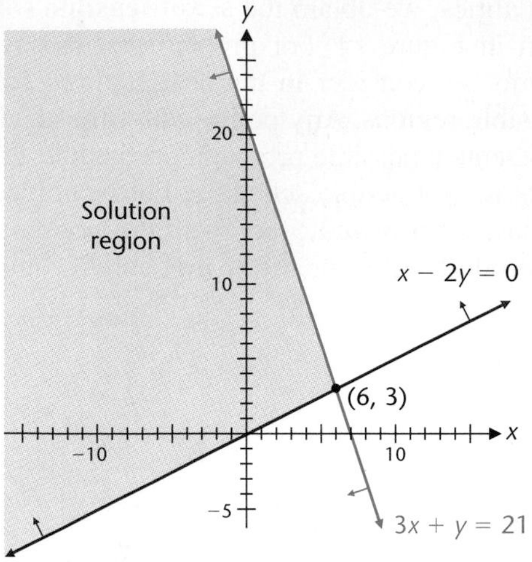

4.

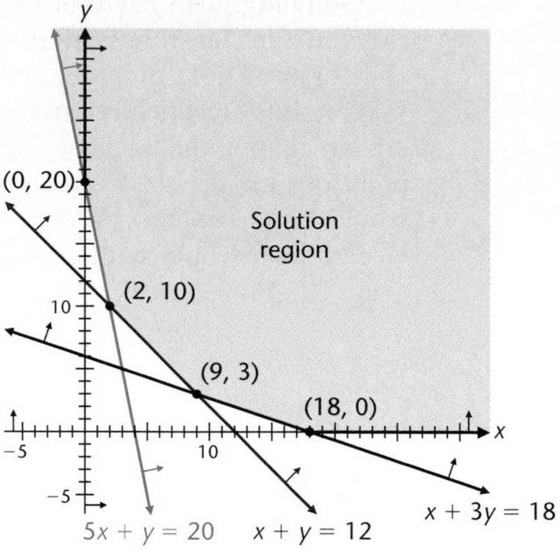

5.

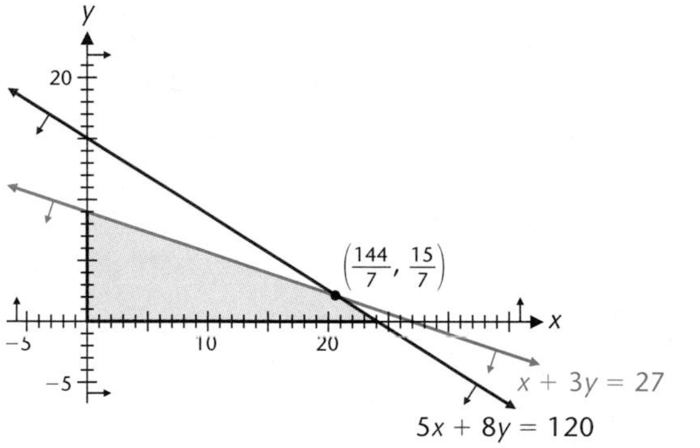

EXERCISE 5-7

A

Graph each inequality in Problems 1–10.

1. $2x - 3y < 6$

2. $3x + 4y < 12$

3. $3x + 2y \geq 18$

4. $3y - 2x \geq 24$

5. $y \leq \frac{2}{3}x + 5$

6. $y \geq \frac{1}{3}x - 2$

7. $y < 8$

8. $x > -5$

9. $-3 \leq y < 2$

10. $-1 < x \leq 3$

In Problems 11–14, match the solution region of each system of linear inequalities with one of the four regions shown in the figure in the next column.

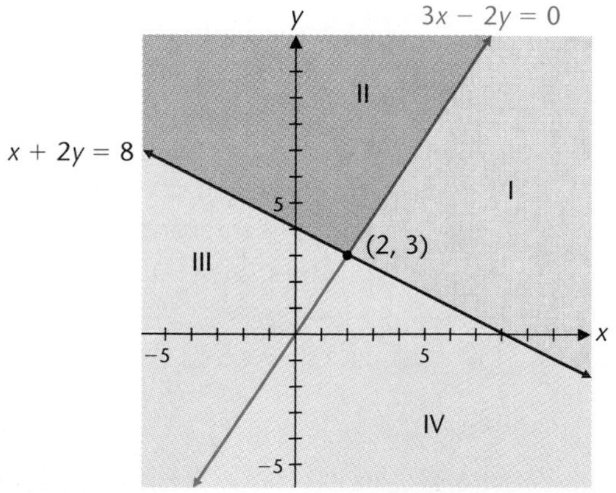

11. $x + 2y \leq 8$
$3x - 2y \geq 0$

12. $x + 2y \geq 8$
$3x - 2y \leq 0$

13. $x + 2y \geq 8$
$3x - 2y \geq 0$

14. $x + 2y \leq 8$
$3x - 2y \leq 0$

In Problems 15–20, solve each system of linear inequalities graphically.

15. $x > 5$
$y \le 6$

16. $x \le 4$
$y \ge 2$

17. $3x + y \ge 6$
$x \le 4$

18. $3x + 4y \le 12$
$y \ge -3$

19. $x - 2y \le 12$
$2x + y \ge 4$

20. $2x + 5y \le 20$
$x - 5y \le -5$

Problems 21–24 require a graphing utility that gives the user the option of shading above or below a graph.

(A) Graph the boundary lines in a standard viewing window and shade the region that contains the points that satisfy each inequality.

(B) Repeat part A, but this time shade the region that contains the points that do not satisfy each inequality (see Explore/Discuss 2)

Explain how you can recognize the solution region in each graph.

21. $x + y \le 5$
$2x - y \le 1$

22. $x - 2y \le 1$
$x + 3y \ge 12$

23. $2x + y \ge 4$
$3x - y \le 7$

24. $3x + y \ge -2$
$x - 2y \ge -6$

B

In Problems 25–28, match the solution region of each system of linear inequalities with one of the four regions shown in the figure below. Identify the corner points of each solution region.

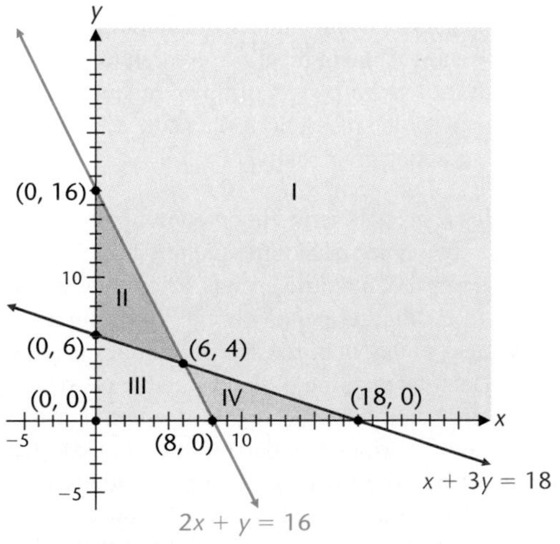

25. $x + 3y \le 18$
$2x + y \ge 16$
$x \ge 0$
$y \ge 0$

26. $x + 3y \le 18$
$2x + y \le 16$
$x \ge 0$
$y \ge 0$

27. $x + 3y \ge 18$
$2x + y \ge 16$
$x \ge 0$
$y \ge 0$

28. $x + 3y \ge 18$
$2x + y \le 16$
$x \ge 0$
$y \ge 0$

In Problems 29–40, solve the systems graphically, and indicate whether each solution region is bounded or unbounded. Find the coordinates of each corner point. Check on a graphing utility.

29. $2x + 3y \le 6$
$x \ge 0$
$y \ge 0$

30. $4x + 3y \le 12$
$x \ge 0$
$y \ge 0$

31. $4x + 5y \ge 20$
$x \ge 0$
$y \ge 0$

32. $5x + 6y \ge 30$
$x \ge 0$
$y \ge 0$

33. $2x + y \le 8$
$x + 3y \le 12$
$x \ge 0$
$y \ge 0$

34. $x + 2y \le 10$
$3x + y \le 15$
$x \ge 0$
$y \ge 0$

35. $4x + 3y \ge 24$
$2x + 3y \ge 18$
$x \ge 0$
$y \ge 0$

36. $x + 2y \ge 8$
$2x + y \ge 10$
$x \ge 0$
$y \ge 0$

37. $2x + y \le 12$
$x + y \le 7$
$x + 2y \le 10$
$x \ge 0$
$y \ge 0$

38. $3x + y \le 21$
$x + y \le 9$
$x + 3y \le 21$
$x \ge 0$
$y \ge 0$

39. $x + 2y \ge 16$
$x + y \ge 12$
$2x + y \ge 14$
$x \ge 0$
$y \ge 0$

40. $3x + y \ge 30$
$x + y \ge 16$
$x + 3y \ge 24$
$x \ge 0$
$y \ge 0$

C

In Problems 41–48, solve the systems graphically, and indicate whether each solution region is bounded or unbounded. Find the coordinates of each corner point.

41. $x + y \le 11$
$5x + y \ge 15$
$x + 2y \ge 12$

42. $4x + y \le 32$
$x + 3y \le 30$
$5x + 4y \ge 51$

43. $3x + 2y \ge 24$
$3x + y \le 15$
$x \ge 4$

44. $3x + 4y \le 48$
$x + 2y \ge 24$
$y \le 9$

45. $x + y \le 10$
$3x + 5y \ge 15$
$3x - 2y \le 15$
$-5x + 2y \le 6$

46. $3x - y \ge 1$
$-x + 5y \ge 9$
$x + y \le 9$
$y \le 5$

47. $16x + 13y \le 119$
$12x + 16y \ge 101$
$-4x + 3y \le 11$

48. $8x + 4y \le 41$
$-15x + 5y \le 19$
$2x + 6y \ge 37$

APPLICATIONS

49. Manufacturing—Resource Allocation. A manufacturing company makes two types of water skis: a trick ski and a slalom ski. The trick ski requires 6 labor-hours for fabricating and 1 labor-hour for finishing. The slalom ski requires 4 labor-hours for fabricating and 1 labor-hour for finishing. The maximum labor-hours available per day for fabricating and finishing are 108 and 24, respectively. If x is the number of trick skis and y is the number of slalom skis produced per day, write a system of inequalities that indicates appropriate restraints on x and y. Find the set of feasible solutions graphically for the number of each type of ski that can be produced.

50. Manufacturing—Resource Allocation. A furniture manufacturing company manufactures dining room tables and chairs. A table requires 8 labor-hours for assembling and 2 labor-hours for finishing. A chair requires 2 labor-hours for assembling and 1 labor-hour for finishing. The maximum labor-hours available per day for assembly and finishing are 400 and 120, respectively. If x is the number of tables and y is the number of chairs produced per day, write a system of inequalities that indicates appropriate restraints on x and y. Find the set of feasible solutions graphically for the number of tables and chairs that can be produced.

★ **51. Manufacturing—Resource Allocation.** Refer to Problem 49. The company makes a profit of \$50 on each trick ski and a profit of \$60 on each slalom ski.

(A) If the company makes 10 trick and 10 slalom skis per day, the daily profit will be \$1,100. Are there other feasible production schedules that will result in a daily profit of \$1,100? How are these schedules related to the graph of the line $50x + 60y = 1{,}100$?

(B) Find a feasible production schedule that will produce a daily profit greater than \$1,100 and repeat part A for this schedule.

(C) Discuss methods for using lines like those in parts A and B to find the largest possible daily profit.

★ **52. Manufacturing—Resource Allocation.** Refer to Problem 50. The company makes a profit of \$50 on each table and a profit of \$15 on each chair.

(A) If the company makes 20 tables and 20 chairs per day, the daily profit will be \$1,300. Are there other feasible production schedules that will result in a daily profit of \$1,300? How are these schedules related to the graph of the line $50x + 15y = 1{,}300$?

(B) Find a feasible production schedule that will produce a daily profit greater than \$1,300 and repeat part A for this schedule.

(C) Discuss methods for using lines like those in parts A and B to find the largest possible daily profit.

53. Nutrition—Plants. A farmer can buy two types of plant food, mix A and mix B. Each cubic yard of mix A contains 20 pounds of phosphoric acid, 30 pounds of nitrogen, and 5 pounds of potash. Each cubic yard of mix B contains 10 pounds of phosphoric acid, 30 pounds of nitrogen, and 10 pounds of potash. The minimum requirements are 460 pounds of phosphoric acid, 960 pounds of nitrogen, and 220 pounds of potash. If x is the number of cubic yards of mix A used and y is the number of cubic yards of mix B used, write a system of inequalities that indicates appropriate restraints on x and y. Find the set of feasible solutions graphically for the amount of mix A and mix B that can be used.

54. Nutrition. A dietitian in a hospital is to arrange a special diet using two foods. Each ounce of food M contains 30 units of calcium, 10 units of iron, and 10 units of vitamin A. Each ounce of food N contains 10 units of calcium, 10 units of iron, and 30 units of vitamin A. The minimum requirements in the diet are 360 units of calcium, 160 units of iron, and 240 units of vitamin A. If x is the number of ounces of food M used and y is the number of ounces of food N used, write a system of linear inequalities that reflects the conditions indicated. Find the set of feasible solutions graphically for the amount of each kind of food that can be used.

55. Sociology. A city council voted to conduct a study on inner-city community problems. A nearby university was contacted to provide sociologists and research assistants. Each sociologist will spend 10 hours per week collecting data in the field and 30 hours per week analyzing data in the research center. Each research assistant will spend 30 hours per week in the field and 10 hours per week in the research center. The minimum weekly labor-hour requirements are 280 hours in the field and 360 hours in the research center. If x is the number of sociologists hired for the study and y is the number of research assistants hired for the study, write a system of linear inequalities that indicates appropriate restrictions on x and y. Find the set of feasible solutions graphically.

56. Psychology. In an experiment on conditioning, a psychologist uses two types of Skinner (conditioning) boxes with mice and rats. Each mouse spends 10 minutes per day in box A and 20 minutes per day in box B. Each rat spends 20 minutes per day in box A and 10 minutes per day in box B. The total maximum time available per day is 800 minutes for box A and 640 minutes for box B. We are interested in the various numbers of mice and rats that can be used in the experiment under the conditions stated. If x is the number of mice used and y is the number of rats used, write a system of linear inequalities that indicates appropriate restrictions on x and y. Find the set of feasible solutions graphically.

Section 5-8 | Linear Programming

- A Linear Programming Problem
- Linear Programming—A General Description
- Application

Several problems in Section 5-7 are related to the general type of problems called *linear programming problems*. Linear programming is a mathematical process that has been developed to help management in decision making, and it has become one of the most widely used and best known tools of management science and industrial engineering. We will use an intuitive graphical approach based on the techniques discussed in Section 5-7 to illustrate this process for problems involving two variables.

The American mathematician George B. Dantzig (1914–) formulated the first linear programming problem in 1947 and introduced a solution technique, called the *simplex method*, that does not rely on graphing and is readily adaptable to computer solutions. Today, it is quite common to use a computer to solve applied linear programming problems involving thousands of variables and thousands of inequalities.

A Linear Programming Problem

We begin our discussion with an example that will lead to a general procedure for solving linear programming problems in two variables.

EXAMPLE 1 **Production Scheduling**

A manufacturer of fiberglass camper tops for pickup trucks makes a compact model and a regular model. Each compact top requires 5 hours from the fabricating department and 2 hours from the finishing department. Each regular top requires 4 hours from the fabricating department and 3 hours from the finishing department. The maximum labor-hours available per week in the fabricating department and the finishing department are 200 and 108, respectively. If the company makes a profit of $40 on each compact top and $50 on each regular top, how many tops of each type should be manufactured each week to maximize the total weekly profit, assuming all tops can be sold? What is the maximum profit?

Solution This is an example of a linear programming problem. To see relationships more clearly, we summarize the manufacturing requirements, objectives, and restrictions in the table:

	Compact Model (labor-hours per top)	Regular Model (labor-hours per top)	Maximum Labor-Hours Available per Week
Fabricating	5	4	200
Finishing	2	3	108
Profit per top	$40	$50	

We now proceed to formulate a *mathematical model* for the problem and then to solve it using graphical methods.

Objective Function

The *objective* of management is to *decide* how many of each camper top model should be produced each week to *maximize* profit. Let

$$\left.\begin{aligned} x &= \text{Number of compact tops produced per week} \\ y &= \text{Number of regular tops produced per week} \end{aligned}\right\} \quad \text{Decision variables}$$

The following function gives the total profit P for x compact tops and y regular tops manufactured each week:

$$P = 40x + 50y \quad \text{Objective function}$$

Mathematically, management needs to decide on values for the **decision variables** (x and y) that achieve its objective, that is, maximizing the **objective function** (profit) $P = 40x + 50y$. It appears that the profit can be made as large as we like by manufacturing more and more tops—or can it?

Constraints

Any manufacturing company, no matter how large or small, has manufacturing limits imposed by available resources, plant capacity, demand, and so forth. These limits are referred to as **problem constraints.**

Fabricating department constraint:

$$\begin{pmatrix}\text{Weekly fabricating} \\ \text{time for } x \\ \text{compact tops}\end{pmatrix} + \begin{pmatrix}\text{Weekly fabricating} \\ \text{time for } y \\ \text{regular tops}\end{pmatrix} \leq \begin{pmatrix}\text{Maximum labor-hours} \\ \text{available per week}\end{pmatrix}$$

$$5x \quad + \quad 4y \quad \leq \quad 200$$

Finishing department constraint:

$$\begin{pmatrix}\text{Weekly finishing} \\ \text{time for } x \\ \text{compact tops}\end{pmatrix} + \begin{pmatrix}\text{Weekly finishing} \\ \text{time for } y \\ \text{regular tops}\end{pmatrix} \leq \begin{pmatrix}\text{Maximum labor-hours} \\ \text{available per week}\end{pmatrix}$$

$$2x \quad + \quad 3y \quad \leq \quad 108$$

Nonnegative constraints: It is not possible to manufacture a negative number of tops; thus, we have the **nonnegative constraints**

$$x \geq 0$$
$$y \geq 0$$

which we usually write in the form

$$x, y \geq 0$$

Mathematical Model

We now have a **mathematical model** for the problem under consideration:

$$\begin{aligned} \text{Maximize} \quad & P = 40x + 50y && \text{Objective function} \\ \text{Subject to} \quad & \left.\begin{aligned} 5x + 4y &\le 200 \\ 2x + 3y &\le 108 \end{aligned}\right\} && \text{Problem constraints} \\ & x, y \ge 0 && \text{Nonnegative constraints} \end{aligned}$$

Graphic Solution

Solving the system of linear inequality constraints **graphically,** as in Section 5-7, we obtain the feasible region for production schedules, as shown in Figure 1.

FIGURE 1

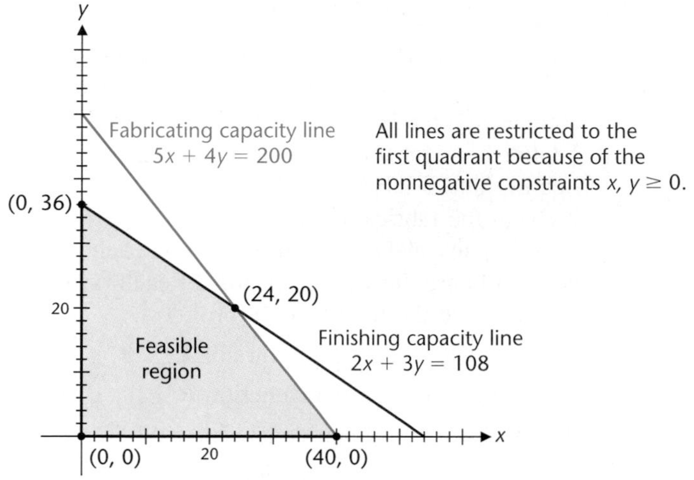

By choosing a production schedule (x, y) from the feasible region, a profit can be determined using the objective function $P = 40x + 50y$. For example, if $x = 24$ and $y = 10$, then the profit for the week is

$$P = 40(24) + 50(10) = \$1{,}460$$

Or if $x = 15$ and $y = 20$, then the profit for the week is

$$P = 40(15) + 50(20) = \$1{,}600$$

The question is, out of all possible production schedules (x, y) from the feasible region, which schedule(s) produces the maximum profit? Such a schedule, if it exists, is called an **optimal solution** to the problem because it produces the maximum value of the objective function and is in the feasible region. It is not practical to use point-by-point checking to find the optimal solution. Even if we consider only points with integer coordinates, there are over 800 such points in the feasible region for this problem. Instead, we use the theory that has been developed to solve linear programming problems. Using advanced techniques, it can be shown that:

If the feasible region is bounded, then one or more of the corner points of the feasible region is an optimal solution to the problem.

The maximum value of the objective function is unique; however, there can be more than one feasible production schedule that will produce this unique value. We will have more to say about this later in this section.

Corner Point (x, y)	Objective Function $P = 40x + 50y$	
(0, 0)	0	
(0, 36)	1,800	
(24, 20)	1,960	Maximum value of P
(40, 0)	1,600	

Since the feasible region for this problem is bounded, at least one of the corner points, (0, 0), (0, 36), (24, 20), or (40, 0), is an optimal solution. To find which one, we evaluate $P = 40x + 50y$ at each corner point and choose the corner point that produces the largest value of P. It is convenient to organize these calculations in a table, as shown in the margin.

Examining the values in the table, we see that the maximum value of P at a corner point is $P = 1{,}960$ at $x = 24$ and $y = 20$. Since the maximum value of P over the entire feasible region must always occur at a corner point, we conclude that the maximum profit is \$1,960 when 24 compact tops and 20 regular tops are produced each week.

MATCHED PROBLEM 1

We now convert the surfboard problem discussed in Section 5-7 into a linear programming problem. A manufacturer of surfboards makes a standard model and a competition model. Each standard board requires 6 labor-hours for fabricating and 1 labor-hour for finishing. Each competition board requires 8 labor-hours for fabricating and 3 labor-hours for finishing. The maximum labor-hours available per week in the fabricating and finishing departments are 120 and 30, respectively. If the company makes a profit of \$40 on each standard board and \$75 on each competition board, how many boards of each type should be manufactured each week to maximize the total weekly profit?

(A) Identify the decision variables.

(B) Write the objective function P.

(C) Write the problem constraints and the nonnegative constraints.

(D) Graph the feasible region, identify the corner points, and evaluate P at each corner point.

(E) How many boards of each type should be manufactured each week to maximize the profit? What is the maximum profit?

Explore/Discuss 1

Refer to Example 1. If we assign the profit function P in $P = 40x + 50y$ a particular value and plot the resulting equation in the coordinate system shown in Figure 1, we obtain a **constant-profit line (isoprofit line).** Every point in the feasible region on this line represents a production schedule that will produce the same profit. Figure 2 shows the constant-profit lines for $P = \$1{,}000$ and $P = \$1{,}500$.

FIGURE 2

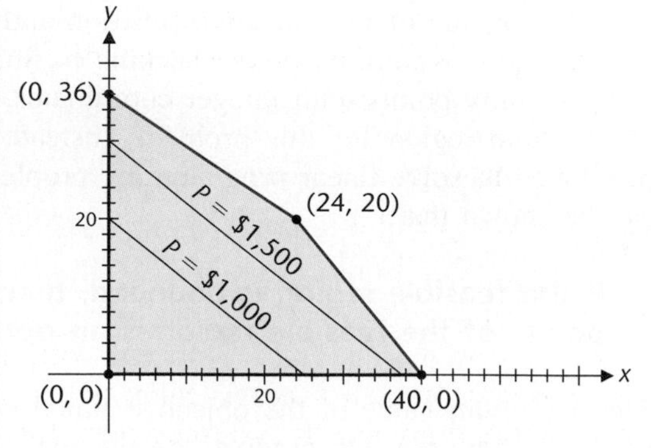

Explore/Discuss 1 continued

(A) How are all the constant-profit lines related?

(B) Place a straightedge along the constant-profit line for $P = \$1{,}000$ and slide it as far as possible in the direction of increasing profit without changing its slope and without leaving the feasible region. Explain how this process can be used to identify the optimal solution to a linear programming problem.

(C) If P is changed to $P = 25x + 75y$, graph the constant-profit lines for $P = \$1{,}000$ and $P = \$1{,}500$, and use a straightedge to identify the optimal solution. Check your answer by evaluating P at each corner point.

(D) Repeat part C for $P = 75x + 25y$.

Linear Programming—A General Description

The linear programming problems considered in Example 1 and Matched Problem 1 were *maximization problems*, where we wanted to maximize profits. The same technique can be used to solve *minimization problems*, where, for example, we may want to minimize costs. Before considering additional examples, we state a few general definitions.

A **linear programming problem** is one that is concerned with finding the **optimal value** (maximum or minimum value) of a linear *objective function* of the form

$$z = ax + by$$

where the *decision variables* x and y are subject to *problem constraints* in the form of linear inequalities and to *nonnegative constraints* $x, y \geq 0$. The set of points satisfying both the problem constraints and the nonnegative constraints is called the *feasible region* for the problem. Any point in the feasible region that produces the optimal value of the objective function over the feasible region is called an *optimal solution.*

Theorem 1 is fundamental to the solving of linear programming problems.

THEOREM 1

FUNDAMENTAL THEOREM OF LINEAR PROGRAMMING

Let S be the feasible region for a linear programming problem, and let $z = ax + by$ be the objective function. If S is bounded, then z has both a maximum and a minimum value on S and each of these occurs at a corner point of S. If S is unbounded, then a maximum or minimum value of z on S may not exist. However, if either does exist, then it must occur at a corner point of S.

We will not consider any problems with unbounded feasible regions in this brief introduction. If a feasible region is bounded, then Theorem 1 provides the basis for the following simple procedure for solving the associated linear programming problem:

SOLUTION OF LINEAR PROGRAMMING PROBLEMS

Step 1. Form a mathematical model for the problem:

(A) Introduce decision variables and write a linear objective function.

(B) Write problem constraints in the form of linear inequalities.

(C) Write nonnegative constraints.

Step 2. Graph the feasible region and find the corner points.

Step 3. Evaluate the objective function at each corner point to determine the optimal solution.

Before considering additional applications, we use this procedure to solve a linear programming problem where the model has already been determined.

EXAMPLE 2

Solving a Linear Programming Problem

Minimize and maximize $z = 5x + 15y$

Subject to
$$x + 3y \le 60$$
$$x + y \ge 10$$
$$x - y \le 0$$
$$x, y \ge 0$$

Solution

This problem is a combination of two linear programming problems—a minimization problem and a maximization problem. Since the feasible region is the same for both problems, we can solve these problems together. To begin, we graph the feasible region S, as shown in Figure 3, and find the coordinates of each corner point.

FIGURE 3

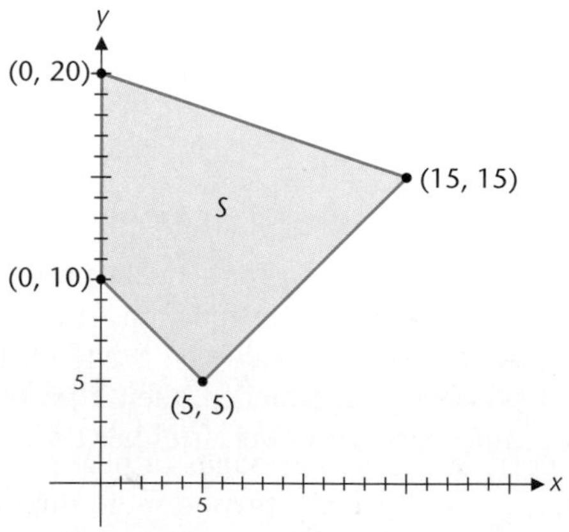

Next, we evaluate the objective function at each corner point, with the results given in the table:

Corner Point (x, y)	Objective Function $z = 5x + 15y$		
(0, 10)	150		
(0, 20)	300	Maximum value	Multiple optimal solutions
(15, 15)	300	Maximum value	
(5, 5)	100	Minimum value	

Examining the values in the table, we see that the minimum value of z on the feasible region S is 100 at (5, 5). Thus, (5, 5) is the optimal solution to the minimization problem. The maximum value of z on the feasible region S is 300, which occurs at (0, 20) and at (15, 15). Thus, the maximization problem has **multiple optimal solutions.** In general,

If two corner points are both optimal solutions of the same type (both produce the same maximum value or both produce the same minimum value) to a linear programming problem, then any point on the line segment joining the two corner points is also an optimal solution of that type.

It can be shown that this is the only time that an optimal value occurs at more than one point.

MATCHED PROBLEM 2

Minimizc and maximize $z = 10x + 5y$

Subject to

$$2x + y \geq 40$$
$$3x + y \leq 150$$
$$2x \quad y \geq 0$$
$$x, y \geq 0$$

Application

Now we consider another application where we must first find the mathematical model and then find its solution.

EXAMPLE 3

Agriculture

A farmer can use two types of plant food, mix A and mix B. The amounts (in pounds) of nitrogen, phosphoric acid, and potash in a cubic yard of each mix are given in the table. Tests performed on the soil in a large field indicate that the field needs at least 840 pounds of potash and at least 350 pounds of nitrogen. The tests also indicate that no more than 630 pounds of phosphoric acid should be added to the field. A cubic yard of mix A costs \$7, and a cubic yard of mix B costs \$9. How many cubic yards of each mix should the farmer add to the field in order to supply the necessary nutrients at minimal cost?

	Pounds Per Cubic Yard	
	Mix A	Mix B
Nitrogen	10	5
Potash	8	24
Phosphoric acid	9	6

Solution Let

x = Number of cubic yards of mix A added to the field
y = Number of cubic yards of mix B added to the field } Decision variables

We form the linear objective function

$$C = 7x + 9y$$

which gives the cost of adding x cubic yards of mix A and y cubic yards of mix B to the field. Using the data in the table and proceeding as in Example 1, we formulate the mathematical model for the problem:

Minimize	$C = 7x + 9y$	Objective function
Subject to	$10x + 5y \geq 350$	Nitrogen constraint
	$8x + 24y \geq 840$	Potash constraint
	$9x + 6y \leq 630$	Phosphoric acid constraint
	$x, y \geq 0$	Nonnegative constraints

Solving the system of constraint inequalities graphically, we obtain the feasible region S shown in Figure 4, and then we find the coordinates of each corner point.

FIGURE 4

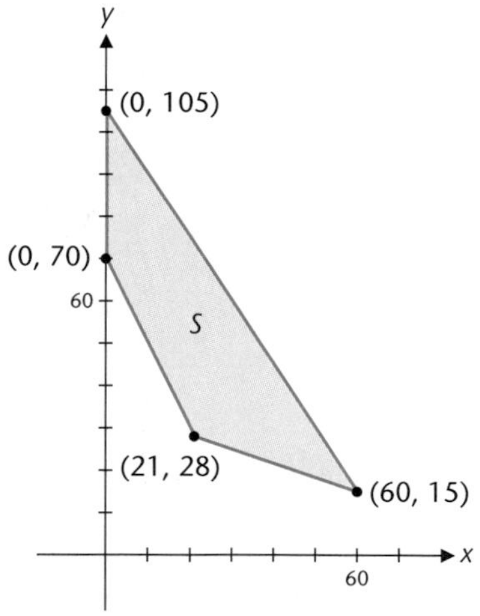

Corner Point (x, y)	Objective Function $C = 7x + 9y$	
(0, 105)	945	
(0, 70)	630	
(21, 28)	399	Minimum value of C
(60, 15)	555	

Next, we evaluate the objective function at each corner point, as shown in the table in the margin.

The optimal value is $C = 399$ at the corner point (21, 28). Thus, the farmer should add 21 cubic yards of mix A and 28 cubic yards of mix B at a cost of \$399. This will result in adding the following nutrients to the field:

Nitrogen:	$10(21) + 5(28)$	= 350 pounds
Potash:	$8(21) + 24(28)$	= 840 pounds
Phosphoric acid:	$9(21) + 6(28)$	= 357 pounds

All the nutritional requirements are satisfied.

MATCHED PROBLEM 3

Repeat Example 3 if the tests indicate that the field needs at least 400 pounds of nitrogen with all other conditions remaining the same.

Answers to Matched Problems

1. (A) x = Number of standard boards manufactured each week
y = Number of competition boards manufactured each week

(B) $P = 40x + 75y$ (C) $6x + 8y \le 120$ Fabricating constraint
$x + 3y \le 30$ Finishing constraint
$x, y \ge 0$ Nonnegative constraints

(D)

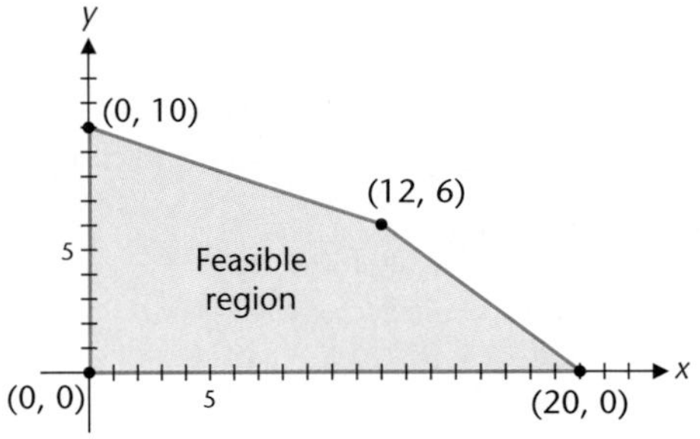

Corner Point (x, y)	Objective Function $P = 40x + 75y$
(0, 0)	0
(0, 10)	750
(12, 6)	930
(20, 0)	800

(E) 12 standard boards and 6 competition boards for a maximum profit of \$930

2. Max $z = 600$ at (30, 60); min $z = 200$ at (10, 20) and (20, 0) (multiple optimal solutions)

3. 27 cubic yards of mix A, 26 cubic yards of mix B; min $C = \$423$

EXERCISE 5-8

A

In Problems 1–4, find the maximum value of each objective function over the feasible region S shown in the figure below.

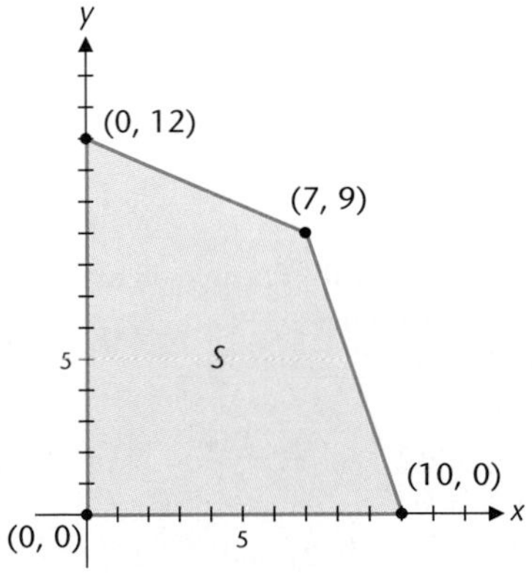

1. $z = x + y$

2. $z = 4x + y$

3. $z = 3x + 7y$

4. $z = 9x + 3y$

In Problems 5–8, find the minimum value of each objective function over the feasible region T shown in the figure below.

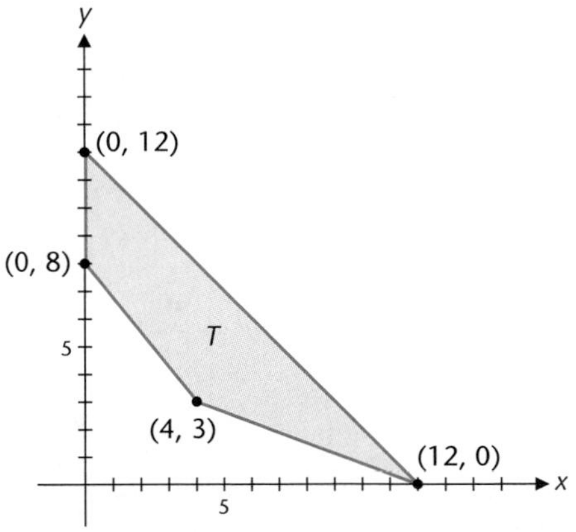

5. $z = 7x + 4y$

6. $z = 7x + 9y$

7. $z = 3x + 8y$

8. $z = 5x + 4y$

B

In Problems 9–22, solve the linear programming problems.

9. Maximize $z = 3x + 2y$
Subject to
$$x + 2y \le 10$$
$$3x + y \le 15$$
$$x, y \ge 0$$

10. Maximize $z = 4x + 5y$
Subject to
$$2x + y \le 12$$
$$x + 3y \le 21$$
$$x, y \ge 0$$

11. Minimize $z = 3x + 4y$
Subject to
$$2x + y \ge 8$$
$$x + 2y \le 10$$
$$x, y \ge 0$$

12. Minimize $z = 2x + y$
Subject to
$$4x + 3y \ge 24$$
$$4x + y \le 16$$
$$x, y \ge 0$$

13. Maximize $z = 3x + 4y$
Subject to
$$x + 2y \le 24$$
$$x + y \le 14$$
$$2x + y \le 24$$
$$x, y \ge 0$$

14. Maximize $z = 5x + 3y$
Subject to
$$3x + y \le 24$$
$$x + y \le 10$$
$$x + 3y \le 24$$
$$x, y \ge 0$$

15. Minimize $z = 5x + 6y$
Subject to
$$x + 4y \ge 20$$
$$4x + y \ge 20$$
$$x + y \le 20$$
$$x, y \ge 0$$

16. Minimize $z = x + 2y$
Subject to
$$2x + 3y \ge 30$$
$$3x + 2y \ge 30$$
$$x + y \le 15$$
$$x, y \ge 0$$

17. Minimize and maximize $z = 25x + 50y$
Subject to
$$x + 2y \le 120$$
$$x + y \ge 60$$
$$x - 2y \ge 0$$
$$x, y \ge 0$$

18. Minimize and maximize $z = 15x + 30y$
Subject to
$$x + 2y \ge 100$$
$$2x - y \le 0$$
$$2x + y \le 200$$
$$x, y \ge 0$$

19. Minimize and maximize $z = 25x + 15y$
Subject to
$$4x + 5y \ge 100$$
$$3x + 4y \le 240$$
$$x \le 60$$
$$y \le 45$$
$$x, y \ge 0$$

20. Minimize and maximize $z = 25x + 30y$
Subject to
$$2x + 3y \ge 120$$
$$3x + 2y \le 360$$
$$x \le 80$$
$$y \le 120$$
$$x, y \ge 0$$

21. Maximize $P = 525x_1 + 478x_2$
Subject to
$$275x_1 + 322x_2 \le 3{,}381$$
$$350x_1 + 340x_2 \le 3{,}762$$
$$425x_1 + 306x_2 \le 4{,}114$$
$$x_1, x_2 \ge 0$$

22. Maximize $P = 300x_1 + 460x_2$
Subject to
$$245x_1 + 452x_2 \le 4{,}181$$
$$290x_1 + 379x_2 \le 3{,}888$$
$$390x_1 + 299x_2 \le 4{,}407$$
$$x_1, x_2 \ge 0$$

C

23. The corner points for the feasible region determined by the problem constraints
$$2x + y \le 10$$
$$x + 3y \le 15$$
$$x, y \ge 0$$
are $O = (0, 0)$, $A = (5, 0)$, $B = (3, 4)$, and $C = (0, 5)$. If $z = ax + by$ and $a, b > 0$, determine conditions on a and b that ensure that the maximum value of z occurs

(A) Only at A

(B) Only at B

(C) Only at C

(D) At both A and B

(E) At both B and C

24. The corner points for the feasible region determined by the problem constraints
$$x + y \ge 4$$
$$x + 2y \ge 6$$
$$2x + 3y \le 12$$
$$x, y \ge 0$$
are $A = (6, 0)$, $B = (2, 2)$, and $C = (0, 4)$. If $z = ax + by$ and $a, b > 0$, determine conditions on a and b that ensure that the minimum value of z occurs

(A) Only at A (B) Only at B

(C) Only at C (D) At both A and B

(E) At both B and C

APPLICATIONS

25. **Resource Allocation.** A manufacturing company makes two types of water skis, a trick ski and a slalom ski. The relevant manufacturing data is given in the table.

(A) If the profit on a trick ski is $40 and the profit on a slalom ski is $30, how many of each type of ski should be manufactured each day to realize a maximum profit? What is the maximum profit?

(B) Discuss the effect on the production schedule and the maximum profit if the profit on a slalom ski decreases to $25 and all other data remains the same.

(C) Discuss the effect on the production schedule and the maximum profit if the profit on a slalom ski increases to $45 and all other data remains the same.

	Trick Ski (labor-hours per ski)	Slalom Ski (labor-hours per ski)	Maximum Labor-Hours Available per Day
Fabricating department	6	4	108
Finishing department	1	1	24

26. **Psychology.** In an experiment on conditioning, a psychologist uses two types of Skinner boxes with mice and rats. The amount of time (in minutes) each mouse and each rat spends in each box per day is given in the table. What is the maximum total number of mice and rats that can be used in this experiment? How many mice and how many rats produce this maximum?

	Mice (minutes)	Rats (minutes)	Max. Time Available per Day (minutes)
Skinner box *A*	10	20	800
Skinner box *B*	20	10	640

27. **Purchasing.** A trucking firm wants to purchase a maximum of 15 new trucks that will provide at least 36 tons of additional shipping capacity. A model *A* truck holds 2 tons and costs $15,000. A model *B* truck holds 3 tons and costs $24,000. How many trucks of each model should the company purchase to provide the additional shipping capacity at minimal cost? What is the minimal cost?

28. **Transportation.** The officers of a high school senior class are planning to rent buses and vans for a class trip. Each bus can transport 40 students, requires 3 chaperones, and costs $1,200 to rent. Each van can transport 8 students, requires 1 chaperone, and costs $100 to rent. The officers want to be able to accommodate at least 400 students with no more than 36 chaperones. How many vehicles of each type should they rent in order to minimize the transportation costs? What are the minimal transportation costs?

★ 29. **Resource Allocation.** A furniture company manufactures dining room tables and chairs. Each table requires 8 hours from the assembly department and 2 hours from the finishing department and contributes a profit of $90. Each chair requires 2 hours from the assembly department and 1 hour from the finishing department and contributes a profit of $25. The maximum labor-hours available each day in the assembly and finishing departments are 400 and 120, respectively.

(A) How many tables and how many chairs should be manufactured each day to maximize the daily profit? What is the maximum daily profit?

(B) Discuss the effect on the production schedule and the maximum profit if the marketing department of the company decides that the number of chairs produced should be at least four times the number of tables produced.

★ 30. **Resource Allocation.** An electronics firm manufactures two types of personal computers, a desktop model and a portable model. The production of a desktop computer requires a capital expenditure of $400 and 40 hours of labor. The production of a portable computer requires a capital expenditure of $250 and 30 hours of labor. The firm has $20,000 capital and 2,160 labor-hours available for production of desktop and portable computers.

(A) What is the maximum number of computers the company is capable of producing?

(B) If each desktop computer contributes a profit of $320 and each portable contributes a profit of $220, how much profit will the company make by producing the maximum number of computers determined in part A? Is this the maximum profit? If not, what is the maximum profit?

31. **Pollution Control.** Because of new federal regulations on pollution, a chemical plant introduced a new process to supplement or replace an older process used in the production of a particular chemical. The older process emitted 20 grams of sulfur dioxide and 40 grams of particulate matter into the atmosphere for each gallon of chemical produced. The new process emits 5 grams of sulfur dioxide and 20 grams of particulate matter for each gallon produced. The company makes a profit of 60¢ per gallon and 20¢ per gallon on the old and new processes, respectively.

(A) If the regulations allow the plant to emit no more than 16,000 grams of sulfur dioxide and 30,000 grams of particulate matter daily, how many gallons of the chemical should be produced by each process to maximize daily profit? What is the maximum daily profit?

(B) Discuss the effect on the production schedule and the maximum profit if the regulations restrict emissions of sulfur dioxide to 11,500 grams daily and all other data remains unchanged.

(C) Discuss the effect on the production schedule and the maximum profit if the regulations restrict emissions of sulfur dioxide to 7,200 grams daily and all other data remains unchanged.

★★ **32. Sociology.** A city council voted to conduct a study on inner-city community problems. A nearby university was contacted to provide a maximum of 40 sociologists and research assistants. Allocation of time and cost per week are given in the table.

(A) How many sociologists and research assistants should be hired to meet the weekly labor-hour requirements and minimize the weekly cost? What is the weekly cost?

(B) Discuss the effect on the solution in part A if the council decides that they should not hire more sociologists than research assistants and all other data remains unchanged.

	Sociologist (labor-hours)	Research Assistant (labor-hours)	Minimum Labor-Hours Needed per Week
Fieldwork	10	30	280
Research center	30	10	360
Cost per week	$500	$300	

★★ **33. Plant Nutrition.** A fruit grower can use two types of fertilizer in her orange grove, brand *A* and brand *B*. The amounts (in pounds) of nitrogen, phosphoric acid, potash, and chlorine in a bag of each mix are given in the table. Tests indicate that the grove needs at least 480 pounds of phosphoric acid, at least 540 pounds of potash, and at most 620 pounds of chlorine. If the grower always uses a combination of bags of brand *A* and brand *B* that will satisfy the constraints of phosphoric acid, potash, and chlorine, discuss the effect that this will have on the amount of nitrogen added to the field.

	Pounds per Bag	
	Brand *A*	Brand *B*
Nitrogen	6	7
Phosphoric acid	2	4
Potash	6	3
Chlorine	3	4

★★ **34. Diet.** A dietitian in a hospital is to arrange a special diet composed of two foods, *M* and *N*. Each ounce of food *M* contains 16 units of calcium, 5 units of iron, 6 units of cholesterol, and 8 units of vitamin A. Each ounce of food *N* contains 4 units of calcium, 25 units of iron, 4 units of cholesterol, and 4 units of vitamin A. The diet requires at least 320 units of calcium, at least 575 units of iron, and at most 300 units of cholesterol. If the dietitian always selects a combination of foods *M* and *N* that will satisfy the constraints for calcium, iron, and cholesterol, discuss the effects that this will have on the amount of vitamin A in the diet.

Chapter 5 Group Activity

Modeling with Systems of Linear Equations

In this group activity we consider two real-world problems that can be solved using systems of linear equations: heat conduction and traffic flow. Both problems involve using a grid and a basic assumption to construct the model (the system of equations). Gauss–Jordan elimination is then used to solve the model. In the heat conduction problem, the solution of the model is easily interpreted in terms of the original problem. The system in the second problem is dependent, and the solution requires a more careful interpretation.

I Heat Conduction

A metal grid consists of four thin metal bars. The end of each bar of the grid is kept at a constant temperature, as shown in Figure 1. We assume that the temperature at each intersection point in the grid is the average of the temperatures at the four adjacent points in the grid (adjacent points are either other intersection points or ends of bars). Thus, the temperature x_1 at the intersection point in the upper left-hand corner of the grid must satisfy

Left Above Right Below

$$x_1 = \tfrac{1}{4}(40 + 0 + x_2 + x_3)$$

Find equations for the temperature at the other three intersection points, and solve the resulting system to find the temperature at each intersection point in the grid.

FIGURE 1

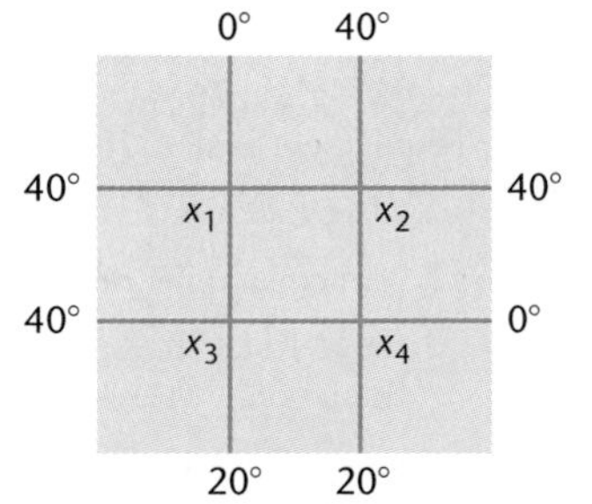

II Traffic Flow

The rush-hour traffic flow for a network of four one-way streets in a city is shown in Figure 2. The numbers next to each street indicate the number of vehicles per hour that enter and leave the network on that street. The variables x_1, x_2, x_3, and x_4 represent the flow of traffic between the four intersections in the network. For a smooth flow of traffic, we assume that the number of vehicles entering each intersection should always equal the number leaving. For example, since 1,500 vehicles enter the intersection of 5th Street and Washington Avenue each hour and $x_1 + x_4$ vehicles leave this intersection, we see that $x_1 + x_4 = 1{,}500$.

CHAPTER 5 GROUP ACTIVITY continued

(A) Find the equations determined by the traffic flow at each of the other three intersections.

(B) Find the solution to the system in part A.

(C) What is the maximum number of vehicles that can travel from Washington Avenue to Lincoln Avenue on 5th Street? What is the minimum number?

(D) If traffic lights are adjusted so that 1,000 vehicles per hour travel from Washington Avenue to Lincoln Avenue on 5th Street, determine the flow around the rest of the network.

FIGURE 2

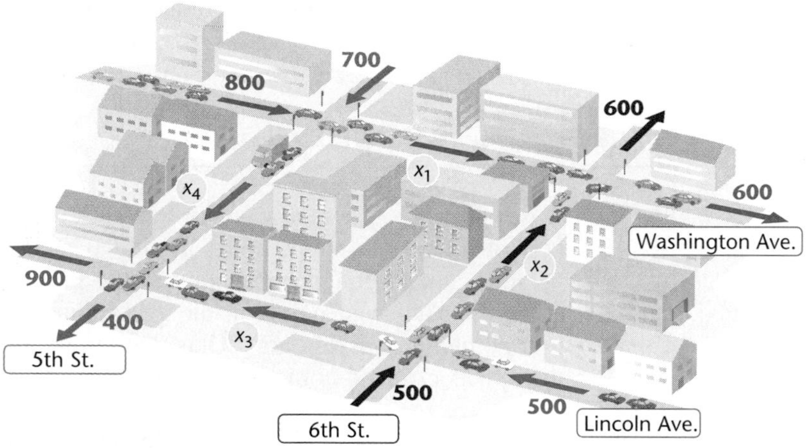

CHAPTER 5 | REVIEW

5-1 SYSTEMS OF LINEAR EQUATIONS IN TWO VARIABLES

A system of two linear equations with two variables is a system of the form

$$\begin{aligned} ax + by &= h \\ cx + dy &= k \end{aligned} \qquad (1)$$

where x and y are variables, a, b, c, and d are real numbers called the **coefficients** of x and y, and h and k are real numbers called the **constant terms** in the equations. The ordered pair of numbers (x_0, y_0) is a **solution** to system (1) if each equation is satisfied by the pair. The set of all such ordered pairs of numbers is called the **solution set** for the system. To **solve** a system is to find its solution set.

In general, a system of linear equations has exactly one solution, no solution, or infinitely many solutions. A system of linear equations is **consistent** if it has one or more solutions and **inconsistent** if no solutions exist. A consistent system is said to be **independent** if it has exactly one solution and **dependent** if it has more than one solution.

Two standard methods for solving system (1) were discussed: **graphing** and **substitution.**

5-2 SYSTEMS OF LINEAR EQUATIONS AND AUGMENTED MATRICES

Two systems of equations are **equivalent** if both have the same solution set. A system of linear equations is transformed into an equivalent system if

1. Two equations are interchanged.
2. An equation is multiplied by a nonzero constant.
3. A constant multiple of another equation is added to a given equation.

These operations form the basis of solution using **elimination by addition.**

The method of solution using elimination by addition is transformed into a more efficient method for larger-scale systems by the introduction of an *augmented matrix*. A **matrix** is a rectangular array of numbers written within brackets. Each number in a matrix is called an **element** of the matrix. If a matrix has m rows and n columns, it is called an $\boldsymbol{m \times n}$ **matrix** (read "m by n matrix"). The expression $m \times n$ is called the **size** of the matrix, and the numbers m and n are called the **dimensions** of the matrix. A matrix with n rows and n columns is called a **square matrix of order $\boldsymbol{n}$.** A matrix with only one column is called a **column matrix,** and a matrix with only one row is called a **row matrix.** The **position** of an element in a matrix is the row and column containing the element. This is usually denoted using **double subscript notation** a_{ij}, where i is the row and j is the column containing the element a_{ij}.

For ease of generalization to larger systems, we change the notation for variables and constants in system (1) to a subscript form:

$$\begin{aligned} a_1x_1 + b_1x_2 &= k_1 \\ a_2x_1 + b_2x_2 &= k_2 \end{aligned} \qquad (2)$$

Associated with each linear system of the form (2), where x_1 and x_2 are variables, is the **augmented matrix** of the system:

Column 1 (C_1)
Column 2 (C_2)
Column 3 (C_3)

$$\left[\begin{array}{cc|c} a_1 & b_1 & k_1 \\ a_2 & b_2 & k_2 \end{array}\right] \begin{array}{l} \leftarrow \text{Row 1 } (R_1) \\ \leftarrow \text{Row 2 } (R_2) \end{array} \qquad (3)$$

Two augmented matrices are **row-equivalent,** denoted by the symbol $\sim$ between the two matrices, if they are augmented matrices of equivalent systems of equations. An augmented matrix is transformed into a row-equivalent matrix if any of the following **row operations** is performed:

1. Two rows are interchanged.
2. A row is multiplied by a nonzero constant.
3. A constant multiple of another row is added to a given row.

The following symbols are used to describe these row operations:

1. $R_i \leftrightarrow R_j$ means "interchange row i with row j."
2. $kR_i \rightarrow R_i$ means "multiply row i by the constant k."
3. $kR_j + R_i \rightarrow R_i$ means "multiply row j by the constant k and add to R_i."

In solving system (2) using row operations, the objective is to transform the augmented matrix (3) into the form

$$\left[\begin{array}{cc|c} 1 & 0 & m \\ 0 & 1 & n \end{array}\right]$$

If this can be done, then (m, n) is the unique solution of system (2). If (3) is transformed into the form

$$\left[\begin{array}{cc|c} 1 & m & n \\ 0 & 0 & 0 \end{array}\right]$$

then system (2) has infinitely many solutions. If (3) is transformed into the form

$$\left[\begin{array}{cc|c} 1 & m & n \\ 0 & 0 & p \end{array}\right] \qquad p \neq 0$$

then system (2) does not have a solution.

5-3 GAUSS–JORDAN ELIMINATION

In Section 5-2 we were actually using *Gauss–Jordan elimination* to solve a system of two equations with two variables. The method generalizes completely for systems with more than two variables, and the number of variables does not have to be the same as the number of equations.

As before, our objective is to start with the augmented matrix of a linear system and transform it using row operations into a simple form where the solution can be read by inspection. The simple form, called the **reduced form,** is achieved if:

1. Each row consisting entirely of 0s is below any row having at least one nonzero element.
2. The leftmost nonzero element in each row is 1.
3. The column containing the leftmost 1 of a given row has 0s above and below the 1.
4. The leftmost 1 in any row is to the right of the leftmost 1 in the preceding row.

A **reduced system** is a system of linear equations that corresponds to a reduced augmented matrix. When a reduced system has more variables than equations and contains no contradictions, the system is dependent and has infinitely many solutions.

The **Gauss–Jordan elimination** procedure for solving a system of linear equations is given in step-by-step form as follows:

Step 1. Choose the leftmost nonzero column, and use appropriate row operations to get a 1 at the top.

Step 2. Use multiples of the row containing the 1 from step 1 to get zeros in all remaining places in the column containing this 1.

Step 3. Repeat step 1 with the **submatrix** formed by (mentally) deleting the row used in step 2 and all rows above this row.

Step 4. Repeat step 2 with the **entire matrix,** including the mentally deleted rows. Continue this process until it is impossible to go further.

If at any point in the above process we obtain a row with all 0s to the left of the vertical line and a nonzero number n to the right, we can stop, since we have a contradiction: $0 = n$, $n \neq 0$. We can then conclude that the system has no solution. If this does not happen and we obtain an augmented matrix in reduced form without any contradictions, the solution can be read by inspection.

5-4 MATRIX OPERATIONS

Two matrices are **equal** if they are the same size and their corresponding elements are equal. The **sum of two matrices** of the same size is a matrix with elements that are the sums of the corresponding elements of the two given matrices. Matrix addition is **commutative** and **associative.** A matrix with all zero elements is called the **zero matrix.** The **negative of a matrix** M**,** denoted $-M$, is a matrix with elements that are the negatives of the elements in M. If A and B are matrices of the same size, then we define **subtraction** as follows: $A - B = A + (-B)$. The **product of a number** k **and a matrix** M**,** denoted by kM, is a matrix formed by multiplying each element of M by k. The

product of a $1 \times n$ row matrix and an $n \times 1$ column matrix is a 1×1 matrix given by

$$\overset{1 \times n}{[a_1 \ a_2 \ \cdots \ a_n]} \overset{n \times 1}{\begin{bmatrix} b_1 \\ b_2 \\ \vdots \\ b_n \end{bmatrix}} = \overset{1 \times 1}{[a_1b_1 + a_2b_2 + \cdots + a_nb_n]}$$

If A is an $m \times p$ matrix and B is a $p \times n$ matrix, then the **matrix product** of A and B, denoted AB, is an $m \times n$ matrix whose element in the ith row and jth column is the real number obtained from the product of the ith row of A and the jth column of B. If the number of columns in A does not equal the number of rows in B, then the matrix product AB is **not defined. Matrix multiplication is not commutative,** and the **zero property does not hold for matrix multiplication.** That is, for matrices A and B, the matrix product AB can be zero without either A or B being the zero matrix.

5-5 INVERSE OF A SQUARE MATRIX

The **identity matrix for multiplication** for the set of all square matrices of order n is the square matrix of order n, denoted by I, with 1s along the **principal diagonal** (from upper left corner to lower right corner) and 0s elsewhere. If M is a square matrix of order n and I is the identity matrix of order n, then

$$IM = MI = M$$

If M is a square matrix of order n and if there exists a matrix M^{-1} (read "M inverse") such that

$$M^{-1}M = MM^{-1} = I$$

then M^{-1} is called the **multiplicative inverse of M** or, more simply, the **inverse of M.** If the augmented matrix $[M \mid I]$ is transformed by row operations into $[I \mid B]$, then the resulting matrix B is M^{-1}. If, however, we obtain all 0s in one or more rows to the left of the vertical line, then M^{-1} does not exist and M is called a **singular matrix.**

5-6 MATRIX EQUATIONS AND SYSTEMS OF LINEAR EQUATIONS

The following properties of matrices are fundamental to the process of solving matrix equations. Assuming all products and sums are defined for the indicated matrices A, B, C, I, and 0, then:

Addition Properties	
Associative:	$(A + B) + C = A + (B + C)$
Commutative:	$A + B = B + A$
Additive Identity:	$A + 0 = 0 + A = A$
Additive Inverse:	$A + (-A) = (-A) + A = 0$
Multiplication Properties	
Associative Property:	$A(BC) = (AB)C$
Multiplicative Identity:	$AI = IA = A$
Multiplicative Inverse:	If A is a square matrix and A^{-1} exists, then $AA^{-1} = A^{-1}A = I$.
Combined Properties	
Left Distributive:	$A(B + C) = AB + AC$
Right Distributive:	$(B + C)A = BA + CA$
Equality	
Addition:	If $A = B$, then $A + C = B + C$.
Left Multiplication:	If $A = B$, then $CA = CB$.
Right Multiplication:	If $A = B$, then $AC = BC$.

A system of linear equations with the same number of variables as equations such as

$$a_{11}x_1 + a_{12}x_2 + a_{13}x_3 = k_1$$
$$a_{21}x_1 + a_{22}x_2 + a_{23}x_3 = k_2$$
$$a_{31}x_1 + a_{32}x_2 + a_{33}x_3 = k_3$$

can be written as the matrix equation

$$\overset{A}{\begin{bmatrix} a_{11} & a_{12} & a_{13} \\ a_{21} & a_{22} & a_{23} \\ a_{31} & a_{32} & a_{33} \end{bmatrix}} \overset{X}{\begin{bmatrix} x_1 \\ x_2 \\ x_3 \end{bmatrix}} = \overset{B}{\begin{bmatrix} k_1 \\ k_2 \\ k_3 \end{bmatrix}}$$

If the inverse of A exists, then the matrix equation has a unique solution given by

$$X = A^{-1}B$$

After multiplying B by A^{-1} from the left, it is easy to read the solution to the original system of equations.

5-7 SYSTEMS OF LINEAR INEQUALITIES

A graph is often the most convenient way to represent the solution of a linear inequality in two variables or of a system of linear inequalities in two variables.

A vertical line divides a plane into **left** and **right half-planes.** A nonvertical line divides a plane into **upper** and **lower half-planes.** Let A, B, and C be real numbers with A and B not both zero, then the **graph of the linear inequality**

$$Ax + By < C \qquad \text{or} \qquad Ax + By > C$$

with $B \neq 0$, is either the upper half-plane or the lower half-plane (but not both) determined by the line $Ax + By = C$. If $B = 0$, then the graph of

$$Ax < C \qquad \text{or} \qquad Ax > C$$

is either the left half-plane or the right half-plane (but not both) determined by the line $Ax = C$. Out of these results follows an easy **step-by-step procedure for graphing a linear inequality in two variables:**

Step 1. Graph $Ax + By = C$ as a broken line if equality is not included in the original statement or as a solid line if equality is included.

Step 2. Choose a test point anywhere in the plane not on the line and substitute the coordinates into the inequality. The origin (0, 0) often requires the least computation.

Step 3. The graph of the original inequality includes the half-plane containing the test point if the inequality is satisfied by that point, or the half-plane not containing that point if the inequality is not satisfied by that point.

We now turn to systems of linear inequalities in two variables. The **solution to a system of linear inequalities in two variables** is the set of all ordered pairs of real numbers that simultaneously satisfy all the inequalities in the system. The graph is called the **solution region.** In many applications the solution region is also referred to as the **feasible region.** To **find the solution region,** we graph each inequality in the system and then take the intersection of all the graphs. A **corner point** of a solution region is a point in the solution region that is the intersection of two boundary lines. A solution region is **bounded** if it can be enclosed within a circle. If it cannot be enclosed within a circle, then it is **unbounded.**

5-8 LINEAR PROGRAMMING

Linear programming is a mathematical process that has been developed to help management in decision making, and it has become one of the most widely used and best-known tools of management science and industrial engineering.

A **linear programming problem** is one that is concerned with finding the **optimal value** (maximum or minimum value) of a linear **objective function** of the form $z = ax + by$, where the **decision variables** x and y are subject to **problem constraints** in the form of linear inequalities and **nonnegative constraints** $x, y \geq 0$. The set of points satisfying both the problem constraints and the nonnegative constraints is called the *feasible region* for the problem. Any point in the feasible region that produces the optimal value of the objective function over the feasible region is called an **optimal solution.** The **fundamental theorem of linear programming** is basic to the solving of linear programming problems: Let S be the feasible region for a linear programming problem, and let $z = ax + by$ be the objective function. If S is bounded, then z has both a maximum and a minimum value on S and each of these occurs at a corner point of S. If S is unbounded, then a maximum or minimum value of z on S may not exist. However, if either does exist, then it must occur at a corner point of S.

Problems with unbounded feasible regions are not considered in this brief introduction. The theorem leads to a simple **step-by-step solution to linear programming problems with a bounded feasible region:**

Step 1. Form a mathematical model for the problem:
- (A) Introduce decision variables and write a linear objective function.
- (B) Write problem constraints in the form of linear inequalities.
- (C) Write nonnegative constraints.

Step 2. Graph the feasible region and find the corner points.

Step 3. Evaluate the objective function at each corner point to determine the optimal solution.

If two corner points are both optimal solutions of the same type (both produce the same maximum value or both produce the same minimum value) to a linear programming problem, then any point on the line segment joining the two corner points is also an optimal solution of that type.

CHAPTER 5 | REVIEW EXERCISES

Work through all the problems in this chapter review and check answers in the back of the book. Answers to all review problems are there, and following each answer is a number in italics indicating the section in which that type of problem is discussed. Where weaknesses show up, review appropriate sections in the text.

A

Solve Problems 1 and 2 by substitution.

1. $y = 4x - 9$; $y = -x + 6$

2. $3x + 2y = 5$; $4x - y = 14$

Solve Problems 3–6 by graphing.

3. $3x - 2y = 8$; $x + 3y = -1$

4. $2x + y = 4$; $-2x + 7y = 9$

5. $3x - 4y \geq 24$

6. $2x + y \leq 2$; $x + 2y \geq -2$

Perform each of the row operations indicated in Problems 7–9 on the following augmented matrix:

$$\left[\begin{array}{cc|c} 1 & -4 & 5 \\ 3 & -6 & 12 \end{array}\right]$$

7. $R_1 \leftrightarrow R_2$

8. $\frac{1}{3}R_2 \rightarrow R_2$

9. $(-3)R_1 + R_2 \rightarrow R_2$

In Problems 10–12, write the linear system corresponding to each reduced augmented matrix and solve.

10. $\left[\begin{array}{cc|c} 1 & 0 & 4 \\ 0 & 1 & -7 \end{array}\right]$

11. $\left[\begin{array}{cc|c} 1 & -1 & 4 \\ 0 & 0 & 1 \end{array}\right]$

12. $\left[\begin{array}{cc|c} 1 & -1 & 4 \\ 0 & 0 & 0 \end{array}\right]$

In Problems 13–21, perform the operations that are defined, given the following matrices:

$$A = \begin{bmatrix} 1 & 2 \\ 3 & 1 \end{bmatrix} \quad B = \begin{bmatrix} 2 & 1 \\ 1 & 1 \end{bmatrix} \quad C = \begin{bmatrix} 2 & 3 \end{bmatrix} \quad D = \begin{bmatrix} 1 \\ 2 \end{bmatrix}$$

13. $A + B$ **14.** $B + D$ **15.** $A - 2B$

16. AB **17.** AC **18.** AD

19. DC **20.** CD **21.** $C + D$

22. Find the inverse of

$$A = \begin{bmatrix} 3 & 2 \\ 4 & 3 \end{bmatrix}$$

23. Write the system

$$3x_1 + 2x_2 = k_1$$
$$4x_1 + 3x_2 = k_2$$

as a matrix equation, and solve using matrix inverse methods for
(A) $k_1 = 3, k_2 = 5$
(B) $k_1 = 7, k_2 = 10$
(C) $k_1 = 4, k_2 = 2$

24. Find the maximum and minimum values of $z = 5x + 3y$ over the feasible region S shown in the figure.

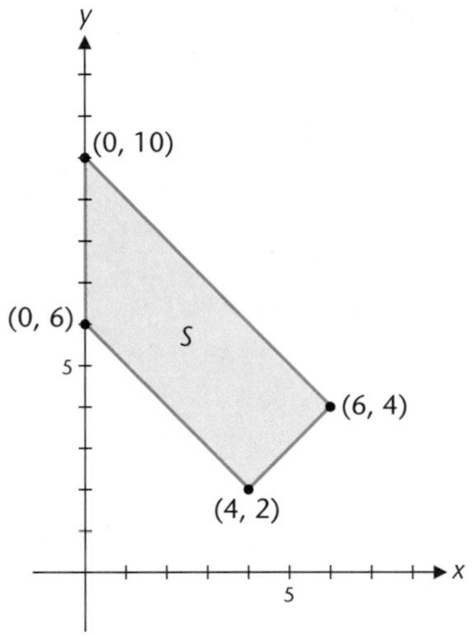

25. Use Gauss–Jordan elimination to solve the system

$$x_1 - x_2 = 4$$
$$2x_1 + x_2 = 2$$

Then write the linear system represented by each augmented matrix in your solution, and solve each of these systems graphically. Discuss the relationship between the solutions of these systems.

B

Solve Problems 26–31 using Gauss–Jordan elimination.

26. $3x_1 + 2x_2 = 3$
$x_1 + 3x_2 = 8$

27. $x_1 + x_2 = 1$
$x_1 - x_3 = -2$
$x_2 + 2x_3 = 4$

28. $x_1 + 2x_2 + 3x_3 = 1$
$2x_1 + 3x_2 + 4x_3 = 3$
$x_1 + 2x_2 + x_3 = 3$

29. $x_1 + 2x_2 - x_3 = 2$
$2x_1 + 3x_2 + x_3 = -3$
$3x_1 + 5x_2 = -1$

30. $x_1 - 2x_2 = 1$
$2x_1 - x_2 = 0$
$x_1 - 3x_2 = -2$

31. $x_1 + 2x_2 - x_3 = 2$
$3x_1 - x_2 + 2x_3 = -3$

In Problems 32–37, perform the operations that are defined, given the following matrices:

$$A = \begin{bmatrix} 2 & -2 \\ 1 & 0 \\ 3 & 2 \end{bmatrix} \quad B = \begin{bmatrix} -1 \\ 2 \\ 3 \end{bmatrix} \quad C = \begin{bmatrix} 2 & 1 & 3 \end{bmatrix}$$

$$D = \begin{bmatrix} 3 & -2 & 1 \\ -1 & 1 & 2 \end{bmatrix} \quad E = \begin{bmatrix} 3 & -4 \\ -1 & 0 \end{bmatrix}$$

32. $A + D$ **33.** $E + DA$ **34.** $DA - 3E$
35. CD **36.** CB **37.** $AD - BC$

38. Find the inverse of

$$A = \begin{bmatrix} 1 & 2 & 3 \\ 2 & 3 & 4 \\ 1 & 2 & 1 \end{bmatrix}$$

39. Write the system

$$x_1 + 2x_2 + 3x_3 = k_1$$
$$2x_1 + 3x_2 + 4x_3 = k_2$$
$$x_1 + 2x_2 + x_3 = k_3$$

as a matrix equation, and solve using matrix inverse methods for
(A) $k_1 = 1, k_2 = 3, k_3 = 3$
(B) $k_1 = 0, k_2 = 0, k_3 = -2$
(C) $k_1 = -3, k_2 = -4, k_3 = 1$

40. Discuss the number of solutions for a system of n equations in n variables if the coefficient matrix
(A) Has an inverse.
(B) Does not have an inverse.

41. If A is a nonzero square matrix of order n satisfying $A^2 = 0$, can A^{-1} exist? Explain.

Solve the systems in Problems 42–44 graphically, and indicate whether each solution region is bounded or unbounded. Find the coordinates of each corner point.

42. $2x + y \le 8$
$2x + 3y \le 12$
$x, y \ge 0$

43. $2x + y \ge 8$
$x + 3y \ge 12$
$x, y \ge 0$

44. $x + y \le 20$
$x + 4y \ge 20$
$x - y \ge 0$

Solve the linear programming problems in Problems 45–47.

45. Maximize $z = 7x + 9y$
Subject to $x + 2y \le 8$
$2x + y \le 10$
$x, y \ge 0$

46. Minimize $z = 5x + 10y$
Subject to $x + y \le 20$
$3x + y \ge 15$
$x + 2y \ge 15$
$x, y \ge 0$

47. Minimize and maximize $z = 5x + 8y$
Subject to $x + 2y \le 20$
$3x + y \le 15$
$x + y \ge 7$
$x, y \ge 0$

C

48. For $n \times n$ matrices A and C and $n \times 1$ column matrices B and X, solve for X assuming all necessary inverses exist:

$$AX - B = CX$$

49. Find the inverse of

$$A = \begin{bmatrix} 4 & 5 & 6 \\ 4 & 5 & -6 \\ 1 & 1 & 1 \end{bmatrix}$$

50. Solve using matrix inverse methods:

$$\begin{aligned} 0.04x_1 + 0.05x_2 + 0.06x_3 &= 360 \\ 0.04x_1 + 0.05x_2 - 0.06x_3 &= 120 \\ x_1 + x_2 + x_3 &= 7{,}000 \end{aligned}$$

51. Solve using Gauss–Jordan elimination:

$$\begin{aligned} x_1 + x_2 + x_3 &= 7{,}000 \\ 0.04x_1 + 0.05x_2 + 0.06x_3 &= 360 \\ 0.04x_1 + 0.05x_2 - 0.06x_3 &= 120 \end{aligned}$$

52. Maximize $z = 30x + 20y$

Subject to

$$\begin{aligned} 1.2x + 0.6y &\le 960 \\ 0.04x + 0.03y &\le 36 \\ 0.2x + 0.3y &\le 270 \\ x, y &\ge 0 \end{aligned}$$

53. Discuss the number of solutions for the system corresponding to the reduced form shown below if
(A) $m \ne 0$ (B) $m = 0$ and $n \ne 0$
(C) $m = 0$ and $n = 0$

$$\left[\begin{array}{ccc|c} 1 & 0 & -3 & 4 \\ 0 & 1 & 2 & 5 \\ 0 & 0 & m & n \end{array}\right]$$

APPLICATIONS

54. **Business.** A container holds 120 packages. Some of the packages weigh $\frac{1}{2}$ pound each, and the rest weigh $\frac{1}{3}$ pound each. If the total contents of the container weigh 48 pounds, how many are there of each type of package? Solve using two-equation–two-variable methods.

★ 55. **Geometry.** Find the dimensions of a rectangle with an area of 48 square meters and a perimeter of 28 meters. Solve using two-equation–two-variable methods.

★ 56. **Diet.** A laboratory assistant wishes to obtain a food mix that contains, among other things, 27 grams of protein, 5.4 grams of fat, and 19 grams of moisture. He has available mixes A, B, and C with the compositions listed in the table. How many grams of each mix should be used to get the desired diet mix? Set up a system of equations and solve using Gauss–Jordan elimination.

Mix	Protein (%)	Fat (%)	Moisture (%)
A	30	3	10
B	20	5	20
C	10	4	10

★★ 57. **Puzzle.** A piggy bank contains 30 coins worth $1.90.
(A) If the bank contains only nickels and dimes, how many coins of each type does it contain?
(B) If the bank contains nickels, dimes, and quarters, how many coins of each type does it contain?

58. **Labor Costs.** A company with manufacturing plants in North and South Carolina has labor-hour and wage requirements for the manufacturing of computer desks and printer stands as given in matrices L and H:

Labor-hour requirements

	Fabricating department	Assembly department	Packaging department	
$L =$	1.7 h	2.4 h	0.8 h	Desk
	0.9 h	1.8 h	0.6 h	Stand

Hourly wages

	North Carolina plant	South Carolina plant	
$H =$	$11.50	$10.00	Fabricating department
	$9.50	$8.50	Assembly department
	$5.00	$4.50	Packaging department

(A) Find the labor cost for producing one printer stand at the South Carolina plant.
(B) Discuss possible interpretations of the elements in the matrix products HL and LH.
(C) If either of the products HL or LH has a meaningful interpretation, find the product and label its rows and columns.

59. **Labor Costs.** The monthly production of computer desks and printer stands for the company in Problem 58 for the months of January and February are given in matrices J and F:

January production

	North Carolina plant	South Carolina plant	
$J =$	1,500	1,650	Desks
	850	700	Stands

February production

	North Carolina plant	South Carolina plant	
$F =$	1,700	1,810	Desks
	930	740	Stands

(A) Find the average monthly production for the months of January and February.
(B) Find the increase in production from January to February.
(C) Find $J\begin{bmatrix}1\\1\end{bmatrix}$ and interpret.

60. Cryptography. The following message was encoded with the matrix B shown below. Decode the message:

25 8 26 24 25 33 21 41 48 41 30 50

21 32 41 52 52 79

$$B = \begin{bmatrix} 1 & 1 & 0 \\ 1 & 0 & 1 \\ 1 & 1 & 1 \end{bmatrix}$$

61. Resource Allocation. A Colorado mining company operates mines at Big Bend and Saw Pit. The Big Bend mine produces ore that is 5% nickel and 7% copper. The Saw Pit mine produces ore that is 3% nickel and 4% copper. How many tons of ore should be produced at each mine to obtain the amounts of nickel and copper listed in the table? Set up a matrix equation and solve using matrix inverses.

	Nickel	Copper
(A)	3.6 tons	5 tons
(B)	3 tons	4.1 tons
(C)	3.2 tons	4.4 tons

★ **62. Resource Allocation.** North Star Sail Loft manufactures regular and competition sails. Each regular sail takes 1 labor-hour to cut and 3 labor-hours to sew. Each competition sail takes 2 labor-hours to cut and 4 labor-hours to sew. There are 140 labor-hours available in the cutting department and 360 labor-hours available in the sewing department.
(A) If the loft makes a profit of \$60 on each regular sail and \$100 on each competition sail, how many sails of each type should the company manufacture to maximize its profit? What is the maximum profit?
(B) An increase in the demand for competition sails causes the profit on a competition sail to rise to \$125. Discuss the effect of this change on the number of sails manufactured and on the maximum profit.
(C) A decrease in the demand for competition sails causes the profit on a competition sail to drop to \$75. Discuss the effect of this change on the number of sails manufactured and on the maximum profit.

★ **63. Nutrition—Animals.** A special diet for laboratory animals is to contain at least 800 units of vitamins, at least 800 units of minerals, and at most 1,300 calories. There are two feed mixes available, mix A and mix B. A gram of mix A contains 5 units of vitamins, 2 units of minerals, and 4 calories. A gram of mix B contains 2 units of vitamins, 4 units of minerals, and 4 calories.
(A) If mix A costs \$0.07 per gram and mix B costs \$0.04 per gram, how many grams of each mix should be used to satisfy the requirements of the diet at minimal cost? What is the minimal cost?
(B) If the price of mix B decreases to \$0.02 per gram, discuss the effect of this change on the solution in part A.
(C) If the price of mix B increases to \$0.15 per gram, discuss the effect of this change on the solution in part A.

ANSWER TO APPLICATION

Let

x = number of bottles of water

y = number of packages of food

The number of people supplied is

$P = 15x + 11y$ Objective function

The decision variables are constrained as follows:

$18x + 9y \le 64{,}800$ Weight constraint

$x + 0.75y \le 4{,}500$ Space constraint

$x, y \ge 0$ Nonnegative constraints

Thus, the mathematical model for this problem is

$$\text{Maximize} \quad P = 15x + 11y$$
$$\text{Subject to} \quad 18x + 9y \le 64{,}800$$
$$x + 0.75y \le 4{,}500$$
$$x, y \ge 0$$

The feasible region is the unshaded region in Figure 1 and the values of the objective function at the corner points are given in the table. The optimal value of the objective function is $P = 66{,}600$ at the corner point (1,800, 3,600). Thus, they should ship 1,800 bottles of water and 3,600 packages of food.

FIGURE 1

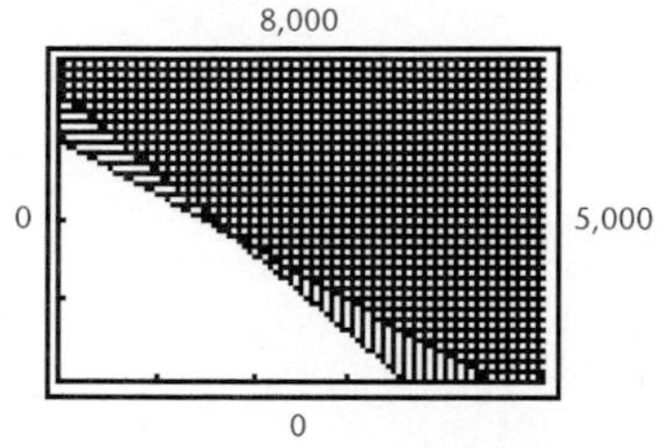

Corner Point (x, y)	Objective Function $P = 15x + 11y$
(0, 0)	0
(0, 6,000)	66,000
(1,800, 3,600)	66,600
(3,600, 0)	54,000

6

Sequences, Series, and Probability

Outline

Application

A reader in the *Ask Marilyn* column by Marilyn vos Savant posed the following problem:

Assume that the probability of giving birth to a boy is the same as the probability of giving birth to a girl. A woman and a man (unrelated) each have two children. At least one of the woman's children is a boy, and the man's first child is a boy. Is the probability that the woman has two boys the same as the probability that the man has two boys?

The lists

1, 4, 9, 16, 25, 36, 49, 64, . . .

and

3, 6, 3, 1, 4, 2, 1, 4, . . .

are examples of sequences. The first sequence exhibits a great deal of regularity. You no doubt recognize it as the sequence of perfect squares. Its terms are increasing, and the differences between terms form a striking pattern. You probably do not recognize the second sequence, whose terms do not suggest an obvious pattern. In fact, the second sequence records the results of repeatedly tossing a single die. Sequences, and the related concept of series, are useful tools in almost all areas of mathematics. In this chapter they play roles in the development of several topics: a method of proof called mathematical induction, techniques for counting, and probability.

Preparing for This Chapter

Before getting started on this chapter, review the following concepts:
Set Notation (Appendix A, Section 1)
Operations on Polynomials (Appendix A, Section 2)
Integer Exponents (Appendix A, Section 5)
Functions (Chapter 1, Section 3)
Set Operations (Appendix A, Section 8)

Section 6-1 | Sequences and Series

- Sequences
- Series

In this section we introduce special notation and formulas for representing and generating sequences and sums of sequences.

Sequences

Consider the function f given by

$$f(n) = 2n - 1 \tag{1}$$

where the domain of f is the set of natural numbers N. Note that

$$f(1) = 1, f(2) = 3, f(3) = 5, \ldots$$

The function f is an example of a sequence. A **sequence** is a function with domain a set of successive integers. However, a sequence is hardly ever represented in the form of equation (1). A special notation for sequences has evolved, which we describe here.

To start, the range value $f(n)$ is usually symbolized more compactly with a symbol such as a_n. Thus, in place of equation (1) we write

$$a_n = 2n - 1$$

The domain is understood to be the set of natural numbers N unless stated to the contrary or the context indicates otherwise. The elements in the range are called **terms of the sequence:** a_1 is the first term, a_2 the second term, and a_n the nth term, or the **general term:**

$$\begin{aligned} a_1 &= 2(1) - 1 = 1 && \text{First term} \\ a_2 &= 2(2) - 1 = 3 && \text{Second term} \\ a_3 &= 2(3) - 1 = 5 && \text{Third term} \\ &\vdots \qquad\quad\;\; \vdots \end{aligned}$$

The ordered list of elements

$$1, 3, 5, \ldots, 2n - 1, \ldots$$

in which the terms of a sequence are written in their natural order with respect to the domain values, is often informally referred to as a sequence. A sequence is also represented in the abbreviated form $\{a_n\}$, where a symbol for the nth term is placed between braces. For example, we can refer to the sequence

$$1, 3, 5, \ldots, 2n - 1, \ldots$$

as the sequence $\{2n - 1\}$.

If the domain of a function is a finite set of successive integers, then the sequence is called a **finite sequence.** If the domain is an infinite set of successive integers, then the sequence is called an **infinite sequence.** The sequence $\{2n - 1\}$ above is an example of an infinite sequence.

Explore/Discuss 1

The sequence $\{2n - 1\}$ is a function whose domain is the set of natural numbers, and so it may be graphed in the same way as any function whose domain and range are sets of real numbers (see Fig. 1).

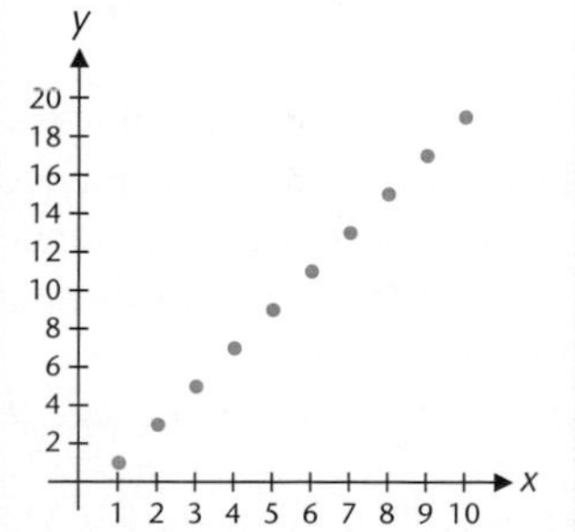

FIGURE 1 Graph of $\{2n - 1\}$.

Explore/Discuss 1 continued

(A) Explain why the graph of the sequence $\{2n - 1\}$ is not continuous.

(B) Explain why the points on the graph of $\{2n - 1\}$ lie on a line. Find an equation for that line.

(C) Graph the sequence $\left\{\dfrac{2n^2 - n + 1}{n}\right\}$. How are the graphs of $\{2n - 1\}$ and $\left\{\dfrac{2n^2 - n + 1}{n}\right\}$ related?

There are several different ways a graphing utility can be used in the study of sequences. Refer to Explore/Discuss 1. Figure 2(a) shows the sequence $\{2n - 1\}$ entered as a function in an equation editor. This produces a continuous graph [Fig. 2(b)] that contains the points in the graph of the sequence (Fig. 1). Figure 2(c) shows the points on the graph of the sequence displayed in a table.

FIGURE 2

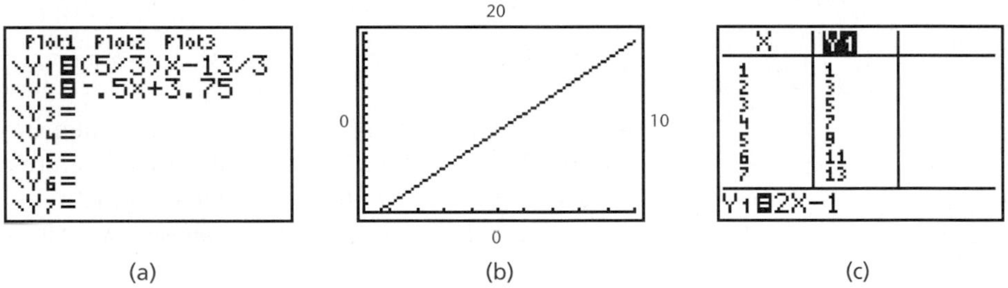

In Figure 3(a), sequence commands are used to store the first and second coordinates of the first 10 points on the graph of the sequence $\{2n - 1\}$ in lists L_1 and L_2, respectively. A statistical plot routine is used to graph these points [Fig. 3(b)], and a statistical editor is used to display the points on the graph [Fig. 3(c)].

FIGURE 3

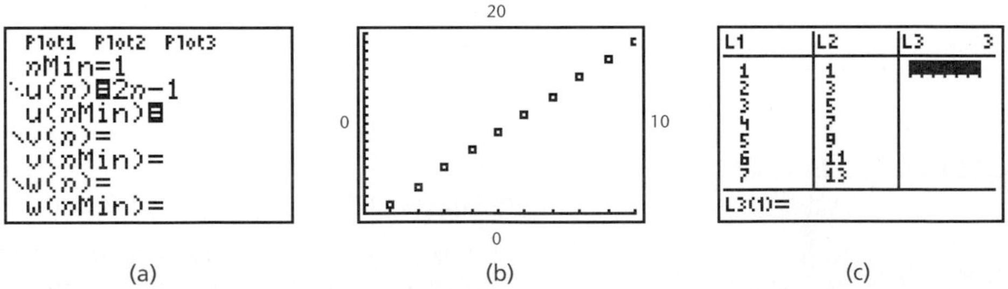

Most graphing utilities can produce the results shown in Figures 2 and 3. The Texas Instruments TI-83 has a special sequence mode that is very useful for studying sequences. Figure 4(a) shows the sequence $\{2n - 1\}$ entered in the sequence editor, Figure 4(b) shows the graph of this sequence, and Figure 4(c) displays the points on the graph in a table.

FIGURE 4

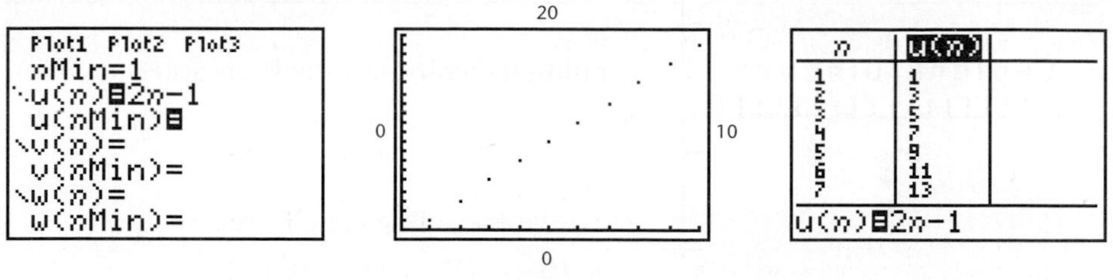

(a) (b) (c)

Examining graphs and displaying values are very helpful activities when working with sequences. Consult your manual to see which of the methods illustrated in Figures 2–4 works on your graphing utility.

Some sequences are specified by a **recursion formula**—that is, a formula that defines each term in terms of one or more preceding terms. The sequence we have chosen to illustrate a recursion formula is a very famous sequence in the history of mathematics called the **Fibonacci sequence.** It is named after the most celebrated mathematician of the thirteenth century, Leonardo Fibonacci from Italy (1180?–1250?).

EXAMPLE 1

Fibonacci Sequence

List the first six terms of the sequence specified by

$$a_1 = 1$$
$$a_2 = 1$$
$$a_n = a_{n-1} + a_{n-2} \qquad n \geq 3$$

Solution

$$a_1 = 1$$
$$a_2 = 1$$
$$a_3 = a_2 + a_1 = 1 + 1 = 2$$
$$a_4 = a_3 + a_2 = 2 + 1 = 3$$
$$a_5 = a_4 + a_3 = 3 + 2 = 5$$
$$a_6 = a_5 + a_4 = 5 + 3 = 8$$

The formula $a_n = a_{n-1} + a_{n-2}$ is a recursion formula that can be used to generate the terms of a sequence in terms of preceding terms. Of course, starting terms a_1 and a_2 must be provided to use the formula. Recursion formulas are particularly suitable for use with calculators and computers (see Problems 57 and 58 in Exercise 6-1).

MATCHED PROBLEM 1

List the first five terms of the sequence specified by

$$a_1 = 4$$
$$a_n = \tfrac{1}{2}a_{n-1} \qquad n \geq 2$$

Explore/Discuss 2

A multiple-choice test question asked for the next term in the sequence:

$$1, 3, 9, \ldots$$

and gave the following choices:

(A) 16 (B) 19 (C) 27

Which is the correct answer?

Compare the first four terms of the following sequences:

(A) $a_n = 3^{n-1}$ (B) $b_n = 1 + 2(n - 1)^2$ (C) $c_n = 8n + \dfrac{12}{n} - 19$

Now which of the choices appears to be correct?

Now we consider the reverse problem. That is, can a sequence be defined just by listing the first three or four terms of the sequence? And can we then use these initial terms to find a formula for the nth term? In general, without other information, the answer to the first question is no. As Explore/Discuss 2 illustrates, many different sequences may start off with the same terms. Simply listing the first three terms, or any other finite number of terms, does not specify a particular sequence. In fact, it can be shown that given any list of m numbers, there are an infinite number of sequences whose first m terms agree with these given numbers.

What about the second question? That is, given a few terms, can we find the general formula for at least one sequence whose first few terms agree with the given terms? The answer to this question is a qualified yes. If we can observe a simple pattern in the given terms, then we may be able to construct a general term that will produce the pattern. The next example illustrates this approach.

EXAMPLE 2 **Finding the General Term of a Sequence**

Find the general term of a sequence whose first four terms are

(A) $5, 6, 7, 8, \ldots$ (B) $2, -4, 8, -16, \ldots$

Solutions

(A) Since these terms are consecutive integers, one solution is $a_n = n$, $n \geq 5$. If we want the domain of the sequence to be all natural numbers, then another solution is $b_n = n + 4$.

(B) Each of these terms can be written as the product of a power of 2 and a power of -1:

$$2 = (-1)^0 2^1$$
$$-4 = (-1)^1 2^2$$
$$8 = (-1)^2 2^3$$
$$-16 = (-1)^3 2^4$$

If we choose the domain to be all natural numbers, then a solution is

$$a_n = (-1)^{n-1} 2^n$$

MATCHED PROBLEM 2

Find the general term of a sequence whose first four terms are

(A) 2, 4, 6, 8, . . . (B) $1, -\frac{1}{2}, \frac{1}{4}, -\frac{1}{8}, \ldots$

In general, there is usually more than one way of representing the nth term of a given sequence. This was seen in the solution of Example 2, part A. However, unless stated to the contrary, we assume the domain of the sequence is the set of natural numbers N.

Explore/Discuss 3

The sequence with general term $b_n = \frac{\sqrt{5}}{5}\left(\frac{1+\sqrt{5}}{2}\right)^n$ is closely related to the Fibonacci sequence. Compute the first 20 terms of both sequences and discuss the relationship. [The first seven values of b_n are shown in Fig. 5(b)].

FIGURE 5

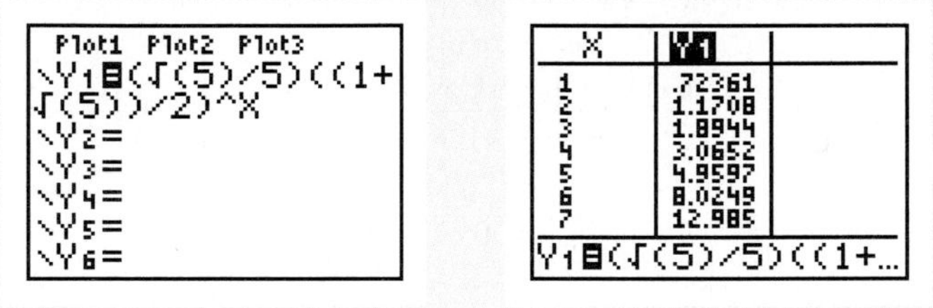

(a) (b)

Series

If $a_1, a_2, a_3, \ldots, a_n, \ldots$ is a sequence, then the expression $a_1 + a_2 + a_3 + \cdots + a_n + \cdots$ is called a **series.** If the sequence is finite, the corresponding series is a **finite series.** If the sequence is infinite, the corresponding series is an **infinite series.** For example,

1, 2, 4, 8, 16	Finite sequence
$1 + 2 + 4 + 8 + 16$	Finite series

We restrict our discussion to finite series in this section.

Series are often represented in a compact form called **summation notation** using the symbol $\sum$, which is a stylized version of the Greek letter sigma. Consider the following examples:

$$\sum_{k=1}^{4} a_k = a_1 + a_2 + a_3 + a_4$$

$$\sum_{k=3}^{7} b_k = b_3 + b_4 + b_5 + b_6 + b_7$$

$$\sum_{k=0}^{n} c_k = c_0 + c_1 + c_2 + \cdots + c_n$$ Domain is the set of integers k satisfying $0 \le k \le n$.

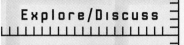

The terms on the right are obtained from the expression on the left by successively replacing the **summing index** k with integers, starting with the first number indicated below $\sum$ and ending with the number that appears above $\sum$. Thus, for example, if we are given the sequence

$$\frac{1}{2}, \frac{1}{4}, \frac{1}{8}, \ldots, \frac{1}{2^n}$$

the corresponding series is

$$\sum_{k=1}^{n} \frac{1}{2^k} = \frac{1}{2} + \frac{1}{4} + \frac{1}{8} + \cdots + \frac{1}{2^n}$$

EXAMPLE 3

Writing the Terms of a Series

Write without summation notation: $\sum_{k=1}^{5} \frac{k-1}{k}$

Solution

$$\sum_{k=1}^{5} \frac{k-1}{k} = \frac{1-1}{1} + \frac{2-1}{2} + \frac{3-1}{3} + \frac{4-1}{4} + \frac{5-1}{5}$$

$$= 0 + \frac{1}{2} + \frac{2}{3} + \frac{3}{4} + \frac{4}{5}$$

MATCHED PROBLEM 3

Write without summation notation: $\sum_{k=0}^{5} \frac{(-1)^k}{2k+1}$

If the terms of a series are alternately positive and negative, it is called an **alternating series.** Example 4 deals with the representation of such a series.

EXAMPLE 4

Writing a Series in Summation Notation

Write the following series using summation notation:

$$1 - \frac{1}{2} + \frac{1}{3} - \frac{1}{4} + \frac{1}{5} - \frac{1}{6}$$

(A) Start the summing index at $k = 1$.

(B) Start the summing index at $k = 0$.

Solutions

(A) $(-1)^{k-1}$ provides the alternation of sign, and $1/k$ provides the other part of each term. Thus, we can write

$$\sum_{k=1}^{6} \frac{(-1)^{k-1}}{k}$$

as can be easily checked.

(B) $(-1)^k$ provides the alternation of sign, and $1/(k+1)$ provides the other part of each term. Thus, we write

$$\sum_{k=0}^{5} \frac{(-1)^k}{k+1}$$

as can be checked.

MATCHED PROBLEM 4

Write the following series using summation notation:

$$1 - \frac{2}{3} + \frac{4}{9} - \frac{8}{27} + \frac{16}{81}$$

(A) Start with $k = 1$. (B) Start with $k = 0$.

Explore/Discuss 4

(A) Find the smallest number of terms of the infinite series

$$1 + \frac{1}{2} + \frac{1}{3} + \cdots + \frac{1}{n} + \cdots$$

that, when added together, give a number greater than 3.

(B) Find the smallest number of terms of the infinite series

$$\frac{1}{2} + \frac{1}{4} + \cdots + \frac{1}{2^n} + \cdots$$

that, when added together, give a number greater than 0.99. Greater than 0.999. Can the sum ever exceed 1? Explain.

Answers to Matched Problems

1. 4, 2, 1, $\frac{1}{2}$, $\frac{1}{4}$ 2. (A) $a_n = 2n$ (B) $a_n = (-1)^{n-1}\left(\frac{1}{2}\right)^{n-1}$ 3. $1 - \frac{1}{3} + \frac{1}{5} - \frac{1}{7} + \frac{1}{9} - \frac{1}{11}$

4. (A) $\sum_{k=1}^{5} (-1)^{k-1}\left(\frac{2}{3}\right)^{k-1}$ (B) $\sum_{k=0}^{4} (-1)^k\left(\frac{2}{3}\right)^k$

EXERCISE 6-1

A

Write the first four terms for each sequence in Problems 1–6.

1. $a_n = n - 2$
2. $a_n = n + 3$
3. $a_n = \dfrac{n-1}{n+1}$
4. $a_n = \left(1 + \dfrac{1}{n}\right)^n$
5. $a_n = (-2)^{n+1}$
6. $a_n = \dfrac{(-1)^{n+1}}{n^2}$
7. Write the eighth term in the sequence in Problem 1.
8. Write the tenth term in the sequence in Problem 2.
9. Write the one-hundredth term in the sequence in Problem 3.
10. Write the two-hundredth term in the sequence in Problem 4.

In Problems 11–16, write each series in expanded form without summation notation.

11. $\sum_{k=1}^{5} k$ **12.** $\sum_{k=1}^{4} k^2$

13. $\sum_{k=1}^{3} \frac{1}{10^k}$ **14.** $\sum_{k=1}^{5} \left(\frac{1}{3}\right)^k$

15. $\sum_{k=1}^{4} (-1)^k$ **16.** $\sum_{k=1}^{6} (-1)^{k+1}k$

B

Write the first five terms of each sequence in Problems 17–26.

17. $a_n = (-1)^{n+1}n^2$ **18.** $a_n = (-1)^{n+1}\left(\frac{1}{2^n}\right)$

19. $a_n = \frac{1}{3}\left(1 - \frac{1}{10^n}\right)$ **20.** $a_n = n[1 - (-1)^n]$

21. $a_n = (-\frac{1}{2})^{n-1}$ **22.** $a_n = (-\frac{3}{2})^{n-1}$

23. $a_1 = 7; a_n = a_{n-1} - 4, n \geq 2$

24. $a_1 = a_2 = 1; a_n = a_{n-1} + a_{n-2}, n \geq 3$

25. $a_1 = 4; a_n = \frac{1}{4}a_{n-1}, n \geq 2$

26. $a_1 = 2; a_n = 2a_{n-1}, n \geq 2$

In Problems 27–38, find the general term of a sequence whose first four terms are given.

27. 4, 5, 6, 7, . . . **28.** −2, −1, 0, 1, . . .

29. 3, 6, 9, 12, . . . **30.** −2, −4, −6, −8, . . .

31. $\frac{1}{2}, \frac{2}{3}, \frac{3}{4}, \frac{4}{5}, \ldots$ **32.** $\frac{1}{2}, \frac{3}{4}, \frac{5}{6}, \frac{7}{8}, \ldots$

33. 1, −1, 1, −1, . . . **34.** 1, −2, 3, −4, . . .

35. −2, 4, −8, 16, . . . **36.** 1, −3, 5, −7, . . .

37. $x, \frac{x^2}{2}, \frac{x^3}{3}, \frac{x^4}{4}, \ldots$ **38.** $x, -x^3, x^5, -x^7, \ldots$

In Problems 39–42, use a graphing utility to graph the first 20 terms of each sequence.

39. $a_n = 1/n$ **40.** $a_n = 2 + \pi n$

41. $a_n = (-0.9)^n$ **42.** $a_1 = -1, a_n = \frac{2}{3}a_{n-1} + \frac{1}{2}$

In Problems 43–48, write each series in expanded form without summation notation.

43. $\sum_{k=1}^{4} \frac{(-2)^{k+1}}{k}$ **44.** $\sum_{k=1}^{5} (-1)^{k+1}(2k-1)^2$

45. $\sum_{k=1}^{3} \frac{1}{k}x^{k+1}$ **46.** $\sum_{k=1}^{5} x^{k-1}$

47. $\sum_{k=1}^{5} \frac{(-1)^{k+1}}{k}x^k$ **48.** $\sum_{k=0}^{4} \frac{(-1)^k x^{2k+1}}{2k+1}$

In Problems 49–56, write each series using summation notation with the summing index k starting at k = 1.

49. $1^2 + 2^2 + 3^2 + 4^2$ **50.** $2 + 3 + 4 + 5 + 6$

51. $\frac{1}{2} + \frac{1}{2^2} + \frac{1}{2^3} + \frac{1}{2^4} + \frac{1}{2^5}$ **52.** $1 - \frac{1}{2} + \frac{1}{3} - \frac{1}{4}$

53. $1 + \frac{1}{2^2} + \frac{1}{3^2} + \cdots + \frac{1}{n^2}$

54. $2 + \frac{3}{2} + \frac{4}{3} + \cdots + \frac{n+1}{n}$

55. $1 - 4 + 9 - \cdots + (-1)^{n+1}n^2$

56. $\frac{1}{2} - \frac{1}{4} + \frac{1}{8} - \cdots + \frac{(-1)^{n+1}}{2^n}$

C

The sequence

$$a_n = \frac{a_{n-1}^2 + M}{2a_{n-1}} \qquad n \geq 2, M \text{ a positive real number}$$

can be used to find $\sqrt{M}$ to any decimal-place accuracy desired. To start the sequence, choose a_1 arbitrarily from the positive real numbers. Problems 57 and 58 are related to this sequence.

57. (A) Find the first four terms of the sequence

$$a_1 = 3 \qquad a_n = \frac{a_{n-1}^2 + 2}{2a_{n-1}} \qquad n \geq 2$$

(B) Compare the terms with $\sqrt{2}$ from a calculator.

(C) Repeat parts A and B letting a_1 be any other positive number, say 1.

58. (A) Find the first four terms of the sequence

$$a_1 = 2 \qquad a_n = \frac{a_{n-1}^2 + 5}{2a_{n-1}} \qquad n \geq 2$$

(B) Find $\sqrt{5}$ with a calculator, and compare with the results of part A.

(C) Repeat parts A and B letting a_1 be any other positive number, say 3.

59. Let $\{a_n\}$ denote the Fibonacci sequence and let $\{b_n\}$ denote the sequence defined by $b_1 = 1, b_2 = 3, b_n = b_{n-1} + b_{n-2}$ for $n \geq 3$. Compute 10 terms of the sequence $\{c_n\}$, where $c_n = b_n/a_n$. Describe the terms of $\{c_n\}$ for large values of n.

60. Define sequences $\{u_n\}$ and $\{v_n\}$ by $u_1 = 1, v_1 = 0, u_n = u_{n-1} + v_{n-1}$ and $v_n = u_{n-1}$ for $n \geq 2$. Find the first 10 terms of each sequence, and explain their relationship to the Fibonacci sequence.

In calculus, it can be shown that

$$e^x = \sum_{k=0}^{\infty} \frac{x^k}{k!} \approx 1 + \frac{x}{1!} + \frac{x^2}{2!} + \frac{x^3}{3!} + \cdots + \frac{x^n}{n!}$$

where the larger n is, the better the approximation. Problems 61 and 62 refer to this series. Note that n!, read "n factorial," is defined by $0! = 1$ *and* $n! = 1 \cdot 2 \cdot 3 \cdot \cdots \cdot n$ *for* $n \in N$.

61. Approximate $e^{0.2}$ using the first five terms of the series. Compare this approximation with your calculator evaluation of $e^{0.2}$.

62. Approximate $e^{-0.5}$ using the first five terms of the series. Compare this approximation with your calculator evaluation of $e^{-0.5}$.

63. Show that $\sum_{k=1}^{n} ca_k = c \sum_{k=1}^{n} a_k$

64. Show that $\sum_{k=1}^{n} (a_k + b_k) = \sum_{k=1}^{n} a_k + \sum_{k=1}^{n} b_k$

Section 6-2 | Mathematical Induction

- Introduction
- Mathematical Induction
- Additional Examples of Mathematical Induction
- Three Famous Problems

Introduction

In common usage, the word "induction" means the generalization from particular cases or facts. The ability to formulate general hypotheses from a limited number of facts is a distinguishing characteristic of a creative mathematician. The creative process does not stop here, however. These hypotheses must then be proved or disproved. In mathematics, a special method of proof called **mathematical induction** ranks among the most important basic tools in a mathematician's toolbox. In this section, mathematical induction will be used to prove a variety of mathematical statements, some new and some that up to now we have just assumed to be true.

We illustrate the process of formulating hypotheses by an example. Suppose we are interested in the sum of the first n consecutive odd integers, where n is a positive integer. We begin by writing the sums for the first few values of n to see if we can observe a pattern:

$$1 = 1 \qquad n = 1$$
$$1 + 3 = 4 \qquad n = 2$$
$$1 + 3 + 5 = 9 \qquad n = 3$$
$$1 + 3 + 5 + 7 = 16 \qquad n = 4$$
$$1 + 3 + 5 + 7 + 9 = 25 \qquad n = 5$$

Is there any pattern to the sums 1, 4, 9, 16, and 25? You no doubt observed that each is a perfect square and, in fact, each is the square of the number of terms in the sum. Thus, the following conjecture seems reasonable:

Conjecture P_n: For each positive integer n,

$$1 + 3 + 5 + \cdots + (2n - 1) = n^2$$

That is, the sum of the first n odd integers is n^2 for each positive integer n.

So far ordinary induction has been used to generalize the pattern observed in the first few cases listed. But at this point conjecture P_n is simply that—a conjecture. How do we prove that P_n is a true statement? Continuing to list specific cases will never provide a general proof—not in your lifetime or all your descendants' lifetimes! Mathematical induction is the tool we will use to establish the validity of conjecture P_n.

Before discussing this method of proof, let's consider another conjecture:

Conjecture Q_n: For each positive integer n, the number $n^2 - n + 41$ is a prime number.

It is important to recognize that a conjecture can be proved false if it fails for only one case. A single case or example for which a conjecture fails is called a **counterexample.** We check the conjecture for a few particular cases in Table 1. From the table, it certainly appears that conjecture Q_n has a good chance of being true. You may want to check a few more cases. If you persist, you will find that conjecture Q_n is true for n up to 41. What happens at $n = 41$?

TABLE 1

n	$n^2 - n + 41$	Prime?
1	41	Yes
2	43	Yes
3	47	Yes
4	53	Yes
5	61	Yes

$$41^2 - 41 + 41 = 41^2$$

which is not prime. Thus, since $n = 41$ provides a counterexample, conjecture Q_n is false. Here we see the danger of generalizing without proof from a few special cases. This example was discovered by Euler (1707–1783).

Explore/Discuss 1

Prove that the following statement is false by finding a counterexample: If $n \geq 2$, then at least one-third of the positive integers less than or equal to n are prime.

Mathematical Induction

We begin by stating the *principle of mathematical induction,* which forms the basis for all our work in this section.

THEOREM 1

PRINCIPLE OF MATHEMATICAL INDUCTION

Let P_n be a statement associated with each positive integer n, and suppose the following conditions are satisfied:

1. P_1 is true.
2. For any positive integer k, if P_k is true, then P_{k+1} is also true.

Then the statement P_n is true for all positive integers n.

Theorem 1 must be read very carefully. At first glance, it seems to say that if we assume a statement is true, then it is true. But that is not the case at all. If the two conditions in Theorem 1 are satisfied, then we can reason as follows:

P_1 is true.	Condition 1
P_2 is true, because P_1 is true.	Condition 2
P_3 is true, because P_2 is true.	Condition 2
P_4 is true, because P_3 is true.	Condition 2
⋮	⋮

Since this chain of implications never ends, we will eventually reach P_n for any positive integer n.

To help visualize this process, picture a row of dominoes that goes on forever (see Fig. 1) and interpret the conditions in Theorem 1 as follows: Condition 1 says that the first domino can be pushed over. Condition 2 says that if the kth domino falls, then so does the $(k + 1)$st domino. Together, these two conditions imply that all the dominoes must fall.

FIGURE 1 Interpreting mathematical induction.

Condition 1: The first domino can be pushed over.
(a)

Condition 2: If the kth domino falls, then so does the $(k + 1)$st.
(b)

Conclusion: All the dominoes will fall.
(c)

Now, to illustrate the process of proof by mathematical induction, we return to the conjecture P_n discussed earlier, which we restate below:

$$P_n\text{: } 1 + 3 + 5 + \cdots + (2n - 1) = n^2 \qquad n \text{ any positive integer}$$

We already know that P_1 is a true statement. In fact, we demonstrated that P_1 through P_5 are all true by direct calculation. Thus, condition 1 in Theorem 1 is satisfied. To show that condition 2 is satisfied, we assume that P_k is a true statement:

$$P_k\text{: } 1 + 3 + 5 + \cdots + (2k - 1) = k^2$$

Now we must show that this assumption implies that P_{k+1} is also a true statement:

$$P_{k+1}\text{: } 1 + 3 + 5 + \cdots + (2k - 1) + (2k + 1) = (k + 1)^2$$

Since we have assumed that P_k is true, we can perform the operations on this equation. Note that the left side of P_{k+1} is the left side of P_k plus $(2k + 1)$. So we start by adding $(2k + 1)$ to both sides of P_k:

$$1 + 3 + 5 + \cdots + (2k - 1) = k^2 \qquad P_k$$

$$1 + 3 + 5 + \cdots + (2k - 1) + (2k + 1) = k^2 + (2k + 1) \qquad \text{Add } 2k + 1 \text{ to both sides.}$$

Factoring the right side of this equation, we have

$$1 + 3 + 5 + \cdots + (2k - 1) + (2k + 1) = (k + 1)^2 \qquad P_{k+1}$$

But this last equation is P_{k+1}. Thus, we have started with P_k, the statement we assumed true, and performed valid operations to produce P_{k+1}, the statement we want to be true. In other words, we have shown that if P_k is true, then P_{k+1} is also true. Since both conditions in Theorem 1 are satisfied, P_n is true for all positive integers n.

Additional Examples of Mathematical Induction

Now we will consider some additional examples of proof by induction. The first is another summation formula. Mathematical induction is the primary tool for proving that formulas of this type are true.

EXAMPLE 1 Proving a Summation Formula

Prove that for all positive integers n

$$\frac{1}{2} + \frac{1}{4} + \frac{1}{8} + \cdots + \frac{1}{2^n} = \frac{2^n - 1}{2^n}$$

Proof State the conjecture:

$$P_n\colon \quad \frac{1}{2} + \frac{1}{4} + \frac{1}{8} + \cdots + \frac{1}{2^n} = \frac{2^n - 1}{2^n}$$

Part 1
Show that P_1 is true.

$$P_1\colon \quad \frac{1}{2} = \frac{2^1 - 1}{2^1}$$

$$= \frac{1}{2}$$

Thus, P_1 is true.

Part 2
Show that if P_k is true, then P_{k+1} is true. It is a good practice to always write out both P_k and P_{k+1} at the beginning of any induction proof to see what is assumed and what must be proved:

$$P_k\colon \quad \frac{1}{2} + \frac{1}{4} + \frac{1}{8} + \cdots + \frac{1}{2^k} = \frac{2^k - 1}{2^k}$$ We assume P_k is true.

$$P_{k+1}\colon \quad \frac{1}{2} + \frac{1}{4} + \frac{1}{8} + \cdots + \frac{1}{2^k} + \frac{1}{2^{k+1}} = \frac{2^{k+1} - 1}{2^{k+1}}$$ We must show that P_{k+1} follows from P_k.

We start with the true statement P_k, add $1/2^{k+1}$ to both sides, and simplify the right side:

$$\frac{1}{2} + \frac{1}{4} + \frac{1}{8} + \cdots + \frac{1}{2^k} = \frac{2^k - 1}{2^k} \qquad P_k$$

$$\begin{aligned}
\frac{1}{2} + \frac{1}{4} + \frac{1}{8} + \cdots + \frac{1}{2^k} + \frac{1}{2^{k+1}} &= \frac{2^k - 1}{2^k} + \frac{1}{2^{k+1}} \\
&= \frac{2^k - 1}{2^k} \cdot \frac{2}{2} + \frac{1}{2^{k+1}} \\
&= \frac{2^{k+1} - 2 + 1}{2^{k+1}} \\
&= \frac{2^{k+1} - 1}{2^{k+1}}
\end{aligned}$$

Thus,

$$\frac{1}{2} + \frac{1}{4} + \frac{1}{8} + \cdots + \frac{1}{2^k} + \frac{1}{2^{k+1}} = \frac{2^{k+1} - 1}{2^{k+1}} \qquad P_{k+1}$$

and we have shown that if P_k is true, then P_{k+1} is true.

Conclusion

Both conditions in Theorem 1 are satisfied. Thus, P_n is true for all positive integers n.

MATCHED PROBLEM 1

Prove that for all positive integers n

$$1 + 2 + 3 + \cdots + n = \frac{n(n + 1)}{2}$$

The next example provides a proof of a law of exponents that previously we had to assume was true. First we redefine a^n for n a positive integer, using a recursion formula:

DEFINITION 1

RECURSIVE DEFINITION OF a^n

For n a positive integer

$$a^1 = a$$
$$a^{n+1} = a^n a \qquad n > 1$$

EXAMPLE 2 **Proving a Law of Exponents**

Prove that $(xy)^n = x^n y^n$ for all positive integers n.

Proof State the conjecture:

$$P_n\text{: } (xy)^n = x^n y^n$$

Part 1

Show that P_1 is true.

$$(xy)^1 = xy \qquad \text{Definition 1}$$
$$= x^1 y^1 \qquad \text{Definition 1}$$

Thus, P_1 is true.

Part 2

Show that if P_k is true, then P_{k+1} is true.

P_k: $(xy)^k = x^k y^k$ Assume P_k is true.

P_{k+1}: $(xy)^{k+1} = x^{k+1} y^{k+1}$ Show that P_{k+1} follows from P_k.

Here we start with the left side of P_{k+1} and use P_k to find the right side of P_{k+1}:

$$\begin{aligned} (xy)^{k+1} &= (xy)^k(xy)^1 && \text{Definition 1} \\ &= x^k y^k xy && \text{Use } P_k\text{: } (xy)^k = x^k y^k. \\ &= (x^k x)(y^k y) && \text{Property of real numbers} \\ &= x^{k+1} y^{k+1} && \text{Definition 1} \end{aligned}$$

Thus, $(xy)^{k+1} = x^{k+1}y^{k+1}$, and we have shown that if P_k is true, then P_{k+1} is true.

Conclusion

Both conditions in Theorem 1 are satisfied. Thus, P_n is true for all positive integers n.

MATCHED PROBLEM 2

Prove that $(x/y)^n = x^n/y^n$ for all positive integers n.

Our last example deals with factors of integers. Before we start, recall that an integer p is *divisible* by an integer q if $p = qr$ for some integer r.

EXAMPLE 3

Proving a Divisibility Property

Prove that $4^{2n} - 1$ is divisible by 5 for all positive integers n.

Proof

Use the definition of divisibility to state the conjecture as follows:

$$P_n\text{: } 4^{2n} - 1 = 5r \qquad \text{for some integer } r$$

Part 1

Show that P_1 is true.

$$P_1\text{: } 4^2 - 1 = 15 = 5 \cdot 3$$

Thus, P_1 is true.

Part 2

Show that if P_k is true, then P_{k+1} is true.

$$\begin{aligned} P_k&\text{: } 4^{2k} - 1 = 5r && \text{for some integer } r && \text{Assume } P_k \text{ is true.} \\ P_{k+1}&\text{: } 4^{2(k+1)} - 1 = 5s && \text{for some integer } s && \text{Show that } P_{k+1} \text{ must follow.} \end{aligned}$$

As before, we start with the true statement P_k:

$$\begin{aligned} 4^{2k} - 1 &= 5r && P_k \\ 4^2(4^{2k} - 1) &= 4^2(5r) && \text{Multiply both sides by } 4^2. \\ 4^{2k+2} - 16 &= 80r && \text{Simplify.} \\ 4^{2(k+1)} - 1 &= 80r + 15 && \text{Add 15 to both sides.} \\ &= 5(16r + 3) && \text{Factor out 5.} \end{aligned}$$

Thus,

$$4^{2(k+1)} - 1 = 5s \qquad P_{k+1}$$

where $s = 16r + 3$ is an integer, and we have shown that if P_k is true, then P_{k+1} is true.

Conclusion

Both conditions in Theorem 1 are satisfied. Thus, P_n is true for all positive integers n.

MATCHED PROBLEM 3

Prove that $8^n - 1$ is divisible by 7 for all positive integers n.

In some cases, a conjecture may be true only for $n \geq m$, where m is a positive integer, rather than for all $n \geq 0$. For example, see Problems 49 and 50 in Exercise 6-2. The principle of mathematical induction can be extended to cover cases like this as follows:

THEOREM 2

EXTENDED PRINCIPLE OF MATHEMATICAL INDUCTION

Let m be a positive integer, let P_n be a statement associated with each integer $n \geq m$, and suppose the following conditions are satisfied:

1. P_m is true.
2. For any integer $k \geq m$, if P_k is true, then P_{k+1} is also true.

Then the statement P_n is true for all integers $n \geq m$.

Three Famous Problems

The problem of determining whether a certain statement about the positive integers is true may be extremely difficult. Proofs may require remarkable insight and ingenuity and the development of techniques far more advanced than mathematical induction. Consider, for example, the famous problems of proving the following statements:

1. **Lagrange's Four Square Theorem, 1772:** Each positive integer can be expressed as the sum of four or fewer squares of positive integers.
2. **Fermat's Last Theorem, 1637:** For $n > 2$, $x^n + y^n = z^n$ does not have solutions in the natural numbers.
3. **Goldbach's Conjecture, 1742:** Every positive even integer greater than 2 is the sum of two prime numbers.

The first statement was considered by the early Greeks and finally proved in 1772 by Lagrange. Fermat's last theorem, defying the best mathematical minds for over 350 years, finally succumbed to a 200-page proof by Prof. Andrew Wiles of Princeton University in 1993. To this date no one has been able to prove or disprove Goldbach's conjecture.

Explore/Discuss 2

(A) Explain the difference between a theorem and a conjecture.

(B) Why is "Fermat's last theorem" a misnomer? Suggest more accurate names for the result.

Answers to Matched Problems

1. Sketch of proof. State the conjecture: P_n: $1 + 2 + 3 + \cdots + n = \dfrac{n(n+1)}{2}$

Part 1. $1 = \dfrac{1(1+1)}{2}$. P_1 is true.

Part 2. Show that if P_k is true, then P_{k+1} is true.

$$1 + 2 + 3 + \cdots + k = \frac{k(k+1)}{2} \qquad P_k$$

$$1 + 2 + 3 + \cdots + k + (k+1) = \frac{k(k+1)}{2} + (k+1)$$

$$= \frac{(k+1)(k+2)}{2} \qquad P_{k+1}$$

Conclusion: P_n is true.

2. Sketch of proof. State the conjecture: P_n: $\left(\dfrac{x}{y}\right)^n = \dfrac{x^n}{y^n}$

Part 1. $\left(\dfrac{x}{y}\right)^1 = \dfrac{x}{y} = \dfrac{x^1}{y^1}$. P_1 is true.

Part 2. Show that if P_k is true, then P_{k+1} is true.

$$\left(\frac{x}{y}\right)^{k+1} = \left(\frac{x}{y}\right)^k\left(\frac{x}{y}\right) = \frac{x^k}{y^k}\left(\frac{x}{y}\right) = \frac{x^kx}{y^ky} = \frac{x^{k+1}}{y^{k+1}}$$

Conclusion: P_n is true.

3. Sketch of proof. State the conjecture: P_n: $8^n - 1 = 7r$ for some integer r

Part 1. $8^1 - 1 = 7 = 7 \cdot 1$. P_1 is true.

Part 2. Show that if P_k is true, then P_{k+1} is true.

$$8^k - 1 = 7r \qquad P_k$$

$$8(8^k - 1) = 8(7r)$$

$$8^{k+1} - 1 = 56r + 7 = 7(8r + 1) = 7s \qquad P_{k+1}$$

Conclusion: P_n is true.

EXERCISE 6-2

A

In Problems 1–4, find the first positive integer n that causes the statement to fail.

1. $(3 + 5)^n = 3^n + 5^n$ **2.** $n < 10$

3. $n^2 = 3n - 2$ **4.** $n^3 + 11n = 6n^2 + 6$

Verify each statement P_n in Problems 5–10 for $n = 1, 2,$ and 3.

5. P_n: $2 + 6 + 10 + \cdots + (4n - 2) = 2n^2$

6. P_n: $4 + 8 + 12 + \cdots + 4n = 2n(n + 1)$

7. P_n: $a^5a^n = a^{5+n}$ **8.** P_n: $(a^5)^n = a^{5n}$

9. P_n: $9^n - 1$ is divisible by 4 **10.** P_n: $4^n - 1$ is divisible by 3

Write P_k and P_{k+1} for P_n as indicated in Problems 11–16.

11. P_n in Problem 5 **12.** P_n in Problem 6

13. P_n in Problem 7
14. P_n in Problem 8
15. P_n in Problem 9
16. P_n in Problem 10

In Problems 17–22, use mathematical induction to prove that each P_n holds for all positive integers n.

17. P_n in Problem 5
18. P_n in Problem 6
19. P_n in Problem 7
20. P_n in Problem 8
21. P_n in Problem 9
22. P_n in Problem 10

B

In Problems 23–26, prove the statement is false by finding a counterexample.

23. If $n > 2$, then any polynomial of degree n has at least one real zero.

24. Any positive integer $n > 7$ can be written as the sum of three or fewer squares of positive integers.

25. If n is a positive integer, then there is at least one prime number p such that $n < p < n + 6$.

26. If a, b, c, and d are positive integers such that $a^2 + b^2 = c^2 + d^2$, then $a = c$ or $a = d$.

In Problems 27–42, use mathematical induction to prove each proposition for all positive integers n, unless restricted otherwise.

27. $2 + 2^2 + 2^3 + \cdots + 2^n = 2^{n+1} - 2$

28. $\frac{1}{2} + \frac{1}{4} + \frac{1}{8} + \cdots + \frac{1}{2^n} = 1 - \left(\frac{1}{2}\right)^n$

29. $1^2 + 3^2 + 5^2 + \cdots + (2n - 1)^2 = \frac{1}{3}(4n^3 - n)$

30. $1 + 8 + 16 + \cdots + 8(n - 1) = (2n - 1)^2$; $n > 1$

31. $1^2 + 2^2 + 3^2 + \cdots + n^2 = \frac{n(n + 1)(2n + 1)}{6}$

32. $1 \cdot 2 + 2 \cdot 3 + 3 \cdot 4 + \cdots + n(n + 1) = \frac{n(n + 1)(n + 2)}{3}$

33. $\frac{a^n}{a^3} = a^{n-3}$; $n > 3$

34. $\frac{a^5}{a^n} = \frac{1}{a^{n-5}}$; $n > 5$

35. $a^m a^n = a^{m+n}$; $m, n \in N$ [*Hint:* Choose m as an arbitrary element of N, and then use induction on n.]

36. $(a^n)^m = a^{mn}$; $m, n \in N$

37. $x^n - 1$ is divisible by $x - 1$; $x \neq 1$ [*Hint:* Divisible means that $x^n - 1 = (x - 1)Q(x)$ for some polynomial $Q(x)$.]

38. $x^n - y^n$ is divisible by $x - y$; $x \neq y$

39. $x^{2n} - 1$ is divisible by $x - 1$; $x \neq 1$

40. $x^{2n} - 1$ is divisible by $x + 1$; $x \neq -1$

41. $1^3 + 2^3 + 3^3 + \cdots + n^3 = (1 + 2 + 3 + \cdots + n)^2$ [*Hint:* See Matched Problem 1 following Example 1.]

42. $$\frac{1}{1 \cdot 2 \cdot 3} + \frac{1}{2 \cdot 3 \cdot 4} + \frac{1}{3 \cdot 4 \cdot 5} + \cdots + \frac{1}{n(n + 1)(n + 2)} = \frac{n(n + 3)}{4(n + 1)(n + 2)}$$

C

In Problems 43–46, suggest a formula for each expression, and prove your hypothesis using mathematical induction, $n \in N$.

43. $2 + 4 + 6 + \cdots + 2n$

44. $\frac{1}{1 \cdot 2} + \frac{1}{2 \cdot 3} + \frac{1}{3 \cdot 4} + \cdots + \frac{1}{n(n + 1)}$

45. The number of lines determined by n points in a plane, no three of which are collinear

46. The number of diagonals in a polygon with n sides

Prove Problems 47–50 true for all integers n as specified.

47. $a > 1 \Rightarrow a^n > 1$; $n \in N$

48. $0 < a < 1 \Rightarrow 0 < a^n < 1$; $n \in N$

49. $n^2 > 2n$; $n \geq 3$

50. $2^n > n^2$; $n \geq 5$

51. Prove or disprove the generalization of the following two facts:

$$3^2 + 4^2 = 5^2$$
$$3^3 + 4^3 + 5^3 = 6^3$$

52. Prove or disprove: $n^2 + 21n + 1$ is a prime number for all natural numbers n.

If $\{a_n\}$ and $\{b_n\}$ are two sequences, we write $\{a_n\} = \{b_n\}$ if and only if $a_n = b_n$, $n \in N$. In Problems 53–56, use mathematical induction to show that $\{a_n\} = \{b_n\}$.

53. $a_1 = 1$, $a_n = a_{n-1} + 2$; $b_n = 2n - 1$

54. $a_1 = 2$, $a_n = a_{n-1} + 2$; $b_n = 2n$

55. $a_1 = 2$, $a_n = 2^2 a_{n-1}$; $b_n = 2^{2n-1}$

56. $a_1 = 2$, $a_n = 3a_{n-1}$; $b_n = 2 \cdot 3^{n-1}$

Section 6-3 Arithmetic and Geometric Sequences

- Arithmetic and Geometric Sequences
- *n*th-Term Formulas
- Sum Formulas for Finite Arithmetic Series
- Sum Formulas for Finite Geometric Series
- Sum Formula for Infinite Geometric Series

For most sequences it is difficult to sum an arbitrary number of terms of the sequence without adding term by term. But particular types of sequences, *arithmetic sequences* and *geometric sequences,* have certain properties that lead to convenient and useful formulas for the sums of the corresponding *arithmetic series* and *geometric series.*

Arithmetic and Geometric Sequences

The sequence 5, 7, 9, 11, 13, . . . , $5 + 2(n - 1)$, . . . , where each term after the first is obtained by adding 2 to the preceding term, is an example of an arithmetic sequence. The sequence 5, 10, 20, 40, 80, . . . , $5\,(2)^{n-1}$, . . . , where each term after the first is obtained by multiplying the preceding term by 2, is an example of a geometric sequence.

DEFINITION 1

ARITHMETIC SEQUENCE

A sequence

$$a_1, a_2, a_3, \ldots, a_n, \ldots$$

is called an **arithmetic sequence,** or **arithmetic progression,** if there exists a constant d, called the **common difference,** such that

$$a_n - a_{n-1} = d$$

That is,

$$a_n = a_{n-1} + d \qquad \text{for every } n > 1$$

DEFINITION 2

GEOMETRIC SEQUENCE

A sequence

$$a_1, a_2, a_3, \ldots, a_n, \ldots$$

is called a **geometric sequence,** or **geometric progression,** if there exists a nonzero constant r, called the **common ratio,** such that

DEFINITION 2 continued

$$\frac{a_n}{a_{n-1}} = r$$

That is,

$$a_n = ra_{n-1} \qquad \text{for every } n > 1$$

Explore/Discuss 1

(A) Graph the arithmetic sequence 5, 7, 9, ...
Describe the graphs of all arithmetic sequences with common difference 2.

(B) Graph the geometric sequence 5, 10, 20, ...
Describe the graphs of all geometric sequences with common ratio 2.

EXAMPLE 1 Recognizing Arithmetic and Geometric Sequences

Which of the following can be the first four terms of an arithmetic sequence? Of a geometric sequence?

(A) 1, 2, 3, 5, . . . (B) $-1, 3, -9, 27, \ldots$
(C) 3, 3, 3, 3, . . . (D) 10, 8.5, 7, 5.5, . . .

Solutions

(A) Since $2 - 1 \neq 5 - 3$, there is no common difference, so the sequence is not an arithmetic sequence. Since $\frac{2}{1} \neq \frac{3}{2}$, there is no common ratio, so the sequence is not geometric either.

(B) The sequence is geometric with common ratio -3, but it is not arithmetic.

(C) The sequence is arithmetic with common difference 0 and it is also geometric with common ratio 1.

(D) The sequence is arithmetic with common difference -1.5, but it is not geometric.

MATCHED PROBLEM 1

Which of the following can be the first four terms of an arithmetic sequence? Of a geometric sequence?

(A) 8, 2, 0.5, 0.125, . . . (B) $-7, -2, 3, 8, \ldots$ (C) 1, 5, 25, 100, . . .

*n*th-Term Formulas

If $\{a_n\}$ is an arithmetic sequence with common difference d, then

$$a_2 = a_1 + d$$
$$a_3 = a_2 + d = a_1 + 2d$$
$$a_4 = a_3 + d = a_1 + 3d$$

This suggests Theorem 1, which can be proved by mathematical induction (see Problem 63 in Exercise 6-3).

THEOREM 1

THE nTH TERM OF AN ARITHMETIC SEQUENCE

$$a_n = a_1 + (n - 1)d \qquad \text{for every } n > 1$$

Similarly, if $\{a_n\}$ is a geometric sequence with common ratio r, then

$$a_2 = a_1 r$$

$$a_3 = a_2 r = a_1 r^2$$

$$a_4 = a_3 r = a_1 r^3$$

This suggests Theorem 2, which can also be proved by mathematical induction (see Problem 69 in Exercise 6-3).

THEOREM 2

THE nTH TERM OF A GEOMETRIC SEQUENCE

$$a_n - a_1 r^{n-1} \qquad \text{for every } n > 1$$

EXAMPLE 2

Finding Terms in Arithmetic and Geometric Sequences

(A) If the first and tenth terms of an arithmetic sequence are 3 and 30, respectively, find the fiftieth term of the sequence.

(B) If the first and tenth terms of a geometric sequence are 1 and 4, find the seventeenth term to three decimal places.

Solutions

(A) First use Theorem 1 with $a_1 = 3$ and $a_{10} = 30$ to find d:

$$a_n = a_1 + (n - 1)d$$

$$a_{10} = a_1 + (10 - 1)d$$

$$30 = 3 + 9d$$

$$d = 3$$

Now find a_{50}:

$$a_{50} = a_1 + (50 - 1)3$$

$$= 3 + 49 \cdot 3$$

$$= 150$$

(B) First let $n = 10$, $a_1 = 1$, $a_{10} = 4$ and use Theorem 2 to find r.

$$a_n = a_1 r^{n-1}$$

$$4 = 1r^{10-1}$$

$$r = 4^{1/9}$$

Now use Theorem 2 again, this time with $n = 17$.

$$a_{17} = a_1 r^{17} = 1\,(4^{1/9})^{17} = 4^{17/9} \approx 13.716$$

MATCHED PROBLEM 2

(A) If the first and fifteenth terms of an arithmetic sequence are -5 and 23, respectively, find the seventy-third term of the sequence.

(B) Find the eighth term of the geometric sequence $\frac{1}{64}, -\frac{1}{32}, \frac{1}{16}, \ldots$.

Sum Formulas for Finite Arithmetic Series

If $a_1, a_2, a_3, \ldots, a_n$ is a finite arithmetic sequence, then the corresponding series $a_1 + a_2 + a_3 + \cdots + a_n$ is called an *arithmetic series*. We will derive two simple and very useful formulas for the sum of an arithmetic series. Let d be the common difference of the arithmetic sequence $a_1, a_2, a_3, \ldots, a_n$ and let S_n denote the sum of the series $a_1 + a_2 + a_3 + \cdots + a_n$.

Then

$$S_n = a_1 + (a_1 + d) + \cdots + [a_1 + (n - 2)d] + [a_1 + (n - 1)d]$$

Reversing the order of the sum, we obtain

$$S_n = [a_1 + (n - 1)d] + [a_1 + (n - 2)d] + \cdots + (a_1 + d) + a_1$$

Adding the left sides of these two equations and corresponding elements of the right sides, we see that

$$\begin{aligned} 2S_n &= [2a_1 + (n - 1)d] + [2a_1 + (n - 1)d] + \cdots + [2a_1 + (n - 1)d] \\ &= n[2a_1 + (n - 1)d] \end{aligned}$$

This can be restated as in Theorem 3.

THEOREM 3

SUM OF AN ARITHMETIC SERIES—FIRST FORM

$$S_n = \frac{n}{2}[2a_1 + (n - 1)d]$$

By replacing $a_1 + (n - 1)d$ with a_n, we obtain a second useful formula for the sum.

THEOREM 4

SUM OF AN ARITHMETIC SERIES—SECOND FORM

$$S_n = \frac{n}{2}(a_1 + a_n)$$

The proof of the first sum formula by mathematical induction is left as an exercise (see Problem 64 in Exercise 6-3).

EXAMPLE 3

Finding the Sum of an Arithmetic Series

Find the sum of the first 26 terms of an arithmetic series if the first term is -7 and $d = 3$.

Solution

Let $n = 26$, $a_1 = -7$, $d = 3$, and use Theorem 3.

$$S_n = \frac{n}{2}[2a_1 + (n - 1)d]$$

$$S_{26} = \tfrac{26}{2}[2(-7) + (26 - 1)3]$$

$$= 793$$

MATCHED PROBLEM 3

Find the sum of the first 52 terms of an arithmetic series if the first term is 23 and $d = -2$.

EXAMPLE 4

Finding the Sum of an Arithmetic Series

Find the sum of all the odd numbers between 51 and 99, inclusive.

Solution

First, use $a_1 = 51$, $a_n = 99$, and Theorem 1 to find n:

$$a_n = a_1 + (n - 1)d$$

$$99 = 51 + (n - 1)2$$

$$n = 25$$

Now use Theorem 4 to find S_{25}:

$$S_n = \frac{n}{2}(a_1 + a_n)$$

$$S_{25} = \tfrac{25}{2}(51 + 99)$$

$$= 1{,}875$$

MATCHED PROBLEM 4

Find the sum of all the even numbers between -22 and 52, inclusive.

EXAMPLE 5

Prize Money

A 16-team bowling league has \$8,000 to be awarded as prize money. If the last-place team is awarded \$275 in prize money and the award increases by the same amount for each successive finishing place, how much will the first-place team receive?

Solution

If a_1 is the award for the first-place team, a_2 is the award for the second-place team, and so on, then the prize money awards form an arithmetic sequence with $n = 16$, $a_{16} = 275$, and $S_{16} = 8{,}000$. Use Theorem 4 to find a_1.

$$S_n = \frac{n}{2}(a_1 + a_n)$$

$$8{,}000 = \tfrac{16}{2}(a_1 + 275)$$

$$a_1 = 725$$

Thus, the first-place team receives \$725.

MATCHED PROBLEM 5

Refer to Example 5. How much prize money is awarded to the second-place team?

Sum Formulas for Finite Geometric Series

If $a_1, a_2, a_3, \ldots, a_n$ is a finite geometric sequence, then the corresponding series $a_1 + a_2 + a_3 + \cdots + a_n$ is called a *geometric series*. As with arithmetic series, we can derive two simple and very useful formulas for the sum of a geometric series. Let r be the common ratio of the geometric sequence $a_1, a_2, a_3, \ldots, a_n$ and let S_n denote the sum of the series $a_1 + a_2 + a_3 + \cdots + a_n$. Then

$$S_n = a_1 + a_1r + a_1r^2 + a_1r^3 + \cdots + a_1r^{n-2} + a_1r^{n-1}$$

Multiply both sides of this equation by r to obtain

$$rS_n = a_1r + a_1r^2 + a_1r^3 + \cdots + a_1r^{n-1} + a_1r^n$$

Now subtract the left side of the second equation from the left side of the first, and the right side of the second equation from the right side of the first to obtain

$$S_n - rS_n = a_1 - a_1r^n$$

$$S_n(1 - r) = a_1 - a_1r^n$$

Thus, solving for S_n, we obtain the following formula for the sum of a geometric series:

THEOREM 5

SUM OF A GEOMETRIC SERIES—FIRST FORM

$$S_n = \frac{a_1 - a_1r^n}{1 - r} \qquad r \neq 1$$

Since $a_n = a_1r^{n-1}$, or $ra_n = a_1r^n$, the sum formula also can be written in the following form:

THEOREM 6

SUM OF A GEOMETRIC SERIES—SECOND FORM

$$S_n = \frac{a_1 - ra_n}{1 - r} \qquad r \neq 1$$

The proof of the first sum formula (Theorem 5) by mathematical induction is left as an exercise (see Problem 70, Exercise 6-3).

If $r = 1$, then

$$S_n = a_1 + a_1(1) + a_1(1^2) + \cdots + a_1(1^{n-1}) = na_1$$

EXAMPLE 6

Finding the Sum of a Geometric Series

Find the sum of the first 20 terms of a geometric series if the first term is 1 and $r = 2$.

Solution Let $n = 20$, $a_1 = 1$, $r = 2$, and use Theorem 5.

$$S_n = \frac{a_1 - a_1r^n}{1 - r}$$

$$= \frac{1 - 1 \cdot 2^{20}}{1 - 2} = 1{,}048{,}575 \qquad \text{Calculation using a calculator}$$

MATCHED PROBLEM 6

Find the sum, to two decimal places, of the first 14 terms of a geometric series if the first term is $\frac{1}{64}$ and $r = -2$.

Sum Formula for Infinite Geometric Series

Consider a geometric series with $a_1 = 5$ and $r = \frac{1}{2}$. What happens to the sum S_n as n increases? To answer this question, we first write the sum formula in the more convenient form

$$S_n = \frac{a_1 - a_1r^n}{1 - r} = \frac{a_1}{1 - r} - \frac{a_1r^n}{1 - r} \tag{1}$$

For $a_1 = 5$ and $r = \frac{1}{2}$,

$$S_n = 10 - 10\left(\frac{1}{2}\right)^n$$

Thus,

$$S_2 = 10 - 10\left(\frac{1}{4}\right)$$

$$S_4 = 10 - 10\left(\frac{1}{16}\right)$$

$$S_{10} = 10 - 10\left(\frac{1}{1{,}024}\right)$$

$$S_{20} = 10 - 10\left(\frac{1}{1{,}048{,}576}\right)$$

It appears that $(\frac{1}{2})^n$ becomes smaller and smaller as n increases and that the sum gets closer and closer to 10.

In general, it is possible to show that, if $|r| < 1$, then r^n will get closer and closer to 0 as n increases. Symbolically, $r^n \to 0$ as $n \to \infty$. Thus, the term

$$\frac{a_1 r^n}{1 - r}$$

in equation (1) will tend to 0 as n increases, and S_n will tend to

$$\frac{a_1}{1 - r}$$

In other words, if $|r| < 1$, then S_n can be made as close to

$$\frac{a_1}{1 - r}$$

as we wish by taking n sufficiently large. Thus, we define the **sum of an infinite geometric series** by the following formula:

DEFINITION 3

SUM OF AN INFINITE GEOMETRIC SERIES

$$S_\infty = \frac{a_1}{1 - r} \qquad |r| < 1$$

If $|r| \geq 1$, an infinite geometric series has no sum.

EXAMPLE 7

Expressing a Repeating Decimal as a Fraction

Represent the repeating decimal $0.454\ 545 \cdots = 0.\overline{45}$ as the quotient of two integers. Recall that a repeating decimal names a rational number and that any rational number can be represented as the quotient of two integers.

Solution

$$0.\overline{45} = 0.45 + 0.0045 + 0.000\,045 + \cdots$$

The right side of the equation is an infinite geometric series with $a_1 = 0.45$ and $r = 0.01$. Thus,

$$S_\infty = \frac{a_1}{1 - r} = \frac{0.45}{1 - 0.01} = \frac{0.45}{0.99} = \frac{5}{11}$$

Hence, $0.\overline{45}$ and $\frac{5}{11}$ name the same rational number. Check the result by dividing 5 by 11.

MATCHED PROBLEM 7

Repeat Example 7 for $0.818\,181 \cdots = 0.\overline{81}$.

EXAMPLE 8

Economy Stimulation

A state government uses proceeds from a lottery to provide a tax rebate for property owners. Suppose an individual receives a \$500 rebate and spends 80% of this, and each of the recipients of the money spent by this individual also spends 80% of what he or she receives, and this process continues without end. According to the **multiplier doctrine** in economics, the effect of the original \$500 tax rebate on the economy is multiplied many times. What is the total amount spent if the process continues as indicated?

Solution

The individual receives \$500 and spends 0.8(500) = \$400. The recipients of this \$400 spend 0.8(400) = \$320, the recipients of this \$320 spend 0.8(320) = \$256, and so on. Thus, the total spending generated by the \$500 rebate is

$$400 + 320 + 256 + \cdots = 400 + 0.8(400) + (0.8)^2(400) + \cdots$$

which we recognize as an infinite geometric series with $a_1 = 400$ and $r = 0.8$. Thus, the total amount spent is

$$S_\infty = \frac{a_1}{1 - r} = \frac{400}{1 - 0.8} = \frac{400}{0.2} = \$2{,}000$$

MATCHED PROBLEM 8

Repeat Example 8 if the tax rebate is \$1,000 and the percentage spent by all recipients is 90%.

Explore/Discuss 2

(A) Find an infinite geometric series with $a_1 = 10$ whose sum is 1,000.

(B) Find an infinite geometric series with $a_1 = 10$ whose sum is 6.

(C) Suppose that an infinite geometric series with $a_1 = 10$ has a sum. Explain why that sum must be greater than 5.

Answers to Matched Problems

1. (A) The sequence is geometric with $r = \frac{1}{4}$, but not arithmetic.
(B) The sequence is arithmetic with $d = 5$, but not geometric.
(C) The sequence is neither arithmetic nor geometric.
2. (A) 139 (B) −2 3. −1,456 4. 570 5. $695 6. −85.33 7. $\frac{9}{11}$ 8. $9,000

EXERCISE 6-3

A

In Problems 1 and 2, determine whether the following can be the first three terms of an arithmetic or geometric sequence, and, if so, find the common difference or common ratio and the next two terms of the sequence.

1. (A) $-11, -16, -21, \ldots$ (B) $2, -4, 8, \ldots$
(C) $1, 4, 9, \ldots$ (D) $\frac{1}{2}, \frac{1}{6}, \frac{1}{18}, \ldots$

2. (A) $5, 20, 100, \ldots$ (B) $-5, -5, -5, \ldots$
(C) $7, 6.5, 6, \ldots$ (D) $512, 256, 128, \ldots$

Let $a_1, a_2, a_3, \ldots, a_n, \ldots$ be an arithmetic sequence. In Problems 3–10, find the indicated qualities.

3. $a_1 = -5, d = 4; a_2 = ?, a_3 = ?, a_4 = ?$
4. $a_1 = -18, d = 3; a_2 = ?, a_3 = ?, a_4 = ?$
5. $a_1 = -3, d = 5; a_{15} = ?, S_{11} = ?$
6. $a_1 = 3, d = 4; a_{22} = ?, S_{21} = ?$
7. $a_1 = 1, a_2 = 5; S_{21} = ?$
8. $a_1 = 5, a_2 = 11; S_{11} = ?$
9. $a_1 = 7, a_2 = 5; a_{15} = ?$
10. $a_1 = -3, d = -4; a_{10} = ?$

Let $a_1, a_2, a_3, \ldots, a_n, \ldots$ be a geometric sequence. In Problems 11–16, find each of the indicated quantities.

11. $a_1 = -6, r = -\frac{1}{2}; a_2 = ?, a_3 = ?, a_4 = ?$
12. $a_1 = 12, r = \frac{2}{3}; a_2 = ?, a_3 = ?, a_4 = ?$
13. $a_1 = 81, r = \frac{1}{3}; a_{10} = ?$
14. $a_1 = 64, r = \frac{1}{2}; a_{13} = ?$
15. $a_1 = 3, a_7 = 2{,}187, r = 3; S_7 = ?$
16. $a_1 = 1, a_7 = 729, r = -3; S_7 = ?$

B

Let $a_1, a_2, a_3, \ldots, a_n, \ldots$ be an arithmetic sequence. In Problems 17–24, find the indicated quantities.

17. $a_1 = 3, a_{20} = 117; d = ?, a_{101} = ?$
18. $a_1 = 7, a_8 = 28; d = ?, a_{25} = ?$
19. $a_1 = -12, a_{40} = 22; S_{40} = ?$
20. $a_1 = 24, a_{24} = -28; S_{24} = ?$
21. $a_1 = \frac{1}{3}, a_2 = \frac{1}{2}; a_{11} = ?, S_{11} = ?$
22. $a_1 = \frac{1}{6}, a_2 = \frac{1}{4}; a_{19} = ?, S_{19} = ?$
23. $a_3 = 13, a_{10} = 55; a_1 = ?$
24. $a_9 = -12, a_{13} = 3; a_1 = ?$

Let $a_1, a_2, a_3, \ldots, a_n, \ldots$ be a geometric sequence. Find each of the indicated quantities in Problems 25–30.

25. $a_1 = 100, a_6 = 1; r = ?$
26. $a_1 = 10, a_{10} = 30; r = ?$
27. $a_1 = 5, r = -2; S_{10} = ?$
28. $a_1 = 3, r = 2; S_{10} = ?$
29. $a_1 = 9, a_4 = \frac{8}{3}; a_2 = ?, a_3 = ?$
30. $a_1 = 12, a_4 = -\frac{4}{9}; a_2 = ?, a_3 = ?$
31. $S_{51} = \sum_{k=1}^{51} (3k + 3) = ?$
32. $S_{40} = \sum_{k=1}^{40} (2k - 3) = ?$
33. $S_7 = \sum_{k=1}^{7} (-3)^{k-1} = ?$
34. $S_7 = \sum_{k=1}^{7} 3^k = ?$
35. Find $g(1) + g(2) + g(3) + \cdots + g(51)$ if $g(t) = 5 - t$.
36. Find $f(1) + f(2) + f(3) + \cdots + f(20)$ if $f(x) = 2x - 5$.
37. Find $g(1) + g(2) + \cdots + g(10)$ if $g(x) = (\frac{1}{2})^x$.
38. Find $f(1) + f(2) + \cdots + f(10)$ if $f(x) = 2^x$.
39. Find the sum of all the even integers between 21 and 135.
40. Find the sum of all the odd integers between 100 and 500.
41. Show that the sum of the first n odd natural numbers is n^2, using approximate formulas from this section.
42. Show that the sum of the first n even natural numbers is $n + n^2$, using appropriate formulas from this section.
43. Find a positive number x so that $-2 + x - 6$ is a three-term geometric series.
44. Find a positive number x so that $6 + x + 8$ is a three-term geometric series.
45. For a given sequence in which $a_1 = -3$ and $a_n = a_{n-1} + 3$, $n > 1$, find a_n in terms of n.

46. For the sequence in Problem 45, find $S_n = \sum_{k=1}^{n} a_k$ in terms of n.

In Problems 47–50, find the least positive integer n such that $a_n < b_n$ by graphing the sequences $\{a_n\}$ and $\{b_n\}$ with a graphing utility. Check your answer by using a graphing utility to display both sequences in table form.

47. $a_n = 5 + 8n$, $b_n = 1.1^n$

48. $a_n = 96 + 47n$, $b_n = 8(1.5)^n$

49. $a_n = 1{,}000\,(0.99)^n$, $b_n = 2n + 1$

50. $a_n = 500 - n$, $b_n = 1.05^n$

In Problems 51–56, find the sum of each infinite geometric series that has a sum.

51. $3 + 1 + \frac{1}{3} + \cdots$

52. $16 + 4 + 1 + \cdots$

53. $2 + 4 + 8 + \cdots$

54. $4 + 6 + 9 + \cdots$

55. $2 - \frac{1}{2} + \frac{1}{8} - \cdots$

56. $21 - 3 + \frac{3}{7} - \cdots$

In Problems 57–62, represent each repeating decimal as the quotient of two integers.

57. $0.\overline{7} = 0.7777\cdots$

58. $0.\overline{5} = 0.5555\cdots$

59. $0.\overline{54} = 0.545\,454\cdots$

60. $0.\overline{27} = 0.272\,727\cdots$

61. $3.\overline{216} = 3.216\,216\,216\cdots$

62. $5.6\overline{3} = 5.636\,363\cdots$

C

63. Prove, using mathemtical induction, that if $\{a_n\}$ is an arithmetic sequence, then

$$a_n = a_1 + (n - 1)d \qquad \text{for every } n > 1$$

64. Prove, using mathematical induction, that if $\{a_n\}$ is an arithmetic sequence, then

$$S_n = \frac{n}{2}[2a_1 + (n - 1)d]$$

65. If in a given sequence, $a_1 = -2$ and $a_n = -3a_{n-1}$, $n > 1$, find a_n in terms of n.

66. For the sequence in Problem 65, find $S_n = \sum_{k=1}^{n} a_k$ in terms of n.

67. Show that $(x^2 + xy + y^2)$, $(z^2 + xz + x^2)$, and $(y^2 + yz + z^2)$ are consecutive terms of an arithmetic progression if x, y, and z form an arithmetic progression. (From U.S.S.R. Mathematical Olympiads, 1955–1956, Grade 9.)

68. Take 121 terms of each arithmetic progression 2, 7, 12, . . . and 2, 5, 8, ... How many numbers will there be in common? (From U.S.S.R. Mathematical Olympiads, 1955–1956, Grade 9.)

69. Prove, using mathematical induction, that if $\{a_n\}$ is a geometric sequence, then

$$a_n = a_1 r^{n-1} \qquad n \in N$$

70. Prove, using mathematical induction, that if $\{a_n\}$ is a geometric sequence, then

$$S_n = \frac{a_1 - a_1 r^n}{1 - r} \qquad n \in N, r \neq 1$$

71. Given the system of equations

$$ax + by = c$$
$$dx + ey = f$$

where a, b, c, d, e, and f is any arithmetic progression with a nonzero constant difference, show that the system has a unique solution.

72. The sum of the first and fourth terms of an arithmetic sequence is 2, and the sum of their squares is 20. Find the sum of the first eight terms of the sequence.

APPLICATIONS

73. Business. In investigating different job opportunities, you find that firm A will start you at \$25,000 per year and guarantee you a raise of \$1,200 each year while firm B will start you at \$28,000 per year but will guarantee you a raise of only \$800 each year. Over a period of 15 years, how much would you receive from each firm?

74. Business. In Problem 73, what would be your annual salary at each firm for the tenth year?

75. Economics. The government, through a subsidy program, distributes \$1,000,000. If we assume that each individual or agency spends 0.8 of what is received, and 0.8 of this is spent, and so on, how much total increase in spending results from this government action?

76. Economics. Due to reduced taxes, an individual has an extra \$600 in spendable income. If we assume that the individual spends 70% of this on consumer goods, that the producers of these goods in turn spend 70% of what they receive on consumer goods, and that this process continues indefinitely, what is the total amount spent on consumer goods?

★ **77. Business.** If \$$P$ is invested at r% compounded annually, the amount A present after n years forms a geometric progression with a common ratio $1 + r$. Write a formula for the amount present after n years. How long will it take a sum of money P to double if invested at 6% interest compounded annually?

★ **78. Population Growth.** If a population of A_0 people grows at the constant rate of r% per year, the population after t years forms a geometric progression with a common ratio $1 + r$. Write a formula for the total population after t years. If the world's population is increasing at the rate of 2% per year, how long will it take to double?

79. Finance. Eleven years ago an investment earned \$7,000 for the year. Last year the investment earned \$14,000. If the earnings from the investment have increased the same amount each year, what is the yearly increase and how much income has accrued from the investment over the past 11 years?

80. Air Temperature. As dry air moves upward, it expands. In so doing, it cools at the rate of about 5°F for each 1,000-foot rise. This is known as the **adiabatic process.**

(A) Temperatures at altitudes that are multiples of 1,000 feet form what kind of a sequence?

(B) If the ground temperature is 80°F, write a formula for the temperature T_n in terms of n, if n is in thousands of feet.

81. Engineering. A rotating flywheel coming to rest rotates 300 revolutions the first minute (see the figure). If in each subsequent minute it rotates two-thirds as many times as in the preceding minute, how many revolutions will the wheel make before coming to rest?

82. Physics. The first swing of a bob on a pendulum is 10 inches. If on each subsequent swing it travels 0.9 as far as on the preceding swing, how far will the bob travel before coming to rest?

83. Food Chain. A plant is eaten by an insect, an insect by a trout, a trout by a salmon, a salmon by a bear, and the bear is eaten by you. If only 20% of the energy is transformed from one stage to the next, how many calories must be supplied by plant food to provide you with 2,000 calories from the bear meat?

★ **84. Genealogy.** If there are 30 years in a generation, how many direct ancestors did each of us have 600 years ago? By *direct* ancestors we mean parents, grandparents, great-grandparents, and so on.

★ **85. Physics.** An object falling from rest in a vacuum near the surface of the Earth falls 16 feet during the first second, 48 feet during the second second, 80 feet during the third second, and so on.

(A) How far will the object fall during the eleventh second?

(B) How far will the object fall in 11 seconds?

(C) How far will the object fall in t seconds?

★ **86. Physics.** In Problem 85, how far will the object fall during:

(A) The twentieth second? (B) The tth second?

★ **87. Bacteria Growth.** A single cholera bacterium divides every $\frac{1}{2}$ hour to produce two complete cholera bacteria. If we start with a colony of A_0 bacteria, how many bacteria will we have in t hours, assuming adequate food supply?

★ **88. Cell Division.** One leukemic cell injected into a healthy mouse will divide into two cells in about $\frac{1}{2}$ day. At the end of the day these two cells will divide again, with the doubling process continuing each $\frac{1}{2}$ day until there are 1 billion cells, at which time the mouse dies. On which day after the experiment is started does this happen?

★★ **89. Astronomy.** Ever since the time of the Greek astronomer Hipparchus, second century B.C., the brightness of stars has been measured in terms of magnitude. The brightest stars, excluding the sun, are classed as magnitude 1, and the dimmest visible to the eye are classed as magnitude 6. In 1856, the English astronomer N. R. Pogson showed that first-magnitude stars are 100 times brighter than sixth-magnitude stars. If the ratio of brightness between consecutive magnitudes is constant, find this ratio. [*Hint:* If b_n is the brightness of an nth-magnitude star, find r for the geometric progression $b_1, b_2, b_3, \ldots,$ given $b_1 = 100b_6$.]

★ **90. Music.** The notes on a piano, as measured in cycles per second, form a geometric progression.

(A) If A is 400 cycles per second and A′, 12 notes higher, is 800 cycles per second, find the constant ratio r.

(B) Find the cycles per second for C, three notes higher than A.

91. **Puzzle.** If you place 1¢ on the first square of a chessboard, 2¢ on the second square, 4¢ on the third, and so on, continuing to double the amount until all 64 squares are covered, how much money will be on the sixty-fourth square? How much money will there be on the whole board?

★ 92. **Puzzle.** If a sheet of very thin paper 0.001 inch thick is torn in half, and each half is again torn in half, and this process is repeated for a total of 32 times, how high will the stack of paper be if the pieces are placed one on top of the other? Give the answer to the nearest mile.

★ 93. **Atmospheric Pressure.** If atmospheric pressure decreases roughly by a factor of 10 for each 10-mile increase in altitude up to 60 miles, and if the pressure is 15 pounds per square inch at sea level, what will the pressure be 40 miles up?

94. **Zeno's Paradox.** Visualize a hypothetical 440-yard oval racetrack that has tapes stretched across the track at the halfway point and at each point that marks the halfway point of each remaining distance thereafter. A runner running around the track has to break the first tape before the second, the second before the third, and so on. From this point of view it appears that he will never finish the race. This famous paradox is attributed to the Greek philosopher Zeno (495–435 B.C.). If we assume the runner runs at 440 yards per minute, the times between tape breakings form an infinite geometric progression. What is the sum of this progression?

95. **Geometry.** If the midpoints of the sides of an equilateral triangle are joined by straight lines, the new figure will be an equilateral triangle with a perimeter equal to half the original. If we start with an equilateral triangle with perimeter 1 and form a sequence of "nested" equilateral triangles proceeding as described, what will be the total perimeter of all the triangles that can be formed in this way?

96. **Photography.** The shutter speeds and f-stops on a camera are given as follows:

Shutter speeds: $1, \frac{1}{2}, \frac{1}{4}, \frac{1}{8}, \frac{1}{15}, \frac{1}{30}, \frac{1}{60}, \frac{1}{125}, \frac{1}{250}, \frac{1}{500}$

f-stops: 1.4, 2, 2.8, 4, 5.6, 8, 11, 16, 22

These are very close to being geometric progressions. Estimate their common ratios.

★★ 97. **Geometry.** We know that the sum of the interior angles of a triangle is 180°. Show that the sums of the interior angles of polygons with 3, 4, 5, 6, . . . sides form an arithmetic sequence. Find the sum of the interior angles for a 21-sided polygon.

Section 6-4 Multiplication Principle, Permutations, and Combinations

- Multiplication Principle
- Factorial
- Permutations
- Combinations

This section introduces some new mathematical tools that are usually referred to as *counting techniques*. In general, a **counting technique** is a mathematical method of determining the number of objects in a set without actually enumerating the objects in the set as 1, 2, 3, ... For example, we can count the number

FIGURE 1

of squares in a checker board (see Fig. 1) by counting 1, 2, 3, . . . , 64. This is enumeration. Or we can note that there are 8 rows with 8 squares in each row. Thus, the total number of squares must be $8 \times 8 = 64$. This is a very simple counting techniquc.

Now consider the problem of assigning telephone numbers. How many different seven-digit telephone numbers can be formed? As we will soon see, the answer is $10^7 = 10{,}000{,}000$, a number that is much too large to obtain by enumeration. Thus, counting techniques are essential tools if the number of elements in a set is very large. The techniques developed in this section will be applied to a brief introduction to probability theory in Section 6-5, and to a famous algebraic formula in Section 6-6.

Multiplication Principle

We start with an example.

EXAMPLE 1 **Combined Outcomes**

Suppose we flip a coin and then throw a single die (see Fig. 2). What are the possible combined outcomes?

Solution To solve this problem, we use a **tree diagram:**

FIGURE 2
Coin and die outcomes.

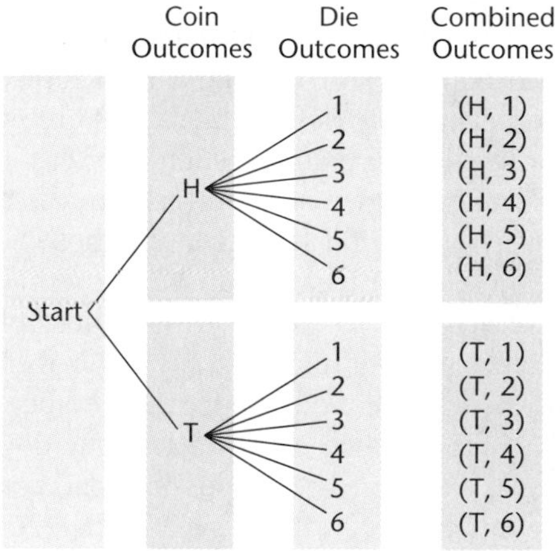

Thus, there are 12 possible combined outcomes—two ways in which the coin can come up followed by six ways in which the die can come up.

MATCHED PROBLEM 1 Use a tree diagram to determine the number of possible outcomes of throwing a single die followed by flipping a coin.

Now suppose you are asked, "From the 26 letters in the alphabet, how many ways can 3 letters appear in a row on a license plate if no letter is repeated?" To try to count the possibilities using a tree diagram would be extremely tedious, to say the least. The following **multiplication principle,** also called the **fundamental counting principle,** enables us to solve this problem easily. In addition, it forms the basis for several other counting techniques developed later in this section.

MULTIPLICATION PRINCIPLE

1. If two operations O_1 and O_2 are performed in order with N_1 possible outcomes for the first operation and N_2 possible outcomes for the second operation, then there are

$$N_1 \cdot N_2$$

possible combined outcomes of the first operation followed by the second.

2. In general, if n operations $O_1, O_2, \ldots, O_n$ are performed in order, with possible number of outcomes $N_1, N_2, \ldots, N_n$, respectively, then there are

$$N_1 \cdot N_2 \cdot \cdots \cdot N_n$$

possible combined outcomes of the operations performed in the given order.

In Example 1, we see that there are two possible outcomes from the first operation of flipping a coin and six possible outcomes from the second operation of throwing a die. Hence, by the multiplication principle, there are $2 \cdot 6 = 12$ possible combined outcomes of flipping a coin followed by throwing a die. Use the multiplication principle to solve Matched Problem 1.

To answer the license plate question, we reason as follows: There are 26 ways the first letter can be chosen. After a first letter is chosen, 25 letters remain; hence there are 25 ways a second letter can be chosen. And after 2 letters are chosen, there are 24 ways a third letter can be chosen. Hence, using the multiplication principle, there are $26 \cdot 25 \cdot 24 = 15{,}600$ possible ways 3 letters can be chosen from the alphabet without allowing any letter to repeat. By not allowing any letter to repeat, earlier selections affect the choice of subsequent selections. If we allow letters to repeat, then earlier selections do not affect the choice in subsequent selections, and there are 26 possible choices for each of the 3 letters. Thus, if we allow letters to repeat, there are $26 \cdot 26 \cdot 26 = 26^3 = 17{,}576$ possible ways the 3 letters can be chosen from the alphabet.

EXAMPLE 2 **Computer-Generated Tests**

Many universities and colleges are now using computer-assisted testing procedures. Suppose a screening test is to consist of 5 questions, and a computer stores 5 equivalent questions for the first test question, 8 equivalent questions for the second, 6 for the third, 5 for the fourth, and 10 for the fifth. How many different 5-question tests can the computer select? Two tests are considered different if they differ in one or more questions.

Solution

O_1:	Select the first question	N_1:	5 ways
O_2:	Select the second question	N_2:	8 ways
O_3:	Select the third question	N_3:	6 ways

O_4: Select the fourth question	N_4: 5 ways	
O_5: Select the fifth question	N_5: 10 ways	

Thus, the computer can generate

$$5 \cdot 8 \cdot 6 \cdot 5 \cdot 10 = 12{,}000 \text{ different tests}$$

MATCHED PROBLEM 2

Each question on a multiple-choice test has 5 choices. If there are 5 such questions on a test, how many different response sheets are possible if only 1 choice is marked for each question?

EXAMPLE 3

Counting Code Words

How many 3-letter code words are possible using the first 8 letters of the alphabet if:

(A) No letter can be repeated? (B) Letters can be repeated?

(C) Adjacent letters cannot be alike?

Solutions

(A) No letter can be repeated.

O_1: Select first letter	N_1: 8 ways	
O_2: Select second letter	N_2: 7 ways	Because 1 letter has been used
O_3: Select third letter	N_3: 6 ways	Because 2 letters have been used

Thus, there are

$$8 \cdot 7 \cdot 6 = 336 \text{ possible code words}$$

(B) Letters can be repeated.

O_1: Select first letter	N_1: 8 ways	
O_2: Select second letter	N_2: 8 ways	Repeats are allowed.
O_3: Select third letter	N_3: 8 ways	Repeats are allowed.

Thus, there are

$$8 \cdot 8 \cdot 8 = 8^3 = 512 \text{ possible code words}$$

(C) Adjacent letters cannot be alike.

O_1: Select first letter	N_1: 8 ways	
O_2: Select second letter	N_2: 7 ways	Cannot be the same as the first

O_3: Select third letter N_3: 7 ways Cannot be the same as the second, but can be the same as the first

Thus, there are

$$8 \cdot 7 \cdot 7 = 392 \text{ possible code words}$$

MATCHED PROBLEM 3

How many 4-letter code words are possible using the first 10 letters of the alphabet under the three conditions stated in Example 3?

Explore/Discuss 1

The postal service of a developing country is choosing a five-character postal code consisting of letters (of the English alphabet) and digits. At least half a million postal codes must be accommodated. Which format would you recommend to make the codes easy to remember?

The multiplication principle can be used to develop two additional counting techniques that are extremely useful in more complicated counting problems. Both of these methods use the factorial function, which we introduce next.

Factorial

For n a natural number, ***n* factorial**—denoted by $n!$—is the product of the first n natural numbers. **Zero factorial** is defined to be 1.

DEFINITION 1

n FACTORIAL

For n a natural number

$$n! = n(n-1) \cdot \cdots \cdot 2 \cdot 1$$

$$1! = 1$$

$$0! = 1$$

It is also useful to note that

THEOREM 1

RECURSION FORMULA FOR *n* FACTORIAL

$$n! = n \cdot (n-1)!$$

EXAMPLE 4

Evaluating Factorials

(A) $4! = 4 \cdot 3! = 4 \cdot 3 \cdot 2! = 4 \cdot 3 \cdot 2 \cdot 1! = 4 \cdot 3 \cdot 2 \cdot 1 = 24$

(B) $5! = 5 \cdot 4 \cdot 3 \cdot 2 \cdot 1 = 120$

(C) $\dfrac{7!}{6!} = \dfrac{7 \cdot \cancel{6!}}{\cancel{6!}} = 7$

(D) $\dfrac{8!}{5!} = \dfrac{8 \cdot 7 \cdot 6 \cdot \cancel{5!}}{\cancel{5!}} = 336$

(E) $\dfrac{9!}{6!3!} = \dfrac{\overset{3}{\cancel{9}} \cdot \overset{4}{\cancel{8}} \cdot 7 \cdot \cancel{6!}}{\cancel{6!}\,\cancel{3} \cdot \cancel{2} \cdot 1} = 84$

MATCHED PROBLEM 4

Find (A) $6!$ (B) $\dfrac{6!}{5!}$ (C) $\dfrac{9!}{6!}$ (D) $\dfrac{10!}{7!3!}$

CAUTION

When reducing fractions involving factorials, don't confuse the single integer n with the symbol $n!$, which represents the product of n consecutive integers.

$$\frac{6!}{3!} \neq 2! \qquad \frac{6!}{3!} = \frac{6 \cdot 5 \cdot 4 \cdot 3!}{3!} = 6 \cdot 5 \cdot 4 = 120$$

Explore/Discuss 2

A student used a calculator to solve Matched Problem 4, as shown in Figure 3. Check these answers. If any are incorrect, explain why and find a correct calculator solution.

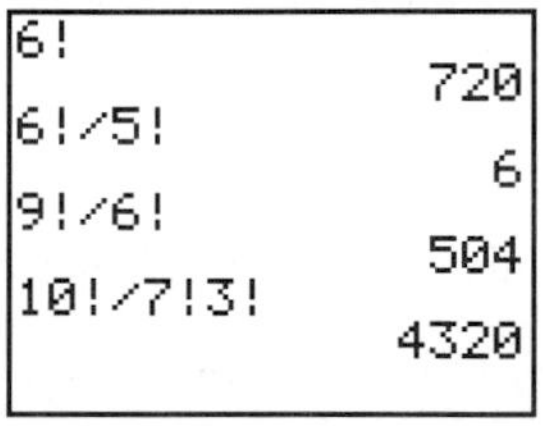

FIGURE 3

It is interesting and useful to note that $n!$ grows very rapidly. Compare the following:

$5! = 120 \qquad 10! = 3{,}628{,}800 \qquad 15! = 1{,}307{,}674{,}368{,}000$

If $n!$ is too large for a calculator to store and display, an error message is displayed. Find the value of n such that your calculator will evaluate $n!$, but not $(n + 1)!$.

Permutations

Suppose 4 pictures are to be arranged from left to right on one wall of an art gallery. How many arrangements are possible? Using the multiplication principle, there are 4 ways of selecting the first picture. After the first picture is selected, there are 3 ways of selecting the second picture. After the first 2 pictures are selected, there are 2 ways of selecting the third picture. And after the first 3 pictures are selected, there is only 1 way to select the fourth. Thus, the number of arrangements possible for the 4 pictures is

$$4 \cdot 3 \cdot 2 \cdot 1 = 4! \qquad \text{or} \qquad 24$$

In general, we refer to a particular arrangement, or **ordering,** of n objects without repetition as a **permutation** of the n objects. How many permutations of n objects are there? From the reasoning above, there are n ways in which the first object can be chosen, there are $n - 1$ ways in which the second object can be chosen, and so on. Applying the multiplication principle, we have Theorem 2.

THEOREM 2

PERMUTATIONS OF n OBJECTS

The number of permutations of n objects, denoted by $P_{n,n}$, is given by

$$P_{n,n} = n \cdot (n-1) \cdot \cdots \cdot 1 = n!$$

Now suppose the director of the art gallery decides to use only 2 of the 4 available pictures on the wall, arranged from left to right. How many arrangements of 2 pictures can be formed from the 4? There are 4 ways the first picture can be selected. After selecting the first picture, there are 3 ways the second picture can be selected. Thus, the number of arrangements of 2 pictures from 4 pictures, denoted by $P_{4,2}$, is given by

$$P_{4,2} = 4 \cdot 3 = 12$$

Or, in terms of factorials, multiplying $4 \cdot 3$ by 1 in the form $2!/2!$, we have

$$P_{4,2} = 4 \cdot 3 = \frac{4 \cdot 3 \cdot 2!}{2!} = \frac{4!}{2!}$$

This last form gives $P_{4,2}$ in terms of factorials, which is useful in some cases.

A **permutation of a set of n objects taken r at a time** is an arrangement of the r objects in a specific order. Thus, reasoning in the same way as in the example above, we find that the number of permutations of n objects taken r at a time, $0 \le r \le n$, denoted by $P_{n,r}$, is given by

$$P_{n,r} = n(n-1)(n-2) \cdot \cdots \cdot (n-r+1)$$

Multiplying the right side of this equation by 1 in the form $(n - r)!/(n - r)!$, we obtain a factorial form for $P_{n,r}$:

$$P_{n,r} = n(n-1)(n-2) \cdot \cdots \cdot (n-r+1)\,\frac{(n-r)!}{(n-r)!}$$

But

$$n(n-1)(n-2) \cdot \cdots \cdot (n-r+1)(n-r)! = n!$$

Hence, we have Theorem 3.

THEOREM 3

PERMUTATION OF n OBJECTS TAKEN r AT A TIME

The number of permutations of n objects taken r at a time is given by

$$P_{n,r} = \underbrace{n(n-1)(n-2) \cdot \cdots \cdot (n-r+1)}_{r \text{ factors}}$$

or

$$P_{n,r} = \frac{n!}{(n-r)!} \qquad 0 \le r \le n$$

Note that if $r = n$, then the number of permutations of n objects taken n at a time is

$$P_{n,n} = \frac{n!}{(n-n)!} = \frac{n!}{0!} = n! \qquad \text{Recall, } 0! = 1.$$

which agrees with Theorem 1, as it should.

The permutation symbol $P_{n,r}$ *also can be denoted by* P^n_r, ${}_nP_r$, or $P(n, r)$. Many calculators use ${}_nP_r$ to denote the function that evaluates the permutation symbol.

EXAMPLE 5 **Selecting Officers**

From a committee of 8 people, in how many ways can we choose a chair and a vice-chair, assuming one person cannot hold more than one position?

Solution We are actually asking for the number of permutations of 8 objects taken 2 at a time—that is, $P_{8,2}$:

$$P_{8,2} = \frac{8!}{(8-2)!} = \frac{8!}{6!} = \frac{8 \cdot 7 \cdot 6!}{6!} = 56$$

MATCHED PROBLEM 5 From a committee of 10 people, in how many ways can we choose a chair, vice-chair, and secretary, assuming one person cannot hold more than one position?

EXAMPLE 6 **Evaluating $P_{n,r}$**

Find the number of permutations of 25 objects taken

(A) 2 at a time (B) 4 at a time (C) 8 at a time

Solution Figure 4 shows the solution on a graphing utility.

FIGURE 4

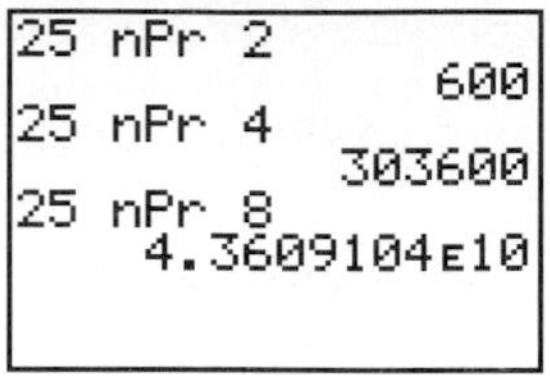

MATCHED PROBLEM 6 Find the number of permutations of 30 objects taken

(A) 2 at a time (B) 4 at a time (C) 6 at a time

Combinations

Now suppose that an art museum owns 8 paintings by a given artist and another art museum wishes to borrow 3 of these paintings for a special show. How many ways can 3 paintings be selected for shipment out of the 8 available? Here, the order of the items selected doesn't matter. What we are actually interested in is how many subsets of 3 objects can be formed from a set of 8 objects. We call such a subset a **combination** of 8 objects taken 3 at a time. The total number of combinations is denoted by the symbol

$$C_{8,3} \quad \text{or} \quad \binom{8}{3}$$

To find the number of combinations of 8 objects taken 3 at a time, $C_{8,3}$, we make use of the formula for $P_{n,r}$ and the multiplication principle. We know that the number of permutations of 8 objects taken 3 at a time is given by $P_{8,3}$, and we have a formula for computing this quantity. Now suppose we think of $P_{8,3}$ in terms of two operations:

O_1: Select a subset of 3 objects (paintings)

N_1: $C_{8,3}$ ways

O_2: Arrange the subset in a given order

N_2: 3! ways

The combined operation, O_1 followed by O_2, produces a permutation of 8 objects taken 3 at a time. Thus,

$$P_{8,3} = C_{8,3} \cdot 3!$$

To find $C_{8,3}$, we replace $P_{8,3}$ in the preceding equation with $8!/(8-3)!$ and solve for $C_{8,3}$:

$$\frac{8!}{(8-3)!} = C_{8,3} \cdot 3!$$

$$C_{8,3} = \frac{8!}{3!(8-3)!} = \frac{8 \cdot 7 \cdot 6 \cdot 5!}{3 \cdot 2 \cdot 1 \cdot 5!} = 56$$

Thus, the museum can make 56 different selections of 3 paintings from the 8 available.

A **combination of a set of *n* objects taken *r* at a time** is an *r*-element subset of the *n* objects. Reasoning in the same way as in the example, the number of combinations of *n* objects taken *r* at a time, $0 \le r \le n$, denoted by $C_{n,r}$, can be obtained by solving for $C_{n,r}$ in the relationship

$$P_{n,r} = C_{n,r} \cdot r!$$

$$C_{n,r} = \frac{P_{n,r}}{r!}$$

$$= \frac{n!}{r!(n-r)!} \qquad P_{n,r} = \frac{n!}{(n-r)!}$$

THEOREM 4

COMBINATION OF *n* OBJECTS TAKEN *r* AT A TIME

The number of combinations of *n* objects taken *r* at a time is given by

$$C_{n,r} = \binom{n}{r} = \frac{P_{n,r}}{r!} = \frac{n!}{r!(n-r)!} \qquad 0 \le r \le n$$

The combination symbols $C_{n,r}$ and $\binom{n}{r}$ also can be denoted by C_r^n, ${}_nC_r$, or $C(n, r)$.

EXAMPLE 7

Selecting Subcommittees

From a committee of 8 people, in how many ways can we choose a subcommittee of 2 people?

Solution

Notice how this example differs from Example 5, where we wanted to know how many ways a chair and a vice-chair can be chosen from a committee of 8 people. In Example 5, ordering matters. In choosing a subcommittee of 2 people, the ordering does not matter. Thus, we are actually asking for the number of combinations of 8 objects taken 2 at a time. The number is given by

$$C_{8,2} = \binom{8}{2} = \frac{8!}{2!(8-2)!} = \frac{8 \cdot 7 \cdot 6!}{2 \cdot 1 \cdot 6!} = 28$$

MATCHED PROBLEM 7

How many subcommittees of 3 people can be chosen from a committee of 8 people?

EXAMPLE 8

Evaluating $C_{n,r}$

Find the number of combinations of 25 objects taken

(A) 2 at a time (B) 4 at a time (C) 8 at a time

Solution

Figure 5 shows the solution on a graphing utility. Compare these results with Example 6.

FIGURE 5

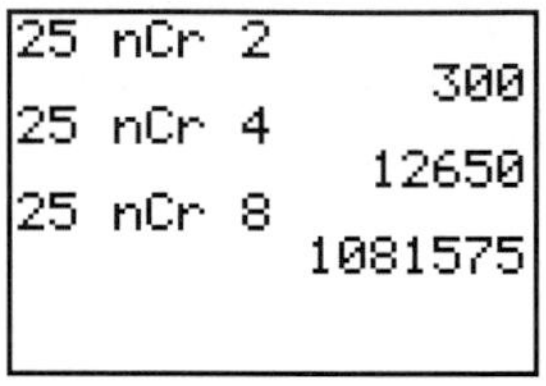

MATCHED PROBLEM 8

Find the number of combinations of 30 objects taken

(A) 2 at a time (B) 4 at a time (C) 6 at a time

Remember: In a permutation, order counts. In a combination, order does not count.

To determine whether a permutation or combination is needed, decide whether rearranging the collection or listing makes a difference. If so, use permutations. If not, use combinations.

Explore/Discuss 3

Each of the following is a selection without repetition. Would you consider the selection to be a combination? A permutation? Discuss your reasoning.

(A) A student checks out three books from the library.

(B) A baseball manager names his starting lineup.

(C) The newly elected president names his cabinet members.

(D) The president selects a delegation of three cabinet members to attend the funeral of a head of state.

(E) An orchestra conductor chooses three pieces of music for a symphony program.

A **standard deck** of 52 cards (see Fig. 6) has four 13-card suits: diamonds, hearts, clubs, and spades. Each 13-card suit contains cards numbered from 2 to 10, a jack, a queen, a king, and an ace. The jack, queen, and king are called **face cards.** Depending on the game, the ace may be counted as the lowest and/or the highest card in the suit. Example 9, as well as other examples and exercises in this chapter, refer to this standard deck.

FIGURE 6
A standard deck of cards.

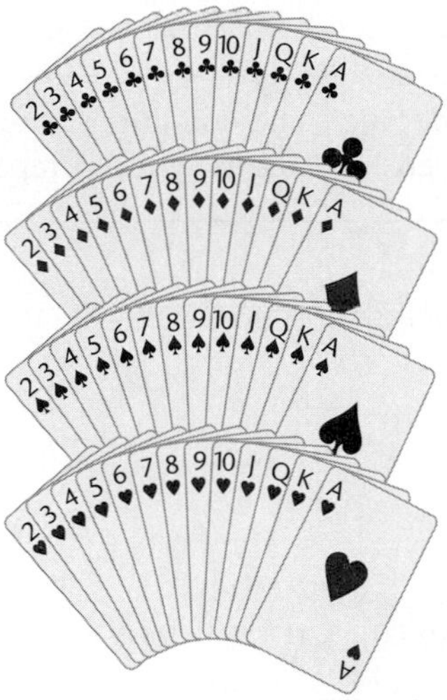

EXAMPLE 9

Counting Card Hands

Out of a standard 52-card deck, how many 5-card hands will have 3 aces and 2 kings?

Solution

O_1: Choose 3 aces out of 4 possible   Order is not important.

N_1: $C_{4,3}$

O_2: Choose 2 kings out of 4 possible   Order is not important.

N_2: $C_{4,2}$

Using the multiplication principle, we have

$$\text{Number of hands} = C_{4,3} \cdot C_{4,2} = 4 \cdot 6 = 24$$

MATCHED PROBLEM 9

From a standard 52-card deck, how many 5-card hands will have 3 hearts and 2 spades?

EXAMPLE 10

Counting Serial Numbers

Serial numbers for a product are to be made using 2 letters followed by 3 numbers. If the letters are to be taken from the first 8 letters of the alphabet with no repeats and the numbers from the 10 digits 0 through 9 with no repeats, how many serial numbers are possible?

Solution

O_1: Choose 2 letters out of 8 available   Order is important.

N_1: $P_{8,2}$

O_2: Choose 3 numbers out of 10 available   Order is important.

N_2: $P_{10,3}$

Using the multiplication principle, we have

$$\text{Number of serial numbers} = P_{8,2} \cdot P_{10,3} = 40{,}320$$

MATCHED PROBLEM 10 Repeat Example 10 under the same conditions, except the serial numbers are now to have 3 letters followed by 2 digits with no repeats.

Answers to Matched Problems

1. 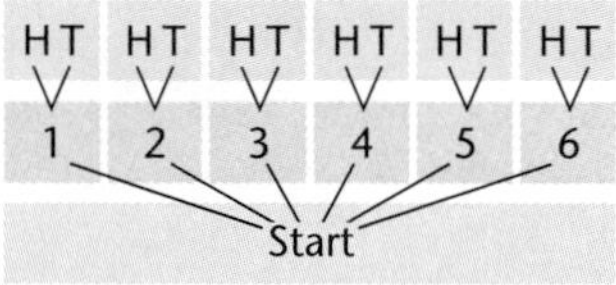

2. 5^5, or 3,125

3. (A) $10 \cdot 9 \cdot 8 \cdot 7 = 5{,}040$ (B) $10 \cdot 10 \cdot 10 \cdot 10 = 10{,}000$ (C) $10 \cdot 9 \cdot 9 \cdot 9 = 7{,}290$

4. (A) 720 (B) 6 (C) 504 (D) 120

5. $P_{10,3} = \dfrac{10!}{(10-3)!} = 720$

6. (A) 870 (B) 657,720 (C) 427,518,000

7. $C_{8,3} = \dfrac{8!}{3!(8-3)!} = 56$

8. (A) 435 (B) 27,405 (C) 593,775

9. $C_{13,3} \cdot C_{13,2} = 22{,}308$

10. $P_{8,3} \cdot P_{10,2} = 30{,}240$

EXERCISE 6-4

A

Evaluate Problems 1–20.

1. $9!$
2. $10!$
3. $11!$
4. $12!$
5. $\dfrac{11!}{8!}$
6. $\dfrac{14!}{12!}$
7. $\dfrac{5!}{2!3!}$
8. $\dfrac{6!}{4!2!}$
9. $\dfrac{7!}{4!(7-4)!}$
10. $\dfrac{8!}{3!(8-3)!}$
11. $\dfrac{7!}{7!(7-7)!}$
12. $\dfrac{8!}{0!(8-0)!}$
13. $P_{5,3}$
14. $P_{4,2}$
15. $P_{52,4}$
16. $P_{52,2}$
17. $C_{5,3}$
18. $C_{4,2}$
19. $C_{52,4}$
20. $C_{52,2}$

In Problems 21 and 22, would you consider the selection to be a combination or a permutation? Explain your reasoning.

21. (A) The recently elected chief executive officer (CEO) of a company named 3 new vice-presidents, of marketing, research, and manufacturing.
    (B) The CEO selected 3 of her vice-presidents to attend the dedication ceremony of a new plant.

22. (A) An individual rented 4 videos from a rental store to watch over a weekend.
    (B) The same individual did some holiday shopping by buying 4 videos, 1 for his father, 1 for his mother, 1 for his younger sister, and 1 for his older brother.

23. A particular new car model is available with 5 choices of color, 3 choices of transmission, 4 types of interior, and 2 types of engine. How many different variations of this model car are possible?

24. A deli serves sandwiches with the following options: 3 kinds of bread, 5 kinds of meat, and lettuce or sprouts. How many different sandwiches are possible, assuming one item is used out of each category?

25. In a horse race, how many different finishes among the first 3 places are possible for a 10-horse race? Exclude ties.

26. In a long-distance foot race, how many different finishes among the first 5 places are possible for a 50-person race? Exclude ties.

27. How many ways can a subcommittee of 3 people be selected from a committee of 7 people? How many ways can a president, vice president, and secretary be chosen from a committee of 7 people?

28. Suppose 9 cards are numbered with the 9 digits from 1 to 9. A 3-card hand is dealt, 1 card at a time. How many hands are possible where:
    (A) Order is taken into consideration?
    (B) Order is not taken into consideration?

29. There are 10 teams in a league. If each team is to play every other team exactly once, how many games must be scheduled?

30. Given 7 points, no 3 of which are on a straight line, how many lines can be drawn joining 2 points at a time?

B

31. How many 4-letter code words are possible from the first 6 letters of the alphabet, with no letter repeated? Allowing letters to repeat?

32. How many 5-letter code words are possible from the first 7 letters of the alphabet, with no letter repeated? Allowing letters to repeat?

33. A combination lock has 5 wheels, each labeled with the 10 digits from 0 to 9. How many opening combinations of 5 numbers are possible, assuming no digit is repeated? Assuming digits can be repeated?

34. A small combination lock on a suitcase has 3 wheels, each labeled with digits from 0 to 9. How many opening combinations of 3 numbers are possible, assuming no digit is repeated? Assuming digits can be repeated?

35. From a standard 52-card deck, how many 5-card hands will have all hearts?

36. From a standard 52-card deck, how many 5-card hands will have all face cards? All face cards, but no kings? Consider only jacks, queens, and kings to be face cards.

37. How many different license plates are possible if each contains 3 letters followed by 3 digits? How many of these license plates contain no repeated letters and no repeated digits?

38. How may 5-digit zip codes are possible? How many of these codes contain no repeated digits?

39. From a standard 52-card deck, how many 7-card hands have exactly 5 spades and 2 hearts?

40. From a standard 52-card deck, how many 5-card hands will have 2 clubs and 3 hearts?

41. A catering service offers 8 appetizers, 10 main courses, and 7 desserts. A banquet chairperson is to select 3 appetizers, 4 main courses, and 2 desserts for a banquet. How many ways can this be done?

42. Three research departments have 12, 15, and 18 members, respectively. If each department is to select a delegate and an alternate to represent the department at a conference, how many ways can this be done?

43. (A) Use a graphing utility to display the sequences $P_{10,0}$, $P_{10,1}, \ldots, P_{10,10}$ and $0!, 1!, \ldots, 10!$ in table form, and show that $P_{10,r} \geq r!$ for $r = 0, 1, \ldots, 10$.

(B) Find all values of r such that $P_{10,r} = r!$

(C) Explain why $P_{n,r} \geq r!$ whenever $0 \leq r \leq n$.

44. (A) How are the sequences $\frac{P_{10,0}}{0!}, \frac{P_{10,1}}{1!}, \ldots, \frac{P_{10,10}}{10!}$ and $C_{10,0}$, $C_{10,1}, \ldots, C_{10,10}$ related?

(B) Use a graphing utility to graph each sequence and confirm the relationship of part A.

C

45. A sporting goods store has 12 pairs of ski gloves of 12 different brands thrown loosely in a bin. The gloves are all the same size. In how many ways can a left-hand glove and a right-hand glove be selected that do not match relative to brand?

46. A sporting goods store has 6 pairs of running shoes of 6 different styles thrown loosely in a basket. The shoes are all the same size. In how many ways can a left shoe and a right shoe be selected that do not match?

47. Eight distinct points are selected on the circumference of a circle.

(A) How many chords can be drawn by joining the points in all possible ways?

(B) How many triangles can be drawn using these 8 points as vertices?

(C) How many quadrilaterals can be drawn using these 8 points as vertices?

48. Five distinct points are selected on the circumference of a circle.

(A) How many chords can be drawn by joining the points in all possible ways?

(B) How many triangles can be drawn using these 5 points as vertices?

49. How many ways can 2 people be seated in a row of 5 chairs? 3 people? 4 people? 5 people?

50. Each of 2 countries sends 5 delegates to a negotiating conference. A rectangular table is used with 5 chairs on each long side. If each country is assigned a long side of the table, how many seating arrangements are possible? [*Hint:* Operation 1 is assigning a long side of the table to each country.]

51. A basketball team has 5 distinct positions. Out of 8 players, how many starting teams are possible if

(A) The distinct positions are taken into consideration?

(B) The distinct positions are not taken into consideration?

(C) The distinct positions are not taken into consideration, but either Mike or Ken, but not both, must start?

52. How many committees of 4 people are possible from a group of 9 people if
(A) There are no restrictions?
(B) Both Juan and Mary must be on the committee?
(C) Either Juan or Mary, but not both, must be on the committee?

53. A 5-card hand is dealt from a standard 52-card deck. Which is more likely: the hand contains exactly 1 king or the hand contains no hearts?

54. A 10-card hand is dealt from a standard 52-card deck. Which is more likely: all cards in the hand are red or the hand contains all four aces?

Section 6-5 Sample Spaces and Probability

- Experiments
- Sample Spaces and Events
- Probability of an Event
- Equally Likely Assumption
- Empirical Probability

This section provides an introduction to probability, a topic that has whole books and courses devoted to it. Probability studies involve many subtle notions, and care must be taken at the beginning to understand the fundamental concepts on which the studies are based. First, we develop a mathematical model for probability studies. Our development, because of space, must be somewhat informal. More formal and precise treatments can be found in books on probability.

Experiments

Our first step in constructing a mathematical model for probability studies is to describe the type of experiments on which probability studies are based. Some types of experiments do not yield the same results, no matter how carefully they are repeated under the same conditions. These experiments are called **random experiments.** Familiar examples of random experiments are flipping coins, rolling dice, observing the frequency of defective items from an assembly line, or observing the frequency of deaths in a certain age group.

Probability theory is a branch of mathematics that has been developed to deal with outcomes of random experiments, both real and conceptual. In the work that follows, the word **experiment** will be used to mean a random experiment.

Sample Spaces and Events

Associated with outcomes of experiments are *sample spaces* and *events*. Our second step in constructing a mathematical model for probability studies is to define these two terms. Set concepts will be useful in this regard.

Consider the experiment, "A single six-sided die is rolled." What outcomes might we observe? We might be interested in the number of dots facing up, or whether the number of dots facing up is an even number, or whether the number of dots facing up is divisible by 3, and so on. The list of possible outcomes appears endless. In general, there is no unique method of analyzing all possible outcomes of an experiment. Therefore, before conducting an experiment, it is important to decide just what outcomes are of interest.

In the die experiment, suppose we limit our interest to the number of dots facing up when the die comes to rest. Having decided what to observe, we make a list of outcomes of the experiment, called *simple events,* such that in each trial of the experiment, one and only one of the results on the list will occur. The set of simple events for the experiment is called a **sample space** for the experiment. The sample space S we have chosen for the die-rolling experiment is

$$S = \{1, 2, 3, 4, 5, 6\}$$

Now consider the outcome, "The number of dots facing up is an even number." This outcome is not a simple event, since it will occur whenever 2, 4, or 6 dots appear, that is, whenever an element in the subset

$$E = \{2, 4, 6\}$$

occurs. Subset E is called a *compound event.* In general, we have the following definition:

DEFINITION 1

EVENT

Given a sample space S for an experiment, we define an **event *E*** to be any subset of S. If an event E has only one element in it, it is called a **simple event.** If event E has more than one element, it is called a **compound event.** We say that **an event *E* occurs** if any of the simple events in E occurs.

EXAMPLE 1

Choosing a Sample Space

A nickel and a dime are tossed. How will we identify a sample space for this experiment?

Solutions

There are a number of possibilities, depending on our interest. We will consider three.

(A) If we are interested in whether each coin falls heads (H) or tails (T), then, using a tree diagram, we can easily determine an appropriate sample space for the experiment:

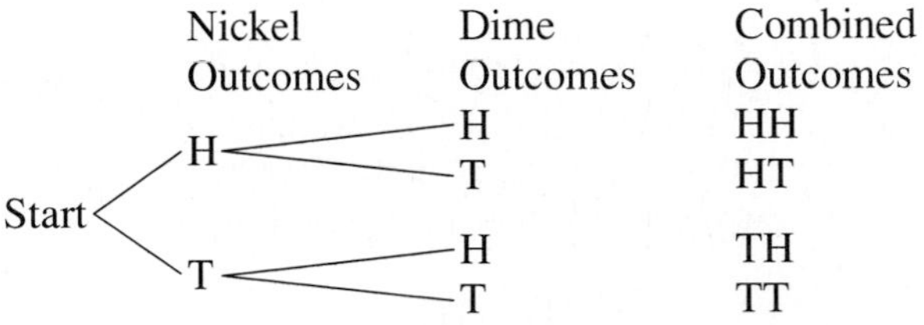

Thus,

$$S_1 = \{HH, HT, TH, TT\}$$

and there are four simple events in the sample space.

(B) If we are interested only in the number of heads that appear on a single toss of the two coins, then we can let

$$S_2 = \{0, 1, 2\}$$

and there are three simple events in the sample space.

(C) If we are interested in whether the coins match (M) or don't match (D), then we can let

$$S_3 = \{M, D\}$$

and there are only two simple events in the sample space.

MATCHED PROBLEM 1

An experiment consists of recording the boy–girl composition of families with 2 children.

(A) What is an appropriate sample space if we are interested in the sex of each child in the order of their births? Draw a tree diagram.

(B) What is an appropriate sample space if we are interested only in the number of girls in a family?

(C) What is an appropriate sample space if we are interested only in whether the sexes are alike (A) or different (D)?

(D) What is an appropriate sample space for all three interests expressed above?

In Example 1, sample space S_1 contains more information than either S_2 or S_3. If we know which outcome has occurred in S_1, then we know which outcome has occurred in S_2 and S_3. However, the reverse is not true. In this sense, we say that S_1 is a more **fundamental sample space** than either S_2 or S_3.

Important Remark: There is no one correct sample space for a given experiment. When specifying a sample space for an experiment, we include as much detail as necessary to answer *all* questions of interest regarding the outcomes of the experiment. If in doubt, include more elements in the sample space rather than fewer.

Now let's return to the 2-coin problem in Example 1 and the sample space

$$S_1 = \{\text{HH, HT, TH, TT}\}$$

Suppose we are interested in the outcome, "Exactly 1 head is up." Looking at S_1, we find that it occurs if either of the two simple events HT or TH occurs.* Thus, to say that the event, "Exactly 1 head is up" occurs is the same as saying the experiment has an outcome in the set

$$E = \{\text{HT, TH}\}$$

This is a subset of the sample space S_1. The event E is a compound event.

---

*Technically, we should write {HT} and {TH}, since there is a logical distinction between an element of a set and a subset consisting of only that element. But we will just keep this in mind and drop the braces for simple events to simplify the notation.

EXAMPLE 2

Rolling Two Dice

Consider an experiment of rolling 2 dice. A convenient sample space that will enable us to answer many questions about interesting events is shown in Figure 1. Let S be the set of all ordered pairs listed in the figure. Note that the simple event (3, 2) is to be distinguished from the simple event (2, 3). The former indicates a 3 turned up on the first die and a 2 on the second, while the latter indicates a 2 turned up on the first die and a 3 on the second. What is the event that corresponds to each of the following outcomes?

(A) A sum of 7 turns up. (B) A sum of 11 turns up.

(C) A sum less than 4 turns up. (D) A sum of 12 turns up.

FIGURE 1
A sample space for rolling two dice.

SECOND DIE

FIRST DIE	⚀	⚁	⚂	⚃	⚄	⚅
⚀	(1, 1)	(1, 2)	(1, 3)	(1, 4)	(1, 5)	(1, 6)
⚁	(2, 1)	(2, 2)	(2, 3)	(2, 4)	(2, 5)	(2, 6)
⚂	(3, 1)	(3, 2)	(3, 3)	(3, 4)	(3, 5)	(3, 6)
⚃	(4, 1)	(4, 2)	(4, 3)	(4, 4)	(4, 5)	(4, 6)
⚄	(5, 1)	(5, 2)	(5, 3)	(5, 4)	(5, 5)	(5, 6)
⚅	(6, 1)	(6, 2)	(6, 3)	(6, 4)	(6, 5)	(6, 6)

Solutions

(A) By "A sum of 7 turns up," we mean that the sum of all dots on both turned-up faces is 7. This outcome corresponds to the event

$$\{(6, 1), (5, 2), (4, 3), (3, 4), (2, 5), (1, 6)\}$$

(B) "A sum of 11 turns up" corresponds to the event

$$\{(6, 5), (5, 6)\}$$

(C) "A sum less than 4 turns up" corresponds to the event

$$\{(1, 1), (2, 1), (1, 2)\}$$

(D) "A sum of 12 turns up" corresponds to the event

$$\{(6, 6)\}$$

MATCHED PROBLEM 2

Refer to the sample space in Example 2 (Fig. 1). What is the event that corresponds to each of the following outcomes?

(A) A sum of 5 turns up.

(B) A sum that is a prime number greater than 7 turns up.

Informally, to facilitate discussion, we often use the terms *event* and *outcome of an experiment* interchangeably. Thus, in Example 2 we might say "the event 'A sum of 11 turns up' " in place of "the outcome 'A sum of 11 turns up,' " or even write

$$E = \text{A sum of 11 turns up} = \{(6, 5), (5, 6)\}$$

Technically speaking, an event is the mathematical counterpart of an outcome of an experiment.

Probability of an Event

The next step in developing our mathematical model for probability studies is the introduction of a *probability function.* This is a function that assigns to an arbitrary event associated with a sample space a real number between 0 and 1, inclusive. We start by discussing ways in which probabilities are assigned to simple events in S.

DEFINITION 2

PROBABILITIES FOR SIMPLE EVENTS

Given a sample space

$$S = \{e_1, e_2, \ldots, e_n\}$$

with n simple events, to each simple event e_i we assign a real number, denoted by $P(e_i)$, that is called the **probability of the event e_i.** These numbers may be assigned in an arbitrary manner as long as the following two conditions are satisfied:

1. $0 \leq P(e_i) \leq 1$
2. $P(e_i) + P(e_2) + \cdots + P(e_n) = 1$ — The sum of the probabilities of all simple events in the sample space is 1.

Any probability assignment that meets conditions 1 and 2 is said to be an **acceptable probability assignment.**

Our mathematical theory does not explain how acceptable probabilities are assigned to simple events. These assignments are generally based on the expected or actual percentage of times a simple event occurs when an experiment is repeated a large number of times. Assignments based on this principle are called *reasonable*.

Let an experiment be the flipping of a single coin, and let us choose a sample space S to be

$$S = \{\text{H}, \text{T}\}$$

If a coin appears to be fair, we are inclined to assign probabilities to the simple events in S as follows:

$$P(\text{H}) = \tfrac{1}{2} \qquad \text{and} \qquad P(\text{T}) = \tfrac{1}{2}$$

These assignments are based on reasoning that, since there are two ways a coin can land, in the long run, a head will turn up half the time and a tail will turn up half the time. These probability assignments are acceptable, since both of the conditions for acceptable probability assignments in Definition 2 are satisfied:

1. $0 \le P(\text{H}) \le 1,\ 0 \le P(\text{T}) \le 1$

2. $P(\text{H}) + P(\text{T}) = \frac{1}{2} + \frac{1}{2} = 1$

But there are other acceptable assignments. Maybe after flipping a coin 1,000 times we find that the head turns up 376 times and the tail turns up 624 times. With this result, we might suspect that the coin is not fair and assign the simple events in the sample space S the probabilities

$$P(\text{H}) = .376 \qquad \text{and} \qquad P(\text{T}) = .624$$

This is also an acceptable assignment. But the probability assignment

$$P(\text{H}) = 1 \qquad \text{and} \qquad P(\text{T}) = 0$$

though acceptable, is not reasonable, unless the coin has 2 heads. The assignment

$$P(\text{H}) = .6 \qquad \text{and} \qquad P(\text{T}) = .8$$

is not acceptable, since $.6 + .8 = 1.4$, which violates condition 2 in Definition 2.

In probability studies, the 0 to the left of the decimal is usually omitted. Thus, we write .8 and not 0.8.

It is important to keep in mind that out of the infinitely many possible acceptable probability assignments to simple events in a sample space, we are generally inclined to choose one assignment over another based on reasoning or experimental results.

Given an acceptable probability assignment for simple events in a sample space S, how do we define the probability of an arbitrary event E associated with S?

DEFINITION 3

PROBABILITY OF AN EVENT *E*

Given an acceptable probability assignment for the simple events in a sample space S, we define the **probability of an arbitrary event *E*,** denoted by $P(E)$, as follows:

1. If E is the empty set, then $P(E) = 0$.

2. If E is a simple event, then $P(E)$ has already been assigned.

3. If E is a compound event, then $P(E)$ is the sum of the probabilities of all the simple events in E.

4. If E is the sample space S, then $P(E) = P(S) = 1$. This is a special case of 3.

EXAMPLE 3

Finding Probabilities of Events

Let's return to Example 1, the tossing of a nickel and dime, and the sample space

$$S = \{\text{HH, HT, TH, TT}\}$$

Since there are four simple outcomes and the coins are assumed to be fair, it appears that each outcome should occur in the long run 25% of the time. Let's assign the same probability of $\frac{1}{4}$ to each simple event in S:

Simple event, e_i	HH	HT	TH	TT
$P(e_i)$	$\frac{1}{4}$	$\frac{1}{4}$	$\frac{1}{4}$	$\frac{1}{4}$

This is an acceptable assignment according to Definition 2 and a reasonable assignment for ideal coins that are perfectly balanced or coins close to ideal.

(A) What is the probability of getting exactly 1 head?

(B) What is the probability of getting at least 1 head?

(C) What is the probability of getting a head or a tail?

(D) What is the probability of getting 3 heads?

Solutions

(A) E_1 = Getting 1 head = {HT, TH}

Since E_1 is a compound event, we use item 3 in Definition 3 and find $P(E_1)$ by adding the probabilities of the simple events in E_1. Thus,

$$P(E_1) = P(\text{HT}) + P(\text{TH}) = \tfrac{1}{4} + \tfrac{1}{4} = \tfrac{1}{2}$$

(B)

$$E_2 = \text{Getting at least 1 head} = \{\text{HH, HT, TH}\}$$

$$P(E_2) = P(\text{HH}) + P(\text{HT}) + P(\text{TH}) = \tfrac{1}{4} + \tfrac{1}{4} + \tfrac{1}{4} = \tfrac{3}{4}$$

(C)

$$E_3 = \{\text{HH, HT, TH, TT}\} = S$$

$$P(E_3) = P(S) = 1 \qquad \tfrac{1}{4} + \tfrac{1}{4} + \tfrac{1}{4} + \tfrac{1}{4} = 1$$

(D)

$$E_3 = \text{Getting 3 heads} = \varnothing \qquad \text{Empty set}$$

$$P(\varnothing) = 0$$

STEPS FOR FINDING PROBABILITIES OF EVENTS

Step 1. Set up an appropriate sample space S for the experiment.

Step 2. Assign acceptable probabilities to the simple events in S.

Step 3. To obtain the probability of an arbitrary event E, add the probabilities of the simple events in E.

The function P defined in steps 2 and 3 is called a **probability function.** The domain of this function is all possible events in the sample space S, and the range is a set of real numbers between 0 and 1, inclusive.

MATCHED PROBLEM 3

Return to Matched Problem 1, recording the boy–girl composition of families with 2 children and the sample space

$$S = \{BB, BG, GB, GG\}$$

Statistics from the U.S. Census Bureau indicate that an acceptable and reasonable probability for this sample space is

Simple event, e_i	BB	BG	GB	GG
$P(e_i)$	.26	.25	.25	.24

Find the probabilities of the following events:

(A) E_1 = Having at least one girl in the family

(B) E_2 = Having at most one girl in the family

(C) E_3 = Having two children of the same sex in the family

Equally Likely Assumption

In tossing a nickel and dime (Example 3), we assigned the same probability, $\frac{1}{4}$, to each simple event in the sample space $S = \{HH, HT, TH, TT\}$. By assigning the same probability to each simple event in S, we are actually making the assumption that each simple event is as likely to occur as any other. We refer to this as an **equally likely assumption.** In general, we have Definition 4.

DEFINITION 4

PROBABILITY OF A SIMPLE EVENT UNDER AN EQUALLY LIKELY ASSUMPTION

If, in a sample space

$$S = \{e_1, e_2, \ldots, e_n\}$$

with n elements, we assume each simple event e_i is as likely to occur as any other, then we assign the probability $1/n$ to each. That is,

$$P(e_i) = \frac{1}{n}$$

Under an equally likely assumption, we can develop a very useful formula for finding probabilities of arbitrary events associated with a sample space S. Consider the following example.

If a single die is rolled and we assume each face is as likely to come up as any other, then for the sample space

$$S = \{1, 2, 3, 4, 5, 6\}$$

we assign a probability of $\frac{1}{6}$ to each simple event, since there are 6 simple events. Then the probability of

$$E = \text{Rolling a prime number} = \{2, 3, 5\}$$

is

$$P(E) = P(2) + P(3) + P(5) = \tfrac{1}{6} + \tfrac{1}{6} + \tfrac{1}{6} = \tfrac{3}{6} = \tfrac{1}{2}$$

Thus, under the assumption that each simple event is as likely to occur as any other, the computation of the probability of the occurrence of any event E in a sample space S is the number of elements in E divided by the number of elements in S.

THEOREM 1

PROBABILITY OF AN ARBITRARY EVENT UNDER AN EQUALLY LIKELY ASSUMPTION

If we assume each simple event in sample space S is as likely to occur as any other, then the probability of an arbitrary event E in S is given by

$$P(E) = \frac{\text{Number of elements in } E}{\text{Number of elements in } S} = \frac{n(E)}{n(S)}$$

EXAMPLE 4

Finding Probabilities of Events

If in rolling 2 dice we assume each simple event in the sample space shown in Figure 1 (page 495) is as likely as any other, find the probabilities of the following events:

(A) E_1 = A sum of 7 turns up

(B) E_2 = A sum of 11 turns up

(C) E_3 = A sum less than 4 turns up

(D) E_4 = A sum of 12 turns up

Solutions

Referring to Figure 1, we see that:

(A) $P(E_1) = \dfrac{n(E_1)}{n(S)} = \dfrac{6}{36} = \dfrac{1}{6}$

(B) $P(E_2) = \dfrac{n(E_2)}{n(S)} = \dfrac{2}{36} = \dfrac{1}{18}$

(C) $P(E_3) = \dfrac{n(E_3)}{n(S)} = \dfrac{3}{36} = \dfrac{1}{12}$

(D) $P(E_4) = \dfrac{n(E_4)}{n(S)} = \dfrac{1}{36}$

MATCHED PROBLEM 4

Under the conditions in Example 4, find the probabilities of the following events:

(A) E_5 = A sum of 5 turns up

(B) E_6 = A sum that is a prime number greater than 7 turns up

Explore/Discuss 1

A box contains 4 red balls and 7 green balls. A ball is drawn at random and then, without replacing the first ball, a second ball is drawn. Discuss whether or not the equally likely assumption would be appropriate for the sample space $S = \{RR, RG, GR, GG\}$.

We now turn to some examples that make use of the counting techniques developed in Section 6-4.

EXAMPLE 5 **Drawing Cards**

In drawing 5 cards from a 52-card deck without replacement, what is the probability of getting 5 spades?

Solution Let the sample space S be the set of all 5-card hands from a 52-card deck. Since the order in a hand does not matter, $n(S) = C_{52,5}$. The event we seek is

$$E = \text{Set of all 5-card hands from 13 spades}$$

Again, the order does not matter and $n(E) = C_{13,5}$. Thus, assuming each 5-card hand is as likely as any other,

$$P(E) = \frac{n(E)}{n(S)} = \frac{C_{13,5}}{C_{52,5}} = \frac{13!/5!8!}{52!/5!47!} = \frac{13!}{5!8!} \cdot \frac{5!47!}{52!} \approx .0005$$

MATCHED PROBLEM 5

In drawing 7 cards from a 52-card deck without replacement, what is the probability of getting 7 hearts?

EXAMPLE 6 **Selecting Committees**

The board of regents of a university is made up of 12 men and 16 women. If a committee of 6 is chosen at random, what is the probability that it will contain 3 men and 3 women?

Solution Let S = Set of all 6-person committees out of 28 people:

$$n(S) = C_{28,6}$$

Let E = Set of all 6-person committees with 3 men and 3 women. To find $n(E)$, we use the multiplication principle and the following two operations:

O_1: Select 3 men out of the 12 available $\quad N_1$: $C_{12,3}$

O_2: Select 3 women out of the 16 available $\quad N_2$: $C_{16,3}$

Thus,

$$n(E) = C_{12,3} \cdot C_{16,3}$$

and

$$P(E) = \frac{n(E)}{n(S)} = \frac{C_{12,3} \cdot C_{16,3}}{C_{28,6}} \approx .327$$

MATCHED PROBLEM 6

What is the probability that the committee in Example 6 will have 4 men and 2 women?

Empirical Probability

In the earlier examples in this section, we made a reasonable assumption about an experiment and used deductive reasoning to assign probabilities. For example, it is reasonable to assume that an ordinary coin will come up heads about as often as it will come up tails. Probabilities determined in this manner are called **theoretical probabilities.** No experiments are ever conducted. But what if the theoretical probabilities are not obvious? Then we assign probabilities to simple events based on the results of actual experiments. Probabilities determined from the results of actually performing an experiment are called **empirical probabilities.** As an experiment is repeated over and over, the percentage of times an event occurs may get closer and closer to a single fixed number. If so, this single fixed number is generally called the **actual probability** of the event.

Explore/Discuss 2

Like a coin, a thumbtack tossed into the air will land in one of two positions, point up or point down [see Fig. 2(a)]. Unlike a coin, we would not expect both events to occur with the same frequency. Indeed, the frequencies of landing point up and point down may well vary from one thumbtack to another [see Fig. 2(b)]. Find two thumbtacks of different sizes and guess which one is likely to land point up more frequently. Then toss each tack 100 times and record the number of times each lands point up. Did the experiment confirm your initial guess?

FIGURE 2

(a) Point up or point down (b) Two different tacks

Suppose when tossing one of the thumbtacks in Explore/Discuss 2, we observe that the tack lands point up 43 times and point down 57 times. Based on this experiment, it seems reasonable to say that for this particular thumbtack

$$P(\text{Point up}) = \frac{43}{100} = .43$$

$$P(\text{Point down}) = \frac{57}{100} = .57$$

Probability assignments based on the results of repeated trials of an experiment are called **approximate empirical probabilities.**

In general, if we conduct an experiment n times and an event E occurs with **frequency** $f(E)$, then the ratio $f(E)/n$ is called the **relative frequency** of the occurrence of event E in n trials. We define the **empirical probability** of E, denoted by $P(E)$, by the number, if it exists, that the relative frequency $f(E)/n$ approaches as n gets larger and larger. Of course, for any particular n, the relative frequency $f(E)/n$ is generally only approximately equal to $P(E)$. However, as n increases in size, we expect the approximation to improve.

DEFINITION 5

EMPIRICAL PROBABILITY

If $f(E)$ is the frequency of event E in n trials, then

$$P(E) \approx \frac{\text{Frequency of occurrence of } E}{\text{Total number of trials}} = \frac{f(E)}{n}$$

If we can also deduce theoretical probabilities for an experiment, then we expect the approximate empirical probabilities to approach the theoretical probabilities. If this does not happen, then we should begin to suspect the manner in which the theoretical probabilities were computed. If $P(E)$ is the theoretical probability of an event E and the experiment is performed n times, then the **expected frequency** of the occurrence of E is $n \cdot P(E)$.

EXAMPLE 7

Finding Approximate Empirical and Theoretical Probabilities

Two coins are tossed 500 times with the following frequencies of outcomes:

2 heads: 121

1 head: 262

0 heads: 117

(A) Compute the approximate empirical probability for each outcome.

(B) Compute the theoretical probability for each outcome.

(C) Compute the expected frequency for each outcome.

Solutions

(A)

$$P(2\text{ heads}) \approx \frac{121}{500} = .242$$

$$P(1\text{ head}) \approx \frac{262}{500} = .524$$

$$P(0\text{ heads}) \approx \frac{117}{500} = .234$$

(B) A sample space of equally likely simple events is $S = \{\text{HH, HT, TH, TT}\}$. Let

$$E_1 = 2 \text{ heads} = \{\text{HH}\}$$
$$E_2 = 1 \text{ head} = \{\text{HT, TH}\}$$
$$E_3 = 0 \text{ heads} = \{\text{TT}\}$$

Then

$$P(E_1) = \frac{n(E_1)}{n(S)} = \frac{1}{4} = .25$$
$$P(E_2) = \frac{n(E_2)}{n(S)} = \frac{2}{4} = .50$$
$$P(E_3) = \frac{n(E_3)}{n(S)} = \frac{1}{4} = .25$$

(C) The expected frequencies are

$$E_1\text{: } 500(.25) = 125$$
$$E_2\text{: } 500(.5) = 250$$
$$E_3\text{: } 500(.25) = 125$$

The actual frequencies obtained from performing the experiment are reasonably close to the expected frequencies. Increasing the number of trials of the experiment would produce even better approximations.

MATCHED PROBLEM 7

One die is rolled 500 times with the following frequencies of outcomes:

Outcome	1	2	3	4	5	6
Frequency	89	83	77	91	72	88

(A) Compute the approximate empirical probability for each outcome.
(B) Compute the theoretical probability for each outcome.
(C) Compute the expected frequency for each outcome.

FIGURE 3
Using a random number generator.

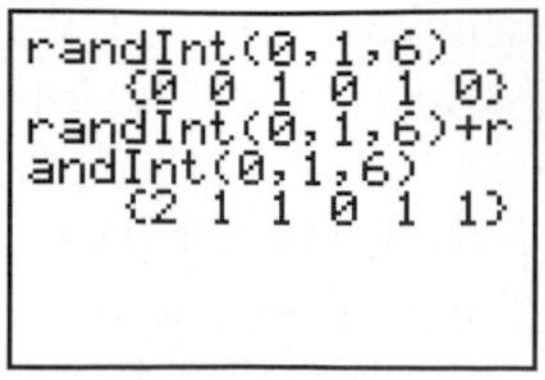

Tossing two coins 500 times is certainly a tedious task and we did not do this to generate the data in Example 7. Instead, we used a random number generator on a graphing utility to simulate this experiment. Specifically, we used the command **randInt(*i*,*k*,*n*)** on a Texas Instruments TI-83, which generates a random sequence of n integers between i and k, inclusively. If we let 0 represent tails and 1 represent heads, then a random sequence of 0s and 1s can be used to represent the outcomes of repeated tosses of one coin (see the first two lines of Fig. 3). Thus, in six tosses, we obtained 2 heads and 4 tails. To simulate tossing two coins, we simply add together two similar statements, as shown in lines three through five of Figure 3. We see that in these six tosses 0 heads occurred once, 1 head occurred four times, and 2 heads occurred once. Of course, to obtain meaningful results, we need to toss the coins many more times. Figure 4(a) shows a command that will simulate 500 tosses of two coins. To determine the frequency of each outcome, we construct a histogram [see Figs. 4(b) and 4(c)] and use the TRACE command to determine the following frequencies [see Figs. 5(a), 5(b), and 5(c)].

FIGURE 4
Simulating 500 tosses of two coins.

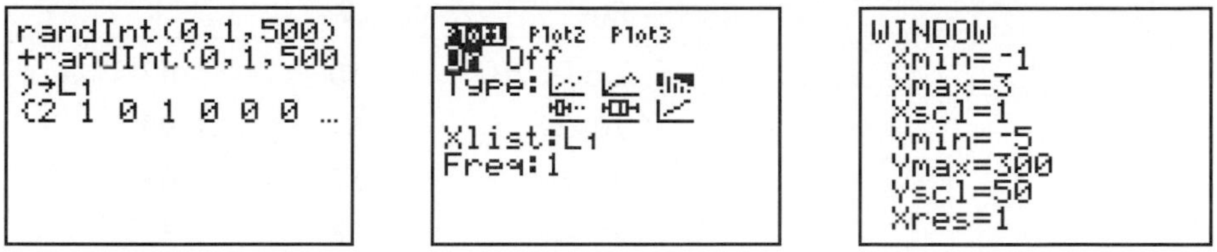

(a) Generating the random numbers (b) Setting up the histogram (c) Selecting the window variables

FIGURE 5
Results of the simulation.

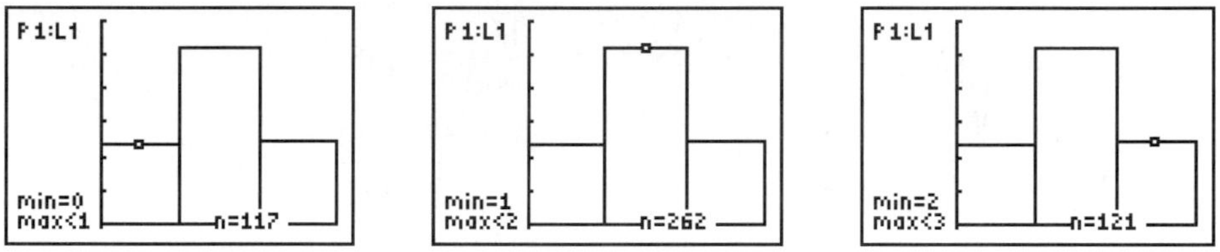

(a) 0 heads: 117 (b) 1 head: 262 (c) 2 heads: 121

If you perform the same simulation on your graphing utility, you are not likely to get exactly the same results. But the approximate empirical probabilities you obtain will be close to the theoretical probabilities.

Explore/Discuss 3

This discussion assumes that your graphing utility has the ability to generate and manipulate sequences of random integers.

(A) As an alternative to using the histogram in Figure 5 to count the outcomes of the sequence of random integers in Figure 4(a), enter the following function and evaluate it for $x = 0$, 1, and 2:

$$y_1 = \text{sum}(\text{seq}(L_1(I)=X,I,1,\dim(L_1)))$$

(B) Simulate the experiment of rolling a single die and compare your empirical results with the results in Matched Problem 7.

EXAMPLE 8

Empirical Probabilities for an Insurance Company

An insurance company selected 1,000 drivers at random in a particular city to determine a relationship between age and accidents. The data obtained are listed in Table 1. Compute the approximate empirical probabilities of the following events for a driver chosen at random in the city:

(A) E_1: being under 20 years old *and* having exactly 3 accidents in 1 year

(B) E_2: being 30–39 years old *and* having 1 or more accidents in 1 year

(C) E_3: having no accidents in 1 year

(D) E_4: being under 20 years old *or* having exactly 3 accidents in 1 year

TABLE 1

	Accidents in 1 Year				
Age	0	1	2	3	Over 3
Under 20	50	62	53	35	20
20–29	64	93	67	40	36
30–39	82	68	32	14	4
40–49	38	32	20	7	3
Over 49	43	50	35	28	24

Solutions

(A) $P(E_1) \approx \frac{35}{1{,}000} = .035$

(B) $P(E_2) \approx \frac{68 + 32 + 14 + 4}{1{,}000} = .118$

(C) $P(E_3) \approx \frac{50 + 64 + 82 + 38 + 43}{1{,}000} = .277$

(D) $P(E_4) \approx \frac{50 + 62 + 53 + 35 + 20 + 40 + 14 + 7 + 28}{1{,}000} = .309$

Notice that in this type of problem, which is typical of many realistic problems, approximate empirical probabilities are the only type we can compute.

MATCHED PROBLEM 8

Referring to Table 1 in Example 8, compute the approximate empirical probabilities of the following events for a driver chosen at random in the city:

(A) E_1: being under 20 years old with no accidents in 1 year

(B) E_2: being 20–29 years old and having fewer than 2 accidents in 1 year

(C) E_3: not being over 49 years old

Approximate empirical probabilities are often used to test theoretical probabilities. Equally likely assumptions may not be justified in reality. In addition to this use, there are many situations in which it is either very difficult or impossible to compute the theoretical probabilities for given events. For example, insurance companies use past experience to establish approximate empirical probabilities to predict future accident rates, baseball teams use batting averages, which are approximate empirical probabilities based on past experience, to predict the future performance of a player, and pollsters use approximate empirical probabilities to predict outcomes of elections.

Answers to Matched Problems

1. (A) $S_1 = \{BB, BG, GB, GG\}$;

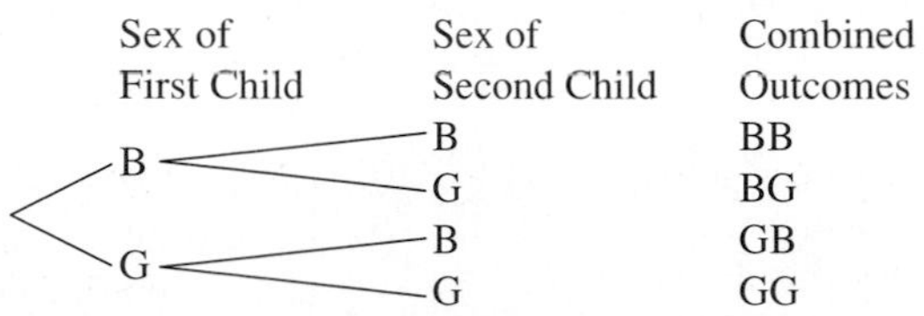

(B) $S_2 = \{0, 1, 2\}$ (C) $S_3 = \{A, D\}$ (D) The sample space in part A.

2. (A) $\{(4, 1), (3, 2), (2, 3), (1, 4)\}$ (B) $\{(6, 5), (5, 6)\}$ **3.** (A) .74 (B) .76 (C) .5

4. (A) $P(E_5) = \frac{1}{9}$ (B) $P(E_6) = \frac{1}{18}$ **5.** $C_{13,7}/C_{52,7} \approx .000013$ **6.** $C_{12,4} \cdot C_{16,2}/C_{28,6} \approx .158$

7. (A) $P(E_1) \approx .178$, $P(E_2) \approx .166$, $P(E_3) \approx .154$, $P(E_4) \approx .182$, $P(E_5) \approx .144$, $P(E_6) \approx .176$
(B) $\frac{1}{6} \approx .167$ for each (C) 83.3 for each

8. (A) $P(E_1) \approx .05$ (B) $P(E_2) \approx .157$ (C) $P(E_3) \approx .82$

EXERCISE 6-5

A

1. How would you interpret $P(E) = 1$?

2. How would you interpret $P(E) = 0$?

3. A spinner can land on 4 different colors: red (R), green (G), yellow (Y), and blue (B). If we do not assume each color is as likely to turn up as any other, which of the following probability assignments have to be rejected, and why?

   (A) $P(R) = .15, P(G) = -.35, P(Y) = .50, P(B) = .70$

   (B) $P(R) = .32, P(G) = .28, P(Y) = .24, P(B) = .30$

   (C) $P(R) = .26, P(G) = .14, P(Y) = .30, P(B) = .30$

4. Under the probability assignments in Problem 3, part C, what is the probability that the spinner will not land on blue?

5. Under the probability assignments in Problem 3, part C, what is the probability that the spinner will land on red or yellow?

6. Under the probability assignments in Problem 3, part C, what is the probability that the spinner will not land on red or yellow?

7. A ski jumper has jumped over 300 feet in 25 out of 250 jumps. What is the approximate empirical probability of the next jump being over 300 feet?

8. In a certain city there are 4,000 youths between 16 and 20 years old who drive cars. If 560 of them were involved in accidents last year, what is the approximate empirical probability of a youth in this age group being involved in an accident this year?

9. Out of 420 times at bat, a baseball player gets 189 hits. What is the approximate empirical probability that the player will get a hit next time at bat?

10. In a medical experiment, a new drug is found to help 2,400 out of 3,000 people. If a doctor prescribes the drug for a particular patient, what is the approximate empirical probability that the patient will be helped?

B

11. A small combination lock on a suitcase has 3 wheels, each labeled with the 10 digits from 0 to 9. If an opening combination is a particular sequence of 3 digits with no repeats, what is the probability of a person guessing the right combination?

12. A combination lock has 5 wheels, each labeled with the 10 digits from 0 to 9. If an opening combination is a particular sequence of 5 digits with no repeats, what is the probability of a person guessing the right combination?

An experiment consists of dealing 5 cards from a standard 52-card deck. In Problems 13–16, what is the probability of being dealt each of the following hands?

13. 5 black cards

14. 5 hearts

15. 5 face cards if aces are considered to be face cards

16. 5 nonface cards if an ace is considered to be 1 and not a face card

17. If 4-digit numbers less than 5,000 are randomly formed from the digits 1, 3, 5, 7, and 9, what is the probability of forming a number divisible by 5? Digits may be repeated; for example, 1,355 is acceptable.

18. If code words of four letters are generated at random using the letters A, B, C, D, E, and F, what is the probability of forming a word without a vowel in it? Letters may be repeated.

19. Suppose 5 thank-you notes are written and 5 envelopes are addressed. Accidentally, the notes are randomly inserted into the envelopes and mailed without checking the addresses. What is the probability that all 5 notes will be inserted into the correct envelopes?

20. Suppose 6 people check their coats in a checkroom. If all claim checks are lost and the 6 coats are randomly returned, what is the probability that all 6 people will get their own coats back?

An experiment consists of rolling 2 fair dice and adding the dots on the 2 sides facing up. Using the sample space shown in Figure 1 (page 495) and assuming each simple event is as likely as any other, find the probabilities of the sums of dots indicated in Problems 21–36.

21. Sum is 2.

22. Sum is 10.

23. Sum is 6.

24. Sum is 8.

25. Sum is less than 5.

26. Sum is greater than 8.

27. Sum is not 7 or 11.

28. Sum is not 2, 4, or 6.

29. Sum is 1.

30. Sum is not 13.

31. Sum is divisible by 3.

32. Sum is divisible by 4.

33. Sum is 7 or 11 (a "natural").

34. Sum is 2, 3, or 12 ("craps").

35. Sum is divisible by 2 or 3.

36. Sum is divisible by 2 and 3.

37. Five thousand people work in a large auto plant. An individual is selected at random and his or her birthday (month and day, not year) is recorded. Set up an appropriate sample space for this experiment and assign acceptable probabilities to the simple events.

38. In a hotly contested three-way race for governor of Minnesota, the leading candidates are running neck-and-neck while the third candidate is receiving half the support of either of the others. Registered voters are chosen at random and are asked for which of the three they are most likely to vote. Set up an appropriate sample space for the random survey experiment and assign acceptable probabilities to the simple events.

39. A pair of dice is rolled 500 times with the following frequencies:

Sum	2	3	4	5	6	7	8	9	10	11	12
Frequency	11	35	44	50	71	89	72	52	36	26	14

(A) Compute the approximate empirical probability for each outcome.

(B) Compute the theoretical probability for each outcome, assuming fair dice.

(C) Compute the expected frequency of each outcome.

(D) Describe how a random number generator could be used to simulate this experiment. If your graphing utility has a random number generator, use it to simulate 500 tosses of a pair of dice and compare your results with part C.

40. Three coins are flipped 500 times with the following frequencies of outcomes:

3 heads: 58	2 heads: 198
1 head: 190	0 heads: 54

(A) Compute the approximate empirical probability for each outcome.

(B) Compute the theoretical probability for each outcome, assuming fair coins.

(C) Compute the expected frequency of each outcome.

(D) Describe how a random number generator could be used to simulate this experiment. If your graphing utility has a random number generator, use it to simulate 500 tosses of three coins and compare your results with part C.

41. (A) Is it possible to get 29 heads in 30 flips of a fair coin? Explain.

(B) If you flip a coin 50 times and get 42 heads, would you suspect that the coin was unfair? Why or why not? If you suspect an unfair coin, what empirical probabilities would you assign to the simple events of the sample space?

42. (A) Is it possible to get 9 double sixes in 12 rolls of a pair of fair dice? Explain.

(B) If you roll a pair of dice 40 times and get 14 double sixes, would you suspect that the dice were unfair? Why or why not? If you suspect loaded dice, what empirical probability would you assign to the event of rolling a double six?

An experiment consists of tossing 3 fair coins, but 1 of the 3 coins has a head on both sides. Compute the probabilities of obtaining the indicated results in Problems 43–48.

43. 1 head

44. 2 heads

45. 3 heads

46. 0 heads

47. More than 1 head

48. More than 1 tail

C

An experiment consists of rolling 2 fair dice and adding the dots on the 2 sides facing up. Each die has 1 dot on two opposite faces, 2 dots on two opposite faces, and 3 dots on two opposite faces. Compute the probabilities of obtaining the indicated sums in Problems 49–56.

49. 2

50. 3

51. 4

52. 5

53. 6

54. 7

55. An odd sum

56. An even sum

An experiment consists of dealing 5 cards from a standard 52-card deck. In Problems 57–64, what is the probability of being dealt the following cards?

57. 5 cards, jacks through aces

58. 5 cards, 2 through 10

59. 4 aces

60. 4 of a kind

61. Straight flush, ace high; that is, 10, jack, queen, king, ace in one suit

62. Straight flush, starting with 2; that is, 2, 3, 4, 5, 6 in one suit

63. 2 aces and 3 queens

64. 2 kings and 3 aces

APPLICATIONS

65. Market Analysis. A company selected 1,000 households at random and surveyed them to determine a relationship between income level and the number of television sets in a home. The information gathered is listed in the table:

	Televisions per Household				
Yearly Income	0	1	2	3	Above 3
Less than \$12,000	0	40	51	11	0
\$12,000–19,999	0	70	80	15	1
\$20,000–39,999	2	112	130	80	12
\$40,000–59,999	10	90	80	60	21
\$60,000 or more	30	32	28	25	20

Compute the approximate empirical probabilities:

(A) Of a household earning \$12,000–\$19,999 per year *and* owning exactly 3 television sets

(B) Of a household earning \$20,000–\$39,999 per year *and* owning more than 1 television set

(C) Of a household earning \$60,000 or more per year *or* owning more than 3 television sets

(D) Of a household not owning 0 television sets

66. Market Analysis. Use the sample results in Problem 65 to compute the approximate empirical probabilities:

(A) Of a household earning \$40,000–\$59,999 per year *and* owning 0 television sets

(B) Of a household earning \$12,000–\$39,999 per year *and* owning more than 2 television sets

(C) Of a household earning less than \$20,000 per year *or* owning exactly 2 television sets

(D) Of a household not owning more than 3 television sets

Section 6-6 Binomial Formula

- Pascal's Triangle
- The Binomial Formula
- Proof of the Binomial Formula

The binomial form

$$(a + b)^n$$

where n is a natural number, appears more frequently than you might expect. It turns out that the coefficients in the expansion are related to probability concepts that we have already discussed.

Pascal's Triangle

Let's begin by expanding $(a + b)^n$ for the first few values of n. We include $n = 0$, which is not a natural number, for reasons of completeness that will become apparent later.

$$\begin{aligned}
(a + b)^0 &= 1\\
(a + b)^1 &= a + b\\
(a + b)^2 &= a^2 + 2ab + b^2\\
(a + b)^3 &= a^3 + 3a^2b + 3ab^2 + b^3
\end{aligned} \tag{1}$$

Explore/Discuss 1

Based on the expansions in equations (1), how many terms would you expect $(a + b)^n$ to have? What is the first term? What is the last term?

Now let's examine just the coefficients of the expansions in equations (1) arranged in a form that is usually referred to as **Pascal's triangle** (see Fig. 1).

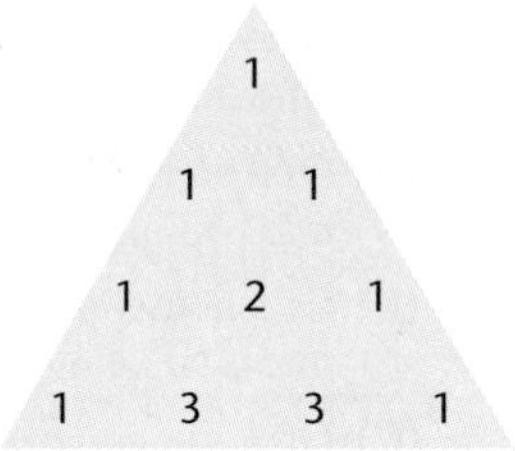

FIGURE 1
Pascal's triangle.

Explore/Discuss 2

Refer to Figure 1.

(A) How is the middle element in the third row related to the elements in the row above it?

(B) How are the two inner elements in the fourth row related to the elements in the row above them?

(C) Based on your observations in parts A and B, make a conjecture about the fifth and sixth rows. Check your conjecture by expanding $(a + b)^4$ and $(a + b)^5$.

Many students find Pascal's triangle a useful tool for determining the coefficients in the expansion of $(a + b)^n$, especially for small values of n. Its major drawback is that to find the elements in a given row, you must write out all the preceding rows. It would be useful to find a formula that would give the coefficients for a binomial expansion directly. Fortunately, such a formula exists—the combination formula $C_{n,r}$ introduced in Section 6-4. Table 1 lists a short program for several popular Texas Instruments calculators that displays the values of the combination formula. You should recognize the output in the table—it is the first 6 lines of Pascal's triangle!

TABLE 1 Combination Formula and Pascal's Triangle

Program PASCAL

TI-82/TI-83

```
Disp "HOW MANY LINES"
Input L
For(N,0,L-1)
N→dim(L1)
For(R,0,N,1)
N nCr R→L1(R+1)
End
Disp L1
End
```

TI-85/TI-86

```
Disp "HOW MANY LINES"
Input L
For(N,0,L-1)
N→dimL L1
For(R,0,N,1)
N nCr R→L1(R+1)
End
Disp L1
End
```

Output

```
                {1}
              {1 1}
            {1 2 1}
          {1 3 3 1}
        {1 4 6 4 1}
  {1 5 10 10 5 1}
               Done
```

The Binomial Formula

When working with binomial expansions, it is customary to use another notation for the combination formula.

DEFINITION 1

COMBINATION FORMULA

For nonnegative integers r and n, $0 \leq r \leq n$,

$$\binom{n}{r} = C_{n,r} = \frac{n!}{r!(n-r)!}$$

The following theorem establishes that the coefficients in a binomial expansion can always be expressed in terms of the combination formula. This is a very important theoretical result and a very practical tool. As we shall see, using this theorem in conjunction with a graphing utility provides a very efficient method for expanding binomials.

THEOREM 1

BINOMIAL FORMULA

For n a positive integer

$$(a+b)^n = \sum_{k=0}^{n} \binom{n}{k} a^{n-k} b^k$$

We defer the proof of Theorem 1 until the end of this section. Since the values of the combination formula are the coefficients in a binomial expansion, it is natural to call them **binomial coefficients.**

EXAMPLE 1 Using the Binomial Formula

Use the binomial formula to expand $(x+y)^6$.

Solution

$$(x + y)^6 = \sum_{k=0}^{6} \binom{6}{k} x^{6-k}y^k$$

$$= \binom{6}{0}x^6 + \binom{6}{1}x^5y + \binom{6}{2}x^4y^2 + \binom{6}{3}x^3y^3 + \binom{6}{4}x^2y^4 + \binom{6}{5}xy^5 + \binom{6}{6}y^6$$

$$= x^6 + 6x^5y + 15x^4y^2 + 20x^3y^3 + 15x^2y^4 + 6xy^5 + y^6$$

FIGURE 2

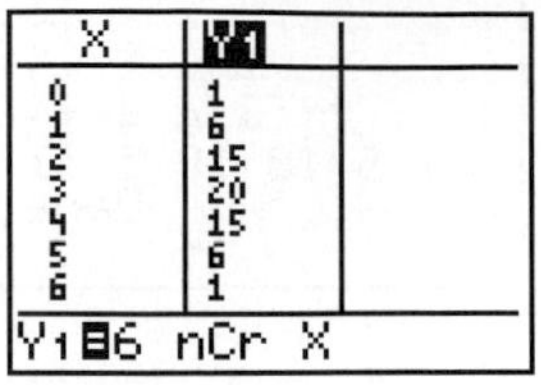

How did we compute the binomial coefficients in the preceding expansion of $(x + y)^6$? There are several ways this can be done. One way is to write out the first 7 lines of Pascal's triangle. Another is to evaluate $\binom{n}{r} = C_{n,r}$ one coefficient at a time, either by hand or on a graphing utility. But the most efficient way is to use the table feature on a graphing utility (see Fig. 2).

MATCHED PROBLEM 1

Use the binomial formula to expand $(x + 1)^5$.

EXAMPLE 2

Using the Binomial Formula

Use the binomial formula to expand $(3p - 2q)^4$.

Solution

$$(3p - 2q)^4 = [(3p) + (-2q)]^4 \qquad a = 3p,\ b = -2q$$

$$= \sum_{k=0}^{4} \binom{4}{k} (3p)^{4-k}(-2q)^k$$

$$= \sum_{k=0}^{4} \binom{4}{k} 3^{4-k}(-2)^k p^{4-k}q^k \qquad \text{See Figure 3.}$$

$$= 81p^4 - 216p^3q + 216p^2q^2 - 96pq^3 + 16q^4$$

FIGURE 3

$y_1 = C_{4,x}3^{4-x}(-2)^x$.

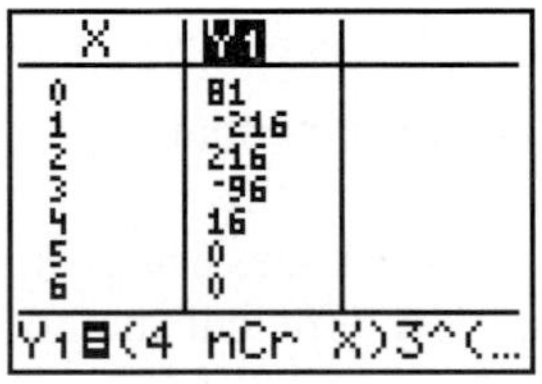

MATCHED PROBLEM 2

Use the binomial formula to expand $(2m - 5n)^3$.

Explore/Discuss 3

(A) Compute each term and also the sum of the alternating series $\binom{6}{0} - \binom{6}{1} + \binom{6}{2} - \cdots + \binom{6}{6}$.

(B) What result about an alternating series can be deduced by letting $a = 1$ and $b = -1$ in the binomial formula?

EXAMPLE 3

Using the Binomial Formula

Find the term containing x^9 in the expansion of $(x + 3)^{14}$.

Solution In the expansion

$$(x + 3)^{14} = \sum_{k=0}^{14} \binom{14}{k} x^{14-k} 3^k$$

the exponent of x is 9 when $k = 5$. Thus, the term containing x^9 is

$$\binom{14}{5} x^9 3^5 = (2{,}002)(243)x^9 = 486{,}486x^9$$

MATCHED PROBLEM 3 Find the term containing y^8 in the expansion of $(2 + y)^{14}$.

EXAMPLE 4 **Using the Binomial Formula**

If the terms in the expansion of $(x - 2)^{20}$ are arranged in decreasing powers of x, find the fourth term and the sixteenth term.

Solution In the expansion of $(a + b)^n$, the exponent of b in the rth term is $r - 1$ and the exponent of a is $n - (r - 1)$. Thus,

Fourth term:

$$\binom{20}{3} x^{17}(-2)^3 = \frac{20 \cdot 19 \cdot 18}{3 \cdot 2 \cdot 1} x^{17}(-8) = -9{,}120x^{17}$$

Sixteenth term:

$$\binom{20}{15} x^5(-2)^{15} = \frac{20 \cdot 19 \cdot 18 \cdot 17 \cdot 16}{5 \cdot 4 \cdot 3 \cdot 2 \cdot 1} x^5(-32{,}768) = -508{,}035{,}072x^5$$

MATCHED PROBLEM 4 If the terms in the expansion of $(u - 1)^{18}$ are arranged in decreasing powers of u, find the fifth term and the twelfth term.

Proof of the Binomial Formula

We now proceed to prove that the binomial formula holds for all natural numbers n using mathematical induction.

Proof State the conjecture.

$$P_n\colon\quad (a + b)^n = \sum_{j=0}^{n} \binom{n}{j} a^{n-j} b^j$$

Part 1

Show that P_1 is true.

$$\sum_{j=0}^{1} \binom{1}{j} a^{1-j} b^j = \binom{1}{0} a + \binom{1}{1} b = a + b = (a + b)^1$$

Thus, P_1 is true.

Part 2

Show that if P_k is true, then P_{k+1} is true.

$$P_k\text{:}\quad (a+b)^k = \sum_{j=0}^{k} \binom{k}{j} a^{k-j}b^j \qquad \text{Assume } P_k \text{ is true.}$$

$$P_{k+1}\text{:}\quad (a+b)^{k+1} = \sum_{j=0}^{k+1} \binom{k+1}{j} a^{k+1-j}b^j \qquad \text{Show } P_{k+1} \text{ is true.}$$

We begin by multiplying both sides of P_k by $(a + b)$:

$$(a+b)^k(a+b) = \left[\sum_{j=0}^{k} \binom{k}{j} a^{k-j}b^j\right](a+b)$$

The left side of this equation is the left side of P_{k+1}. Now we multiply out the right side of the equation and try to obtain the right side of P_{k+1}:

$$\begin{aligned}
(a+b)^{k+1} &= \left[\binom{k}{0}a^k + \binom{k}{1}a^{k-1}b + \binom{k}{2}a^{k-2}b^2 + \cdots + \binom{k}{k}b^k\right](a+b) \\
&= \left[\binom{k}{0}a^{k+1} + \binom{k}{1}a^kb + \binom{k}{2}a^{k-1}b^2 + \cdots + \binom{k}{k}ab^k\right] \\
&\quad + \left[\binom{k}{0}a^kb + \binom{k}{1}a^{k-1}b^2 + \cdots + \binom{k}{k-1}ab^k + \binom{k}{k}b^{k+1}\right] \\
&= \binom{k}{0}a^{k+1} + \left[\binom{k}{0} + \binom{k}{1}\right]a^kb + \left[\binom{k}{1} + \binom{k}{2}\right]a^{k-1}b^2 + \cdots \\
&\quad + \left[\binom{k}{k-1} + \binom{k}{k}\right]ab^k + \binom{k}{k}b^{k+1}
\end{aligned}$$

We now use the following facts (the proofs are left as exercises; see Problems 59–61, Exercise 6-6).

$$\binom{k}{r-1} + \binom{k}{r} = \binom{k+1}{r} \qquad \binom{k}{0} = \binom{k+1}{0} \qquad \binom{k}{k} = \binom{k+1}{k+1}$$

to rewrite the right side as

$$\begin{aligned}
&\binom{k+1}{0}a^{k+1} + \binom{k+1}{1}a^kb + \binom{k+1}{2}a^{k-1}b^2 + \cdots \\
&\qquad + \binom{k+1}{k}ab^k + \binom{k+1}{k+1}b^{k+1} = \sum_{j=0}^{k+1}\binom{k+1}{j}a^{k+1-j}b^j
\end{aligned}$$

Since the right side of the last equation is the right side of P_{k+1}, we have shown that P_{k+1} follows from P_k.

Conclusion

P_n is true. That is, the binomial formula holds for all positive integers n.

Answers to Matched Problems

1. $x^5 + 5x^4 + 10x^3 + 10x^2 + 5x + 1$ **2.** $8m^3 - 60m^2n + 150mn^2 - 125n^3$ **3.** $192{,}192y^8$ **4.** $3{,}060u^{14}$; $-31{,}824u^7$

EXERCISE 6-6

A

In Problems 1–8, use Pascal's triangle to evaluate each expression.

1. $\binom{5}{3}$ **2.** $\binom{6}{4}$

3. $\binom{4}{2}$ **4.** $\binom{7}{5}$

5. $C_{6,3}$ **6.** $C_{5,3}$

7. $C_{7,4}$ **8.** $C_{4,3}$

In Problems 9–16, use a graphing utility to evaluate each expression.

9. $\binom{9}{3}$ **10.** $\binom{10}{6}$

11. $\binom{12}{10}$ **12.** $\binom{13}{8}$

13. $\binom{17}{13}$ **14.** $\binom{20}{16}$

15. $\binom{50}{4}$ **16.** $\binom{50}{45}$

Expand Problems 17–28 using the binomial formula.

17. $(m + n)^3$ **18.** $(x + 2)^3$ **19.** $(2x - 3y)^3$

20. $(3u + 2v)^3$ **21.** $(x - 2)^4$ **22.** $(x - y)^4$

23. $(m + 3n)^4$ **24.** $(3p - q)^4$ **25.** $(2x - y)^5$

26. $(2x - 1)^5$ **27.** $(m + 2n)^6$ **28.** $(2x - y)^6$

B

In Problems 29–38, find the term of the binomial expansion containing the given power of x.

29. $(x + 1)^7$; x^4 **30.** $(x + 1)^8$; x^5 **31.** $(2x - 1)^{11}$; x^6

32. $(3x + 1)^{12}$; x^7 **33.** $(2x + 3)^{18}$; x^{14} **34.** $(3x - 2)^{17}$; x^5

35. $(x^2 - 1)^6$; x^8 **36.** $(x^2 - 1)^9$; x^7 **37.** $(x^2 + 1)^9$; x^{11}

38. $(x^2 + 1)^{10}$; x^{14}

In Problems 39–46, find the indicated term in each expansion if the terms of the expansion are arranged in decreasing powers of the first term in the binomial.

39. $(u + v)^{15}$; seventh term

40. $(a + b)^{12}$; fifth term

41. $(2m + n)^{12}$; eleventh term

42. $(x + 2y)^{20}$; third term

43. $[(w/2) - 2]^{12}$; seventh term

44. $(x - 3)^{10}$; fourth term

45. $(3x - 2y)^8$; sixth term

46. $(2p - 3q)^7$; fourth term

C

In Problems 47–50, use the binomial formula to expand and simplify the difference quotient

$$\frac{f(x + h) - f(x)}{h}$$

for the indicated function f. Discuss the behavior of the simplified form as h approaches 0.

47. $f(x) = x^3$ **48.** $f(x) = x^4$

49. $f(x) = x^5$ **50.** $f(x) = x^6$

In Problems 51–54, use a graphing utility to graph each sequence and to display it in table form.

51. Find the number of terms of the sequence

$$\binom{20}{0}, \binom{20}{1}, \binom{20}{2}, \ldots, \binom{20}{20}$$

that are greater than one-half of the largest term.

52. Find the number of terms of the sequence

$$\binom{40}{0}, \binom{40}{1}, \binom{40}{2}, \ldots, \binom{40}{40}$$

that are greater than one-half of the largest term.

53. (A) Find the largest term of the sequence $a_0, a_1, a_2, \ldots, a_{10}$ to three decimal places, where

$$a_k = \binom{10}{k}(0.6)^{10-k}(0.4)^k$$

(B) According to the binomial formula, what is the sum of the series $a_0 + a_1 + a_2 + \cdots + a_{10}$?

54. (A) Find the largest term of the sequence $a_0, a_1, a_2, \ldots, a_{10}$ to three decimal places, where

$$a_k = \binom{10}{k}(0.3)^{10-k}(0.7)^k.$$

(B) According to the binomial formula, what is the sum of the series $a_0 + a_1 + a_2 + \cdots + a_{10}$?

C

55. Evaluate $(1.01)^{10}$ to four decimal places, using the binomial formula. [*Hint:* Let $1.01 = 1 + 0.01$.]

56. Evaluate $(0.99)^6$ to four decimal places, using the binomial formula.

57. Show that: $\binom{n}{r} = \binom{n}{n-r}$

58. Show that: $\binom{n}{0} = \binom{n}{n}$

59. Show that: $\binom{k}{r-1} + \binom{k}{r} = \binom{k+1}{r}$

60. Show that: $\binom{k}{0} = \binom{k+1}{0}$

61. Show that: $\binom{k}{k} = \binom{k+1}{k+1}$

62. Show that: $\binom{n}{r}$ is given by the recursion formula

$$\binom{n}{r} = \frac{n-r+1}{r}\binom{n}{r-1}$$

where $\binom{n}{0} = 1$.

63. Write $2^n = (1+1)^n$ and expand, using the binomial formula to obtain

$$2^n = \binom{n}{0} + \binom{n}{1} + \binom{n}{2} + \cdots + \binom{n}{n}$$

Chapter 6 | Group Activity

Sequences Specified by Recursion Formulas

The recursion formula $a_n = 5a_{n-1} - 6a_{n-2}$, together with the initial values $a_1 = 4$, $a_2 = 14$, specifies the sequence $\{a_n\}$ whose first several terms are 4, 14, 46, 146, 454, 1394, ... The sequence $\{a_n\}$ is neither arithmetic nor geometric. Nevertheless, because it satisfies a simple recursion formula, it is possible to obtain an nth-term formula for $\{a_n\}$ that is analogous to the nth-term formulas for arithmetic and geometric sequences. Such an nth-term formula is valuable because it allows us to estimate a term of a sequence without computing all the preceding terms.

If the geometric sequence $\{r^n\}$ satisfies the recursion formula above, then $r^n = 5r^{n-1} - 6r^{n-2}$. Dividing by r^{n-2} leads to the quadratic equation $r^2 - 5r + 6 = 0$, whose solutions are $r = 2$ and $r = 3$. Now it is easy to check that the geometric sequences $\{2^n\} = 2, 4, 8, 16, \ldots$ and $\{3^n\} = 3, 9, 27, 81, \ldots$ satisfy the recursion formula. Therefore, any sequence of the form $\{u2^n + v3^n\}$, where u and v are constants, will satisfy the same recursion formula.

We now find u and v so that the first two terms of $\{u2^n + v3^n\}$ are $a_1 = 4$, $a_2 = 14$. Letting $n = 1$ and $n = 2$ we see that u and v must satisfy the following linear system:

$$2u + 3v = 4$$

$$4u + 9v = 14$$

Solving the system gives $u = -1$, $v = 2$. Therefore, an nth-term formula for the original sequence is $a_n = (-1)2^n + (2)3^n$.

Note that the nth-term formula was obtained by solving a quadratic equation and a system of two linear equations in two variables.

(A) Compute $(-1)2^n + (2)3^n$ for $n = 1, 2, \ldots, 6$, and compare with the terms of $\{a_n\}$.

(B) Estimate the one-hundredth term of $\{a_n\}$.

(C) Show that any sequence of the form $\{u2^n + v3^n\}$, where u and v are constants, satisfies the recursion formula $a_n = 5a_{n-1} - 6a_{n-2}$.

(D) Find an nth-term formula for the sequence $\{b_n\}$ that is specified by $b_1 = 5$, $b_2 = 55$, $b_n = 3b_{n-1} + 4b_{n-2}$.

(E) Find an nth-term formula for the Fibonacci sequence.

(F) Find an nth-term formula for the sequence $\{c_n\}$ that is specified by $c_1 = -3$, $c_2 = 15$, $c_3 = 99$, $c_n = 6c_{n-1} - 3c_{n-2} - 10c_{n-3}$. (Since the recursion formula involves the three terms which precede c_n, our method will involve the solution of a cubic equation and a system of three linear equations in three variables.)

CHAPTER 6 | REVIEW

6-1 SEQUENCES AND SERIES

A **sequence** is a function with the domain a set of successive integers. The symbol a_n, called the ***n*th term,** or **general term,** represents the range value associated with the domain value n. Unless specified otherwise, the domain is understood to be the set of natural numbers. A **finite sequence** has a finite domain, and an **infinite sequence** has an infinite domain. A **recursion formula** defines each term of a sequence in terms of one or more of the preceding terms. For example, the **Fibonacci sequence** is defined by $a_n = a_{n-1} + a_{n-2}$ for $n \ge 3$, where $a_1 = a_2 = 1$. If $a_1, a_2, \ldots, a_n, \ldots$ is a sequence, then the expression $a_1 + a_2 + \cdots + a_n + \cdots$ is called a **series.** A finite sequence produces a **finite series,** and an infinite sequence produces an **infinite series.** Series can be represented using the summation notation:

$$\sum_{k=m}^{n} a_k = a_m + a_{m+1} + \cdots + a_n$$

where k is called the **summing index.** If the terms in the series are alternately positive and negative, the series is called an **alternating series.**

6-2 MATHEMATICAL INDUCTION

A wide variety of statements can be proven using the **principle of mathematical induction:** Let P_n be a statement associated with each positive integer n and suppose the following conditions are satisfied:

1. P_1 is true.
2. For any positive integer k, if P_k is true, then P_{k+1} is also true.

Then the statement P_n is true for all positive integers n.

To use mathematical induction to prove statements involving laws of exponents, it is convenient to state a **recursive definition of a^n:**

$$a^1 = a \qquad \text{and} \qquad a^{n+1} = a^n a \qquad \text{for any integer } n > 1$$

To deal with conjectures that may be true only for $n \ge m$, where m is a positive integer, we use the **extended principle of mathematical induction:** Let m be a positive integer, let P_n be a statement associated with each integer $n \ge m$, and suppose the following conditions are satisfied:

1. P_m is true.
2. For any integer $k \ge m$, if P_k is true, then P_{k+1} is also true.

Then the statement P_n is true for all integers $n \ge m$.

6-3 ARITHMETIC AND GEOMETRIC SEQUENCES

A sequence is called an **arithmetic sequence,** or **arithmetic progression,** if there exists a constant d, called the **common difference,** such that

$$a_n - a_{n-1} = d \qquad \text{or} \qquad a_n = a_{n-1} + d \qquad \text{for every } n > 1$$

The following formulas are useful when working with arithmetic sequences and their corresponding series:

$a_n = a_1 + (n-1)d$ — ***n*th-Term Formula**

$S_n = \dfrac{n}{2}[2a_1 + (n-1)d]$ — **Sum Formula—First Form**

$S_n = \dfrac{n}{2}(a_1 + a_n)$ — **Sum Formula—Second Form**

A sequence is called a **geometric sequence,** or a **geometric progression,** if there exists a nonzero constant r, called the **common ratio,** such that

$$\frac{a_n}{a_{n-1}} = r \qquad \text{or} \qquad a_n = ra_{n-1} \qquad \text{for every } n > 1$$

The following formulas are useful when working with geometric sequences and their corresponding series:

$a_n = a_1 r^{n-1}$ — ***n*th-Term Formula**

$S_n = \dfrac{a_1 - a_1 r^n}{1 - r}$ $\quad r \ne 1$ — **Sum Formula—First Form**

$S_n = \dfrac{a_1 - ra_n}{1 - r}$ $\quad r \ne 1$ — **Sum Formula—Second Form**

$S_\infty = \dfrac{a_1}{1 - r}$ $\quad |r| < 1$ — **Sum of an Infinite Geometric Series**

6-4 MULTIPLICATION PRINCIPLE, PERMUTATIONS, AND COMBINATIONS

Given a sequence of operations, **tree diagrams** are often used to list all the possible combined outcomes. To count the number of combined outcomes without actually listing them, we use the **multiplication principle:**

1. If operations O_1 and O_2 are performed in order with N_1 possible outcomes for the first operation and N_2 possible outcomes for the second operation, then there are

   $N_1 \cdot N_2$

   possible outcomes of the first operation followed by the second.
2. In general, if n operations $O_1, O_2, \ldots, O_n$ are performed in order, with possible number of outcomes $N_1, N_2, \ldots, N_n$, respectively, then there are

   $N_1 \cdot N_2 \cdot \cdots \cdot N_n$

   possible combined outcomes of the operations performed in the given order.

A particular arrangement or ordering of n objects without repetition is called a **permutation.** The number of permutations of n objects is given by

$$P_{n,n} = n \cdot (n-1) \cdot \cdots \cdot 1 = n!$$

and the number of permutations of n objects taken r at a time is given by

$$P_{n,r} = \frac{n!}{(n-r)!} \qquad 0 \le r \le n$$

A **combination of a set of n elements taken r at a time** is an r-element subset of the n objects. The number of combinations of n objects taken r at a time is given by

$$C_{n,r} = \binom{n}{r} = \frac{P_{n,r}}{r!} = \frac{n!}{r!(n-r)!} \qquad 0 \le r \le n$$

In a permutation, order is important. In a combination, order is not important.

6-5 SAMPLE SPACES AND PROBABILITY

The outcomes of an experiment are called **simple events** if one and only one of these results will occur in each trial of the experiment. The set of all simple events is called the **sample space.** Any subset of the sample space is called an **event.** An event is a **simple event** if it has only one element in it and a **compound event** if it has more than one element in it. We say that **an event E occurs** if any of the simple events in E occurs. A sample space S_1 is **more fundamental** than a second sample space S_2 if knowledge of which event occurs in S_1 tells us which event in S_2 occurs, but not conversely.

Given a sample space $S = \{e_1, e_2, \dots, e_n\}$ with n simple events, to each simple event e_i we assign a real number, denoted by $P(e_i)$, that is called the **probability of the event e_i** and satisfies:

1. $0 \le P(e_i) \le 1$
2. $P(e_1) + P(e_2) + \cdots + P(e_n) = 1$

Any probability assignment that meets conditions 1 and 2 is said to be an **acceptable probability assignment.**

Given an acceptable probability assignment for the simple events in a sample space S, the **probability of an arbitrary event E** is defined as follows:

1. If E is the empty set, then $P(E) = 0$.
2. If E is a simple event, then $P(E)$ has already been assigned.
3. If E is a compound event, then $P(E)$ is the sum of the probabilities of all the simple events in E.
4. If E is the sample space S, then $P(E) = P(S) = 1$.

If each of the simple events in a sample space $S = \{e_1, e_2, \dots, e_n\}$ with n simple events is **equally likely** to occur, then we assign the probability $1/n$ to each. If E is an arbitrary event in S, then

$$P(E) = \frac{\text{Number of elements in } E}{\text{Number of elements in } S} = \frac{n(E)}{n(S)}$$

If we conduct an experiment n times and event E occurs with **frequency** $f(E)$, then the ratio $f(E)/n$ is called the **relative frequency** of the occurrence of event E in n trials. As n increases, $f(E)/n$ usually approaches a number that is called the **empirical probability** $P(E)$. Thus, $f(E)/n$ is used as an **approximate empirical probability** for $P(E)$.

If $P(E)$ is the theoretical probability of an event E and the experiment is performed n times, then the **expected frequency** of the occurrence of E is $n \cdot P(E)$.

The command **randInt(*i*,*k*,*n*)** on a Texas Instruments TI-83 generates a random sequence of n integers between i and k, inclusively, that can be used to simulate repeated trials of experiments.

6-6 BINOMIAL FORMULA

Pascal's triangle is a triangular array of coefficients for the expansion of the binomial $(a + b)^n$, where n is a positive integer. New notation for the combination formula is

$$\binom{n}{r} = C_{n,r} = \frac{n!}{r!(n-r)!}$$

For n a positive integer, the **binomial formula** is

$$(a + b)^n = \sum_{k=0}^{n} \binom{n}{k} a^{n-k} b^k$$

The numbers $\binom{n}{k}$, $0 \le k \le n$, are called **binomial coefficients.**

CHAPTER 6 | REVIEW EXERCISES

Work through all the problems in this chapter review and check answers in the back of the book. Answers to all review problems are there, and following each answer is a number in italics indicating the section in which that type of problem is discussed. Where weaknesses show up, review appropriate sections in the text.

A

1. Determine whether each of the following can be the first three terms of a geometric sequence, an arithmetic sequence, or neither.
   (A) $16, -8, 4, \dots$ (B) $5, 7, 9, \dots$
   (C) $-8, -5, -2, \dots$ (D) $2, 3, 5, \dots$
   (E) $-1, 2, -4, \dots$

In Problems 2–5:
(A) Write the first four terms of each sequence.
(B) Find a_{10}. *(C) Find S_{10}.*

2. $a_n = 2n + 3$
3. $a_n = 32(\frac{1}{2})^n$
4. $a_1 = -8;\ a_n = a_{n-1} + 3,\ n \ge 2$
5. $a_1 = -1;\ a_n = (-2)a_{n-1},\ n \ge 2$
6. Find S_∞ in Problem 3.

Evaluate Problems 7–10.

7. $6!$
8. $\dfrac{22!}{19!}$
9. $\dfrac{7!}{2!(7-2)!}$
10. $C_{6,2}$ and $P_{6,2}$
11. A single die is rolled and a coin is flipped. How many combined outcomes are possible? Solve
    (A) By using a tree diagram
    (B) By using the multiplication principle

12. How many seating arrangements are possible with 6 people and 6 chairs in a row? Solve by using the multiplication principle.

13. Solve Problem 12 using permutations or combinations, whichever is applicable.

14. In a single deal of 5 cards from a standard 52-card deck, what is the probability of being dealt 5 clubs?

15. Betty and Bill are members of a 15-person ski club. If the president and treasurer are selected by lottery, what is the probability that Betty will be president and Bill will be treasurer? A person cannot hold more than one office.

16. A drug has side effects for 50 out of 1,000 people in a test. What is the approximate empirical probability that a person using the drug will have side effects?

Verify Problems 17–19 for $n = 1, 2,$ and 3.

17. P_n: $5 + 7 + 9 + \cdots + (2n + 3) = n^2 + 4n$

18. P_n: $2 + 4 + 8 + \cdots + 2^n = 2^{n+1} - 2$

19. P_n: $49^n - 1$ is divisible by 6

In Problems 20–22, write P_k and P_{k+1}.

20. For P_n in Problem 17

21. For P_n in Problem 18

22. For P_n in Problem 19

23. Either prove the statement is true or prove it is false by finding a counterexample: If n is a positive integer, then the sum of the series $1 + \frac{1}{2} + \frac{1}{3} + \cdots + \frac{1}{n}$ is less than 4.

B

Write Problems 24 and 25 without summation notation, and find the sum.

24. $S_{10} = \sum_{k=1}^{10} (2k - 8)$

25. $S_7 = \sum_{k=1}^{7} \frac{16}{2^k}$

26. $S_\infty = 27 - 18 + 12 + \cdots = ?$

27. Write

$$S_n = \frac{1}{3} - \frac{1}{9} + \frac{1}{27} + \cdots + \frac{(-1)^{n+1}}{3^n}$$

using summation notation, and find S_∞.

28. Someone tells you that the following approximate empirical probabilities apply to the sample space $\{e_1, e_2, e_3, e_4\}$: $P(e_1) \approx .1$, $P(e_2) \approx -.2$, $P(e_3) \approx .6$, $P(e_4) \approx 2$. There are three reasons why P cannot be a probability function. Name them.

29. Six distinct points are selected on the circumference of a circle. How many triangles can be formed using these points as vertices?

30. In an arithmetic sequence, $a_1 = 13$ and $a_7 = 31$. Find the common difference d and the fifth term a_5.

31. How many 3-letter code words are possible using the first 8 letters of the alphabet if no letter can be repeated? If letters can be repeated? If adjacent letters cannot be alike?

32. Two coins are flipped 1,000 times with the following frequencies:

2 heads:	210
1 head:	480
0 heads:	310

(A) Compute the empirical probability for each outcome.
(B) Compute the theoretical probability for each outcome.
(C) Using the theoretical probabilities computed in part B, compute the expected frequency of each outcome, assuming fair coins.

33. From a standard deck of 52 cards, what is the probability of obtaining a 5-card hand:
(A) Of all diamonds?
(B) Of 3 diamonds and 2 spades?
Write answers in terms of $C_{n,r}$ or $P_{n,r}$, as appropriate. Do not evaluate.

34. A group of 10 people includes one married couple. If 4 people are selected at random, what is the probability that the married couple is selected?

35. A spinning device has three numbers, 1, 2, 3, each as likely to turn up as the other. If the device is spun twice, what is the probability that:
(A) The same number turns up both times?
(B) The sum of the numbers turning up is 5?

36. Use the formula for the sum of an infinite geometric series to write $0.727\,272 \cdots = 0.\overline{72}$ as the quotient of two integers.

37. Solve the following problems using $P_{n,r}$ or $C_{n,r}$, as appropriate:
(A) How many 3-digit opening combinations are possible on a combination lock with 6 digits if the digits cannot be repeated?
(B) Suppose 5 tennis players have made the finals. If each of the 5 players is to play every other player exactly once, how many games must be scheduled?

Evaluate Problems 38–40.

38. $\frac{20!}{18!(20 - 18)!}$

39. $\binom{16}{12}$

40. $\binom{11}{11}$

41. Expand $(x - y)^5$ using the binomial formula.

42. Find the term containing x^6 in the expansion of $(x + 2)^9$.

43. If the terms in the expansion of $(2x - y)^{12}$ are arranged in descending powers of x, find the tenth term.

Establish each statement in Problems 44–46 for all natural numbers, using mathematical induction.

44. P_n in Problem 17

45. P_n in Problem 18

46. P_n in Problem 19

In Problems 47 and 48, find the smallest positive integer n such that $a_n < b_n$ by graphing the sequences $\{a_n\}$ and $\{b_n\}$ with a graphing utility. Check your answer by using a graphing utility to display both sequences in table form.

47. $a_n = C_{50,n}$, $b_n = 3^n$

48. $a_1 = 100$, $a_n = 0.99a_{n-1} + 5$, $b_n = 9 + 7n$

C

49. How many different families with 5 children are possible, excluding multiple births, where the sex of each child in the order of their birth is taken into consideration? How many families are possible if the order pattern is not taken into account?

50. A free-falling body travels $g/2$ feet in the first second, $3g/2$ feet during the next second, $5g/2$ feet the next, and so on. Find the distance fallen during the twenty-fifth second and the total distance fallen from the start to the end of the twenty-fifth second.

51. How many ways can 2 people be seated in a row of 4 chairs?

52. Expand $(x + i)^6$, where i is the imaginary unit, using the binomial formula.

53. If 3 people are selected from a group of 7 men and 3 women, what is the probability that at least 1 woman is selected?

54. Three fair coins are tossed 1,000 times with the following frequencies of outcomes:

Number of heads	0	1	2	3
Frequency	120	360	350	170

(A) What is the approximate empirical probability of obtaining 2 heads?
(B) What is the theoretical probability of obtaining 2 heads?
(C) What is the expected frequency of obtaining 2 heads?

Prove that each statement in Problems 55–59 holds for all positive integers, using mathematical induction.

55. $\sum_{k=1}^{n} k^3 = \left(\sum_{k=1}^{n} k\right)^2$

56. $x^{2n} - y^{2n}$ is divisible by $x - y$, $x \neq y$

57. $\dfrac{a^n}{a^m} = a^{n-m}$; $n > m$, n, m positive integers

58. $\{a_n\} = \{b_n\}$, where $a_n = a_{n-1} + 2$, $a_1 = -3$, $b_n = -5 + 2n$

59. $(1!)1 + (2!)2 + (3!)3 + \cdots + (n!)n = (n + 1)! - 1$ (From U.S.S.R. Mathematical Olympiads, 1955–1956, Grade 10.)

APPLICATIONS

60. Loan Repayment. You borrow \$7,200 and agree to pay 1% of the unpaid balance each month for interest. If you decide to pay an additional \$300 each month to reduce the unpaid balance, how much interest will you pay over the 24 months it will take to repay this loan?

61. Economics. Due to reduced taxes, an individual has an extra \$2,400 in spendable income. If we assume that the individual spends 75% of this on consumer goods, and the producers of those consumer goods in turn spend 75% on consumer goods, and that this process continues indefinitely, what is the total amount (to the nearest dollar) spent on consumer goods?

62. Compound Interest. If \$500 is invested at 6% compounded annually, the amount A present after n years forms a geometric sequence with common ratio $1 + 0.06 = 1.06$. Use a geometric sequence formula to find the amount A in the account (to the nearest cent) after 10 years. After 20 years.

63. Transportation. A distribution center A wishes to distribute its products to 5 different retail stores, B, C, D, E, and F, in a city. How many different route plans can be constructed so that a single truck can start from A, deliver to each store exactly once, and then return to the center?

64. Market Analysis. A videocassette company selected 1,000 persons at random and surveyed them to determine a relationship between age of purchaser and annual videocassette purchases. The results are given in the table.

	Cassettes Purchased Annually				
Age	0	1	2	Above 2	Totals
Under 12	60	70	30	10	170
12–18	30	100	100	60	290
19–25	70	110	120	30	330
Over 25	100	50	40	20	210
Totals	260	330	290	120	1,000

Find the empirical probability that a person selected at random
(A) Is over 25 *and* buys exactly 2 cassettes annually.
(B) Is 12–18 years old *and* buys more than 1 cassette annually.
(C) Is 12–18 years old *or* buys more than 1 cassette annually.

★ **65. Quality Control.** Twelve precision parts, including 2 that are substandard, are sent to an assembly plant. The plant manager selects 4 at random and will return the whole shipment if 1 or more of the sample are found to be substandard. What is the probability that the shipment will be returned?

ANSWER TO APPLICATION

Since the man's first child was a boy, the equally likely sample space of all possible two-child families for the man is $S = \{BB, BG\}$ and the probability that he has two boys is $\frac{1}{2}$. Since the woman has at least one boy, the equally likely sample space of all possible two-child families for the woman is $S = \{BB, BG, GB\}$ and the probability that she has two boys is $\frac{1}{3}$.

www.mhhe.com/barnett

Additional Topics in Analytic Geometry

7

Outline

Application

A solar cooker in the shape of a paraboloid generates high temperatures by reflecting the rays of sunlight that enter the cooker into the focus of the paraboloid. If a solar cooker is 20 inches wide and 10 inches deep, how far from the bottom of the cooker is the focus of the paraboloid?

Analytic geometry, a union of geometry and algebra, enables us to analyze certain geometric concepts algebraically and to interpret certain algebraic relationships geometrically. Our two main concerns center around graphing algebraic equations and finding equations of useful geometric figures. We have discussed a number of topics in analytic geometry, such as straight lines and circles, in earlier chapters. In this chapter we discuss additional analytic geometry topics: conic sections and translation of axes.

René Descartes (1596–1650), the French philosopher–mathematician, is generally recognized as the founder of analytic geometry.

Preparing for This Chapter

Before getting started on this chapter, review the following concepts:

Graphs and Transformations (Chapter 1, Section 5)

Cartesian Coordinate System (Chapter 1, Section 1)

Linear Functions (Chapter 2, Section 1)

Quadratic Functions (Chapter 2, Section 3)

Quadratic Equations (Chapter 2, Section 5)

Equation Solving Techniques (Chapter 2, Section 6)

Asymptotes (Chapter 3, Section 4)

Section 7-1 Conic Sections; Parabola

- Conic Sections
- Definition of a Parabola
- Drawing a Parabola
- Standard Equations and Their Graphs
- Applications

In this section we introduce the general concept of a conic section and then discuss the particular conic section called a *parabola*. In the next two sections we will discuss two other conic sections called *ellipses* and *hyperbolas*.

Conic Sections

In Section 2-1 we found that the graph of a first-degree equation in two variables,

$$Ax + By = C \tag{1}$$

where A and B are not both 0, is a straight line, and every straight line in a rectangular coordinate system has an equation of this form. What kind of graph will a second-degree equation in two variables,

$$Ax^2 + Bxy + Cy^2 + Dx + Ey + F = 0 \tag{2}$$

where A, B, and C are not all 0, yield for different sets of values of the coefficients? The graphs of equation (2) for various choices of the coefficients are plane curves obtainable by intersecting a cone* with a plane, as shown in Figure 1. These curves are called **conic sections.**

FIGURE 1
Conic sections.

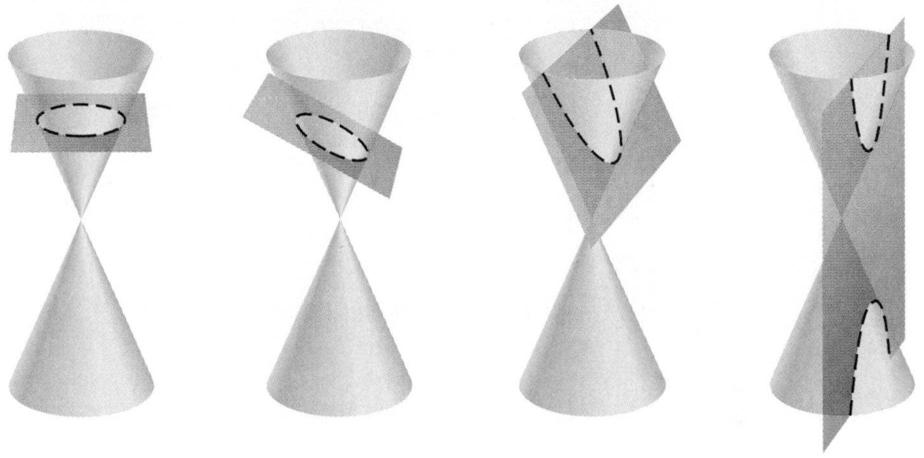

If a plane cuts clear through one nappe, then the intersection curve is called a **circle** if the plane is perpendicular to the axis and an **ellipse** if the plane is not perpendicular to the axis. If a plane cuts only one nappe, but does not cut clear through, then the intersection curve is called a **parabola.** Finally, if a plane cuts through both nappes, but not through the vertex, the resulting intersection curve is called a **hyperbola.** A plane passing through the vertex of the cone produces a **degenerate conic**—a point, a line, or a pair of lines.

Conic sections are very useful and are readily observed in your immediate surroundings: wheels (circle), the path of water from a garden hose (parabola), some serving platters (ellipses), and the shadow on a wall from a light surrounded by a cylindrical or conical lamp shade (hyperbola) are some examples (see Fig. 2). We will discuss many applications of conics throughout the remainder of this chapter.

FIGURE 2
Examples of conics.

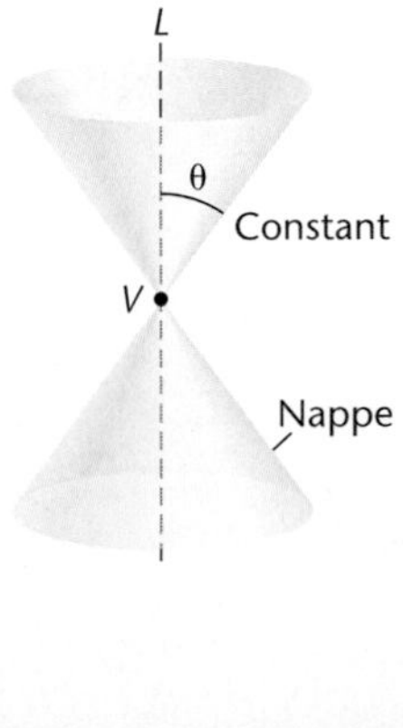

*Starting with a fixed line L and a fixed point V on L, the surface formed by all straight lines through V making a constant angle θ with L is called a **right circular cone.** The fixed line L is called the **axis** of the cone, and V is its **vertex.** The two parts of the cone separated by the vertex are called **nappes.**

A definition of a conic section that does not depend on the coordinates of points in any coordinate system is called a **coordinate-free definition.** In Section 1-1 we gave a coordinate-free definition of a circle and developed its standard equation in a rectangular coordinate system. In this and the next two sections we will give coordinate-free definitions of a parabola, ellipse, and hyperbola, and we will develop standard equations for each of these conics in a rectangular coordinate system.

Definition of a Parabola

The following definition of a parabola does not depend on the coordinates of points in any coordinate system:

DEFINITION 1

PARABOLA

A **parabola** is the set of all points in a plane equidistant from a fixed point F and a fixed line L in the plane. The fixed point F is called the **focus,** and the fixed line L is called the **directrix.** A line through the focus perpendicular to the directrix is called the **axis,** and the point on the axis halfway between the directrix and focus is called the **vertex.**

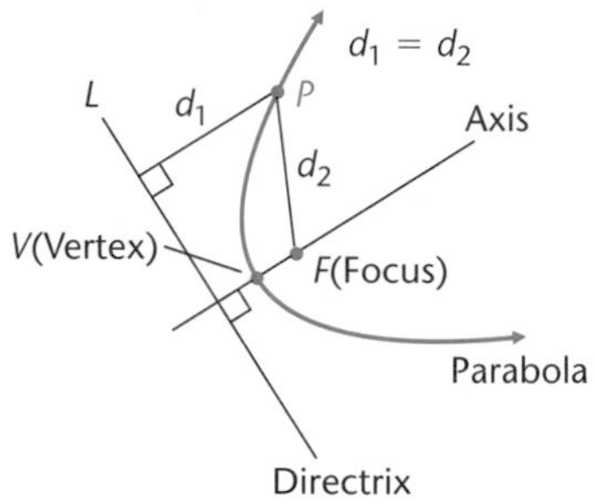

Drawing a Parabola

Using the definition, we can draw a parabola with fairly simple equipment—a straightedge, a right-angle drawing triangle, a piece of string, a thumbtack, and a pencil. Referring to Figure 3, tape the straightedge along the line AB and place the thumbtack above the line AB. Place one leg of the triangle along the straightedge as indicated, then take a piece of string the same length as the other leg, tie one end to the thumbtack, and fasten the other end with tape at C on the triangle. Now press the string to the edge of the triangle, and keeping the string taut, slide the triangle along the straightedge. Since DE will always equal DF, the resulting curve will be part of a parabola with directrix AB lying along the straightedge and focus F at the thumbtack.

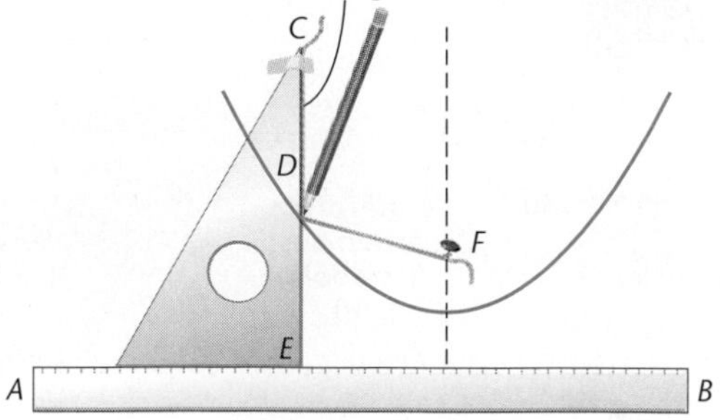

FIGURE 3
Drawing a parabola.

Explore/Discuss 1

The line through the focus F that is perpendicular to the axis of a parabola intersects the parabola in two points G and H. Explain why the distance from G to H is twice the distance from F to the directrix of the parabola.

Standard Equations and Their Graphs

Using the definition of a parabola and the distance-between-two-points formula

$$d = \sqrt{(x_2 - x_1)^2 + (y_2 - y_1)^2} \tag{3}$$

we can derive simple standard equations for a parabola located in a rectangular coordinate system with its vertex at the origin and its axis along a coordinate axis. We start with the axis of the parabola along the x axis and the focus at $F(a, 0)$. We locate the parabola in a coordinate system as in Figure 4 and label key lines and points. This is an important step in finding an equation of a geometric figure in a coordinate system. Note that the parabola opens to the right if $a > 0$ and to the left if $a < 0$. The vertex is at the origin, the directrix is $x = -a$, and the coordinates of M are $(-a, y)$.

FIGURE 4
Parabola with center at the origin and axis the x axis.

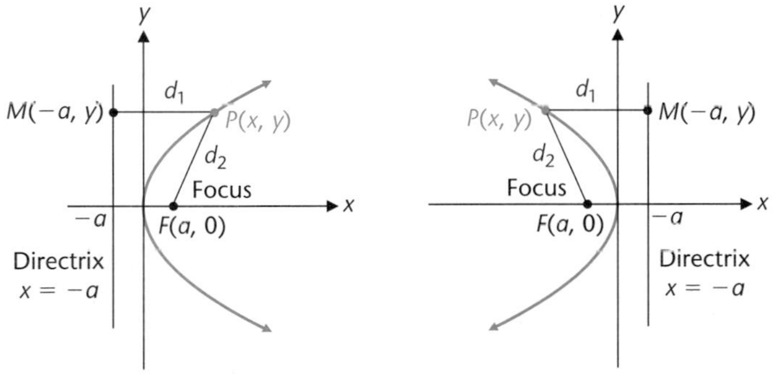

$a > 0$, focus on positive x axis
(a)

$a < 0$, focus on negative x axis
(b)

The point $P(x, y)$ is a point on the parabola if and only if

$$\begin{aligned} d_1 &= d_2 \\ d(P, M) &= d(P, F) \\ \sqrt{(x + a)^2 + (y - y)^2} &= \sqrt{(x - a)^2 + (y - 0)^2} && \text{Use equation (3).} \\ (x + a)^2 &= (x - a)^2 + y^2 && \text{Square both sides.} \\ x^2 + 2ax + a^2 &= x^2 - 2ax + a^2 + y^2 && \text{Simplify.} \\ y^2 &= 4ax \end{aligned} \tag{4}$$

Equation (4) is the standard equation of a parabola with vertex at the origin, axis the x axis, and focus at $(a, 0)$.

Now we locate the vertex at the origin and focus on the y axis at $(0, a)$. Looking at Figure 5, we note that the parabola opens upward if $a > 0$ and downward if $a < 0$. The directrix is $y = -a$, and the coordinates of N are $(x, -a)$. The point $P(x, y)$ is a point on the parabola if and only if

$$d_1 = d_2$$

$$d(P, N) = d(P, F)$$

$$\sqrt{(x - x)^2 + (y + a)^2} = \sqrt{(x - 0)^2 + (y - a)^2} \quad \text{Use equation (3).}$$

$$(y + a)^2 = x^2 + (y - a)^2 \quad \text{Square both sides.}$$

$$y^2 + 2ay + a^2 = x^2 + y^2 - 2ay + a^2 \quad \text{Simplify.}$$

$$x^2 = 4ay \tag{5}$$

Equation (5) is the standard equation of a parabola with vertex at the origin, axis the y axis, and focus at $(0, a)$.

FIGURE 5
Parabola with center at the origin and axis the y axis.

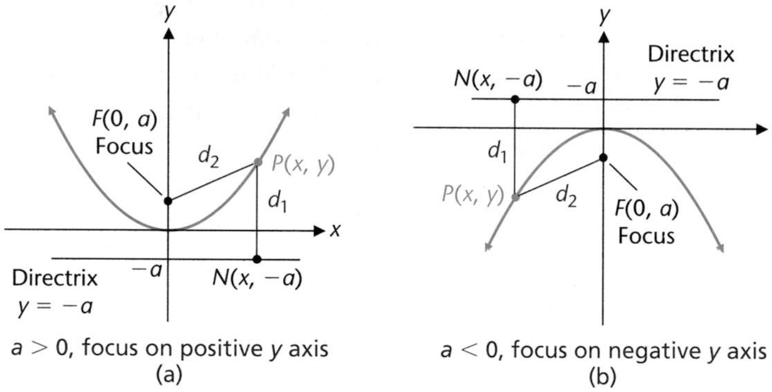

We summarize these results for easy reference in Theorem 1.

THEOREM 1

STANDARD EQUATIONS OF A PARABOLA WITH VERTEX AT (0, 0)

1. $y^2 = 4ax$
   Vertex: $(0, 0)$
   Focus: $(a, 0)$
   Directrix: $x = -a$
   Symmetric with respect to the x axis
   Axis the x axis

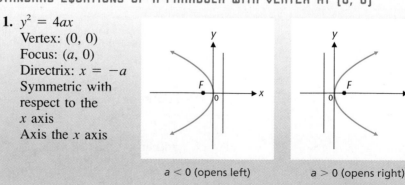

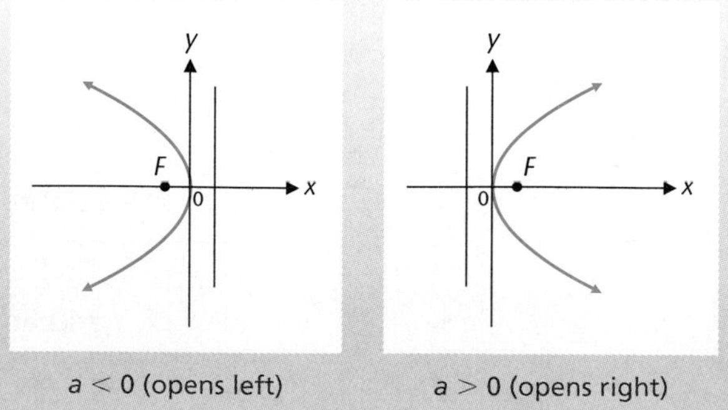

THEOREM 1 continued

2. $x^2 = 4ay$
Vertex: (0, 0)
Focus: (0, a)
Directrix: $y = -a$
Symmetric with respect to the y axis
Axis the y axis

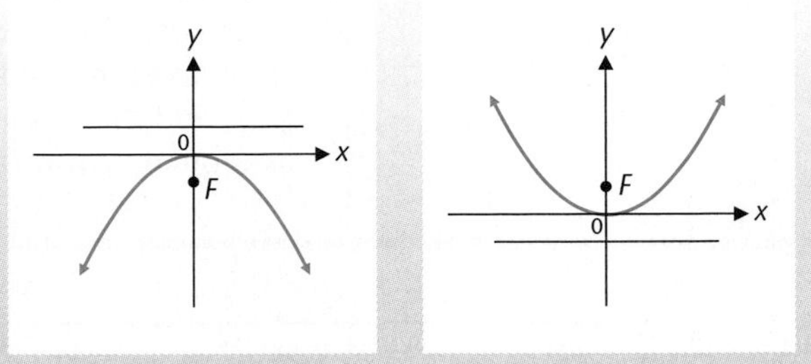

EXAMPLE 1 Graphing $x^2 = 4ay$

Graph $x^2 = -16y$, and locate the focus and directrix.

Solution To graph $x^2 = -16y$, it is convenient to assign y values that make the right side a perfect square, and solve for x. Note that y must be 0 or negative for x to be real. Since the coefficient of y is negative, a must be negative, and the parabola opens downward (Fig. 6).

x	0	± 4	± 8
y	0	-1	-4

Focus: $x^2 = -16y = 4\overset{a}{(-4)}y$

$F(0, a) = F(0, -4)$

Directrix: $y = -a$

$= -(-4) = 4$

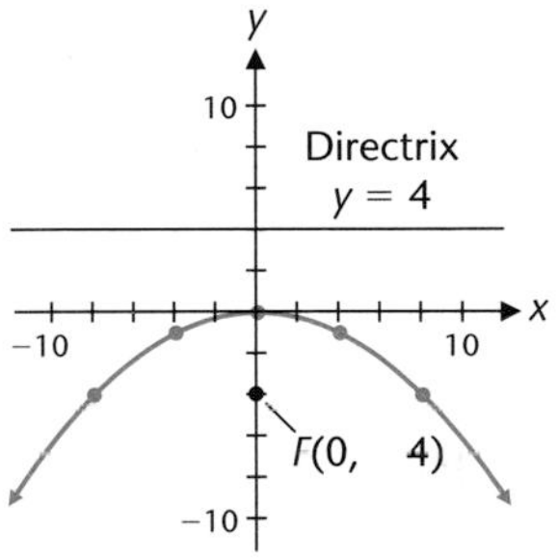

FIGURE 6
$x^2 = -16y$.

MATCHED PROBLEM 1

Graph $y^2 = -8x$, and locate the focus and directrix.

Remark To graph the equation $x^2 = -16y$ of Example 1 on a graphing utility, we first solve the equation for y and then graph the function $y = -\frac{1}{16}x^2$. If that same approach is used to graph the equation $y^2 = -8x$ of Matched Problem 1, then $y = \pm\sqrt{-8x}$, and there are two functions to graph. The graph of $y = \sqrt{-8x}$ is the upper half of the parabola, and the graph of $y = -\sqrt{-8x}$ is the lower half (see Fig. 7).

FIGURE 7

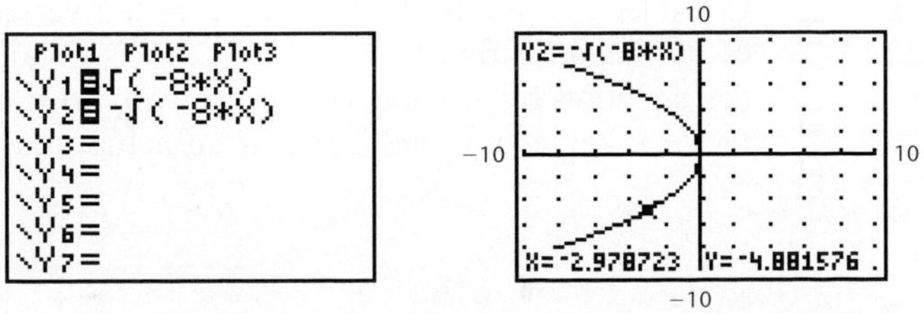

CAUTION

A common error in making a quick sketch of $y^2 = 4ax$ or $x^2 = 4ay$ is to sketch the first with the y axis as its axis and the second with the x axis as its axis. The graph of $y^2 = 4ax$ is symmetric with respect to the x axis, and the graph of $x^2 = 4ay$ is symmetric with respect to the y axis, as a quick symmetry check will reveal.

EXAMPLE 2 Finding the Equation of a Parabola

(A) Find the equation of a parabola having the origin as its vertex, the y axis as its axis, and $(-10, -5)$ on its graph.

(B) Find the coordinates of its focus and the equation of its directrix.

Solutions

(A) The parabola is opening down and has an equation of the form $x^2 = 4ay$. Since $(-10, -5)$ is on the graph, we have

$$\begin{aligned} x^2 &= 4ay \\ (-10)^2 &= 4a(-5) \\ 100 &= -20a \\ a &= -5 \end{aligned}$$

Thus, the equation of the parabola is

$$\begin{aligned} x^2 &= 4(-5)y \\ &= -20y \end{aligned}$$

(B) Focus: $x^2 = -20y = 4(\overset{a}{-5})y$

$$F(0, a) = F(0, -5)$$

Directrix:
$$\begin{aligned} y &= -a \\ &= -(-5) \\ &= 5 \end{aligned}$$

MATCHED PROBLEM 2

(A) Find the equation of a parabola having the origin as its vertex, the x axis as its axis, and $(4, -8)$ on its graph.

(B) Find the coordinates of its focus and the equation of its directrix.

Explore/Discuss 2

Consider the graph of an equation in the variables x and y. The equation of its magnification by a factor $k > 0$ is obtained by replacing x and y in the equation by x/k and y/k, respectively. (Of course, a magnification by a factor k between 0 and 1 means an actual reduction in size.)

Explore/Discuss 2 continued

(A) Show that the magnification by a factor 3 of the circle with equation $x^2 + y^2 = 1$ has equation $x^2 + y^2 = 9$.

(B) Explain why every circle with center at (0, 0) is a magnification of the circle with equation $x^2 + y^2 = 1$.

(C) Find the equation of the magnification by a factor 3 of the parabola with equation $x^2 = y$. Graph both equations.

(D) Explain why every parabola with vertex (0, 0) that opens upward is a magnification of the parabola with equation $x^2 = y$.

Applications

Parabolic forms are frequently encountered in the physical world. Suspension bridges, arch bridges, microphones, symphony shells, satellite antennas, radio and optical telescopes, radar equipment, solar furnaces, and searchlights are only a few of many items that utilize parabolic forms in their design.

Figure 8(a) illustrates a parabolic reflector used in all reflecting telescopes—from 3- to 6-inch home type to the 200-inch research instrument on Mount Palomar in California. Parallel light rays from distant celestial bodies are reflected to the focus off a parabolic mirror. If the light source is the sun, then the parallel rays are focused at F and we have a solar furnace. Temperatures of over 6,000°C have been achieved by such furnaces. If we locate a light source at F, then the rays in Figure 8(a) reverse, and we have a spotlight or a searchlight. Automobile headlights can use parabolic reflectors with special lenses over the light to diffuse the rays into useful patterns.

Figure 8(b) shows a suspension bridge, such as the Golden Gate Bridge in San Francisco. The suspension cable is a parabola. It is interesting to note that a free-hanging cable, such as a telephone line, does not form a parabola. It forms another curve called a *catenary.*

Figure 8(c) shows a concrete arch bridge. If all the loads on the arch are to be compression loads (concrete works very well under compression), then using physics and advanced mathematics, it can be shown that the arch must be parabolic.

FIGURE 8 Uses of parabolic forms.

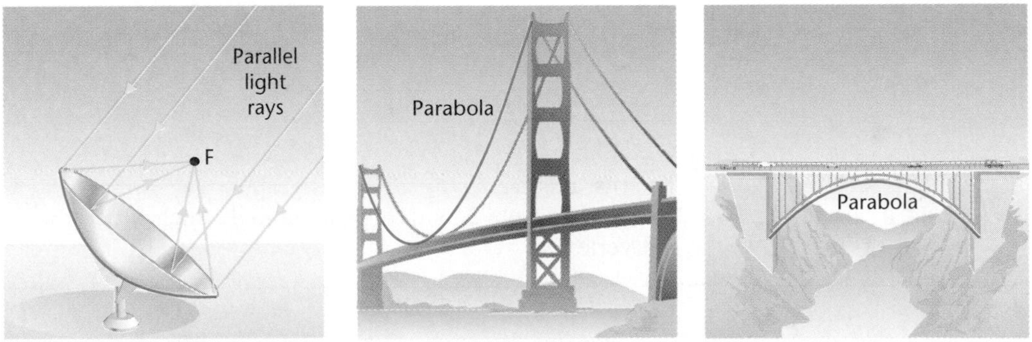

Parabolic reflector (a)

Suspension bridge (b)

Arch bridge (c)

EXAMPLE 3

Parabolic Reflector

A **paraboloid** is formed by revolving a parabola about its axis. A spotlight in the form of a paraboloid 5 inches deep has its focus 2 inches from the vertex. Find, to one decimal place, the radius R of the opening of the spotlight.

Solution

Step 1. Locate a parabolic cross section containing the axis in a rectangular coordinate system, and label all known parts and parts to be found. This is a very important step and can be done in infinitely many ways. Since we are in charge, we can make things simpler for ourselves by locating the vertex at the origin and choosing a coordinate axis as the axis. We choose the y axis as the axis of the parabola with the parabola opening upward (see Fig. 9).

FIGURE 9

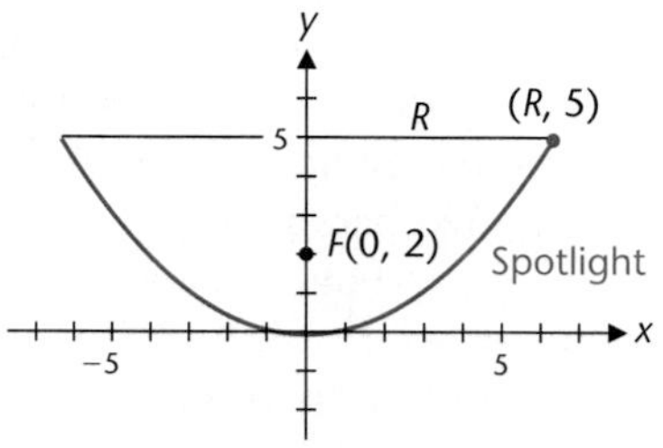

Step 2. Find the equation of the parabola in the figure. Since the parabola has the y axis as its axis and the vertex at the origin, the equation is of the form

$$x^2 = 4ay$$

We are given $F(0, a) = F(0, 2)$; thus, $a = 2$, and the equation of the parabola is

$$x^2 = 8y$$

Step 3. Use the equation found in step 2 to find the radius R of the opening. Since $(R, 5)$ is on the parabola, we have

$$R^2 = 8(5)$$
$$R = \sqrt{40} \approx 6.3 \text{ inches}$$

MATCHED PROBLEM 3

Repeat Example 3 with a paraboloid 12 inches deep and a focus 9 inches from the vertex.

Answers to Matched Problems

1. Focus: $(-2, 0)$
Directrix: $x = 2$

x	0	-2
y	0	± 4

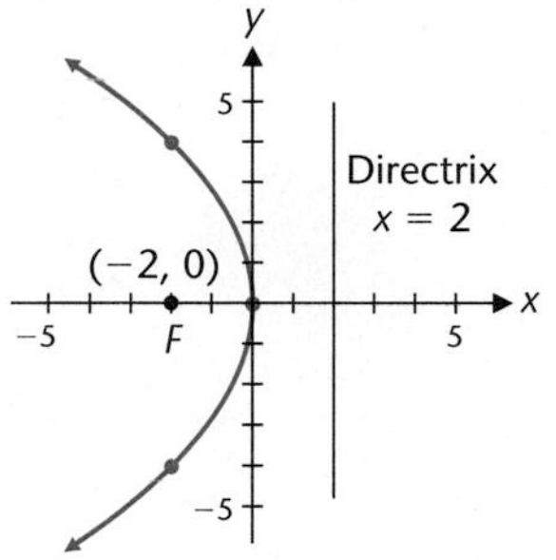

2. (A) $y^2 = 16x$ (B) Focus: $(4, 0)$; Directrix: $x = -4$ 3. $R = 20.8$ in.

EXERCISE 7-1

A

In Problems 1–12, graph each equation, and locate the focus and directrix. Check by graphing on a graphing utility.

1. $y^2 = 4x$
2. $y^2 = 8x$
3. $x^2 = 8y$
4. $x^2 = 4y$
5. $y^2 = -12x$
6. $y^2 = -4x$
7. $x^2 = -4y$
8. $x^2 = -8y$
9. $y^2 = -20x$
10. $x^2 = -24y$
11. $x^2 = 10y$
12. $y^2 = 6x$

Find the coordinates to two decimal places of the focus for each parabola in Problems 13–18.

13. $y^2 = 39x$
14. $x^2 = 58y$
15. $x^2 = -105y$
16. $y^2 = -93x$
17. $y^2 = -77x$
18. $x^2 = -205y$

B

In Problems 19–26, find the equation of a parabola with vertex at the origin, axis the x or y axis, and

19. Directrix $y = -3$
20. Directrix $y = 4$
21. Focus $(0, -7)$
22. Focus $(0, 5)$
23. Directrix $x = 6$
24. Directrix $x = -9$
25. Focus $(2, 0)$
26. Focus $(-4, 0)$

In Problems 27–32, find the equation of the parabola having its vertex at the origin, its axis as indicated, and passing through the indicated point.

27. y axis; $(4, 2)$
28. x axis; $(4, 8)$
29. x axis; $(-3, 6)$
30. y axis; $(-5, 10)$
31. y axis; $(-6, -9)$
32. x axis; $(-6, -12)$

In Problems 33–36, find the first-quadrant points of intersection for each system of equations to three decimal places.

33. $x^2 = 4y$
$y^2 = 4x$
34. $y^2 = 3x$
$x^2 = 3y$
35. $y^2 = 6x$
$x^2 = 5y$
36. $x^2 = 7y$
$y^2 = 2x$

37. Consider the parabola with equation $x^2 = 4ay$.
(A) How many lines through $(0, 0)$ intersect the parabola in exactly one point? Find their equations.
(B) Find the coordinates of all points of intersection of the parabola with the line through $(0, 0)$ having slope $m \neq 0$.

38. Find the coordinates of all points of intersection of the parabola with equation $x^2 = 4ay$ and the parabola with equation $y^2 = 4bx$.

39. The line segment AB through the focus in the figure is called a **focal chord** of the parabola. Find the coordinates of A and B.

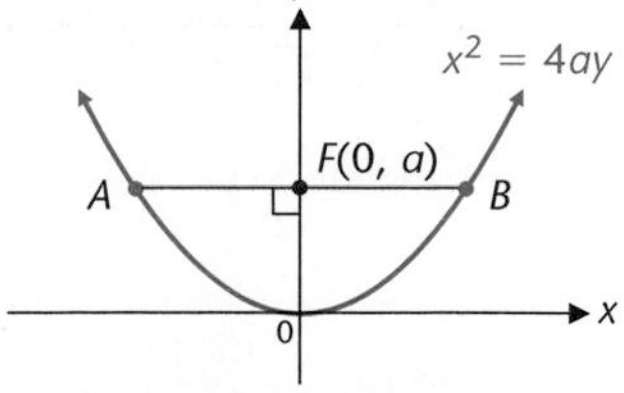

40. The line segment AB through the focus in the figure is called a **focal chord** of the parabola. Find the coordinates of A and B.

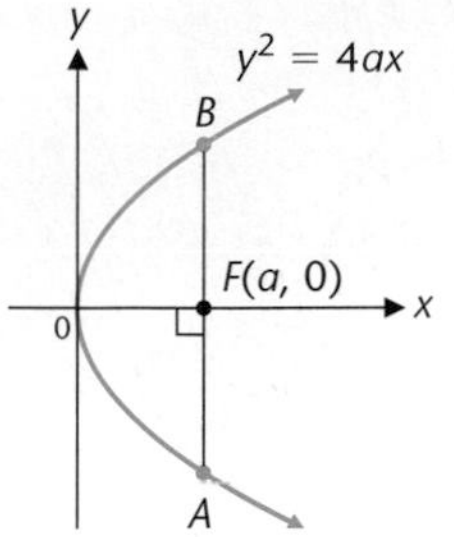

C

In Problems 41–44, use the definition of a parabola and the distance formula to find the equation of a parabola with

41. Directrix $y = -4$ and focus $(2, 2)$

42. Directrix $y = 2$ and focus $(-3, 6)$

43. Directrix $x = 2$ and focus $(6, -4)$

44. Directrix $x = -3$ and focus $(1, 4)$

In Problems 45–48, find the coordinates of all points of intersection to two decimal places.

45. $x^2 = 8y, y = 5x + 4$

46. $x^2 = 3y, 7x + 4y = 11$

47. $x^2 = -8y, y^2 = -5x$

48. $y^2 = 6x, 2x - 9y = 13$

APPLICATIONS

49. **Engineering.** The parabolic arch in the concrete bridge in the figure must have a clearance of 50 feet above the water and span a distance of 200 feet. Find the equation of the parabola after inserting a coordinate system with the origin at the vertex of the parabola and the vertical y axis (pointing upward) along the axis of the parabola.

50. **Astronomy.** The cross section of a parabolic reflector with 6-inch diameter is ground so that its vertex is 0.15 inch below the rim (see the figure).

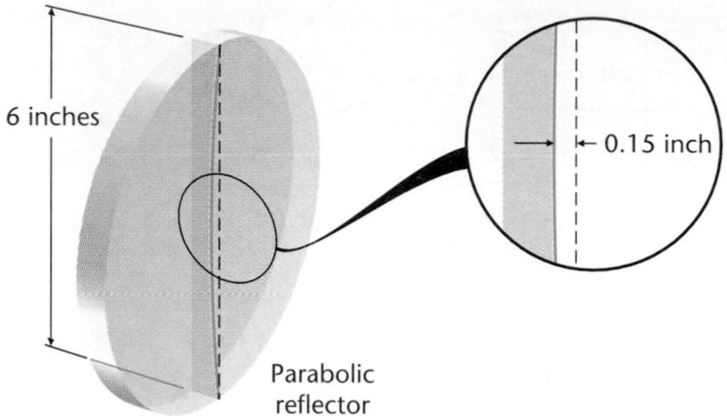

(A) Find the equation of the parabola after inserting an xy coordinate system with the vertex at the origin, the y axis (pointing upward) the axis of the parabola.

(B) How far is the focus from the vertex?

51. **Space Science.** A designer of a 200-foot-diameter parabolic electromagnetic antenna for tracking space probes wants to place the focus 100 feet above the vertex (see the figure).

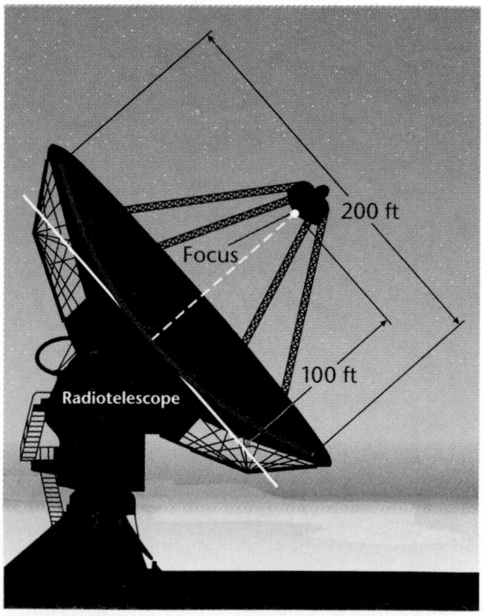

(A) Find the equation of the parabola using the axis of the parabola as the y axis (up positive) and vertex at the origin.

(B) Determine the depth of the parabolic reflector.

52. Signal Light. A signal light on a ship is a spotlight with parallel reflected light rays (see the figure). Suppose the parabolic reflector is 12 inches in diameter and the light source is located at the focus, which is 1.5 inches from the vertex.

(A) Find the equation of the parabola using the axis of the parabola as the x axis (right positive) and vertex at the origin.

(B) Determine the depth of the parabolic reflector.

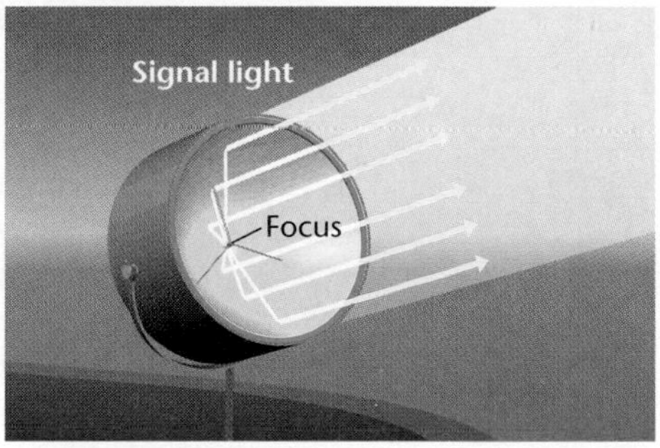

Section 7-2 Ellipse

- Definition of an Ellipse
- Drawing an Ellipse
- Standard Equations and Their Graphs
- Applications

We start our discussion of the ellipse with a coordinate-free definition. Using this definition, we show how an ellipse can be drawn and we derive standard equations for ellipses specially located in a rectangular coordinate system.

Definition of an Ellipse

The following is a coordinate-free definition of an ellipse:

DEFINITION 1

ELLIPSE

An **ellipse** is the set of all points P in a plane such that the sum of the distances of P from two fixed points in the plane is constant. Each of the fixed points, F' and F, is called a **focus,** and together they are called **foci.** Referring to the figure, the line segment $V'V$ through the foci is the **major axis.** The perpendicular bisector $B'B$ of the major axis is the **minor axis.** Each end of the major axis, V' and V, is called a **vertex.** The midpoint of the line segment $F'F$ is called the **center** of the ellipse.

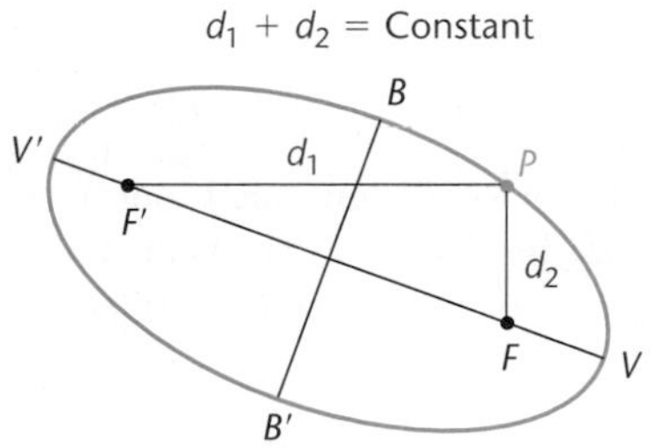

Drawing an Ellipse

An ellipse is easy to draw. All you need is a piece of string, two thumbtacks, and a pencil or pen (see Fig. 1). Place the two thumbtacks in a piece of cardboard. These form the foci of the ellipse. Take a piece of string longer than the distance between the two thumbtacks—this represents the constant in the definition—and tie each end to a thumbtack. Finally, catch the tip of a pencil under the string and move it while keeping the string taut. The resulting figure is by definition an ellipse. Ellipses of different shapes result, depending on the placement of thumbtacks and the length of the string joining them.

FIGURE 1
Drawing an ellipse.

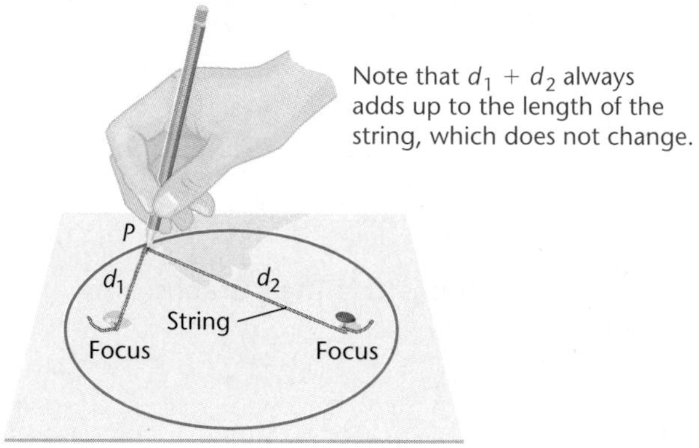

Standard Equations and Their Graphs

Using the definition of an ellipse and the distance-between-two-points formula, we can derive standard equations for an ellipse located in a rectangular coordinate system. We start by placing an ellipse in the coordinate system with the foci on the x axis equidistant from the origin at $F'(-c, 0)$ and $F(c, 0)$, as in Figure 2.

FIGURE 2
Ellipse with foci on x axis.

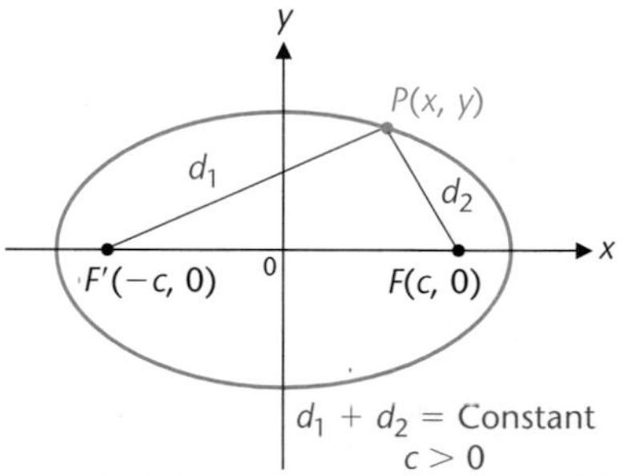

For reasons that will become clear soon, it is convenient to represent the constant sum $d_1 + d_2$ by $2a$, $a > 0$. Also, the geometric fact that the sum of the lengths of any two sides of a triangle must be greater than the third side can be applied to Figure 2 to derive the following useful result:

$$\begin{aligned} d(F', P) + d(P, F) &> d(F', F) \\ d_1 + d_2 &> 2c \\ 2a &> 2c \\ a &> c \end{aligned} \qquad (1)$$

We will use this result in the derivation of the equation of an ellipse, which we now begin.

Referring to Figure 2, the point $P(x, y)$ is on the ellipse if and only if

$$\begin{aligned} d_1 + d_2 &= 2a \\ d(P, F') + d(P, F) &= 2a \\ \sqrt{(x + c)^2 + (y - 0)^2} + \sqrt{(x - c)^2 + (y - 0)^2} &= 2a \end{aligned}$$

After eliminating radicals and simplifying, a good exercise for you, we obtain

$$(a^2 - c^2)x^2 + a^2y^2 = a^2(a^2 - c^2) \tag{2}$$

$$\frac{x^2}{a^2} + \frac{y^2}{a^2 - c^2} = 1 \tag{3}$$

Dividing both sides of equation (2) by $a^2(a^2 - c^2)$ is permitted, since neither a^2 nor $a^2 - c^2$ is 0. From equation (1), $a > c$; thus $a^2 > c^2$ and $a^2 - c^2 > 0$. The constant a was chosen positive at the beginning.

To simplify equation (3) further, we let

$$b^2 = a^2 - c^2 \qquad b > 0 \tag{4}$$

to obtain

$$\frac{x^2}{a^2} + \frac{y^2}{b^2} = 1 \tag{5}$$

From equation (5) we see that the x intercepts are $x = \pm a$ and the y intercepts are $y = \pm b$. The x intercepts are also the vertices. Thus,

Major axis length = $2a$

Minor axis length = $2b$

To see that the major axis is longer than the minor axis, we show that $2a > 2b$. Returning to equation (4),

$$\begin{aligned} b^2 &= a^2 - c^2 && a, b, c > 0 \\ b^2 + c^2 &= a^2 \\ b^2 &< a^2 && \text{Definition of} < \\ b^2 - a^2 &< 0 \\ (b - a)(b + a) &< 0 \\ b - a &< 0 && \text{Since } b + a \text{ is positive, } b - a \text{ must be negative.} \\ b &< a \\ 2b &< 2a \\ 2a &> 2b \end{aligned}$$

$$\begin{pmatrix}\text{Length of}\\ \text{major axis}\end{pmatrix} > \begin{pmatrix}\text{Length of}\\ \text{minor axis}\end{pmatrix}$$

If we start with the foci on the y axis at $F(0, c)$ and $F'(0, -c)$ as in Figure 3, instead of on the x axis as in Figure 2, then, following arguments similar to those used for the first derivation, we obtain

$$\frac{x^2}{b^2} + \frac{y^2}{a^2} = 1 \qquad a > b \tag{6}$$

where the relationship among a, b, and c remains the same as before:

$$b^2 = a^2 - c^2 \tag{7}$$

The center is still at the origin, but the major axis is now along the y axis and the minor axis is along the x axis.

FIGURE 3
Ellipse with foci on y axis.

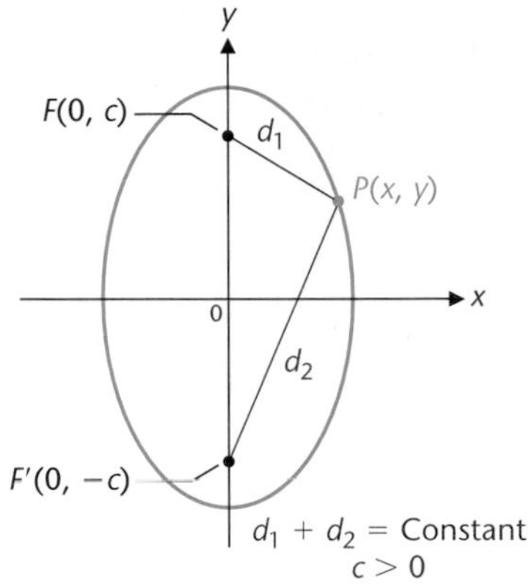

To sketch graphs of equations of the form of equations (5) or (6) is an easy matter. We find the x and y intercepts and sketch in an appropriate ellipse. Since replacing x with $-x$ or y with $-y$ produces an equivalent equation, we conclude that the graphs are symmetric with respect to the x axis, y axis, and origin. If further accuracy is required, additional points can be found with the aid of a calculator and the use of symmetry properties.

Given an equation of the form of equations (5) or (6), how can we find the coordinates of the foci without memorizing or looking up the relation $b^2 = a^2 - c^2$? There is a simple geometric relationship in an ellipse that enables us to get the same result using the Pythagorean theorem. To see this relationship, refer to Figure 4(a). Then, using the definition of an ellipse and $2a$ for the constant sum, as we did in deriving the standard equations, we see that

$$\begin{aligned} d + d &= 2a \\ 2d &= 2a \\ d &= a \end{aligned}$$

Thus,

The length of the line segment from the end of a minor axis to a focus is the same as half the length of a major axis.

This geometric relationship is illustrated in Figure 4(b). Using the Pythagorean theorem for the triangle in Figure 4(b), we have

$$b^2 + c^2 = a^2$$

or

$$b^2 = a^2 - c^2 \qquad \text{Equations (4) and (7)}$$

or

$$c^2 = a^2 - b^2 \qquad \text{Useful for finding the foci, given } a \text{ and } b$$

Thus, we can find the foci of an ellipse given the intercepts a and b simply by using the triangle in Figure 4(b) and the Pythagorean theorem.

FIGURE 4
Geometric relationships.

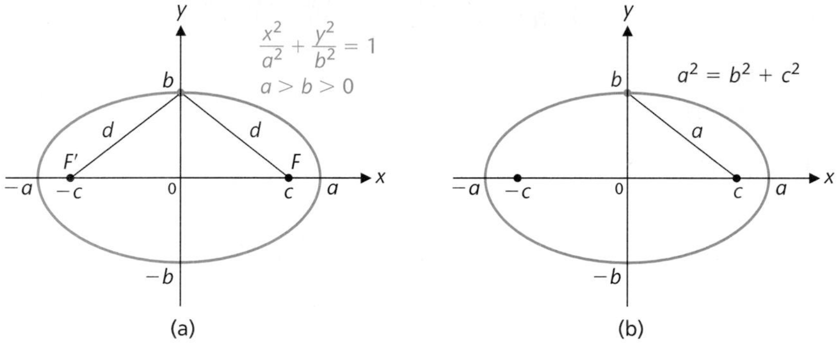

We summarize all of these results for convenient reference in Theorem 1.

THEOREM 1

STANDARD EQUATIONS OF AN ELLIPSE WITH CENTER AT (0, 0)

1. $\frac{x^2}{a^2} + \frac{y^2}{b^2} = 1 \qquad a > b > 0$

x intercepts: $\pm a$ (vertices)

y intercepts: $\pm b$

Foci: $F'(-c, 0)$, $F(c, 0)$

$$c^2 = a^2 - b^2$$

Major axis length $= 2a$

Minor axis length $= 2b$

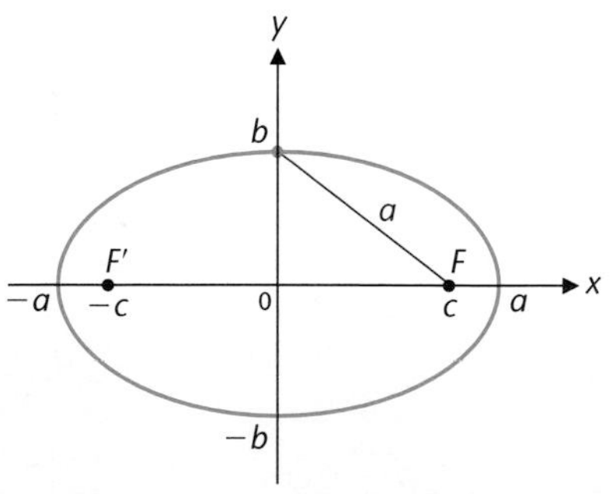

THEOREM 1 continued

2. $\dfrac{x^2}{b^2} + \dfrac{y^2}{a^2} = 1 \qquad a > b > 0$

x intercepts: $\pm b$

y intercepts: $\pm a$ (vertices)

Foci: $F'(0, -c)$, $F(0, c)$

$$c^2 = a^2 - b^2$$

Major axis length $= 2a$

Minor axis length $= 2b$

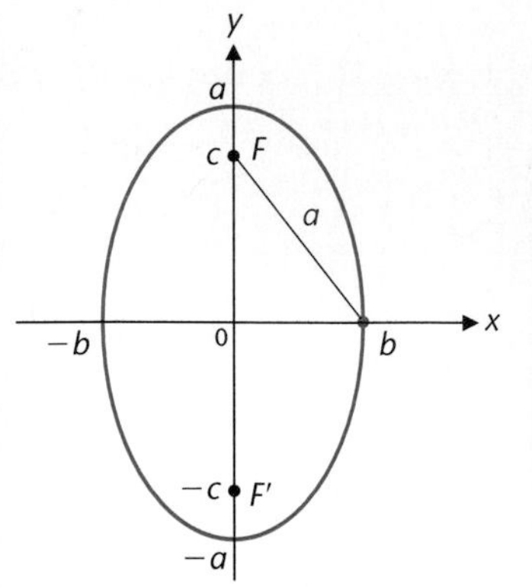

[*Note:* Both graphs are symmetric with respect to the x axis, y axis, and origin. Also, the major axis is always longer than the minor axis.]

Explore/Discuss 1

The line through a focus F of an ellipse that is perpendicular to the major axis intersects the ellipse in two points G and H. For each of the two standard equations of an ellipse with center (0, 0), find an expression in terms of a and b for the distance from G to H.

EXAMPLE 1

Graphing Ellipses

Sketch the graph of each equation, find the coordinates of the foci, and find the lengths of the major and minor axes. Check by graphing on a graphing utility.

(A) $9x^2 + 16y^2 = 144$ (B) $2x^2 + y^2 = 10$

Solutions

(A) First, write the equation in standard form by dividing both sides by 144:

$$9x^2 + 16y^2 = 144$$

$$\frac{9x^2}{144} + \frac{16y^2}{144} = \frac{144}{144}$$

$$\frac{x^2}{16} + \frac{y^2}{9} = 1 \qquad a^2 = 16 \text{ and } b^2 = 9$$

Locate the intercepts:

x intercepts: ± 4

y intercepts: ± 3

FIGURE 5
$9x^2 + 16y^2 = 144.$

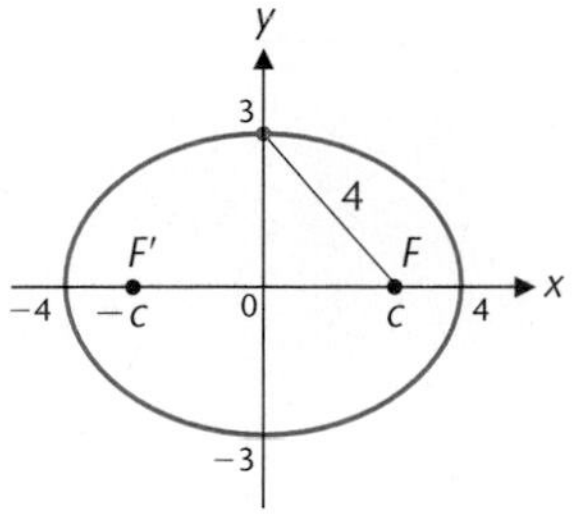

and sketch in the ellipse, as shown in Figure 5.

$$\begin{aligned} \text{Foci:}\quad c^2 &= a^2 - b^2 \\ &= 16 - 9 \\ &= 7 \\ c &= \sqrt{7} \qquad c \text{ is positive.} \end{aligned}$$

Thus, the foci are $F'(-\sqrt{7}, 0)$ and $F(\sqrt{7}, 0)$.

$$\text{Major axis length} = 2(4) = 8$$
$$\text{Minor axis length} = 2(3) = 6$$

To check the graph on a graphing utility, we solve the original equation for y:

FIGURE 6
$y_1 = \sqrt{(144 - 9x^2)/16};$
$y_2 = -\sqrt{(144 - 9x^2)/16}.$

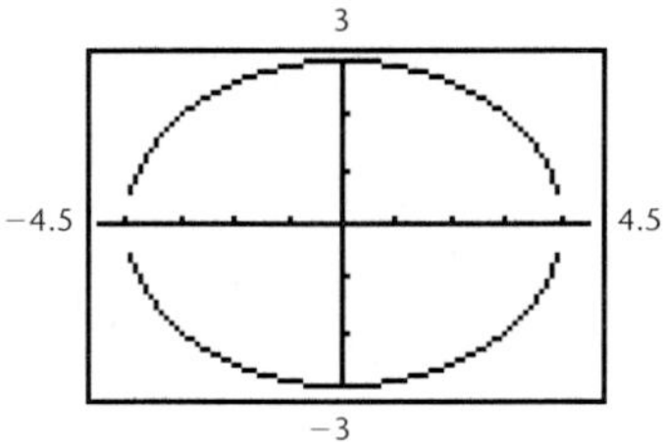

$$\begin{aligned} 9x^2 + 16y^2 &= 144 \\ y^2 &= (144 - 9x^2)/16 \\ y &= \pm\sqrt{(144 - 9x^2)/16} \end{aligned}$$

This produces the two functions whose graphs are shown in Figure 6. Notice that we used a squared viewing window to avoid distorting the shape of the ellipse. Also note the gaps in the graph at ± 4. This is a common occurrence in graphs involving the square root function.

(B) Write the equation in standard form by dividing both sides by 10:

$$2x^2 + y^2 = 10$$

$$\boxed{\frac{2x^2}{10} + \frac{y^2}{10} = \frac{10}{10}}$$

$$\frac{x^2}{5} + \frac{y^2}{10} = 1 \qquad a^2 = 10 \text{ and } b^2 = 5$$

Locate the intercepts:

FIGURE 7
$2x^2 + y^2 = 10.$

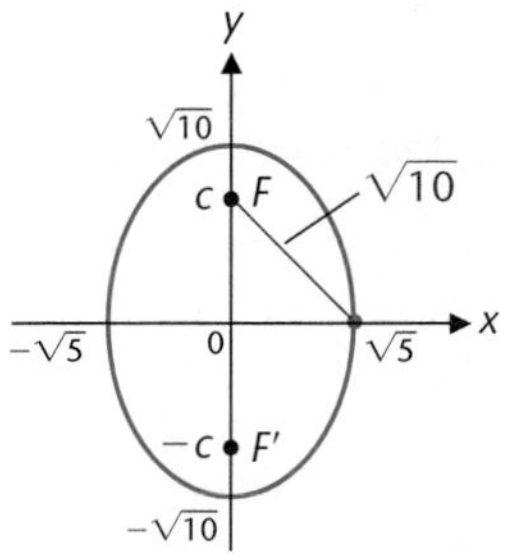

$$x \text{ intercepts: } \pm\sqrt{5} \approx \pm 2.24$$
$$y \text{ intercepts: } \pm\sqrt{10} \approx \pm 3.16$$

and sketch in the ellipse, as shown in Figure 7.

$$\begin{aligned} \text{Foci:}\quad c^2 &= a^2 - b^2 \\ &= 10 - 5 \\ &= 5 \\ c &= \sqrt{5} \end{aligned}$$

FIGURE 8
$y_1 = \sqrt{10 - 2x^2}$;
$y_2 = -\sqrt{10 - 2x^2}$.

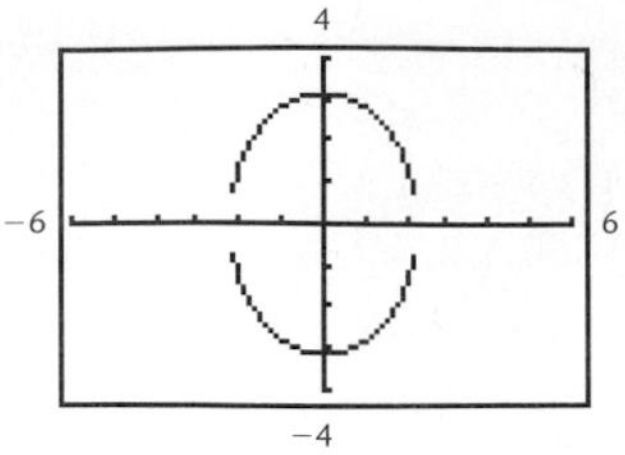

Thus, the foci are $F'(0, -\sqrt{5})$ and $F(0, \sqrt{5})$.

$$\text{Major axis length} = 2\sqrt{10} \approx 6.32$$

$$\text{Minor axis length} = 2\sqrt{5} \approx 4.47$$

Figure 8 shows a check of the graph.

MATCHED PROBLEM 1

Sketch the graph of each equation, find the coordinates of the foci, and find the lengths of the major and minor axes. Check by graphing on a graphing utility.

(A) $x^2 + 4y^2 = 4$ (B) $3x^2 + y^2 = 18$

EXAMPLE 2

Finding the Equation of an Ellipse

Find an equation of an ellipse in the form

$$\frac{x^2}{M} + \frac{y^2}{N} = 1 \qquad M, N > 0$$

if the center is at the origin, the major axis is along the y axis, and

(A) Length of major axis = 20
Length of minor axis = 12

(B) Length of major axis = 10
Distance of foci from center = 4

Solutions

(A) Compute x and y intercepts and make a rough sketch of the ellipse, as shown in Figure 9.

FIGURE 9
$\dfrac{x^2}{36} + \dfrac{y^2}{100} = 1.$

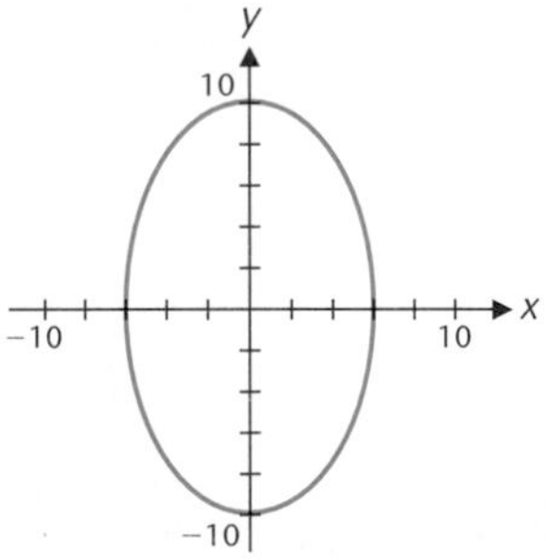

$$\frac{x^2}{b^2} + \frac{y^2}{a^2} = 1$$

$$a = \frac{20}{2} = 10 \qquad b = \frac{12}{2} = 6$$

$$\frac{x^2}{36} + \frac{y^2}{100} = 1$$

(B) Make a rough sketch of the ellipse, as shown in Figure 10; locate the foci and y intercepts, then determine the x intercepts using the special triangle relationship discussed earlier.

FIGURE 10
$\dfrac{x^2}{9} + \dfrac{y^2}{25} = 1.$

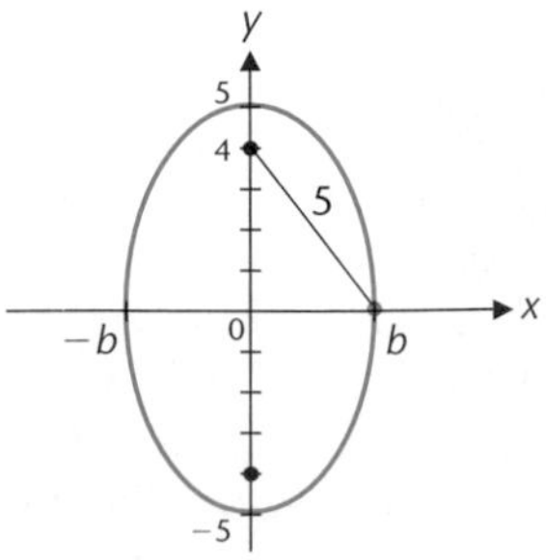

$$\frac{x^2}{b^2} + \frac{y^2}{a^2} = 1$$

$$a = \frac{10}{2} = 5 \qquad b^2 = 5^2 - 4^2 = 25 - 16 = 9$$

$$b = 3$$

$$\frac{x^2}{9} + \frac{y^2}{25} = 1$$

MATCHED PROBLEM 2

Find an equation of an ellipse in the form

$$\frac{x^2}{M} + \frac{y^2}{N} = 1 \qquad M, N > 0$$

if the center is at the origin, the major axis is along the x axis, and

(A) Length of major axis = 50
Length of minor axis = 30

(B) Length of minor axis = 16
Distance of foci from center = 6

Explore/Discuss 2

Consider the graph of an equation in the variables x and y. The equation of its magnification by a factor $k > 0$ is obtained by replacing x and y in the equation by x/k and y/k, respectively.

(A) Find the equation of the magnification by a factor 3 of the ellipse with equation $(x^2/4) + y^2 = 1$. Graph both equations.

(B) Give an example of an ellipse with center (0, 0) with $a > b$ that is not a magnification of $(x^2/4) + y^2 = 1$.

(C) Find the equations of all ellipses that are magnifications of $(x^2/4) + y^2 = 1$.

Applications

You are no doubt aware of many occurrences and uses of elliptical forms: orbits of satellites, planets, and comets; shapes of galaxies; gears and cams; some airplane wings, boat keels, and rudders; tabletops; public fountains; and domes in buildings are a few examples (see Fig. 11).

FIGURE 11
Uses of elliptical forms.

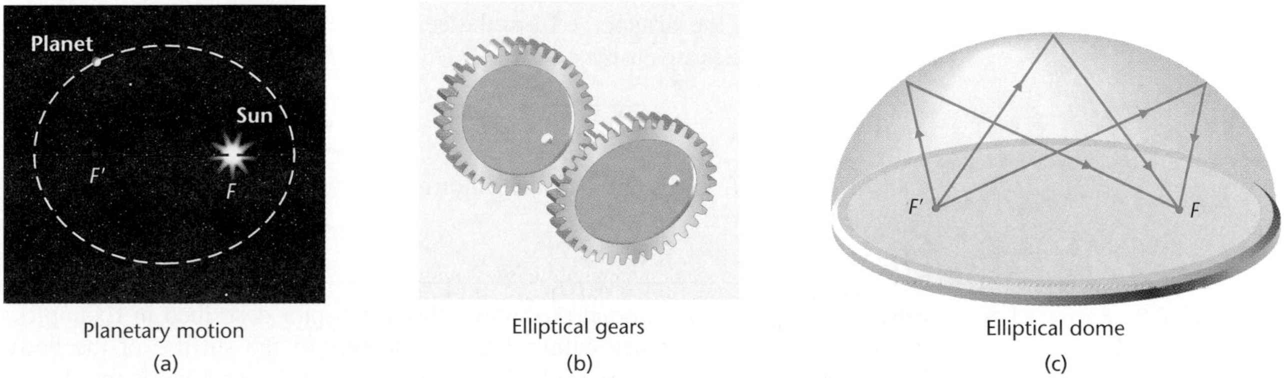

Johannes Kepler (1571–1630), a German astronomer, discovered that planets move in elliptical orbits, with the sun at a focus, and not in circular orbits as had been thought before [Fig. 11(a)]. Figure 11(b) shows a pair of elliptical gears with pivot points at foci. Such gears transfer constant rotational speed to variable rotational speed, and vice versa. Figure 11(c) shows an elliptical dome. An interesting property of such a dome is that a sound or light source at one focus will reflect off the dome and pass through the other focus. One of the chambers in the Capitol

Building in Washington, D.C., has such a dome, and is referred to as a whispering room because a whispered sound at one focus can be easily heard at the other focus.

A fairly recent application in medicine is the use of elliptical reflectors and ultrasound to break up kidney stones. A device called a lithotripter is used to generate intense sound waves that break up the stone from outside the body, thus avoiding surgery. To be certain that the waves do not damage other parts of the body, the reflecting property of the ellipse is used to design and correctly position the lithotripter.

EXAMPLE 3

Medicinal Lithotripsy

A lithotripter is formed by rotating the portion of an ellipse below the minor axis around the major axis (see Fig. 12). The lithotripter is 20 centimeters wide and 16 centimeters deep. If the ultrasound source is positioned at one focus of the ellipse and the kidney stone at the other, then all the sound waves will pass through the kidney stone. How far from the kidney stone should the point V on the base of the lithotripter be positioned to focus the sound waves on the kidney stone? Round the answer to one decimal place.

FIGURE 12 Lithotripter.

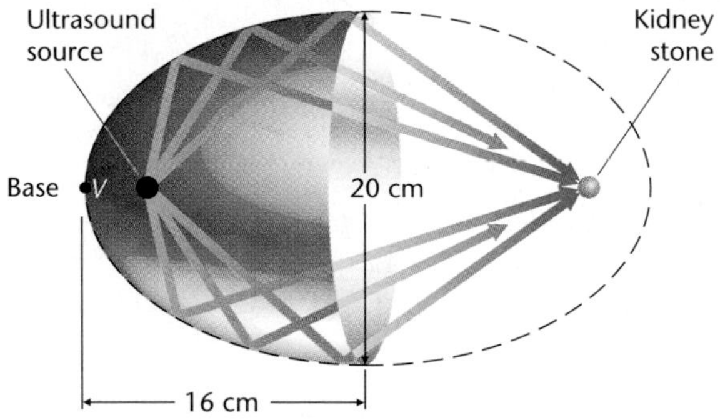

Solution

From the figure, we see that $a = 16$ and $b = 10$ for the ellipse used to form the lithotripter. Thus, the distance c from the center to either the kidney stone or the ultrasound source is given by

$$c = \sqrt{a^2 - b^2} = \sqrt{16^2 - 10^2} = \sqrt{156} \approx 12.5$$

and the distance from the base of the lithotripter to the kidney stone is $16 + 12.5 = 28.5$ centimeters.

MATCHED PROBLEM 3

Since lithotripsy is an external procedure, the lithotripter described in Example 3 can be used only on stones within 12.5 centimeters of the surface of the body. Suppose a kidney stone is located 14 centimeters from the surface. If the diameter is kept fixed at 20 centimeters, how deep must a lithotripter be to focus on this kidney stone? Round answer to one decimal place.

Answers to Matched Problems

1. (A)

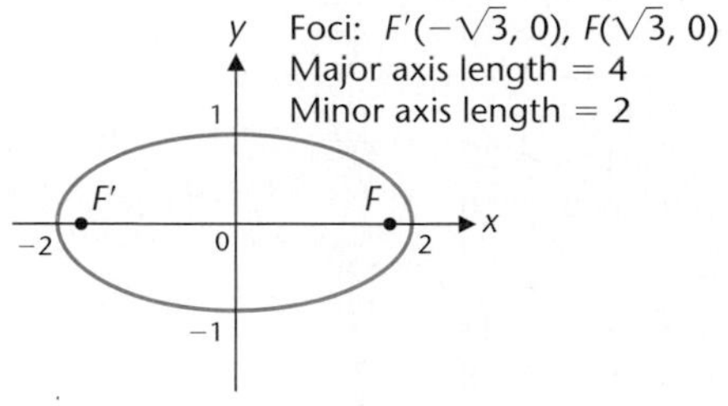

(B)

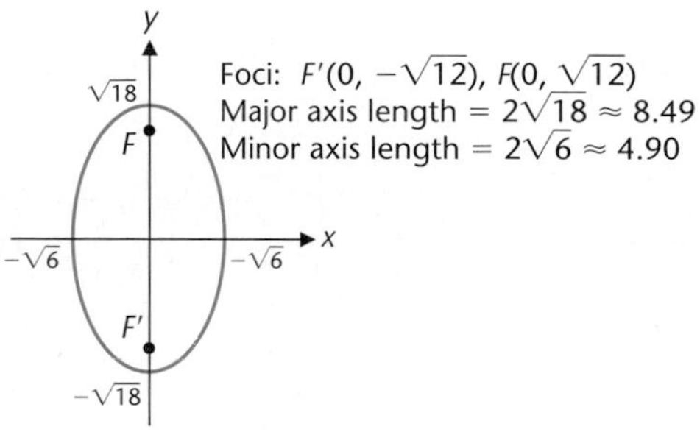

2. (A) $\dfrac{x^2}{625} + \dfrac{y^2}{225} = 1$ (B) $\dfrac{x^2}{100} + \dfrac{y^2}{64} = 1$ **3.** 17.2 centimeters

---

EXERCISE 7-2

A

In Problems 1–6, sketch a graph of each equation, find the coordinates of the foci, and find the lengths of the major and minor axes. Check by graphing on a graphing utility.

1. $\dfrac{x^2}{25} + \dfrac{y^2}{4} = 1$ **2.** $\dfrac{x^2}{9} + \dfrac{y^2}{4} = 1$

3. $\dfrac{x^2}{4} + \dfrac{y^2}{25} = 1$ **4.** $\dfrac{x^2}{4} + \dfrac{y^2}{9} = 1$

5. $x^2 + 9y^2 = 9$ **6.** $4x^2 + y^2 = 4$

In Problems 7–10, match each equation with one of graphs (a)–(d).

7. $9x^2 + 16y^2 = 144$ **8.** $16x^2 + 9y^2 = 144$

9. $4x^2 + y^2 = 16$ **10.** $x^2 + 4y^2 = 16$

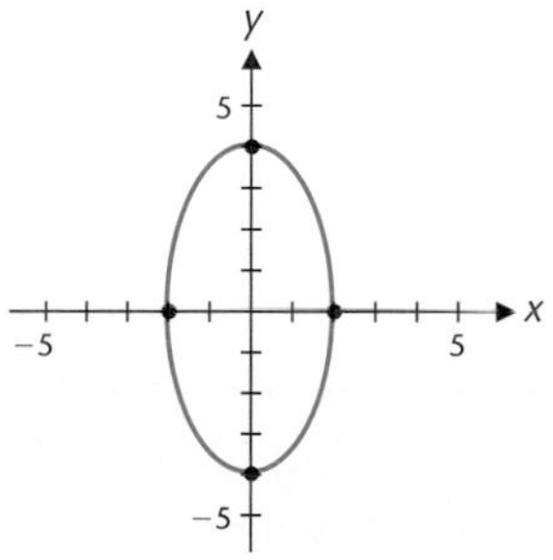

(a)

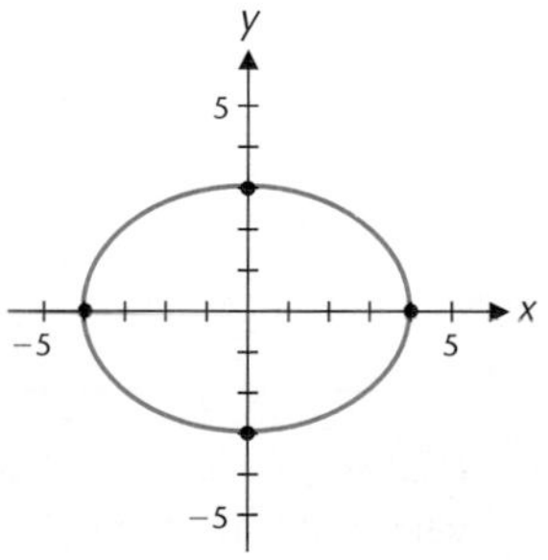

(b)

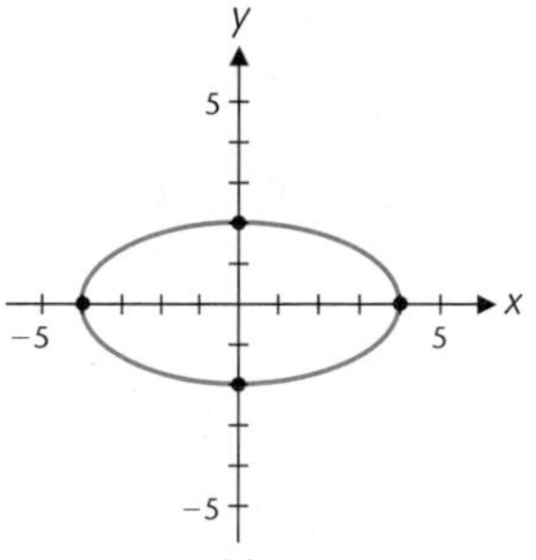

(c)

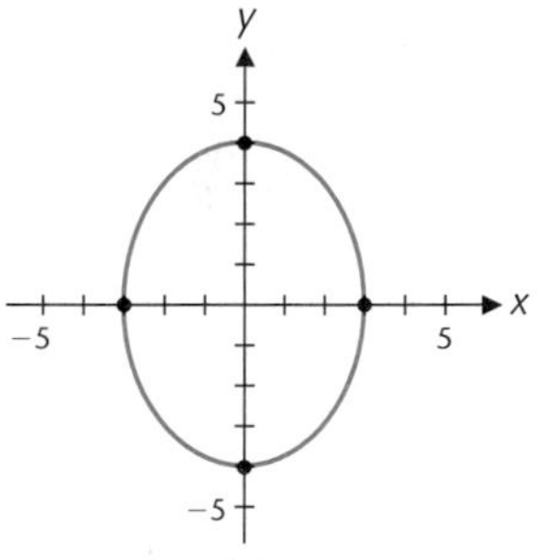

(d)

B

In Problems 11–16, sketch a graph of each equation, find the coordinates of the foci, and find the lengths of the major and minor axes. Check by graphing on a graphing utility.

11. $25x^2 + 9y^2 = 225$

12. $16x^2 + 25y^2 = 400$

13. $2x^2 + y^2 = 12$

14. $4x^2 + 3y^2 = 24$

15. $4x^2 + 7y^2 = 28$

16. $3x^2 + 2y^2 = 24$

In Problems 17–28, find an equation of an ellipse in the form

$$\frac{x^2}{M} + \frac{y^2}{N} = 1 \qquad M, N > 0$$

if the center is at the origin, and

17. The graph is

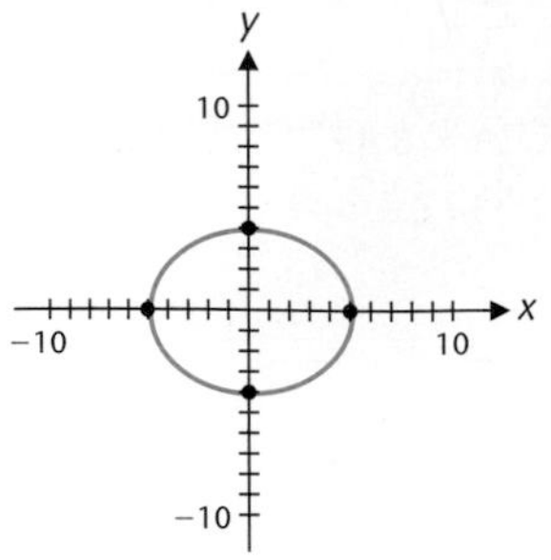

18. The graph is

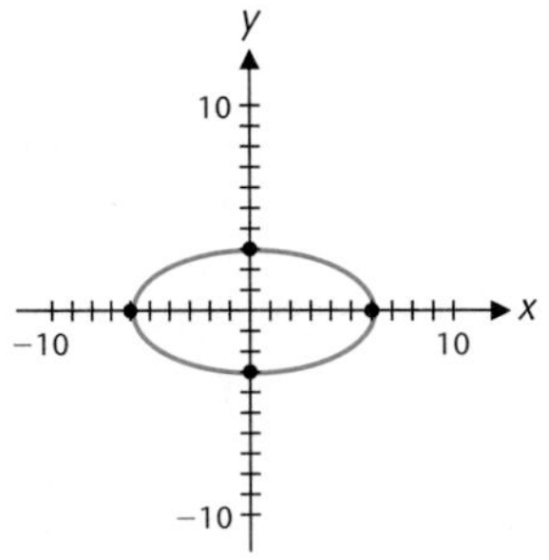

19. The graph is

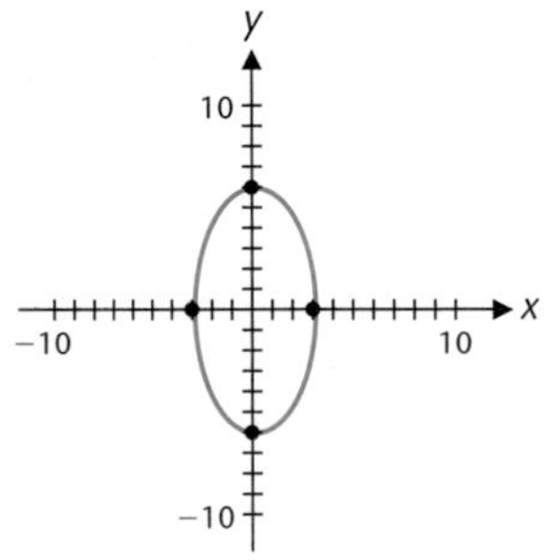

20. The graph is

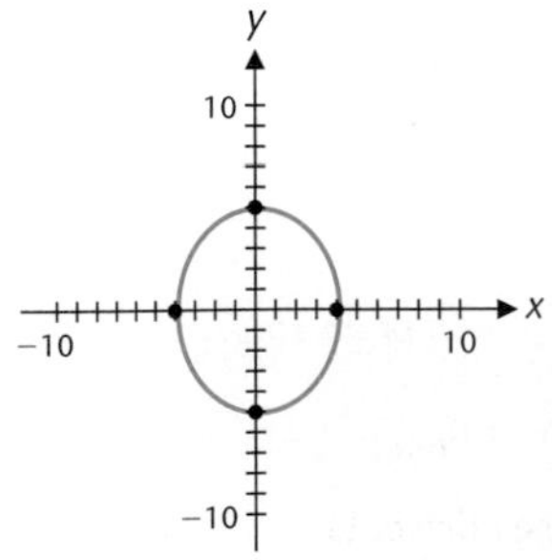

21. Major axis on x axis
Major axis length $= 10$
Minor axis length $= 6$

22. Major axis on x axis
Major axis length $= 14$
Minor axis length $= 10$

23. Major axis on y axis
Major axis length $= 22$
Minor axis length $= 16$

24. Major axis on y axis
Major axis length $= 24$
Minor axis length $= 18$

25. Major axis on x axis
Major axis length $= 16$
Distance of foci from center $= 6$

26. Major axis on y axis
Major axis length $= 24$
Distance of foci from center $= 10$

27. Major axis on y axis
Minor axis length $= 20$
Distance of foci from center $= \sqrt{70}$

28. Major axis on x axis
Minor axis length $= 14$
Distance of foci from center $= \sqrt{200}$

29. Explain why an equation whose graph is an ellipse does not define a function.

30. Consider all ellipses having $(0, \pm 1)$ as the ends of the minor axis. Describe the connection between the elongation of the ellipse and the distance from a focus to the origin.

In Problems 31–38, find all points of intersection. Round any approximate values to three decimal places.

31. $16x^2 + 25y^2 = 400$
$2x - 5y = 10$

32. $25x^2 + 16y^2 = 400$
$5x + 8y = 20$

33. $25x^2 + 16y^2 = 400$
$25x^2 - 36y = 0$

34. $16x^2 + 25y^2 = 400$
$3x^2 - 20y = 0$

35. $5x^2 + 2y^2 = 63$
$2x - y = 0$

36. $3x^2 + 4y^2 = 57$
$x - 2y = 0$

37. $2x^2 + 3y^2 = 33$
$x^2 - 8y = 0$

38. $3x^2 + 2y^2 = 43$
$x^2 - 12y = 0$

39. Find an equation of the set of points in a plane, each of whose distance from (2, 0) is one-half its distance from the line $x = 8$. Identify the geometric figure.

40. Find an equation of the set of points in a plane, each of whose distance from (0, 9) is three-fourths its distance from the line $y = 16$. Identify the geometric figure.

In Problems 41–44, find the coordinates of all points of intersection to two decimal places.

41. $x^2 + 3y^2 = 20,\ 4x + 5y = 11$

42. $8x^2 + 35y^2 = 3{,}600,\ x^2 = -25y$

43. $50x^2 + 4y^2 = 1{,}025,\ 9x^2 + 2y^2 = 300$

44. $2x^2 + 7y^2 = 95,\ 13x^2 + 6y^2 = 63$

APPLICATIONS

45. Engineering. The semielliptical arch in the concrete bridge in the figure must have a clearance of 12 feet above the water and span a distance of 40 feet. Find the equation of the ellipse after inserting a coordinate system with the center of the ellipse at the origin and the major axis on the x axis. The y axis points up, and the x axis points to the right. How much clearance above the water is there 5 feet from the bank?

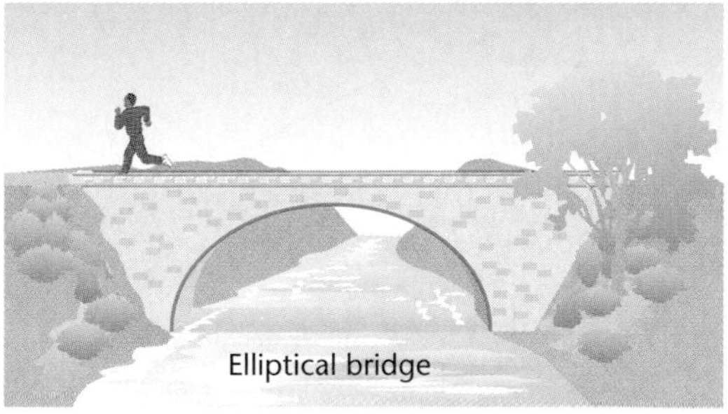

46. Design. A 4 × 8 foot elliptical tabletop is to be cut out of a 4 × 8 foot rectangular sheet of teak plywood (see the figure). To draw the ellipse on the plywood, how far should the foci be located from each edge and how long a piece of string must be fastened to each focus to produce the ellipse (see Fig. 1 in the text)? Compute the answer to two decimal places.

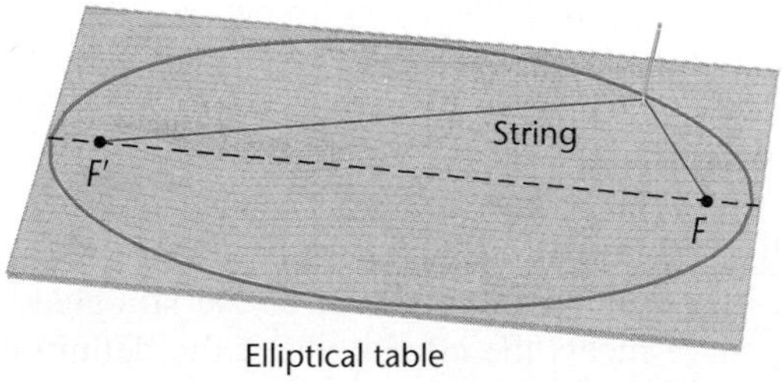

★ **47. Aeronautical Engineering.** Of all possible wing shapes, it has been determined that the one with the least drag along the trailing edge is an ellipse. The leading edge may be a straight line, as shown in the figure. One of the most famous planes with this design was the World War II British Spitfire. The plane in the figure has a wingspan of 48.0 feet.

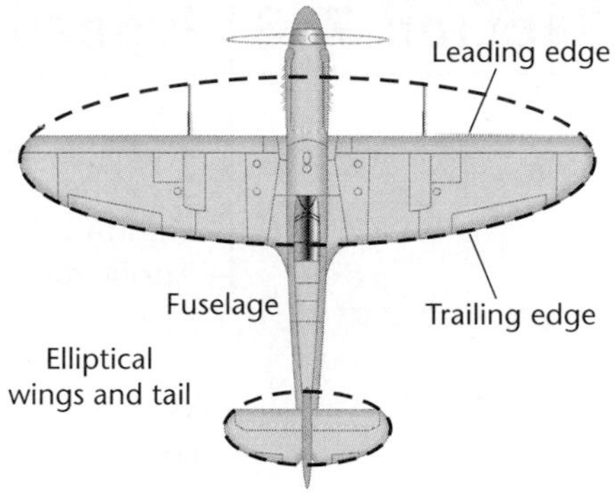

(A) If the straight-line leading edge is parallel to the major axis of the ellipse and is 1.14 feet in front of it, and if the leading edge is 46.0 feet long (including the width of the fuselage), find the equation of the ellipse. Let the x axis lie along the major axis (positive right), and let the y axis lie along the minor axis (positive forward).

(B) How wide is the wing in the center of the fuselage (assuming the wing passes through the fuselage)?

Compute quantities to three significant digits.

★ **48. Naval Architecture.** Currently, many high-performance racing sailboats use elliptical keels, rudders, and main sails for the same reasons stated in Problem 47—less drag along the trailing edge. In the accompanying figure, the ellipse containing the keel has a 12.0-foot major axis. The straight-line leading edge is parallel to the major axis of the ellipse and 1.00 foot in front of it. The chord is 1.00 foot shorter than the major axis.

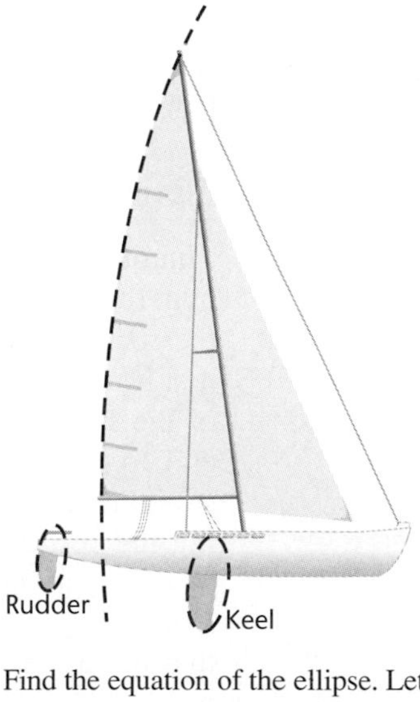

(A) Find the equation of the ellipse. Let the y axis lie along the minor axis of the ellipse, and let the x axis lie along the major axis, both with positive direction upward.

(B) What is the width of the keel, measured perpendicular to the major axis, 1 foot up the major axis from the bottom end of the keel?

Compute quantities to three significant digits.

Section 7-3 Hyperbola

- Definition of a Hyperbola
- Drawing a Hyperbola
- Standard Equations and Their Graphs
- Applications

As before, we start with a coordinate-free definition of a hyperbola. Using this definition, we show how a hyperbola can be drawn and we derive standard equations for hyperbolas specially located in a rectangular coordinate system.

Definition of a Hyperbola

The following is a coordinate-free definition of a hyperbola:

DEFINITION 1

HYPERBOLA

A **hyperbola** is the set of all points P in a plane such that the absolute value of the difference of the distances of P to two fixed points in the plane is a positive constant. Each of the fixed points, F' and F, is called a **focus.** The intersection points V' and V of the line through the foci and the two branches of the hyperbola are called **vertices,** and each is called a **vertex.** The line segment $V'V$ is called the **transverse axis.** The midpoint of the transverse axis is the **center** of the hyperbola.

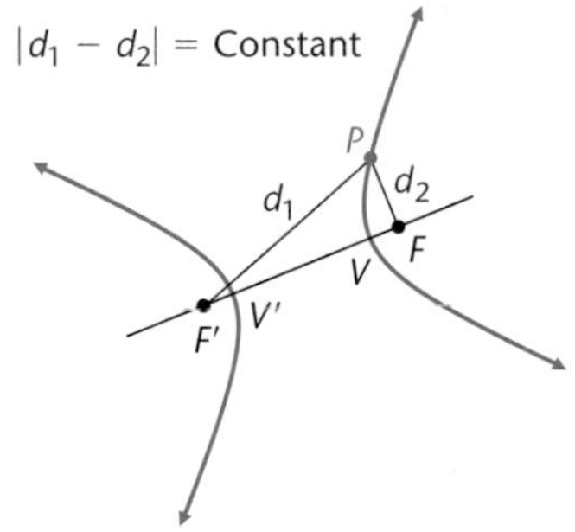

Drawing a Hyperbola

Thumbtacks, a straightedge, string, and a pencil are all that are needed to draw a hyperbola (see Fig. 1). Place two thumbtacks in a piece of cardboard—these form the foci of the hyperbola. Rest one corner of the straightedge at the focus F' so that it is free to rotate about this point. Cut a piece of string shorter than the length of the straightedge, and fasten one end to the straightedge corner A and the other end to the thumbtack at F. Now push the string with a pencil up against the straightedge at B. Keeping the string taut, rotate the straightedge about F', keeping the corner at F'. The resulting curve will be part of a hyperbola. Other parts of the hyperbola can be drawn by changing the position of the straightedge and string. To see that the resulting curve meets the conditions of the definition, note that the difference of the distances BF' and BF is

$$\begin{aligned} BF' - BF &= BF' + BA - BF - BA \\ &= AF' - (BF + BA) \\ &= \begin{pmatrix}\text{Straightedge}\\ \text{length}\end{pmatrix} - \begin{pmatrix}\text{String}\\ \text{length}\end{pmatrix} \\ &= \text{Constant} \end{aligned}$$

FIGURE 1
Drawing a hyperbola.

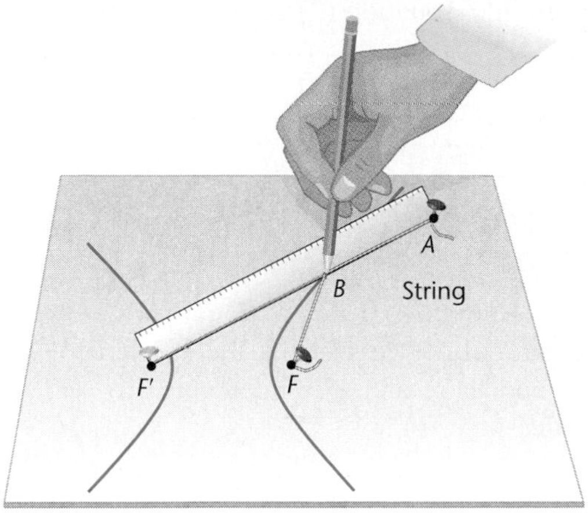

Standard Equations and Their Graphs

FIGURE 2
Hyperbola with foci on the x axis.

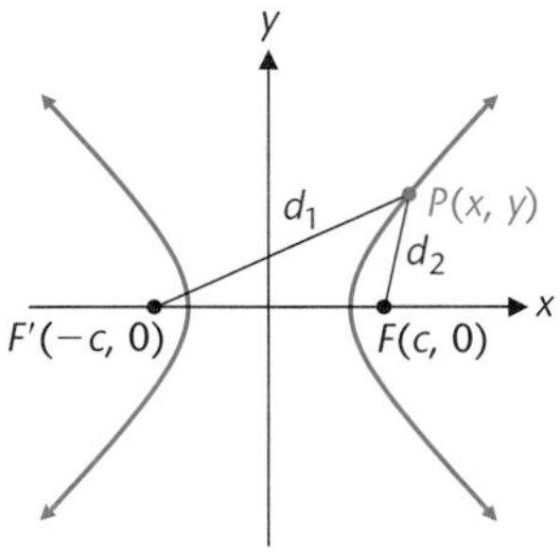

$c > 0$
$|d_1 - d_2|$ = Positive constant

Using the definition of a hyperbola and the distance-between-two-points formula, we can derive the standard equations for a hyperbola located in a rectangular coordinate system. We start by placing a hyperbola in the coordinate system with the foci on the x axis equidistant from the origin at $F'(-c, 0)$ and $F(c, 0)$, $c > 0$, as in Figure 2.

Just as for the ellipse, it is convenient to represent the constant difference by $2a$, $a > 0$. Also, the geometric fact that the difference of two sides of a triangle is always less than the third side can be applied to Figure 2 to derive the following useful result:

$$\begin{aligned}|d_1 - d_2| &< 2c \\ 2a &< 2c \\ a &< c\end{aligned} \tag{1}$$

We will use this result in the derivation of the equation of a hyperbola, which we now begin.

Referring to Figure 2, the point $P(x, y)$ is on the hyperbola if and only if

$$\begin{aligned}|d_1 - d_2| &= 2a \\ |d(P, F') - d(P, F)| &= 2a \\ \left|\sqrt{(x + c)^2 + y^2} - \sqrt{(x - c)^2 + y^2}\right| &= 2a\end{aligned}$$

After eliminating radicals and absolute value signs by appropriate use of squaring and simplifying, another good exercise for you, we have

$$(c^2 - a^2)x^2 - a^2y^2 = a^2(c^2 - a^2) \tag{2}$$

$$\frac{x^2}{a^2} - \frac{y^2}{c^2 - a^2} = 1 \tag{3}$$

Dividing both sides of equation (2) by $a^2(c^2 - a^2)$ is permitted, since neither a^2 nor $c^2 - a^2$ is 0. From equation (1), $a < c$; thus, $a^2 < c^2$ and $c^2 - a^2 > 0$. The constant a was chosen positive at the beginning.

To simplify equation (3) further, we let

$$b^2 = c^2 - a^2 \qquad b > 0 \tag{4}$$

to obtain

$$\frac{x^2}{a^2} - \frac{y^2}{b^2} = 1 \tag{5}$$

From equation (5) we see that the x intercepts, which are also the vertices, are $x = \pm a$ and there are no y intercepts. To see why there are no y intercepts, let $x = 0$ and solve for y:

$$\frac{0^2}{a^2} - \frac{y^2}{b^2} = 1$$

$$y^2 = -b^2$$

$$y = \pm\sqrt{-b^2} \qquad \text{An imaginary number}$$

FIGURE 3
Hyperbola with foci on the y axis.

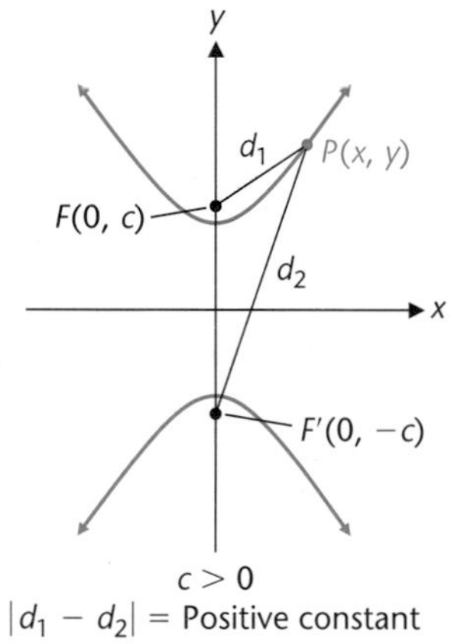

If we start with the foci on the y axis at $F'(0, -c)$ and $F(0, c)$ as in Figure 3, instead of on the x axis as in Figure 2, then, following arguments similar to those used for the first derivation, we obtain

$$\frac{y^2}{a^2} - \frac{x^2}{b^2} = 1 \tag{6}$$

where the relationship among a, b, and c remains the same as before:

$$b^2 = c^2 - a^2 \tag{7}$$

The center is still at the origin, but the transverse axis is now on the y axis.

As an aid to graphing equation (5), we solve the equation for y in terms of x, another good exercise for you, to obtain

$$y = \pm\frac{b}{a}x\sqrt{1 - \frac{a^2}{x^2}} \tag{8}$$

As x changes so that $|x|$ becomes larger, the expression $1 - (a^2/x^2)$ within the radical approaches 1. Hence, for large values of $|x|$, equation (5) behaves very much like the lines

$$y = \pm\frac{b}{a}x \tag{9}$$

These lines are **asymptotes** for the graph of equation (5). The hyperbola approaches these lines as a point $P(x, y)$ on the hyperbola moves away from the origin (see Fig. 4). An easy way to draw the asymptotes is to first draw the rectangle as in Figure 4, then extend the diagonals. We refer to this rectangle as the **asymptote rectangle.**

FIGURE 4
Asymptotes.

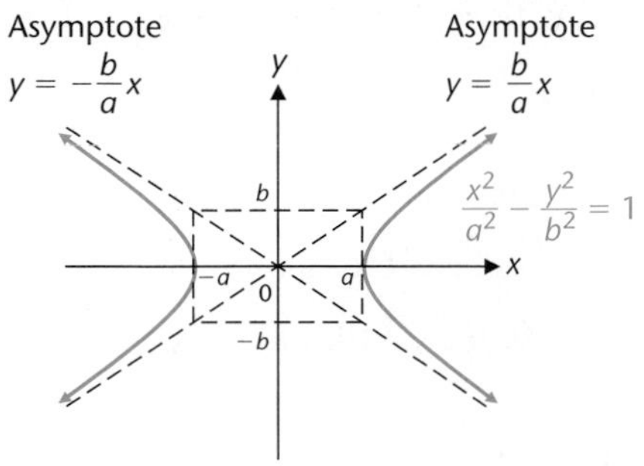

Starting with equation (6) and proceeding as we did for equation (5), we obtain the asymptotes for the graph of equation (6):

$$y = \pm\frac{a}{b}x \tag{10}$$

The perpendicular bisector of the transverse axis, extending from one side of the asymptote rectangle to the other, is called the **conjugate axis** of the hyperbola.

Given an equation of the form of equations (5) or (6), how can we find the coordinates of the foci without memorizing or looking up the relation $b^2 = c^2 - a^2$? Just as with the ellipse, there is a simple geometric relationship in a hyperbola that enables us to get the same result using the Pythagorean theorem. To see this relationship, we rewrite $b^2 = c^2 - a^2$ in the form

$$c^2 = a^2 + b^2 \tag{11}$$

Note in the figures in Theorem 1 below that the distance from the center to a focus is the same as the distance from the center to a corner of the asymptote rectangle. Stated in another way:

A circle, with center at the origin, that passes through all four corners of the asymptote rectangle also passes through all foci of hyperbolas with asymptotes determined by the diagonals of the rectangle.

We summarize all the preceding results in Theorem 1 for convenient reference.

THEOREM 1

STANDARD EQUATIONS OF A HYPERBOLA WITH CENTER AT (0, 0)

1. $\frac{x^2}{a^2} - \frac{y^2}{b^2} = 1$

x intercepts: $\pm a$ (vertices)

y intercepts: none

Foci: $F'(-c, 0)$, $F(c, 0)$

$$c^2 = a^2 + b^2$$

Transverse axis length $= 2a$

Conjugate axis length $= 2b$

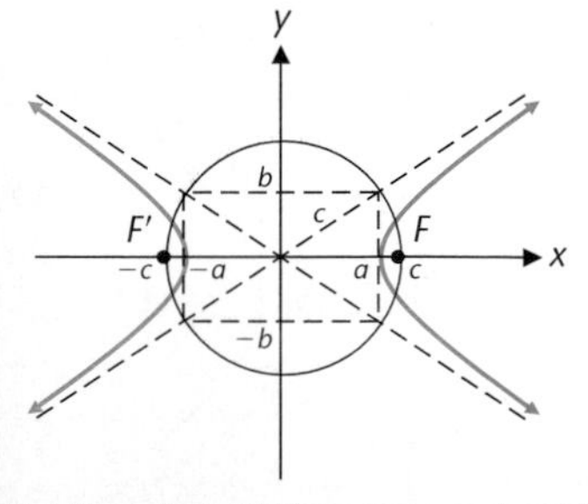

THEOREM 1 continued

2. $\dfrac{y^2}{a^2} - \dfrac{x^2}{b^2} = 1$

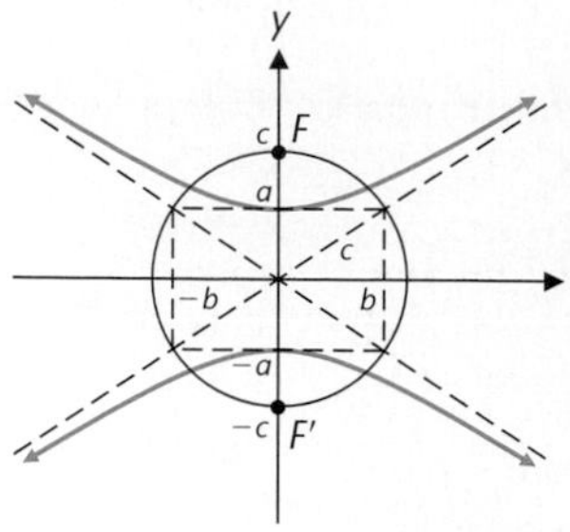

x intercepts: none

y intercepts: $\pm a$ (vertices)

Foci: $F'(0, -c)$, $F(0, c)$

$$c^2 = a^2 + b^2$$

Transverse axis length $= 2a$

Conjugate axis length $= 2b$

[*Note:* Both graphs are symmetric with respect to the x axis, y axis, and origin.]

Explore/Discuss 1

The line through a focus F of a hyperbola that is perpendicular to the transverse axis intersects the hyperbola in two points G and H. For each of the two standard equations of a hyperbola with center (0, 0), find an expression in terms of a and b for the distance from G to H.

EXAMPLE 1

Graphing Hyperbolas

Sketch the graph of each equation, find the coordinates of the foci, and find the lengths of the transverse and conjugate axes. Check by graphing on a graphing utility.

(A) $9x^2 - 16y^2 = 144$ (B) $16y^2 - 9x^2 = 144$ (C) $2x^2 - y^2 = 10$

Solutions

(A) First, write the equation in standard form by dividing both sides by 144:

$$9x^2 - 16y^2 = 144$$

$$\frac{x^2}{16} - \frac{y^2}{9} = 1 \qquad a^2 = 16 \text{ and } b^2 = 9$$

Locate x intercepts, $x = \pm 4$; there are no y intercepts. Sketch the asymptotes using the asymptote rectangle, then sketch in the hyperbola (Fig. 5).

FIGURE 5
$9x^2 - 16y^2 = 144.$

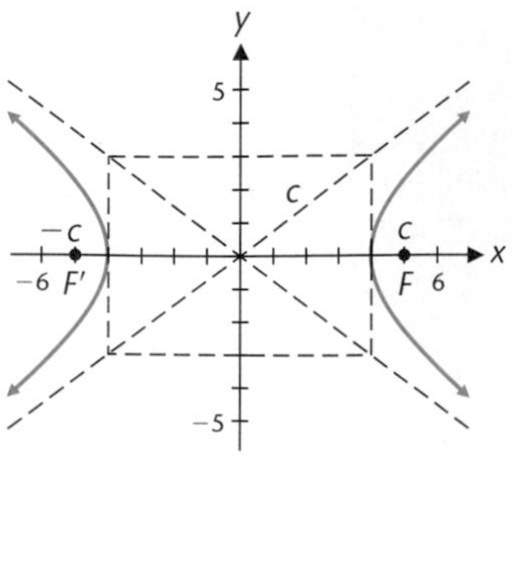

$$\begin{aligned} \text{Foci:}\quad c^2 &= a^2 + b^2 \\ &= 16 + 9 \\ &= 25 \\ c &= 5 \end{aligned}$$

Thus, the foci are $F'(-5, 0)$ and $F(5, 0)$.

Transverse axis length $= 2(4) = 8$

Conjugate axis length $= 2(3) = 6$

FIGURE 6
$y_1 = \sqrt{(9x^2 - 144)/16}$;
$y_2 = -\sqrt{(9x^2 - 144)/16}$.

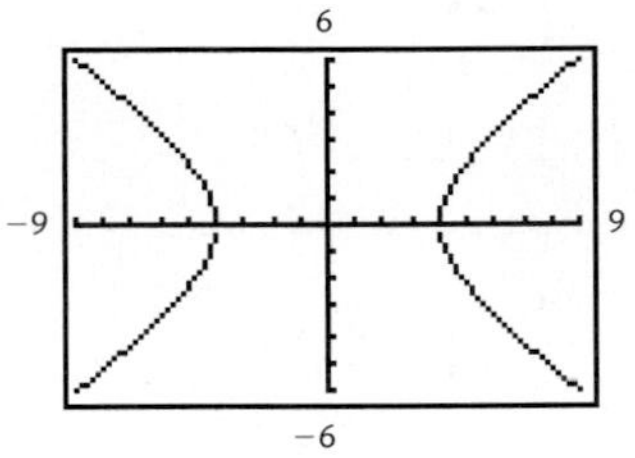

To check the graph on a graphing utility, we solve the original equation for y:

$$9x^2 - 16y^2 = 144$$
$$y^2 = (9x^2 - 144)/16$$
$$y = \pm\sqrt{(9x^2 - 144)/16}$$

This produces two functions whose graphs are shown in Figure 6.

(B) $16y^2 - 9x^2 = 144$

$$\frac{y^2}{9} - \frac{x^2}{16} = 1 \qquad a^2 = 9 \text{ and } b^2 = 16$$

FIGURE 7
$16y^2 - 9x^2 = 144$.

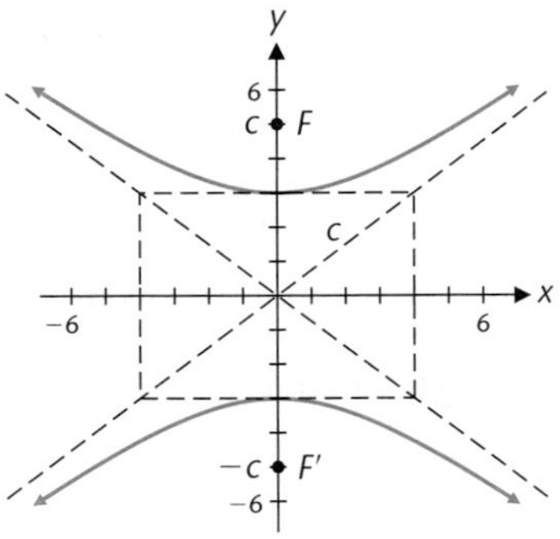

Locate y intercepts, $y = \pm 3$; there are no x intercepts. Sketch the asymptotes using the asymptote rectangle, then sketch in the hyperbola (Fig. 7). It is important to note that the transverse axis and the foci are on the y axis.

$$\text{Foci:}\quad c^2 = a^2 + b^2$$
$$= 9 + 16$$
$$= 25$$
$$c = 5$$

Thus, the foci are $F'(0, -5)$ and $F(0, 5)$.

$$\text{Transverse axis length} = 2(3) = 6$$
$$\text{Conjugate axis length} = 2(4) = 8$$

A check of the graph is shown in Figure 8.

FIGURE 8
$y_1 = \sqrt{(9x^2 + 144)/16}$;
$y_2 = -\sqrt{(9x^2 + 144)/16}$.

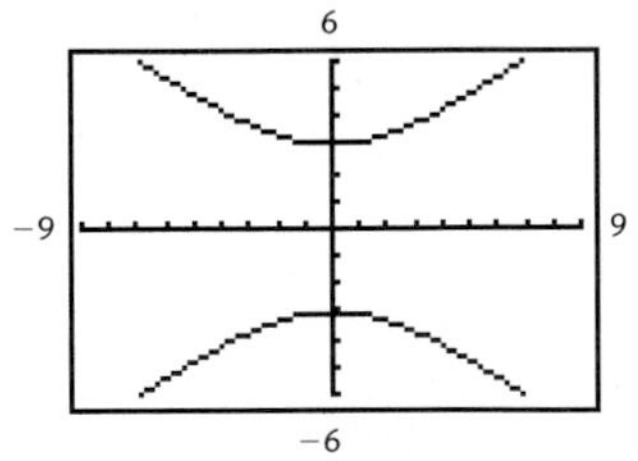

(C) $2x^2 - y^2 = 10$

$$\frac{x^2}{5} - \frac{y^2}{10} = 1 \qquad a^2 = 5 \text{ and } b^2 = 10$$

FIGURE 9
$2x^2 - y^2 = 10$.

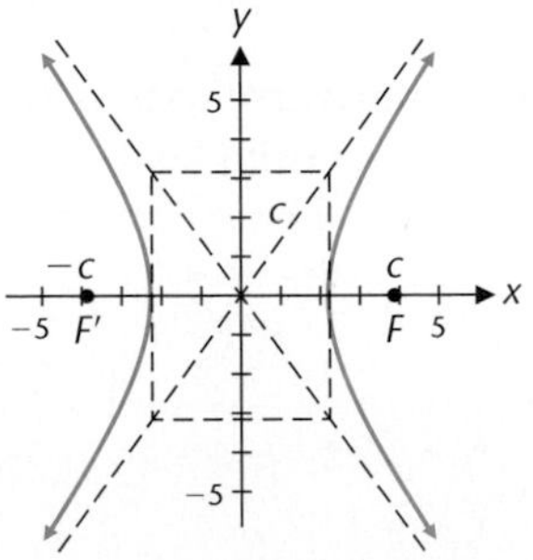

Locate x intercepts, $x = \pm\sqrt{5}$; there are no y intercepts. Sketch the asymptotes using the asymptote rectangle, then sketch in the hyperbola (Fig. 9).

$$\text{Foci:}\quad c^2 = a^2 + b^2$$
$$= 5 + 10$$
$$= 15$$
$$c = \sqrt{15}$$

FIGURE 10
$y_1 = \sqrt{2x^2 - 10}$;
$y_2 = -\sqrt{2x^2 - 10}$.

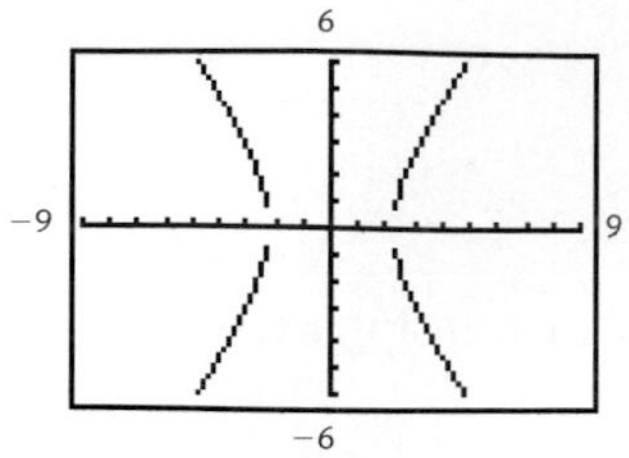

Thus, the foci are $F'(-\sqrt{15}, 0)$ and $F(\sqrt{15}, 0)$.

$$\text{Transverse axis length} = 2\sqrt{5} \approx 4.47$$

$$\text{Conjugate axis length} = 2\sqrt{10} \approx 6.32$$

A check of the graph is shown in Figure 10.

MATCHED PROBLEM 1

Sketch the graph of each equation, find the coordinates of the foci, and find the lengths of the transverse and conjugate axes. Check by graphing on a graphing utility.

(A) $16x^2 - 25y^2 = 400$ (B) $25y^2 - 16x^2 = 400$ (C) $y^2 - 3x^2 = 12$

Hyperbolas of the form

$$\frac{x^2}{M} - \frac{y^2}{N} = 1 \qquad \text{and} \qquad \frac{y^2}{N} - \frac{x^2}{M} = 1 \qquad M, N > 0$$

are called **conjugate hyperbolas.** In Example 1 and Matched Problem 1, the hyperbolas in parts A and B are conjugate hyperbolas—they share the same asymptotes.

CAUTION

When making a quick sketch of a hyperbola, it is a common error to have the hyperbola opening up and down when it should open left and right, or vice versa. The mistake can be avoided if you first locate the intercepts accurately.

EXAMPLE 2

Finding the Equation of a Hyperbola

Find an equation of a hyperbola in the form

$$\frac{y^2}{M} - \frac{x^2}{N} = 1 \qquad M, N > 0$$

if the center is at the origin, and:

(A) Length of transverse axis is 12
Length of conjugate axis is 20

(B) Length of transverse axis is 6
Distance of foci from center is 5

Solutions

(A) Start with

$$\frac{y^2}{a^2} - \frac{x^2}{b^2} = 1$$

and find a and b:

$$a = \frac{12}{2} = 6 \quad \text{and} \quad b = \frac{20}{2} = 10$$

Thus, the equation is

$$\frac{y^2}{36} - \frac{x^2}{100} = 1$$

(B) Start with

$$\frac{y^2}{a^2} - \frac{x^2}{b^2} = 1$$

and find a and b:

$$a = \frac{6}{2} = 3$$

FIGURE 11
Asymptote rectangle.

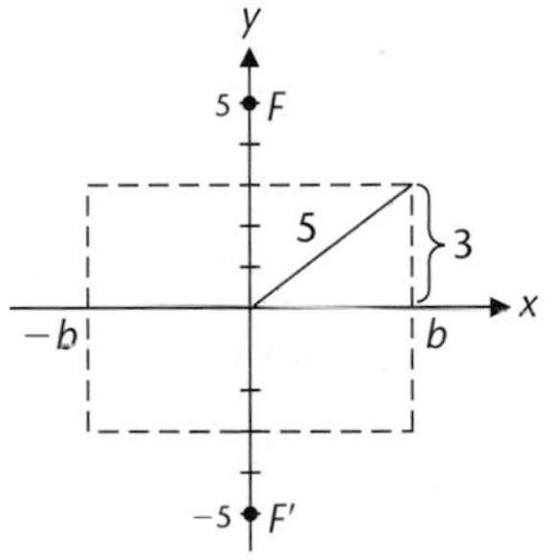

To find b, sketch the asymptote rectangle (Fig. 11), label known parts, and use the Pythagorean theorem:

$$b^2 = 5^2 - 3^2$$
$$= 16$$
$$b = 4$$

Thus, the equation is

$$\frac{y^2}{9} - \frac{x^2}{16} = 1$$

MATCHED PROBLEM 2

Find an equation of a hyperbola in the form

$$\frac{x^2}{M} - \frac{y^2}{N} = 1 \qquad M, N > 0$$

if the center is at the origin, and:

(A) Length of transverse axis is 50
Length of conjugate axis is 30

(B) Length of conjugate axis is 12
Distance of foci from center is 9

Explore/Discuss 2

(A) Does the line with equation $y = x$ intersect the hyperbola with equation $x^2 - (y^2/4) = 1$? If so, find the coordinates of all intersection points.

Explore/Discuss 2 continued

(B) Does the line with equation $y = 3x$ intersect the hyperbola with equation $x^2 - (y^2/4) = 1$? If so, find the coordinates of all intersection points.

(C) For which values of m does the line with equation $y = mx$ intersect the hyperbola $\frac{x^2}{a^2} - \frac{y^2}{b^2} = 1$? Find the coordinates of all intersection points.

Applications

You may not be aware of the many important uses of hyperbolic forms. They are encountered in the study of comets; the loran system of navigation for pleasure boats, ships, and aircraft; sundials; capillary action; nuclear cooling towers; optical and radiotelescopes; and contemporary architectural structures. The TWA building at Kennedy Airport is a *hyperbolic paraboloid,* and the St. Louis Science Center Planetarium is a *hyperboloid.* With such structures, thin concrete shells can span large spaces [see Fig. 12(a)]. Some comets from outer space occasionally enter the sun's gravitational field, follow a hyperbolic path around the sun (with the sun as a focus), and then leave, never to be seen again [Fig. 12(b)]. The next example illustrates the use of hyperbolas in navigation.

FIGURE 12
Uses of hyperbolic forms.

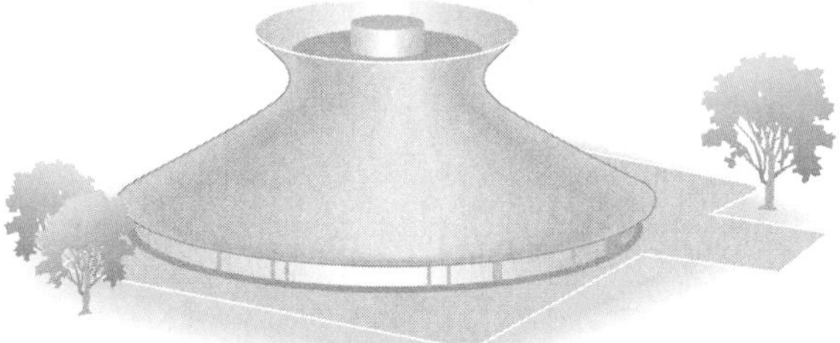

St. Louis Planetarium
(a)

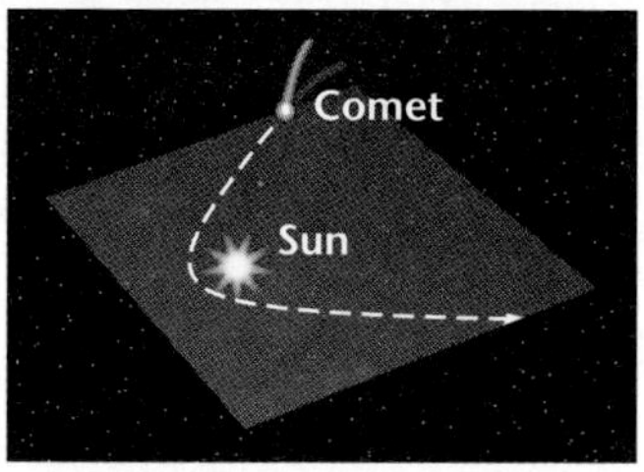

Comet around sun
(b)

EXAMPLE 3

Navigation

A ship is traveling on a course parallel to and 60 miles from a straight shoreline. Two transmitting stations, S_1 and S_2, are located 200 miles apart on the shoreline (see Fig. 13). By timing radio signals from the stations, the ship's navigator determines that the ship is between the two stations and 50 miles closer to S_2 than to S_1. Find the distance from the ship to each station. Round answers to one decimal place.

FIGURE 13
$d_1 - d_2 = 50.$

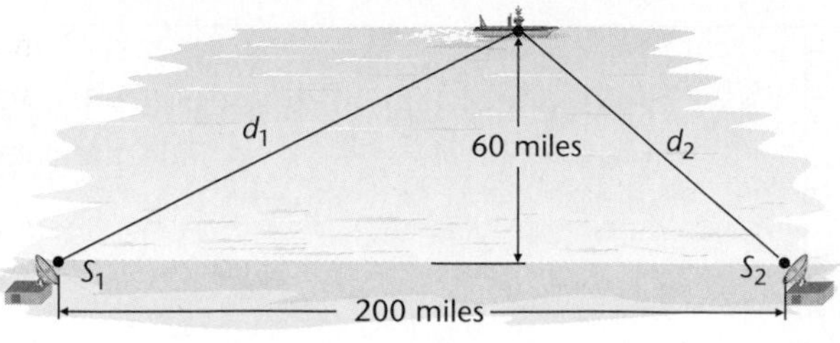

Solution If d_1 and d_2 are the distances from the ship to S_1 and S_2, respectively, then $d_1 - d_2 = 50$ and the ship must be on the hyperbola with foci at S_1 and S_2 and fixed difference 50, as illustrated in Figure 14. In the derivation of the equation of a hyperbola, we represented the fixed difference as $2a$. Thus, for the hyperbola in Figure 14 we have:

FIGURE 14

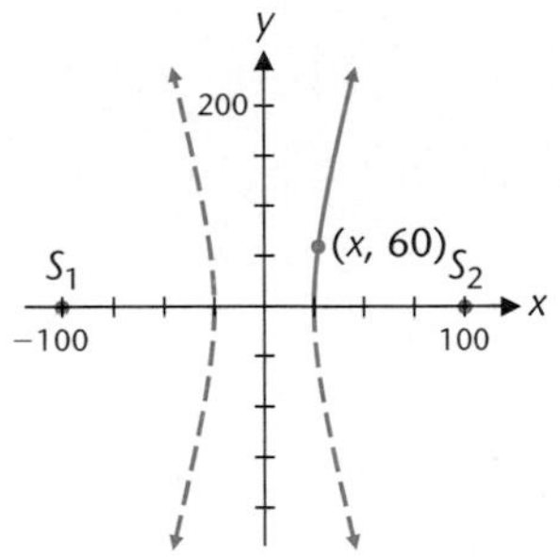

$$c = 100$$
$$a = \tfrac{1}{2}(50) = 25$$
$$b = \sqrt{100^2 - 25^2} = \sqrt{9{,}375}$$

The equation for this hyperbola is

$$\frac{x^2}{625} - \frac{y^2}{9{,}375} = 1$$

Substitute $y = 60$ and solve for x (see Fig. 14):

$$\frac{x^2}{625} - \frac{60^2}{9{,}375} = 1$$
$$\frac{x^2}{625} = \frac{3{,}600}{9{,}375} + 1$$
$$x^2 = 625\,\frac{3{,}600 + 9{,}375}{9{,}375}$$
$$= 865$$

Thus, $x = \sqrt{865} \approx 29.41$. (The negative square root is discarded, since the ship is closer to S_2 than to S_1.)

Distance from ship to S_1	Distance from ship to S_2
$d_1 = \sqrt{(29.41 + 100)^2 + 60^2}$	$d_2 = \sqrt{(29.41 - 100)^2 + 60^2}$
$= \sqrt{20{,}346.9841}$	$= \sqrt{8{,}582.9841}$
≈ 142.6	≈ 92.6 miles

Notice that the difference between these two distances is 50, as it should be.

MATCHED PROBLEM 3 Repeat Example 3 if the ship is 80 miles closer to S_2 than to S_1.

Example 3 illustrates a simplified form of the loran (LOng RAnge Navigation) system. In practice, three transmitting stations are used to send out signals simultaneously (Fig. 15), instead of the two used in Example 3. A computer onboard a ship will record these signals and use them to determine the differences of the distances that the ship is to S_1 and S_2, and to S_2 and S_3.

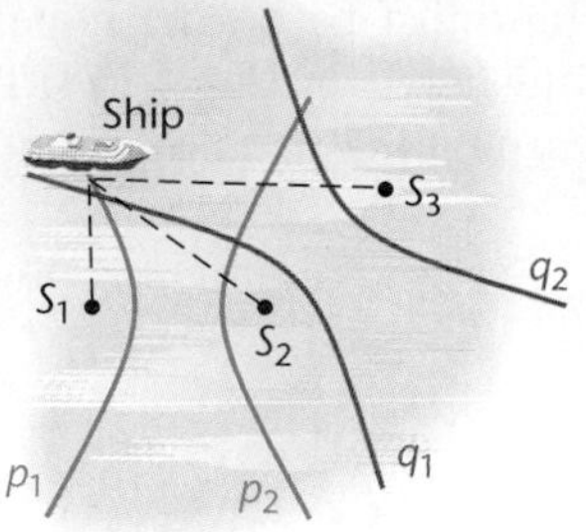

FIGURE 15
Loran navigation.

Plotting all points so that these distances remain constant produces two branches, p_1 and p_2, of a hyperbola with foci S_1 and S_2, and two branches, q_1 and q_2, of a hyperbola with foci S_2 and S_3. It is easy to tell which branches the ship is on by comparing the signals from each station. The intersection of a branch of each hyperbola locates the ship and the computer expresses this in terms of longitude and latitude.

Answers to Matched Problems

1. (A)

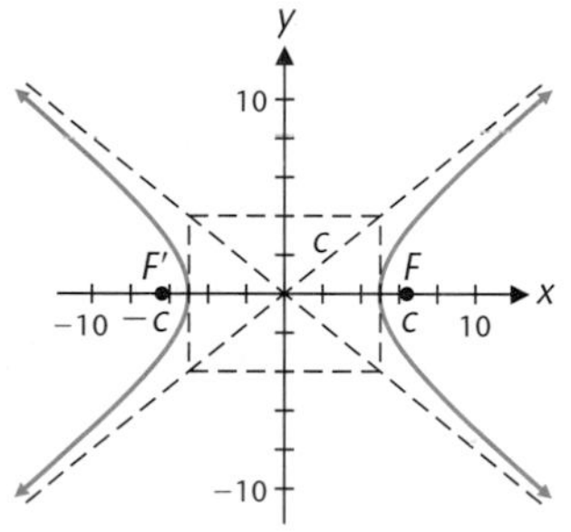

$$\frac{x^2}{25} - \frac{y^2}{16} = 1$$

Foci: $F'(-\sqrt{41}, 0)$, $F(\sqrt{41}, 0)$
Transverse axis length = 10
Conjugate axis length = 8

(B)

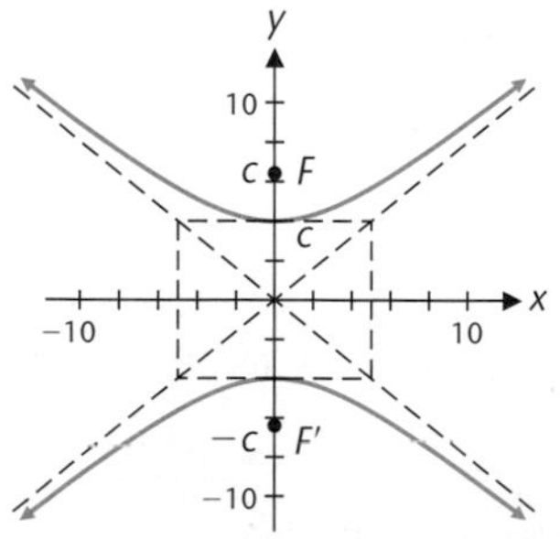

$$\frac{y^2}{16} - \frac{x^2}{25} = 1$$

Foci: $F'(0, -\sqrt{41})$, $F(0, \sqrt{41})$
Transverse axis length = 8
Conjugate axis length = 10

(C)

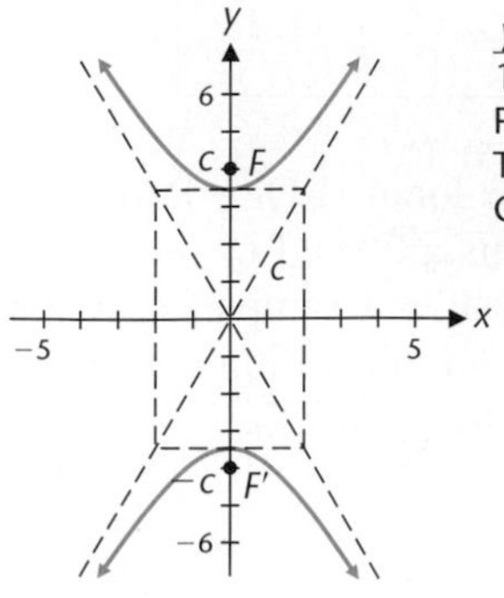

$$\frac{y^2}{12} - \frac{x^2}{4} = 1$$

Foci: $F'(0, -4)$, $F(0, 4)$
Transverse axis length $= 2\sqrt{12} \approx 6.93$
Conjugate axis length = 4

2. (A) $\frac{x^2}{625} - \frac{y^2}{225} = 1$ (B) $\frac{x^2}{45} - \frac{y^2}{36} = 1$ **3.** $d_1 = 159.5$ miles, $d_2 = 79.5$ miles

EXERCISE 7-3

A

In Problems 1–4, match each equation with one of graphs (a)–(d).

1. $x^2 - y^2 = 1$

2. $y^2 - x^2 = 1$

3. $y^2 - x^2 = 4$

4. $x^2 - y^2 = 4$

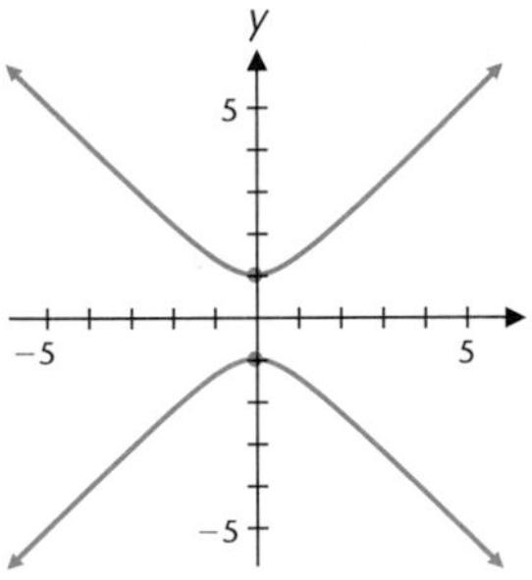

(a)

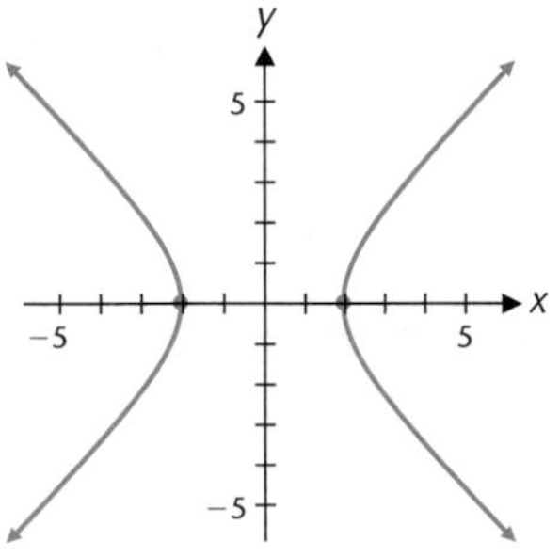

(b)

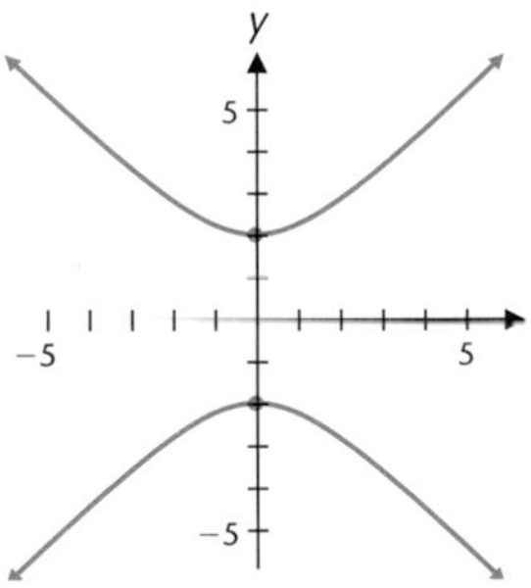

(c)

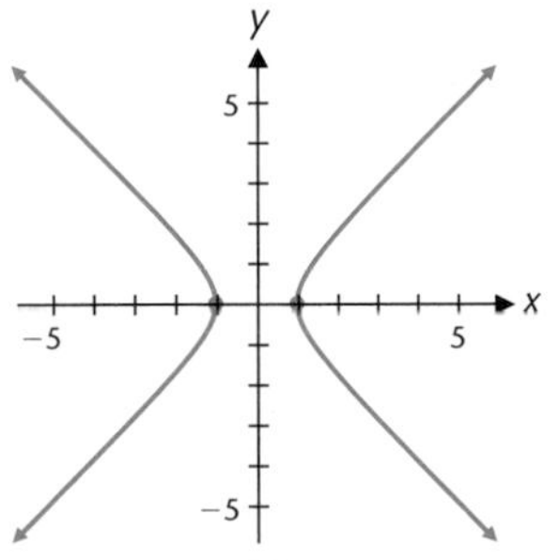

(d)

Sketch a graph of each equation in Problems 5–12, find the coordinates of the foci, and find the lengths of the transverse and conjugate axes. Check by graphing on a graphing utility.

5. $\frac{x^2}{9} - \frac{y^2}{4} = 1$

6. $\frac{x^2}{9} - \frac{y^2}{25} = 1$

7. $\frac{y^2}{4} - \frac{x^2}{9} = 1$

8. $\frac{y^2}{25} - \frac{x^2}{9} = 1$

9. $4x^2 - y^2 = 16$

10. $x^2 - 9y^2 = 9$

11. $9y^2 - 16x^2 = 144$

12. $4y^2 - 25x^2 = 100$

B

Sketch a graph of each equation in Problems 13–16, find the coordinates of the foci, and find the lengths of the transverse and conjugate axes. Check by graphing on a graphing utility.

13. $3x^2 - 2y^2 = 12$

14. $3x^2 - 4y^2 = 24$

15. $7y^2 - 4x^2 = 28$

16. $3y^2 - 2x^2 = 24$

In Problems 17–28, find an equation of a hyperbola in the form

$$\frac{x^2}{M} - \frac{y^2}{N} = 1 \quad \textit{or} \quad \frac{y^2}{N} - \frac{x^2}{M} = 1 \quad M, N > 0$$

if the center is at the origin, and

17. The graph is

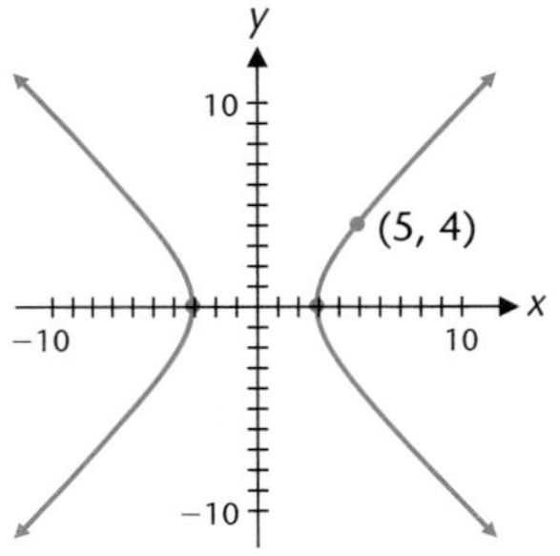

18. The graph is

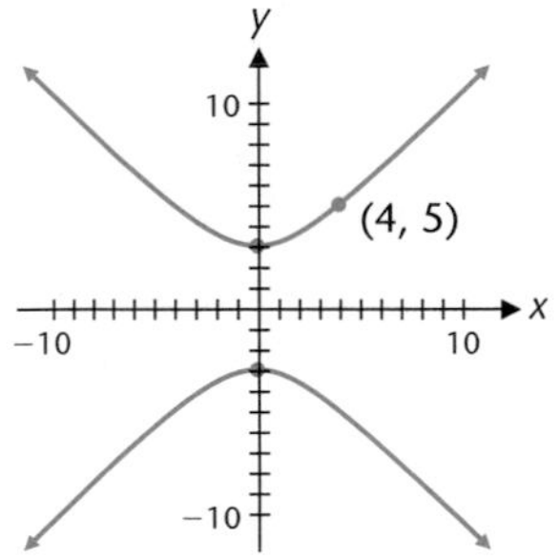

19. The graph is

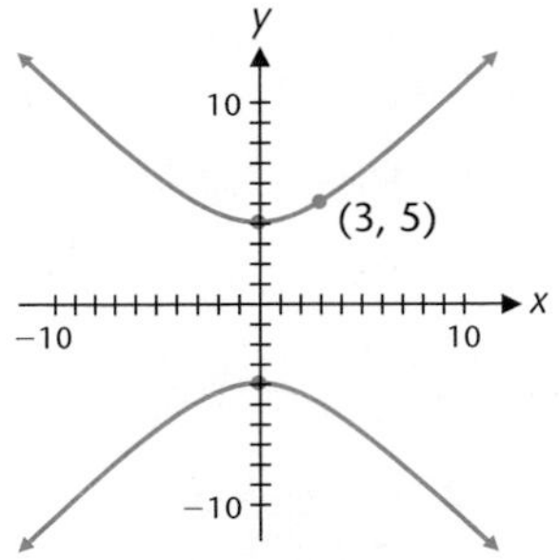

20. The graph is

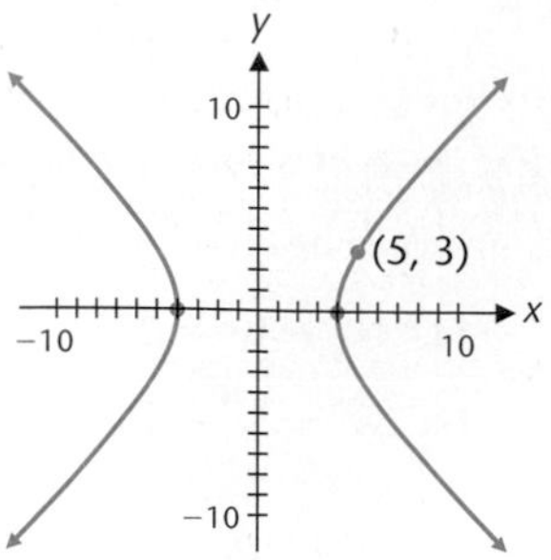

21. Transverse axis on x axis
Transverse axis length $= 14$
Conjugate axis length $= 10$

22. Transverse axis on x axis
Transverse axis length $= 8$
Conjugate axis length $= 6$

23. Transverse axis on y axis
Transverse axis length $= 24$
Conjugate axis length $= 18$

24. Transverse axis on y axis
Transverse axis length $= 16$
Conjugate axis length $= 22$

25. Transverse axis on x axis
Transverse axis length $= 18$
Distance of foci from center $= 11$

26. Transverse axis on x axis
Transverse axis length $= 16$
Distance of foci from center $= 10$

27. Conjugate axis on x axis
Conjugate axis length $= 14$
Distance of foci from center $= \sqrt{200}$

28. Conjugate axis on x axis
Conjugate axis length $= 10$
Distance of foci from center $= \sqrt{70}$

29. (A) How many hyperbolas have center at (0, 0) and a focus at (1, 0)? Find their equations.

(B) How many ellipses have center at (0, 0) and a focus at (1, 0)? Find their equations.

(C) How many parabolas have center at (0, 0) and focus at (1, 0)? Find their equations.

30. How many hyperbolas have the lines $y = \pm 2x$ as asymptotes? Find their equations.

In Problems 31–38, find the coordinates of all points of intersection. Round any approximate values to three decimal places.

31. $3y^2 - 4x^2 = 12$
$y^2 + x^2 = 25$

32. $y^2 - x^2 = 3$
$y^2 + x^2 = 5$

33. $2x^2 + y^2 = 24$
$x^2 - y^2 = -12$

34. $2x^2 + y^2 = 17$
$x^2 - y^2 = -5$

35. $y^2 - x^2 = 9$
$2y - x = 8$

36. $y^2 - x^2 = 4$
$y - x = 6$

37. $y^2 - x^2 = 4$
$y^2 + 2x^2 = 36$

38. $y^2 - x^2 = 1$
$2y^2 + x^2 = 16$

C

Eccentricity. *Problems 39 and 40 below and Problems 39 and 40 in Exercise 7-2 are related to a property of conics called* ***eccentricity,*** *which is denoted by a positive real number E. Parabolas, ellipses, and hyperbolas all can be defined in terms of E, a fixed point called a focus, and a fixed line not containing the focus called a directrix as follows: The set of points in a plane each of whose distance from a fixed point is E times its distance from a fixed line is an ellipse if $0 < E < 1$, a parabola if $E = 1$, and a hyperbola if $E > 1$.*

39. Find an equation of the set of points in a plane each of whose distance from (3, 0) is three-halves its distance from the line $x = \frac{4}{3}$. Identify the geometric figure.

40. Find an equation of the set of points in a plane each of whose distance from (0, 4) is four-thirds its distance from the line $y = \frac{9}{4}$. Identify the geometric figure.

In Problems 41–44, find the coordinates of all points of intersection to two decimal places.

41. $2x^2 - 3y^2 = 20,\ 7x + 15y = 10$

42. $y^2 - 3x^2 = 8,\ x^2 = -\dfrac{y}{3}$

43. $24y^2 - 18x^2 = 175,\ 90x^2 + 3y^2 = 200$

44. $8x^2 - 7y^2 = 58,\ 4y^2 - 11x^2 = 45$

APPLICATIONS

45. Architecture. An architect is interested in designing a thin-shelled dome in the shape of a hyperbolic paraboloid, as shown in Figure (a). Find the equation of the hyperbola located in a coordinate system [Fig. (b)] satisfying the indicated conditions. How far is the hyperbola above the vertex 6 feet to the right of the vertex? Compute the answer to two decimal places.

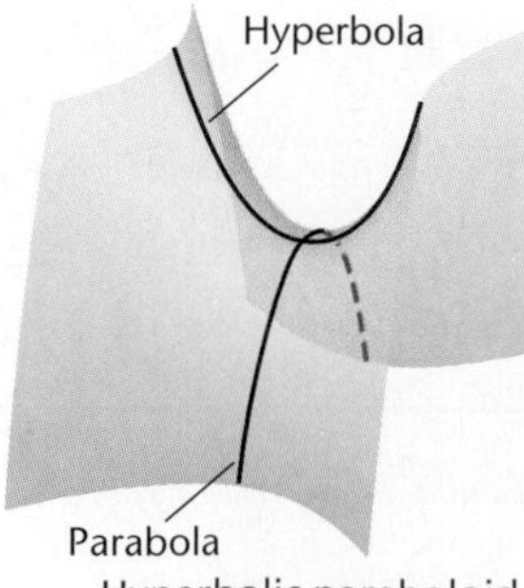

Hyperbolic paraboloid
(a)

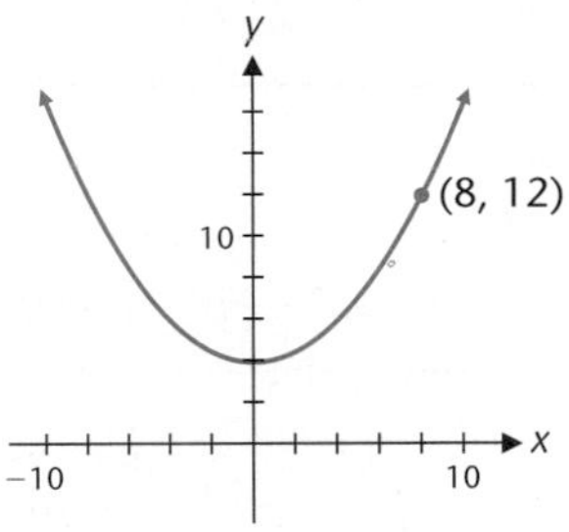

Hyperbola part of dome
(b)

46. Nuclear Power. A nuclear cooling tower is a **hyperboloid,** that is, a hyperbola rotated around its conjugate axis, as shown in Figure (a). The equation of the hyperbola in Figure (b) used to generate the hyperboloid is

$$\frac{x^2}{100^2} - \frac{y^2}{150^2} = 1$$

Nuclear cooling tower
(a)

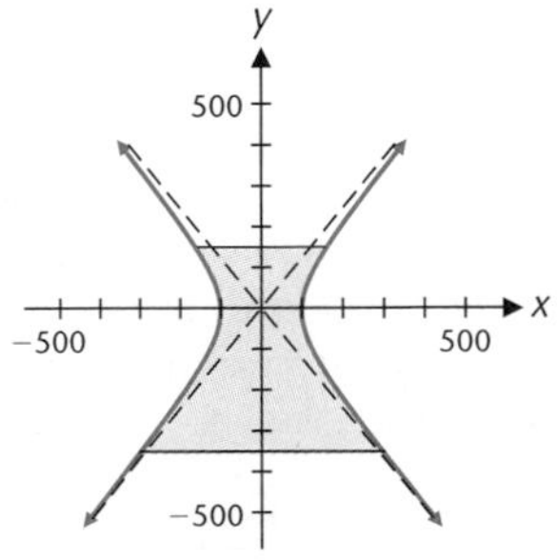

Hyperbola part of dome
(b)

If the tower is 500 feet tall, the top is 150 feet above the center of the hyperbola, and the base is 350 feet below the center, what is the radius of the top and the base? What is the radius of the smallest circular cross section in the tower? Compute answers to 3 significant digits.

47. Space Science. In tracking space probes to the outer planets, NASA uses large parabolic reflectors with diameters equal to two-thirds the length of a football field. Needless to say, many design problems are created by the weight of these reflectors. One weight problem is solved by using a hyperbolic reflector sharing the parabola's focus to reflect the incoming electromagnetic waves to the other focus of the hyperbola where receiving equipment is installed (see the figure).

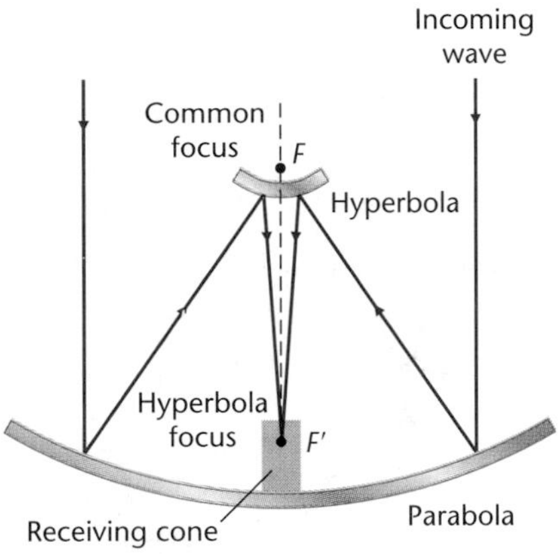

(a)

(b)

For the receiving antenna shown in the figure, the common focus F is located 120 feet above the vertex of the parabola, and focus F' (for the hyperbola) is 20 feet above the vertex. The vertex of the reflecting hyperbola is 110 feet above the vertex for the parabola. Introduce a coordinate system by using the axis of the parabola as the y axis (up positive), and let the x axis pass through the center of the hyperbola (right positive). What is the equation of the reflecting hyperbola? Write y in terms of x.

Section 7-4 Translation of Axes

In the last three sections we found standard equations for parabolas, ellipses, and hyperbolas located with their axes on the coordinate axes and centered relative to the origin. What happens if we move conics away from the origin while keeping their axes parallel to the coordinate axes? We will show that we can obtain new standard equations that are special cases of the equation $Ax^2 + Cy^2 + Dx + Ey + F = 0$, where A and C are not both zero. The basic mathematical tool used in this endeavor is *translation of axes*. The usefulness of translation of axes is not limited to graphing conics, however. Translation of axes can be put to good use in many other graphing situations.

Translation of Axes

A **translation of coordinate axes** occurs when the new coordinate axes have the same direction as and are parallel to the original coordinate axes. To see how coordinates in the original system are changed when moving to the translated system, and vice versa, refer to Figure 1.

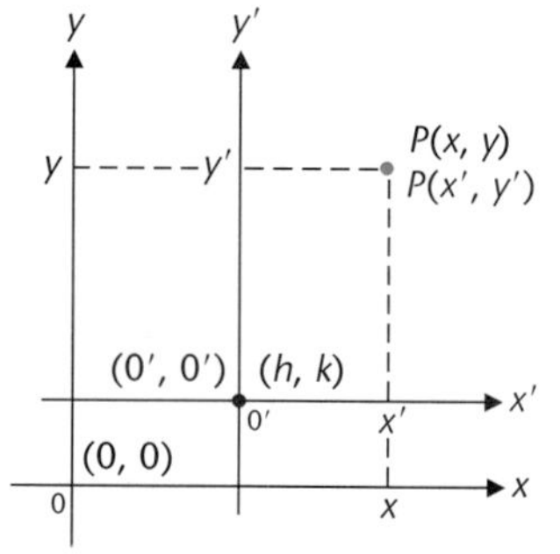

FIGURE 1
Translation of coordinates.

A point P in the plane has two sets of coordinates: (x, y) in the original system and (x', y') in the translated system. If the coordinates of the origin of the translated system are (h, k) relative to the original system, then the old and new coordinates are related as given in Theorem 1.

THEOREM 1

TRANSLATION FORMULAS

1. $x = x' + h$
   $y = y' + k$

2. $x' = x - h$
   $y' = y - k$

It can be shown that these formulas hold for (h, k) located anywhere in the original coordinate system.

EXAMPLE 1

Equation of a Curve in a Translated System

A curve has the equation

$$(x - 4)^2 + (y + 1)^2 = 36$$

If the origin is translated to (4, −1), find the equation of the curve in the translated system and identify the curve.

Solution

Since $(h, k) = (4, -1)$, use translation formulas

$$x' = x - h = x - 4$$
$$y' = y - k = y + 1$$

to obtain, after substitution,

$$x'^2 + y'^2 = 36$$

This is the equation of a circle of radius 6 with center at the new origin. The coordinates of the new origin in the original coordinate system are (4, −1) (Fig. 2). Note that this result agrees with our general treatment of the circle in Section 1-1.

FIGURE 2
$(x - 4)^2 + (y + 1)^2 = 36.$

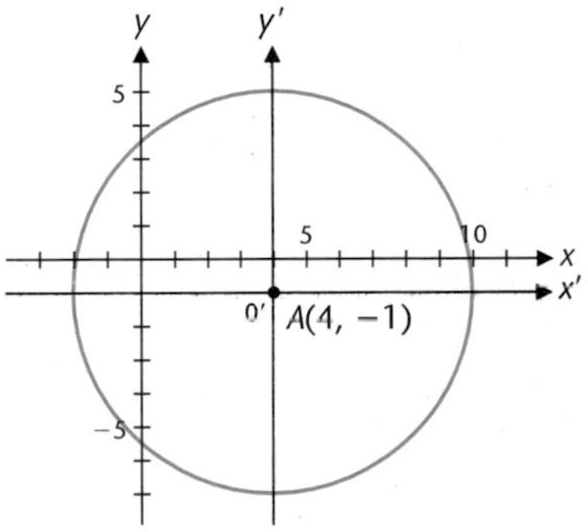

MATCHED PROBLEM 1

A curve has the equation $(y + 2)^2 = 8(x - 3)$. If the origin is translated to (3, −2), find an equation of the curve in the translated system and identify the curve.

Standard Equations of Translated Conics

We now proceed to find standard equations of conics translated away from the origin. We do this by first writing the standard equations found in earlier sections in the $x'y'$ coordinate system with $0'$ at (h, k). We then use translation equations to find the standard forms relative to the original xy coordinate system. The equations of translation in all cases are

$$x' = x - h$$
$$y' = y - k$$

For parabolas we have

$$x'^2 = 4ay' \qquad (x - h)^2 = 4a(y - k)$$
$$y'^2 = 4ax' \qquad (y - k)^2 = 4a(x - h)$$

For circles we have

$$x'^2 + y'^2 = r^2 \qquad (x - h)^2 + (y - k)^2 = r^2$$

For ellipses we have for $a > b > 0$

$$\frac{x'^2}{a^2} + \frac{y'^2}{b^2} = 1 \qquad \frac{(x - h)^2}{a^2} + \frac{(y - k)^2}{b^2} = 1$$
$$\frac{x'^2}{b^2} + \frac{y'^2}{a^2} = 1 \qquad \frac{(x - h)^2}{b^2} + \frac{(y - k)^2}{a^2} = 1$$

For hyperbolas we have

$$\frac{x'^2}{a^2} - \frac{y'^2}{b^2} = 1 \qquad \frac{(x - h)^2}{a^2} - \frac{(y - k)^2}{b^2} = 1$$
$$\frac{y'^2}{a^2} - \frac{x'^2}{b^2} = 1 \qquad \frac{(y - k)^2}{a^2} - \frac{(x - h)^2}{b^2} = 1$$

Table 1 summarizes these results with appropriate figures and some properties discussed earlier.

Graphing Equations of the Form $Ax^2 + Cy^2 + Dx + Ey + F = 0$

It can be shown that the graph of

$$Ax^2 + Cy^2 + Dx + Ey + F = 0 \qquad (1)$$

where A and C are not both zero, is a conic or a degenerate conic or that there is no graph. If we can transform equation (1) into one of the standard forms in Table 1, then we will be able to identify its graph and sketch it rather quickly. The process of completing the square discussed in Section 2-3 will be our primary tool in accomplishing this transformation. A couple of examples should help make the process clear.

EXAMPLE 2

Graphing a Translated Conic

Transform

$$y^2 - 6y - 4x + 1 = 0 \qquad (2)$$

into one of the standard forms in Table 1. Identify the conic and graph it. Check by graphing on a graphing utility.

TABLE 1 Standard Equations for Translated Conics

Parabolas

$$(x - h)^2 = 4a(y - k)$$

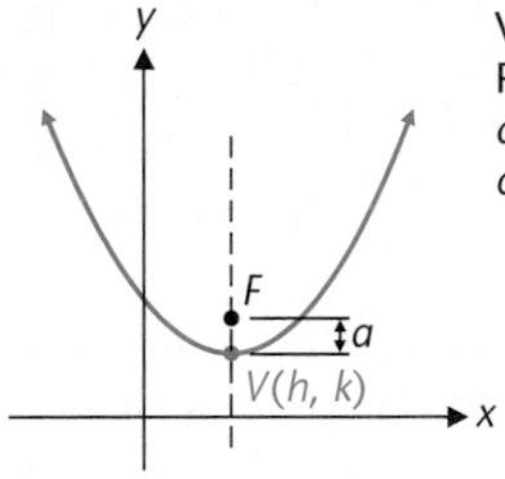

Vertex (h, k)
Focus $(h, k + a)$
$a > 0$ opens up
$a < 0$ opens down

$$(y - k)^2 = 4a(x - h)$$

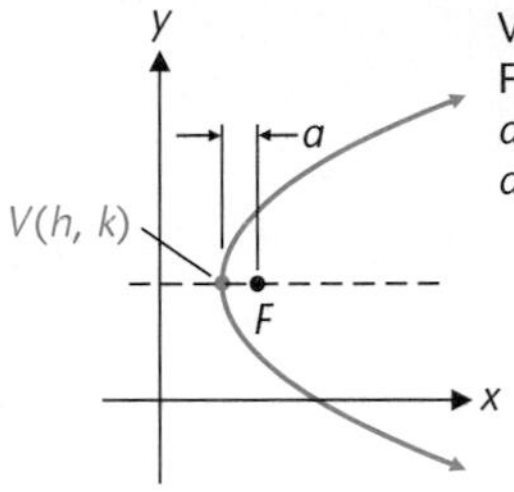

Vertex (h, k)
Focus $(h + a, k)$
$a < 0$ opens left
$a > 0$ opens right

Circles

$$(x - h)^2 + (y - k)^2 = r^2$$

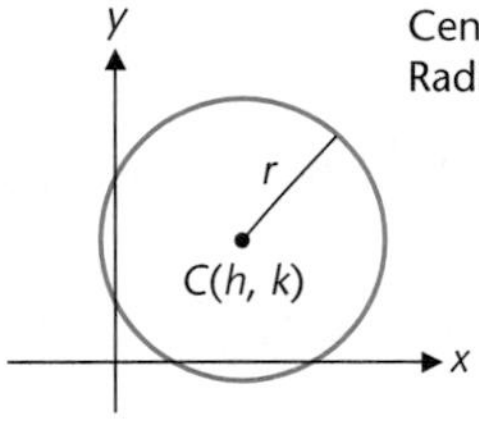

Center (h, k)
Radius r

Ellipses

$$\frac{(x - h)^2}{a^2} + \frac{(y - k)^2}{b^2} = 1 \qquad a > b > 0 \qquad \frac{(x - h)^2}{b^2} + \frac{(y - k)^2}{a^2} = 1$$

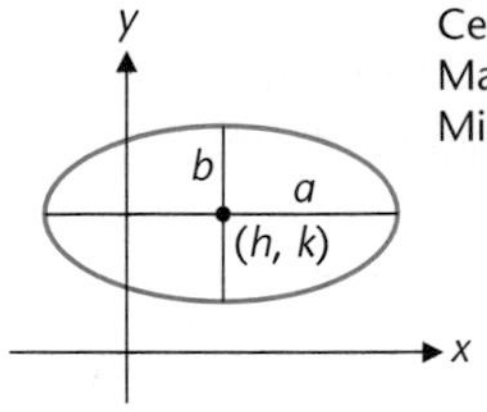

Center (h, k)
Major axis $2a$
Minor axis $2b$

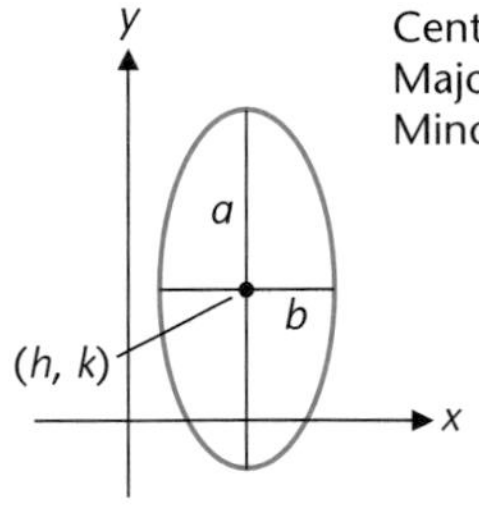

Center (h, k)
Major axis $2a$
Minor axis $2b$

Hyperbolas

$$\frac{(x - h)^2}{a^2} - \frac{(y - k)^2}{b^2} = 1$$

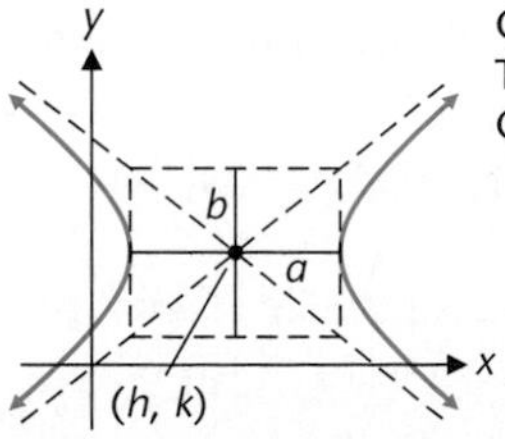

Center (h, k)
Transverse axis $2a$
Conjugate axis $2b$

$$\frac{(y - k)^2}{a^2} - \frac{(x - h)^2}{b^2} = 1$$

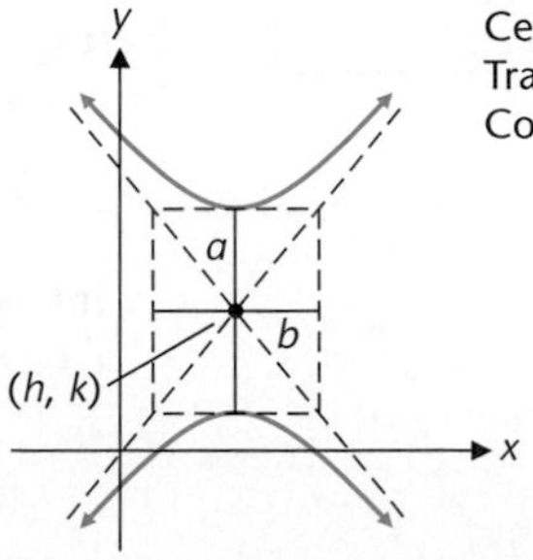

Center (h, k)
Transverse axis $2a$
Conjugate axis $2b$

Solution

Step 1. Complete the square in equation (2) relative to each variable that is squared—in this case y:

$$y^2 - 6y - 4x + 1 = 0$$

$$y^2 - 6y \qquad = 4x - 1$$

$$y^2 - 6y \quad + 9 = 4x + 8 \qquad \text{Add 9 to both sides to complete the square on the left side.}$$

$$(y - 3)^2 = 4(x + 2) \tag{3}$$

From Table 1 we recognize equation (3) as an equation of a parabola opening to the right with vertex at $(h, k) = (-2, 3)$.

Step 2. Find the equation of the parabola in the translated system with origin $0'$ at $(h, k) = (-2, 3)$. The equations of translation are read directly from equation (3):

$$x' = x + 2$$

$$y' = y - 3$$

FIGURE 3
$y^2 - 6y - 4x + 1 = 0.$

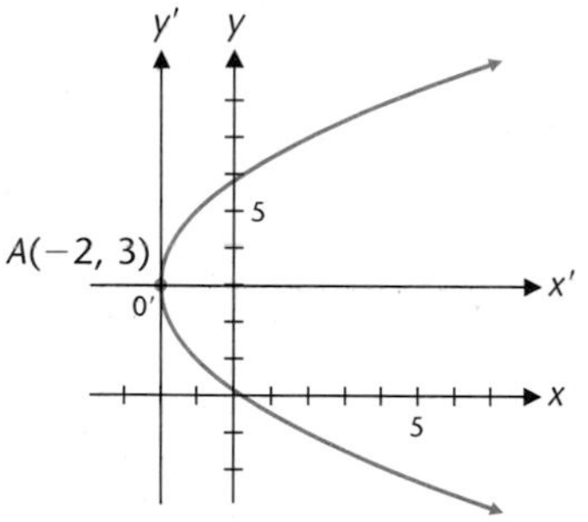

Making these substitutions in equation (3) we obtain

$$y'^2 = 4x' \tag{4}$$

the equation of the parabola in the $x'y'$ system.

Step 3. Graph equation (4) in the $x'y'$ system following the process discussed in Section 7-1. The resulting graph is the graph of the original equation relative to the original xy coordinate system (Fig. 3).

To check the graph in Figure 3 on a graphing utility, we can solve either equation (2) or equation (3) for y. Choosing equation (2) has the added benefit of providing a check of the derivation of equation (3).

$$y^2 - 6y - 4x + 1 = 0 \qquad \text{Quadratic equation with } a = 1,\ b = -6, \text{ and } c = -4x + 1$$

$$y = \frac{6 \pm \sqrt{36 - 4(1)(-4x + 1)}}{2(1)}$$

$$= \frac{6 \pm \sqrt{32 + 16x}}{2}$$

$$= 3 \pm 2\sqrt{2 + x} \tag{5}$$

FIGURE 4
$y_1 = 3 + 2\sqrt{2 + x}$;
$y_2 = 3 - 2\sqrt{2 + x}$.

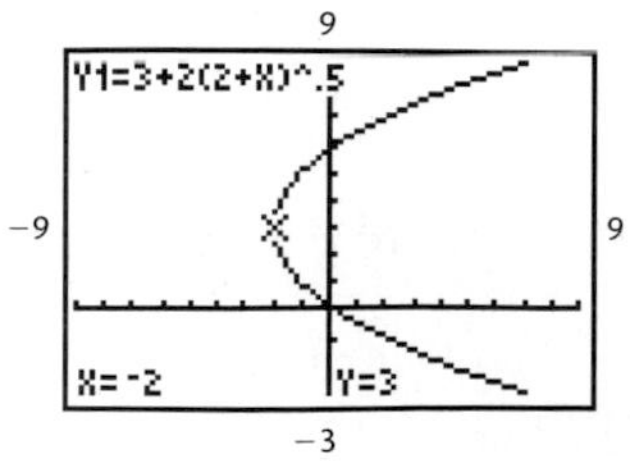

Figure 4 shows the graph of the two functions determined by equation (5) and the vertex of the parabola.

MATCHED PROBLEM 2

Transform

$$x^2 + 4x + 4y - 12 = 0$$

into one of the standard forms in Table 1. Identify the conic and graph it.

EXAMPLE 3

Graphing a Translated Conic

Transform

$$9x^2 - 4y^2 - 36x - 24y - 35 = 0$$

into one of the standard forms in Table 1. Identify the conic and graph it. Find the coordinates of any foci relative to the original system. Check by graphing on a graphing utility.

Solution

Step 1. Complete the square relative to both x and y.

$$\begin{aligned}
9x^2 - 4y^2 - 36x - 24y - 36 &= 0 \\
9x^2 - 36x \qquad - 4y^2 - 24y \qquad &= 36 \\
9(x^2 - 4x \qquad) - 4(y^2 + 6y \qquad) &= 36 \\
9(x^2 - 4x + 4) - 4(y^2 + 6y + 9) &= 36 + 36 - 36 \\
9(x - 2)^2 - 4(y + 3)^2 &= 36 \\
\frac{(x - 2)^2}{4} - \frac{(y + 3)^2}{9} &= 1
\end{aligned}$$

From Table 1 we recognize the last equation as an equation of a hyperbola opening left and right with center at $(h, k) = (2, -3)$.

Step 2. Find the equation of the hyperbola in the translated system with origin $0'$ at $(h, k) = (2, -3)$. The equations of translation are read directly from the last equation in step 1:

$$x' = x - 2$$
$$y' = y + 3$$

Making these substitutions, we obtain

$$\frac{x'^2}{4} - \frac{y'^2}{9} = 1$$

the equation of the hyperbola in the $x'y'$ system.

Step 3. Graph the equation obtained in step 2 in the $x'y'$ system following the process discussed in Section 7-3. The resulting graph is the graph of the original equation relative to the original xy coordinate system (Fig. 5).

FIGURE 5
$9x^2 - 4y^2 - 36x - 24y - 36 = 0$.

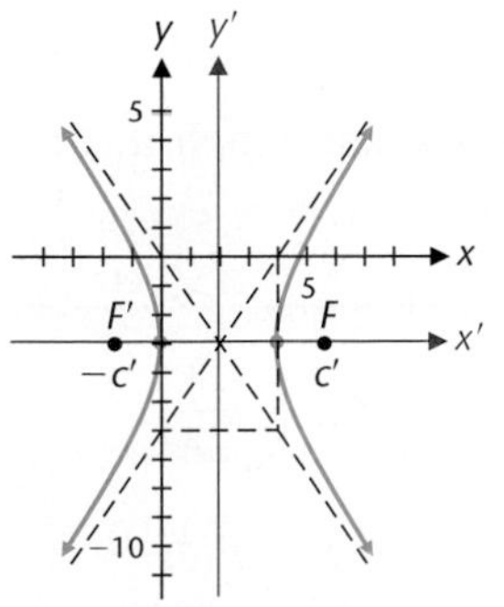

Step 4. Find the coordinates of the foci. To find the coordinates of the foci in the original system, first find the coordinates in the translated system:

$$c'^2 = 2^2 + 3^2 = 13$$
$$c' = \sqrt{13}$$
$$-c' = -\sqrt{13}$$

Thus, the coordinates in the translated system are

$$F'(-\sqrt{13}, 0) \quad \text{and} \quad F(\sqrt{13}, 0)$$

Now, use

$$x = x' + h = x' + 2$$
$$y = y' + k = y' - 3$$

to obtain

$$F'(-\sqrt{13} + 2, -3) \quad \text{and} \quad F(\sqrt{13} + 2, -3)$$

as the coordinates of the foci in the original system.

To check the graph in Figure 5, we return to the original equation and use the quadratic formula to solve for y:

FIGURE 6
$y_1 = -3 + 1.5\sqrt{x^2 - 4x}$;
$y_2 = -3 - 1.5\sqrt{x^2 - 4x}$.

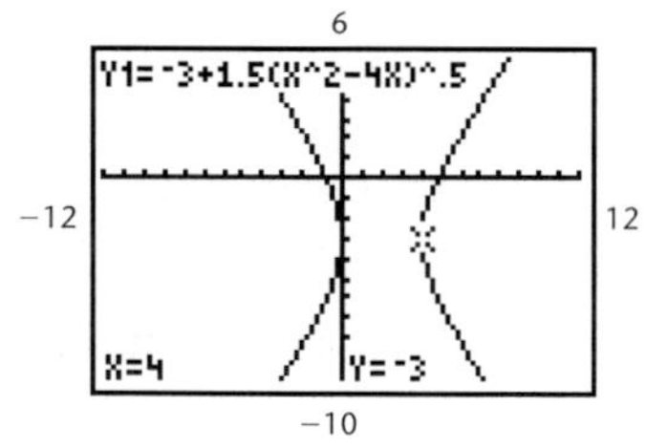

$$9x^2 - 4y^2 - 36x - 24y - 36 = 0 \quad \text{Write in the form } ay^2 + by + c = 0.$$
$$4y^2 + 24y + (-9x^2 + 36x + 36) = 0$$
$$y = \frac{-24 \pm \sqrt{24^2 - 4(4)(-9x^2 + 36x + 36)}}{8}$$
$$= -3 \pm 1.5\sqrt{x^2 - 4x} \qquad (6)$$

The two functions determined by equation (6) are graphed in Figure 6.

MATCHED PROBLEM 3

Transform

$$9x^2 + 16y^2 + 36x - 32y - 92 = 0$$

into one of the standard forms in Table 1. Identify the conic and graph it. Find the coordinates of any foci relative to the original system.

Explore/Discuss 1

If $A \neq 0$ and $C \neq 0$, show that the translation of axes $x' = x + \dfrac{D}{2A}$, $y' = y + \dfrac{E}{2C}$ transforms the equation $Ax^2 + Cy^2 + Dx + Ey + F = 0$ into an equation of the form $Ax'^2 + Cy'^2 = K$.

Finding Equations of Conics

We now reverse the problem: Given certain information about a conic in a rectangular coordinate system, find its equation.

EXAMPLE 4 Finding the Equation of a Translated Conic

Find the equation of a hyperbola with vertices on the line $x = -4$, conjugate axis on the line $y = 3$, length of the transverse axis $= 4$, and length of the conjugate axis $= 6$.

Solution Locate the vertices, asymptote rectangle, and asymptotes in the original coordinate system [Fig. 7(a)], then sketch the hyperbola and translate the origin to the center of the hyperbola [Fig. 7(b)].

FIGURE 7

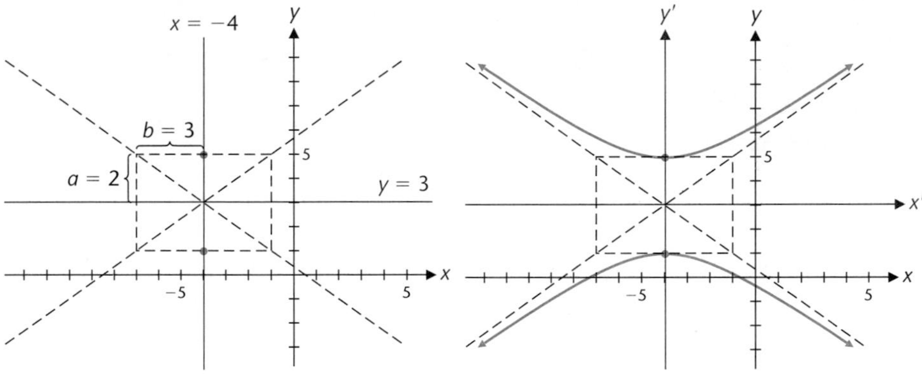

(a) Asymptote rectangle (b) Hyperbola

Next write the equation of the hyperbola in the translated system:

$$\frac{y'^2}{4} - \frac{x'^2}{9} = 1$$

The origin in the translated system is at $(h, k) = (-4, 3)$, and the translation formulas are

$$x' = x - h = x - (-4) = x + 4$$

$$y' = y - k = y - 3$$

Thus, the equation of the hyperbola in the original system is

$$\frac{(y - 3)^2}{4} - \frac{(x + 4)^2}{9} = 1$$

or, after simplifying and writing in the form of equation (1),

$$4x^2 - 9y^2 + 32x + 54y + 19 = 0$$

MATCHED PROBLEM 4 Find the equation of an ellipse with foci on the line $x = 4$, minor axis on the line $y = -3$, length of the major axis $= 8$, and length of the minor axis $= 4$.

Explore/Discuss 2

Use the strategy of completing the square to transform each equation to an equation in an $x'y'$ coordinate system. Note that the equation you obtain is not one of the standard forms in Table 1; instead, it is either the equation of a degenerate conic or the equation has no solution. If the solution set of the equation is not empty, graph it and identify the graph (a point, a line, two parallel lines, or two interesting lines).

(A) $x^2 + 2y^2 - 2x + 16y + 33 = 0$
(B) $4x^2 - y^2 - 24x - 2y + 35 = 0$
(C) $y^2 - 2y - 15 = 0$
(D) $5x^2 + y^2 + 12y + 40 = 0$
(E) $x^2 - 18x + 81 = 0$

Answers to Matched Problems

1. $y'^2 = 8x'$; a parabola
2. $(x + 2)^2 = -4(y - 4)$; a parabola

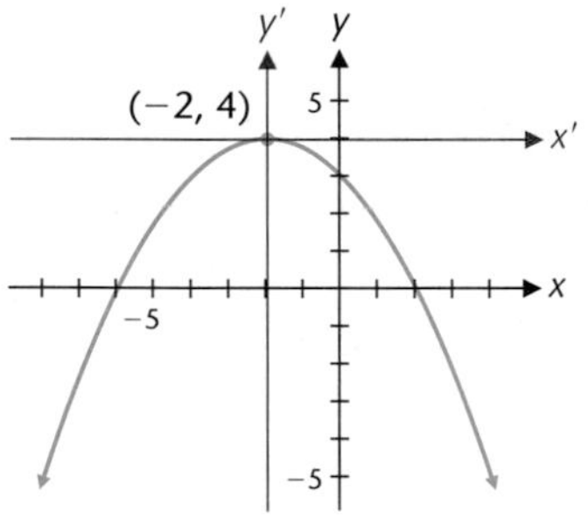

3. $\dfrac{(x + 2)^2}{16} + \dfrac{(y - 1)^2}{9} = 1$; ellipse   Foci: $F'(-\sqrt{7} - 2, 1)$, $F(\sqrt{7} - 2, 1)$

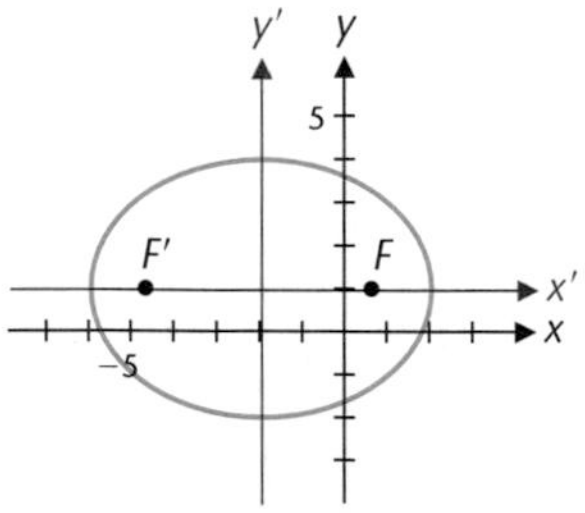

4. $\dfrac{(x - 4)^2}{4} + \dfrac{(y + 3)^2}{16} = 1$, or $4x^2 + y^2 - 32x + 6y + 57 = 0$

EXERCISE 7-4

A

In Problems 1–8:
(A) Find translation formulas that translate the origin to the indicated point (h, k).
(B) Write the equation of the curve for the translated system.
(C) Identify the curve.

1. $(x - 3)^2 + (y - 5)^2 = 81$; $(3, 5)$
2. $(x - 3)^2 = 8(y + 2)$; $(3, -2)$
3. $\dfrac{(x + 7)^2}{9} + \dfrac{(y - 4)^2}{16} = 1$; $(-7, 4)$

4. $(x + 2)^2 + (y + 6)^2 = 36; (-2, -6)$

5. $(y + 9)^2 = 16(x - 4); (4, -9)$

6. $\dfrac{(y - 9)^2}{10} - \dfrac{(x + 5)^2}{6} = 1; (-5, 9)$

7. $\dfrac{(x + 8)^2}{12} + \dfrac{(y + 3)^2}{8} = 1; (-8, -3)$

8. $\dfrac{(x + 7)^2}{25} - \dfrac{(y - 8)^2}{50} = 1; (-7, 8)$

In Problems 9–14:

(A) Write each equation in one of the standard forms listed in Table 1.

(B) Identify the curve.

9. $16(x - 3)^2 - 9(y + 2)^2 = 144$

10. $(y + 2)^2 - 12(x - 3) = 0$

11. $6(x + 5)^2 + 5(y + 7)^2 = 30$

12. $12(y - 5)^2 - 8(x - 3)^2 = 24$

13. $(x + 6)^2 + 24(y - 4) = 0$

14. $4(x - 7)^2 + 7(y - 3)^2 = 28$

B

In Problems 15–22, transform each equation into one of the standard forms in Table 1. Identify the curve and graph it.

15. $4x^2 + 9y^2 - 16x - 36y + 16 = 0$

16. $16x^2 + 9y^2 + 64x + 54y + 1 = 0$

17. $x^2 + 8x + 8y = 0$

18. $y^2 + 12x + 4y - 32 = 0$

19. $x^2 + y^2 + 12x + 10y + 45 = 0$

20. $x^2 + y^2 - 8x - 6y = 0$

21. $-9x^2 + 16y^2 - 72x - 96y - 144 = 0$

22. $16x^2 - 25y^2 - 160x = 0$

23. If $A \neq 0$, $C = 0$, and $E \neq 0$, find h and k so that the translation of axes $x = x' + h$, $y = y' + k$ transforms the equation $Ax^2 + Cy^2 + Dx + Ey + F = 0$ into one of the standard forms of Table 1.

24. If $A = 0$, $C \neq 0$, and $D \neq 0$, find h and k so that the translation of axes $x = x' + h$, $y = y' + k$ transforms the equation $Ax^2 + Cy^2 + Dx + Ey + F = 0$ into one of the standard forms of Table 1.

In Problems 25–36, use the given information to find the equation of each conic. Express the answer in the form $Ax^2 + Cy^2 + Dx + Ey + F = 0$ with integer coefficients and $A > 0$.

25. A parabola with vertex at (2, 5), axis the line $x = 2$, and passing through the point $(-2, 1)$.

26. A parabola with vertex at $(4, -1)$, axis the line $y = -1$, and passing through the point (2, 3).

27. An ellipse with major axis on the line $y = -3$, minor axis on the line $x = -2$, length of major axis = 8, and length of minor axis = 4.

28. An ellipse with major axis on the line $x = -4$, minor axis on the line $y = 1$, length of major axis = 4, and length of minor axis = 2.

29. An ellipse with vertices $(4, -7)$ and (4, 3) and foci $(4, -6)$ and (4, 2).

30. An ellipse with vertices $(-3, 1)$ and (7, 1) and foci $(-1, 1)$ and (5, 1).

31. A hyperbola with transverse axis on the line $x = 2$, length of transverse axis = 4, conjugate axis on the line $y = 3$, and length of conjugate axis = 2.

32. A hyperbola with transverse axis on the line $y = -5$, length of transverse axis = 6, conjugate axis on the line $x = 2$, and length of conjugate axis = 6.

33. An ellipse with the following graph:

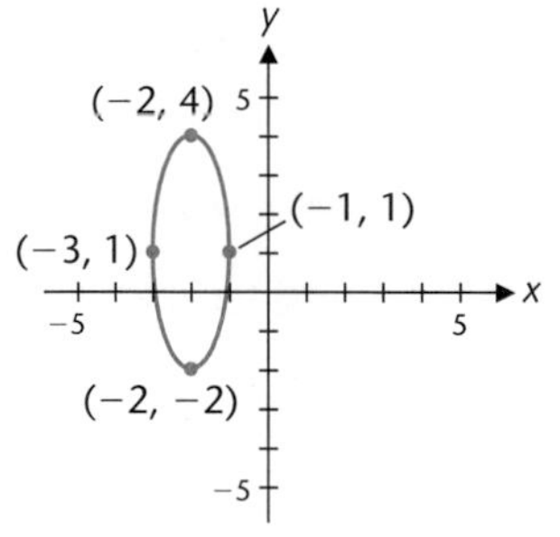

34. An ellipse with the following graph:

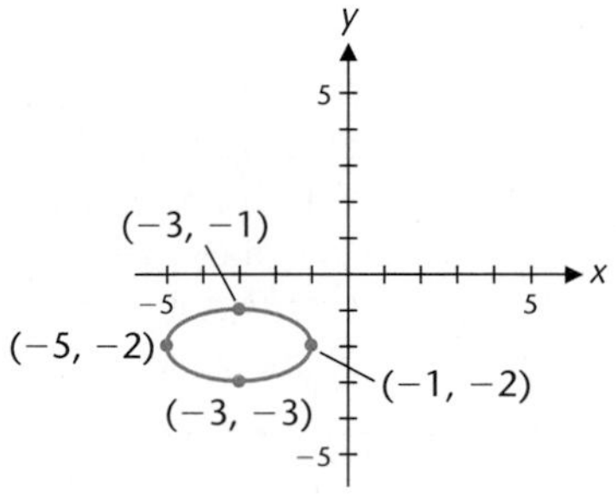

35. A hyperbola with the following graph:

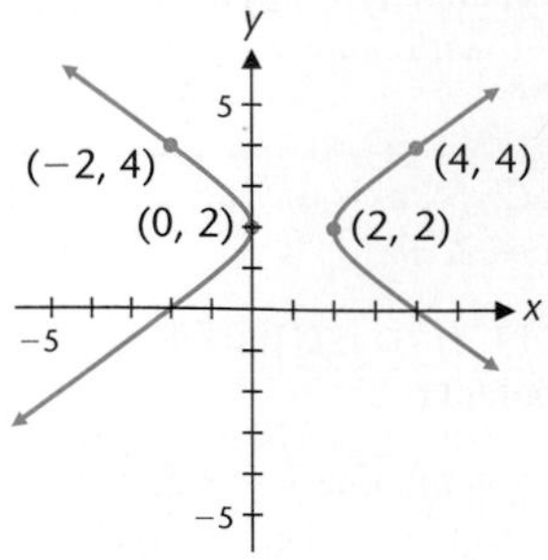

36. A hyperbola with the following graph:

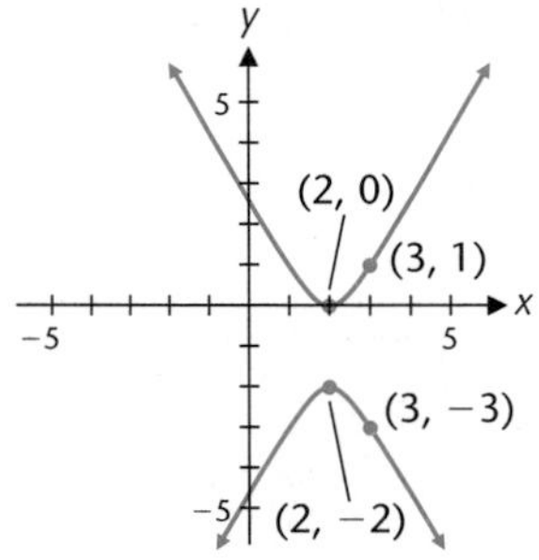

C

In Problems 37–42, find the coordinates of any foci relative to the original coordinate system.

37. Problem 15

38. Problem 16

39. Problem 17

40. Problem 18

41. Problem 21

42. Problem 22

In Problems 43–46, find the coordinates of all points of intersection to two decimal places.

43. $3x^2 - 5y^2 + 7x - 2y + 11 = 0, 6x + 4y = 15$

44. $8x^2 + 3y^2 - 14x + 17y - 39 = 0, 5x - 11y = 23$

45. $7x^2 - 8x + 5y - 25 = 0, x^2 + 4y^2 + 4x - y - 12 = 0$

46. $4x^2 - y^2 - 24x - 2y + 35 = 0, 2x^2 + 6y^2 - 3x - 34 = 0$

Section 7-5 | Parametric Equations

- Parametric Equations and Plane Curves
- Projectile Motion

Parametric Equations and Plane Curves

FIGURE 1
Graph of $x = t + 1, y = t^2 - 2t, -\infty < t < \infty$.

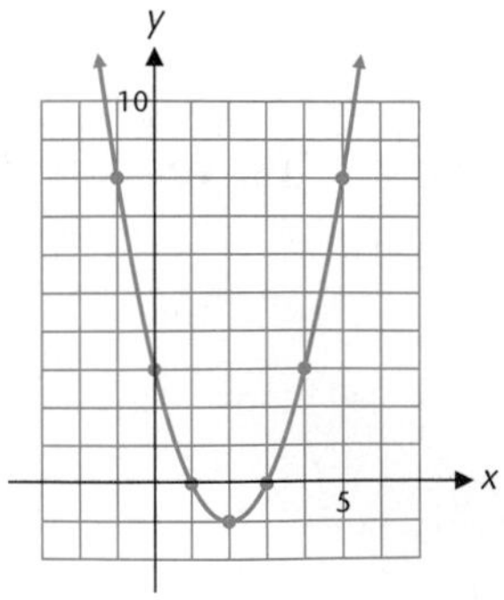

Consider the two equations

$$x = t + 1 \qquad -\infty < t < \infty \qquad (1)$$
$$y = t^2 - 2t$$

Each value of t determines a value of x, a value of y, and hence, an ordered pair (x, y). To graph the set of ordered pairs (x, y) determined by letting t assume all real values, we construct Table 1 listing selected values of t and the corresponding values of x and y. Then we plot the ordered pairs (x, y) and connect them with a continuous curve, as shown in Figure 1. The variable t is called a *parameter* and does not appear on the graph. Equations (1) are called *parametric equations* because both x and y are expressed in terms of the parameter t. The graph of the ordered pairs (x, y) is called a *plane curve*.

TABLE 1

t	0	1	2	3	4	−1	−2
x	1	2	3	4	5	0	−1
y	0	−1	0	3	8	3	8

Parametric equations can also be graphed on a graphing utility. Figure 2(a) shows the Parametric mode selected on a Texas Instruments TI-83 calculator. Figure 2(b) shows the equation editor with the parametric equations in (1) entered as x_{1T} and y_{1T}. In Figure 2(c), notice that there are three new window variables, Tmin, Tmax, and Tstep, that must be entered by the user.

FIGURE 2
Graphing parametric equations on a graphing utility.

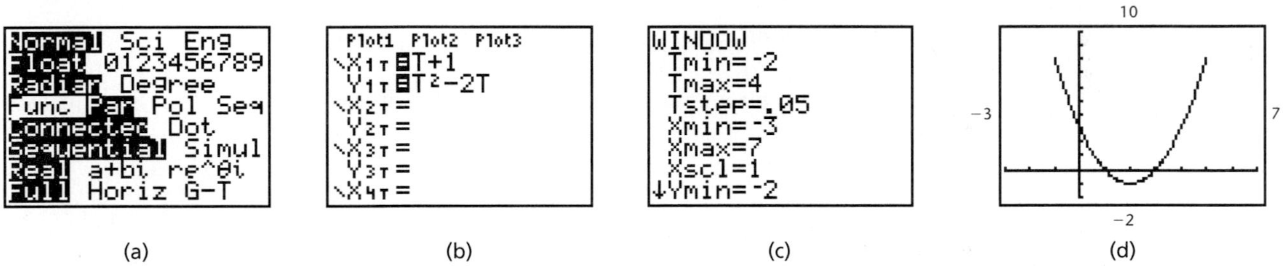

Explore/Discuss 1

(A) Consult the manual for your graphing utility and reproduce Figure 2(a).

(B) Discuss the effect of using different values for Tmin and Tmax. Try Tmin = −1 and −3. Try Tmax = 3 and 5.

(C) Discuss the effect of using different values for Tstep. Try Tstep = 1, 0.1, and 0.01.

In some cases it is possible to eliminate the parameter by solving one of the equations for t and substituting into the other. In the example just considered, solving the first equation for t in terms of x, we have

$$t = x - 1$$

Then, substituting the result into the second equation, we obtain

$$\begin{aligned} y &= (x - 1)^2 - 2(x - 1) \\ &= x^2 - 4x + 3 \end{aligned}$$

We recognize this as the equation of a parabola, as we would guess from Figure 1.

In other cases, it may not be easy or possible to eliminate the parameter to obtain an equation in just x and y. For example, for

$$\begin{aligned} x &= t + \log t \\ y &= t - e^t \end{aligned} \qquad t > 0$$

you will not find it possible to solve either equation for t in terms of functions we have considered.

Is there more than one parametric representation for a plane curve? The answer is yes. In fact, there is an unlimited number of parametric representations for the same plane curve. The following are two additional representations of the parabola in Figure 1.

$$\begin{aligned} x &= t + 3 \\ y &= t^2 + 2t \end{aligned} \qquad -\infty < t < \infty \tag{2}$$

$$\begin{aligned} x &= t \\ y &= t^2 - 4t + 3 \end{aligned} \qquad -\infty < t < \infty \tag{3}$$

The concepts introduced in the preceding discussion are summarized in Definition 1.

DEFINITION 1

PARAMETRIC EQUATIONS AND PLANE CURVES

A **plane curve** is the set of points (x, y) determined by the **parametric equations**

$$x = f(t)$$
$$y = g(t)$$

where the **parameter** t varies over an interval I and the functions f and g are both defined on the interval I.

Why are we interested in parametric representations of plane curves? It turns out that this approach is more general than using equations with two variables as we have been doing. In addition, the approach generalizes to curves in three- and higher-dimensional spaces. Other important reasons for using parametric representations of plane curves will be brought out in the discussion and examples that follow.

EXAMPLE 1

Eliminating the Parameter

Eliminate the parameter and identify the plane curve given parametrically by

$$\begin{aligned} x &= \sqrt{t} \\ y &= \sqrt{9 - t} \end{aligned} \qquad 0 \le t \le 9 \tag{4}$$

Solution To eliminate the parameter t, we solve each equation (4) for t:

$$x = \sqrt{t} \qquad y = \sqrt{9 - t}$$

$$x^2 = t \qquad y^2 = 9 - t$$

$$t = 9 - y^2$$

Equating the last two equations, we have

$$x^2 = 9 - y^2$$

$$x^2 + y^2 = 9 \qquad \text{A circle of radius 3 centered at (0, 0)}$$

Thus, the graph of the parametric equations in equation (4) is the quarter of the circle of radius 3 centered at the origin that lies in the first quadrant (Fig. 3).

FIGURE 3

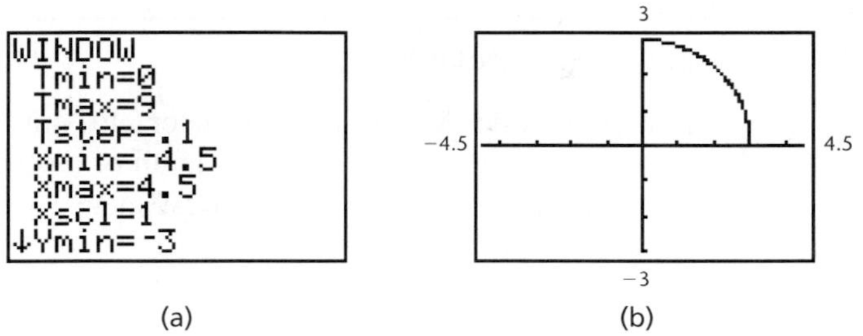

MATCHED PROBLEM 1 Eliminate the parameter and identify the plane curve given parametrically by $x = \sqrt{4 - t}$, $y = -\sqrt{t}$, $0 \le t \le 4$.

Projectile Motion

Newton's laws and advanced mathematics can be used to determine the **path of a projectile.** If v_0 is the vertical speed of the projectile, h_0 is the horizontal speed, and a_0 is the initial altitude of the projectile (see Fig. 4), then, neglecting air resistance, the path of the projectile is given by

$$\begin{aligned} x &= h_0 t \\ y &= a_0 + v_0 t - 4.9t^2 \end{aligned} \qquad 0 \le t \le b \qquad (5)$$

FIGURE 4
Projectile motion.

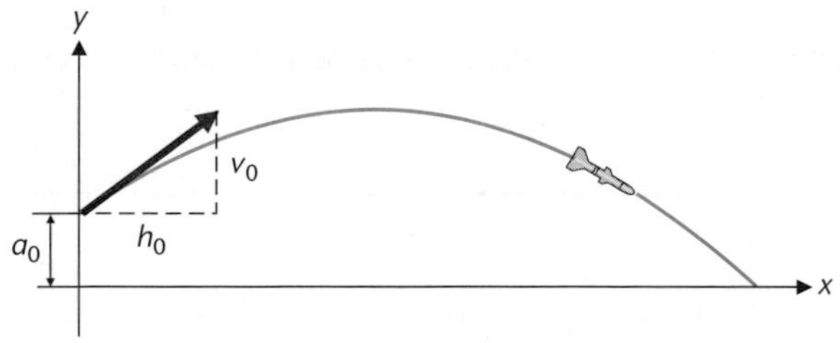

The parameter t represents time in seconds, and x and y are distances measured in meters. Solving the first equation in equations (5) for t in terms of x, substituting into the second equation, and simplifying produces the following equation:

$$y = a_0 + \frac{v_0}{h_0}x - \frac{4.9}{h_0^2}x^2 \qquad (6)$$

You should verify this by supplying the omitted details.

We recognize equation (6) as a parabola. This equation in x and y describes the path the projectile follows but tells us little else about its flight. On the other hand, the parametric equations (5) not only determine the path of the projectile but also tell us where it is at any time t. Furthermore, using concepts from physics and calculus, the parametric equations can be used to determine the velocity and acceleration of the projectile at any time t. This illustrates another advantage of using parametric representations of plane curves.

EXAMPLE 2 Projectile Motion

An automobile drives off a 50 meter cliff traveling at 25 meters per second (see Fig. 5). When (to the nearest tenth of a second) will the automobile strike the ground? How far (to the nearest meter) from the base of the cliff is the point of impact?

FIGURE 5

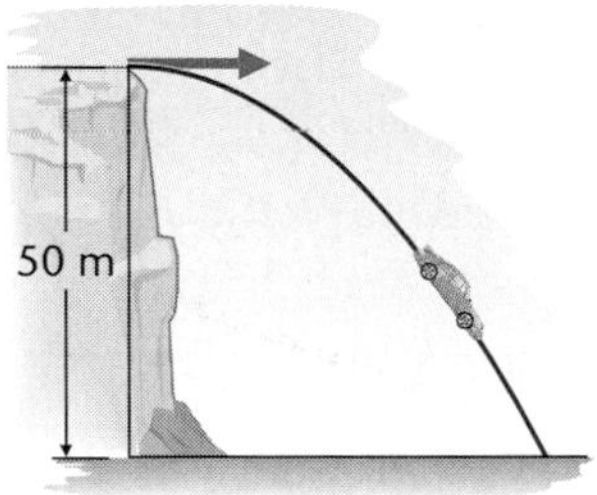

Solution At the instant the automobile leaves the cliff, the vertical speed is 0, the horizontal speed is 25 meters per second, and the altitude is 50 meters. Substituting these values in equations (5), the parametric equations for the path of the automobile are

$$x = 25t$$
$$y = 50 - 4.9t^2$$

The automobile strikes the ground when $y = 0$. Using the parametric equation for y, we have

$$y = 50 - 4.9t^2 = 0$$
$$-4.9t^2 = -50$$
$$t = \sqrt{\frac{-50}{-4.9}} \approx 3.2 \text{ seconds}$$

The distance from the base of the cliff is the same as the value of x. Substituting $t = 3.2$ in the first parametric equation, the distance from the base of the cliff at the point of impact is $x = 25(3.2) = 80$ meters.

MATCHED PROBLEM 2

A gardener is holding a hose in a horizontal position 1.5 meters above the ground. Water is leaving the hose at a speed of 5 meters per second. What is the distance (to the nearest tenth of a meter) from the gardener's feet to the point where the water hits the gound?

Answers to Matched Problems

1. The quarter of the circle of radius 2 centered at the origin that lies in the fourth quadrant.
2. 2.8 meters

EXERCISE 7-5

A

1. If $x = t^2$ and $y = t^2 - 2$, then $y = x - 2$. Discuss the differences between the graph of the parametric equations and the graph of the line $y = x - 2$.

2. If $x = t^2$ and $y = t^4 - 2$, then $y = x^2 - 2$. Discuss the differences between the graph of the parametric equations and the graph of the parabola $y = x^2 - 2$.

In Problems 3–12, the interval for the parameter is the whole real line. For each pair of parametric equations, eliminate the parameter t and find an equation for the curve in terms of x and y. Identify and graph the curve.

3. $x = -t$, $y = 2t - 2$
4. $x = t$, $y = t + 1$
5. $x = -t^2$, $y = 2t^2 - 2$
6. $x = t^2$, $y = t^2 + 1$
7. $x = 3t$, $y = -2t$
8. $x = 2t$, $y = t$
9. $x = \frac{1}{4}t^2$, $y = t$
10. $x = 2t$, $y = t^2$
11. $x = \frac{1}{4}t^4$, $y = t^2$
12. $x = 2t^2$, $y = t^4$

B

In Problems 13–20, obtain an equation in x and y by eliminating the parameter. Identify the curve.

13. $x = t - 2$, $y = 4 - 2t$

14. $x = t - 1$, $y = 2t + 2$

15. $x = t - 1$, $y = \sqrt{t}$, $t \geq 0$

16. $x = \sqrt{t}$, $y = t + 1$, $t \geq 0$

17. $x = \sqrt{t}$, $y = 2\sqrt{16 - t}$, $0 \leq t \leq 16$

18. $x = -3\sqrt{t}$, $y = \sqrt{25 - t}$, $0 \leq t \leq 25$

19. $x = -\sqrt{t + 1}$, $y = -\sqrt{t - 1}$, $t \geq 1$

20. $x = \sqrt{2 - t}$, $y = -\sqrt{4 - t}$, $t \leq 2$

21. If $A \neq 0$, $C = 0$, and $E \neq 0$, find parametric equations for $Ax^2 + Cy^2 + Dx + Ey + F = 0$. Identify the curve.

22. If $A = 0$, $C \neq 0$, and $D \neq 0$, find parametric equations for $Ax^2 + Cy^2 + Dx + Ey + F = 0$. Identify the curve.

C

In Problems 23–28, the interval for the parameter is the entire real line. Obtain an equation in x and y by eliminating the parameter and identify the curve.

23. $x = \sqrt{t^2 + 1}$, $y = \sqrt{t^2 + 9}$

24. $x = \sqrt{t^2 + 4}$, $y = \sqrt{t^2 + 1}$

25. $x = \dfrac{2}{\sqrt{t^2 + 1}}$, $y = \dfrac{2t}{\sqrt{t^2 + 1}}$

26. $x = \dfrac{3t}{\sqrt{t^2 + 1}}$, $y = \dfrac{3}{\sqrt{t^2 + 1}}$

27. $x = \dfrac{8}{t^2 + 4}$, $y = \dfrac{4t}{t^2 + 4}$

28. $x = \dfrac{4t}{t^2 + 1}$, $y = \dfrac{4t^2}{t^2 + 1}$

29. Consider the following two pairs of parametric equations:
1. $x_1 = t$, $y_1 = e^t$, $-\infty < t < \infty$
2. $x_2 = e^t$, $y_2 = t$, $-\infty < t < \infty$

(A) Graph both pairs of parametric equations in a squared viewing window and discuss the relationship between the graphs.

(B) Eliminate the parameter and express each equation as a function of x. How are these functions related?

30. Consider the following two pairs of parametric equations:

1. $x_1 = t$, $y_1 = \log t$, $t > 0$

2. $x_2 = \log t$, $y_2 = t$, $t > 0$

(A) Graph both pairs of parametric equations in a squared viewing window and discuss the relationship between the graphs.

(B) Eliminate the parameter and express each equation as a function of x. How are these functions related?

APPLICATIONS

31. **Projectile Motion.** An airplane flying at an altitude of 1,000 meters is dropping medical supplies to hurricane victims on an island. The path of the plane is horizontal, the speed is 125 meters per second, and the supplies are dropped at the instant the plane crosses the shoreline. How far inland (to the nearest meter) will the supplies land?

32. **Projectile Motion.** One stone is dropped vertically from the top of a tower 40 meters high. A second stone is thrown horizontally from the top of the tower with a speed of 30 meters per second. How far apart (to the nearest tenth of a meter) are the stones when they land?

Chapter 7 Group Activity

Focal Chords

Many of the applications of the conic sections are based on their reflective or focal properties. One of the interesting algebraic properties of the conic sections concerns their focal chords.

If a line through a focus F contains two points G and H of a conic section, then the line segment GH is called a **focal chord.** Let $G(x_1, y_1)$ and $H(x_2, y_2)$ be points on the graph of $x^2 = 4ay$ such that GH is a focal chord. Let u denote the length of GF and v the length of FH (see Fig. 1).

FIGURE 1

Focal chord GH of the parabola $x^2 = 4ay$.

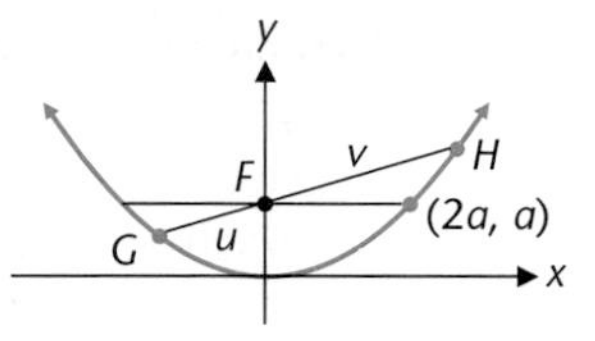

(A) Use the distance formula to show that $u = y_1 + a$.

(B) Show that G and H lie on the line $y - a = mx$, where $m = (y_2 - y_1)/(x_2 - x_1)$.

(C) Solve $y - a = mx$ for x and substitute in $x^2 = 4ay$, obtaining a quadratic equation in y. Explain why $y_1y_2 = a^2$.

(D) Show that $\frac{1}{u} + \frac{1}{v} = \frac{1}{a}$.

(E) Show that $u + v - 4a = \frac{(u - 2a)^2}{u - a}$. Explain why this implies that $u + v \geq 4a$, with equality if and only if $u = v = 2a$.

(F) Which focal chord is the shortest? Is there a longest focal chord?

(G) Is $\frac{1}{u} + \frac{1}{v}$ a constant for focal chords of the ellipse? For focal chords of the hyperbola? Obtain evidence for your answers by considering specific examples.

CHAPTER 7 | REVIEW

7-1 CONIC SECTIONS; PARABOLA

The plane curves obtained by intersecting a right circular cone with a plane are called **conic sections.** If the plane cuts clear through one nappe, then the intersection curve is called a **circle** if the plane is perpendicular to the axis and an **ellipse** if the plane is not perpendicular to the axis. If a plane cuts only one nappe, but does not cut clear through, then the intersection curve is called a **parabola.** If a plane cuts through both nappes, but not through the vertex, the resulting intersection curve is called a **hyperbola.** A plane passing through the vertex of the cone produces a **degenerate conic**—a point, a line, or a pair of lines. The figure illustrates the four nondegenerate conics.

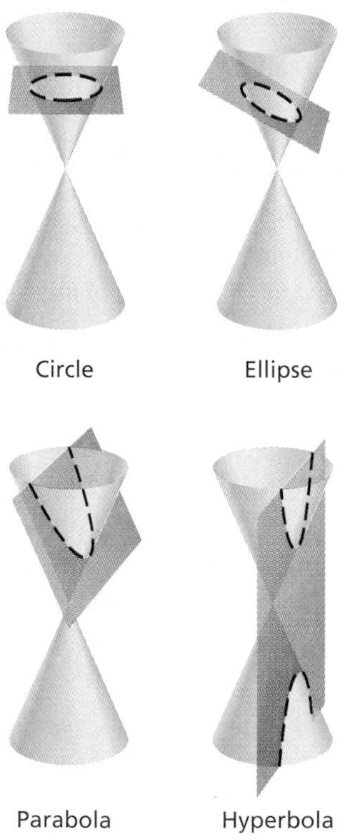

The graph of

$$Ax^2 + Bxy + Cy^2 + Dx + Ey + F = 0$$

where A, B, and C are not all 0, is a conic.

The following is a coordinate-free definition of a parabola:

Parabola. A *parabola* is the set of all points in a plane equidistant from a fixed point F and a fixed line L in the plane. The fixed point F is called the **focus,** and the fixed line L is called the **directrix.** A line through the focus perpendicular to the directrix is called the **axis,** and the point on the axis halfway between the directrix and focus is called the **vertex.**

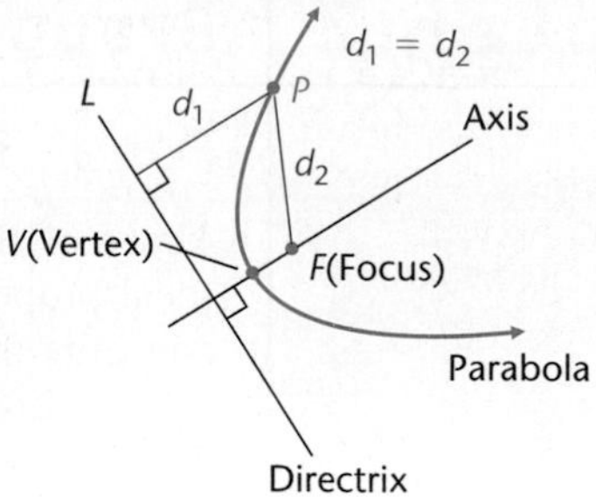

From the definition of a parabola, we can obtain the following standard equations:

Standard Equations of a Parabola with Vertex at (0, 0)

1. $y^2 = 4ax$
Vertex: (0, 0)
Focus: $(a, 0)$
Directrix: $x = -a$
Symmetric with respect to the x axis
Axis the x axis

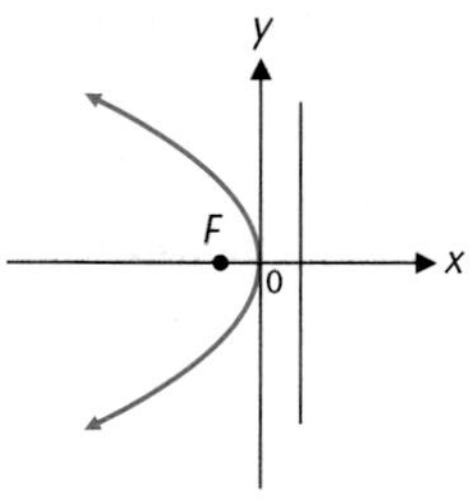

$a < 0$ (opens left)

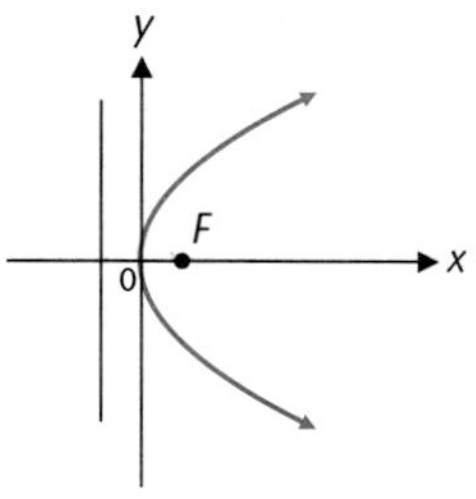

$a > 0$ (opens right)

2. $x^2 = 4ay$
Vertex: (0, 0)
Focus: $(0, a)$
Directrix: $y = -a$
Symmetric with respect to the y axis
Axis the y axis

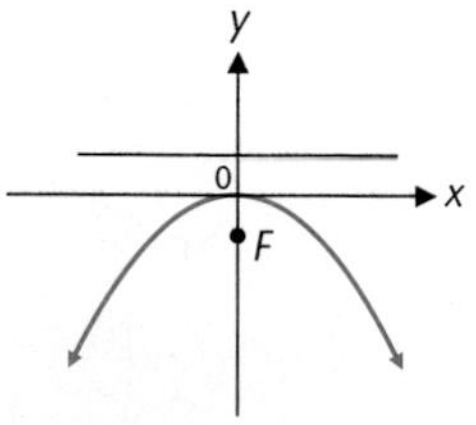

$a < 0$ (opens down)

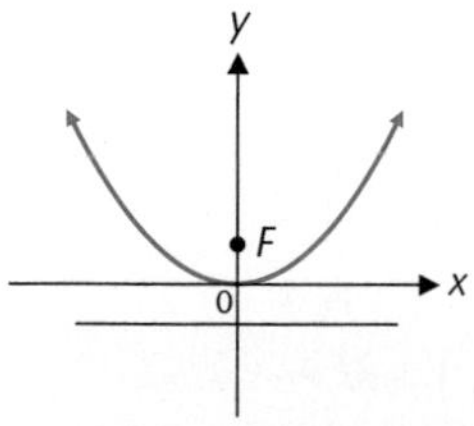

$a > 0$ (opens up)

7-2 ELLIPSE

The following is a coordinate-free definition of an ellipse:

Ellipse. An *ellipse* is the set of all points P in a plane such that the sum of the distances of P from two fixed points in the plane is constant. Each of the fixed points, F' and F, is called a *focus*, and together they are called **foci.** Referring to the figure, the line segment $V'V$ through the foci is the **major axis.** The perpendicular bisector $B'B$ of the major axis is the **minor axis.** Each end of the major axis, V' and V, is called a *vertex*. The midpoint of the line segment $F'F$ is called the **center** of the ellipse.

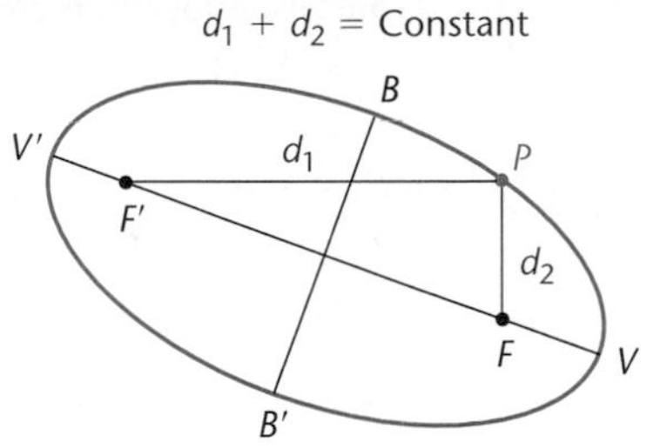

From the definition of an ellipse, we can obtain the following standard equations:

Standard Equations of an Ellipse with Center at (0, 0)

1. $\dfrac{x^2}{a^2} + \dfrac{y^2}{b^2} = 1 \qquad a > b > 0$

x intercepts: $\pm a$ (vertices)
y intercepts: $\pm b$
Foci: $F'(-c, 0)$, $F(c, 0)$

$c^2 = a^2 - b^2$

Major axis length $= 2a$
Minor axis length $= 2b$

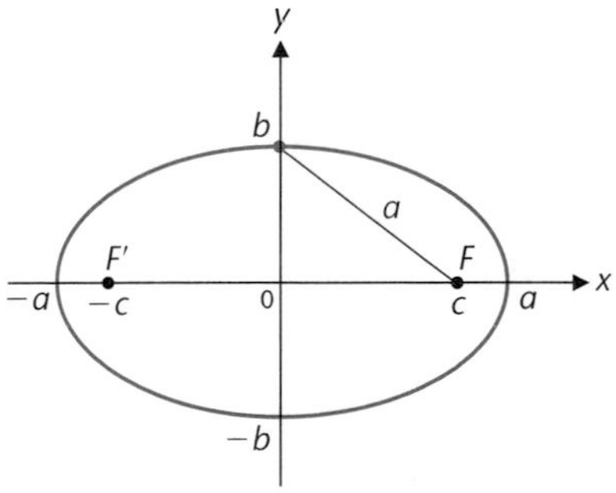

2. $\dfrac{x^2}{b^2} + \dfrac{y^2}{a^2} = 1 \qquad a > b > 0$

x intercepts: $\pm b$
y intercepts: $\pm a$ (vertices)
Foci: $F'(0, -c)$, $F(0, c)$

$c^2 = a^2 - b^2$

Major axis length $= 2a$
Minor axis length $= 2b$

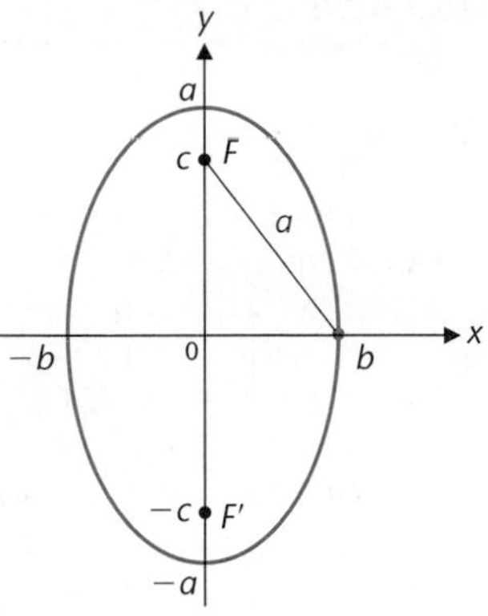

[*Note:* Both graphs are symmetric with respect to the x axis, y axis, and origin. Also, the major axis is always longer than the minor axis.]

7-3 HYPERBOLA

The following is a coordinate-free definition of a hyperbola:

Hyperbola. A *hyperbola* is the set of all points P in a plane such that the absolute value of the difference of the distances of P to two fixed points in the plane is a positive constant. Each of the fixed points, F' and F, is called a *focus*. The intersection points V' and V of the line through the foci and the two branches of the hyperbola are called **vertices,** and each is called a *vertex*. The line segment $V'V$ is called the **transverse axis.** The midpoint of the transverse axis is the *center* of the hyperbola.

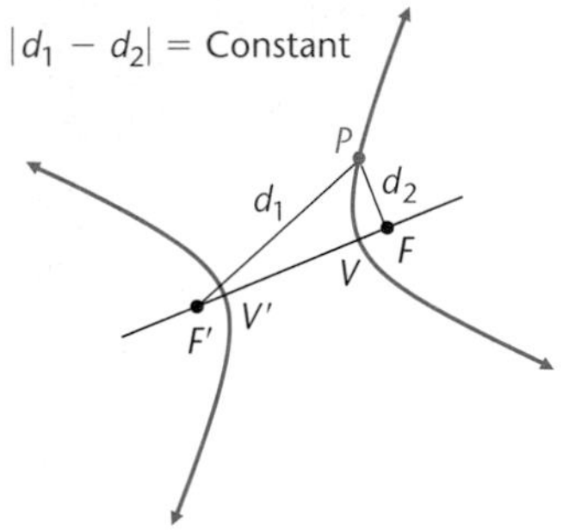

From the definition of a hyperbola, we can obtain the following standard equations:

Standard Equations of a Hyperbola with Center at (0, 0)

1. $\dfrac{x^2}{a^2} - \dfrac{y^2}{b^2} = 1$

x intercepts: $\pm a$ (vertices)
y intercepts: none

Foci: $F'(-c, 0)$, $F(c, 0)$

$c^2 = a^2 + b^2$

Transverse axis length $= 2a$
Conjugate axis length $= 2b$

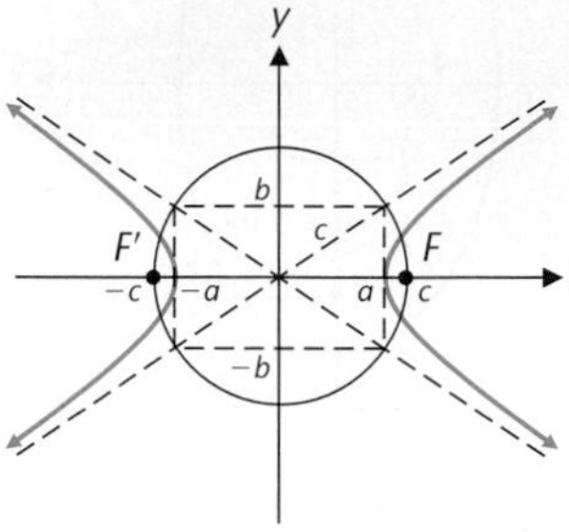

2. $\dfrac{y^2}{a^2} - \dfrac{x^2}{b^2} = 1$

x intercepts: none
y intercepts: $\pm a$ (vertices)
Foci: $F'(0, -c)$, $F(0, c)$

$c^2 = a^2 + b^2$

Transverse axis length $= 2a$
Conjugate axis length $= 2b$

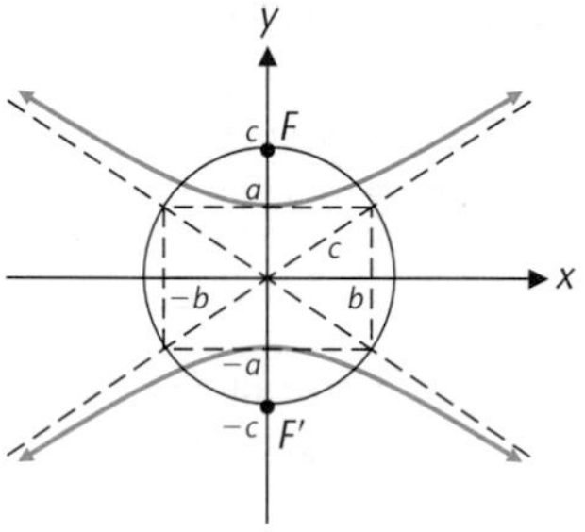

[*Note:* Both graphs are symmetric with respect to the x axis, y axis, and origin.]

7-4 TRANSLATION OF AXES

In the last three sections we found standard equations for parabolas, ellipses, and hyperbolas located with their axes on the coordinate axes and centered relative to the origin. We now move the conics away from the origin while keeping their axes parallel to the coordinate axes. In this process we obtain new standard equations that are special cases of the equation $Ax^2 + Cy^2 + Dx + Ey + F = 0$, where A and C are not both zero. The basic mathematical tool used is *translation of axes*.

A **translation of coordinate axes** occurs when the new coordinate axes have the same direction as and are parallel to the original coordinate axes. **Translation formulas** are as follows:

1. $x = x' + h$
   $y = y' + k$

2. $x' = x - h$
   $y' = y - k$

where (h, k) are the coordinates of the origin $0'$ relative to the original system.

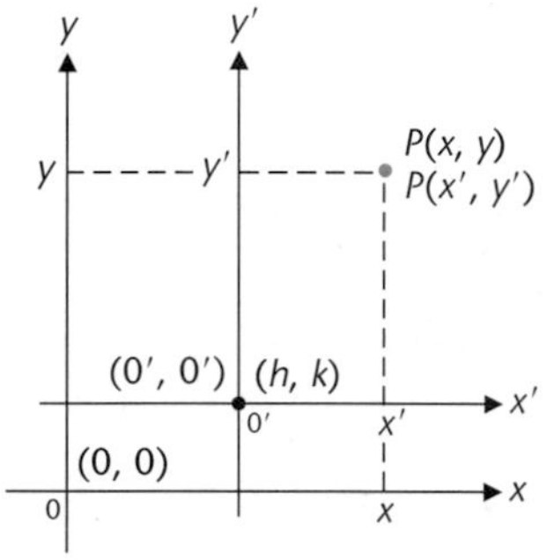

Table 1 lists the standard equations for translated conics.

7-5 PARAMETRIC EQUATIONS

A **plane curve** is the set of points (x, y) given by the **parametric equations**

$x = f(t)$ and $y = g(t)$

where the **parameter** t varies over an interval I.

The **path of a projectile** with an initial vertical speed v_0, an initial horizontal speed h_0, and an initial altitude a_0 is given by

$x = h_0$ and $y = a_0 + v_0 t - 4.9t^2, 0 \le t \le b$

or, after eliminating the parameter t, by

$$y = a_0 + \frac{v_0}{h_0}x - \frac{4.9}{h_0^2}x^2$$

where t is time in seconds and x and y are distances in meters.

TABLE 1 Standard Equations for Translated Conics

Parabolas

$(x - h)^2 = 4a(y - k)$

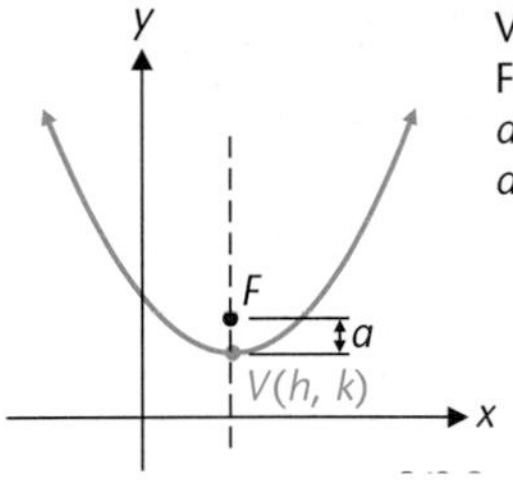

Vertex (h, k)
Focus $(h, k + a)$
$a > 0$ opens up
$a < 0$ opens down

$(y - k)^2 = 4a(x - h)$

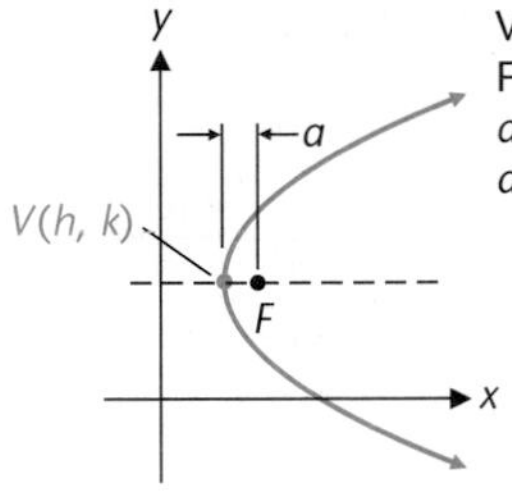

Vertex (h, k)
Focus $(h + a, k)$
$a < 0$ opens left
$a > 0$ opens right

Circles

$(x - h)^2 + (y - k)^2 = r^2$

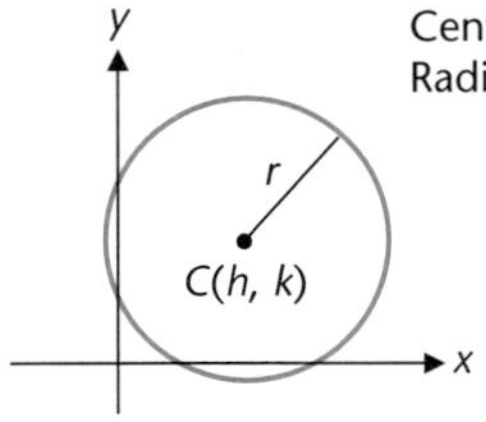

Center (h, k)
Radius r

Ellipses

$a > b > 0$

$$\frac{(x - h)^2}{a^2} + \frac{(y - k)^2}{b^2} = 1$$

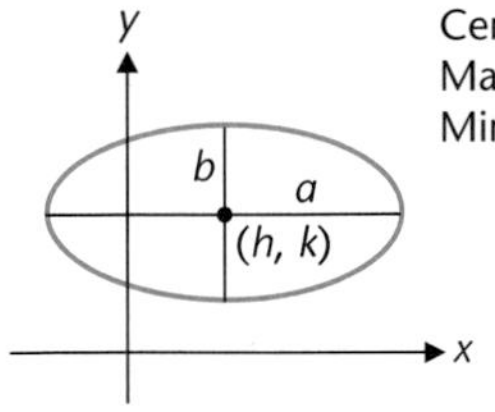

Center (h, k)
Major axis $2a$
Minor axis $2b$

$$\frac{(x - h)^2}{b^2} + \frac{(y - k)^2}{a^2} = 1$$

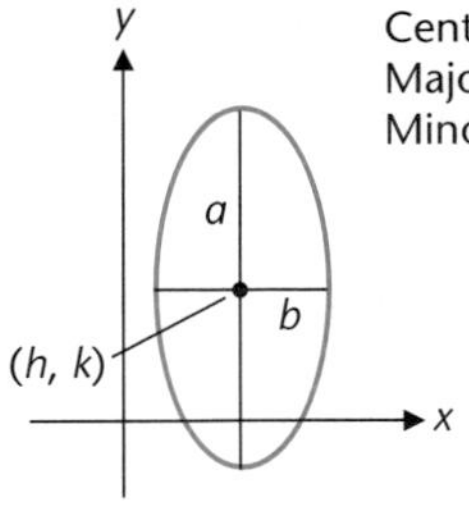

Center (h, k)
Major axis $2a$
Minor axis $2b$

Hyperbolas

$$\frac{(x - h)^2}{a^2} - \frac{(y - k)^2}{b^2} = 1$$

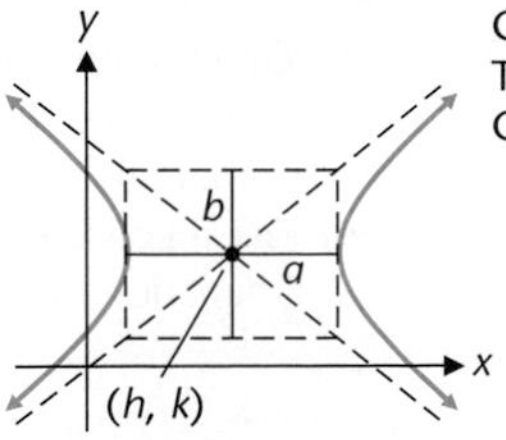

Center (h, k)
Transverse axis $2a$
Conjugate axis $2b$

$$\frac{(y - k)^2}{a^2} - \frac{(x - h)^2}{b^2} = 1$$

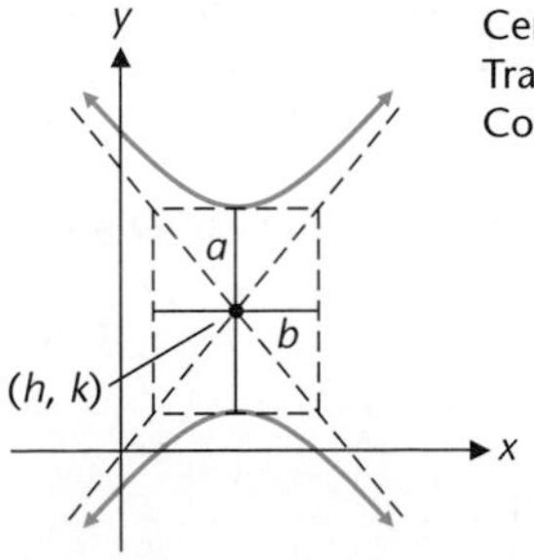

Center (h, k)
Transverse axis $2a$
Conjugate axis $2b$

CHAPTER 7 | REVIEW EXERCISES

Work through all the problems in this chapter review and check answers in the back of the book. Answers to all review problems are there, and following each answer is a number in italics indicating the section in which that type of problem is discussed. Where weaknesses show up, review appropriate sections in the text.

A

In Problems 1–3, graph each equation and locate foci. Locate the directrix for any parabolas. Find the lengths of major, minor, transverse, and conjugate axes where applicable.

1. $9x^2 + 25y^2 = 225$ **2.** $x^2 = -12y$

3. $25y^2 - 9x^2 = 225$

In Problems 4–6:
(A) Write each equation in one of the standard forms listed in Table 1 of the review.
(B) Identify the curve.

4. $4(y + 2)^2 - 25(x - 4)^2 = 100$

5. $(x + 5)^2 + 12(y + 4) = 0$

6. $16(x - 6)^2 + 9(y - 4)^2 = 144$

7. If $x = t^4 + 1$ and $y = t^2$, then $x = y^2 + 1$. Discuss the difference between the graph of the parametric equations and the graph of the parabola $x = y^2 + 1$.

B

8. Find the equation of the parabola having its vertex at the origin, its axis the x axis, and $(-4, -2)$ on its graph.

9. Find an equation of an ellipse in the form

$$\frac{x^2}{M} + \frac{y^2}{N} = 1 \qquad M, N > 0$$

if the center is at the origin, the major axis is on the y axis, the minor axis length is 6, and the distance of the foci from the center is 4.

10. Find an equation of a hyperbola in the form

$$\frac{y^2}{M} - \frac{x^2}{N} = 1 \qquad M, N > 0$$

if the center is at the origin, the conjugate axis length is 8, and the foci are 5 units from the center.

11. Plot the curve given parametrically by

$$x = -t^2$$
$$y = -\tfrac{1}{2}t^2 + 1$$

Obtain an equation in x and y by eliminating the parameter, and identify the curve.

CIC *In Problems 12–14, graph each system of equations in the same coordinate system and find the coordinates of any points of intersection.*

12. $x^2 + 4y^2 = 32$
$x + 2y = 0$

13. $16x^2 + 25y^2 = 400$
$16x^2 - 45y = 0$

14. $x^2 + y^2 = 10$
$16x^2 + y^2 = 25$

In Problems 15–17, transform each equation into one of the standard forms in Table 1 in the review. Identify the curve and graph it.

15. $16x^2 + 4y^2 + 96x - 16y + 96 = 0$

16. $x^2 - 4x - 8y - 20 = 0$

17. $4x^2 - 9y^2 + 24x - 36y - 36 = 0$

18. Use a graphing utility to graph $x^2 = y$ and $x^2 = 50y$ in the viewing window $-10 \le x, y \le 10$. Find m so that the graph of $x^2 = y$ in the viewing window $-m \le x, y \le m$, has the same appearance as the graph of $x^2 = 50y$ in $-10 \le x, y \le 10$. Explain.

C

19. Use the definition of a parabola and the distance formula to find the equation of a parabola with directrix $x = 6$ and focus at $(2, 4)$.

20. Find an equation of the set of points in a plane each of whose distance from $(4, 0)$ is twice its distance from the line $x = 1$. Identify the geometric figure.

21. Find an equation of the set of points in a plane each of whose distance from $(4, 0)$ is two-thirds its distance from the line $x = 9$. Identify the geometric figure.

In Problems 22–24, find the coordinates of any foci relative to the original coordinate system.

22. Problem 15 **23.** Problem 16 **24.** Problem 17

In Problems 25 and 26, the interval for the parameter is the entire real line. Obtain an equation in x and y by eliminating the parameter and identify the curve.

25. $x = 2^t, y = 2^{-t}$

26. $x = \dfrac{3t}{\sqrt{t^2 + 4}}, y = \dfrac{4}{\sqrt{t^2 + 4}}$

27. Consider the following two pairs of parametric equations:
1. $x_1 = t, y_1 = 2^t, -\infty < t < \infty$
2. $x_2 = 2^t, y_2 = t, -\infty < t < \infty$

(A) Graph both pairs of parametric equations in a squared viewing window and discuss the relationship between the graphs.
(B) Eliminate the parameter and express each equation as a function of x. How are these functions related?

28. Find, to two decimal places, the coordinates of all points of intersection of $x^2 - 3y^2 + 9x + 7y - 22 = 0$ and $4x^2 + 5x + 10y - 53 = 0$.

APPLICATIONS

29. Communications. A parabolic satellite television antenna has a diameter of 8 feet and is 1 foot deep. How far is the focus from the vertex?

30. Engineering. An elliptical gear is to have foci 8 centimeters apart and a major axis 10 centimeters long. Letting the x axis lie along the major axis (right positive) and the y axis lie along the minor axis (up positive), write the equation of the ellipse in the standard form

$$\frac{x^2}{a^2} + \frac{y^2}{b^2} = 1$$

31. Space Science. A hyperbolic reflector for a radiotelescope (such as that illustrated in Problem 47, Exercise 7-3) has the equation

$$\frac{y^2}{40^2} - \frac{x^2}{30^2} = 1$$

If the reflector has a diameter of 30 feet, how deep is it? Compute the answer to three significant digits.

CUMULATIVE REVIEW EXERCISE FOR CHAPTERS 5, 6, AND 7

Work through all the problems in this cumulative review and check answers in the back of the book. Answers to all review problems are there, and following each answer is a number in italics indicating the section in which that type of problem is discussed. Where weaknesses show up, review appropriate sections in the text.

A

1. Solve using substitution or elimination by addition:

$$3x - 5y = 11$$
$$2x + 3y = 1$$

2. Solve by graphing: $2x - y = -4$
$$3x + y = -1$$

3. Solve by substitution or elimination by addition:

$$x^2 + y^2 = 2$$
$$2x - y = 1$$

4. Solve by graphing: $3x + 5y \le 15$
$$x, y \ge 0$$

5. Find the maximum and minimum value of $z = 2x + 3y$ over the feasible region S:

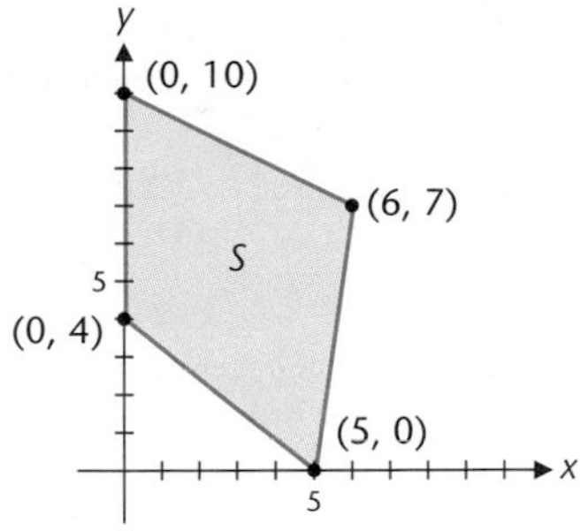

6. Perform the operations that are defined, given the following matrices:

$$M = \begin{bmatrix} 2 & 1 \\ 1 & -3 \end{bmatrix} \quad N = \begin{bmatrix} 1 & 2 \\ -1 & 3 \end{bmatrix}$$

$$P = \begin{bmatrix} 1 & 2 \end{bmatrix} \quad Q = \begin{bmatrix} -1 \\ 2 \end{bmatrix}$$

(A) $M - 2N$ (B) $P + Q$ (C) PQ
(D) MN (E) PN (F) QM

7. Write the linear system corresponding to each augmented matrix and solve:

(A) $\left[\begin{array}{cc|c} 1 & 0 & 3 \\ 0 & 1 & -4 \end{array}\right]$ (B) $\left[\begin{array}{cc|c} 1 & -2 & 3 \\ 0 & 0 & 0 \end{array}\right]$

(C) $\left[\begin{array}{cc|c} 1 & -2 & 3 \\ 0 & 0 & 1 \end{array}\right]$

8. Given the system: $x_1 + x_2 = 3$
$$-x_1 + x_2 = 5$$

(A) Write the augmented matrix for the system.
(B) Transform the augmented matrix into reduced form.
(C) Write the solution to the system.

9. Given the system: $x_1 - 3x_2 = k_1$
$$2x_1 - 5x_2 = k_2$$

(A) Write the system as a matrix equation of the form $AX = B$.
(B) Find the inverse of the coefficient matrix A.
(C) Use A^{-1} to find the solution for $k_1 = -2$ and $k_2 = 1$.
(D) Use A^{-1} to find the solution for $k_1 = 1$ and $k_2 = -2$.

10. Use Gauss–Jordan elimination to solve the system

$$x_1 + 3x_2 = 10$$
$$2x_1 - x_2 = -1$$

Then write the linear system represented by each augmented matrix in your solution, and solve each of these systems graphically. Discuss the relationship between the solutions of these systems.

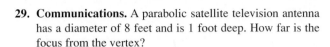

11. Solve graphically to two decimal places.

$$\begin{aligned} -2x + 3y &= 7 \\ 3x + 4y &= 18 \end{aligned}$$

12. Determine whether each of the following can be the first three terms of an arithmetic sequence, a geometric sequence, or neither.

(A) 20, 15, 10, . . . (B) 5, 25, 125, . . .
(C) 5, 25, 50, . . . (D) 27, −9, 3, . . .
(E) −9, −6, −3, . . .

In Problems 13–15:
(A) Write the first four terms of each sequence.
(B) Find a_8. *(C) Find S_8.*

13. $a_n = 2 \cdot 5^n$ **14.** $a_n = 3n - 1$

15. $a_1 = 100$; $a_n = a_{n-1} - 6$, $n \geq 2$

16. Evaluate each of the following:

(A) $8!$ (B) $\dfrac{32!}{30!}$ (C) $\dfrac{9!}{3!(9-3)!}$

17. Evaluate each of the following:

(A) $\dbinom{7}{2}$ (B) $C_{7,2}$ (C) $P_{7,2}$

In Problems 18–20, graph each equation and locate foci. Locate the directrix for any parabolas. Find the lengths of major, minor, transverse, and conjugate axes where applicable.

18. $25x^2 - 36y^2 = 900$ **19.** $25x^2 + 36y^2 = 900$

20. $25x^2 - 36y = 0$

21. A coin is flipped three times. How many combined outcomes are possible? Solve:
(A) By using a tree diagram
(B) By using the multiplication principle

22. How many ways can 4 distinct books be arranged on a shelf? Solve:
(A) By using the multiplication principle
(B) By using permutations or combinations, whichever is applicable

23. In a single deal of 3 cards from a standard 52-card deck, what is the probability of being dealt 3 diamonds?

24. Each of the 10 digits 0 through 9 is printed on 1 of 10 different cards. Four of these cards are drawn in succession without replacement. What is the probability of drawing the digits 4, 5, 6, and 7 by drawing 4 on the first draw, 5 on the second draw, 6 on the third draw, and 7 on the fourth draw? What is the probability of drawing the digits 4, 5, 6, and 7 in any order?

25. A thumbtack lands point down in 38 out of 100 tosses. What is the approximate empirical probability of the tack landing point up?

26. Plot the curve given parametrically by

$$\begin{aligned} x &= 2t + 3 \\ y &= 4t + 5 \end{aligned}$$

Obtain an equation in x and y by eliminating the parameter, and identify the curve.

Verify Problems 27 and 28 for $n = 1, 2,$ and 3.

27. P_n: $1 + 5 + 9 + \cdots + (4n - 3) = n(2n - 1)$

28. P_n: $n^2 + n + 2$ is divisible by 2

In Problems 29 and 30, write P_k and P_{k+1}.

29. For P_n in Problem 27 **30.** For P_n in Problem 28

B

Solve Problems 31–33 using Gauss–Jordan elimination.

31. $$\begin{aligned} x_1 + 2x_2 - x_3 &= 3 \\ x_2 + x_3 &= -2 \\ 2x_1 + 3x_2 + x_3 &= 0 \end{aligned}$$

32. $$\begin{aligned} x_1 + x_2 - x_3 &= 2 \\ 4x_2 + 6x_3 &= -1 \\ 6x_2 + 9x_3 &= 0 \end{aligned}$$

33. $$\begin{aligned} x_1 - 2x_2 + x_3 &= 1 \\ 3x_1 - 2x_2 - x_3 &= -5 \end{aligned}$$

34. Given $M = \begin{bmatrix} 1 & 2 & -1 \end{bmatrix}$ and $N = \begin{bmatrix} 1 \\ -1 \\ 2 \end{bmatrix}$, find

(A) MN (B) NM

35. Given

$$L = \begin{bmatrix} 2 & -1 & 0 \\ 1 & 2 & 1 \end{bmatrix} \quad M = \begin{bmatrix} 1 & 2 \\ -1 & 0 \\ 1 & 1 \end{bmatrix} \quad N = \begin{bmatrix} 2 & 1 \\ -1 & 0 \end{bmatrix}$$

find, if defined: (A) $LM - 2N$ (B) $ML + N$

36. Solve graphically and indicate whether the solution region is bounded or unbounded. Find the coordinates of each corner point.

$$\begin{aligned} 3x + 2y &\geq 12 \\ x + 2y &\geq 8 \\ x, y &\geq 0 \end{aligned}$$

37. Solve the linear programming problem:

Maximize $z = 4x + 9y$

Subject to
$$\begin{aligned} x + 2y &\leq 14 \\ 2x + y &\leq 16 \\ x, y &\geq 0 \end{aligned}$$

38. Given the system
$$\begin{aligned} x_1 + 4x_2 + 2x_3 &= k_1 \\ 2x_1 + 6x_2 + 3x_3 &= k_2 \\ 2x_1 + 5x_2 + 2x_3 &= k_3 \end{aligned}$$

(A) Write the system as a matrix equation in the form $AX = B$.
(B) Find the inverse of the coefficient matrix A.
(C) Use A^{-1} to solve the system when $k_1 = -1$, $k_2 = 2$, and $k_3 = 1$.
(D) Use A^{-1} to solve the system when $k_1 = 2$, $k_2 = 0$, and $k_3 = -1$.

39. Find the equation of the parabola having its vertex at the origin, its axis the y axis, and $(2, -8)$ on its graph.

40. Find an equation of an ellipse in the form

$$\frac{x^2}{M} + \frac{y^2}{N} = 1 \qquad M, N > 0$$

if the center is at the origin, the major axis is the x axis, the major axis length is 10, and the distance of the foci from the center is 3.

41. Find an equation of a hyperbola in the form

$$\frac{x^2}{M} - \frac{y^2}{N} = 1 \qquad M, N > 0$$

if the center is at the origin, the transverse axis length is 16, and the distance of the foci from the center is $\sqrt{89}$.

42. Write $\sum_{k=1}^{5} k^k$ without summation notation and find the sum.

43. Write the series $\frac{2}{2!} - \frac{2^2}{3!} + \frac{2^3}{4!} - \frac{2^4}{5!} + \frac{2^5}{6!} - \frac{2^6}{7!}$ using summation notation with the summation index k starting at $k = 1$.

44. Find S_∞ for the geometric series $108 - 36 + 12 - 4 + \cdots$.

45. How many 4-letter code words are possible using the first 6 letters of the alphabet if no letter can be repeated? If letters can be repeated? If adjacent letters cannot be alike?

46. A basketball team with 12 members has two centers. If 5 players are selected at random, what is the probability that both centers are selected? Express the answer in terms of $C_{n,r}$ or $P_{n,r}$, as appropriate, and evaluate.

47. A single die is rolled 1,000 times with the frequencies of outcomes shown in the table below:

Number of dots facing up	1	2	3	4	5	6
Frequency	160	155	195	180	140	170

(A) What is the approximate empirical probability that the number of dots showing is divisible by 3?
(B) What is the theoretical probability that the number of dots showing is divisible by 3?

48. Let $a_n = 100(0.9)^n$ and $b_n = 10 + 0.03n$. Find the least positive integer n such that $a_n < b_n$ by graphing the sequences $\{a_n\}$ and $\{b_n\}$ with a graphing utility. Check your answer by using a graphing utility to display both sequences in table form.

49. Given the parametric equations of a plane curve

$$x = 2 + t^2$$
$$y = -3 + t$$

obtain an equation in x and y by eliminating the parameter. Identify the curve.

50. Evaluate each of the following:

(A) $P_{25,5}$ (B) $C(25, 5)$ (C) $\binom{25}{20}$

51. Expand $(a + \frac{1}{2}b)^6$ using the binomial formula.

52. Find the fifth and the eighth terms in the expansion of $(3x - y)^{10}$.

Establish each statement in Problems 53 and 54 for all positive integers using mathematical induction.

53. P_n in Problem 27 54. P_n in Problem 28

55. Find the sum of all the odd integers between 50 and 500.

56. Use the formula for the sum of an infinite geometric series to write $2.\overline{45} = 2.454\,545\cdots$ as the quotient of two integers.

57. Let $a_k = \binom{30}{k}(0.1)^{30-k}(0.9)^k$ for $k = 0, 1, \ldots, 30$. Use a graphing utility to find the largest term of the sequence $\{a_k\}$ and the number of terms that are greater than 0.01.

In Problems 58–60, use a translation of coordinates to transform each equation into a standard equation for a nondegenerate conic. Identify the curve and graph it.

58. $4x + 4y - y^2 + 8 = 0$

59. $x^2 + 2x - 4y^2 - 16y + 1 = 0$

60. $4x^2 - 16x + 9y^2 + 54y + 61 = 0$

61. How many 9-digit zip codes are possible? How many of these have no repeated digits?

62. Find, to two decimal places, the coordinates of all points of intersection of $5x^2 + 2y^2 - 7x + 8y - 48 = 0$ and $e^x - e^{-x} - 2y = 0$.

63. Use mathematical induction to prove that the following statement holds for all positive integers:

$$P_n\text{:} \quad \frac{1}{1\cdot 3} + \frac{1}{3\cdot 5} + \frac{1}{5\cdot 7} + \cdots + \frac{1}{(2n-1)(2n+1)} = \frac{n}{2n+1}$$

64. Three-digit numbers are randomly formed from the digits 1, 2, 3, 4, and 5. What is the probability of forming an even number if digits cannot be repeated? If digits can be repeated?

C

65. Use the binomial formula to expand $(x - 2i)^6$, where i is the imaginary unit.

66. Use the definition of a parabola and the distance formula to find the equation of a parabola with directrix $y = 3$ and focus $(6, 1)$.

67. An ellipse has vertices $(\pm 4, 0)$ and foci $(\pm 2, 0)$. Find the y intercepts.

68. A hyperbola has vertices $(2, \pm 3)$ and foci $(2, \pm 5)$. Find the length of the conjugate axis.

69. Discuss the number of solutions for the system corresponding to the reduced form shown below if
(A) $m = 0$ and $n = 0$ (B) $m = 0$ and $n \neq 0$
(C) $m \neq 0$

$$\left[\begin{array}{ccc|c} 1 & 0 & -5 & 2 \\ 0 & 1 & 3 & 6 \\ 0 & 0 & m & n \end{array}\right]$$

70. If a square matrix A satisfies the equation $A^2 = A$, find A. Assume that A^{-1} exists.

71. Which of the following augmented matrices are in reduced form?

$$L = \left[\begin{array}{ccc|c} 1 & 0 & 0 & 2 \\ 0 & 1 & 0 & 0 \\ 0 & 0 & 1 & -1 \end{array}\right] \qquad M = \left[\begin{array}{ccc|c} 1 & 0 & 3 & 3 \\ 0 & 1 & -2 & 2 \\ 0 & 0 & 0 & 0 \end{array}\right]$$

$$N = \left[\begin{array}{cc|c} 0 & 0 & 0 \\ 1 & 0 & 2 \\ 0 & 1 & -3 \end{array}\right] \qquad P = \left[\begin{array}{cccc|c} 1 & 2 & 0 & 2 & -2 \\ 0 & 0 & 1 & 3 & 1 \end{array}\right]$$

72. Seven distinct points are selected on the circumference of a circle. How many triangles can be formed using these seven points as vertices?

73. Given the parametric equations of a plane curve

$$x = e^{2t} - 4$$
$$y = 1 - e^t$$

obtain an equation in x and y by eliminating the parameter. Identify the curve.

74. Use mathematical induction to prove that $2^n < n!$ for all integers $n > 3$.

75. Use mathematical induction to show that $\{a_n\} = \{b_n\}$, where $a_1 = 3$, $a_n = 2a_{n-1} - 1$ for $n > 1$, and $b_n = 2^n + 1$, $n \geq 1$.

76. Find an equation of the set of points in the plane each of whose distance from (1, 4) is three times its distance from the x axis. Write the equation in the form $Ax^2 + Cy^2 + Dx + Ey + F = 0$, and identify the curve.

77. A box of 12 lightbulbs contains 4 defective bulbs. If 3 bulbs are selected at random, what is the probability of selecting at least one defective bulb?

APPLICATIONS

78. **Finance.** An investor has \$12,000 to invest. If part is invested at 8% and the rest in a higher-risk investment at 14%, how much should be invested at each rate to produce the same yield as if all had been invested at 10%?

79. **Economics.** The government, through a subsidy program, distributes \$2,000,000. If we assume that each individual or agency spends 75% of what it receives, and 75% of this is spent, and so on, how much total increase in spending results from this government action?

80. **Diet.** In an experiment involving mice, a zoologist needs a food mix that contains, among other things, 23 grams of protein, 6.2 grams of fat, and 16 grams of moisture. She has on hand mixes of the following compositions: Mix A contains 20% protein, 2% fat, and 15% moisture; mix B contains 10% protein, 6% fat, and 10% moisture; and mix C contains 15% protein, 5% fat, and 5% moisture. How many grams of each mix should be used to get the desired diet mix?

81. **Purchasing.** A soft-drink distributor has budgeted \$300,000 for the purchase of 12 new delivery trucks. If a model A truck costs \$18,000, a model B truck costs \$22,000, and a model C truck costs \$30,000, how many trucks of each model should the distributor purchase to use exactly all the budgeted funds?

82. **Engineering.** An automobile headlight contains a parabolic reflector with a diameter of 8 inches. If the light source is located at the focus, which is 1 inch from the vertex, how deep is the reflector?

83. **Architecture.** A sound whispered at one focus of a whispering chamber can be easily heard at the other focus. Suppose that a cross section of this chamber is a semielliptical arch which is 80 feet wide and 24 feet high (see the figure). How far is each focus from the center of the arch? How high is the arch above each focus?

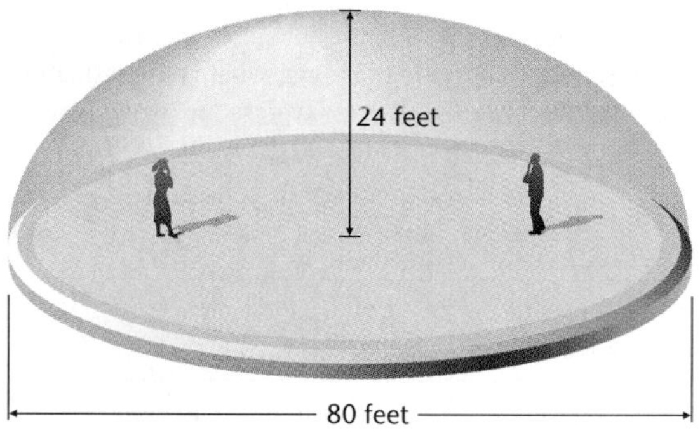

84. **Manufacturing.** A manufacturer makes two types of day packs, a standard model and a deluxe model. Each standard model requires 0.5 labor-hour from the fabricating department and 0.3 labor-hour from the sewing department. Each deluxe model requires 0.5 labor-hour from the fabricating department and 0.6 labor-hour from the sewing department. The maximum number of labor-hours available per week in the fabricating department and the sewing department are 300 and 240, respectively.
(A) If the profit on a standard day pack is \$8 and the profit on a deluxe day pack is \$12, how many of each type of pack should be manufactured each day to realize a maximum profit? What is the maximum profit?
(B) Discuss the effect on the production schedule and the maximum profit if the profit on a standard day pack decreases by \$3 and the profit on a deluxe day pack increases by \$3.
(C) Discuss the effect on the production schedule and the maximum profit if the profit on a standard day pack increases by \$3 and the profit on a deluxe day pack decreases by \$3.

85. **Averaging Tests.** A teacher has given four tests to a class of five students and stored the results in the following matrix:

$$\begin{array}{r} \\ \text{Ann} \\ \text{Bob} \\ \text{Carol} \\ \text{Dan} \\ \text{Eric} \end{array} \begin{array}{c} \text{Tests} \\ \begin{array}{cccc} 1 & 2 & 3 & 4 \end{array} \\ \begin{bmatrix} 78 & 84 & 81 & 86 \\ 91 & 65 & 84 & 92 \\ 95 & 90 & 92 & 91 \\ 75 & 82 & 87 & 91 \\ 83 & 88 & 81 & 76 \end{bmatrix} \end{array} = M$$

Discuss methods of matrix multiplication that the teacher can use to obtain the indicated information in parts A–C below. In each case, state the matrices to be used and then perform the necessary multiplications.

(A) The average on all four tests for each student, assuming that all four tests are given equal weight
(B) The average on all four tests for each student, assuming that the first three tests are given equal weight and the fourth is given twice this weight
(C) The class average on each of the four tests

86. **Political Science.** A random survey of 1,000 residents in a state produced the following results:

	Party Affiliation			
Age	**Democrat**	**Republican**	**Independent**	**Totals**
Under 30	130	80	40	250
30–39	120	90	20	230
40–49	70	80	20	170
50–59	50	60	10	120
Over 59	90	110	30	230
Totals	460	420	120	1,000

Find the empirical probability that a person selected at random:

(A) Is under 30 *and* a Democrat
(B) Is under 40 *and* a Republican
(C) Is over 59 *or* is an Independent

ANSWER TO APPLICATION

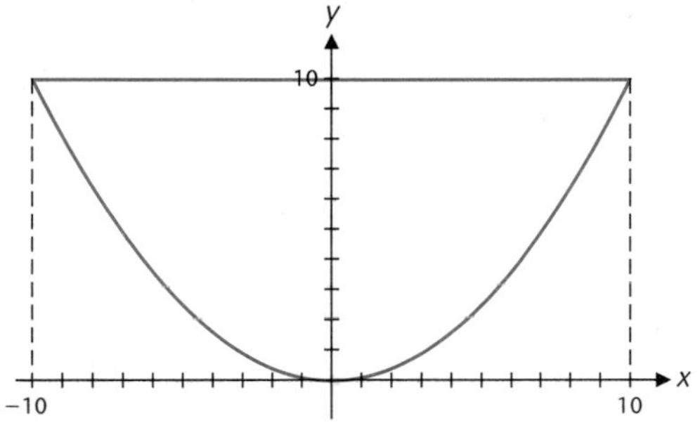

FIGURE 1

The paraboloid is formed by rotating a parabola about its axis (see Fig. 1). If the vertex of the parabola is placed at the origin and the axis along the positive y axis, then the equation of the parabola has the form $x^2 = 4ay$ and the focus is located at the point $(0, a)$. Since the parabola passes through the point (10, 10), we have

$$\begin{aligned} x^2 &= 4ay \\ 10^2 &= 4a(10) \\ a &= 2.5 \end{aligned}$$

Thus, the focus is 2.5 inches above the bottom of the cooker.

APPENDIX A

A Basic Algebra Review

Outline

Algebra is often referred to as "generalized arithmetic." In arithmetic we deal with the basic arithmetic operations of addition, subtraction, multiplication, and division performed on specific numbers. In algebra we continue to use all that we know in arithmetic, but, in addition, we reason and work with symbols that represent one or more numbers. In this appendix we review some important basic algebraic operations usually studied in earlier courses. The material can be studied systematically before commencing with the rest of the book or reviewed as needed.

Section A-1 Algebra and Real Numbers

- Sets
- The Set of Real Numbers
- The Real Number Line
- Basic Real Number Properties
- Further Properties

The rules for manipulating and reasoning with symbols in algebra depend, in large measure, on properties of the real numbers. In this section we look at some of the important properties of this number system. To make our discussions here and elsewhere in the text clearer and more precise, we first introduce a few useful notions about sets.

Sets

George Cantor (1845–1918) developed a theory of sets as an outgrowth of his studies on infinity. His work has become a milestone in the development of mathematics.

Our use of the word "set" will not differ appreciably from the way it is used in everyday language. Words such as "set," "collection," "bunch," and "flock" all convey the same idea. Thus, we think of a **set** as a collection of objects with the important property that we can tell whether any given object is or is not in the set.

Each object in a set is called an **element,** or **member,** of the set. Symbolically,

$a \in A$	means	"a is an element of set A"	$3 \in \{1, 3, 5\}$
$a \notin A$	means	"a is not an element of set A"	$2 \notin \{1, 3, 5\}$

Capital letters are often used to represent sets and lowercase letters to represent elements of a set.

A set is **finite** if the number of elements in the set can be counted and **infinite** if there is no end in counting its elements. A set is **empty** if it contains no elements. The empty set is also called the **null** set and is denoted by $\varnothing$. It is important to observe that the empty set is *not* written as $\{\varnothing\}$.

A set is usually described in one of two ways—by **listing** the elements between braces, { }, or by enclosing within braces a **rule** that determines its elements. For example, if D is the set of all numbers x such that $x^2 = 4$, then using the listing method we write

$$D = \{-2, 2\} \quad \text{Listing method}$$

or using the rule method, we write

$$D = \{x \mid x^2 = 4\} \quad \text{Rule method}$$

Note that in the rule method, the vertical bar | represents "such that," and the entire symbolic form $\{x \mid x^2 = 4\}$ is read, "The set of all x such that $x^2 = 4$."

The letter x introduced in the rule method is a *variable*. In general, a **variable** is a symbol that is used as a placeholder for the elements of a set with two or more elements. This set is called the **replacement set** for the variable. A **constant,** on the other hand, is a symbol that names exactly one object. The symbol "8" is a constant, since it always names the number eight.

If each element of set A is also an element of set B, we say that A is a **subset** of set B, and we write

$$A \subset B \qquad \{1, 5\} \subset \{1, 3, 5\}$$

Note that the definition of a subset allows a set to be a subset of itself.

Since the empty set $\varnothing$ has no elements, every element of $\varnothing$ is also an element of any given set. Thus, the empty set is a subset of every set. For example,

$$\varnothing \subset \{1, 3, 5\} \quad \text{and} \quad \varnothing \subset \{2, 4, 6\}$$

If two sets A and B have exactly the same elements, the sets are said to be **equal,** and we write

$$A = B \qquad \{4, 2, 6\} = \{6, 4, 2\}$$

Notice that the order of listing elements in a set does not matter.

We can now begin our discussion of the real number system. Additional set concepts will be introduced as needed.

The Set of Real Numbers

The real number system is the number system you have used most of your life. Informally, a **real number** is any number that has a decimal representation. Table 1 on the next page describes the set of real numbers and some of its important subsets. Figure 1 illustrates how these sets of numbers are related to each other.

FIGURE 1
Real numbers and important subsets.

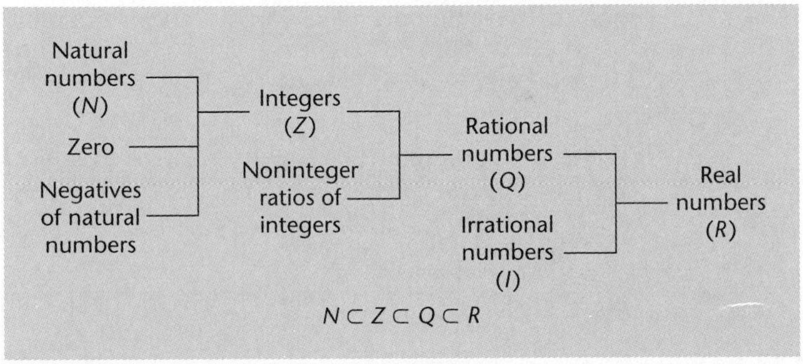

TABLE 1 The Set of Real Numbers

Symbol	Name	Description	Examples
N	Natural numbers	Counting numbers (also called positive integers)	$1, 2, 3, \ldots$
Z	Integers	Natural numbers, their negatives, and 0	$\ldots, -2, -1, 0, 1, 2, \ldots$
Q	Rational numbers	Numbers that can be represented as a/b, where a and b are integers and $b \neq 0$; decimal representations are repeating or terminating	$-4, 0, 1, 25, \frac{-3}{5}, \frac{2}{3}, 3.67, -0.33\overline{3},$* $5.2727\overline{27}$
I	Irrational numbers	Numbers that can be represented as nonrepeating and nonterminating decimal numbers	$\sqrt{2}, \pi, \sqrt[3]{7}, 1.414213\ldots, 2.71828182\ldots$
R	Real numbers	Rational numbers and irrational numbers	

*The overbar indicates that the number (or block of numbers) repeats indefinitely.

The Real Number Line

A one-to-one correspondence exists between the set of real numbers and the set of points on a line. That is, each real number corresponds to exactly one point, and each point to exactly one real number. A line with a real number associated with each point, and vice versa, as in Figure 2, is called a **real number line,** or simply a **real line.** Each number associated with a point is called the **coordinate** of the point. The point with coordinate 0 is called the **origin.** The arrow on the right end of the line indicates a positive direction. The coordinates of all points to the right of the origin are called **positive real numbers,** and those to the left of the origin are called **negative real numbers.** The real number 0 is neither positive nor negative.

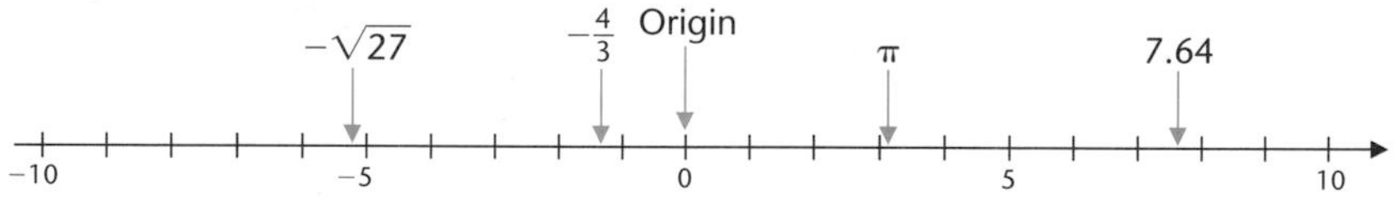

FIGURE 2
A real number line.

Basic Real Number Properties

We now take a look at some of the basic properties of real numbers. (See the box on page A-5.)

You are already familiar with the **commutative properties** for addition and multiplication. They indicate that the order in which the addition or multiplication of two numbers is performed doesn't matter. For example,

$$4 + 5 = 5 + 4 \qquad \text{and} \qquad 4 \cdot 5 = 5 \cdot 4$$

Is there a commutative property relative to subtraction or division? That is, does $x - y = y - x$ or does $x \div y = y \div x$ for all real numbers x and y (division by 0 excluded)? The answer is no, since, for example,

$$7 - 5 \neq 5 - 7 \qquad \text{and} \qquad 6 \div 3 \neq 3 \div 6$$

BASIC PROPERTIES OF THE SET OF REAL NUMBERS

Let R be the set of real numbers, and let x, y, and z be arbitrary elements of R.

Addition Properties

Closure:	$x + y$ is a unique element in R.
Associative:	$(x + y) + z = x + (y + z)$
Commutative:	$x + y = y + x$
Identity:	0 is the additive identity; that is, $0 + x = x + 0 = x$ for all x in R, and 0 is the only element in R with this property.
Inverse:	For each x in R, $-x$ is its unique additive inverse; that is, $x + (-x) = (-x) + x = 0$, and $-x$ is the only element in R relative to x with this property.

Multiplication Properties

Closure:	xy is a unique element in R.
Associative:	$(xy)z = x(yz)$
Commutative:	$xy = yx$
Identity:	1 is the multiplicative identity; that is, for x in R, $(1)x = x(1) = x$, and 1 is the only element in R with this property.
Inverse:	For each x in R, $x \neq 0$, $1/x$ is its unique multiplicative inverse; that is, $x(1/x) = (1/x)x = 1$, and $1/x$ is the only element in R relative to x with this property.

Combined Property

Distributive:	$x(y + z) = xy + xz \qquad (x + y)z = xz + yz$

When computing

$$2 + 5 + 3 \qquad \text{or} \qquad 2 \cdot 5 \cdot 3$$

why don't we need parentheses to indicate which two numbers are to be added or multiplied first? The answer is to be found in the **associative properties.** These properties allow us to write

$$(2 + 5) + 3 = 2 + (5 + 3) \qquad \text{and} \qquad (2 \cdot 5) \cdot 3 = 2 \cdot (5 \cdot 3)$$

so it doesn't matter how we group numbers relative to either operation. Is there an associative property for subtraction or division? The answer is no, since, for example,

$$(8 - 4) - 2 \neq 8 - (4 - 2) \quad \text{and} \quad (8 \div 4) \div 2 \neq 8 \div (4 \div 2)$$

Evaluate both sides of these equations to see why.

CONCLUSION

Relative to addition, **commutativity** and **associativity** permit us to change the order of addition at will and insert or remove parentheses as we please. The same is true for multiplication, but not for subtraction and division.

What number added to a given number will give that number back again? What number times a given number will give that number back again? The answers are 0 and 1, respectively. Because of this, 0 and 1 are called the **identity elements** for the real numbers. Hence, for any real numbers x and y,

$7 + 0 = 7 \qquad 0 + (x + y) = x + y$ 0 is the additive identity.

$1 \cdot 6 = 6 \qquad 1(x + y) = x + y$ 1 is the multiplicative identity.

We now consider **inverses.** For each real number x, there is a unique real number $-x$ such that $x + (-x) = 0$. The number $-x$ is called the **additive inverse** of x, or the **negative** of x. For example, the additive inverse of 4 is -4, since $4 + (-4) = 0$. The additive inverse of -4 is $-(-4) = 4$, since $-4 + [-(-4)] = 0$. It is important to remember:

$-x$ is not necessarily a negative number; it is positive if x is negative and negative if x is positive.

For each nonzero real number x there is a unique real number $1/x$ such that $x(1/x) = 1$. The number $1/x$ is called the **multiplicative inverse** of x, or the **reciprocal** of x. For example, the multiplicative inverse of 7 is $\frac{1}{7}$, since $7(\frac{1}{7}) = 1$. Also note that 7 is the multiplicative inverse of $\frac{1}{7}$. The number 0 has no multiplicative inverse.

We now turn to a real number property that involves both multiplication and addition. Consider the two computations:

$$3(4 + 2) = 3(6) = 18$$

$$3(4) + 3(2) = 12 + 6 = 18$$

Thus,

$$3(4 + 2) = 3(4) + 3(2)$$

and we say that multiplication by 3 *distributes* over the sum $(4 + 2)$. In general, multiplication **distributes** over addition in the real number system. Two more illustrations are

$$2(x + y) = 2x + 2y \qquad (3 + 5)x = 3x + 5x$$

EXAMPLE 1

Using Real Number Properties

Which real number property justifies the indicated statement?

Statement	Property Illustrated
(A) $(7x)y = 7(xy)$	Associative $(\cdot)$
(B) $a(b + c) = (b + c)a$	Commutative $(\cdot)$
(C) $(2x + 3y) + 5y = 2x + (3y + 5y)$	Associative $(+)$
(D) $(x + y)(a + b) = (x + y)a + (x + y)b$	Distributive
(E) If $a + b = 0$, then $b = -a$.	Inverse $(+)$

MATCHED PROBLEM 1

Which real number property justifies the indicated statement?

(A) $4 + (2 + x) = (4 + 2) + x$ (B) $(a + b) + c = c + (a + b)$

(C) $3x + 7x = (3 + 7)x$ (D) $(2x + 3y) + 0 = 2x + 3y$

(E) If $ab = 1$ and $a \neq 0$, then $b = 1/a$.

Further Properties

Subtraction and division can be defined in terms of addition and multiplication, respectively:

DEFINITION 1

SUBTRACTION AND DIVISION

For all real numbers a and b,

Subtraction: $a - b = a + (-b)$ $\quad (-5) - (-3) = (-5) + (3) = -2$

Division: $b\overline{)a} = a \div b = \dfrac{a}{b} = a\left(\dfrac{1}{b}\right) \quad b \neq 0 \quad 3 \div 2 = 3\left(\dfrac{1}{2}\right)$

Thus, to subtract b from a, add the negative of b to a. To divide a by b, multiply a by the reciprocal of b. Note that division by 0 is not defined, since 0 does not have a reciprocal. It is important to remember:

Division by 0 is never allowed.

The following properties of negatives can be proved using the preceding properties and definitions.

THEOREM 1

PROPERTIES OF NEGATIVES

For all real numbers a and b,

1. $-(-a) = a$
2. $(-a)b = -(ab) = a(-b) = -ab$
3. $(-a)(-b) = ab$
4. $(-1)a = -a$
5. $\frac{-a}{b} = -\frac{a}{b} = \frac{a}{-b} \qquad b \neq 0$
6. $\frac{-a}{-b} = -\frac{-a}{b} = -\frac{a}{-b} = \frac{a}{b} \qquad b \neq 0$

We now state an important theorem involving 0.

THEOREM 2

ZERO PROPERTIES

For all real numbers a and b,

1. $a \cdot 0 = 0$
2. $ab = 0$ if and only if $a = 0$ or $b = 0$ or both

EXAMPLE 2

Using Negative and Zero Properties

Which real number property or definition justifies each statement?

Statement	**Property or Definition Illustrated**
(A) $3 - (-2) = 3 + [-(-2)] = 5$	Subtraction (Definition 1 and Theorem 1, part 1)
(B) $-(-2) = 2$	Negatives (Theorem 1, part 1)
(C) $-\frac{-3}{2} = \frac{3}{2}$	Negatives (Theorem 1, part 6)
(D) $\frac{5}{-2} = -\frac{5}{2}$	Negatives (Theorem 1, part 5)
(E) If $(x - 3)(x + 5) = 0$, then either $x - 3 = 0$ or $x + 5 = 0$.	Zero (Theorem 2, part 2)

MATCHED PROBLEM 2

Which real number property or definition justifies each statement?

(A) $\frac{3}{5} = 3\left(\frac{1}{5}\right)$ (B) $(-5)(2) = -(5 \cdot 2)$ (C) $(-1)3 = -3$

(D) $\frac{-7}{9} = -\frac{7}{9}$ (E) If $x + 5 = 0$, then $(x - 3)(x + 5) = 0$.

Explore/Discuss 1

In general, a set of numbers is closed under an operation if performing the operation on numbers in the set always produces another number in the set. For example, the real numbers are closed under addition, multiplication, subtraction, and division, excluding division by 0. Replace each ? in the following tables with T (true) or F (false), and illustrate each false statement with an example. (See Table 1 for the definitions of the sets N, Z, I, Q, and R.)

	Closed under Addition	Closed under Multiplication
N	?	?
Z	?	?
Q	?	?
I	?	?
R	T	T

	Closed under Subtraction	Closed under Division*
N	?	?
Z	?	?
Q	?	?
I	?	?
R	T	T

*Excluding division by 0.

Recall that the quotient $a \div b$, $b \neq 0$, written in the form a/b is called a **fraction.** The quantity a is called the **numerator** and the quantity b is the **denominator.**

THEOREM 3

FRACTION PROPERTIES

For all real numbers a, b, c, d, and k (division by 0 excluded),

1. $\frac{a}{b} = \frac{c}{d}$ if and only if $ad = bc$

   $\frac{4}{6} = \frac{6}{9}$ since $4 \cdot 9 = 6 \cdot 6$

2. $\frac{ka}{kb} = \frac{a}{b}$

   $\frac{7 \cdot 3}{7 \cdot 5} = \frac{3}{5}$

3. $\frac{a}{b} \cdot \frac{c}{d} = \frac{ac}{bd}$

   $\frac{3}{5} \cdot \frac{7}{8} = \frac{3 \cdot 7}{5 \cdot 8}$

4. $\frac{a}{b} \div \frac{c}{d} = \frac{a}{b} \cdot \frac{d}{c}$

   $\frac{2}{3} \div \frac{5}{7} = \frac{2}{3} \cdot \frac{7}{5}$

5. $\frac{a}{b} + \frac{c}{b} = \frac{a + c}{b}$

   $\frac{3}{6} + \frac{5}{6} = \frac{3 + 5}{6}$

6. $\frac{a}{b} - \frac{c}{b} = \frac{a - c}{b}$

   $\frac{7}{8} - \frac{3}{8} = \frac{7 - 3}{8}$

7. $\frac{a}{b} + \frac{c}{d} = \frac{ad + bc}{bd}$

   $\frac{2}{3} + \frac{3}{5} = \frac{2 \cdot 5 + 3 \cdot 3}{3 \cdot 5}$

Answers to Matched Problems

1. (A) Associative (+) (B) Commutative (+) (C) Distributive (D) Identity (+) (E) Inverse (·)
2. (A) Division (Definition 1) (B) Negatives (Theorem 1, part 2) (C) Negatives (Theorem 1, part 4) (D) Negatives (Theorem 1, part 5) (E) Zero (Theorem 2, part 1)

EXERCISE A-1

All variables represent real numbers.

A

In Problems 1–8, indicate true (T) or false (F).

1. $4 \in \{3, 4, 5\}$

2. $6 \in \{2, 4, 6\}$

3. $3 \notin \{3, 4, 5\}$

4. $7 \notin \{2, 4, 6\}$

5. $\{1, 2\} \subset \{1, 3, 5\}$

6. $\{2, 6\} \subset \{2, 4, 6\}$

7. $\{7, 3, 5\} \subset \{3, 5, 7\}$

8. $\{7, 3, 5\} = \{3, 5, 7\}$

In Problems 9–14, replace each question mark with an appropriate expression that will illustrate the use of the indicated real number property.

9. Commutative property (+): $x + 7 = ?$

10. Commutative property (·): $uv = ?$

11. Associative property (·): $x(yz) = ?$

12. Associative property (+): $3 + (7 + y) = ?$

13. Identity property (+): $0 + 9m = ?$

14. Identity property (·): $1(u + v) = ?$

In Problems 15–26, each statement illustrates the use of one of the following properties or definitions. Indicate which one.

Commutative (+, ·)	*Subtraction*
Associative (+, ·)	*Division*
Distributive	*Negatives (Theorem 1)*
Identity (+, ·)	*Zero (Theorem 2)*
Inverse (+, ·)	

15. $x + ym = x + my$

16. $7(3m) = (7 \cdot 3)m$

17. $7u + 9u = (7 + 9)u$

18. $-\frac{u}{-v} = \frac{u}{v}$

19. $(-2)(\frac{1}{-2}) = 1$

20. $8 - 12 = 8 + (-12)$

21. $w + (-w) = 0$

22. $5 \div (-6) = 5(\frac{1}{-6})$

23. $3(xy + z) + 0 = 3(xy + z)$

24. $ab(c + d) = abc + abd$

25. $\frac{-x}{-y} = \frac{x}{y}$

26. $(x + y) \cdot 0 = 0$

B

Write each set in Problems 27–32 using the listing method; that is, list the elements between braces. If the set is empty, write ∅.

27. $\{x \mid x$ is an even integer between -3 and $5\}$

28. $\{x \mid x$ is an odd integer between -4 and $6\}$

29. $\{x \mid x$ is a letter in "status"$\}$

30. $\{x \mid x$ is a letter in "consensus"$\}$

31. $\{x \mid x$ is a month starting with B$\}$

32. $\{x \mid x$ is a month with 32 days$\}$

In Problems 33–40, each statement illustrates the use of one of the following properties or definitions. Indicate which one.

Commutative (+, ·)	*Subtraction*
Associative (+, ·)	*Division*
Distributive	*Negatives (Theorem 1)*
Identity (+, ·)	*Zero (Theorem 2)*
Inverse (+, ·)	

33. $(3x + 5) + 7 = 7 + (3x + 5)$

34. $(5x)(7y) = 5[x(7y)]$

35. $(3x + 2) + (x + 5) = 3x + [2 + (x + 5)]$

36. $(x + 3)(x + 5) = (x + 3)x + (x + 3)5$

37. $x(x - y) + y(x - y) = (x + y)(x - y)$

38. $\frac{-7}{-(m - n)} = \frac{7}{m - n}$

39. $(2x - 3)(x + 5) = 0$ if and only if $2x - 3 = 0$ or $x + 5 = 0$.

40. If $x(3x - 7) = 0$, then either $x = 0$ or $3x - 7 = 0$.

41. If $ab = 0$, does either a or b have to be 0?

42. If $ab = 1$, does either a or b have to be 1?

43. Indicate which of the following are true:

(A) All natural numbers are integers.

(B) All real numbers are irrational.

(C) All rational numbers are real numbers.

44. Indicate which of the following are true:

(A) All integers are natural numbers.

(B) All rational numbers are real numbers.

(C) All natural numbers are rational numbers.

45. Give an example of a rational number that is not an integer.

46. Give an example of a real number that is not a rational number.

47. Given the sets of numbers N (natural numbers), Z (integers), Q (rational numbers), and R (real numbers), indicate to which set(s) each of the following numbers belongs:

(A) -3 (B) 3.14 (C) π (D) $\frac{2}{3}$

48. Given the sets of numbers N, Z, Q, and R (see Problem 47), indicate to which set(s) each of the following numbers belongs:

(A) 8 (B) $\sqrt{2}$ (C) -1.414 (D) $\frac{-5}{2}$

In Problems 49 and 50, use a calculator to express each number as a decimal fraction to the capacity of your calculator (refer to the user's manual for your calculator). Observe the repeating decimal representation of the rational numbers and the apparent nonrepeating decimal representation of the irrational numbers.

49. (A) $\frac{8}{9}$ (B) $\frac{3}{11}$ (C) $\sqrt{5}$ (D) $\frac{11}{8}$

50. (A) $\frac{13}{6}$ (B) $\sqrt{21}$ (C) $\frac{7}{16}$ (D) $\frac{29}{111}$

51. Indicate true (T) or false (F), and for each false statement find real number replacements for a and b that will provide a counterexample. For all real numbers a and b:

(A) $a + b = b + a$ (B) $a - b = b - a$

(C) $ab = ba$ (D) $a \div b = b \div a$

52. Indicate true (T) or false (F), and for each false statement find real number replacements for a, b, and c that will provide a counterexample. For all real numbers a, b, and c:

(A) $(a + b) + c = a + (b + c)$

(B) $(a - b) - c = a - (b - c)$

(C) $a(bc) = (ab)c$

(D) $(a \div b) \div c = a \div (b \div c)$

C

53. If $A = \{1, 2, 3, 4\}$ and $B = \{2, 4, 6\}$, find

(A) $\{x \mid x \in A \text{ or } x \in B\}$

(B) $\{x \mid x \in A \text{ and } x \in B\}$

54. If $F = \{-2, 0, 2\}$ and $G = \{-1, 0, 1, 2\}$, find

(A) $\{x \mid x \in F \text{ or } x \in G\}$

(B) $\{x \mid x \in F \text{ and } x \in G\}$

55. If $c = 0.151\,515\ldots$, then $100c = 15.1515\ldots$ and

$$\begin{aligned} 100c - c &= 15.1515\ldots - 0.151\,515\ldots \\ 99c &= 15 \\ c &= \tfrac{15}{99} = \tfrac{5}{33} \end{aligned}$$

Proceeding similarly, convert the repeating decimal 0.090909 . . . into a fraction. (All repeating decimals are rational numbers, and all rational numbers have repeating decimal representations.)

56. Repeat Problem 55 for 0.181 818. . . .

57. To see how the distributive property is behind the mechanics of long multiplication, compute each of the following and compare:

Long Multiplication	Use of the Distributive Property
23	$23 \cdot 12$
$\times$ 12	$= 23(2 + 10)$
	$= 23 \cdot 2 + 23 \cdot 10 =$

58. For a and b real numbers, justify each step using a property in this section.

Statement	Reason
1. $(a + b) + (-a) = (-a) + (a + b)$	**1.**
2. $= [(-a) + a] + b$	**2.**
3. $= 0 + b$	**3.**
4. $= b$	**4.**

Section A-2 | Polynomials: Basic Operations

- Natural Number Exponents
- Polynomials
- Combining Like Terms
- Addition and Subtraction
- Multiplication
- Combined Operations
- Application

In this section we review the basic operations on *polynomials*, a mathematical form encountered frequently throughout mathematics. We start the discussion with a brief review of natural number exponents. Integer and rational exponents and their properties will be discussed in detail in subsequent sections.

Natural Number Exponents

The definition of a **natural number exponent** is given below.

DEFINITION 1

NATURAL NUMBER EXPONENT

For n a natural number and a any real number,

$$a^n = \underbrace{a \cdot a \cdot \cdots \cdot a}_{n \text{ factors of } a} \qquad 2^4 = \underbrace{2 \cdot 2 \cdot 2 \cdot 2}_{4 \text{ factors of } 2}$$

Also, the **first property of exponents** is stated as follows:

THEOREM 1

FIRST PROPERTY OF EXPONENTS

For any natural numbers m and n, and any real number a,

$$a^m a^n = a^{m+n} \qquad (3x^5)(2x^7) = \boxed{= (3 \cdot 2)x^{5+7}} = 6x^{12}$$

Polynomials

Algebraic expressions are formed by using constants and variables and the algebraic operations of addition, subtraction, multiplication, division, raising to powers, and taking roots. Some examples are

$$\sqrt[3]{x^3 + 5} \qquad 5x^4 + 2x^2 - 7$$

$$x + y - 7 \qquad (2x - y)^2$$

$$\frac{x - 5}{x^2 + 2x - 5} \qquad 1 + \frac{1}{1 + \frac{1}{x}}$$

An algebraic expression involving only the operations of addition, subtraction, multiplication, and raising to natural number powers on variables and constants is called a **polynomial.** Some examples are

$$2x - 3 \qquad 4x^2 - 3x + 7$$

$$x - 2y \qquad 5x^3 - 2x^2 - 7x + 9$$

$$5 \qquad x^2 - 3xy + 4y^2$$

$$0 \qquad x^3 - 3x^2y + xy^2 + 2y^7$$

In a polynomial, a variable cannot appear in a denominator, as an exponent, or within a radical. Accordingly, a **polynomial in one variable** x is constructed by adding or subtracting constants and terms of the form ax^n, where a is a real number and n is a natural number. A **polynomial in two variables** x and y is constructed by adding and subtracting constants and terms of the form ax^my^n, where a is a real number and m and n are natural numbers. Polynomials in three or more variables are defined in a similar manner.

Polynomial forms can be classified according to their *degree.* If a term in a polynomial has only one variable as a factor, then the **degree of that term** is the power of the variable. If two or more variables are present in a term as factors, then the **degree of the term** is the sum of the powers of the variables. The **degree of a polynomial** is the degree of the nonzero term with the highest degree in the polynomial. Any nonzero constant is defined to be a **polynomial of degree 0.** The number 0 is also a polynomial but is not assigned a degree.

EXAMPLE 1 Polynomials and Nonpolynomials

(A) Polynomials in one variable:

$$x^2 - 3x + 2 \qquad 6x^3 - \sqrt{2}x - \tfrac{1}{3}$$

(B) Polynomials in several variables:

$$3x^2 - 2xy + y^2 \qquad 4x^3y^2 - \sqrt{3}xy^2z^5$$

(C) Nonpolynomials:

$$\sqrt{2x} - \frac{3}{x} + 5 \qquad \frac{x^2 - 3x + 2}{x - 3} \qquad \sqrt{x^2 - 3x + 1}$$

(D) The degree of the first term in $6x^3 - \sqrt{2}x - \frac{1}{3}$ is 3, the degree of the second term is 1, the degree of the third term is 0, and the degree of the whole polynomial is 3.

(E) The degree of the first term in $4x^3y^2 - \sqrt{3}xy^2$ is 5, the degree of the second term is 3, and the degree of the whole polynomial is 5.

MATCHED PROBLEM 1

(A) Which of the following are polynomials?

$$3x^2 - 2x + 1 \qquad \sqrt{x - 3} \qquad x^2 - 2xy + y^2 \qquad \frac{x - 1}{x^2 + 2}$$

(B) Given the polynomial $3x^5 - 6x^3 + 5$, what is the degree of the first term? The second term? The whole polynomial?

(C) Given the polynomial $6x^4y^2 - 3xy^3$, what is the degree of the first term? The second term? The whole polynomial?

In addition to classifying polynomials by degree, we also call a single-term polynomial a **monomial,** a two-term polynomial a **binomial,** and a three-term polynomial a **trinomial.**

$\frac{5}{2}x^2y^3$	Monomial
$x^3 + 4.7$	Binomial
$x^4 - \sqrt{2}x^2 + 9$	Trinomial

Combining Like Terms

We start with a word about *coefficients.* A constant in a term of a polynomial, including the sign that precedes it, is called the **numerical coefficient,** or simply, the **coefficient,** of the term. If a constant doesn't appear, or only a + sign appears, the coefficient is understood to be 1. If only a − sign appears, the coefficient is understood to be −1. Thus, given the polynomial

$$2x^4 - 4x^3 + x^2 - x + 5 \quad 2x^4 + (-4)x^3 + 1x^2 + (-1)x + 5$$

the coefficient of the first term is 2, the coefficient of the second term is −4, the coefficient of the third term is 1, the coefficient of the fourth term is −1, and the coefficient of the last term is 5.

At this point, it is useful to state two additional distributive properties of real numbers that follow from the distributive properties stated in Section A-1.

ADDITIONAL DISTRIBUTIVE PROPERTIES

1. $a(b - c) = (b - c)a = ab - ac$
2. $a(b + c + \cdots + f) = ab + ac + \cdots + af$

Two terms in a polynomial are called **like terms** if they have exactly the same variable factors to the same powers. The numerical coefficients may or may not be the same. Since constant terms involve no variables, all constant terms are like terms. If a polynomial contains two or more like terms, these terms can be combined into a single term by making use of distributive properties. Consider the following example:

$$\begin{aligned} 5x^3y - 2xy - x^3y - 2x^3y &= 5x^3y - x^3y - 2x^3y - 2xy \\ &= (5x^3y - x^3y - 2x^3y) - 2xy \\ &= (5 - 1 - 2)x^3y - 2xy \\ &= 2x^3y - 2xy \end{aligned}$$

It should be clear that free use has been made of the real number properties discussed earlier. The steps done in the dashed box are usually done mentally, and the process is quickly mechanized as follows:

Like terms in a polynomial are combined by adding their numerical coefficients.

EXAMPLE 2

Simplifying Polynomials

Remove parentheses and combine like terms.

(A) $2(3x^2 - 2x + 5) + (x^2 + 3x - 7)$

$= 2(3x^2 - 2x + 5) + 1(x^2 + 3x - 7)$ Think

$= 6x^2 - 4x + 10 + x^2 + 3x - 7$

$= 7x^2 - x + 3$

(B) $(x^3 - 2x - 6) - (2x^3 - x^2 + 2x - 3)$

$= 1(x^3 - 2x - 6) + (-1)(2x^3 - x^2 + 2x - 3)$ Think — Be careful with the sign here.

$= x^3 - 2x - 6 - 2x^3 + x^2 - 2x + 3$

$= -x^3 + x^2 - 4x - 3$

(C) $[3x^2 - (2x + 1)] - (x^2 - 1) = [3x^2 - 2x - 1] - (x^2 - 1)$ Remove inner parentheses first.

$= 3x^2 - 2x - 1 - x^2 + 1$

$= 2x^2 - 2x$

MATCHED PROBLEM 2

Remove parentheses and combine like terms.

(A) $3(u^2 - 2v^2) + (u^2 + 5v^2)$

(B) $(m^3 - 3m^2 + m - 1) - (2m^3 - m + 3)$

(C) $(x^3 - 2) - [2x^3 - (3x + 4)]$

Addition and Subtraction

Addition and subtraction of polynomials can be thought of in terms of removing parentheses and combining like terms, as illustrated in Example 2. Horizontal and vertical arrangements are illustrated in the next two examples. You should be able to work either way, letting the situation dictate the choice.

EXAMPLE 3

Adding Polynomials

Add.

$$x^4 - 3x^3 + x^2, \quad -x^3 - 2x^2 + 3x, \quad \text{and} \quad 3x^2 - 4x - 5$$

Solution Add horizontally.

$$(x^4 - 3x^3 + x^2) + (-x^3 - 2x^2 + 3x) + (3x^2 - 4x - 5)$$
$$= x^4 - 3x^3 + x^2 - x^3 - 2x^2 + 3x + 3x^2 - 4x - 5$$
$$= x^4 - 4x^3 + 2x^2 - x - 5$$

Or vertically, by lining up like terms and adding their coefficients.

$$\begin{array}{rrrrr} x^4 & - 3x^3 & + x^2 & & \\ & - x^3 & - 2x^2 & + 3x & \\ & & 3x^2 & - 4x & - 5 \\ \hline x^4 & - 4x^3 & + 2x^2 & - x & - 5 \end{array}$$

MATCHED PROBLEM 3 Add horizontally and vertically.

$$3x^4 - 2x^3 - 4x^2, \quad x^3 - 2x^2 - 5x, \quad \text{and} \quad x^2 + 7x - 2$$

EXAMPLE 4 **Subtracting Polynomials**

Subtract.

$$4x^2 - 3x + 5 \quad \text{from} \quad x^2 - 8$$

Solution

$$(x^2 - 8) - (4x^2 - 3x + 5) \quad \text{or} \quad \begin{array}{rrr} x^2 & & - 8 \\ -4x^2 & + 3x & - 5 \\ \hline -3x^2 & + 3x & - 13 \end{array} \quad \leftarrow \text{Change signs and add.}$$
$$= x^2 - 8 - 4x^2 + 3x - 5$$
$$= -3x^2 + 3x - 13$$

MATCHED PROBLEM 4 Subtract.

$$2x^2 - 5x + 4 \quad \text{from} \quad 5x^2 - 6$$

CAUTION When you use a horizontal arrangement to subtract a polynomial with more than one term, you must enclose the polynomial in parentheses. Thus, to subtract $2x + 5$ from $4x - 11$, you must write

$$4x - 11 - (2x + 5) \quad \text{and not} \quad 4x - 11 - 2x + 5$$

Multiplication

Multiplication of algebraic expressions involves the extensive use of distributive properties for real numbers, as well as other real number properties.

EXAMPLE 5

Multiplying Polynomials

Multiply.

$$(2x - 3)(3x^2 - 2x + 3)$$

Solution

$$\begin{aligned}(2x - 3)(3x^2 - 2x + 3) &= 2x(3x^2 - 2x + 3) - 3(3x^2 - 2x + 3)\\ &= 6x^3 - 4x^2 + 6x - 9x^2 + 6x - 9\\ &= 6x^3 - 13x^2 + 12x - 9\end{aligned}$$

or, using a vertical arrangement,

$$\begin{array}{llll} & 3x^2 & - 2x & + 3\\ & 2x & - 3 & \\ \hline 6x^3 & - 4x^2 & + 6x & \\ & - 9x^2 & + 6x & - 9\\ \hline 6x^3 & - 13x^2 & + 12x & - 9\end{array}$$

MATCHED PROBLEM 5

Multiply.

$$(2x - 3)(2x^2 + 3x - 2)$$

Thus, to multiply two polynomials, multiply each term of one by each term of the other, and combine like terms.

Products of certain binomial factors occur so frequently that it is useful to develop procedures that will enable us to write down their products by inspection. To find the product $(2x - 1)(3x + 2)$, we will use the popular **FOIL method.** We multiply each term of one factor by each term of the other factor as follows:

$$\begin{array}{lcccc} & \text{F} & \text{O} & \text{I} & \text{L}\\ & \text{First product} & \text{Outer product} & \text{Inner product} & \text{Last product}\\ & \downarrow & \downarrow & \downarrow & \downarrow\\ (2x - 1)(3x + 2) = & 6x^2 & + 4x & - 3x & - 2\end{array}$$

The inner and outer products are like terms and hence combine into one term. Thus,

$$(2x - 1)(3x + 2) = 6x^2 + x - 2$$

To speed up the process, we combine the inner and outer product mentally.

Products of certain binomial factors occur so frequently that it is useful to remember formulas for their products. The following formulas are easily verified by multiplying the factors on the left using the FOIL method:

SPECIAL PRODUCTS

1. $(a - b)(a + b) = a^2 - b^2$
2. $(a + b)^2 = a^2 + 2ab + b^2$
3. $(a - b)^2 = a^2 - 2ab + b^2$

Explore/Discuss 1

(A) Explain the relationship between special product formula 1 and the areas of the rectangles in the figures.

$$(a - b)(a + b) \qquad = \qquad a^2 - b^2$$

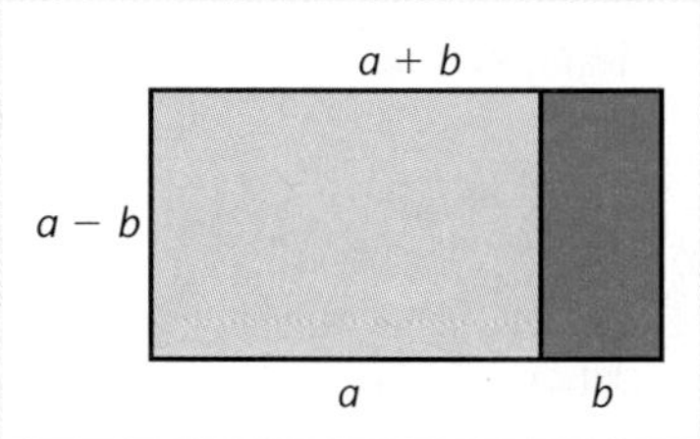

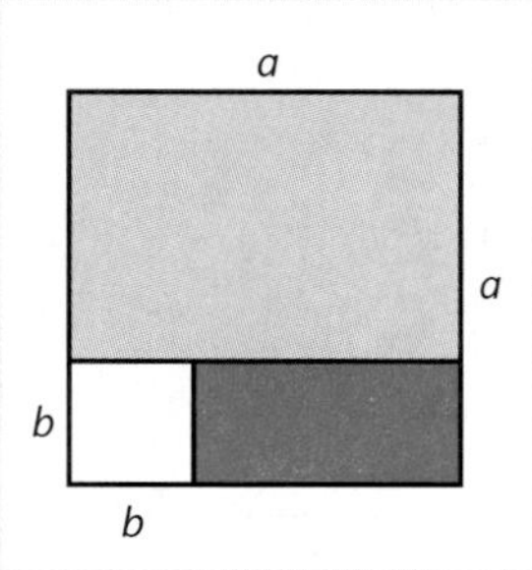

(B) Construct similar figures to provide geometric interpretations for special product formulas 2 and 3.

EXAMPLE 6

Multiplying Binomials

Multiply.

(A) $(2x - 3y)(5x + 2y) = 10x^2 + 4xy - 15xy - 6y^2$

$= 10x^2 - 11xy - 6y^2$

(B) $(3a - 2b)(3a + 2b) = (3a)^2 - (2b)^2 = 9a^2 - 4b^2$

(C) $(5x - 3)^2 = (5x)^2 - 2(5x)(3) + 3^2 = 25x^2 - 30x + 9$

(D) $(m + 2n)^2 = m^2 + 4mn + 4n^2$

MATCHED PROBLEM 6

Multiply.

(A) $(4u - 3v)(2u + v)$ (B) $(2xy + 3)(2xy - 3)$

(C) $(m + 4n)(m - 4n)$ (D) $(2u - 3v)^2$ (E) $(6x + y)^2$

CAUTION

Remember to include the sum of the inner and outer terms when using the FOIL method to square a binomial. That is,

$$(x + 3)^2 \neq x^2 + 9 \qquad (x + 3)^2 = x^2 + 6x + 9$$

Combined Operations

We now consider several examples that use all the operations just discussed. Before considering these examples, it is useful to summarize order-of-operation conventions pertaining to exponents, multiplication and division, and addition and subtraction.

ORDER OF OPERATIONS

1. Simplify inside the innermost grouping first, then the next innermost, and so on.

$$\begin{aligned} 2[3 - (x - 4)] &= 2[3 - x + 4] \\ &= 2(7 - x) = 14 - 2x \end{aligned}$$

2. Unless grouping symbols indicate otherwise, apply exponents before multiplication or division is performed.

$$2(x - 2)^2 = 2(x^2 - 4x + 4) = 2x^2 - 8x + 8$$

3. Unless grouping symbols indicate otherwise, perform multiplication and division before addition and subtraction. In either case, proceed from left to right.

$$5 - 2(x - 3) = 5 - 2x + 6 = 11 - 2x$$

EXAMPLE 7

Combined Operations

Perform the indicated operations and simplify.

(A)
$$\begin{aligned} 3x - \{5 - 3[x - x(3 - x)]\} &= 3x - \{5 - 3[x - 3x + x^2]\} \\ &= 3x - \{5 - 3[-2x + x^2]\} \\ &= 3x - \{5 + 6x - 3x^2\} \\ &= 3x - 5 - 6x + 3x^2 \\ &= 3x^2 - 3x - 5 \end{aligned}$$

(B)
$$\begin{aligned} (x - 2y)(2x + 3y) - (2x + y)^2 &= 2x^2 + 3xy - 4xy - 6y^2 \\ &\quad - (4x^2 + 4xy + y^2) \\ &= 2x^2 - xy - 6y^2 - 4x^2 - 4xy - y^2 \\ &= -2x^2 - 5xy - 7y^2 \end{aligned}$$

(C) $(2m + 3n)^3 = (2m + 3n)(2m + 3n)^2$
$= (2m + 3n)(4m^2 + 12mn + 9n^2)$
$= 8m^3 + 24m^2n + 18mn^2 + 12m^2n + 36mn^2 + 27n^3$
$= 8m^3 + 36m^2n + 54mn^2 + 27n^3$

MATCHED PROBLEM 7

Perform the indicated operations and simplify.

(A) $2t - \{7 - 2[t - t(4 + t)]\}$ (B) $(u - 3v)^2 - (2u - v)(2u + v)$

(C) $(4x - y)^3$

Application

EXAMPLE 8

Volume of a Cylindrical Shell

A plastic water pipe with a hollow center is 100 inches long, 1 inch thick, and has an inner radius of x inches (see the figure). Write an algebraic expression in terms of x that represents the volume of the plastic used to construct the pipe. Simplify the expression. [*Recall:* The volume V of a right circular cylinder of radius r and height h is given by $V = \pi r^2 h$.]

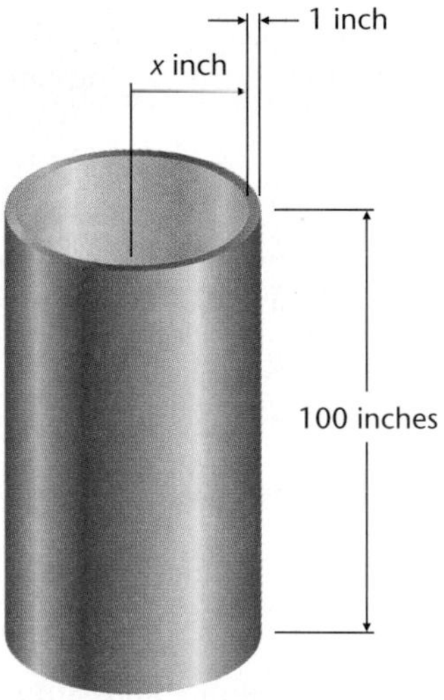

Solution A right circular cylinder with a hollow center is called a **cylindrical shell.** The volume of the shell is equal to the volume of the cylinder minus the volume of the hole. Since the radius of the hole is x inches and the pipe is 1 inch thick, the radius of the cylinder is $x + 1$ inches. Thus, we have

$$\begin{pmatrix}\text{Volume of}\\ \text{shell}\end{pmatrix} = \begin{pmatrix}\text{Volume of}\\ \text{cylinder}\end{pmatrix} - \begin{pmatrix}\text{Volume of}\\ \text{hole}\end{pmatrix}$$

$$\begin{aligned}\text{Volume} &= \pi(x + 1)^2\,100 - \pi x^2 100\\ &= 100\pi(x^2 + 2x + 1) - 100\pi x^2\\ &= 100\pi x^2 + 200\pi x + 100\pi - 100\pi x^2\\ &= 200\pi x + 100\pi\end{aligned}$$

MATCHED PROBLEM 8

A plastic water pipe is 200 inches long, 2 inches thick, and has an outer radius of x inches. Write an algebraic expression in terms of x that represents the volume of the plastic used to construct the pipe. Simplify the expression.

Answers to Matched Problems

1. (A) $3x^2 - 2x + 1$, $x^2 - 2xy + y^2$ (B) 5, 3, 5 (C) 6, 4, 6
2. (A) $4u^2 - v^2$ (B) $-m^3 - 3m^2 + 2m - 4$ (C) $-x^3 + 3x + 2$
3. $3x^4 - x^3 - 5x^2 + 2x - 2$ **4.** $3x^2 + 5x - 10$ **5.** $4x^3 - 13x + 6$
6. (A) $8u^2 - 2uv - 3v^2$ (B) $4x^2y^2 - 9$ (C) $m^2 - 16n^2$ (D) $4u^2 - 12uv + 9v^2$ (E) $36x^2 + 12xy + y^2$
7. (A) $-2t^2 - 4t - 7$ (B) $-3u^2 - 6uv + 10v^2$ (C) $64x^3 - 48x^2y + 12xy^2 - y^3$
8. Volume $= 200\pi x^2 - 200\pi(x - 2)^2 = 800\pi x - 800\pi$

EXERCISE A-2

A

Problems 1–8 refer to the following polynomials:
(a) $2x^3 - 3x^2 + x + 5$ *(b)* $2x^2 + x - 1$ *(c)* $3x - 2$

1. What is the degree of (a)? **2.** What is the degree of (b)?

3. Add (a) and (b). **4.** Add (b) and (c).

5. Subtract (b) from (a). **6.** Subtract (c) from (b).

7. Multiply (a) and (c). **8.** Multiply (b) and (c).

In Problems 9–28, perform the indicated operations and simplify.

9. $2(x - 1) + 3(2x - 3) - (4x - 5)$

10. $2(u - 1) - (3u + 2) - 2(2u - 3)$

11. $2y - 3y[4 - 2(y - 1)]$ **12.** $4a - 2a[5 - 3(a + 2)]$

13. $(m - n)(m + n)$ **14.** $(a + b)(a - b)$

15. $(4t - 3)(t - 2)$ **16.** $(3x - 5)(2x + 1)$

17. $(3x + 2y)(x - 3y)$ **18.** $(2x - 3y)(x + 2y)$

19. $(2m - 7)(2m + 7)$ **20.** $(3y + 2)(3y - 2)$

21. $(6x - 4y)(5x + 3y)$ **22.** $(3m + 7n)(2m - 5n)$

23. $(3x - 2y)(3x + 2y)$ **24.** $(4m + 3n)(4m - 3n)$

25. $(4x - y)^2$ **26.** $(3u + 4v)^2$

27. $(a + b)(a^2 - ab + b^2)$ **28.** $(a - b)(a^2 + ab + b^2)$

B

In Problems 29–42, perform the indicated operations and simplify.

29. $2x - 3\{x + 2[x - (x + 5)] + 1\}$

30. $m - \{m - [m - (m - 1)]\}$

31. $2\{3[a - 4(1 - a)] - (5 - a)\}$

32. $5b - 3\{-[2 - 4(2b - 1)] + 2(2 - 3b)\}$

33. $(2x^2 + x - 2)(x^2 - 3x + 5)$

34. $(x^2 - 2xy + y^2)(x^2 + 2xy + y^2)$

35. $(h^2 + hk + k^2)(h^2 - hk + k^2)$

36. $(n^2 + 2n + 1)(n^2 - 4n - 3)$

37. $(2x - 1)^2 - (3x + 2)(3x - 2)$

38. $(3a - b)(3a + b) - (2a - 3b)^2$

39. $(m - 3n)(m + 8n) + (m + 6n)(m + 4n)$

40. $(y - 2)(y + 1) + (y - 3)(y + 4)$

41. $(2m - n)^3$ **42.** $(3a + 2b)^3$

C

Problems 43–50 are calculus-related. Perform the indicated operations and simplify.

43. $5(x + h) - 4 - (5x - 4)$ **44.** $6(x + h) + 2 - (6x + 2)$

45. $3(x + h)^2 + 2(x + h) - (3x^2 + 2x)$

46. $4(x + h)^2 - 5(x + h) - (4x^2 - 5x)$

47. $-2(x + h)^2 - 3(x + h) + 7 - (-2x^2 - 3x + 7)$

48. $-(x + h)^2 + 4(x + h) - 9 - (-x^2 + 4x - 9)$

49. $(x + h)^3 - x^3$

50. $2(x + h)^2 + 3(x + h) - (2x^2 + 3x)$

51. Subtract the sum of the first two polynomials from the sum of the last two: $3m^2 - 2m + 5$, $4m^2 - m$, $3m^2 - 3m - 2$, $m^3 + m^2 + 2$

52. Subtract the sum of the last two polynomials from the sum of the first two: $2x^2 - 4xy + y^2$, $3xy - y^2$, $x^2 - 2xy - y^2$, $-x^2 + 3xy - 2y^2$

C

In Problems 53–56, perform the indicated operations and simplify.

53. $2(x-2)^3 - (x-2)^2 - 3(x-2) - 4$

54. $(2x-1)^3 - 2(2x-1)^2 + 3(2x-1) + 7$

55. $-3x\{x[x - x(2-x)] - (x+2)(x^2-3)\}$

56. $2\{(x-3)(x^2-2x+1) - x[3 - x(x-2)]\}$

57. Show by example that, in general, $(a+b)^2 \neq a^2 + b^2$. Discuss possible conditions on a and b that would make this a valid equation.

58. Show by example that, in general, $(a-b)^2 \neq a^2 - b^2$. Discuss possible conditions on a and b that would make this a valid equation.

59. If you are given two polynomials, one of degree m and the other of degree n, $m > n$, what is the degree of the sum?

60. What is the degree of the product of the two polynomials in Problem 59?

61. How does the answer to Problem 59 change if the two polynomials can have the same degree?

62. How does the answer to Problem 60 change if the two polynomials can have the same degree?

APPLICATIONS

63. Geometry. The width of a rectangle is 5 centimeters less than its length. If x represents the length, write an algebraic expression in terms of x that represents the perimeter of the rectangle. Simplify the expression.

64. Geometry. The length of a rectangle is 8 meters more than its width. If x represents the width of the rectangle, write an algebraic expression in terms of x that represents its area. Change the expression to a form without parentheses.

★ **65. Coin Problem.** A parking meter contains nickels, dimes, and quarters. There are 5 fewer dimes than nickels, and 2 more quarters than dimes. If x represents the number of nickels, write an algebraic expression in terms of x that represents the value of all the coins in the meter in cents. Simplify the expression.

★ **66. Coin Problem.** A vending machine contains dimes and quarters only. There are 4 more dimes than quarters. If x represents the number of quarters, write an algebraic expression in terms of x that represents the value of all the coins in the vending machine in cents. Simplify the expression.

67. Packaging. A spherical plastic container for designer wristwatches has an inner radius of x centimeters (see the figure). If the plastic shell is 0.3 centimeters thick, write an algebraic expression in terms of x that represents the volume of the plastic used to construct the container. Simplify the expression. [*Recall:* The volume V of a sphere of radius r is given by $V = \frac{4}{3}\pi r^3$.]

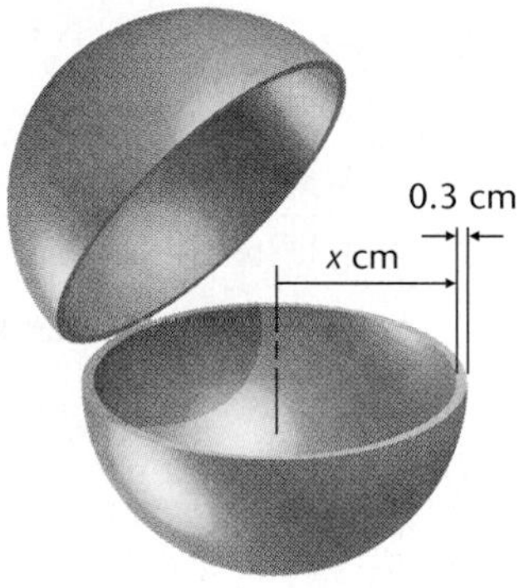

68. Packaging. A cubical container for shipping computer components is formed by coating a metal mold with polystyrene. If the metal mold is a cube with sides x centimeters long and the polystyrene coating is 2 centimeters thick, write an algebraic expression in terms of x that represents the volume of the polystyrene used to construct the container. Simplify the expression. [*Recall:* The volume V of a cube with sides of length t is given by $V = t^3$.]

Section A-3 Polynomials: Factoring

- Factoring—What Does It Mean?
- Common Factors and Factoring by Grouping
- Factoring Second-Degree Polynomials
- More Factoring

Factoring—What Does It Mean?

A **factor of a number** is one of two or more numbers whose product is the given number. Similarly, a **factor of an algebraic expression** is one of two or more algebraic expressions whose product is the given algebraic expression. For example,

$30 = 2 \cdot 3 \cdot 5$ — 2, 3, and 5 are each factors of 30.

$x^2 - 4 = (x - 2)(x + 2)$ — $(x - 2)$ and $(x + 2)$ are each factors of $x^2 - 4$.

The process of writing a number or algebraic expression as the product of other numbers or algebraic expressions is called **factoring.** We start our discussion of factoring with the positive integers.

An integer such as 30 can be represented in a factored form in many ways. The products

$$6 \cdot 5 \qquad (\tfrac{1}{2})(10)(6) \qquad 15 \cdot 2 \qquad 2 \cdot 3 \cdot 5$$

all yield 30. A particularly useful way of factoring positive integers greater than 1 is in terms of *prime* numbers.

DEFINITION 1

PRIME AND COMPOSITE NUMBERS

An integer greater than 1 is **prime** if its only positive integer factors are itself and 1. An integer greater than 1 that is not prime is called a **composite number.** The integer 1 is neither prime nor composite.

Examples of prime numbers: 2, 3, 5, 7, 11, 13

Examples of composite numbers: 4, 6, 8, 9, 10, 12

Explore/Discuss 1

In the array below, cross out all multiples of 2, except 2 itself. Then cross out all multiples of 3, except 3 itself. Repeat this for each integer in the array that has not yet been crossed out. Describe the set of numbers that remains when this process is completed.

1	2	3	4	5	6	7	8	9	10	11	12	13	14	15	16	17	18	19	20
21	22	23	24	25	26	27	28	29	30	31	32	33	34	35	36	37	38	39	40
41	42	43	44	45	46	47	48	49	50	51	52	53	54	55	56	57	58	59	60
61	62	63	64	65	66	67	68	69	70	71	72	73	74	75	76	77	78	79	80
81	82	83	84	85	86	87	88	89	90	91	92	93	94	95	96	97	98	99	100

This process is referred to as the **sieve of Eratosthenes.** (Eratosthenes was a Greek mathematician and astronomer who was a contemporary of Archimedes, circa 200 B.C.)

A composite number is said to be **factored completely** if it is represented as a product of prime factors. The only factoring of 30 given above that meets this condition is $30 = 2 \cdot 3 \cdot 5$.

EXAMPLE 1 **Factoring a Composite Number**

Write 60 in completely factored form.

Solution

$$60 = 6 \cdot 10 = 2 \cdot 3 \cdot 2 \cdot 5 = 2^2 \cdot 3 \cdot 5$$

or

$$60 = 5 \cdot 12 = 5 \cdot 4 \cdot 3 = 2^2 \cdot 3 \cdot 5$$

or

$$60 = 2 \cdot 30 = 2 \cdot 2 \cdot 15 = 2^2 \cdot 3 \cdot 5$$

MATCHED PROBLEM 1

Write 180 in completely factored form.

Notice in Example 1 that we end up with the same prime factors for 60 irrespective of how we progress through the factoring process. This illustrates an important property of integers:

THEOREM 1

THE FUNDAMENTAL THEOREM OF ARITHMETIC

Each integer greater than 1 is either prime or can be expressed uniquely, except for the order of factors, as a product of prime factors.

We can also write polynomials in completely factored form. A polynomial such as $2x^2 - x - 6$ can be written in factored form in many ways. The products

$$(2x + 3)(x - 2) \qquad 2(x^2 - \tfrac{1}{2}x - 3) \qquad 2(x + \tfrac{3}{2})(x - 2)$$

all yield $2x^2 - x - 6$. A particularly useful way of factoring polynomials is in terms of prime polynomials.

DEFINITION 2

PRIME POLYNOMIALS

A polynomial of degree greater than 0 is said to be **prime** relative to a given set of numbers if: (1) all of its coefficients are from that set of numbers; and (2) it cannot be written as a product of two polynomials, excluding 1 and itself, having coefficients from that set of numbers.

Relative to the set of integers:

$x^2 - 2$ is prime
$x^2 - 9$ is not prime, since $x^2 - 9 = (x - 3)(x + 3)$

[*Note:* The set of numbers most frequently used in factoring polynomials is the set of integers.]

A nonprime polynomial is said to be **factored completely relative to a given set of numbers** if it is written as a product of prime polynomials relative to that set of numbers.

Our objective in this section is to review some of the standard factoring techniques for polynomials with integer coefficients. In Chapter 3 we treated in detail the topic of factoring polynomials of higher degree with arbitrary coefficients.

Common Factors and Factoring by Grouping

The next example illustrates the use of the distributive properties in factoring.

EXAMPLE 2 **Factoring Out Common Factors**

Factor out, relative to the integers, all factors common to all terms.

(A) $2x^3y - 8x^2y^2 - 6xy^3$ (B) $2x(3x - 2) - 7(3x - 2)$

Solutions

(A) $2x^3y - 8x^2y^2 - 6xy^3 = (2xy)x^2 - (2xy)4xy - (2xy)3y^2$

$= 2xy(x^2 - 4xy - 3y^2)$

(B) $2x(3x - 2) - 7(3x - 2) = 2x(3x - 2) - 7(3x - 2)$

$= (2x - 7)(3x - 2)$

MATCHED PROBLEM 2

Factor out, relative to the integers, all factors common to all terms.

(A) $3x^3y - 6x^2y^2 - 3xy^3$ (B) $3y(2y + 5) + 2(2y + 5)$

EXAMPLE 3 CU **Factoring Out Common Factors**

Factor completely relative to the integers:

$$4(2x + 7)(x - 3)^2 + 2(2x + 7)^2(x - 3)$$

Solution

$$\begin{aligned} &4(2x + 7)(x - 3)^2 + 2(2x + 7)^2(x - 3) \\ &\quad = 2(2x + 7)(x - 3)[2(x - 3) + (2x + 7)] \\ &\quad = 2(2x + 7)(x - 3)(2x - 6 + 2x + 7) \\ &\quad = 2(2x + 7)(x - 3)(4x + 1) \end{aligned}$$

MATCHED PROBLEM 3

Factor completely relative to the integers.

$$4(2x + 5)(3x + 1)^2 + 6(2x + 5)^2(3x + 1)$$

Some polynomials can be factored by first grouping terms in such a way that we obtain an algebraic expression that looks something like Example 2, part B. We can then complete the factoring by the method used in that example.

EXAMPLE 4 **Factoring by Grouping**

Factor completely, relative to the integers, by grouping.

(A) $3x^2 - 6x + 4x - 8$ (B) $wy + wz - 2xy - 2xz$
(C) $3ac + bd - 3ad - bc$

Solutions (A) $3x^2 - 6x + 4x - 8$

$= (3x^2 - 6x) + (4x - 8)$ Group the first two and last two terms.

$= 3x(x - 2) + 4(x - 2)$ Remove common factors from each group.

$= (3x + 4)(x - 2)$ Factor out the common factor $(x - 2)$.

(B) $wy + wz - 2xy - 2xz$

$= (wy + wz) - (2xy + 2xz)$ Group the first two and last two terms—be careful of signs.

$= w(y + z) - 2x(y + z)$ Remove common factors from each group.

$= (w - 2x)(y + z)$ Factor out the common factor $(y + z)$.

(C) $3ac + bd - 3ad - bc$

In parts A and B the polynomials are arranged in such a way that grouping the first two terms and the last two terms leads to common factors. In this problem neither the first two terms nor the last two terms have a common factor. Sometimes rearranging terms will lead to a factoring by grouping. In this case, we interchange the second and fourth terms to obtain a problem comparable to part B, which can be factored as follows:

$$\begin{aligned} 3ac - bc - 3ad + bd &= (3ac - bc) - (3ad - bd) \\ &= c(3a - b) - d(3a - b) \\ &= (c - d)(3a - b) \end{aligned}$$

MATCHED PROBLEM 4 Factor completely, relative to the integers, by grouping.

(A) $2x^2 + 6x + 5x + 15$ (B) $2pr + ps - 6qr - 3qs$
(C) $6wy - xz - 2xy + 3wz$

Factoring Second-Degree Polynomials

We now turn our attention to factoring second-degree polynomials of the form

$$2x^2 - 5x - 3 \quad \text{and} \quad 2x^2 + 3xy - 2y^2$$

into the product of two first-degree polynomials with integer coefficients. The following example will illustrate an approach to the problem.

EXAMPLE 5 **Factoring Second-Degree Polynomials**

Factor each polynomial, if possible, using integer coefficients.

(A) $2x^2 + 3xy - 2y^2$ (B) $x^2 - 3x + 4$ (C) $6x^2 + 5xy - 4y^2$

Solutions

(A) $2x^2 + 3xy - 2y^2 = (2x + \underset{?}{\overset{\uparrow}{\ }}y)(x - \underset{?}{\overset{\uparrow}{\ }}y)$ Put in what we know. Signs must be opposite. (We can reverse this choice if we get $-3xy$ instead of $+3xy$ for the middle term.)

Now, what are the factors of 2 (the coefficient of y^2)?

2	
$1 \cdot 2$	$(2x + y)(x - 2y) = 2x^2 - 3xy - 2y^2$
$2 \cdot 1$	$(2x + 2y)(x - y) = 2x^2 - 2y^2$

The first choice gives us $-3xy$ for the middle term—close, but not there—so we reverse our choice of signs to obtain

$$2x^2 + 3xy - 2y^2 = (2x - y)(x + 2y)$$

(B) $x^2 - 3x + 4 = (x - \quad)(x - \quad)$ Signs must be the same because the third term is positive and must be negative because the middle term is negative.

4	
$2 \cdot 2$	$(x - 2)(x - 2) = x^2 - 4x + 4$
$1 \cdot 4$	$(x - 1)(x - 4) = x^2 - 5x + 4$
$4 \cdot 1$	$(x - 4)(x - 1) = x^2 - 5x + 4$

No choice produces the middle term; hence $x^2 - 3x + 4$ is not factorable using integer coefficients.

(C) $6x^2 + 5xy - 4y^2 = (\underset{?}{\overset{\uparrow}{\ }}x + \underset{?}{\overset{\uparrow}{\ }}y)(\underset{?}{\overset{\uparrow}{\ }}x - \underset{?}{\overset{\uparrow}{\ }}y)$

The signs must be opposite in the factors, because the third term is negative. We can reverse our choice of signs later if necessary. We now write all factors of 6 and of 4:

6	4
$2 \cdot 3$	$2 \cdot 2$
$3 \cdot 2$	$1 \cdot 4$
$1 \cdot 6$	$4 \cdot 1$
$6 \cdot 1$	

and try each choice on the left with each on the right—a total of 12 combinations that give us the first and last terms in the polynomial $6x^2 + 5xy - 4y^2$. The question is: Does any combination also give us the middle term, $5xy$? After trial and error and, perhaps, some educated guessing among the choices, we find that $3 \cdot 2$ matched with $4 \cdot 1$ gives us the correct middle term. Thus,

$$6x^2 + 5xy - 4y^2 = (3x + 4y)(2x - y)$$

If none of the 24 combinations (including reversing our sign choice) had produced the middle term, then we would conclude that the polynomial is not factorable using integer coefficients.

MATCHED PROBLEM 5

Factor each polynomial, if possible, using integer coefficients.

(A) $x^2 - 8x + 12$ (B) $x^2 + 2x + 5$

(C) $2x^2 + 7xy - 4y^2$ (D) $4x^2 - 15xy - 4y^2$

More Factoring

The factoring formulas listed below will enable us to factor certain polynomial forms that occur frequently.

SPECIAL FACTORING FORMULAS

1. $u^2 + 2uv + v^2 = (u + v)^2$	**Perfect Square**
2. $u^2 - 2uv + v^2 = (u - v)^2$	**Perfect Square**
3. $u^2 - v^2 = (u - v)(u + v)$	**Difference of Squares**
4. $u^3 - v^3 = (u - v)(u^2 + uv + v^2)$	**Difference of Cubes**
5. $u^3 + v^3 = (u + v)(u^2 - uv + v^2)$	**Sum of Cubes**

The formulas in the box can be established by multiplying the factors on the right.

CAUTION

Note that we did not list a special factoring formula for the sum of two squares. In general,

$$u^2 + v^2 \neq (au + bv)(cu + dv)$$

for any choice of real number coefficients a, b, c, and d.

EXAMPLE 6

Using Special Factoring Formulas

Factor completely relative to the integers.

(A) $x^2 + 6xy + 9y^2$ (B) $9x^2 - 4y^2$ (C) $8m^3 - 1$ (D) $x^3 + y^3z^3$

Solutions

(A) $x^2 + 6xy + 9y^2 = x^2 + 2(x)(3y) + (3y)^2 = (x + 3y)^2$

(B) $9x^2 - 4y^2 = (3x)^2 - (2y)^2 = (3x - 2y)(3x + 2y)$

(C)
$$\begin{aligned} 8m^3 - 1 &= (2m)^3 - 1^3 \\ &= (2m - 1)[(2m)^2 + (2m)(1) + 1^2] \\ &= (2m - 1)(4m^2 + 2m + 1) \end{aligned}$$

(D)
$$\begin{aligned} x^3 + y^3z^3 &= x^3 + (yz)^3 \\ &= (x + yz)(x^2 - xyz + y^2z^2) \end{aligned}$$

MATCHED PROBLEM 6

Factor completely relative to the integers.

(A) $4m^2 - 12mn + 9n^2$ (B) $x^2 - 16y^2$ (C) $z^3 - 1$ (D) $m^3 + n^3$

Explore/Discuss 2

(A) Verify the following factor formulas for $u^4 - v^4$:

$$\begin{aligned} u^4 - v^4 &= (u - v)(u + v)(u^2 + v^2) \\ &= (u - v)(u^3 + u^2v + uv^2 + v^3) \end{aligned}$$

(B) Discuss the pattern in the following formulas:

$$\begin{aligned} u^2 - v^2 &= (u - v)(u + v) \\ u^3 - v^3 &= (u - v)(u^2 + uv + v^2) \\ u^4 - v^4 &= (u - v)(u^3 + u^2v + uv^2 + v^3) \end{aligned}$$

(C) Use the pattern you discovered in part B to write similar formulas for $u^5 - v^5$ and $u^6 - v^6$. Verify your formulas by multiplication.

We complete this section by considering factoring that involves combinations of the preceding techniques as well as a few additional ones. Generally speaking,

When asked to factor a polynomial, we first take out all factors common to all terms, if they are present, and then proceed as above until all factors are prime.

EXAMPLE 7 **Combining Factoring Techniques**

Factor completely relative to the integers.

(A) $18x^3 - 8x$ (B) $x^2 - 6x + 9 - y^2$ (C) $4m^3n - 2m^2n^2 + 2mn^3$

(D) $2t^4 - 16t$ (E) $2y^4 - 5y^2 - 12$

Solutions

(A) $18x^3 - 8x = 2x(9x^2 - 4)$

$= 2x(3x - 2)(3x + 2)$

(B) $x^2 - 6x + 9 - y^2$

$= (x^2 - 6x + 9) - y^2$ Group the first three terms.

$= (x - 3)^2 - y^2$ Factor $x^2 - 6x + 9$.

$= [(x - 3) - y][(x - 3) + y]$ Difference of squares

$= (x - 3 - y)(x - 3 + y)$

(C) $4m^3n - 2m^2n^2 + 2mn^3 = 2mn(2m^2 - mn + n^2)$

(D) $2t^4 - 16t = 2t(t^3 - 8)$

$= 2t(t - 2)(t^2 + 2t + 4)$

(E) $2y^4 - 5y^2 - 12 = (2y^2 + 3)(y^2 - 4)$

$= (2y^2 + 3)(y - 2)(y + 2)$

MATCHED PROBLEM 7

Factor completely relative to the integers.

(A) $3x^3 - 48x$ (B) $x^2 - y^2 - 4y - 4$ (C) $3u^4 - 3u^3v - 9u^2v^2$

(D) $3m^4 - 24mn^3$ (E) $3x^4 - 5x^2 + 2$

Answers to Matched Problems

1. $2^2 \cdot 3^2 \cdot 5$ 2. (A) $3xy(x^2 - 2xy - y^2)$ (B) $(3y + 2)(2y + 5)$ 3. $2(2x + 5)(3x + 1)(12x + 17)$
4. (A) $(2x + 5)(x + 3)$ (B) $(p - 3q)(2r + s)$ (C) $(3w - x)(2y + z)$
5. (A) $(x - 2)(x - 6)$ (B) Not factorable using integers (C) $(2x - y)(x + 4y)$ (D) $(4x + y)(x - 4y)$
6. (A) $(2m - 3n)^2$ (B) $(x - 4y)(x + 4y)$ (C) $(z - 1)(z^2 + z + 1)$ (D) $(m + n)(m^2 - mn + n^2)$
7. (A) $3x(x - 4)(x + 4)$ (B) $(x - y - 2)(x + y + 2)$ (C) $3u^2(u^2 - uv - 3v^2)$
(D) $3m(m - 2n)(m^2 + 2mn + 4n^2)$ (E) $(3x^2 - 2)(x - 1)(x + 1)$

EXERCISE A-3

A

In Problems 1–8, factor out, relative to the integers, all factors common to all terms.

1. $6x^4 - 8x^3 - 2x^2$
2. $6m^4 - 9m^3 - 3m^2$
3. $10x^3y + 20x^2y^2 - 15xy^3$
4. $8u^3v - 6u^2v^2 + 4uv^3$
5. $5x(x + 1) - 3(x + 1)$
6. $7m(2m - 3) + 5(2m - 3)$
7. $2w(y - 2z) - x(y - 2z)$
8. $a(3c + d) - 4b(3c + d)$

In Problems 9–16, factor completely relative to integers.

9. $x^2 - 2x + 3x - 6$
10. $2y^2 - 6y + 5y - 15$
11. $6m^2 + 10m - 3m - 5$
12. $5x^2 - 40x - x + 8$
13. $2x^2 - 4xy - 3xy + 6y^2$
14. $3a^2 - 12ab - 2ab + 8b^2$

15. $8ac + 3bd - 6bc - 4ad$ **16.** $3pr - 2qs - qr + 6ps$

In Problems 17–28, factor completely relative to the integers. If a polynomial is prime relative to the integers, say so.

17. $2x^2 + 5x - 3$ **18.** $3y^2 - y - 2$

19. $x^2 - 4xy - 12y^2$ **20.** $u^2 - 2uv - 15v^2$

21. $x^2 + x - 4$ **22.** $m^2 - 6m - 3$

23. $25m^2 - 16n^2$ **24.** $w^2x^2 - y^2$

25. $x^2 + 10xy + 25y^2$ **26.** $9m^2 - 6mn + n^2$

27. $u^2 + 81$ **28.** $y^2 + 16$

B

In Problems 29–44, factor completely relative to the integers. If a polynomial is prime relative to the integers, say so.

29. $6x^2 + 48x + 72$ **30.** $4z^2 - 28z + 48$

31. $2y^3 - 22y^2 + 48y$ **32.** $2x^4 - 24x^3 + 40x^2$

33. $16x^2y - 8xy + y$ **34.** $4xy^2 - 12xy + 9x$

35. $6s^2 + 7st - 3t^2$ **36.** $6m^2 - mn - 12n^2$

37. $x^3y - 9xy^3$ **38.** $4u^3v - uv^3$

39. $3m^3 - 6m^2 + 15m$ **40.** $2x^3 - 2x^2 + 8x$

41. $m^3 + n^3$ **42.** $r^3 - t^3$

43. $c^3 - 1$ **44.** $a^3 + 1$

Problems 45–50 are calculus-related. Factor completely relative to the integers.

45. $6(3x - 5)(2x - 3)^2 + 4(3x - 5)^2(2x - 3)$

46. $2(x - 3)(4x + 7)^2 + 8(x - 3)^2(4x + 7)$

47. $5x^4(9 - x)^4 - 4x^5(9 - x)^3$

48. $3x^4(x - 7)^2 + 4x^3(x - 7)^3$

49. $2(x + 1)(x^2 - 5)^2 + 4x(x + 1)^2(x^2 - 5)$

50. $4(x - 3)^3(x^2 + 2)^3 + 6x(x - 3)^4(x^2 + 2)^2$

In Problems 51–56, factor completely relative to the integers. In polynomials involving more than three terms, try grouping the terms in various combinations as a first step. If a polynomial is prime relative to the integers, say so.

51. $(a - b)^2 - 4(c - d)^2$ **52.** $(x + 2)^2 - 9y^2$

53. $2am - 3an + 2bm - 3bn$

54. $15ac - 20ad + 3bc - 4bd$

55. $3x^2 - 2xy - 4y^2$ **56.** $5u^2 + 4uv - 2v^2$

C

In Problems 57–72, factor completely relative to the integers. In polynomials involving more than three terms, try grouping the terms in various combinations as a first step. If a polynomial is prime relative to the integers, say so.

57. $x^3 - 3x^2 - 9x + 27$ **58.** $x^3 - x^2 - x + 1$

59. $a^3 - 2a^2 - a + 2$ **60.** $t^3 - 2t^2 + t - 2$

61. $4(A + B)^2 - 5(A + B) - 6$

62. $6(x - y)^2 + 23(x - y) - 4$

63. $m^4 - n^4$ **64.** $y^4 - 3y^2 - 4$

65. $s^4t^4 - 8st$ **66.** $27a^2 + a^5b^3$

67. $m^2 + 2mn + n^2 - m - n$

68. $y^2 - 2xy + x^2 - y + x$

69. $18a^3 - 8a(x^2 + 8x + 16)$

70. $25(4x^2 - 12xy + 9y^2) - 9a^2b^2$

71. $x^4 + 2x^2 + 1 - x^2$ **72.** $a^4 + 2a^2b^2 + b^4 - a^2b^2$

APPLICATIONS

73. Construction. A rectangular open-topped box is to be constructed out of 20-inch-square sheets of thin cardboard by cutting x-inch squares out of each corner and bending the sides up as indicated in the figure. Express each of the following quantities as a polynomial in both factored and expanded form.

(A) The area of cardboard after the corners have been removed.

(B) The volume of the box.

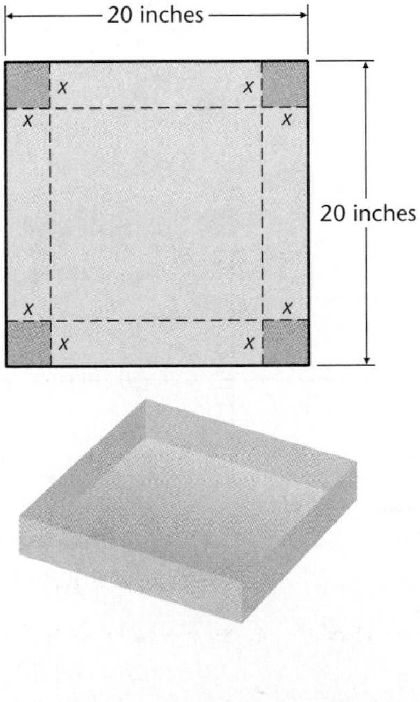

74. Construction. A rectangular open-topped box is to be constructed out of 9- by 16-inch sheets of thin cardboard by cutting x-inch squares out of each corner and bending the sides up. Express each of the following quantities as a polynomial in both factored and expanded form.

(A) The area of cardboard after the corners have been removed.

(B) The volume of the box.

Section A-4 | Rational Expressions: Basic Operations

- Reducing to Lowest Terms
- Multiplication and Division
- Addition and Subtraction
- Compound Fractions

We now turn our attention to fractional forms. A quotient of two algebraic expressions, division by 0 excluded, is called a **fractional expression.** If both the numerator and denominator of a fractional expression are polynomials, the fractional expression is called a **rational expression.** Some examples of rational expressions are the following (recall, a nonzero constant is a polynomial of degree 0):

$$\frac{x-2}{2x^2-3x+5} \qquad \frac{1}{x^4-1} \qquad \frac{3}{x} \qquad \frac{x^2+3x-5}{1}$$

In this section we discuss basic operations on rational expressions, including multiplication, division, addition, and subtraction.

Since variables represent real numbers in the rational expressions we are going to consider, the properties of real number fractions summarized in Section A-1 play a central role in much of the work that we will do.

Even though not always explicitly stated, we always assume that variables are restricted so that division by 0 is excluded.

Reducing to Lowest Terms

We start this discussion by restating the **fundamental property of fractions** (from Theorem 3 in Section A-1):

FUNDAMENTAL PROPERTY OF FRACTIONS

If a, b, and k are real numbers with $b, k \neq 0$, then

$$\frac{ka}{kb} = \frac{a}{b} \qquad \frac{2 \cdot 3}{2 \cdot 4} = \frac{3}{4} \qquad \frac{(x-3)2}{(x-3)x} = \frac{2}{x}$$
$$x \neq 0, x \neq 3$$

Using this property from left to right to eliminate all common factors from the numerator and the denominator of a given fraction is referred to as **reducing**

a fraction to lowest terms. We are actually dividing the numerator and denominator by the same nonzero common factor.

Using the property from right to left—that is, multiplying the numerator and the denominator by the same nonzero factor—is referred to as **raising a fraction to higher terms.** We will use the property in both directions in the material that follows.

We say that a rational expression is **reduced to lowest terms** if the numerator and denominator do not have any factors in common. Unless stated to the contrary, factors will be relative to the integers.

EXAMPLE 1

Reducing Rational Expressions

Reduce each rational expression to the lowest terms.

(A) $$\frac{x^2 - 6x + 9}{x^2 - 9} = \frac{(x-3)^2}{(x-3)(x+3)}$$

Factor numerator and denominator completely. Divide numerator and denominator by $(x - 3)$; this is a valid operation as long as $x \neq 3$ and $x \neq -3$.

$$= \frac{x-3}{x+3}$$

(B) $$\frac{x^3 - 1}{x^2 - 1} = \frac{\overset{1}{\cancel{(x-1)}}(x^2 + x + 1)}{\underset{1}{\cancel{(x-1)}}(x+1)}$$

Dividing numerator and denominator by $(x - 1)$ can be indicated by drawing lines through both $(x - 1)$s and writing the resulting quotients, 1s.

$$= \frac{x^2 + x + 1}{x + 1}$$

$x \neq -1$ and $x \neq 1$

MATCHED PROBLEM 1

Reduce each rational expression to lowest terms.

(A) $\dfrac{6x^2 + x - 2}{2x^2 + x - 1}$ (B) $\dfrac{x^4 - 8x}{3x^3 - 2x^2 - 8x}$

EXAMPLE 2

Reducing a Rational Expression

Reduce the following rational expression to lowest terms.

$$\frac{6x^5(x^2+2)^2 - 4x^3(x^2+2)^3}{x^8} = \frac{2x^3(x^2+2)^2[3x^2 - 2(x^2+2)]}{x^8}$$

$$= \frac{2\overset{1}{\cancel{x^3}}(x^2+2)^2(x^2-4)}{\underset{x^5}{\cancel{x^8}}}$$

$$= \frac{2(x^2+2)^2(x-2)(x+2)}{x^5}$$

MATCHED PROBLEM 2

Reduce the following rational expression to lowest terms.

$$\frac{6x^4(x^2+1)^2 - 3x^2(x^2+1)^3}{x^6}$$

CAUTION

Remember to always factor the numerator and denominator first, then divide out any *common factors*. Do not indiscriminately eliminate *terms* that appear in both the numerator and the denominator. For example,

$$\frac{2x^3 + y^2}{y^2} \neq \frac{2x^3 + \overset{1}{\cancel{y^2}}}{\underset{1}{\cancel{y^2}}} = 2x^3 + 1$$

Since the term y^2 is not a factor of the numerator, it cannot be eliminated. In fact, $(2x^3 + y^2)/y^2$ is already reduced to lowest terms.

Multiplication and Division

Since we are restricting variable replacements to real numbers, multiplication and division of rational expressions follow the rules for multiplying and dividing real number fractions (Theorem 3 in Section A-1).

MULTIPLICATION AND DIVISION

If a, b, c, and d are real numbers with $b, d \neq 0$, then:

1. $\dfrac{a}{b} \cdot \dfrac{c}{d} = \dfrac{ac}{bd}$ $\qquad \dfrac{2}{3} \cdot \dfrac{x}{x-1} = \dfrac{2x}{3(x-1)}$
2. $\dfrac{a}{b} \div \dfrac{c}{d} = \dfrac{a}{b} \cdot \dfrac{d}{c} \qquad c \neq 0$ $\qquad \dfrac{2}{3} \div \dfrac{x}{x-1} = \dfrac{2}{3} \cdot \dfrac{x-1}{x}$

Explore/Discuss 1

Write a verbal description of the process of multiplying two fractions. Do the same for the quotient of two fractions.

EXAMPLE 3

Multiplying and Dividing Rational Expressions

Perform the indicated operations and reduce to lowest terms.

(A) $$\frac{10x^3y}{3xy + 9y} \cdot \frac{x^2 - 9}{4x^2 - 12x} = \frac{\overset{5x^2}{\cancel{10x^3y}}}{\underset{3\cdot 1}{\cancel{3y}(\cancel{x+3})}} \cdot \frac{\overset{1\cdot 1}{(\cancel{x-3})(\cancel{x+3})}}{\underset{2\cdot 1}{\cancel{4x}(\cancel{x-3})}} = \frac{5x^2}{6}$$

Factor numerators and denominators; then divide any numerator and any denominator with a like common factor.

(B) $$\frac{4 - 2x}{4} \div (x - 2) = \frac{\overset{1}{\cancel{2}}(2 - x)}{\underset{2}{\cancel{4}}} \cdot \frac{1}{x - 2}$$

$x - 2$ is the same as $\frac{x - 2}{1}$.

$$= \frac{2 - x}{2(x - 2)} = \frac{\overset{-1}{-(\cancel{x-2})}}{\underset{1}{2(\cancel{x-2})}}$$

$b - a = -(a - b)$, a useful change in some problems.

$$= -\frac{1}{2}$$

(C) $$\frac{2x^3 - 2x^2y + 2xy^2}{x^3y - xy^3} \div \frac{x^3 + y^3}{x^2 + 2xy + y^2}$$

$$= \frac{\overset{2}{\cancel{2x}}\overset{1}{(\cancel{x^2 - xy + y^2})}}{\underset{y}{\cancel{xy}}\underset{1}{(\cancel{x + y})}(x - y)} \cdot \frac{\overset{1}{\cancel{(x + y)^2}}}{\underset{1}{(\cancel{x + y})}\underset{1}{(\cancel{x^2 - xy + y^2})}}$$

$$= \frac{2}{y(x - y)}$$

MATCHED PROBLEM 3

Perform the indicated operations and reduce to lowest terms.

(A) $\frac{12x^2y^3}{2xy^2 + 6xy} \cdot \frac{y^2 + 6y + 9}{3y^3 + 9y^2}$ (B) $(4 - x) \div \frac{x^2 - 16}{5}$

(C) $\frac{m^3 + n^3}{2m^2 + mn - n^2} \div \frac{m^3n - m^2n^2 + mn^3}{2m^3n^2 - m^2n^3}$

Addition and Subtraction

Again, because we are restricting variable replacements to real numbers, addition and subtraction of rational expressions follow the rules for adding and subtracting real number fractions (Theorem 3 in Section A-1).

ADDITION AND SUBTRACTION

For a, b, and c real numbers with $b \neq 0$:

1. $\dfrac{a}{b} + \dfrac{c}{b} = \dfrac{a + c}{b} \qquad \dfrac{x}{x - 3} + \dfrac{2}{x - 3} = \dfrac{x + 2}{x - 3}$

2. $\dfrac{a}{b} - \dfrac{c}{b} = \dfrac{a - c}{b} \qquad \dfrac{x}{2xy^2} - \dfrac{x - 4}{2xy^2} = \dfrac{x - (x - 4)}{2xy^2}$

Thus, we add rational expressions with the same denominators by adding or subtracting their numerators and placing the result over the common denominator. If the denominators are not the same, we raise the fractions to higher terms, using the fundamental property of fractions to obtain common denominators, and then proceed as described.

Even though any common denominator will do, our work will be simplified if the least common denominator (LCD) is used. Often, the LCD is obvious, but if it is not, the steps in the box describe how to find it.

THE LEAST COMMON DENOMINATOR (LCD)

The LCD of two or more rational expressions is found as follows:

1. Factor each denominator completely.
2. Identify each different prime factor from all the denominators.
3. Form a product using each different factor to the highest power that occurs in any one denominator. This product is the LCD.

EXAMPLE 4 Adding and Subtracting Rational Expressions

Combine into a single fraction and reduce to lowest terms.

(A) $\dfrac{3}{10} + \dfrac{5}{6} - \dfrac{11}{45}$ (B) $\dfrac{4}{9x} - \dfrac{5x}{6y^2} + 1$

(C) $\dfrac{x + 3}{x^2 - 6x + 9} - \dfrac{x + 2}{x^2 - 9} - \dfrac{5}{3 - x}$

Solutions (A) To find the LCD, factor each denominator completely:

$$\left.\begin{aligned} 10 &= 2 \cdot 5 \\ 6 &= 2 \cdot 3 \\ 45 &= 3^2 \cdot 5 \end{aligned}\right\} \text{LCD} = 2 \cdot 3^2 \cdot 5 = 90$$

Now use the fundamental property of fractions to make each denominator 90:

$$\frac{3}{10} + \frac{5}{6} - \frac{11}{45} = \frac{9 \cdot 3}{9 \cdot 10} + \frac{15 \cdot 5}{15 \cdot 6} - \frac{2 \cdot 11}{2 \cdot 45}$$

$$= \frac{27}{90} + \frac{75}{90} - \frac{22}{90}$$

$$= \frac{27 + 75 - 22}{90} = \frac{80}{90} = \frac{8}{9}$$

(B) $\left.\begin{aligned} 9x &= 3^2x \\ 6y^2 &= 2 \cdot 3y^2 \end{aligned}\right\}$ LCD $= 2 \cdot 3^2xy^2 = 18xy^2$

$$\frac{4}{9x} - \frac{5x}{6y^2} + 1 = \frac{2y^2 \cdot 4}{2y^2 \cdot 9x} - \frac{3x \cdot 5x}{3x \cdot 6y^2} + \frac{18xy^2}{18xy^2}$$

$$= \frac{8y^2 - 15x^2 + 18xy^2}{18xy^2}$$

(C) $$\frac{x+3}{x^2 - 6x + 9} - \frac{x+2}{x^2 - 9} - \frac{5}{3 - x} = \frac{x+3}{(x-3)^2} - \frac{x+2}{(x-3)(x+3)} + \frac{5}{x-3}$$

Note: $-\dfrac{5}{3-x} = -\dfrac{5}{-(x-3)} = \dfrac{5}{x-3}$

We have again used the fact that $a - b = -(b - a)$.

The LCD $= (x-3)^2(x+3)$. Thus,

$$\frac{(x+3)^2}{(x-3)^2(x+3)} - \frac{(x-3)(x+2)}{(x-3)^2(x+3)} + \frac{5(x-3)(x+3)}{(x-3)^2(x+3)}$$

$$= \frac{(x^2 + 6x + 9) - (x^2 - x - 6) + 5(x^2 - 9)}{(x-3)^2(x+3)}$$

Be careful of sign errors here.

$$= \frac{x^2 + 6x + 9 - x^2 + x + 6 + 5x^2 - 45}{(x-3)^2(x+3)}$$

$$= \frac{5x^2 + 7x - 30}{(x-3)^2(x+3)}$$

MATCHED PROBLEM 4

Combine into a single fraction and reduce to lowest terms.

(A) $\dfrac{5}{28} - \dfrac{1}{10} + \dfrac{6}{35}$ (B) $\dfrac{1}{4x^2} - \dfrac{2x+1}{3x^3} + \dfrac{3}{12x}$

(C) $\dfrac{y-3}{y^2 - 4} - \dfrac{y+2}{y^2 - 4y + 4} - \dfrac{2}{2 - y}$

Explore/Discuss 2

What is the value of $\dfrac{\dfrac{16}{4}}{2}$?

What is the result of entering 16 ÷ 4 ÷ 2 on a calculator?

What is the difference between 16 ÷ (4 ÷ 2) and (16 ÷ 4) ÷ 2?

How could you use fraction bars to distinguish between these two cases when writing $\dfrac{\dfrac{16}{4}}{2}$?

Compound Fractions

A fractional expression with fractions in its numerator, denominator, or both is called a **compound fraction.** It is often necessary to represent a compound fraction as a **simple fraction**—that is (in all cases we will consider), as the quotient of two polynomials. The process does not involve any new concepts. It is a matter of applying old concepts and processes in the right sequence. We will illustrate two approaches to the problem, each with its own merits, depending on the particular problem under consideration.

EXAMPLE 5 **Simplifying Compound Fractions**

Express as a simple fraction reduced to lowest terms.

$$\frac{\dfrac{2}{x} - 1}{\dfrac{4}{x^2} - 1}$$

Solution *Method 1.* Multiply the numerator and denominator by the LCD of all fractions in the numerator and denominator—in this case, x^2. (We are multiplying by $1 = x^2/x^2$).

$$\frac{x^2\left(\dfrac{2}{x} - 1\right)}{x^2\left(\dfrac{4}{x^2} - 1\right)} = \boxed{\frac{x^2\dfrac{2}{x} - x^2}{x^2\dfrac{4}{x^2} - x^2}} = \frac{2x - x^2}{4 - x^2} = \frac{x\overset{1}{\cancel{(2 - x)}}}{(2 + x)\underset{1}{\cancel{(2 - x)}}}$$

$$= \frac{x}{2 + x}$$

Method 2. Write the numerator and denominator as single fractions. Then treat as a quotient.

$$\frac{\dfrac{2}{x}-1}{\dfrac{4}{x^2}-1} = \frac{\dfrac{2-x}{x}}{\dfrac{4-x^2}{x^2}} = \frac{2-x}{x} \div \frac{4-x^2}{x^2} = \frac{\overset{1}{\cancel{2-x}}}{\underset{1}{\cancel{x}}} \cdot \frac{\overset{x}{\cancel{x^2}}}{(\underset{1}{\cancel{2-x}})(2+x)}$$

$$= \frac{x}{2+x}$$

MATCHED PROBLEM **5**

Express as a simple fraction reduced to lowest terms. Use the two methods described in Example 5.

$$\frac{1+\dfrac{1}{x}}{x-\dfrac{1}{x}}$$

EXAMPLE **6**

Simplifying Compound Fractions

Express as a simple fraction reduced to lowest terms.

$$\frac{\dfrac{y}{x^2}-\dfrac{x}{y^2}}{\dfrac{y}{x}-\dfrac{x}{y}}$$

Solution

Using the first method described in Example 5, we have

$$\frac{x^2y^2\left(\dfrac{y}{x^2}-\dfrac{x}{y^2}\right)}{x^2y^2\left(\dfrac{y}{x}-\dfrac{x}{y}\right)} = \boxed{\frac{x^2y^2\dfrac{y}{x^2}-x^2y^2\dfrac{x}{y^2}}{x^2y^2\dfrac{y}{x}-x^2y^2\dfrac{x}{y}}} = \frac{y^3-x^3}{xy^3-x^3y} = \frac{(\overset{1}{\cancel{y-x}})(y^2+xy+x^2)}{xy(\underset{1}{\cancel{y-x}})(y+x)}$$

$$= \frac{y^2+xy+x^2}{xy(y+x)}$$

MATCHED PROBLEM **6**

Express as a simple fraction reduced to lowest terms. Use the first method described in Example 5.

$$\frac{\dfrac{a}{b}-\dfrac{b}{a}}{\dfrac{a}{b}+2+\dfrac{b}{a}}$$

Answers to Matched Problems

1. (A) $\frac{3x+2}{x+1}$ (B) $\frac{x^2+2x+4}{3x+4}$ 2. $\frac{3(x^2+1)^2(x+1)(x-1)}{x^4}$ 3. (A) $2x$ (B) $\frac{-5}{x+4}$ (C) mn

4. (A) $\frac{1}{4}$ (B) $\frac{3x^2-5x-4}{12x^3}$ (C) $\frac{2y^2-9y-6}{(y-2)^2(y+2)}$ 5. $\frac{1}{x-1}$ 6. $\frac{a-b}{a+b}$

EXERCISE A-4

A

In Problems 1–20, perform the indicated operations and reduce answers to lowest terms. Represent any compound fractions as simple fractions reduced to lowest terms.

1. $\left(\frac{d^5}{3a} \div \frac{d^2}{6a^2}\right) \cdot \frac{a}{4d^3}$

2. $\frac{d^5}{3a} \div \left(\frac{d^2}{6a^2} \cdot \frac{a}{4d^3}\right)$

3. $\frac{2y}{18} - \frac{-1}{28} - \frac{y}{42}$

4. $\frac{x^2}{12} + \frac{x}{18} - \frac{1}{30}$

5. $\frac{3x+8}{4x^2} - \frac{2x-1}{x^3} - \frac{5}{8x}$

6. $\frac{4m-3}{18m^3} + \frac{3}{4m} - \frac{2m-1}{6m^2}$

7. $\frac{2x^2+7x+3}{4x^2-1} \div (x+3)$

8. $\frac{x^2-9}{x^2-3x} \div (x^2-x-12)$

9. $\frac{m+n}{m^2-n^2} \div \frac{m^2-mn}{m^2-2mn+n^2}$

10. $\frac{x^2-6x+9}{x^2-x-6} \div \frac{x^2+2x-15}{x^2+2x}$

11. $\frac{1}{a^2-b^2} + \frac{1}{a^2+2ab+b^2}$

12. $\frac{3}{x^2-1} - \frac{2}{x^2-2x+1}$

13. $m - 3 - \frac{m-1}{m-2}$

14. $\frac{x+1}{x-1} - 1$

15. $\frac{5}{x-3} - \frac{2}{3-x}$

16. $\frac{3}{a-1} - \frac{2}{1-a}$

17. $\frac{2}{y+3} - \frac{1}{y-3} + \frac{2y}{y^2-9}$

18. $\frac{2x}{x^2-y^2} + \frac{1}{x+y} - \frac{1}{x-y}$

19. $\frac{1-\frac{y^2}{x^2}}{1-\frac{y}{x}}$

20. $\frac{1+\frac{3}{x}}{x-\frac{9}{x}}$

B

Problems 21–26 are calculus-related. Reduce each fraction to lowest terms.

21. $\frac{6x^3(x^2+2)^2 - 2x(x^2+2)^3}{x^4}$

22. $\frac{4x^4(x^2+3) - 3x^2(x^2+3)^2}{x^6}$

23. $\frac{2x(1-3x)^3 + 9x^2(1-3x)^2}{(1-3x)^6}$

24. $\frac{2x(2x+3)^4 - 8x^2(2x+3)^3}{(2x+3)^8}$

25. $\frac{-2x(x+4)^3 - 3(3-x^2)(x+4)^2}{(x+4)^6}$

26. $\frac{3x^2(x+1)^3 - 3(x^3+4)(x+1)^2}{(x+1)^6}$

In Problems 27–40, perform the indicated operations and reduce answers to lowest terms. Represent any compound fractions as simple fractions reduced to lowest terms.

27. $\frac{y}{y^2-y-2} - \frac{1}{y^2+5y-14} - \frac{2}{y^2+8y+7}$

28. $\frac{x^2}{x^2+2x+1} + \frac{x-1}{3x+3} - \frac{1}{6}$

29. $\frac{9-m^2}{m^2+5m+6} \cdot \frac{m+2}{m-3}$

30. $\frac{2-x}{2x+x^2} \cdot \frac{x^2+4x+4}{x^2-4}$

31. $\frac{x+7}{ax-bx} + \frac{y+9}{by-ay}$

32. $\frac{c+2}{5c-5} - \frac{c-2}{3c-3} + \frac{c}{1-c}$

33. $\frac{x^2-16}{2x^2+10x+8} \div \frac{x^2-13x+36}{x^3+1}$

34. $\left(\frac{x^3-y^3}{y^3} \cdot \frac{y}{x-y}\right) \div \frac{x^2+xy+y^2}{y^2}$

35. $\dfrac{x^2 - xy}{xy + y^2} \div \left(\dfrac{x^2 - y^2}{x^2 + 2xy + y^2} \div \dfrac{x^2 - 2xy + y^2}{x^2y + xy^2}\right)$

36. $\left(\dfrac{x^2 - xy}{xy + y^2} \div \dfrac{x^2 - y^2}{x^2 + 2xy + y^2}\right) \div \dfrac{x^2 - 2xy + y^2}{x^2y + xy^2}$

37. $\left(\dfrac{x}{x^2 - 16} - \dfrac{1}{x + 4}\right) \div \dfrac{4}{x + 4}$

38. $\left(\dfrac{3}{x - 2} - \dfrac{1}{x + 1}\right) \div \dfrac{x + 4}{x - 2}$

39. $\dfrac{1 + \dfrac{2}{x} - \dfrac{15}{x^2}}{1 + \dfrac{4}{x} - \dfrac{5}{x^2}}$

40. $\dfrac{\dfrac{x}{y} - 2 + \dfrac{y}{x}}{\dfrac{x}{y} - \dfrac{y}{x}}$

Problems 41–44 are calculus-related. Perform the indicated operations and reduce answers to lowest terms. Represent any compound fractions as simple fractions reduced to lowest terms.

41. $\dfrac{\dfrac{1}{x + h} - \dfrac{1}{x}}{h}$

42. $\dfrac{\dfrac{1}{(x + h)^2} - \dfrac{1}{x^2}}{h}$

43. $\dfrac{\dfrac{(x + h)^2}{x + h + 2} - \dfrac{x^2}{x + 2}}{h}$

44. $\dfrac{\dfrac{2x + 2h + 3}{x + h} - \dfrac{2x + 3}{x}}{h}$

In Problems 45–52, imagine that the indicated "solutions" were given to you by a student whom you were tutoring in this class.

(A) Is the solution correct? If the solution is incorrect, explain what is wrong and how it can be corrected.

(B) Show a correct solution for each incorrect solution.

45. $\dfrac{x^2 + 5x + 4}{x + 4} = \dfrac{x^2 + 5x}{x} = x + 5$

46. $\dfrac{x^2 - 2x - 3}{x - 3} = \dfrac{x^2 - 2x}{x} = x - 2$

47. $\dfrac{(x + h)^2 - x^2}{h} = (x + 1)^2 - x^2 = 2x + 1$

48. $\dfrac{(x + h)^3 - x^3}{h} = (x + 1)^3 - x^3 = 3x^2 + 3x + 1$

49. $\dfrac{x^2 - 2x}{x^2 - x - 2} + x - 2 = \dfrac{x^2 - 2x + x - 2}{x^2 - x - 2} = 1$

50. $\dfrac{2}{x - 1} - \dfrac{x + 3}{x^2 - 1} = \dfrac{2x + 2 - x - 3}{x^2 - 1} = \dfrac{1}{x + 1}$

51. $\dfrac{2x^2}{x^2 - 4} - \dfrac{x}{x - 2} = \dfrac{2x^2 - x^2 - 2x}{x^2 - 4} = \dfrac{x}{x + 2}$

52. $x + \dfrac{x - 2}{x^2 - 3x + 2} = \dfrac{x + x - 2}{x^2 - 3x + 2} = \dfrac{2}{x - 2}$

C

In Problems 53–56, perform the indicated operations and reduce answers to lowest terms. Represent any compound fractions as simple fractions reduced to lowest terms.

53. $\dfrac{y - \dfrac{y^2}{y - x}}{1 + \dfrac{x^2}{y^2 - x^2}}$

54. $\dfrac{\dfrac{s^2}{s - t} - s}{\dfrac{t^2}{s - t} + t}$

55. $2 - \dfrac{1}{1 - \dfrac{2}{a + 2}}$

56. $1 - \dfrac{1}{1 - \dfrac{1}{1 - \dfrac{1}{x}}}$

In Problems 57 and 58, a, b, c, and d represent real numbers.

57. (A) Prove that d/c is the multiplicative inverse of c/d $(c, d \neq 0)$.

(B) Use part A to prove that

$$\frac{a}{b} \div \frac{c}{d} = \frac{a}{b} \cdot \frac{d}{c} \qquad b, c, d \neq 0$$

58. Prove that

$$\frac{a}{b} + \frac{c}{b} = \frac{a + c}{b} \qquad b \neq 0$$

Section A-5 | Integer Exponents

- Integer Exponents
- Scientific Notation

The French philosopher/mathematician René Descartes (1596–1650) is generally credited with the introduction of the very useful exponent notation "x^n." This notation as well as other improvements in algebra may be found in his *Geometry,* published in 1637.

In Section A-2 we introduced the natural number exponent as a short way of writing a product involving the same factors. In this section we will expand the meaning of exponent to include all integers so that exponential forms of the following types will all have meaning:

$$7^5 \qquad 5^{-4} \qquad 3.14^0$$

Integer Exponents

Definition 1 generalizes exponent notation to include 0 and negative integer exponents.

DEFINITION 1

a^n, n AN INTEGER AND a A REAL NUMBER

1. For n a positive integer,

$$a^n = \underbrace{a \cdot a \cdot \cdots \cdot a}_{n \text{ factors of } a} \qquad 3^5 = 3 \cdot 3 \cdot 3 \cdot 3 \cdot 3$$

2. For $n = 0$,

$$a^0 = 1 \quad a \neq 0 \qquad 132^0 = 1$$

0^0 is not defined

3. For n a negative integer,

$$a^n = \frac{1}{a^{-n}} \quad a \neq 0 \qquad 7^{-3} = \frac{1}{7^{-(-3)}} = \frac{1}{7^3}$$

Note: In general, it can be shown that for *all* integers n

$$a^{-n} = \frac{1}{a^n} \qquad a^{-5} = \frac{1}{a^5} \qquad a^{-(-3)} = \frac{1}{a^{-3}}$$

EXAMPLE 1

Using the Definition of Integer Exponents

Write each part as a decimal fraction or using positive exponents.

(A) $(u^3v^2)^0 = 1 \quad u \neq 0, \quad v \neq 0$

(B) $10^{-3} = \dfrac{1}{10^3} = \dfrac{1}{1{,}000} = 0.001$

(C) $x^{-8} = \dfrac{1}{x^8}$

(D) $\dfrac{x^{-3}}{y^{-5}} = \dfrac{x^{-3}}{1} \cdot \dfrac{1}{y^{-5}} = \dfrac{1}{x^3} \cdot \dfrac{y^5}{1} = \dfrac{y^5}{x^3}$

MATCHED PROBLEM 1

Write parts A–D as decimal fractions and parts E and F with positive exponents.

(A) 636^0 (B) $(x^2)^0 \quad x \neq 0$ (C) 10^{-5}

(D) $\dfrac{1}{10^{-3}}$ (E) $\dfrac{1}{x^{-4}}$ (F) $\dfrac{u^{-7}}{v^{-3}}$

The basic properties of integer exponents are summarized in Theorem 1. The proof of this theorem involves *mathematical induction*, which is discussed in Chapter 9.

THEOREM 1

PROPERTIES OF INTEGER EXPONENTS

For n and m integers and a and b real numbers,

1. $a^m a^n = a^{m+n}$ $\qquad a^5 a^{-7} = a^{5+(-7)} = a^{-2}$
2. $(a^n)^m = a^{mn}$ $\qquad (a^3)^{-2} = a^{(-2)3} = a^{-6}$
3. $(ab)^m = a^m b^m$ $\qquad (ab)^3 = a^3 b^3$
4. $\left(\dfrac{a}{b}\right)^m = \dfrac{a^m}{b^m} \quad b \neq 0$ $\qquad \left(\dfrac{a}{b}\right)^4 = \dfrac{a^4}{b^4}$
5. $\dfrac{a^m}{a^n} = \begin{cases} a^{m-n} \\ \dfrac{1}{a^{n-m}} \end{cases} \quad a \neq 0$ $\qquad \dfrac{a^3}{a^{-2}} = a^{3\ (\ 2)} = a^5$ $\qquad \dfrac{a^3}{a^{-2}} = \dfrac{1}{a^{-2-3}} = \dfrac{1}{a^{-5}}$

Explore/Discuss 1

Property 1 in Theorem 1 can be expressed verbally as follows:

> To find the product of two exponential forms with the same base, add the exponents and use the same base.
>
> Express the other properties in Theorem 1 verbally. Decide which you find easier to remember, a formula or a verbal description.

EXAMPLE 2

Using Exponent Properties

Simplify using exponent properties, and express answers using positive exponents only.*

(A) $(3a^5)(2a^{-3}) = (3 \cdot 2)(a^5a^{-3}) = 6a^2$

(B) $\dfrac{6x^{-2}}{8x^{-5}} = \dfrac{3x^{-2-(-5)}}{4} = \dfrac{3x^3}{4}$

(C) $-4y^3 - (-4y)^3 = -4y^3 - (-4)^3y^3 = -4y^3 - (-64)y^3$

$= -4y^3 + 64y^3 = 60y^3$

MATCHED PROBLEM 2

Simplify using exponent properties, and express answers using positive exponents only.

(A) $(5x^{-3})(3x^4)$ (B) $\dfrac{9y^{-7}}{6y^{-4}}$ (C) $2x^4 - (-2x)^4$

CAUTION

Be careful when using the relationship $a^{-n} = \dfrac{1}{a^n}$.

$ab^{-1} \neq \dfrac{1}{ab}$ $\quad ab^{-1} = \dfrac{a}{b}$ and $(ab)^{-1} = \dfrac{1}{ab}$

$\dfrac{1}{a+b} \neq a^{-1} + b^{-1}$ $\quad \dfrac{1}{a+b} = (a+b)^{-1}$ and $\dfrac{1}{a} + \dfrac{1}{b} = a^{-1} + b^{-1}$

Do not confuse properties 1 and 2 in Theorem 1.

$a^3a^4 \neq a^{3\cdot 4}$ $\quad a^3a^4 = a^{3+4} = a^7$ property 1, Theorem 1

$(a^3)^4 \neq a^{3+4}$ $\quad (a^3)^4 = a^{3\cdot 4} = a^{12}$ property 2, Theorem 1

From the definition of negative exponents and the five properties of exponents, we can easily establish the following properties, which are used very frequently when dealing with exponent forms.

*By "simplify" we mean eliminate common factors from numerators and denominators and reduce to a minimum the number of times a given constant or variable appears in an expression. We ask that answers be expressed using positive exponents only in order to have a definite form for an answer. Sometimes we will encounter situations where we will want negative exponents in a final answer.

THEOREM 2

FURTHER EXPONENT PROPERTIES

For a and b any real numbers and m, n, and p any integers (division by 0 excluded),

1. $(a^m b^n)^p = a^{pm} b^{pn}$
2. $\left(\frac{a^m}{b^n}\right)^p = \frac{a^{pm}}{b^{pn}}$
3. $\frac{a^{-n}}{b^{-m}} = \frac{b^m}{a^n}$
4. $\left(\frac{a}{b}\right)^{-n} = \left(\frac{b}{a}\right)^n$

Proof

We prove properties 1 and 4 in Theorem 2 and leave the proofs of 2 and 3 to you.

1. $(a^m b^n)^p = (a^m)^p (b^n)^p$ property 3, Theorem 1
$= a^{pm} b^{pn}$ property 2, Theorem 1

4. $\left(\frac{a}{b}\right)^{-n} = \frac{a^{-n}}{b^{-n}}$ property 4, Theorem 1
$= \frac{b^n}{a^n}$ property 3, Theorem 2
$= \left(\frac{b}{a}\right)^n$ property 4, Theorem 1

EXAMPLE 3

Using Exponent Properties

Simplify using exponent properties, and express answers using positive exponents only.

(A) $(2a^{-3}b^2)^{-2} = 2^{-2}a^6b^{-4} = \frac{a^6}{4b^4}$

(B) $\left(\frac{a^3}{b^5}\right)^{-2} = \frac{a^{-6}}{b^{-10}} = \frac{b^{10}}{a^6}$ or $\left(\frac{a^3}{b^5}\right)^{-2} = \left(\frac{b^5}{a^3}\right)^2 = \frac{b^{10}}{a^6}$

(C) $\frac{4x^{-3}y^{-5}}{6x^{-4}y^3} = \frac{2x^{-3-(-4)}}{3y^{3-(-5)}} = \frac{2x}{3y^8}$

(D) $\left(\frac{m^{-3}m^3}{n^{-2}}\right)^{-2} = \left(\frac{m^{-3+3}}{n^{-2}}\right)^{-2} = \left(\frac{m^0}{n^{-2}}\right)^{-2} = \left(\frac{1}{n^{-2}}\right)^{-2} = \frac{1}{n^4}$

(E) $(x + y)^{-3} = \frac{1}{(x + y)^3}$

MATCHED PROBLEM 3

Simplify using exponent properties, and express answers using positive exponents only.

(A) $(3x^4y^{-3})^{-2}$ (B) $\left(\dfrac{x^2}{y^4}\right)^{-3}$ (C) $\dfrac{6m^{-2}n^3}{15m^{-1}n^{-2}}$

(D) $\left(\dfrac{x^{-3}}{y^4y^{-4}}\right)^{-3}$ (E) $\dfrac{1}{(a-b)^{-2}}$

In simplifying exponent forms there is often more than one sequence of steps that will lead to the same result (see Example 3, part B). Use whichever sequence of steps makes sense to you.

EXAMPLE 4

Simplifying a Compound Fraction

Express as a simple fraction reduced to lowest terms.

$$\frac{x^{-2}-y^{-2}}{x^{-1}+y^{-1}} = \frac{\dfrac{1}{x^2}-\dfrac{1}{y^2}}{\dfrac{1}{x}+\dfrac{1}{y}} = \frac{x^2y^2\left(\dfrac{1}{x^2}-\dfrac{1}{y^2}\right)}{x^2y^2\left(\dfrac{1}{x}+\dfrac{1}{y}\right)}$$

$$= \frac{y^2-x^2}{xy^2+x^2y} = \frac{(y-x)\overset{1}{\cancel{(y+x)}}}{xy\underset{1}{\cancel{(y+x)}}}$$

$$= \frac{y-x}{xy}$$

MATCHED PROBLEM 4

Express as a simple fraction reduced to lowest terms.

$$\frac{x-x^{-1}}{1-x^{-2}}$$

Scientific Notation

Scientific work often involves the use of very large numbers or very small numbers. For example, the average cell contains about 200,000,000,000,000 molecules, and the diameter of an electron is about 0.000 000 000 0004 centimeter. It is generally troublesome to write and work with numbers of this type in standard decimal form. The two numbers written here cannot even be entered into most calculators as they are written. With exponents now defined for all integers, it is possible to express any decimal form as the product of a number between 1 and 10 and an integer power of 10; that is, in the form

$$a \times 10^n \qquad 1 \le a < 10,\ n \text{ an integer, } a \text{ in decimal form}$$

A number expressed in this form is said to be in **scientific notation.**

EXAMPLE 5

Scientific Notation

Each number is written in scientific notation:

$$7 = 7 \times 10^0 \qquad 0.5 = 5 \times 10^{-1}$$

$$720 = 7.2 \times 10^2 \qquad 0.08 = 8 \times 10^{-2}$$

$$6{,}430 = 6.43 \times 10^3 \qquad 0.000\ 32 = 3.2 \times 10^{-4}$$

$$5{,}350{,}000 = 5.35 \times 10^6 \qquad 0.000\ 000\ 0738 = 7.38 \times 10^{-8}$$

Can you discover a rule relating the number of decimal places the decimal point is moved to the power of 10 that is used?

$$7{,}320{,}000 = 7.320\ 000. \times 10^6 = 7.32 \times 10^6$$

6 places left
Positive exponent

$$0.000\ 000\ 54 = 0.000\ 000\ 5.4 \times 10^{-7} = 5.4 \times 10^{-7}$$

7 places right
Negative exponent

MATCHED PROBLEM 5

(A) Write each number in scientific notation: 430; 23,000; 345,000,000; 0.3; 0.0031; 0.000 000 683.

(B) Write in standard decimal form: 4×10^3; 5.3×10^5; 2.53×10^{-2}; 7.42×10^{-6}.

Most calculators express very large and very small numbers in scientific notation. Consult the manual for your calculator to see how numbers in scientific notation are entered in your calculator. Some common methods for displaying scientific notation on a calculator are shown below.

Number Represented	Typical Scientific Calculator Display	Typical Graphing Calculator Display
$5.427\ 493 \times 10^{-17}$	5.427493 −17	5.427493E−17
$2.359\ 779 \times 10^{12}$	2.359779 12	2.359779E12

EXAMPLE 6

Using Scientific Notation on a Calculator

Write each number in scientific notation; then carry out the computations using your calculator. (Refer to the user's manual accompanying your calculator for the procedure.) Express the answer to three significant digits in scientific notation.

$$\frac{325{,}100{,}000{,}000}{0.000\ 000\ 000\ 000\ 0871} = \frac{3.251 \times 10^{11}}{8.71 \times 10^{-14}}$$

$= \boxed{3.732491389\text{E}24}$ Calculator display

$= 3.73 \times 10^{24}$ To three significant digits

FIGURE 1

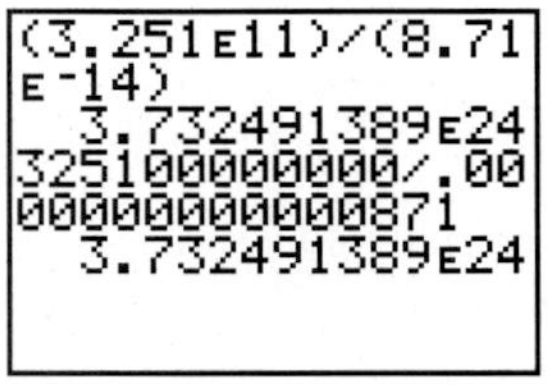

Figure 1 shows two solutions to this problem on a graphing calculator. In the first solution, we entered the numbers in scientific notation, and in the second, we used standard decimal notation. Although the multiple-line screen display on a graphing calculator allows us to enter very long standard decimals, scientific notation is usually more efficient and less prone to errors in data entry. Furthermore, as Figure 1 shows, the calculator uses scientific notation to display the answer, regardless of the manner in which the numbers are entered.

MATCHED PROBLEM 6

Repeat Example 6 for

$$\frac{0.000\ 000\ 006\ 932}{62{,}600{,}000{,}000}$$

EXAMPLE 7

Measuring Time with an Atomic Clock

An atomic clock that counts the radioactive emissions of cesium is used to provide a precise definition of a second. One second is defined to be the time it takes cesium to emit 9,192,631,770 cycles of radiation. How many of these cycles will occur in 1 hour? Express the answer to five significant digits in scientific notation.

Solution

$(9{,}192{,}631{,}770)(60^2) = \boxed{3.309347437\text{E}13}$

$= 3.3093 \times 10^{13}$

MATCHED PROBLEM 7

Refer to Example 7. How many of these cycles will occur in 1 year? Express the answer to five significant digits in scientific notation.

Answers to Matched Problems

1. (A) 1 (B) 1 (C) 0.000 01 (D) 1,000 (E) x^4 (F) v^3/u^7 **2.** (A) $15x$ (B) $3/(2y^3)$ (C) $-14x^4$
3. (A) $y^6/(9x^8)$ (B) y^{12}/x^6 (C) $2n^5/(5m)$ (D) x^9 (E) $(a-b)^2$ **4.** x
5. (A) 4.3×10^2; 2.3×10^4; 3.45×10^8; 3×10^{-1}; 3.1×10^{-3}; 6.83×10^{-7} (B) 4,000; 530,000; 0.0253; 0.000 007 42
6. 1.11×10^{-19} **7.** 2.8990×10^{17}

EXERCISE A-5

All variables are restricted to prevent division by 0.

A

Simplify Problems 1–16, and write the answers using positive exponents only.

1. $y^{-5}y^5$ **2.** x^3x^{-3} **3.** $(2x^2)(3x^3)(x^4)$

4. $(2x^5)(3x^7)(4x^2)$ **5.** $(3x^3y^{-2})^2$ **6.** $(2cd^2)^{-3}$

7. $\left(\frac{ab^3}{c^2d}\right)^4$ **8.** $\left(\frac{x^2y}{2w^2}\right)^3$ **9.** $\frac{10^{23}\cdot 10^{-11}}{10^{-3}\cdot 10^{-2}}$

10. $\frac{10^{-13}\cdot 10^{-4}}{10^{-21}\cdot 10^3}$ **11.** $\frac{4x^{-2}y^{-3}}{2x^{-3}y^{-1}}$ **12.** $\frac{2a^6b^{-2}}{16a^{-3}b^2}$

13. $\left(\frac{n^{-3}}{n^{-2}}\right)^{-2}$ **14.** $\left(\frac{x^{-1}}{x^{-8}}\right)^{-1}$

15. $\frac{8\times 10^3}{2\times 10^{-5}}$ **16.** $\frac{18\times 10^{12}}{6\times 10^{-4}}$

Write the numbers in Problems 17–22 in scientific notation.

17. 32,250,000 **18.** 4,930

19. 0.085 **20.** 0.017

21. 0.000 000 0729 **22.** 0.000 592

In Problems 23–28, write each number in standard decimal form.

23. 5×10^{-3} **24.** 4×10^{-4} **25.** 2.69×10^7

26. 6.5×10^9 **27.** 5.9×10^{-10} **28.** 6.3×10^{-6}

B

Simplify Problems 29–42, and write the answers using positive exponents only. Write compound fractions as simple fractions.

29. $\frac{27x^{-5}x^5}{18y^{-6}y^2}$ **30.** $\frac{32n^5n^{-8}}{24m^{-7}m^7}$ **31.** $\left(\frac{x^4y^{-1}}{x^{-2}y^3}\right)^2$

32. $\left(\frac{m^{-2}n^3}{m^4n^{-1}}\right)^2$ **33.** $\left(\frac{2x^{-3}y^2}{4xy^{-1}}\right)^{-2}$ **34.** $\left(\frac{6mn^{-2}}{3m^{-1}n^2}\right)^{-3}$

35. $\left[\left(\frac{u^3v^{-1}w^{-2}}{u^{-2}v^{-2}w}\right)^{-2}\right]^2$ **36.** $\left[\left(\frac{x^{-2}y^3t}{x^{-3}y^{-2}t^2}\right)^2\right]^{-1}$

37. $(x+y)^{-2}$ **38.** $(a^2-b^2)^{-1}$ **39.** $\frac{1+x^{-1}}{1-x^{-2}}$

40. $\frac{1-x}{x^{-1}-1}$ **41.** $\frac{x^{-1}-y^{-1}}{x-y}$ **42.** $\frac{u+v}{u^{-1}+v^{-1}}$

CL **43.** $-3(x^3+3)^{-4}(3x^2)$

CL **44.** $-2(x^2+3x)^{-3}(2x+3)$

45. What is the result of entering 2^{3^2} on a calculator?

46. Refer to Problem 45. What is the difference between $2^{(3^2)}$ and $(2^3)^2$? Which agrees with the value of 2^{3^2} obtained with a calculator?

47. If $n = 0$, then property 1 in Theorem 1 implies that $a^ma^0 = a^{m+0} = a^m$. Explain how this helps motivate the definition of a^0.

48. If $m = -n$, then property 1 in Theorem 1 implies that $a^{-n}a^n = a^0 = 1$. Explain how this helps motivate the definition of a^{-n}.

CL *Problems 49–54 are calculus-related. Write each problem in the form $ax^p + bx^q$ or $ax^p + bx^q + cx^r$, where a, b, and c are real numbers and p, q, and r are integers. For example,*

$$\frac{2x^4-3x^2+1}{2x^3} = \frac{2x^4}{2x^3}-\frac{3x^2}{2x^3}+\frac{1}{2x^3}$$
$$= x - \tfrac{3}{2}x^{-1} + \tfrac{1}{2}x^{-3}$$

49. $\frac{4x^2-12}{2x}$ **50.** $\frac{6x^3+9x}{3x^3}$

51. $\frac{5x^3-2}{3x^2}$ **52.** $\frac{7x^5-x^2}{4x^5}$

53. $\frac{2x^3-3x^2+x}{2x^2}$ **54.** $\frac{3x^4-4x^2-1}{4x^3}$

Evaluate Problems 55–58 to three significant digits using scientific notation where appropriate and a calculator.

55. $\frac{(32.7)(0.000\ 000\ 008\ 42)}{(0.0513)(80{,}700{,}000{,}000)}$

56. $\frac{(4{,}320)(0.000\ 000\ 000\ 704)}{(835)(635{,}000{,}000{,}000)}$

57. $\frac{(5{,}760{,}000{,}000)}{(527)(0.000\ 007\ 09)}$

58. $\frac{0.000\ 000\ 007\ 23}{(0.0933)(43{,}700{,}000{,}000)}$

In Problems 59–64, use a calculator to evaluate each of the following problems to five significant digits. (Read the instruction book accompanying your calculator.)

59. $(23.8)^8$ **60.** $(-302)^7$

61. $(-302)^{-7}$ **62.** $(23.8)^{-8}$

63. $(9{,}820{,}000{,}000)^3$ **64.** $(0.000\ 000\ 000\ 482)^{-4}$

C

Simplify Problems 65–70, and write the answers using positive exponents only. Write compound fractions as simple fractions.

65. $\dfrac{12(a+2b)^{-3}}{6(a+2b)^{-8}}$

66. $\dfrac{4(x-3)^{-4}}{8(x-3)^{-2}}$

67. $\dfrac{xy^{-2}-yx^{-2}}{y^{-1}-x^{-1}}$

68. $\dfrac{b^{-2}-c^{-2}}{b^{-3}-c^{-3}}$

69. $\left(\dfrac{x^{-1}}{x^{-1}-y^{-1}}\right)^{-1}$

70. $\left[\dfrac{u^{-2}-v^{-2}}{(u^{-1}-v^{-1})^2}\right]^{-1}$

APPLICATIONS

71. Earth Science. If the mass of the earth is approximately 6.1×10^{27} grams and each gram is 2.2×10^{-3} pound, what is the mass of the earth in pounds?

72. Biology. In 1929 Vernadsky, a biologist, estimated that all the free oxygen of the earth weighs 1.5×10^{21} grams and that it is produced by life alone. If 1 gram is approximately 2.2×10^{-3} pound, what is the weight of the free oxygen in pounds?

73. Computer Science. If a computer can perform a single operation in 10^{-10} second, how many operations can it perform in 1 second? In 1 minute? Compute answers to three significant digits.

★ **74. Computer Science.** If electricity travels in a computer circuit at the speed of light (1.86×10^5 miles per second), how far will electricity travel in the superconducting computer (see Problem 73) in the time it takes it to perform one operation? (Size of circuits is a critical problem in computer design.) Give the answer in miles, feet, and inches (1 mile = 5,280 feet). Compute answers to two significant digits.

75. Economics. If in 1986 individuals in the United States paid about $349,000,000,000 in federal income taxes and the population was about 242,000,000, estimate to three significant digits the average amount of tax paid per person. Write your answer in scientific notation and in standard decimal form.

76. Economics. If the gross national product (GNP) was about $4,240,000,000,000 in the United States in 1986 and the population was about 242,000,000, estimate to three significant digits the GNP per person. Write your answer in scientific notation and in standard decimal form.

Section A-6 | Rational Exponents

- Roots of Real Numbers
- Rational Exponents

We now know what symbols such as 3^5, 2^{-3}, and 7^0 mean; that is, we have defined a^n, where n is any integer and a is a real number. But what do symbols such as $4^{1/2}$ and $7^{2/3}$ mean? In this section we will extend the definition of exponent to the rational numbers. Before we can do this, however, we need a precise knowledge of what is meant by "a root of a number."

Roots of Real Numbers

Perhaps you recall that a **square root** of a number b is a number c such that $c^2 = b$, and a **cube root** of a number b is a number d such that $d^3 = b$.

What are the square roots of 9?

3 is a square root of 9, since $3^2 = 9$

-3 is a square root of 9, since $(-3)^2 = 9$

Thus, 9 has two real square roots, one the negative of the other.

What are the cube roots of 8?

2 is a cube root of 8, since $2^3 = 8$

And 2 is the only real number with this property. In general,

DEFINITION 1

DEFINITION OF AN *n*TH ROOT

For a natural number n and a and b real numbers,

a is an nth root of b if $a^n = b$ 3 is a fourth root of 81, since $3^4 = 81$.

Explore/Discuss 1

Is -4 a cube root of -64?

Is either 8 or -8 a square root of -64?

Can you find any real number b with the property that $b^2 = -64$? [*Hint:* Consider the sign of b^2 for $b > 0$ and $b < 0$.]

How many real square roots of 4 exist? Of 5? Of -9? How many real fourth roots of 5 exist? Of -5? How many real cube roots of 27 are there? Of -27? The following important theorem (which we state without proof) answers these questions.

THEOREM 1

NUMBER OF REAL *n*TH ROOTS OF A REAL NUMBER *b*

	n even	*n* odd
b positive	Two real nth roots	One real nth root
	-3 and 3 are both fourth roots of 81	2 is the only real cube root of 8
b negative	No real nth root	One real nth root
	-9 has no real square roots	-2 is the only real cube root of -8

Thus, 4 and 5 have two real square roots each, and -9 has none. There are two real fourth roots of 5 and none for -5. And 27 and -27 have one real cube root each. What symbols do we use to represent these roots? We turn to this question now.

Rational Exponents

If all exponent properties are to continue to hold even if some of the exponents are rational numbers, then

$$(5^{1/3})^3 = 5^{3/3} = 5 \qquad \text{and} \qquad (7^{1/2})^2 = 7^{2/2} = 7$$

Since Theorem 1 states that the number 5 has one real cube root, it seems reasonable to use the symbol $5^{1/3}$ to represent this root. On the other hand, Theorem

1 states that 7 has two real square roots. Which real square root of 7 does $7^{1/2}$ represent? We answer this question in the following definition.

DEFINITION 2

$b^{1/n}$, PRINCIPAL nTH ROOT

For n a natural number and b a real number,

$b^{1/n}$ is the **principal nth root of b**

defined as follows:

1. If n is even and b is positive, then $b^{1/n}$ represents the positive nth root of b.

   $16^{1/2} = 4$ not -4 and 4.

   $-16^{1/2} = -4$ $-16^{1/2}$ and $(-16)^{1/2}$ are not the same.

2. If n is even and b is negative, then $b^{1/n}$ does not represent a real number. (More will be said about this case later.)

   $(-16)^{1/2}$ is not real.

3. If n is odd, then $b^{1/n}$ represents the real nth root of b (there is only one).

   $32^{1/5} = 2$ $(-32)^{1/5} = -2$

4. $0^{1/n} = 0$ $0^{1/9} = 0$ $0^{1/6} = 0$

EXAMPLE 1

Principal nth Roots

(A) $9^{1/2} = 3$

(B) $-9^{1/2} = -3$ Compare parts B and C.

(C) $(-9)^{1/2}$ is not a real number. (D) $27^{1/3} = 3$

(E) $(-27)^{1/3} = -3$ (F) $0^{1/7} = 0$

MATCHED PROBLEM 1

Find each of the following:

(A) $4^{1/2}$ (B) $-4^{1/2}$ (C) $(-4)^{1/2}$

(D) $8^{1/3}$ (E) $(-8)^{1/3}$ (F) $0^{1/8}$

How should a symbol such as $7^{2/3}$ be defined? If the properties of exponents are to hold for rational exponents, then $7^{2/3} = (7^{1/3})^2$; that is, $7^{2/3}$ must represent the square of the cube root of 7. This leads to the following general definition:

DEFINITION 3

$b^{m/n}$ AND $b^{-m/n}$, RATIONAL NUMBER EXPONENT

For m and n natural numbers and b any real number (except b cannot be negative when n is even),

$$b^{m/n} = (b^{1/n})^m \quad \text{and} \quad b^{-m/n} = \frac{1}{b^{m/n}}$$

$$4^{3/2} = (4^{1/2})^3 = 2^3 = 8 \qquad 4^{-3/2} = \frac{1}{4^{3/2}} = \frac{1}{8} \qquad (-4)^{3/2} \text{ is not real.}$$

$$(-32)^{3/5} = [(-32)^{1/5}]^3 = (-2)^3 = -8$$

We have now discussed $b^{m/n}$ for all rational numbers m/n and real numbers b. It can be shown, though we will not do so, that all five properties of exponents listed in Theorem 1 in Section A-5 continue to hold for rational exponents as long as we avoid even roots of negative numbers. With the latter restriction in effect, the following useful relationship is an immediate consequence of the exponent properties:

THEOREM 2

RATIONAL EXPONENT PROPERTY

For m and n natural numbers and b any real number (except b cannot be negative when n is even),

$$b^{m/n} = \begin{cases} (b^{1/n})^m \\ (b^m)^{1/n} \end{cases} \qquad 8^{2/3} = \begin{cases} (8^{1/3})^2 \\ (8^2)^{1/3} \end{cases}$$

Explore/Discuss 2

Find the contradiction in the following chain of equations:

$$-1 = (-1)^{2/2} = [(-1)^2]^{1/2} = 1^{1/2} = 1 \qquad (1)$$

Where did we try to use Theorem 2? Why was this not correct?

The three exponential forms in Theorem 2 are equal as long as only real numbers are involved. But if b is negative and n is even, then $b^{1/n}$ is not a real number and Theorem 2 does not necessarily hold, as illustrated in Explore/Discuss 2. One way to avoid this difficulty is to assume that m and n have no common factors.

EXAMPLE 2

Using Rational Exponents

Simplify, and express answers using positive exponents only. All letters represent positive real numbers.

(A) $8^{2/3} = (8^{1/3})^2 = 2^2 = 4$ or $8^{2/3} = (8^2)^{1/3} = 64^{1/3} = 4$

(B) $(-8)^{5/3} = [(-8)^{1/3}]^5 = (-2)^5 = -32$

(C) $(3x^{1/3})(2x^{1/2}) = 6x^{1/3+1/2} = 6x^{5/6}$

(D) $\left(\dfrac{4x^{1/3}}{x^{1/2}}\right)^{1/2} = \dfrac{4^{1/2}x^{1/6}}{x^{1/4}} = \dfrac{2}{x^{1/4-1/6}} = \dfrac{2}{x^{1/12}}$

(E) $(u^{1/2} - 2v^{1/2})(3u^{1/2} + v^{1/2}) = 3u - 5u^{1/2}v^{1/2} - 2v$

MATCHED PROBLEM **2**

Simplify, and express answers using positive exponents only. All letters represent positive real numbers.

(A) $9^{3/2}$ (B) $(-27)^{4/3}$ (C) $(5y^{3/4})(2y^{1/3})$ (D) $(2x^{-3/4}y^{1/4})^4$

(E) $\left(\dfrac{8x^{1/2}}{x^{2/3}}\right)^{1/3}$ (F) $(2x^{1/2} + y^{1/2})(x^{1/2} - 3y^{1/2})$

EXAMPLE **3**

Evaluating Rational Exponential Forms with a Calculator

Evaluate to four significant digits using a calculator. (Refer to the instruction book for your particular calculator to see how exponential forms are evaluated.)

(A) $11^{3/4}$ (B) $3.1046^{-2/3}$ (C) $(0.000\ 000\ 008\ 437)^{3/11}$

Solutions

(A) First change $\frac{3}{4}$ to the standard decimal form 0.75; then evaluate $11^{0.75}$ using a calculator.

$$11^{3/4} = 6.040$$

(B) $3.1046^{-2/3} = 0.4699$

(C) $(0.000\ 000\ 008\ 437)^{3/11} = (8.437 \times 10^{-9})^{3/11}$

$= 0.006\ 281$

MATCHED PROBLEM **3**

Evaluate to four significant digits using a calculator.

(A) $2^{3/8}$ (B) $57.28^{-5/6}$ (C) $(83{,}240{,}000{,}000)^{5/3}$

EXAMPLE **4** C|/

Simplifying Fractions Involving Rational Exponents

Write the following expression as a simple fraction reduced to lowest terms and without negative exponents:

$$\frac{(1 + x^2)^{1/2}(2x) - x^2(\frac{1}{2})(1 + x^2)^{-1/2}(2x)}{1 + x^2}$$

Solution

The negative exponent indicates the presence of a fraction in the numerator. Multiply numerator and denominator by $(1 + x^2)^{1/2}$ to eliminate the negative exponent and simplify.

$$\frac{(1 + x^2)^{1/2}(2x) - x^2(\frac{1}{2})(1 + x^2)^{-1/2}(2x)}{1 + x^2} \cdot \frac{(1 + x^2)^{1/2}}{(1 + x^2)^{1/2}}$$

$$= \frac{2x(1 + x^2) - x^3}{(1 + x^2)^{3/2}} = \frac{2x + 2x^3 - x^3}{(1 + x^2)^{3/2}} = \frac{2x + x^3}{(1 + x^2)^{3/2}}$$

$$= \frac{x(2 + x^2)}{(1 + x^2)^{3/2}}$$

MATCHED PROBLEM 4

Write the following expression as a simple fraction reduced to lowest terms and without negative exponents:

$$\frac{x^2(\frac{1}{2})(1 + x^2)^{-1/2}(2x) - (1 + x^2)^{1/2}(2x)}{x^4}$$

Answers to Matched Problems

1. (A) 2 (B) -2 (C) Not real (D) 2 (E) -2 (F) 0
2. (A) 27 (B) 81 (C) $10y^{13/12}$ (D) $16y/x^3$ (E) $2/x^{1/18}$ (F) $2x - 5x^{1/2}y^{1/2} - 3y$
3. (A) 1.297 (B) 0.034 28 (C) 1.587×10^{18} **4.** $-(2 + x^2)/[x^3(1 + x^2)^{1/2}]$

EXERCISE A-6

All variables represent positive real numbers unless otherwise stated.

A

In Problems 1–12, evaluate each expression that results in a rational number.

1. $16^{1/2}$ **2.** $64^{1/3}$

3. $16^{3/2}$ **4.** $16^{3/4}$

5. $-36^{1/2}$ **6.** $32^{3/5}$

7. $(-36)^{1/2}$ **8.** $(-32)^{3/5}$

9. $(\frac{4}{25})^{3/2}$ **10.** $(\frac{8}{27})^{2/3}$

11. $9^{-3/2}$ **12.** $8^{-2/3}$

Simplify Problems 13–20, and express answers using positive exponents only.

13. $y^{1/5}y^{2/5}$ **14.** $x^{1/4}x^{3/4}$

15. $d^{2/3}d^{-1/3}$ **16.** $x^{1/4}x^{-3/4}$

17. $(y^{-8})^{1/16}$ **18.** $(x^{-2/3})^{-6}$

19. $(8x^3y^{-6})^{1/3}$ **20.** $(4u^{-2}v^4)^{1/2}$

B

Simplify Problems 21–30, and express answers using positive exponents only.

21. $\left(\frac{a^{-3}}{b^4}\right)^{1/12}$ **22.** $\left(\frac{m^{-2/3}}{n^{-1/2}}\right)^{-6}$ **23.** $\left(\frac{4x^{-2}}{y^4}\right)^{-1/2}$

24. $\left(\frac{w^4}{9x^{-2}}\right)^{-1/2}$ **25.** $\left(\frac{8a^{-4}b^3}{27a^2b^{-3}}\right)^{1/3}$ **26.** $\left(\frac{25x^5y^{-1}}{16x^{-3}y^{-5}}\right)^{1/2}$

27. $\frac{8x^{-1/3}}{12x^{1/4}}$ **28.** $\frac{6a^{3/4}}{15a^{-1/3}}$ **29.** $\left(\frac{a^{2/3}b^{-1/2}}{a^{1/2}b^{1/2}}\right)^2$

30. $\left(\frac{x^{-1/3}y^{1/2}}{x^{-1/4}y^{1/3}}\right)^6$

In Problems 31–38, multiply, and express answers using positive exponents only.

31. $2m^{1/3}(3m^{2/3} - m^6)$ **32.** $3x^{3/4}(4x^{1/4} - 2x^8)$

33. $(a^{1/2} + 2b^{1/2})(a^{1/2} - 3b^{1/2})$

34. $(3u^{1/2} - v^{1/2})(u^{1/2} - 4v^{1/2})$

35. $(2x^{1/2} - 3y^{1/2})(2x^{1/2} + 3y^{1/2})$

36. $(5m^{1/2} + n^{1/2})(5m^{1/2} - n^{1/2})$

37. $(x^{1/2} + 2y^{1/2})^2$ **38.** $(3x^{1/2} - y^{1/2})^2$

In Problems 39–46, evaluate to four significant digits using a calculator. (Refer to the instruction book for your calculator to see how exponential forms are evaluated.)

39. $15^{5/4}$ **40.** $22^{3/2}$ **41.** $103^{-3/4}$

42. $827^{-3/8}$ **43.** $2.876^{8/5}$ **44.** $37.09^{7/3}$

45. $(0.000\,000\,077\,35)^{-2/7}$ **46.** $(491{,}300{,}000{,}000)^{7/4}$

Problems 47–50 illustrate common errors involving rational exponents. In each case, find numerical examples that show that the left side is not always equal to the right side.

47. $(x + y)^{1/2} \neq x^{1/2} + y^{1/2}$ **48.** $(x^3 + y^3)^{1/3} \neq x + y$

49. $(x + y)^{1/3} \neq \dfrac{1}{(x + y)^3}$ **50.** $(x + y)^{-1/2} \neq \dfrac{1}{(x + y)^2}$

C *Problems 51–56 are calculus-related. Write each problem in the form $ax^p + bx^q$, where a and b are real numbers and p and q are rational numbers. For example,*

$$\frac{2x^{1/3} + 4}{4x} = \frac{2x^{1/3}}{4x} + \frac{4}{4x} = \frac{1}{2}x^{1/3-1} + x^{-1}$$

$$= \frac{1}{2}x^{-2/3} + x^{-1}$$

51. $\dfrac{12x^{1/2} - 3}{4x^{1/2}}$ **52.** $\dfrac{x^{2/3} + 2}{2x^{1/3}}$ **53.** $\dfrac{3x^{2/3} + x^{1/2}}{5x}$

54. $\dfrac{2x^{3/4} + 3x^{1/3}}{3x}$ **55.** $\dfrac{x^2 - 4x^{1/2}}{2x^{1/3}}$ **56.** $\dfrac{2x^{1/3} - x^{1/2}}{4x^{1/2}}$

C

In Problems 57–60, m and n represent positive integers. Simplify and express answers using positive exponents.

57. $(a^{3/n}b^{3/m})^{1/3}$ **58.** $(a^{n/2}b^{n/3})^{1/n}$

59. $(x^{m/4}y^{n/3})^{-12}$ **60.** $(a^{m/3}b^{n/2})^{-6}$

61. If possible, find a real value of x such that

(A) $(x^2)^{1/2} \neq x$ (B) $(x^2)^{1/2} = x$ (C) $(x^3)^{1/3} \neq x$

62. If possible, find a real value of x such that

(A) $(x^2)^{1/2} \neq -x$ (B) $(x^2)^{1/2} = -x$ (C) $(x^3)^{1/3} = -x$

63. If n is even and b is negative, then $b^{1/n}$ is not real. If m is odd, n is even, and b is negative, is $(b^m)^{1/n}$ real?

64. If we assume that m is odd and n is even, is it possible that one of $(b^{1/n})^m$ and $(b^m)^{1/n}$ is real and the other is not?

C *Problems 65–68 are calculus-related. Simplify by writing each expression as a simple fraction reduced to lowest terms and without negative exponents.*

65. $\dfrac{(2x - 1)^{1/2} - (x + 2)(\frac{1}{2})(2x - 1)^{-1/2}(2)}{2x - 1}$

66. $\dfrac{(x - 1)^{1/2} - x(\frac{1}{2})(x - 1)^{-1/2}}{x - 1}$

67. $\dfrac{2(3x - 1)^{1/3} - (2x + 1)(\frac{1}{3})(3x - 1)^{-2/3}(3)}{(3x - 1)^{2/3}}$

68. $\dfrac{(x + 2)^{2/3} - x(\frac{2}{3})(x + 2)^{-1/3}}{(x + 2)^{4/3}}$

APPLICATIONS

69. Economics. The number of units N of a finished product produced from the use of x units of labor and y units of capital for a particular Third-World country is approximated by

$$N = 10x^{3/4}y^{1/4} \quad \text{Cobb-Douglas equation}$$

Estimate how many units of a finished product will be produced using 256 units of labor and 81 units of capital.

70. Economics. The number of units N of a finished product produced by a particular automobile company where x units of labor and y units of capital are used is approximated by

$$N = 50x^{1/2}y^{1/2} \quad \text{Cobb-Douglas equation}$$

Estimate how many units will be produced using 256 units of labor and 144 units of capital.

71. Braking Distance. R. A. Moyer of Iowa State College found, in comprehensive tests carried out on 41 wet pavements, that the braking distance d (in feet) for a particular automobile traveling at v miles per hour was given approximately by

$$d = 0.0212v^{7/3}$$

Approximate the braking distance to the nearest foot for the car traveling on wet pavement at 70 miles per hour.

72. Braking Distance. Approximately how many feet would it take the car in Problem 71 to stop on wet pavement if it were traveling at 50 miles per hour? (Compute answer to the nearest foot.)

Section A-7 Radicals

- From Rational Exponents to Radicals, and Vice Versa
- Properties of Radicals
- Simplifying Radicals
- Sums and Differences
- Products
- Rationalizing Operations

What do the following algebraic expressions have in common?

$$2^{1/2} \qquad 2x^{2/3} \qquad \frac{1}{x^{1/2} + y^{1/2}}$$

$$\sqrt{2} \qquad 2\sqrt[3]{x^2} \qquad \frac{1}{\sqrt{x} + \sqrt{y}}$$

Each vertical pair represents the same quantity, one in rational exponent form and the other in *radical form.* There are occasions when it is more convenient to work with radicals than with rational exponents, or vice versa. In this section we see how the two forms are related and investigate some basic operations on radicals.

From Rational Exponents to Radicals, and Vice Versa

We start this discussion by defining an ***n*th-root radical.**

DEFINITION 1

$\sqrt[n]{b}$, *n*TH-ROOT RADICAL

For n a natural number greater than 1 and b a real number, we define $\sqrt[n]{b}$ to be the **principal *n*th root of *b*** (see Definition 2 in Section A-6); that is,

$$\sqrt[n]{b} = b^{1/n}$$

If $n = 2$, we write $\sqrt{b}$ in place of $\sqrt[2]{b}$

$$\sqrt{25} = 25^{1/2} = 5 \qquad \sqrt[5]{32} = 32^{1/5} = 2$$

$$-\sqrt{25} = -25^{1/2} = -5 \qquad \sqrt[5]{-32} = (-32)^{1/5} = -2$$

$\sqrt{-25}$ is not real. $\qquad \sqrt[4]{0} = 0^{1/4} = 0$

The symbol $\sqrt{\ }$ is called a **radical,** n is called the **index,** and b is called the **radicand.**

As stated above, it is often an advantage to be able to shift back and forth between rational exponent forms and radical forms. The following relationships, which are direct consequences of Definition 1 and Theorem 2 in Section A-6, are useful in this regard:

RATIONAL EXPONENT/RADICAL CONVERSIONS

For m and n positive integers ($n > 1$), and b not negative when n is even,

$$b^{m/n} = \begin{cases} (b^m)^{1/n} = \sqrt[n]{b^m} \\ (b^{1/n})^m = (\sqrt[n]{b})^m \end{cases} \qquad 2^{2/3} = \begin{cases} \sqrt[3]{2^2} \\ (\sqrt[3]{2})^2 \end{cases}$$

Note: Unless stated to the contrary, all variables in the rest of the discussion are restricted so that all quantities involved are real numbers.

Explore/Discuss 1

In each of the following, evaluate both radical forms:

$$16^{3/2} = \sqrt{16^3} = (\sqrt{16})^3$$
$$27^{2/3} = \sqrt[3]{27^2} = (\sqrt[3]{27})^2$$

Which radical conversion form is easier to use if you are performing the calculations by hand?

EXAMPLE 1 Rational Exponents/Radical Conversions

Change from rational exponent form to radical form.

(A) $x^{1/7} = \sqrt[7]{x}$

(B) $(3u^2v^3)^{3/5} = \sqrt[5]{(3u^2v^3)^3}$ or $(\sqrt[5]{3u^2v^3})^3$ The first is usually preferred.

(C) $y^{-2/3} = \dfrac{1}{y^{2/3}} = \dfrac{1}{\sqrt[3]{y^2}}$ or $\sqrt[3]{y^{-2}}$ or $\sqrt[3]{\dfrac{1}{y^2}}$

Change from radical form to rational exponent form.

(D) $\sqrt[5]{6} = 6^{1/5}$ (E) $-\sqrt[3]{x^2} = -x^{2/3}$ (F) $\sqrt{x^2 + y^2} = (x^2 + y^2)^{1/2}$

MATCHED PROBLEM 1

Change from rational exponent form to radical form.

(A) $u^{1/5}$ (B) $(6x^2y^5)^{2/9}$ (C) $(3xy)^{-3/5}$

Change from radical form to rational exponent form.

(D) $\sqrt[4]{9u}$ (E) $-\sqrt[7]{(2x)^4}$ (F) $\sqrt[3]{x^3 + y^3}$

Properties of Radicals

The process of changing and simplifying radical expressions is aided by the introduction of several properties of radicals that follow directly from exponent properties considered earlier.

THEOREM 1

PROPERTIES OF RADICALS

For n a natural number greater than 1, and x and y positive real numbers,

1. $\sqrt[n]{x^n} = x \qquad \sqrt[3]{x^3} = x$
2. $\sqrt[n]{xy} = \sqrt[n]{x}\sqrt[n]{y} \qquad \sqrt[5]{xy} = \sqrt[5]{x}\sqrt[5]{y}$
3. $\sqrt[n]{\dfrac{x}{y}} = \dfrac{\sqrt[n]{x}}{\sqrt[n]{y}} \qquad \sqrt[4]{\dfrac{x}{y}} = \dfrac{\sqrt[4]{x}}{\sqrt[4]{y}}$

EXAMPLE 2

Simplifying Radicals

Simplify.

(A) $\sqrt[5]{(3x^2y)^5} = 3x^2y$

(B) $\sqrt{10}\sqrt{5} = \sqrt{50} = \sqrt{25 \cdot 2} = \sqrt{25}\sqrt{2} = 5\sqrt{2}$

(C) $\sqrt[3]{\dfrac{x}{27}} = \dfrac{\sqrt[3]{x}}{\sqrt[3]{27}} = \dfrac{\sqrt[3]{x}}{3} \quad$ or $\quad \dfrac{1}{3}\sqrt[3]{x}$

MATCHED PROBLEM 2

Simplify.

(A) $\sqrt[7]{(u^2 + v^2)^7}$ (B) $\sqrt{6}\sqrt{2}$ (C) $\sqrt[3]{\dfrac{x^2}{8}}$

CAUTION

In general, properties of radicals can be used to simplify terms raised to powers, not sums of terms raised to powers. Thus, for x and y positive real numbers,

$$\sqrt{x^2 + y^2} \neq \sqrt{x^2} + \sqrt{y^2} = x + y$$

but

$$\sqrt{x^2 + 2xy + y^2} = \sqrt{(x + y)^2} = x + y$$

Simplifying Radicals

The properties of radicals provide us with the means of changing algebraic expressions containing radicals to a variety of equivalent forms. One form that is often useful is a *simplified form.* An algebraic expression that contains radicals is said to be in **simplified form** if all four of the conditions listed in the following definition are satisfied.

DEFINITION 2

SIMPLIFIED (RADICAL) FORM

1. No radicand (the expression within the radical sign) contains a factor to a power greater than or equal to the index of the radical.
   For example, $\sqrt{x^5}$ violates this condition.
2. No power of the radicand and the index of the radical have a common factor other than 1.
   For example, $\sqrt[6]{x^4}$ violates this condition.
3. No radical appears in a denominator.
   For example, $y/\sqrt{x}$ violates this condition.
4. No fraction appears within a radical.
   For example, $\sqrt{\frac{3}{5}}$ violates this condition.

EXAMPLE 3

Finding Simplified Form

Express radicals in simplified form.

(A) $\sqrt{12x^3y^5z^2} = \sqrt{(4x^2y^4z^2)(3xy)}$ Condition 1 is not met.

$= \sqrt{(2xy^2z)^2(3xy)}$ $x^{pm}y^{pn} = (x^my^n)^p$

$= \sqrt{(2xy^2z)^2}\sqrt{3xy}$ $\sqrt[n]{xy} = \sqrt[n]{x}\sqrt[n]{y}$

$= 2xy^2z\sqrt{3xy}$ $\sqrt[n]{x^n} = x$

(B) $\sqrt[3]{6x^2y}\sqrt[3]{4x^5y^2} = \sqrt[3]{(6x^2y)(4x^5y^2)}$ $\sqrt[n]{x}\sqrt[n]{y} = \sqrt[n]{xy}$

$= \sqrt[3]{24x^7y^3}$

$= \sqrt[3]{(8x^6y^3)(3x)}$ Condition 1 is not met.

$= \sqrt[3]{(2x^2y)^3(3x)}$ $x^{pm}y^{pn} = (x^my^n)^p$

$- \sqrt[3]{(2x^2y)^3}\sqrt[3]{3x}$ $\sqrt[n]{xy} = \sqrt[n]{x}\sqrt[n]{y}$

$= 2x^2y\sqrt[3]{3x}$ $\sqrt[n]{x^n} = x$

(C) $\sqrt[6]{16x^4y^2} = [(4x^2y)^2]^{1/6}$ Condition 2 is not met.

$= (4x^2y)^{2/6}$ Note the convenience of using rational exponents.

$= (4x^2y)^{1/3}$

$= \sqrt[3]{4x^2y}$

(D) $\sqrt[3]{\sqrt{27}} = [(3^3)^{1/2}]^{1/3}$

$$= (3^3)^{1/6} = 3^{3/6} \quad = 3^{1/2} = \sqrt{3}$$

MATCHED PROBLEM 3

Express radicals in simplified form.

(A) $\sqrt{18x^5y^2z^3}$ (B) $\sqrt[4]{27a^3b^3}\sqrt[4]{3a^5b^3}$ (C) $\sqrt[9]{8x^6y^3}$ (D) $\sqrt{\sqrt[3]{4}}$

Sums and Differences

Algebraic expressions involving radicals often can be simplified by adding and subtracting terms that contain exactly the same radical expressions. We proceed in essentially the same way as we do when we combine like terms in polynomials. The distributive property of real numbers plays a central role in this process.

EXAMPLE 4

Combining Like Terms

Combine as many terms as possible.

(A) $5\sqrt{3} + 4\sqrt{3} \quad = (5 + 4)\sqrt{3} \quad = 9\sqrt{3}$

(B) $2\sqrt[3]{xy^2} - 7\sqrt[3]{xy^2} \quad = (2 - 7)\sqrt[3]{xy^2} \quad = -5\sqrt[3]{xy^2}$

(C) $3\sqrt{xy} - 2\sqrt[3]{xy} + 4\sqrt{xy} - 7\sqrt[3]{xy} \quad = 3\sqrt{xy} + 4\sqrt{xy} - 2\sqrt[3]{xy} - 7\sqrt[3]{xy}$

$$= 7\sqrt{xy} - 9\sqrt[3]{xy}$$

MATCHED PROBLEM 4

Combine as many terms as possible.

(A) $6\sqrt{2} + 2\sqrt{2}$ (B) $3\sqrt[5]{2x^2y^3} - 8\sqrt[5]{2x^2y^3}$

(C) $5\sqrt[3]{mn^2} - 3\sqrt{mn} - 2\sqrt[3]{mn^2} + 7\sqrt{mn}$

Products

We will now consider several types of special products that involve radicals. The distributive property of real numbers plays a central role in our approach to these problems.

EXAMPLE 5

Multiplication with Radical Forms

Multiply and simplify.

(A) $\sqrt{2}(\sqrt{10} - 3) = \sqrt{2}\sqrt{10} - \sqrt{2} \cdot 3 = \sqrt{20} - 3\sqrt{2} = 2\sqrt{5} - 3\sqrt{2}$

(B) $(\sqrt{2} - 3)(\sqrt{2} + 5) = \sqrt{2}\sqrt{2} - 3\sqrt{2} + 5\sqrt{2} - 15$

$$= 2 + 2\sqrt{2} - 15$$

$$= 2\sqrt{2} - 13$$

(C) $(\sqrt{x}-3)(\sqrt{x}+5) = \sqrt{x}\sqrt{x} - 3\sqrt{x} + 5\sqrt{x} - 15$
$= x + 2\sqrt{x} - 15$

(D) $(\sqrt[3]{m} + \sqrt[3]{n^2})(\sqrt[3]{m^2} - \sqrt[3]{n}) = \sqrt[3]{m^3} + \sqrt[3]{m^2n^2} - \sqrt[3]{mn} - \sqrt[3]{n^3}$
$= m - \sqrt[3]{mn} + \sqrt[3]{m^2n^2} - n$

MATCHED PROBLEM 5

Multiply and simplify.

(A) $\sqrt{3}(\sqrt{6}-4)$ (B) $(\sqrt{3}-2)(\sqrt{3}+4)$

(C) $(\sqrt{y}-2)(\sqrt{y}+4)$ (D) $(\sqrt[3]{x^2} - \sqrt[3]{y^2})(\sqrt[3]{x} + \sqrt[3]{y})$

Rationalizing Operations

We now turn to algebraic fractions involving radicals in the denominator. Eliminating a radical from a denominator is referred to as **rationalizing the denominator.** To rationalize the denominator, we multiply the numerator and denominator by a suitable factor that will rationalize the denominator—that is, will leave the denominator free of radicals. This factor is called a **rationalizing factor.** The following special products are of use in finding some rationalizing factors (see Example 6, parts C and D):

$$(a-b)(a+b) = a^2 - b^2 \quad (1)$$

$$(a-b)(a^2+ab+b^2) = a^3 - b^3 \quad (2)$$

$$(a+b)(a^2-ab+b^2) = a^3 + b^3 \quad (3)$$

Explore/Discuss 2

Use special products in equations (1) to (3) to find a rationalizing factor for each of the following:

(A) $\sqrt{a}-\sqrt{b}$ (B) $\sqrt{a}+\sqrt{b}$ (C) $\sqrt[3]{a}-\sqrt[3]{b}$ (D) $\sqrt[3]{a}+\sqrt[3]{b}$

EXAMPLE 6

Rationalizing Denominators

Rationalize denominators.

(A) $\dfrac{3}{\sqrt{5}}$ (B) $\sqrt[3]{\dfrac{2a^2}{3b^2}}$ (C) $\dfrac{\sqrt{x}+\sqrt{y}}{3\sqrt{x}-2\sqrt{y}}$ (D) $\dfrac{1}{\sqrt[3]{m}+2}$

Solutions

(A) $\sqrt{5}$ is a rationalizing factor for $\sqrt{5}$, since $\sqrt{5}\sqrt{5} = \sqrt{5^2} = 5$. Thus, we multiply the numerator and denominator by $\sqrt{5}$ to rationalize the denominator:

$$\frac{3}{\sqrt{5}} = \frac{3\sqrt{5}}{\sqrt{5}\sqrt{5}} = \frac{3\sqrt{5}}{5}$$

(B) $\sqrt[3]{\dfrac{2a^2}{3b^2}} = \dfrac{\sqrt[3]{2a^2}}{\sqrt[3]{3b^2}} = \dfrac{\sqrt[3]{2a^2}\sqrt[3]{3^2b}}{\sqrt[3]{3b^2}\sqrt[3]{3^2b}} = \dfrac{\sqrt[3]{2 \cdot 3^2a^2b}}{\sqrt[3]{3^3b^3}} = \dfrac{\sqrt[3]{18a^2b}}{3b}$

(C) The special product in equation (1) suggests that if we multiply the denominator $3\sqrt{x} - 2\sqrt{y}$ by $3\sqrt{x} + 2\sqrt{y}$, we will obtain the difference of two squares and the denominator will be rationalized.

$$\frac{\sqrt{x} + \sqrt{y}}{3\sqrt{x} - 2\sqrt{y}} = \frac{(\sqrt{x} + \sqrt{y})(3\sqrt{x} + 2\sqrt{y})}{(3\sqrt{x} - 2\sqrt{y})(3\sqrt{x} + 2\sqrt{y})}$$

$$= \frac{3\sqrt{x^2} + 2\sqrt{xy} + 3\sqrt{xy} + 2\sqrt{y^2}}{(3\sqrt{x})^2 - (2\sqrt{y})^2}$$

$$= \frac{3x + 5\sqrt{xy} + 2y}{9x - 4y}$$

(D) The special product in equation (3) suggests that if we multiply the denominator $\sqrt[3]{m} + 2$ by $(\sqrt[3]{m})^2 - 2\sqrt[3]{m} + 2^2$, we will obtain the sum of two cubes and the denominator will be rationalized.

$$\frac{1}{\sqrt[3]{m} + 2} = \frac{1[(\sqrt[3]{m})^2 - 2\sqrt[3]{m} + 2^2]}{(\sqrt[3]{m} + 2)[(\sqrt[3]{m})^2 - 2\sqrt[3]{m} + 2^2]}$$

$$= \frac{\sqrt[3]{m^2} - 2\sqrt[3]{m} + 4}{(\sqrt[3]{m})^3 + 2^3}$$

$$= \frac{\sqrt[3]{m^2} - 2\sqrt[3]{m} + 4}{m + 8}$$

MATCHED PROBLEM 6

Rationalize denominators.

(A) $\dfrac{6}{\sqrt{2x}}$ (B) $\dfrac{10x^3}{\sqrt[3]{4x}}$ (C) $\dfrac{\sqrt{x} + 2}{2\sqrt{x} + 3}$ (D) $\dfrac{1}{1 - \sqrt[3]{y}}$

Answers to Matched Problems

1. (A) $\sqrt[5]{u}$ (B) $\sqrt[9]{(6x^2y^5)^2}$ or $(\sqrt[9]{6x^2y^5})^2$ (C) $1/\sqrt[5]{(3xy)^3}$ (D) $(9u)^{1/4}$ (E) $-(2x)^{4/7}$ (F) $(x^3 + y^3)^{1/3}$

2. (A) $u^2 + v^2$ (B) $2\sqrt{3}$ (C) $(\sqrt[3]{x^2})/2$ or $\frac{1}{2}\sqrt[3]{x^2}$

3. (A) $3x^2yz\sqrt{2xz}$ (B) $3a^2b\sqrt[4]{b^2} = 3a^2b\sqrt{b}$ (C) $\sqrt[3]{2x^2y}$ (D) $\sqrt[3]{2}$

4. (A) $8\sqrt{2}$ (B) $-5\sqrt[5]{2x^2y^3}$ (C) $3\sqrt[3]{mn^2} + 4\sqrt{mn}$

5. (A) $3\sqrt{2} - 4\sqrt{3}$ (B) $2\sqrt{3} - 5$ (C) $y + 2\sqrt{y} - 8$ (D) $x + \sqrt[3]{x^2y} - \sqrt[3]{xy^2} - y$

6. (A) $\dfrac{3\sqrt{2x}}{x}$ (B) $5x^2\sqrt[3]{2x^2}$ (C) $\dfrac{2x + \sqrt{x} - 6}{4x - 9}$ (D) $\dfrac{1 + \sqrt[3]{y} + \sqrt[3]{y^2}}{1 - y}$

EXERCISE A-7

Unless stated to the contrary, all variables are restricted so that all quantities involved are real numbers.

A

In Problems 1–8, change to radical form. Do not simplify.

1. $m^{2/3}$ 2. $n^{4/5}$ 3. $6x^{3/5}$

4. $7y^{2/5}$ 5. $(4xy^3)^{2/5}$ 6. $(7x^2y)^{5/7}$

7. $(x + y)^{1/2}$ 8. $x^{1/2} + y^{1/2}$

In Problems 9–16, change to rational exponent form. Do not simplify.

9. $\sqrt[5]{b}$ 10. $\sqrt{c}$ 11. $5\sqrt[4]{x^3}$

12. $7m\sqrt[5]{n^2}$ 13. $\sqrt[5]{(2x^2y)^3}$ 14. $\sqrt[9]{(3m^4n)^2}$

15. $\sqrt[3]{x} + \sqrt[3]{y}$ 16. $\sqrt[3]{x + y}$

In Problems 17–32, write in simplified form.

17. $\sqrt[3]{-8}$ 18. $\sqrt[3]{-27}$ 19. $\sqrt{9x^8y^4}$

20. $\sqrt{16m^4y^8}$ 21. $\sqrt[4]{16m^4n^8}$ 22. $\sqrt[5]{32a^{15}b^{10}}$

23. $\sqrt{8a^3b^5}$ 24. $\sqrt{27m^2n^7}$ 25. $\sqrt[3]{2^4x^4y^7}$

26. $\sqrt[4]{2^4x^5y^8}$ 27. $\sqrt[4]{m^2}$ 28. $\sqrt[10]{n^6}$

29. $\sqrt[5]{\sqrt[3]{xy}}$ 30. $\sqrt{\sqrt[4]{5x}}$ 31. $\sqrt[3]{9x^2}\sqrt[3]{9x}$

32. $\sqrt{2x}\sqrt{8xy}$

In Problems 33–40, rationalize denominators, and write in simplified form.

33. $\dfrac{1}{\sqrt{5}}$ 34. $\dfrac{1}{\sqrt{7}}$ 35. $\dfrac{6x}{\sqrt{3x}}$

36. $\dfrac{12y^2}{\sqrt{6y}}$ 37. $\dfrac{2}{\sqrt{2} - 1}$ 38. $\dfrac{4}{\sqrt{6} - 2}$

39. $\dfrac{\sqrt{2}}{\sqrt{6} + 2}$ 40. $\dfrac{\sqrt{2}}{\sqrt{10} - 2}$

B

In Problems 41–52, write in simplified form.

41. $x\sqrt[5]{3^6x^7y^{11}}$ 42. $2a\sqrt[3]{8a^8b^{13}}$ 43. $\dfrac{\sqrt{32m^7n^9}}{2mn}$

44. $\dfrac{\sqrt[5]{32u^{12}v^8}}{uv}$ 45. $\sqrt[6]{a^4(b - a)^2}$ 46. $\sqrt[8]{3^6(u + v)^6}$

47. $\sqrt[3]{\sqrt[4]{a^9b^3}}$ 48. $\sqrt{\sqrt[6]{x^8y^6}}$

49. $\sqrt[3]{2x^2y^4}\sqrt[3]{3x^5y}$ 50. $\sqrt[4]{4m^5n}\sqrt[4]{6m^3n^4}$

51. $\sqrt[3]{a^3 + b^3}$ 52. $\sqrt{x^2 + y^2}$

In Problems 53–64, rationalize denominators and write in simplified form.

53. $\dfrac{\sqrt{2m}\sqrt{5}}{\sqrt{20m}}$ 54. $\dfrac{\sqrt{6}\sqrt{8c}}{\sqrt{18c}}$ 55. $\dfrac{4a^3b^2}{\sqrt[3]{2ab^2}}$

56. $\dfrac{8x^3y^5}{\sqrt[3]{4x^2y}}$ 57. $\sqrt[4]{\dfrac{3y^3}{4x^3}}$ 58. $\sqrt[5]{\dfrac{4x^2}{16y^3}}$

59. $\dfrac{3\sqrt{y}}{2\sqrt{y} - 3}$ 60. $\dfrac{5\sqrt{x}}{3 - 2\sqrt{x}}$ 61. $\dfrac{2\sqrt{5} + 3\sqrt{2}}{5\sqrt{5} + 2\sqrt{2}}$

62. $\dfrac{3\sqrt{2} - 2\sqrt{3}}{3\sqrt{3} - 2\sqrt{2}}$ 63. $\dfrac{x^2}{\sqrt{x^2 + 9} - 3}$ 64. $\dfrac{-y^2}{2 - \sqrt{y^2 + 4}}$

CI *Problems 65–68 are calculus-related. Rationalize the numerators; that is, perform operations on the fractions that eliminate radicals from the numerators. (This is a particularly useful operation in some problems in calculus.)*

65. $\dfrac{\sqrt{t} - \sqrt{x}}{t - x}$ 66. $\dfrac{\sqrt{x} - \sqrt{y}}{\sqrt{x} + \sqrt{y}}$

67. $\dfrac{\sqrt{x + h} - \sqrt{x}}{h}$ 68. $\dfrac{\sqrt{2 + h} + \sqrt{2}}{h}$

In Problems 69–80, evaluate to four significant digits using a calculator. (Read the instruction booklet accompanying your calculator for the process required to evaluate $\sqrt[n]{x}$.)

69. $\sqrt{0.049\ 375}$ 70. $\sqrt{306.721}$

71. $\sqrt[5]{27.0635}$ 72. $\sqrt[8]{0.070\ 144}$

73. $\sqrt[7]{0.000\ 000\ 008\ 066}$ 74. $\sqrt[12]{6{,}423{,}000{,}000{,}000}$

75. $\sqrt[3]{7 + \sqrt[3]{7}}$ 76. $\sqrt[5]{4 + \sqrt[5]{4}}$

77. $\sqrt[3]{\sqrt[4]{2}}$ and $\sqrt[12]{2}$ 78. $\sqrt[3]{\sqrt{5}}$ and $\sqrt[6]{5}$

79. $\dfrac{1}{\sqrt[3]{4}}$ and $\dfrac{\sqrt[3]{2}}{2}$ 80. $\dfrac{1}{\sqrt[3]{5}}$ and $\dfrac{\sqrt[3]{25}}{5}$

C

For what real numbers are Problems 81–84 true?

81. $\sqrt{x^2} = -x$ 82. $\sqrt{x^2} = x$

83. $\sqrt[3]{x^3} = x$ 84. $\sqrt[3]{x^3} = -x$

In Problems 85 and 86, evaluate each expression on a calculator and determine which pairs have the same value. Verify these results algebraically.

85. (A) $\sqrt{3} + \sqrt{5}$ (B) $\sqrt{2 + \sqrt{3}} + \sqrt{2 - \sqrt{3}}$
(C) $1 + \sqrt{3}$ (D) $\sqrt[3]{10 + 6\sqrt{3}}$
(E) $\sqrt{8 + \sqrt{60}}$ (F) $\sqrt{6}$

86. (A) $2\sqrt[3]{2} + \sqrt{5}$ (B) $\sqrt{8}$
(C) $\sqrt{3} + \sqrt{7}$ (D) $\sqrt{3 + \sqrt{8}} + \sqrt{3 - \sqrt{8}}$
(E) $\sqrt{10 + \sqrt{84}}$ (F) $1 + \sqrt{5}$

In Problems 87–90, rationalize denominators.

87. $\dfrac{1}{\sqrt[3]{a} - \sqrt[3]{b}}$ **88.** $\dfrac{1}{\sqrt[3]{m} + \sqrt[3]{n}}$

89. $\dfrac{1}{\sqrt{x} - \sqrt{y} + \sqrt{z}}$ **90.** $\dfrac{1}{\sqrt{x} + \sqrt{y} - \sqrt{z}}$

[*Hint for Problem 89:* Start by multiplying numerator and denominator by $(\sqrt{x} - \sqrt{y}) - \sqrt{z}$.]

C *Problems 91 and 92 are calculus-related. Rationalize numerators.*

91. $\dfrac{\sqrt[3]{x + h} - \sqrt[3]{x}}{h}$ **92.** $\dfrac{\sqrt[3]{t} - \sqrt[3]{x}}{t - x}$

93. Show that $\sqrt[kn]{x^{km}} = \sqrt[n]{x^m}$ for k, m, and n natural numbers greater than 1.

94. Show that $\sqrt[m]{\sqrt[n]{x}} = \sqrt[mn]{x}$ for m and n natural numbers greater than 1.

APPLICATIONS

95. Physics—Relativistic Mass. The mass M of an object moving at a velocity v is given by

$$M = \frac{M_0}{\sqrt{1 - \dfrac{v^2}{c^2}}}$$

where M_0 = mass at rest and c = velocity of light. The mass of an object increases with velocity and tends to infinity as the velocity approaches the speed of light. Show that M can be written in the form

$$M = \frac{M_0 c\sqrt{c^2 - v^2}}{c^2 - v^2}$$

96. Physics—Pendulum. A simple pendulum is formed by hanging a bob of mass M on a string of length L from a fixed support (see the figure). The time it takes the bob to swing from right to left and back again is called the **period** T and is given by

$$T = 2\pi\sqrt{\frac{L}{g}}$$

where g is the gravitational constant. Show that T can be written in the form

$$T = \frac{2\pi\sqrt{gL}}{g}$$

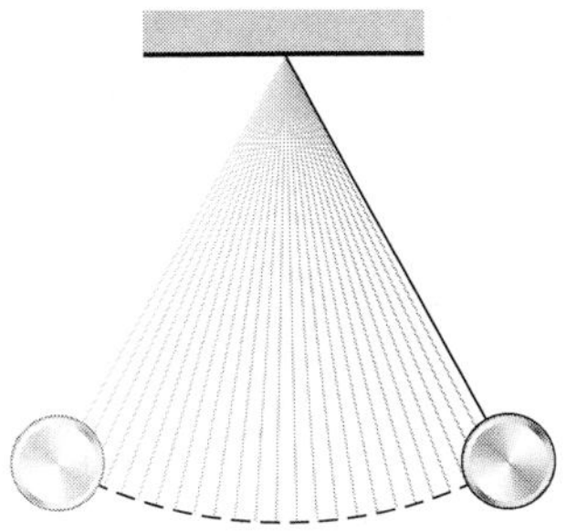

Section A-8 | Linear Equations and Inequalities

- Equations
- Solving Linear Equations
- Inequality Relations and Interval Notation
- Solving Linear Inequalities

Equations

An **algebraic equation** is a mathematical statement that relates two algebraic expressions involving at least one variable. Some examples of equations with x as a variable are

$$3x - 2 = 7 \qquad \frac{1}{1 + x} = \frac{x}{x - 2}$$

$$2x^2 - 3x + 5 = 0 \qquad \sqrt{x + 4} = x - 1$$

The **replacement set,** or **domain,** for a variable is defined to be the set of numbers that are permitted to replace the variable.

ASSUMPTION

ON DOMAINS OF VARIABLES

Unless stated to the contrary, we assume that the domain for a variable is the set of those real numbers for which the algebraic expressions involving the variable are real numbers.

For example, the domain for the variable x in the expression

$$2x - 4$$

is R, the set of all real numbers, since $2x - 4$ represents a real number for all replacements of x by real numbers. The domain of x in the equation

$$\frac{1}{x} = \frac{2}{x - 3}$$

is the set of all real numbers except 0 and 3. These values are excluded because the left member is not defined for $x = 0$ and the right member is not defined for $x = 3$. The left and right members represent real numbers for all other replacements of x by real numbers.

The **solution set** for an equation is defined to be the set of elements in the domain of the variable that make the equation true. Each element of the solution set is called a **solution,** or **root,** of the equation. To **solve an equation** is to find the solution set for the equation.

Knowing what we mean by the solution set of an equation is one thing; finding it is another. To this end we introduce the idea of equivalent equations. Two equations are said to be **equivalent** if they both have the same solution set for a given replacement set. A basic technique for solving equations is to perform operations on equations that produce simpler equivalent equations, and to continue the process until an equation is reached whose solution is obvious.

Application of any of the properties of equality given in Theorem 1 will produce equivalent equations.

THEOREM 1

PROPERTIES OF EQUALITY

For a, b, and c any real numbers,

1. If $a = b$, then $a + c = b + c$. — **Addition Property**
2. If $a = b$, then $a - c = b - c$. — **Subtraction Property**
3. If $a = b$, then $ca = cb$, $c \neq 0$. — **Multiplication Property**
4. If $a = b$, then $\frac{a}{c} = \frac{b}{c}$, $c \neq 0$. — **Division Property**
5. If $a = b$, then either may replace the other in any statement without changing the truth or falsity of the statement. — **Substitution Property**

Solving Linear Equations

We now turn our attention to methods of solving *first-degree,* or *linear, equations* in one variable.

DEFINITION 1

LINEAR EQUATION IN ONE VARIABLE

Any equation that can be written in the form

$$ax + b = 0 \qquad a \neq 0 \qquad \textbf{Standard Form}$$

where a and b are real constants and x is a variable, is called a **linear,** or **first-degree, equation** in one variable.

$5x - 1 = 2(x + 3)$ is a linear equation, since it can be written in the standard form $3x - 7 = 0$.

EXAMPLE 1

Solving a Linear Equation

Solve $5x - 9 = 3x + 7$ and check.

Solution

We use the properties of equality to transform the given equation into an equivalent equation whose solution is obvious.

$5x - 9 = 3x + 7$	Original equation
$5x - 9 + 9 = 3x + 7 + 9$	Add 9 to both sides.
$5x = 3x + 16$	Combine like terms.
$5x - 3x = 3x + 16 - 3x$	Subtract $3x$ from both sides.
$2x = 16$	Combine like terms.
$\frac{2x}{2} = \frac{16}{2}$	Divide both sides by 2.
$x = 8$	Simplify.

The solution set for this last equation is obvious:

Solution set: $\{8\}$

And since the equation $x = 8$ is equivalent to all the preceding equations in our solution, $\{8\}$ is also the solution set for all these equations, including the original equation. [*Note:* If an equation has only one element in its solution set, we generally use the last equation (in this case, $x = 8$) rather than set notation to represent the solution.]

Check

$5x - 9 = 3x + 7$	Original equation
$5(8) - 9 \stackrel{?}{=} 3(8) + 7$	Substitute $x = 8$.
$40 - 9 \stackrel{?}{=} 24 + 7$	Simplify each side.
$31 \stackrel{\checkmark}{=} 31$	A true statement

MATCHED PROBLEM 1 Solve $7x - 10 = 4x + 5$ and check.

EXAMPLE 2 Solving a Linear Equation

Solve $3x - 2(2x - 5) = 2(x + 3) - 8$ and check.

Solution

$3x - 2(2x - 5) = 2(x + 3) - 8$	Original equation
$3x - 4x + 10 = 2x + 6 - 8$	Clear parentheses.
$-x + 10 = 2x - 2$	Combine like terms.
$-3x = -12$	Subtract $2x$ and 10 from both sides.
$x = 4$	Divide both sides by -3.

Check

$$3x - 2(2x - 5) = 2(x + 3) - 8$$

$$3(4) - 2[2(4) - 5] \stackrel{?}{=} 2[(4) + 3] - 8$$

$$6 \stackrel{\checkmark}{=} 6$$

MATCHED PROBLEM 2 Solve $2(3 - x) - (3x + 1) = 8 - 2(x + 2)$ and check.

Inequality Relations and Interval Notation

The above mathematical forms involve the **inequality,** or **order, relation**—that is, "less than" and "greater than" relations. Just as we use = to replace the words "is equal to," we use the **inequality symbols** < and > to represent "is less than" and "is greater than," respectively.

While it probably seems obvious to you that

$$2 < 4 \qquad 5 > 0 \qquad 25{,}000 > 1$$

are true, it may not seem as obvious that

$$-4 < -2 \qquad 0 > -5 \qquad -25{,}000 < -1$$

To make the inequality relation precise so that we can interpret it relative to all real numbers, we need a precise definition of the concept.

DEFINITION 2

a < *b* AND *b* > *a*

For a and b real numbers, we say that ***a* is less than *b*** or ***b* is greater than *a*** and write

$$a < b \qquad \text{or} \qquad b > a$$

if there exists a positive real number p such that $a + p = b$ (or equivalently, $b - a = p$).

We certainly expect that if a positive number is added to *any* real number, the sum is larger than the original. That is essentially what the definition states.

When we write

$$a \le b$$

we mean $a < b$ or $a = b$ and say ***a* is less than or equal to *b*.** When we write

$$a \ge b$$

we mean $a > b$ or $a = b$ and say ***a* is greater than or equal to *b*.**

The inequality symbols $<$ and $>$ have a very clear geometric interpretation on the real number line. If $a < b$, then a is to the left of b; if $c > d$, then c is to the right of d (Fig. 1).

FIGURE 1

$a < b, c > d.$

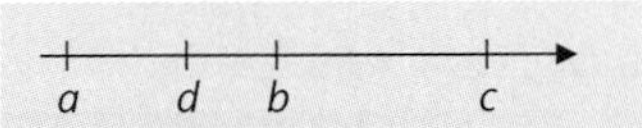

It is an interesting and useful fact that for any two real numbers a and b, either $a < b$, or $a > b$, or $a = b$. This is called the **trichotomy property** of real numbers.

The double inequality $a < x \le b$ means that $x > a$ and $x \le b$; that is, x is between a and b, including b but not including a. The set of all real numbers x satisfying the inequality $a < x \le b$ is called an **interval** and is represented by $(a, b]$. Thus,

$$(a, b] = \{x \mid a < x \le b\}^*$$

The number a is called the **left endpoint** of the interval, and the symbol (indicates that a is not included in the interval. The number b is called the **right endpoint** of the interval, and the symbol] indicates that b is included in the interval. Other types of intervals of real numbers are shown in Table 1.

TABLE 1 Interval Notation

Interval Notation	Inequality Notation	Line Graph	Type
$[a, b]$	$a \le x \le b$		Closed
$[a, b)$	$a \le x < b$		Half-open
$(a, b]$	$a < x \le b$		Half-open
(a, b)	$a < x < b$		Open
$[b, \infty)$	$x \ge b$	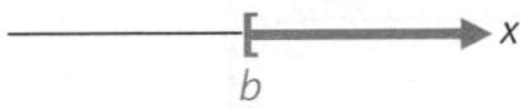	Closed
(b, ∞)	$x > b$	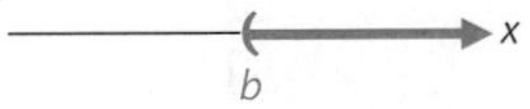	Open
$(-\infty, a]$	$x \le a$	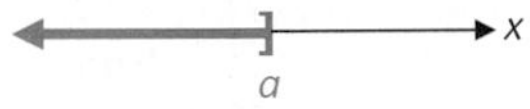	Closed
$(-\infty, a)$	$x < a$	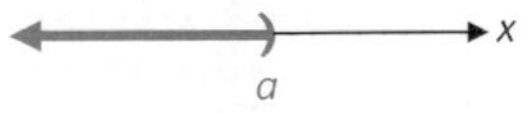	Open

*In general, $\{x \mid P(x)\}$ represents the set of all x such that statement $P(x)$ is true. To express this set verbally, just read the vertical bar as "such that."

Note that the symbol ∞, read "infinity," used in Table 1 is not a numeral. When we write $[b, \infty)$, we are simply referring to the interval starting at b and continuing indefinitely to the right. We would never write $[b, \infty]$ or $b \le x \le \infty$, since ∞ cannot be used as an endpoint of an interval. The interval $(-\infty, \infty)$ represents the set of real numbers R, since its graph is the entire real number line.

CAUTION

It is important to note that

$$5 > x \ge -3 \qquad \text{is equivalent to } [-3, 5) \text{ and not to } (5, -3]$$

In interval notation, the smaller number is always written to the left. Thus, it may be useful to rewrite the inequality as $-3 \le x < 5$ before rewriting it in interval notation.

EXAMPLE 3

Graphing Intervals and Inequalities

Write each of the following in inequality notation and graph on a real number line:

(A) $[-2, 3)$ (B) $(-4, 2)$ (C) $[-2, \infty)$ (D) $(-\infty, 3)$

Solutions

(A) $-2 \le x < 3$

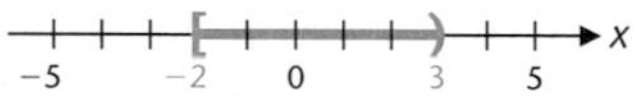

(B) $-4 < x < 2$

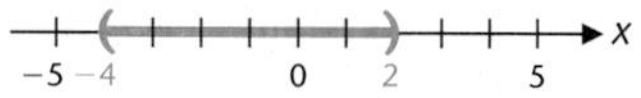

(C) $x \ge -2$

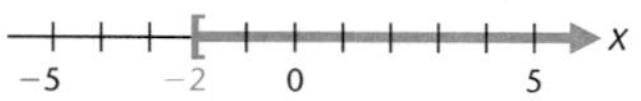

(D) $x < 3$

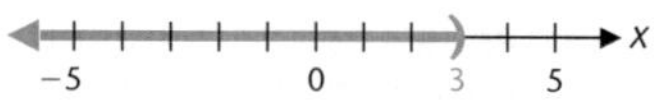

MATCHED PROBLEM 3

Write each of the following in interval notation and graph on a real number line:

(A) $-3 < x \le 3$ (B) $2 \ge x \ge -1$ (C) $x > 1$ (D) $x \le 2$

Explore/Discuss 1

Example 3, part C, shows the graph of the inequality $x \ge -2$. What is the graph of $x < -2$? What is the corresponding interval? Describe the relationship between these sets.

Since intervals are sets of real numbers, the set operations of *union* and *intersection* are often useful when working with intervals. The **union** of sets A and B, denoted by $A \cup B$, is the set formed by combining all the elements of A and all the elements of B. The **intersection** of sets A and B, denoted by $A \cap B$, is the set of elements of A that are also in B. Symbolically:

DEFINITION 3

UNION AND INTERSECTION

Union: $A \cup B = \{x \mid x \text{ is in } A \text{ or } x \text{ is in } B\}$

$\{1, 2, 3\} \cup \{2, 3, 4, 5\} = \{1, 2, 3, 4, 5\}$

Intersection: $A \cap B = \{x \mid x \text{ is in } A \text{ and } x \text{ is in } B\}$

$\{1, 2, 3\} \cap \{2, 3, 4, 5\} = \{2, 3\}$

EXAMPLE 4

Graphing Unions and Intersections of Intervals

If $A = [-2, 3]$, $B = (1, 6)$, and $C = (4, \infty)$, graph the indicated sets and write as a single interval, if possible.

(A) $A \cup B$ and $A \cap B$ (B) $A \cup C$ and $A \cap C$

Solutions

(A)

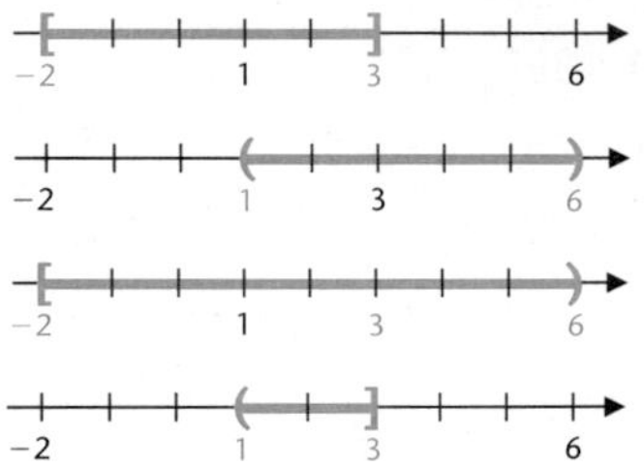

$A = [-2, 3]$

$B = (1, 6)$

$A \cup B = [-2, 6)$

$A \cap B = (1, 3]$

(B)

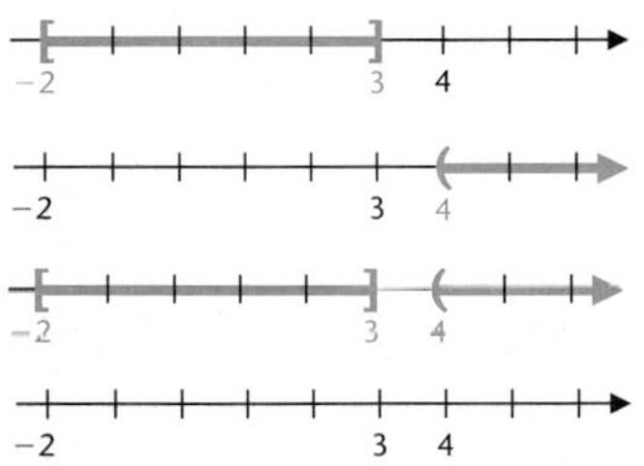

$A = [-2, 3]$

$C = (4, \infty)$

$A \cup C = [-2, 3] \cup (4, \infty)$

$A \cap C = \varnothing$

MATCHED PROBLEM 4

If $D = [-4, 1)$, $E = (-1, 3]$, and $F = [2, \infty)$, graph the indicated sets and write as a single interval, if possible.

(A) $D \cup E$ (B) $D \cap E$ (C) $E \cup F$ (D) $E \cap F$

Explore/Discuss 2

Replace ? with < or > in each of the following:

(A) $-1 \ ? \ 3$ and $2(-1) \ ? \ 2(3)$

(B) $-1 \ ? \ 3$ and $-2(-1) \ ? \ -2(3)$

(C) $12 \ ? \ -8$ and $\dfrac{12}{4} \ ? \ \dfrac{-8}{4}$

(D) $12 \ ? \ -8$ and $\dfrac{12}{-4} \ ? \ \dfrac{-8}{-4}$

Based on these examples, describe verbally the effect of multiplying both sides of an inequality by a number.

Solving Linear Inequalities

We now turn to the problem of solving linear inequalities in one variable, such as

$$2(2x + 3) < 6(x - 2) + 10 \qquad \text{and} \qquad -3 < 2x + 3 \le 9$$

The **solution set** for an inequality is the set of all values of the variable that make the inequality a true statement. Each element of the solution set is called a **solution** of the inequality. To **solve an inequality** is to find its solution set. Two inequalities are **equivalent** if they have the same solution set for a given replacement set. Just as with equations, we perform operations on inequalities that produce simpler equivalent inequalities, and continue the process until an inequality is reached whose solution is obvious. The properties of inequalities given in Theorem 2 can be used to produce equivalent inequalities.

THEOREM 2

INEQUALITY PROPERTIES

For a, b, and c any real numbers,

1. If $a < b$ and $b < c$, then $a < c$. — **Transitive Property**
2. If $a < b$, then $a + c < b + c$. — **Addition Property**
   $-2 < 4 \qquad -2 + 3 < 4 + 3$
3. If $a < b$, then $a - c < b - c$. — **Subtraction Property**
   $-2 < 4 \qquad -2 - 3 < 4 - 3$
4. If $a < b$ and c is positive, then $ca < cb$.
   $-2 < 4 \qquad 3(-2) < 3(4)$
5. If $a < b$ and c is negative, then $ca > cb$.
   $-2 < 4 \qquad (-3)(-2) > (-3)(4)$

   (4 and 5: **Multiplication Property (Note difference between 4 and 5.)**)
6. If $a < b$ and c is positive, then $\frac{a}{c} < \frac{b}{c}$.
   $-2 < 4 \qquad \frac{-2}{2} < \frac{4}{2}$
7. If $a < b$ and c is negative, then $\frac{a}{c} > \frac{b}{c}$.
   $-2 < 4 \qquad \frac{-2}{-2} > \frac{4}{-2}$

   (6 and 7: **Division Property (Note difference between 6 and 7.)**)

Similar properties hold if each inequality sign is reversed, or if $<$ is replaced with $\le$ and $>$ is replaced with $\ge$. Thus, we find that we can perform essentially the same operations on inequalities that we perform on equations. When working with inequalities, however, we have to be particularly careful of the use of the multiplication and division properties.

The order of the inequality reverses if we multiply or divide both sides of an inequality statement by a negative number.

Explore/Discuss 3

Properties of equality are easily summarized. We can add, subtract, multiply, or divide both sides of an equation by any nonzero real number to produce an equivalent equation. Write a similar summary for the properties of inequalities.

Now let's see how the inequality properties are used to solve linear inequalities. Several examples will illustrate the process.

EXAMPLE 5 **Solving a Linear Inequality**

Solve $2(2x + 3) - 10 < 6(x - 2)$ and graph.

Solution

$2(2x + 3) - 10 < 6(x - 2)$

$4x + 6 - 10 < 6x - 12$ Simplify left and right sides.

$4x - 4 < 6x - 12$

$4x - 4 + 4 < 6x - 12 + 4$ Addition property

$4x < 6x - 8$

$4x - 6x < 6x - 8 - 6x$ Subtraction property

$-2x < -8$

$\frac{-2x}{-2} > \frac{-8}{-2}$ Division property—note that order reverses since -2 is negative.

$x > 4$ or $(4, \infty)$ Solution set

Graph of solution set

MATCHED PROBLEM 5

Solve $3(x - 1) \geq 5(x + 2) - 5$ and graph.

EXAMPLE 6 **Solving a Double Inequality**

Solve $-1 \leq 3 - 4x < 11$ and graph.

Solution

We proceed as before, except we try to isolate x in the middle with a coefficient of 1.

$-1 \leq 3 - 4x < 11$

$-1 - 3 \leq 3 - 4x - 3 < 11 - 3$ Subtract 3 from each member.

$$-4 \le -4x < 8$$

$$\frac{-4}{-4} \ge \frac{-4x}{-4} > \frac{8}{-4}$$ Divide each number by −4 and reverse each inequality.

$1 \ge x > -2$ or $-2 < x \le 1$ or $(-2, 1]$

MATCHED PROBLEM 6 Solve $-5 < 10 - 3x \le 10$ and graph.

Answers to Matched Problems

1. $x = 5$ **2.** $x = \frac{1}{3}$

3. (A) $(-3, 3]$
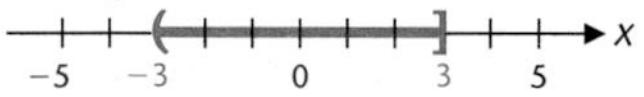

(B) $[-1, 2]$
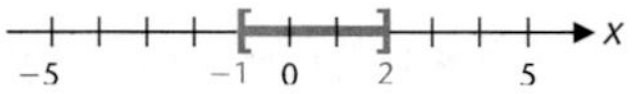

(C) $(1, \infty)$
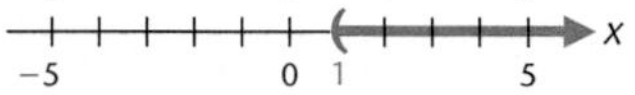

(D) $(-\infty, 2]$
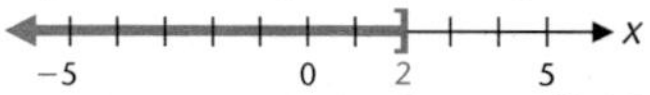

4. (A)

$D \cup E = [-4, 3]$

(B)

$D \cap E = (-1, 1)$

(C)

$E \cup F = (-1, \infty)$

(D)
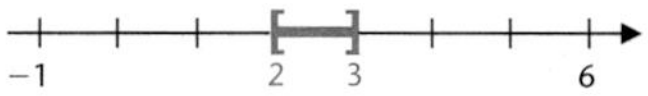

$E \cap F = [2, 3]$

5. $x \le -4$ or $(-\infty, -4]$
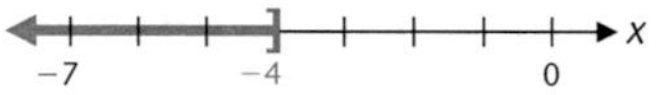

6. $5 > x \ge 0$ or $0 \le x < 5$ or $[0, 5)$

EXERCISE A-8

A

Solve Problems 1–6.

1. $x + 5 = 12$ **2.** $x - 9 = -2$

3. $2s - 7 = -2$ **4.** $7 - 3t = 1$

5. $2m + 8 = 5m - 7$ **6.** $3y + 5 = 6y - 10$

In Problems 7–12, rewrite in inequality notation and graph on a real number line.

7. $[-8, 7]$ **8.** $(-4, 8)$

9. $[-6, 6)$ **10.** $(-3, 3]$

11. $[-6, \infty)$ **12.** $(-\infty, 7)$

In Problems 13–18, rewrite in interval notation and graph on a real number line.

13. $-2 < x \le 6$ **14.** $-5 \le x \le 5$

15. $-7 < x < 8$ **16.** $-4 \le x < 5$

17. $x \le -2$ **18.** $x > 3$

In Problems 19–22, write in interval and inequality notation.

19.
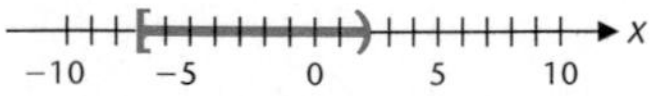

20.
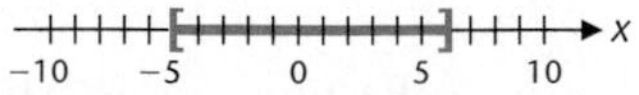

21.
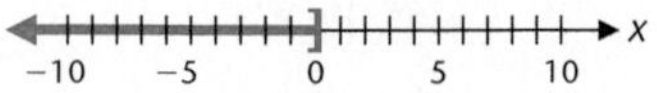

22.
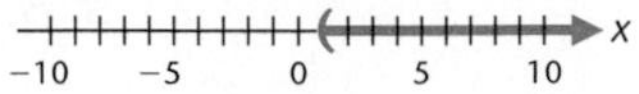

In Problems 23–32, solve and graph.

23. $7x - 8 < 4x + 7$

24. $4x + 8 \geq x - 1$

25. $3 - x \geq 5(3 - x)$

26. $2(x - 3) + 5 < 5 - x$

27. $\frac{N}{-2} > 4$

28. $\frac{M}{-3} \leq -2$

29. $3 - m < 4(m - 3)$

30. $2(1 - u) \geq 5u$

31. $-2 - \frac{B}{4} \leq \frac{1 + B}{3}$

32. $\frac{y - 3}{4} - 1 > \frac{y}{2}$

B

Solve Problems 33–38.

33. $3 - \frac{2x - 3}{3} = \frac{5 - x}{2}$

34. $\frac{x - 2}{3} + 1 = \frac{x}{7}$

35. $0.1(x - 7) + 0.05x = 0.8$

36. $0.4(x + 5) - 0.3x = 17$

37. $0.3x - 0.04(x + 1) = 2.04$

38. $0.02x - 0.5(x - 2) = 5.32$

In Problems 39–50, graph the indicated set and write as a single interval, if possible.

39. $(-5, 5) \cup [4, 7]$

40. $(-5, 5) \cap [4, 7]$

41. $[-1, 4) \cap (2, 6]$

42. $[-1, 4) \cup (2, 6]$

43. $(-\infty, 1) \cup (-2, \infty)$

44. $(-\infty, 1) \cap (2, \infty)$

45. $(-\infty, -1) \cup [3, 7)$

46. $(1, 6] \cup [9, \infty)$

47. $[2, 3] \cup (1, 5)$

48. $[2, 3] \cap (1, 5)$

49. $(-\infty, 4) \cup (-1, 6]$

50. $(-3, 2) \cup [0, \infty)$

In Problems 51–60, solve and graph.

51. $\frac{q}{7} - 3 > \frac{q - 4}{3} + 1$

52. $\frac{p}{3} - \frac{p - 2}{2} \leq \frac{p}{4} - 4$

53. $\frac{2x}{5} - \frac{1}{2}(x - 3) \leq \frac{2x}{3} - \frac{3}{10}(x + 2)$

54. $\frac{2}{3}(x + 7) - \frac{x}{4} > \frac{1}{2}(3 - x) + \frac{x}{6}$

55. $-4 \leq \frac{9}{5}x + 32 \leq 68$

56. $-1 \leq \frac{2}{3}A + 5 \leq 11$

57. $16 < 7 - 3x \leq 31$

58. $-1 \leq 9 - 2x < 5$

59. $-6 < -\frac{3}{5}(1 - x) \leq 4$

60. $15 \leq 7 - \frac{2}{5}x \leq 21$

C

61. Indicate true (T) or false (F):

(A) If $p > q$ and $m > 0$, then $mp < mq$.

(B) If $p < q$ and $m < 0$, then $mp > mq$.

(C) If $p > 0$ and $q < 0$, then $p + q > q$.

62. Assume that $m > n > 0$; then

$$mn > n^2$$
$$mn - m^2 > n^2 - m^2$$
$$m(n - m) > (n + m)(n - m)$$
$$m > n + m$$
$$0 > n$$

But it was assumed that $n > 0$. Find the error.

Prove each inequality property in Problems 63–66, given a, b, and c are arbitrary real numbers.

63. If $a < b$, then $a + c < b + c$

64. If $a < b$, then $a - c < b - c$.

65. (A) If $a < b$ and c is positive, then $ca < cb$.

(B) If $a < b$ and c is negative, then $ca > cb$.

66. (A) If $a < b$ and c is positive, then $\frac{a}{c} < \frac{b}{c}$.

(B) If $a < b$ and c is negative, then $\frac{a}{c} > \frac{b}{c}$.

Appendix A | Group Activity

Rational Number Representations

The set of real numbers can be partitioned into two disjoint subsets, the set of rational numbers and the set of irrational numbers. Rational numbers can be represented two ways: as a/b, where a and b are integers and $b \neq 0$, and as terminating or repeating decimal expansions. The irrational numbers can be represented as nonrepeating and nonterminating decimal expansions. In this activity, we want to explore the relationship between the two different methods for representing rational numbers.

Consider the rational number $r = a/b$, where a and b are integers with no common factors and $b \neq 0$.

1. If $b = 10^n$ for a positive integer n, what kind of decimal expansion will r have?

2. If $b = 2^m 5^n$ for positive integers m and n, what kind of decimal expansion will r have?

3. If r has a terminating decimal expansion, show that r can be expressed in the form a/b, where $b = 2^m 5^n$.

4. Find a and b for the following repeating expansions (the overbar indicates the repeating block):

(A) $0.\overline{63}$ (B) $0.\overline{486}$ (C) $0.\overline{846\ 153}$

5. Find a and b for $r = 0.1\overline{9}$ and then find a terminating decimal expansion for r.

APPENDIX A | REVIEW

A-1 ALGEBRA AND REAL NUMBERS

A **set** is a collection of objects called **elements** or **members** of the set. Sets are usually described by **listing** the elements or by stating a **rule** that determines the elements. A set may be **finite** or **infinite.** A set with no elements is called the **empty set** or the **null set** and is denoted $\varnothing$. A **variable** is a symbol that represents unspecified elements from a **replacement set.** A **constant** is a symbol for a single object. If each element of Set A is also in set B, we say A is a **subset** of B and write $A \subset B$.

Real numbers:

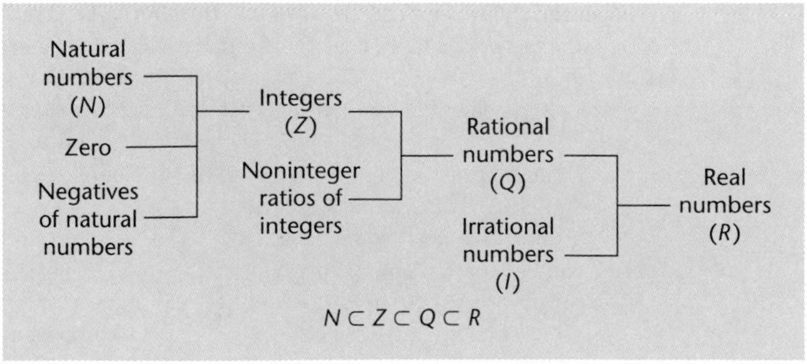

Real number line:

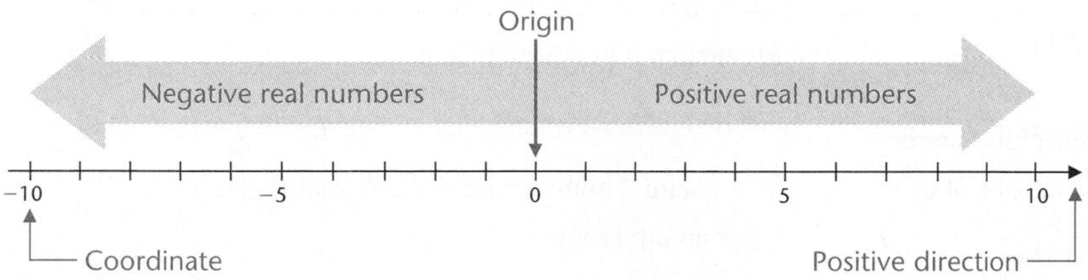

Basic real number properties include **associative properties:** $x + (y + z) = (x + y) + z$ and $x(yz) = (xy)z$; **commutative properties:** $x + y = y + x$ and $xy = yx$; **identities:** $0 + x = x + 0 = x$ and $(1)x = x(1) = x$; **inverses:** $-x$ is the additive inverse or **negative** of x and, if $x \neq 0$, $1/x$ is the multiplicative inverse or **reciprocal** of x; and **distributive property:** $x(y + z) = xy + xz$. **Subtraction** is defined by $a - b = a + (-b)$ and **division** by $a/b = a(1/b)$. Division by 0 is never allowed. Additional properties include **properties of negatives:**

1. $-(-a) = a$
2. $(-a)b = -(ab) = a(-b) = -ab$
3. $(-a)(-b) = ab$
4. $(-1)a = -a$
5. $\frac{-a}{b} = -\frac{a}{b} = \frac{a}{-b} \quad b \neq 0$
6. $\frac{-a}{-b} = -\frac{-a}{b} = -\frac{a}{-b} = \frac{a}{b} \quad b \neq 0$

Zero properties:

1. $a \cdot 0 = 0$.
2. $ab = 0$ if and only if $a = 0$ or $b = 0$ or both.

and **fraction properties** (division by 0 excluded):

1. $\frac{a}{b} = \frac{c}{d}$ if and only if $ad = bc$
2. $\frac{ka}{kb} = \frac{a}{b}$
3. $\frac{a}{b} \cdot \frac{c}{d} = \frac{ac}{bd}$
4. $\frac{a}{b} \div \frac{c}{d} = \frac{a}{b} \cdot \frac{d}{c}$
5. $\frac{a}{b} + \frac{c}{b} = \frac{a + c}{b}$
6. $\frac{a}{b} - \frac{c}{b} = \frac{a - c}{b}$
7. $\frac{a}{b} + \frac{c}{d} = \frac{ad + bc}{bd}$

A-2 POLYNOMIALS: BASIC OPERATIONS

For n and m natural numbers and a any real number:

$$a^n = a \cdot a \cdot \cdots \cdot a \text{ (} n \text{ factors of } a\text{)} \quad \text{and} \quad a^m a^n = a^{m+n}$$

An **algebraic expression** is formed by using constants and variables and the operations of addition, subtraction, multiplication, division, raising to powers, and taking roots. A **polynomial** is an algebraic expression formed by adding and subtracting constants and terms of the form ax^n (one variable), ax^ny^m (two variables), and so on. The **degree of a term** is the sum of the powers of all variables in the term, and the **degree of a polynomial** is the degree of the nonzero term with highest degree in the polynomial. Polynomials with one, two, or three terms are called **monomials, binomials,** and **trinomials,** respectively. **Like terms** have exactly the same variable factors to the same powers and can be combined by adding their **coefficients.** Polynomials can be **added, subtracted,** and **multiplied** by repeatedly applying the distributive property and combining like terms. The **FOIL method** is used to multiply two binomials. **Special products** obtained using FOIL method are

1. $(a - b)(a + b) = a^2 - b^2$
2. $(a - b)^2 = a^2 - 2ab + b^2$
3. $(a + b)^2 = a^2 + 2ab + b^2$

A-3 POLYNOMIALS: FACTORING

A number or algebraic expression is **factored** if it is expressed as a product of other numbers or algebraic expressions, which are called **factors.** An integer greater than 1 is a **prime number** if its only positive integer factors are itself and 1, and a **composite number** otherwise. Each composite number can be **factored uniquely into a product of prime numbers.** A polynomial is **prime** relative to a given set of numbers (usually the set of integers) if (1) all its coefficients are from that set of numbers, and (2) it cannot be written as a product of two polynomials, excluding 1 and itself, having coefficients from that set of numbers. A nonprime polynomial is **factored completely relative to a given set of numbers** if it is written as a product of prime polynomials relative to that set of numbers. **Common factors** can be factored out by applying the distributive properties. **Grouping** can be used to identify common factors.

Second-degree polynomials can be factored by trial and error. The following special factoring formulas are useful:

1. $u^2 + 2uv + v^2 = (u + v)^2$ Perfect Square
2. $u^2 - 2uv + v^2 = (u - v)^2$ Perfect Square
3. $u^2 - v^2 = (u - v)(u + v)$ Difference of Squares
4. $u^3 - v^3 = (u - v)(u^2 + uv + v^2)$ Difference of Cubes
5. $u^3 + v^3 = (u + v)(u^2 - uv + v^2)$ Sum of Cubes

There is no factoring formula relative to the real numbers for $u^2 + v^2$.

A-4 RATIONAL EXPRESSIONS: BASIC OPERATIONS

A **fractional expression** is the ratio of two algebraic expressions, and a **rational expression** is the ratio of two polynomials. The rules for adding, subtracting, multiplying, and dividing real number fractions (see Section A-1 in this review) all extend to fractional expressions with the understanding that **variables are always restricted to exclude division by zero.** Fractions can be **reduced to lowest terms** or **raised to higher terms** by using the fundamental property of fractions:

$$\frac{ka}{kb} = \frac{a}{b} \quad \text{with } b, k \neq 0$$

A rational expression is **reduced to lowest terms** if the numerator and denominator do not have any factors in common relative to the integers. The **least common denominator** (LCD) is useful for adding and subtracting fractions with different denominators and for reducing **compound fractions** to **simple fractions.**

A-5 INTEGER EXPONENTS

$a^n = a \cdot a \cdot \cdots \cdot a$ (n factors of a) for n a positive integer, $a^0 = 1$ ($a \neq 0$), and $a^n = 1/a^{-n}$ for n a negative integer ($a \neq 0$). 0^0 is not defined.

Properties of integer exponents (division by 0 excluded):

1. $a^m a^n = a^{m+n}$
2. $(a^n)^m = a^{mn}$
3. $(ab)^m = a^m b^m$
4. $\left(\dfrac{a}{b}\right)^m = \dfrac{a^m}{b^m}$
5. $\dfrac{a^m}{a^n} = a^{m-n} = \dfrac{1}{a^{n-m}}$

Further exponent properties (division by 0 excluded):

1. $(a^m b^n)^p = a^{pm} b^{pn}$
2. $\left(\dfrac{a^m}{b^n}\right)^p = \dfrac{a^{pm}}{b^{pn}}$
3. $\dfrac{a^{-n}}{b^{-m}} = \dfrac{b^m}{a^n}$
4. $\left(\dfrac{a}{b}\right)^{-n} = \left(\dfrac{b}{a}\right)^n$

Scientific notation:

$$a \times 10^n \qquad 1 \leq a < 10$$

n an integer, a in decimal form.

A-6 RATIONAL EXPONENTS

For n a natural number and a and b real numbers:

a is an nth root of b if $a^n = b$

The **principal nth root** of b is denoted by $b^{1/n}$. If n is odd, b has one real nth root that is the principal nth root. If n is even and $b > 0$, b has two real nth roots and the positive nth root is the principal nth root. If n is even and $b < 0$, b has no real nth roots.

Rational number exponents (even roots of negative numbers excluded):

$$b^{m/n} = (b^{1/n})^m = (b^m)^{1/n} \quad \text{and} \quad b^{-m/n} = \frac{1}{b^{m/n}}$$

A-7 RADICALS

An **nth root radical** is defined by $\sqrt[n]{b} = b^{1/n}$, where $b^{1/n}$ is the principal nth root of b, $\sqrt{\ }$ is a **radical,** n is the **index,** and b is the **radicand.** Rational exponents and radicals are related by

$$b^{m/n} = (b^m)^{1/n} = \sqrt[n]{b^m} = (b^{1/n})^m = (\sqrt[n]{b})^m$$

Properties of radicals ($x > 0$, $y > 0$):

1. $\sqrt[n]{x^n} = x$
2. $\sqrt[n]{xy} = \sqrt[n]{x}\sqrt[n]{y}$
3. $\sqrt[n]{\dfrac{x}{y}} = \dfrac{\sqrt[n]{x}}{\sqrt[n]{y}}$

A radical is in **simplified form** if:

1. No radicand contains a factor to a power greater than or equal to the index of the radical.
2. No power of the radicand and the index of the radical have a common factor other than 1.
3. No radical appears in a denominator.
4. No fraction appears within a radical.

Algebraic fractions containing radicals are **rationalized** by multiplying numerator and denominator by a **rationalizing factor** often determined by using a special product formula.

A-8 LINEAR EQUATIONS AND INEQUALITIES

A **solution** or **root** of an equation is a number in the **domain** or **replacement set** of the variable that when substituted for the variable makes the equation a true statement. Two equations are **equivalent** if they have the same **solution set.** The **properties of equality** are used to solve equations:

1. If $a = b$, then $a + c = b + c$. Addition Property
2. If $a = b$, then $a - c = b - c$. Subtraction Property
3. If $a = b$, then $ca = cb$, $c \neq 0$. Multiplication Property
4. If $a = b$, then $\dfrac{a}{c} = \dfrac{b}{c}$, $c \neq 0$. Division Property

5. If $a = b$, then either may replace the other in any statement without changing the truth or falsity of statement. — Substitution Property

An equation that can be written in the **standard form** $ax + b = 0$, $a \neq 0$, is a **linear** or **first-degree equation.**

The inequality symbols $<, >, \leq, \geq$ are used to express **inequality relations. Line graphs, interval notation,** and the set operations of **union** and **intersection** are used to describe inequality relations. A **solution** of a linear inequality in one variable is a value of the variable that makes the inequality a true statement. Two inequalities are **equivalent** if they have the same **solution set. Inequality properties** are used to solve inequalities:

1. If $a < b$ and $b < c$, then $a < c$. — Transitive Property
2. If $a < b$, then $a + c < b + c$. — Addition Property
3. If $a < b$, then $a - c < b - c$. — Subtraction Property
4. If $a < b$ and $c > 0$, then $ca < cb$.
5. If $a < b$ and $c < 0$, then $ca > cb$. } Multiplication Property
6. If $a < b$ and $c > 0$, then $\frac{a}{c} < \frac{b}{c}$.
7. If $a < b$ and $c < 0$, then $\frac{a}{c} > \frac{b}{c}$. } Division Property

The order of an inequality reverses if we multiply or divide both sides of an inequality statement by a negative number.

APPENDIX A | REVIEW EXERCISES

Work through all the problems in this review and check answers in the back of the book. Answers to all review problems are there, and following each answer is a number in italics indicating the section in which that type of problem is discussed. Where weaknesses show up, review appropriate sections in the text.

A

1. For $A = \{1, 2, 3, 4, 5\}$, $B = \{1, 2, 4\}$, and $C = \{4, 1, 2\}$, indicate true (T) or false (F):
(A) $3 \in A$ (B) $5 \notin C$ (C) $B \in A$
(D) $B \subset A$ (E) $B \neq C$ (F) $A \subset B$

2. Replace each question mark with an appropriate expression that will illustrate the use of the indicated real number property:
(A) Commutative $(\cdot)$: $x(y + z) = ?$
(B) Associative $(+)$: $2 + (x + y) = ?$
(C) Distributive: $(2 + 3)x = ?$

Problems 3–7 refer to the following polynomials:
(a) $3x - 4$ (b) $x + 2$ (c) $3x^2 + x - 8$ (d) $x^3 + 8$

3. Add all four.
4. Subtract the sum of (a) and (c) from the sum of (b) and (d).
5. Multiply (c) and (d).
6. What is the degree of (d)?
7. What is the coefficient of the second term in (c)?

In Problems 8–11, perform the indicated operations and simplify.

8. $5x^2 - 3x[4 - 3(x - 2)]$
9. $(3m - 5n)(3m + 5n)$
10. $(2x + y)(3x - 4y)$
11. $(2a - 3b)^2$

In Problems 12–14, write each polynomial in a completely factored form relative to the integers. If the polynomial is prime relative to the integers, say so.

12. $9x^2 - 12x + 4$
13. $t^2 - 4t - 6$
14. $6n^3 - 9n^2 - 15n$

In Problems 15–18, perform the indicated operations and reduce to lowest terms. Represent all compound fractions as simple fractions reduced to lowest terms.

15. $\dfrac{2}{5b} - \dfrac{4}{3a^3} - \dfrac{1}{6a^2b^2}$
16. $\dfrac{3x}{3x^2 - 12x} + \dfrac{1}{6x}$
17. $\dfrac{y - 2}{y^2 - 4y + 4} \div \dfrac{y^2 + 2y}{y^2 + 4y + 4}$
18. $\dfrac{u - \dfrac{1}{u}}{1 - \dfrac{1}{u^2}}$

Simplify Problems 19–24, and write answers using positive exponents only. All variables represent positive real numbers.

19. $6(xy^3)^5$
20. $\dfrac{9u^8v^6}{3u^4v^8}$
21. $(2 \times 10^5)(3 \times 10^{-3})$
22. $(x^{-3}y^2)^{-2}$
23. $u^{5/3}u^{2/3}$
24. $(9a^4b^{-2})^{1/2}$
25. Change to radical form: $3x^{2/5}$.
26. Change to rational exponent form: $-3\sqrt[3]{(xy)^2}$

Simplify Problems 27–31, and express answers in simplified form. All variables represent positive real numbers.

27. $3x\sqrt[3]{x^5y^4}$
28. $\sqrt{2x^2y^5}\sqrt{18x^3y^2}$
29. $\dfrac{6ab}{\sqrt{3a}}$
30. $\dfrac{\sqrt{5}}{3 - \sqrt{5}}$
31. $\sqrt[8]{y^6}$

Solve Problems 32 and 33 for x.

32. $3x - 2 = x + 6$
33. $4x - 5 < 2x + 7$

B

34. Write using the listing method:
$\{x \mid x \text{ is an odd integer between } -4 \text{ and } 2\}$

In Problems 35–40, each statement illustrates the use of one of the following real number properties or definitions. Indicate which one.

Commutative $(+, \cdot)$	*Identity $(+, \cdot)$*
Division	*Associative $(+, \cdot)$*
Inverse $(+, \cdot)$	*Zero*
Distributive	*Subtraction*
Negatives	

35. $(-3) - (-2) = (-3) + [-(-2)]$

36. $3y + (2x + 5) = (2x + 5) + 3y$

37. $(2x + 3)(3x + 5) = (2x + 3)3x + (2x + 3)5$

38. $3 \cdot (5x) = (3 \cdot 5)x$

39. $\dfrac{a}{-(b - c)} = -\dfrac{a}{b - c}$ **40.** $3xy + 0 = 3xy$

41. Indicate true (T) or false (F):
(A) An integer is a rational number and a real number.
(B) An irrational number has a repeating decimal representation.

42. Give an example of an integer that is not a natural number.

43. Given the algebraic expressions:
(a) $2x^2 - 3x + 5$ (b) $x^2 - \sqrt{x - 3}$
(c) $x^{-3} + x^{-2} - 3x^{-1}$ (d) $x^2 - 3xy - y^2$
(A) Identify all second-degree polynomials.
(B) Identify all third-degree polynomials.

In Problems 44–48, perform the indicated operations and simplify.

44. $(2x - y)(2x + y) - (2x - y)^2$

45. $(m^2 + 2mn - n^2)(m^2 - 2mn - n^2)$

46. $5(x + h)^2 - 7(x + h) - (5x^2 - 7x)$

47. $-2x\{(x^2 + 2)(x - 3) - x[x - x(3 - x)]\}$

48. $(x - 2y)^3$

In Problems 49–55, write in a completely factored form relative to the integers.

49. $(4x - y)^2 - 9x^2$ **50.** $2x^2 + 4xy - 5y^2$

51. $6x^3y + 12x^2y^2 - 15xy^3$ **52.** $(y - b)^2 - y + b$

53. $3x^3 + 24y^3$ **54.** $y^3 + 2y^2 - 4y - 8$

55. $2x(x - 4)^3 + 3x^2(x - 4)^2$

In Problems 56–60, perform the indicated operations and reduce to lowest terms. Represent all compound fractions as simple fractions reduced to lowest terms.

56. $\dfrac{3x^2(x + 2)^2 - 2x(x + 2)^3}{x^4}$

57. $\dfrac{m - 1}{m^2 - 4m + 4} + \dfrac{m + 3}{m^2 - 4} + \dfrac{2}{2 - m}$

58. $\dfrac{y}{x^2} \div \left(\dfrac{x^2 + 3x}{2x^2 + 5x - 3} \div \dfrac{x^3y - x^2y}{2x^2 - 3x + 1}\right)$

59. $\dfrac{1 - \dfrac{1}{1 + \dfrac{x}{y}}}{1 - \dfrac{1}{1 - \dfrac{x}{y}}}$ **60.** $\dfrac{a^{-1} - b^{-1}}{ab^{-2} - ba^{-2}}$

61. Check the following solution. If it is wrong, explain what is wrong and how it can be corrected, and then show a correct solution.

$$\frac{x^2 + 2x}{x^2 + x - 2} + x + 2 = \frac{x^2 + 3x + 2}{x^2 + x - 2} = \frac{x + 1}{x - 1}$$

In Problems 62–67, perform the indicated operations, simplify and write answers using positive exponents only. All variables represent positive real numbers.

62. $\left(\dfrac{8u^{-1}}{2^2u^2v^0}\right)^{-2}\left(\dfrac{u^{-5}}{u^{-3}}\right)^3$ **63.** $\dfrac{5^0}{3^2} + \dfrac{3^{-2}}{2^{-2}}$

64. $\left(\dfrac{27x^2y^{-3}}{8x^{-4}y^3}\right)^{1/3}$ **65.** $(a^{-1/3}b^{1/4})(9a^{1/3}b^{-1/2})^{3/2}$

66. $(x^{1/2} + y^{1/2})^2$ **67.** $(3x^{1/2} - y^{1/2})(2x^{1/2} + 3y^{1/2})$

68. Convert to scientific notation and simplify:

$$\frac{0.000\ 000\ 000\ 52}{(1{,}300)(0.000\ 002)}$$

Evaluate Problems 69–76 to four significant digits using a calculator.

69. $\dfrac{(20{,}410)(0.000\ 003\ 477)}{0.000\ 000\ 022\ 09}$ **70.** 0.1347^5

71. $(-60.39)^{-3}$ **72.** $82.45^{8/3}$

73. $(0.000\ 000\ 419\ 9)^{2/7}$ **74.** $\sqrt[5]{0.006\ 604}$

75. $\sqrt[3]{3 + \sqrt{2}}$ **76.** $\dfrac{2^{-1/2} - 3^{-1/2}}{2^{-1/3} + 3^{-1/3}}$

In Problems 77–85, perform the indicated operations and express answers in simplified form. All radicands represent positive real numbers.

77. $-2x\sqrt[5]{3^6x^7y^{11}}$ **78.** $\dfrac{2x^2}{\sqrt[3]{4x}}$ **79.** $\sqrt[5]{\dfrac{3y^2}{8x^2}}$

80. $\sqrt[9]{8x^6y^{12}}$ **81.** $\sqrt{\sqrt[3]{4x^4}}$

82. $(2\sqrt{x} - 5\sqrt{y})(\sqrt{x} + \sqrt{y})$

83. $\dfrac{3\sqrt{x}}{2\sqrt{x} - \sqrt{y}}$ **84.** $\dfrac{2\sqrt{u} - 3\sqrt{v}}{2\sqrt{u} + 3\sqrt{v}}$ **85.** $\dfrac{y^2}{\sqrt{y^2 + 4} - 2}$

86. Rationalize the numerator: $\dfrac{\sqrt{t} - \sqrt{5}}{t - 5}$.

87. Write in the form $ax^p + bx^q$, where a and b are real numbers and p and q are rational numbers:

$$\frac{4\sqrt{x} - 3}{2\sqrt{x}}$$

Solve Problems 88 and 89 for x.

88. $1.6x + 1 = 1.9 - 0.9x$ **89.** $-3 \le 5 - 2x < 9$

C

90. Write the repeating decimal 0.545454 . . . in the form a/b reduced to lowest terms, where a and b are positive integers. Is the number rational or irrational?

91. If $M = \{-4, -3, 2\}$ and $N = \{-3, 0, 2\}$, find
(A) $\{x \mid x \in M$ or $x \in N\}$
(B) $\{x \mid x \in M$ and $x \in N\}$

92. Evaluate $x^2 - 4x + 1$ for $x = 2 - \sqrt{3}$.

93. Simplify $x(2x - 1)(x + 3) - (x - 1)^3$.

94. Factor completely with respect to the integers.

$$4x(a^2 - 4a + 4) - 9x^3$$

95. Evaluate each expression on a calculator and determine which pairs have the same value. Verify these results algebraically.
(A) $\sqrt{3 + \sqrt{5}} + \sqrt{3 - \sqrt{5}}$
(B) $\sqrt{4 + \sqrt{15}} + \sqrt{4 - \sqrt{15}}$
(C) $\sqrt{10}$

In Problems 96–99, simplify and express answers using positive exponents only (m is an integer greater than 1).

96. $\dfrac{8(x-2)^{-3}(x+3)^2}{12(x-2)^{-4}(x+3)^{-2}}$ **97.** $\left(\dfrac{a^{-2}}{b^{-1}} + \dfrac{b^{-2}}{a^{-1}}\right)^{-1}$

98. $(x^{1/3} - y^{1/3})(x^{2/3} + x^{1/3}y^{1/3} + y^{2/3})$

99. $\left(\dfrac{x^{m^2}}{x^{2m-1}}\right)^{1/(m-1)} \quad m > 1$

100. Rationalize the denominator: $\dfrac{1}{1 - \sqrt[3]{x}}$.

101. Rationalize the numerator: $\dfrac{\sqrt[3]{t} - \sqrt[3]{5}}{t - 5}$.

102. Write in simplified form: $\sqrt[n+1]{x^{n^2}x^{2n+1}} \quad n > 0$

APPLICATIONS

103. Construction. A circular fountain in a park includes a concrete wall that is 3 feet high and 2 feet thick (see the figure). If the inner radius of the wall is x feet, write an algebraic expression in terms of x that represents the volume of the concrete used to construct the wall. Simplify the expression.

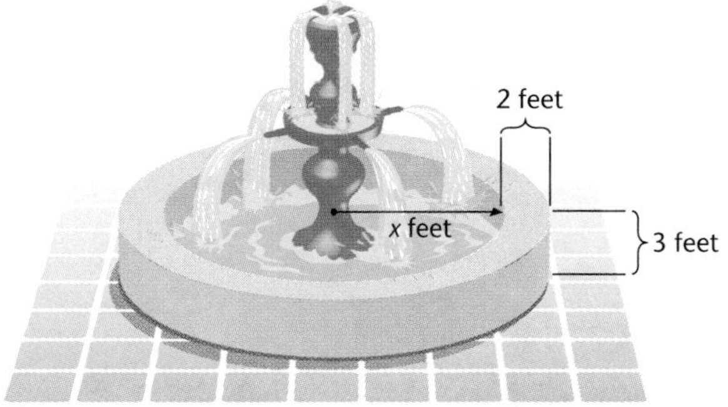

104. Energy Consumption. In 1984 the total amount of energy consumed in the United States was equivalent to 2,257,000,000,000 kilograms of coal (consumption of all forms of energy is expressed in terms of coal for purposes of comparison) and the population was about 235,000,000. Estimate to three significant digits the average energy consumption per person. Write your answer in scientific notation and in standard decimal notation.

105. Economics. The number of units N produced by a petroleum company from the use of x units of capital and y units of labor is approximated by

$$N = 20x^{1/2}y^{1/2}$$

(A) Estimate the number of units produced by using 1,600 units of capital and 900 units of labor.
(B) What is the effect on production if the number of units of capital and labor are doubled to 3,200 units and 1,800 units, respectively?
(C) What is the effect on production of doubling the units of labor and capital at any production level?

106. Electric Circuit. If three electric resistors with resistances R_1, R_2, and R_3 are connected in parallel, then the total resistance R for the circuit shown in the figure is given by

$$R = \frac{1}{\dfrac{1}{R_1} + \dfrac{1}{R_2} + \dfrac{1}{R_3}}$$

Represent this compound fraction as a simple fraction.

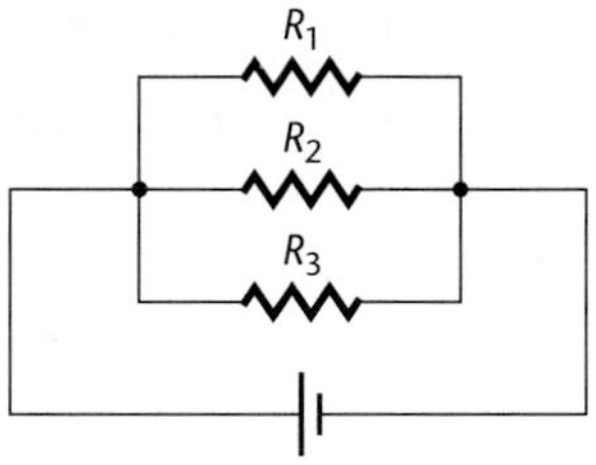

★ **107. Construction.** A box with a hinged lid is to be made out of a piece of cardboard that measures 16 by 30 inches. Six squares, x inches on a side, will be cut from each corner and the middle, and then the ends and sides will be folded up to form the box and its lid (see the figure). Express each of the following quantities as a polynomial in both factored and expanded form.
(A) The area of cardboard after the corners have been removed.
(B) The volume of the box.

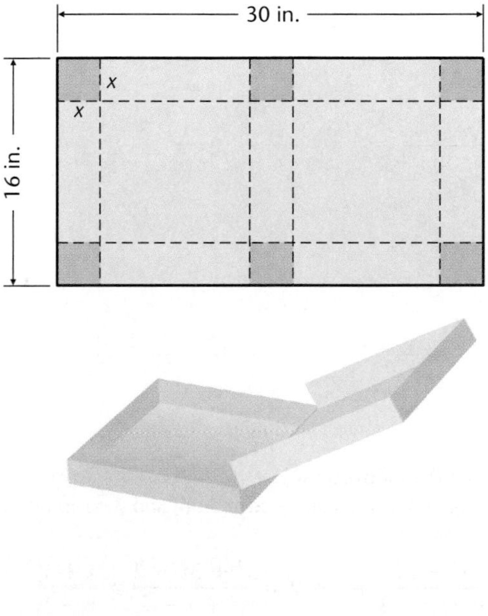

APPENDIX

B

Partial Fractions

Outline

Basic Theorems

Partial Fraction Decomposition

You have now had considerable experience combining two or more rational expressions into a single rational expression. For example, problems such as

$$\frac{2}{x+5} + \frac{3}{x-4} = \frac{2(x-4) + 3(x+5)}{(x+5)(x-4)} = \frac{5x+7}{(x+5)(x-4)}$$

should seem routine. Frequently in more advanced courses, particularly in calculus, it is advantageous to be able to reverse this process—that is, to be able to express a rational expression as the sum of two or more simpler rational expressions called **partial fractions.** As is often the case with reverse processes, the process of decomposing a rational expression into partial fractions is more difficult than combining rational expressions. Basic to the process is the factoring of polynomials, so the topics discussed in Section 3-2 can be put to effective use.

We confine our attention to rational expressions of the form $P(x)/D(x)$, where $P(x)$ and $D(x)$ are polynomials with real coefficients. In addition, we assume that the degree of $P(x)$ is less than the degree of $D(x)$. If the degree of $P(x)$ is greater than or equal to that of $D(x)$, we have only to divide $P(x)$ by $D(x)$ to obtain

$$\frac{P(x)}{D(x)} = Q(x) + \frac{R(x)}{D(x)}$$

where the degree of $R(x)$ is less than that of $D(x)$. For example,

$$\frac{x^4 - 3x^3 + 2x^2 - 5x + 1}{x^2 - 2x + 1} = x^2 - x - 1 + \frac{-6x + 2}{x^2 - 2x + 1}$$

If the degree of $P(x)$ is less than that of $D(x)$, then $P(x)/D(x)$ is called a **proper fraction.**

Basic Theorems

Our task now is to establish a systematic way to decompose a proper fraction into the sum of two or more partial fractions. The following three theorems take care of the problem completely. Theorems 1 and 3 are stated without proof.

THEOREM 1

EQUAL POLYNOMIALS

Two polynomials are equal to each other if and only if the coefficients of terms of like degree are equal.

For example, if

Equate the constant terms.

$$(A + 2B)x + B = 5x - 3$$

Equate the coefficients of x.

then

$$B = -3 \quad \text{Substitute } B = -3 \text{ into the second equation to solve for } A.$$

$$A + 2B = 5$$

$$A + 2(-3) = 5$$

$$A = 11$$

Explore/Discuss 1

If

$$x + 5 = A(x + 1) + B(x - 3) \tag{1}$$

is a polynomial identity (that is, both sides represent the same polynomial), then equating coefficients produces the system

$$1 = A + B \quad \text{Equating coefficients of } x$$

$$5 = A - 3B \quad \text{Equating constant terms}$$

(A) Solve this system graphically.

(B) For an alternate method of solution, substitute $x = 3$ in equation (1) to find A and then substitute $x = -1$ in equation (1) to find B. Explain why this method is valid.

THEOREM 2

LINEAR AND QUADRATIC FACTOR THEOREM

For a polynomial with real coefficients, there always exists a complete factoring involving only linear and/or quadratic factors with real coefficients where the linear and quadratic factors are prime relative to the real numbers.

That Theorem 2 is true can be seen as follows: From earlier theorems in Chapter 3, we know that an nth-degree polynomial $P(x)$ has n zeros and n linear factors. The real zeros of $P(x)$ correspond to linear factors of the form $(x - r)$, where r is a real number. Since $P(x)$ has real coefficients, the imaginary zeros occur in conjugate pairs. Thus, the imaginary zeros correspond to pairs of factors of the form $[x - (a + bi)]$ and $[x - (a - bi)]$, where a and b are real numbers. Multiplying these two imaginary factors, we have

$$[x - (a + bi)][x - (a - bi)] = x^2 - 2ax + a^2 + b^2$$

This quadratic polynomial with real coefficients is a factor of $P(x)$. Thus, $P(x)$ can be factored into a product of linear factors and quadratic factors, all with real coefficients.

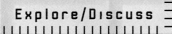

Partial Fraction Decomposition

We are now ready to state Theorem 3, which forms the basis for partial fraction decomposition.

THEOREM 3

PARTIAL FRACTION DECOMPOSITION

Any proper fraction $P(x)/D(x)$ reduced to lowest terms can be decomposed into the sum of partial fractions as follows:

1. If $D(x)$ has a nonrepeating linear factor of the form $ax + b$, then the partial fraction decomposition of $P(x)/D(x)$ contains a term of the form

$$\frac{A}{ax + b} \qquad A \text{ a constant}$$

2. If $D(x)$ has a k-repeating linear factor of the form $(ax + b)^k$, then the partial fraction decomposition of $P(x)/D(x)$ contains k terms of the form

$$\frac{A_1}{ax + b} + \frac{A_2}{(ax + b)^2} + \cdots + \frac{A_k}{(ax + b)^k} \qquad A_1, A_2, \ldots, A_k \text{ constants}$$

3. If $D(x)$ has a nonrepeating quadratic factor of the form $ax^2 + bx + c$, which is prime relative to the real numbers, then the partial fraction decomposition of $P(x)/D(x)$ contains a term of the form

$$\frac{Ax + B}{ax^2 + bx + c} \qquad A, B \text{ constants}$$

4. If $D(x)$ has a k-repeating quadratic factor of the form $(ax^2 + bx + c)^k$, where $ax^2 + bx + c$ is prime relative to the real numbers, then the partial fraction decomposition of $P(x)/D(x)$ contains k terms of the form

$$\frac{A_1x + B_1}{ax^2 + bx + c} + \frac{A_2x + B_2}{(ax^2 + bx + c)^2} + \cdots + \frac{A_kx + B_k}{(ax^2 + bx + c)^k}$$

$$A_1, \ldots, A_k, B_1, \ldots, B_k \text{ constants}$$

Let's see how the theorem is used to obtain partial fraction decompositions in several examples.

EXAMPLE 1 Nonrepeating Linear Factors

Decompose into partial fractions: $\dfrac{5x + 7}{x^2 + 2x - 3}$.

Solution We first try to factor the denominator. If it can't be factored in the real numbers, then we can't go any further. In this example, the denominator factors, so we apply part 1 from Theorem 3:

$$\frac{5x + 7}{(x - 1)(x + 3)} = \frac{A}{x - 1} + \frac{B}{x + 3} \tag{2}$$

To find the constants A and B, we combine the fractions on the right side of equation (2) to obtain

$$\frac{5x + 7}{(x - 1)(x + 3)} = \frac{A(x + 3) + B(x - 1)}{(x - 1)(x + 3)}$$

Since these fractions have the same denominator, their numerators must be equal. Thus

$$5x + 7 = A(x + 3) + B(x - 1) \tag{3}$$

We could multiply the right side and find A and B by using Theorem 1, but in this case it is easier to take advantage of the fact that equation (3) is an identity—that is, it must hold for all values of x. In particular, we note that if we let $x = 1$, then the second term of the right side drops out and we can solve for A:

$$5 \cdot 1 + 7 = A(1 + 3) + B(1 - 1)$$
$$12 = 4A$$
$$A = 3$$

Similarly, if we let $x = -3$, the first term drops out and we find

$$-8 = -4B$$
$$B = 2$$

Hence,

$$\frac{5x + 7}{x^2 + 2x - 3} = \frac{3}{x - 1} + \frac{2}{x + 3} \tag{4}$$

as can easily be checked by adding the two fractions on the right.

MATCHED PROBLEM 1 Decompose into partial fractions: $\dfrac{7x + 6}{x^2 + x - 6}$.

Explore/Discuss 2

A graphing utility can also be used to check a partial fraction decomposition. To check Example 1, we graph the left and right sides of equation (4) in a graphing utility (Fig. 1). Discuss how the TRACE feature on the graphing utility can be used to check that the graphing utility is displaying two identical graphs.

FIGURE 1

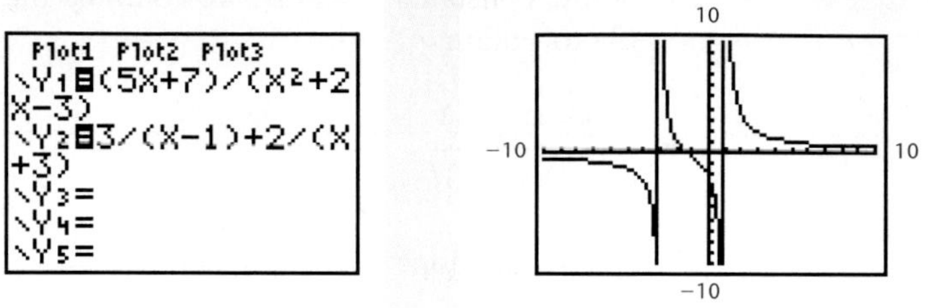

EXAMPLE 2 Repeating Linear Factors

Decompose into partial fractions: $\dfrac{6x^2 - 14x - 27}{(x + 2)(x - 3)^2}$.

Solution Using parts 1 and 2 from Theorem 3, we write

$$\frac{6x^2 - 14x - 27}{(x + 2)(x - 3)^2} = \frac{A}{x + 2} + \frac{B}{x - 3} + \frac{C}{(x - 3)^2}$$

$$= \frac{A(x - 3)^2 + B(x + 2)(x - 3) + C(x + 2)}{(x + 2)(x - 3)^2}$$

Thus, for all x,

$$6x^2 - 14x - 27 = A(x - 3)^2 + B(x + 2)(x - 3) + C(x + 2)$$

If $x = 3$, then If $x = -2$, then

$$-15 = 5C \qquad 25 = 25A$$
$$C = -3 \qquad A = 1$$

There are no other values of x that will cause terms on the right to drop out. Since any value of x can be substituted to produce an equation relating A, B, and C, we let $x = 0$ and obtain

$$-27 = 9A - 6B + 2C \qquad \text{Substitute } A = 1 \text{ and } C = -3.$$
$$-27 = 9 - 6B - 6$$
$$B = 5$$

Thus,

$$\frac{6x^2 - 14x - 27}{(x + 2)(x - 3)^2} = \frac{1}{x + 2} + \frac{5}{x - 3} - \frac{3}{(x - 3)^2}$$

MATCHED PROBLEM 2

Decompose into partial fractions: $\dfrac{x^2 + 11x + 15}{(x - 1)(x + 2)^2}$.

EXAMPLE 3

Nonrepeating Linear and Quadratic Factors

Decompose into partial fractions: $\dfrac{5x^2 - 8x + 5}{(x - 2)(x^2 - x + 1)}$.

Solution

First, we see that the quadratic in the denominator can't be factored further in the real numbers. Then, we use parts 1 and 3 from Theorem 3 to write

$$\frac{5x^2 - 8x + 5}{(x - 2)(x^2 - x + 1)} = \frac{A}{x - 2} + \frac{Bx + C}{x^2 - x + 1}$$

$$= \frac{A(x^2 - x + 1) + (Bx + C)(x - 2)}{(x - 2)(x^2 - x + 1)}$$

Thus, for all x,

$$5x^2 - 8x + 5 = A(x^2 - x + 1) + (Bx + C)(x - 2)$$

If $x = 2$, then

$$9 = 3A$$

$$A = 3$$

If $x = 0$, then, using $A = 3$, we have

$$5 = 3 - 2C$$

$$C = -1$$

If $x = 1$, then, using $A = 3$ and $C = -1$, we have

$$2 = 3 + (B - 1)(-1)$$

$$B = 2$$

Hence,

$$\frac{5x^2 - 8x + 5}{(x - 2)(x^2 - x + 1)} = \frac{3}{x - 2} + \frac{2x - 1}{x^2 - x + 1}$$

MATCHED PROBLEM 3

Decompose into partial fractions: $\dfrac{7x^2 - 11x + 6}{(x - 1)(2x^2 - 3x + 2)}$.

EXAMPLE 4

Repeating Quadratic Factors

Decompose into partial fractions: $\dfrac{x^3 - 4x^2 + 9x - 5}{(x^2 - 2x + 3)^2}$.

Solution

Since $x^2 - 2x + 3$ can't be factored further in the real numbers, we proceed to use part 4 from Theorem 3 to write

$$\frac{x^3 - 4x^2 + 9x - 5}{(x^2 - 2x + 3)^2} = \frac{Ax + B}{x^2 - 2x + 3} + \frac{Cx + D}{(x^2 - 2x + 3)^2}$$

$$= \frac{(Ax + B)(x^2 - 2x + 3) + Cx + D}{(x^2 - 2x + 3)^2}$$

Thus, for all x,

$$x^3 - 4x^2 + 9x - 5 = (Ax + B)(x^2 - 2x + 3) + Cx + D$$

Since the substitution of carefully chosen values of x doesn't lead to the immediate determination of A, B, C, or D, we multiply and rearrange the right side to obtain

$$x^3 - 4x^2 + 9x - 5 = Ax^3 + (B - 2A)x^2 + (3A - 2B + C)x + (3B + D)$$

Now we use Theorem 1 to equate coefficients of terms of like degree:

$$A = 1$$
$$B - 2A = -4$$
$$3A - 2B + C = 9$$
$$3B + D = -5$$

$$1x^3 \qquad -4x^2 \qquad +9x \qquad -5$$
$$Ax^3 + (B - 2A)x^2 + (3A - 2B + C)x + (3B + D)$$

From these equations we easily find that $A = 1$, $B = -2$, $C = 2$, and $D = 1$. Now we can write

$$\frac{x^3 - 4x^2 + 9x - 5}{(x^2 - 2x + 3)^2} = \frac{x - 2}{x^2 - 2x + 3} + \frac{2x + 1}{(x^2 - 2x + 3)^2}$$

MATCHED PROBLEM 4

Decompose into partial fractions: $\dfrac{3x^3 - 6x^2 + 7x - 2}{(x^2 - 2x + 2)^2}$.

Answers to Matched Problems

1. $\dfrac{4}{x - 2} + \dfrac{3}{x + 3}$ **2.** $\dfrac{3}{x - 1} - \dfrac{2}{x + 2} + \dfrac{1}{(x + 2)^2}$ **3.** $\dfrac{2}{x - 1} + \dfrac{3x - 2}{2x^2 - 3x + 2}$ **4.** $\dfrac{3x}{x^2 - 2x + 2} + \dfrac{x - 2}{(x^2 - 2x + 2)^2}$

EXERCISE B

A

In Problems 1–10, find constants A, B, C, and D so that the right side is equal to the left.

1. $\dfrac{7x-14}{(x-4)(x+3)}=\dfrac{A}{x-4}+\dfrac{B}{x+3}$

2. $\dfrac{9x+21}{(x+5)(x-3)}=\dfrac{A}{x+5}+\dfrac{B}{x-3}$

3. $\dfrac{17x-1}{(2x-3)(3x-1)}=\dfrac{A}{2x-3}+\dfrac{B}{3x-1}$

4. $\dfrac{x-11}{(3x+2)(2x-1)}=\dfrac{A}{3x+2}+\dfrac{B}{2x-1}$

5. $\dfrac{3x^2+7x+1}{x(x+1)^2}=\dfrac{A}{x}+\dfrac{B}{x+1}+\dfrac{C}{(x+1)^2}$

6. $\dfrac{x^2-6x+11}{(x+1)(x-2)^2}=\dfrac{A}{x+1}+\dfrac{B}{x-2}+\dfrac{C}{(x-2)^2}$

7. $\dfrac{3x^2+x}{(x-2)(x^2+3)}=\dfrac{A}{x-2}+\dfrac{Bx+C}{x^2+3}$

8. $\dfrac{5x^2-9x+19}{(x-4)(x^2+5)}=\dfrac{A}{x-4}+\dfrac{Bx+C}{x^2+5}$

9. $\dfrac{2x^2+4x-1}{(x^2+x+1)^2}=\dfrac{Ax+B}{x^2+x+1}+\dfrac{Cx+D}{(x^2+x+1)^2}$

10. $\dfrac{3x^3-3x^2+10x-4}{(x^2-x+3)^2}=\dfrac{Ax+B}{x^2-x+3}+\dfrac{Cx+D}{(x^2-x+3)^2}$

B

In Problems 11–22, decompose into partial fractions.

11. $\dfrac{-x+22}{x^2-2x-8}$

12. $\dfrac{-x-21}{x^2+2x-15}$

13. $\dfrac{3x-13}{6x^2-x-12}$

14. $\dfrac{11x-11}{6x^2+7x-3}$

15. $\dfrac{x^2-12x+18}{x^3-6x^2+9x}$

16. $\dfrac{5x^2-36x+48}{x(x-4)^2}$

17. $\dfrac{5x^2+3x+6}{x^3+2x^2+3x}$

18. $\dfrac{6x^2-15x+16}{x^3-3x^2+4x}$

19. $\dfrac{2x^3+7x+5}{x^4+4x^2+4}$

20. $\dfrac{-5x^2+7x-18}{x^4+6x^2+9}$

21. $\dfrac{x^3-7x^2+17x-17}{x^2-5x+6}$

22. $\dfrac{x^3+x^2-13x+11}{x^2+2x-15}$

C

In Problems 23–30, decompose into partial fractions.

23. $\dfrac{4x^2+5x-9}{x^3-6x-9}$

24. $\dfrac{4x^2-8x+1}{x^3-x+6}$

25. $\dfrac{x^2+16x+18}{x^3+2x^2-15x-36}$

26. $\dfrac{5x^2-18x+1}{x^3-x^2-8x+12}$

27. $\dfrac{-x^2+x-7}{x^4-5x^3+9x^2-8x+4}$

28. $\dfrac{-2x^3+12x^2-20x-10}{x^4-7x^3+17x^2-21x+18}$

29. $\dfrac{4x^5+12x^4-x^3+7x^2-4x+2}{4x^4+4x^3-5x^2+5x-2}$

30. $\dfrac{6x^5-13x^4+x^3-8x^2+2x}{6x^4-7x^3+x^2+x-1}$

APPENDIX C

Significant Digits

Most calculations involving problems of the real world deal with figures that are only approximate. It therefore seems reasonable to assume that a final answer should not be any more accurate than the least accurate figure used in the calculation. This is an important point, since calculators tend to give the impression that greater accuracy is achieved than is warranted.

Suppose we wish to compute the length of the diagonal of a rectangular field from measurements of its sides of 237.8 meters and 61.3 meters. Using the Pythagorean theorem and a calculator, we find

$$\begin{aligned} d &= \sqrt{237.8^2 + 61.3^2} \\ &= 245.573\ 878\ldots \end{aligned}$$

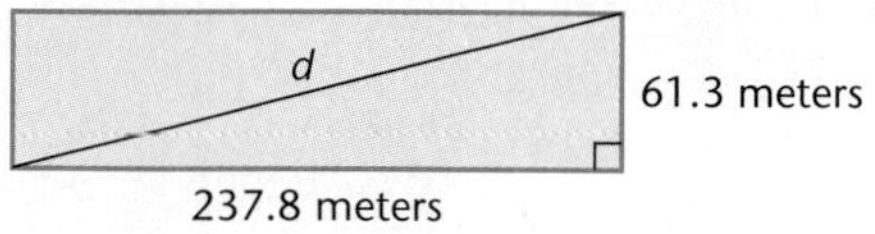

The calculator answer suggests an accuracy that is not justified. What accuracy is justified? To answer this question, we introduce the idea of *significant digits*.

Whenever we write a measurement such as 61.3 meters, we assume that the measurement is accurate to the last digit written. Thus, the measurement 61.3 meters indicates that the measurement was made to the nearest tenth of a meter. That is, the actual width is between 61.25 meters and 61.35 meters. In general, the digits in a number that indicate the accuracy of the number are called **significant digits.** If all the digits in a number are nonzero, then they are all significant. Thus, the measurement 61.3 meters has three significant digits, and the measurement 237.8 meters has four significant digits.

What are the significant digits in the number 7,800? The accuracy of this number is not clear. It could represent a measurement with any of the following accuracies:

Between 7,750 and 7,850	Correct to the hundreds place
Between 7,795 and 7,805	Correct to the tens place
Between 7,799.5 and 7,800.5	Correct to the units place

In order to give a precise definition of significant digits that resolves this ambiguity, we use scientific notation.

DEFINITION 1

SIGNIFICANT DIGITS

If a number x is written in scientific notation as

$$x = a \times 10^n \qquad 1 \le a < 10,\ n \text{ an integer}$$

then the number of significant digits in x is the number of digits in a.

Thus,

7.8×10^3	has two significant digits
7.80×10^3	has three significant digits
7.800×10^3	has four significant digits

All three of these measurements have the same decimal representation (7,800), but each represents a different accuracy.

Definition 1 tells us how to write a number so that the number of significant digits is clear, but it does not tell us how to interpret the accuracy of a number that is not written in scientific notation. We will use the following convention for numbers that are written as decimal fractions:

SIGNIFICANT DIGITS IN DECIMAL FRACTIONS

The number of significant digits in a number with no decimal point is found by counting the digits from left to right, starting with the first digit and ending with the last *nonzero* digit.

The number of significant digits in a number containing a decimal point is found by counting the digits from left to right, starting with the first *nonzero* digit and ending with the last digit.

Applying this rule to the number 7,800, we conclude that this number has two significant digits. If we want to indicate that it has three or four significant digits, we must use scientific notation. The significant digits in the following numbers are underlined:

$\underline{70{,}007}$ $\underline{82}{,}000$ $\underline{5.600}$ $0.000\underline{8}$ $0.000\ \underline{830}$

In calculations involving multiplication, division, powers, and roots, we adopt the following convention:

ROUNDING CALCULATED VALUES

The result of a calculation is rounded to the same number of significant digits as the number used in the calculation that has the least number of significant digits.

Thus, in computing the length of the diagonal of the rectangular field shown earlier, we write the answer rounded to three significant digits because the width has three significant digits and the length has four significant digits:

$d = 246$ meters Three significant digits

One Final Note: In rounding a number that is exactly halfway between a larger and a smaller number, we use the convention of making the final result even.

EXAMPLE 1

Rounding Numbers

Round each number to three significant digits.

(A) 43.0690 (B) 48.05 (C) 48.15 (D) $8.017\ 632 \times 10^{-3}$

Solutions

(A) 43.1

(B) 48.0 } Use the convention of making the digit before the

(C) 48.2 } 5 even if it is odd, or leaving it alone if it is even.

(D) 8.02×10^{-3}

MATCHED PROBLEM 1

Round each number to three significant digits.

(A) 3.1495 (B) 0.004 135 (C) 32,450 (D) $4.314\,764\,09 \times 10^{12}$

Answers to Matched Problems

1. (A) 3.15 (B) 0.004 14 (C) 32,400 (D) 4.31×10^{12}

Circle

R = Radius

D = Diameter

$D = 2R$

$A = \pi R^2 = \frac{1}{4}\pi D^2$ Area

$C = 2\pi R = \pi D$ Circumference

$\frac{C}{D} = \pi$ For all circles

$\pi \approx 3.141\ 59$

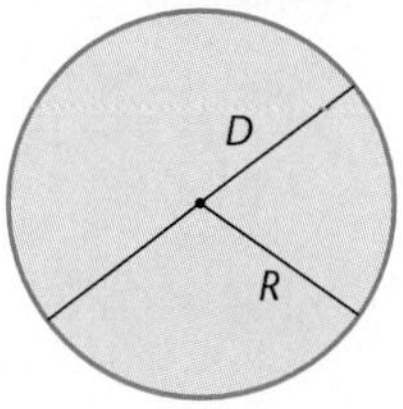

Rectangular Solid

$V = abc$ Volume

$T = 2ab + 2ac + 2bc$ Total surface area

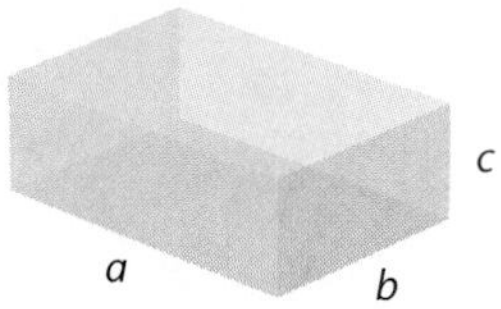

Right Circular Cylinder

R = Radius of base

h = Height

$V = \pi R^2 h$ Volume

$S = 2\pi Rh$ Lateral surface area

$T = 2\pi R(R + h)$ Total surface area

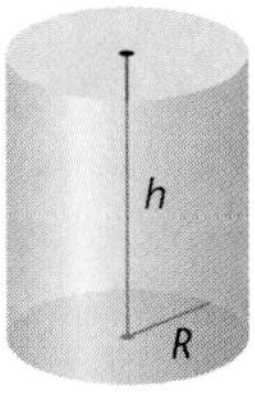

Right Circular Cone

R = Radius of base

h = Height

s = Slant height

$V = \frac{1}{3}\pi R^2 h$ Volume

$S = \pi Rs = \pi R\sqrt{R^2 + h^2}$ Lateral surface area

$T = \pi R(R + s) = \pi R(R + \sqrt{R^2 + h^2})$ Total surface area

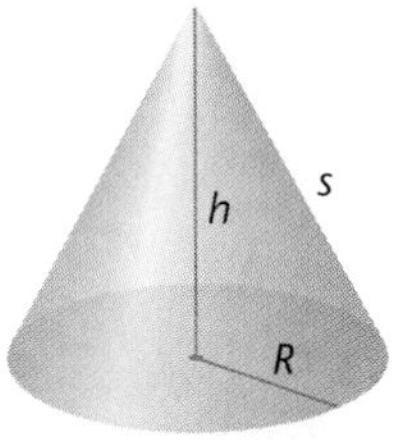

Sphere

R = Radius

D = Diameter

$D = 2R$

$V = \frac{4}{3}\pi R^3 = \frac{1}{6}\pi D^3$ Volume

$S = 4\pi R^2 = \pi D^2$ Surface area

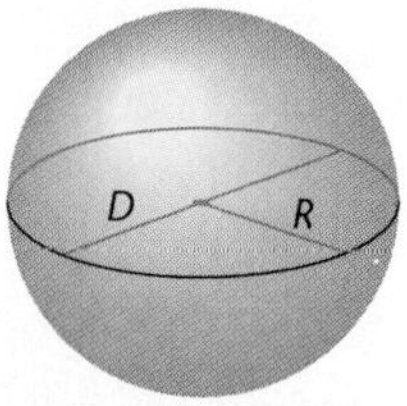

Answers

CHAPTER 1 Exercise 1-1

1.

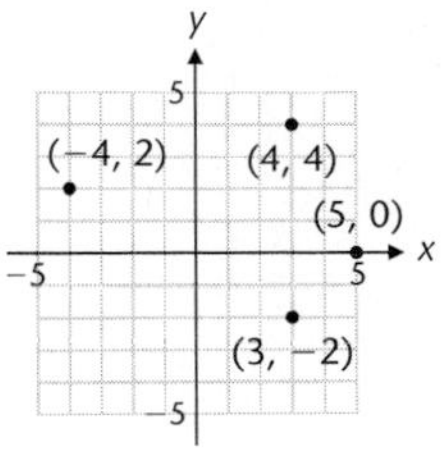

3.

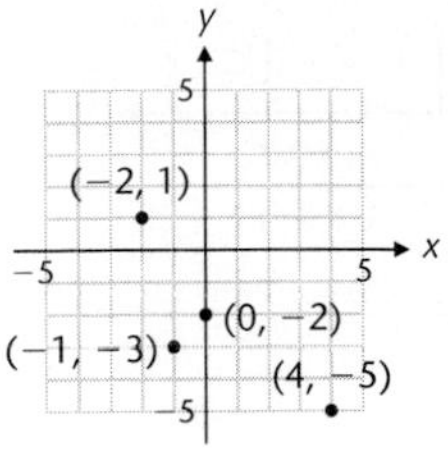

5. $A(2, 4)$, $B(3, -1)$, $C(-4, 0)$, $D(-5, 2)$

7. $A(-3, -3)$, $B(0, 4)$, $C(-3, 2)$, $D(5, -1)$ **9.** $\sqrt{145}$ **11.** $\sqrt{68}$ **13.** $x^2 + y^2 = 49$ **15.** $(x - 2)^2 + (y - 3)^2 = 36$

17. $(x + 4)^2 + (y - 1)^2 = 7$ **19.** $(x + 3)^2 + (y + 4)^2 = 2$

21.

x	y
−3	−2
−2	−1
−1	0
0	1
1	2
2	3
3	4

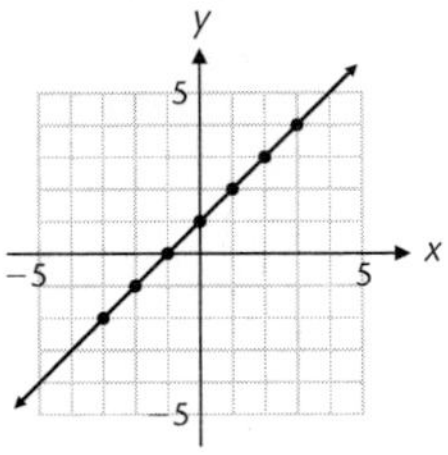

23.

x	y
−3	4
−2	−1
−1	−4
0	−5
1	−4
2	−1
3	4

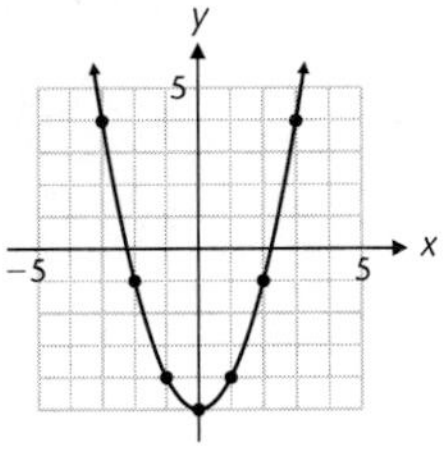

25.

x	y
−3	−4.5
−2	−1
−1	1.5
0	3
1	3.5
2	3
3	1.5

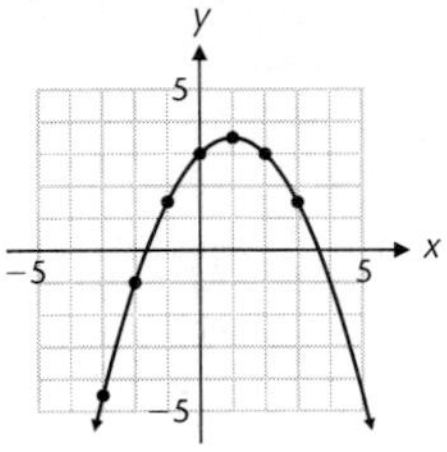

27. (A) 6 (B) −5 (C) −1 (D) 8 (E) −5 (F) 5

29. (A) 6 (B) 4 (C) 4 (D) 8 (E) −8, 0, 6 (F) −7, −2, 7 **31.** A right triangle **33.** 18.11

35.

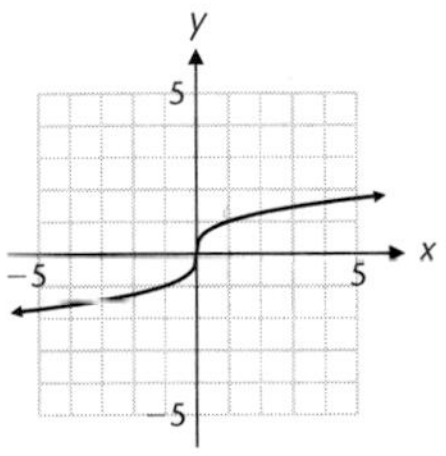

37.

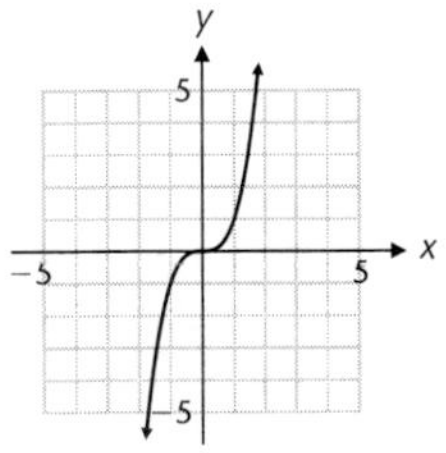

39.

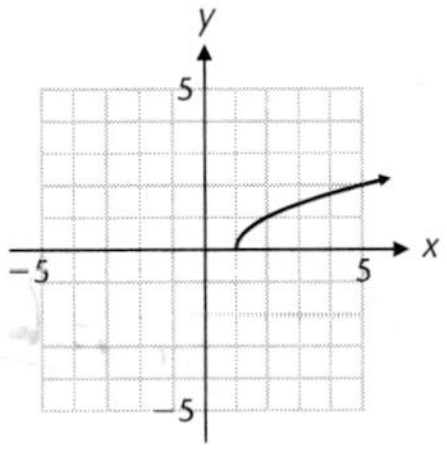

41.

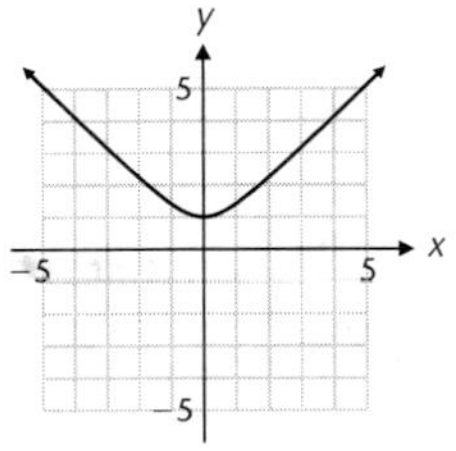

43.

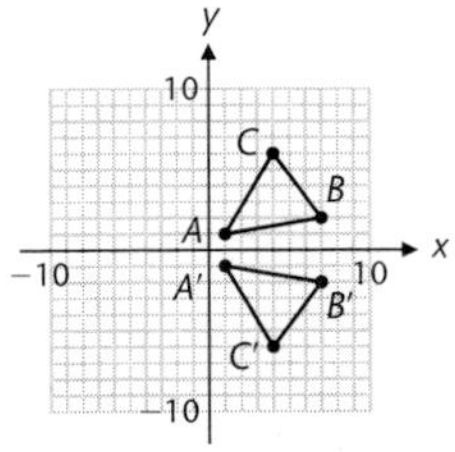

45.

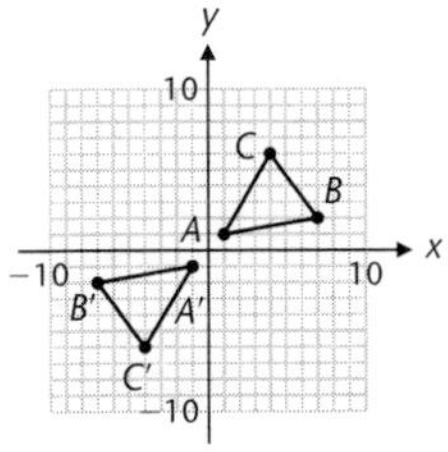

49. $(x - 4)^2 + (y - 2)^2 = 34$ **51.** $(x - 2)^2 + (y - 2)^2 = 50$

53. (A) 3,000 cases (B) Demand decreases by 400 cases (C) Demand increases by 600 cases

55. (A) 53°F (B) 68°F at 3 P.M. (C) 1 A.M., 7 A.M., 11 P.M.

57.

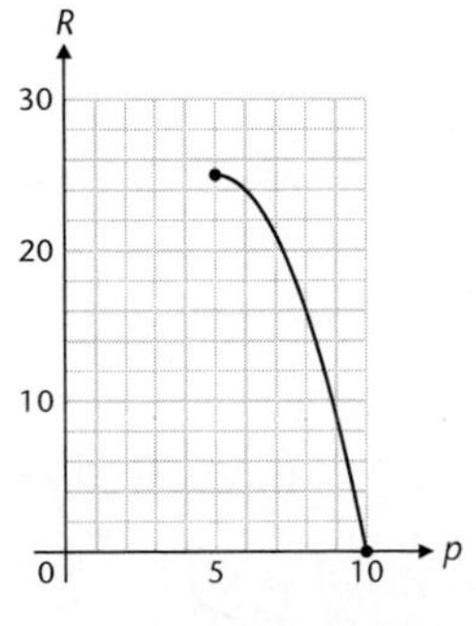

59.

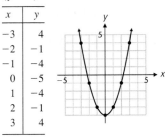

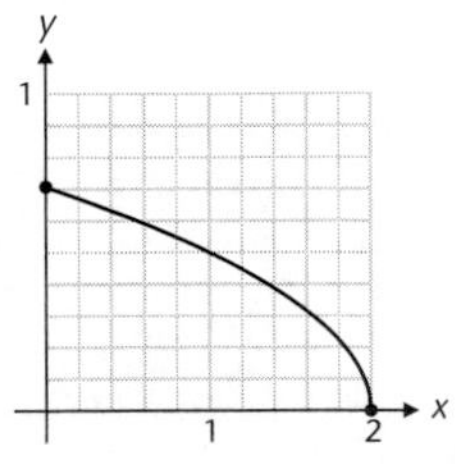

Exercise 1-2

1. Yes **3.** No **5.** Yes **7.** (A) Xmin = −7, Xmax = 6, Ymin = −9, Ymax = 14

9.

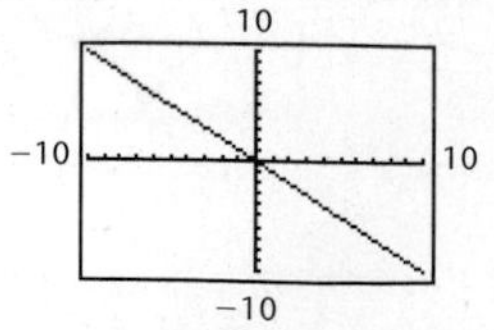

11.

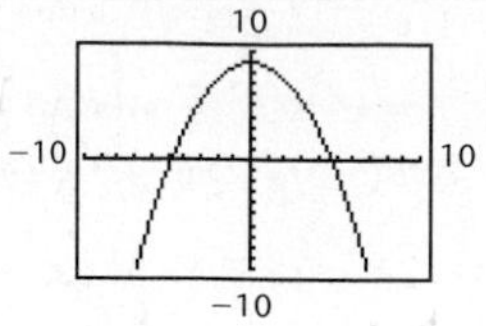

13.

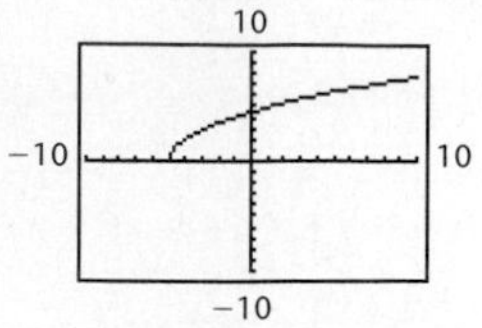

15.

x	−2	0	2	4	6
y	−8	4	8	4	−8

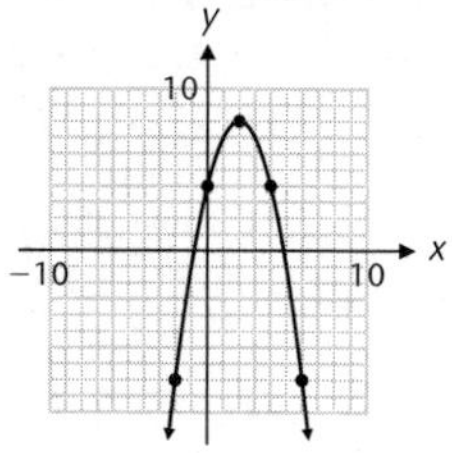

17.

x	−5	−3	−1	1	3	5
y	0	4	5.7	6.9	8	8.9

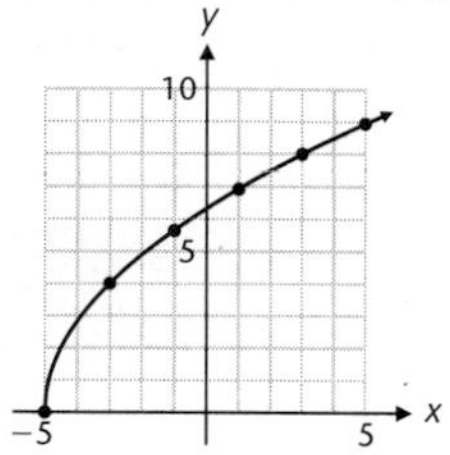

19.

x	−3	−1	1	3	5
y	10.5	−2.5	4.5	7.5	−17.5

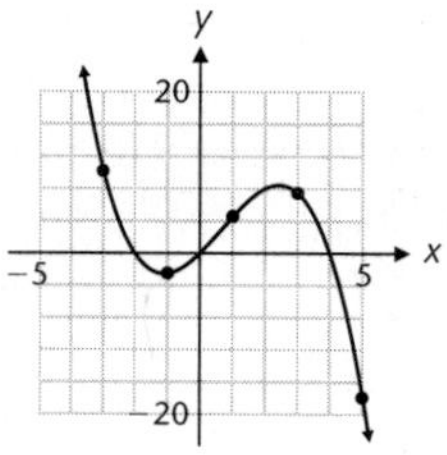

21. (A) −6.37 (B) 0.63 (C) −1.63

23. (A) 0.92 (B) −3.93 (C) −2.09

25.

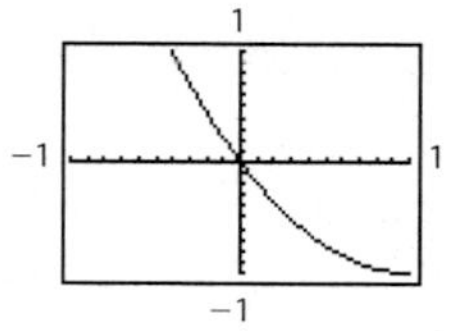

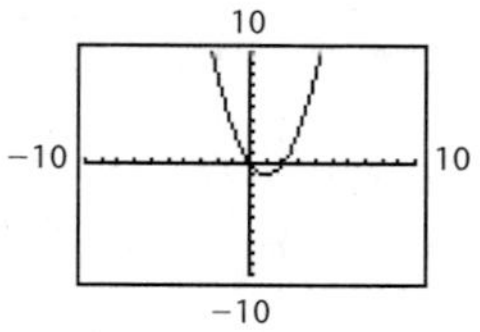

Best view

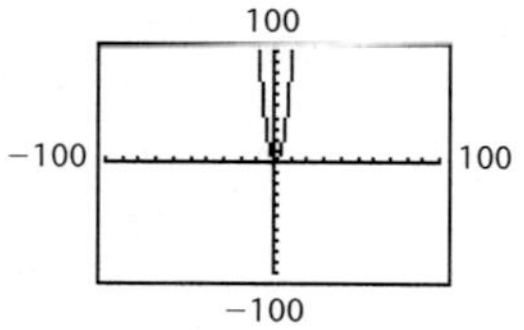

27.

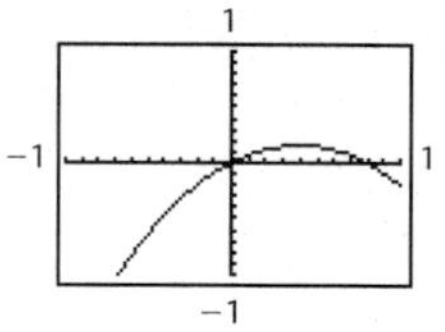

Best view

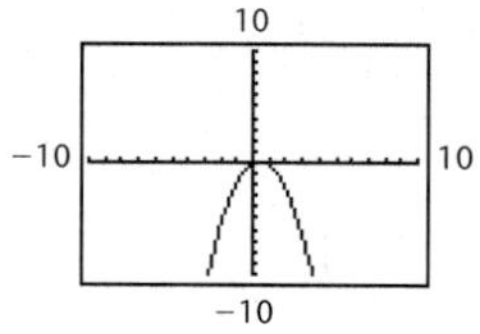

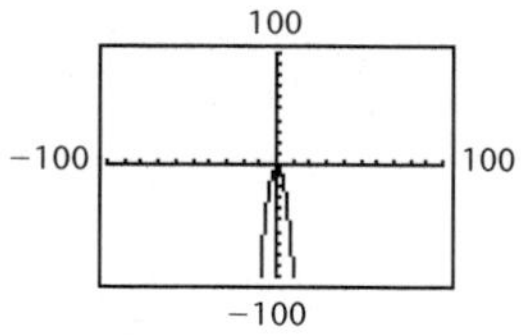

29. −0.57, 0.76, 2.31 **31.** ±6.32, ±2.24

33. 1.4142 **39.** 9.5 sec

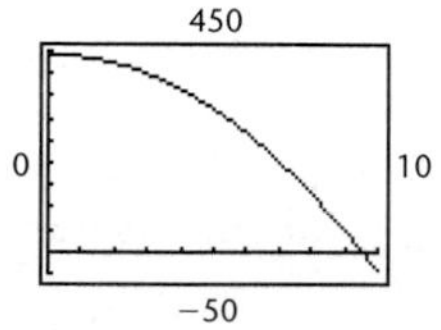

41. A 0.89 in. square or a 2.40 in. square can be cut out. Dimensions for smaller square: 0.89 in. × 9.23 in. × 6.73 in.; dimensions for larger square: 2.40 in. × 6.20 in. × 3.70 in.

43. A 0.93 in. square or a 3.92 in. square can be cut out. Dimensions for smaller square: 0.93 in. × 10.14 in. × 10.61 in.; dimensions for larger square: 3.92 in. × 4.16 in. × 6.12 in.

45. (A)

x	17,800	15,600	13,600
y	20	25	30

(B) Demand decreases 2,000 cases (C) Demand increases 2,200 cases

47. (A)

y	20	25	30
R	356,000	390,000	408,000

(B) Revenue increases $18,000 (C) Revenue decreases $34,000

(D) The company should raise the price $5 to increase the revenue.

Exercise 1-3

1. A function **3.** Not a function **5.** A function **7.** A function; domain = {2, 3, 4, 5}; range = {4, 6, 8, 10} **9.** Not a function **11.** A function; domain = {0, 1, 2, 3, 4, 5}; range = {1, 2} **13.** A function **15.** Not a function **17.** Not a function **19.** -8 **21.** -6 **23.** 1 **25.** 10 **27.** $-\frac{30}{17}$ **29.** 3 **31.** Domain: all real numbers or $(-\infty, \infty)$ **33.** Domain: all real numbers except 4 or $(-\infty, 4) \cup (4, \infty)$ **35.** Domain: $x \geq 0$ or $[0, \infty)$ **37.** Domain: $x \leq 0$ or $(-\infty, 0]$ **39.** Domain: all real numbers except -1 and 1 or $(-\infty, -1) \cup (-1, 1) \cup (1, \infty)$ **41.** Domain: $x \geq 0$, $x \neq 5$ or $[0, 5) \cup (5, \infty)$ **43.** $f(x) = 2x - 3$ **45.** $f(x) = 4x^2 - 2x + 9$ **47.** 3 **49.** $-6 - h$ **51.** $11 - 2h$ **53.** $g(x) = 3x + 1$ **55.** $F(x) = \dfrac{x}{8 + \sqrt{x}}$ **57.** Function f multiplies the domain element by 2 and then subtracts the product of 3 and the square of the domain element. **59.** Function F takes the square root of the sum of the fourth power of the domain element and 9. **61.** $f(x) = 2x^2 - 4x + 6$ **63.** $m(x) = 4x - 3\sqrt{x} + 9$ **65.** (A) 3 (B) 3 **67.** (A) $2x + h$ (B) $x + a$ **69.** (A) $-6x - 3h + 9$ (B) $-3x - 3a + 9$ **71.** (A) $3x^2 + 3xh + h^2$ (B) $x^2 + ax + a^2$ **73.** 2 **75.** (A) $s(0) = 0$, $s(1) = 16$, $s(2) = 64$, $s(3) = 144$ (B) $64 + 16h$ (C) Value of expression tends to 64; this number appears to be the speed of the object at the end of 2 s.

77. (A)

x	0	5,000	10,000	15,000	20,000	25,000	30,000
$B(x)$	212	203	194	185	176	167	158

(B) The boiling point drops 9°F for each 5,000 foot increase in altitude.

79. The rental charges are \$20 per day plus \$0.25 per mile driven.

81. (A)

t	0	1	2	3	4
Sales	5.9	6.5	7.7	8.6	9.7
$S(t)$	5.7	6.7	7.7	8.6	9.6

(B)

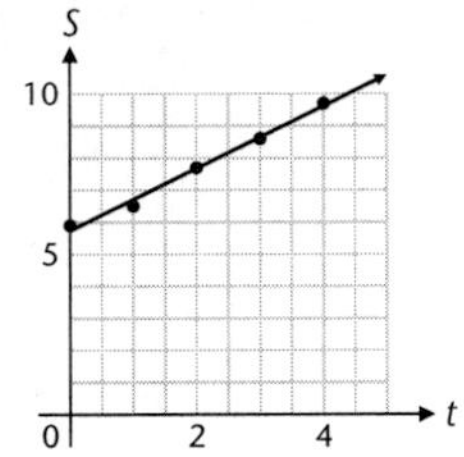

(C) \$10.6 billion, \$17.4 billion

83. (A)

r (R&D)	0.66	0.75	0.85	0.99	1.1
Sales	5.9	6.5	7.7	8.6	9.7
$S(r)$	5.9	6.6	7.5	8.7	9.7

(B)

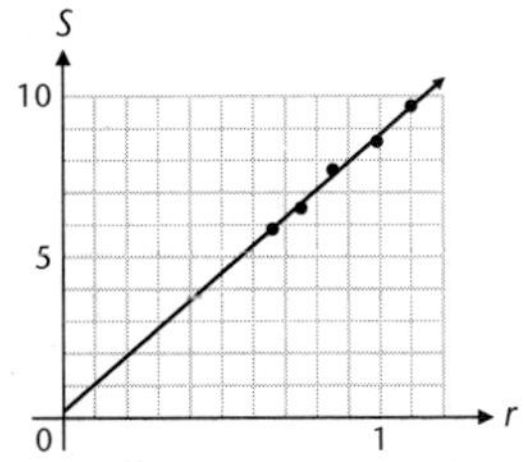

(C) \$13.1 billion, \$17.4 billion

Exercise 1-4

1. (A) $[-4, 4)$ (B) $[-3, 3)$ (C) 0 (D) 0 (E) $[-4, 4)$ (F) None (G) None (H) None **3.** (A) $(-\infty, \infty)$ (B) $[-4, \infty)$ (C) -3, 1 (D) -3 (E) $[-1, \infty)$ (F) $(-\infty, -1]$ (G) None (H) None **5.** (A) $(-\infty, 2) \cup (2, \infty)$ (The function is not defined at $x = 2$.) (B) $(-\infty, -1) \cup [1, \infty)$ (C) None (D) 1 (E) None (F) $(-\infty, -2] \cup (2, \infty)$ (G) $[-2, 2)$ (H) $x = 2$ **7.** Increasing: $[-2, 10]$; decreasing: $[-10, -2]$ **9.** Increasing: $[-3, 10]$; constant: $[-10, -3]$ **11.** Decreasing: $[-4, 3]$; constant: $[-10, -4]$, $[3, 10]$ **13.** Increasing: $[1, 10]$; decreasing: $[-10, -3]$; constant: $[-3, 1]$ **15.** Increasing: $[-4, 0]$, $[4, 10]$; decreasing: $[-10, -4]$, $[0, 4]$

17. One possible answer:

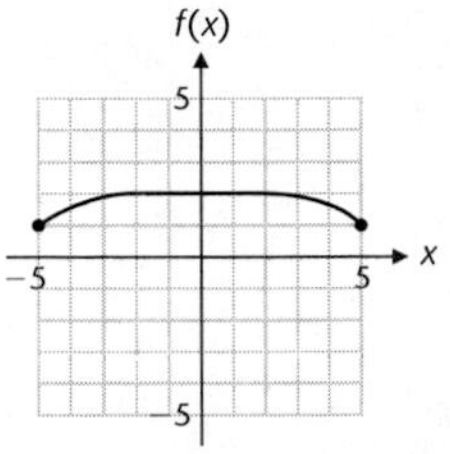

19. One possible answer:

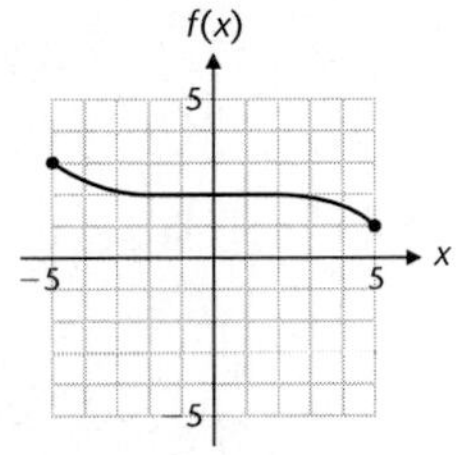

21. One possible answer:

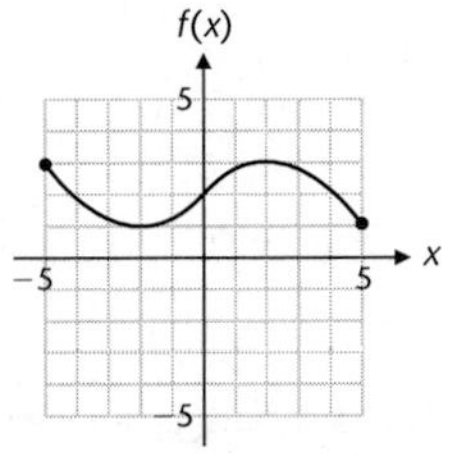

23. $f(3.6) \approx -5.9$ is a local minimum; x intercepts: 0.7, 6.5 **25.** $f(3.1) \approx 6.1$ is a local maximum; x intercepts: 0, 12.25 **27.** $f(0.8) \approx -7.1$ is a local minimum; x intercepts: -0.7, 3.1

29. Domain: $[-1, 1]$; range: $[0, 1]$

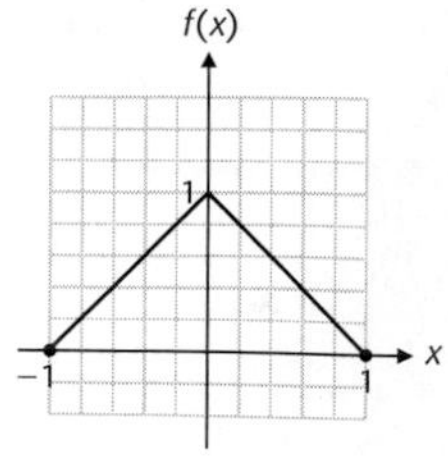

31. Domain: $[-3, -1) \cup (-1, 2]$; range: $\{-2, 4\}$ (a set, not an interval); discontinuous at $x = -1$

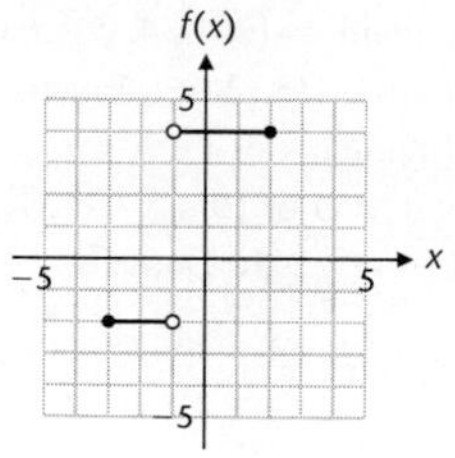

33. Domain: all real numbers; range: all real numbers; discontinuous at $x = -1$

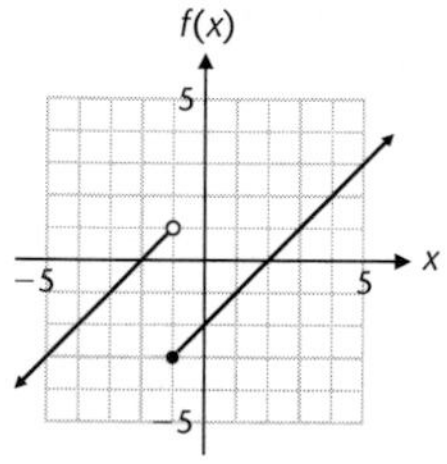

35. Domain: $(-\infty, 0) \cup (0, \infty)$; discontinuous at $x = 0$; range: $(-\infty, -1) \cup (1, \infty)$

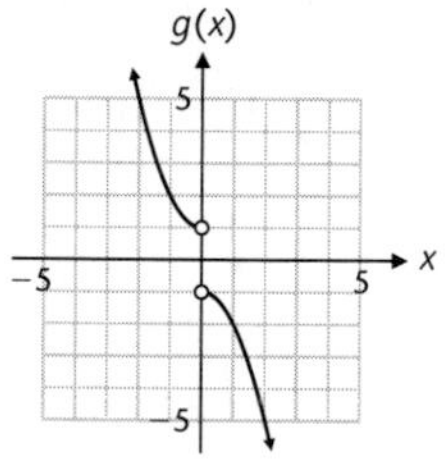

37. The graph of f decreases on $[-10, -2.15]$ to a local minimum value, $f(-2.15) \approx -36.62$, and then increases on $[-2.15, 10]$.

39. The graph of h increases on $[-10, -4.64]$ to a local maximum value, $h(-4.64) \approx 281.93$, decreases on $[-4.64, 5.31]$ to a local minimum value, $h(5.31) \approx -211.41$, and then increases on $[5.31, 10]$.

41. The graph of p decreases on $[-10, -3.77]$ to a local minimum value, $p(-3.77) \approx 0$, increases on $[-3.77, 0.50]$ to a local maximum value, $p(0.50) = 18.25$, decreases on $[0.50, 4.77]$ to a local minimum value, $p(4.77) \approx 0$, and then increases on $[4.77, 10]$.

43. One possible answer:

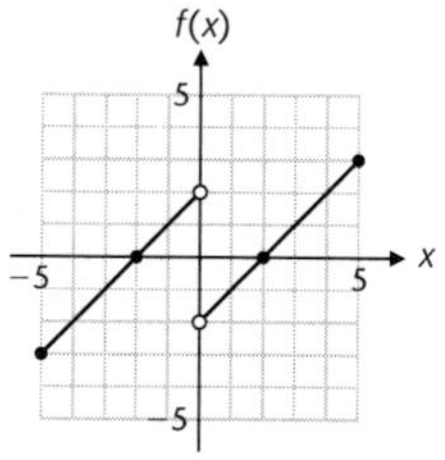

45. One possible answer:

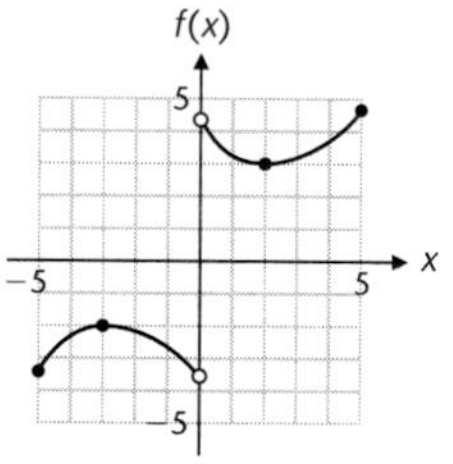

47. One possible answer:

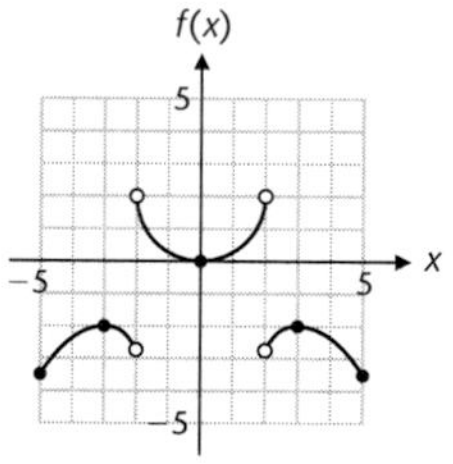

49. Domain: all real numbers except $x = 2$; range: $\{-5, 5\}$ (a set, not an interval); discontinuous at $x = 2$

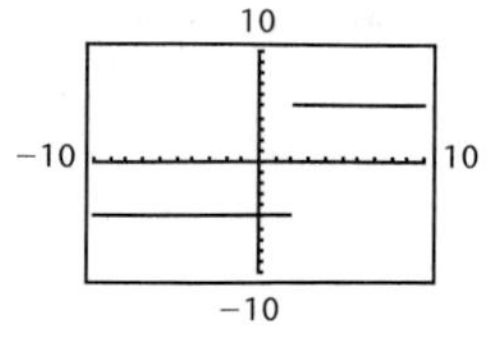

51. Domain: all real numbers except $x = 1$; range: $(-\infty, -3) \cup (5, \infty)$; discontinuous at $x = 1$

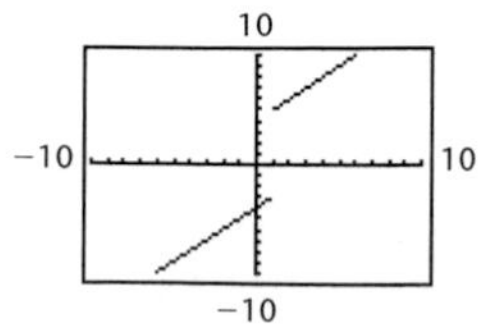

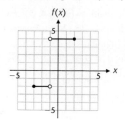

53. Domain: all real numbers except $x = 3$; range: $(0, \infty)$; discontinuous at $x = 3$

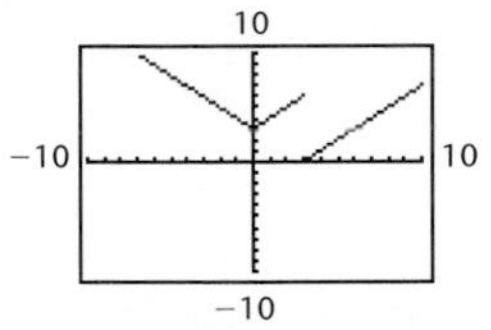

55. Domain: all real numbers; range: all integers; discontinuous at the even integers

$$f(x) = \begin{cases} \vdots & & \vdots \\ -2 & \text{if} & -4 \le x < -2 \\ -1 & \text{if} & -2 \le x < 0 \\ 0 & \text{if} & 0 \le x < 2 \\ 1 & \text{if} & 2 \le x < 4 \\ 2 & \text{if} & 4 \le x < 6 \\ \vdots & & \vdots \end{cases}$$

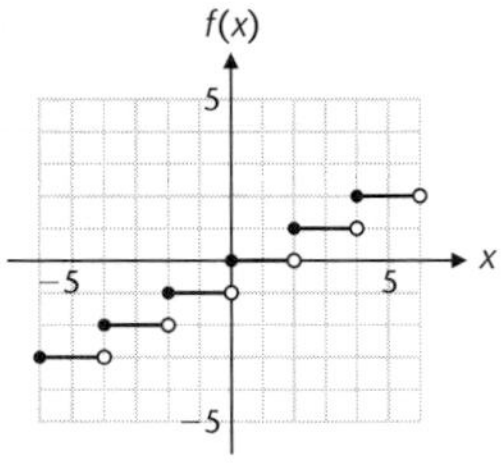

57. Domain: all real numbers; range: all integers; discontinuous at rational numbers of the form $\frac{k}{3}$, where k is an integer.

$$f(x) = \begin{cases} \vdots & & \vdots \\ -2 & \text{if} & -\frac{2}{3} \le x < -\frac{1}{3} \\ -1 & \text{if} & -\frac{1}{3} \le x < 0 \\ 0 & \text{if} & 0 \le x < \frac{1}{3} \\ 1 & \text{if} & \frac{1}{3} \le x < \frac{2}{3} \\ 2 & \text{if} & \frac{2}{3} \le x < 1 \\ \vdots & & \vdots \end{cases}$$

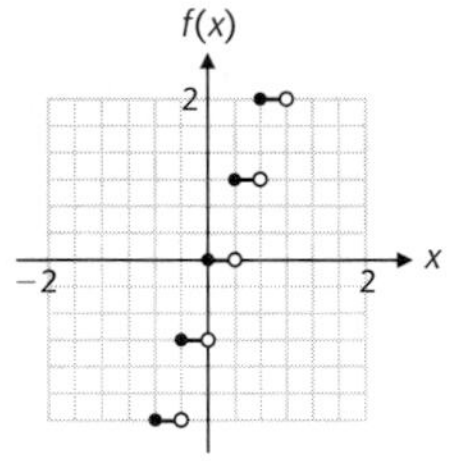

59. Domain: all real numbers; range: $[0, 1)$; discontinuous at all integers

$$f(x) = \begin{cases} \vdots & & \vdots \\ x + 2 & \text{if} & -2 \le x < -1 \\ x + 1 & \text{if} & -1 \le x < 0 \\ x & \text{if} & 0 \le x < 1 \\ x - 1 & \text{if} & 1 \le x < 2 \\ x - 2 & \text{if} & 2 \le x < 3 \\ \vdots & & \vdots \end{cases}$$

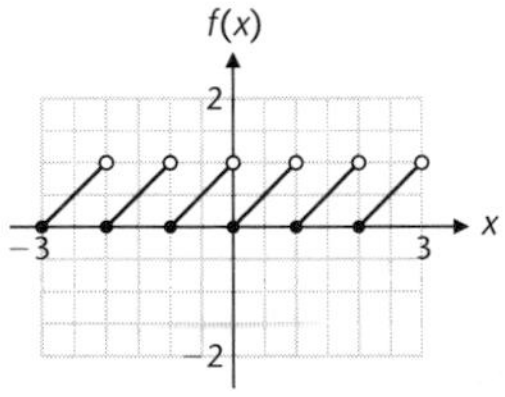

61. (A) One possible answer:

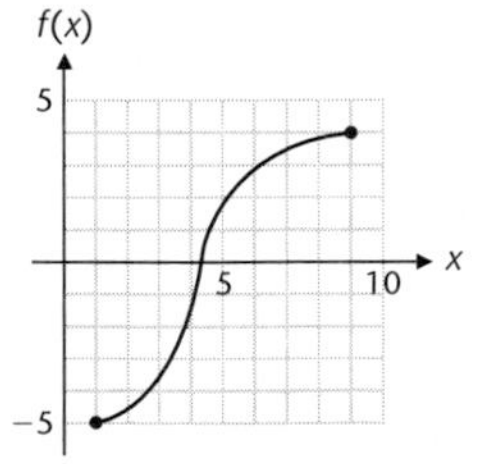

(B) The graph must cross the x axis exactly once.

63. (A) One possible answer:

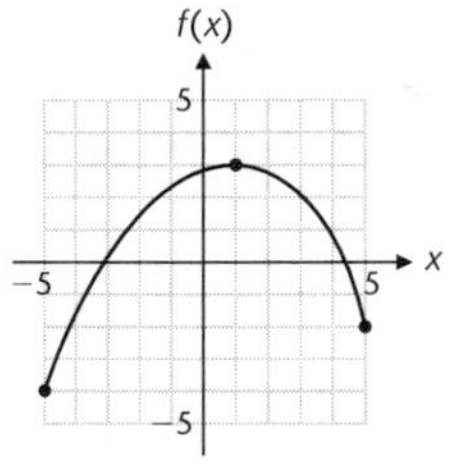

(B) The graph must cross the x axis at least twice. There is no upper limit on the number of times it can cross the x axis.

65. (A) One possible answer:

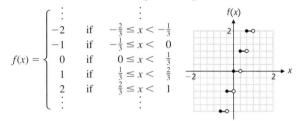

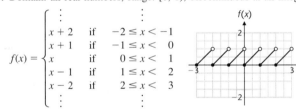

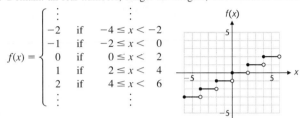

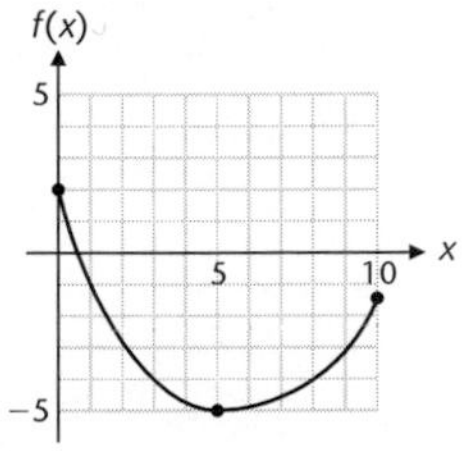

(B) The graph can cross the axis 0, 1, or 2 times.

67. (A) $C(x) = \begin{cases} 15 & 0 < x \le 1 \\ 18 & 1 < x \le 2 \\ 21 & 2 < x \le 3 \\ 24 & 3 < x \le 4 \\ 27 & 4 < x \le 5 \\ 30 & 5 < x \le 6 \end{cases}$

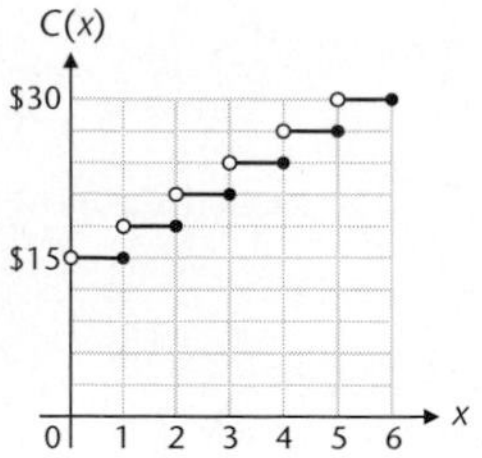

(B) No, since $f(x) \ne C(x)$ at $x = 1, 2, 3, 4, 5$, or 6

69. $E(x) = \begin{cases} 200 & \text{if} \quad 0 \le x \le 3{,}000 \\ 80 + 0.04x & \text{if} \quad 3{,}000 < x < 8{,}000 \\ 180 + 0.04x & \text{if} \quad 8{,}000 \le x \end{cases}$

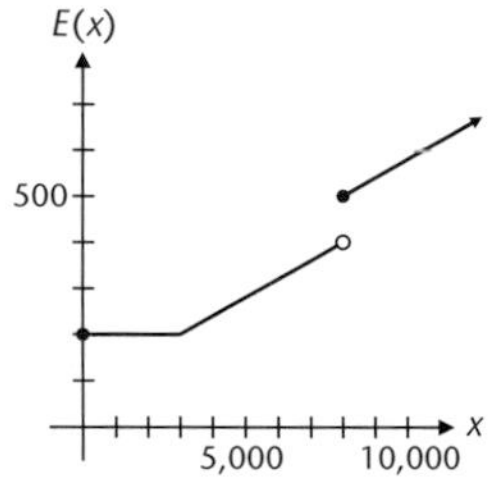

$E(5{,}750) = \$310$; $E(9{,}200) = \$548$ discontinuous at $x = 8{,}000$

71.

$$\left.\begin{array}{rlll} f(4) &= 10[\![0.5 + 0.4]\!] &= 10(0) &= 0 \\ f(-4) &= 10[\![0.5 - 0.4]\!] &= 10(0) &= 0 \\ f(6) &= 10[\![0.5 + 0.6]\!] &= 10(1) &= 10 \\ f(-6) &= 10[\![0.5 - 0.6]\!] &= 10(-1) &= -10 \\ f(24) &= 10[\![0.5 + 2.4]\!] &= 10(2) &= 20 \\ f(25) &= 10[\![0.5 + 2.5]\!] &= 10(3) &= 30 \\ f(247) &= 10[\![0.5 + 24.7]\!] &= 10(25) &= 250 \\ f(-243) &= 10[\![0.5 - 24.3]\!] &= 10(-24) &= -240 \\ f(-245) &= 10[\![0.5 - 24.5]\!] &= 10(-24) &= -240 \\ f(-246) &= 10[\![0.5 - 24.6]\!] &= 10(-25) &= -250 \end{array}\right\} f \text{ rounds numbers to the tens place}$$

73. $f(x) = [\![100x + 0.5]\!]/100$

75. (A) Estimated maximum revenue is \$25,650 when 900 car seats are sold.
(B) The maximum revenue is \$25,714.28 when 857 car seats are sold.

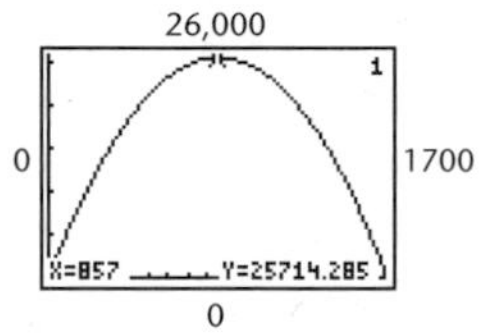

77. (A) The estimated maximum volume is 648 in.^{3} when the side of the cutout square is 3 in.
(B) The maximum volume is approximately 654.98 in.^{3} when the side of the cutout square is approximately 3.39 in.

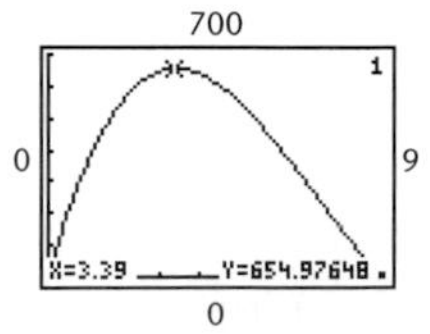

79. (A) The estimated minimal cost is \$301,000 when the land portion of the pipe is 15 miles long.
(B) The minimal cost is approximately \$300,000 when the land portion of the pipe is approximately 13.6 miles long.

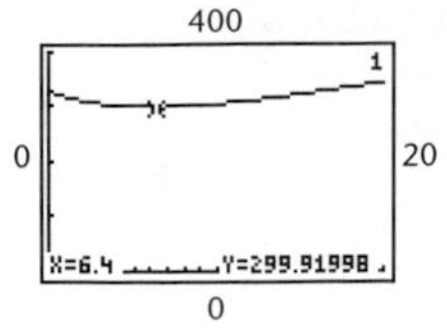

81. (A)

x	28	30	32	34	36
Mileage	45	52	55	51	47
$f(x)$	45.3	51.8	54.2	52.4	46.5

(B)

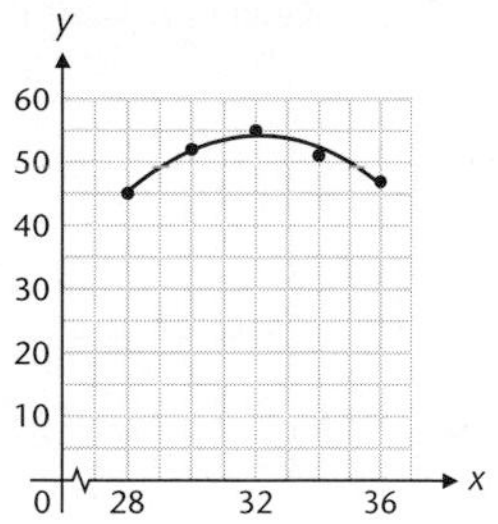

(C) $f(31) \approx 53.50$ thousand miles; $f(35) \approx 49.95$ thousand miles

Exercise 1-5

1. Odd **3.** Even **5.** Neither **7.** Even **9.** Neither **11.**

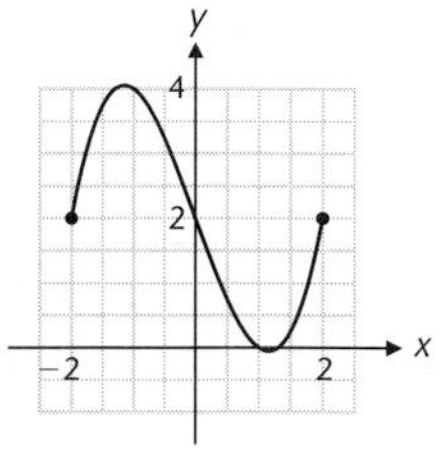

13.

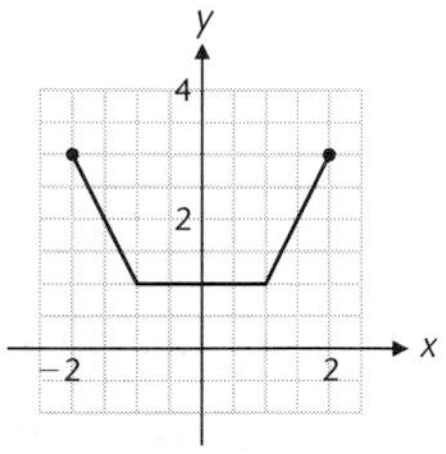

15.

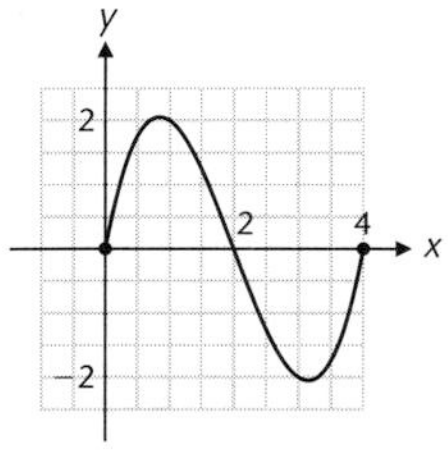

17.

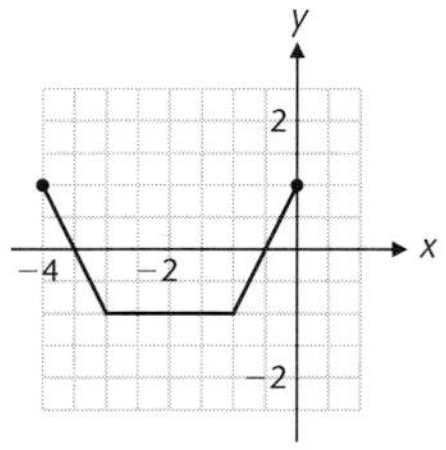

19.

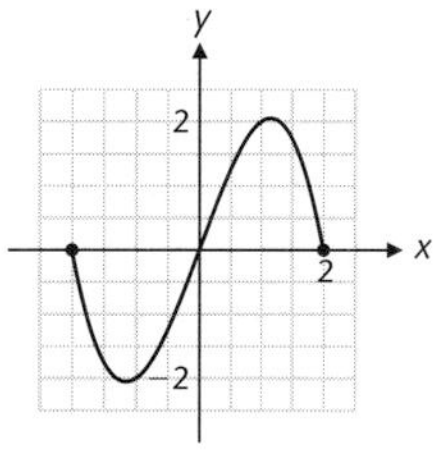

21.

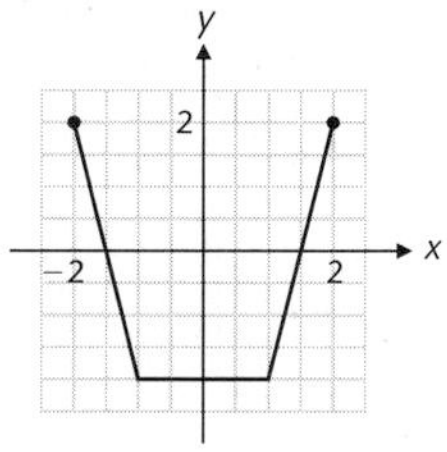

23. The graph of $y = x^2$ is shifted 2 units to the right; $y = (x - 2)^2$. **25.** The graph of $y = x^3$ is shifted down 2 units; $y = x^3 - 2$.
27. The graph of $y = |x|$ is contracted by a factor of 0.25; $y = 0.25|x|$.
29. The graph of $y = x^3$ is reflected in the x axis (or the y axis); $y = -x^3$. **31.** $g(x) = \sqrt[3]{x + 4} - 5$ **33.** $g(x) = -0.5(6 + \sqrt{x})$
35. $g(x) = -2(x + 4)^2 - 2$ **37.** The graph of $y = x^2$ is shifted 7 units left and 9 units up.
39. The graph of $y = |x|$ is shifted 8 units right and reflected in the x axis.
41. The graph of $y = \sqrt{x}$ is reflected in the x axis and shifted 3 units up.
43. The graph of $y = x^2$ is expanded by a factor of 4 and reflected in the x axis. **45.** $y = |x + 2| + 2$
47. $y = 4 - \sqrt{x}$ **49.** $y = 4 - (x - 1)^2$ **51.** $y = 0.5(x - 3)^3 + 1$ **53.** Reversing the order does not change the result.
55. Reversing the order can change the result. **57.** Reversing the order can change the result.

59.

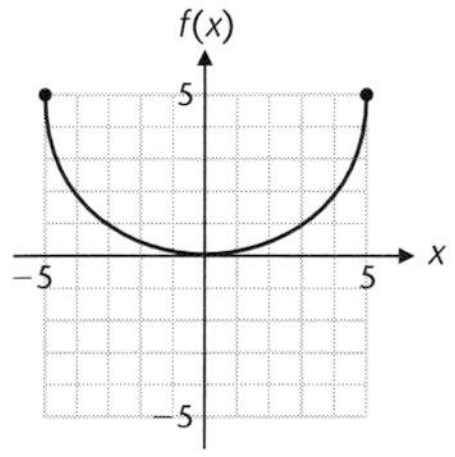

61.

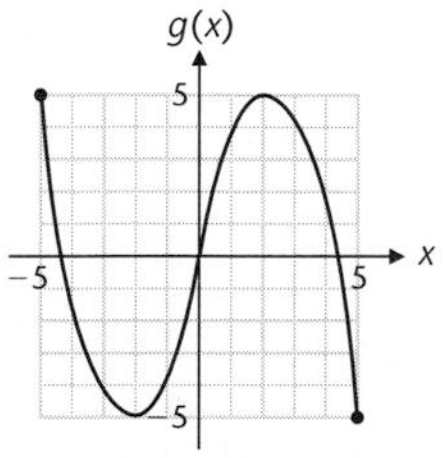

63. Conclusion: any function can be written as the sum of two other functions, one even and the other odd.

65. Graph of $f(x)$

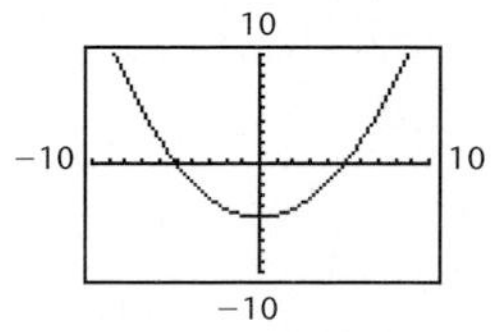

Graph of $|f(x)|$

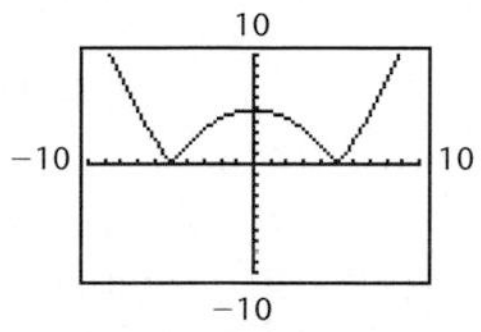

Graph of $-|f(x)|$

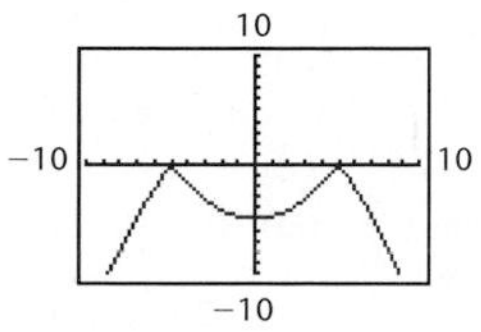

67. Graph of $f(x)$ Graph of $|f(x)|$ Graph of $-|f(x)|$

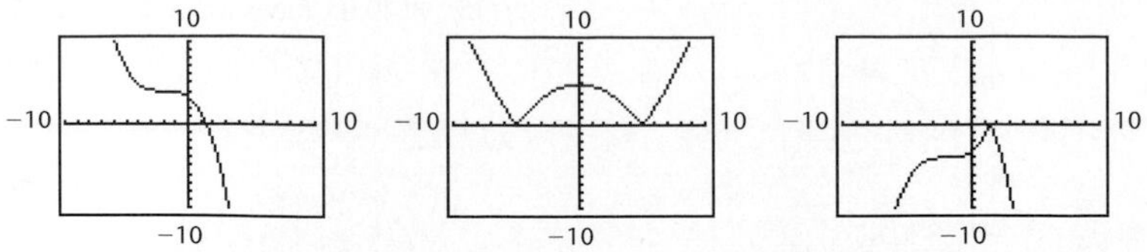

69. The graph of $y = |f(x)|$ is the same as the graph of $y = f(x)$ whenever $f(x) \geq 0$ and is the reflection of the graph of $y = f(x)$ with respect to the x axis whenever $f(x) < 0$.

71.

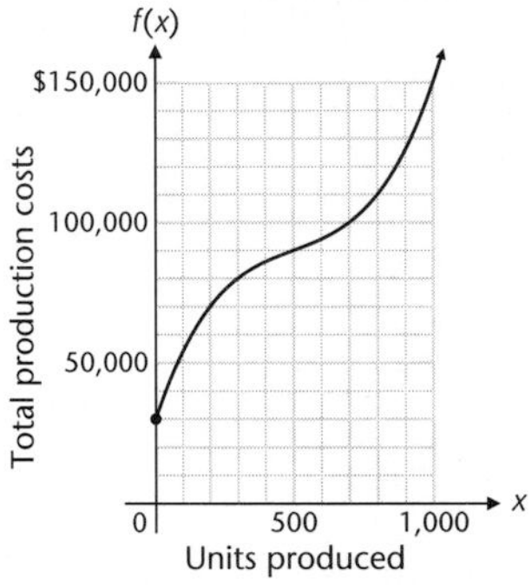

73. Each graph is a vertical translation of the graph of $y = 0.004(x - 10)^3$.

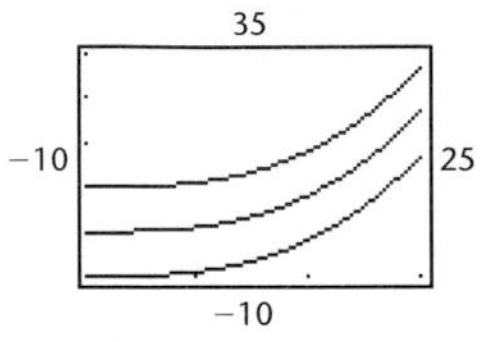

75. Each graph is a contraction followed by a vertical translation of the graph of $y = x^2$.

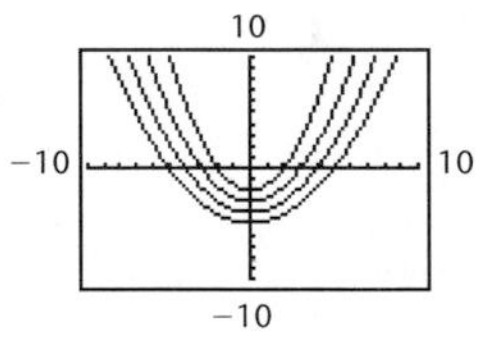

77. Each graph is a portion of the graph of a horizontal translation followed by an expansion (except for $C = 8$) of the graph of $y = t^2$.

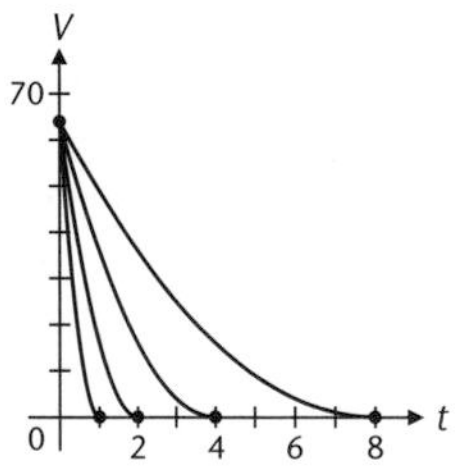

Chapter 1 Review

1. *(1-1)*

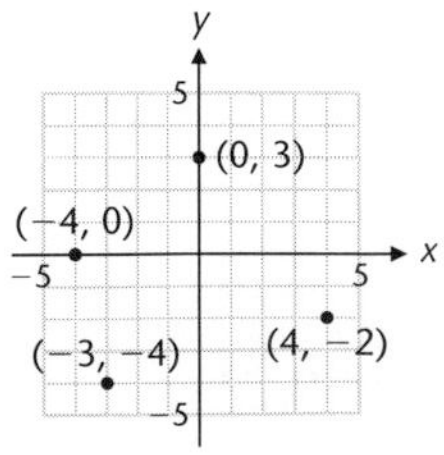

2. (A) $x^2 + y^2 = 7$ (B) $(x - 3)^2 + (y + 2)^2 = 7$ *(1-1)*

3. Xmin = −4, Xmax = 9, Ymin = −6, Ymax = 7 *(1-2)*

4. (A) Function; domain = {1, 2, 3}; range = {1, 4, 9} (B) Not a function
(C) Function; domain = {−2, −1, 0, 1, 2}; range = {2} *(1-3)*

5. (A) Not a function (B) A function (C) A function (D) Not a function *(1-3)* **6.** (A) −1 (B) 24 (C) 0 (D) 0 *(1-3)*

7. (A) Odd (B) Even (C) Neither *(1-5)* **8.** $f(-4) = 4, f(0) = -4, f(3) = 0, f(5)$ is not defined *(1-1, 1-3, 1-4)*

9. $x = -2, x = 1$ *(1-1, 1-3, 1-4)* **10.** Domain: $[-4, 5)$; range: $[-4, 4]$ *(1-4)* **11.** Increasing: $[0, 5)$; decreasing: $[-4, 0]$ *(1-4)*

12. $x = 0$ *(1-4)* **13.** *(1-5)*

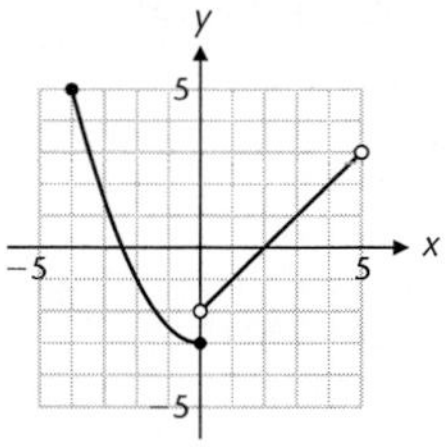

14. *(1-5)*

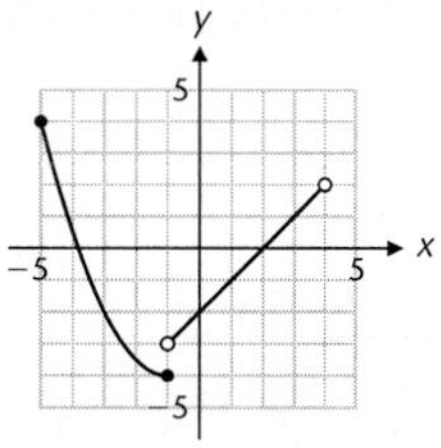

15. *(1-5)*

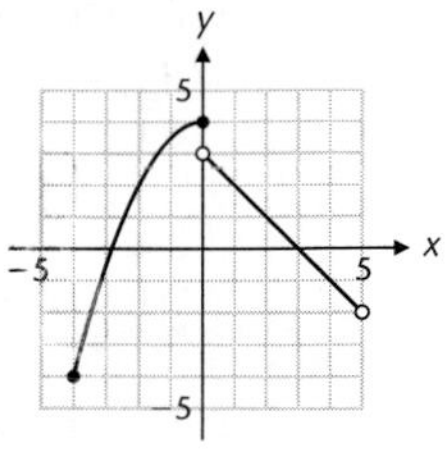

16. *(1-5)*

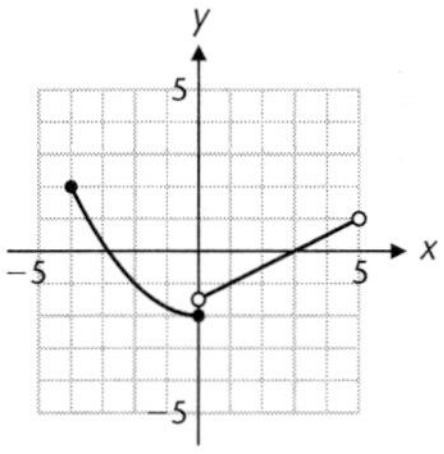

17. (A) g (B) m (C) n (D) f *(1-5)* **18.** (A) 0 (B) 1 (C) 2 (D) 0 *(1-4)*

19. (A) $-2, 0$ (B) $-1, 1$ (C) No solution (D) $x = 3$ and $x < -2$ *(1-4)* **20.** Domain $= (-\infty, \infty)$; range $= (-3, \infty)$ *(1-4)*
21. $[-2, -1], [1, \infty)$ *(1-4)* **22.** $[-1, 1)$ *(1-4)* **23.** $(-\infty, -2)$ *(1-4)* **24.** $x = -2, x = 1$ *(1-4)*
25. (A)

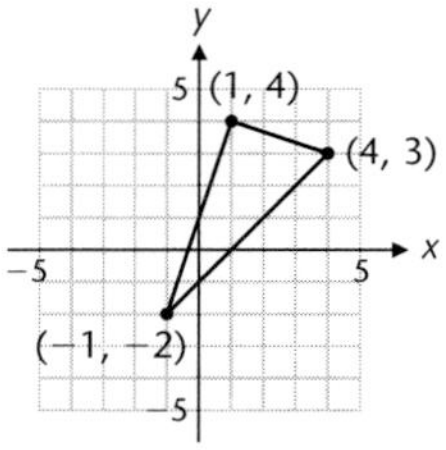

(B) 16.56 (C) A right triangle *(1-1)*

26. (A)

x	0	1	2	3	4	5	6
y	−4	1	4	5	4	1	−4

(B) *(1-2)*

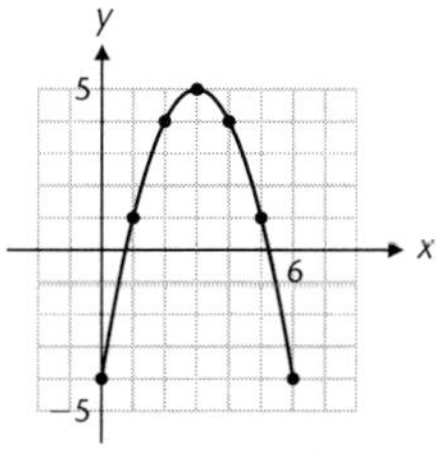

27. (A) All real numbers (B) All real numbers except $t = 5$ (C) $w \geq 0$ or $[0, \infty)$ *(1-3)*
28. $5 + 2h$ *(1-3)* **29.** $f(x) = 4x^3 - \sqrt{x}$ *(1-3)*
30. The function f multiplies the square of the domain element by 3, adds 4 times the domain element, and then subtracts 6. *(1-3)*
31. $c = 1.3, f(1.3) = 5.2$; x intercepts: $x = 0, x = 3.3$ *(1-4)*
32. (A)

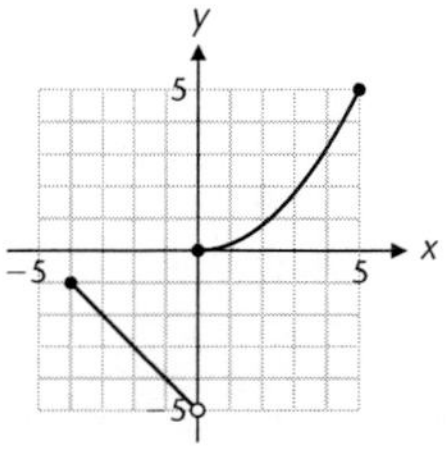

(B) Domain: $[-4, 5]$; range: $(-5, -1] \cup [0, 5]$ (C) $x = 0$

(D) Decreasing on $[-4, 0)$; increasing on $[0, 5]$ *(1-4)*
33. The graph of f increases on $[-10, -4.47]$ to a local maximum value, $f(-4.47) \approx 22.89$, decreases on $[-4.47, 4.47]$ to a local minimum value, $f(4.47) \approx -12.89$, and then increases on $[4.47, 10]$. *(1-4)*
34. (A) Reflected in x axis (B) Shifted down 3 units (C) Shifted left 3 units *(1-5)*
35. (A) $-(x - 2)^2 + 4$ (B) $4 - 4\sqrt{x}$ *(1-5)*

36. $g(x) = 8 - 3|x - 4|$ *(1-5)*

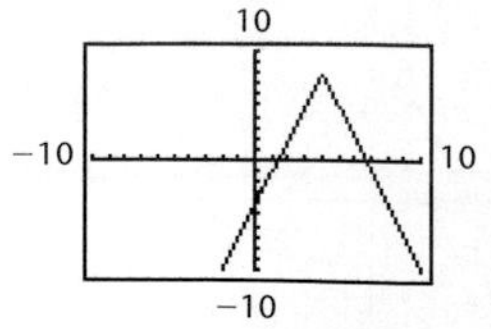

37. Center: $(-2, 2)$, radius: 2 *(1-1)*

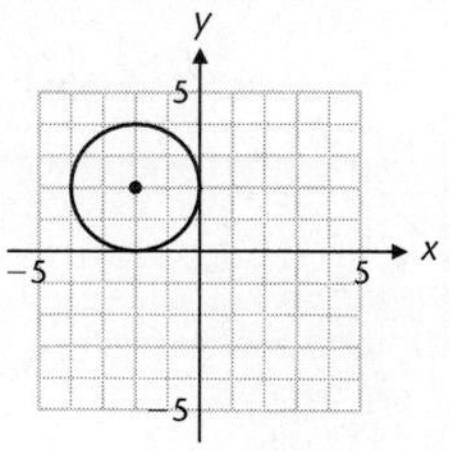

38. Domain: $x \geq 0$, $x \neq 9$ or $[0, 9) \cup (9, \infty)$ *(1-3)*

39. (A) One possible answer:

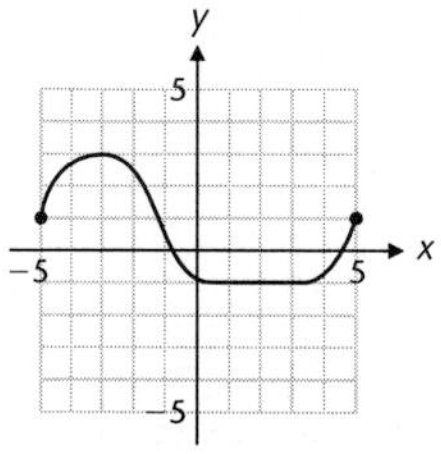

(B) One possible answer: *(1-4)*

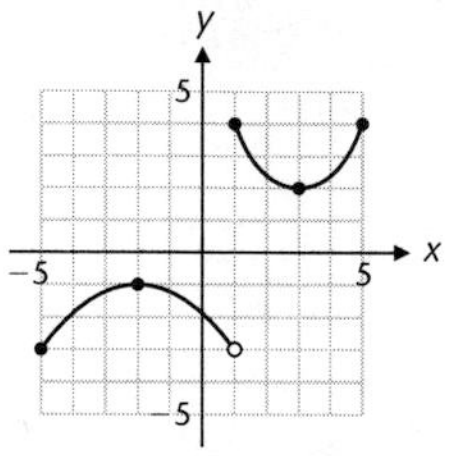

40. $g(t) = 2t^2 - 4t + 5$ *(1-3)*

41. Domain: all real numbers except $x = 2$; range: $y > -3$ or $(-3, \infty)$; discontinuous at $x = 2$ *(1-4)*

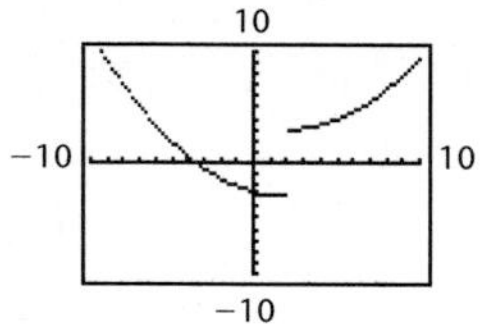

42. (A)

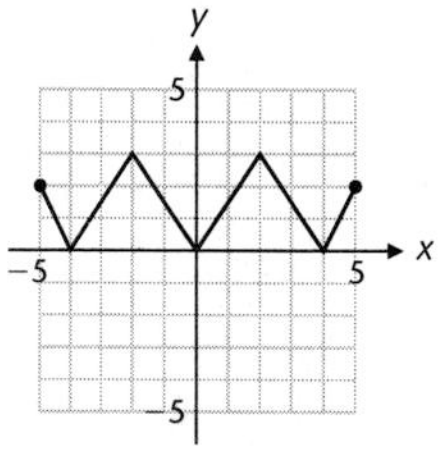

(B) *(1-5)*

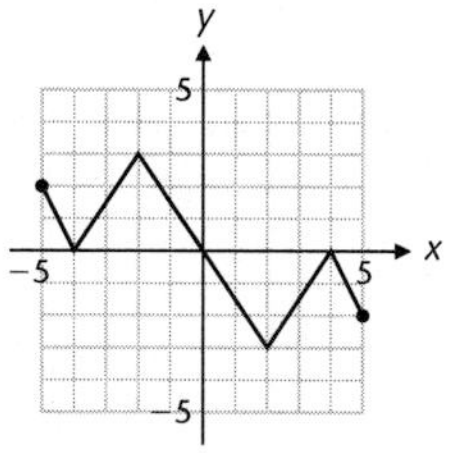

43. $(x - 4)^2 + (y + 3)^2 = 34$ *(1-1)* **44.** (A) $6x - 5 + 3h$ (B) $3x + 3a - 5$ *(1-3)*

45. (A) The graph must cross the x axis exactly once. (B) The graph can cross the x axis at most once. *(1-4)*

46. (A) $f(x) = \begin{cases} 2 & \text{for } -3 < x \leq -2 \\ 1 & \text{for } -2 < x \leq -1 \\ 0 & \text{for } -1 < x < 1 \\ 1 & \text{for } 1 \leq x < 2 \\ 2 & \text{for } 2 \leq x < 3 \end{cases}$ (B)

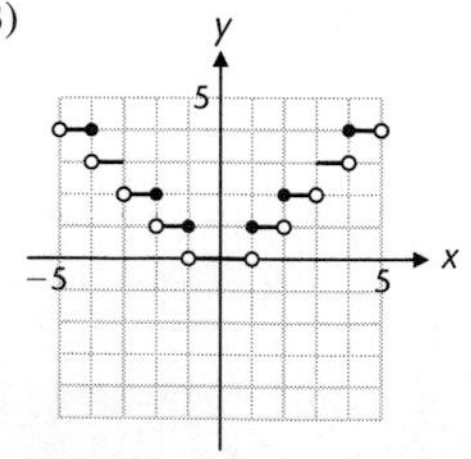

(C) Range: nonnegative integers

(D) Discontinuous at all integers except 0 (E) Even *(1-4, 1-5)*

47. (A) 3,800 bottles (B) The demand decreases by 400 bottles. (C) The demand increases by 500 bottles. *(1-1)*

48. (A) 3,970 bottles (B) The demand decreases by 490 bottles. (C) The demand increases by 520 bottles. *(1-2)*

49. (A)

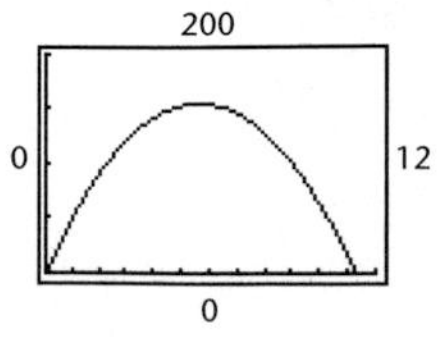

(B) 11.3 sec (C) 155 m *(1-2, 1-4)*

50. (A) The estimated maximum volume is 10,472 in.^{3} when the flap is 7 in. wide.

(B) The maximum volume is approximately 10,480 in.^{3} when the flap is 6.8 in. wide. *(1-4)*

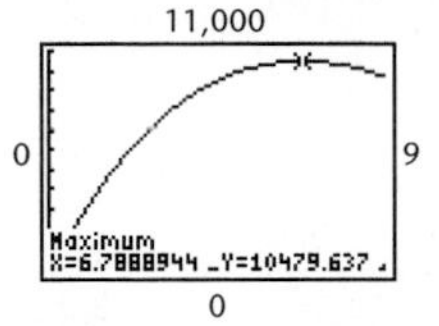

51. (A)

x	0	5	10	15	20
Consumption	309	276	271	255	233
$f(x)$	303	286	269	252	234

(B)

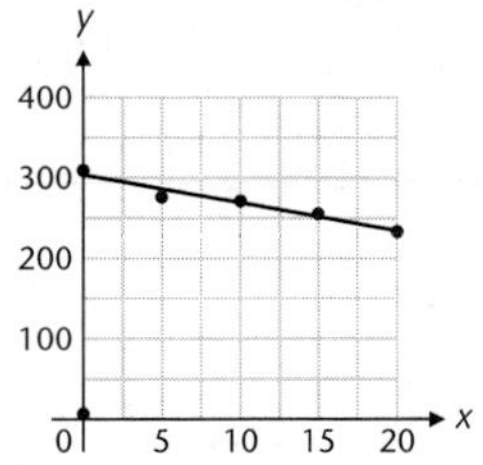

(C) 217 in 1995, 200 in 2000

(D) $y = -3.46x + 303.4$, which is the same as the modeling function f.

(E) Per capita egg consumption is dropping about 17 eggs every 5 years. *(1-3)*

52. (A)

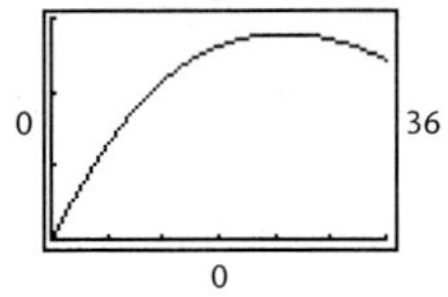

(B) The function increases on [0, 24.8] to a local maximum of 2.8 cc/sec, and then decreases on [24.8, 36]. *(1-4)*

53. (A) $C(x) = \begin{cases} 0.49x & \text{for } 0 \le x < 36 \\ 0.44x & \text{for } 36 \le x < 72 \\ 0.39x & \text{for } 72 \le x \end{cases}$

(B) Discontinuous at $x = 36$ and $x = 72$ *(1-4)*

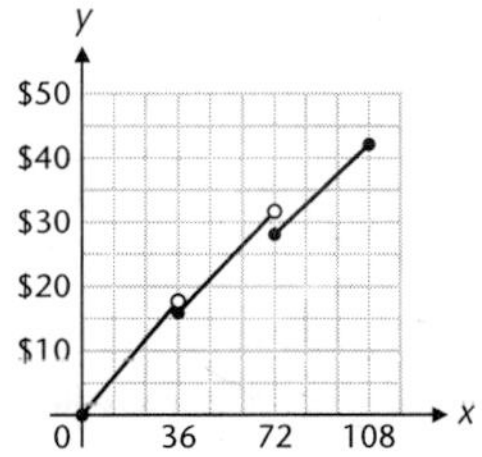

54. (A) 0 (B) 1 (C) 2 (D) 0 (E) 1 (F) 0 *(1-5)*

CHAPTER 2 Exercise 2-1

1. x intercept $= -2$; y intercept $= 2$; slope $= 1$ **3.** x intercept $= -2$; y intercept $= -4$; slope $= -2$
5. x intercept $= 3$; y intercept $= -1$; slope $= \frac{1}{3}$ **7.** Not linear **9.** Linear **11.** Linear **13.** Linear **15.** Not linear
17. x intercept: $\frac{20}{3}$; y intercept: 4; slope: $-\frac{3}{5}$ **19.** x intercept: 0; y intercept: 0; slope: $-\frac{3}{4}$ **21.** x intercept: $\frac{15}{2}$; y intercept: -5; slope: $\frac{2}{3}$

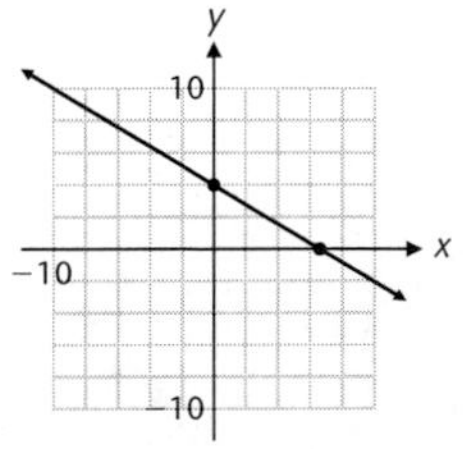

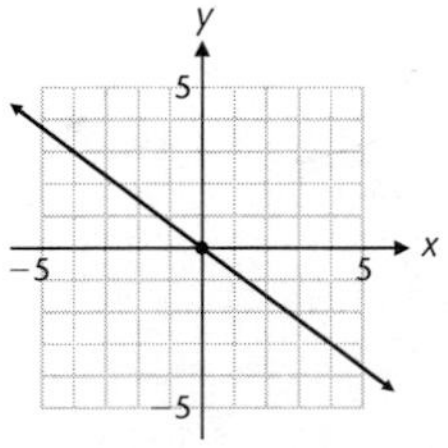

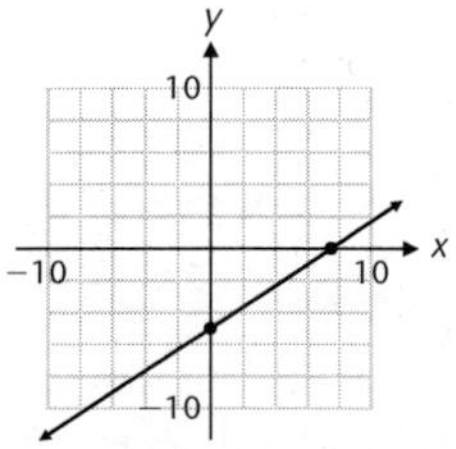

23. x intercept: -4; y intercept: 8; slope: 2 **25.** x intercept: -3; y intercept: none; slope is not defined

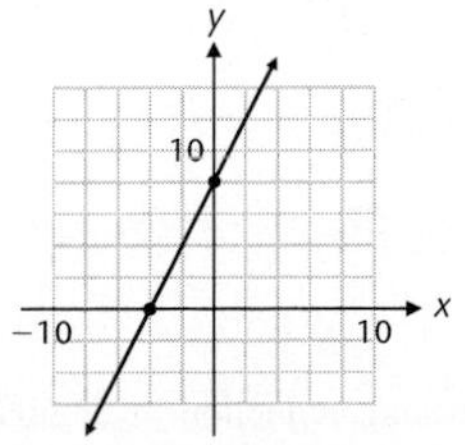

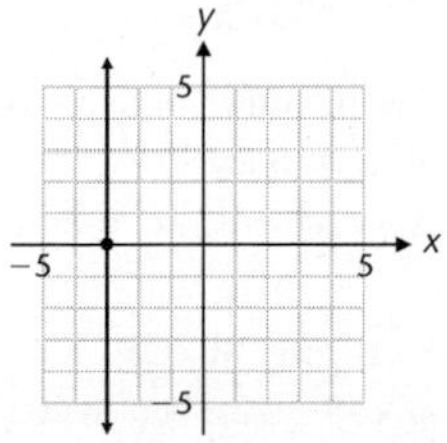

27. x intercept: none; y intercept: 3.5; slope: 0 **29.** $y = x$ **31.** $y = -\frac{2}{3}x - 4$ **33.** $y = -3x + 4$ **35.** $y = -\frac{2}{5}x + 2$

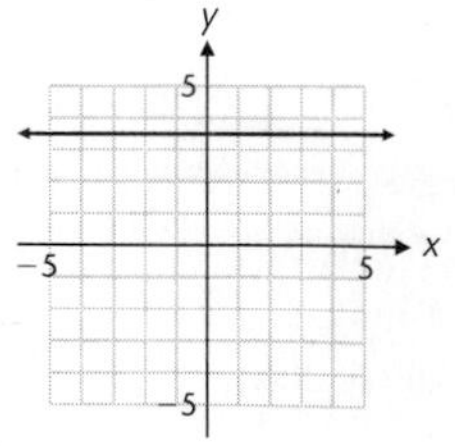

37. $y = -2x + 8$ **39.** $y = -\frac{4}{3}x + \frac{8}{3}$ **41.** $y = 4$ **43.** $x = 4$ **45.** $y = \frac{3}{4}x + 3$ **47.** $3x - y = -13$
49. $3x - y = 9$ **51.** $x = 2$ **53.** $x = 3$ **55.** $3x - 2y = 15$ **57.** $3x - y = 4$ **61.** slope $AB = -\frac{3}{4}$ = slope DC
63. (slope AB) (slope BC) $= (-\frac{3}{4})(\frac{4}{3}) = -1$
65. $6x + 8y = -9$ **67.** $3x + 4y = 25$ **69.** $x - y = 10$ **71.** $232 = 5x - 12y$

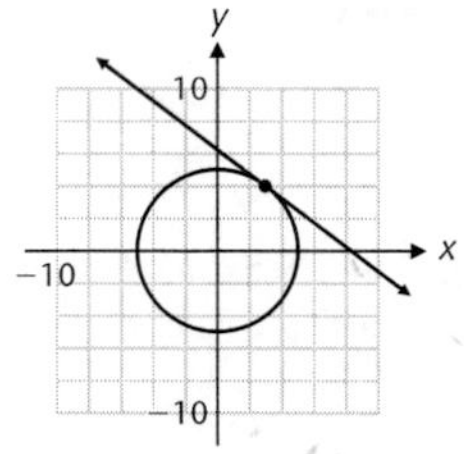

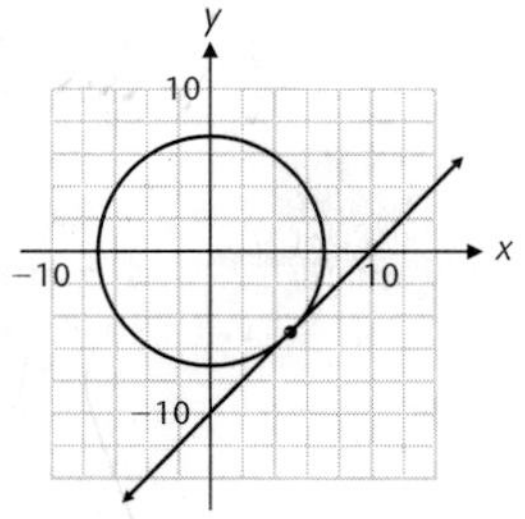

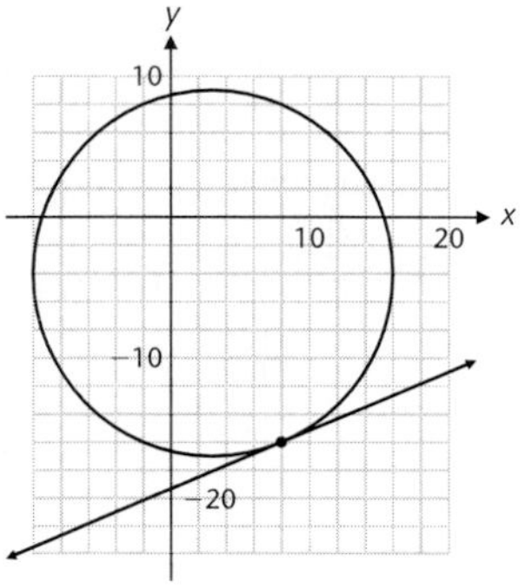

73. (A)

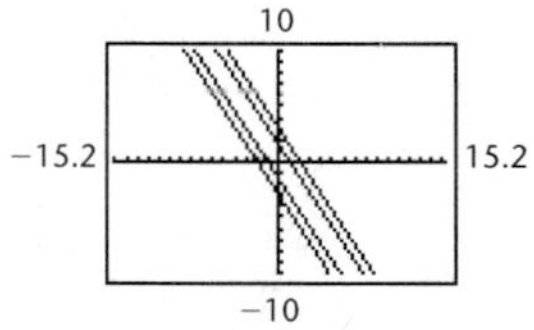

(B) Varying C produces a family of parallel lines. **75.** The function g is never linear.

79. (A) $F = \frac{9}{5}C + 32$ (B) 68°F, 30°C (C) $\frac{9}{5}$ **81.** (A) $V = -1{,}600t + 8{,}000, 0 \le t \le 5$ (B) $V = \$3{,}200$ (C) −1,600
83. (A) $T = -5A + 70, A \ge 0$ (B) A = 14,000 feet (C) Slope = −5; the temperature changes −5°F for each 1,000 foot rise in altitude.
85. (A) $h = 1.13t + 12.8$ (B) t = 32.9 hours
87. (A) $R = 0.00152C - 0.159, C \ge 210$ (B) $R = 0.236$
(C) Slope = 0.00152; coronary risk increases 0.00152 per unit increase in cholesterol above the 210 cholesterol level.
89. (A) $C(x) = 128x + 375$
(B) The fixed costs are \$375, the variable costs are $128x$, and the cost of producing an additional surfboard is \$128. (C) \$3,571
91. (A) $f(x) = -1.2x + 97.4$ (B) 64%, 32% (C) 1979

Exercise 2-2

1. $x = c, x = f$ **3.** $x = b, x = e$ **5.** (c, f) **7.** $(-\infty, b] \cup [e, \infty)$ **9.** 18 **11.** 9 **13.** $\frac{11}{2}$ or 5.5
15. $x < 5$ or $(-\infty, 5)$;

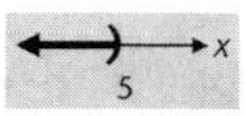

17. $t > 2$ or $(2, \infty)$;

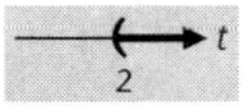

19. $-2 < t \le 3$ or $(-2, 3]$;

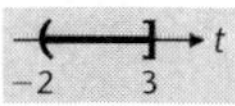

21. $y = 2, 8$;

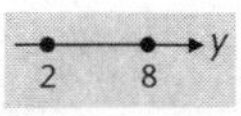

23. $-2 \le t \le 0.8$; $[-2, 0.8]$ **25.** 3 **27.** $-30 \le x < 18$ or $[-30, 18)$ **29.** $t = 1.5, 4.5$

31. $x < -7.5$ or $x > 1.25$; $(-\infty, -7.5) \cup (1.25, \infty)$ **33.** $x = 3.4$ **35.** $-5.5 < x < -2.5$ or $3.5 < x < 6.5$ **37.** $d = \dfrac{a_n - a_1}{n - 1}$

39. $f = \dfrac{d_1d_2}{d_2 + d_1}$ **41.** $a = \dfrac{A - 2bc}{2b + 2c}$ **43.** $x = \dfrac{5y + 3}{2 - 3y}$
45. The graphs are identical for $x \ge 0$. For $x < 0$, each is the reflection of the other in the x axis.
47. (A) and (C), $a > 0$ and $b > 0$, or $a < 0$ and $b < 0$ (B) and (D), $a > 0$ and $b < 0$, or $a < 0$ and $b > 0$
49. > **51.** > **53.** $(2.9, 3) \cup (3, 3.1)$;

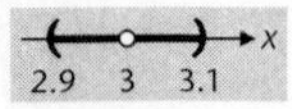

55. $(c - d, c) \cup (c, c + d)$;

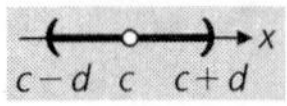

57. ±1 **59.** \$19,750

61. $|A - 12.436| < 0.001$, $(12.435, 12.437)$ **63.** (A) $x > 40{,}625$ (B) $x = 40{,}625$
65. (B) $x > 52{,}000$ (C) Raise wholesale price \$3.50 to \$66.50. **67.** $|N - 2.37| \le 0.005$ **69.** \$2,060 ≤ benefit reduction ≤ \$3,560

71. 16°C to 27°C **73.** (A) $T = 30 + 25(x - 3)$ (B) 330°C (C) $9.8 \le x \le 13.8$ (from 9.8 km to 13.8 km)

Exercise 2-3

1. $f(x) = (x - 2)^2 + 1$ **3.** $h(x) = -(x + 1)^2$ **5.** $m(x) = (x - 2)^2 - 3$
7. The graph of f is the graph of $y = x^2$ shifted to the right 2 units and up 1 unit.
9. The graph of h is the graph of $y = x^2$ reflected in the x axis and shifted to the left 1 unit.
11. The graph of m is the graph of $y = x^2$ shifted to the right 2 units and down 3 units. **13.** k **15.** m **17.** h

19.

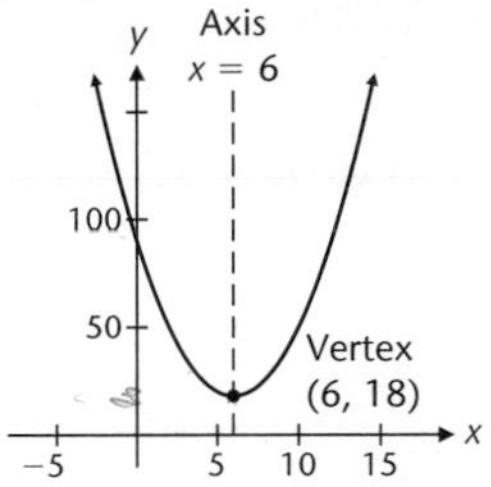

21.

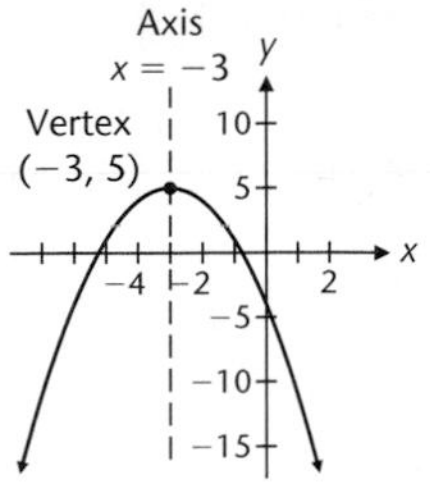

23.

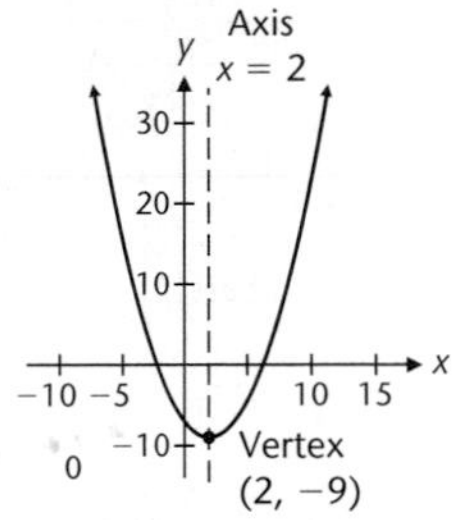

25. Increasing: $[2.25, \infty)$; decreasing: $(-\infty, 2.25]$; range: $[4.75, \infty)$ **27.** Increasing: $(-\infty, 2.2]$; decreasing: $[2.2, \infty)$; range: $(-\infty, 60.4]$
29. $y = 2x^2 - 4x - 2$ **31.** $y = -0.5x^2 - x + 3.5$ **33.** $y = -2x^2 + 16x - 24$ **35.** $y = -0.5x^2 - 4x + 4$
37. $y = 5x^2 + 50x + 100$ **43.** Center: (3, 2); radius: 7 **45.** Center: (−4, 1); radius: 5 **49.** $y = 2x - 1$

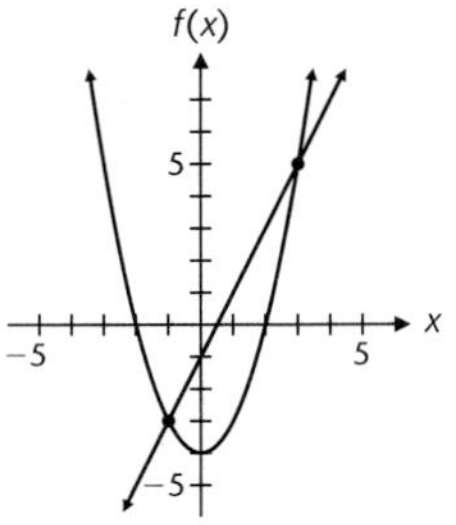

51. (A) $1 + h$ (B)

h	slope $= 1 + h$
1	2
0.1	1.1
0.01	1.01
0.001	1.001

; The slope seems to be approaching 1.

53. The minimum product is −225 for the numbers 15 and −15. There is no maximum product.
55. (A) $A(x) = 5{,}000 + 50x - x^2$, $0 \le x \le 100$ (B) $x = 25$ (C) 75 ft × 75 ft
57. (A) $d(t) = 176t - 16t^2$, $0 \le t \le 11$ (B) 1.68 sec; 9.32 sec
59. (A) $h(x) = -0.14x^2 + 14$, $0 \le x \le 10$ (B) No (C) 11.76 ft (D) 7.56 ft
61. (A)

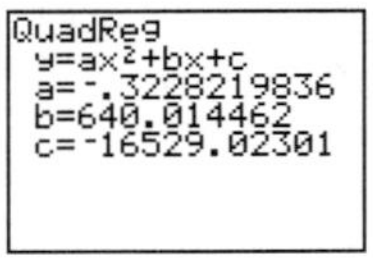

(B)

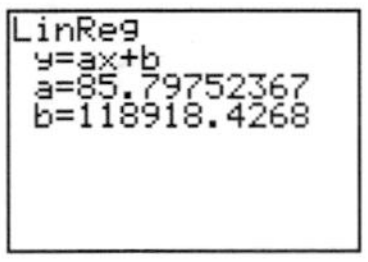

(C) 295 mowers; 1,422 mowers **63.** (A)

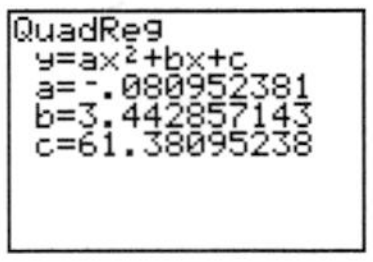

(B) 2003

Exercise 2-4

1. $7 + 5i$ **3.** $5 + 3i$ **5.** $2 + 4i$ **7.** $5 + 9i$ **9.** $4 - 3i$ **11.** -24 or $-24 + 0i$ **13.** $-12 - 6i$ **15.** $15 - 3i$
17. $-4 - 33i$ **19.** 65 or $65 + 0i$ **21.** $\frac{2}{5} - \frac{1}{5}i$ **23.** $\frac{3}{13} + \frac{11}{13}i$ **25.** $5 + 3i$ **27.** $7 - 5i$ **29.** $-3 + 2i$ **31.** $8 + 25i$
33. $\frac{5}{7} - \frac{2}{7}i$ **35.** $\frac{2}{13} + \frac{3}{13}i$ **37.** $-\frac{2}{5}i$ or $0 - \frac{2}{5}i$ **39.** $\frac{3}{2} - \frac{1}{2}i$ **41.** $-6i$ or $0 - 6i$ **45.** $i^{18} = -1$, $i^{32} = 1$, $i^{67} = -i$
47. $x = 3$, $y = -2$ **49.** $x = 2$, $y = 3$ **51.** $0.6 + 1.2i$ **53.** $1.5 + 0.5i$ **57.** $(a + c) + (b + d)i$
59. $a^2 + b^2$ or $(a^2 + b^2) + 0i$ **61.** $(ac - bd) + (ad + bc)i$ **63.** $i^{4k} = (i^4)^k = (i^2 \cdot i^2)^k = [(-1)(-1)]^k = 1^k = 1$
67. $3 - i$, $-3 + i$ **71.** (1) Definition of addition; (2) Commutative (+) property for R; (3) Definition of addition

Exercise 2-5

1. $u = 0, 2$ **3.** $y = \frac{2}{3}$ (double root) **5.** $x = \frac{3}{2}, 4$ **7.** $m = \pm 2\sqrt{3}$ **9.** $x = \pm 5i$ **11.** $y = \pm\frac{4}{3}$ **13.** $x = \pm\frac{5i}{2}$ or $\pm\frac{5}{2}i$

15. $n = -2, -8$ **17.** $d = 3 \pm 2i$ **19.** $x = 5 \pm 2\sqrt{7}$ **21.** $x = 2 \pm 2i$ **23.** $x = \frac{2 \pm \sqrt{2}}{2}$ **25.** $x = \frac{1}{5} \pm \frac{3}{5}i$

27. $(-5, 2)$, $-5 < x < 2$,
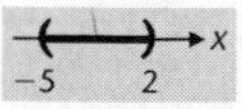

29. $(-\infty, 3) \cup (7, \infty)$, $x < 3$ or $x > 7$,
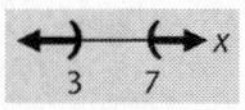

31. $[0, 8]$, $0 \le x \le 8$,
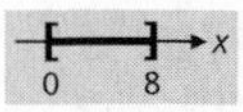

33. $[-5, 0]$, $-5 \le x \le 0$,
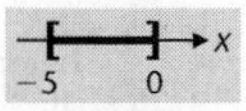

35. $x = 3 \pm 2\sqrt{3}$ **37.** $y = \frac{3 \pm \sqrt{3}}{2}$ **39.** $x = \frac{1 \pm \sqrt{7}}{3}$ **41.** $x = -\frac{5}{4}, \frac{2}{3}$

43. $x = \frac{3 \pm \sqrt{13}}{2}$ **45.** $t = \sqrt{\frac{2s}{g}}$ **47.** $I = \frac{E + \sqrt{E^2 - 4RP}}{2R}$ **49.** $x < 0.48$ or $x > 1.35$ **51.** $-1.05 \le x \le 0.63$

53. $(-\infty, -3] \cup [3, \infty)$ **55.** $(-\infty, -2] \cup [1.5, \infty)$ **57.** $(3 - \sqrt{5}, 3 + \sqrt{5})$

59. If $c < 4$ there are two distinct real roots, if $c = 4$ there is one real double root, and if $c > 4$ there are two distinct imaginary roots.

61. If $a > 0$, the solution set is $(-\infty, r_1) \cup (r_2, \infty)$. If $a < 0$, the solution set is (r_1, r_2).

63. If $a > 0$, the solution set is R, the set of real numbers. If $a < 0$, the solution set is $\{r\}$.

65. $x^2 \ge 0$ **67.** $x = -i, -2i$ **69.** $x = \sqrt{2} - i, -\sqrt{2} - i$ **71.** $x = 1, -\frac{1}{2} \pm \frac{1}{2}i\sqrt{3}$

77. The $\pm$ in front still yields the same two numbers even if a is negative. **79.** 8, 13 **81.** 12, 14

83. $33 < x < 122$ **85.** At 8:06 A.M. **87.** 2.19 ft

89. (A) $A(w) = 400w - 2w^2$, $0 \le w \le 200$ (B) $50 \le w \le 150$
(C) No, the maximum cross-sectional area is 20,000 square feet when $w = 100$ feet.

91. 52 mi

Exercise 2-6

1. T **3.** F **5.** F **7.** $x = 22$ **9.** $n = 8$ **11.** No solution **13.** $x = 0, 4$ **15.** $y = \pm 2, \pm i\sqrt{2}$

17. $x = \frac{1}{2}i$ **19.** $x = \frac{1}{8}, -8$ **21.** $m = 3, -2, \frac{1}{2} \pm \frac{\sqrt{7}}{2}i$ **23.** No solution **25.** $y = 1$ **27.** $x = 2$ **29.** $x = -\frac{3}{2} + \frac{1}{2}i$

31. $n = -\frac{3}{4}, \frac{1}{5}$ **33.** $y = \pm 3, \pm 1$ **35.** $y = 1, 16$ **37.** $m = 3, 7, 2, 8$ **39.** $x = -2$ **41.** $y = \pm\sqrt{\frac{3 \pm \sqrt{3}}{2}}$ (four roots)

43. $m = 9, 16$ **45.** $t = 4, 81$ **47.** $x = -4, 39{,}596$ **49.** $x = \left(\frac{4}{5 \pm \sqrt{17}}\right)^5 \approx 0.016203, 1974.98$

51. 13.1 in. by 9.1 in. **53.** 1.65 ft or 3.65 ft

Chapter 2 Review

1. Slope: $-\frac{3}{2}$ *(2-1)* **2.** $2x + 3y = 12$ *(2-1)* **3.** $y = -\frac{2}{3}x + 2$ *(2-1)*

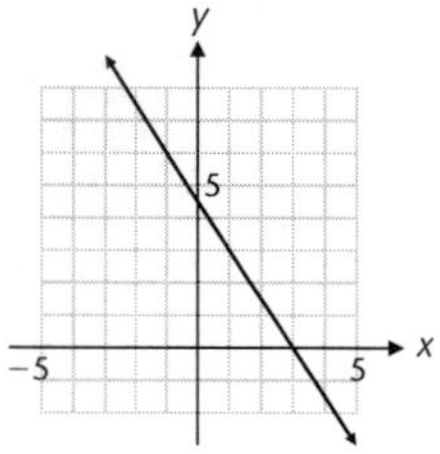

4. Vertical: $x = -3$, slope not defined; horizontal: $y = 4$, slope $= 0$ *(2-1)* **5.** (A) $x = 21$ (B) $x = \frac{30}{11}$ *(2-2)*

6. (A) $f(x) = -(x + 1)^2 + 4$ (B) It is the same as the graph of $y = x^2$ reflected in the x axis, shifted left 1 unit, and up 4 units.
(C) $x = -3, 1$ *(2-3, 2-5)*

7. (A) $f(x) = (x - \frac{3}{2})^2 - \frac{17}{4}$ (B) It is the same as the graph of $y = x^2$ shifted right $\frac{3}{2}$ units, and down $\frac{17}{4}$ units.
(C) $x = \frac{3 \pm \sqrt{17}}{2}$ *(2-3, 2-5)*

8. (A) $3 - 6i$ (B) $15 + 3i$ (C) $2 + i$ *(2-4)* **9.** $x = 2$ *(2-1, 2-5)* **10.** $x = \pm\frac{\sqrt{14}}{2}$ *(2-6)* **11.** $x = 0, 2$ *(2-5)*

12. $x = \frac{1}{2}, 3$ *(2-5)* **13.** $m = -\frac{1}{2} \pm \frac{\sqrt{3}}{2}i$ *(2-5)* **14.** $y = \frac{3 \pm \sqrt{33}}{4}$ *(2-5)* **15.** $x = 2, 3$ *(2-6)*

16. $x \ge 1$; $[1, \infty)$
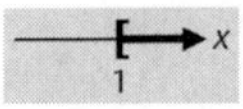

(2-2) **17.** $(-5, 4)$; $-5 < x < 4$;
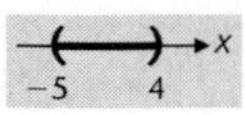

(2-2)

18. $x < -2$ or $x > 6$; $(-\infty, -2) \cup (6, \infty)$
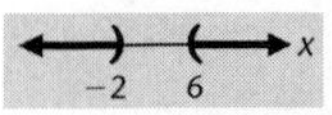

(2-5)

20. $3x + 2y = -6$ *(2-1)* **21.** (A) $y = -2x - 3$ (B) $y = \frac{1}{2}x + 2$ *(2-1)* **22.** $-14 < y < -4$ *(2-2)*

23. $x \le 2.5$ or $5.5 \le x$ *(2-2)* **24.** $-1 \le m \le 2$ *(2-2)*

25. (A) $(-\infty, 2]$ (B) $[-1 - \sqrt{5}, -1 + \sqrt{5}]$ (C) $(-\infty, 3 - \sqrt{3}) \cup (3 + \sqrt{3}, \infty)$ *(2-2, 2-5)*

26. (A)

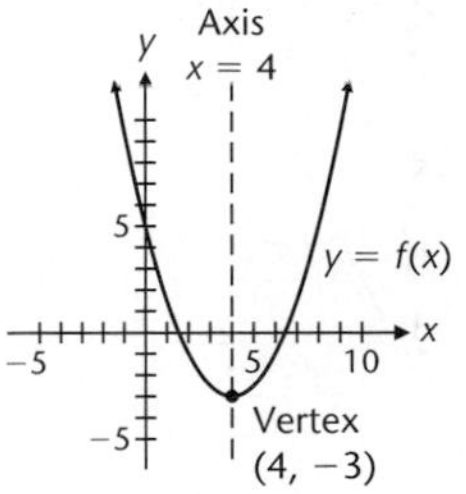

(B) Increasing: $[4, \infty)$; decreasing $(-\infty, 4]$; range: $[-3, \infty)$ *(2-3)*

27. $g(x) = 2x + 2$; $f(x) = -0.5x^2 + x + 1.5$ *(2-1, 2-3)* **28.** (A) $5 + 4i$ (B) $-i$ *(2-4)*

29. (A) $-1 + i$ (B) $\frac{4}{13} - \frac{7}{13}i$ (C) $\frac{5}{2} - 2i$ *(2-4)* **30.** $x = \dfrac{-5 \pm \sqrt{5}}{2}$ *(2-5)* **31.** $u = 1 \pm i\sqrt{2}$ *(2-6)* **32.** $\frac{1}{2} - \frac{3}{2}i$ *(2-6)*

33. $x = -\frac{27}{8}, 64$ *(2-6)* **34.** $m = \pm 3i, \pm 2$ *(2-6)* **35.** $y = \frac{9}{4}, 3$ *(2-6)* **38.** $x < 1.98$ *(2-2)* **39.** $x \le -0.67$ or $x \ge 3.07$ *(2-5)*

40. If $c < 9$ there are two distinct real roots, if $c = 9$ there is one real double root, and if $c > 9$ there are two distinct imaginary roots. *(2-5)*

41. $M = \dfrac{P}{1 - dt}$ *(2-2)* **42.** $I = \dfrac{E \pm \sqrt{E^2 - 4PR}}{2R}$ *(2-5)* **43.** True for all real b and all negative a *(2-2)* **44.** $\dfrac{a}{b}$ is less than 1. *(2-2)*

45. $6 - d < x < 6 + d$, $x \ne 6$, $(6 - d, 6) \cup (6, 6 + d)$

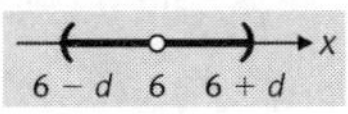

(2-2) **46.** 1 *(2-4)* **47.** Perpendicular *(2-1)*

48. Center: (2, 1); radius: $2\sqrt{2}$ *(2-3)* **49.** $y = -x + 7$

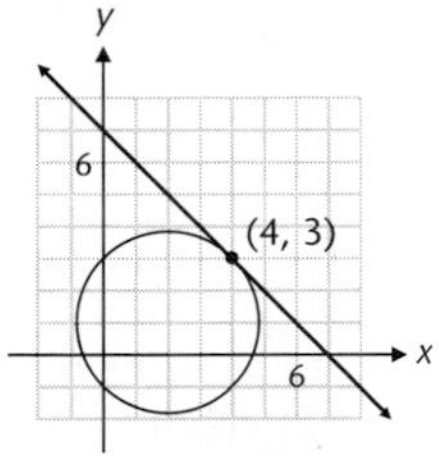

(1-1, 2-1) **50.** $x = 1, 243$ *(2-6)*

51. $x = -1; \dfrac{1 \pm i\sqrt{3}}{2}$ *(2-5)* **52.** (A) 4,750 calculators; \$7,437.50 (B) 2,614 or 6,886 calculators (C) None *(2-3)*

53. 3,240 or 9,260 calculators *(2-3)* **54.** Profit: $3{,}240 < x < 9{,}260$; loss: $0 \le x < 3{,}240$ or $x > 9{,}260$ *(2-3)*

55. (A) $V = -1{,}250t + 12{,}000$ (B) $V = \$5{,}750$ *(2-1)* **56.** (A) $R = 1.6C$ (B) $R = \$168$ *(2-1)*

57. $E(x) = \begin{cases} 200 & \text{if } 0 \le x \le 3{,}000 \\ 0.1x - 100 & \text{if } x > 3{,}000 \end{cases}$; $E(2{,}000) = 200$, $E(5{,}000) = 400$ *(2-1)*

58. (A) $A(x) = 60x - \frac{3}{2}x^2$ (B) $0 < x < 40$ (C) $x = 20$, $y = 15$ *(2-3)*

59. (A) $H = 0.7(220 - A)$ (B) $H = 140$ beats per minute (C) $A = 40$ years old *(2-1)*

60. 20 cm by 24 cm *(2-5)* **61.** $B = 14.58$ ft or 6.58 ft *(2-6)* **62.** 6.6 ft *(2-3)*

63. (A)

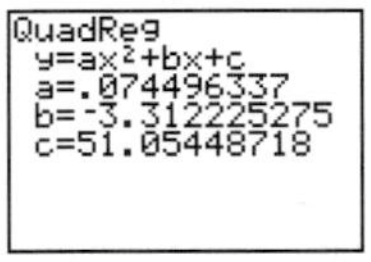

(B) 2005 *(2-3)* **64.** (A)

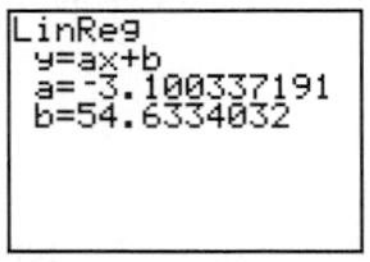

(B) 45.33% *(2-1)*

Cumulative Review Exercise for Chapters 1 and 2

1. (A)

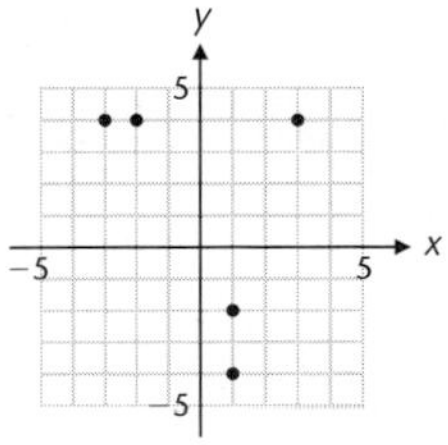

(B) Xmin = −3, Xmax = 3, Ymin = −4, Ymax = 4 (C) No *(1-1, 1-2, 1-3)*

2. (A) $2\sqrt{5}$ (B) $y = 2x - 4$ (C) $y = -\frac{1}{2}x + \frac{17}{2}$ (D) $(x - 3)^2 + (y - 2)^2 = 20$ (E) *(1-1, 2-1)*

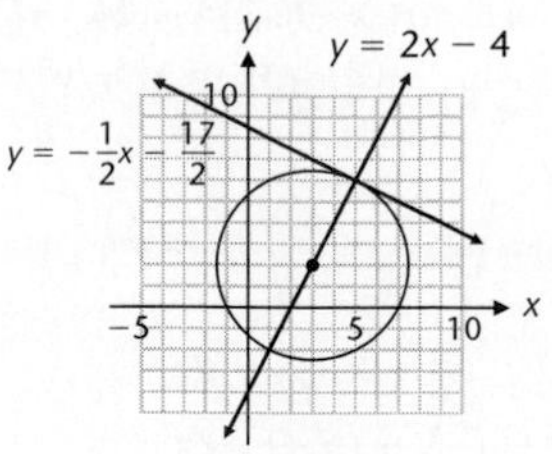

3. Slope: $\frac{2}{3}$; y intercept: -2; x intercept: 3 *(2-1)* **4.** (A) 2 (B) 4 (C) $-\frac{2}{5}$ *(1-3)*

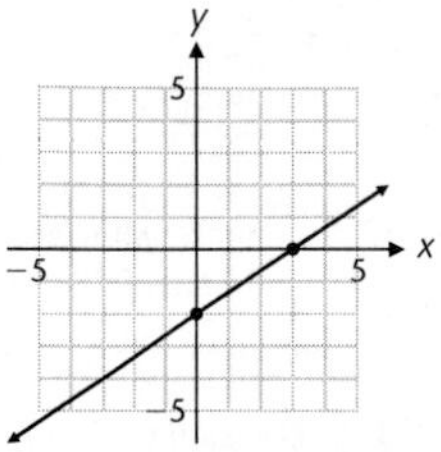

5. (A) Expanded by a factor of 2 (B) Shifted right 2 units (C) Shifted down 2 units *(1-5)*

6. Domain: $[-2, 3]$; range: $[-1, 2]$ *(1-3)* **7.** Neither *(1-5)*

8. (A)

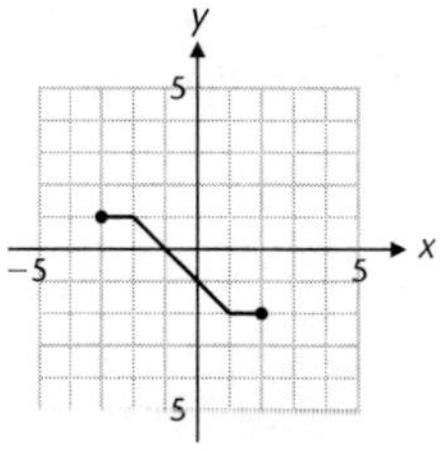

(B)

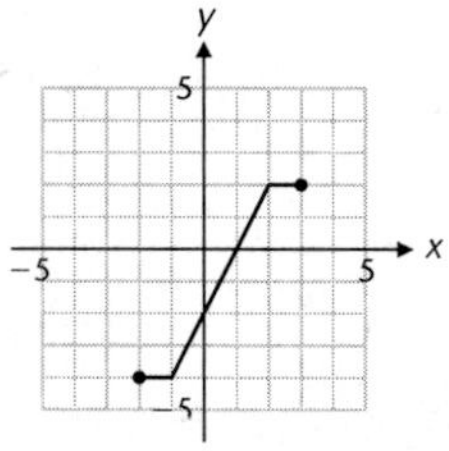

(1-5) **9.** $x = \frac{5}{2}$ *(2-2)* **10.** $x = 0, -4$ *(2-5)*

11. $x = \pm\sqrt{5}$ *(2-5)* **12.** $x = 3 \pm \sqrt{7}$ *(2-5)* **13.** $x = 3$ *(2-6)* **14.** $y \geq 5$; $[5, \infty)$

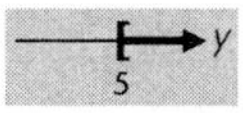

(2-2)

15. $-5 < x < 9$; $(-5, 9)$

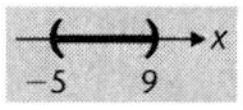

(2-2) **16.** $x \leq -5$ or $x \geq 2$; $(-\infty, -5] \cup [2, \infty)$

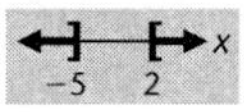

(2-5)

17. (A) $f(x) = (x - 2)^2 - 5$ (B) It is the same as the graph of $y = x^2$ shifted to the right 2 units and down 5 units. (C) $x = 2 \pm \sqrt{5}$ *(2-3, 2-5)*

18. (A) $7 - 10i$ (B) $23 + 7i$ (C) $1 - i$ *(2-4)*

19. (A) All real numbers $(-\infty, \infty)$ (B) $\{-2\} \cup [1, \infty)$ (C) 1 (D) $[-3, -2]$ and $[2, \infty)$ (E) $-2, 2$ *(1-4)*

20. (A) $y = -\frac{3}{2}x - 8$ (B) $y = \frac{2}{3}x + 5$ *(2-1)*

21. Range: $[-9, \infty)$; min $f(x) = f\left(-\dfrac{b}{2a}\right) = -9$; y intercept: $f(0) = -8$; x intercepts: $x = 4$ and $x = -2$ *(2-3)*

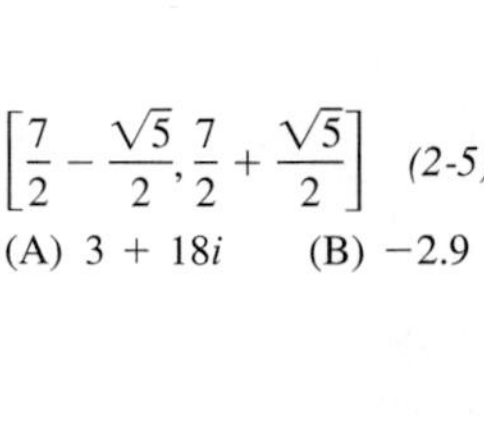

22. $x < \frac{3}{2}$ or $x > 3$; $(-\infty, \frac{3}{2}) \cup (3, \infty)$

(2-2) **23.** $\frac{2}{3} \leq m \leq 2$; $[\frac{2}{3}, 2]$

(2-2) **24.** $(-\infty, \frac{9}{7})$ *(2-2)*

25. $\left[\dfrac{7}{2} - \dfrac{\sqrt{5}}{2}, \dfrac{7}{2} + \dfrac{\sqrt{5}}{2}\right]$ *(2-5)* **26.** (A) $0 + 0i$ or 0 (B) $\frac{6}{5}$ (C) $i^{35} = i^{32}i^3 = (i^4)^8(-i) = 1^8(-i) = -i$ *(2-4)*

27. (A) $3 + 18i$ (B) $-2.9 + 10.7i$ (C) $-4 - 6i$ *(2-4)*

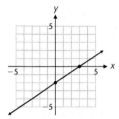

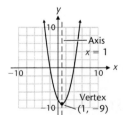

28. Domain: all real numbers; range: $(-\infty, -1) \cup [1, \infty)$; discontinuous at: $x = 0$ *(1-4)*

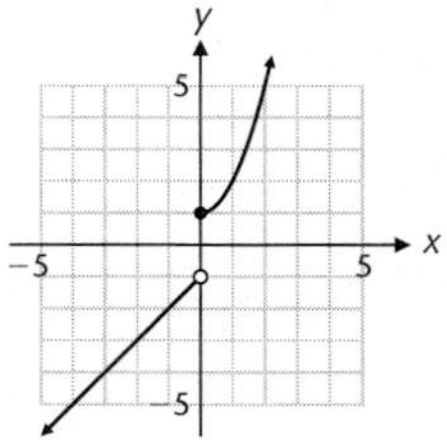

29. Center: $(3, -1)$; radius: $\sqrt{10}$ *(1-1, 2-3)*

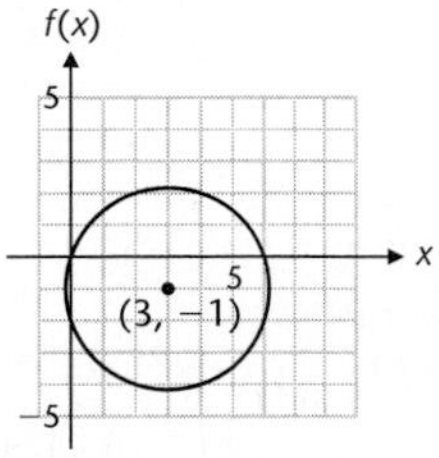

30. The graph of $y = |x|$ is contracted by $\frac{1}{2}$, reflected in the x axis, shifted two units to the right and three units up; $y = -\frac{1}{2}|x - 2| + 3$. *(1-5)*
31. $y = (x + 2)^2 - 3$ *(2-3)* **32.** $y = 3 \pm i\sqrt{5}$ *(2-6)* **33.** $x = \frac{27}{8}, -\frac{1}{8}$ *(2-6)* **34.** $u = \pm 2i, \pm\sqrt{3}$ *(2-6)* **35.** $t = \frac{9}{4}$ *(2-6)*
36. $x = \frac{4}{3}i$ *(2-4)* **37.** $-18.36 \le x < 16.09$ or $[-18.36, 16.09)$ *(2-2)* **38.** $-5.68, 1.23$ *(2-5)*
39. If $b < -2$ or $b > 2$ there are two distinct real roots; if $b = -2$ or $b = 2$, there is one real double root; and if $-2 < b < 2$ there are two distinct imaginary roots. *(2-5)*
42. $-5 - 2h$ *(1-3)* **43.** $y = 2\sqrt[3]{x+1} - 1$ *(1-5)*
44. (A) $h = \dfrac{A - 2\pi r^2}{2\pi r}$ (B) $r = -\dfrac{h}{2} + \sqrt{\dfrac{h^2}{4} + \dfrac{A}{2\pi}}$ The negative root is discarded since r must be positive. *(2-2, 2-5)* **45.** 0 *(2-4)*
46. All a and b such that $a < b$. *(2-2)* **47.** $\dfrac{a^2 - b^2}{a^2 + b^2} + \dfrac{2ab}{a^2 + b^2}i$ *(2-4)* **48.** $x = \dfrac{\sqrt{2} \pm i}{3}$ *(2-5)* **49.** $x = \pm 2i, \pm 3i$ *(2-6)*
50. $x = -1.5 + 0.5i$ *(2-6)* **51.** $x = \left(\dfrac{6}{1 \pm \sqrt{13}}\right)^5$ *(2-6)* **53.** (A) $x - 3 + 0.5h$ (B) $0.5x + 0.5a - 3$ *(1-3)*

55. $f(x) = \begin{cases} -2x & \text{if } x < -2 \\ 4 & \text{if } -2 \le x \le 2 \\ 2x & \text{if } x > 2 \end{cases}$ Domain: all real numbers; range: $[4, \infty)$ *(1-4)*

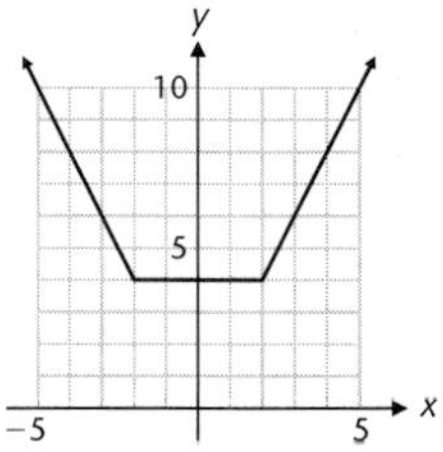

56. $f(x) = \begin{cases} \vdots & & \vdots \\ 2x + 2 & \text{if} & -1 \le x < -\frac{1}{2} \\ 2x + 1 & \text{if} & -\frac{1}{2} \le x < 0 \\ 2x & \text{if} & 0 \le x < \frac{1}{2} \\ 2x - 1 & \text{if} & \frac{1}{2} \le x < 1 \\ 2x - 2 & \text{if} & 1 \le x < \frac{3}{2} \\ 2x - 3 & \text{if} & \frac{3}{2} \le x < 2 \\ \vdots & & \vdots \end{cases}$

Domain: all real numbers; range: $[0, 1)$; discontinuous at $x = k/2$, k an integer *(1-4)*

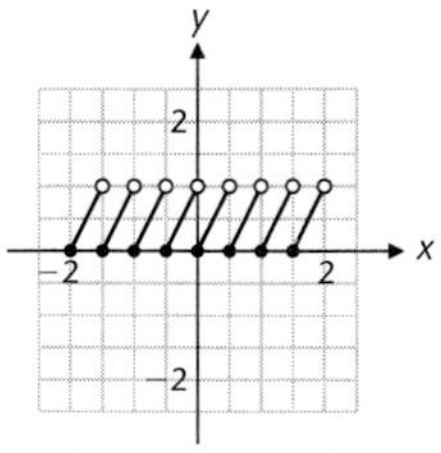

57. $x = 2, -1 \pm i\sqrt{3}$ *(2-5)* **58.** $x = 8{,}800$ books *(2-2)* **59.** $|p - 200| \le 10$ *(2-2)*
60. (A) Profit: $\$5.5 < p < \$8$ or $(\$5.5, \$8)$ (B) Loss: $\$0 \le p < \$5.5$ or $p > \$8$. $[\$0, \$5.5) \cup (\$8, \infty)$ *(2-3)*
61. 40 mi from A to B and 75 mi from B to C or 75 mi from A to B and 40 mi from B to C *(2-5)*
62. $x = -900(3.29) + 4{,}571$; 1,610 bottles *(2-2)*

63. $C(x) = \begin{cases} 0.06x & \text{if } 0 \le x \le 60 \\ 0.05x + 0.6 & \text{if } 60 < x \le 150 \\ 0.04x + 2.1 & \text{if } 150 < x \le 300 \\ 0.03x + 5.1 & \text{if } 300 < x \end{cases}$ *(1-4)*

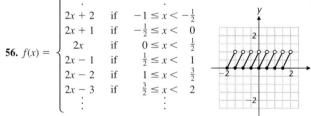

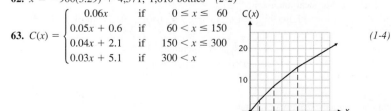

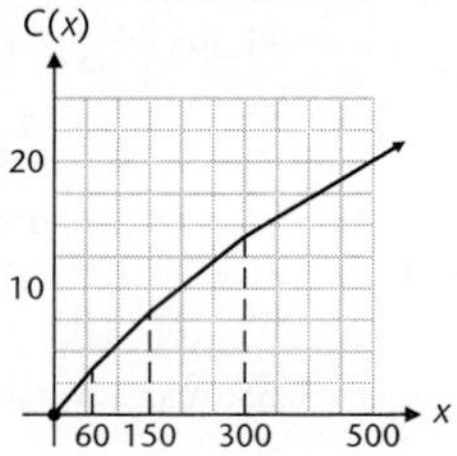

64. (A) $A(x) = 80x - 2x^2$ (B) $0 < x < 40$ (C) 20 feet by 40 feet *(1-4, 2-3)*

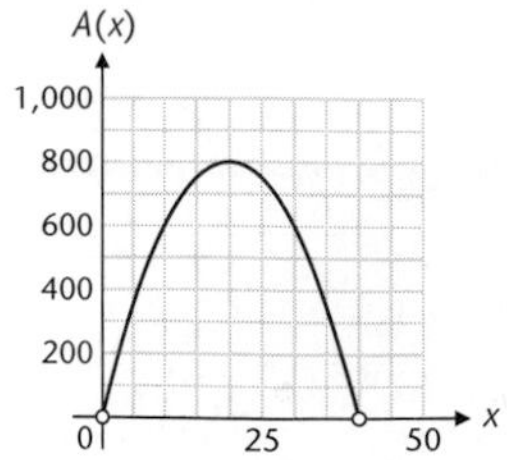

65. (A) $f(1) = 1; f(2) = 0; f(3) = 1; f(4) = 0$ (B) $f(n) = \begin{cases} 1 & \text{if } n \text{ is an odd integer} \\ 0 & \text{if } n \text{ is an even integer} \end{cases}$ *(1-4)*

66. (A) 30,000 bushels (B) The demand decreases to 20,000 bushels. (C) The demand increases to 40,000 bushels.

(E)

q	20	25	30	35	40
p	340	332	325	320	315

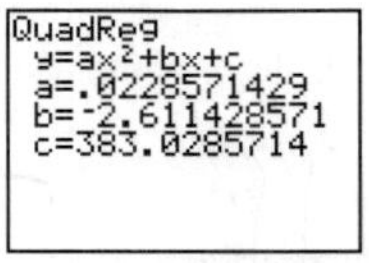

(1-1, 1-2, 2-1)

67. (A)
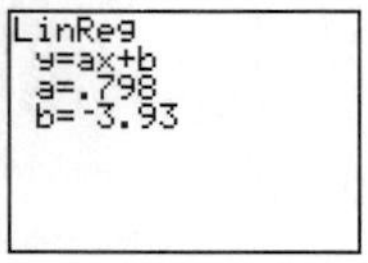

(B)
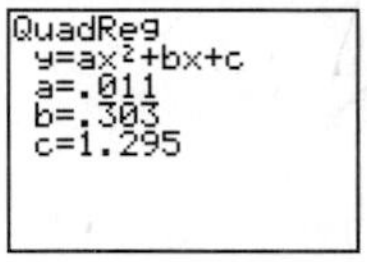

(C) The quadratic model is a better fit.

MPH	Linear Model	Quadratic Model
15	8.0	8.3
20	12.0	11.8
25	16.0	15.7
30	20.0	20.3

(D) Range at 17 MPH: 19 hours and 320 miles; Range at 28 MPH: 10 hours and 280 miles *(2-1, 2-3)*

CHAPTER 3 Exercise 3-1

1. c **3.** d **5.** h **7.** h, k **9.** $2m + 1, R = 0$ **11.** $4x - 5, R = 11$ **13.** $x^2 + x + 1, R = 0$

15. $2y^2 - 5y + 13, R = -27$ **17.** $\frac{x^2 + 3x - 7}{x - 2} = x + 5 + \frac{3}{x - 2}$ **19.** $\frac{4x^2 + 10x - 9}{x + 3} = 4x - 2 - \frac{3}{x + 3}$

21. $\frac{2x^3 - 3x + 1}{x - 2} = 2x^2 + 4x + 5 + \frac{11}{x - 2}$ **23.** 4 **25.** 3 **27.** -6 **29.** $3x^3 - 3x^2 + 3x - 4, R = 0$

31. $x^4 - x^3 + x^2 - x + 1, R = 0$ **33.** $3x^3 - 7x^2 + 21x - 67, R = 200$ **35.** $2x^5 - 3x^4 - 15x^3 + 2x + 10, R = 0$

37. $4x^3 - 6x - 2, R = 2$ **39.** $4x^2 - 2x - 4, R = 0$ **41.** $3x^3 - 0.8x^2 + 1.68x - 2.328, R = 0.0688$

43. $3x^4 - 0.4x^3 + 5.32x^2 - 4.256x - 3.5952, R = -0.12384$

45. (A) $P(x) \to \infty$ as $x \to \infty$ and $P(x) \to -\infty$ as $x \to -\infty$; three intercepts and two local extrema
(B) x intercepts: $-0.86, 1.68, 4.18$; local maximum: $P(0.21) \approx 6.21$; local minimum: $P(3.12) \approx -6.06$

47. (A) $P(x) \to -\infty$ as $x \to \infty$ and $P(x) \to \infty$ as $x \to -\infty$; three intercepts and two local extrema
(B) x intercept: 4.47; local minimum: $P(-0.12) \approx 4.94$; local maximum: $P(2.79) \approx 17.21$

49. (A) $P(x) \to \infty$ as $x \to \infty$ and as $x \to -\infty$; four intercepts and three local extrema
(B) x intercepts: none; local minimum: $P(-1.87) \approx 5.81$; local maximum: $P(-0.28) \approx 12.43$; local minimum: $P(1.41) \approx 4.59$

51. $P(x) = x^3$ **53.** No such polynomial exists. **55.** $2x^2 - 3x + 2, R = 0$ **57.** $x^2 + (-3 + i)x - 3i, R = 0$

59. (A) -5 (B) $-40i$ (C) 0 (D) 0

61. x intercepts: $-12.69, -0.72, 4.41$; local maximum: $P(2.07) \approx 96.07$; local minimum: $P(-8.07) \approx -424.07$

63. x intercepts: $-16.06, 0.50, 15.56$; local maximum: $P(-9.13) \approx 65.86$; local minimum: $P(9.13) \approx -55.86$

65. x intercepts: $-16.15, -2.53, 1.56, 14.12$; local minimum: $P(-11.68) \approx -1{,}395.99$; local maximum: $P(-0.50) \approx 95.72$; local minimum: $P(9.92) \approx -1{,}140.27$

67. x intercepts: 1, 1.09; local minimum: $P(1.05) \approx -0.20$; local maximum: $P(6.01) \approx 605.01$; local minimum: $P(10.94) \approx 9.70$

69. (A) In both cases the coefficient of x is a_2, the constant term $a_2r + a_1$, and the remainder is $(a_2r + a_1)r + a_0$.
(B) The remainder expanded is $a_2r^2 + a_1r + a_0 = P(r)$.

71. $P(-2) = 81$; $P(1.7) = 6.2452$ or 6.2

73. (A) $R(x) = 0.0004x^3 - x^2 + 569x$ (B) 364 air conditioners; price: \$258; max revenue: \$93,911

75. (A) $V(x) = (1 + 2x)(2 + 2x)(4 + 2x) - 8$ (B) 0.097 ft

77. (A)

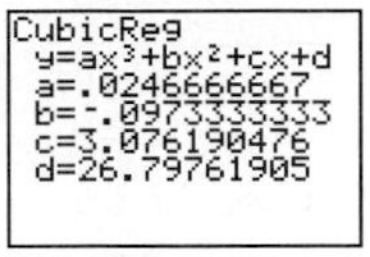

(B) \$1,072.8 billion

79. (A)

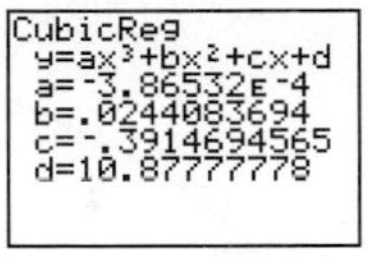

(B) 7.5 marriages per 1,000 population

Exercise 3-2

1. -8 (multiplicity 3), 6 (multiplicity 2); degree of $P(x)$ is 5 **3.** -4 (multiplicity 3), 3 (multiplicity 2); -1; degree of $P(x)$ is 6 **5.** $P(x) = (x - 3)^2(x + 4)$; degree 3 **7.** $P(x) = (x + 7)^3[x - (-3 + \sqrt{2})][x - (-3 - \sqrt{2})]$; degree 5 **9.** $P(x) = [x - (2 - 3i)][x - (2 + 3i)](x + 4)^2$; degree 4 **11.** $(x + 2)(x - 1)(x - 3)$; degree 3 **13.** $(x + 2)^2(x - 1)^2$; degree 4 **15.** $(x + 3)(x + 2)\,x(x - 1)(x - 2)$; degree 5 **17.** Yes **19.** Yes **21.** $\pm 1, \pm 2, \pm 3, \pm 6$ **23.** $\pm 1, \pm 2, \pm 4, \pm\frac{1}{3}, \pm\frac{2}{3}, \pm\frac{4}{3}$ **25.** $\pm 1, \pm 3, \pm\frac{1}{2}, \pm\frac{3}{2}, \pm\frac{1}{3}, \pm\frac{1}{4}, \pm\frac{3}{4}, \pm\frac{1}{6}, \pm\frac{1}{12}$ **27.** $P(x) = (x + 4)^2(x + 1)$ **29.** $P(x) = (x - 1)(x + 1)(x - i)(x + i)$ **31.** $P(x) = (2x - 1)[x - (4 + 5i)][x - (4 - 5i)]$ **33.** $\frac{1}{2}, 1 \pm \sqrt{2}$ **35.** -2 (double), $\pm\sqrt{5}$ **37.** $\pm 2, 1 \pm \sqrt{2}$ **39.** $\pm 1, \frac{3}{2}, \pm i$ **41.** $2, 3, -5$ **43.** $0, 2, -\frac{2}{5}, \frac{1}{2}$ **45.** 2 (double), $\frac{1}{2} \pm \frac{1}{2}\sqrt{3}$ **47.** -1 (double), $-\frac{1}{3}, 2 \pm i$ **49.** $P(x) = (x + 2)(3x + 2)(2x - 1)$ **51.** $P(x) = (x + 4)[x - (1 + \sqrt{2})][x - (1 - \sqrt{2})]$ **53.** $P(x) = (x - 2)(x + 1)(2x + 1)(2x - 1)$ **55.** $x^2 - 8x + 41$ **57.** $x^2 - 6x + 25$ **59.** $x^2 - 2ax + a^2 + b^2$ **61.** -1 and $3 + i$ **63.** $5i$ and 3 **65.** $2 + i, 2 - i, \sqrt{2}, -\sqrt{2}$ **67.** $\sqrt{6}$ is a zero of $P(x) = x^2 - 6$, but $P(x)$ has no rational zeros. **69.** $\sqrt[3]{5}$ is a zero of $P(x) = x^3 - 5$, but $P(x)$ has no rational zeros. **71.** $\frac{1}{3}, 6 \pm 2\sqrt{3}$ **73.** $\frac{3}{2}, -\frac{5}{2}, \pm 4i$ **75.** $\frac{3}{2}$ (double), $4 \pm \sqrt{6}$ **77.** (A) 3 (B) $-\frac{1}{2} + \frac{\sqrt{3}}{2}i$ and $-\frac{1}{2} - \frac{\sqrt{3}}{2}i$

79. maximum of n; minimum of 1

81. No, since $P(x)$ is not a polynomial with real coefficients (the coefficient of x is the imaginary number $2i$).

83. 2 feet **85.** 0.5×0.5 inches or 1.59×1.59 inches

Exercise 3-3

1. There is at least one x intercept in each of the intervals $(-5, -1)$, $(-1, 3)$, and $(5, 8)$.

3. There is at least one x intercept in each of the intervals $(-6, -4)$, $(-4, 0)$, $(2, 4)$, and $(4, 7)$.

5. Zeros in $(0, 1)$, $(3, 4)$, and $(4, 5)$ **7.** Zeros in $(-3, -2)$, $(-2, -1)$, and $(1, 2)$ **9.** Upper bound: 2; lower bound: -2 **11.** Upper bound: 3; lower bound: -2 **13.** Upper bound: 2; lower bound: -3

15. (A) Upper bound: 4; lower bound: -2; real zeros in $(-2, -1)$, $(0, 1)$, and $(3, 4)$ (B) 5 intervals, 3.2

17. (A) Upper bound: 3; lower bound: -2; real zero in $(-2, -1)$ (B) 6 intervals, -1.4

19. (A) Upper bound: 4; lower bound: -3; real zeros in $(-3, -2)$, $(-1, 0)$, $(1, 2)$, and $(3, 4)$ (B) 4 intervals, 3.1

21. (A) Upper bound: 3; lower bound: -2; real zeros in $(-2, -1)$ and $(-1, 0)$ (B) 5 intervals, -0.5

23. (A) Upper bound: 3; lower bound: -1 (B) 2.25 **25.** (A) Upper bound: 3; lower bound: -4 (B) -3.51, 2.12

27. (A) Upper bound: 2; lower bound: -3 (B) -2.09, 0.75, 1.88 **29.** (A) Upper bound: 1; lower bound: -1 (B) 0.83

35. -1.83 (double zero); 3.83 (double zero) **37.** -1.24 (double zero); 2 (simple zero); 3.24 (double zero)

39. -0.22 (double zero); 2 (simple zero); 2.22 (double zero) **41.** (A) Upper bound: 30; lower bound: -10 (B) -1.29, 0.31, 24.98

43. (A) Upper bound: 30; lower bound: -40 (B) -36.53, -2.33, 2.40, 24.46

45. (A) Upper bound: 20; lower bound: -10 (B) -7.47, 14.03

47. (A) Upper bound: 30; lower bound: -20 (B) -17.66, 2.5 (double zero), 22.66

49. (A) Upper bound: 40; lower bound: -40 (B) -30.45, 9.06, 39.80 **51.** $x^4 - 3x^2 - 2x + 4 = 0$; (1, 1) and (1.659, 2.752)

53. $4x^3 - 84x^2 + 432x - 600 = 0$; 2.319 in. or 4.590 in. **55.** $x^3 - 15x^2 + 30 = 0$; 1.490 ft

Exercise 3-4

1. $g(x)$ **3.** $h(x)$ **5.** Domain: $(-\infty, -1) \cup (-1, \infty)$; x intercept: 2 **7.** Domain: $(-\infty, -4) \cup (-4, 4) \cup (4, \infty)$; x intercepts: $-1, 1$ **9.** Domain: $(-\infty, -3) \cup (-3, 4) \cup (4, \infty)$; x intercepts: $-2, 3$ **11.** Domain: all real numbers; x intercept: 0

13. Vertical asymptote: $x = 4$; horizontal asymptote: $y = 2$ **15.** Vertical asymptote: $x = -4$, $x = 4$; horizontal asymptote: $y = \frac{2}{3}$

17. No vertical asymptotes; horizontal asymptote: $y = 0$ **19.** Vertical asymptotes: $x = -1$, $x = \frac{5}{3}$; no horizontal asymptote

21.

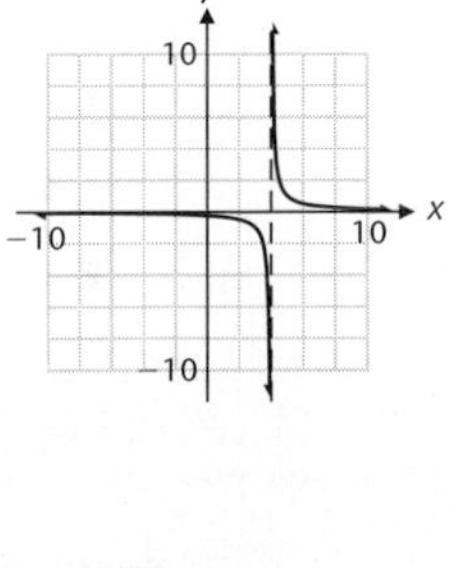

23.

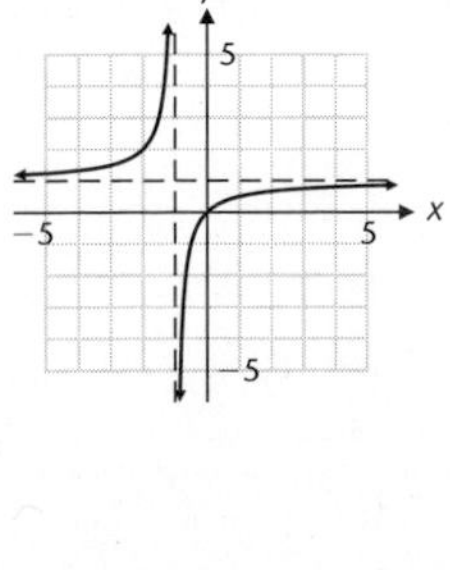

25.

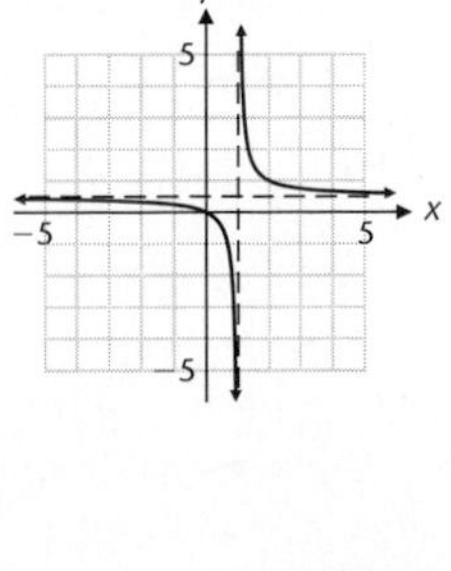

27.

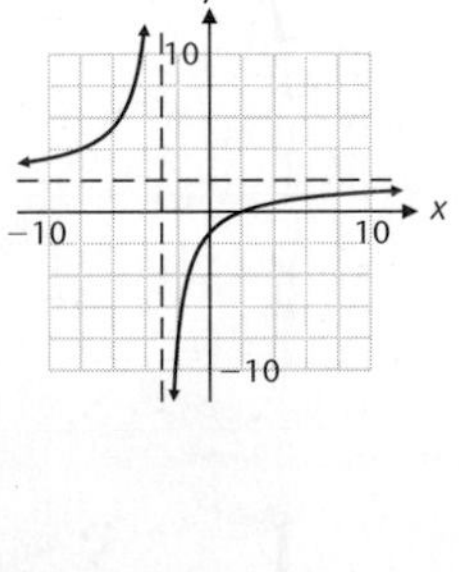

29.

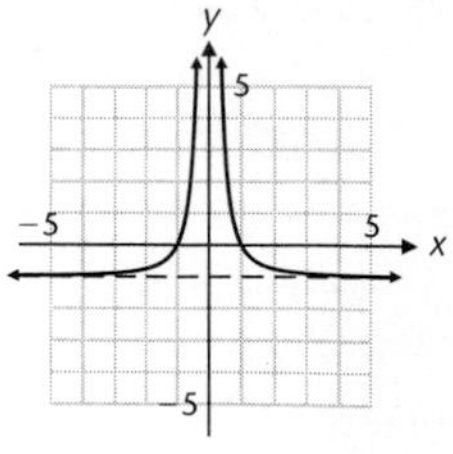

31.

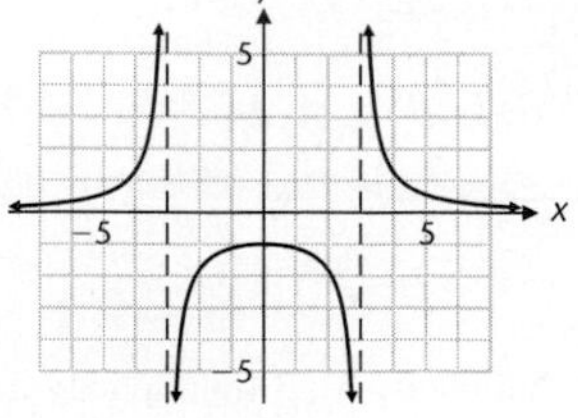

33.

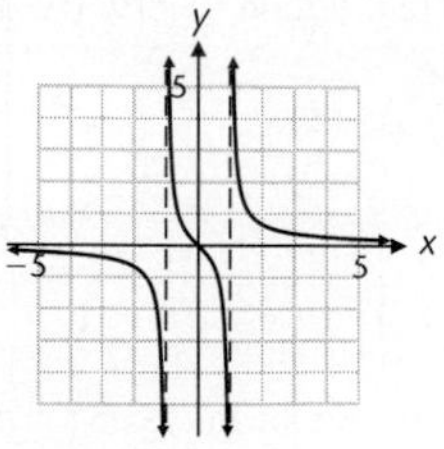

35.

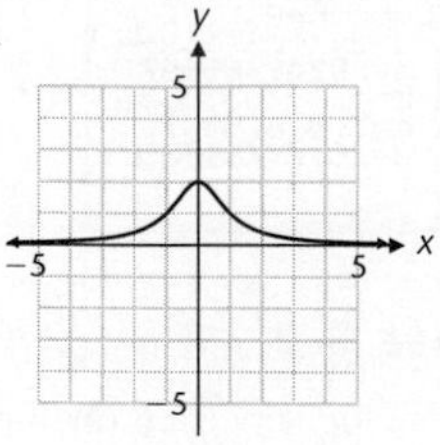

37.

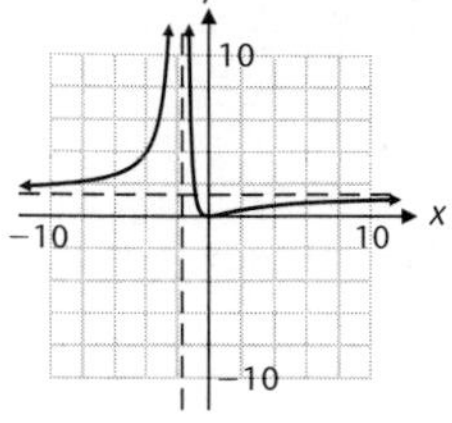

39.

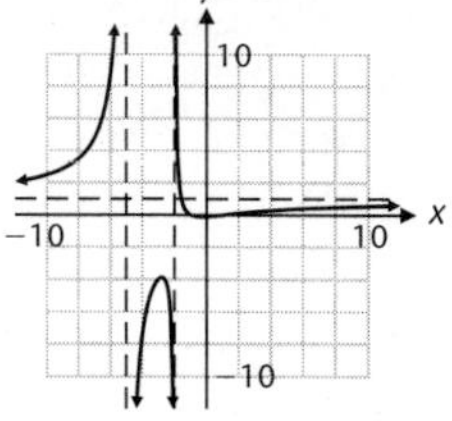

41. The maximum number of x intercepts is 2 and the minimum number is 0. For example, $\dfrac{x^2 - 1}{x^2}$ has two x intercepts and $\dfrac{x^2 + 1}{x^2}$ has none.

43. Vertical asymptote: $x = 1$; oblique asymptote: $y = 2x + 2$ **45.** Oblique asymptote: $y = x$

47. Vertical asymptote: $x = 0$; oblique asymptote: $y = 2x - 3$

49. $f(x) \to 5$ as $x \to \infty$ and $f(x) \to -5$ as $x \to -\infty$; the lines $y = 5$ and $y = -5$ are horizontal asymptotes.

51. $f(x) \to 4$ as $x \to \infty$ and $f(x) \to -4$ as $x \to -\infty$; the lines $y = 4$ and $y = -4$ are horizontal asymptotes.

53.

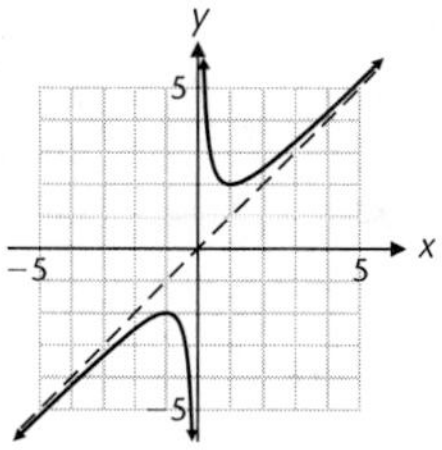

55.

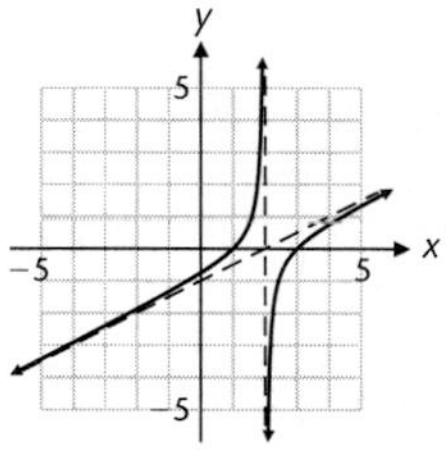

57.

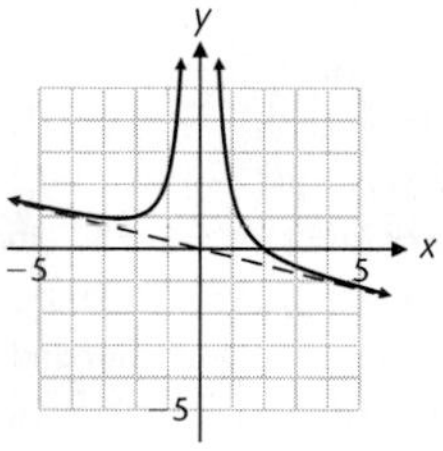

59. Let $p(x) = x^2 - 1$; $[f(x) - p(x)] \to 0$ as $x \to \infty$ and as $x \to -\infty$ **61.** $p(x) = x^3 + x$; $[f(x) - p(x)] \to 0$ as $x \to \infty$ and as $x \to -\infty$

63. Domain: $x \neq 2$, or $(-\infty, 2) \cup (2, \infty)$; $f(x) = x + 2$

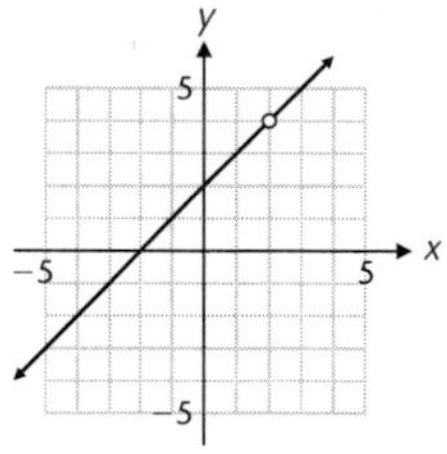

65. Domain: $x \neq 2, -2$ or $(-\infty, -2) \cup (-2, 2) \cup (2, \infty)$; $r(x) = \dfrac{1}{x - 2}$

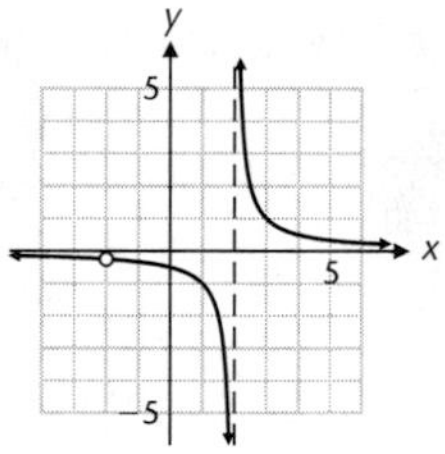

67. As $t \to \infty$, $N \to 50$

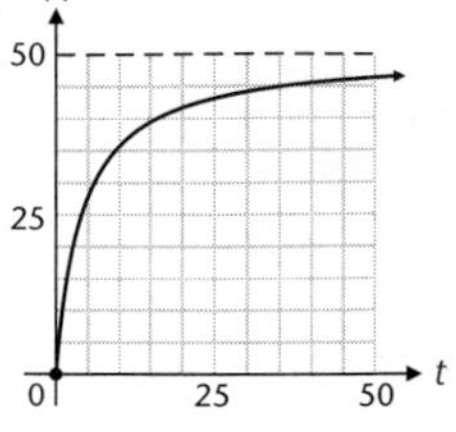

69. As $t \to \infty$, $N \to 5$

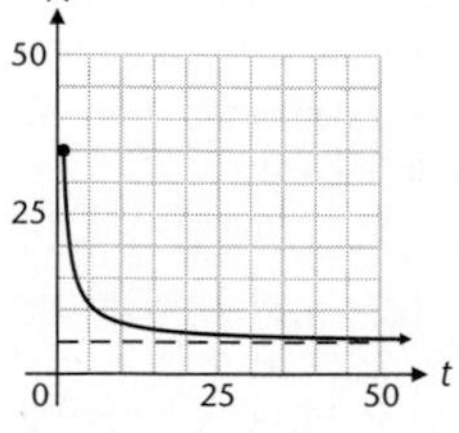

71. (A) $\overline{C}(n) = 25n + 175 + \dfrac{2{,}500}{n}$

(B) 10 yr

(C)

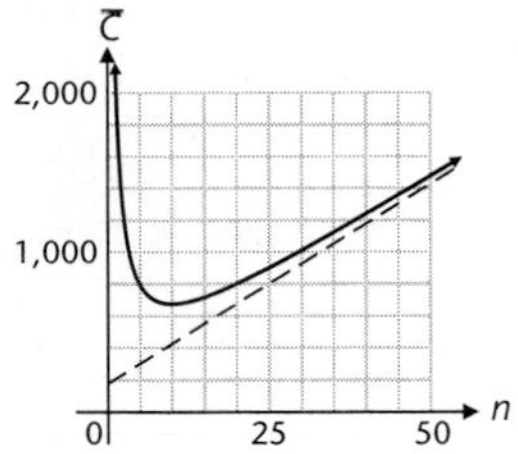

73. (A) $L(x) = 2x + \dfrac{450}{x} = \dfrac{2x^2 + 450}{x}$

(B) $(0, \infty)$ (C) 15 ft by 15 ft

(D)

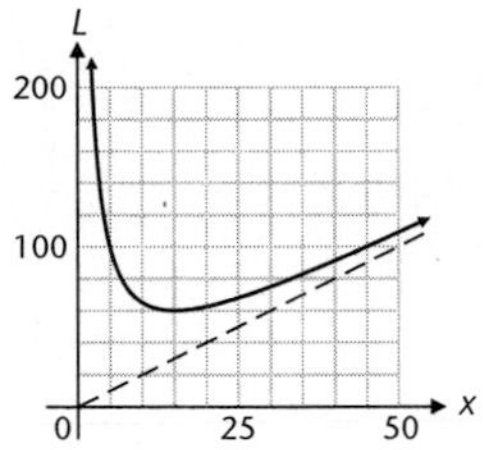

Chapter 3 Review

1. $2x^3 + 3x^2 - 1 = (x + 2)(2x^2 - x + 2) - 5$ *(3-1)* **2.** $P(3) = -8$ *(3-1, 3-2)* **3.** 2, -4, -1 *(3-2)* **4.** $1 - i$ is a zero. *(3-2)*

5. (A) $P(x) = (x + 2)x(x - 2) = x^3 - 4x$ (B) $P(x) \to \infty$ as $x \to \infty$ and $P(x) \to -\infty$ as $x \to -\infty$ *(3-1, 3-2)*

6. Lower bound: -2, -1; upper bound: 4 *(3-3)* **7.** $P(1) = -5$ and $P(2) = 1$ are of opposite sign. *(3-2)*

8. $\pm 1, \pm 2, \pm 3, \pm 6$ *(3-2)* **9.** -1, 2, 3 *(3-2)*

10. (A) Domain is $(-\infty, -4) \cup (-4, \infty)$; x intercept is $\frac{3}{2}$. (B) Domain is $(-\infty, -2) \cup (-2, 3) \cup (3, \infty)$; x intercept is 0. *(3-4)*

11. (A) Horizontal asymptote: $y = 2$; vertical asymptote: $x = -4$

(B) Horizontal asymptote: $y = 0$; vertical asymptotes: $x = -2$, $x = 3$ *(3-4)*

12. (A) The graph of $P(x)$ has three x intercepts and two turning points; $P(x) \to \infty$ as $x \to \infty$ and $P(x) \to -\infty$ as $x \to -\infty$ (B) 3.53 *(3-1, 3-3)*

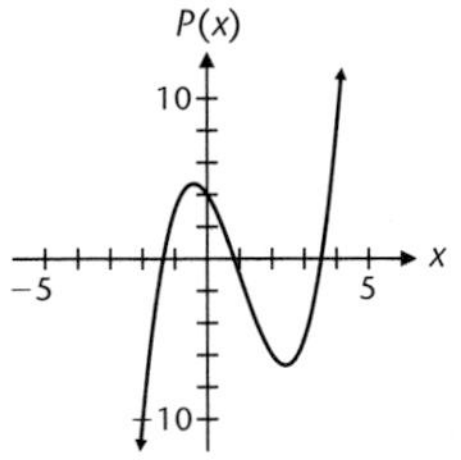

13. $Q(x) = 8x^3 - 12x^2 - 16x - 8$, $R = 5$; $P(\frac{1}{4}) = 5$ *(3-1)* **14.** -4 *(3-1)* **15.** $P(x) = [x - (1 + \sqrt{2})][x - (1 - \sqrt{2})]$ *(3-2)*

16. Yes, since $P(-1) = 0$, $x - (-1) = x + 1$ must be a factor. *(3-2)* **17.** 4, $-\frac{1}{2}$, -2 *(3-2)* **18.** $(x - 4)(2x + 1)(x + 2)$ *(3-2)*

19. No rational zeros *(3-2)* **20.** -1, $\dfrac{1}{2}$, and $\dfrac{1 \pm i\sqrt{3}}{2}$ *(3-2)* **21.** $(x + 1)(2x - 1)\left(x - \dfrac{1 + i\sqrt{3}}{2}\right)\left(x - \dfrac{1 - i\sqrt{3}}{2}\right)$ *(3-2)*

22. (A) -0.24 (double zero); 2 (simple zero); 4.24 (double zero)

(B) -0.24 can be approximated with a maximum routine; 2 can be approximated with the bisection; 4.24 can be approximated with a minimum routine. *(3-3)*

23. (A) Upper bound: 7; lower bound: -5 (B) Four intervals (C) -4.67, 6.62 *(3-3)*

24. (A) Domain is $(-\infty, -1) \cup (-1, \infty)$; x intercept: $x = 1$; y intercept: $y = -\frac{1}{2}$.

(B) Vertical asymptote: $x = -1$; horizontal asymptote: $y = \frac{1}{2}$

(C) *(3-4)*

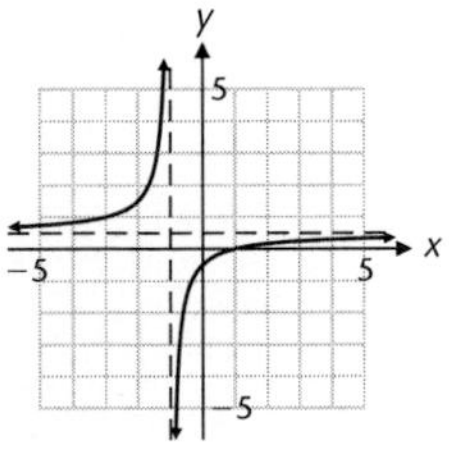

25. $P(x) = [x^2 + (1 + i)x + (3 + 2i)][x - (1 + i)] + 3 + 5i$ *(3-1)* **26.** $P(x) = (x + \frac{1}{2})^2 (x + 3)(x - 1)^3$. The degree is 6. *(3-2)*

27. $P(x) = (x + 5)[x - (2 - 3i)][x - (2 + 3i)]$. The degree is 3. *(3-2)* **28.** $\frac{1}{2}$, ± 2, $1 \pm \sqrt{2}$ *(3-2)*

29. $(x - 2)(x + 2)(2x - 1)[x - (1 - \sqrt{2})][x - (1 + \sqrt{2})]$ *(3-2)*

30. zeros: 0.91, 1; local minimum: $P(-8.94) \approx 9.70$; local maximum: $P(-4.01) \approx 605.01$; local minimum: $P(0.95) \approx -0.20$ *(3-1)*

31. Since $P(x)$ changes sign three times, the minimal degree is 3. *(3-3)*

32. $P(x) = a(x - r)(x^2 - 2x + 5)$ and since the constant term, $-5ar$, must be an integer, r must be a rational number. *(3-2)*

33. (A) 3 (B) $-\frac{3}{2} \pm \frac{3i\sqrt{3}}{2}$ *(3-2)* **34.** (A) Upper bound: 30; lower bound: −30 (B) −23.54, 21.57 *(3-3)*

35.

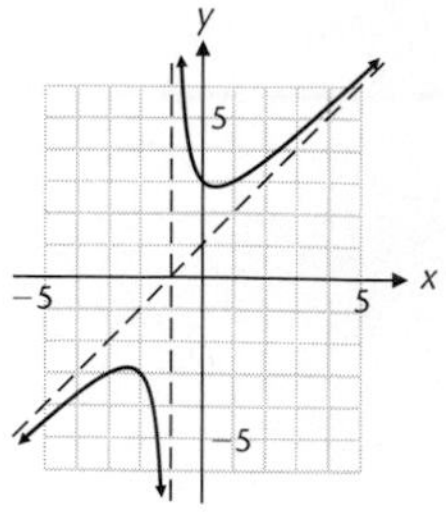

(3-4) **36.** $y = 2$ and $y = -2$ *(3-4)* **37.** $2x^3 - 32x + 48 = 0$, 4×12 ft or 5.211×9.211 ft *(3-2)*

38. $x^3 + 27x^2 - 729 = 0$, 4.789 ft *(3-3)* **39.** $4x^3 - 70x^2 + 300x - 300 = 0$, 1.450 in. or 4.465 in. *(3-3)*

40. $x^4 - 7x^2 - 2x + 8 = 0$, $(-2, 4)$, $(-1.562, 2.440)$, $(1, 1)$, $(2.562, 6.564)$ *(3-2)*

41. (A)

```
CubicReg
 y=ax³+bx²+cx+d
 a=-.0102791694
 b=.8486894107
 c=-2.575453575
 d=221.667258
```

(B) 339 refrigerators (C) 36 ads *(3-1)* **42.** (B)

```
CubicReg
 y=ax³+bx²+cx+d
 a=-.0048619529
 b=.5893160173
 c=-22.74548148
 d=341.7396941
```

(C) 25 yr, 41 yr, 55 yr *(3-1)*

CHAPTER 4 Exercise 4-1

1.

x	−3	−2	−1	0	1	2	3
$(f+g)(x)$	3	3	1	−1	−3	−3	−3

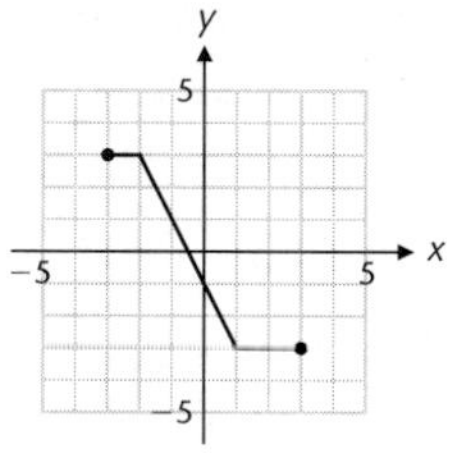

3. −2 **5.** 1 **7.** −2 **9.** 3

11. $(f+g)(x) = 5x + 1$; $(f-g)(x) = 3x - 1$; $(fg)(x) = 4x^2 + 4x$; $\left(\frac{f}{g}\right)(x) = \frac{4x}{x+1}$; domain $f + g, f - g, fg = (-\infty, \infty)$; domain of $f/g = (-\infty, -1) \cup (-1, \infty)$

13. $(f+g)(x) = 3x^2 + 1$; $(f-g)(x) = x^2 - 1$; $(fg)(x) = 2x^4 + 2x^2$; $\left(\frac{f}{g}\right)(x) = \frac{2x^2}{x^2+1}$; domain of each function: $(-\infty, \infty)$

15. $(f+g)(x) = x^2 + 3x + 4$; $(f-g)(x) = -x^2 + 3x + 6$; $(fg)(x) = 3x^3 + 5x^2 - 3x - 5$; $\left(\frac{f}{g}\right)(x) = \frac{3x+5}{x^2-1}$; domain $f + g, f - g, fg$: $(-\infty, \infty)$; domain of f/g: $(-\infty, -1) \cup (-1, 1) \cup (1, \infty)$

17. $(f \circ g)(x) = (x^2 - x + 1)^3$; domain: $(-\infty, \infty)$; $(g \circ f)(x) = x^6 - x^3 + 1$; domain: $(-\infty, \infty)$

19. $(f \circ g)(x) = |2x + 4|$; domain: $(-\infty, \infty)$; $(g \circ f)(x) = 2|x + 1| + 3$; domain: $(-\infty, \infty)$

21. $(f \circ g)(x) = (2x^3 + 4)^{1/3}$; domain: $(-\infty, \infty)$; $(g \circ f)(x) = 2x + 4$; domain: $(-\infty, \infty)$

23. $(f \circ g)(x) = (g \circ f)(x) = x$; symmetric with respect to the line $y = x$

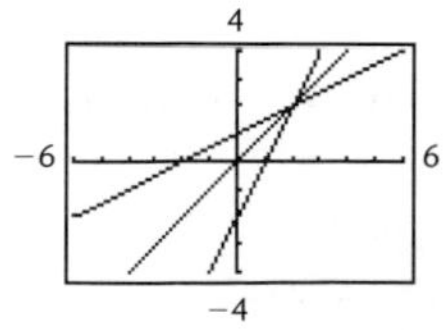

25. $(f \circ g)(x) = (g \circ f)(x) = x$; symmetric with respect to the line $y = x$

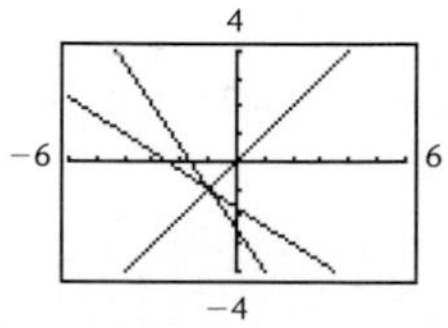

27. $(f+g)(x) = \sqrt{2-x} + \sqrt{x+3}$; $(f-g)(x) = \sqrt{2-x} - \sqrt{x+3}$; $(fg)(x) = \sqrt{6 - x - x^2}$; $\left(\frac{f}{g}\right)(x) = \sqrt{\frac{2-x}{x+3}}$

The domain of the functions $f + g$, $f - g$, and fg is $[-3, 2]$. The domain of $\frac{f}{g}$ is $(-3, 2]$.

29. $(f+g)(x) = 2\sqrt{x} - 2$; $(f-g)(x) = 6$; $(fg)(x) = x - 2\sqrt{x} - 8$; $\left(\frac{f}{g}\right)(x) = \frac{\sqrt{x}+2}{\sqrt{x}-4}$

The domain of $f + g$, $f - g$, and fg is $[0, \infty)$. Domain of $\frac{f}{g} = [0, 16) \cup (16, \infty)$.

31. $(f+g)(x) = \sqrt{x^2+x-6} + \sqrt{7+6x-x^2}$; $(f-g)(x) = \sqrt{x^2+x-6} - \sqrt{7+6x-x^2}$; $(fg)(x) = \sqrt{-x^4+5x^3+19x^2-29x-42}$; $\left(\frac{f}{g}\right)(x) = \sqrt{\frac{x^2+x-6}{7+6x-x^2}}$. The domain of the functions $f+g$, $f-g$, and fg is $[2, 7]$. The domain of $\frac{f}{g}$ is $[2, 7)$.

33. $(f \circ g)(x) = \sqrt{x-4}$; domain: $[4, \infty)$; $(g \circ f)(x) = \sqrt{x} - 4$; domain: $[0, \infty)$

35. $(f \circ g)(x) = \frac{1}{x} + 2$; domain: $(-\infty, 0) \cup (0, \infty)$; $(g \circ f)(x) = \frac{1}{x+2}$; domain: $(-\infty, -2) \cup (-2, \infty)$

37. $(f \circ g)(x) = \frac{1}{|x-1|}$; domain: $(-\infty, 1) \cup (1, \infty)$; $(g \circ f)(x) = \frac{1}{|x|-1}$; domain of $g \circ f$: $(-\infty, -1) \cup (-1, 1) \cup (1, \infty)$

39. (e) **41.** (a) **43.** (c) **45.** $g(x) = 2x - 7$; $f(x) = x^4$; $h(x) = (f \circ g)(x)$ **47.** $g(x) = 4 + 2x$; $f(x) = x^{1/2}$; $h(x) = (f \circ g)(x)$

49. $f(x) = x^7$; $g(x) = 3x - 5$; $h(x) = (g \circ f)(x)$ **51.** $f(x) = x^{-1/2}$; $g(x) = 4x + 3$; $h(x) = (g \circ f)(x)$

57. $(f+g)(x) = 2x$; $(f-g)(x) = \frac{2}{x}$; $(fg)(x) = x^2 - \frac{1}{x^2}$; $\left(\frac{f}{g}\right)(x) = \frac{x^2+1}{x^2-1}$

The domain of $f+g$, $f-g$, and fg is $(-\infty, 0) \cup (0, \infty)$. The domain of $\frac{f}{g}$ is $(-\infty, -1) \cup (-1, 0) \cup (0, 1) \cup (1, \infty)$.

59. $(f+g)(x) = 2$; $(f-g)(x) = \frac{-2x}{|x|}$; $(fg)(x) = 0$; $\left(\frac{f}{g}\right)(x) = 0$

The domain of $f+g$, $f-g$, and fg is $(-\infty, 0) \cup (0, \infty)$; domain of $\frac{f}{g}$ is $(0, \infty)$.

61. $(f \circ g)(x) = \sqrt{4-x^2}$; domain of $f \circ g$ is $[-2, 2]$; $(g \circ f)(x) = 4 - x$; domain of $g \circ f$ is $(-\infty, 4]$.

63. $(f \circ g)(x) = \frac{6x-10}{x}$; domain of $f \circ g$ is $(-\infty, 0) \cup (0, 2) \cup (2, \infty)$; $(g \circ f)(x) = \frac{x+5}{5-x}$; domain of $g \circ f$ is $(-\infty, 0) \cup (0, 5) \cup (5, \infty)$.

65. $(f \circ g)(x) = \sqrt{16-x^2}$; domain of $f \circ g$ is $[-4, 4]$; $(g \circ f)(x) = \sqrt{34-x^2}$; domain of $g \circ f$ is $[-5, 5]$.

67. $(f \circ g)(x) = \sqrt{2+x}$; domain of $f \circ g$ is $[-2, 3]$; the first graph is correct.

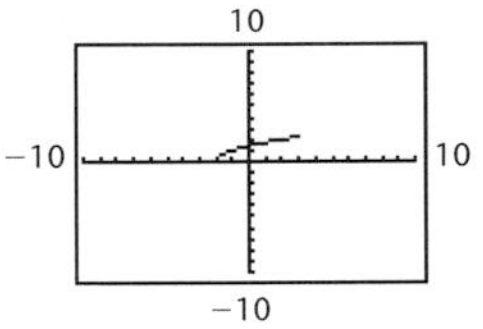

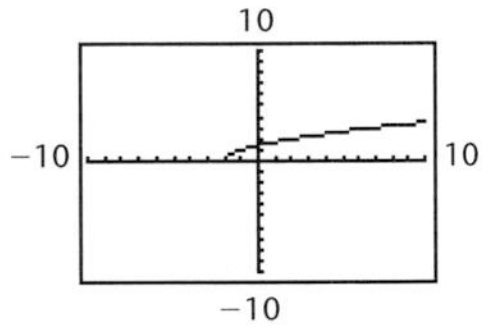

69. $(f \circ g)(x) = \sqrt{x^2+1}$; domain of $f \circ g$ is $(-\infty, -2] \cup [2, \infty)$; the first graph is correct.

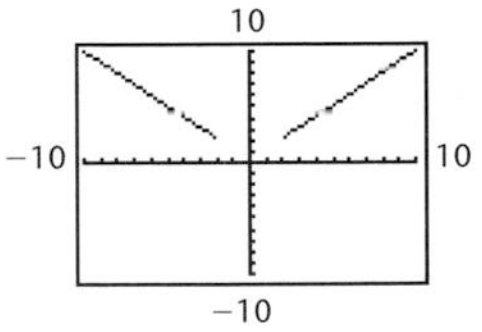

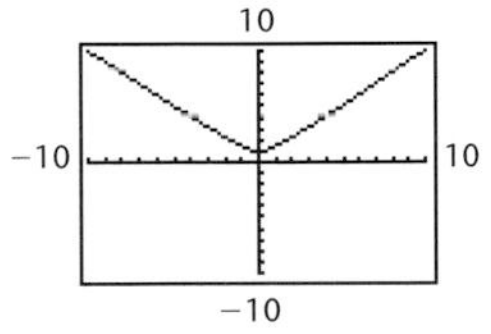

71. $(f \circ g)(x) = \sqrt{16-x^2}$; domain of $f \circ g$ is $[-3, 3]$; the first graph is correct.

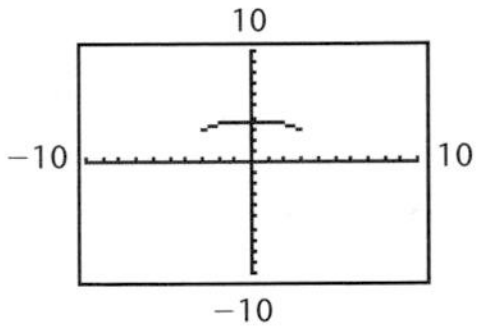

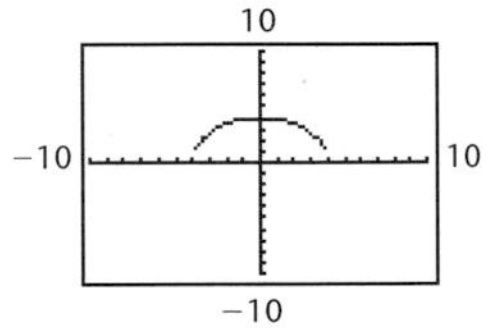

73. $P(p) = -70{,}000 + 6{,}000p - 200p^2$ **75.** $V(t) = 0.016\pi t^{2/3}$ **77.** (A) $r(h) = \frac{1}{2}h$ (B) $V(h) = \frac{1}{12}\pi h^3$ (C) $V(t) = \frac{0.125}{12}\pi t^{3/2}$

Exercise 4-2

1. One-to-one **3.** Not one-to-one **5.** One-to-one **7.** Not one-to-one **9.** One-to-one **11.** Not one-to-one
13. One-to-one **15.** One-to-one **17.** One-to-one **19.** Not one-to-one **21.** One-to-one
23. One-to-one **25.** Not one-to-one **27.** Not one-to-one **29.** One-to-one

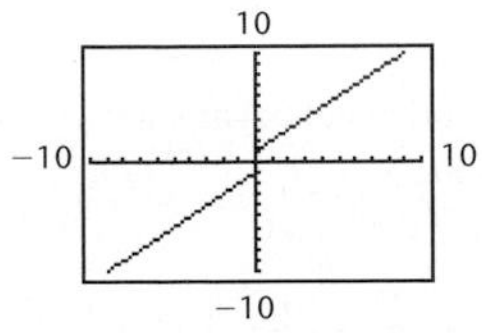

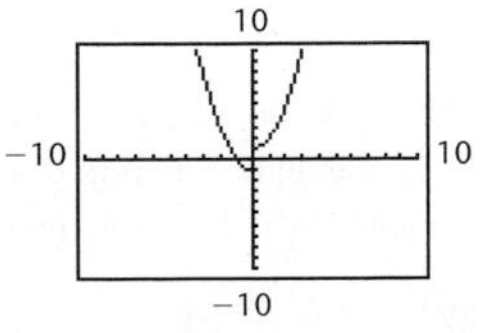

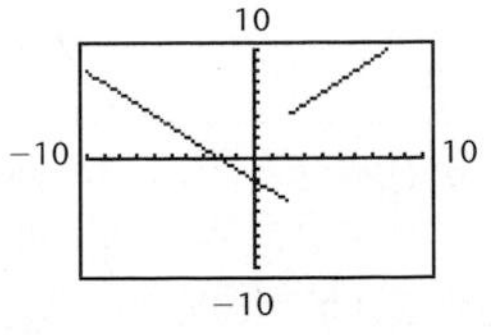

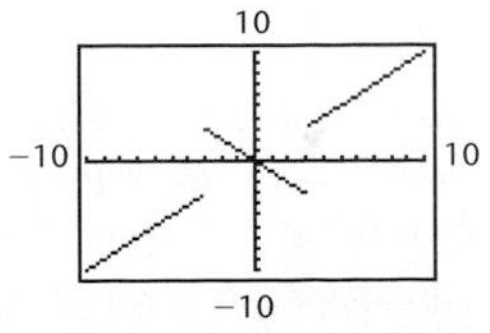

31. Range of $f^{-1} = [-4, 4]$; domain of $f^{-1} = [1, 5]$

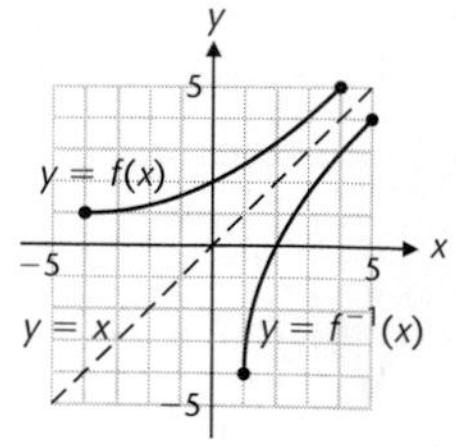

33. Range of $f^{-1} = [-5, 3]$; domain of $f^{-1} = [-3, 5]$

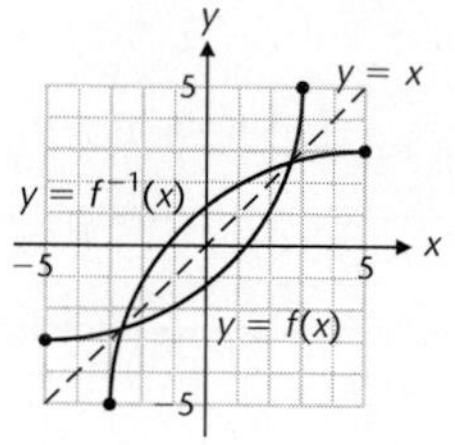

35.

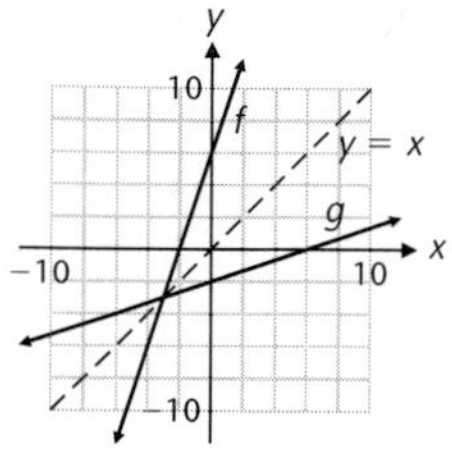

37.

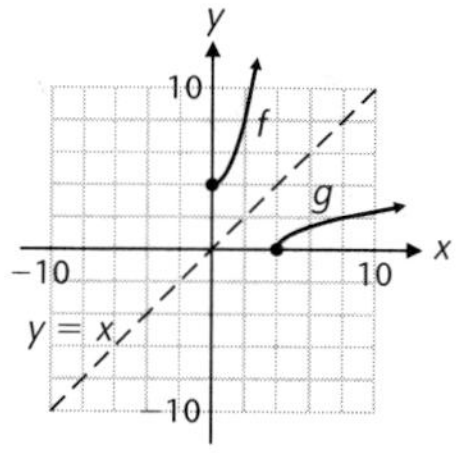

39.

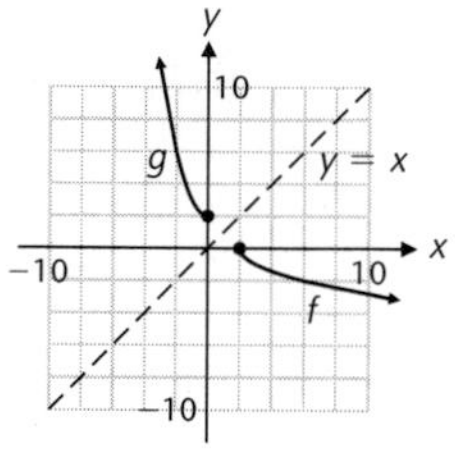

41. $f^{-1}(x) = \frac{1}{3}x$ **43.** $f^{-1}(x) = \dfrac{x+3}{4}$ **45.** $f^{-1}(x) = 10x - 6$ **47.** $f^{-1}(x) = \dfrac{x+2}{x}$ **49.** $f^{-1}(x) = \dfrac{2x}{1-x}$ **51.** $f^{-1}(x) = \dfrac{4x+5}{3x-2}$

53. $f^{-1}(x) = \sqrt[3]{x} - 1$ **55.** $f^{-1}(x) = (4 - x)^5 - 2$ **57.** $f^{-1}(x) = 16 - 4x^2, x \geq 0$ **59.** $f^{-1}(x) = (3 - x)^2 + 2, x \leq 3$

61. The x intercept of f is the y intercept of f^{-1} and the y intercept of f is the x intercept of f^{-1}.

63. $f^{-1}(x) = 1 + \sqrt{x - 2}$ **65.** $f^{-1}(x) = -1 - \sqrt{x + 3}$

67. $f^{-1}(x) = \sqrt{9 - x^2}$; domain of $f^{-1} = [-3, 0]$; range of $f^{-1} = [0, 3]$

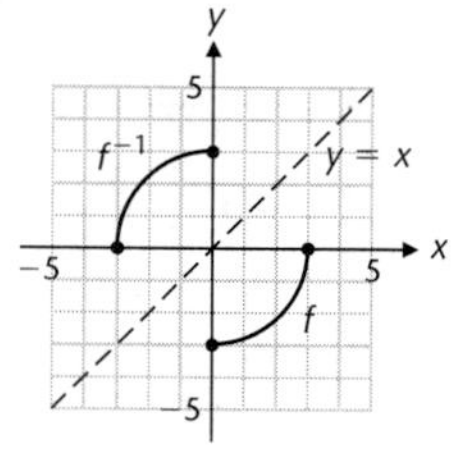

69. $f^{-1}(x) = -\sqrt{9 - x^2}$; domain: $f^{-1} = [0, 3]$; range of $f^{-1} = [-3, 0]$

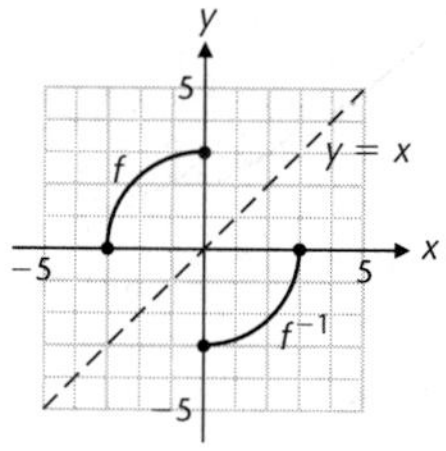

71. $f^{-1}(x) = \sqrt{2x - x^2}$; domain of $f^{-1} = [1, 2]$; range of $f^{-1} = [0, 1]$

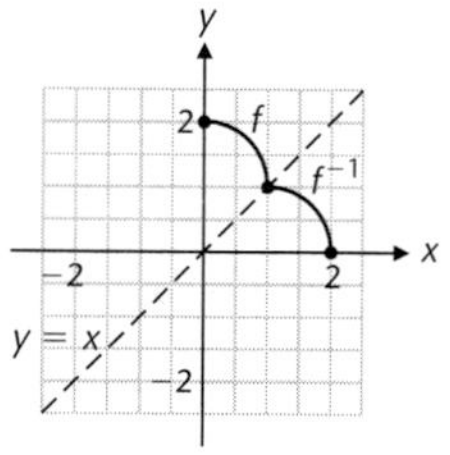

73. $f^{-1}(x) = -\sqrt{2x - x^2}$; domain of $f^{-1} = [0, 1]$; range of $f^{-1} = [-1, 0]$

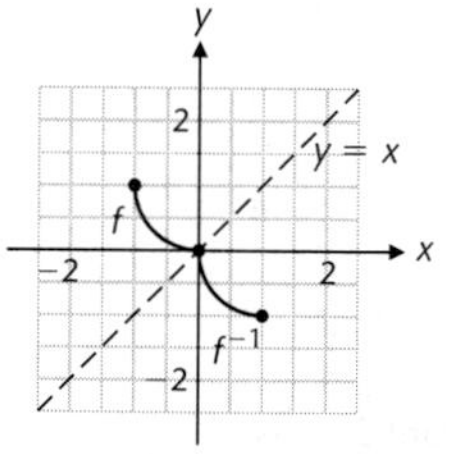

75. $f^{-1}(x) = \dfrac{x - b}{a}$ **77.** $a = 1$ and $b = 0$ or $a = -1$ and b arbitrary. **81.** (A) $f^{-1}(x) = 2 - \sqrt{x}$ (B) $f^{-1}(x) = 2 + \sqrt{x}$

83. (A) $f^{-1}(x) = 2 - \sqrt{4 - x^2}, 0 \leq x \leq 2$ (B) $f^{-1}(x) = 2 + \sqrt{4 - x^2}, 0 \leq x \leq 2$

85. (A) $200 \leq q \leq 1{,}000$ (B) $p = \dfrac{15{,}000}{q} - 5$; domain: $200 \leq q \leq 1{,}000$; range: $10 \leq p \leq 70$

87. (A) $r = 1.25w + 2.45$ (C) $w = 0.8r - 1.96$

Exercise 4-3

1. (A) g (B) n (C) f (D) m **3.** 16.24 **5.** 5.047 **7.** 4.469 **9.** Increasing; y intercept: 1; horizontal asymptote: $y = 0$

11. Decreasing; y intercept: 1; horizontal asymptote: $y = 0$ **13.** Increasing; y intercept: -1; horizontal asymptote: $y = 0$

15. Increasing; y intercept: 5; horizontal asymptote: $y = 0$ **17.** Increasing; y intercept: 22; horizontal asymptote: $y = -5$

19. 10^{2x+3} **21.** 3^{2x-1} **23.** $\dfrac{4^{3xz}}{5^{3yz}}$ **25.** $x = 2$ **27.** $x = -1, 3$ **29.** $x = \frac{2}{3}$ **31.** $x = -2$ **33.** $x = \frac{1}{2}$ **35.** $x = \frac{1}{2}, 1$

37. $a = 1$ or $a = -1$

41.
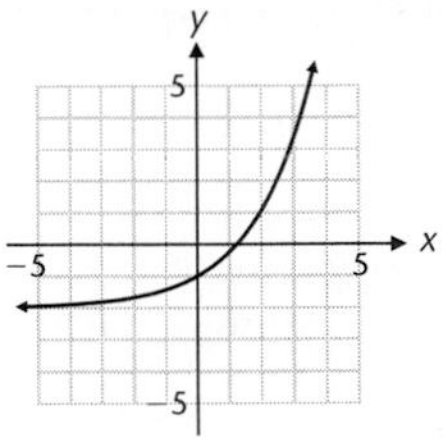

43.
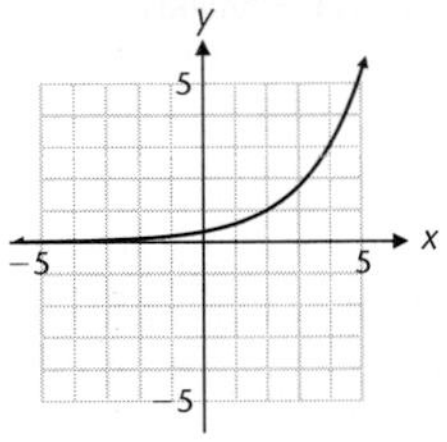

45.
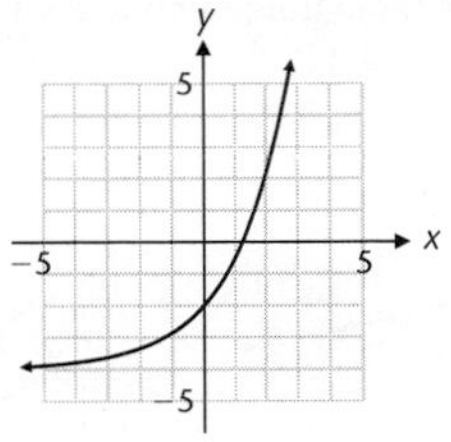

47.
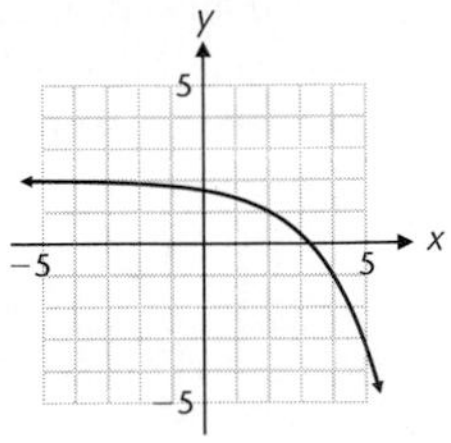

49.
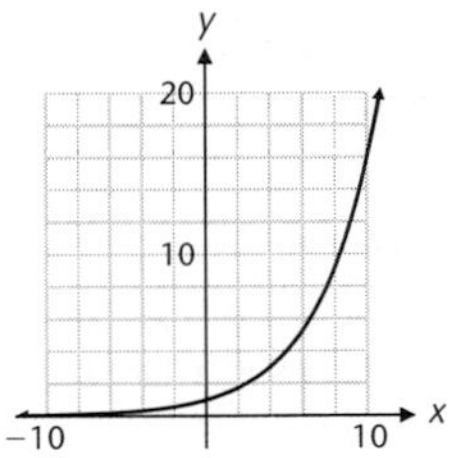

51.
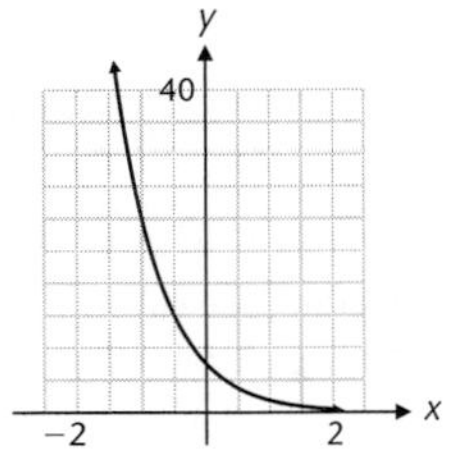

53.
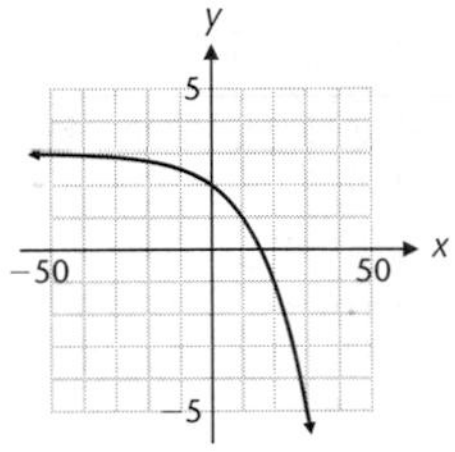

55.
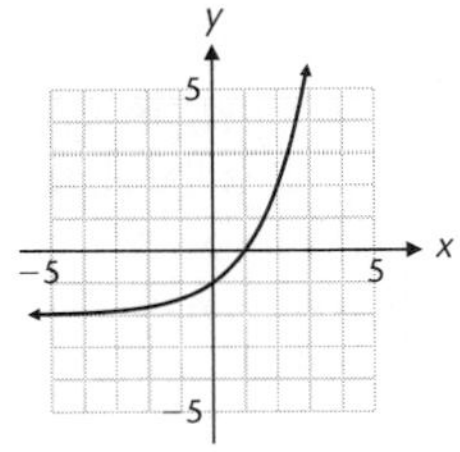

57. $6^{2x} - 6^{-2x}$ **59.** 4 **61.** Local maximum: $m(0.91) \approx 2.67$; x intercept: -0.55; horizontal asymptote: $y = 2$

63. Local minimum: $f(0) = 1$; no x intercepts; no horizontal asymptotes **65.**

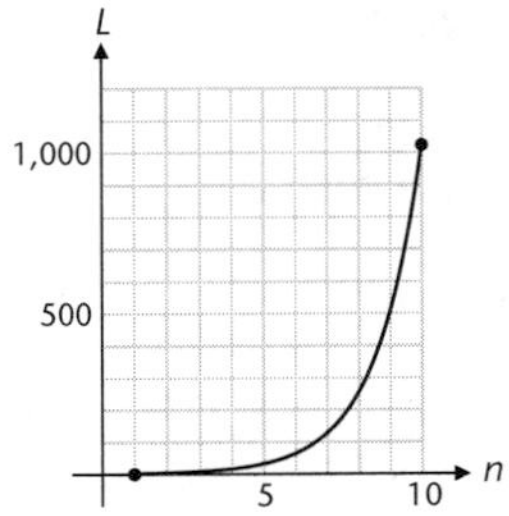

67. (A) 76 flies (B) 570 flies **69.** (A) 19 pounds (B) 7.9 pounds **71.** (A) $4,225.92 (B) $12,002.75 **73.** $9,841

75. Yes, after 6,217 days **77.** No **79.**

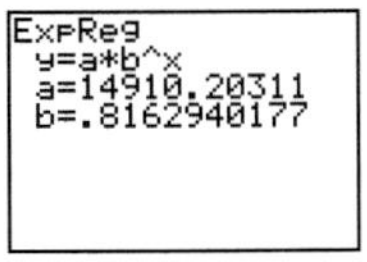

Estimated purchase price: $14,910; estimated value after 10 years: $1,959

Exercise 4-4

1. (A) m (B) f (C) n (D) g **3.** 7.524 **5.** 1.649 **7.** 15.15 **9.** e^{-x} **11.** e^{3x} **13.** e^{3x-1} **15.** (B) e

17. Decreasing; no x intercept; y intercept: -1; horizontal asymptote: x axis

19. Increasing; no x intercept; y intercept: 10; horizontal asymptote: x axis

21. Decreasing; no x intercept; y intercept: 100; horizontal asymptote: x axis

23. Increasing; x intercept: -0.69; y intercept: 1; horizontal asymptote: $y = 2$

25. Increasing; x intercept: 0.23; y intercept: -1; horizontal asymptote: $y = -2$

27. The graph of g is the same as the graph of f shifted to the right 2 units.

29. The graph of g is the same as the graph of f shifted upward 2 units.

31. The graph of g is the same as the graph of f reflected in the y axis, shifted to the left 2 units, and expanded vertically.

33. $\dfrac{e^{-2x}(-2x - 3)}{x^4}$ **35.** $2e^{2x} + 2e^{-2x}$ **37.** $2e^{2x}$ **39.** $x = 0$ **41.** $x = 0, 5$

43. No local extrema; no x intercept; y intercept: 2.14; horizontal asymptote: $y = 2$

45. Local minimum: $m(0) = 1$; no x intercepts; y intercept: 1; no horizontal asymptotes

47. Local maximum: $s(0) = 1$; no x intercepts; y intercept: 1; horizontal asymptote: x axis

49. No local extrema; no x intercept; y intercept: 50; horizontal asymptotes: x axis and $y = 200$ **51.** $f(x) \to 2.7183 \approx e$ as $x \to 0$

53.
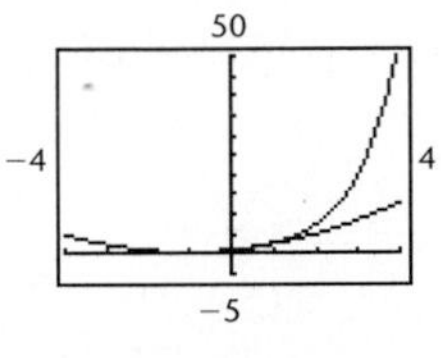

55.
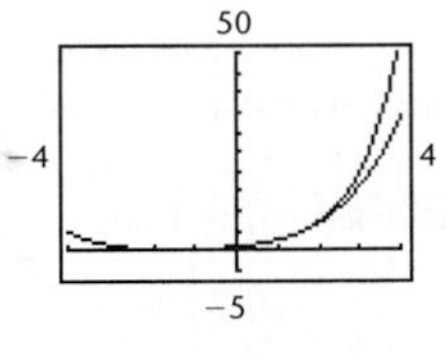

57. As $x \to \infty$, $f_n(x) \to 0$; the line $y = 0$ is a horizontal asymptote. As $x \to -\infty$, $f_1(x) \to -\infty$ and $f_3(x) \to -\infty$, while $f_2(x) \to \infty$. As $x \to -\infty$, $f_n(x) \to \infty$ if n is even and $f_n(x) \to -\infty$ if n is odd.

59. 7.1 billion **61.** 2006 **63.**

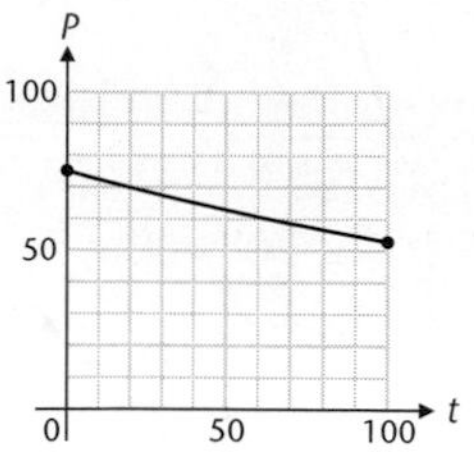

65. (A) 62% (B) 39% **67.** (A) $10,691.81 (B) $36,336.69

69. Gill Savings: $1,230.60; Richardson S & L: $1,231.00; U.S.A. Savings: $1,229.03 **71.** $12,197.09

73. (A) 15 million (B) 30 million

75. (A) 12 boards; 18 boards (B) 8 days (C) N approaches 40, the upper limit for the number of boards an average employee can produce

77. $T = 50°F$ **79.** q approaches 0.0009 coulombs, the upper limit for the charge on the capacitor

81. (A) 25 deer; 37 deer (B) 10 years (C) N approaches 100 deer, the upper limit for the number of deer the island can support

Exercise 4-5

1. $81 = 3^4$ **3.** $0.001 = 10^{-3}$ **5.** $3 = 81^{1/4}$ **7.** $16 = (\frac{1}{2})^{-4}$ **9.** $\log_{10} 0.0001 = -4$ **11.** $\log_4 8 = \frac{3}{2}$ **13.** $\log_{32} \frac{1}{2} = -\frac{1}{5}$

15. $\log_{49} 7 = \frac{1}{2}$ **17.** 0 **19.** 1 **21.** 4 **23.** $\log_{10} 10^{-2} = -2$ **25.** $\frac{1}{3}$ **27.** $\sqrt{x}$ **29.** x^2 **31.** $x = 2^2 = 4$

33. $y = 2$ **35.** $b = 4$ **37.** b is any positive real number except 1 **39.** $x = 2$ **41.** $y = -2$ **43.** $b = 100$

45. $2 \log_b u + 7 \log_b v$ **47.** $\frac{2}{3} \log_b m - \frac{1}{2} \log_b n$ **49.** $\log_b u - \log_b v - \log_b w$ **51.** $-2 \log_b a$ **53.** $\frac{1}{3} \log_b (x^2 - y^2)$

55. $\frac{1}{3} \log_b N - 2 \log_b p - 3 \log_b q$ **57.** $\frac{1}{4}(2 \log_b x + 3 \log_b y - \frac{1}{2} \log_b z)$ **59.** $\log_b \dfrac{x^2}{y}$ **61.** $\log_b \dfrac{w}{xy}$ **63.** $\log_b \dfrac{x^3 y^2}{z^{1/4}}$

65. $\log_b \left(\dfrac{u^{1/2}}{v^2}\right)^5$ **67.** $\log_b \sqrt[5]{x^2 y^3}$ **69.** $5 \log_b (x + 3) + 2 \log_b (2x - 7)$ **71.** $7 \log_b (x + 10) - 2 \log_b (1 + 10x)$

73. $2 \log_b x - \frac{1}{2} \log_b (x + 1)$ **75.** $2 \log_b x + \log_b (x + 5) + \log_b (x - 4)$ **77.** $x = 4$ **79.** $x = \frac{1}{3}$ **81.** $x = \frac{8}{7}$ **83.** $x = 2$

85. $x = 2$ **87.** 3.40 **89.** −0.92 **91.** 3.30 **93.** 0.23 **95.** −0.05

97. (A)

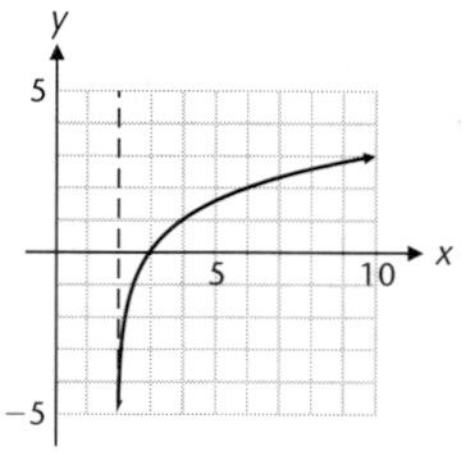

(B) $f^{-1}(x) = 2^x + 2$

99. (A)

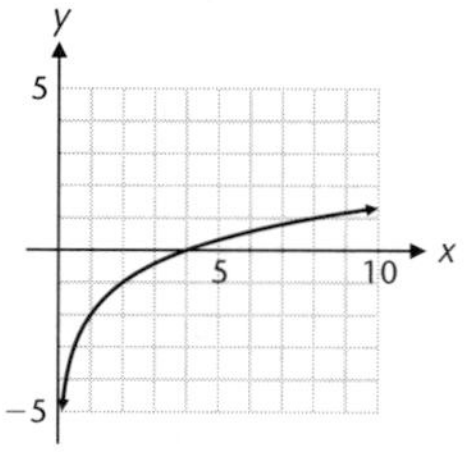

(B) $f^{-1}(x) = 2^{x+2}$

101. (A)

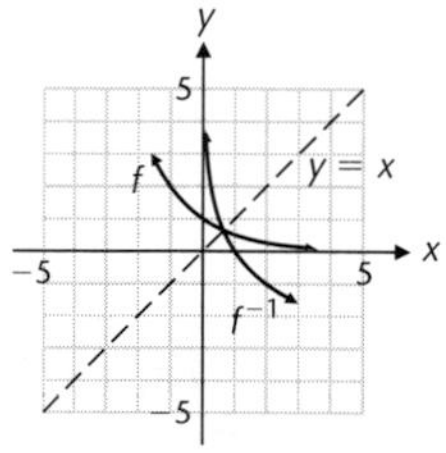

(B) Domain $f = (-\infty, \infty) =$ range f^{-1}; range $f = (0, \infty) =$ domain f^{-1}

(C) $f^{-1}(x) = \log_{1/2} x = -\log_2 x$

103. $f^{-1}(x) = \frac{1}{3}[1 + \log_5 (x - 4)]$ **105.** $g^{-1}(x) = \frac{1}{5}(e^{x/3} + 2)$ **107.** The reflection is not a function since $y = 3^{x^2}$ is not one-to-one.

109. $x = 100e^{-0.08t}$

Exercise 4-6

1. 4.9177 **3.** −2.8419 **5.** 3.7623 **7.** −2.5128 **9.** 200,800 **11.** 0.0006648 **13.** 47.73 **15.** 0.6760

17. 4.959 **19.** 7.861 **21.** 3.301 **23.** 4.561 **25.** $x = 12.725$ **27.** −25.715 **29.** $x = 1.1709 \times 10^{32}$ **31.** 4.2672×10^{-7}

33. $f^{-1}(x) = e^{x/2} - 2$ **35.** $f^{-1}(x) = e^{(x+3)/4}$ **37.** Domain: $(-\infty, \infty)$; range: $[-2, \infty)$; x intercepts: ± 2.53; y intercept: -2; no asymptotes

39. Domain: $(-1, 1)$; range: $(-\infty, 1]$; x intercepts: ± 0.80; y intercept: 1; vertical asymptotes: $x = \pm 1$

41. The inequality sign in the last step reverses because $\log \frac{1}{3}$ is negative.

43. (B) Domain $= (1, \infty)$; range $= (-\infty, \infty)$ **45.** (0.90, −0.11), (38.51, 3.65) **47.** (6.41, 1.86), (93.35, 4.54)

49. 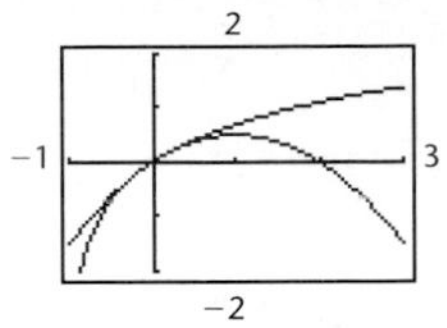

51. 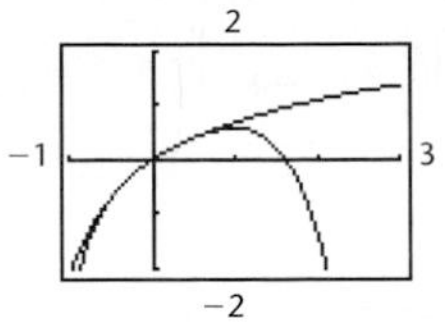

53. (A) 0 decibels (B) 120 decibels **55.** 30 decibels **57.** 8.6

59. 1,000 times as powerful **61.** 7.67 km/s **63.** (A) 8.3, basic (B) 3.0, acidic **65.** 6.3×10^{-6} moles per liter

Exercise 4-7

1. 1.46 **3.** 0.321 **5.** 1.29 **7.** 3.50 **9.** 1.80 **11.** 2.07 **13.** 20 **15.** $x = 5$ **17.** $x = \frac{11}{9}$ **19.** 14.2

21. −1.83 **23.** 11.7 **25.** ±1.21 **27.** $x = 5$ **29.** $2 + \sqrt{3}$ **31.** $\frac{1 + \sqrt{89}}{4}$ **33.** $1, e^2, e^{-2}$ **35.** $x = e^e$

37. $x = 100, 0.1$ **39.** (B) 2 **41.** (B) −1.252, 1.707 **43.** 3.6776 **45.** −1.6094 **47.** −1.7372 **49.** $r = \frac{1}{t} \ln \frac{A}{P}$

51. $I = I_0(10^{D/10})$ **53.** $I = I_0[10^{(6-M)/2.5}]$ **55.** $t = -\frac{L}{R} \ln \left(1 - \frac{RI}{E}\right)$ **57.** $x = \ln (y \pm \sqrt{y^2 - 1})$ **59.** $x = \frac{1}{2} \ln \frac{1 + y}{1 - y}$

61. 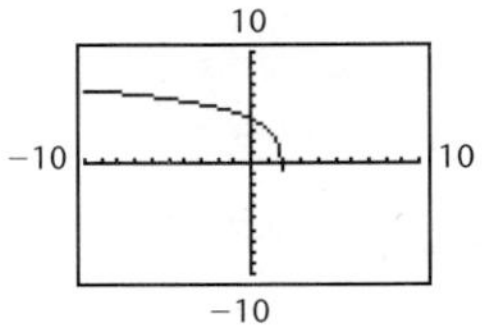

63. 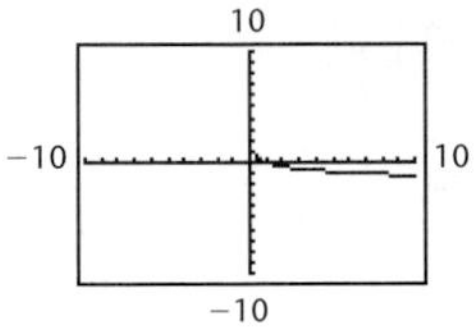

65. 0.38 **67.** 0.55 **69.** 0.57 **71.** 0.85 **73.** 0.43

75. 0.27 **77.** $n = 5$ years to the nearest year **79.** $r = 0.0916$ or 9.16% **81.** (A) $m = 6$ (B) 100 times brighter
83. $t = 35$ years to the nearest year **85.** $t = 18{,}600$ years old **87.** $t = 7.52$ seconds
89. $k = 0.40$, $t = 2.9$ hours **91.** (A) 1996: 123.0 bush./acre; 2010: 141.4 bush./acre

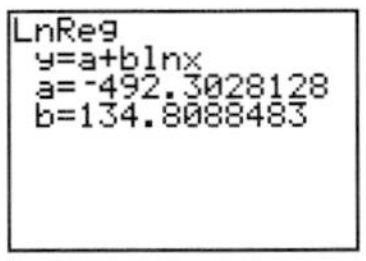

Chapter 4 Review

1.

x	−3	−2	−1	0	1	2	3
$(f - g)(x)$	7	5	3	1	1	1	1

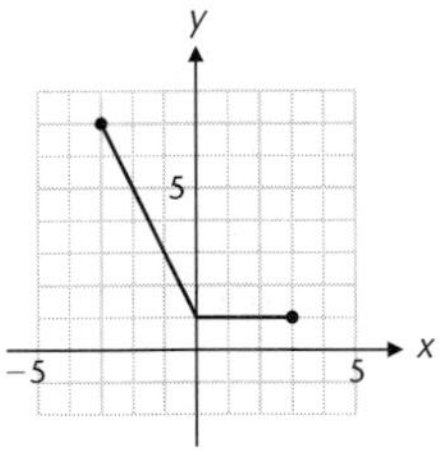

(4-1) **2.** 2 *(4-1)* **3.** 1 *(4-1)* **4.** 0 *(4-1)*

5. 2 *(4-1)* **6.** No *(4-2)* **7.** Yes *(4-2)* **8.** (A) m (B) f (C) n (D) g *(4-2, 4-3, 4-5)* **9.** $\log m = n$ *(4-5)*
10. $\ln x = y$ *(4-5)* **11.** $x = 10^y$ *(4-5)* **12.** $y = e^x$ *(4-5)* **13.** 7^{2x} *(4-3)* **14.** e^{2x^2} *(4-3)* **15.** $x = 8$ *(4-5)*
16. $x = 5$ *(4-5)* **17.** $x = 3$ *(4-5)* **18.** $x = 1.24$ *(4-6)* **19.** $x = 11.9$ *(4-6)* **20.** $x = 0.984$ *(4-6)* **21.** $x = 103$ *(4-6)*
22. 1.145 *(4-5)* **23.** Not defined *(4-5)* **24.** 2.211 *(4-5)* **25.** 11.59 *(4-5)*
26. (A) $(f/g)(x) = (x^2 - 4)/(x + 3)$; domain of $f/g = (-\infty, -3) \cup (-3, \infty)$
(B) $(g/f)(x) = (x + 3)/(x^2 - 4)$; domain of $g/f = (-\infty, -2) \cup (-2, 2) \cup (2, \infty)$
(C) $(f \circ g)(x) = x^2 + 6x + 5$; domain of $f \circ g = (-\infty, \infty)$ (D) $(g \circ f)(x) = x^2 - 1$; domain of $g \circ f = (-\infty, \infty)$ *(4-1)*
27. $x = 4$ *(4-5)* **28.** $x = 2$ *(4-5)* **29.** $x = 3, -1$ *(4-4)* **30.** $x = 1$ *(4-3)* **31.** $x = 3, -3$ *(4-4)* **32.** $x = -2$ *(4-5)*
33. $x = \frac{1}{3}$ *(4-5)* **34.** $x = 64$ *(4-5)* **35.** $x = e$ *(4-5)* **36.** $x = 33$ *(4-5)* **37.** $x = 1$ *(4-5)* **38.** $x = 41.8$ *(4-3)*
39. $x = 1.95$ *(4-5)* **40.** $x = 0.0400$ *(4-5)* **41.** $x = -6.67$ *(4-5)* **42.** $x = 1.66$ *(4-5)* **43.** $x = 2.32$ *(4-7)*
44. $x = 3.92$ *(4-7)* **45.** $x = 92.1$ *(4-7)* **46.** $x = 2.11$ *(4-7)* **47.** $x = 0.881$ *(4-7)*
48. (A) $(f \circ g)(x) = \sqrt{|x|} - 8$; $(g \circ f)(x) = |\sqrt{x} - 8|$
(B) Domain of $f \circ g$ is the set of all real numbers. The domain of $(g \circ f)$ is $[0, \infty)$. *(4-1)*
49. (A) $(f \circ g)(x) = e^{\ln(x+1)-1}$; $(g \circ f)(x) = \ln (e^{x-1} + 1)$ (B) Domain of $f \circ g = (-1, \infty)$; Domain of $g \circ f = (-\infty, \infty)$ *(4-1, 4-4, 4-6)*
50. Functions f, h, F, and H are one-to-one *(4-2)*

51. $f^{-1}(x) = (x + 7)/3$;
domain of f^{-1} = range of $f^{-1} = (-\infty, \infty)$ *(4-2)*

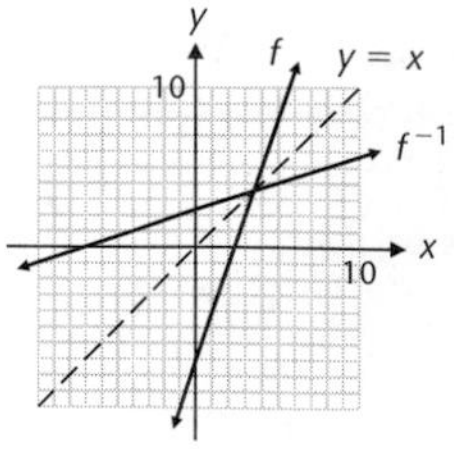

52. $f^{-1}(x) = x^2 + 1$;
domain of $f^{-1} = [0, \infty)$; range of $f^{-1} = [1, \infty)$ *(4-2)*

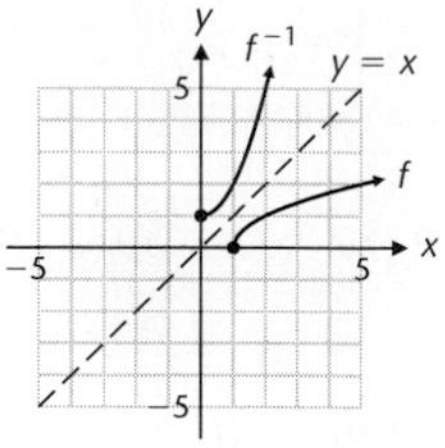

53. $f^{-1}(x) = \sqrt{x + 1}$;
domain of $f^{-1} = [-1, \infty)$; range of $f^{-1} = [0, \infty)$ *(4-2)*

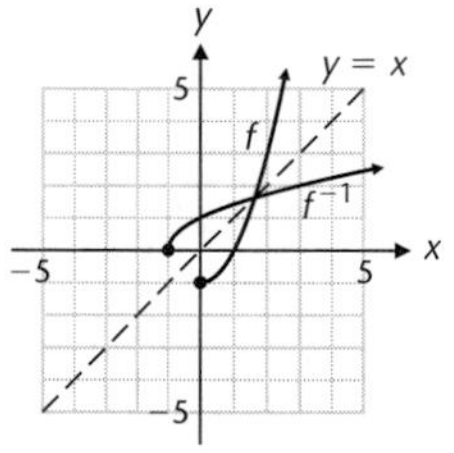

54. $f^{-1}(x) = \ln(x - 1)$;
domain of $f^{-1} = (1, \infty)$; range of $f^{-1} = (-\infty, \infty)$ *(4-2)*

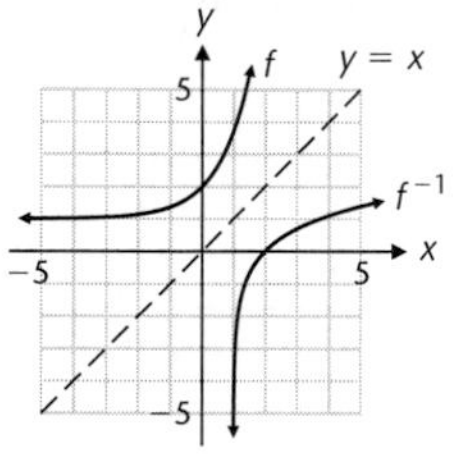

55. $f^{-1}(x) = 1 + e^{x/2}$; domain of $f^{-1} = (-\infty, \infty)$; range of $f^{-1} = (1, \infty)$ *(4-2)* **56.** $x = 300$ *(4-7)* **57.** $x = 2$ *(4-7)*

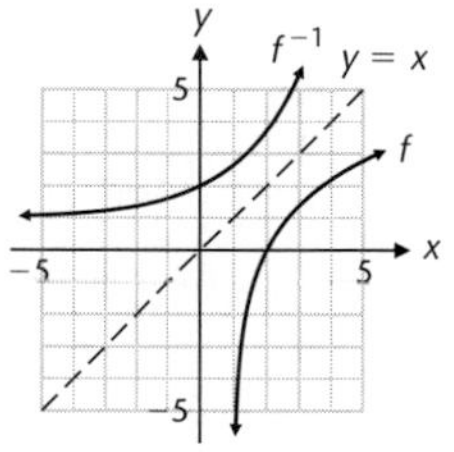

58. $x = 1$ *(4-7)* **59.** $x = \dfrac{3 + \sqrt{13}}{2}$ *(4-7)* **60.** $x = 1, 10^3, 10^{-3}$ *(4-7)* **61.** $x = 10^e$ *(4-7)* **62.** $e^{-x} - 1$ *(4-4)*

63. $2 - 2e^{-2x}$ *(4-4)* **64.** Domain = $(-\infty, \infty)$; range = $(0, \infty)$; y intercept: 0.5; horizontal asymptote: $y = 0$ *(4-3)*

65. Domain = $(-\infty, \infty)$; range = $(0, \infty)$; y intercept: 10; horizontal asymptote: $y = 0$ *(4-4)*

66. Domain = $(1, \infty)$; range = $(-\infty, \infty)$; x intercept: 2; vertical asymptote: $x = 1$ *(4-6)*

67. Domain = $(-\infty, \infty)$; range = $(0, 100)$; y intercept: 25; horizontal asymptotes: $y = 0$ and $y = 100$ *(4-4)*

68. $y = -e^x$; $y = e^{-x}$ or $y = \dfrac{1}{e^x}$ or $y = \left(\dfrac{1}{e}\right)^x$ *(4-5)*

69. (A) $y = e^{-x/3}$ is decreasing while $y = 4 \ln(x + 1)$ is increasing without bound. (B) 0.258 *(4-5)* **70.** 0.018, 2.187 *(4-5)*

71. (1.003, 0.010), (3.653, 4.502) *(4-6)* **72.** (A) $(fg)(x) = x^2\sqrt{1 - x}$; domain is $(-\infty, 1]$ (B) $\left(\dfrac{f}{g}\right)(x) = \dfrac{x^2}{\sqrt{1 - x}}$; domain of $\dfrac{f}{g}$ is $(-\infty, 1)$
(C) $(f \circ g)(x) = 1 - x$; domain of $f \circ g$ is $(-\infty, 1]$ (D) $(g \circ f)(x) = \sqrt{1 - x^2}$; domain of $g \circ f$ is $[-1, 1]$ *(4-1)*

73. $I = I_0(10^{D/10})$ *(4-7)* **74.** $x = \pm\sqrt{-2 \ln(\sqrt{2\pi}y)}$ *(4-7)* **75.** $I = I_0(e^{-kx})$ *(4-7)*

76. $n = -\dfrac{\ln(1 - \frac{P_i}{r})}{\ln(1 + i)}$ *(4-7)* **77.** $f^{-1}(x) = \dfrac{3x + 2}{x - 1}$ *(4-2)* **78.** $f^{-1}(x) = \ln(x + \sqrt{x^2 + 1})$ *(4-2, 4-7)* **79.** $y = ce^{-5t}$ *(4-5, 4-7)*

80. Domain $f = (0, \infty)$ = Range f^{-1}
Range $f = (-\infty, \infty)$ = Domain f^{-1} *(4-2, 4-5)*

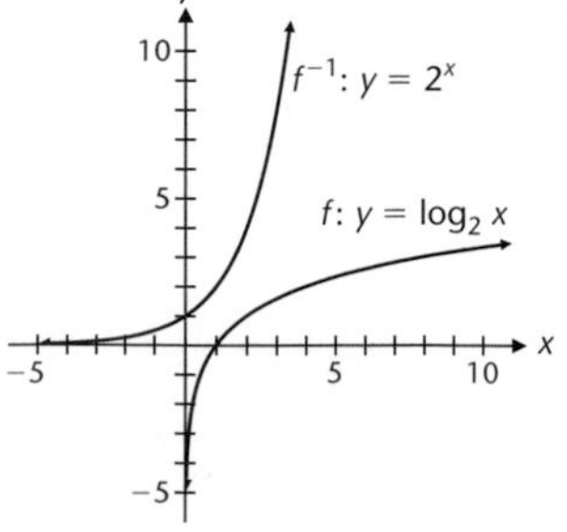

81. If $\log_1 x = y$, then we would have to have $1^y = x$; that is, $1 = x$ for arbitrary positive x, which is impossible. *(4-5)*

83. (A) [1, 3] (B) $q = d^{-1}(p) = 4{,}500/p - 500$; domain = [1, 3]; range = [1,000, 4,000] *(4-2)*
84. $P(p) = -14{,}000 + 700p - 10p^2$ *(4-1)* **85.** $t = 23.4$ years *(4-7)* **86.** $t = 23.1$ years *(4-7)* **87.** $t = 37{,}100$ years *(4-7)*
88. (A) $N = 2^{2t}$ (or $N = 4^t$) (B) $t = 15$ days *(4-7)* **89.** $A = 1.1 \times 10^{26}$ dollars *(4-4)*
90. (A)

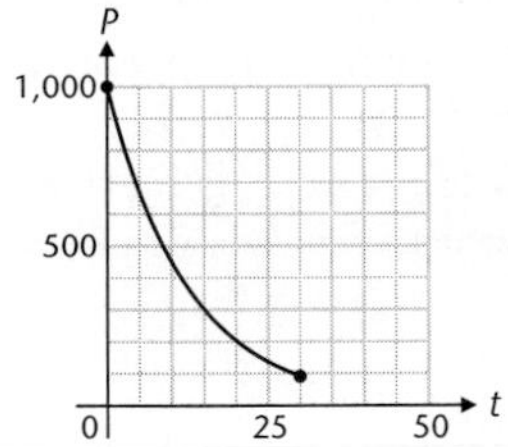

(B) 0 *(4-4)* **91.** $M = 6.6$ *(4-6)* **92.** $E = 10^{16.85}$ or 7.08×10^{16} joules *(4-6)*

93. The level of the louder sound is 50 decibels more. *(4-6)* **94.** $k = 0.00942$, $d = 489$ feet *(4-4)* **95.** $t = 3$ years *(4-7)*
96. (A) 1996: \$207 billion; 2010: \$886 billion (B) Midway through 2004 *(4-3)*

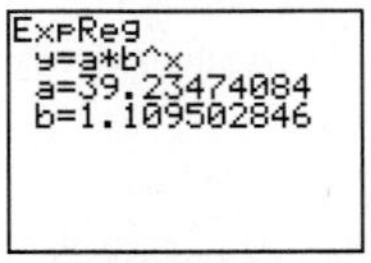

97. (A) 1996: 1,724 million bush.; 2010: 2,426 million bush. *(4-7)*

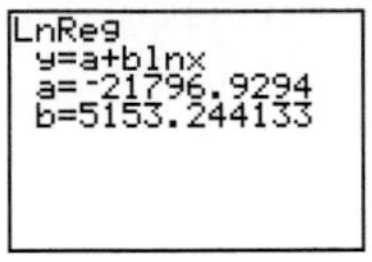

Cumulative Review Exercise for Chapters 3 and 4

1. (A) $P(x) = (x + 1)^2(x - 1)(x - 2)$ (B) $P(x) \to \infty$ as $x \to \infty$ and as $x \to -\infty$ *(3-1)*
2. (A) m (B) g (C) n (D) f *(4-1, 4-3)* **3.** $3x^3 + 5x^2 - 18x - 3 = (x + 3)(3x^2 - 4x - 6) + 15$ *(3-1)*
4. $-2, 3, 5$ *(3-2)* **5.** $P(1) = -5$ and $P(2) = 5$ are of opposite sign. *(3-3)* **6.** $1, 2, -4$ *(3-2)*
7. (A) $x = \log y$ (B) $x = e^y$ *(4-6)* **8.** (A) $8e^{3x}$ (B) e^{5x} *(4-4)* **9.** (A) 9 (B) 4 (C) $\frac{1}{2}$ *(4-5)*
10. (A) 0.371 (B) 11.4 (C) 0.0562 (D) 15.6 *(4-6)* **11.** $(f \circ g)(x) = \dfrac{x}{3 - x}$; domain: $(-\infty, 0) \cup (0, 3) \cup (3, \infty)$ *(4-1)*
12. $f^{-1}(x) = \dfrac{x - 5}{2}$ or $\dfrac{1}{2}x - \dfrac{5}{2}$ *(4-2)* **13.** $f(x) = 3 \ln x - \sqrt{x}$ *(4-5)*
14. The function f multiplies the base e raised to power of one-half the domain element by 100 and then subtracts 50. *(4-4)*
15. (A) Domain: $x \neq -2$; x intercept: $x = -4$; y intercept: $y = 4$
(B) Vertical asymptote: $x = -2$; horizontal asymptote: $y = 2$
(C) *(3-4)*

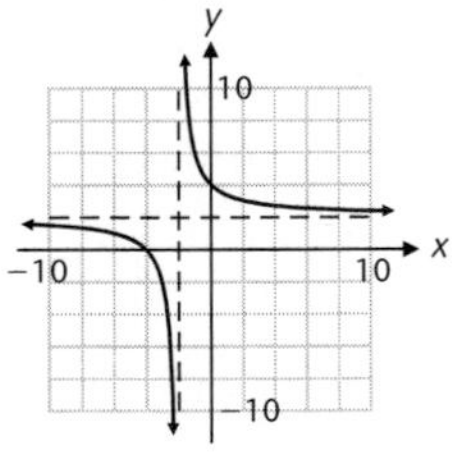

16. (A) $f^{-1}(x) = x^2 - 4$; domain: $x \geq 0$
(B) Domain of $f = [-4, \infty) =$ Range of f^{-1}
Range of $f = [0, \infty) =$ Domain of f^{-1}
(C) *(4-2)*

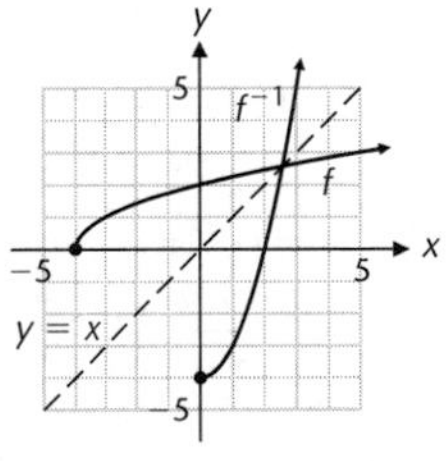

17. (A) and (C) *(4-2)* **18.** $P(\frac{1}{2}) = \frac{5}{2}$ *(3-2)* **19.** (B) *(3-1)*
20. (A) The graph of $P(x)$ has four x intercepts and three turning points; $P(x) \to \infty$ as $x \to \infty$ and as $x \to -\infty$

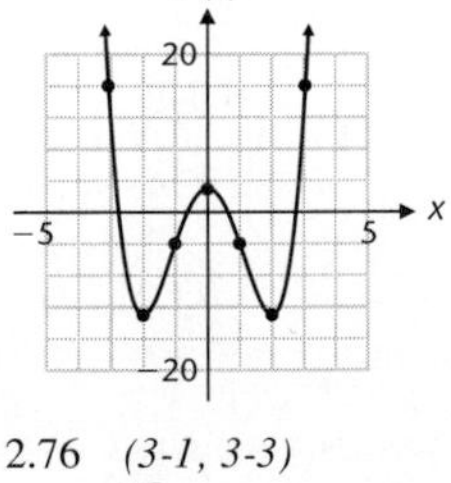

(B) 2.76 *(3-1, 3-3)*

21. (A) -0.56 (double zero); 2 (simple zero); 3.56 (double zero)
(B) -0.56 can be approximated with a maximum routine; 2 can be approximated with the bisection; 3.56 can be approximated with a minimum routine *(3-3)*

22. (A) Upper bound: 4; lower bound: -6 (B) four intervals (C) -5.68, 3.80 *(3-3)* **23.** 3, $1 \pm \frac{1}{2}i$ *(3-2)*

24. $P(x) = (x + 1)(x + 4)(x^2 - 3) = (x + 1)(x + 4)(x - \sqrt{3})(x + \sqrt{3})$; the four zeros are $-1, -4, \pm\sqrt{3}$. *(3-2)* **25.** $x = 4, -2$ *(4-4)*

26. $\frac{1}{2}, -1$ *(4-5)* **27.** $x = 2.5$ *(4-5)* **28.** $x = 10$ *(4-5)* **29.** $x = \frac{1}{27}$ *(4-7)* **30.** $x = 5$ *(4-7)* **31.** $x = 7$ *(4-7)*

32. $x = 5$ *(4-6)* **33.** $x = e^{0.1}$ *(4-7)* **34.** $x = 1, e^{0.5}$ *(4-7)* **35.** $x = 3.38$ *(4-6)* **36.** $x = 4.26$ *(4-6)* **37.** $x = 2.32$ *(4-7)*

38. $x = 3.67$ *(4-7)* **39.** $x = 0.549$ *(4-7)* **40.** Domain: $(-\infty, \infty)$; range: $(0, \infty)$; y intercept: 3; horizontal asymptote: $y = 0$ *(4-3)*

41. Domain: $(-\infty, 2)$; range: $(-\infty, \infty)$; x intercept: 1; y intercept: ln 2; vertical asymptote: $x = 2$ *(4-6)*

42. Domain: $(-\infty, \infty)$; range: $(0, \infty)$; y intercept: 100; horizontal asymptote: $y = 0$ *(4-4)*

43. Domain: $(-\infty, \infty)$; range: $(-\infty, 3)$; x intercept: -0.41; y intercept: 1; horizontal asymptote: $y = 3$ *(4-4)*

44. Domain: $(-\infty, \infty)$; range: $(0, 3)$; y intercept: 2; horizontal asymptotes: $y = 0$ and $y = 3$ *(4-4)*

45. A reflection in the x axis transforms the graph of $y = \ln x$ into the graph of $y = -\ln x$. A reflection in the y axis transforms the graph of $y = \ln x$ into the graph of $y = \ln(-x)$. *(4-5)*

46. (A) For $x > 0$, $y = e^{-x}$ decreases from 1 to 0 while $\ln x$ increases from $-\infty$ to ∞. Consequently, the graphs can intersect at exactly one point.
(B) 1.31 *(4-5)*

47. (A) Domain g: $[-2, 2]$ (B) $\left(\frac{f}{g}\right)(x) = \frac{x^2}{\sqrt{4 - x^2}}$; domain of f/g is $(-2, 2)$ (C) $(f \circ g)(x) = 4 - x^2$; domain of $f \circ g$ is $[-2, 2]$ *(4-1)*

48. (A) $f^{-1}(x) = 1 + \sqrt{x + 4}$
(B) Domain of f^{-1} is $[-4, \infty)$.
Range of f^{-1} = Domain of f is $[1, \infty)$
(C) *(4-2)*

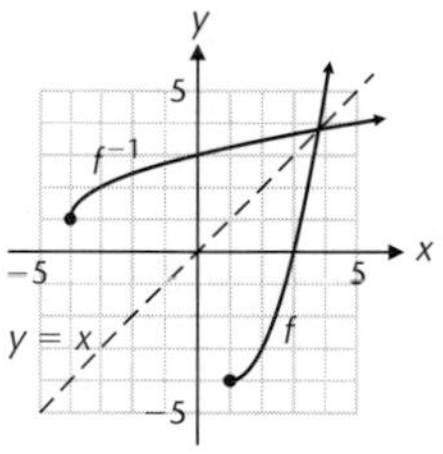

49. Vertical asymptote: $x = -2$;
oblique asymptote: $y = x + 2$ *(3-4)*

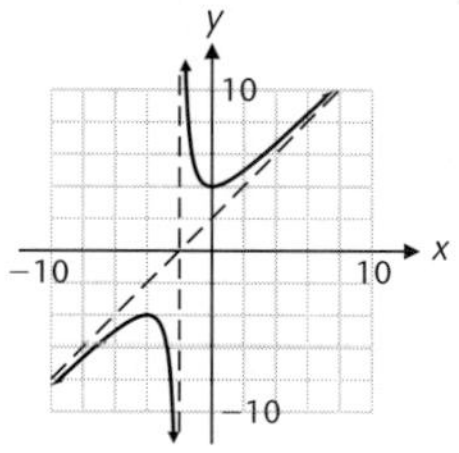

50. Zeros: 2.97, 3; local minimum: $P(2.98) \approx -0.02$; local maximum: $P(7.03) \approx 264.03$; local minimum: $P(10.98) \approx 15.98$ *(3-1)*

51. $P(x) = (x + 1)^2x^3(x - 3 - 5i)(x - 3 + 5i)$; degree 7 *(3-2)*

52. Yes, for example, $P(x) = (x + i)(x - i)(x + \sqrt{2})(x - \sqrt{2}) = x^4 - x^2 - 2$ *(3-2)*

53. (A) Upper bound: 20; lower bound: -30 (B) $-26.68, -6.22, 7.23, 16.67$ *(3-3)*

54. 2, -1 (double), and $2 \pm i\sqrt{2}$; $P(x) = (x - 2)(x + 1)^2(x - 2 - i\sqrt{2})(x - 2 + i\sqrt{2})$ *(3-2)*

55. -2 (double), -1.88, 0.35, 1.53 *(3-3)*

56. (A) $f^{-1}(x) = e^{x/3} + 2$ (B) Domain of $f = (2, \infty)$ = Range of f^{-1}. Range of $f = (-\infty, \infty)$ = Domain of f^{-1} *(4-1, 4-7)*
(C) *(4-1, 4-7)*

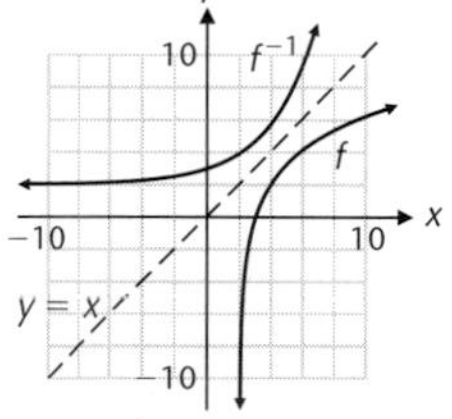

57. $n = \frac{\ln(1 + \frac{Ai}{P})}{\ln(1 + i)}$ *(4-7)* **58.** $y = Ae^{5x}$ *(4-7)* **59.** $x = \ln(y + \sqrt{y^2 + 2})$ *(4-7)* **60.** $V(t) = 0.036\pi t^{3/4}$ *(4-1)*

61. (A) $v = -2{,}000t + 20{,}000$ (B) $t = -0.0005v + 10$ *(4-2)*

62. $x = 2$ feet and $y = 2$ feet, or $x = 1.28$ feet and $y = 4.88$ feet *(3-2)* **63.** 1.79 feet by 3.35 feet *(3-3)*

64. (A) 46.8 million (B) 103 million *(4-3)* **65.** $t = 10.2$ years *(4-7)* **66.** $t = 9.90$ years *(4-7)*

67. 63.1 times as powerful *(4-6)* **68.** $I = 6.31 \times 10^{-4}$ W/m^2 *(4-6)* **69.** (A) 79.3 (B) 75.4 (C) 77.8 (D) 79.5 *(3-1, 4-3)*

70. Cubic regression *(3-1, 4-3)*

CHAPTER 5 Exercise 5-1

1. b, no solution **3.** d, (1, −3) **5.** (5, 2) **7.** (2, −3) **9.** No solution (parallel lines) **11.** $x = 8, y = 19$
13. $x = 6, y = 2$ **15.** $x = 2, y = -1$ **17.** $x = 5, y = 2$ **19.** $m = 1, n = -2/3$ **21.** $x = 2{,}500, y = 200$
23. $u = 1.1$ and $v = 0.3$ **25.** $x = -5/4, y = 5/3$ **27.** (1.12, 2.41) **29.** (−2.24, −3.31) **31.** The system has no solution.
33. $q = x + y - 5, p = 3x + 2y - 12$ **35.** $x = \frac{dh - bk}{ad - bc}, y = \frac{ak - ch}{ad - bc}, ad - bc \neq 0$ **37.** Airspeed = 330 mph; wind rate = 90 mph
39. 2.475 km **41.** 40 ml of 50% solution and 60 ml of 80% solution **43.** \$7,200 invested at 10% and \$4,800 invested at 15%
45. Mexico plant: 75 hours; Taiwan plant: 50 hours **47.** Mix *A*: 80 g; mix *B*: 60 g
49. (A) Supply: 143 T-shirts; demand: 611 T-shirts
(B) Supply: 714 T-shirts; demand: 389 T-shirts
(C) Equilibrium price: \$6.36; equilibrium quantity: 480 T-shirts
(D)

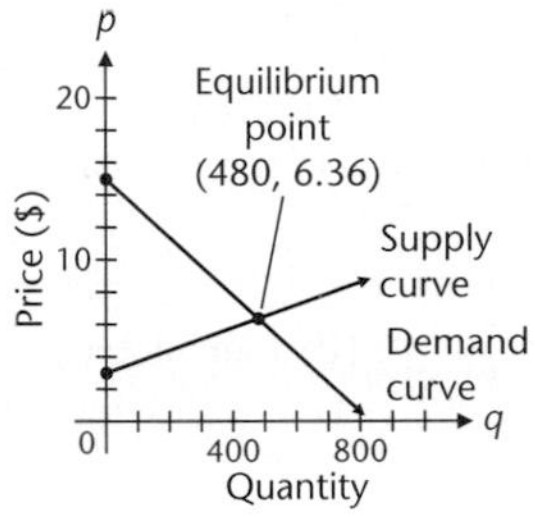

51. (A) $p = 0.001q + 0.15$ (B) $p = -0.002q + 1.89$ (C) Equilibrium price = \$0.73; equilibrium quantity = 580 bushels
53. (A) $a = 196, b = -16$ (B) 196 ft (C) 3.5 sec **55.** 40 sec, 24 sec, 120 mi

Exercise 5-2

1. (2, −1) **3.** (3, −1) **5.** 2 × 3, 1 × 3 **7.** *C* **9.** *B* **11.** −2, −6 **13.** −2, 6, 0

15. $\left[\begin{array}{cc|c} 4 & -6 & -8 \\ 1 & -3 & 2 \end{array}\right]$ **17.** $\left[\begin{array}{cc|c} -4 & 12 & -8 \\ 4 & -6 & -8 \end{array}\right]$ **19.** $\left[\begin{array}{cc|c} 1 & -3 & 2 \\ 8 & -12 & -16 \end{array}\right]$ **21.** $\left[\begin{array}{cc|c} 1 & -3 & 2 \\ 0 & 6 & -16 \end{array}\right]$ **23.** $\left[\begin{array}{cc|c} 1 & -3 & 2 \\ 2 & 0 & -12 \end{array}\right]$

25. $\left[\begin{array}{cc|c} 1 & -3 & 2 \\ 3 & -3 & -10 \end{array}\right]$ **27.** $\frac{1}{3}R_2 \to R_2$ **29.** $6R_1 + R_2 \to R_2$ **31.** $\frac{1}{3}R_2 + R_1 \to R_1$

33. $x_1 = 4, x_2 = 3$; each pair of lines has the same intersection point.

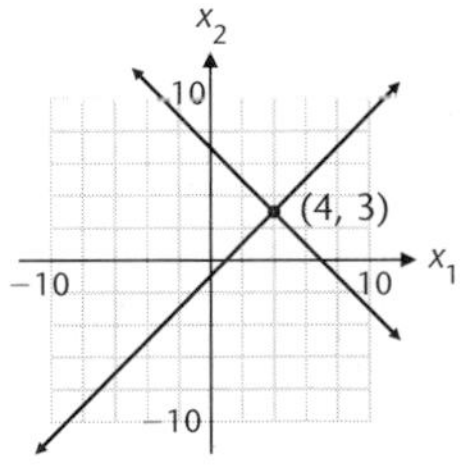

$x_1 + x_2 = 7$
$x_1 - x_2 = 1$

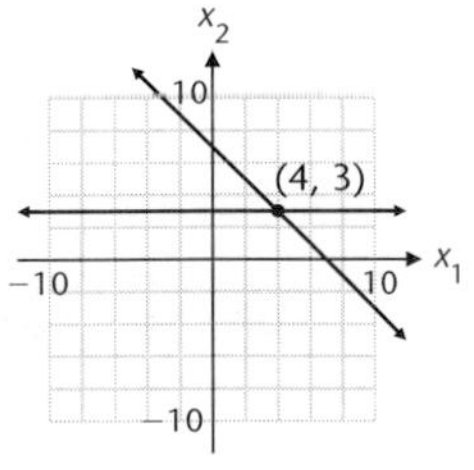

$x_1 + x_2 = 7$
$-2x_2 = -6$

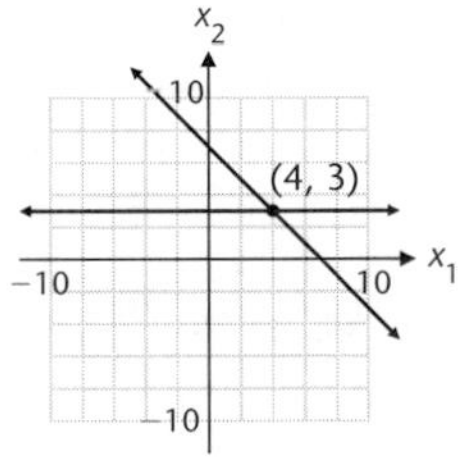

$x_1 + x_2 = 7$
$x_2 = 3$

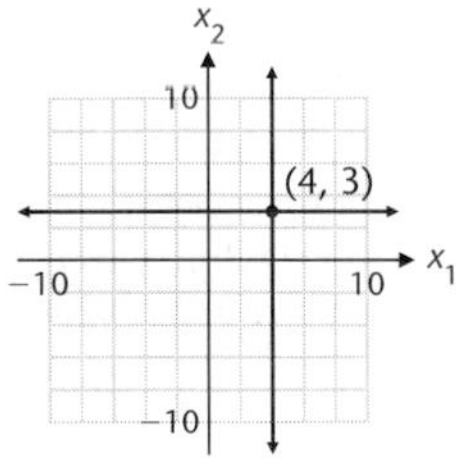

$x_1 = 4$
$x_2 = 3$

35. $x_1 = 2$ and $x_2 = 1$ **37.** $x_1 = 2$ and $x_2 = 4$ **39.** No solution **41.** $x_1 = 1$ and $x_2 = 4$
43. Infinitely many solutions for any real number s, $x_2 = s, x_1 = 2s - 3$
45. Infinitely many solutions for any real number s, $x_2 = s, x_1 = \frac{1}{2}s + \frac{1}{2}$
47. (A) (−24, 20) (B) (6, −4) (C) No solution **49.** (−23.125, 7.8125) **51.** (3.225, −6.9375)
53. 25 32¢ stamps, 50 23¢ stamps **55.** \$107,500 in bond *A* and \$92,500 in bond *B*
57. 30 liters of 20% solution and 70 liters of 80% solution **59.** 200 g of mix *A* and 80 g of mix *B*
61. Base price = \$17.95, surcharge = \$2.45 per pound **63.** 5,720 pounds of the robust blend and 6,160 pounds of the mild blend

Exercise 5-3

1. Reduced form **3.** Not reduced form; $R_2 \leftrightarrow R_3$ **5.** Not reduced form; $\frac{1}{3}R_2 \to R_2$ **7.** Reduced form

9. Not reduced form; $3R_2 + R_1 \to R_1$ **11.** $x_1 = -2, x_2 = 3, x_3 = 0$

13. $x_1 = 2t + 3$, $x_2 = -t - 5$, $x_3 = t$ is the solution for t any real number **15.** No solution
17. $x_1 = 2s + 3t - 5$, $x_2 = s$, $x_3 = -3t + 2$, $x_4 = t$ is the solution, for s and t any real numbers

19. $\left[\begin{array}{cc|c} 1 & 0 & -7 \\ 0 & 1 & 3 \end{array}\right]$ **21.** $\left[\begin{array}{ccc|c} 1 & 0 & 0 & -5 \\ 0 & 1 & 0 & 4 \\ 0 & 0 & 1 & -2 \end{array}\right]$ **23.** $\left[\begin{array}{ccc|c} 1 & 0 & 2 & -\frac{5}{3} \\ 0 & 1 & -2 & \frac{1}{3} \\ 0 & 0 & 0 & 0 \end{array}\right]$ **25.** $x_1 = -2$, $x_2 = 3$, and $x_3 = 1$

27. $x_1 = 0$, $x_2 = -2$, and $x_3 = 2$ **29.** $x_1 = 2t + 3$, $x_2 = t - 2$, $x_3 = t$, t any real number **31.** $x_1 = 1$, $x_2 = 2$ **33.** No solution
35. $x_1 = 2t + 4$, $x_2 = t + 1$, $x_3 = t$, t any real number **37.** $x_1 = s + 2t - 1$, $x_2 = s$, $x_3 = t$, s and t any real numbers
39. No solution **41.** $x_1 = 2.5t - 4$, $x_2 = t$, $x_3 = -5$ for t any real number **43.** $x_1 = 1$, $x_2 = -2$, $x_3 = 1$
45. (A) Dependent with two parameters (B) Dependent with one parameter (C) Independent (D) Impossible
51. $x_1 = 2s - 3t + 3$, $x_2 = s + 2t + 2$, $x_3 = s$, $x_4 = t$, s and t any real numbers **53.** $x_1 = -0.5$, $x_2 = 0.2$, $x_3 = 0.3$, $x_4 = -0.4$
55. $x_1 = 2s - 1.5t + 1$, $x_2 = s$, $x_3 = -t + 1.5$, $x_4 = 0.5t - 0.5$, $x_5 = t$ for s and t any real numbers **57.** $x_1 = 4$, $x_2 = 1$
59. $x_1 = -1.4$, $x_2 = 4.8$, $x_3 = 4$
63. $x_1 = (3t - 100)$ 15¢ stamps, $x_2 = (145 - 4t)$ 20¢ stamps, $x_3 = t$ 35¢ stamps, where $t = 34, 35$, or 36
65. $x_1 = (6t - 24)$, 500-cc containers of 10% solution, $x_2 = (48 - 8t)$, 500-cc containers of 20% solution, $x_3 = t$, 1000-cc containers of 50% solution
67. $a = 3$, $b = 2$, $c = 1$ **69.** $a = -2$, $b = -4$, and $c = -20$
71. (A) $x_1 = 20$ one-person boats, $x_2 = 220$ two-person boats, $x_3 = 100$ four-person boats
(B) $x_1 = (t - 80)$ one-person boats, $x_2 = (-2t + 420)$ two-person boats, $x_3 = t$ four-person boats, $80 \le t \le 210$, t an integer
(C) No solution; no production schedule will use all the work-hours in all departments.
73. (A) $x_1 = 8$ oz food A, $x_2 = 2$ oz food B, $x_3 = 4$ oz food C
(B) No solution (C) $x_1 = 8$ oz food A, $x_2 = -2t + 10$ oz food B, $x_3 = t$ ounces food C, $0 \le t \le 5$
75. $10 - t$ barrels of mix A, $t - 5$ barrels of mix B, $25 - 2t$ barrels of mix C, and t barrels of mix D, where t is an integer satisfying $5 \le t \le 10$
77. $x_1 = 10$ hours company A, $x_2 = 15$ hours company B

Exercise 5-4

1. $\begin{bmatrix} 0 & 2 \\ 2 & -1 \end{bmatrix}$ **3.** $\begin{bmatrix} -1 & 6 \\ -4 & 3 \\ 1 & -1 \end{bmatrix}$ **5.** Not defined **7.** $\begin{bmatrix} 2 & 3 & -5 \\ 5 & -5 & 7 \end{bmatrix}$ **9.** $\begin{bmatrix} 20 & -10 & 30 \\ 0 & -40 & 50 \end{bmatrix}$ **11.** $[10]$ **13.** $\begin{bmatrix} 5 \\ -3 \end{bmatrix}$

15. $\begin{bmatrix} 2 & 4 \\ 1 & -5 \end{bmatrix}$ **17.** $\begin{bmatrix} 1 & -5 \\ -2 & -4 \end{bmatrix}$ **19.** $[-14]$ **21.** $\begin{bmatrix} -20 & 10 \\ -12 & 6 \end{bmatrix}$ **23.** $[11]$ **25.** $\begin{bmatrix} 3 & -2 & -4 \\ 6 & -4 & -8 \\ -9 & 6 & 12 \end{bmatrix}$ **27.** Not defined

29. $\begin{bmatrix} -6 & 7 & -11 \\ 4 & 18 & -4 \end{bmatrix}$ **31.** $\begin{bmatrix} -3 & 6 & 8 \\ -18 & 12 & 10 \\ 4 & 6 & 24 \end{bmatrix}$ **33.** $\begin{bmatrix} 5 & -11 & 15 \\ 4 & -7 & 3 \\ 0 & 10 & 4 \end{bmatrix}$ **35.** $\begin{bmatrix} -0.2 & 1.2 \\ 2.6 & -0.6 \\ -0.2 & 2.2 \end{bmatrix}$ **37.** $\begin{bmatrix} -31 & 16 \\ 61 & -25 \\ -3 & 77 \end{bmatrix}$ **39.** Not defined

41. $\begin{bmatrix} -2 & 25 & -15 \\ 26 & -25 & 45 \\ -2 & 45 & -25 \end{bmatrix}$ **43.** $\begin{bmatrix} -26 & -15 & -25 \\ -4 & -18 & 4 \\ 2 & 43 & -19 \end{bmatrix}$ **45.** $B^n \to \begin{bmatrix} 0.25 & 0.75 \\ 0.25 & 0.75 \end{bmatrix}$, $AB^n \to [0.25 \ 0.75]$ **47.** $a = -1$, $b = 1$, $c = 3$, $d = -5$

49. $x = 1$, $y = 2$ **51.** $x = -5$, $y = 4$ **53.** $a = 3$, $b = 1$, $c = 1$, $d = -2$ **55.** All are true

57.

	Guitar	Banjo
Materials	$33	$26
Labor	$57	$77

59.

Markup

	Basic car	Air	AM/FM radio	Cruise control
Model *A*	$3,330	$77	$42	$27
Model *B*	$2,125	$93	$95	$50
Model *C*	$1,270	$113	$121	$52

61. (A) $11.80 (B) $30.30 (C) MN gives the labor costs per boat at each plant. (D)

$MN =$

	Plant I	Plant II
One-person boat	$11.80	$13.80
Two-person boat	$18.50	$21.60
Four-person boat	$26.00	$30.30

63. (A) $A^2 = \begin{bmatrix} 0 & 0 & 2 & 0 & 0 \\ 1 & 0 & 0 & 0 & 1 \\ 0 & 1 & 0 & 2 & 0 \\ 1 & 0 & 0 & 0 & 1 \\ 0 & 0 & 1 & 0 & 0 \end{bmatrix}$;

There is one way to travel from Baltimore to Atlanta with one intermediate connection; there are two ways to travel from Atlanta to Chicago with one intermediate connection. In general, the elements in A^2 indicate the number of different ways to travel from the *i*th city to the *j*th city with one intermediate connection.

(B) $A^3 = \begin{bmatrix} 2 & 0 & 0 & 0 & 2 \\ 0 & 1 & 0 & 2 & 0 \\ 0 & 0 & 3 & 0 & 0 \\ 0 & 1 & 0 & 2 & 0 \\ 1 & 0 & 0 & 0 & 1 \end{bmatrix}$;

There is one way to travel from Denver to Baltimore with two intermediate connections; there are two ways to travel from Atlanta to El Paso with two intermediate connections. In general, the elements in A^3 indicate the number of different ways to travel from the *i*th city to the *j*th city with two intermediate connections.

(C) $A + A^2 + A^3 + A^4 = \begin{bmatrix} 2 & 3 & 2 & 5 & 2 \\ 1 & 1 & 4 & 2 & 1 \\ 4 & 1 & 3 & 2 & 4 \\ 1 & 1 & 4 & 2 & 1 \\ 1 & 1 & 1 & 3 & 1 \end{bmatrix}$;

It is possible to travel from any origin to any destination with at most 3 intermediate connections.

65. (A) \$3,550 (B) \$6,000 (C) *NM* gives the total cost per town.

(D) $NM = \begin{bmatrix} \$3{,}550 \\ \$6{,}000 \end{bmatrix}$ (Cost/town) Berkeley, Oakland

(E) $[1 \quad 1]N = [3{,}000 \quad 1{,}300 \quad 13{,}000]$ (Telephone call, House call, Letter)

(F) $N\begin{bmatrix} 1 \\ 1 \\ 1 \end{bmatrix} = \begin{bmatrix} 6{,}500 \\ 10{,}800 \end{bmatrix}$ (Total contacts) Berkeley, Oakland

67. (A) $\begin{bmatrix} 0 & 0 & 1 & 1 & 1 & 0 \\ 1 & 0 & 0 & 1 & 1 & 0 \\ 0 & 1 & 0 & 1 & 0 & 0 \\ 0 & 0 & 0 & 0 & 0 & 1 \\ 0 & 0 & 1 & 1 & 0 & 1 \\ 1 & 1 & 1 & 0 & 0 & 0 \end{bmatrix}$ (B) $\begin{bmatrix} 0 & 1 & 2 & 3 & 1 & 2 \\ 1 & 0 & 2 & 3 & 2 & 2 \\ 1 & 1 & 0 & 2 & 1 & 1 \\ 1 & 1 & 1 & 0 & 0 & 1 \\ 1 & 2 & 2 & 2 & 0 & 2 \\ 2 & 2 & 2 & 3 & 2 & 0 \end{bmatrix}$ (C) $BC = \begin{bmatrix} 9 \\ 10 \\ 6 \\ 4 \\ 9 \\ 11 \end{bmatrix}$ where $C = \begin{bmatrix} 1 \\ 1 \\ 1 \\ 1 \\ 1 \\ 1 \end{bmatrix}$

(D) Frank, Bart, Aaron and Elvis (tie), Charles, Dan

Exercise 5-5

1. $\begin{bmatrix} 2 & -3 \\ 4 & 5 \end{bmatrix}$ **3.** $\begin{bmatrix} 2 & -3 \\ 4 & 5 \end{bmatrix}$ **5.** $\begin{bmatrix} -2 & 1 & 3 \\ 2 & 4 & -2 \\ 5 & 1 & 0 \end{bmatrix}$ **7.** $\begin{bmatrix} -2 & 1 & 3 \\ 2 & 4 & -2 \\ 5 & 1 & 0 \end{bmatrix}$ **9.** Yes **11.** No **13.** Yes **15.** No **17.** Yes

19. $\begin{bmatrix} 4 & 1 \\ -1 & 0 \end{bmatrix}$ **21.** $\begin{bmatrix} 3 & -2 \\ -1 & 1 \end{bmatrix}$ **23.** $\begin{bmatrix} 7 & -3 \\ -2 & 1 \end{bmatrix}$ **25.** $\begin{bmatrix} -3 & -4 & 2 \\ -2 & -2 & 1 \\ 2 & 3 & -1 \end{bmatrix}$ **27.** $\begin{bmatrix} 2 & -1 & -1 \\ -1 & 1 & 1 \\ -2 & 1 & 2 \end{bmatrix}$ **29.** Does not exist

31. $\begin{bmatrix} 5 & -3 \\ -3 & 2 \end{bmatrix}$ **33.** $\begin{bmatrix} 1 & 0 & 1 \\ \frac{1}{2} & \frac{1}{2} & 1 \\ 2 & 1 & 4 \end{bmatrix}$ **35.** Does not exist **37.** $\begin{bmatrix} -9 & -15 & 10 \\ 4 & 5 & -4 \\ -1 & -1 & 1 \end{bmatrix}$

39. M^{-1} exists if and only if all the elements on the main diagonal are nonzero. **41.** In both parts, $A^{-1} = A$ and $A^2 = I$
43. In both parts, $(A^{-1})^{-1} = A$

45. (A) $(AB)^{-1} = \begin{bmatrix} 29 & -41 \\ -12 & 17 \end{bmatrix}$, $A^{-1}B^{-1} = \begin{bmatrix} 23 & -33 \\ -16 & 23 \end{bmatrix}$, $B^{-1}A^{-1} = \begin{bmatrix} 29 & -41 \\ -12 & 17 \end{bmatrix}$

(B) $(AB)^{-1} = \begin{bmatrix} 0.7 & -0.1 \\ -1.8 & 0.4 \end{bmatrix}$, $A^{-1}B^{-1} = \begin{bmatrix} 0.1 & 0 \\ -0.4 & 1 \end{bmatrix}$, $B^{-1}A^{-1} = \begin{bmatrix} 0.7 & -0.1 \\ -1.8 & 0.4 \end{bmatrix}$

47. 14 5 195 74 97 37 181 67 49 18 121 43 103 41 **49.** GREEN EGGS AND HAM
51. 21 56 55 25 58 46 97 94 48 75 45 58 63 45 59 48 64 80 44 69 68 104 123 72 127 **53.** LYNDON BAINES JOHNSON

Exercise 5-6

1. $\begin{aligned} 2x_1 - x_2 &= 3 \\ x_1 + 3x_2 &= -2 \end{aligned}$ **3.** $\begin{aligned} -2x_1 + x_3 &= 3 \\ x_1 + 2x_2 + x_3 &= -4 \\ x_2 - x_3 &= 2 \end{aligned}$

5. $\begin{bmatrix} 4 & -3 \\ 1 & 2 \end{bmatrix} \begin{bmatrix} x_1 \\ x_2 \end{bmatrix} = \begin{bmatrix} 2 \\ 1 \end{bmatrix}$ **7.** $\begin{bmatrix} 1 & -2 & 1 \\ -1 & 1 & 0 \\ 2 & 3 & 1 \end{bmatrix} \begin{bmatrix} x_1 \\ x_2 \\ x_3 \end{bmatrix} = \begin{bmatrix} -1 \\ 2 \\ -3 \end{bmatrix}$ **9.** $x_1 = -8$ and $x_2 = 2$ **11.** $x_1 = 0$ and $x_2 = 4$

13. $x_1 = 3, x_2 = -2$ **15.** $x_1 = 11, x_2 = 4$ **17.** (A) $x_1 = -3, x_2 = 2$ (B) $x_1 = -1, x_2 = 2$ (C) $x_1 = -8, x_2 = 3$
19. (A) $x_1 = 17, x_2 = -5$ (B) $x_1 = 7, x_2 = -2$ (C) $x_1 = 24, x_2 = -7$
21. (A) $x_1 = 1, x_2 = 0, x_3 = 0$ (B) $x_1 = -1, x_2 = 0, x_3 = 1$ (C) $x_1 = 4, x_2 = 1, x_3 = -3$
23. (A) $x_1 = 0, x_2 = 2, x_3 = 4$ (B) $x_1 = -2, x_2 = 2, x_3 = 0$ (C) $x_1 = 6, x_2 = -2, x_3 = -6$
25. $x_1 = 2t + 2.5, x_2 = t$, t any real number **27.** No solution **29.** $x_1 = 13t + 3, x_2 = 8t + 1, x_3 = t$, t any real number
31. $X = (A - B)^{-1}C$ $[X \neq C(A - B)^{-1}]$ **33.** $X = (A + I)^{-1}C$ **35.** $X = (A + B)^{-1}(C + D)$
37. (A) $x_1 = 1, x_2 = 0$ (B) $x_1 = -2{,}000, x_2 = 1{,}000$ (C) $x_1 = 2{,}001, x_2 = -1{,}000$
39. (A) Concert 1: 6,000 \$4 tickets and 4,000 \$8 tickets; concert 2: 5,000 \$4 tickets and 5,000 \$8 tickets; concert 3: 3,000 \$4 tickets and 7,000 \$8 tickets (B) No (C) Between \$40,000 and \$80,000
41. (A) $I_1 = 4, I_2 = 6, I_3 = 2$ (B) $I_1 = 3, I_2 = 7, I_3 = 4$ (C) $I_1 = 7, I_2 = 8, I_3 = 1$
43. (A) $a = 1, b = 0, c = -3$ (B) $a = -2, b = 5, c = 1$ (C) $a = 11, b = -46, c = 43$
45. (A) Diet 1: 60 oz mix A and 80 oz mix B; diet 2: 20 oz mix A and 60 oz mix B; diet 3: 0 oz mix A and 100 oz mix B (B) No

Exercise 5-7

1.
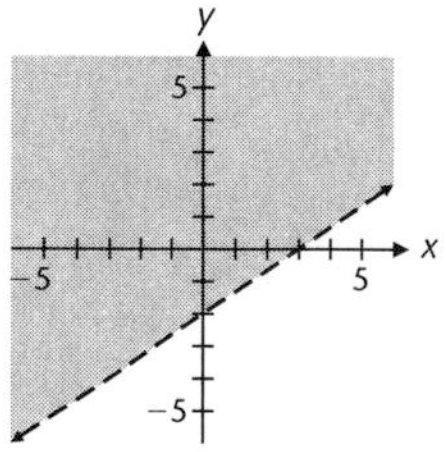

3.
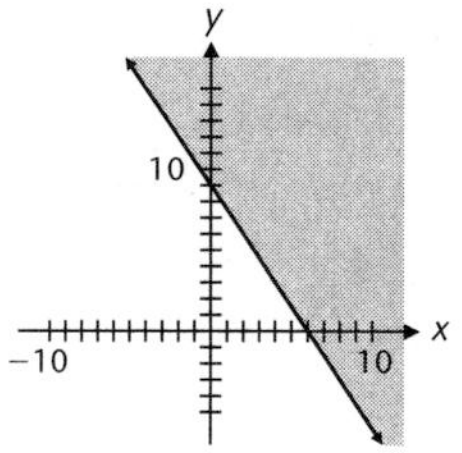

5.
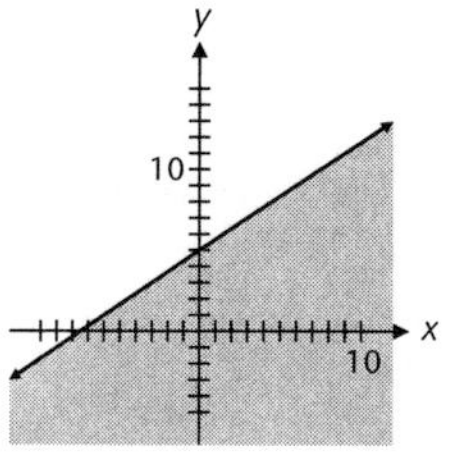

7.
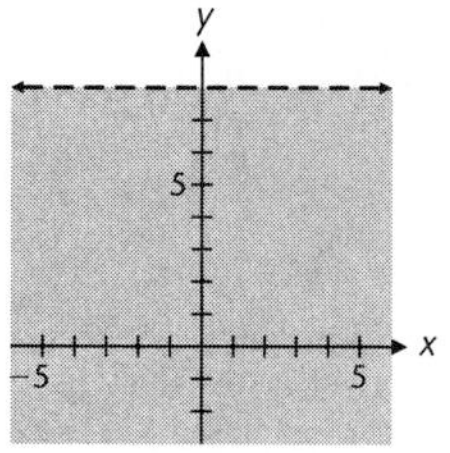

9.
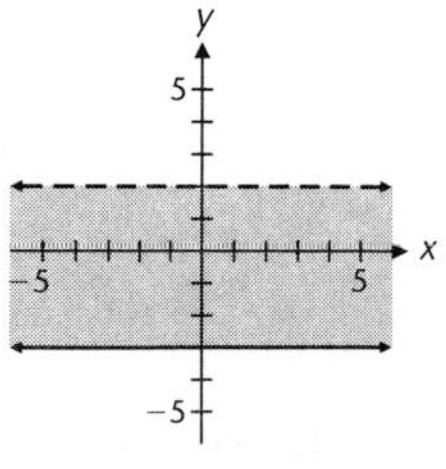

11. Region IV **13.** Region I

15.
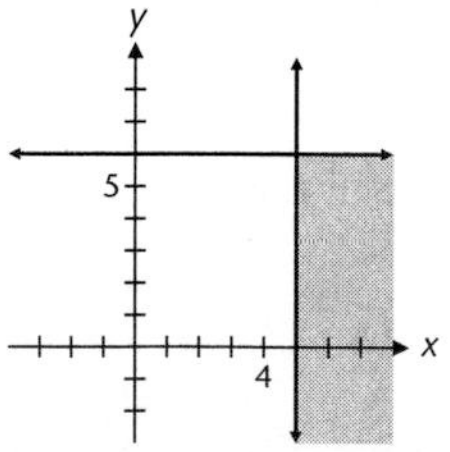

17.
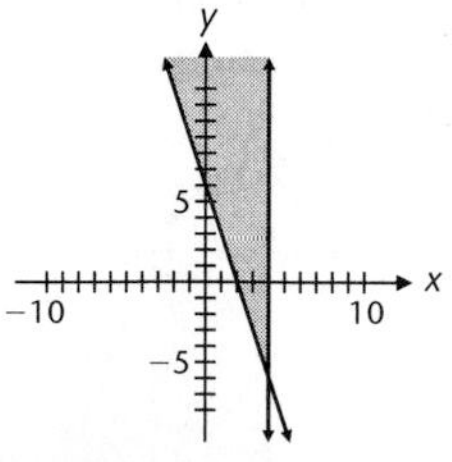

19.
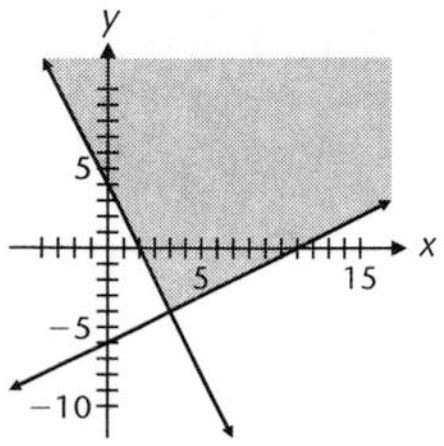

21. (A) Solution region is the double-shaded region. (B) Solution region is the unshaded region.

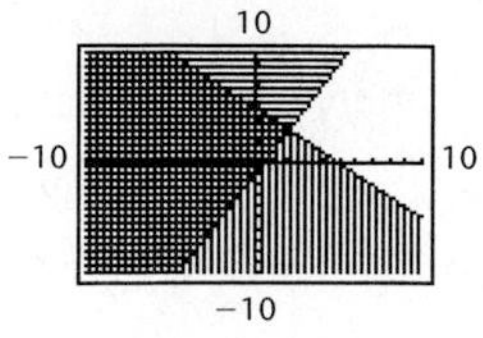

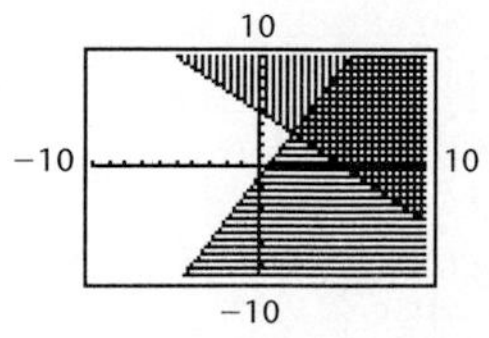

23. (A) Solution region is the double-shaded region.

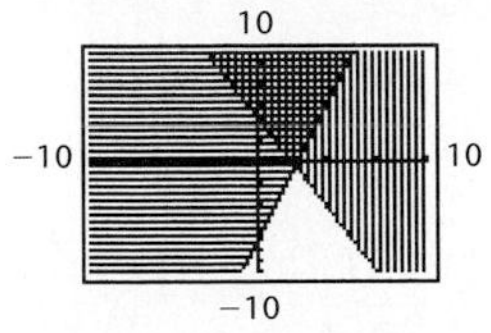

(B) Solution region is the unshaded region.

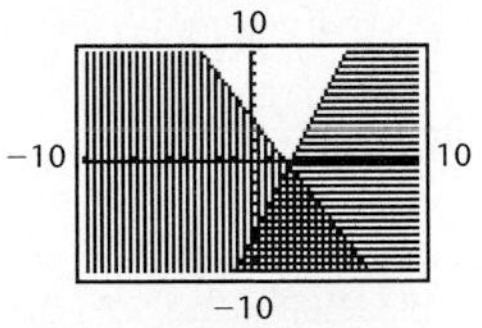

25. Region IV; corner points are (6, 4), (8, 0), and (18, 0)

27. Region I; corner points are (0, 16), (6, 4), and (18, 0)

29.

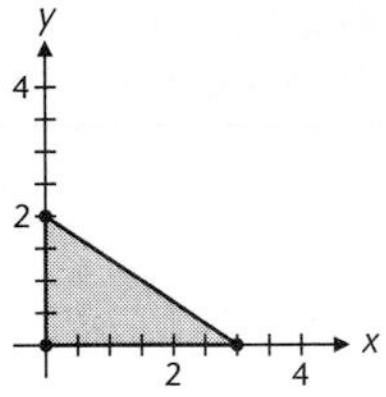

Corner points: (0, 0), (0, 2), (3, 0); bounded

31.

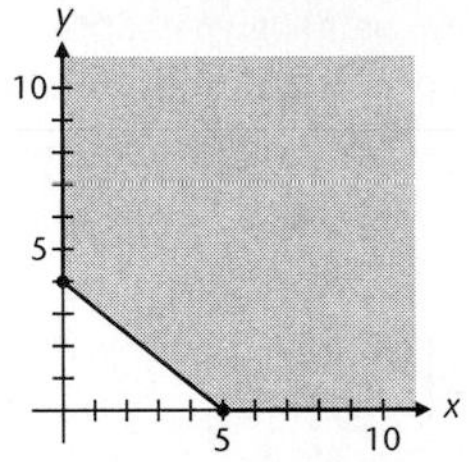

Corner points: (0, 4) and (5, 0); unbounded

33.

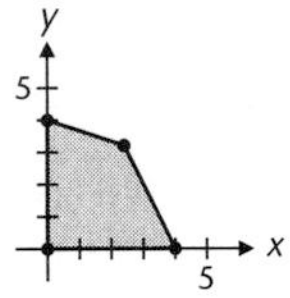

Corner points: (0, 4), (0, 0), $\left(\frac{12}{5}, \frac{16}{5}\right)$, (4, 0); bounded

35.

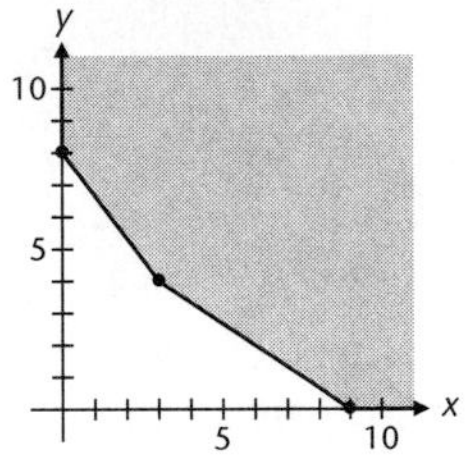

Corner points: (9, 0), (0, 8), and (3, 4); unbounded

37.

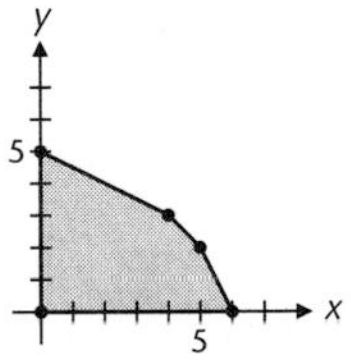

Corner points: (6, 0), (4, 3), (5, 2), (0, 0), and (0, 5); bounded

39.

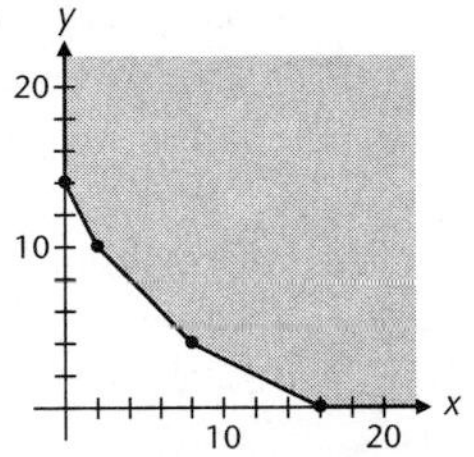

Corner points: (0, 14), (2, 10), (8, 4), (16, 0); unbounded

41.

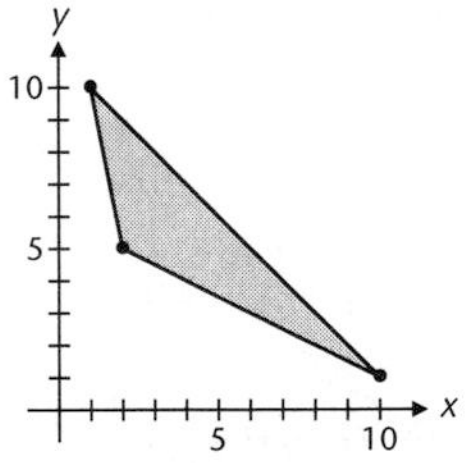

Corner points: (2, 5), (10, 1), (1, 10); bounded

43. The feasible region is empty.

45.

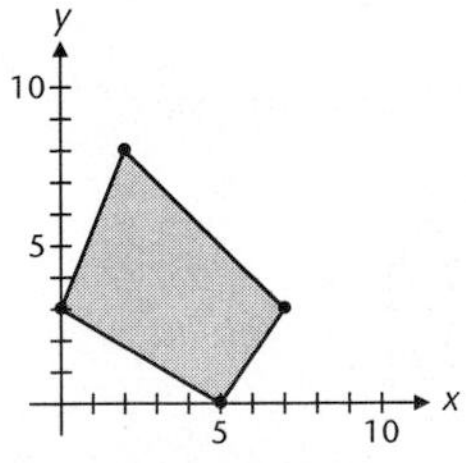

Corner points: (0, 3), (5, 0), (7, 3), (2, 8); bounded

47.

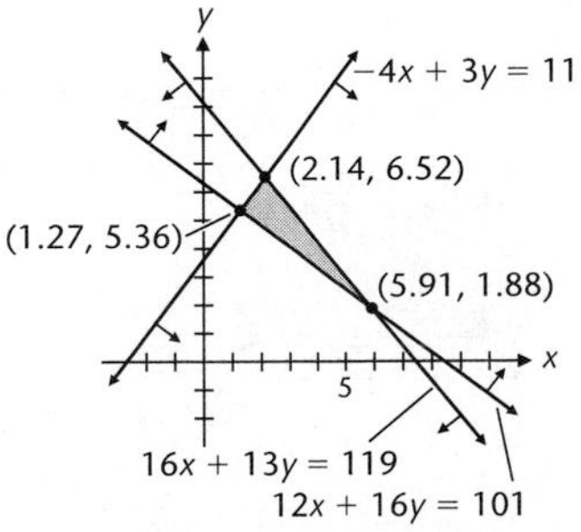

Corner points: (1.27, 5.36), (2.14, 6.52), (5.91, 1.88); bounded

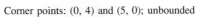

49. $6x + 4y \le 108$
$x + y \le 24$
$x \ge 0$
$y \ge 0$

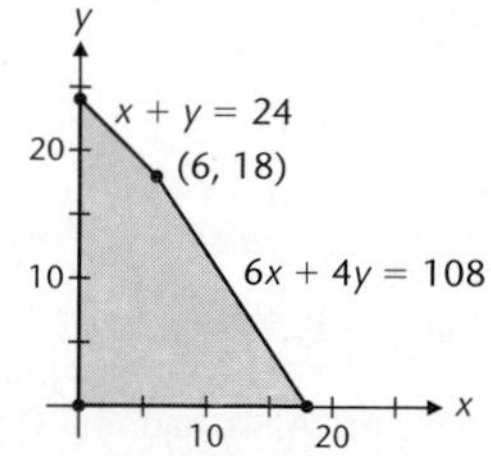

51. (A) All production schedules in the feasible region that are on the graph of $50x + 60y = 1{,}100$ will result in a profit of \$1,100.
(B) There are many possible choices. For example, producing 5 trick and 15 slalom skis will produce a profit of \$1,150. The graph of the line $50x + 60y = 1{,}150$ includes all the production schedules in the feasible region that result in a profit of \$1,150.

53. $20x + 10y \ge 460$
$30x + 30y \ge 960$
$5x + 10y \ge 220$
$x \ge 0$
$y \ge 0$

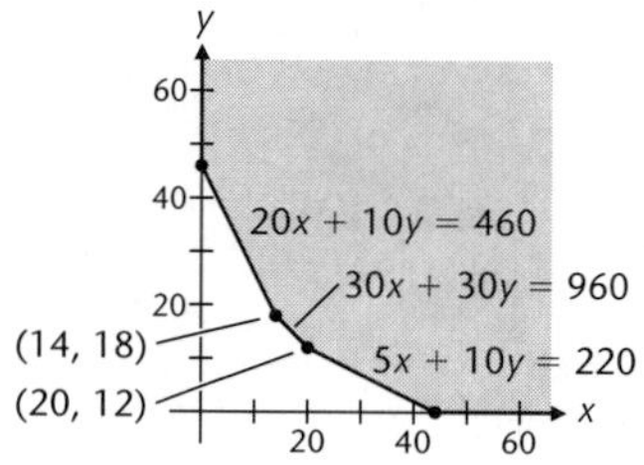

55. $10x + 30y \ge 280$
$30x + 10y \ge 360$
$x \ge 0$
$y \ge 0$

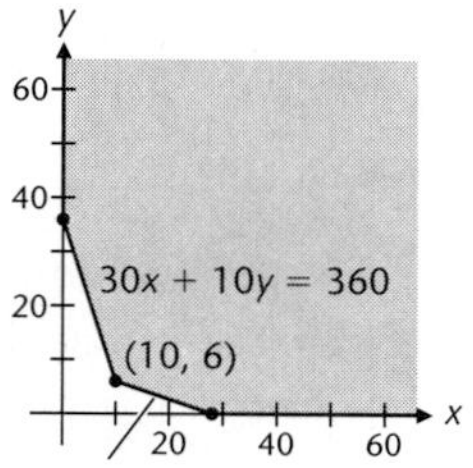

Exercise 5-8

1. Maximum value of z on S is 16 at (7, 9).
3. Maximum value of z on S is 84 at both (0, 12) and (7, 9).
5. Minimum value of z on S is 32 at (0, 8).
7. Minimum value of z on S is 36 at both (12, 0) and (4, 3).
9. Maximum value of z on S is 18 at (4, 3).
11. Minimum value of z on S is 12 at (4, 0).
13. Maximum value of z on S is 52 at (4, 10).
15. Minimum value of z on S is 44 at (4, 4).
17. The minimum value of z on S is 1,500 at (60, 0). The maximum value of z on S is 3,000 at (60, 30) and (120, 0) (multiple optimal solutions).
19. The minimum value of z on S is 300 at (0, 20). The maximum value of z on S is 1,725 at (60, 15).
21. Max $P = 5{,}507$ at $x_1 = 6.62$ and $x_2 = 4.25$
23. (A) $a > 2b$ (B) $\frac{1}{3}b < a < 2b$ (C) $a < \frac{1}{3}b$ or $b > 3a$ (D) $a = 2b$ (E) $b = 3a$
25. (A) 6 trick skis, 18 slalom skis; \$780 (B) The maximum profit decreases to \$720 when 18 trick and so slalom skis are produced.
(C) The maximum profit increases to \$1,080 when no trick and 24 slalom skis are produced.
27. 9 model A trucks and 6 model B trucks to realize the minimum cost of \$279,000
29. (A) 40 tables, 40 chairs; \$4,600 (B) The maximum profit decreases to \$3,800 when 20 tables and 80 chairs are produced.
31. (A) Max P = \$450 when 750 gallons are produced using the old process exclusively.
(B) The maximum profit decreases to \$380 when 400 gal are produced using the old process and 700 gal using the new process.
(C) The maximum profit decreases to \$288 when 1,440 gal are produced using the new process exclusively.
33. The nitrogen will range from a minimum of 940 lb when 40 bags of brand A and 100 bags of brand B are used to a maximum of 1,190 lb when 140 bags of brand A and 50 bags of brand B are used.

Chapter 5 Review

1. $x = 3, y = 3$ *(5-1)*
2. $x = 3, y = -2$ *(5-1)*
3. $x = 2, y = -1$ *(5-1)*
4. $x = 1.1875, y = 1.625$ *(5-1)*

5. *(5-7)*

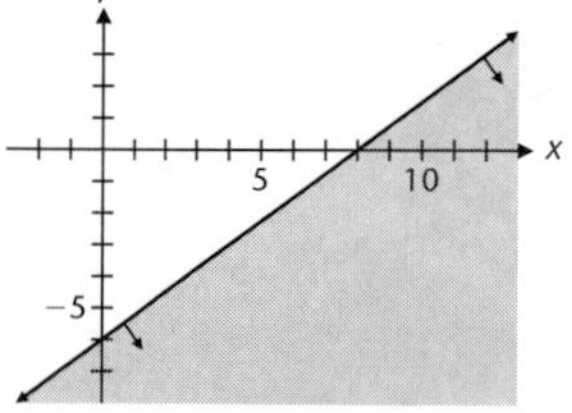

6. *(5-7)*

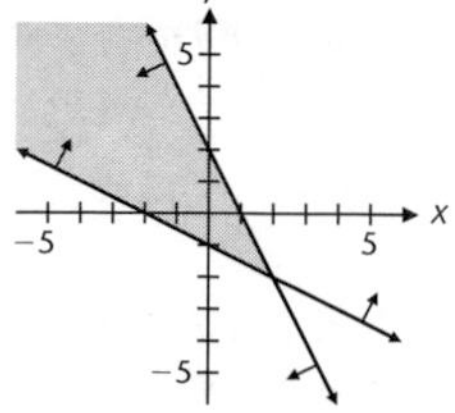

7. $\left[\begin{array}{cc|c} 3 & -6 & 12 \\ 1 & -4 & 5 \end{array}\right]$ *(5-2)* **8.** $\left[\begin{array}{cc|c} 1 & -4 & 5 \\ 1 & -2 & 4 \end{array}\right]$ *(5-2)* **9.** $\left[\begin{array}{cc|c} 1 & -4 & 5 \\ 0 & 6 & -3 \end{array}\right]$ *(5-2)*

10. $x_1 = 4$
$x_2 = -7$
The solution is $(4, -7)$ *(5-3)*

11. $x_1 - x_2 = 4$
$0 = 1$
No solution *(5-3)*

12. $x_1 - x_2 = 4$
$x_1 = t + 4$, $x_2 = t$ is the solution, for t any real number *(5-3)*

13. $\begin{bmatrix} 3 & 3 \\ 4 & 2 \end{bmatrix}$ *(5-4)* **14.** Not defined *(5-4)* **15.** $\begin{bmatrix} -3 & 0 \\ 1 & -1 \end{bmatrix}$ *(5-4)* **16.** $\begin{bmatrix} 4 & 3 \\ 7 & 4 \end{bmatrix}$ *(5-4)* **17.** Not defined *(5-4)*

18. $\begin{bmatrix} 5 \\ 5 \end{bmatrix}$ *(5-4)* **19.** $\begin{bmatrix} 2 & 3 \\ 4 & 6 \end{bmatrix}$ *(5-4)* **20.** $[8]$ *(5-4)* **21.** Not defined *(5-4)* **22.** $\begin{bmatrix} 3 & -2 \\ -4 & 3 \end{bmatrix}$ *(5-5)*

23. (A) $x_1 = -1, x_2 = 3$ (B) $x_1 = 1, x_2 = 2$ (C) $x_1 = 8, x_2 = -10$ *(5-6)*

24. The maximum value of z on S is 42 at (6, 4). The minimum value of z on S is 18 at (0, 6). *(5-8)*

25. $x_1 = 2, x_2 = -2$; each pair of lines has the same intersection point. *(5-3)*

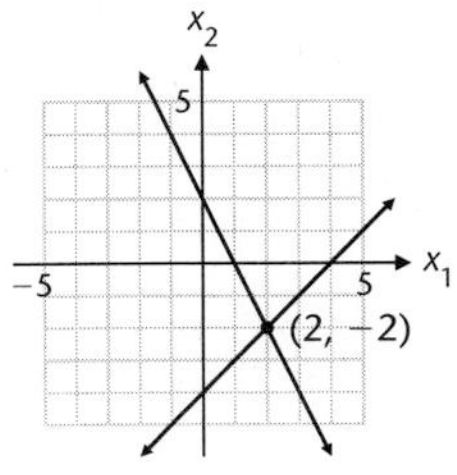

$x_1 - x_2 = 4$
$2x_1 + x_2 = 2$

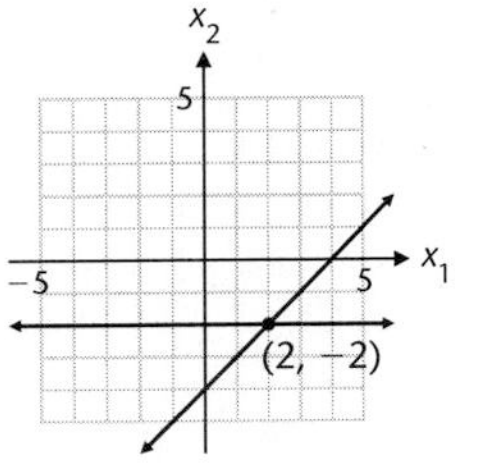

$x_1 - x_2 = 4$
$3x_2 = -6$

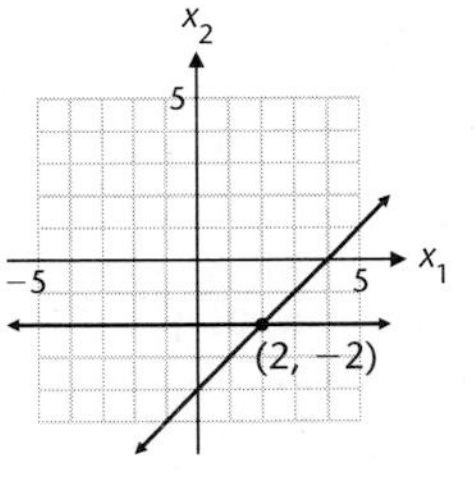

$x_1 - x_2 = 4$
$x_2 = -2$

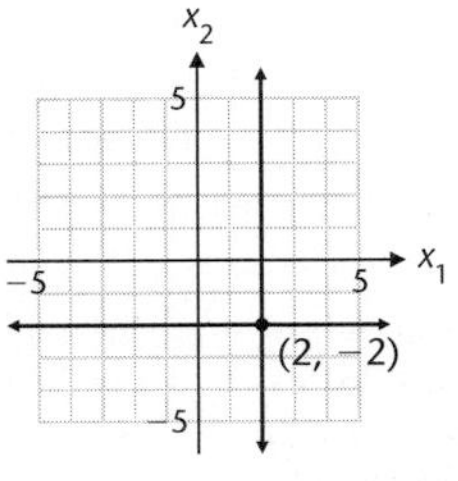

$x_1 = 2$
$x_2 = -2$

26. $x_1 = -1, x_2 = 3$ *(5-3)* **27.** $x_1 = -1, x_2 = 2, x_3 = 1$ *(5-3)* **28.** $x_1 = 2, x_2 = 1, x_3 = -1$ *(5-3)*

29. $x_1 = -5t - 12, x_2 = 3t + 7, x_3 = t$ is a solution for every real number t. There are infinitely many solutions. *(5-3)*

30. No solution *(5-3)*

31. $x_1 = -\frac{3}{7}t - \frac{4}{7}, x_2 = \frac{5}{7}t + \frac{9}{7}, x_3 = t$ is a solution for every real number t. There are infinitely many solutions. *(5-3)*

32. Not defined *(5-4)* **33.** $\begin{bmatrix} 10 & -8 \\ 4 & 6 \end{bmatrix}$ *(5-4)* **34.** $\begin{bmatrix} -2 & 8 \\ 8 & 6 \end{bmatrix}$ *(5-4)* **35.** Not defined *(5-4)* **36.** $[9]$ *(5-4)*

37. $\begin{bmatrix} 10 & -5 & 1 \\ -1 & -4 & -5 \\ 1 & -7 & -2 \end{bmatrix}$ *(5-4)* **38.** $\begin{bmatrix} -\frac{5}{2} & 2 & -\frac{1}{2} \\ 1 & -1 & 1 \\ \frac{1}{2} & 0 & -\frac{1}{2} \end{bmatrix}$ or $\frac{1}{2}\begin{bmatrix} -5 & 4 & -1 \\ 2 & -2 & 2 \\ 1 & 0 & -1 \end{bmatrix}$ *(5-5)*

39. (A) $x_1 = 2, x_2 = 1, x_3 = -1$ (B) $x_1 = 1, x_2 = -2, x_3 = 1$ (C) $x_1 = -1, x_2 = 2, x_3 = -2$ *(5-6)*

40. (A) A unique solution (B) Either no solution or an infinite number *(5-6)* **41.** No *(5-5)*

42. Corners: (0, 4), (0, 0), (4, 0), and (3, 2); bounded *(5-7)*

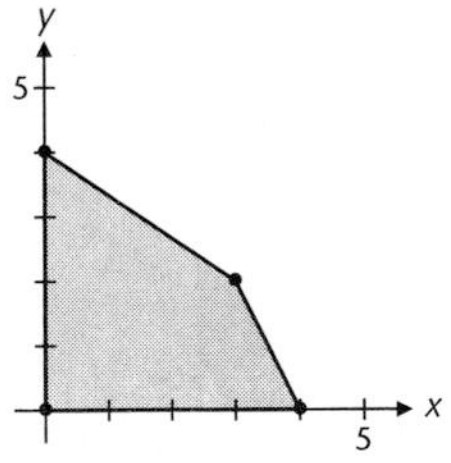

43. Corners: (0, 8), (12, 0), and $(\frac{12}{5}, \frac{16}{5})$; unbounded *(5-7)*

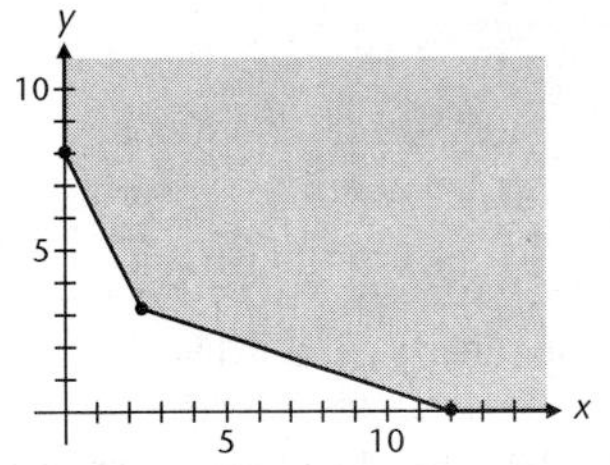

44. Corners: (4, 4), (10, 10), (20, 0); bounded *(5-7)*

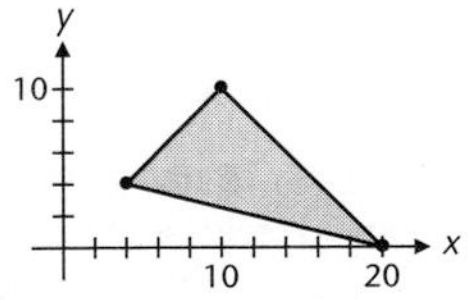

45. The maximum value of z on S is 46 at (4, 2). *(5-8)*

46. The minimum value of z on S is 75 at (3, 6) and (15, 0) (multiple optimal solutions). *(5-8)*

47. The minimum value of z on S is 44 at (4, 3). The maximum value of z on S is 82 at (2, 9). *(5-8)* **48.** $X = (A - C)^{-1}B$ *(5-6)*

49. $\begin{bmatrix} -\frac{11}{12} & -\frac{1}{12} & 5 \\ \frac{10}{12} & \frac{2}{12} & -4 \\ \frac{1}{12} & -\frac{1}{12} & 0 \end{bmatrix}$ or $\frac{1}{12}\begin{bmatrix} -11 & -1 & 60 \\ 10 & 2 & -48 \\ 1 & -1 & 0 \end{bmatrix}$ *(5-5)* **50.** $x_1 = 1{,}000, x_2 = 4{,}000, x_3 = 2{,}000$ *(5-6)*

51. $x_1 = 1000, x_2 = 4000, x_3 = 2000$ *(5-3)* **52.** The maximum value of z on S is 26,000 at (600, 400). *(5-8)*

53. (A) A unique solution (B) No solution (C) An infinite number of solutions *(5-3)*

54. 48 $\frac{1}{2}$-lb packages and 72 $\frac{1}{3}$-lb packages *(5-1)* **55.** 6 m by 8 m *(5-1)*
56. $x_1 = 40$ g mix A, $x_2 = 60$ g mix B, $x_3 = 30$ g mix C *(5-3)*
57. (A) $x_1 = 22$ nickels, $x_2 = 8$ dimes (B) $x_1 = 3t + 22$ nickels, $x_2 = 8 - 4t$ dimes, $x_3 = t$ quarters, $t = 0, 1,$ or 2 *(5-3)*
58. (A) \$27 (B) Elements in LH give the total cost of manufacturing each product at each plant.
(C)
$$LH = \begin{array}{c} \text{N.C.} \quad \text{S.C.} \\ \begin{bmatrix} \$46.35 & \$41.00 \\ \$30.45 & \$27.00 \end{bmatrix} \begin{array}{l} \text{Desk} \\ \text{Stands} \end{array} \end{array} \quad (5\text{-}4)$$
59. (A) $\begin{bmatrix} 1{,}600 & 1{,}730 \\ 890 & 720 \end{bmatrix}$ (B) $\begin{bmatrix} 200 & 160 \\ 80 & 40 \end{bmatrix}$ (C) $\begin{bmatrix} 3{,}150 \\ 1{,}550 \end{bmatrix} \begin{array}{l} \text{Desks} \\ \text{Stands} \end{array}$
This matrix represents the total production of each item in January. *(5-4)*
60. GRAPHING UTILITY *(5-5)*
61. (A) 60 tons of ore must be produced at Big Bend, 20 tons of ore at Saw Pit.
(B) 30 tons of ore must be produced at Big Bend, 50 tons of ore at Saw Pit.
(C) 40 tons of ore must be produced at Big Bend, 40 tons of ore at Saw Pit. *(5-6)*
62. (A) Maximum profit is $P = \$7{,}800$ when 80 regular and 30 competition sails are produced.
(B) The maximum profit increases to \$8,750 when 70 competition and no regular sails are produced.
(C) The maximum profit decreases to \$7,200 when no competition and 120 regular sales are produced. *(5-8)*
63. (A) The minimum cost is $C = \$13$ when 100 g of mix A and 150 g of mix B are used.
(B) The minimum cost decreases to \$9 when 50 g of mix A and 275 g of mix B are used.
(C) The minimum cost increases to \$28.75 when 250 g of mix A and 75 g of mix B are used. *(5-8)*

CHAPTER 6 Exercise 6-1

1. $-1, 0, 1, 2$ **3.** $0, \frac{1}{3}, \frac{1}{2}, \frac{3}{5}$ **5.** $4, -8, 16, -32$ **7.** 6 **9.** $\dfrac{99}{101}$ **11.** $1 + 2 + 3 + 4 + 5$ **13.** $\dfrac{1}{10} + \dfrac{1}{100} + \dfrac{1}{1{,}000}$
15. $-1 + 1 - 1 + 1$ **17.** $1, -4, 9, -16, 25$ **19.** $0.3, 0.33, 0.333, 0.3333, 0.33333$ **21.** $1, -\frac{1}{2}, \frac{1}{4}, -\frac{1}{8}, \frac{1}{16}$
23. $7, 3, -1, -5, -9$ **25.** $4, 1, \frac{1}{4}, \frac{1}{16}, \frac{1}{64}$ **27.** $a_n = n + 3$ **29.** $a_n = 3n$ **31.** $a_n = \dfrac{n}{n+1}$
33. $a_n = (-1)^{n+1}$ **35.** $a_n = (-2)^n$ **37.** $a_n = \dfrac{x^n}{n}$

39.
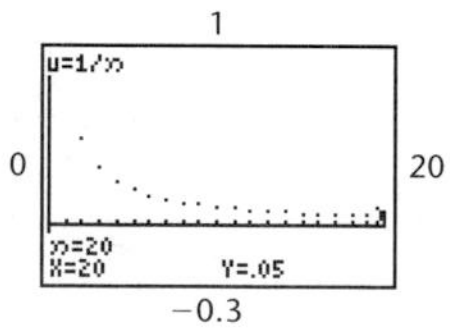

41.
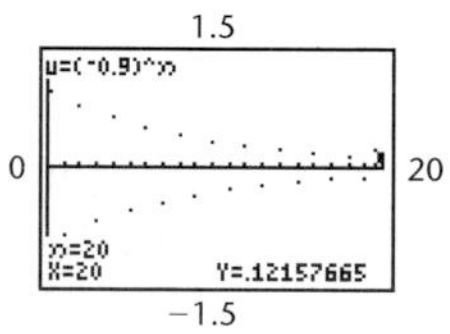

43. $\frac{4}{1} - \frac{8}{2} + \frac{16}{3} - \frac{32}{4}$ **45.** $x^2 + \dfrac{x^3}{2} + \dfrac{x^4}{3}$

47. $x - \dfrac{x^2}{2} + \dfrac{x^3}{3} - \dfrac{x^4}{4} + \dfrac{x^5}{5}$ **49.** $\sum_{k=1}^{4} k^2$ **51.** $\sum_{k=1}^{5} \dfrac{1}{2^k}$ **53.** $\sum_{k=1}^{n} \dfrac{1}{k^2}$ **55.** $\sum_{k=1}^{n} (-1)^{k+1}k^2$
57. (A) 3, 1.83, 1.46, 1.415 (B) Calculator $\sqrt{2} = 1.4142135\ldots$ (C) $a_1 = 1$; 1, 1.5, 1.417, 1.414
59. The values of c_n are approximately 2.236 (i.e., $\sqrt{5}$) for large values of n.
61. $e^{0.2} = 1.2214000$; $e^{0.2} = 1.2214028$ (calculator—direct evaluation)

Exercise 6-2

1. Fails at $n = 2$ **3.** Fails at $n = 3$ **5.** P_1: $2 = 2 \cdot 1^2$; P_2: $2 + 6 = 2 \cdot 2^2$; P_3: $2 + 6 + 10 = 2 \cdot 3^2$
7. P_1: $a^5a^1 = a^{5+1}$; P_2: $a^5a^2 = a^5(a^1a) = (a^5a)a = a^6a = a^7 = a^{5+2}$; P_3: $a^5a^3 = a^5(a^2a) = a^5(a^1a)a = [(a^5a)a]a = a^8 = a^{5+3}$
9. P_1: $9^1 - 1 = 8$ is divisible by 4; P_2: $9^2 - 1 = 80$ is divisible by 4; P_3: $9^3 - 1 = 728$ is divisible by 4
11. P_k: $2 + 6 + 10 + \cdots + (4k - 2) = 2k^2$; P_{k+1}: $2 + 6 + 10 + \cdots + (4k - 2) + (4k + 2) = 2(k + 1)^2$
13. P_k: $a^5a^k = a^{5+k}$; P_{k+1}: $a^5a^{k+1} = a^{5+k+1}$ **15.** P_k: $9^k - 1 = 4r$ for some integer r; P_{k+1}: $9^{k+1} - 1 = 4s$ for some integer s
23. $n = 4$, $p(x) = x^4 + 1$ **25.** $n = 23$ **43.** P_n: $2 + 4 + 6 + \cdots + 2n = n(n + 1)$
45. $1 + 2 + 3 + \cdots + (n - 1) = \dfrac{n(n-1)}{2}$, $n \geq 2$ **51.** $3^4 + 4^4 + 5^4 + 6^4 \neq 7^4$

Exercise 6-3

1. (A) Arithmetic with $d = -5$; $-26, -31$ (B) Geometric with $r = -2$; $-16, 32$ (C) Neither (D) Geometric with $r = \frac{1}{3}$; $\frac{1}{54}, \frac{1}{162}$
3. $a_2 = -1$; $a_3 = 3$; $a_4 = 7$ **5.** $a_{15} = 67$; $S_{11} = 242$ **7.** $S_{21} = 861$ **9.** $a_{15} = -21$ **11.** $a_2 = 3$; $a_3 = -\frac{3}{2}$; $a_4 = \frac{3}{4}$

13. $a_{10} = \frac{1}{243}$ **15.** $S_7 = 3{,}279$ **17.** $d = 6$; $a_{101} = 603$ **19.** $S_{40} = 200$ **21.** $a_{11} = 2$; $S_{11} = \frac{77}{6}$ **23.** $a_1 = 1$
25. $r = 0.398$ **27.** $S_{10} = -1{,}705$ **29.** $a_2 = 6$; $a_3 = 4$ **31.** $S_{51} = 4{,}131$ **33.** $S_7 = 547$ **35.** $-1{,}071$ **37.** $\frac{1{,}023}{1{,}024}$
39. 4,446 **43.** $x = 2\sqrt{3}$ **45.** $a_n = -3 + (n - 1)3$ or $3n - 6$ **47.** 66 **49.** 133 **51.** $S_\infty = \frac{9}{2}$ **53.** no sum
55. $S_\infty = \frac{8}{5}$ **57.** $\frac{7}{9}$ **59.** $\frac{6}{11}$ **61.** $3\frac{8}{37}$ or $\frac{119}{37}$ **65.** $a_n = (-2)(-3)^{n-1}$ **67.** *Hint:* $y = x + d$, $z = x + 2d$ **71.** $x = -1$, $y = 2$
73. Firm *A*: \$501,000; firm *B*: \$504,000 **75.** \$4,000,000 **77.** $P(1 + r)^n$; approximately 12 yr **79.** \$700 per year; \$115,500
81. 900 **83.** 1,250,000 **85.** (A) 336 ft (B) 1,936 ft (C) $16t^2$ ft **87.** $A = A_0 2^{2t}$ **89.** $r = 10^{-0.4} = 0.398$
91. 9.22×10^{16} dollars; 1.845×10^{17} dollars **93.** 0.0015 pounds per square inch **95.** 2 **97.** 3,420°

Exercise 6-4

1. 362,880 **3.** 39,916,800 **5.** 990 **7.** 10 **9.** 35 **11.** 1 **13.** 60 **15.** 6,497,400 **17.** 10 **19.** 270,725
23. $5 \cdot 3 \cdot 4 \cdot 2 = 120$ **25.** $P_{10,3} = 10 \cdot 9 \cdot 8 = 720$ **27.** $C_{7,3} = 35$ subcommittees; $P_{7,3} = 210$ **29.** $C_{10,2} = 45$
31. No repeats: $6 \cdot 5 \cdot 4 \cdot 3 = 360$; with repeats: $6 \cdot 6 \cdot 6 \cdot 6 = 1{,}296$
33. No repeats: $10 \cdot 9 \cdot 8 \cdot 7 \cdot 6 = 30{,}240$; with repeats: $10 \cdot 10 \cdot 10 \cdot 10 \cdot 10 = 100{,}000$ **35.** $C_{13,5} = 1{,}287$
37. $26 \cdot 26 \cdot 26 \cdot 10 \cdot 10 \cdot 10 = 17{,}576{,}000$ possible license plates; no repeats: $26 \cdot 25 \cdot 24 \cdot 10 \cdot 9 \cdot 8 = 11{,}232{,}000$
39. $C_{13,5} \cdot C_{13,2} = 100{,}386$ **41.** $C_{8,3} \cdot C_{10,4} \cdot C_{7,2} = 246{,}960$
43. (B) $r = 0, 10$ (C) Each is the product of r consecutive integers, the largest of which is n for $P_{n,r}$ and r for $r!$.
45. $12 \cdot 11 = 132$ **47.** (A) $C_{8,2} = 28$ (B) $C_{8,3} = 56$ (C) $C_{8,4} = 70$
49. Two people: $5 \cdot 4 = 20$; three people: $5 \cdot 4 \cdot 3 = 60$; four people: $5 \cdot 4 \cdot 3 \cdot 2 = 120$; five people: $5 \cdot 4 \cdot 3 \cdot 2 \cdot 1 = 120$
51. (A) $P_{8,5} = 6{,}720$ (B) $C_{8,5} = 56$ (C) $C_{2,1} \cdot C_{6,4} = 30$
53. There are $C_{4,1} \cdot C_{48,4} = 778{,}320$ hands which contain exactly one king, and $C_{39,5} = 575{,}757$ hands containing no hearts, so the former is more likely.

Exercise 6-5

1. Occurrence of *E* is certain
3. (A) No probability can be negative (B) $P(R) + P(G) + P(Y) + P(B) \neq 1$ (C) Is an acceptable probability assignment.
5. $P(R) + P(Y) = .56$ **7.** .1 **9.** .45 **11.** $P(E) = \frac{n(E)}{n(S)} = \frac{1}{720} \approx .0014$ **13.** $\frac{C_{26,5}}{C_{52,5}} \approx .025$ **15.** $\frac{C_{16,5}}{C_{52,5}} \approx .0017$
17. $P(E) = \frac{n(E)}{n(S)} = \frac{50}{250} = .2$ **19.** $P(E) = \frac{n(E)}{n(S)} = \frac{1}{120} \approx .008$ **21.** $\frac{1}{36}$ **23.** $\frac{5}{36}$ **25.** $\frac{1}{6}$ **27.** $\frac{7}{9}$ **29.** 0 **31.** $\frac{1}{3}$ **33.** $\frac{2}{9}$
35. $\frac{2}{3}$ **37.** $S = \{1, 2, 3, \ldots, 365\}$; $P(e_i) = 1/365$
39. (A) $P(2) \approx .022$, $P(3) \approx .07$, $P(4) \approx .088$, $P(5) \approx .1$, $P(6) \approx .142$, $P(7) \approx .178$, $P(8) \approx .144$, $P(9) \approx .104$, $P(10) \approx .072$, $P(11) \approx .052$, $P(12) \approx .028$
(B) $P(2) = \frac{1}{36}$, $P(3) = \frac{2}{36}$, $P(4) = \frac{3}{36}$, $P(5) = \frac{4}{36}$, $P(6) = \frac{5}{36}$, $P(7) = \frac{6}{36}$, $P(8) = \frac{5}{36}$, $P(9) = \frac{4}{36}$, $P(10) = \frac{3}{36}$, $P(11) = \frac{2}{36}$, $P(12) = \frac{1}{36}$
(C)

Sum	2	3	4	5	6	7	8	9	10	11	12
Expected frequency	13.9	27.8	41.7	55.6	69.4	83.3	69.4	55.6	41.7	27.8	13.9

43. $\frac{1}{4}$ **45.** $\frac{1}{4}$ **47.** $\frac{3}{4}$ **49.** $\frac{1}{9}$ **51.** $\frac{1}{3}$ **53.** $\frac{1}{9}$ **55.** $\frac{4}{9}$ **57.** $\frac{C_{16,5}}{C_{52,5}} \approx .00168$ **59.** $\frac{48}{C_{52,5}} \approx .000\,0185$
61. $\frac{4}{C_{52,5}} \approx .000\,0015$ **63.** $\frac{C_{4,2} \cdot C_{4,3}}{C_{52,5}} \approx .000\,009$ **65.** (A) .015 (B) .222 (C) .169 (D) .958

Exercise 6-6

1. 10 **3.** 6 **5.** 20 **7.** 35 **9.** 84 **11.** 66 **13.** 2,380 **15.** 230,300 **17.** $m^3 + 3m^2n + 3mn^2 + n^3$
19. $8x^3 - 36x^2y + 54xy^2 - 27y^3$ **21.** $x^4 - 8x^3 + 24x^2 - 32x + 16$ **23.** $m^4 + 12m^3n + 54m^2n^2 + 108mn^3 + 81n^4$
25. $32x^5 - 80x^4y + 80x^3y^2 - 40x^2y^3 + 10xy^4 - y^5$ **27.** $m^6 + 12m^5n + 60m^4n^2 + 160m^3n^3 + 240m^2n^4 + 192mn^5 + 64n^6$
29. $35x^4$ **31.** $-29{,}586x^6$ **33.** $4{,}060{,}938{,}240x^{14}$ **35.** $15x^8$ **37.** Does not exist **39.** $5{,}005u^9v^6$ **41.** $264m^2n^{10}$
43. $924w^6$ **45.** $-48{,}384x^3y^5$ **47.** $3x^2 + 3xh + h^2$; approaches $3x^3$ **49.** $5x^4 + 10x^3h + 10x^2h^2 + 5xh^3 + h^4$; approaches $5x^4$
51. 5 **53.** (A) $a_4 = 0.251$ (B) 1 **55.** 1.1046

Chapter 6 Review

1. (A) Geometric (B) Arithmetic (C) Arithmetic (D) Neither (E) Geometric *(6-1, 6-3)*
2. (A) 5, 7, 9, 11 (B) $a_{10} = 23$ (C) $S_{10} = 140$ *(6-1, 6-3)* **3.** (A) 16, 8, 4, 2 (B) $a_{10} = \frac{1}{32}$ (C) $S_{10} = 31\frac{31}{32}$ *(6-1, 6-3)*
4. (A) −8, −5, −2, 1 (B) $a_{10} = 19$ (C) $S_{10} = 55$ *(6-1, 6-3)*
5. (A) −1, 2, −4, 8 (B) $a_{10} = 512$ (C) $S_{10} = 341$ *(6-1, 6-3)* **6.** $S_\infty = 32$ *(6-3)* **7.** 720 *(6-4)*

8. $\frac{22 \cdot 21 \cdot 20 \cdot 19!}{19!} = 9{,}240$ *(6-4)* **9.** 21 *(6-4)* **10.** $C_{6,2} = 15$; $P_{6,2} = 30$ *(6-5)*

11. (A) 12 combined outcomes: (B) $6 \cdot 2 = 12$ *(6-5)*

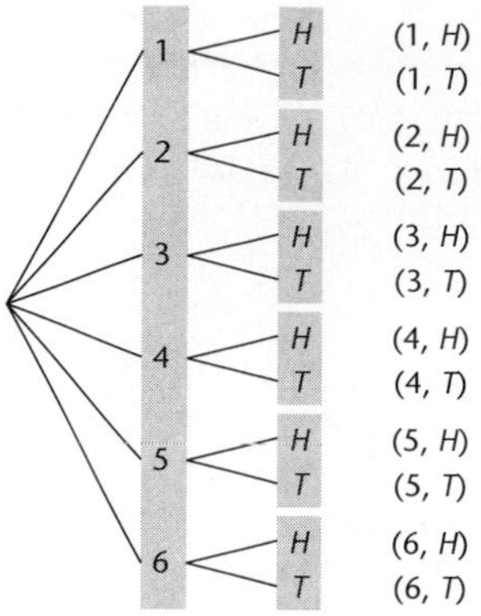

12. $6 \cdot 5 \cdot 4 \cdot 3 \cdot 2 \cdot 1 = 720$ *(6-5)* **13.** $P_{6,6} = 6! = 720$ *(6-5)* **14.** $\frac{C_{13,5}}{C_{52,5}} \approx .0005$ *(6-5)* **15.** $\frac{1}{P_{15,2}} \approx .0048$ *(6-5)*

16. .05 *(6-5)* **17.** P_1: $5 = 1^2 + 4 \cdot 1 = 5$; P_2: $5 + 7 = 2^2 + 4 \cdot 2$; P_3: $5 + 7 + 9 = 3^2 + 4 \cdot 3$ *(6-2)*

18. P_1: $2 = 2^{1+1} - 2$; P_2: $2 + 4 = 2^{2+1} - 2$; P_3: $2 + 4 + 8 = 2^{3+1} - 2$ *(6-2)*

19. P_1: $49^1 - 1 = 48$ is divisible by 6; P_2: $49^2 - 1 = 2{,}400$ is divisible by 6; P_3: $49^3 - 1 = 117{,}648$ is divisible by 6 *(6-2)*

20. P_k: $5 + 7 + 9 + \cdots + (2k + 3) = k^2 + 4k$; P_{k+1}: $5 + 7 + 9 + \cdots + (2k + 3) + (2k + 5) = (k + 1)^2 + 4(k + 1)$ *(6-2)*

21. P_k: $2 + 4 + 8 + \cdots + 2^k = 2^{k+1} - 2$; P_{k+1}: $2 + 4 + 8 + \cdots + 2^k + 2^{k+1} = 2^{k+2} - 2$ *(6-2)*

22. P_k: $49^k - 1 = 6r$ for some integer r; P_{k+1}: $49^{k+1} - 1 = 6s$ for some integer s *(6-2)* **23.** $n = 31$ is a counterexample *(6-2)*

24. $S_{10} = (-6) + (-4) + (-2) + 0 + 2 + 4 + 6 + 8 + 10 + 12 = 30$ *(6-3)* **25.** $S_7 = 8 + 4 + 2 + 1 + \frac{1}{2} + \frac{1}{4} + \frac{1}{8} = 15\frac{7}{8}$ *(6-3)*

26. $S_\infty = \frac{81}{5}$ *(6-3)* **27.** $S_n = \sum_{k=1}^{n} \frac{(-1)^{k+1}}{3^k}$; $S_\infty = \frac{1}{4}$ *(6-3)*

28. The probability of an event cannot be negative, but $P(e_2)$ is given as negative. The sum of the probabilities of the simple events must be 1, but it is given as 2.5. The probability of an event cannot be greater than 1, but $P(e_4)$ is given as 2. *(6-5)*

29. $C_{6,3} = 20$ *(6-4)* **30.** $d = 3$, $a_5 = 25$ *(6-3)* **31.** 336; 512; 392 *(6-4)*

32. (A) $P(\text{2 heads}) = .21$; $P(\text{1 head}) = .48$; $P(\text{0 heads}) = .31$ (B) $P(E_1) = .25$; $P(E_2) = .5$; $P(E_3) = .25$
(C) 2 heads = 250; 1 head = 500; 0 heads = 250 *(6-5)*

33. (A) $\frac{C_{13,5}}{C_{52,5}}$ (B) $\frac{C_{13,3} \cdot C_{13,2}}{C_{52,5}}$ *(6-5)* **34.** $\frac{C_{8,2}}{C_{10,4}} = \frac{2}{15}$ *(6-5)* **35.** (A) $\frac{1}{3}$ (B) $\frac{2}{9}$ *(6-5)* **36.** $\frac{8}{11}$ *(6-3)*

37. (A) $P_{6,3} = 120$ (B) $C_{5,2} = 10$ *(6-4)* **38.** 190 *(6-6)* **39.** 1,820 *(6-6)* **40.** 1 *(6-6)*

41. $x^5 - 5x^4y + 10x^3y^2 - 10x^2y^3 + 5xy^4 - y^5$ *(6-6)* **42.** $672x^6$ *(6-6)* **43.** $-1{,}760x^3y^9$ *(6-6)* **47.** 29 *(6-6)* **48.** 26 *(6-1)*

49. $2 \cdot 2 \cdot 2 \cdot 2 \cdot 2 = 32$; 6 *(6-4)* **50.** $\frac{49g}{2}$ ft; $\frac{625g}{2}$ ft *(6-3)* **51.** 12 *(6-4)*

52. $x^6 + 6ix^5 - 15x^4 - 20ix^3 + 15x^2 + 6ix - 1$ *(6-6)* **53.** $1 - \frac{C_{7,3}}{C_{10,3}} = \frac{17}{24}$ *(6-5)* **54.** (A) .350 (B) $\frac{3}{8} = .375$ (C) 375 *(6-5)*

60. \$900 *(6-3)* **61.** \$7,200 *(6-3)* **62.** \$895.42; \$1,603.57 *(6-3)* **63.** $P_{5,5} = 120$ *(6-4)*

64. (A) .04 (B) .16 (C) .54 *(6-5)* **65.** $1 - \frac{C_{10,4}}{C_{12,4}} \approx .576$ *(6-5)*

CHAPTER 7 Exercise 7-1

1.

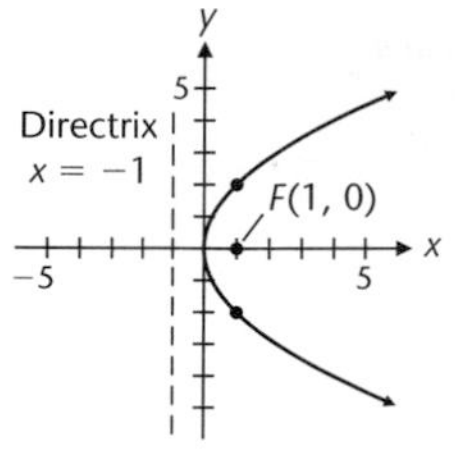

3.

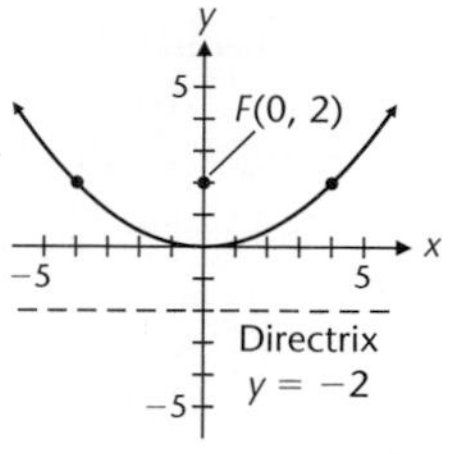

5.

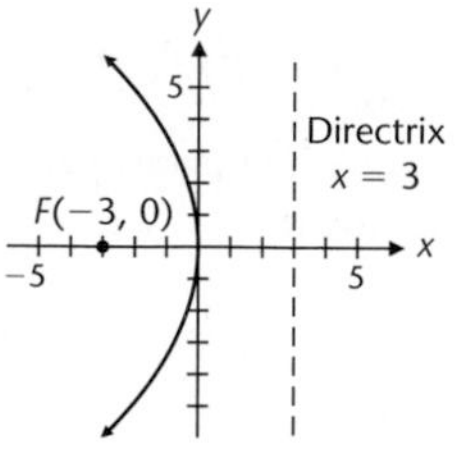

7.

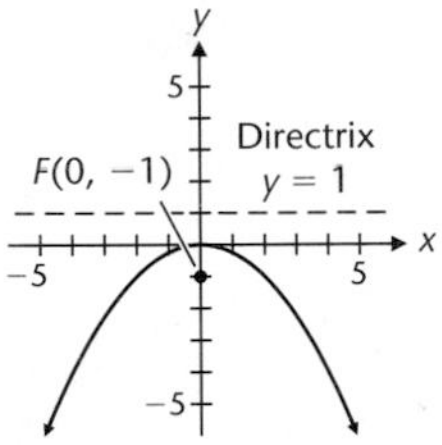

9.

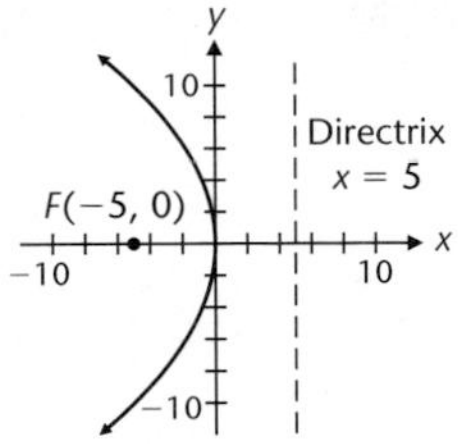

11.

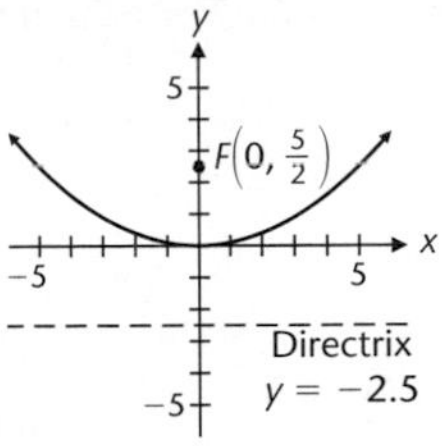

13. (9.75, 0) **15.** (0, −26.25) **17.** (−19.25, 0) **19.** $x^2 = 12y$ **21.** $x^2 = -28y$ **23.** $y^2 = -24x$ **25.** $y^2 = 8x$
27. $x^2 = 8y$ **29.** $y^2 = -12x$ **31.** $x^2 = -4y$ **33.**

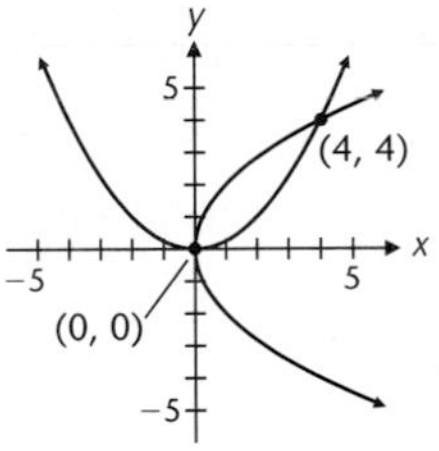

35.

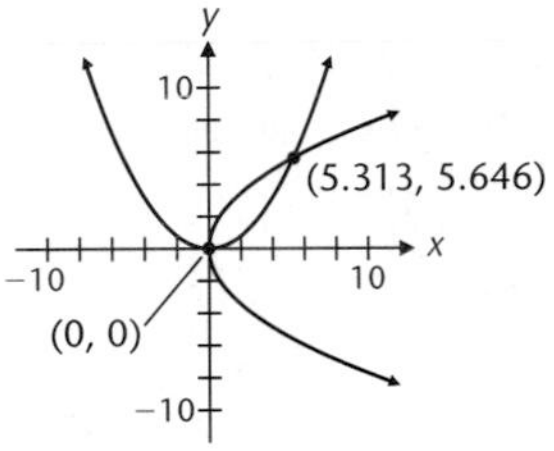

37. (A) 2; $x = 0$ and $y = 0$ (B) (0, 0), $(4am, 4am^2)$ **39.** $A(-2a, a)$, $B(2a, a)$ **41.** $x^2 - 4x - 12y - 8 = 0$
43. $y^2 + 8y - 8x + 48 = 0$ **45.** (−0.78, 0.08), (40.78, 207.92) **47.** (−6.84, −5.85), (0, 0) **49.** $x^2 = -200y$
51. (A) $y = 0.0025x^2$, $-100 \le x \le 100$ (B) 25 ft

Exercise 7-2

1. Foci: $F'(-\sqrt{21}, 0)$, $F(\sqrt{21}, 0)$; major axis length = 10; minor axis length = 4

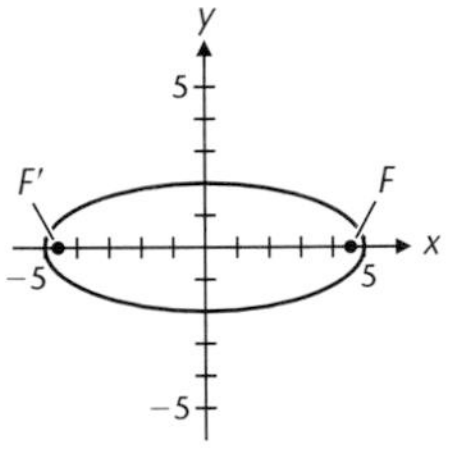

3. Foci: $F'(0, -\sqrt{21})$, $F(0, \sqrt{21})$; major axis length = 10; minor axis length = 4

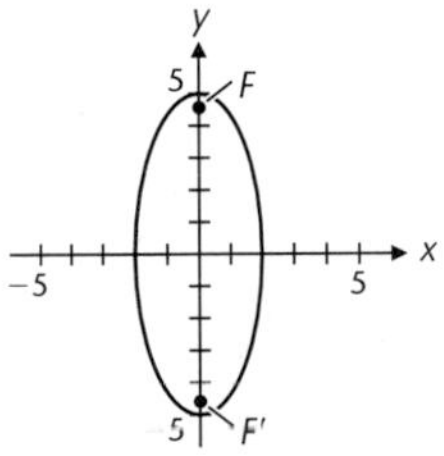

5. Foci: $F'(-\sqrt{8}, 0)$, $F(\sqrt{8}, 0)$; major axis length = 6; minor axis length = 2 **7.** (b) **9.** (a)

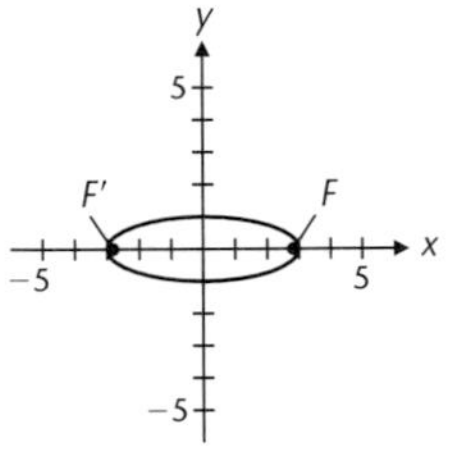

11. Foci: $F'(0, -4)$, $F(0, 4)$; major axis length = 10; minor axis length = 6

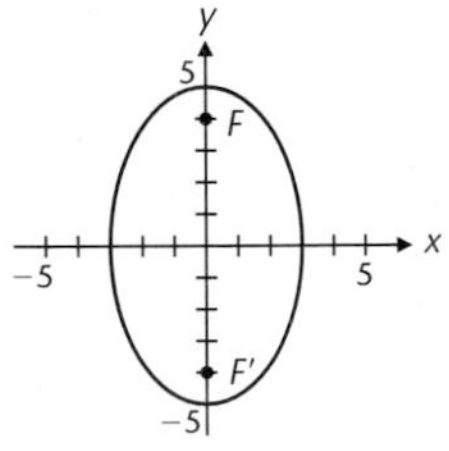

13. Foci: $F'(0, -\sqrt{6})$, $F(0, \sqrt{6})$; major axis length = $2\sqrt{12} \approx 6.93$; minor axis length = $2\sqrt{6} \approx 4.90$

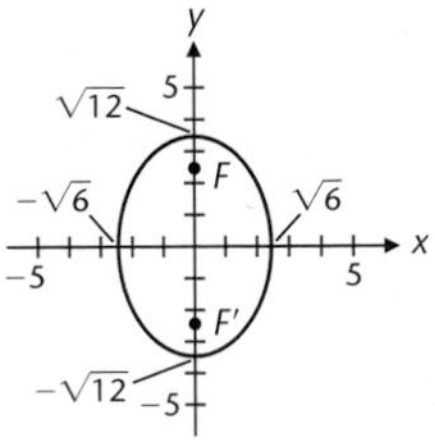

15. Foci: $F'(-\sqrt{3}, 0)$, $F(\sqrt{3}, 0)$; major axis length $= 2\sqrt{7} \approx 5.29$; minor axis length $= 4$

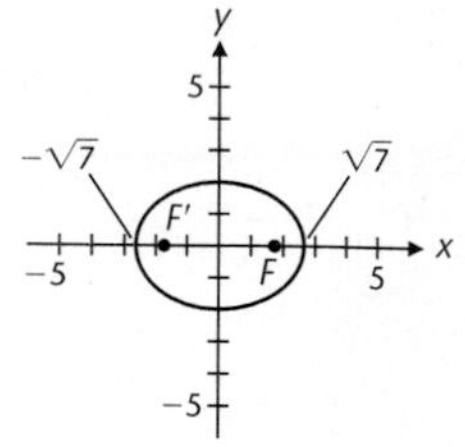

17. $\frac{x^2}{25} + \frac{y^2}{16} = 1$ **19.** $\frac{x^2}{9} + \frac{y^2}{36} = 1$ **21.** $\frac{x^2}{25} + \frac{y^2}{9} = 1$ **23.** $\frac{x^2}{64} + \frac{y^2}{121} = 1$ **25.** $\frac{x^2}{64} + \frac{y^2}{28} = 1$ **27.** $\frac{x^2}{100} + \frac{y^2}{170} = 1$

29. It does not pass the vertical line test. **31.** $(5, 0)$, $(-3, -3.2)$ **33.** $(-2.4, 4)$, $(2.4, 4)$ **35.** $(2.201, 4.403)$, $(-2.201, -4.403)$

37. $(3.565, 1.589)$, $(-3.565, 1.589)$ **39.** $\frac{x^2}{16} + \frac{y^2}{12} = 1$: ellipse **41.** $(-0.46, 2.57)$, $(4.08, -1.06)$ **43.** $(\pm 3.64, \pm 9.50)$

45. $\frac{x^2}{400} + \frac{y^2}{144} = 1$; 7.94 ft approximately **47.** (A) $\frac{x^2}{576} + \frac{y^2}{15.9} = 1$ (B) 5.13 ft

Exercise 7-3

1. (d) **3.** (c)

5. Foci: $F'(-\sqrt{13}, 0)$, $F(\sqrt{13}, 0)$; transverse axis length $= 6$; conjugate axis length $= 4$

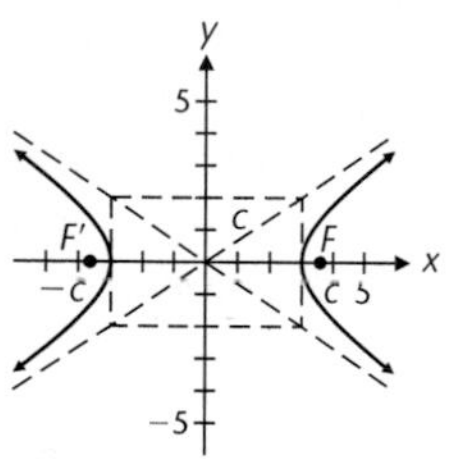

7. Foci: $F'(0, -\sqrt{13})$, $F(0, \sqrt{13})$; transverse axis length $= 4$; congugate axis length $= 6$

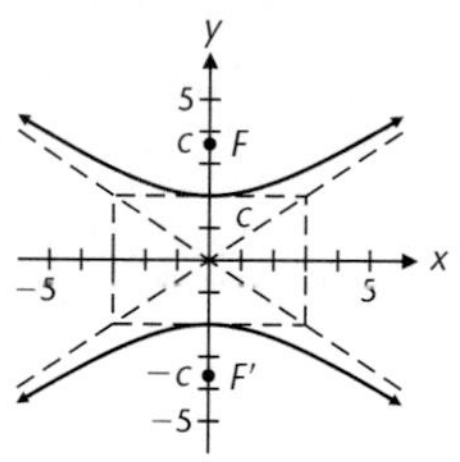

9. Foci: $F'(-\sqrt{20}, 0)$, $F(\sqrt{20}, 0)$; transverse axis length $= 4$; conjugate axis length $= 8$

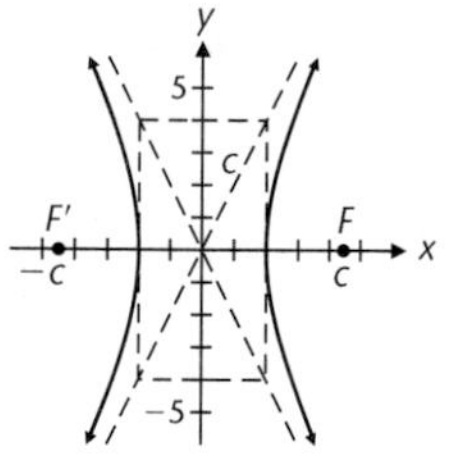

11. Foci: $F'(0, -5)$, $F(0, 5)$; transverse axis length $= 8$; conjugate axis length $= 6$

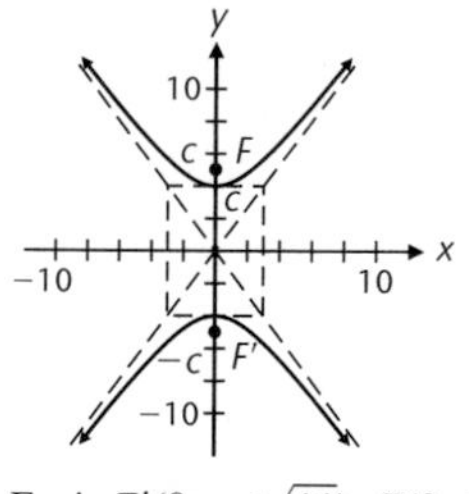

13. Foci: $F'(-\sqrt{10}, 0)$, $F(\sqrt{10}, 0)$; transverse axis length $= 4$; conjugate axis length $= 2\sqrt{6} \approx 4.90$

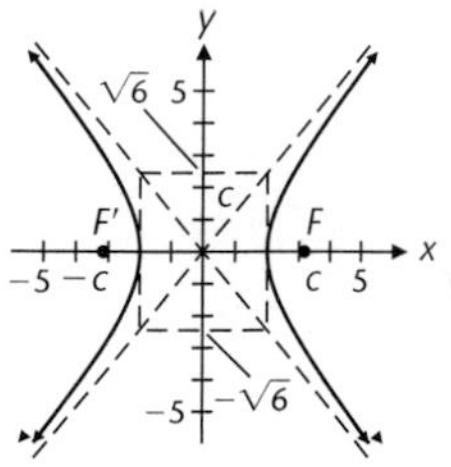

15. Foci: $F'(0, -\sqrt{11})$, $F(0, \sqrt{11})$; transverse axis length $= 4$; conjugate axis length $= 2\sqrt{7} \approx 5.29$

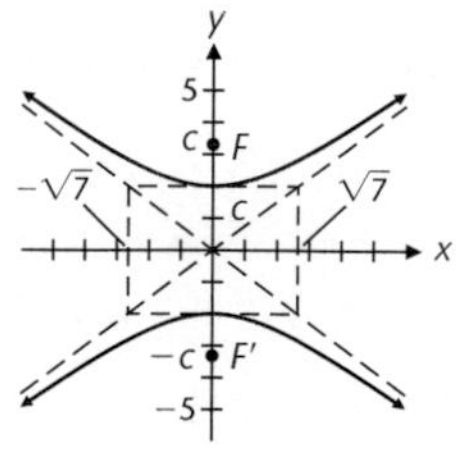

17. $\frac{x^2}{9} - \frac{y^2}{9} = 1$ **19.** $\frac{y^2}{16} - \frac{x^2}{16} = 1$ **21.** $\frac{x^2}{49} - \frac{y^2}{25} = 1$ **23.** $\frac{y^2}{144} - \frac{x^2}{81} = 1$ **25.** $\frac{x^2}{81} - \frac{y^2}{40} = 1$ **27.** $\frac{y^2}{151} - \frac{x^2}{49} = 1$

29. (A) Infinitely many; $\frac{x^2}{a^2} - \frac{y^2}{1 - a^2} = 1$ $(0 < a < 1)$ (B) Infinitely many; $\frac{x^2}{a^2} + \frac{y^2}{a^2 - 1} = 1$ $(a > 1)$ (C) One; $y^2 = 4x$

31. $(-3, -4), (-3, 4), (3, -4), (3, 4)$ **33.** $(-2, -4), (-2, 4), (2, -4), (2, 4)$ **35.** $(-1.389, 3.306), (6.722, 7.361)$

37. $(3.266, 3.830), (-3.266, 3.830), (-3.266, -3.830), (3.266, -3.830)$ **39.** $\frac{x^2}{4} - \frac{y^2}{5} = 1$; hyperbola **41.** $(-4.73, 2.88), (3.35, -0.90)$

43. $(\pm 1.39, \pm 2.96)$ **45.** $\frac{y^2}{16} - \frac{x^2}{8} = 1$; 5.38 ft above vertex **47.** $y = \frac{4}{3}\sqrt{x^2 + 30^2}$

Exercise 7-4

1. (A) $x' = x - 3$; $y' = y - 5$ (B) $x'^2 + y'^2 = 81$ (C) Circle

3. (A) $x' = x + 7$, $y' = y - 4$ (B) $\frac{x'^2}{9} + \frac{y'^2}{16} = 1$ (C) Ellipse

5. (A) $x' = x - 4$, $y' = y + 9$ (B) $y'^2 = 16x'$ (C) Parabola

7. (A) $x' = x + 8$, $y' = y + 3$ (B) $\frac{x'^2}{12} + \frac{y'^2}{8} = 1$ (C) Ellipse **9.** (A) $\frac{(x-3)^2}{9} - \frac{(y+2)^2}{16} = 1$ (B) Hyperbola

11. (A) $\frac{(x+5)^2}{5} + \frac{(y+7)^2}{6} = 1$ (B) Ellipse **13.** (A) $(x + 6)^2 = -24(y - 4)$ (B) Parabola

15. $\frac{(x-2)^2}{9} + \frac{(y-2)^2}{4} = 1$; ellipse **17.** $(x + 4)^2 = -8(y - 2)$; parabola

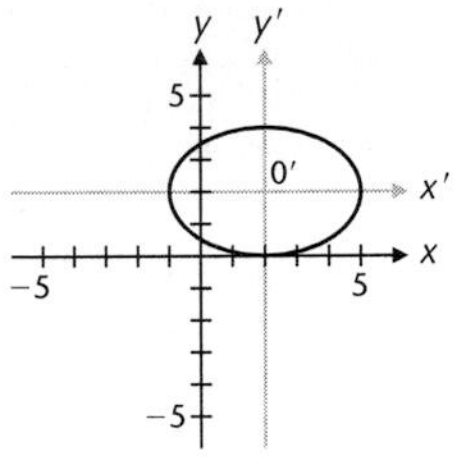

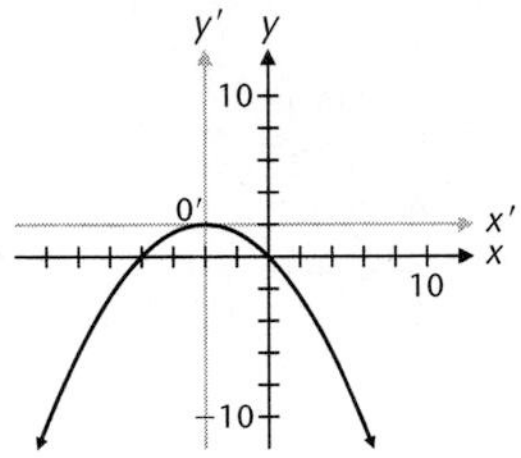

19. $(x + 6)^2 + (y + 5)^2 = 16$; circle **21.** $\frac{(y-3)^2}{9} - \frac{(x+4)^2}{16} = 1$; hyperbola

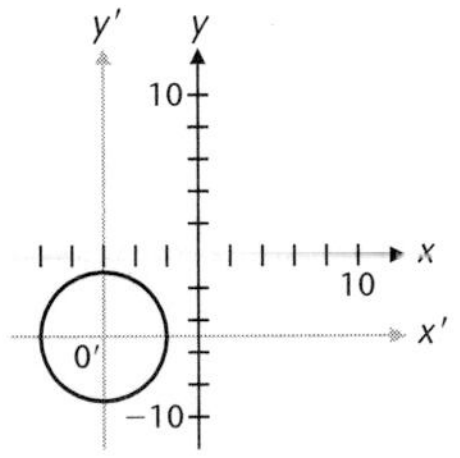

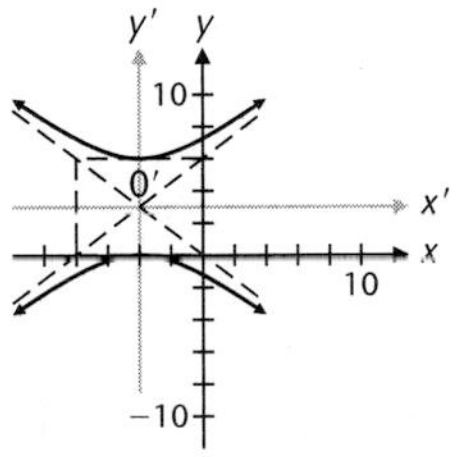

23. $h = \frac{-D}{2A}$, $k = \frac{D^2 - 4AF}{4AE}$ **25.** $x^2 - 4x + 4y - 16 = 0$ **27.** $x^2 + 4y^2 + 4x + 24y + 24 = 0$

29. $25x^2 + 9y^2 - 200x + 36y + 211 = 0$ **31.** $4x^2 - y^2 - 16x + 6y + 11 = 0$ **33.** $9x^2 + y^2 + 36x - 2y + 28 = 0$

35. $x^2 - 2y^2 - 2x + 8y - 8 = 0$ **37.** $F'(-\sqrt{5} + 2, 2)$ and $F(\sqrt{5} + 2, 2)$ **39.** $F(-4, 0)$ **41.** $F'(-4, -2)$, $F(-4, 8)$

43. $(1.18, 1.98), (6.85, -6.52)$ **45.** $(-1.72, -1.87), (-0.99, 2.06)$

Exercise 7-5

3. $y = -2x - 2$; straight line **5.** $y = -2x - 2$, $x \le 0$; a ray (part of a straight line)

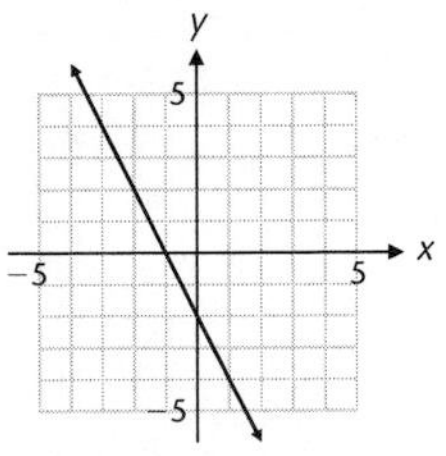

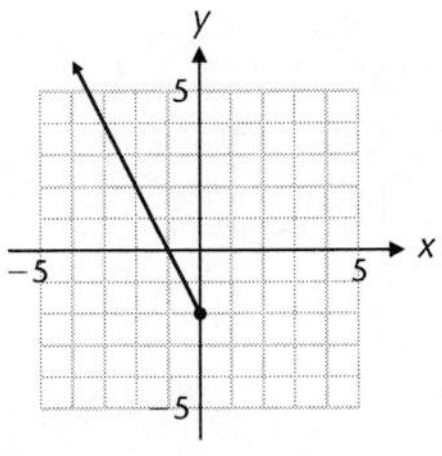

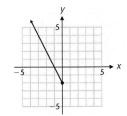

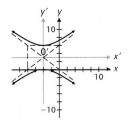

7. $y = -\frac{2}{3}x$; straight line

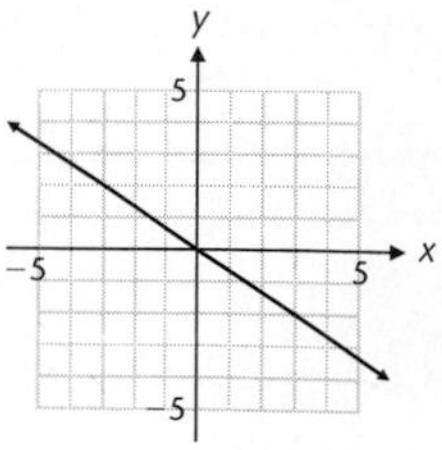

9. $y^2 = 4x$; parabola

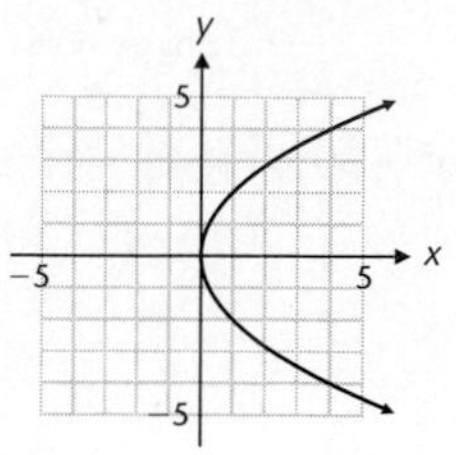

11. $y^2 = 4x$, $y \geq 0$; parabola (upper half)

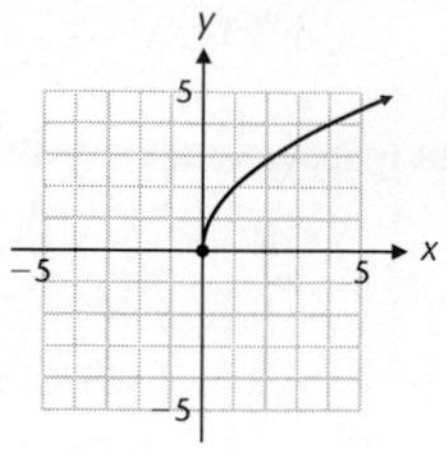

13. $y = -2x$; line **15.** $y^2 = x + 1$, $y \geq 0$, $x \geq -1$; parabola (upper half)
17. $4x^2 + y^2 = 64$, $0 \leq x \leq 4$, $0 \leq y \leq 8$; ellipse (first quadrant portion)
19. $x^2 - y^2 = 2$, $x \leq -\sqrt{2}$, $y \leq 0$; hyperbola (third quadrant portion) **21.** $x = t$, $y = \dfrac{At^2 + Dt + F}{-E}$, $-\infty < t < \infty$; parabola
23. $y^2 - x^2 = 1$, $x \geq 1$, $y \geq \sqrt{2}$; part of a hyperbola **25.** $x^2 + y^2 = 4$, $0 < x \leq 2$, $-2 < y < 2$; semicircle (excluding the end points)
27. $x^2 + y^2 = 2x$, $x \neq 0$ or $(x - 1)^2 + y^2 = 1$, $x \neq 0$; circle (note hole at origin) **31.** 1,786 m

Chapter 7 Review

1. Foci: $F'(-4, 0)$, $F(4, 0)$; major axis length = 10; minor axis length = 6 *(7-2)*

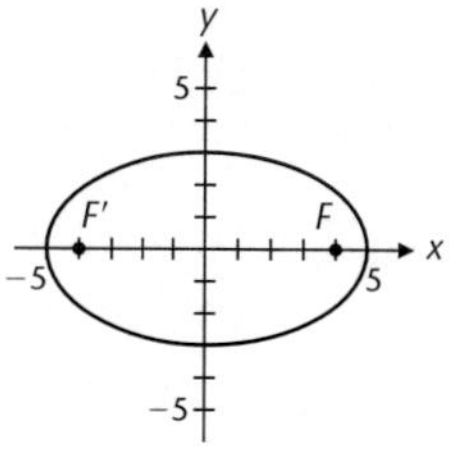

2. *(7-1)*

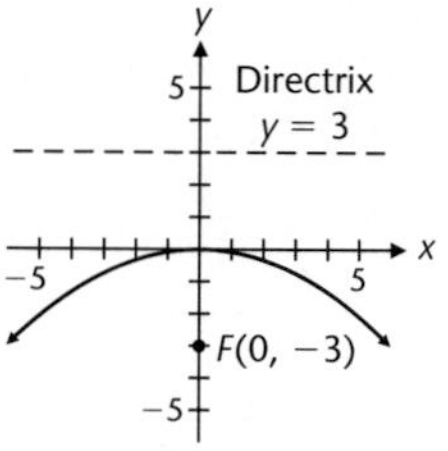

3. Foci: $F'(0, -\sqrt{34})$, $F(0, \sqrt{34})$; transverse axis length = 6; conjugate axis length = 10 *(7-3)*

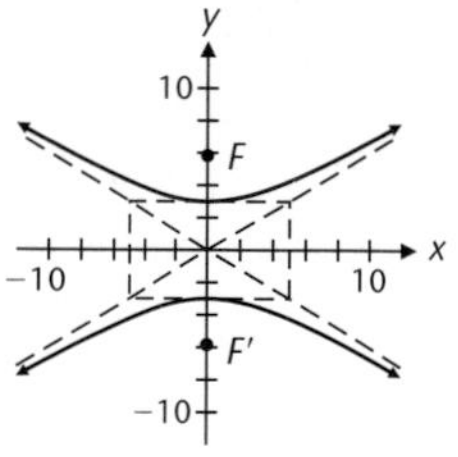

4. (A) $\dfrac{(y + 2)^2}{25} - \dfrac{(x - 4)^2}{4} = 1$ (B) Hyperbola *(7-4)*

5. (A) $(x + 5)^2 = -12(y + 4)$ (B) Parabola *(7-4)* **6.** (A) $\dfrac{(x - 6)^2}{9} + \dfrac{(y - 4)^2}{16} = 1$ (B) Ellipse *(7-4)*

8. $y^2 = -x$ *(7-1)* **9.** $\dfrac{x^2}{9} + \dfrac{y^2}{25} = 1$ *(7-2)* **10.** $\dfrac{y^2}{9} - \dfrac{x^2}{16} = 1$ *(7-3)*

11. $y = \frac{1}{2}x + 1$, $x \leq 0$; a ray (part of a straight line) *(7-5)*

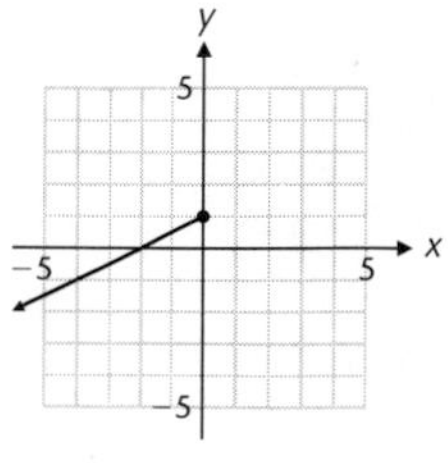

12. *(7-2)*

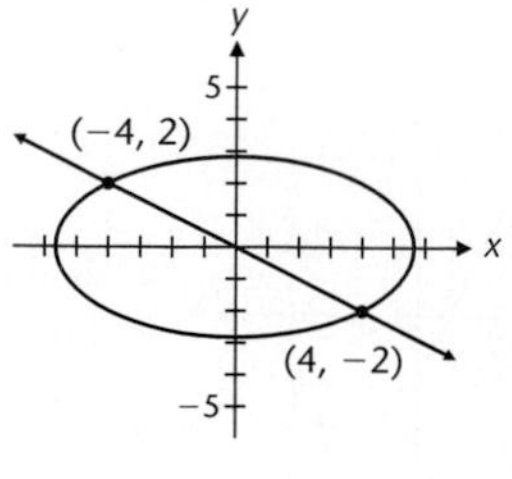

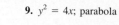

13. (7-1, 7-2)

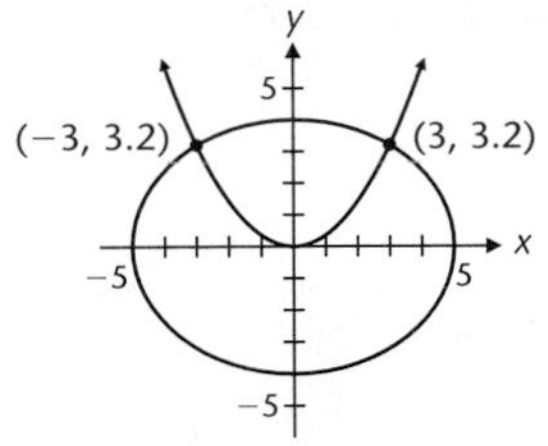

14. (7-2)

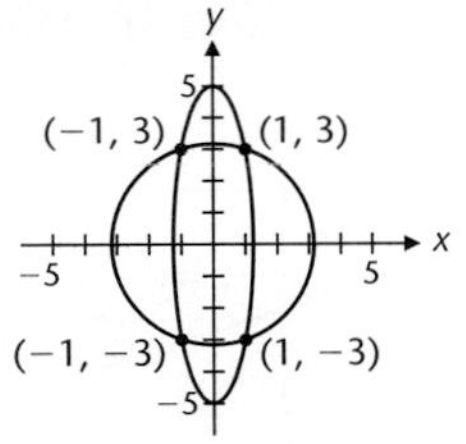

15. $\frac{(x+3)^2}{4} + \frac{(y-2)^2}{16} = 1$; ellipse (7-4)

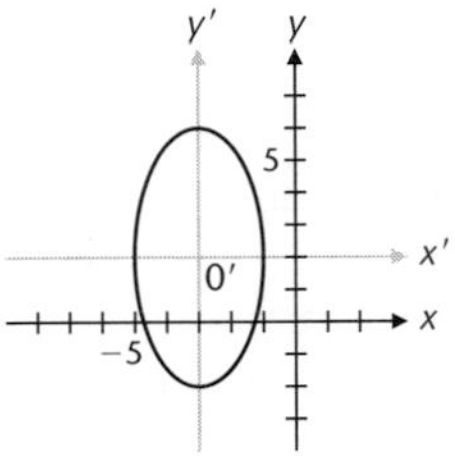

16. $(x-2)^2 = 4(2)(y+3)$; parabola (7-4)

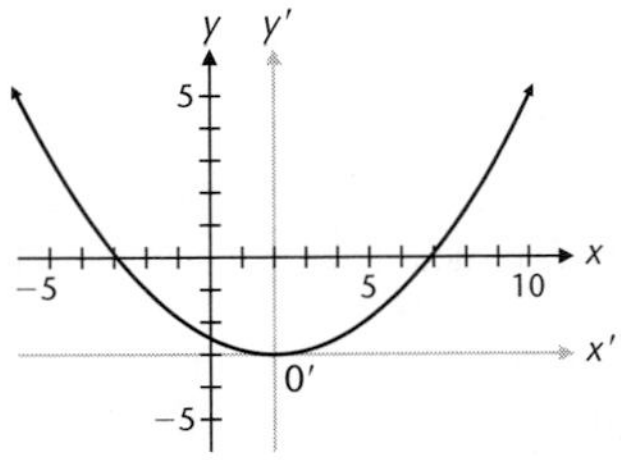

17. $\frac{(x+3)^2}{9} - \frac{(y+2)^2}{4} = 1$; hyperbola (7-4)

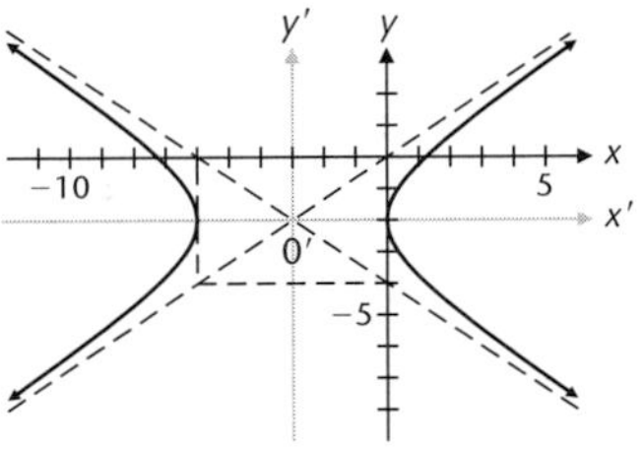

18. $m = 0.2$; $x^2 = 50y$ is a magnification by a factor 50 of $x^2 = y$ (7-1)

19. $(y-4)^2 = -8(x-4)$ or $y^2 - 8y + 8x - 16 = 0$ (7-1) **20.** $\frac{x^2}{4} - \frac{y^2}{12} = 1$; hyperbola (7-3) **21.** $\frac{x^2}{36} + \frac{y^2}{20} = 1$; ellipse (7-2)

22. $F'(-3, -\sqrt{12} + 2)$ and $F(-3, \sqrt{12} + 2)$ (7-4) **23.** $F(2, -1)$ (7-4) **24.** $F'(-\sqrt{13} - 3, -2)$ and $F(\sqrt{13} - 3, -2)$ (7-4)

25. $y = \frac{1}{x}$, $x > 0$; $y > 0$; hyperbola (one branch) (7-5)

26. $4x^2 + 9y^2 = 36$, $-3 < x < 3$, $0 < y \le 2$; ellipse (upper half, excluding the end points) (7-5) **28.** (2.09, 2.50), (3.67, −1.92) (7-4)

29. 4 ft (7-1) **30.** $\frac{x^2}{5^2} + \frac{y^2}{3^2} = 1$ (7-2) **31.** 4.72 ft deep (7-3)

Cumulative Review Exercise for Chapters 5, 6, and 7

1. (2, −1) (5-1, 5-2) **2.** (−1, 2) (5-1) **3.** $(-\frac{1}{5}, -\frac{7}{5})$, (1, 1) (5-1, 5-2) **4.** (5-2)

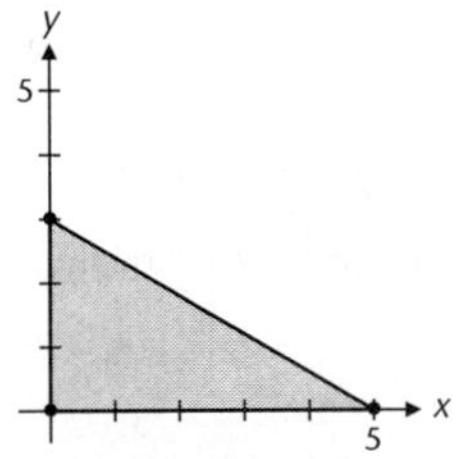

5. The minimum value of z on S is 10 at (5, 0). The maximum value of z on S is 33 at (6, 7). (5-7)

6. (A) $\begin{bmatrix} 0 & -3 \\ 3 & -9 \end{bmatrix}$ (B) Not defined (C) $[3]$ (D) $\begin{bmatrix} 1 & 7 \\ 4 & -7 \end{bmatrix}$ (E) $[-1 \quad 8]$ (F) Not defined (5-4)

7. (A) $x_1 = 3$, $x_2 = -4$ (B) $x_1 = 2t + 3$, $x_2 = t$ is a solution for every real number t (C) No solution (5-2)

8. (A) $\left[\begin{array}{rr|r} 1 & 1 & 3 \\ -1 & 1 & 5 \end{array}\right]$ (B) $\left[\begin{array}{rr|r} 1 & 0 & -1 \\ 0 & 1 & 4 \end{array}\right]$ (C) $x_1 = -1$, $x_2 = 4$ (5-2, 5-3)

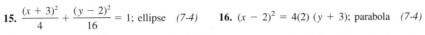

9. (A) $\begin{bmatrix} 1 & -3 \\ 2 & -5 \end{bmatrix} \begin{bmatrix} x_1 \\ x_2 \end{bmatrix} = \begin{bmatrix} k_1 \\ k_2 \end{bmatrix}$ (B) $A^{-1} = \begin{bmatrix} -5 & 3 \\ -2 & 1 \end{bmatrix}$ (C) $x_1 = 13, x_2 = 5$ (D) $x_1 = -11, x_2 = -4$ *(5-6)*

10. $x_1 = 1, x_2 = 3$; each pair of lines has the same intersection point.

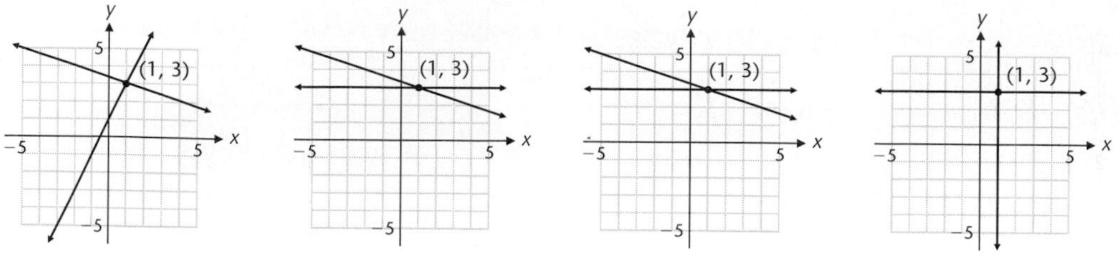

$x_1 + 3x_2 = 10$, $2x_1 - x_2 = -1$ | $x_1 + 3x_2 = 10$, $-7x_2 = -21$ | $x_1 + 3x_2 = 10$, $x_2 = 3$ | $x_1 = 1$, $x_2 = 3$ *(5-1, 5-3)*

11. (1.53, 3.35) *(5-1)* **12.** (A) Arithmetic (B) Geometric (C) Neither (D) Geometric (E) Arithmetic *(6-3)*

13. (A) 10, 50, 250, 1,250 (B) $a_8 = 781{,}250$ (C) $S_8 = 976{,}560$ *(6-3)*

14. (A) 2, 5, 8, 11 (B) $a_8 = 23$ (C) $S_8 = 100$ *(6-3)* **15.** (A) 100, 94, 88, 82 (B) $a_8 = 58$ (C) $S_8 = 632$ *(6-3)*

16. (A) 40,320 (B) 992 (C) 84 *(6-4)* **17.** (A) 21 (B) 21 (C) 42 *(6-4, 6-5)*

18. Foci: $F'(-\sqrt{61}, 0)$, $F(\sqrt{61}, 0)$; transverse axis length = 12; conjugate axis length = 10 *(7-3)*

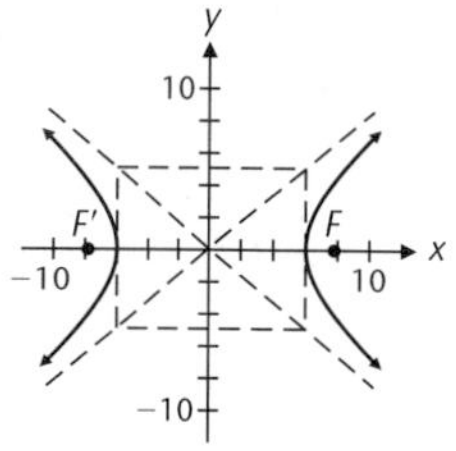

19. Foci: $F'(-\sqrt{11}, 0)$, $F(\sqrt{11}, 0)$; major axis length = 12; minor axis length = 10 *(7-2)*

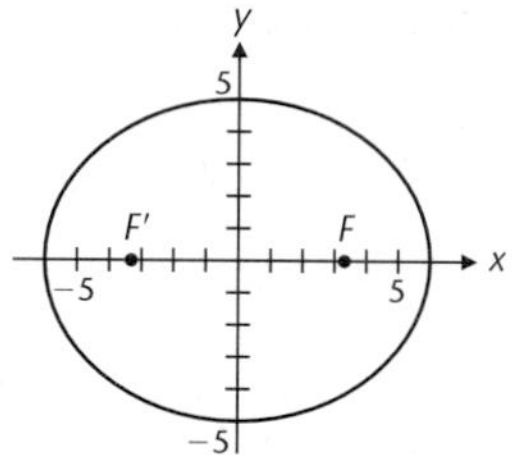

20. *(7-1)*

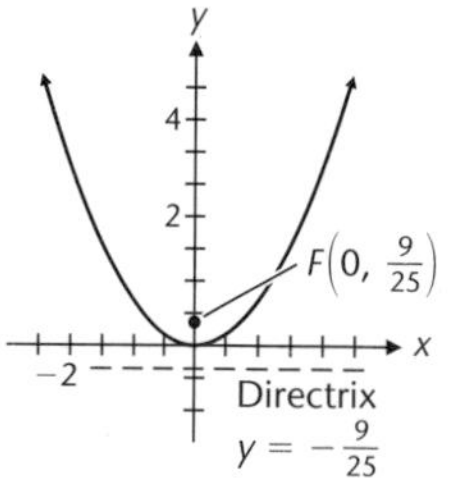

21. (A) 8 combined outcomes:

HHH
HHT
HTH
HTT
THH
THT
TTH
TTT

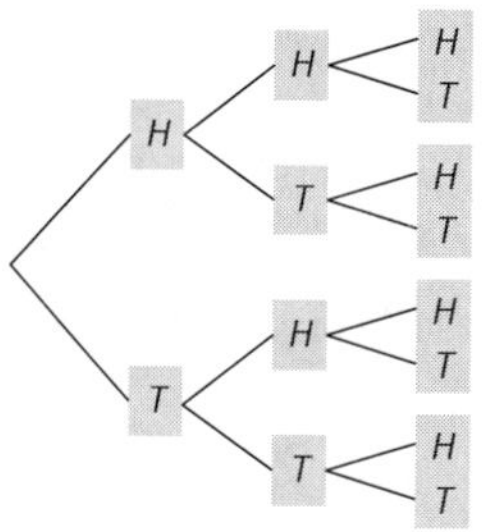

(B) $2 \cdot 2 \cdot 2 = 8$ *(6-5)*

22. (A) $4 \cdot 3 \cdot 2 \cdot 1 = 24$ (B) $P_{4,4} = 4! = 24$ *(6-5)* **23.** $\dfrac{C_{13,3}}{C_{52,3}} \approx .0129$ *(6-5)* **24.** $\dfrac{1}{P_{10,4}} \approx .0002$; $\dfrac{1}{C_{10,4}} \approx .0048$ *(6-5)*

25. .62 *(6-5)* **26.** $y = 2x - 1$; straight line *(7-5)*

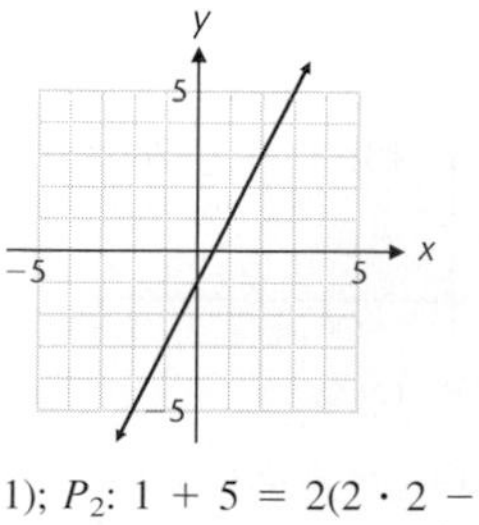

27. P_1: $1 = 1(2 \cdot 1 - 1)$; P_2: $1 + 5 = 2(2 \cdot 2 - 1)$; P_3: $1 + 5 + 9 = 3(2 \cdot 3 - 1)$ *(6-2)*

28. P_1: $1^2 + 1 + 2 = 4$ is divisible by 2; P_2: $2^2 + 2 + 2 = 8$ is divisible by 2; P_3: $3^2 + 3 + 2 = 14$ is divisible by 2 *(6-2)*

29. P_k: $1 + 5 + 9 + \cdots + (4k - 3) = k(2k - 1)$; P_{k+1}: $1 + 5 + 9 + \cdots + (4k - 3) + (4k + 1) = (k + 1)(2k + 1)$ *(6-2)*

30. P_k: $k^2 + k + 2 = 2r$ for some integer r; P_{k+1}: $(k + 1)^2 + (k + 1) + 2 = 2s$ for some integer s *(6-2)*

31. $x_1 = 1, x_2 = 0, x_3 = -2$ *(5-3)* **32.** No solution *(5-3)*

33. $x_1 = t - 3, x_2 = t - 2, x_3 = t$ is a solution for every real number t *(5-3)*

34. (A) $[-3]$ (B) $\begin{bmatrix} 1 & 2 & -1 \\ -1 & -2 & 1 \\ 2 & 4 & -2 \end{bmatrix}$ *(5-4)* **35.** (A) $\begin{bmatrix} -1 & 2 \\ 2 & 3 \end{bmatrix}$ (B) Not defined *(5-4)*

36. Unbounded *(5-7)* **37.** 63 *(5-8)*

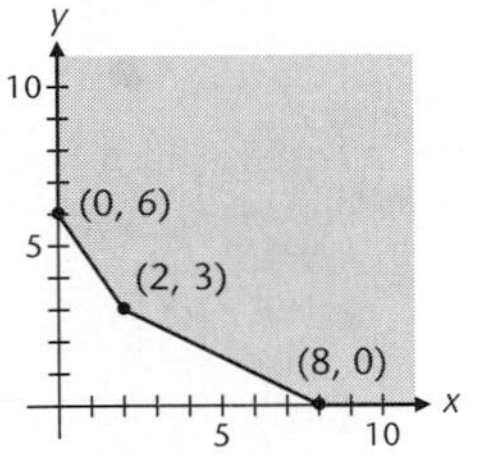

38. (A) $\begin{bmatrix} 1 & 4 & 2 \\ 2 & 6 & 3 \\ 2 & 5 & 2 \end{bmatrix}\begin{bmatrix} x_1 \\ x_2 \\ x_3 \end{bmatrix} = \begin{bmatrix} k_1 \\ k_2 \\ k_3 \end{bmatrix}$ (B) $A^{-1} = \begin{bmatrix} -3 & 2 & 0 \\ 2 & -2 & 1 \\ -2 & 3 & -2 \end{bmatrix}$ (C) $x_1 = 7, x_2 = -5, x_3 = 6$ (D) $x_1 = -6, x_2 = 3, x_3 = -2$ *(5-6)*

39. $y = -2x^2$ *(7-1)* **40.** $\dfrac{x^2}{25} + \dfrac{y^2}{16} = 1$ *(7-2)* **41.** $\dfrac{x^2}{64} - \dfrac{y^2}{25} = 1$ *(7-3)* **42.** $1 + 4 + 27 + 256 + 3{,}125 = 3{,}413$ *(6-1)*

43. $\displaystyle\sum_{k=1}^{6} (-1)^{k+1}\frac{2^k}{(k+1)!}$ *(6-1)* **44.** 81 *(6-3)* **45.** 360; 1,296; 750 *(6-4)* **46.** $\dfrac{C_{10,3}}{C_{12,5}} = \dfrac{5}{33} = .\overline{15}$ *(6-5)*

47. (A) .365 (B) $\frac{1}{3}$ *(6-5)* **48.** $n = 22$ *(6-3)* **49.** $x = y^2 + 6y + 11$; parabola *(7-5)*

50. (A) 6,375,600 (B) 53,130 (C) 53,130 *(6-4, 6-6)* **51.** $a^6 + 3a^5b + \frac{15}{4}a^4b^2 + \frac{5}{2}a^3b^3 + \frac{15}{16}a^2b^4 + \frac{3}{16}ab^5 + \frac{1}{64}b^6$ *(6-6)*

52. $153{,}090x^6y^4$; $-3{,}240x^3y^7$ *(6-6)* **55.** 61,875 *(6-3)* **56.** $\frac{27}{11}$ *(6-3)* **57.** $a_{22} = 0.236$; 8 terms *(6-6)*

58. $4(x + 3) = (y - 2)^2$; parabola *(7-1)* **59.** $\dfrac{(y + 2)^2}{4} - \dfrac{(x + 1)^2}{16} = 1$; hyperbola *(7-3)*

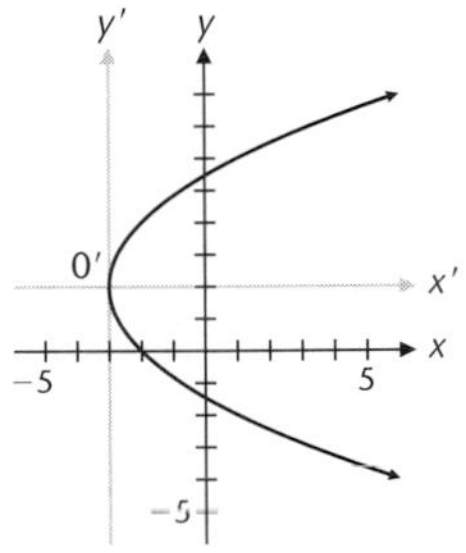

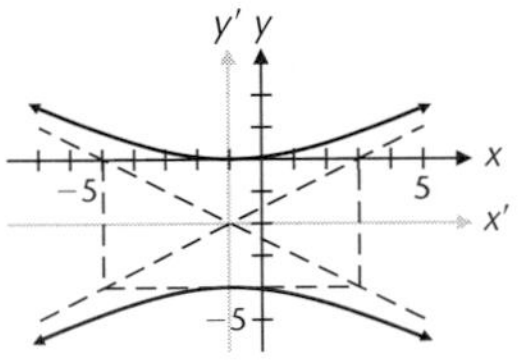

60. $\dfrac{(x - 2)^2}{9} + \dfrac{(y + 3)^2}{4} = 1$; ellipse *(7-2)* **61.** 10^9; 3,628,800 zip codes *(6-1)* **62.** $(-2.26, -4.72)$, $(1.85, 3.09)$ *(7-4)*

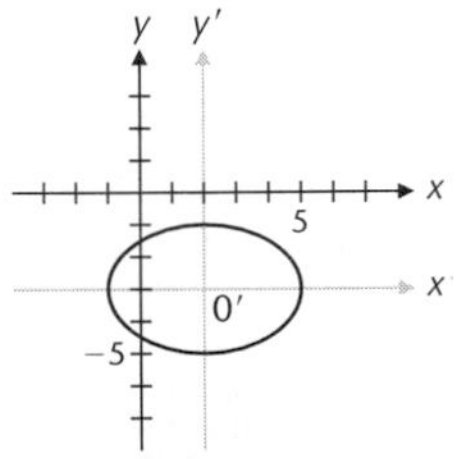

64. $\frac{2}{5}; \frac{2}{5}$ *(6-5)* **65.** $x^6 - 12ix^5 - 60x^4 + 160ix^3 + 240x^2 - 192ix - 64$ *(6-6)* **66.** $x^2 - 12x + 4y + 28 = 0$ *(7-1)*

67. $\pm 2\sqrt{3}$ *(7-2)* **68.** 8 *(7-3)* **69.** (A) Infinite number of solutions (B) No solution (C) Unique solution *(5-3)*

70. $A = I$, the $n \times n$ identity *(5-5)* **71.** L, M, and P *(5-3)* **72.** $C_{7,3} = 35$ *(6-4)*

73. $x + 4 = (y - 1)^2$, $y < 1$; lower half of a parabola (excluding the vertex) *(7-5)*

76. $x^2 - 8y^2 - 2x - 8y + 17 = 0$; hyperbola *(7-3)* **77.** $1 - \dfrac{C_{8,3}}{C_{12,3}} = \dfrac{41}{55} = .7\overline{45}$ *(6-5)*

78. \$8,000 at 8% and \$4,000 at 14% *(5-1, 5-2)* **79.** \$6,000,000 *(6-3)*

80. $x_1 = 60$ g mix A, $x_2 = 50$ g mix B, $x_3 = 40$ g mix C *(5-3)*

81. 1 model A truck, 6 model B trucks, and 5 model C trucks; or 3 model A trucks, 3 model B trucks, and 6 model C trucks; or 5 model A trucks and 7 model C trucks. *(5-3)*

82. 4 in. *(7-1)* **83.** 32 ft, 14.4 ft *(7-2)*

84. (A) Manufacturing 400 standard and 200 deluxe day packs produces a maximum weekly profit of \$5,600.
(B) The maximum weekly profit increases to \$6,000 when 0 standard and 400 deluxe day packs are manufactured.
(C) The maximum weekly profit increases to \$6,600 when 600 standard and 0 deluxe day packs are manufactured. *(5-8)*

85. (A) $M\begin{bmatrix}0.25\\0.25\\0.25\\0.25\end{bmatrix} = \begin{bmatrix}82.25\\83\\92\\83.75\\82\end{bmatrix}\begin{matrix}\text{Ann}\\\text{Bob}\\\text{Carol}\\\text{Dan}\\\text{Eric}\end{matrix}$ (B) $M\begin{bmatrix}0.2\\0.2\\0.2\\0.4\end{bmatrix} = \begin{bmatrix}83\\84.8\\91.8\\85.2\\80.8\end{bmatrix}\begin{matrix}\text{Ann}\\\text{Bob}\\\text{Carol}\\\text{Dan}\\\text{Eric}\end{matrix}$

(C) Class averages: Test 1, Test 2, Test 3, Test 4
$[0.2\ \ 0.2\ \ 0.2\ \ 0.2\ \ 0.2]M = [84.4\ \ 81.8\ \ 85\ \ 87.2]$ *(5-4)*

86. (A) .13 (B) .17 (C) .32 *(6-5)*

APPENDIX A Exercise A-1

1. T **3.** F **5.** F **7.** T **9.** $7 + x$ **11.** $(xy)z$ **13.** $9m$ **15.** Commutative (·) **17.** Distributive
19. Inverse (·) **21.** Inverse (+) **23.** Identity (+) **25.** Negatives **27.** $\{-2, 0, 2, 4\}$ **29.** $\{a, s, t, u\}$ **31.** $\varnothing$
33. Commutative (+) **35.** Associative (+) **37.** Distributive **39.** Zero **41.** Yes **43.** (A) True (B) False (C) True
45. $\frac{3}{5}$ and -1.43 are two examples of infinitely many **47.** (A) Z, Q, R (B) Q, R (C) R (D) Q, R
49. (A) 0.888 888 88 . . . (B) 0.272 727 27 . . . (C) 2.236 067 97 . . . (D) 1.375 000 00 . . .
51. (A) T (B) F; for example $3 - 5 \neq 5 - 3$. (C) T (D) F; for example $9 \div 3 \neq 3 \div 9$.
53. (A) $\{1, 2, 3, 4, 6\}$ (B) $\{2, 4\}$ **55.** $\frac{1}{11}$ **57.**

$$\begin{array}{r} 23\\ \underline{12}\\ 46\\ \underline{230}\\ 276\end{array} \qquad \begin{aligned} 23 \cdot 12 &= 23(2 + 10)\\ &= 23 \cdot 2 + 23 \cdot 10\\ &= 46 + 230\\ &= 276\end{aligned}$$

Exercise A-2

1. 3 **3.** $2x^3 - x^2 + 2x + 4$ **5.** $2x^3 - 5x^2 + 6$ **7.** $6x^4 - 13x^3 + 9x^2 + 13x - 10$ **9.** $4x - 6$ **11.** $6y^2 - 16y$
13. $m^2 - n^2$ **15.** $4t^2 - 11t + 6$ **17.** $3x^2 - 7xy - 6y^2$ **19.** $4m^2 - 49$ **21.** $30x^2 - 2xy - 12y^2$ **23.** $9x^2 - 4y^2$
25. $16x^2 - 8xy + y^2$ **27.** $a^3 + b^3$ **29.** $-x + 27$ **31.** $32a - 34$ **33.** $2x^4 - 5x^3 + 5x^2 + 11x - 10$
35. $h^4 + h^2k^2 + k^4$ **37.** $-5x^2 - 4x + 5$ **39.** $2m^2 + 15mn$ **41.** $8m^3 - 12m^2n + 6mn^2 - n^3$ **43.** $5h$
45. $6xh + 3h^2 + 2h$ **47.** $-4xh - 2h^2 - 3h$ **49.** $3x^2h + 3xh^2 + h^3$ **51.** $m^3 - 3m^2 - 5$ **53.** $2x^3 - 13x^2 + 25x - 18$
55. $9x^3 - 9x^2 - 18x$ **57.** $(1 + 1)^2 \neq 1^2 + 1^2$; either a or b must be zero **59.** m **61.** Now the degree is less than or equal to m.
63. Perimeter $= 4x - 10$ **65.** Value $= 40x - 125$ **67.** Volume $= 1.2\pi x^2 + 0.36\pi x + 0.036\pi$

Exercise A-3

1. $2x^2(3x^2 - 4x - 1)$ **3.** $5xy(2x^2 + 4xy - 3y^2)$ **5.** $(x + 1)(5x - 3)$ **7.** $(y - 2z)(2w - x)$ **9.** $(x - 2)(x + 3)$
11. $(3m + 5)(2m - 1)$ **13.** $(x - 2y)(2x - 3y)$ **15.** $(2c - d)(4a - 3b)$ **17.** $(2x - 1)(x + 3)$ **19.** $(x - 6y)(x + 2y)$
21. Prime **23.** $(5m + 4n)(5m - 4n)$ **25.** $(x + 5y)^2$ **27.** Prime **29.** $6(x + 2)(x + 6)$ **31.** $2y(y - 3)(y - 8)$
33. $y(4x - 1)^2$ **35.** $(3s - t)(2s + 3t)$ **37.** $xy(x - 3y)(x + 3y)$ **39.** $3m(m^2 - 2m + 5)$ **41.** $(m + n)(m^2 - mn + n^2)$
43. $(c - 1)(c^2 + c + 1)$ **45.** $2(3x - 5)(2x - 3)(12x - 19)$ **47.** $9x^4(9 - x)^3(5 - x)$ **49.** $2(x + 1)(x^2 - 5)(3x + 5)(x - 1)$
51. $[(a - b) - 2(c - d)][(a - b) + 2(c - d)]$ **53.** $(2m - 3n)(a + b)$ **55.** Prime **57.** $(x - 3)^2(x + 3)$
59. $(a - 2)(a + 1)(a - 1)$ **61.** $[4(A + B) + 3][(A + B) - 2]$ **63.** $(m - n)(m + n)(m^2 + n^2)$
65. $st(st - 2)[s^2t^2 + 2st + 4]$ **67.** $(m + n)(m + n - 1)$ **69.** $2a[3a - 2(x + 4)][3a + 2(x + 4)]$
71. $(x^2 - x + 1)(x^2 + x + 1)$ **73.** (A) $4(10 - x)(10 + x) = 400 - 4x^2$ (B) $4x(10 - x)^2 = 400x - 80x^2 + 4x^3$

Exercise A-4

1. $\frac{a^2}{2}$ **3.** $\frac{22y + 9}{252}$ **5.** $\frac{x^2 + 8}{8x^3}$ **7.** $\frac{1}{2x - 1}$ **9.** $\frac{1}{m}$ **11.** $\frac{2a}{(a + b)^2(a - b)}$ **13.** $\frac{m^2 - 6m + 7}{m - 2}$ **15.** $\frac{7}{x - 3}$
17. $\frac{3}{y + 3}$ **19.** $\frac{x + y}{x}$ **21.** $\frac{4(x^2 + 2)^2(x + 1)(x - 1)}{x^3}$ **23.** $\frac{x(2 + 3x)}{(1 - 3x)^4}$ **25.** $\frac{(x + 1)(x - 9)}{(x + 4)^4}$ **27.** $\frac{y + 3}{(y - 2)(y + 7)}$ **29.** -1
31. $\frac{7y - 9x}{xy(a - b)}$ **33.** $\frac{x^2 - x + 1}{2(x - 9)}$ **35.** $\frac{(x - y)^2}{y^2(x + y)}$ **37.** $\frac{1}{x - 4}$ **39.** $\frac{x - 3}{x - 1}$ **41.** $\frac{-1}{x(x + h)}$ **43.** $\frac{x^2 + hx + 4x + 2h}{(x + h + 2)(x + 2)}$

45. (A) Incorrect (B) $x + 1$ **47.** (A) Incorrect (B) $2x + h$ **49.** (A) Incorrect (B) $\frac{x^2 - 2}{x + 1}$ **51.** (A) Correct

53. $\frac{-x(x + y)}{y}$ **55.** $\frac{a - 2}{a}$

Exercise A-5

1. 1 **3.** $6x^9$ **5.** $\frac{9x^6}{y^4}$ **7.** $\frac{a^4b^{12}}{c^8d^4}$ **9.** 10^{17} **11.** $\frac{2x}{y^2}$ **13.** n^2 **15.** 4×10^8 **17.** 3.225×10^7 **19.** 8.5×10^{-2}

21. 7.29×10^{-8} **23.** 0.005 **25.** 26,900,000 **27.** 0.000 000 000 59 **29.** $\frac{3y^4}{2}$ **31.** $\frac{x^{12}}{y^8}$ **33.** $\frac{4x^8}{y^6}$ **35.** $\frac{w^{12}}{u^{20}v^4}$

37. $\frac{1}{(x + y)^2}$ **39.** $\frac{x}{x - 1}$ **41.** $\frac{-1}{xy}$ **43.** $\frac{-9x^2}{(x^3 + 3)^4}$ **45.** 64 **49.** $2x - 6x^{-1}$ **51.** $\frac{5}{3}x - \frac{2}{3}x^{-2}$ **53.** $x - \frac{3}{2} + \frac{1}{2}x^{-1}$

55. 6.65×10^{-17} **57.** 1.54×10^{12} **59.** 1.0295×10^{11} **61.** -4.3647×10^{-18} **63.** 9.4697×10^{29} **65.** $2(a + 2b)^5$

67. $\frac{x^2 + xy + y^2}{xy}$ **69.** $\frac{y - x}{y}$ **71.** 1.3×10^{25} lb **73.** 10^{10} or 10 billion, 6×10^{11} or 600 billion

75. 1.44×10^3 dollars per person; $1,440 per person

Exercise A-6

1. 4 **3.** 64 **5.** −6 **7.** $(-36)^{1/2}$ is not a real number **9.** $\frac{8}{125}$ **11.** $\frac{1}{27}$ **13.** $y^{3/5}$ **15.** $d^{1/3}$

17. $\frac{1}{y^{1/2}}$ **19.** $\frac{2x}{y^2}$ **21.** $\frac{1}{a^{1/4}b^{1/3}}$ **23.** $\frac{xy^2}{2}$ **25.** $\frac{2b^2}{3a^2}$ **27.** $\frac{2}{3x^{7/12}}$ **29.** $\frac{a^{1/3}}{b^2}$ **31.** $6m - 2m^{19/3}$ **33.** $a - a^{1/2}b^{1/2} - 6b$

35. $4x - 9y$ **37.** $x + 4x^{1/2}y^{1/2} + 4y$ **39.** 29.52 **41.** 0.03093 **43.** 5.421 **45.** 107.6

47. $x = y = 1$ is one of many choices. **49.** $x = y = 1$ is one of many choices. **51.** $3 - \frac{3}{4}x^{-1/2}$ **53.** $\frac{3}{5}x^{-1/3} + \frac{1}{5}x^{-1/2}$

55. $\frac{1}{2}x^{5/3} - 2x^{1/6}$ **57.** $a^{1/n}b^{1/m}$ **59.** $\frac{1}{x^{3m}y^{4n}}$ **61.** (A) $x = -2$, for example (B) $x = 2$, for example (C) Not possible

63. No **65.** $\frac{x - 3}{(2x - 1)^{3/2}}$ **67.** $\frac{4x - 3}{(3x - 1)^{4/3}}$ **69.** 1,920 units **71.** 428 ft

Exercise A-7

1. $\sqrt[3]{m^2}$ or $(\sqrt[3]{m})^2$ (first preferred) **3.** $6\sqrt[5]{x^3}$ **5.** $\sqrt[5]{(4xy^3)^2}$ **7.** $\sqrt{x + y}$ **9.** $b^{1/5}$ **11.** $5x^{3/4}$ **13.** $(2x^2y)^{3/5}$

15. $x^{1/3} + y^{1/3}$ **17.** −2 **19.** $3x^4y^2$ **21.** $2mn^2$ **23.** $2ab^2\sqrt{2ab}$ **25.** $2xy^2\sqrt[3]{2xy}$ **27.** $\sqrt{m}$ **29.** $\sqrt[15]{xy}$

31. $3x\sqrt[3]{3}$ **33.** $\frac{\sqrt{5}}{5}$ **35.** $2\sqrt{3x}$ **37.** $2\sqrt{2} + 2$ **39.** $\sqrt{3} - \sqrt{2}$ **41.** $3x^2y^2\sqrt[3]{3x^2y}$ **43.** $n\sqrt[4]{2m^3n}$ **45.** $\sqrt[3]{a^2(b - a)}$

47. $\sqrt[4]{a^3b}$ **49.** $x^2y\sqrt[3]{6xy^2}$ **51.** In simplified form **53.** $\frac{\sqrt{2}}{2}$ or $\frac{1}{2}\sqrt{2}$ **55.** $2a^2b\sqrt[3]{4a^2b}$ **57.** $\frac{\sqrt[4]{12xy^3}}{2x}$ or $\frac{1}{2x}\sqrt[4]{12xy^3}$

59. $\frac{6y + 9\sqrt{y}}{4y - 9}$ **61.** $\frac{38 + 11\sqrt{10}}{117}$ **63.** $\sqrt{x^2 + 9} + 3$ **65.** $\frac{1}{\sqrt{t} + \sqrt{x}}$ **67.** $\frac{1}{\sqrt{x + h} + \sqrt{x}}$ **69.** 0.2222 **71.** 1.934

73. 0.069 79 **75.** 2.073 **77.** Both are 1.059 **79.** Both are 0.6300 **81.** $x \le 0$ **83.** All real numbers

85. A and E, B and F, C and D **87.** $\frac{\sqrt[3]{a^2} + \sqrt[3]{ab} + \sqrt[3]{b^2}}{a - b}$ **89.** $\frac{(\sqrt{x} - \sqrt{y} - \sqrt{z})[(x + y - z) + 2\sqrt{xy}]}{(x + y - z)^2 - 4xy}$

91. $\frac{1}{\sqrt[3]{(x + h)^2} + \sqrt[3]{x(x + h)} + \sqrt[3]{x^2}}$ **93.** $\sqrt[kn]{x^{km}} = (x^{km})^{1/kn} = x^{km/kn} = x^{m/n} = \sqrt[n]{x^m}$

Exercise A-8

1. $x = 7$ **3.** $s = 2.5$ **5.** $m = 5$ **7.** $-8 \le x \le 7$; 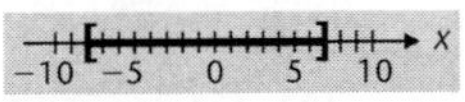 **9.** $-6 \le x < 6$; 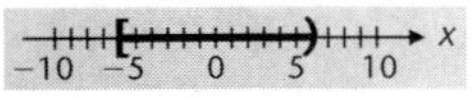

11. $x \ge -6$; 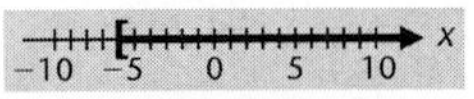 **13.** $(-2, 6]$; 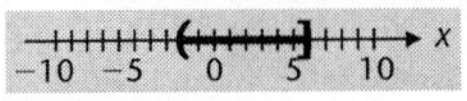 **15.** $(-7, 8)$; 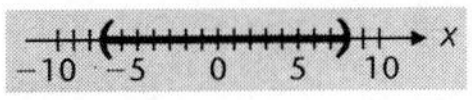

17. $(-\infty, -2]$; 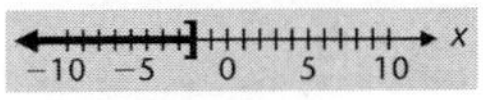 **19.** $[-7, 2)$; $-7 \le x < 2$ **21.** $(-\infty, 0]$; $x \le 0$ **23.** $x < 5$; $(-\infty, 5)$; 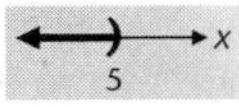

25. $x \ge 3$; $[3, \infty)$; 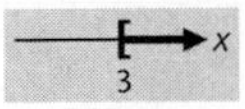 **27.** $N < -8$; $(-\infty, -8)$; 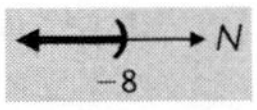 **29.** $m > 3$; $(3, \infty)$; 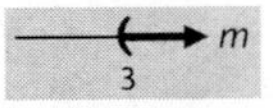

31. $B \ge -4$; $[-4, \infty)$; 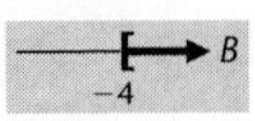 **33.** 9 **35.** 10 **37.** 8 **39.** $(-5, 7]$; $-5 < x \le 7$; 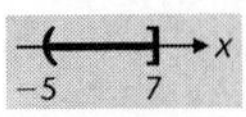

41. $(2, 4)$; $2 < x < 4$; 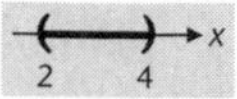

43. $(-\infty, \infty)$; $-\infty < x < \infty$; 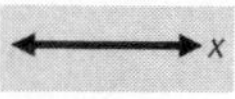

45. $(-\infty, -1) \cup [3, 7)$; $x < -1$ or $3 \le x < 7$; 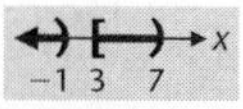

47. $(1, 5)$; $1 < x < 5$; 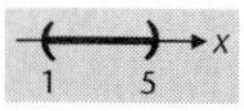

49. $(-\infty, 6]$; $x \le 6$; 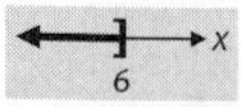

51. $q < -14$; $(-\infty, -14)$; 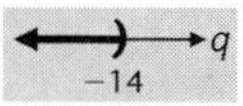

53. $x \ge 4.5$; $[4.5, \infty)$; 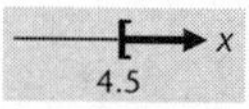

55. $-20 \le x \le 20$; $[-20, 20]$; 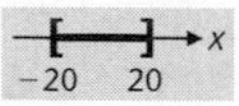

57. $-8 \le x < -3$; $[-8, -3)$; 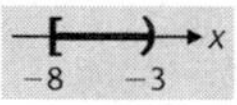

59. $-14 < x \le 11$; $(-14, 11]$; 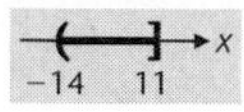

61. (A) F (B) T (C) T

Appendix A Review

1. (A) T (B) T (C) F (D) T (E) F (F) F *(A-1)* **2.** (A) $(y + z)x$ (B) $(2 + x) + y$ (C) $2x + 3x$ *(A-1)* **3.** $x^3 + 3x^2 + 5x - 2$ *(A-2)* **4.** $x^3 - 3x^2 - 3x + 22$ *(A-2)* **5.** $3x^5 + x^4 - 8x^3 + 24x^2 + 8x - 64$ *(A-2)* **6.** 3 *(A-2)* **7.** 1 *(A-2)* **8.** $14x^2 - 30x$ *(A-2)* **9.** $9m^2 - 25n^2$ *(A-2)* **10.** $6x^2 - 5xy - 4y^2$ *(A-2)* **11.** $4a^2 - 12ab + 9b^2$ *(A-2)* **12.** $(3x - 2)^2$ *(A-3)* **13.** Prime *(A-3)* **14.** $3n(2n - 5)(n + 1)$ *(A-3)* **15.** $\dfrac{12a^3b - 40b^2 - 5a}{30a^3b^2}$ *(A-4)* **16.** $\dfrac{7x - 4}{6x(x - 4)}$ *(A-4)* **17.** $\dfrac{y + 2}{y(y - 2)}$ *(A-4)* **18.** u *(A-4)* **19.** $6x^5y^{15}$ *(A-5)* **20.** $\dfrac{3u^4}{v^2}$ *(A-5)* **21.** 6×10^2 *(A-5)* **22.** $\dfrac{x^6}{y^4}$ *(A-5)* **23.** $u^{7/3}$ *(A-6)* **24.** $\dfrac{3a^2}{b}$ *(A-6)* **25.** $3\sqrt[5]{x^2}$ *(A-7)* **26.** $-3(xy)^{2/3}$ *(A-7)* **27.** $3x^2y\sqrt[3]{x^2y}$ *(A-7)* **28.** $6x^2y^3\sqrt{xy}$ *(A-7)* **29.** $2b\sqrt{3a}$ *(A-7)* **30.** $\dfrac{3\sqrt{5} + 5}{4}$ *(A-7)* **31.** $\sqrt[4]{y^3}$ *(A-7)* **32.** $x = 4$ *(A-8)* **33.** $x < 6$; $(-\infty, 6)$ *(A-8)* **34.** $\{-3, -1, 1\}$ *(A-1)* **35.** Subtraction *(A-1)* **36.** Commutative (+) *(A-1)* **37.** Distributive *(A-1)* **38.** Associative (·) *(A-1)* **39.** Negatives *(A-1)* **40.** Identity (+) *(A-1)* **41.** (A) T (B) F *(A-1)* **42.** 0 and −3 are two examples of infinitely many *(A-1)* **43.** (A) a and d (B) None *(A-2)* **44.** $4xy - 2y^2$ *(A-2)* **45.** $m^4 - 6m^2n^2 + n^4$ *(A-2)* **46.** $10xh + 5h^2 - 7h$ *(A-2)* **47.** $2x^3 - 4x^2 + 12x$ *(A-2)* **48.** $x^3 - 6x^2y + 12xy^2 - 8y^3$ *(A-2)* **49.** $(x - y)(7x - y)$ *(A-3)* **50.** Prime *(A-3)* **51.** $3xy(2x^2 + 4xy - 5y^2)$ *(A-3)* **52.** $(y - b)(y - b - 1)$ *(A-3)* **53.** $3(x + 2y)(x^2 - 2xy + 4y^2)$ *(A-3)* **54.** $(y - 2)(y + 2)^2$ *(A-3)* **55.** $x(x - 4)^2(5x - 8)$ *(A-3)* **56.** $\dfrac{(x + 2)^2(x - 4)}{x^3}$ *(A-4)* **57.** $\dfrac{2m}{(m - 2)^2(m + 2)}$ *(A-4)* **58.** $\dfrac{y^2}{x}$ *(A-4)* **59.** $\dfrac{x - y}{x + y}$ *(A-4)* **60.** $\dfrac{-ab}{a^2 + ab + b^2}$ *(A-4, A-5)* **61.** Incorrect; correct final form is $\dfrac{x^2 + 2x - 2}{x - 1}$ *(A-4)* **62.** $\frac{1}{4}$ *(A-5)* **63.** $\frac{5}{9}$ *(A-5)* **64.** $\dfrac{3x^2}{2y^2}$ *(A-6)* **65.** $\dfrac{27a^{1/6}}{b^{1/2}}$ *(A-6)* **66.** $x + 2x^{1/2}y^{1/2} + y$ *(A-6)* **67.** $6x + 7x^{1/2}y^{1/2} - 3y$ *(A-6)* **68.** 2×10^{-7} *(A-5)* **69.** 3.213×10^6 *(A-5)* **70.** 4.434×10^{-5} *(A-5)* **71.** -4.541×10^{-6} *(A-5)* **72.** 128,800 *(A-6)* **73.** 0.01507 *(A-6)* **74.** 0.3664 *(A-7)* **75.** 1.640 *(A-7)* **76.** 0.08726 *(A-6)* **77.** $-6x^2y^2\sqrt[5]{3x^2y}$ *(A-7)* **78.** $x\sqrt[3]{2x^2}$ *(A-7)* **79.** $\dfrac{\sqrt[5]{12x^3y^2}}{2x}$ *(A-7)* **80.** $y\sqrt[3]{2x^2y}$ *(A-7)* **81.** $\sqrt[3]{2x^2}$ *(A-7)* **82.** $2x - 3\sqrt{xy} - 5y$ *(A-7)* **83.** $\dfrac{6x + 3\sqrt{xy}}{4x - y}$ *(A-7)* **84.** $\dfrac{4u - 12\sqrt{uv} + 9v}{4u - 9v}$ *(A-7)* **85.** $\sqrt{y^2 + 4} + 2$ *(A-7)* **86.** $\dfrac{1}{\sqrt{t} + \sqrt{5}}$ *(A-7)* **87.** $2 - \frac{3}{2}x^{-1/2}$ *(A-7)* **88.** $x = 0.36$ *(A-8)* **89.** $-2 < x \le 4$; $(-2, 4]$ *(A-8)* **90.** $\frac{6}{11}$; rational *(A-1)* **91.** (A) $\{-4, -3, 0, 2\}$ (B) $\{-3, 2\}$ *(A-1)* **92.** 0 *(A-7)* **93.** $x^3 + 8x^2 - 6x + 1$ *(A-2)* **94.** $x(2a + 3x - 4)(2a - 3x - 4)$ *(A-3)* **95.** All three have the same value. *(A-7)* **96.** $\frac{2}{3}(x - 2)(x + 3)^4$ *(A-5)* **97.** $\dfrac{a^2b^2}{a^3 + b^3}$ *(A-5)* **98.** $x - y$ *(A-6)* **99.** x^{m-1} *(A-6)* **100.** $\dfrac{1 + \sqrt[3]{x} + \sqrt[3]{x^2}}{1 - x}$ *(A-7)* **101.** $\dfrac{1}{\sqrt[3]{t^2} + \sqrt[3]{5t} + \sqrt[3]{25}}$ *(A-7)* **102.** x^{n+1} *(A-7)* **103.** Volume $= 12\pi x + 12\pi$ ft^3 *(A-2)*

104. $9.60 \times 10^3 = 9{,}600$ kg per person *(A-5)*

105. (A) 24,000 units (B) Production doubles to 48,000 units
(C) At any production level, doubling the units of capital and labor doubles production. *(A-6)*

106. $R = \dfrac{R_1R_2R_3}{R_2R_3 + R_1R_3 + R_1R_2}$ *(A-4)*

107. (A) $A = 480 - 6x^2 = 6(80 - x^2)$ (B) $V = x(16 - 2x)(15 - 1.5x) = 240x - 54x^2 + 3x^3$ *(A-3)*

APPENDIX B Exercise B-1

1. $A = 2, B = 5$ **3.** $A = 7, B = -2$ **5.** $A = 1, B = 2, C = 3$ **7.** $A = 2, B = 1, C = 3$

9. $A = 0, B = 2, C = 2, D = -3$ **11.** $\frac{-4}{x+2} + \frac{3}{x-4}$ **13.** $\frac{3}{3x+4} - \frac{1}{2x-3}$ **15.** $\frac{2}{x} - \frac{1}{x-3} - \frac{3}{(x-3)^2}$

17. $\frac{2}{x} + \frac{3x-1}{x^2+2x+3}$ **19.** $\frac{2x}{x^2+2} + \frac{3x+5}{(x^2+2)^2}$ **21.** $x - 2 + \frac{3}{x-2} - \frac{2}{x-3}$ **23.** $\frac{2}{x-3} + \frac{2x+5}{x^2+3x+3}$

25. $\frac{2}{x-4} - \frac{1}{x+3} + \frac{3}{(x+3)^2}$ **27.** $\frac{2}{x-2} - \frac{3}{(x-2)^2} - \frac{2x}{x^2-x+1}$ **29.** $x + 2 - \frac{2}{x+2} + \frac{1}{2x-1} + \frac{x-1}{2x^2-x+1}$

Index

Inequalities and Intervals (A-8)

$a < b$	a is less than b
$a \le b$	a is less than or equal to b
$a > b$	a is greater than b
$a \ge b$	a is greater than or equal to b
(a, b)	Open interval; $\{x \mid a < x < b\}$
$(a, b]$	Half-open interval; $\{x \mid a < x \le b\}$
$[a, b)$	Half-open interval; $\{x \mid a \le x < b\}$
$[a, b]$	Closed interval; $\{x \mid a \le x \le b\}$

Quadratic Formula (2-5)

If $ax^2 + bx + c = 0$, $a \neq 0$, then

$$x = \frac{-b \pm \sqrt{b^2 - 4ac}}{2a}$$

Rectangular Coordinates (1-1, 2-1)

(x_1, y_1)	Coordinates of point P_1
$d = \sqrt{(x_2 - x_1)^2 + (y_2 - y_1)^2}$	Distance between $P_1(x_1, y_1)$ and $P_2(x_2, y_2)$
$\left(\frac{x_1 + x_2}{2}, \frac{y_1 + y_2}{2}\right)$	Midpoint of line joining P_1 and P_2
$m = \frac{y_2 - y_1}{x_2 - x_1}$, $x_1 \neq x_2$	Slope of line through P_1 and P_2

Arithmetic Sequence (6-3)

$a_1, a_2, \ldots, a_n, \ldots$

$a_n - a_{n-1} = d$ Common difference

$a_n = a_1 + (n - 1)d$ nth-term formula

$S_n = a_1 + \cdots + a_n = \frac{n}{2}[2a_1 + (n - 1)d]$ Sum of n terms

$S_n = \frac{n}{2}(a_1 + a_n)$

Geometric Sequence (6-3)

$a_1, a_2, \ldots, a_n, \ldots$

$\frac{a_n}{a_{n-1}} = r$ Common ratio

$a_n = a_1 r^{n-1}$ nth-term formula

$S_n = a_1 + \cdots + a_n = \frac{a_1 - a_1 r^n}{1 - r}$, $r \neq 1$ Sum of n terms

$S_n = \frac{a_1 - r a_n}{1 - r}$, $r \neq 1$

$S_\infty = a_1 + a_2 + \cdots = \frac{a_1}{1 - r}$, $|r| < 1$ Sum of infinitely many terms

Permutations and Combinations (6-4, 6-6)

$n! = n(n - 1) \cdots 2 \cdot 1$, $n \in N$ n factorial

$0! = 1$

For $0 \le r \le n$,

$P_{n,r} = \frac{n!}{(n - r)!}$ Permutation

$C_{n,r} = \binom{n}{r} = \frac{n!}{r!(n - r)!}$ Combination

$(a + b)^n = \sum_{k=0}^{n} \binom{n}{k} a^{n-k} b^k$, $n \ge 1$ Binomial formula

Circle (1-1)

$(x - h)^2 + (y - k)^2 = r^2$ Center at (h, k); radius r

$x^2 + y^2 = r^2$ Center at $(0, 0)$; radius r

Parabola (2-3, 7-1)

$y^2 = 4ax$, $a > 0$, opens right; $a < 0$, opens left
Focus: $(a, 0)$; Directrix: $x = -a$;
Axis: x axis

$x^2 = 4ay$, $a > 0$, opens up; $a < 0$, opens down
Focus: $(0, a)$; Directrix: $y = -a$;
Axis: y axis

Ellipse (7-2)

$\frac{x^2}{a^2} + \frac{y^2}{b^2} = 1$, $a > b > 0$
Foci: $F'(-c, 0)$, $F(c, 0)$; $c^2 = a^2 - b^2$

$\frac{x^2}{b^2} + \frac{y^2}{a^2} = 1$, $a > b > 0$
Foci: $F'(0, -c)$, $F(0, c)$; $c^2 = a^2 - b^2$

Hyperbola (7-3)

$\frac{x^2}{a^2} - \frac{y^2}{b^2} = 1$ Foci: $F'(-c, 0)$, $F(c, 0)$; $c^2 = a^2 + b^2$

$\frac{y^2}{a^2} - \frac{x^2}{b^2} = 1$ Foci: $F'(0, -c)$, $F(0, c)$; $c^2 = a^2 + b^2$

Translation Formulas (7-4)

$x = x' + h, y = y' + k;$ $x' = x - h, y' = y - k$
New origin (h, k)